WASTEWATER RECLAMATION AND REUSE

WATER QUALITY MANAGEMENT LIBRARY
VOLUME 10

WASTEWATER RECLAMATION AND REUSE

EDITED BY

Takashi Asano, Ph.D., P.E.

LIBRARY EDITORS

W. W. ECKENFELDER, D.Sc., P.E.   J. F. MALINA, JR., Ph.D., P.E., D.E.E.   J. W. PATTERSON, Ph.D.

TECHNOMIC
PUBLISHING CO., INC.
LANCASTER · BASEL

Water Quality Management Library—Volume 10
a TECHNOMIC publication

Technomic Publishing Company, Inc.
851 New Holland Avenue, Box 3535
Lancaster, Pennsylvania 17604 U.S.A.

Printed in the United States of America
10 9 8 7 6 5 4 3 2 1

Main entry under title:
  Water Quality Management Library—Volume 10 / Wastewater Reclamation and Reuse

A Technomic Publishing Company book
Bibliography: p.
Includes index p. 1501

Library of Congress Catalog Card No. 97-62550
ISBN No. 1-56676-620-6 (Volume 10)
ISBN No. 1-56676-660-5 (11-Volume Set)

Table of Contents

    FRANK J. LOGE, JEANNIE L. DARBY, GEORGE TCHOBANOGLOUS
    and DAVID SCHWARTZEL

WILLIAM D. JOHNSON and JOHN R. PARNELL

Historical Background   1037

Foreword

IN 1992 the United States National Committee of IAWQ (International Association on Water Quality) organized eight specialty courses offered in conjunction with the 1992 IAWQ Biennial Conference in Washington, D.C. Designed for the practicing engineer, the specialty courses covered critical topics in environmental quality management water pollution control, wastewater treatment, toxicity reduction, and residuals management. These courses were compiled in an eight-volume series as the Water Quality Management Library. Experts from the United States and many countries contributed their expertise and experience to the preparation of these state-of-the-art texts.

The success of this series prompted the editors to expand the series to include volumes on such timely topics as water reuse, non-point source control and aeration and oxygen transfer. Additional volumes are presently being considered. In addition to the new topics, in order to keep pace with this rapidly developing field, many of the original volumes have been updated to reflect current advances in the field. In addition to providing an up-to-date technical reference for the practicing engineer and scientist as in the first series, these volumes will provide a text for continuing education courses and workshops.

The Water Quality Management Library should provide a unique reference source for professional and education libraries.

W. WESLEY ECKENFELDER
JOSEPH F. MALINA, JR.
JAMES W. PATTERSON

Preface

FOR the last quarter century, a repeated thesis has been that the treatment of municipal and industrial wastewater provides a water of such quality that it should not be wasted but put to beneficial use. This conviction, coupled with the increasing frequency of water shortages and the high costs of water development and environmental protection, has provided an impetus for considering wastewater reclamation, recycling, and reuse in many parts of the world. A few examples of the nonpotable water reuse applications are agricultural and landscape irrigation, industrial cooling, toilet flushing in large office buildings, groundwater recharge, and water for aesthetics and environmental purposes. While direct potable use is, at present, limited to extreme situations, highly treated reclaimed water is being seriously evaluated as a raw water source for potable purposes. Reclaimed water is, after all, a water resource created right at the doorstep of the urban environment where water resources are needed the most and priced the highest. Furthermore, reclaimed water provides a dependable source of water even in drought years because the generation of urban wastewater is affected little by drought.

Today, technically proven wastewater treatment or water purification processes exist to provide water of almost any quality desired. However, the effective integration of water and reclaimed wastewater still requires close examination of public health issues, infrastructure and facilities planning, wastewater treatment plant siting, treatment process reliability, economic and financial analyses, and water utility management. Whether wastewater reuse can be implemented will depend upon careful economic considerations, potential uses for the reclaimed water, stringency of waste discharge requirements, public health considerations, and public policy emphasizing water conservation, rather than development of new water resources. Thus, wastewater reclamation and reuse have a rightful place in the integrated management of water resources and play an important role in optimal planning and more efficient management and use of water resources, now and in the future.

In this book, I have made a determined attempt to assemble, analyze, and review the various aspects of wastewater reclamation, recycling, and reuse in most parts of the world. The contributing authors come, not only from the United States, but also from Brazil, France, Israel, Japan, Portugal, Spain, South Africa, Tunisia, and the United Kingdom. They are all leading authorities in their fields and are involved in various aspects of wastewater reuse from fundamental research to public education, and together, they provide a global perspective and valuable experiences.

I am indeed grateful to all the contributors, for their professional collaboration and support. I am also thankful to Professor W. Wesley Eckenfelder who initially convinced me to edit this book, some three years ago. Finally, the successful completion of this book owes much to Susan G. Farmer, Managing Editor, Book Division, and Kimberly J. Martin, Copy Editor, Technomic Publishing Co., Inc. whose patience and expert guidance contributed to making *Wastewater Reclamation and Reuse* a reality.

TAKASHI ASANO
DAVIS, CALIFORNIA

List of Contributors

AVNER ADIN, Graduate School of Applied Science, Division of Environmental Sciences, The Hebrew University of Jerusalem, Jerusalem 91904, Israel

TAKASHI ASANO, Department of Civil and Environmental Engineering, University of California at Davis, Davis, CA 95616

AKISSA BAHRI, Centre de Recherche du Génie Rural, Laboratoire de Chimie des Eaux-Sols-Boues, Ministere de L'Agriculture, B.P. 10, Ariana 2080, Tunisia

DANIEL E. BLACKSON, Water Reclamation Facility, Arizona Public Service Company, Palo Verde Nuclear Generating Station, P.O. Box 52034, Phoenix, AZ 85072-2034

JEAN BONTOUX, Département Sciences de l'Environnement et Santé Publique, Faculté de Pharmacie, Universite de Montepellier I, av. Ch. Flahault - 34060 Montepellier Cedex, France

LAURENT BONTOUX, European Commission, Institute for Prospective Technological Studies, W.T.C., Isla de la Cartuja s/n, E-41092 Sevilla, Spain

SUSAN M. BRADFORD, Orange County Water District, 10500 Ellis Avenue, P.O. Box 8300, Fountain Valley, CA 92728-8300

ANDREW C. CHANG, Department of Soil and Environmental Sciences, University of California at Riverside, Riverside, CA 92521

CHING-LIN CHEN, Wastewater Research Section, County Sanitation Districts of Los Angeles County, 1955 Workman Mill Road, Whittier, CA 90601-1400

GORDON COLOGNE, Justice (ret.), California Court of Appeal. 48511 Via Amistad, La Quinta, CA 92253

ROBERT C. COOPER, School of Public Health, University of California at

Berkeley, Berkeley, CA 94720 and BioVir Laboratories, Inc., 685 Stone Road, Benicia, CA 94510

ROBIN CORT, Parsons Engineering Science, Inc., 2101 Webster Street, Suite 700, Oakland, CA 94612

GERARD COURTOIS, Direction Départementale des Affaires Sanitaires et Sociales, Montpellier, 615, boulevard Antigone - 34000 Montpellier Cedex, France

RONALD W. CRITES, Brown and Caldwell, 9616 Micron Avenue, Suite 600, Sacramento, CA 95827

JAMES CROOK, Black & Veatch, 230 Congress Street, Suite 802, Boston, MA 02110

JEANNIE L. DARBY, Department of Civil and Environmental Engineering, University of California at Davis, Davis, CA 95616

RICHARD A. DIAMOND, Irvine Ranch Water District, 15600 Sand Canyon Avenue, Irvine, CA 92618

DON M. EISENBERG, Eisenberg, Olivieri & Associates, 1410 Jackson Street, Oakland, CA 94612

DAVE FERGUSON, Irvine Ranch Water District, 15600 Sand Canyon Avenue, Irvine, CA 92618

NAOYUKI FUNAMIZU, Department of Sanitary and Environmental Engineering, Faculty of Engineering, Hokkaido University, Kita-13, Nishi 8, Kita-ku, Sapporo 060, Japan

CHARLES P. GERBA, Department of Soil, Water, and Environmental Science, University of Arizona, Tucson, AZ 85721

MARK B. GINGRAS, Irvine Ranch Water District, 15600 Sand Canyon Avenue, Irvine, CA 92618

JOAN GLADSTONE, Gladstone International, Lakeshore Towers, 18101 Von Karman Avenue, suite 1280, Irvine, CA 92715

G. L. GROBLER, Department of Water Affairs and Forestry, Private Bag X313, Pretoria 0001, South Africa

EARLE C. HARTLING, County Sanitation Districts of Los Angeles County, 1955 Workman Mill Road, P.O. Box 4998, Whittier, CA 90607

JOE HAWORTH, County Sanitation Districts of Los Angeles County, 1955 Workman Mill Road, Whittier, CA 90601

GREGORY K. HERR, Irvine Ranch Water District, 15600 Sand Canyon Avenue, Irvine, CA 92618

IVANILDO HESPANHOL, Department of Hydraulic and Sanitary Engineering, Escola Politécnica da Universidade de São Paulo, P.O. Box 61548, 05508-900 São Paulo, S.P. Brazil

THOMAS R. HOLLIMAN, Engineering and Planning, Long Beach Water Department, 1800 East Wardlow Road, Long Beach, CA 90807-4994

ROBERT S. JAQUES, Monterey Regional Water Pollution Control Agency, # 5 Harris Court, Building D, Monterey, CA 93940

JOHNNIE JOHANNESSEN, Irvine Ranch Water District, 15600 Sand Canyon Avenue, Irvine, CA 92618

WILLIAM D. JOHNSON, Public Utilities Department, City of St. Petersburg, 1635 3rd Avenue North, P.O. Box 2842, St. Petersburg, FL 33731

JILL KANZLER, Gladstone International, Lakeshore Towers, 18101 Von Karman Avenue, Suite 1280, Irvine, CA 92715

SARA M. KATZ, Katz & Associates, 4275 Executive Square, Suite 530, La Jolla, CA 92037-1477

LINDA KELLY, Unified Sewerage Agency, 155 North First Avenue, #270. Hillsboro, OR 97124

KAREN S. KUBICK, Utilities Commission, City & County of San Francisco, 1990 Newcomb Avenue, San Francisco, CA 94124-1617

JIH-FEN KUO, County Sanitation Districts of Los Angeles County, 1955 Workman Mill Road, Whittier, CA 90601-1400

WILLIAM C. LAUER, American Water Works Association, 6666 W. Quincy Avenue, Denver, CO 80235

AUDREY D. LEVINE, Department of Civil and Environmental Engineering, Utah State University, Utah Water Research Laboratory, Logan, Utah 84322-8200

FRANK J. LOGE, Department of Civil and Environmental Engineering, University of California at Davis, Davis, CA 95616

PETER M. MACLAGGAN, WateReuse Association of California, 4021 Liggett Drive, San Diego, CA 92106

DUNCAN MARA, Department of Civil Engineering, University of Leeds, Leeds LS2 9JT United Kingdom

BROCK MCEWEN, CH2M Hill, P.O. Box 241325, Denver, CO 80224-9325

ROBERT R. MCVICKER, Irvine Ranch Water District, 15600 Sand Canyon Avenue, Irvine, CA 92618

RICHARD A. MILLS, Office of Water Recycling, California State Water Resources Control Board, 2014 T Street, P.O. Box 944212, Sacramento, CA 94244-2120

WILLIAM R. MILLS, JR., Orange County Water District, 10500 Ellis Avenue, P.O. Box 8300, Fountain Valley, CA 92728-8300

MARIA HELENA F. MARECOS DO MONTE, Ministerio das Obras Publicas, Transportes e Comunicaçoes, Laboratório Nacional de Engenharia Civil, Avenida do Brasil, 101, 1799 Lisboa Codex, Portugal

JERALD L. MORELAND, Water Reclamation Facility, Arizona Public Service

Company, Palo Verde Nuclear Generating Station, P.O. Box 52034, Phoenix, AZ 85072-2034

MARGARET H. NELLOR, County Sanitation Districts of Los Angeles County, 1955 Workman Mill Road, P.O. Box 4998, Whittier, CA 90607

BONNIE NIXON, Public Affairs Management, 101 The Embarcadero, Suite 210, San Francisco, CA 94105

P. E. ODENDAAL, Water Research Commission, P.O. Box 824, Pretoria 0001, South Africa

ADAM W. OLIVIERI, Eisenberg, Olivieri & Associates, 1410 Jackson Street, Oakland, CA 94612

GIDEON ORON, The Jacob Blaustein Institute for Desert Research, Ben-Gurion University of the Negev, Sede Boker Campus 84990, Israel

ALBERT L. PAGE, Department of Soil and Environmental Sciences, University of California at Riverside, Riverside, CA 92521

JOHN R. PARNELL, Public Utilities Department, City of St. Petersburg, 1635 3rd Avenue North, P.O. Box 2842, St. Petersburg, FL 33731

JOHN J. PARSONS, Irvine Ranch Water District, 15600 Sand Canyon Avenue, Irvine, CA 92618

SHERWOOD C. REED, Environmental Engineering Consultant, 50 Butternut Road, Norwich, VT 05055

DAVID RICHARD, Nolte and Associates, Inc., 1750 Creekside Oaks Drive, Suite 200, Sacramento, CA 95833

MARTIN RIGBY, Orange County Water District, 10500 Ellis Avenue, P.O. Box 8300, Fountain Valley, CA 92728-8300

STEPHEN E. ROGERS, Camp Dresser & McKee Inc., 1331 17th Street, Suite 1200, Denver, CO 80202

RICHARD H. SAKAJI, Division of Drinking Water and Environmental Management, California Department of Health Services, 2151 Berkeley Way, Berkeley, CA 94704

DAVID T. SCHWARTZEL, Trojan Technologies Inc., 3020 Gore Road, London, Ontario, Canada N5V 4T7

BAHMAN SHEIKH, Parsons Engineering Science, Inc., 2101 Webster Street, Suite 700, Oakland, CA 94612

JAMES F. STAHL, County Sanitation Districts of Los Angeles County, 1955 Workman Mill Road, Whittier, CA 90601-1400

GEORGE TCHOBANOGLOUS, Department of Civil and Environmental Engineering, University of California at Davis, Davis, CA 95616

PATRICIA A. TENNYSON, Katz & Associates, 4275 Executive Square, Suite 530, La Jolla, CA 92037-1477

KENNETH A. THOMPSON, Irvine Ranch Water District, 15600 Sand Canyon Avenue, Irvine, CA 92618

JOYCE WEGNER-GWIDT, Irvine Ranch Water District, 15600 Sand Canyon Avenue, Irvine, CA 92618

MICHAEL P. WEHNER, Orange County Water District, 10500 Ellis Avenue, P.O. Box 8300, Fountain Valley, CA 92728-8300

J. L. J. VAN DER WESTHUIZEN, Department of Water Affairs and Forestry, Private Bag X313, Pretoria 0001, South Africa

MARK WILF, Hydranautics, 401 Jones Road, Oceanside, CA 92054-1216

MARYLYNN V. YATES, Department of Soil & Environmental Sciences, University of California at Riverside, Riverside, CA 92521

RONALD E. YOUNG, Irvine Ranch Water District, 15600 Sand Canyon Avenue, Irvine, CA 92618

CHAPTER 1

Wastewater Reclamation, Recycling, and Reuse: An Introduction

INTRODUCTION

THE concept of deriving beneficial uses from treated municipal and industrial wastewater coupled with increasing pressures on water resources has prompted the emergence of wastewater reclamation, recycling, and reuse as integral components of water resource management. The inherent benefits associated with reclaiming treated wastewater for supplemental applications prior to discharge or disposal include preservation of higher quality water resources, environmental protection, and economic advantages. A major catalyst for the evolution of wastewater reclamation, recycling, and reuse has been the need to provide alternative water sources to satisfy water requirements for irrigation, industry, urban non-potable and potable water applications due to unprecedented growth and development in many regions of the world. Water shortages, particularly during periods of drought, have necessitated stricter control measures on rates of water consumption and development of alternative water sources.

Advances in the effectiveness and reliability of wastewater treatment technologies have improved the capacity to produce reclaimed wastewater that can serve as a supplemental water source in addition to meeting water quality protection and pollution abatement requirements. In developing countries, particularly those in arid parts of the world, reliable low-cost

Takashi Asano, Department of Civil and Environmental Engineering, University of California at Davis, Davis, CA 95616; Audrey D. Levine, Department of Civil and Environmental Engineering, Utah Water Research Laboratory, Utah State University, Logan, UT 84322-8200.

1

technologies are needed for acquiring new water supplies and protecting existing water sources from pollution. The implementation of wastewater reclamation, recycling, and reuse promotes the preservation of limited water resources in conjunction with water conservation and watershed protection programs. In the planning and implementation of water reclamation and reuse, the intended water reuse applications dictate the extent of wastewater treatment required, the quality of the finished water, and the method of distribution and application.

The purpose of this chapter is to provide an introduction to the major elements of wastewater reclamation, recycling, and reuse. The topics included are: (1) an overview of wastewater reclamation, recycling, and reuse, (2) a survey of concepts associated with planning wastewater reclamation and reuse systems, (3) a summary of technologies and treatment systems appropriate for wastewater reclamation, and (4) a discussion of water quality factors associated with specific wastewater reuse applications. Examples of water reuse applications are provided and cross-referenced to appropriate chapters of this book.

WASTEWATER REUSE TERMINOLOGY

The terminology currently used in wastewater reuse engineering is derived from sanitary and environmental engineering practice. The water potentially available for reuse includes municipal and industrial wastewater, agricultural return flows, and stormwater. Of these, return flows from agricultural irrigation and stormwater are usually collected and reused without further treatment.

Wastewater reclamation involves the treatment or processing of wastewater to make it reusable, and *wastewater reuse* or *water reuse* is the beneficial use of the treated water. Reclamation and reuse of water frequently require water conveyance facilities for delivering the reclaimed water and may require intermittent storage of the reclaimed water prior to reuse. In contrast to reuse, *wastewater recycling* or *water recycling* normally involves only one use or user, and the effluent from the user is captured and redirected back into that use scheme. In this context, the term *wastewater recycling* is applied predominantly to industrial applications such as in the steam-electric, manufacturing, and minerals industries.

The use of reclaimed wastewater where there is a direct link from the treatment system to the reuse application is termed *direct reuse*. Direct reuse provides water for agricultural and landscape irrigation, industrial applications, urban applications, and dual water systems. *Indirect reuse* includes mixing, dilution, and dispersion of reclaimed wastewater by discharge into an impoundment, receiving water, or groundwater aquifer prior to reuse such as in groundwater recharge. Indirect reuse, through discharge of a treated effluent to a receiving water for assimilation and withdrawal downstream, while important, does not normally constitute *planned reuse.* For example, the diversion of water from a river downstream of a discharge of treated wastewater constitutes an incidental or *unplanned reuse.* The topics covered in this book are related to deliberate or *planned reuse* as described above.

Unplanned indirect wastewater reuse, through effluent disposal to streams and groundwater basins, has been an accepted practice throughout the world for centuries. Communities that are situated at the end of major waterways have a long history of producing potable water from river water sources that have circulated through multiple cycles of withdrawal, treatment, and discharge. Examples of unplanned indirect wastewater reuse include: New Orleans, Louisiana, U.S.A. (the Mississippi River), London, England (the River Thames), the cities and towns in the Rhine River Valley, Germany, and Osaka, Japan (the Yodo River). It has been estimated that more than 80% of the water in the Santa Ana River in Southern California originates from wastewater discharged by upstream municipal wastewater treatment (Argo and Rigby, 1986). Similarly, river beds or percolation ponds may recharge underlying groundwater aquifers with wastewater dominated water, which in turn, is withdrawn by down-gradient communities for domestic water supplies.

Indirect water reuse can also be planned. For example, the City of London derives about 20% of its drinking water supply from the River Lea, a tributary of the River Thames. Upstream of London, the City of Stevenage discharges treated wastewater into the River Lea. To protect the City of London's water supply, the City of Stevenage is required to remove nitrate as an integral component of wastewater treatment (Dean and Lund, 1981). Another example of indirect reuse is an artificial groundwater recharge program in Los Angeles County, California where reclaimed water has provided a source of recharge water since 1962 (Nellor, Baird, and Smyth, 1985; Nellor et al., 1985). A maximum of 50% of the total water applied to surface spreading areas during a 12-month period can be supplied by reclaimed wastewater (State of California, 1989).

It is also important to differentiate between *potable* and *non-potable* reuse applications. *Potable water reuse* refers to the use of highly treated reclaimed water to augment drinking water supplies. Although direct potable reuse is limited to extreme cases, it consists of incorporating reclaimed water into a potable water supply system, without relinquishing control over the resource. Indirect potable reuse includes an intermediate step in which reclaimed water is mixed with surface or groundwater sources prior to drinking water treatment. *Non-potable water reuse* includes all water applications other than drinking water supplies. Currently, on an international scale, direct non-potable water reuse comprises the dominant mode of wastewater reuse for supplementing public water supplies for such uses as landscape and agricultural irrigation. Detailed examples of non-potable reuse applications are provided in Chapters 16, 17, 18, and 19. Strategies for indirect potable reuse are discussed in more detail in Chapter 27 by McEwen.

THE ROLE OF WASTEWATER RECLAMATION, RECYCLING, AND REUSE IN THE WATER CYCLE

The inclusion of planned wastewater reclamation, recycling, and reuse in water resource management systems reflects the application of complementary developments in technology, health risk understanding, and public acceptance to mitigate limitations imposed by the increasing scarcity of water resources. As the link between wastewater, reclaimed water, and water reuse has become better defined, increasingly smaller recycle loops can be developed. The hydrologic cycle is a conceptual model of the continuous transport of water in the environment. The water cycle consists of fresh and saline surface water resources, subsurface groundwater, water associated with various landuse functions, and atmospheric water vapor. Many subcycles to the large-scale hydrologic cycle exist, including the engineered transport of water, such as an aqueduct. Wastewater reclamation, recycling, and reuse have become significant components of the hydrologic cycle in urban, industrial, and agricultural areas. An overview of the cycling of water from surface and groundwater resources to water treatment facilities, irrigation, municipal, and industrial applications, and to wastewater reclamation and reuse facilities is shown in Figure 1.1

Water reuse may involve a completely controlled ''pipe-to-pipe'' system with an intermittent storage step, or it may include blending of reclaimed water with natural water either indirectly through surface water supplies or groundwater recharge or directly in an engineered system. The major pathways of water reuse are depicted in Figure 1.1 and include groundwater recharge, irrigation, industrial use, and surface water replenishment. Surface water replenishment and groundwater recharge also occur through natural drainage and through infiltration of irrigation water and stormwater runoff. The potential use of reclaimed wastewater for potable water treatment is also shown, although this application is reserved for extreme situations. The quantity of water transferred via each pathway depends on the watershed characteristics, climatic and geohydrological factors, the degree of water utilization for various purposes, and the degree of direct or indirect water reuse.

The water used or reused for agricultural and landscape irrigation includes agricultural, residential, commercial, and municipal applications.

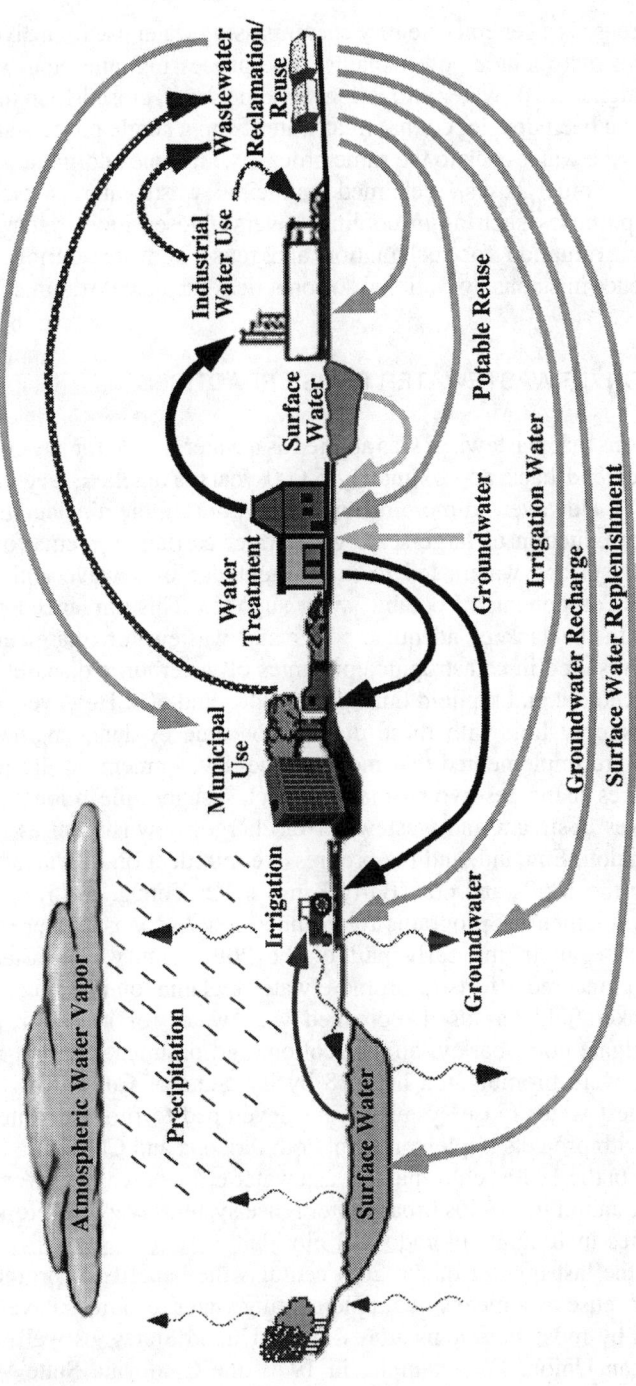

Figure 1.1. The role of engineered treatment, reclamation, and reuse facilities in the cycling of water through the hydrologic cycle.

Industrial reuse is a general category encompassing water use for a diversity of industries that include power plants, food processing, and other industries with high rates of water utilization. In some cases, closed-loop recycle systems have been developed that treat water from a single process stream and recycle the water back to the same process with some additional make-up water. In other cases, reclaimed municipal wastewater is used for industrial purposes such as in cooling towers. Closed-loop systems are also under evaluation for reclamation and reuse of water during long-duration space missions by National Aeronautics and Space Administration (NASA).

EVOLUTION OF WASTEWATER REUSE PRACTICES

Indications that wastewater was applied as a water source for agricultural irrigation extend back approximately 5,000 years (Angelakis and Spyridakis, 1995). However, in more recent history, during the nineteenth century, the introduction of large-scale wastewater carriage systems for discharge into surface waters led to inadvertent use of sewage and other effluents as components of potable water supplies. This unplanned reuse, coupled with the lack of adequate water and wastewater treatment and sanitation, resulted in catastrophic epidemics of waterborne diseases such as Asiatic cholera and typhoid during the 1840s and 50s. However, when the water supply link with these diseases became evident, engineering solutions were implemented that included the development of alternative water sources using reservoirs and aqueduct systems, the relocating of water intakes upstream and wastewater discharges downstream as in the case of London, England, and the progressive introduction of water filtration during the 1850s and 60s (Barty-King, 1992; Young, 1985).

The development of programs for planned reuse of wastewater within the U.S.A. began in the early part of the 20th century. The State of California pioneered efforts to promote water reclamation and reuse. The City of Bakersfield has used reclaimed wastewater for irrigation since 1912 to irrigate corn, barley, alfalfa, cotton, and pasture. The first reuse regulations were promulgated in 1918 by the State of California. Some of the earliest water reuse systems were developed to provide water for irrigation with projects implemented in both Arizona and California in the late 1920s. In the 1940s, chlorinated wastewater effluent was used for steel processing, and in the 1960s urban water reuse systems were developed in several states including Colorado and Florida.

During the last quarter of the 20th century, the benefits of promoting wastewater reuse as a means of supplementing water resources have been recognized by most state legislatures in the United States, as well as by the European Union. For example, in 1970, the California State Water

Code stated that "it is the intention of the Legislature that the State undertake all possible steps to encourage development of water reclamation facilities so that reclaimed water is available to help meet the growing water requirements of the State" (State of California, 1978). In the same context, the European Communities Commission Directive (91/271/EEC) declared that "treated wastewater shall be reused whenever appropriate. Disposal routes shall minimize the adverse effects on the environment" (European Communities Commission Directive, 1991).

Currently, in the late 1990s, increased interest in wastewater reuse in many parts of the world is occurring in response to growing pressures for high quality, dependable water supplies by agriculture, industry, and the public. Today, technically proven wastewater treatment and engineered purification processes exist to produce water of almost any quality desired. Thus, wastewater reclamation, recycling, and reuse have evolved to become an integral factor in fostering the optimal planning and efficient use of water resources in many countries.

Milestone events that had significant impacts on the evolution of wastewater reclamation, recycling, and reuse are itemized on the timeline in Figure 1.2. Microbiological advances in the late 19th century precipitated the "Great Sanitary Awakening" (Fair and Geyer, 1954) and the advent of disinfection processes. The development of the activated sludge process in 1904 was a significant step towards advancement of wastewater treatment and pollution control, and the development of biological treatment systems. In 1918, the California State Board of Public Health adopted its first regulations addressing the use of sewage for irrigation.

Technological advances in physical, chemical, and biological processing of water and wastewater during the middle of the 20th century led to the "Era of Wastewater Reclamation, Recycling, and Reuse." Since the 1960s, intensive research efforts, fueled by regulatory pressures and water shortages, have provided valuable insight into health risks and reliable treatment system design concepts for water reuse engineering. In 1965, the Israeli Ministry of Health issued regulations to allow the reuse of secondary effluents for crop irrigation with the exclusion of vegetable crops that are eaten uncooked. In 1968, extensive research on direct potable reuse was conducted in Windhoek, Namibia. In the U.S.A., a milestone event was the passage of the Federal Water Pollution Control Act of 1972 (PL 92-500) "to restore and maintain the chemical, physical, and biological integrity of the Nation's waters" with the ultimate goal of zero discharge of pollutants into navigable, "fishable and swimmable waters." During the 1970s and 80s, several comprehensive research and demonstration projects were conducted to evaluate the potential for nonpotable and potable reuse with a major emphasis on quantifying health risks and defining reliable treatment and technological requirements. These research efforts

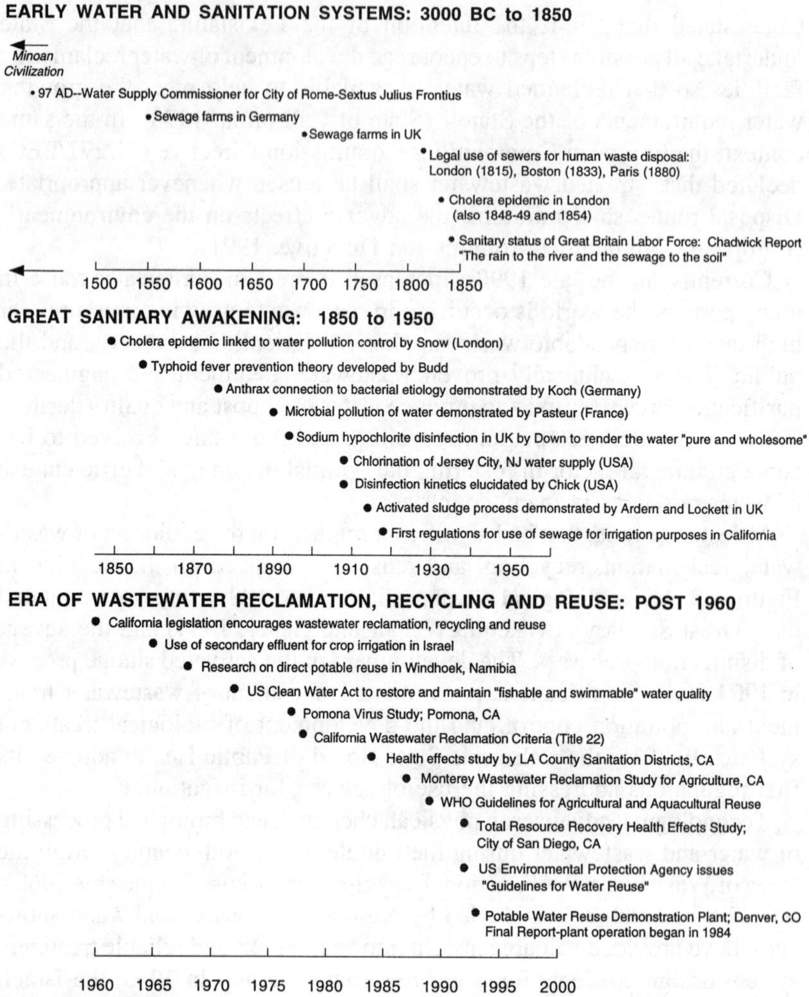

EARLY WATER AND SANITATION SYSTEMS: 3000 BC to 1850

Minoan Civilization

• 97 AD--Water Supply Commissioner for City of Rome-Sextus Julius Frontius

    • Sewage farms in Germany

        • Sewage farms in UK

            • Legal use of sewers for human waste disposal:
            London (1815), Boston (1833), Paris (1880)

                • Cholera epidemic in London
                (also 1848-49 and 1854)

                • Sanitary status of Great Britain Labor Force: Chadwick Report
                "The rain to the river and the sewage to the soil"

    1500   1550   1600   1650   1700   1750   1800   1850

GREAT SANITARY AWAKENING: 1850 to 1950

● Cholera epidemic linked to water pollution control by Snow (London)

    ● Typhoid fever prevention theory developed by Budd

        ● Anthrax connection to bacterial etiology demonstrated by Koch (Germany)

        ● Microbial pollution of water demonstrated by Pasteur (France)

            ● Sodium hypochlorite disinfection in UK by Down to render the water "pure and wholesome"

            ● Chlorination of Jersey City, NJ water supply (USA)

            ● Disinfection kinetics elucidated by Chick (USA)

                ● Activated sludge process demonstrated by Ardern and Lockett in UK

                ● First regulations for use of sewage for irrigation purposes in California

    1850   1870   1890   1910   1930   1950

ERA OF WASTEWATER RECLAMATION, RECYCLING AND REUSE: POST 1960

    ● California legislation encourages wastewater reclamation, recycling and reuse

        ● Use of secondary effluent for crop irrigation in Israel

        ● Research on direct potable reuse in Windhoek, Namibia

            ● US Clean Water Act to restore and maintain "fishable and swimmable" water quality

            ● Pomona Virus Study; Pomona, CA

                ● California Wastewater Reclamation Criteria (Title 22)

                ● Health effects study by LA County Sanitation Districts, CA

                    ● Monterey Wastewater Reclamation Study for Agriculture, CA

                    ● WHO Guidelines for Agricultural and Aquacultural Reuse

                        ● Total Resource Recovery Health Effects Study;
                        City of San Diego, CA

                            ● US Environmental Protection Agency issues
                            "Guidelines for Water Reuse"

                                ● Potable Water Reuse Demonstration Plant; Denver, CO
                                Final Report-plant operation began in 1984

    1960   1965   1970   1975   1980   1985   1990   1995   2000

Figure 1.2. Milestone events in the evolution of wastewater reclamation, recycling, and reuse.

have resulted in increased implementation of wastewater reuse projects in various regions and the evolution of new options for wastewater reuse throughout the world.

The continued testing and implementation of treatment systems and new applications have helped to overcome many technical barriers to wastewater reuse projects. Improvements in treatment process reliability, health risk assessment, and public confidence in reuse systems in conjunction with increasing water demands and pollution control requirements have promoted

TABLE 1.1. Summary of Major Elements of Wastewater Reuse Planning.

Planning Phase	Objective of Planning
Assess wastewater treatment and disposal needs	Evaluate quantity of wastewater available for reuse and disposal options
Assess water supply and demand	Evaluate dominant water use patterns
Analyze market for reclaimed water	Identify potential users of reclaimed water and associated water quantity and quality requirements
Conduct engineering and economic analyses	Determine treatment and distribution system requirements for potential users of reclaimed water
Develop implementation plan with financial analysis	Develop strategies, schedule, and financing options for implementation of project

the integration of water reuse into water resources management strategies throughout the world. It is important to recognize that public acceptance of water reuse projects is vital to the future of wastewater reclamation, recycling, and reuse; the consequences of poor public perception could jeopardize future projects involving the use of reclaimed wastewater.

WASTEWATER REUSE PLANNING

The impetus for wastewater reclamation, recycling, and reuse projects is to address needs for water pollution control, to augment water supplies through replenishment of groundwater or surface water resources, or to provide an alternative water source for some applications thereby preserving higher quality water resources. To optimize the net benefits from implementation of wastewater reuse, a well-designed, integrated planning process is essential. Conceptual level planning for wastewater reuse typically involves definition of the project, cost estimation, and identification of a potential reclaimed water market. This preliminary planning provides insight into the viability of wastewater reuse and is a prelude to detailed planning. Effective project planning provides a unique opportunity to achieve multiple objectives by unifying wastewater reclamation efforts with water use requirements. The major elements of wastewater reuse planning are summarized in Table 1.1 and key components of the planning process are discussed. Detailed information on the planning of wastewater reclamation, recycling, and reuse projects is found in Chapter 2 by Mills and Asano.

PROJECT STUDY AREA

Delineation of the project study area is a critical planning issue. The project study area should encompass all locations within a practical distance of the project that can potentially benefit from reuse of reclaimed wastewater. Because water supplies are often derived from water resources outside of the project sponsor's jurisdictional boundaries, it is prudent to consider expanding the scope of the project beyond the immediate area to obtain a comprehensive assessment of short-term and long-term water resources requirements. For example, the most serious impacts from overdrafted groundwater basins may be manifested in communities beyond the local area. Thus, implementing wastewater reuse in the project area could yield water supply savings for other water users within the watershed.

It is important to identify all potential reclaimed water applications appropriate for the study area. As a starting point, water use patterns associated with landscape irrigation, industrial cooling, and irrigation of food crops should be characterized. On the basis of water use categories, health and water pollution control regulatory authorities should be consulted to determine specific wastewater treatment requirements, on-site facilities modifications (e.g., backflow prevention devices), and use area controls (e.g., no irrigation in areas of direct human contact). Projects designed with the primary purpose of water supply can usually be operated more flexibly if alternate disposal, such as stream discharge, is available to dispose of effluent that cannot be reused. In other cases intermittent storage may be necessary to accommodate seasonal variations in water usage rates.

If land application is to be used for treatment and disposal of wastewater, candidate sites should be identified where water can be applied at high rates and low cost. Unless the system is designed with alternative wastewater disposal methods, the users will have to make a long-term commitment to accept the treated effluent and may not have full control over the quantities of water delivered. If users cannot be found to accept treated effluent on a voluntary contractual basis, the wastewater agency itself will have to purchase wastewater application sites and apply the reclaimed water or lease the land to a private farmer.

MARKET ANALYSIS

After determining the water quantity and quality requirements, it is necessary to conduct a market analysis to identify potential users of reclaimed water. Access to records of water retailers can be especially helpful. It is important to review several years of actual water use records to avoid misinterpretation of data from unusually wet or dry precipitation years. Potential users should be contacted and the reuse sites visited to determine potential site problems or on-site water system modifications needed to accommodate the use of reclaimed water. These factors have cost implications which must be assessed in the planning stage. The concerns, needs, and financial expectations of users must be identified. It is advisable to meet with potential users to disseminate information, address concerns, and provide access to technical experts to respond to questions.

MONETARY ANALYSES

While technical, environmental, and social factors are considered in project planning, monetary factors tend to control the ultimate decisions of whether and how to implement a wastewater reuse project. Monetary analyses fall into two categories: economic analysis and financial analysis. The economic analysis focuses on the value of the resources invested in a project to construct and operate it, measured in monetary terms and computed in terms of present value estimations. On the other hand, the financial analysis is based on the market value of goods and services at the time of sale, incorporating subsidies or monetary transfers which may exist. The objective of conducting economic analyses of wastewater reuse projects is to quantify impacts on society, whereas financial analyses are targeted on the local ability to raise money from project revenues, government grants, loans, and bonds to pay for the project (Asano and Mills, 1990; State of California, 1979).

The overriding goal of the economic analysis is to answer the question: *should* a reuse project be constructed? Equally important, however, is the question: *can* a reuse project be constructed? Both orientations are necessary. However, only wastewater reuse projects which are viable in the economic context should be given further consideration for a financial analysis (Engineering-Science, 1987; U.S. Environmental Protection Agency, 1986).

OTHER PLANNING FACTORS

In addition to monetary aspects, other important planning factors for a wastewater reuse project include system engineering and minimizing potential public health and environmental impacts associated with the project. A wastewater reuse project can be viewed as a small-scale water supply project. Appropriate levels of wastewater treatment are needed along with reclaimed water storage and supplemental or backup fresh water supply to accommodate variations in relative rates of reclaimed water production and water utilization.

Another important planning distinction is in the design approach. Design of water treatment systems is usually based on providing adequate water to meet a projected demand, whereas wastewater reuse system design stems from the quantity of wastewater available. The wastewater supply rate is balanced with the reclaimed water demand rate until the economic optimum is met. For example, landscape irrigation demand in California is seasonal. However, wastewater production is relatively constant year-round. Reclaimed water supply may be sufficient to meet annual demands, but only if seasonal storage is provided. Seasonal storage, however, is costly and, sometimes impossible to site, particularly in urban settings. Another option is to include fewer users in the system such that the peak demands can be met entirely by the reclaimed water supply without seasonal storage. This approach, however, could result in the waste of as much as 40% of the available reclaimed water. Another option is to consider providing supplemental fresh water to meet peak demands and/or to provide an emergency backup water supply during periods of treatment plant upset or equipment failure. These design decisions influence the need for equipment redundancy in the reclaimed water system, overall system reliability, and cost. A thorough analysis of treatment, storage, and distribution system trade-offs is necessary to optimize system planning. A more detailed discussion of planning issues is provided in Chapter 2.

RECLAIMED WASTEWATER QUALITY CHARACTERIZATION

Water quality characterization is necessary to evaluate the biological and chemical safety of using reclaimed wastewater for various applications and the effectiveness of individual treatment technologies. The water quality parameters that are used to evaluate reclaimed wastewater are based on current practice in water and wastewater treatment. A summary of relevant water quality monitoring parameters is given in Table 1.2. Municipal wastewater treatment systems are typically designed to meet water quality objectives based on biochemical oxygen demand (BOD_5), total suspended solids (TSS), total or fecal coliform, nutrient levels (nitrogen and phosphorus), and chlorine residual. Potable water quality monitoring parameters include coliform organisms, turbidity, dissolved minerals, disinfection by-products, and specific inorganic and organic contaminants. Recently there has been increased emphasis in developing monitoring tools for detection of microbial pathogens including *Giardia lambia, Cryptosporidium parvum*, and viruses in potable water supplies (Fox and Lytle, 1994; Gerba et al., 1996; Rose, Darbin, and Gerba, 1988; Rose and Gerba, 1991). Particle size analysis has also been proposed as a water quality monitoring tool (LeChevallier and Norton, 1992; Levine, Tchobanoglous, and Asano, 1985).

TABLE 1.2. Summary of Major Parameters Used to Characterize Reclaimed Wastewater Quality.

Parameter	Significance in Wastewater Reclamation	Approximate Range in Treated Wastewater	Treatment Goal in Reclaimed Wastewater[a]
Organic indicators			
BOD_5	Organic substrate for microbial or algal growth	10–30 mg/L	<1 to 10 mg/L
TOC	Measure of organic carbon	1–20 mg/L	<1 to 10 mg/L
Measurement of particulate matter			
Total suspended solids (TSS)	Measure of particles in wastewater can be related to microbial contamination, turbidity. Can interfere with disinfection effectiveness	<1 to 30 mg/L	<1 to 10 mg/L
Turbidity	Measure of particles in wastewater; can be correlated to TSS	1 to 30 NTU	0.1 to 10 NTU
Pathogenic organisms	Measure of microbial health risks due to enteric viruses, pathogenic bacteria and protozoa.	Coliform organisms: <1 to 10^4/100 mL Other pathogens: Controlled by treatment technology	<1 to 2,000/mL
Nutrients			
Nitrogen	Nutrient source for irrigation; can also contribute to microbial growth	10 to 30 mg/L	<1 to 30 mg/L
Phosphorus	Nutrient source for irrigation; can also contribute to microbial growth	0.1 to 30 mg/L	<1 to 20 mg/L

[a]Treatment goal depends on specific wastewater reuse application.

Where reclaimed wastewater is used for applications that have potential human exposure routes, the major acute health risks are associated with exposure to biological agents including bacterial pathogens, helminths, protozoa, and enteric viruses (Asano et al., 1992). From a public health and process control perspective, enteric viruses are the most critical group of pathogenic organisms in the developed world due to the possibility of infection from exposure to low doses and the lack of routine, cost-effective methods for detection and quantification of viruses. In addition, treatment systems that can remove viruses effectively will most likely be effective for control of other pathogenic organisms. Details on microbiological considerations in wastewater reclamation and reuse and public health concerns are provided in Chapter 10 by Gerba and Yates and in Chapter 11 by Cooper and Olivieri.

OVERVIEW OF WASTEWATER TREATMENT TECHNOLOGY

The effective treatment of wastewater to meet water quality objectives for water reuse applications and to protect public health is a critical element of water reuse systems. Municipal wastewater treatment consists of a combination of physical, chemical, and biological processes and operations to remove solids, organic matter, pathogens, metals, and sometimes nutrients from wastewater. General terms used to describe different degrees of treatment, in order of increasing treatment level, are preliminary, primary, secondary, tertiary, and advanced treatment. A disinfection step for control of pathogenic organisms is often the final treatment step prior to distribution or storage of reclaimed wastewater. Wastewater reclamation, recycling, and reuse treatment systems are derived from applying technologies used for conventional wastewater treatment and drinking water treatment. The goal in designing a wastewater reclamation and reuse system is to develop an integrated cost-effective treatment scheme that is capable of reliably meeting water quality objectives.

The degree of treatment required in individual water treatment and wastewater reclamation facilities varies according to the specific reuse application and associated water quality requirements. The simplest treatment systems involve solid/liquid separation processes and disinfection whereas more complex treatment systems involve combinations of physical, chemical, and biological processes employing multiple barrier treatment approaches for contaminant removal. An overview of the major technologies that are appropriate for wastewater reclamation and reuse systems is given in Table 1.3. More detailed discussions on treatment processes for water reuse can be found in Chapter 5 by Adin.

TABLE 1.3. Overview of Representative Unit Processes and Operations Used in Wastewater Reclamation.

Process	Description	Application
Solid/liquid separation		
Sedimentation	Gravity sedimentation of particulate matter, chemical floc, and precipitates from suspension by gravity settling	Removal particles from wastewater that are larger than about 30 μm. Typically used as primary treatment and downstream of secondary biological processes.
Filtration	Particle removal by passing water through sand or other porous medium	Removal of particles from wastewater that are larger than about 3 μm. Typically used downstream of sedimentation (conventional treatment), or following coagulation/flocculation.
Biological treatment		
Aerobic biological treatment	Biological metabolism of wastewater by microorganisms in an aeration basin or biofilm (trickling filter) process.	Removal of dissolved and suspended organic matter from wastewater
Oxidation pond	Ponds with 2 to 3 feet of water depth for mixing and sunlight penetration	Reduction of suspended solids, BOD, pathogenic bacteria, and ammonia from wastewater.
Biological nutrient removal	Combination of aerobic, anoxic, and anaerobic processes to optimize conversion of organic and ammonia nitrogen to molecular nitrogen (N_2) and removal of phosphorus	Reduction of nutrient content of reclaimed wastewater
Disinfection	The inactivation of pathogenic organisms using oxidizing chemicals, ultraviolet light, caustic chemicals, heat, or physical separation processes (e.g., membranes)	Protection of public health by removal of pathogenic organisms

15

TABLE 1.3. (continued).

Process	Description	Application
Advanced treatment		
Activated carbon	Process by which contaminants are physically absorbed onto the surface of activated carbon	Removal of hydrophobic organic compounds
Air stripping	Transfer of ammonia and other volatile constituents from water to air	Removal of ammonia nitrogen and some volatile organics from wastewater
Ion exchange	Exchange of ions between an exchange resin and water using a flow through reactor	Effective for removal of cations such as calcium, magnesium, iron, ammonium, and anions such as nitrate
Chemical coagulation and Precipitation	Use of aluminum or iron salts, polyelectrolytes, and/or ozone to promote destabilization of colloidal particles from reclaimed wastewater and precipitation of phosphorus.	Formation of phosphorus precipitates and flocculation of particles for removal by sedimentation and filtration
Lime treatment	The use of lime to precipitate cations and metals from solution	Used to reduce scale forming potential of water, precipitate phosphorus, and modify pH
Membrane filtration	Microfiltration, nanofiltration, and ultrafiltration	Removal of particles and microorganisms from water
Reverse osmosis	Membrane system to separate ions from solution based on reversing osmotic pressure differentials	Removal of dissolved salts and minerals from solution; also effective for pathogen removal

PRIMARY TREATMENT

Primary treatment refers to the initial processing of wastewater for removal of particulate matter. In conventional wastewater treatment facilities, primary treatment includes screening and comminution for removal of large solids, grit removal, and sedimentation. Conventional primary treatment is effective for removal of particulate matter larger than about 50 μm from wastewater. In general, about 50% of the suspended solids and 25 to 50% of the BOD_5 are removed from the untreated wastewater by primary treatment processes (Metcalf & Eddy, 1991). Nutrients, hydrophobic constituents, metals, and microorganisms that are associated with particulates in wastewater can also be removed by primary treatment processes. About 10 to 20% of the organic nitrogen and about 10% of the phosphorus are removed by conventional primary treatment. The removal efficiency of primary treatment processes can be increased by incorporating coagulation/flocculation upstream of gravity sedimentation and/or by using filtration downstream of gravity sedimentation. For most wastewater reuse applications, primary treatment alone does not provide adequate treatment to meet water quality objectives.

SECONDARY TREATMENT

Secondary treatment systems include an array of biological treatment processes coupled with solid/liquid separation. Biological processes are engineered to provide effective microbiological metabolism of organic substrates dissolved or suspended in wastewater. The microbial biomass interacts with wastewater using a suspended growth or a fixed film process. Examples of suspended growth processes include activated sludge processes, aerated lagoons, and stabilization ponds. Examples of engineered fixed-film processes are trickling filters, rotating biological contactors, and other biofilm bioreactors. A portion of the degradable organic material in wastewater provides energy and nutrients to support microbial growth while the remainder is microbially oxidized to carbon dioxide, water, and other end products.

Conventional biological treatment systems consists of an aerobic biological reactor coupled with secondary sedimentation to remove and concentrate biomass produced from conversion of wastewater organic constituents. The effluent from conventional secondary processes contains levels of suspended solids and BOD_5 ranging from about 10 to 30 mg/L. Depending on process operation, from 10 to 50% of the organic nitrogen is removed during conventional secondary treatment and phosphorus is converted to phosphate (PO_4^{-3}). Biosolids produced during secondary treatment are treated by aerobic or anaerobic digestion, composting, or other solids processing technologies. Some removal of pathogens, trace elements, and dissolved organic contaminants occurs in conjunction with biological treatment and physical separation.

The design of a secondary process depends on the required capacity of the treatment facility, the treatment objectives, and the need for nutrient removal and advanced treatment. Aerated lagoons and stabilization ponds can operate effectively without upstream primary treatment and are often used in small-scale, decentralized facilities where adequate land area is available. Activated sludge and biofilm systems are normally designed to operate downstream of primary treatment and are applicable for large- or small-scale facilities. There are multiple combinations and variations of biological processes that can be designed to achieve alternative levels of performance for removal of suspended solids, biodegradable organic constituents, and nutrients.

For many wastewater reclamation and reuse systems, secondary treatment can provide adequate removal of organic matter from wastewater. Frequently, secondary treatment is supplemented by filtration for additional removal of particles and disinfection. Details on smallscale systems and natural systems for wastewater reuse can be found in Chapter 3 by Tchobanoglous, Reed, and Crites. The effectiveness of waste stabilization ponds and reservoirs for producing reclaimed water for agricultural irrigation is discussed in Chapter 4 by Mara.

TERTIARY TREATMENT

In general, tertiary treatment refers to additional removal of colloidal and suspended solids by chemical coagulation and granular medium filtration. Advanced treatment refers to more complete removal of specific constituents, such as ammonia or nitrate removal by ion exchange or total dissolved solids (TDS) removal by reverse osmosis. Tertiary and advanced treatment processes are normally applied downstream of biological treatment processes (e.g., activated sludge, trickling filtration). More detailed discussions are found in Chapters 5, 7, 8, 20, 21, 22, 23, 30, and 31.

Chemical Coagulation and Flocculation

Coagulation/flocculation processes involve the addition of chemicals to wastewater to promote aggregation of particles for improved solid/liquid separation by sedimentation and filtration. Inorganic coagulant chemicals are metallic salts such as alum (aluminum sulfate), ferric chloride, or ferric sulfate. The salts hydrolyze in water and react with particle surfaces resulting in particle destabilization. Organic polyelectrolytes are also used in conjunction with inorganic coagulant chemicals to improve process effectiveness. The dosages of coagulant chemicals used depend on the characteristics of the wastewater and the process design and range from 1 to 50 mg/L of inorganic coagulants and 0.5 to 10 mg/L of organic polyelectrolytes. Ozonation of water can serve to improve coagulation effectiveness.

Flocculation is a process used downstream of coagulation to aggregate destabilized particles into flocs that are of a size range amenable to removal by gravity sedimentation or filtration. Particle flocculation is accomplished by passing the water through a mixing system that promotes interparticle collisions and particle aggregation. If coagulation/flocculation is used directly upstream of filtration, the filtration system is defined as "direct-filtration." A process sequence in which coagulation/flocculation is followed by sequential sedimentation and filtration is defined as "conventional treatment." Chemical costs are a major operating expense associated with coagulation/flocculation and therefore, careful control of chemical dosage is important.

Granular-Medium Filtration

Filtration is a solid/liquid separation process that is effective for removal of suspended particles larger than about 3 μm. Wastewater is passed through a column of granular media and particles are removed by impaction, interception, and physical straining. As particles accumulate in the filter media, headloss through the filter increases. When the headloss reaches a terminal value or there is a breakthrough of turbidity, filters are cleaned using a combination of water backwash and air scour. To optimize filter run-time and avoid rapid increases in filter headloss, filtration is most effective if the particle concentration (measured as TSS) applied to the filter is below about 20 mg/L. Filtration can be used downstream of primary sedimentation (primary effluent filtration), or downstream of secondary sedimentation (tertiary filtration). A variety of filter designs are available that vary in media type (single, dual, or multi-media), operating mode (upflow, downflow or crossflow; constant rate or constant head-declining rate), cleaning mode (continuous or intermittent; water backwash with or without air scour or chemical amendment).

Because pathogenic organisms are associated with particles, filtration is effective at reducing pathogen concentration in wastewater streams and provides an excellent pre-treatment for disinfection. Filtration is stipulated as a required treatment process in many wastewater reuse applications to remove particulate matter that can compromise disinfection effectiveness. If water is to be treated by activated carbon, ion exchange, or reverse osmosis, filtration is used as a pretreatment to reduce the particulate loading on these processes and improve their overall effectiveness. More details on the application of filtration in water reuse are provided in Chapter 6 by Chen, Kuo, and Stahl.

Disinfection

Disinfection is an essential treatment component for almost all wastewater reclamation and reuse applications. The objective of disinfection processes is to destroy pathogenic organisms. The major groups of pathogenic organisms include waterborne bacteria, viruses, amoebic cysts, and protozoa including *Giardia lambia,* and *Cryptosporidium parvum.*

A disinfection step is typically one of the final treatment processes in a wastewater treatment system. Chemical disinfection practices are based on addition of a strong oxidizing chemical such as chlorine, ozone, hydrogen peroxide, or bromine. Oxidation chemicals, particularly ozone, can also be effective in reduction of odor and color in wastewater and in improving the biodegradability of organic constituents. Ultraviolet radiation is an alternative process that can achieve disinfection. Other methods for reduction of microbial content of reclaimed wastewater include exposing pathogenic organisms to alkaline environments as in lime treatment. Alternatively, physical methods can be designed for removal of microorganisms such as granular medium or membrane filtration systems.

The most common type of disinfection system in wastewater reclamation is chlorine disinfection at typical dosages ranging from 5 to 20 mg/L with a maximum of two hours of contact time. Because chlorine can have negative impacts on irrigated crops, the residual chlorine in the treated wastewater is controlled. Dechlorination, if necessary, is applied after adequate chlorine contact time has been achieved. Sulfur dioxide or other reducing agents are used for dechlorination. Activated carbon adsorption is also effective for removal of residual chlorine.

Ultraviolet (UV) disinfection has earned a reputation as a viable alternative to chemical disinfection processes in wastewater reclamation and reuse applications (Darby, Snider, and Tchobanoglous, 1993). The performance of UV disinfection is influenced by water turbidity, suspended solids, and the UV lamp intensity. The age of the lamp and the fouling characteristics of the wastewater tend to have negative impacts on lamp intensity and effectiveness. The presence of particulate matter in wastewater can interfere with the effectiveness of UV disinfection by shielding pathogenic organisms from UV irradiation. Filtration is employed upsteam of UV systems to reduce the concentration of particulates and improve the disinfection effectiveness. More details on UV disinfection systems are provided in Chapter 8.

Nutrient Removal

The need for nutrient removal depends on the ultimate fate of the treated wastewater. The primary nutrients of concern are nitrogen and phosphorus. For many irrigation applications, nutrients can be beneficial for crop growth. However, excess nutrients in wastewater can foster the growth of algae in reservoirs, streams, and storage facilities. Nutrient removal treatment systems can be designed to remove nitrogen and/or phosphorus from wastewater. In untreated wastewater, nitrogen can exist in dissolved or particulate form and in several oxidation states. Ammonia and organic nitrogen are the dominant forms of nitrogen associated with untreated wastewater. During biological treatment, organic nitrogen is converted to ammonia nitrogen and provides a nitrogen source for microbial growth. In consort with microbial growth, some of the nitrogen may be microbially oxidized to nitrite and nitrate in aerobic processes. The biological conversion of nitrogen to nitrate is known as nitrification. Nitrate can be converted to molecular nitrogen (N_2) by a biological process known as denitrification in the absence of molecular oxygen. Molecular nitrogen is released to the atmosphere. The combined effect of nitrification-denitrification is nitrogen removal from the wastewater. Biological nitrogen removal is designed using sequential aerobic-anoxic processes and adequate organic substrate to optimize growth of nitrifying and denitrifying microbial consortia. Alternatively, nitrogen can be removed by air stripping, chemical oxidation, or ion exchange processes.

Phosphorus removal is accomplished by converting soluble phosphorus to particulate phosphorus that can be removed by sedimentation and/or filtration. Particulate phosphorus can be formed through chemical precipitation as calcium phosphate using lime treatment or using iron or aluminum salts to form iron or aluminum phosphate precipitates. Biological phosphorus removal can be accomplished by operating a biological treatment system with alternating anaerobic and aerobic conditions that promote uptake of orthophosphate, polyphosphate, and organic phosphate by microbial cells. Separation of microbial cells containing phosphorus results in phosphorus removal from the liquid stream. In many cases, biological removal of nitrogen and phosphorus are coupled in a biological nutrient removal treatment system.

Ion exchange is another option for nitrogen removal. A cation exchange process can be used to remove ammonium (NH_4^+); alternatively anion exchange processes can be used to remove nitrate (NO_3^-) and nitrite (NO_2^-). An example of using ion exchange for ammonium removal is the Upper Occoquan Sewage Authority's Advanced Water Treatment Plant in Northern Virginia that discharges reclaimed wastewater to the principal water supply (Robbins, 1985). The treatment train consists of headworks, grit removal, primary clarification, aeration, secondary clarification, flocculation, chemical clarification, first stage recarbonation clarifier, second stage recarbonation clarifier, ballast pond, multimedia filtration, carbon column, ion exchange (final filtration), and chlorination. When the plant is operating in the nitrogen removal mode, the ion exchange columns, which exchange sodium ions for ammonium ions to remove ammonia, function as final filters to remove particulates from granular activated carbon effluent.

Membrane Processes

Membrane processes include microfiltration, ultrafiltration, nanofiltration, reverse osmosis, and electrodialysis. Microfiltration is effective for removal of particles and can be cost competitive with conventional granular medium filtration. Removal of macromolecules and particles larger than about 0.1 μm can be achieved using ultrafiltration whereas nanofiltration and reverse osmosis are applied for removal of dissolved ions from liquids. While membranes have multiple applications, the useful life of a membrane is dependent on avoiding conditions that will cause fouling, scaling, or chemical interactions. The success of membrane process operation is dependent on appropriate pretreatment. A key issue is to prevent biological growth on membrane surfaces that can result in membrane fouling. Also, the use of strong oxidizing chemicals can damage membrane integrity. Pretreatment options include filtration for coarse particle removal, scale control, and chemical addition. Post treatment includes stabilization of the water to prevent corrosion. More detailed discussions on membrane processes are found in Chapter 7 by Wilf.

Adsorption

Activated carbon adsorption is effective in removing hydrophobic organic compounds from surface and groundwater sources. Compounds with low water solubilities such as organic solvents and chlorinated organic solvents are adsorbable. Water soluble compounds and larger compounds are removed more effectively by oxidation or ultrafiltration. In most cases, testing is necessary (isotherm evaluation, dynamic adsorption testing) to determine the applicability of activated carbon to meet a specific treatment objective. Activated carbon adsorption can be used in conjunction with air stripping for removal of vapor-phase volatile compounds. Specific applications of adsorption in wastewater reclamation are found in Chapter 28 by Lauer and Rogers.

STORAGE

Water reuse for some applications, such as irrigation, is needed on a seasonal basis. Storage represents an important intermediate step between wastewater treatment and water reuse. Storage acts to equalize flow variations and to balance the production of reclaimed water with the utilization of water. Another benefit of storage is the additional residence time and treatment afforded. Options for storage of reclaimed water include above ground reservoirs or tanks or a below ground storage system. In above ground storage systems, additional biological degradation of organic matter occurs as well as reduction of ammonia and phosphate content. Variable results have been observed for pathogenic organism removal. In some cases, post-storage treatment may be required because of algal growth and increases in turbidity. More detailed discussions are found in Chapter 9 by Holliman.

WASTEWATER REUSE APPLICATIONS

To provide a framework for evaluating wastewater reuse, it is important to correlate major water use patterns with potential water reuse applications. On the basis of water quantity, irrigation use, consisting of both agriculture and landscape applications, is projected to account for 54% of total freshwater withdrawals by the year 2000 in the U.S.A. The second major user of reclaimed water is industry, primarily for cooling and process needs. However, industrial uses vary greatly and additional wastewater treatment beyond secondary treatment is usually required. Thus, the effective integration of wastewater reuse into water resource management is based on the quantity of water required for a specific application and the associated water quality requirements.

Significant regional and seasonal variations in water use patterns exist. For example, in urban areas, industrial, commercial, and nonpotable urban water requirements account for the major water demand. For agricultural applications in arid and semi-arid regions, irrigation is the dominant component of water demand. Water requirements for irrigation applications tend to vary seasonally whereas industrial water needs are more consistent. The feasibility of water reuse for a given watershed is limited by the degree to which reclaimed wastewater could augment existing water supplies through substitution of water in commercial, industrial, and agricultural applications.

The purpose of this section is to provide an overview of water reuse applications and water quality requirements including health aspects of wastewater reuse. The applications for municipal wastewater reuse parallel major water use applications as discussed above. An overview of the major categories of wastewater reuse is given in Table 1.4. These categories are arranged according to current and projected volumes of reclaimed wastewater. Treatment goals are based on effluent quality and the applica-

TABLE 1.4. Categories of Municipal Wastewater Reuse.[a]

Category of Wastewater Reuse	Treatment Goals	Example Applications
Urban use		
Unrestricted	Secondary, filtration, disinfection BOD_5: ≤10 mg/L; Turbidity: ≤2NTU Fecal coliform: ND[b]/100 mL Cl_2 residual: 1 mg/L; pH 6 to 9	Landscape irrigation: Parks, playgrounds, school yards; Fire protection; Construction; Ornamental fountains; Impoundments; In-building uses: toilet flushing, air conditioning
Restricted access irrigation	Secondary and disinfection BOD_5: ≤30 mg/L; TSS: ≤30 mg/L Fecal coliform: ≤200/100 mL Cl_2 residual: 1 mg/L; pH 6 to 9	Irrigation of areas where public access is infrequent and controlled Golf courses; Cemeteries; Residential; Greenbelts
Agricultural irrigation		
Food crops	Secondary, filtration, disinfection BOD_5: ≤10 mg/L; Turbidity: ≤2NTU Fecal coliform: ND/100 mL Cl_2 residual: 1 mg/L; pH 6 to 9	Crops grown for human consumption and consumed uncooked
Non-food crops and food crops consumed after processing	Secondary, disinfection BOD_5: ≤30 mg/L; TSS: ≤30 mg/L Fecal coliform: ≤200/100 mL Cl_2 residual: 1 mg/L; pH 6 to 9	Fodder, fiber, seed crops, pastures, commercial nurseries, sod farms commercial aquaculture
Recreational use		
Unrestricted	Secondary, filtration, disinfection BOD_5: ≤10 mg/L; Turbidity: ≤2NTU Fecal coliform: ND/100 mL Cl_2 residual: 1 mg/L; pH 6 to 9	No limitations on body-contact: lakes and ponds used for swimming, snowmaking
Restricted	Secondary, disinfection BOD_5: ≤30 mg/L TSS: ≤30 mg/L Fecal coliform: ≤200/100 mL Cl_2 residual: 1 mg/L; pH 6 to 9	Fishing, boating, and other non-contact recreational activities

TABLE 1.4. (continued).

Category of Wastewater Reuse	Treatment Goals	Example Applications
Environmental enhancement	Site specific treatment levels comparable to unrestricted urban uses Dissolved oxygen; pH Coliform organisms; Nutrients	Use of reclaimed wastewater to create artificial wetlands, enhance natural wetlands and sustain stream flows
Groundwater recharge	Site specific	Groundwater replenishment Salt water intrusion control Subsidence control
Industrial reuse	Secondary and disinfection BOD$_5$: ≤30 mg/L; TSS: ≤30 mg/L Fecal coliform: ≤200/100 mL	Cooling-system make-up water, process waters, boiler feed water, construction activities and washdown waters
Potable reuse	Safe drinking water requirements	Blending with municipal water supply Pipe to pipe supply

[a] Adapted from U.S. Environmental Protection Agency, 1992.
[b] Not detected.

25

tion of specific technologies. For most applications, effective secondary treatment is a prerequisite to production of high-quality reclaimed water.

Over the past two decades, the degree to which reclaimed wastewater is treated and recovered for applications listed in Table 4 has increased throughout the world. The primary incentives for implementation of water reuse are related to the need for augmentation of water supplies or control of water pollution. By reducing the quantity of treated wastewater discharged to surface waters, effluent requirements tend to be more favorable, particularly with respect to nutrient removal. Thus, water reuse is becoming an economic alternative for treatment facilities discharging into small streams.

The increased implementation of wastewater reuse projects in various regions has facilitated the evolution of new reuse alternatives. As treatment systems and applications are tested and design parameters are developed, technical barriers to reuse projects are reduced. Improvements in treatment process reliability, health risk assessment, and confidence in reuse systems coupled with increasing water demands and pollution control requirements have led to a general increase in the number of reuse projects.

Geographic, climatic, and economic factors dictate the degree and form of wastewater reclamation and reuse in different regions. In agricultural regions, irrigation is a dominant reuse application. In arid regions, such as California and Arizona in the U.S.A., groundwater recharge is a major reuse objective either to replenish existing groundwater resources or to mitigate salt water intrusion in coastal areas. Industrial reuse of water varies with industries and locations. In contrast to the arid or semi-arid regions of the world where irrigation comprises a major beneficial use of reclaimed wastewater, wastewater reuse in Japan is dominated by non-potable urban uses such as toilet flushing, industrial use, and stream restoration and flow augmentation.

In Japan, several factors have contributed to increased implementation of wastewater reclamation and reuse. The vulnerability of the fresh water supply during drought years or in the aftermath of earthquakes or other catastrophic events has become evident in recent years. To increase the dependability and capacity of water resources throughout Japan, the Japanese Government instituted a multi-faceted program including the development of new water supply reservoirs, implementation of water conservation measures in large metropolitan areas, and wastewater reclamation and reuse where practical (Asano, Maeda, and Takaki, 1996).

A comparison of water reuse applications for two locations in the U.S.A. (Florida and California) and Japan is given in Figure 1.3. The specific applications reflect the water balance associated with each region and other local constraints. Agricultural and landscape irrigation, groundwater recharge, and industrial reuse account for the majority of reused water.

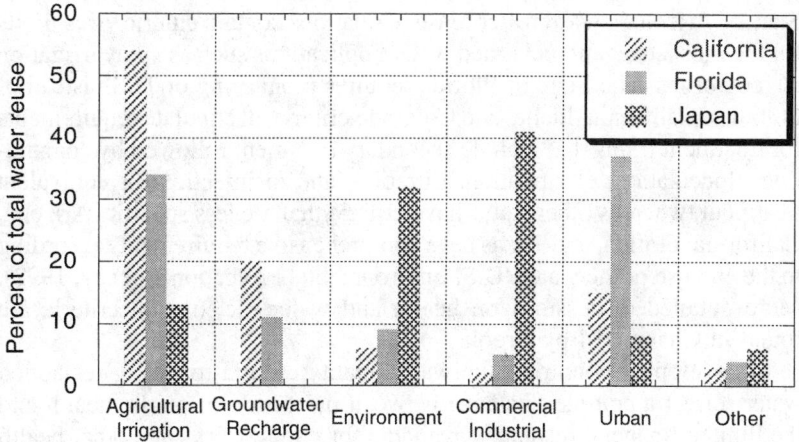

Figure 1.3. Comparison of distribution of reclaimed water applications in California, Florida, and Japan (adapted from Asano et al., 1996; Wright and Missimer, 1995).

HEALTH AND REGULATORY REQUIREMENTS

The potential health risks associated with wastewater reclamation and reuse are related to the extent of direct exposure to the reclaimed water and the adequacy, effectiveness, and reliability of the treatment system. The goal of each water reuse project is to protect public health without unnecessarily discouraging wastewater reclamation and reuse. Regulatory approaches stipulate water quality standards in conjunction with requirements for treatment, sampling, and monitoring. To minimize health risks and aesthetic problems, tight controls are imposed on the delivery and use of reclaimed water after it leaves the treatment facility. Since major issues surrounding wastewater reclamation and reuse are often related to health, considerable research efforts have been directed towards protection of public health.

In 1992, the US Environmental Protection Agency issued ''Guidelines for Water Reuse'' (U.S. Environmental Protection Agency, 1992), however the specific criteria for wastewater reclamation in the U.S.A. are developed by individual states, often in conjunction with regulations on land treatment and disposal of wastewater. Some of the major differences among the approaches taken by individual states are associated with the degree of specificity provided in the rules. Also, some discrepancies exist from place to place in terms of monitoring and treatment requirements. For example, the State of California bases microbial water quality assessment on total coliform whereas many other states require fecal coliform testing. The

State of Arizona's wastewater reuse regulations contain enteric virus limits for the most stringent reclaimed water applications such as spray irrigation of food crops. The State of Florida requires monitoring of TSS instead of turbidity. California, Idaho, and Colorado criteria all stipulate requirements for treatment trains that include secondary treatment followed by coagulation, flocculation, clarification, filtration, and disinfection, or equivalent treatment, whereas other states have criteria that are less specific. Arizona, California, Florida, and Texas have comprehensive requirements according to the end use of the water (U.S. Environmental Protection Agency, 1992). More detailed discussions on health and water reclamation criteria are found in Chapter 14 by Crook.

In developing countries, the water quality criteria for using reclaimed water reflect a complex balance between protection of public health and the limited financial resources available for public works and other health delivery systems. In many cases, engineered sewage collection systems and wastewater treatment are non-existent and reclaimed wastewater often provides an essential water resource and fertilizer source. A major concern for the use of wastewater for irrigation is control of enteric helminths such as hookworm, ascaris, trichuris, and under certain circumstances, the beef tapeworm. In this context, it is necessary to provide protection from exposure to pathogens by preventing direct consumption of crops irrigated with untreated wastewater.

The degree of treatment required and the extent of monitoring necessary depend on the specific application. In general, irrigation systems are categorized according to the potential degree of human exposure. A higher degree of treatment is required for irrigation of crops that are consumed uncooked, or use of reclaimed water for irrigation of locations that are likely to have frequent human contact. To illustrate alternative regulatory practices governing the use of reclaimed wastewater for irrigation, the major microbiological quality guidelines by the World Health Organization (World Health Organization, 1989) and the State of California's current Wastewater Reclamation Criteria (State of California, 1978) are compared in Table 1.5. More details on these regulations can be found in the Appendices.

The WHO guidelines emphasize that a series of stabilization ponds is necessary to meet microbial water quality requirements. In contrast, the California criteria stipulate conventional biological wastewater treatment followed by tertiary treatment including filtration and chlorine disinfection to produce effluent that is virtually pathogen-free. Microbiological monitoring requirements also vary. The WHO guidelines also require monitoring of intestinal nematodes; whereas the California criteria rely on treatment systems and monitoring of the total coliform density for assessment of microbiological quality.

TABLE 1.5. Comparison of Microbiological Quality Guidelines and Criteria for Irrigation by the World Health Organization (World Health Organization, 1989) and the State of California's Current Wastewater Reclamation Criteria (State of California, 1978).

Category	Reuse Conditions	Intestinal Nematodes[a]	Fecal or Total[b] Coliforms	Wastewater Treatment Requirements
WHO	Irrigation of crops likely to be eaten uncooked, sports fields, public parks	<1/L	<1,000/100 mL	A series of stabilization ponds or equivalent treatment
WHO	Landscape irrigation where there is public access, such as hotels	<1/L	<200/100 mL	Secondary treatment followed by disinfection
Calif.	Spray and surface irrigation of food crops, high exposure landscape irrigation such as parks	No standard recommended	<2.2/100 mL[b]	Secondary treatment followed by filtration and disinfection
WHO	Irrigation of cereal crops, industrial crops, fodder crops, pasture, and trees	<1/L	No standard recommended	Stabilization ponds with 8–10 day retention or equivalent removal
Calif.	Irrigation of pasture for milking animals, landscape impoundment	No standard recommended	<23/100 mL[b]	Secondary treatment followed by disinfection

[a] Intestinal nematodes (Ascaris and Trichuris species and hookworms) are expressed as the arithmetic mean number of eggs per liter during the irrigation period.
[b] California Wastewater Reclamation Criteria is expressed as the median number of total coliforms per 100 mL, as determined from the bacteriological results of the last 7 days for which analyses have been completed.

29

WATER REUSE FOR IRRIGATION

Approximately 67% of the world-wide water demand and about 40% of the annual water use in the U.S.A. is associated with agricultural production; thus, it is an excellent candidate for reclaimed water (Solley et al., 1983, National Research Council, 1996). Since the 1960s, the total world acreage that is irrigated has almost doubled. In many regions, there is a shortage of fresh water resources that can be used for irrigation. The use of reclaimed wastewater for irrigation water provides a vital resource to enhance agricultural productivity. In addition to providing a low cost water source, other side benefits include increases in crop yields, decreased reliance on chemical fertilizers, and increased protection against frost damage. In arid regions, such as Egypt, Israel, and Tunisia, water reuse provides an essential water source for agricultural production. This topic is covered in more detail in Chapter 16 by Oron and Chapter 19 by Bahri.

In the U.S.A., the practice of crop irrigation began in the early 1900s with the settlement of the arid western states in response to the limited availability of rainfall. In the humid eastern states, irrigation is used to supplement natural rainfall to increase the number of plantings per year and yield of crops, and to reduce the risk of crop failures during drought periods. Irrigation also is used to maintain recreational lands such as parks and golf courses (Solley et al., 1983). In recent years, improvements in the efficiency of irrigation systems have been implemented including efforts to reduce water losses, replacement of surface irrigation with sprinkler and microirrigation systems to reduce surface runoff, and efforts to develop irrigation schedules appropriate for crops and regional climatic variations (National Research Council, 1996).

The development of reliable wastewater treatment systems to produce irrigation water and ensure production of agricultural crops in consort with protection of public health is a direct result of two milestone studies that were conducted in California during the 1970s and 1980s: the Pomona Virus Study and the Monterey Wastewater Reclamation Study for Agriculture (Engineering-Science, 1987; Sanitation District of Los Angeles County, 1977; Sheikh et al. 1990). Both studies demonstrated conclusively that virtually pathogen-free effluents could be produced from municipal wastewater via tertiary treatment and extended disinfection with chlorine. A major result of these studies was the sound scientific demonstration that even food crops that are consumed uncooked could be successfully irrigated with reclaimed municipal wastewater without adverse environmental or health effects. More details on these studies are found in Chapters 6 and 17.

Currently, reclaimed water is widely used for agricultural and landscape irrigation throughout the world. The use of reclaimed wastewater has been successful for irrigation of a wide array of crops. Reported increases in crop yields range from 10 to 30% (Asano and Pettygrove, 1987; Deis, Scherzinger, and Thornton, 1986; Jackson and Cross, 1993). The design approach for irrigation with reclaimed municipal wastewater depends upon whether emphasis is placed on wastewater treatment or on water supply. The term *slow rate* land treatment refers to the use of land application for wastewater treatment and disposal, while the terms *agricultural reuse* and *wastewater irrigation* imply that reclaimed water is used for irrigation as a substitute for fresh water (Metcalf & Eddy, 1991). A significant use of reclaimed water in recent years is irrigation of golf courses, turfgrass, and landscaped areas in urban environments. Preapplication treatment of wastewater is necessary to protect public health, prevent nuisance conditions during application and storage, and prevent damage to crops and soils.

The nutrients in reclaimed municipal wastewater provide fertilizer value to crop or landscape production but, in certain instances, are in excess of plant needs and cause problems related to excessive vegetative growth, delayed or uneven maturity, or reduced quality. Nutrients in reclaimed water which are important to agriculture and landscape management include nitrogen, potassium, zinc, boron, and sulfur. Application rates are based on providing nutrients at levels that are beneficial but not in excess. Frequently, the nitrogen content of wastewater limits the rate of application.

Reclaimed wastewater is applied by surface, subsurface, drip, or sprinkler irrigation methods, depending on the specific situation. A key issue is to minimize exposure of workers to the reclaimed water. Drip irrigation systems are totally enclosed and provide additional worker protection compared to spray systems. Clogging problems with sprinkler and drip irrigation systems have been reported, particularly with effluents from primary sedimentation and oxidation ponds. Microbial growth, high concentrations of algae, and suspended solids can cause plugging of sprinklers or supply lines. The most frequent clogging problems occur with drip irrigation systems. More detailed discussions are found in Chapter 16 by Oron.

EVALUATION OF IRRIGATION WATER QUALITY

The feasibility of using reclaimed water for irrigation is evaluated based on several factors including: salinity, trace elements, and water infiltration rates, and other water quality criteria. Salinity can influence the soil osmotic potential, specific ion toxicity, and result in degradation of soil physical conditions (Ayers and Westcott, 1985; Pettygrove and Asano, 1985). Excess salinity results in salt accumulation in the crop root zone that leads to a loss in yield. Plant damage can result from excess salinity. The best way to avoid salinity problems is to ensure a net downward flux of water and salt through the root zone. Under such conditions, adequate drainage is needed to allow water and salt to migrate below the root zone. Long-term use of reclaimed water for irrigation is not generally possible without adequate drainage. Long-term soil exposure to reclaimed water results in higher levels of nitrogen and phosphorus, while potassium, calcium, magnesium, and sodium tend to be more variable (Ayers and Westcott, 1985).

Irrigation water that percolates through and below the root zone transports a portion of the accumulated salts from the upper root zone. Salts leached from the upper root zone accumulate to some extent in the lower part but eventually are moved below the root zone by sufficient leaching. Consequently, salinity tends to increase with depth resulting in a threefold higher average soil salinity compared to the salinity of the applied water. Crops respond to average salinity of the root zone (Pettygrove and Asano, 1985). The fraction of applied water that passes through the entire rooting depth and percolates below is called the *leaching fraction* (LF).

$$LF = \frac{\text{Depth of water leached below the root zone}}{\text{Depth of water applied at the surface}} \qquad (1)$$

The amount of salt that accumulates in the root zone is inversely proportional to the LF. For reclaimed water irrigation, it is desirable to achieve an LF above 0.5. If the salinity of irrigation water (EC_W) and the leaching fraction are known, that salinity of the drainage water that percolates below the root zone can be estimated by using Equation 1-2.

$$EC_{dw} = \frac{EC_w}{LF} \qquad (2)$$

where EC_{dw} is the electrical conductivity of the drainage water percolating below the root zone which is equal to salinity of soil-water, Ec_{sw}.

Water infiltration problems tend to occur in about the top four inches (10 cm) of the soil and are related to the structural stability of the surface soil. Depending on the relative pH and mineral composition of the soil and the applied water, minerals may precipitate in the soil from the applied water, be leached from soils by applied water, or minimal interaction will result. Changes in soil permeability that result from applied water influence the effective infiltration rate. An important parameter used to evaluate soil/water interactions is the *sodium adsorption ratio* (SAR):

$$SAR = \frac{(Na^+)}{\sqrt{\dfrac{(Ca^{+2}) + (Na^+)}{2}}} \qquad (3)$$

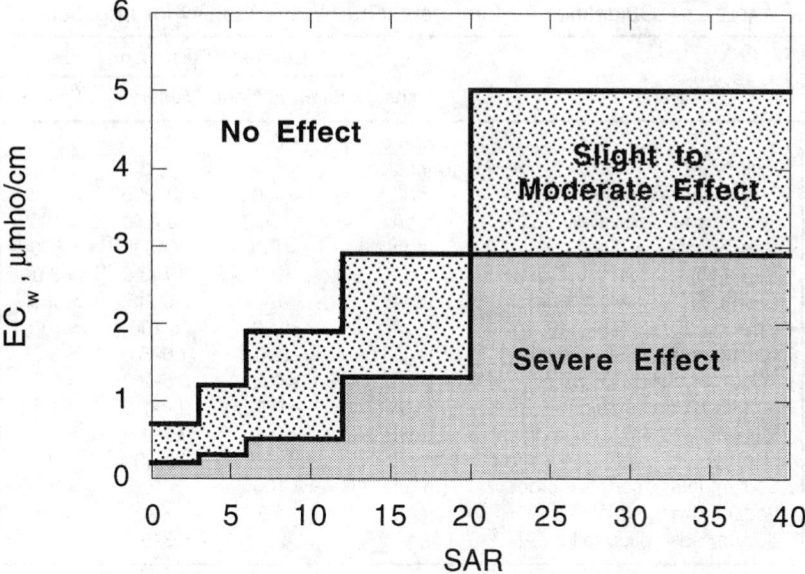

Figure 1.4. Influence of irrigation water EC_w and SAR levels on soil permeability (adapted from Ayers and Westcott, 1985).

where the concentrations of sodium and calcium are expressed in milliequivalents per liter (meq/L). The calcium levels in reclaimed wastewater are high enough that leaching of calcium from the surface soil does not occur to any significant extent. However, high sodium levels can be a major concern in planning irrigation projects with reclaimed water.

At a given sodium adsorption ratio, the infiltration rate is proportional to salinity. Adverse effects of excess sodium include impaired soil permeability. The SAR can be used in conjunction with the salinity [electrical conductivity (EC_w)] of the applied irrigation water to assess potential permeability problems as shown in Figure 1.4. For SAR levels below about 5, and EC_w levels above about 1, minimal effect on soil permeability will result for application of reclaimed wastewater. However, for SAR levels above 20 and EC_w levels below 3, severe effects on soil permeability are likely unless additional precautions are taken (Ayers and Westcott, 1985).

Sodium, chloride, and boron are soluble constituents in reclaimed wastewater that can interfere with plant growth. Sodium in irrigation water can affect soil structure as well as reduce soil aeration. The source of boron in reclaimed wastewater is household detergents or discharges from industrial plants. Chloride and sodium also increase during domestic water usage, especially where water softeners are used. When excessive residual chlorine (above 5 mg/L) is present in reclaimed wastewater due to chlorine

TABLE 1.6. Guidelines for Interpretation of Water Quality for Irrigation[a].

Parameter	Units	Degree of Restriction of Use		
		Slight to None	Moderate	Severe
Salinity, EC_w	dS/m or μmhos/cm	< 0.7	0.7–3.0	> 3.0
Total dissolved solids, TDS	mg/L	< 450	450–2000	> 2000
Total suspended solids, TSS	mg/L	< 50	50–100	> 100
Bicarbonate, (HCO_3)	mg/L	< 90	90–500	> 500
Boron (B)	mg/L	< 0.7	0.7–3.0	> 3.0
Chloride (Cl^-), sensitive crops	mg/L	< 140	140–350	> 350
Chloride (Cl^-), sprinklers	mg/L	< 100	> 100	> 100
Chlorine (Cl_2), total residual	mg/L	< 1.0	1.0–5.0	> 5.0
Hydrogen Sulfide (H_2S)	mg/L	< 0.5	0.5–2.0	> 2.0
Iron (Fe), drip irrigation	mg/L	< 0.1	0.1–1.5	> 1.5
Manganese (Mn), drip irrigation	mg/L	< 0.1	0.1–1.5	> 1.5
Nitrogen (N), total	mg/L	< 5	5–30	> 30
Sodium (Na^+), sensitive crops	mg/L	< 100	> 100	> 100
Sodium (Na^+), sprinklers	mg/L	< 70	> 70	> 70
Sodium adsorption ratio, SAR	$(meq/L)^{0.5}$	< 3	3–9	> 9

[a]After Ayers and Westcott, 1985.

disinfection, severe plant damage can occur if reclaimed water is sprayed directly on foliage. For sensitive crops, specific ion toxicity is difficult to correct without changing the crop or the water supply. The problem is usually accentuated by severe (hot and dry) climatic conditions due to high evapotranspiration rates. Recommended guidelines to prevent specific ion toxicity from irrigation water are given in Table 1.6. If the infiltration rate is greatly reduced, it may be impossible to supply the crop or landscape plant with enough water for adequate growth. In addition, irrigation systems with reclaimed municipal wastewater are often located on less desirable soils or those already having soil permeability and management problems, which accentuate the problem (Ayers and Westcott, 1985; Suarez, 1981).

WATER REUSE FOR INDUSTRIAL APPLICATIONS

About 25% of worldwide water demand is related to industrial applications (Crook et al. 1994; Metcalf & Eddy, 1991). The cost-effectiveness of using reclaimed water for industrial purposes depends on the distance the water must be transported between the reclamation facility and the point of use. In addition, the availability and cost of alternative water sources influences the degree to which reclaimed water is used.

Water recycling has been implemented successfully in several industries, and in other cases, reclaimed municipal wastewater has been used as an external water source for industrial applications. Alternative approaches for industrial recycling and reuse include reuse of municipal wastewater for an industrial process, cascading use of industrial process water between successive processes within an industry, and agricultural reuse of industrial plant effluent. Among these approaches, three categories of industrial water use are of particular interest because they are high volume uses with excellent prospects for using reclaimed municipal wastewater: (1) recirculating cooling tower makeup, (2) once-through cooling, and (3) process water. Other industrial recycling applications include commercial laundries that can recover heat, detergent and water; car and truck washing establishments; pulp and paper industries; steel production; textiles; electroplating and semiconductor industries; boiler-feed water; and water for stack gas scrubbing.

One of the earliest industrial wastewater reclamation and reuse programs was adopted in Japan in 1951 to provide water for a paper manufacturing mill from nearby Mikawashima Wastewater Treatment Plant in Tokyo (Maeda, Nakada, Kawamoto, and Ikeda, 1995). In that case, a higher quality water was produced from treated wastewater effluent than was available from surface water sources and saltwater intrusion coupled with land subsidence had compromised the availability of groundwater. In addition to paper mills, reclaimed wastewater is used as process or cooling water for manufacturing requirements, power plants, iron and steel production, and carpet dyeing. The mining industry has also adopted the use of reclaimed wastewater for cooling, transport, or process water. Wastewater reclamation and reuse for cooling tower applications are covered in more detail in Chapter 24.

Water Quality Factors

Water quality requirements tend to be industry specific because, in many cases, if the chemistry of reclaimed water differs from other source waters, industrial process performance can be impacted. Water quality concerns for industrial water applications include scaling, corrosion, biological growth, fouling, and foaming. TDS, ammonia, and metals can increase corrosion rates. In addition, protection of worker safety may be necessary to minimize exposure to aerosols containing toxic volatile organic compounds or biological pathogens. A summary of water quality issues unique to industrial applications is given in Table 1.7.

Cooling Tower Make-up Water

For industries such as electric power generating stations, oil refining, and many types of chemical and metal plants, one-quarter to more than

TABLE 1.7. **Summary of Water Quality Issues of Importance for Industrial Water Reuse.**

Water Quality Parameter	Industrial Reuse Concern	Treatment Alternatives
Residual organics	Bacterial growth, microbial fouling on surfaces, foaming in process waters	Carbon adsorption, ion exchange
Ammonia	Forms combined chlorines with lower disinfection effectiveness, causes corrosion, promotes microbial growth	Nitrification, nutrient removal, ion exchange, air stripping
Phosphorus	Scale formation, algal growth, biofouling of process equipment	Biological nutrient removal, chemical precipitation, ion exchange
TSS	Deposition in materials, microbial growth	Filtration, microfiltration
TDS	Corrosion, scale formation	Blending, reverse osmosis
Dissolved minerals: Calcium, magnesium, iron, and silica	Scale formation	Softening, ion exchange, reverse osmosis

one-half of a facility's water use may be cooling tower make-up water. Because a cooling tower normally operates as a closed-loop system isolated from the process, it can be viewed as a separate water system with its own specific set of water quality requirements which are largely independent of the particular industry (State of California, 1979).

There are significant variations among large industrial cooling systems. Once-through-non-contact cooling is often used at large power facilities or refineries near the ocean, whereas direct contact cooling is used when inert material being processed, as in quenching in the primary metals industry. Non-contact recirculating cooling is used at large inland industries with limited water resources.

Once-through-cooling transfers process heat to water which is then wasted. Recirculating systems go one step further and transfer the heat from the warmed water to the air so that the water can be re-used to absorb process heat. The heat is transferred from water to air primarily through evaporation. Cool, dry outside air is pulled into the sides of the tower, at a rate of 600–750 m^3 air per m^3 of recirculating water, up through and out the top by a rotating fan. Warmed water from an industrial heat exchanger is pumped into the top of the tower and allowed to descend through the upflowing air stream. Packing inside the tower breaks the water into droplets to allow efficient airwater contact. A fraction of the water evaporates and leaves the tower as vapor. The cooled water collects at the base of the tower, and is again available for use in process (Burger, 1979; State of California, 1980).

Industrial cooling tower operations face four water quality problems: (1) scaling, (2) corrosion, (3) biological growth, and (4) fouling in heat exchanger and condensers. In most cases, disinfected secondary effluent is supplied to non-critical, once-through cooling. For recirculating cooling tower operation, additional treatment generally includes lime clarification, alum precipitation, or ion exchange. Internal chemical treatment involves the addition of acid for pH control, biocide, scale, and biofoul inhibitors. In many cases, the water quality requirements for the use of reclaimed municipal wastewater as a cooling water source are identical to constraints associated with the use of fresh water. More detailed discussions on the use of reclaimed municipal wastewater as cooling water in a nuclear power plant can be found in Chapter 24 by Blackson and Moreland.

WATER REUSE FOR URBAN APPLICATIONS

Another application for reuse of reclaimed wastewater is to satisfy secondary water requirements in urban areas. The development of dual distribution systems (one line for domestic water supply and the other for reclaimed water) is a growing practice worldwide, particularly in areas with high rates of urban water usage. Other urban water reclamation options include subpotable uses, such as for recreational lakes, parks and playgrounds, and toilet flushing. Examples of sub-potable applications include: water hazards on golf courses; ponds and lakes for swimming; fishing and boating; and creation of wetlands as wildlife habitats.

Urban water reuse systems are derived from secondary treatment processes supplemented with sand filtration and high-dose chlorination. For stream restoration and flow augmentation, chemical coagulation, filtration, and ozonation are effective for maintaining acceptable aesthetic water quality in conjunction with control of foam and prevention of excessive hatches of waterborne insects, such as mosquitoes. The reclaimed water quality must be appropriate for support of indigenous flora and fauna. Increased nutrient levels in the reclaimed wastewater may give rise to algal blooms that can cause eutrophic conditions and non-aesthetic conditions in ponds and lakes. Ponds used for fishing must be examined for constituents that can bioaccumulate in fish and biomagnify in the food chain. Wetlands provide water quality enhancement by natural assimilation processes as long as the quality of the applied reclaimed wastewater is suitable for plant growth. As with landscape irrigation, the salinity of the water is an important consideration.

Dual Distribution Systems

Dual distribution systems segregate the potable water supply from the nonpotable system. Thus, a high quality system can be provided for drinking water and in-building applications, and reclaimed water is provided for use in irrigation and fire protection (American Water Works Association, 1994). Water distribution lines for reclaimed water parallel pipelines used for domestic water. When retrofitting an existing system, the high cost of installing a second distribution system can often render a project economically unfeasible; however if the dual system is installed as part of a new development the economics are more favorable.

Dual distribution systems can be developed in two ways. One approach is to construct a city-wide system in which the wastewater is returned to a municipal wastewater treatment plant for processing before being redistributed to the population. The second approach is using small-scale individual systems where ''gray water'' from sinks, bathtubs, showers, and other sources of dilute non-fecal wastewater is redistributed to the system. As with the applications of reclaimed water discussed earlier, protection of public health is a prime concern in implementation of dual distribution system water reuse.

In the United States, many cities have invested in dual distribution systems to provide water for residential irrigation and reserve higher quality water for potable requirements. It is interesting to note that dual distribution systems have been used for almost 70 years; the first U.S.A. system was installed in 1926 to provide secondary water for Grand Canyon Village in Arizona. More recently, the Irvine Ranch Water District (IRWD) in Southern California initiated the use of a dual distribution system to supply reclaimed wastewater for secondary uses in highrise office buildings. In addition, a dual distribution system is available for all new developments within the district. It is anticipated that about 13 mgd (49,000 m³/d) of reclaimed water will be provided to IRWD customers by the year 2000. Details on the IRWD can be found in Chapter 21 by Young et al.

Project Apricot is an example of a dual distribution system in Altamonte Springs, Florida (Jackson and Cross, 1993). The reclaimed water is delivered to properties through an underground system entirely separate from the potable water system. The reclaimed water is used for car washing, lawn irrigation, or irrigation of vegetables that require peeling or cooking prior to consumption. Other uses are fire protection in certain areas of the city, fountains, and waterfalls and toilet flushing in commercial buildings. The plant was designed by modifying an existing wastewater treatment facility. With an average daily capacity of 12.5 mgd (0.55 m³/s). The plant has a peak daily capacity of 25 mgd (1.10 m³/s). With the exception of nitrogen and phosphorus, which are beneficial nutrients in irrigation systems, the reclaimed water meets current drinking water standards. The reclaimed water is odorless, non-staining, and has a turbidity level below 1 NTU. Fecal coliform and virus levels are consistently below detectable limits.

A critical component of dual distribution systems is the prevention of cross connections between potable and nonpotable water lines (American Water Works Association, 1994). Frequently, local plumbing codes specify methods for identifying potable and nonpotable water pipelines. In general, water supply lines must be clearly marked as to whether they carry domestic or reclaimed water. Another preventive measure is the addition of a harmless coloring agent to the reclaimed water supply to alert users of its source. Safety measures that have been used successfully for nonpotable water reuse applications include (1) complete segregation of potable storage and distribution systems, (2) use of color coded labels to distinguish potable and nonpotable pipes, (3) backflow prevention devices, (4) periodic testing with tracer dyes to detect the occurrence of cross contamination in potable supply lines, and (5) irrigation during off hours to further minimize the potential for human contact.

Applications of Urban Wastewater Reclamation and Reuse in Japan

In Japan, several urban water reuse systems have been implemented successfully. Reclaimed wastewater is used in more than 650 locations for toilet flushing within business buildings and apartment complexes. Dual distribution systems are also used to provide reclaimed wastewater for irrigation of ornamental gardens and water sources for streams in parks and playgrounds. Similar, but smaller scale projects have been implemented in Hong Kong and Singapore (Asano et al., 1996). A summary of wastewater reclamation and reuse in the Tokyo Metropolitan District is given in Table 1.8. The largest consumption of reclaimed wastewater is for stream augmentation and industrial water.

Reclaimed wastewater is used in the Shinjuku District as a supplemental in-building water source. Treated wastewater from the Ochiai Municipal Wastewater Treatment Plant is pumped to a reclaimed wastewater storage and distribution system that allows for supplemental disinfection. The disinfected reclaimed water is distributed to 19 high-rise buildings for toilet flushing. The total length of the reclaimed water distribution pipe network is 2.9 km (1.8 miles). Each building has a reclaimed water tank, and the reclaimed water is pumped to a roof-top tank for distribution to maintain 0.5 kg/cm² (7 psi) pressure within each building (Maeda et al., 1995).

Another urban reuse application in Japan provided a mechanism to

TABLE 1.8. Comparison of Annual Quantity of Wastewater Reclaimed for Various Applications in the Tokyo Metropolitan Districts[a].

Reclaimed Water Use	Volume, 1,000 m³/year
Stream augmentation	12,370
Industrial water	8,835
Toilet flushing	970
Refuse incineration plant	386
Passenger train washing	111
Dust control	6

[a]After Maeda et al., 1995; Asano et al., 1996.

restore the Nobidome Stream. Due to a headwater diversion project that began in 1976, the stream had degraded to the quality of a combined open sewer and refuse dump. The streamflow was augmented using 15,000 m³/d (4 mgd) of filtered secondary effluent with partial phosphorus removal from the Tama-Joryu Wastewater Treatment Plant. Additional treatment was added in 1989 and included chemical coagulation (10–15 mg/L polyalminium chloride) and ozonation (5–10 mg/L) processes were added in 1989 for color and odor control. The net effect of the project was a dramatic improvement in water quality (Asano et al., 1996).

WATER REUSE FOR GROUNDWATER RECHARGE

Groundwater aquifers provide natural mechanisms for storage and subsurface transmission of reclaimed water. The use of groundwater recharge as a water reuse application can satisfy multiple objectives. Groundwater recharge with reclaimed wastewater is an approach to wastewater reuse that results in the planned augmentation of groundwater resources (Bouwer, 1991; Crook, Asano, and Nellor, 1990; National Research Council, 1994). The purposes of artificial recharge of groundwater include: (1) arresting the decline of groundwater levels due to excessive groundwater withdrawals, (2) protection of coastal aquifers against saltwater intrusion from the ocean, and (3) to store surface water, including flood or other surplus water, and reclaimed wastewater for future use. Groundwater recharge is also incidentally achieved in land treatment and disposal of municipal and industrial wastewater via percolation and infiltration.

There are several advantages to storing water underground: (1) the cost of artificial recharge may be less than the cost of equivalent surface reservoirs, (2) the aquifer serves as an eventual distribution system and may eliminate the need for surface pipelines or canals, (3) water stored in surface reservoirs is subject to evaporation, potential taste and odor problems due to algae and other aquatic productivity, and to pollution;

these may be avoided by underground storage, (4) suitable sites for surface reservoirs may not be available or environmentally acceptable, and (5) the inclusion of groundwater recharge in a water reuse project may also provide psychological and aesthetic secondary benefits as a result of the transition between reclaimed water and groundwater. This aspect is particularly significant when a possibility exists in the wastewater reclamation and reuse plan to augment substantial portions of potable water supplies. Because of the importance of groundwater recharge in the future of wastewater reclamation, recycling, and reuse, this subject is covered in more detail in Chapters 20 and 23.

Techniques of Groundwater Recharge

There are two options by which groundwater can be recharged with reclaimed wastewater: (1) surface spreading or percolation, and (2) direct injection. In surface spreading, reclaimed municipal wastewater percolates from spreading basins through the unsaturated zone. The advantage of groundwater recharge by surface spreading is that groundwater supplies may be replenished in the vicinity of metropolitan and agricultural areas where groundwater overdraft is severe, with the added benefits of the treatment effect of soils and transporting facilities of aquifers. In direct injection, treated wastewater is pumped under pressure directly into the groundwater zone, usually into a well-confined aquifer. Groundwater recharge by direct injection is practiced, in most cases, where groundwater is deep or where the topography or existing land use makes surface spreading impractical or too expensive. This method of groundwater recharge is particularly effective in creating freshwater barriers in coastal aquifers against saltwater intrusion.

Both in surface spreading and in direct injection, locating the extraction wells as great a distance as possible from the spreading basins or the injection wells increases the flow path length and residence time of the recharged groundwater. The separation of injection and withdrawal in-space and in-time contribute to the mixing of the recharged water and the other aquifer contents, and the loss of identity of the recharged water originated from wastewater. Groundwater recharge operations in Los Angeles County are discussed in Chapter 20 by Hartling and Nellor. Groundwater recharge at the Orange County Water District in California is presented in Chapter 23 by Mills et al.

Proposed Groundwater Recharge Regulations

The use of reclaimed wastewater for groundwater recharge, particularly in groundwater basins that serve as sources of domestic water supply is associated with a broad spectrum of health concerns (National Research Council, 1994). Water extracted from a groundwater basin for domestic use must be of acceptable physical, chemical, microbiological, and radiological quality. The major concerns governing the acceptability of groundwater recharge projects are that adverse health effects could result from the introduction of pathogens or trace amounts of toxic chemicals into groundwater that is eventually consumed by the public. In light of uncertainties over potential long-term health effects from exposure to trace levels of organic and inorganic contaminants, it is important to ensure the absence of toxic compounds in recharged groundwater. A source control program to limit discharges of harmful constituents to the sewer system must be an integral part of any recharge project. Extreme caution is warranted because of the difficulty in restoring a groundwater basin once it is contaminated. Additional cost would be incurred if groundwater quality changes resulting from recharge necessitated the treatment of extracted groundwater and/or the development of additional water sources.

The level of municipal wastewater treatment necessary to produce reclaimed water suitable for groundwater recharge depends upon the groundwater quality objectives, hydrogeologic characteristics of the groundwater basin, and the amount of reclaimed water and percentage of reclaimed water applied. The major considerations are: (1) the total quantity and type of water available for recharge on an annual basis, (2) the size of the groundwater basin and probability of dilution with natural groundwaters, (3) soil types, (4) depth to groundwater, (5) method of recharge, and (6) the length of time the reclaimed water is retained in the basin prior to withdrawal for domestic use. These factors must be evaluated in establishing criteria for groundwater recharge with reclaimed wastewater (Odendaal and Hattingh, 1988; State of California, 1990).

In the United States, federal requirements for groundwater recharge in the context of wastewater reuse have not been established to date. In general, wastewater reuse requirements for groundwater recharge are regulated by individual state agencies on a case-by-case determination. In the 1990s, wastewater reuse regulations were proposed in California targeted at groundwater recharge applications. The proposed regulations reflect a cautious attitude toward short-term and long-term health concerns (State of California, 1992). The regulations rely on a combination of controls intended to maintain a microbiologically and chemically safe groundwater recharge operation. No single method of control is universally effective for control of the transmission and transport of contaminants of concern into and through the environment. Therefore, a combination of source control, wastewater treatment processes, treatment standards, recharge methods, recharge area, extraction well proximity, and monitoring wells are all specified (Hultquist, Sakaji, and Asano, 1991).

The method by which reclaimed wastewater is applied for groundwater recharge and the "project category" identify a set of conditions that constitute an acceptable project. An equivalent level of perceived risk is inherent in each project category when all conditions are met and enforced. The main concerns governing the acceptability of groundwater recharge projects with reclaimed municipal wastewater are that adverse health effects could result from the introduction of pathogens or trace amounts of toxic chemicals into groundwater that is eventually consumed by the public. A summary of the proposed State of California regulations is presented in Table 1.9.

Microbiological Considerations

Of the known waterborne pathogens, enteric viruses have been considered most critical in wastewater reuse because of the possibility of contracting disease from exposure to relatively low doses and the difficulty of routine examination of reclaimed wastewater for their presence. Thus, essentially virus-free effluent via the full treatment process (primary/secondary, coagulation/flocculation, clarification, filtration, and disinfection) is deemed necessary for reclaimed wastewater applications with higher potential exposures, e.g., spray irrigation of food crops eaten uncooked, or most of groundwater recharge applications (Project Categories I, II, and IV in Table 1.9).

The wastewater treatment requirements shown in Table 1.9 are designed to provide assurance that reclaimed water is essentially pathogen-free prior to extraction from the groundwater. The passage of treated wastewater through an unsaturated zone of at least 10 ft (3 m) provides additional removal of organic constituents and pathogens in treated effluents (Fujita, Ding, and Reinhard, 1996; Quanrud et al., 1996; Wilson et al., 1995). At infiltration rates below 16 ft/day (5 m/day) in sands and sandy loams, the rates of virus removal can be correlated to infiltration rates and approximately 99.2% removal can be achieved during passage through 10 ft (3 m) depth soils. The total removal of enteric viruses by a combination of treatment, unsaturated zone infiltration, and horizontal separation (retention time in groundwater) is estimated to be between 13 and 17 log reduction (Asano and Mujeriego, 1988; State of California, 1989).

Trace Organics Removal

The proposed groundwater recharge regulations are designed to control the concentrations of enteric viruses, organics of municipal wastewater origin as well as anthropogenic chemicals that have an impact on health when present in trace amounts. Thus, the purpose of the dilution require-

TABLE 1.9. Proposed Requirements for Groundwater Recharge with Reclaimed Municipal Wastewater (State of California, 1992).

Treatment and Recharge Site Requirements	Project Category			
	Surface Spreading			Direct Injection
	I	II	III	IV
Level of wastewater treatment:				
Primary/secondary	X	X	X	X
Filtration	X	X		X
Organics removal	X			X
Disinfection	X	X	X	X
Max. percent reclaimed wastewater in extracted well water	50	20	20	50
Depth to groundwater ft (m) at initial percolation rate of:				
2 in/min (50 mm/min)	3 (1)	3 (1)	6 (2)	NA[a]
3 in/min (80 mm/min)	6 (2)	6 (2)	15 (5)	NA[a]
Retention time underground (months)	6	6	12	12
Horizontal separation[b] ft (m)	150 (50)	150 (50)	300 (100)	300 (100)

[a] Not applicable.
[b] From the edge of the groundwater recharge operation to nearest potable water supply well.

ments and the organics removal specified in project categories I and IV in Table 1.9 is to limit the average concentrations of unregulated organics in extracted groundwater affected by the groundwater recharge operation. The concentration of unregulated and unidentified trace organics is of great concern since other constituents and specific organics are dealt with through the established maximum contaminant levels and action levels developed by the California Department of Health Services. A variety of approaches have been used to characterize and quantify trace organics in wastewater (Fujita et al., 1996; Levine, Tchobanoglous, and Asano, 1985; Manka and Rebhun, 1982; Quanrud et al., 1996). Significant progress has been made using gas chromatography/mass spectrometry, nuclear magnetic resonance, and functional separation of hydrophobic, hydrophilic, and volatile constituents (Fujita et al., 1996). Currently, the total organic carbon (TOC) level is used to provide a benchmark of the organic content and to determine the removal efficiency of treatment systems in practice.

Dilution of reclaimed wastewater with surface water or native groundwater is effective for reducing the levels of trace contaminants within the recharged water. When reclaimed water makes up more than 20% of the water reaching any extraction well for potable water supply, treatment to remove organics must be provided. However, because TOC is a nonspecific measure of organic content, it is not possible to establish a TOC standard that protects public health. As a control measure, the maximum TOC level in extracted water must be below 1 mg/L. In addition, the maximum amount of reclaimed wastewater in an extraction well is restricted to 20–50% for surface spreading (Category I), and 0–50% for direct injection (Category IV).

The recommended reclaimed water TOC levels for surface spreading and direct injection systems are compared in Figure 1.5 as a function of the percent of reclaimed water at the extraction well. As shown, a higher degree of TOC removal is necessary for direct injection systems. In fact, direct injection projects would have to achieve a 70% TOC reduction to compensate for the lack of unsaturated zone in the overall soil-aquifer treatment capability (State of California, 1992).

Inorganic Chemicals

Inorganic chemicals, with the exception of nitrogen in its various forms, are adequately controlled by meeting all maximum contaminant limits (MCLs) in the reclaimed wastewater. By limiting the concentration of total nitrogen in the reclaimed water, detrimental health effects such as methemoglobinemia can be controlled. If adequate nitrogen removal cannot be achieved by treatment processes or passage through unsaturated zone, treat-

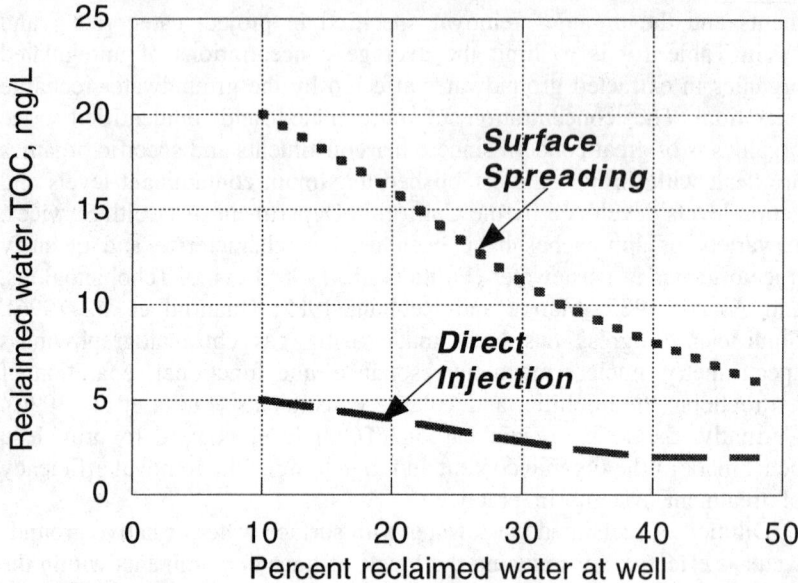

Figure 1.5. Maximum TOC concentrations in reclaimed wastewater to achieve 1 mg/L in extracted well water.

ment can be provided at the wellhead to meet the allowable total nitrogen concentration of 10 mg/L as N (Odendaal and Hattingh, 1988).

EVOLUTION OF POTABLE WATER REUSE

Increasing problems of water shortage, coupled with advancements in technology and treatment reliability have provided a sound basis for development of wastewater reclamation and reuse applications. Currently, direct nonpotable water reuse is the dominant planned reuse venue for supplementing public water supplies.

Health and safety concerns have cultivated a cautious attitude toward crossing the threshold from nonpotable to potable reuse. Nevertheless, some communities are developing or implementing plans for direct or indirect potable reuse where no readily attainable options exist for supplemental freshwater supplies. Although the quantities involved in potable reuse are small, the technological and public health interests are the greatest and hence considerable debate has been directed toward formulation of potable reuse criteria and standards (Asano and Tchobanoglous, 1995; Okun, 1996).

Planned indirect potable reuse systems in operation today include such groundwater recharge operations as in the Whittier Narrows groundwater recharge site in Los Angeles County, California (Nellor et al., 1985; Nellor et al., 1985) and in El Paso, Texas, where reclaimed waters are injected into groundwater aquifers (Knorr et al., 1988). Another example is the Occoquan Reservoir in northern Virginia (Robbins, 1984). Effluent from the 15 mgd (0.65 m³/s) Manassas advanced wastewater treatment plant operated by the Upper Occoquan Sewage Authority is directly discharged into the Occoquan Reservoir, a principal drinking water reservoir for more than 660,000 people. More discussions on Windhoek, Namibia are found in Chapter 25 and other examples are found in Chapter 27 by McEwen.

MICROBIAL HEALTH RISK ESTIMATION FOR WASTEWATER REUSE

Despite a long history of wastewater reuse in many parts of the world, question of *safety* of wastewater reuse is still difficult to define and *acceptable* health risks have been hotly debated. When treated municipal wastewater effluents are used in urban environments, considerable health concerns may be justified when there is a strong likelihood of direct human contact. The control of enteric viruses is a major health concern, particularly in industrialized countries with high health standards. In this section, a comparative assessment of the safety of wastewater reuse is discussed based on work by Asano and associates (Asano, Richard, Crites, and Tchobanoglous, 1992; Tanaka, Asano, Schroeder, and Tchobanoglous, 1998) on enteric virus risk assessment. Other chapters dealing with health risk assessment include Chapters 11,12, and 15.

ENTERIC VIRUS CONCENTRATION IN WASTEWATER

To evaluate health risks associated with exposure to enteric viruses, data from two different sources were used: virus levels in unchlorinated activated sludge effluents and virus levels in chlorinated tertiary filtration effluents. The database was obtained from reports published by water and wastewater agencies in California. The enteric virus data included 424 unchlorinated secondary effluent samples in which 283 samples (67%) were virus positive and 814 chlorinated tertiary (filtered) effluent samples with 7 positive samples (1%). Quantifying the virus concentration [expressed as viral unit (vu) per liter] in the treated effluent is the first step for estimating the risk of virus infection due to exposure to reclaimed municipal wastewater. The statistical model used was the lognormal distribution and the goodness-of-fit of the hypothesized distribution was evaluated using the nonparametric Kolmogorov-Smirnov test.

Virus concentrations vary over a wide range in unchlorinated secondary effluents. The geometric mean values of unchlorinated secondary effluent samples ranged from 0.0002 to 2.3 vu/L and 90 percentile concentrations ranged from 0.34 to 29 vu/L. It should be noted that detection limits for quantifying virus concentrations are about 0.01 vu/L.

VIRUS RISK ANALYSIS

To assess potential risks associated with the use of reclaimed wastewater, exposure scenarios were developed for landscape irrigation, spray irrigation of food crops, recreational impoundments, and groundwater recharge (Asano et al., 1992). The beta-distributed probability model based on Haas' work was chosen for use in risk calculations because it best represented the frequency distribution of virus infection. Infectious models based on echovirus 12, and poliovirus 1 and 3 were used by Asano (Asano et al., 1992), and the rotavirus model based on Rose and Gerba was used in the study by (Tanaka et al., 1993).

Results of the annual risk calculations are shown in Table 1.10, using virus concentrations of 0.01 vu/L and 1.11 vu/L from the chlorinated tertiary filtration effluents which are a reasonable estimate of the limit of detection for enteric viruses and the maximum concentration found in tertiary effluents in California.

The estimates of risk of infection presented in Table 1.10 show the range of risks associated with annual exposures encountered in different wastewater reuse situations. The overall probability of infection due to ingestion of viruses is a combination of virus removal and inactivation by wastewater treatment, die-off in the environment, and dose-response. For each exposure scenario presented, the range of risks covers two to three orders of magnitude. This reflects the differences in infectivity among different viruses. For groundwater recharge with reclaimed wastewater, with an effluent virus concentration of 0.01 vu/L, the annual risk of infection ranges from 5×10^{-10} to 5×10^{-11}. When virus concentration is increased to 1.11 vu/L which is the maximum virus concentration found in the chlorinated tertiary effluent, the risk of infection increased by two to three order of magnitude (6×10^{-8} and 5×10^{-9}). Similar trends are noted in the other exposure scenarios.

Of the remaining three categories, the highest risk associated with use of reclaimed wastewater is for recreational impoundments where water contact sports such as swimming take place. In all cases, regardless of the initial virus concentration, the use of reclaimed wastewater for unrestricted recreational impoundments results in the most extensive potential exposure to enteric viruses. The relatively high probability of infection is attributed to the fact that the risk calculations did not account for dilution or virus die-off in the environment to evaluate a worst case scenario.

TABLE 1.10. Annual Risk of Contracting at Least One Infection from Exposure to Reclaimed Wastewater at Two Different Enteric Virus Concentrations (Asano et al., 1992).

Virus	Exposure Scenarios			
	Landscape Irrigation for Golf Courses	Spray Irrigation for Food Crops	Unrestricted Recreational Impoundments	Groundwater Recharge
Maximum enteric virus concentration of 1.11 vu/L in chlorinated tertiary effluent				
Echovirus 12	1E-03	4E-06	7E-02	6E-08
Poliovirus 1	3E-05	2E-07	3E-03	5E-09
Poliovirus 3	3E-02	1E-04	8E-01	2E-08
Minimum enteric virus concentration of 0.01 vu/L in chlorinated tertiary effluent[a]				
Echovirus 12	9E-06	4E-08	7E-04	5E-10
Poliovirus 1	3E-07	1E-09	2E-05	5E-11
Poliovirus 3	2E-04	1E-06	2E-02	2E-10

[a] The limit of detection.

49

Landscape irrigation for golf courses posed the second highest level of exposure to reclaimed wastewater, and spray irrigation of food crops ranked third being two orders of magnitude lower in relative risk. The lower risks of infection associated with spray irrigation of food crops and groundwater recharge can be attributed to environmental factors such as use area controls. In both cases, the exposure scenarios considered virus die-off in the environment. These risk estimates, however, do not account for the variability of enteric viruses in the environment. Seasonal fluctuations in the endemic virus populations will affect the quantity and species present in the wastewater at any given time (see Chapter 11).

Daily and seasonal water quality variations, flow fluctuations, and process variability in wastewater treatment can result in variations in effluent quality. Based on existing database, it is reasonable to assume chlorinated tertiary effluent can produce a virus free effluent 99% of the time. Thus, the permit limit of virus-free effluent is expected to be exceeded 1% of the time, or three to four times a year. The question in risk management in wastewater reuse, then, becomes one of determining whether or not the presence of enteric viruses in the concentration range of 2 to 111 vu/100 L in approximately 1% of tertiary-treated reclaimed wastewater poses a significant risk to public health (Asano et al., 1992). Further research is needed to link treatment process reliability with risk reduction in wastewater reclamation and reuse.

The goal of essentially virus-free reclaimed wastewater contained in California's *Wastewater Reclamation Criteria* should not be interpreted to mean that the practice of using such water is risk-free. As Table 1.10 clearly shows there is always some risk of infection due to exposure to reclaimed wastewater. However, this does not mean that the practice of wastewater reclamation and reuse is unsafe. The *safety* of wastewater reclamation and reuse practice is defined by the acceptable level of risks developed by the regulatory agencies responsible for risk management and endorsed by the public. More detailed discussions are contained in Chapter 14 by Crook and Chapter 15 by Sakaji and Funamizu.

FUTURE OF WATER REUSE

Significant progress has been made with respect to developing sound technical approaches to producing a high quality and reliable water source from reclaimed wastewater. Continued research and demonstration efforts will result in additional progress in the development of water reuse applications. Some key topics include: assessment of health risks associated with trace contaminants in reclaimed water; improved monitoring techniques to evaluate microbiological quality; optimization of treatment trains; improved removal of wastewater particles to increase disinfection effectiveness; the application of membrane processes in production of reclaimed water; the effect of reclaimed water storage systems on water quality; evaluation of the fate of microbiological, chemical, and organic contaminants in reclaimed water; and the long-term sustainability of soil-aquifer treatment systems. A key to promoting the implementation of water reuse is the continued development of cost-effective treatment systems.

To date the major emphasis on wastewater reclamation and reuse has been for nonpotable applications such as agricultural and landscape irrigation, industrial cooling, and in-building applications such as toilet flushing. While direct potable reuse of reclaimed municipal wastewater is, at present, limited to extreme situations, it has been argued that there should be a single water quality standard for potable water. If reclaimed water can meet this standard, it should be acceptable regardless of the source of water. While indirect potable reuse by groundwater recharge or surface water augmentation has gained support, some concerns still remain regarding trace organics, treatment and reuse reliability, and particularly, public acceptance.

Considering that modern wastewater reclamation and reuse started in the 1960s (see Figure 1.2), remarkable progress has been made. A cautious and judicious approach is warranted to avoid potential health consequences that could result if a water reuse project is not successful. In addition, the importance of public confidence cannot be underestimated. This book provides state-of-the-art information on current and evolving wastewater reclamation and reuse practices. It is clear that reclaimed wastewater is a viable water resource. Continued research and development will provide a sound scientific basis for crossing the threshold to direct potable reuse, when necessary. As technology continues to advance and the reliability of wastewater reuse systems is demonstrated, applications for wastewater reclamation, recycling, and reuse will continue to expand as a vital element in integrated water resources management twenty-first century.

SUMMARY AND CONCLUSIONS

The successful implementation of wastewater reuse options into a water resources management program requires careful planning; economic and financial analyses; and effective design, operation and management of wastewater reclamation, storage, and distribution facilities. Technologies for wastewater reclamation and purification have developed to the point where it is technically feasible to produce water of almost any quality and advances continue to be made. Current water reclamation strategies incorporate multiple measures to minimize health and environmental risks associated with various reuse applications. A combination of source control, advanced treatment process flowschemes, and other engineering controls provides a sound basis for increased implementation of water reuse applications. The feasibility of producing reclaimed water of a specified quality to fulfill multiple water use objectives is now a reality due to the progressive evolution of technologies and risk assessment procedures. However, the ultimate decision to harvest reclaimed wastewater is dependent on economic, regulatory, and public policy factors reflecting the demand and need for dependable water supply.

REFERENCES

American Water Works Association (1994). Dual Water Systems. *AWWA Manual, M24.* Denver, Colorado.

Angelakis, A. N. and Spyridakis, S. V. (1995). The Status of Water Resources in Minoan Times: A Preliminary Study. in A. N. Angelakis, A. Issar, and O. K. Davis (Eds.), *Diachronic Climatic Impacts on Water Resources in Mediterranean Region.* Springer-Verlag, Heidelberg, Germany.

Argo, D. G. and Rigby, M. G. (1986). Water Reuse—What's the Big Deal? In *Proceedings of the 1986 Conference,* American Water Works Association. Denver, Colorado.

Asano, T., Leong, L. Y. C., Rigby, M. G., and Sakaji, R. H. (1992). Evaluation of the California Wastewater Reclamation Criteria Using Enteric Virus Monitoring Data. *Water Science and Technology, 26*(7–8), 1513–1524.

Asano, T. and Levine, A. D. (1995). Wastewater Reclamation, Recycling, and Reuse: Past, Present and Future. *Water Science and Technology, 33*(10–11), 1–14.

Asano, T., Maeda, M., and Takaki, M. (1996). Wastewater Reclamation and Reuse in Japan: Overview and Implementation Examples. *Water Science and Technology, 34*(11), 219–226.

Asano, T. and Mills, R. A. (1990). Planning and Analysis for Water Reuse Projects. *Journal of the American Water Works Association,* 38–47.

Asano, T. and Mujeriego, R. (1988). Pretreatment for Wastewater Reclamation and Reuse. *Pretreatment in Chemical Water and Wastewater Treatment,* edited by H. H. Hahn and R. Klute, Springer-Verlag, Berlin, Germany.

Asano, T. and Pettygrove, G. S. (1987). Using Reclaimed Municipal Wastewater for Irrigation. *California Agriculture, 41*(3 and 4), 15–18.

Asano, T., Richard, D., Crites, R. W., and Tchobanoglous, G. (1992). Evolution of Tertiary Treatment Requirements in California. *Water Environment and Technology, 4*(2), 36–41.

Asano, T. and Tchobanoglous, G. (1995). Drinking Repurified Wastewater. *Journal of Environmental Engineering (ASCE), 121*(8).

Ayers, R. S. and Westcott, D. W. (1985). *Water Quality for Agriculture.* FAO Irrigation and Drainage Paper 29 Rev. 1, Food and Agriculture, Organization of the United Nations, Rome, Italy.

Barty-King, H. (1992). *Water the Book, An Illustrated History of Water Supply and Wastewater in the United Kingdom.* Quiller Press Limited, London, England.

Bouwer, H. (1991). Role of Groundwater Recharge in Treatment and Storage of Wastewater for Reuse. *Water Science and Technology, 24*(9), 295–302.

Burger, R. (1979). *Cooling Tower Technology: Maintenance, Upgrading and Rebuilding.* Cooling Tower Institute, Houston, Texas.

City of San Diego (1992). *Total Resource Recovery Project Health Effects Study.* Western Consortium for Public Health, San Diego, California.

Crook, J., Asano, T., and Nellor, M. (1990). Groundwater Recharge with Reclaimed Water in California. *Water Environment and Technology, 49*(August).

Crook, J., Okum, D. A., Pincince, A. B., and Camp Dresser and McKee, I. (1994). *Water Reuse.* (No. Project 92-WRE-1). Water Environment Research Foundation, Alexandria, Virginia.

Darby, J. L., Snider, K. E., and Tchobanoglous, G. (1993). Ultraviolet Disinfection for Wastewater Reclamation and Reuse Subject to Restrictive Standards. *Water Environment Research, 65*(3), 169–180.

Dean, R. B. and Lund, E. (1981). *Water Reuse: Problems and Solutions.* Academic Press, New York, New York.

Deis, G. C., Scherzinger, D., and Thornton, J. R. (1986). Reclaimed Water Irrigation to Eliminate Summer Discharge. *Journal Water Pollution Control Federation, 58*(11), 1034–1038.

Engineering-Science (1987). *Monterey Wastewater Reclamation Study for Agriculture.* Monterey Regional Water Pollution Control Agency, Pacific Grove, California.

European Communities Commission Directive (1991). Council Directive Regarding the Treatment of Urban Wastewater (91/271). *Official Journal of the European Communities, No. L of the 91.5.30,* 40–50, Brussels, Belgium.

Fair, G. M. and Geyer, J. C. (1954). *Water Supply and Wastewater Disposal.* John Wiley and Sons, New York, New York.

Fox, K. R. and Lytle, D. A. (1994). *Cryptosporidium:* The Milwaukee Experience and Relevant Research. In *American Water Works Association Conference,* pp. 505–516, June 19–23. New York, New York.

Fujita, Y., Ding, W.-H., and Reinhard, M. (1996). Identification of Wastewater Dissolved Organic Carbon Characteristics in Reclaimed Wastewater and Recharged Groundwater. *Water Environment Research, 68*(5), 867–876.

Gerba, C. P., Rose, J. B., Haas, C. N., and Crabtree, K. D. (1996). Waterborne Rotavirus: A Risk Assessment. *Water Research, 30*(12), 2929–2940.

Haas, C. N. (1983). Estimation of Risk Due to Low Doses of Microorganisms: A Comparison of Alternative Methodologies. *American Journal of Epidemiology, 118*(4), 573.

Hultquist, R. H., Sakaji, R. H., and Asano, T. (1991). Proposed California Regulations for Groundwater Recharge with Reclaimed Municipal Wastewater. In *American Society of Civil Engineers Environmental Engineering Proceedings,* 759–764. *Specialty Conference/EE Div/ASCE.* New York, New York.

Isaacson, M., et al. (1987). *Studies of Health Aspects of Water Reclamation During 1974–1983 in Windhoek, South West Africa/Namibia.* (No. WRC-Report No. 38/1/87). Water Research Commission, Pretoria, South Africa.

Isaacson, M. and Sayed, A. R. (1988). Human Consumption of Reclaimed Water—The Namibian Experience. In *Water Reuse Symposium IV, Implementing Water Reuse, 1047, AWWA Research Foundation,* (p. 1047). Denver, Colorado.

Jackson, J. L. and Cross, P. (1993). Citrus Trees Blossom with Reclaimed Water. *Water Environment and Technology, 5*(2), 27–28.

Knorr, D. B., et al. (1988). Wastewater Treatment and Groundwater Recharge: A Learning Experience at El Paso, TX. In *Implementing Water Reuse, Proceedings of the Water Reuse Symposium IV, 211, AWWA Research Foundation.* Denver, Colorado.

Lauer, W. C. and Rogers, S. E. (1994). The Demonstration of Direct Potable Water Reuse: Denver's Pioneer Project. In *1994 Water Reuse Symposium Proceedings,* February 27–March 2, (pp. 779–798). Dallas, Texas.

LeChevallier, M. W. and Norton, W. D. (1992). Examining Relationships Between Particle Count and *Giardia, Cryptosporidium,* and Turbidity. *Journal of the American Water Works Association, 84*(12), 54–60.

Levine, A. D., Tchobanoglous, G., and Asano, T. (1985). Characterization of the Size Distribution of Contaminants in Wastewater: Treatment and Reuse Implications. *Journal of the Water Pollution Control Federation, 57*(7), 805–816.

Maeda, M., Nakada, K., Kawamoto, K., and Ikeda, M. (1995). Area-Wide Use of Reclaimed Water in Tokyo, Japan. In *The Proceedings of the Second International Symposium on Wastewater Reclamation and Reuse, Book 1.* International Association on Water Quality, October 17–20, Iraklio, Greece.

Manka, R. and Rebhun, M. (1982). Organic Groups and Molecular Weight Distribution in Tertiary Effluents and Renovated Waters. *Water Research, 15,* 399–403.

Metcalf & Eddy (1991). *Wastewater Engineering: Treatment, Disposal, and Reuse.* McGraw-Hill, Inc., New York, New York.

Metzler, D. F. et al. (1958). Emergency Use of Reclaimed Water for Potable Supply at Chanute, Kansas. *Journal of the American Water Works Association, 50*(8), 1021.

National Academy Press (1984). *The Potomac Estuary Experimental Water Treatment Plant.* A Review of the U.S. Army Corps of Engineers, Evaluation of the Operation, Maintenance and Performance of the Experimental Estuary Water Treatment Plant, Washington, D.C.

National Research Council (1996). *A New Era for Irrigation.* National Academy Press, Washington, DC.

National Research Council (1994). *Ground Water Recharge Using Waters of Impaired Quality.* National Academy Press, Washington, DC.

National Research Council (1982). *Quality Criteria for Water Reuse.* Panel on Quality Criteria for Water Reuse, National Academy Press, Washington, DC.

Nellor, M. H., Baird, R. B., and Smyth, J. R. (1985). Health Aspects of Groundwater Recharge. In T. Asano (Eds.), *Artificial Recharge of Groundwater.* Butterworth Publishers, Boston, Massachusetts.

Nellor, M. H., et al. (1985). Health Effects of Indirect Potable Water Reuse. *Journal of the American Water Works Association, 77*(7), 88.

Odendaal, P. E. and Hattingh, W. H. (1988). The Status of Potable Reuse Research in South Africa. In *Proceedings of Water Reuse Symposium IV, Implementing Water Reuse,* 1339, AWWA Research Foundation, Denver, Colorado.

Odendaal, P. E. and Vuuren, L. R. V. (1980). Reuse of Wastewater in South Africa— Research and Application. In *Proceedings of Water Reuse Symposium, Water Reuse— From Research to Application,* Vol. 2, 886, AWWA Research Foundation, 2 (p. 886). Denver, Colorado.

Okun, D. A. (1996). A Preference for Nonpotable Urban Water Reuse Over Drinking "Repurified Wastewater." *Journal of Environmental Engineering.*

Pettygrove, G. S. and Asano, T. (Eds.). (1985). *Irrigation with Reclaimed Municipal Wastewater—A Guidance Manual.* Lewis Publishers, Inc, Chelsea, Michigan.

Quanrud, D. M., Arnold, R. G., Wilson, L. G., Gordon, H. J., Graham, D. W., and Amy, G. L. (1996). Fate of Organics During Column Studies of Soil Aquifer Treatment. *Journal of Environmental Engineering,* 314–321.

Robbins, Jr., M. H. (1985). Operational and Maintenance of the UOSA Water Reclamation Plant (Upper Occoquan Sewage Authority of Virginia). *Journal Water Pollution Control Federation, 57*(12), 1122–1127.

Rose, J. B., Darbin, H., and Gerba, C. P. (1988). Correlations of Protozoa, *Cryptosporidium* and *Giardia* with Water Quality Variables in a Watershed. *Water Science and Technology, 20*(11/12), 271–276.

Rose, J. B. and Gerba, C. P. (1991). Use of Risk Assessment for Development of Microbial Standards. *Water Science and Technology, 24*(2), 29–34.

Sanitation Districts of Los Angeles County (1977). *Pomona Virus Study—Final Report.* State Water Resources Control Board, Sacramento, California.

Sheikh, B., Cort, R. P., Kirkpatrick, W. R., Jaques, R. S., and Asano, T. (1990). Monterey Wastewater Reclamation Study for Agriculture. *Research Journal of the Water Pollution Control Federation, 62*(3).

Solley, W. B., et al. (1983). *Estimated Use of Water in the United States in 1980.* Geological Survey Circular 1001, U.S. Geological Survey.

State of California (1978). *Wastewater Reclamation Criteria, An Excerpt from the California Code of Regulations, Title 22, Division 4.* Environmental Health, Department of Health Services, Berkeley, California.

State of California (1979). *Interim Guidelines for Economic and Financial Analyses of Water Reclamation Projects.* Prepared by Ernst and Ernst for State Water Resources Control Board, February, Sacramento, California.

State of California (1979). *Industrial Water Recycling.* Office of Water Recycling, March Sacramento, California.

State of California (1980). *Evaluation of Industrial Cooling Systems Using Reclaimed Municipal Wastewater: Applications for Potential Users.* California State Water Resources Control Board Office of Water Recycling, November, Sacramento, California.

State of California (1989). *Policy and Guidelines for Ground Water Recharge with Reclaimed Municipal Wastewater.* Sacramento, CA: Draft Edition, Department of Health Services and State Water Resources Control Board, Sacramento, California.

State of California (1990). *California Municipal Wastewater Reclamation in 1987.* State Water Resources Control Board, Office of Water Recycling, June, Sacramento, California.

State of California (1992). *Proposed Guidelines for Groundwater Recharge with Reclaimed Municipal Wastewater, and Background Information on Proposed Guidelines for Groundwater Recharge with Reclaimed Municipal Wastewater.* Interagency Water Reclamation Coordinating Committee and the Groundwater Recharge Committee, Sacramento, California.

Suarez, D. L. (1981). Relation Between pH_c and Sodium Adsorption Ratio (SAR) and Alternative Method of Estimating SAR of Soil or Drainage Waters. *Soil Science Society of America Journal, 45,* 469.

Tanaka, H., Asano, T., Schroeder, E. D., and Tchobanoglous, G. (1998). Estimating the Safety of Wastewater Reclamation and Reuse Using Enteric Virus Monitoring Data. Water Environment Research, Vol. 70, No. 1, pp. 39–51, January/February 1998.

U.S. Environmental Protection Agency (1986). *Environmental Regulations and Technology, The National Pretreatment Program.* (No. EPA/625/10-86/005). Office of Water Enforcement and Permits, Office of Water, July, Washington, DC.

U.S. Environmental Protection Agency (1992). *Guidelines for Water Reuse Manual.* (No. EPA/625/R-92/004), Washington, DC.

Wilson, L. G., Amy, G. L., Gerba, C. P., Gordon, H., Johnson, B., and Miller, J. (1995). Water Quality Changes During Soil Aquifer Treatment of Tertiary Effluent. *Water Environment Research, 67*(3), 371–376.

World Health Organization (1980). *Health Aspects of Treated Sewage Re-Use.* Report on a WHO Seminar, EURO Reports and Studies 42, Regional Office for Europe, Copenhagen, Denmark.

World Health Organization (1989). *Health Guidelines for the Use of Wastewater in Agriculture and Aquaculture.* Report of a WHO Scientific Group, Geneva, Switzerland.

Wright, R. R. and Missimer, T. M. (1995). Reuse. The U.S. Experience and Trend Direction. *Desalination and Water Reuse, 5/3,* 28–34.

Young, D. D. (1985). Reuse Via Rivers for Water Supply in Reuse of Sewage Effluent. In *Proceedings of the International Symposium Organized by the Institution of Civil Engineers,* October 30–31, Thomas Telford, London, U.K.

CHAPTER 2

Planning and Analysis of Water Reuse Projects

T HE scope of this chapter is the planning and analysis of water reuse projects. The objective is to provide a framework for planning and a discussion of issues that will assist in identifying most of the potential problems related to water reuse project planning and implementation.

General concepts of planning for water reclamation and reuse projects are discussed first. Then planning phases and the factors to be evaluated in each phase are described. Some engineering and marketing issues are also discussed. Finally, economic and financial analysis of water reuse projects are explained in detail. These analyses are essential in project planning and the thorough understanding of the principles are crucial. There are many factors that must be considered in project planning. Some are addressed in detail in other chapters. This chapter will provide a framework to integrate these factors into a systematic approach of project analysis.

OBJECTIVES OF WATER REUSE

Water reclamation and reuse can serve several possible objectives. The most prominent can be summarized as to provide:

(1) An incidental secondary benefit to the disposal of wastewater, primarily crop production by irrigation with effluent
(2) A water supply to displace the need for other sources of water

Richard A. Mills, Office of Water Recycling, California State Water Resources Control Board, 2014 T St., Sacramento, CA 94244-2120; Takashi Asano, Department of Civil and Environmental Engineering, University of California at Davis, Davis, CA 95616.

(3) A cost-effective means of environmentally sound treatment and disposal of wastewater
(4) A water supply to generate regional economic development
(5) A water supply for environmental enhancement

The first objective is perhaps the one most commonly encountered in the USA until the early decades of the 20th century. Wastewater may be put to use because of its convenient availability. A farmer may convert from dry land to irrigated farming because effluent is available nearby, producing higher, dependable yields and increasing the farmer's income. If there is enough value to the water, a wastewater agency may sell the reclaimed water, providing revenue to offset the costs of wastewater treatment and disposal. This is a common scenario in rural California or urban areas in agricultural regions [1]. While this objective is a water supply function, such projects are not approached from a water resources perspective. They are analyzed more as a financial opportunity for a wastewater agency or a prospective user of reclaimed water.

In recent decades the second objective has become the most critical driving force in the USA and many other countries, as described in Chapter 1. As freshwater resources become scarce or more expensive, reclaimed water may be recognized as an alternative water supply source to meet water demands in existence or projected in the future.

However, the third objective of pollution control has historically played an important role in water reclamation and reuse. For example, when effluent comprises a high percentage of stream flow that is a potable water source, expensive technology may be the only effective means of sufficiently treating wastewater before discharge. Land application of wastewater may be possible with cheaper technology and yield a useful by-product. Water reuse can be the cost-effective pollution control alternative.

Where economic development of a region is limited by water resources, water reuse may be promoted specifically for the economic benefits from the use of the reclaimed water. While the analysis of an economic development project should include alternative water supplies besides reclaimed water, the analysis would include measurement of net income generated by the economic activity stemming from the new water supply.

An example of environmental enhancement would be the restoration of a wetlands with a reclaimed water supply. Such a project would include an analysis of the optimum location and design of the wetlands as well as the optimum source of water.

Reflecting the experience and primary motivations in the USA, the emphasis of this chapter will be on the second and third objectives listed above, that is, water reuse planned as a water supply to offset alternative water development and as a cost-effective means of municipal wastewater pollution control.

MULTIPLE-PURPOSE PLANNING

Because of the single-purpose nature of many governmental agencies and programs, most water reclamation projects have been developed with one of the five objectives in mind. Unlike other wastewater treatment and disposal alternatives, however, water reclamation and reuse provides opportunities to address several problems with a potential solution. However, unless there is a multi-objective planning approach, many benefits of water reuse will be ignored, resulting in a pollution control or water supply alternative that would not be the optimal solution.

Setting the planning objective at the outset is necessary to establish the criteria for ultimate selection. It is also necessary to establish the scope of project analysis. Will the scope be confined to pollution control alternatives and their cost-effectiveness? Will water supply be the focus of the data gathering and analysis? It is important to understand that water reuse is not pursued as a purpose in itself. Water reuse is not an objective. It is a means to an objective. It is important, therefore, to outline the objectives and pursue the analysis accordingly. In addition, a multi-objective planning approach with respect to water reclamation should be strongly considered.

Multiple-purpose planning is best pursued if the participants in the planning process include all those that would have a stake in the outcome. Water supply and wastewater agencies are the two key participants. While one agency or another can take a planning lead, if several agencies can see a long-term benefit, they will be willing to contribute to the planning process and share in project financing. Without a team approach, the feasibility of a water reuse project can be severely impaired.

Thus, set the objectives, develop a team of agencies and interest groups to participate in the planning process, and analyze water reuse as one of several alternatives that would meet the objectives. Encourage a multi-objective approach. If water reuse is the cost-effective alternative, the groundwork will have been established for all relevant participants to willingly support and finance project implementation.

PLANNING CONCEPTS

The approach to planning taken in this chapter is to conduct a cost-effectiveness analysis to arrive at the optimum or recommended project. "Cost-effectiveness analysis" is defined here as an analysis to determine which alternative system will result in the minimum total resources costs over time to meet project objectives. Nonmonetary factors (e.g., social and environmental) are accounted for descriptively in the analysis to determine their significance and impacts.

One of the primary criteria in public works decision-making is to reduce monetary costs. Reflecting this, the most cost-effective alternative is the alternative determined to have the lowest present worth or equivalent annual value without overriding adverse nonmonetary costs and to realize at least identical minimum benefits as defined in project objectives. Alternatives analyzed should include taking no action and, as appropriate, nonstructural alternatives [2].

A common weakness in water reclamation planning is to take a singular viewpoint. This viewpoint is usually that of the agency proposing the project. The success of a water reclamation project involves the support of the public at large and the willingness of water users to accept reclaimed water. A project may also involve cooperation of several agencies. The cost-effectiveness must be conducted with these several viewpoints in mind.

In identifying the benefits of a project and its alternatives, the comparison must be with and without the project rather than before and after the project [3]. Some outcomes may occur after implementation of a project that would have occurred without the project in any case. One must separate the outcomes that are dependent on implementation of a project in order to assess the project's costs and benefits.

An example of a common fallacy of before and after analysis of water reclamation is to ignore indirect reuse that is occurring after the discharge of treated effluent into a river. Replacing indirect reuse downstream with direct reuse in the vicinity of a wastewater treatment plant may be a shift of benefits of the water supply from one location to another, rather than the creation of new benefits by a water reclamation project.

PLANNING PHASES

It is helpful to have a systematic and phased approach to planning to evaluate project feasibility. A simplified strategy for project planning is illustrated in Figure 2.1. The alternatives formulation and analysis will involve seven major feasibility criteria [3]:

(1) Engineering feasibility
(2) Economic feasibility
(3) Financial feasibility
(4) Institutional feasibility
(5) Environmental impact
(6) Social impact and public acceptance
(7) Market feasibility

The first six criteria are common to all water resources projects. While estimating projected water demand is an important element of water re-

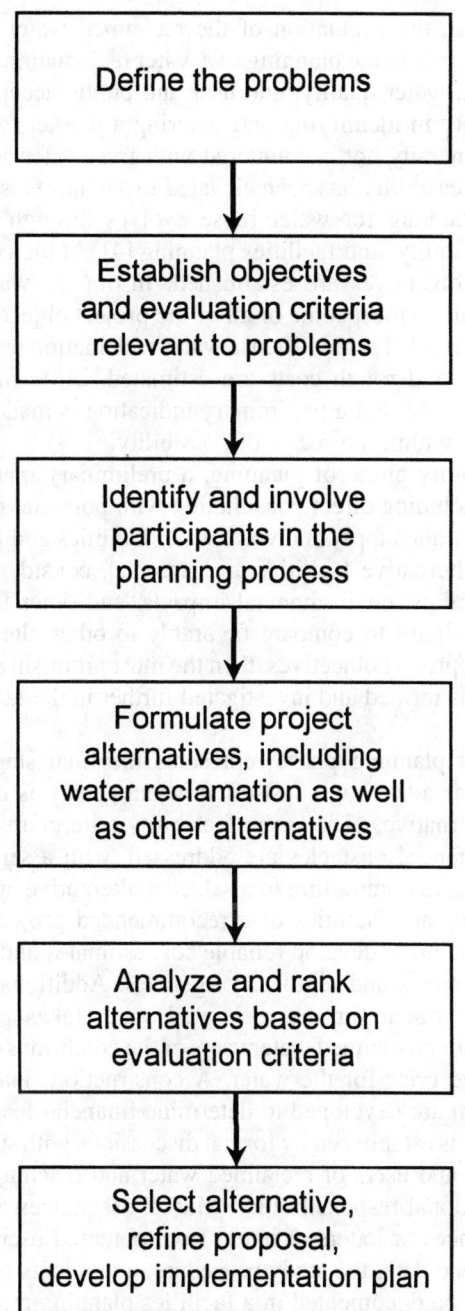

Figure 2.1. Planning strategy for water reuse projects.

sources planning, the evaluation of the reclaimed water market plays a more significant role in the planning of a water reclamation project. Factors of public health, water quality, and user and public acceptance create an added complexity in identifying and securing a market for the reclaimed water that is generally not encountered with freshwater sources. For this reason, market feasibility has been elevated to a major feasibility criterion.

Typically, planning for water reuse evolves through three phases—conceptual, feasibility, and facilities planning [4]. At the conceptual level, overall project objectives are established. Insofar as water reclamation appears to be able to meet some or all of the project objectives, a potential reclaimed water market is identified, a water reclamation project alternative is sketched out, and rough costs are estimated. Little investigation has occurred at this point, but a preliminary indication is made whether water reclamation lies within the realm of possibility.

At the feasibility phase of planning, a preliminary market assessment is performed, including direct consultation with potential reclaimed water users. Existing water supply and wastewater facilities and long term needs are assessed. Alternative facilities are screened, considering economics, technical constraints, environmental impacts, and other factors. If water reclamation continues to compare favorably to other alternatives for accomplishing the project objectives, then the most promising water reclamation alternative is refined and investigated further in the facilities planning phase.

The facilities planning phase represents the final stage of planning. During this phase a thorough cost-effectiveness analysis is conducted for all potential alternatives. The seven feasibility criteria are investigated in more detail. Potential obstacles are addressed, with a serious attempt to resolve them before committing to a selected alternative and commencing design. All necessary facilities of a recommended project are identified with sufficient detail to develop reliable cost estimates and seek approvals from funding sources and regulatory agencies. Additional refinement of the market assessment for the reclaimed water takes place, including informing potential reclaimed water users of the conditions of water service and the probable price for the water. A construction financing plan and revenue program are developed to determine financial feasibility. Institutional feasibility is established by formal discussions with suppliers, wholesalers, retailers, and users of reclaimed water and reaching agreement on legal and operational responsibilities. Market assurances, such as mandatory use ordinances or letters of intent from potential users, are obtained.

All of the basic data, the study procedure, and results of the feasibility analyses should be documented in a facilities planning report. Putting the information into a structured format that is readily accessible to review forces the project proponent to analyze a project more carefully. A report is the best basis for a thorough review by the public as well as the many project participants and funding and regulatory agencies.

PLANNING AREA DETERMINATION

One of the important first steps in a planning study is to define the project study area. The typical approach is to equate the study area with the project sponsor's jurisdictional boundaries. For a wastewater management agency, the agency boundaries may cover the wastewater collection service area. For a water supplier, the boundaries may be the potable water service area. Such boundaries are often not optimum for a study area for water reuse. The study area for the reclaimed water market should be the area that can feasibly be served from the source of the reclaimed water, the wastewater treatment plant. The appropriate project study area may cross jurisdictional lines between water and wastewater agencies and between adjacent water purveyors. Often wholesale water agencies cover broad service areas and can be a logical lead agency for a water reclamation planning study. In any case, cooperation and joint participation of agencies should be sought to be able to define a logical study area that will provide the maximum flexibility in formulating optimum facilities alternatives for the distribution and use of the reclaimed water.

While the project study area is suited for identifying a market and formulating facilities alternatives, limiting the scope of analysis to this area will prevent a complete evaluation of the potential benefits or costs of a water reclamation project. Water reuse may relieve water supplies or offset new water developments such that other areas outside of the project study area are benefitted. Removing effluent from a receiving water may have beneficial water quality effects. On the other hand, there may be incidental indirect reuse occurring downstream that will be harmed by planned reuse in the study area. Evaluating water reclamation within the context of the overall objectives of the project, such as meeting pollution control requirements or serving water supply needs, must involve analysis of effects beyond the immediate study area.

RECLAIMED WATER MARKET ASSESSMENT

A key task in planning a water reclamation project is to find potential users or customers of reclaimed water which are capable of using and are willing to use the water. The market assessment involves the gathering of background information on the constraints applicable to various categories of reclaimed water use and detailed data on each potential user or use site. The background information and user data will provide a basis for determining whether a potential user is capable of using the reclaimed water. This information will also help the potential user to decide whether it is willing to use the water. Willingness is also dependent on whether the reclaimed water will be marketed on a voluntary or mandatory basis. A discussion later on market assurances will address these marketing approaches.

In addition to identifying potential users, the market assessment provides much of the data needed for formulating project alternatives. Information about the users is needed to determine facility location and capacities, design criteria, reclaimed water pricing policy, financial feasibility, the amount and sources of freshwater displaced, and the institutional framework for the project.

Whether a user is capable of using reclaimed water depends on the quality of effluent available and its suitability for the type of use involved. Willingness to use reclaimed water depends on whether use is voluntary, and, if so, on how well reclaimed water competes with freshwater with respect to cost, quality, and convenience. It is essential to have a thorough knowledge of the water supply context of the users, especially if reclaimed water is to be marketed on a voluntary basis [4].

Gathering of background information involves the following steps:

(1) Create an inventory of potential users in the study area and locate them on a map. Group the users by types of use. Cooperation of retail water agencies can be very helpful in this task.

(2) Determine public health-related requirements by consulting regulatory agencies. Such requirements will determine the levels of treatment for the various types of use and application requirements that will apply on the sites of use, e.g., backflow prevention devices to protect the potable water supply, irrigation methods that are acceptable, use-area controls to prevent ponding or runoff of reclaimed water, practices to protect workers or the public having contact with the water.

(3) Determine water quality regulatory requirements to prevent nuisance or water quality problems, such as restrictions to protect groundwater quality.

(4) Determine water quality needs of various types of use, such as industrial cooling or irrigation of various crops. Government farm advisors may be helpful in this regard.

(5) Identify the wholesale and retail water agencies serving the study area. Collect data from them on current and projected freshwater supply prices that would be applicable to the reclaimed water users. Also, collect data on the quality of freshwater being provided.

(6) Identify the sources of the reclaimed water and estimate the probable

TABLE 2.1. **Information Required for Reclaimed Water Market Survey.**

1. Specific potential uses of reclaimed water
2. Location of user
3. Recent historical and future quantity needs (because of fluctuations in water demands, at least three years' of past use data should be collected)
4. Timing of needs (seasonal, daily, and hourly water demand variations)
5. Water quality needs
6. Water pressure needs
7. Reliability needs regarding availability and quality of reclaimed water, that is, how susceptible is the user to interruptions in water supply or fluctuations in water quality
8. Needs of the user regarding the disposal of any residual reclaimed water after use
9. Identification of on-site treatment or plumbing retrofit facilities needed to accept reclaimed water
10. Internal capital investment and possible operation and maintenance costs for on-site facilities needed to accept reclaimed water
11. Needed monetary savings on reclaimed water to recover on-site costs or desired pay-back period and rate of return on on-site investments
12. Present source of water, present water retailer if the water is purchased, cost of present source of water
13. When user would be prepared to begin using reclaimed water
14. Future land use trends that could eliminate reclaimed water use, such as conversion of farm lands to urban development
15. For undeveloped future potential sites, the year in which water demand is expected to begin, current status and schedule of development
16. After informing user of potential project conditions, a preliminary indication of the willingness of user to accept reclaimed water

quality of the reclaimed water after treatment to the level or levels under evaluation. Determine what types of use would be permitted at the various levels of treatment based on public health requirements and requirements suitable for various usages, such as industrial or agricultural uses.

(7) Conduct a survey of the identified potential reclaimed water users to obtain detailed and more accurate data for evaluating each users' capability and willingness to use reclaimed water. The types of data that should be collected on each user are shown in Table 2.1. While most of these data must be obtained directly from the user, some of these data may be assessed from the background information obtained from other sources.

(8) Inform potential users of applicable regulatory restrictions, probable quality of reclaimed water at various levels of treatment compared to freshwater sources, reliability of the reclaimed water supply, projected reclaimed water and freshwater prices. Determine on a preliminary basis the willingness of the potential user to accept reclaimed water.

Steps 1 through 6 are conducted simultaneously and are addressed, at least on a preliminary basis, before Steps 7 and 8. To some extent these steps may be carried out iteratively, returning to them during the course of planning to refine the data or survey additional potential users. The first phase of data collection on users may concentrate on larger water users and the core information of type and annual quantity of use. As facilities alternatives take shape, more detailed information will be collected and smaller users will be identified. Potential users may be contacted more than once as planning progresses to gather more information from the users and to provide the users with more information on the potential project. Group presentations with potential users may be useful for disseminating information and providing technical experts to answer questions.

WATER SUPPLY BACKGROUND

If reclaimed water use will displace the existing use of other sources of water or offset the development of alternative new water sources, it is essential to have a thorough understanding of the freshwater supplies. This knowledge is necessary to obtain acceptance of reclaimed water users and cooperation of affected water suppliers and to determine the net water savings and economic and financial feasibility. The information will be collected from the market survey of users as well as consultation with local and regional water suppliers.

From the viewpoint of the reclaimed water user, reclaimed water will be compared with other sources of water in terms of quality, dependability, safety, and cost. The planner must identify the current or future sources of freshwater that each user is or would be using if reclaimed water were not available. It must not be assumed that because a user is located within the boundaries of a water supplier, this user purchases water from the supplier. Many users have their own sources of supply, especially wells. The independent well user may see quite different water supply costs than customers of the water supplier.

On the other hand, the water supplier may be comparing water reclamation costs and revenues to the supplier's current freshwater supplies, future water supply developments, or purchases from wholesale water suppliers. Wholesale and retail water supply service areas and the potential users that rely upon each of the suppliers must be identified. The reclaimed water planner must be familiar with the freshwater sources of supply and the major facilities for capturing, treating, and distributing these supplies. With this knowledge, it can be evaluated how serving each user will offset a particular freshwater demand, not only reducing the costs of providing the freshwater but also reducing the associated revenue. In some cases the reduction in revenues will exceed the reduction in costs, resulting in a financial loss to the freshwater purveyor. Costs and benefits must be evaluated from the viewpoint of each wholesale and retail water supplier. With an understanding of how each agency will be impacted, a plan can be developed to share the water reclamation costs and revenues equitably between the water suppliers and wastewater agencies, helping to ensure full cooperation.

When gathering cost data, it is important to distinguish between fixed and variable costs. The use of reclaimed water will reduce the variable costs of a freshwater supply, but not the fixed costs, such as debt service on existing facilities.

Retail water suppliers are also useful sources of data on water quality, freshwater prices, and water demands of water users. Water suppliers can often easily provide actual water use records of individual users. Obtaining several years of records can help ensure that planners are not misled by data from unusually wet or dry years.

ENGINEERING ISSUES

Often, reclaimed water system design is approached in the same way as conventional potable water system design. However, special issues arise from the water quality, reliability, variation in supply and demand, and other differences between reclaimed and fresh water. Engineering or technical issues for a water reclamation project generally fall into the following categories:

(1) Water quality
(2) Public health protection
(3) Wastewater treatment alternatives
(4) Storage and distribution system siting and design
(5) On-site conversions at water use sites, such as potable and reclaimed water plumbing separation
(6) Matching of supply and demand for reclaimed water
(7) Supplemental and backup water supplies

Many aspects of these issues will be addressed in other chapters. A few will be highlighted here.

BALANCING SUPPLY AND DEMAND

One of the major engineering considerations is the balancing of reclaimed water supply with the demand. Wastewater flows vary by time of day, day of the week, and perhaps by season. It is likewise with water demands. Unfortunately, these variations usually do not match. For example, a seasonal variation is illustrated in Figure 2.2. Reclaimed water supply is fairly uniform in this example, uninfluenced by significant stormwater inflows or seasonal groundwater infiltration into the sewer system. On the other hand, a dry growing season can result in a pronounced seasonal peak in demand. Seasonal wastewater flows can be influenced by:

(1) Groundwater infiltration
(2) Stormwater inflow
(3) Seasonal population influx due to tourism
(4) Seasonal industrial activity such as vegetable canning

Seasonal water demands for reclaimed water are characteristic of irrigation uses and use in the food processing industry.

An hourly variation is illustrated in Figure 2.3. Wastewater flows are influenced by domestic, commercial, and industrial activities during the day. In this example, simplified but typical of the Southwest USA, landscape irrigation demands are at their peak at night, whereas industrial water demands may be uniform throughout the day. There may also be long-term trends, such as population growth, that will change the base wastewater supply over time.

Before analyzing the options for balancing supply and demand, some overall system constraints must be understood. Is the objective to remove a certain quantity of flow from discharge to comply with water quality requirements? Is the objective to reclaim whatever quantity of demand is cost-effective? The options for disposal of unused wastewater effluent must also be considered. If there are no restrictions on the discharge of unused effluent, then there is freedom to adjust the reclaimed water demands by connecting more or less users to the system.

Each situation must be analyzed to determine whether there are periods when demand exceeds the available reclaimed water supply and, if so, how to address the problem. There are many alternatives and it is important not to impose constraints that preclude an optimum cost-effective project. The alternatives of balancing supply and demand are the following:

(1) Adjust the available supply by adjusting the flow capacity of the reclaimed water system components or by adding additional levels of wastewater treatment.
(2) Adjust the level of demand by connecting more or less reclaimed water users to the system.
(3) Adjust the quantity of seasonal or daily operational storage.
(4) Provide a supplemental water supply to augment the reclaimed water during peak demands.

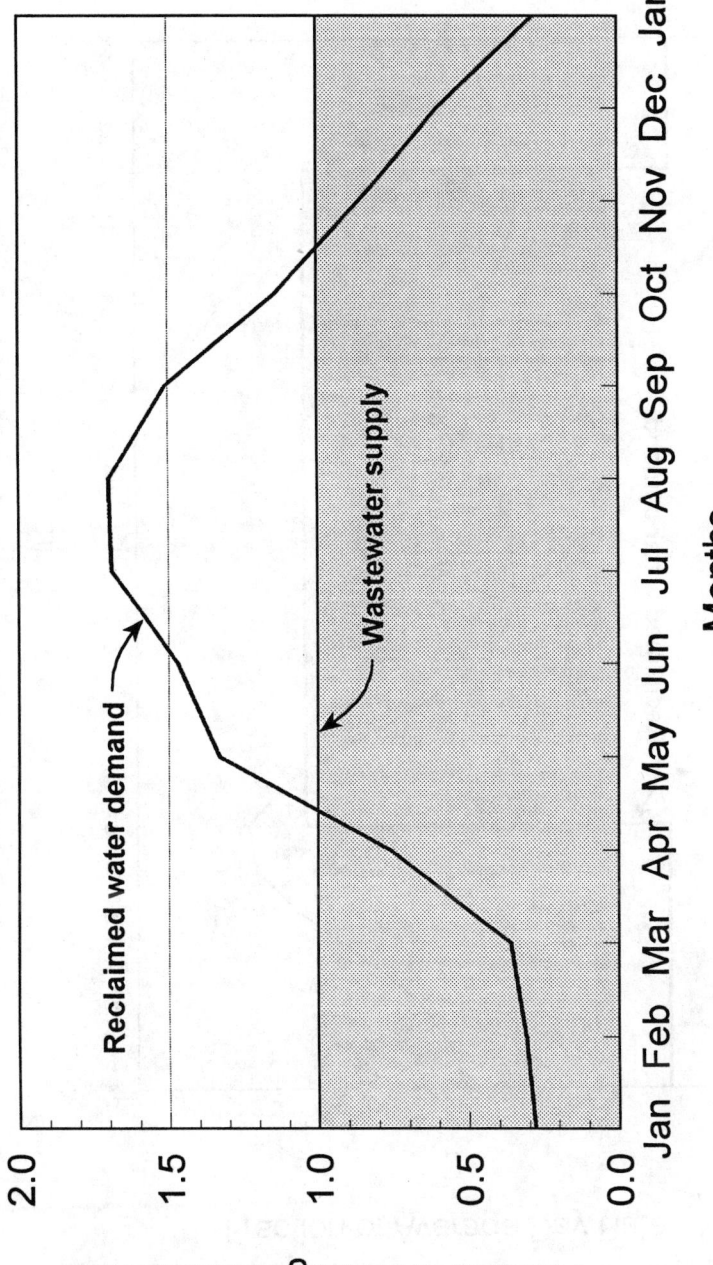

Figure 2.2. Relationship between supply and irrigation demand of reclaimed water (seasonal basis).

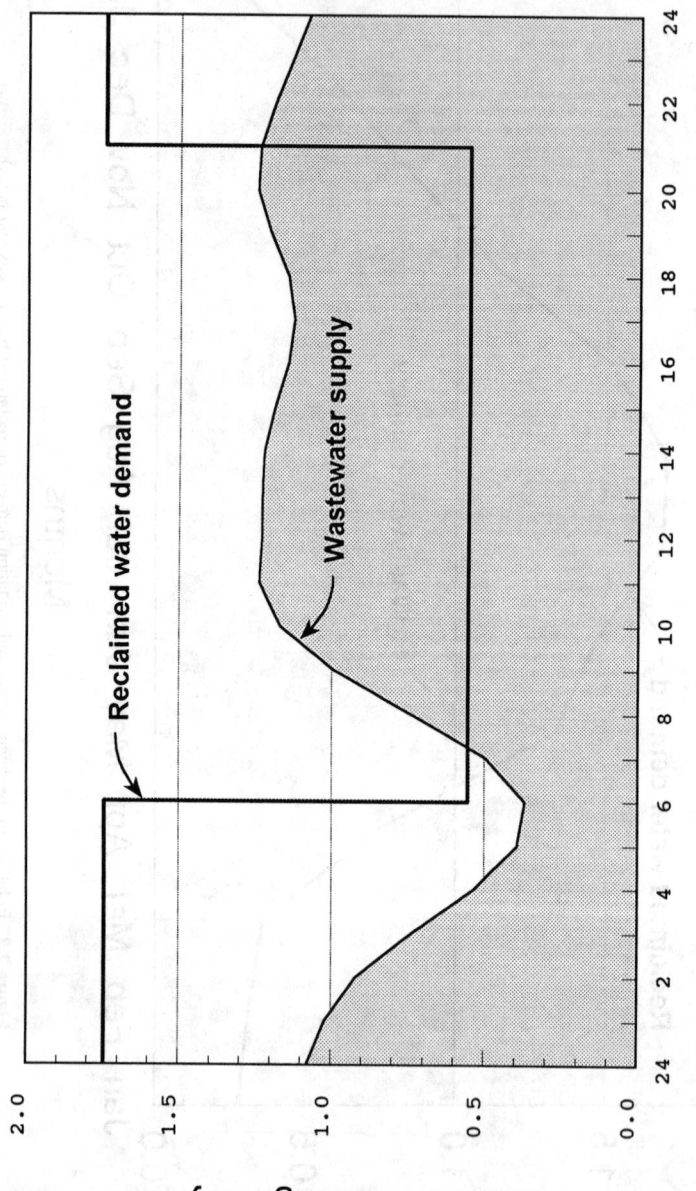

Figure 2.3. Relationship between supply and irrigation demand of reclaimed water (hourly basis).

Regarding the first alternative, the baseline reclaimed water supply rate is the total wastewater flow at the treatment level acceptable for discharge to a receiving water. However, this level of treatment may be insufficient for certain types of uses. Adding levels of treatment for reclamation and reuse will expand the potential uses of reclaimed water and the potential demand. The capacity of the higher levels of treatment does not have to match the baseline treatment for discharge and disposal, so adjusting this capacity adjusts the available supply for certain uses.

It is possible to adjust the level of demand by connecting users such that the demand never exceeds the supply. However, this would mean that wastewater flows in low demand periods will be lost. In order to capture all flows, seasonal storage may be necessary as well as a potential market for the reclaimed water. However, seasonal storage is costly, and in urban settings it may be impossible to locate a suitable site. Instead of providing seasonal storage, adding a supplemental supply can be a mechanism to add more baseline demand to the system to use more reclaimed water in the non-peak demand periods and use supplemental water to meet peak demands. The development of a reclaimed water system may open up an opportunity to use other nonpotable water supplies in the region. The reclaimed water system may be suitable for use of a nonpotable groundwater, such as groundwater containing excessive nitrates. The optimum solution will be determined by the economics, as will be discussed later in the chapter.

SUPPLEMENTAL AND BACKUP WATER SUPPLIES

Reliable water service may be an important factor in the success of a water reclamation project. Consideration should be given to the necessity or desirability of introducing other sources of water supply into the reclaimed water distribution system and providing an emergency backup water supply. In addition to using a potable or other freshwater supply as an emergency backup, a freshwater supply can be used as a supplemental supply to meet peak water demands in the reclaimed water system.

Blending reclaimed water with potable or nonpotable water can be a cost-effective means of meeting peak water demands in lieu of designing treatment and distribution system components to deliver 100 percent reclaimed water. Instantaneous peak water demands may occur in the aggregate only a few hours or days over a year's period. The total amount of supplemental water added during these peak periods may be less than 5 percent of total deliveries for a year, yet this may result in substantial capital cost savings.

From the perspective of drought protection, a reclaimed water supply can be a more reliable water supply than one dependent on precipitation.

On the other hand, the quality of a raw wastewater can be quite variable and less amenable to treatment to a uniform and reliable quality. Wastewater treatment can be adversely affected by unregulated introduction of toxic substances into the sewer system, excessive stormwater inflows, or poor control of biological treatment systems. Thus, there will be periods when a wastewater treatment plant will not be able to produce an acceptable reclaimed water.

The planner should estimate the frequency and duration of lapses in reclaimed water supply. These lapses may be acceptable if the types of uses can tolerate the lapses, would suffer only minor damage, or can rebound from damage without permanent loss. Otherwise, consideration may have to be given to providing a backup water supply or emergency storage of reclaimed water. Which of these two alternatives is chosen will depend on the accessibility of a backup supply and the costs. In urban areas where potable water supply pipelines are usually in close proximity to the reclaimed water distribution system, use of the potable supply as a backup supply is usually the cost-effective choice.

There are several methods of introducing the alternative water supply into the system. One method would be to bypass the reclaimed water distribution system and make emergency connections at the sites of each user. Some users may have groundwater wells that can be used for the reclaimed water uses. The water meter vaults at each use site can be designed to feed either reclaimed water or potable water by switching pipes in an emergency. Appropriate backflow prevention devices may be necessary to prevent residual reclaimed water in plumbing from draining into a potable water system.

If potable water is used as a backup or supplemental supply, the most reliable protection of the potable system from contamination is to introduce the potable water into a reservoir or pump station wet well of the reclaimed water system, leaving an air gap between the potable water discharge and the surface of the reservoir. This will prevent a suction of any reclaimed water into the potable water supply. Introducing the backup or supplemental supply at wet wells or reservoirs can include automatic controls to respond whenever the reclaimed water supply is detected not to meet demands.

PLANNING AND DESIGN TIME HORIZONS

There are several time horizons that are factors in facilities planning, design, economic evaluation, and financial analyses.

(1) *Planning period* is the total period for which facilities needs will be assessed and alternatives will be evaluated for cost-effectiveness.

(2) *Design life* is the period in which a phase of a component of facilities to be constructed is expected to reach capacity.

(3) *Useful life* is the estimated period of time during which a facility or component of a facility will be operated before replacement or abandonment. The useful life is usually equivalent to the period during which a facility is capable of performing its function. However, in some cases a facility will cease being useful even though it is still functional, in which case its useful life will be shorter than its operable life. The useful life may be shorter or longer than the planning period.

(4) *Financing period* is the period for meeting debt obligations or required paybacks for undertaking a project. This period may also be shorter or longer than the planning period.

The alternatives that are conceived during planning should all meet the planning objectives for the planning period. Each alternative should incorporate an evaluation of phasing of construction during the planning period. While the ultimate design for each alternative should serve the needs for the entire planning period, the initial phase to be constructed may serve only a portion of the planning period.

The optimum period for the design life of each project component is dependent on the degree of certainty for predicting future needs, the useful life of the component, the practical ability to add facility expansions, and the economy of scale related to the component. Present worth cost analyses, discussed later in this chapter, can be used to arrive at the optimum design life by testing different phasing intervals for certain types of facilities. However, if there is relative uncertainty in projecting growth in demand, either in terms of quantities or geographic direction, shorter design lives for the initial phase are desirable. Based on the experience observed for many water reuse projects, projecting future reclaimed water demand is a very uncertain task [5].

INSTITUTIONAL FEASIBILITY

Water reuse projects involve the interaction between, as a minimum, a supplier of reclaimed water and the user. More typically, several entities are involved. Separate entities may be responsible for collection of wastewater, treatment of wastewater, distribution of wastewater, wholesale supply of water, and retail distribution of water. All of these entities may have to cooperate and establish areas of responsibility for the successful operation of a water reclamation project.

The Walnut Valley Water District, located in southern California, represents a typical situation. The District operates a water reclamation project that delivers water not only within its service area, but also within an adjacent retail water district. To understand the economics and financial feasibility of its project, it was necessary to understand the water supply infrastructure from the sources of supply to the end water user. A chain of water supply agencies are involved, as illustrated in Figure 2.4.

Reclaimed water, on the other hand, reaches the district through another chain of agencies, as shown in Figure 2.5. An agreement between the Los Angeles County Sanitation Districts and the City of Pomona entitled the city to all of the effluent from the wastewater treatment plant. An interagency agreement had to be negotiated between Walnut Valley Water District and its supplier of reclaimed water, the City of Pomona. Another agreement was executed between the district and Rowland Water District so users located within Rowland Water District could participate in the project. In this situation, Walnut Valley Water District constructed and operates all facilities for the distribution of reclaimed water. Rowland Water District, while it does not operate the facilities, purchases the reclaimed water from Walnut Valley, reads the water meters, and bills the reclaimed water users within its territory.

From a regulatory standpoint, permits or approvals may be necessary from authorities governing water quality, water supply, public health, agricultural affairs, or water rate setting. Regulatory agencies should be involved early in the planning process.

MARKET ASSURANCES

During the market assessment phase of planning, a list is developed of potential users capable of using reclaimed water. While this list can be used to begin shaping alternative plans to serve these users with reclaimed water, at some point an assessment must be made of the degree of willingness these users have to participate in a project. There are a multitude of reasons that potential reclaimed water users may resist participation. The more frequently encountered reasons are listed in Table 2.2. These concerns can be overcome only by becoming familiar with the user's needs through the market assessment and educating the user about the nature of the water reclamation project.

There are various means of measuring the degree of willingness. During planning, the options in order of strength are usually:

(1) Oral interview
(2) Letter of interest signed by potential user, indicating general awareness of potential project and expressing interest in participation
(3) Letter of intent signed by potential user, incorporating key understand-

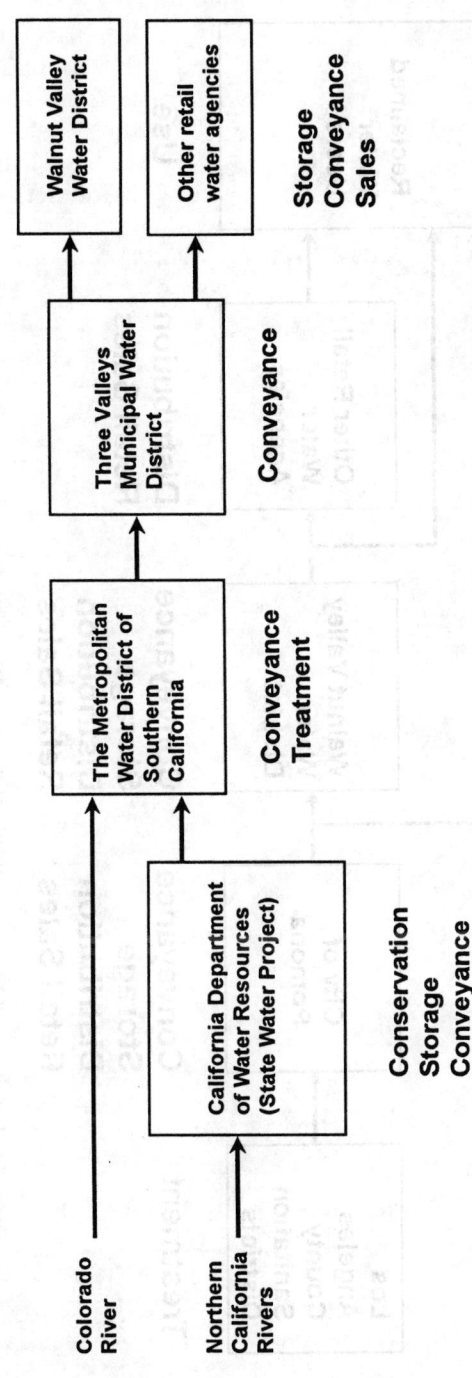

Figure 2.4. Potable water system structure serving Walnut Valley Water District (adapted from Reference [6]).

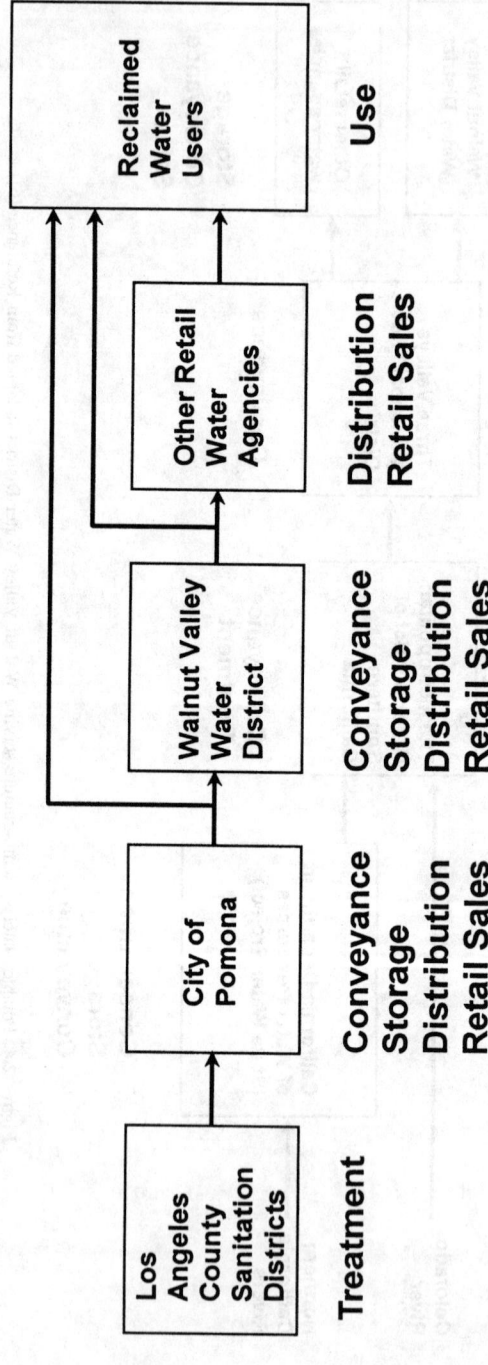

Figure 2.5. Institutional relationships for Walnut Valley Water District Water Reclamation Project (adapted from Reference [6]).

TABLE 2.2. **Customer Concerns about Use of Reclaimed Water.**

1. Price of reclaimed water in relation to freshwater costs is too high
2. Inability to finance on-site conversion costs
3. Concern over water quality
4. Inability to prevent worker contact
5. Possibility of employee union objections
6. Lack of reliable reclaimed water supply
7. Water supply costs are too insignificant to tolerate added inconvenience of reclaimed water
8. Liability

ings of the conditions of reclaimed water service, such as quantity of supply, water pressure, responsibilities for making on-site conversions, dates of service, and estimated price of reclaimed water

Before a commitment is made to commence design of a recommended project, a letter of intent is recommended from each user identified with the project. The letter contains many of the details of service and financial responsibilities so that it is clear that the user has been fully educated about the project and, with this understanding, fully intends to participate in the project. However, before construction commences, experience has demonstrated that these letters are insufficient assurances that users will participate in the project [5]. Stronger, legally enforceable measures are usually called for.

If a user resists signing a letter of intent during planning or a contract during the design phase of a project, this is usually an indication that there are outstanding concerns that have not been resolved. These concerns may seem inconsequential to the project planner, and the planner may anticipate that they can be overcome. However, if they cannot be overcome before beginning construction, the project planner is risking a major investment. Experience has shown that although potential users quite easily express positive interest in using reclaimed water, they often will resist a long-term commitment or refuse to use the reclaimed water after the facilities are built. Negotiations for firm commitments should be completed before project construction. This sometimes laborious process can be an educational exercise to win the support and confidence of potential customers and bring out hidden issues much earlier in the project development process [4].

There is little room for optimism regarding user participation in water reuse projects. A survey of 16 water reclamation and reuse projects in California found that only one quarter were delivering the amount of reclaimed water that was expected before project design was initiated. One quarter of this sample were delivering 50 percent or less of the planned amount. As part of the same survey, data on elapsed time for design were collected for 28 projects. One third had not gone to construction after over 2 years of design, including one having surpassed 5 years after commencing design. Nearly all of the projects experiencing these poor records of deliveries and long implementation periods suffered from problems that need to be addressed in planning of water reclamation projects. The problems included using unreliable data for estimating water demand quantities and encountering institutional difficulties, but most frequently failing to obtain agreement with potential users to purchase reclaimed water [5]. Legally enforceable mechanisms are needed to ensure user participation.

The ultimate measure of success of a water reclamation project is the delivery of reclaimed water in the quantities projected during planning during the life of the project. To ensure that a project will achieve this, there must be some assurance that an adequate reclaimed market exists. Examples of market assurances are listed below:

(1) Land ownership: The agency may own the land upon which the reclaimed water will be applied.

(2) Land lease: The agency may lease the land upon which the reclaimed water will be applied.

(3) User contract: The reclaimed water may be sold or provided to a water user under the provisions of a legally enforceable contract between the user and the reclaimed water supplier.

(4) Mandatory reclaimed water use ordinance: A legal mandate may be imposed by the retail water purveyor that water users are obligated to accept reclaimed water in lieu of another source of water available from the purveyor.

(5) Broad water rights authority: National, provincial, or regional governments may have authority to restrict the rights to use of freshwater if reclaimed water is available.

(6) Sale by reclaimed water use permit or informal arrangement: Reclaimed water may be sold or provided to a water user under provision of a permit or informal arrangement without any long-term obligations.

The mechanisms above are generally ranked in order of the strongest to weakest form of market assurance, though ordinances and water rights authority have the potential to be very strong, depending on the willingness of the agencies or governments to exercise their authorities and on the ease with which the authorities can be exercised.

The project proponent has to decide whether to take a voluntary or mandatory approach to marketing the reclaimed water. The choice of mechanism depends on the local situation and project purpose. The voluntary approaches can include land ownership if willing sellers are

TABLE 2.3. **Desirable Provisions of Reclaimed Water User Contracts.**

1.	Contract duration: terms and conditions for termination
2.	Reclaimed water characteristics: source, quality, and pressure
3.	Quantity of reclaimed water demand
4.	Flow variation of reclaimed water demand (hours and season of use): as needed by user or allowed by supplier
5.	Reliability of supply: potential lapses in supply and back-up supply provisions
6.	Commencement of use: when user can or will begin use
7.	Specific areas and conditions for reuse
8.	Financial arrangement: pricing of reclaimed water and payment for facilities, land lease costs
9.	Ownership of facilities and rights-of-way
10.	Responsibility for operation and maintenance
11.	Notification of problems: obligation of user to notify reclaimed water purveyor of violations of regulatory requirements; obligation of purveyor to notify user of water quality violations or supply interruptions
12.	Liability: for example, the user contract should indemnify the reuse agency from damages caused by use of reclaimed water due to no fault of the reuse agency
13.	Operations plan: operations or practices of the purveyor and the user to ensure monitoring and safe use of reclaimed water; obligation of user to observe regulatory requirements
14.	Violations of contract provisions: definition of what constitutes a violation, specification of penalties and remedies
15.	Inspection of on-site reuse facilities: the right of entry into the user's premises to inspect the reuse facilities during construction and operation

Source: Adapted from Reference [7].

found for the land or land lease if a land owner willingly leases the land to an agency for land application of reclaimed water. If willing landowners are not found, an agency may use the legal powers of eminent domain, taking a mandatory approach. These two mechanisms are most common when the project purpose is for treatment or disposal of treated wastewater.

A user contract is considered a voluntary mechanism in which the reclaimed water purveyor and the user mutually agree upon a long-term, legally binding commitment for the sale or transfer of reclaimed water. The contract should contain provisions specifying the responsibilities, obligations, conditions of reclaimed water service, and liability of the two parties. A list of common provisions is given in Table 2.3.

A mandatory use ordinance is a rule or law of local governmental bodies, such as water districts, mandating the use of reclaimed water in lieu of alternative water supplies, usually potable water. An ordinance, which may have different names under different jurisdictions, has legal effect only if the agency declaring the ordinance is the retail supplier of both the reclaimed water and the alternative supply that a user would have

TABLE 2.4. **Desirable Provisions for Mandatory Use Ordinance.**

1. Specification of the types of use of water for which reclaimed water must be used
2. Specification of the conditions under which reclaimed water must be used or new development must be plumbed for future reclaimed water use
3. Procedure for determining which water users are required to either convert to reclaimed water service or be plumbed to accept reclaimed water upon new water service
4. Procedure to provide notice to potential users that they are subject to the ordinance and specification that the notice include information about the project, the responsibilities of the users under the ordinance, the price of the reclaimed water, and description of the on-site retrofit facilities requirements
5. Procedure for request by the users for a waiver
6. A penalty for noncompliance with the ordinance. Example penalties are discontinuance of freshwater service or a freshwater rate surcharge of 50 percent of the freshwater rate

Source: Adapted from Reference [8].

to depend on. If an agency has control over both supplies, it can impose restrictions on the use of freshwater or fines if freshwater is used instead of reclaimed water. However, a wholesale agency does not have a direct relationship with the water user; so it has limited legal jurisdiction. However, wholesale water suppliers can encourage retail water agencies to mandate the use of reclaimed water by imposing fines or restrictions on the sale of freshwater to retail agencies.

Desirable provisions to include in a mandatory use ordinance are listed in Table 2.4. A mandatory use ordinance can apply to existing development that is currently using freshwater. It can also apply to new development by requiring dual plumbing or dual water distribution systems during new construction. Because reclaimed water service may not be cost-effective or feasible in certain geographic areas, the ordinance should specify the service areas governed by the ordinance. There are always special circumstances that prevent the use of reclaimed water, such as sites that would require expensive conversion (retrofit) costs or sites containing sensitive ornamental plants. A mandatory use ordinance should have a formal procedure to apply for, review, and approve a waiver.

The State of California has a ''waste and unreasonable use of water'' doctrine incorporated into its constitution. As an extension of this, the state legislature has created laws that require the use of reclaimed water in lieu of potable water under certain conditions that make allowances for public health protection, reasonable costs, and suitable water quality. However, the exercise of this state authority is cumbersome, requiring a quasi-judicial proceeding. While it is a strong mechanism, it is not in the control of the local water purveyor and has been applied in only a few circumstances.

If a reclamation project is intended to meet a pollution control objective, then the success of the project may be dependent on having users to accept all of the effluent from the wastewater treatment plant. Alternative effluent disposal options may not be incorporated in the recommended project or be acceptable to pollution control authorities. The facilities plan should identify a reclaimed water market for all present and future projected flows and describe the facilities needed to serve that market, even if the facilities would be constructed in future phases. Strong market assurances are necessary for the life of the facility to prevent violations of pollution restrictions. The pollution control agency can purchase land or obtain long term leases to exercise direct control over the ability to dispose of effluent in land application. Long term user contracts are also commonly used.

FACILITIES PLAN COMPONENTS

A facilities plan or facilities planning report is the main project documentation that is available to the public, decision-making authorities, regulatory authorities, funding agencies, and potential users. While environmental impact reports, feasibility reports, market assessment reports, or financial feasibility reports provide valuable information, their narrow scope omit discussion of many important issues and analyses performed during the project planning. Without a single report to tie all of the various analyses together, it is very difficult for independent parties to see the continuity between the various analyses and the final recommended project.

A suggested outline for a facilities plan that includes water reclamation alternatives is shown in Table 2.5. All of the items shown have been relevant at one time or another in a water reclamation and reuse project. Thus, although all of the factors shown do not deserve an in-depth analysis for every project, each item should at least be considered. The overall level of detail should be commensurate with the size and complexity of the proposed project. Although the emphasis on the wastewater or water supply aspects will vary depending on whether a project is single or multiple purpose, the nature of wastewater reclamation and reuse is such that both aspects must at least be considered.

Even if it is determined that a wastewater reclamation and reuse project is not feasible at the conclusion of a study, it is still advisable to publish the information and data collected and the analyses performed to arrive at this conclusion. Wastewater reclamation and reuse is good public policy in appropriate situations and public interest in wastewater reuse will continue as long as water supply needs are perceived to be critical. Documentation of even unsuccessful reuse planning is helpful in responding to public inquiry and in developing future planning efforts.

TABLE 2.5. **Wastewater Reclamation and Reuse Facilities Plan Outline.**

1. Study area characteristics: geography, geology, climate, groundwater basins, surface waters, land use, and population growth
2. Water supply characteristics and facilities: agency jurisdictions, sources and qualities of supplies, description of major facilities and existing capacities, water use trends, future facilities needs, groundwater management and problems, present and future freshwater costs, subsidies, and customer prices
3. Wastewater characteristics and facilities: agency jurisdictions, description of major facilities, quantity and quality of treated effluent, seasonal and hourly flow and quality variations, future facilities needs, need for source control of constituents affecting reuse, and description of existing reuse (users, quantities, contractual and pricing agreements)
4. Treatment requirements for discharge and reuse and other restrictions: health- and water quality-related requirements, user-specific water quality requirements, and use-area controls
5. Reclaimed water market assessment: description of market analysis procedures, inventory of potential reclaimed water users and results of user survey
6. Project alternative analysis: planning and design assumptions; evaluation of the full array of alternatives to achieve the water supply, pollution control, or other project objectives; preliminary screening of alternatives based on feasibility criteria; selection of limited alternatives for more detailed review, including one or more reclamation alternatives and at least one base alternative that does not involve reclamation for comparison; for each alternative, presentation of capital and operation and maintenance costs, engineering feasibility, economic analyses, financial analyses, energy analysis, water quality effects, public and market acceptance, water rights effects, environmental and social effects; and comparison of alternatives and selection, including consideration of the following alternatives:
   a. water reclamation alternatives: levels of treatment, treatment processes, pipeline route alternatives, alternative markets based on different levels of treatment and service areas, storage alternatives
   b. freshwater or other water supply alternatives to reclaimed water
   c. water pollution control alternatives to water reclamation
   d. no project alternative
7. Recommended plan: description of proposed facilities, preliminary design criteria, projected cost, list of potential users and commitments, quantity and variation of reclaimed water demand in relation to supply, reliability of supply and need for supplemental or backup water supply, implementation plan, and operational plan
8. Construction financing plan and revenue program: sources and timing of funds for design and construction; pricing policy of reclaimed water; cost allocation between water supply benefits and pollution control purpose; projection of future reclaimed water use, freshwater prices, reclamation project costs, unit costs, unit prices, total revenue, subsidies, sunk costs and indebtedness; and analysis of sensitivity to changed conditions

Source: Adapted from References [4,7].

GENERAL CONCEPTS OF ECONOMIC AND FINANCIAL ANALYSES

As cost-effectiveness analysis is defined in this chapter, monetary costs play a key role in project decision-making. Monetary analyses fall into two categories: economic analyses and financial analyses. While the two categories sound similar and are often used interchangeably, the two analyses have different roles as defined in water resources economics [3]. Economic feasibility, established through an economic analysis, is a test of whether the total benefits that result with a project exceed those which would accrue without the project by an amount greater than the project cost. Procedures have been developed by economists to calculate benefits and costs on a common basis so a comparison can be made between project alternatives. On the other hand, a financial analysis is intended to determine whether there is the ability to finance the construction of a project and raise the revenues for debt service and project operation. Economic justification and financial feasibility do not necessarily go hand in hand. As a matter of sound public policy, economic feasibility should have a key role in alternative selection and determining cost-effectiveness. Financial analysis is a tool to determine whether a project can be implemented rather than to measure the net benefits of a project. Expressed in simpler terms [9], an economic analysis addresses the question, *Should* a project be constructed? A financial analysis addresses the question, *Can* a project be constructed?

Before entering a detailed discussion of economic and financial analyses, some general concepts of the time value of money will be discussed. Of course, any standard engineering economics textbook should be consulted for a fuller and more theoretical discussion of the material in this and the following sections.

TIME VALUE OF MONEY

It is generally understood that money has a time value, that is, a dollar is worth more to us today than the promise of receiving a dollar a year from now. This is the basis for charging interest for the use of money. This time value presents a dilemma in the evaluation of projects. How do we compare the costs for two different projects that have different streams of capital and operations costs over the lives of the projects? A simple addition of the costs does not reflect the value we place on the time value of money.

Interest rate factors have been developed to allow us to calculate the value of a monetary transaction made at one time to the value at another time. The interest rate is the measure of the time value of money. The rate is influenced by inflation, tax rates, and the risk in an investment. However, even assuming a condition without inflation, taxes, or risk, there is still some expectation of a rate of return on the use of money. Which rate to use in an analysis depends on the purpose of the analysis, as will be discussed later.

The formula for a simple interest rate calculation is

$$I = (P)(i)(n)$$

where

I = earned interest
P = present amount or principal
i = interest rate per time period
n = number of time periods

To compute the total amount accrued at the end of the investment period, the following formula applies to a simple interest rate calculation:

$$F = P(1 + (i)(n))$$

where

F = future sum of money.

The time period commonly used as a basis for interest rates is one year. Thus, i is the interest rate per annum and n is the number of years. The formula works as well for fractions of a year. For example, the amount due after 7 months on a loan of $1000 borrowed at a rate of 5 percent per annum would be

$$F = \$1000(1 + (0.05)(\frac{7 \text{ months}}{12 \text{ months}})) = \$1029.17$$

For periods of longer than one year, compound interest is more common than simple interest. At the end of each year, the accrued interest is added to the amount due and the interest for the following year is calculated on the balance due, not the original principal. To illustrate this concept, assume that the $1000 loan described above is extended to 2 years. The balance at the end of the first year is

$$F_1 = \$1000(1 + (0.05)(1)) = \$1050.00$$

At the end of year 2, the balance is

$$F_2 = \$1050(1 + (0.05)(1)) = \$1102.50$$

The general formula for compound interest is

$$F = P(1 + i)^n$$

The *compound amount factor (F/P)* is $(1 + i)^n$. Interest tables are published that list this factor for various interest rates and numbers of years.

The *present worth factor* can be used to answer the question, how much should be invested now to yield a goal of \$1000 in 2 years at 5 percent interest? The formula is derived from the compound interest formula above.

$$P = \frac{F}{(1 + i)^n}$$

$$\text{Present worth factor } (P/F) = \frac{1}{(1 + i)^n}$$

$$P = \$1000(\text{Present worth factor}) = \$1000(\frac{1}{(1 + 0.05)^2}) = \$907.03$$

The formulas above deal with single transactions occurring at the beginning and ending of a time period. Another common financial situation is a repeating transaction, such as a loan resulting in repeating monthly repayments or the million dollar lottery prize paid out in equal annual amounts. Factors can be derived for computations dealing with these repetitive transactions, or *annuities*. If the transactions are equal amounts, it is called a uniform series. There are four uniform series compound interest factors commonly used:

$$\text{Sinking fund factor } (A/F) = \frac{i}{(1 + i)^n - 1}$$

$$\text{Capital recovery factor } (A/P) = \frac{i(1 + i)^n}{(1 + i)^n - 1}$$

$$\text{Series compound amount factor } (F/A) = \frac{(1 + i)^n - 1}{i}$$

$$\text{Series present worth factor } (P/A) = \frac{(1 + i)^n - 1}{i(1 + i)^n}$$

In each of these factors, the periodic or annuity amount, A, is assumed to occur at the end of each time period.

To illustrate the meaning and use of each of these factors, four examples are provided.

Example 1

A pump having a ten year life will cost $10,000 to replace. How much funds must be set aside at the end of each year into a savings account earning 5 percent interest to yield $10,000 in ten years?

$$A = F(\text{Sinking fund factor}) = (\$10,000)(\frac{0.05}{(1 + 0.05)^{10} - 1}) = \$795.05$$

where

A = annuity amount.

Example 2

A loan is approved for $100,000. It must be repaid in monthly install-ments for 30 years at an interest rate of 6 percent. What must the monthly repayment be?

This problem has an added complication in that the interest rate is expressed in a per annum amount, but the period between repayments is only monthly. The interest rate must be converted to a monthly equivalent:

$$\text{Monthly interest rate} = \frac{0.06/\text{year}}{12 \text{ months/year}} = 0.005$$

$$n = \text{Number of payments} = (12/\text{year})(30 \text{ years}) = 360$$

$$A = P(\text{Capital recovery factor}) = (\$100,000)(\frac{0.005(1 + 0.005)^{360}}{(1 + 0.005)^{360} - 1})$$

$$= \$599.55$$

Example 3

A deposit of $100 is made at the end of each month into a savings account earning 6 percent interest. What will be the account balance in 10 years?

$$\text{Monthly interest rate} = \frac{0.06/\text{year}}{12 \text{ months/year}} = 0.005$$

n = Number of deposits = (12/year)(10 years) = 120

$F = A$(Series compound amount factor) = ($100)$\dfrac{(1 + 0.005)^{120} - 1}{0.005}$

= $16,387.93

Example 4

College expenses are expected to be $15,000 per year for six years. How much money must be set aside in an account earning 5 percent per annum such that $15,000 can be drawn for six years, depleting the account at the end?

This solution is simplified by assuming that the expenses accumulate until the end of each year and are paid in lump sums from the savings account. In planning studies it is common to assume annual expenses and revenues occur in lump sums at the end of annual periods. If this assumption is inappropriate, adjustments must be made in the calculations.

$P = A$(Series present worth factor) = ($15,000)$\dfrac{(1 + 0.05)^{6} - 1}{0.05(1 + 0.05)^{6}}$

= $76,135.38

Additional explanation and examples of usage of these factors, as well as additional factors to fit more specialized situations, can be found in engineering economics textbooks [10]. The application of some of these factors in economic analyses will be illustrated in a later section.

INFLATION

Inflation must be considered in estimating costs used for economic and financial analyses. Inflation and, less commonly, deflation are changes in the price levels for goods and services. General inflation, as it affects prices in the overall economy, represent changes in the buying power of money. Differential inflation, that is, the difference in the inflation rate for a particular product as compared to the general inflation rate, may represent a change in the scarcity of an item or the efficiency of its production.

Cost estimates used in planning are often based on the actual experience of past projects. Cost indices are the usual tool for tracking past inflation and adjusting historical costs into current costs. The Consumer Price Index is the common measure of general inflation. Because of the differential inflation, specialized indices have been developed for many industries or types of construction activities. For public works projects, the *Engineering News-Record* magazine publishes the Construction Cost Index (ENRCCI), which is commonly used for wastewater and water supply projects, including water reclamation projects. It is, therefore, important to understand the proper use of this index.

As a general rule, a cost at one point in time can be converted to an actual cost at another point in time using the following formula:

$$\text{Cost}_1 = (\text{Cost}_0)\frac{\text{Index}_1}{\text{Index}_0}$$

where

Cost_0 = the source cost at time 0 to be adjusted
Cost_1 = adjusted cost at time 1
Index_0 = cost index at time 0
Index_1 = cost index at time 1

Because different geographic regions experience different rates of inflation, the ENRCCI is computed for 20 cities in the United States. In addition, these indices, published monthly, are averaged to produce the U.S. 20-city average. One of the common errors in use of the ENRCCI is to use it to convert the cost in one geographic area to a cost in another. Each city index is tracked independently and can be used to measure cost changes over time but not differentials between cities. Annually in an issue in March *Engineering News-Record* publishes a historical summary of indices and an explanation of the use of the index [11].

Example 5

In December 1996 a community proposed to add direct filtration tertiary treatment to its 5 million gallon per day secondary wastewater treatment plant to produce reclaimed water suitable for irrigating school grounds. The capital cost of upgrading treatment was estimated to be $2,525,000 in May 1991 dollars [12]. Construction is expected to occur March 1999 through December 1999, to be ready to cross the bridge into the 21st century. Estimate the capital cost of the treatment plant construction.

Cost indices record historical inflation. Projections of costs into the future must be based on assumptions of future inflation. One approach to estimate the cost is to adjust the cost to December 1996 dollars using the 20-City Average ENRCCI, then forecast the cost based on assumed inflation. Based on recent inflation rates, a rate of 3 percent per annum will be assumed from December 1996 to the midpoint of construction.

Adjusted capital cost through December 1996 =

$$= (\$2,525,000) \frac{\text{ENRCCI@Dec 1996}}{\text{ENRCCI@May 1991}} = (\$2,525,000) \frac{5730}{4801}$$

$$= \$3,013,591$$

Midpoint of construction = August 1999

Inflation rates are compounded. The compound amount factor will be used to adjust the cost from December 1996 through December 1998 and the simple interest factor will be used to complete the adjustment through August 1999.

Adjusted cost through August 1999 =

$$= (\$3,013,591)(\text{F/P}, 3\%, 2 \text{ years})(\text{Simple interest factor}, 3\%, 8 \text{ months})$$
$$= (\$3,013,591)(1 + 0.03)^2 (1 + (0.03)(8/12))$$
$$= \$3,261,000$$

Inflation rates represent changes in *actual* costs over time. General inflation does not indicate a change in *real* costs, that is, changes in the actual investment of resources and labor in the production of goods and services. Planning involves the estimation of costs occurring in the future. Actual costs or real costs can be used in monetary analyses, depending on the purpose of the analysis. Whether to adjust future costs for inflation will be discussed in later sections on economic and financial analyses.

COMPARISON OF ALTERNATIVES BY PRESENT WORTH ANALYSIS

A present worth analysis is a technique to be able to compare two alternative actions with different cash flows to determine which alternative has the lowest cost over time. It involves translating future cash flow streams into a single present lump sum called the *present worth* or *present value.* The technique relies on the concept of the time value of money, that a dollar earned in the future is worth less to us than a dollar earned today, because we can put it to use today by either investing it or spending it for an immediate benefit. Future cash flows are converted to present worth amounts by use of the present worth factor (*P/F*).

The interest rate used in this analysis is called the *discount rate.* The appropriate discount rate depends on the purpose of the analysis and the time value of money within the context of the purpose. The appropriate discount rate also depends upon whether actual (inflated) costs or real (constant dollar) costs are used in the analysis. Market interest rates are influenced by rates of inflation and tend to rise as the rate of inflation rises. Market value interest rates should not be used for a discount rate if real costs are used in the analysis [10]. During periods of inflation, the discount rate should be less than the market rate when real costs are used in the analysis.

Example 6

An agency is constructing a water reclamation project that will cost $5 million. The agency has three financing options: A) borrowing funds from a private lender at an interest rate of 8 percent, B) receiving a federal grant of 25 percent of the cost and borrowing the remainder from a private lender, and C) receiving a state loan at an interest rate of 4 percent. All borrowing options have a repayment period of 20 years. Which financing option is the lowest cost?

From the agency's perspective, an appropriate discount rate would be the interest rate it would have to pay to finance the project on its own, that is, the private lending rate of 8 percent per annum.

Option A:
The cash flow for this option is the annual repayments for 20 years computed using the capital recovery factor (A/P):

$$A = (\$5,000,000)\ \frac{0.08(1 + 0.08)^{20}}{(1 + 0.08)^{20} - 1} = \$509,261/\text{year}$$

$$\text{Present worth} = (\$509,261/\text{year})(P/A) = (509,261)\ \frac{(1 + 0.08)^{20} - 1}{0.08(1 + 0.08)^{20}}$$

$$= \$5,000,000$$

Option B:
Amount borrowed = \$5,000,000 − 0.25(\$5,000,000) = \$3,750,000

$$\text{Repayments} = A = (\$3,750,000)\ \frac{0.08(1 + 0.08)^{20}}{(1 + 0.08)^{20} - 1} = \$381,946/\text{year}$$

$$\text{Present worth} = (\$381,946/\text{year})(P/A) = (381,946)\ \frac{(1 + 0.08)^{20} - 1}{0.08(1 + 0.08)^{20}}$$

$$= \$3,750,000$$

Option C:

$$\text{Repayments} = A = (\$5,000,000) = \frac{0.04(1 + 0.04)^{20}}{(1 + 0.04)^{20} - 1} = \$367,909/\text{year}$$

$$\text{Present worth} = (\$367,909/\text{year})(P/A) = (367,909)\frac{(1 + 0.08)^{20} - 1}{0.08(1 + 0.08)^{20}}$$

$$= \$3,612,182$$

Option C has the lowest present worth and is, therefore, the lowest cost option for the agency. The equivalent subsidy of this low interest loan can be calculated from the present worth:

$$\text{Subsidy} = \frac{\$5,000,000 - \$3,612,182}{\$5,000,000} = 28 \text{ percent}$$

ECONOMIC ANALYSES

Economic analyses are performed to determine whether project alternatives have a net benefit in monetary terms and to rank alternatives in terms of relative benefit. For public works projects, such as water reclamation projects, an attempt should be made to quantify all costs and benefits associated with a project, not just the costs that will be experienced by the project sponsor. The objective is to determine the project alternative that will achieve the highest net benefit to the public as a whole. In a cost-effectiveness analysis, this alternative would be recommended for implementation unless other nonmonetary factors were over-riding. The time horizon of the economic analysis is the planning period. This period may not coincide with the financing period, that is, the period for debt repayment.

MEASUREMENT OF COSTS IN ECONOMIC ANALYSES

An attempt is made to compare projects on a real cost basis that will represent the relative investment of material and labor resources in each alternative. Because the rate of inflation represents a change in the value of money rather than a change in the investment of social resources, real costs in constant dollars rather than actual costs are used in estimating future costs for an economic analysis. However, because there is a time value of money aside from the inflation factor, a present worth analysis is used to compare alternatives on an equivalent basis.

Future costs are estimated in constant dollars, that is, the dollar value of materials and labor at a chosen reference point. If there is a basis for assuming that certain goods are inflating at a different rate from common inflation, it is permissible to adjust costs of those goods by the differential inflation rate, that is, the difference between the two inflation rates. The reference point is usually a date close to the period of the analysis or the date when a project is expected to be constructed. Cost indices can be used to adjust costs to a common current date. However, if the reference point is the period of construction, then an assumption of inflation between the present and the future construction date will have to be made and all costs adjusted to that common reference point.

MEASUREMENT OF BENEFITS IN ECONOMIC ANALYSES

Benefits are measured in terms of the effectiveness of actions or projects in achieving their stated goals [3]. The two most common goals of water reclamation projects are the production of a usable water supply and the reduction of pollution in the aquatic environment. Benefits can be measured by the market value of project outputs. However, the valuation of the output of a cleaner environment is very difficult. Furthermore, while water has a market value, the price of water to customers, the transfer of water does not occur within a free market economic environment. The price of water is often set to recover the costs of production rather than to reflect its worth in the market place. The price of water is generally less than the cost of development of new sources of water.

An alternative method of measuring project benefits is through the cost of producing the same outputs in an alternative manner [3]. This is the most common approach in public works projects. Within an array of alternatives that achieve the project goals equally effectively, the least cost alternative is the recommended project. The benefits are the cost of the second-best alternative. Inherent in this approach is that the second-best alternative would in fact be built if the recommended project were not built. One can always find a more expensive project to compare to, but to provide a valid analysis, the second-best alternative must be a practicable one.

Using the alternative costs as the basis of comparison for projects, each alternative cost is computed in the present worth using common assumptions for each alternative, such as an equivalent planning period and discount rate. Because the present worth incorporates all capital and operational costs over a time span, the present worth is also called the *life cycle cost*.

There are two common derivations from the present worth cost: equivalent annual cost and unit cost. The equivalent annual cost is a uniform annual cost that has a present worth equal to the present worth of the project. It is derived as follows:

Equivalent annual cost = (Present worth)(A/P)

where the series compound amount factor, A/P, is based on the same discount rate and time period as the present worth.

The unit cost approach is valuable when there are alternative approaches to achieving a given output but in different quantities of the output. Alternative water supply projects may not yield the same amounts of water. Comparison on the basis of cost per unit of water yield may be the only means of cost comparison. The derivation of unit cost will be illustrated by example.

BASIC ASSUMPTIONS IN ECONOMIC ANALYSES

The preferred basis for expressing costs in an economic analysis is real costs in constant dollars. Thus, the discount rate should be a rate that is not dominated by inflation influences. Long-term government borrowing rates averaged over several years are often used as a basis for discount rates. However, the issue is complex and economics references should be consulted for more background [3,10,13].

Sunk costs are expenditures that occurred in the past. They should not be allowed to influence decisions on future actions. Likewise, continuing debt payments for past investments can generally be considered a sunk cost because the debt obligation will remain regardless of which future action is taken.

The time reference point for economic analyses is usually the beginning of project operation. All costs and benefits for each alternative through the end of the planning period should be identified. The present worth of costs and benefits is computed to this point in time. Design and construction costs, occurring before this reference point, must be brought forward to this point in time.

All costs stemming from project alternatives should be considered in the economic analysis. There is a tendency to ignore costs that may be essential for project implementation but not a responsibility of the entity performing the feasibility study. For example, on-site conversion costs, also called retrofit costs, are necessary for users presently using freshwater to convert to reclaimed water. There are installation of dual plumbing or special signage warning against improper use of reclaimed water, for example. The project proponent may not be intending to pay for these costs. Nevertheless, they are a part of the project and should be calculated in the economic analysis.

If a first phase of a project is dependent upon future phases for fulfilment of reclaimed water delivery projections, then the facilities plan should identify the market, facilities, and associated costs of all dependent phases.

It is common, but inappropriate, to attribute an output to a project without identifying all of the facilities and costs that must be added in future to realize this yield.

REPLACEMENT COSTS AND SALVAGE VALUES

Facilities that will reach their useful lives during the planning period will need replacement. The replacement costs must be included in the economic and financial analyses. If a useful life exceeds the planning period and if the facility is expected to remain in use, the facility will have a salvage value at the end of the planning period.

There are various formulas for depreciation used for taxing purposes and private sector financing. For the purposes of an economic analysis, straight-line depreciation from the first day of facility operation is assumed for determining salvage value.

Example 7

A small reclaimed water distribution system will cost $1,000,000. It consists of a pump station structure costing $200,000, pumps costing $150,000, and pipelines costing $650,000. The useful lives are 20 years for the structure, 15 years for the pumps, and 50 years for the pipelines. What is the salvage value at the end of a 20 year planning period?

The pumps will be replaced in 15 years.

The remaining lives of the facilities at the end of the planning period will be:

pump station structure = Useful life − Planning period =
                        20 years − 20 years = 0 years
replacement pumps = 15 years + 15 years − 20 years = 10 years
pipelines = 50 years − 20 years = 30 years

Salvage values:

Salvage value = (Initial cost)(Remaining life)/(Useful life)
pump station structure = ($200,000)(0/20) = $0
replacement pumps = ($150,000)(10/15) = $100,000
pipelines = ($650,000)(30/50) = $390,000

Total salvage value = $100,000 + $390,000 = $490,000

COMPUTATION OF ECONOMIC COST

To illustrate the various concepts described above for economic analyses, Example 8 is provided.

Example 8

A water reclamation and reuse project is proposed with a capital cost of $1.5 million, including design and construction costs. The breakdown of capital costs is shown in Table 2.6. Design and construction are each expected to take a year, beginning in 1998. Deliveries of reclaimed water will be 200 acre-feet during the first year of operation, 270 acre-feet during the second year, and 450 acre-feet each year thereafter (1 acre-foot equals 1233.5 cubic meters). Deliveries include 100 acre-feet/year to industrial users, beginning in the first year, and the remaining to landscape irrigation users. Because of the increased salinity of the reclaimed water as compared to freshwater, each acre-foot of reclaimed water replaces only 0.8 acre-foot of freshwater. Operation and maintenance (O&M) costs will be $40,000 during the first year, $60,000 the second year, and $85,000 each year thereafter. The fertilizer value of the nutrients in the reclaimed water is $40/acre-feet. The water district will be responsible for the financing and construction of the treatment and distribution facilities. The reclaimed water users will be responsible for paying for and making the necessary plumbing changes on their sites to accept reclaimed water. The useful life of pump facilities is 15 years, waste treatment facilities 20 years, and pipelines and on-site plumbing 50 years. All costs given are in current dollars. Assuming a 20-year planning period, what is the economic cost of the project expressed as present worth and unit cost?

A summary of capital costs is given in Table 2.6. Also in this table is a calculation of the salvage value of these costs at the end of the 20-year planning period.

The annual costs and reclaimed water deliveries are presented in Table 2.7. The present worth of net costs and unit cost is calculated in the table.

$$\text{Present worth} = \$2,151,861$$

The computation of unit cost, as shown in Table 2.7, involves a computation of the present worth of reclaimed water delivered. A commodity, the same as money, has a time value. The delivery of a unit of water today has more value to us than a promise of its delivery in the future. The same concepts of present worth can be applied to such commodities as to money. To obtain a valid unit cost, this must be taken into consideration.

TABLE 2.6. Example 8: Capital Costs, Salvage Values, Years of Occurrence.

Item	Year of Occurrence	Capital Cost, $	Useful Life, y	Salvage Value at end of 2014, $[a]
I. Construction				
Advanced waste treatment facilities	1999	600,000	20	0
Pump facilities	1999	50,000	15	33,333
Distribution pipelines	1999	500,000	50	300,000
On-site plumbing conversions	1999	100,000	50	60,000
Total construction cost		1,250,000		393,333
II. Design	1998	150,000		0
III. Services during construction	1999	100,000		0
IV. Total project costs		1,500,000		393,333

[a] Assumed straight line depreciation for 20 y planning period. Example: for Distribution pipelines, ($500,000) * (50y − 20y)/50y = $300,000.

Unit cost of reclaimed / water delivered $=\dfrac{\text{(Present worth cost)}}{\text{(Present worth of reclaimed water delivered)}}=$
= \$456/acre-foot

Because the amount of reclaimed water deliveries is not equivalent to freshwater replaced, calculations for freshwater replaced are also presented. The unit cost in relation to freshwater replaced would be the appropriate basis of comparison to alternative freshwater projects.

Unit cost of / freshwater replaced $=\dfrac{\text{(Present worth cost)}}{\text{(Present worth of freshwater replaced)}}=$
= \$570/acre-foot

PROJECT OPTIMIZATION

Economic analyses can be used not only to identify the least cost alternative within a group of distinct alternatives to achieve a project objective, but also to optimize a given alternative with respect to size and features. Optimization involves the use of marginal cost analysis. The *marginal cost* is the cost associated with adding a particular project feature or adding an increment of size to a project or component of project. It is cost-effective to add the project feature or increase the size if the marginal cost is less than the associated increase in benefits.

Each increment of additional demand added to the reclaimed water system is associated with potential marginal costs of additional wastewater treatment, additional pipelines, additional daily or seasonal storage, and operational costs of treatment and pumping. An example of when marginal cost analysis should be used is in the consideration of whether to upgrade from secondary to tertiary level of treatment in order to deliver more reclaimed water. A certain market for a hypothetical project may have been identified that could use 1.0 million m^3 per year of reclaimed water of secondary quality. Because of fewer restrictions on the use of tertiary effluent, the potential market for tertiary quality reclaimed water could be 1.5 million m^3 per year. There is a base cost for this project using secondary effluent, consisting of treatment and distribution facilities. To upgrade treatment to tertiary and to make use of this effluent, there are marginal costs of the added treatment level, the expansion of the capacity of the secondary and tertiary treatment to serve additional users, and the expansion of the pipeline distribution system to reach the additional users that could use only tertiary effluent. If the basis of project justification is the cost of alternative freshwater supplies, then the upgrade to tertiary

TABLE 2.7. Example 8: Total Economic Cost.[a]

A	B	C	D	E	F	G	H
	Reclaimed Water Deliveries, AF						
Year	Industrial Use	Landscape Irrigation Use[b]	Total	Fresh Water Replaced, AF[c]	Design Period Costs, $[d]	Construction & Related Costs, $[d]	Operation, Maintenance, & Replacement Costs, $[e]
1998					150,000		
1999						1,350,000	
2000	100	100	200	160			40,000
2001	100	170	270	216			60,000
2002	100	350	450	360			85,000
2003	100	350	450	360			85,000
2004	100	350	450	360			85,000
2005	100	350	450	360			85,000
2006	100	350	450	360			85,000
2007	100	350	450	360			85,000
2008	100	350	450	360			85,000
2009	100	350	450	360			85,000
2010	100	350	450	360			85,000
2011	100	350	450	360			85,000
2012	100	350	450	360			85,000
2013	100	350	450	360			85,000
2014	100	350	450	360			135,000
2015	100	350	450	360			85,000
2016	100	350	450	360			85,000
2017	100	350	450	360			85,000
2018	100	350	450	360			85,000
2019	100	350	450	360			85,000
Total					150,000	1,350,000	1,680,000

Unit cost of fresh water replaced = (Present worth of costs)/(Present worth of fresh water relaced) = 570 $/AF
Unit cost of reclaimed water delivered = (Present worth of costs)/(Present worth of reclaimed water deliveries) = 456 $/AF

TABLE 2.7. (continued).

A Year	I Fertilizer Credit, $[f]	J Salvage Value, $[d]	K Total Net Cost, $	L Present Worth Factor at 6% Discount Rate[g]	M Present Worth of Net Cost, $[h]	N Present Worth of Freshwater Replaced, AF[i]	O Present Worth of Reclaimed Water Deliveries, AF[j]
1998			150,000	1.06000	159,000	0	0
1999			1,350,000	1.00000	1,350,000	0	0
2000	(4,000)		36,000	0.94340	33,962	151	189
2001	(6,800)		53,200	0.89000	47,348	192	240
2002	(14,000)		71,000	0.83962	59,613	302	378
2003	(14,000)		71,000	0.79209	56,239	285	356
2004	(14,000)		71,000	0.74726	53,055	269	336
2005	(14,000)		71,000	0.70496	50,052	254	317
2006	(14,000)		71,000	0.66506	47,219	239	299
2007	(14,000)		71,000	0.62741	44,546	226	282
2008	(14,000)		71,000	0.59190	42,025	213	266
2009	(14,000)		71,000	0.55839	39,646	201	251
2010	(14,000)		71,000	0.52679	37,402	190	237
2011	(14,000)		71,000	0.49697	35,285	179	224
2012	(14,000)		71,000	0.46884	33,288	169	211
2013	(14,000)		71,000	0.44230	31,403	159	199
2014	(14,000)		121,000	0.41727	50,489	150	188
2015	(14,000)		71,000	0.39365	27,949	142	177
2016	(14,000)		71,000	0.37136	26,367	134	167
2017	(14,000)		71,000	0.35034	24,874	126	158
2018	(14,000)		71,000	0.33051	23,466	119	149
2019	(14,000)	(393,333)	(322,333)	0.31180	(100,505)	112	140
Total	(262,800)	(393,333)	(322,333)		2,172,724	3,812	4,765

Unit cost of fresh water replaced = (Present worth of costs)/(Present worth of fresh water relaced) = 570 $/AF
Unit cost of reclaimed water delivered = (Present worth of costs)/(Present worth of reclaimed water deliveries) = 456 $/AF

[a]AF = acre-foot. 1 acre-foot = 1233.5 cubic meters. Col = column; [b]Calculated, Col D – Col B; [c]Freshwater replaced equals 80 percent of reclaimed water deliveries: Col E = (Col D)*0.8; [d]See Table 2.6; [e]This column includes the pump replacement cost after 15 years of useful life; [f]Calculated, (Col C)*($40/AF); [g]Present worth factor = $1/[(1 + i)^n]$, where $i = 0.06$, n = years from beginning of operation. All costs or benefits are assumed to occur at end of year; [h]Column K multiplied by Column L; [i](Col E)*(Col L); [j](Col D)*(Col L).

99

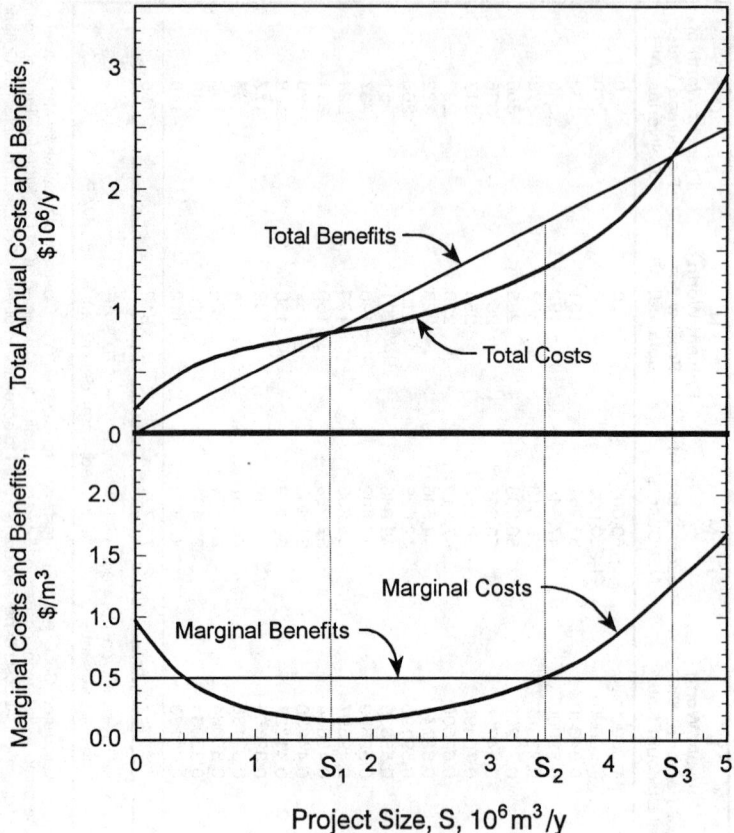

Figure 2.6. Project size optimization.

treatment would be justified if the marginal costs were less than providing 0.5 million m³ per year from new freshwater development.

Another common example where marginal cost analysis should be applied is in determining the geographic extent of a reclaimed water distribution system. To reach each new user of reclaimed water, there are marginal costs of additional pipeline to the user and expanded capacity of treatment, pumping facilities, and pipelines to serve the additional water demand.

A common fallacy in project justification is to use total or average costs and benefits rather than marginal costs and benefits in cost analyses. Total and average costs tend to mask project components or sizes that are not cost-effective. This is illustrated in Figure 2.6. These hypothetical curves show typical trends of costs for water reclamation projects. The benefits shown in this illustration are the least-cost alternative freshwater development costs, which are assumed to have a constant marginal unit cost within the range of the reclamation project size. The cost curves, shown as smooth curves, in reality would be irregular steps representing the addition of individual users or groups of users and the costs associated with each discrete addition.

The marginal cost curve in the lower half of the figure illustrates the initial effect of economy of scale, drawing down the unit cost as the project becomes larger. However, as the distribution becomes more extensive to reach users progressively farther away from the wastewater treatment plant or to reach users with progressively smaller water demands, the marginal costs begin to rise to a point of being uneconomical to serve. If total costs alone are compared, as is done in calculating benefit-cost ratios or in comparing average unit costs, any project within the size range between S_1 and S_3 appears to be cost-effective. However, when marginal costs are analyzed, it is seen that any project addition beyond S_2 has a marginal cost greater than the marginal benefit. It is not cost-effective to build a project larger than S_2. While any project within the range of S_1 and S_2 is justified, the optimum size would be at S_2, where the maximum net benefit is achieved. Making decisions based on benefit-cost ratios or average unit costs may lead to uneconomic, oversized projects.

INFLUENCE OF SUBSIDIES

The objective of the economic analysis is to identify the project that has the least true cost or the most net benefits. Subsidies, such as grants or rebates received from external sources, should not be incorporated into the analysis. Subsidies would lower the cost curves shown in Figure 2.6, falsely giving the impression that a given project alternative has a greater net benefit. Subsidies incorporated into analyses for project selection or optimization result in oversized, if not entirely uneconomic, projects.

Subsidies can come in many forms. For water reclamation projects, the most common forms are grants or below-market interest rate loans for capital financing and operational rebates based on reclaimed water usage or deliveries. Because subsidies may apply to only certain categories of costs, such as capital costs, they can sometimes influence local decisions contrary to the broader public interest. A high cost alternative characterized by high capital costs but low operational costs may be favored by an agency if there is a subsidy for capital costs but not operational costs. Unless there are overriding nonmonetary benefits to the public at large, subsidies should not bias local decisions in favor of more expensive project alternatives.

Subsidies should be a tool to encourage actions that are beneficial to the public at large. After the most cost-effective project is identified considering the economic analysis and other nonmonetary factors, subsidies should be a tool to make this alternative more desirable, affordable, or capable of being implemented. From a local viewpoint, water reclamation may appear to be financially more expensive than alternative water supplies, even though from an economic analysis it can be shown to be the least cost water supply. Subsidies, usually from higher levels of government or regional water supply or wastewater agencies, can lower the cost to facilitate the implementation of the most cost-effective project. Subsidies rightfully belong in the financial feasibility analyses, discussed later in this chapter.

FINANCIAL ANALYSES

The financial analysis addresses the question of whether a water reuse project is financially feasible. It must be addressed from perspectives of all project participants, and, in some cases, external parties that may experience financial effects of a project. There are the viewpoints of the project sponsor, other suppliers and distributors of the reclaimed water, and the reclaimed water users. However, even freshwater users that are not participants in a water reuse project may see their water rates rise or fall because of the use of reclaimed water. Each participant or affected party expects to remain financially the same or better off by the use of reclaimed water in lieu of freshwater.

Financial analyses for water reuse projects generally fall into two categories: *construction financing plans* and *revenue programs.* The options for financing of capital costs are addressed in the construction financing plan. The sources of revenue during the operation of a project to cover debt service and operation and maintenance costs are analyzed in the revenue program. Because capital financing affects annual costs, the two analyses are related.

While the methods of financing were disregarded in the economic analysis, they are an important consideration in financial feasibility analysis. Different sources of capital funds have different interest rates or repayment constraints to consider. Subsidies of various kinds may be available to make the economically justified alternative more attractive. It is appropriate in financial analyses to incorporate subsidies and actual borrowing interest rates in the computation of debt service and annual costs. The computation of debt service should reflect actual repayment periods, regardless of the planning period or useful life of facilities.

In establishing reclaimed water rates, it may be necessary to consider on-going costs of facilities existing prior to the project under investigation. Debt service on existing facilities, treated as sunk costs in economic analyses, could be given consideration in financial analyses. Existing costs may have to be melded with the costs of new facilities in establishing water rates. However, care must be taken not to attribute sunk costs as a cost of a new project and to treat sunk costs the same in comparison of all alternatives.

Because reclaimed water use may lower the use of freshwater, the freshwater rates may be influenced. The existing freshwater infrastructure will have to be financed by a lower base of freshwater sales, which could cause freshwater rates to rise. On the other hand, the use of reclaimed water may offset the purchase of more expensive freshwater. It is best to approach reclaimed water as another component of water supply for the community as a whole, rather than as a separate water system to be financed independently. Reclaimed water rates should be set to reflect the value of the reclaimed water relative to freshwater. Revenues from reclaimed water sales may need to be transferred to offset freshwater revenue losses or freshwater revenue may be appropriately used to subsidize water reuse to create an equitable sharing of the costs and benefits of the reclaimed water supply.

Within the reclaimed water study area, potential reclaimed water users may obtain their water supplies from different sources at different costs. In performing the financial analysis, it is important to consider what users are presently paying for water and would be paying in the future. The use of average water prices for users may be appropriate for preliminary feasibility analyses, but in the final analysis the situation of each potential user must be considered to judge the financial feasibility from the viewpoint of the user.

The costs of on-site conversions from freshwater to reclaimed water use often are inappropriately ignored. The assumption may be that these costs are to be borne by the users and are of no concern to the project proponent. However, these costs will clearly be an issue of user acceptance of reclaimed water and must be considered early in the marketing analysis and later in the financial analysis. To encourage user participation, the project proponent may elect to pay for on-site costs, to loan the funds to the user at a subsidized rate, or to reduce the reclaimed water rates in the first few years of the project operation to offset the costs of conversion.

Inflation, which is generally ignored in the economic analysis, should be considered in financial analyses insofar as it can be estimated realistically.

SOURCES OF REVENUE AND PRICING OF RECLAIMED WATER

There are a variety of sources of revenue to cover the costs of a water reuse project. An obvious possible source is payment for the reclaimed water by users. Insofar as water reclamation serves as a means of treatment and disposal of wastewater, sewer use charges may be justified as a source of funds to cover reclamation costs. As noted earlier, revenues from freshwater sales may be used to offset reclaimed water costs. Taxes on real estate have been used. Subsidies from other governmental sources or wholesale water suppliers may be available.

Revenue from reclaimed water users can be collected in various forms, including water service connection fees, fixed monthly or annual charges, and variable charges based on quantities of water use. When other water sources are available, the upper boundary for the reclaimed water rate is the competing freshwater rate. A discount from this upper boundary is common to provide an incentive to use reclaimed water. The discount reflects some added costs for the users resulting from on-site conversions or poorer water quality. The lower boundary is providing the water for free or even paying users to accept reclaimed water. This kind of incentive may be appropriate when water reclamation and reuse is an essential wastewater treatment and disposal scheme in which there is no flexibility of users to refuse to take the water. To the extent that users realize a genuine benefit from the use of reclaimed water, prices should reflect this benefit, even if they exceed the costs of the water reclamation and reuse project. Excess revenues can be used to offset other water supply or wastewater management costs to benefit the community as a whole.

Example 9

For the project described in Example 8, perform a financial feasibility analysis. Capital costs will be financed with a 20-year loan at 8 percent annual interest. Operation and maintenance costs are predicted to increase with inflation at an annual rate of 4 percent. The costs in Tables 2.6 and 2.7 were adjusted to dollars in midpoint of construction, 1999. All of the reclaimed water customers currently purchase potable water from the water district proposing the water reclamation project. The district currently obtains potable water from two sources: groundwater pumping at a variable cost of $150/acre-foot and purchased imported water at a variable cost of $650/acre-foot, expressed in 1999 dollars. The imported water cost is expected to increase at a rate of 4 percent annually. Estimate the potential price range to charge for the reclaimed water if the project is to be self-supporting.

From the water district's viewpoint, financial feasibility is contingent upon being able to generate sufficient revenue to recover the water reuse project costs, taking into consideration other financial impacts on the district. From the reclaimed water customers' viewpoints, the expectation is that reclaimed water will cost no more than potable water costs. Calculations to address these perspectives are shown in Tables 2.8 and 2.9.

TABLE 2.8. Example 9: Net Annual Cost.[a]

A	B	C	D	E	F	G	H
	Total Reclaimed	Fresh Water		Operation, Maintenance & Replacement			Total
	Water Deliveries,	Replaced, AF[b]	Debt Service, $		Inflation		Project
Year	AF[b]			Base Year $[b]	Factor at 4%	Adjusted $[c]	Cost, $[d]
1998							
1999							
2000	200	160	154,001	40,000	1.04000	41,600	195,601
2001	270	216	154,001	60,000	1.08160	64,896	218,897
2002	450	360	154,001	85,000	1.12486	95,613	249,614
2003	450	360	154,001	85,000	1.16986	99,438	253,439
2004	450	360	154,001	85,000	1.21665	103,415	257,416
2005	450	360	154,001	85,000	1.26532	107,552	261,553
2006	450	360	154,001	85,000	1.31593	111,854	265,855
2007	450	360	154,001	85,000	1.36857	116,328	270,329
2008	450	360	154,001	85,000	1.42331	120,982	274,982
2009	450	360	154,001	85,000	1.48024	125,821	279,821
2010	450	360	154,001	85,000	1.53945	130,854	284,854
2011	450	360	154,001	85,000	1.60103	136,088	290,088
2012	450	360	154,001	85,000	1.66507	141,531	295,532
2013	450	360	154,001	85,000	1.73168	147,192	301,193
2014	450	360	154,001	135,000	1.80094	243,127	397,128
2015	450	360	154,001	85,000	1.87298	159,203	313,204
2016	450	360	154,001	85,000	1.94790	165,572	319,572
2017	450	360	154,001	85,000	2.02582	172,194	326,195
2018	450	360	154,001	85,000	2.10685	179,082	333,083
2019	450	360	154,001	85,000	2.19112	186,245	340,246
Total			3,080,011			2,648,589	5,728,600

(continued)

TABLE 2.8. (continued).

A	I	J	K	L	M	N
	Avoided Freshwater Purchases			Lost Freshwater Revenue		
Year	Adjusted Unit Cost, $/AF[e]	Total Cost, $[f]	Potable Price Escalation Factor at 2%	Adjusted Potable Water Price, $/AF[g]	Total Lost Revenue, $	Net Cost, $[h]
1998						
1999						
2000	676	108,160	1.02000	408	65,280	152,721
2001	703	151,857	1.04040	416	89,891	156,930
2002	731	263,218	1.06121	424	152,814	139,210
2003	760	273,747	1.08243	433	155,870	135,562
2004	791	284,697	1.10408	442	158,988	131,707
2005	822	296,085	1.12616	450	162,167	127,635
2006	855	307,928	1.14869	459	165,411	123,337
2007	890	320,245	1.17166	469	168,719	118,803
2008	925	333,055	1.19509	478	172,093	114,020
2009	962	346,377	1.21899	488	175,535	108,979
2010	1,001	360,232	1.24337	497	179,046	103,668
2011	1,041	374,642	1.26824	507	182,627	98,074
2012	1,082	389,627	1.29361	517	186,279	92,184
2013	1,126	405,212	1.31948	528	190,005	85,986
2014	1,171	421,421	1.34587	538	193,805	169,512
2015	1,217	438,278	1.37279	549	197,681	72,607
2016	1,266	455,809	1.40024	560	201,635	65,398
2017	1,317	474,041	1.42825	571	205,667	57,821
2018	1,369	493,003	1.45681	583	209,781	49,861
2019	1,424	512,723	1.48595	594	213,976	41,500
Total		7,010,355			3,427,271	2,145,515

[a]AF = acre-feet, 1 acre-foot = 1233.5 cubic meters; [b]Refer to Table 2.7; [c](Col E) * (Col F); [d](Col D) + (Col G); [f](Col C) * (Col I); [g]$650 * (Col G); [g]$400 * (Col K); [h](Col H) − (Col J) + (Col M).

TABLE 2.9. Example 9: Reclaimed Water Price Range.[a]

A Year	B Total Reclaimed Water Deliveries, AF[b]	C Net Cost, $[b]	D Adjusted Potable Water Price, $/AF[b]	E Minimum Reclaimed Water Price, $/AF[c]	F Maximum Reclaimed Water Price, $/AF[d]	G Pricing Margin Available, $/AF[e]
1998						
1999						
2000	200	152,721	408	764	326	(437)
2001	270	156,930	416	581	333	(248)
2002	450	139,210	424	309	340	30
2003	450	135,562	433	301	346	45
2004	450	131,707	442	293	353	61
2005	450	127,635	450	284	360	77
2006	450	123,337	459	274	368	93
2007	450	118,803	469	264	375	111
2008	450	114,020	478	253	382	129
2009	450	108,979	488	242	390	148
2010	450	103,668	497	230	398	168
2011	450	98,074	507	218	406	188
2012	450	92,184	517	205	414	209
2013	450	85,986	528	191	422	231
2014	450	169,512	538	377	431	54
2015	450	72,607	549	161	439	278
2016	450	65,398	560	145	448	303
2017	450	57,821	571	128	457	329
2018	450	49,861	583	111	466	355
2019	450	41,500	594	92	476	383
Total		2,145,515				

[a]AF = acre-feet. 1 acre-foot = 1233.5 cubic meters; [b]Refer to Table 2.8.; [c](Col C)/(Col B);; [d]Because 1 AF of reclaimed water replaces only 0.8 AF of freshwater, maximum price = (Col D) * 0.8.; [e](Col F) − (Col E).

A. Project costs:

(1) Debt service: The repayment period will be assumed to begin upon completion of construction. The loan principal will be computed as of that date. All costs will be assumed to occur at the end of the year, so one year of interest will have accrued for loan disbursements made to cover design costs during 1998.

$$\text{Loan principal} = (\$150,000)(1.08) + (\$1,350,000) = \$1,512,000$$

$$\text{Capital recovery factor} = \frac{i(1 + i)^n}{(1 + i)^n - 1} = \frac{(0.08)(1 + 0.08)^{20}}{(1 + 0.08)^{20} - 1} = 0.101852$$

$$\text{Debt service} = (\$1,512,000)(0.101852) = \$154,001$$

(2) Operation, maintenance, and replacement costs: In the economic analysis in Example 8, all costs are given in constant dollars. In a financial analysis, where actual cash flow is of relevance, inflated costs must be estimated. Thus, operation, maintenance, and replacement costs given in Table 2.7 must be inflated using the assumed 4 percent rate. The inflation factor, shown in Column F of Table 2.8, is based on the compound amount factor.

(3) External financial impacts: As a result of use of reclaimed water, there are financial impacts external to the project itself that should be taken into consideration. For example, the district will avoid variable costs associated with supplying potable water to the reclaimed water customers. Typically, the district would reduce the most expensive source of supply, the imported water purchase, at $650/acre-foot, as reflected in Columns I and J in Table 2.8. However, the district will also lose the revenue from the sale of potable water, as shown in Columns L and M. An external cost possibly borne by the customers is the on-site conversion cost. In this example this cost, shown in Table 2.6, is assumed to be part of the project cost borne by the water district.

(4) Net project cost: The net project cost is shown in Column N in Table 2.8 and Column D in Table 2.9.

B. Reclaimed water pricing:

(1) Calculations related to reclaimed water pricing are shown in Table 2.9. The hypothetical minimum price is the unit cost to recover the net cost of the project, Column E. From the customers' perspective, the maximum reclaimed water price is the price that would have been paid for potable water. If on-site conversion costs were the responsibility of the customers, an allowance for this would have to

be considered in setting reclaimed water prices. A further consideration is that it is assumed that customers will have to purchase more reclaimed water than they would have purchased of potable water due to the quality difference. (This could be needed, for example, because of a reduction in the number of cycles through an industrial cooling tower or an increase in irrigation flows to leach salts from the plant root zone.) The maximum reclaimed water price is shown in Column F.

(2) The pricing margin, that is, the spread between the minimum and maximum possible reclaimed water prices, is shown in Column G. It is indicated that there is not any margin available during the first two years. Unless a subsidy can be obtained to cover a negative cash flow in these years of project operation, this project would not be feasible. However, a case can be made to raise additional potable water sales revenue to provide this subsidy. In the later years, excess reclaimed water can be used to cover potable system costs.

SENSITIVITY ANALYSIS AND CONSERVATIVE ASSUMPTIONS

The tendency for engineers to be ''conservative'' in design criteria can be misleading for certain planning criteria. Estimating water demands on the high side is not appropriate for evaluating economic justification and financial feasibility of a project and leads to false expectations with respect to costs as well as project yields. It is advisable to report estimated project yield as a range. A sensitivity analysis should be performed testing the effects of high and low project deliveries on economic cost and financial feasibility. Depending on the degree of uncertainty in estimating demands for reclaimed water and the level of market assurances obtained, the worst case scenario in terms of project cost would be minimum deliveries representing potential water users in existence that are willing to execute a binding agreement to accept reclaimed water. The best case of maximum deliveries would include demands such as future undeveloped land or users located outside the project proponent's boundaries where institutional agreements have not been formalized. It should be made clear that the best case scenario is associated with the highest degree of risk of failing expectations.

CONCLUSION

A systematic planning approach to water reuse projects will provide a framework for defining project objectives, identifying the issues to be addressed and potential problems that may be encountered, formulating facilities alternatives, and analyzing the alternatives in a realistic way. The resulting project may not be as grand as the initial optimism may have envisioned, but it will have a greater chance of meeting expectations. A cautious approach is especially appropriate for agencies getting involved in water reclamation and reuse for the first time. After initial experience with a smaller project, additional phases can be approached with background knowledge and with a number of issues, such as institutional arrangements, already resolved. Shortcuts can be taken in planning to build a project faster, but they will not lead to satisfied reclaimed water users and an optimum, cost-effective project.

ACKNOWLEDGEMENT

Nick Kontos reviewed the manuscript of this chapter and provided valuable comments.

REFERENCES

1   Office of Water Recycling, California State Water Resources Control Board. 1990. *California Municipal Wastewater Reclamation in 1987*. [Sacramento, California], June 1990.

2   Office of the Federal Register, National Archives and Records Administration. 1991. *Code of Federal Regulations, Protection of Environment, 40, Parts 1 to 51*. Washington: U.S. Government Printing Office, Title 40, Part 35, Subpart E, Appendix A, pp. 539–546.

3   James, L. D., and Lee, R. R. 1971. *Economics of Water Resources Planning*. New York: McGraw-Hill Book Company, pp. 161–162.

4   Asano, T., and Mills, R. A. 1990. "Planning and Analysis for Water Reuse Projects," *Journal American Water Works Association,* 82(1):38–47.

5   Mills, R. A. and Asano, T. 1996. "A Retrospective Assessment of Water Reclamation Projects," *Water Science and Technology,* 33(10–11):59–70.

6   Bales, R. C., Biederman, E. M. and Arant, G. 1979. "Reclaimed Water Distribution System Planning—Walnut Valley, California," *Proceedings Water Reuse Symposium: Water Reuse—From Research to Application.* American Water Works Association Research Foundation, Denver, Colorado.

7   Task Force on Water Reuse. 1989. *Water Reuse, Manual of Practice SM-3,* Second Edition. Alexandria, Virginia: Water Pollution Control Federation, p. 115.

8   Office of Water Recycling, California State Water Resources Control Board. 1994. "Water Reclamation Loan Program Guidelines." [Sacramento, California], June 1994.

9   Mills, R. A., and Asano, T. 1986/87. "The Economic Benefits of Using Reclaimed Water," *Journal of Freshwater,* 10:14–15.

10   Riggs, J. L., and West, T. M. 1986. *Engineering Economics,* Third Edition. New York: McGraw-Hill Book Company.

11 Grogan, T. 1996. "A New Era of Stability," *Engineering News-Record,* 236(12):71–77.

12 Richard, D., Asano, T. and Tchobanoglous, G. November 1992. *The Cost of Wastewater Reclamation in California.* Davis, California: Department of Civil and Environmental Engineering, University of California, Davis.

13 U.S. Bureau of Reclamation. 1990. "Bureau of Reclamation, Change in Discount Rate for Water Resources Planning," *Federal Register,* 55(15):2265.

Wastewater Reclamation and Reuse in Small and Decentralized Wastewater Management Systems

T HE importance of wastewater reclamation in the field of water resources management is now acknowledged widely, as discussed
throughout this book. Similarly, the important role of wastewater reclamation in maintaining a sustainable environment is also gaining wider acceptance. While the principal focus of this book is on the reclamation and
reuse of wastewater from centralized wastewater treatment systems, it
must be recognized that, at the present time, more than sixty million people
in the United States live in homes that are not connected to a centralized
wastewater collection and treatment system. In the early 1970s, with the
passage of the Clean Water Act, it was thought that it would only be a
matter of time before sewerage facilities would be available to almost all
residents of the continental United States. Now, more than 25 years later,
it is recognized that complete sewerage of our country may never be
possible, due to both geographic and economic reasons. Given that complete sewerage will not be possible in the foreseeable future, and that
increasing demands are being made on freshwater supplies, it is clear that
decentralized wastewater management, in which reuse will be an important
element, is of great importance in developing long-term strategies for the
management of our environment.

The purposes of this chapter are fourfold: (1) to examine the role of decentralized systems in wastewater management, with specific reference to reclamation and reuse, (2) to review appropriate technologies for wastewater

George Tchobanoglous, Department of Civil and Environmental Engineering, University of California at Davis, Davis, CA 95616; Ronald W. Crites, Brown and Caldwell,
9616 Micron Ave., Suite 600, Sacramento, CA 95827; Sherwood C. Reed, Environmental Engineering Consultant, 50 Butternut Rd., Norwich, VT 05055.

treatment and reuse for individual and other DWM applications, (3) to compare the performance of selected decentralized and centralized wastewater treatment technologies, and (4) to illustrate representative reclamation and reuse applications in small and decentralized wastewater management.

ROLE OF DECENTRALIZED WASTEWATER MANAGEMENT

Decentralized wastewater management (DWM) may be defined as the collection, treatment, and disposal/reuse of wastewater from individual homes, clusters of homes, or isolated communities, industries, or institutional facilities (Tchobanoglous, 1995). In decentralized systems both the solid and liquid fractions of the wastewater are retained near their point of origin, although the liquid portion may be transported to a centralized point for further treatment and reuse. Because it will not be possible, for the reasons cited above, to provide centralized sewerage facilities for all of the residents of the United States now or in the future, the focus of the field of wastewater management is beginning to change from the construction and management of regional centralized sewerage systems to small, decentralized, and satellite wastewater treatment facilities. As more emphasis is placed on DWM, the opportunities for localized reclamation/reuse will increase. Thus, in addition to the implementation of large scale reclamation projects, it will become important, in the future, to develop appropriate strategies and operating agencies for the management of localized reclamation and projects.

WASTEWATER REUSE APPLICATIONS IN DWM SYSTEMS

The most common uses of reclaimed wastewater from centralized reclamation systems worldwide, as discussed in Chapter 1, have been for agricultural and landscape irrigation. Recently, groundwater recharge and indirect potable reuse have received considerable attention in the United States. The repurification project in San Diego, CA, in which it is proposed to blend repurified wastewater with local runoff and imported water in a local water supply storage reservoir, is an example of such a project (Montgomery/Watson and NBS Lowry, 1994). In DWM systems, however, the most common reuse application will be for agricultural and landscape irrigation. Other uses will include toilet flushing in complete recycle systems.

APPROPRIATE WASTEWATER TECHNOLOGIES FOR INDIVIDUAL AND SMALL SYSTEMS

To protect the environment and to maximize reuse opportunities, discharge requirements for treated wastewater for small dischargers are now the same as those for large dischargers. The challenge is to be able to provide the required level of treatment in decentralized systems, subject to serious economic constraints. Alternative wastewater management technologies for unsewered areas for the collection and transport of wastewater, wastewater treatment, and the disposal and/or reuse, are reported in Table 3.1. Specific technologies are discussed in the following section.

The most common system used for the management of wastewater from individual dwellings and community facilities in unsewered locations consists of a septic tank for the partial treatment of the wastewater, and a subsurface disposal field for final treatment and disposal of the septic tank effluent. The most significant change that has occurred in the past 15 years, in the implementation of DWM systems, is the development of new technology and the reapplication of old technology. Many of these new technologies have been significant with respect to the development of reuse opportunities. Some of these technologies are highlighted in the following discussion. Additional details on technologies suitable for small and decentralized systems may be found in Crites and Tchobanoglous (1998).

PRETREATMENT

Two noteworthy improvements in pretreatment include the development of watertight structurally sound septic tanks and the effluent filter vault (see Figure 3.1). By eliminating the entry of extraneous flows into the septic tank, downstream treatment units such as intermittent and recirculating packed bed filters have become feasible. Use of the effluent filter vault to eliminate the discharge of untreated solids has brought about significant improvements in the quality of the effluent discharged from septic tanks (see Table 3.2). In addition to improving the performance of septic tanks, the development of the effluent filter vault is significant because it has made the use of small high-head (e.g., 100 m, 300 ft) multi-stage well pumps feasible, for use with pressure sewers, and for pressure dosing packed-bed filters and soil absorption systems.

INTERMITTENT AND RECIRCULATING PACKED BED FILTERS

Because conventional disposal fields cannot be used in some locations, many alternative systems have been developed. The most successful of these include intermittent sand filters (ISF) and recirculating granular-medium filters (RGF) (see Figure 3.2). Intermittent sand filters have become quite popular for single family residences, in all parts of the

TABLE 3.1. Technologies Used for Decentralized Wastewater Management Systems.

Collection and Transport of Wastewater	Wastewater Treatment and/or Containment	Wastewater Reuse/Disposal
House drains and building sewers	Primary treatment	Landscape irrigation
Conventional gravity sewers	Septic tank (individual, home, home cluster, and community)	Surface application
Pressure sewers (with non grinder pumps)	Imhoff tank	Drip application
Pressure sewers (with grinder pumps)	Advanced primary treatment	Subsurface soil disposal
Small-diameter variable-slope gravity sewers	Septic tank with attached growth reactor element	Leachfields
Vacuum sewers	Secondary treatment	Conventional
Combinations of the above	Aerobic/anaerobic units	Shallow trench (pressure dosed)
	Anaerobic units	Shallow sand-filled
	Two stage anaerobic/aerobic units with disinfection (used in Japan)	Drip application
	Intermittent packed bed filter (sand, foam, textile chips, crushed glass)	Seepage beds
	Recirculating packed bed filter	Mound systems
	Peat filter	Evapotranspiration/percolation beds
	Constructed wetland	Evaporation systems
	Recycle treatment systems	Evapotranspiration bed
	Toilet flushing	Evaporation pond
	Landscape watering and toilet flushing	Constructed wetlands
	Onsite containment	Discharge to water bodies
	Privy	Constructed wetlands
	Holding tank	Lakes
		Streams/rivers
		Combinations of the above

116

TABLE 3.2. Typical Data on the Expected Effluent Wastewater Characteristics from a Residential Septic Tank without and with an Effluent Filter Vault.[a]

Constituent (1)	Typical Complete Mix Value[b] (mg/L) (2)	Concentration (mg/L)					
		Without Effluent Filter			With Effluent Filter		
		Range (3)	Typical without Ground up Kitchen Waste (4)	Typical with Ground up Kitchen Waste (5)	Range (6)	Typical without Ground up Kitchen Waste (7)	Typical with Ground up Kitchen Waste (8)
BOD_5	450	150–250	180	190	100–140	130	140
COD	1,050	250–500	345	400	160–300	250	300
TSS	503	40–140	80	85	20–55	30	30
NH_3 as N	41.2	30–50	40	44	30–50	40	44
Org. N as N	29.1	20–40	28	31	20–40	28	31
TKN as N	70.4	50–90	68	75	50–90	68	75
Org. P as P	6.5	4–8	6	6	4–8	6	6
Inorg. P as P	10.8	8–12	10	10	8–12	10	10
Total P as P	17.3	12–20	16	16	12–20	16	16
Oil and grease	164	20–50	25	30	10–20	15	20

[a]From Crites and Tchobanoglous, 1998.
[b]Concentration if waste constituents were mixed completely.

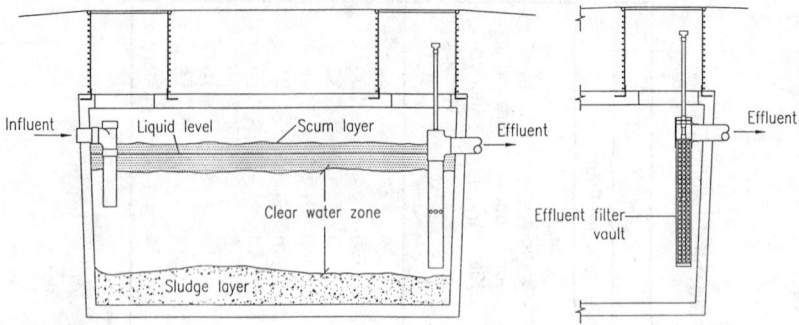

Figure 3.1. Water-tight septic tank with effluent filter (Courtesy Orenco Systems, Inc.).

United States, because of their excellent performance, reliability, and relatively low cost. The effluent from an intermittent sand filter and from recirculating gravel filters is of such high quality (see Table 3.3) that it can be used in a variety of reuse applications, including drip irrigation. Recirculating granular-medium filters are used for larger flows. Typical design criteria for ISF and RGF are reported in Table 3.4.

BIOLOGICAL TREATMENT UNITS

A variety of anaerobic, aerobic, and combined anaerobic/aerobic treatment systems have also been developed. The use of aerobic biological treatment systems without the availability of a responsible management agency is, however, not recommended. For enhanced removal of nitrogen, septic tanks with integral trickling filter or absorbent highly porous plastic elements have been developed (Orenco Systems, Inc., 1997; Jowett and McMaster, 1994).

The packed bed absorbent biofilter (see Figure 3.3) is also significant in that a high quality effluent is produced at loading rates of 20 to 40 times greater than those used for intermittent sand filters (Jowett and McMaster, 1994). The effluent from the absorbent biofilter is suitable for application by drip irrigation.

MEMBRANE TREATMENT SYSTEMS

In some membrane systems, currently being developed and tested for the treatment of septic tank effluent, all of the wastewater would be used

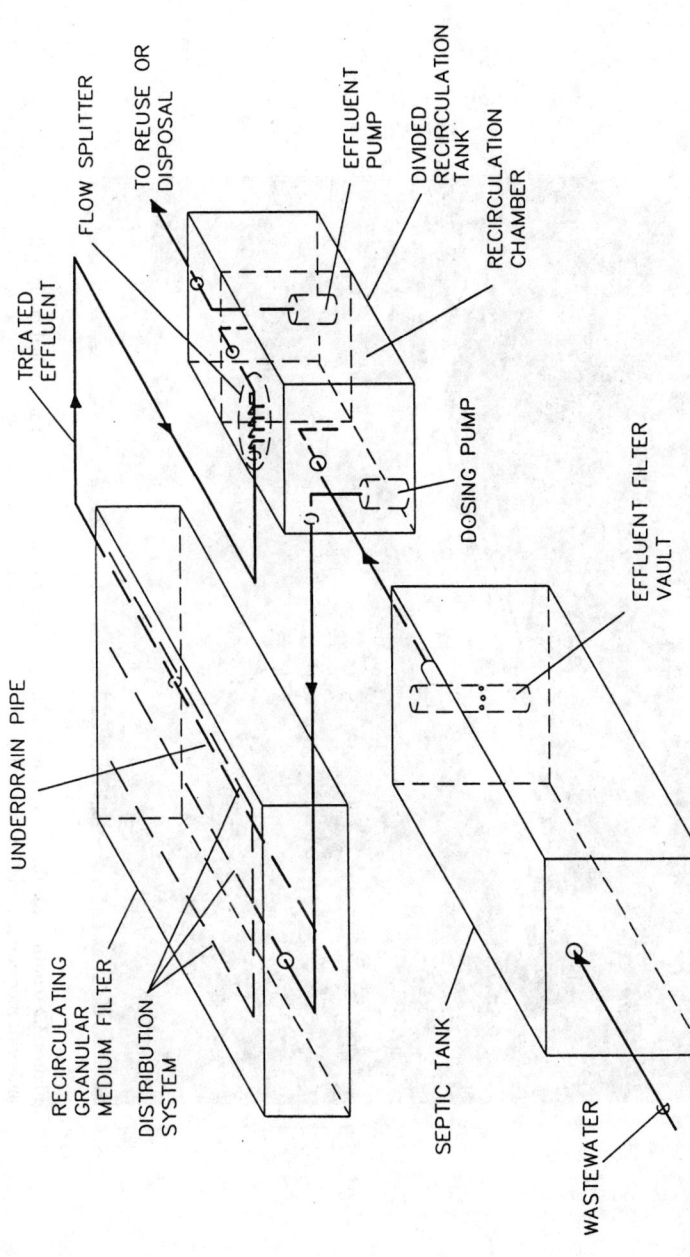

FLOW SPLITTER

TREATED EFFLUENT

TO REUSE OR DISPOSAL

EFFLUENT PUMP

DIVIDED RECIRCULATION TANK

RECIRCULATION CHAMBER

UNDERDRAIN PIPE

RECIRCULATING GRANULAR MEDIUM FILTER

DISTRIBUTION SYSTEM

DOSING PUMP

EFFLUENT FILTER VAULT

SEPTIC TANK

WASTEWATER

Figure 3.2. Views of recirculating granular medium packed bed filters: (a) schematic of recirculating granular medium filter (from Crites and Tchobanoglous, 1998).

119

Figure 3.2 (continued). (b) View of recirculating granular medium filter for individual home with cover and orifice shields removed for inspection.

TABLE 3.3. Summary of ISF Performance with Respect to the Removal of BOD and Nitrogen.[a]

Study and Location	BOD$_5$			Total nitrogen		
	Influent (mg/L)	Effluent (mg/L)	Percent Removal	Influent (mg/L)	Effluent (mg/L)	Percent Removal
Grantham[b] (1949), FL	148	14	90	37	32	14
Furman (1955), FL	57	4.8	92	30	16	47
Sauer (1977), WI	123	9	93	nd[c]	nd	nd
Ronayne (1984), OR	217	3.2	98	58	30	48
Effert[d] (1988), OH	127	4	97	42	38	10
Stinson Beach (1987–1990), CA	203	11	94	57	41	28
Nor (1991), Davis, CA	82	0.5	99	14	7.2	47
Town of Paradise (1992), CA	148	6	96	38	19	50
Stinson Beach (1996–1997), CA	180	<5[e]	97	60	35	42

[a]From Crites and Tchobanoglous, 1998.
[b]Relatively high hydraulic loadings (1.7 to 4 gal/ft^2•d) and shallow beds (18 to 30 in).
[c]nd = no data.
[d]High hydraulic loading rates (2 to 10 gal/ft^2•d).
[e]Values less than 5 mg/L reported as <5 by laboratory.

121

TABLE 3.4. Design Criteria for Intermittent Sand and Recirculating Gravel Filters.

Design Factor	Unit	ISF		RGF	
		Range	Typical	Range	Typical
Filter medium					
Material		Washed, durable sand		Washed, durable gravel	
Effective size	mm	0.25–0.75	0.35	1–5	3
Depth	mm	450–900	600	450–900	600
Uniformity coeff.	U.C.	<4	3.5	<2	2.0
Underdrain					
Type		Slotted or perforated drain pipe		Slotted or perforated drain pipe	
Size	mm	75–100	100	75–100	100
Slope	%	0–0.1	0	0–0.1	0
Pressure distribution					
Pipe size	mm	25–50	38	25–50	38
Orifice size	mm	3–6	3	3–6	3
Head on orifice	m	1–2	1.6	1–2	1.6
Lateral spacing	m	0.5–1.2	0.6	0.5–1.2	0.6
Orifice spacing	m	0.5–1.2	0.6	0.5–1.2	0.6
Design parameters					
Hydraulic loading[a]	mm/d	16–48	40	120–240	200
Organic loading	kg BOD/m²·d	0.0025–0.01	<0.005	0.01–0.04	<0.001
Dosing frequency	times/d	12–72	18		
Recirculation ratio	unitless			3:1–6:1	4:1
Dosing frequency	min/30 min			1–10	5
Dosing tank vol	days flow	0.5–1.0	0.5	0.5–1.0	0.5
Filter temp	°C		>4		>4

aBased on peak flow, with a peaking factor of 2.5.

Figure 3.3. View of absorbent porous plastic medium used in the Waterloo biofilter.

locally. Residual solids remaining in the retentate from the membrane processing step would be returned to the septic tank for further treatment and long-term thickening.

SHALLOW PRESSURE-DOSED DISPOSAL TRENCHES

Final treatment and disposal of the effluent from a septic tank or other treatment unit is accomplished, most commonly, by means of subsurface-soil absorption. A soil absorption system, commonly known as a leachfield, consists of a series of trenches [0.9 to 1.5 m (3 to 5 ft)] filled with a porous medium (e.g., gravel). Effluent from the septic tank is applied to the disposal field by intermittent gravity flow, or by periodic dosing using a pump or a dosing siphon. Unfortunately, conventional trench designs fail to take maximum advantage of the treatment capabilities of the soil because they are typically located below the region of maximum bacterial activity and soil carbon. New trench designs now involve the use of very shallow trenches with no porous medium (see Figure 3.4). The use of shallow trenches enhances (1) the treatment of the effluent with respect to the removal of BOD, TSS, phosphorus, and nitrogen and (2) the reuse opportunities for septic tank effluent. It is interesting to note that the use of such shallow trenches was recommended in an early Public Health Bulletin in 1915 (Lumsden et al., 1915).

DRIP IRRIGATION

Drip irrigation technology has advanced over the years to where nonclog emitters are available for both surface and subsurface uses. Sand filter and other high quality effluent can be used in drip irrigation of landscape and other crops. Modern emitters are designed with a turbulent flow path to minimize clogging from suspended solids. They are also treated with a herbicide to protect them from root intrusion.

Drip emitters typically operate at a flowrate of 1 to 2 gal/hr with 0.06 to 0.07 in (1.5 to 1.8 mm) diameter openings. Drip irrigation systems usually require 15 to 25 lb/in^2 pressure. The ability to flush the lines and to apply periodic doses of chlorine for control of clogging from bacterial growth is also provided. A typical onsite drip irrigation system, as shown in Figure 3.5, consists of emitter lines placed on 2 ft (0.6 m) centers with a 2 ft (0.6 m) emitter spacing. This spacing is typical for sandy and loamy soils. Closer spacings of 15 to 18 in (0.4 to 0.45 m) are used for clay soils where lateral movement of water is restricted. The emitter lines are placed at depths of 6 to 10 in (150 to 250 mm). The recommended application rate is 0.125 gal/ft^2•d.

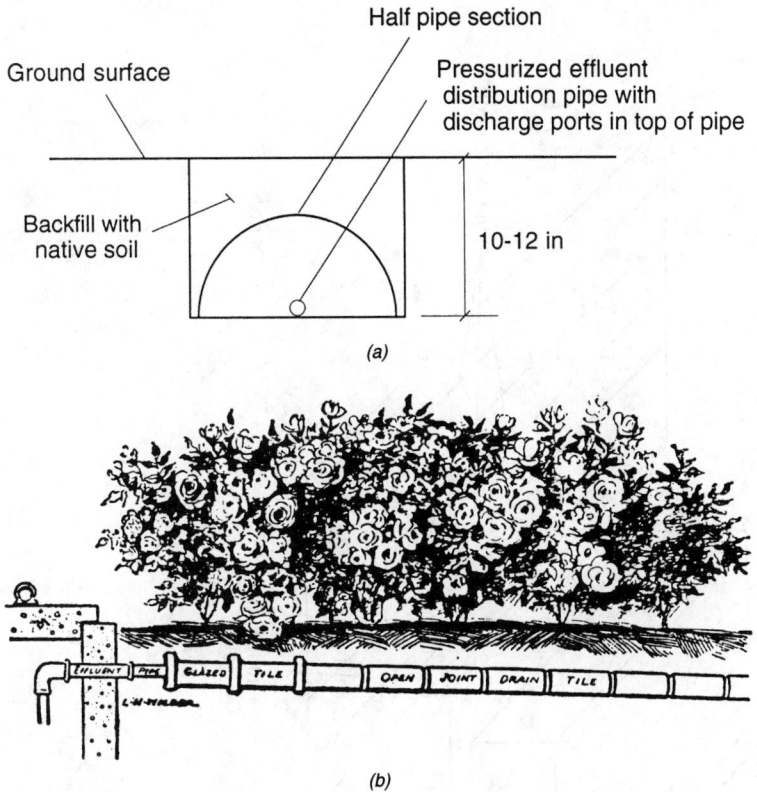

Half pipe section

Ground surface

Pressurized effluent
distribution pipe with
discharge ports in top of pipe

Backfill with
native soil

10-12 in

(a)

(b)

Figure 3.4. Shallow trenches for the reuse and disposal of septic tank effluent: (a) current embodiment, and (b) recommended shallow placement (circa 1915, Lumsden et al., 1915).

WASTEWATER TECHNOLOGIES FOR CLUSTER AND COMMUNITY SYSTEMS AND ISOLATED COMMERCIAL FACILITIES

The treatment component of cluster and community systems and for other isolated commercial facilities will vary with the size of the installation and the final disposition of the effluent.

CLUSTER AND COMMUNITY SYSTEMS

Typically, a large septic tank will be used for a cluster of homes. Imhoff tanks, commonly used in the past, are making a comeback in modified forms. In some communities, septic tanks may be used for the separation of settleable solids and oil and grease. Recirculating granular-medium filters are used in conjunction with septic tanks where a higher level of

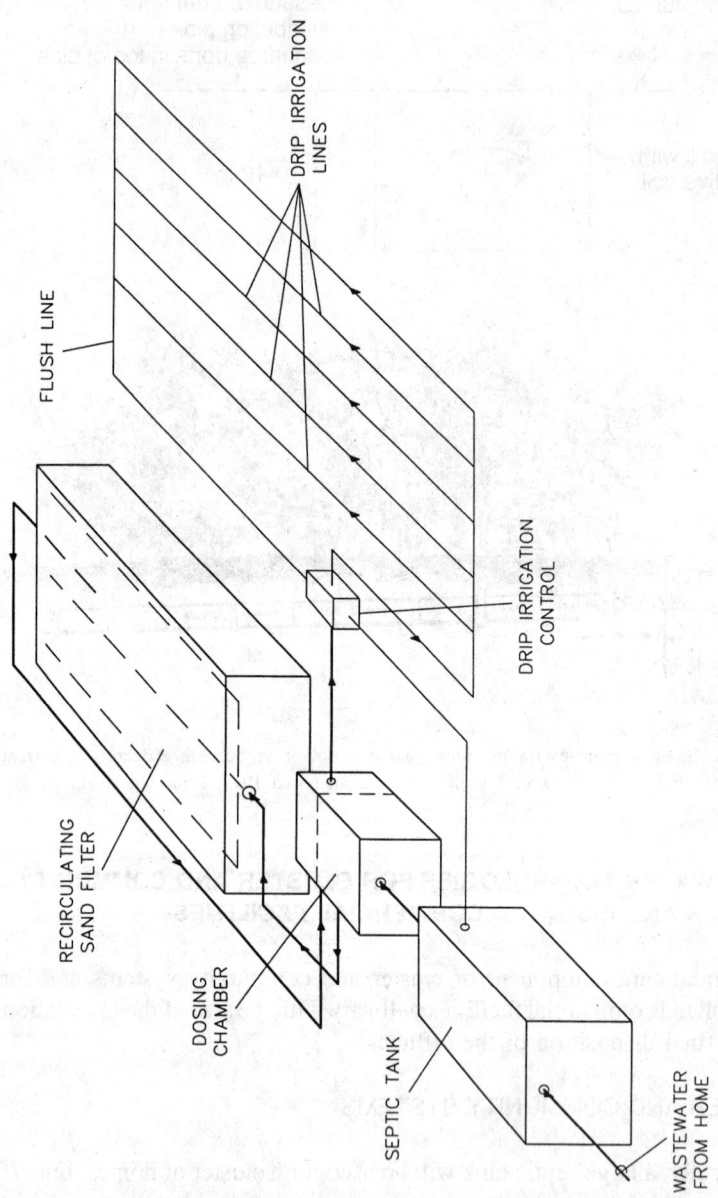

Figure 3.5. Schematic for drip irrigation system for individual residence (Crites and Tchobanoglous, 1998).

FLUSH LINE

DRIP IRRIGATION LINES

DRIP IRRIGATION CONTROL

RECIRCULATING SAND FILTER

DOSING CHAMBER

SEPTIC TANK

WASTEWATER FROM HOME

treatment is required. Pre-engineered and constructed "package" plants, and individually designed plants, are used where the flows are larger.

The methods used for effluent disposal will also vary with the size of the system and the local reuse opportunities. For small installations serving a cluster of homes, effluent disposal has been accomplished most commonly using disposal fields and drip irrigation in greenbelts and recreational areas. As the size of the system increases, the methods used for the disposal of effluent are, as shown in Table 3.1, essentially the same as those used for larger systems. Disinfection, where required, is by means of chemical addition (e.g., chlorine and related compounds) or ultraviolet (UV) radiation. With the high quality effluent obtained with packed bed filters, treated effluent from cluster systems is being reused increasingly for landscape irrigation.

SYSTEMS FOR ISOLATED COMMERCIAL FACILITIES

Complete recycle systems have been developed for isolated commercial buildings where the treated effluent is used for toilet and urinal flushing or for landscape irrigation. Where an acceptable onsite disposal system cannot be installed, holding tanks can be employed. One such unit involves three treatment steps: (1) the solids in the wastewater are collected and treated aerobically, (2) the effluent from the biological treatment unit is then passed through a self-cleaning ultrafiltration step where residual organics, microorganisms, and suspended solids are removed, and (3) in the final step the effluent is passed through an activated carbon column for polishing (see Figure 3.6). The material removed in the ultrafiltration step is returned to the first processing step for further treatment. The effluent from the carbon filters is disinfected with ozone before it is reused for flush water. Although such processes are expensive, they have been used for office buildings located in unsewered areas, and where water for domestic use is in short supply.

PERFORMANCE EXPECTATIONS FOR APPROPRIATE TECHNOLOGIES

Performance expectations for various technologies are of importance with respect to their use in reclamation and reuse applications. Treatment levels achievable with various combinations of unit operations and processes used for wastewater treatment, are reported in Table 3.5. Treatment technologies for individual and small systems are reported in the last four lines of Table 3.5. What is interesting to note, is that many of the treatment systems used for individual homes, cluster systems, and small systems

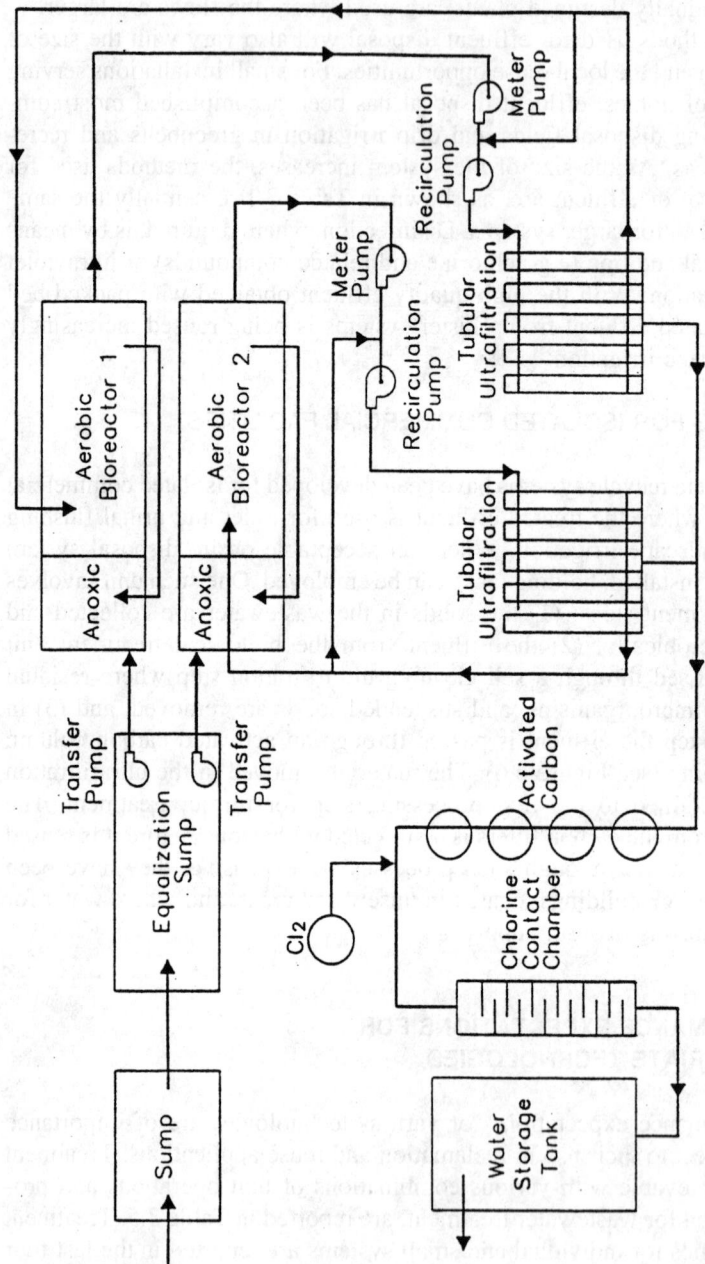

Figure 3.6. Schematic diagram of complete recycle system for 25,000 to 50,000 gal/d used at The Water Gardens, Santa Monica, CA (adapted from Jordan and Senthilnathan, 1996).

TABLE 3.5. Treatment Levels Achievable with Various Combinations of Unit Operations and Processes Used for Wastewater Treatment.[a]

Treatment Process	Typical Effluent Quality (mg/L) except Turbidity (NTU)						
	SS	BOD$_5$	COD	Total N	NH$_3$-N	PO$_4$-P	Turb
Primary sedimentation + primary effluent filtration	20–40	50–100	100–200	20–30	15–25	6–10	20–30
Activated sludge + granular medium filtration	4–6	<5–10	30–70	15–35	15–25	4–10	0.3–5
Activated sludge + granular medium filtration + carbon adsorption	<5	<5	5–20	15–30	15–25	4–10	0.3–3
Activated sludge/nitrification, single stage	10–25	5–15	20–45	20–30	1–5	6–10	5–15
Activated sludge/nitrification/denitrification separate stages	10–25	5–15	20–35	5–10	1–2	6–10	5–15
Metal salt addition to activated sludge	10–20	10–20	30–70	15–30	15–25	<2	5–10
Metal salt addition to activated sludge + nitrification/denitrification + filtration	<5–10	<5–10	20–30	3–5	1–2	<1	0.3–2
Mainstream biological phosphorus removal[b]	10–20	5–15	20–35	15–25	5–10	<2	5–10
Mainstream biological nitrogen and phosphorus removal[b] + filtration	<10	<5	20–30	<5	<2	<2	0.3–2
Activated sludge + granular medium filtration + carbon adsorption + reverse osmosis	<1	<1	5–10	<2	<2	<1	0.01–1
Activated sludge/nitrification-denitrification and phosphorus removal + granular medium filtration + carbon adsorption + reverse osmosis	<1	<1	2–8	<0.1–0.5	<0.1–0.5	<0.1–0.5	0.01–1
Septic tank with effluent filter vault	25–40	80–120	120–260	40–80	30–60	8–12	10–20
Septic tank with internal trickling filter	20–40	40–60	60–100	10–20	8–16	8–12	8–20
Septic tank with effluent filter vault + intermittent sand filtration	0–5	0–5	10–40	4–10	0–2	6–10	0–2
Septic tank + absorbent biofilter	5–15	5–15	30–80	10–20	8–16	6–10	1–2

[a]Adapted, in part, from Tchobanoglous and Burton, 1991.
[b]Removal process occurs in the main flowstream as opposed to sidestream treatment.

129

perform at a level equal to or better than comparable large scale centralized systems. Where effluent disinfection is required, the use of chlorine and related compounds and conventional low pressure UV radiation are used. Clearly, technology is now available for the production of high quality water from wastewater, regardless of system size.

REPRESENTATIVE DWM SYSTEMS DESIGNED FOR REUSE APPLICATIONS

In the past, effluent disposal was the principal concern with DWM systems. However, with the advent of new technologies, the reuse of treated effluent has become an integral part of the design of DWM systems.

INDIVIDUAL HOME REUSE SYSTEM

With the development and use of both intermittent and recirculating packed bed filters for the treatment of septic tank effluent and the development of improved drip irrigation systems effluent reuse has now become feasible. The most common application is for irrigation of lawns and individual plants.

STINSON BEACH WATER DISTRICT

Since its inception as a summer community north of San Francisco in the late 1800s, onsite systems have been used for the treatment and disposal of liquid wastewater in Stinson Beach. The Stinson Beach Water District formed an Onsite Management District in 1978, eschewed an expensive centralized collection and treatment system, and has managed over 650 onsite systems since then. Instead of a centralized collection system that would have dewatered the western slope of the hills above the community, onsite systems continue to recharge the shallow aquifer, sustaining the growth of trees and shrubs (see Figure 3.7). Effluent from intermittent sand filters is also used for irrigation of plants and ground cover in individual yards.

SMALL RESIDENTIAL DEVELOPMENT

Stonehurst is a 47 lot subdivision located near the City of Martinez, CA, approximately 25 miles east of San Francisco. The subdivision is located in a hilly, rural area without a wastewater collection system (see Figure 3.8). Although the subdivision was approved originally for the use of conventional septic tank/leachfield systems, the developer chose to

Figure 3.7. View of Stinson Beach, CA looking north along Highway One, showing variation from beach front to hillside terrain.

131

Figure 3.8. View of Stonehurst residential development showing rolling terrain.

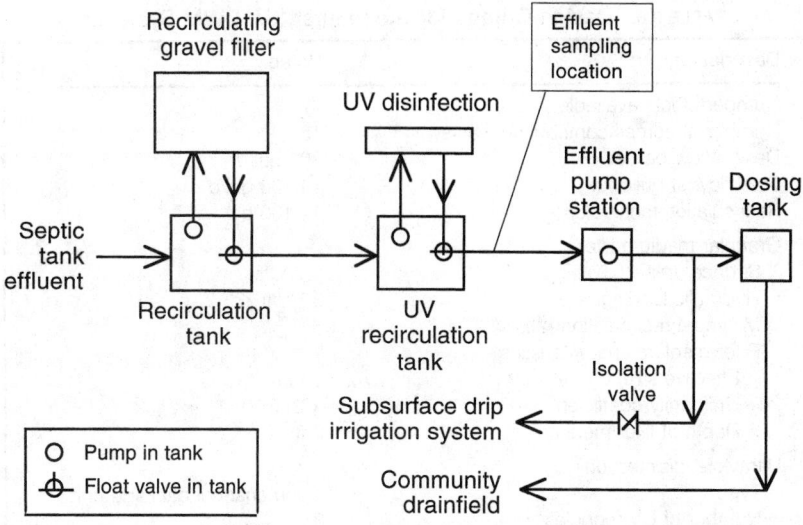

Figure 3.9. Flow diagram for Stonehurst wastewater treatment system.

implement a community wastewater collection, treatment, and reuse system. The rationale for this approach was to free the homeowners from concerns about possible leachfield failure, and to provide for the beneficial reuse of the wastewater generated and treated at the subdivision.

Wastewater Management System

The wastewater management system that was designed and constructed incorporates a number of innovative technologies. The system includes the use of 1500 gal, two-compartment, water-tight septic tanks with screened effluent filter vaults, high-head effluent pumps, small diameter variable grade sewers, and two pressure sewers. The principal elements of the wastewater treatment system include: a recirculating granular-medium filter for treatment, and a UV system for disinfection. A schematic flow diagram of the treatment system is shown in Figure 3.9. Design criteria for the treatment plant and disposal/reuse facilities are summarized in Table 3.6. A subsurface drip irrigation system has been installed for reuse of the treated effluent. Non-clog drip emitters are buried in short lengths of gravel-filled PVC pipe. The park is landscaped with drought-tolerant trees and shrubs. Three drip emitters are provided for each tree and one for each shrub. At buildout, reclaimed water will provide 100 percent of the landscaped water demands at the one acre park. In the wet season, reclaimed water not needed for

TABLE 3.6. **Design Criteria for Stonehurst, CA DWM System.**[a]

Description	Value
Number of lots available	47
Number of homes contributing wastewater	15
Design flow per lot	300 gal/d
Total flow at buildout	14,100 gal/d
Recirculation tank volume	14,100 gal
Granular medium filter	
Surface area	4,700 ft^2
Hydraulic loading rate	3 gal/ft$^2 \cdot$d
Minimum recirculation ratio[b]	5:1
Filter medium characteristics	
Effective size, d_{10}	3 mm
Uniformity coefficient	<2
Depth of filter medium	2 ft
Ultraviolet disinfection	
Type	Open channel package unit
Number of UV modules	3
Lamps per UV module	2
Minimum UV dose	95 mW$\cdot$s/cm^2
Community drainfield	
Trench dimensions	24 in wide × 36 in deep
Total trench length	5,000 linear ft
Total drainfield area	2.5 ac
Design loading rate	2.8 gal/lineal ft$\cdot$d
Drip irrigation system	
Emitter type	Non-clog, buried
Reuse area	1 acre park
Reuse vegetation	Landscape, trees, and shrubs

[a]From Crites et al., 1997.
[b]Based on wastewater flow to treatment plant.

irrigation purposes, is disposed of automatically in the pressure-dosed community drainfield.

Performance Data

Summary performance data for the treatment system for the 28 month period from June 1994 through September 1996 are presented in Table 3.7. As reported in Table 3.7, both monthly and quarterly values are reported. The reported monthly values for BOD$_5$ are based on an average of at least two samples per month. Monthly values reported for TSS, COD, pH, and total coliform are based on an average of at least four samples per month. The reported quarterly values are based on grab samples. The measured data for the individual constituents are reported in terms of the

TABLE 3.7. Performance Data for Stonehurst Wastewater Treatment System for the 28-Month Period from June 1994 through September 1996. Effluent Samples Taken Following Recirculating Gravel Filter and UV Disinfection as Shown in Figure 3.5.

Constituent[a]	Unit	Sample Reporting Period[b]	Total Number of Samples to Date	Observed Values[c]			
				Range	Arithmetic Mean	Geometric Mean	Median
BOD5	mg/L	Monthly	56	0- < 5		<5	<5
COD	mg/L	Monthly	120	1.0–18.0		4.7	5.0
TSS	mg/L	Monthly	118	2.0–15.0		5.0	4.9
pH	Unitless	Monthly	120	6.96–8.65	—[d]	—[d]	7.61
Total coliform	MPN/100 mL	Monthly	118	<2–12.5	—[d]	—[d]	<2
NH4	mg/L	Quarterly	9	0–15	—[e]	—[e]	0.0
NO3	mg/L	Quarterly	9	3.55–37.0		12.5	12.2
TKN	mg/L	Quarterly	9	0–3	—[e]	—[e]	0.4
Oil and grease	mg/L	Quarterly	9	0–12	—[e]	—[e]	0.0
TDS	mg/L	Quarterly	9	340–770	630	—[e]	656
EC	µmhos/cm	Quarterly	9	433–1,200	894	—[e]	1,000

[a]TDS = dissolved solids, EC = electrical conductivity.
[b]Monthly values for BOD5 are based on an average of two samples per month. Monthly values for TSS, COD, pH, and total coliform are based on an average of at least four samples per month.
[c]Arithmetic or geometric means are reported depending on the distribution of the monthly values.
[d]Mean values cannot be reported for pH which is a logarithmic function and total coliform which is based on a Poisson distribution.
[e]Unable to define nature of distribution, because of the number of zero values reported.

135

range of values observed, the arithmetic or geometric mean, and the median. Either the arithmetic or geometric mean are reported depending on the nature of the statistical distribution of the monthly or quarterly data. As shown, all of the values reported for BOD were either reported as 0 or less than 5. As a result, both the mean (assumed to be normally distributed) and median values are reported as less than 5. The reported monthly values for TSS and COD were found to be log-normally distributed (i.e., skewed). Because of the number of zero values reported, it was not possible to define the nature of distribution for the TKN, NH_4, and oil and grease values. The TDS and conductivity values were found to be normally distributed. Additional details may be found in Crites et al. (1997).

ARCATA MARSH AND WILDLIFE SANCTUARY

The City of Arcata, CA reuses 8,700 m^3/d (2.3 Mgal/d) of effluent from their constructed treatment wetlands in a 12.5 ha (31 ac) enhancement marsh (see Figure 3.10). The marsh is the home or rest stop for over 200 species of shorebirds, waterfowl, raptors, and migratory birds. Over 150,000 people a year use the area for passive recreation, bird-watching, or scientific study. Further water quality improvement occurs in the enhancement marsh in terms of nitrogen (Gearheart and Finney, 1996). The City of Arcata, by avoiding a regional sewerage plan in the 1970s, serves as a good example of resource conservation and water reuse. Satellite treatment plants can be used to provide the treated effluent for similar applications.

SMALL SATELLITE TREATMENT SYSTEMS

Satellite treatment is another form of DWM. Water reuse is expected to increase in the future for both centralized and decentralized wastewater management systems. For decentralized systems and developing areas on the periphery of existing communities, satellite reclamation plants will continue to increase in number. Satellite plants either as stand-alone facilities, as planned for Paradise, CA, or connected to downstream collection systems (as in Los Angeles County for their upstream reclamation plants), will employ a variety of technologies. Recirculating packed bed filters and constructed wetlands are often cost-effective for small flows. Improvements in technology, such as for membrane systems, will continue and, as a result, more water reuse will occur because of the lower costs of reclamation.

WATER GARDENS PROJECT, SANTA MONICA, CA

Recycling of reclaimed water from office buildings has been practiced in Japan and in the United States. One example is The Water Gardens

Figure 3.10. Views of the Arcata, CA, enhancement marsh.

137

Figure 3.10 (continued).

TABLE 3.8. Water Gardens Reclamation and
Reuse System Operating Data.[a]

Item	Unit	Value
Design flow	gal/d	20,000
Initial year of operation		1992
Influent BOD$_5$	mg/L	600
Effluent BOD$_5$	mg/L	5
Influent TSS	mg/L	600
Effluent TSS	mg/L	5
Influent nitrogen	mg/L	150
Effluent nitrogen	mg/L	4
Effluent nitrogen limit	mg/L	10
Effluent turbidity	NTU	0.15
Effluent turbidity limit	NTU	2
Effluent total coliform	MPN/100 mL	<2
Effluent total coliform limit	MPN/100 mL	<2

[a]From Jordan and Senthilnathan, 1996.

development, a 4-multistory office building complex in Santa Monica, CA. Wastewater collected from the offices is reclaimed and reused for landscape irrigation, water features, and, ultimately, urinal and toilet flushing. The treatment processes include pretreatment, biological oxidation, membrane filtration (ultrafiltration), activated carbon, ultraviolet radiation, and chlorination. The flow diagram is shown in Figure 3.6. At full building occupancy, approximately 26,000 gal/d (98 m³/d) of wastewater from the sinks, showers, toilets, and urinals will pass through a 26,000 gal trash trap to remove large solids and then into a 60,000 gal equalization and emergency storage reservoir. Biological treatment is accomplished by a 20,000 gal/d (76 m³/d) activated sludge system. Solids are separated from the treated liquid by means of membrane filters, allowing large concentrations of mixed liquor suspended solids (20,000 to 30,000 mg/L) to be maintained in the reactor. By eliminating the clarifiers, the system requires much less area than conventional activated sludge.

Treated wastewater is pumped through tubular cross-flow membrane filters, that are composed of synthetic organic polymers, that have an average pore size of approximately 0.005 microns. The permeate from the ultrafiltration units is then passed through granular activated carbon adsorption for color removal and a combination of UV and chlorine disinfection. The performance data for operations from 1992 through 1996 are presented in Table 3.8. During that period the system has never been out of compliance (Jordan and Senthilnathan, 1996). The Water Gardens system was manufactured by Thetford Systems (now owned by Zenon Municipal Systems) and was called the Cycle-Let system. Thetford installed approximately 27 systems between 1974 and 1982, mostly for small commercial facilities on the east coast of the United States.

FUTURE OF WASTEWATER REUSE IN DWM SYSTEMS

Based on what has occurred during the past 25 years and the current economic situation (first quarter 1997), it can be concluded that it will not be possible to provide conventional sewerage facilities to all of the population in the foreseeable future. Thus, it is clear that decentralized systems will become more important in the future as long-term strategies are developed to protect the environment. The use of new equipment and technologies has made it possible to produce a high quality effluent suitable for most reuse applications. With the development of membrane technology it will be possible to develop additional reuse opportunities. With the improved effluent quality now achievable, reuse has become a viable alternative for DWM systems.

REFERENCES

Crites, R., C. Lekven, S. Wert, and G. Tchobanoglous (1997) "A Decentralized Wastewater System for a Small Residential Development in California," *Small Flow Journal,* Vol. 3, Issue 1.

Crites, R. and G. Tchobanoglous (1998) *Small and Decentralized Wastewater Management Systems,* McGraw-Hill Book Company, New York, NY.

Gearheart, R. A. and B. A. Finney (1996) Criteria for Design of Free Surface Constructed Wetlands Based Upon a Coupled Ecological and Water Quality Model, Presented at the *Fifth International Conference on Wetland Systems for Water Pollution Control,* Vienna, Austria.

Jordan, E. J. and P. R. Senthilnathan (1996) Advanced Wastewater Treatment With Integrated Membrane BioSystems, Paper presented at *AIChE 1996 Spring National Meeting,* New Orleans, LA.

Jowett, E. C. and M. M. McMaster (1994) "A New Single-Pass Aerobic Biofilter For On-Site Wastewater Treatment," *Environmental Science & Engineering,* Vol. 6, No. 6.

Lumsden, L. L., Stiles, C. W., and Freeman, A. W. (1915) *Safe Disposal of Human Excreta at Unsewered Homes,* Public Health Bulletin No. 68, United States Public Health Service, Government Printing Office, Washington, D.C.

Montgomery/Watson and NBS Lowry, (1994) "Water Repurification Feasibility Study," Report prepared for San Diego Water Authority, San Diego, CA.

Orenco Systems, Inc., (1997) Equipment Catalog.

Tchobanoglous, G. (1995) "Decentralized Systems for Wastewater Management," Paper presented at *The 24th Annual WEAO Technical Symposium,* Toronto, Canada.

Tchobanoglous, G. and Angelakis, A. (1996) "Appropriate Technologies for Wastewater Treatment and Reuse Including Water Resources Management In Greece," In Angelakis, A., Asano, T., Diamadopoulos, E.,and Tchobanoglous, G. eds. (1995) *Water Science & Technology,* vol. 33, nos. 10–11, pp. 15–24.

CHAPTER 4

Waste Stabilization Ponds and Wastewater Storage and Treatment Reservoirs: The Low-Cost Production of Microbiologically Safe Effluents for Agricultural and Aquacultural Reuse

WASTE STABILIZATION PONDS

WASTE stabilization ponds (WSP) are shallow man-made basins into which wastewater flows and from which, after a retention time of many days (rather than several hours in conventional treatment processes), a well treated effluent is discharged. WSP systems comprise a series of anaerobic, facultative and maturation ponds, or two or more such series in parallel. In essence, anaerobic and facultative ponds are designed for BOD removal and maturation ponds for pathogen removal, although some BOD removal occurs in maturation ponds and some pathogen removal in anaerobic and facultative ponds. The final effluent quality depends largely on the size and number of maturation ponds. The functions and modes of operation of these three different types of pond are described in later sections.

ADVANTAGES OF WSP

The advantages of WSP systems, which can be summarized as *simplicity, low cost* and *high efficiency,* are as follows:

(1) *Simplicity:* WSP are simple to construct: earth moving is the principal activity; other civil works are minimal—preliminary treatment, inlets and outlets, pond embankment protection and, if necessary, pond lining.

Duncan Mara, School of Civil Engineering, University of Leeds, Leeds LS2 9JT, England.

They are also simple to operate and maintain: routine tasks comprise cutting the embankment grass, removing scum and any floating vegetation from the pond surface, keeping the inlets and outlets clear, and repairing any damage to the embankments. Only unskilled, but carefully supervised, labor is needed for pond operation and maintenance.

(2) *Low cost:* Because of their simplicity, WSP are much cheaper than other wastewater treatment processes. There is no need for expensive electro-mechanical equipment (with its attendant problems, in developing countries, of foreign exchange and spare parts), nor for a high annual consumption of electrical energy. The latter point is well illustrated by the following data from the United States (where one third of all wastewater treatment plants are WSP systems) for a flow of 1 million US gallons per day (3780 m³/d) [1]:

Treatment Process	Energy Consumption (kWh/yr)
Activated sludge	1,000,000
Aerated lagoons	800,000
Biodiscs	120,000
Waste stabilization ponds	nil

The cost advantages of WSP were analysed in detail by Arthur in a World Bank Technical Paper [2]. Arthur compared four treatment processes—trickling filters, aerated lagoons, oxidation ditches and WSP, all designed to produce the same quality of final effluent, and he found that WSP systems were the cheapest treatment process at land costs of US$ 50,000–150,000 (1983 $) per hectare, depending on the discount rate used (5–15 percent). These figures are much higher than most land costs, and so land costs are unlikely to be a factor operating against the selection of WSP for wastewater treatment (but, of course, land availability may be).

(3) *High efficiency:* BOD removals >90 percent are readily obtained in a series of well designed ponds. The removal of suspended solids is less, due to the presence of algae in the final effluent (but, since algae are very different to the suspended solids in conventional secondary effluents, this is not cause for alarm: indeed the European directive on urban wastewater treatment [3] permits WSP effluents to contain up to 150 mg suspended solids/L, and it also allows sample filtration prior to BOD analysis to remove the algae). Total nitrogen removal is 70–90 percent, and total phosphorus removal 30–50 percent.

WSP are particularly efficient in removing excreted pathogens, whereas in contrast all other treatment processes are very inefficient at this and

TABLE 4.1. Removals of Excreted Pathogens Achieved by Waste Stabilization Ponds and Conventional Treatment Processes.

Excreted Pathogens	Removal in WSP	Removal in Conventional Treatment
Bacteria	up to 6 log units[a]	1–2 log units
Viruses	up to 4 log units	1–2 log units
Protozoan cysts	100%	90–99%
Helminth eggs	100%	90–99%

[a]1 log unit = 90 percent removal; 2 = 99 percent; 3 = 99.9 percent, and so on.

require a tertiary treatment process, such as chlorination (with all its inherent operational and environmental problems), to achieve the destruction of fecal bacteria. Activated sludge plants may, if operating very well, achieve a 99 percent removal of fecal coliform bacteria: this might, at first inspection, appear very impressive, but in fact it only represents a reduction from 10^8 per 100 ml to 10^6 per 100 ml (that is, almost nothing). A properly designed series of WSP, on the other hand, can easily reduce fecal coliform numbers from 10^8 per 100 ml to below the WHO guideline value for unrestricted irrigation of no more than 1000 per 100 ml [4], which is a removal of 99.999 percent (or 5 $\log_{10}$ units). WSP can also easily achieve the WHO guideline value for restricted irrigation of no more than 1 intestinal nematode egg per liter. (The rationale for these WHO guideline values is discussed in the next section.) A general comparison between WSP and conventional treatment processes for the removal of excreted pathogens is shown in Table 4.1; detailed information is given in Feachem et al. [5].

Industrial Wastes

WSP are also extremely robust: due to their long hydraulic retention time, they can withstand both organic and hydraulic shock loads. They can also cope with high levels of heavy metals, up to 60 mg/L [6], so they can treat a wide variety of industrial wastewaters that would be too toxic for other treatment processes. Strong wastewaters from agro-industrial processes (for example, abattoirs, food canneries, dairies) are easily treated in WSP.

RATIONALE FOR THE WHO GUIDELINE VALUES FOR WASTEWATER REUSE

As noted above, WHO [4] has recommended the following guidelines for the microbiological quality of treated wastewater used for crop irrigation (see also Appendix III, p. 1491):

(1) Restricted irrigation: $\not\geq$ 1 intestinal nematode egg per liter
(2) Unrestricted irrigation: as (1), plus $\not\geq$ 1000 fecal coliform bacteria per 100 ml

The nematological guideline was introduced as it has been shown [7] that both farm workers and crop consumers were at highest risk from infection due to the human intestinal nematodes (the human roundworm, *Ascaris lumbricoides;* the human whipworm, *Trichuris trichiura;* and the human hookworms, *Ancylostoma duodenale* and *Necator americanus*).

The fecal coliform guideline is based on an appraisal of *actual* health risk, rather than *potential* health risk which is the basis of more stringent fecal coliform requirements—for example, those used in California [8]. The WHO guideline of no more than 1000 fecal coliforms per 100 ml has not been without its critics [9], but recent appraisals of the available evidence [7] and of requirements for the microbiological quality of foods (including those eaten raw) [10] indicate that it is nonetheless appropriate.

WHO [4] recognized that WSP were a most appropriate wastewater treatment process for achieving compliance with both its nematode and fecal coliform guidelines.

ANAEROBIC PONDS

Function

Anaerobic ponds are 2–5 m deep and receive such a high organic loading (usually >100 g BOD/m^3d, equivalent to >3000 kg/ha d for a depth of 3 m) that they contain no dissolved oxygen and no algae (although occasionally a thin film of mainly *Chlamydomonas* may be seen at the surface). They function much like open septic tanks, and their primary function is BOD removal. They work extremely well in warm climates: a properly designed and not significantly underloaded anaerobic pond will achieve around 60 percent BOD removal at 20°C and as much as 75 percent at 25°C. Retention times are short: for wastewater with a BOD of up to 300 mg/L, 1 day is sufficient at temperatures >20°C.

Designers have in the past been too afraid to incorporate anaerobic ponds in case they cause odor. Hydrogen sulfide, formed mainly by the anaerobic reduction of sulfate by sulfate-reducing bacteria such as *Desulfovibrio,* is the principal potential source of odor. However, in aqueous solution hydrogen sulfide is present as either dissolved hydrogen sulfide gas (H_2S) or the bisulfide ion (HS^-), with the sulfide ion (S^{2-}) only really being formed in significant quantities at high pH. At the pH values normally found in well designed anaerobic ponds (around 7.5), most of the sulfide is present as the odorless bisulfide ion. Odor is only caused by escaping

TABLE 4.2. Design Loadings on, and BOD Removals in, Anaerobic Ponds at Various Temperatures (*T*).

Temperature (°C)	Volumetric BOD Loading (g/m³d)	Percentage BOD Removal
≤10	100	40
10–20	20*T*–100	2*T* + 20
≥20	300	60[a]

[a]70 at ≥25°C.

hydrogen sulfide molecules as they seek to achieve a partial pressure in the air above the pond which is in equilibrium with their concentration in it (Henry's law). Thus, for any given total sulfide concentration, the greater the proportion of sulfide present as HS^-, the lower the release of H_2S. Odor is *not* a problem if the recommended design loadings (Table 4.2) are not exceeded and if the sulfate concentration in the raw wastewater is less than 500 mg SO_4^{2-}/L [11]. A small amount of sulfide is beneficial as it is lethal to *Vibrio cholerae* [12], and it reacts with heavy metals to form insoluble metal sulfides which precipitate out, but concentrations of 50–150 mg/L can inhibit methanogenesis [13].

Design

Anaerobic ponds are designed on the basis of volumetric BOD loading (λ_v, g/m³d), which is given by:

$$\lambda_v = L_iQ/V_a \qquad (1)$$

where

L_i = influent BOD, mg/L (= g/m³)
Q = flow, m³/d
V_a = anaerobic pond volume, m³

The permissible design value of λ_v increases with temperature, but there are too few reliable data to permit the development of a suitable design equation, and the recommendations given in Table 4.2 should be used [14]. These values were based on earlier recommendations [15] that λ_v should lie between 100 and 400 g/m³d, the former in order to maintain anaerobic conditions and the latter to avoid odor release. However, in Table 4.2 the upper limit for design is set at 300 g/m³d in order to provide an adequate margin of safety with respect to odor. This is appropriate for

TABLE 4.3. BOD Removals in Anaerobic Ponds in Northeast Brazil at 25°C.

Pond No.	Retention Time (days)	Volumetric BOD Loading (g/m³d)	BOD Removal (%)
1	0.8	306	76
2	1.0	215	76
3	1.9	129	80
4	2.0	116	75
5	4.0	72	68
6	6.8	35	74

normal domestic or municipal wastewaters which contain less than 500 mg SO_4^{2-}/L.

Once a value of λ_v has been selected, the anaerobic pond volume is then calculated from Equation (1). The mean hydraulic retention time in the pond (θ_a, d) is determined from:

$$\theta_a = V_a/Q \tag{2}$$

Performance

Table 4.2 gives values for BOD removal at various temperatures: these are conservative design values. In practise higher values are obtained. Table 4.3 gives percentage removals obtained in anaerobic ponds at various volumetric loadings in northeast Brazil at 25°C [16,17]. This shows the extremely high efficiency of anaerobic ponds in warm climates. Indeed, as noted by Marais [18], "Pretreatment in anaerobic ponds is so advantageous that the first consideration in the design of a series of ponds should always include the possibility of anaerobic treatment."

FACULTATIVE PONDS

Function

Facultative ponds are designed for BOD removal on the basis of a relatively low surface loading (100–400 kg BOD/ha d) to permit the development of a healthy algal population as the oxygen for BOD removal by the pond bacteria is mostly generated by algal photosynthesis. Due to the algae facultative ponds are colored dark green, although they may occasionally appear red or pink (especially when slightly overloaded) due to the presence of anaerobic purple sulfide-oxidising photosynthetic bacteria. The algae that predominate in the turbid waters of facultative ponds are the motile genera, such as *Chlamydomonas, Pyrobotrys* and *Euglena,* as these can optimise their position in the pond in relation to incident light intensity and temperature more easily than non-motile forms (such as *Chlorella,* although this is fairly common in facultative ponds). The concentration of algae in a healthy facultative pond depends on loading and temperature, but it is usually in the range 500–2000 μg chlorophyll *a* per liter.

As a result of the photosynthetic activities of the pond algae, there is a diurnal variation in the concentration of dissolved oxygen. After sunrise, the dissolved oxygen level gradually rises to a maximum in the mid-afternoon, after which it falls to a minimum during the night. The position of the oxypause (the depth at which the dissolved oxygen concentration reaches zero) similarly changes, as does the pH since at peak algal activity carbonate and bicarbonate ions react to provide more carbon dioxide for the algae, so leaving an excess of hydroxyl ions with the result that the pH can rise to above 9, which rapidly kills fecal bacteria.

Design

Although there are several methods available for designing facultative ponds [19], it is recommended that they be designed on the basis of surface BOD loading (λ_s, kg/ha d), which is given by:

$$\lambda_s = 10 \, L_i Q / A_f \tag{3}$$

where

A_f = facultative pond area, m^2.

The permissible design value of λ_s increases with temperature (T, °C). The earliest relationship between λ_s and T is that given by McGarry and Pescod [20], but their value of λ_s is the *maximum* that can be applied to a facultative pond before it fails (that is, becomes anaerobic). Their relationship, which is therefore an *envelope of failure,* is:

$$\lambda_s = 60 \, (1.099)^T \tag{4}$$

Mara [21] recommended the following equation for design in the temperature range 8–35°C:

$$\lambda_s = 350 \, (1.107 - 0.002T)^{T-25} \tag{5}$$

At temperatures below 8°C a design loading of 80 kg/ha day should be used [22].

Once a suitable value of λ_s has been selected, the pond area is calculated from Equation (3) and its retention time (θ_f, d) from:

$$\theta_f = A_f D / Q_m \qquad (6)$$

where

D = pond depth, m (range: 1–2 m)
Q_m = mean flow, m³/day

The mean flow is the mean of the influent and effluent flows (Q_f and Q_e), the latter being the former less evaporation and seepage. If seepage is negligible, then Equation (6) becomes:

$$\theta_f = A_f D / [\tfrac{1}{2}(Q_i + Q_e)] \qquad (7)$$

Since $Q_e = Q_i - 0.001 A_f e$ (where e is the evaporation rate, mm/d), Equation (7) becomes:

$$\theta_f = 2A_f D / (2Q_i - 0.001 A_f e) \qquad (8)$$

BOD removal in facultative ponds is usually in the range 70–80 percent based on unfiltered samples (that is, including the BOD exerted by the algae), and above 90 percent based on filtered samples.

Helminth eggs, which can number up to 2000 per liter of wastewater depending on the endemicity of intestinal nematode infections, are removed by sedimentation and thus most egg removal occurs in the anaerobic and facultative ponds. If the final effluent is to be used for restricted irrigation, then it is necessary to ensure that it complies with the WHO guideline of no more than one egg per liter [4]. At this stage in the design it is sensible to check whether the facultative pond effluent contains ≤1 egg per liter and so is suitable for restricted irrigation. However, depending on the number of helminth eggs present in the raw wastewater and the retention times in the anaerobic and facultative ponds, it may be necessary to incorporate a maturation pond to ensure that the final effluent contains at most only one egg per liter. Analysis of egg removal data from ponds in Brazil, India and Kenya has yielded the following relationship which is equally valid for anaerobic, facultative and maturation ponds [23]:

$$R = 100 \, [1 - 0.14 \, \exp(-0.38\theta)] \qquad (9)$$

where

R = percentage egg removal
θ = retention time, d

The equation corresponding to the lower 95 percent confidence limit of Equation (9) is:

$$R = 100\ [1 = 9.41\ \exp(-0.49\theta + 0.00850^2)] \tag{10}$$

Equation (10) is recommended for use in design; it is applied sequentially to each pond in the series, so that the number of eggs in the final effluent can be determined.

MATURATION PONDS

Function

A series of maturation ponds receives the effluent from the facultative pond, and the size and number of maturation ponds is governed mainly by the required bacteriological quality of the final effluent; for unrestricted irrigation this is ≤1000 fecal coliforms per 100 ml [4]. Maturation ponds usually show less vertical biological and physicochemical stratification than facultative ponds, and are well oxygenated throughout the day. Their algal population is thus much more diverse than that of facultative ponds, with non-motile genera tending to be more common; algal diversity increases from pond to pond along the series.

The primary function of maturation ponds is the removal of excreted pathogens, and this is extremely efficient in a properly designed series of ponds. Maturation ponds achieve only a small removal of BOD, usually around 10–25 percent in each pond.

Design

The method of Marais [24] is generally used to design a pond series for fecal coliform removal. This assumes that fecal coliform removal can be modelled by first order kinetics in a completely mixed reactor. The resulting equation for a single pond is thus:

$$N_e = N_i/(1 + k_T\theta) \tag{11}$$

where

N_e = number of FC per 100 ml of effluent
N_i = number of FC per 100 ml of influent
k_T = first order rate constant for FC removal, d^{-1}
θ = retention time, d

For a series of anaerobic, facultative and maturation ponds, Equation (11) becomes:

$$N_e = N_i/[(1 + k_T\theta_a)(1 + k_T\theta_f)(1 + k_T\theta_m)^n)] \qquad (12)$$

where N_e and N_i now refer to the numbers of FC per 100 ml of the final effluent and raw wastewater respectively; the subscripts a, f and m refer to the anaerobic, facultative and maturation ponds; and n is the number of maturation ponds.

It is assumed in Equation (12) that all the maturation ponds are equally sized: this is the most efficient configuration [24], but may not be topographically possible (in which case the last term of the denominator in Equation (12) is replaced by $[(1 + k_T\theta_{m1}) + (1 + k_T\theta_{m2}) + \ldots + (1 + k_T\theta_{mn})]$).

The value of k_T is highly temperature dependent. Marais found that [24]:

$$k_T = 2.6 \, (1.19)^{T-20} \qquad (13)$$

Thus k_T changes by 19 percent for every change in temperature of 1°C.

Maturation ponds require careful design to ensure that their FC removal follows that given by Equations (12) and (13). If they are suboptimally loaded, then their FC removal performance may be correspondingly suboptimal, as was found in Kenya [25,26].

Examination of Equation (12) shows that it contains two unknowns, θ_m and n, since by this stage of the design process θ_a and θ_f will have been calculated, N_i measured or estimated, N_e set (at, for example, 1000 per 100 ml for unrestricted irrigation) and k_T calculated from Equation [13]. The best approach to solving Equation (12) is to calculate the values of θ_m corresponding to $n = 1, 2, 3$ etc. and then adopt the following rules to select the most appropriate combination of θ_m and n:

$$\theta_m \not> \theta_f$$

$$\theta_m \not< \theta_m^{min}$$

where θ_m^{min} is the minimum acceptable retention time in a maturation pond.

This is introduced to minimise hydraulic shortcircuiting and prevent algal washout; a value of 3 days is recommended [24].

The remaining pairs of θ_m and n, together with the pair θ_m^{min} and $\tilde{n}$, where $\tilde{n}$ is the first value of n for which θ_m is less than θ_m^{min}, are then compared and the one with the least product selected as this will give the least land area requirements. A check must be made on the BOD loading on the first maturation pond: this must not be higher than that on the preceding facultative pond, and it is preferable that it is significantly lower, for example no more than 75 percent of that on the preceding facultative pond. The loading on the first maturation pond is calculated on the assumption that 70–80 percent of the BOD (depending on temperature) has been removed in the preceding anaerobic and facultative pond. Thus, for 80 percent removal at 20°C, for example:

$$\lambda_s(m1) = 10\ (0.2\ L_i)\ Q/A_{m1} \tag{14}$$

or, since $Q\theta_{m1} = A_{m1}D_m$:

$$\lambda_s(m1) = 10\ (0.2\ L_i)\ D_{m1}/\theta_{m1} \tag{15}$$

Taking $\lambda_s(m1)$ as $0.75\ \lambda_s(f)$, rearrangement of Equation (15) gives the following equation for the minimum retention time in the first maturation pond:

$$\theta_{m1} = (8/3)(L_iD)/\lambda_s(f) \tag{16}$$

The area of each maturation pond is calculated from a rearrangement of Equation (8):

$$A_m = 2Q_i\theta_m/(2D + 0.001e\theta_m) \tag{17}$$

Performance

The very high efficiency of a series of WSP which includes maturation ponds is shown in Table 4.4, which is based on results from northeast Brazil at 25°C [16]. Fecal coliforms were reduced to 30 per 100 ml after a total retention time of 29 days, and intestinal nematode eggs to 1 per liter after only 12 days.

Table 4.5, also based on results from northeast Brazil at 25°C [27], shows that fecal bacterial pathogens such as *Salmonella* spp. and *Campylo-bacter* spp. were absent when the FC count was 7000 per 100 ml. It was also found that *Vibrio cholerae* was totally removed in a series of WSP

TABLE 4.4. **Performance of a Series of Five WSP in Northeast Brazil at 25°C.**

	Retention Time (days)	BOD (mg/L)	SS (mg/L)	FC (per 100 ml)	Nematode Eggs (per liter)
Raw wastewater	—	240	305	4.6×10^7	804
Effluent of:					
Anaerobic pond	6.8[a]	63	56	2.9×10^6	29
Facultative pond	5.5	45	74	3.2×10^5	1
1st maturation pond	5.5	25	61	2.4×10^4	0
2nd maturation pond	5.5	19	43	450	0
3rd maturation pond	5.8	17	45	30	0

[a] Similar performance was obtained at a 0.8 day retention time (see Table 4.3). More recent work has shown that retention times in the facultative and maturation ponds can also be reduced to 3–4 days at 25°C without any significant performance loss [37].

when the FC count was 60,000 per 100 ml [12]. These results lend much credence to the safety of the WHO guideline value of ≤1000 FC per 100 ml for unrestricted irrigation (see also [10]).

POND TREATMENT PRIOR TO AQUACULTURE

Pond treatment prior to crop irrigation is relatively straightforward, as shown previously. However, a slightly different approach is required for fish culture [28]. Anaerobic and facultative ponds are used as described previously, and the facultative pond effluent is discharged into fishponds. These are designed on the basis of surface nitrogen loading, and then a check is made to see that they do not contain more than 1000 FC per 100 ml, which is the WHO guideline for aquacultural reuse [4].

The optimum nitrogen loading on the fishpond is 4 kg total N per ha per day [28]. Too much nitrogen causes too high an algal biomass, with the resultant risk of deoxygenation at night and consequent fish kills; and too little nitrogen results in too low an algal biomass and consequently low fish yields. In order to apply a loading of 4 kg total N/ha day on the fishpond, it is first necessary to determine the removal of total nitrogen in the facultative pond (none occurs in the anaerobic pond). Reed's equation is used [29]:

$$C_e = C_i \exp\{-[0.0064 \, (1.039)^{T-20}][\theta_f + 60.6 \, (\text{pH} - 6.6)]\} \quad (18)$$

TABLE 4.5. Geometric Mean Bacterial and Viral Numbers[a] and in Raw Wastewater and the Effluents of Five Waste Stabilization Ponds in Series in Northeast Brazil at 26°C.

Organism	RW[b]	A	F	M1	M2	M3
Faecal coliforms	2×10^7	4×10^6	8×10^5	2×10^5	3×10^4	7×10^3
Campylobacters	70	20	0.2	0	0	0
Salmonellae	20	8	0.1	0.002	0.01	0
Enteroviruses	1×10^4	6×10^3	1×10^3	400	50	9
Rotaviruses	800	200	70	30	10	3

[a]Bacterial numbers per 100 ml, viral numbers per 10 liters.
[b]RW, raw wastewater; A was an anaerobic pond with a mean hydraulic retention time of 1 day; F and M1–M3 were a facultative pond and maturation ponds, respectively, each with a retention time of 5 days.

153

where

C_e = total N concentration in the facultative pond effluent, mg/L
C_i = total N concentration in the raw wastewater, mg/L
T = temperature, °C
θ_f = retention time in the facultative pond, days

The pH value to be used in Equation (18) may be estimated from:

$$pH = 7.3\ \exp(0.0005A) \tag{19}$$

where

A = alkalinity, mg $CaCO_3$ per liter.

The fishpond area is now determined from Equation (3) with $\lambda_s = 4$ kg total N/ha d; and its retention time (θ_{fp}, days) from Equation (8), using a depth of 1 m. The fecal coliform concentration in the fishpond (N_{fp}, per 100 ml) is calculated from the following version of Equation (12):

$$N_{fp} = N_i/(1 + k_T\theta_a)(1 + k_T\theta_f)(1 + k\theta_{fp}) \tag{20}$$

If N_{fp} is not less than 1000, θ_{fp} is increased until it is and the fishpond area recalculated.

Fish yields of up to 13 tonnes per ha per year can be obtained with careful management [28]. This figure includes a loss of 25 percent resulting from mortality, consumption by fish-eating birds and poaching.

WASTEWATER STORAGE AND TREATMENT RESERVOIRS

Wastewater storage and treatment reservoirs (WSTR), also called effluent storage reservoirs, are especially useful in arid and semi-arid areas. They were developed in Israel to store the effluent from a WSP system during the period (8 months in Israel) when it is not required for irrigation [30]. It is thus a method of conserving effluent so that, during the irrigation season, a greater area of land can be irrigated. Current Israeli practice is to treat the wastewater in an anaerobic pond and discharge its effluent into a single 5–15 m deep WSTR with an 8-month retention time. This is perfectly satisfactory, as the WSTR effluent is only used to irrigate cotton (*i.e.* for restricted irrigation) and so this usage complies with the WHO guideline for restricted irrigation since any helminth eggs settle out in the anaerobic pond and the WSTR.

TABLE 4.6. Management Strategy for Three Batch-fed Sequential Wastewater Storage and Treatment Stabilization Reservoirs in Parallel for an Irrigation Season of 6 Months.[a]

Month	WSTR 1	WSTR 2	WSTR 3
January	Fill (1)[b]	USE	Rest
February	Fill (1)	USE	Rest
March	Fill ($^1/_2$)	Fill ($^1/_2$)	USE
April	Fill ($^1/_2$)	Fill ($^1/_2$)	USE
May	Fill ($^1/_2$)	Fill ($^1/_2$)	Empty
June	Fill ($^1/_2$)	Fill ($^1/_2$)	Empty
July	Rest	Fill (1)	Empty
August	Rest	Fill (1)	Empty
September	Rest	Rest	Fill (1)
October	Rest	Rest	Fill (1)
November	USE	Rest	Fill (1)
December	USE	Rest	Fill (1)

[a]The hot season is assumed to be January–February, so WSTR 3 has the minimum rest period of 2 months in these months; the other WSTRs have rest periods of 4 months.
[b]Proportion of flow diverted to each WSTR.

If the WSTR effluent is to be used for unrestricted irrigation, then it should contain ≤ 1000 FC per 100 ml, which the above single WSTR cannot achieve, at least not during the irrigation season [31]. Instead several sequential batch-fed WSTR in parallel are required [32]. These receive anaerobic pond effluent and are each operated on a sequential cycle of fill, rest and use, with fecal coliform die-off to <1000 per 100 ml occurring during the fill and rest periods. Table 4.6 illustrates this sequence with three WSTR in parallel for an irrigation season of six months. Similar strategies can be readily developed for irrigation seasons of different lengths [32].

Recent research in northeast Brazil [33] has shown that sequential batch-fed WSTR are very efficient at removing fecal coliforms: at temperatures of 25°C die-off to <1000 per 100 ml throughout the whole reservoir depth of 6 m occurred 3 weeks into the rest phase. WSTR were found to behave much like deep facultative ponds [34] with an algal biomass of around 500 μg chlorophyll *a* per liter (as with WSP, such algal concentrations in WSTR effluents are beneficial for crop irrigation as the algae act as slow-release fertilisers in the soil). The much greater depth of WSTR (5–15 m, compared with 1–2 m for WSP) reduces evaporative losses; in northeast Brazil such losses amounted to under 14% of the inflow to a 6 m deep WSTR during a 4-month rest phase in the hottest part of the year (25–27°C), with a corresponding increase in electrical conductivity to 160 mS/m. Wastewaters of such conductivity have been successfully used to irrigate local cash crops, including lettuce [35].

WSTR are a very flexible system of wastewater treatment and storage. Juanico [36] details several arrangements, including two WSTR in series, with effluent from the first being used for restricted irrigation and that from the second, for unrestricted irrigation. An alternative ''hybrid'' WSP-WSTR system is to treat the wastewater in anaerobic and facultative ponds, the effluent from the latter being discharged into a WSTR during the non-irrigation season, but used for restricted irrigation during the irrigation season when the WSTR contents are used for unrestricted irrigation [33].

REFERENCES

1 Middlebrooks, E. J. et al. 1982. *Wastewater Stabilization Lagoon Design, Performance and Upgrading.* New York, NY: Macmillan Publishing Co.

2 Arthur, J. P. 1983. Notes on the Design and Operation of Waste Stabilization Ponds in Warm Climates of Developing Countries (Technical Paper No. 7). Washington, DC: The World Bank.

3 Council of the European Communities. 1991. ''Council Directive of 21 May 1991 concerning Urban Waste Water Treatment,'' *Official Journal of the European Communities,* L135/40-52 (30 May).

4 World Health Organization. 1989. *Health Guidelines for the Use of Wastewater in Agriculture and Aquaculture* (Technical Report Series No. 778). Geneva: WHO.

5 Feachem, R. G. et al. 1983. *Sanitation and Disease: Health Aspects of Excreta and Wastewater Management.* Chichester, England: John Wiley and Sons Ltd.

6 Moshe, M. et al. 1972. ''Effect of Industrial Wastes on Oxidation Pond Performance,'' *Water Research,* 6: 1165–1171.

7 Shuval, H. I. et al. 1986. Wastewater Irrigation in Developing Countries: Health Effects and Technical Solutions (Technical Paper No. 51). Washington, DC: The World Bank.

8 California Department of Health Services. 1978. *Wastewater Reclamation Criteria* (California Administrative Code, Title 22, Division 4, Environmental Health). Berkeley, CA: CDHS.

9 Shelef, G. 1991. ''Wastewater Reclamation and Water Resources Management,'' *Water Science and Technology,* 24(9): 251–265.

10 Mara, D. D. 1995. ''Faecal Coliforms—Everywhere (But Not a Cell to Drink),'' *Water Quality International,* (3): 29–30.

11 Gloyna, E. F. and Espino, E. 1969. ''Sulfide Production in Waste Stabilization Ponds,'' *Journal of the Sanitary Engineering Division, American Society of Civil Engineers,* 95: 607–628.

12 Oragui, J. I. et al. 1993. ''*Vibrio cholerae* O1 (El Tor) Removal in Waste Stabilization Ponds in Northeast Brazil,'' *Water Research,* 27: 727–728.

13 Pfeffer, J. T. 1970. *Proceedings of the Second International Symposium on Waste Treatment Lagoons,* June 23–25, 1970, Kansas City, Missouri, pp. 310–320.

14 Mara, D. D. and Pearson, H. W. 1986. ''Artificial Freshwater Environments: Waste Stabilization Ponds,'' *Biotechnology, Volume 8,* Schoenborn, W. ed. Weinheim, Germany: VCH Verlagsgesellschaft, pp. 177–206.

15 Meiring, P. G. et al. 1968. A Guide to the Use of Pond Systems in South Africa for

the Purification of Raw and Partially Treated Sewage (CSIR Special Report WAT34). Pretoria, South Africa: National Institute for Water Research.

16   Silva, S. A. 1982. On the Treatment of Domestic Wastewater in Waste Stabilization Ponds in Northeast Brazil (Ph.D. Thesis). Dundee, Scotland: University of Dundee, Department of Civil Engineering.

17   Mara, D. D. and Mills, S. W. 1994. "Who's Afraid of Anaerobic Ponds?" *Water Quality International,* (2): 34–36.

18   Marais, G.v.R. 1970. *Proceedings of the Second International Symposium on Waste Treatment Lagoons,* June 23–25, Kansas City, Missouri, pp. 15–46.

19   Mara, D. D. 1976. *Sewage Treatment in Hot Climates.* Chichester, England: John Wiley and Sons Ltd.

20   McGarry, M. G. and Pescod, M. B. 1970. *Proceedings of the Second International Symposium on Waste Treatment Lagoons,* June 23–25, Kansas City, Missouri, pp. 114–132.

21   Mara, D. D. 1987. "Waste Stabilization Ponds: Problems and Controversies," *Water Quality International,* (1): 20–22.

22   CEMAGREF and Agences de l'Eau, 1997. *Le Lagunage Naturel: les Leçons de 15 Ans de Pratique en France.* Lyon, France: Centre National du Machinisme Agricole, du Génie Rural, des Eaux et des Forêts.

23   Ayres, R. M. et al. 1992. "A Design Equation for Human Intestinal Nematode Egg Removal in Waste Stabilization Ponds," *Water Research,* 26: 863–865.

24   Marais, G.v.R. 1974. "Faecal Bacterial Kinetics in Waste Stabilization Ponds," *Journal of the Environmental Engineering Division, American Society of Civil Engineers,* 100: 119–139.

25   Mara, D. D. 1992. *Waste Stabilization Ponds: A Design Manual for Eastern Africa.* Leeds, England: Lagoon Technology International Ltd.

26   Mills, S. W. et al. 1992. "Efficiency of Faecal Bacterial Removal in Waste Stabilization Ponds in Kenya," *Water Science and Technology,* 26 (7/8): 1739–1748.

27   Oragui, J. I. et al. 1987. "The Removal of Excreted Bacteria and Viruses in Deep Waste Stabilization Ponds in Northeast Brazil," *Water Science and Technology,* 19 (Rio): 569–573.

28   Mara, D. D. et al. 1993. "A Rational Approach to the Design of Wastewater-fed Fishponds," *Water Research,* 27: 1797–1799.

29   Reed, S. C. 1985. "Nitrogen Removal in Wastewater Stabilization Ponds," *Journal of the Water Pollution Control Federation,* 57: 39–41.

30   Juanico, M. and Shelef, G. 1991. "The Performance of Stabilization Reservoirs as a Function of Design and Operation Parameters," *Water Science and Technology,* 23 (Kyoto): 1509–1516.

31   Liran, A. et al. 1994. "Coliform Removal in a Stabilization Reservoir for Wastewater Irrigation in Israel," *Water Research,* 28: 1305–1314.

32   Mara, D. D. and Pearson, H. W. 1992. "Sequential Batch-fed Effluent Storage Reservoirs: A New Concept of Wastewater Treatment Prior to Unrestricted Crop Irrigation," *Water Science and Technology,* 26(7/8): 1459–1464.

33   Mara, D. D. et al. 1996. Sequential Batch-fed Effluent Storage Reservoirs (Final Report, ODA Research Scheme R5678). Leeds, England: University of Leeds, Department of Civil Engineering.

34   de Oliveira, R. 1990. The Performance of Deep Waste Stabilization Ponds in Northeast Brazil (Ph.D. Thesis). Leeds, England: University of Leeds, Department of Civil Engineering.

35 Ayres, R. M. 1992. On the Removal of Nematode Eggs in Waste Stabilization Ponds and Consequent Potential Health Risks from Effluent Reuse (Ph. D. Thesis). Leeds, England: University of Leeds, Department of Civil Engineering.

36 Juanico, M. 1995. "The Effect of the Operational Regime on the Performance of Wastewater Storage Reservoirs." Paper presented at the *3rd IAWQ International Specialist Conference on Waste Stabilization Ponds: Technology and Applications,* João Pessoa, Brazil, 27–31 March.

37 Pearson, H. W. et al. 1995. "The Influence of Pond Geometry and Configuration on Facultative and Maturation Pond Performance and Efficiency," *Water Science and Technology,* 31(12): 129–139.

CHAPTER 5

Physicochemical Mechanisms in Treatment Processes for Water Reuse

INTRODUCTION

TERTIARY or advanced wastewater treatment is a requirement in some countries prior to disposal to a water body or land, and is one of the most important demands for wastewater reuse. Common tertiary or advanced wastewater treatment schemes include chemical coagulation or flocculation of secondary clarifier effluent followed by sedimentation, filtration and disinfection.

The quality of a biologically treated wastewater, i.e., secondary effluents, is generally insufficient for application to variety of users besides agriculture, e.g., industrial cooling towers, process water for textile and paper factories, municipal drip irrigation systems and recreational facilities. Effluent quality varies from one place to another, a good example could be suspended solids (SS) contents of 20–30 mg/L, residual organics content ≥ 30 mg/L DOC and nitrates content ≥ 100 mg/L. While the removal of nitrates may be based on a modification of biological processes, the removal of residual organics and SS is normally done by a tertiary treatment.

In case of in-line filtration, i.e. contact filtration, equipment and chemical costs are minimized. Microflocculation followed by filtration, i.e. direct filtration, can be highly efficient in colloid removal, however, it is only partly efficient in soluble organics removal. The use of flocculants—commonly aluminum sulfate—has both economical and environmental

Avner Adin, Graduate School of Applied Science, Division of Environmental Sciences, The Hebrew University of Jerusalem, Jerusalem 91904, Israel.

impacts which are undesirable. Flocculation with iron salts may become an attractive replacement, both environmentally and economically. A combined process of granular filtration and membrane filtration could optimize the use of flocculants for simultaneous removal of SS and organics.

High-technology treatment plants, however, have no place in low-income countries because they are too expensive, too complicated, and too dependent on well-trained operators and reliable maintenance and electric power. Too many high-tech plants that have been constructed in such countries under some international program are now sitting idle. Chemically enhanced settling followed by filtration through soil provides an attractive solution: simple, inexpensive, effective, and implementable for smaller towns (and, perhaps, for mega-cities too). The low concentration of suspended solids after settling favors such filtration, while the high concentration of dissolved organic compounds enhances removal of nitrogen and synthetic organic compounds. Chemically enhanced settling can be retrofitted to existing treatment plants and is very useful at overloaded treatment plants, enabling overflow rates to be increased significantly.

Drip irrigation systems are water efficient systems often incorporated into Israeli agricultural practice. Investigations have demonstrated that drip irrigation systems are dependent on virtually particle free water since even colloidal particles can clog dripper pathways (Adin and Sacks, 1991). Research work and field observations concerning the performance of drip irrigation systems, utilizing either fresh water or wastewater effluents, indicate that the causes of clogging of low rate applicators may be divided into three main categories: (1) suspended matter; (2) chemical precipitation; and (3) bacterial growth (Nakayama et al., 1977). Clogging is actually a combination of some of the aforementioned factors: algae, clay and corrosion products entrapped within a biological mass cemented with $CaCo_3$ precipitate. It is now clear, however, that the major clogging problems are caused by the presence of suspended particles in the irrigation water (Bucks, 1982; Dasberg, 1985; Adin, 1987).

Filtration prevents immediate clogging by large particles. It also plays an important role in protection from smaller particles that cause gradual clogging. Two main categories of filters may be defined on the basis of filtration mechanisms: (1) strainers, in which straining is the main filtration mechanism (also referred to as mechanical or surface filtration), e.g., screen filters; and (2) granular filters, also deep bed, rapid or high rate filters, e.g., sand filters. Here the suspended particle size is smaller than the filter grain size and the removal is dominated by physical-chemical mechanisms. Proper design of filters for water reuse in agriculture can only occur through a more complete understanding of filtration mechanisms.

Particle size distribution (PSD) can serve as a powerful tool for understanding filter behavior and discrepancies in case of filtration under various conditions. It may be applicable for advanced wastewater filter design. PSD should be taken into consideration in filtration models, rather than relying on a single representative particle diameter of questionable value. Removal mechanisms of particulates in granular beds are markedly affected by the particle size. Yao et al. (1971) showed that particles of 1–2 μm in size have minimal opportunity for removal, since transport mechanisms of these particles within the filter bed are less efficient.

There is a large variability in the concentration and properties of the suspended solids in secondary effluents, depending on the treatment applied prior to the filtration system. Tebbutt (1971) reported that the concentration of suspended solids in effluents may change by a factor of 2 within an hour. Biological flocs from secondary effluents appear to be stronger and more resistant to sheer forces than are chemical flocs found in water treatment, whereas flocs in chlorinated effluents are smaller, lighter and more fragile than those in unchlorinated effluents (Hsiung, 1980).

This chapter is not a design manual. At the outset it discusses the application of particle size analysis to better understanding and improved design of suspended solids removal processes in relation to water reuse. Later on, the mechanisms and models of chemically enhanced sedimentation and of the various filtration processes (screening, high rate, slow rate, membrane) with and without chemical pretreatment are presented. An extensive reference list provides the reader with an access to basic as well as advanced developments in understanding the field of chemical/physical treatment for the reuse of water.

PARTICLE CHARACTERISTICS IN WASTEWATER EFFLUENTS

PARTICLE SIZE AND VOLUME DISTRIBUTIONS

PSD in Wastewater

It has been demonstrated that particle size distribution (PSD) in wastewater effluents relates to a power law function (Adin and Elimelech, 1989; Alon and Adin, 1994) given by the following equation, for particles larger than 1 micrometer:

$$dN/d(d_p) = A\ (d_p)^{-\beta} \tag{1}$$

where N is the number of particles in the size interval of the particle, d_p is the average particle size of the interval and A and β are empirical constants. Each group size is ruled by different mechanism. Particles smaller than about one micrometer are transported by diffusion. Larger particles are transported by gravity. Transport of relatively large particles may also be dominated by interception, or they may be retained by interstitial straining.

Tchobanoglous and Eliassen (1970) suggested that PSD (particle size distribution) in an activated sludge effluent was bimodal. On the other hand, PSD data for aqueous particulates larger than 1 μm in many freshwater and wastewater systems are known to be modeled with a two-parameter power law distribution function (Kavanaugh, 1980) where the exponent provides an estimate of particulates contribution by size to the total particulate number, the surface area, the volume and the light scattering coefficient. When β values exceed 3 it indicates that the smaller size fraction dominates and vise versa. Adin et al. (1989) did not observe any bimodal distribution, not in activated sludge effluents nor in trickling filters, aerated lagoons or wastewater reservoir effluents. The distribution in those effluents was either of the power-law or the exponential type.

Most particle size analyzers are equipped with detectors based on light blockage or electric field modification or laser scattering principles. The PSD characterization methodology requires some elaboration. Since wastewater effluents are highly turbid (up to hundreds turbidity units) and particle number density is high, the testing water samples often have to be diluted. Most researchers use distilled water for dilution and are not aware of the importance of choosing a suitable dilution technique for obtaining reliable information about particle size distribution in wastewater.

Testing the effect of different dilution factors upon PSD, resulted in change of the particle number density as compared to initial dilution figures (Adin et al., 1989). At high degree of dilution (>40 times) the normalized PSD in all ranges increased sharply. In spite of the differences in the absolute numbers of particles at different dilutions, particle distribution in percent over different sizes of particles remained nearly unchanged. It could be concluded from the dilution study that (a) dilution with distilled water affects particle size distribution significantly and does not produce any stable and uniform image at different dilutions for any kind of wastewater, while (b) dilution with wastewater filtrates yields a stable and uniform picture with PSD virtually constant over a wide range of dilution factors (5–20), and (c) over 20 times dilution, the results are unstable and it is not advisable to work in that range.

Particle Size and Volume Distributions

PSD for different types of effluents at different dates are plotted in Figure 5.1 in terms of cumulative particle number frequency vs particle size (Adin et al., 1989). For example, curve no. 8 indicates that in that

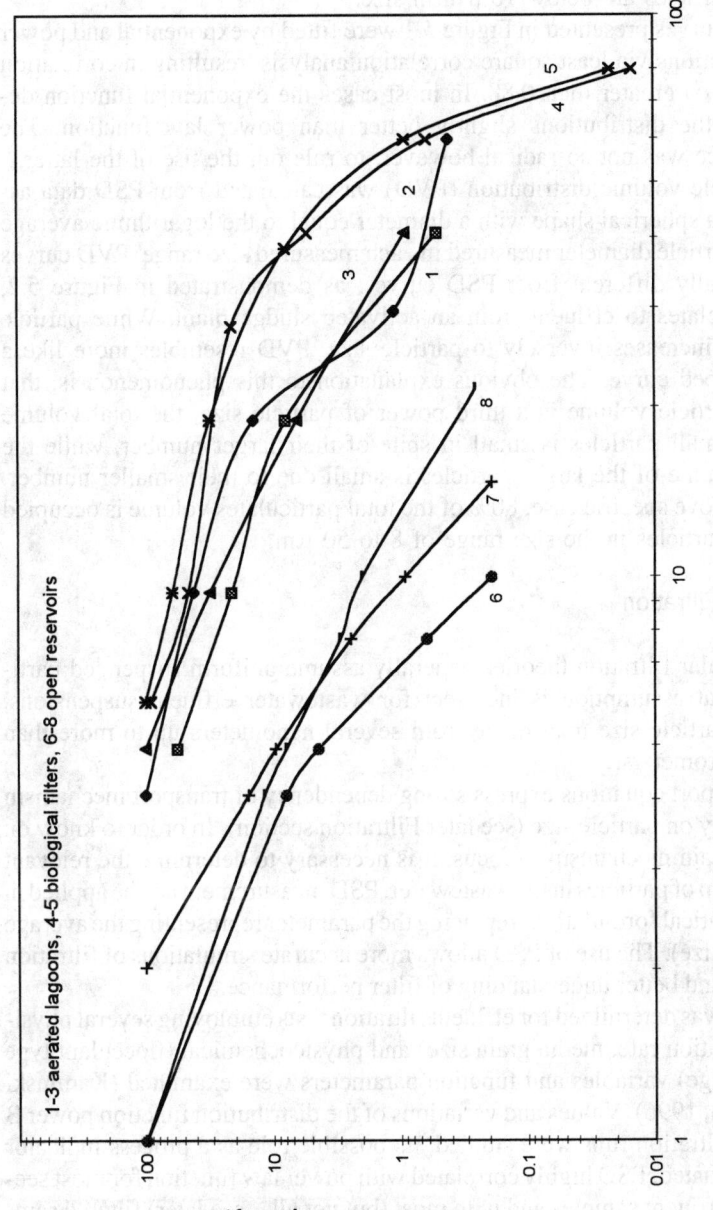

Figure 5.1. PSD for different effluents (after Adin et al., 1989).

1-3 aerated lagoons, 4-5 biological filters, 6-8 open reservoirs

specific sample 90% of all particles are below 5μm in size whereas 99% of all particles are below 10 μm in size.

The curves presented in Figure 5.1 were fitted by exponential and power law functions via least square correlation analysis, resulting in correlation factors (r^2) greater than 0.88. In most cases the exponential function described the distributions slightly better than power law function. The difference was not so radical however, to rule out the use of the latter.

Particle volume distribution (PVD) was calculated from PSD data assuming a spherical shape with a diameter equal to the logarithmic average of the particle diameter measured in each measured size range. PVD curves look totally different from PSD curves, as demonstrated in Figure 5.2, which relates to effluent from an activated sludge plant. While particle number increases inversely to particle size, PVD resembles more like a bell shaped curve. The obvious explanation to this phenomenon is, that since particle volume is a third power of particle size, the total volume of the small particles is small in spite of their larger number, while the total volume of the larger particles is small due to their smaller number. In the above specific case, 80% of the total particulates volume is occupied by the particles in the size range of 8 to 50 μm.

PSD in Filtration

Granular filtration theories generally assume uniform suspended particles. That assumption is incorrect for wastewater effluent suspensions, where particle size may range from several nanometers up to more than 100 micrometers.

Transport equations express strong dependency of transport mechanism efficiency on particle size (see later Filtration section). In order to know on which main mechanism to focus, it is necessary to determine the relevant size group of particles in the wastewater. PSD measurement can be applied in mathematical formulation (replacing the parameter representing the average particle size). The use of PSD allows more accurate simulations of filtration models and better understanding of filter performance.

PSD was determined for effluent filtration tests employing several physical (filtration rate, media grain size) and physicochemical (flocculant type and dosage) variables and function parameters were examined (Kaminski and Adin, 1996). Values and variations of the distribution function power β during filtration runs were studied, its possible role as a process indicator was evaluated. PSD highly correlated with power law function for most secondary effluent samples and with most (but not all—see later) filtered samples. A typical influent distribution consisted of 20,000, 14,000 and 1,000 particles per 1 ml for particle size ranges of 5–10 μm, 10–15 μm and 30–40 μm respectively. Less than 100 particles/ml larger than 40 μm were counted.

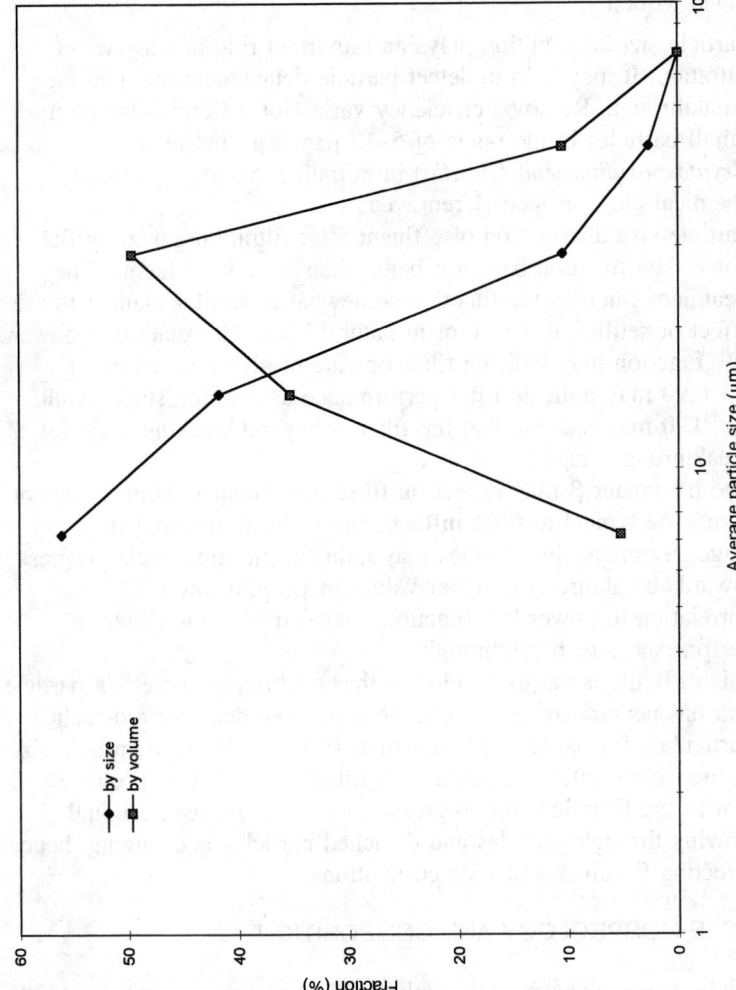

Figure 5.2. PSD versus PVD for activated sludge effluent (after Adin and Alon, 1993).

PVD was calculated from PSD data, assuming equivalent spherical particles. Total particle volume (TPV) contained in the filter influent was in the order of 10^{-4} to 10^{-5} cc/cc. After filtration, a decrease of approximately 1 order of magnitude in TPV was observed.

Several conclusions are made in relation to particle size from the above mentioned works:

- Particle size distribution plays an important role in wastewater filtration. It may help to detect particle detachment and particle breakthrough. Removal efficiency varies for different size groups: Small particles in the range of 5–10 μm (e.g., between *Cryptosporidium* and *Giardia*) in initialization stage, with no chemical aids, are poorly removed.
- Particle size distribution of effluent after filtration generally fits power law function behavior better than for raw effluents. The treatment smooths the function somewhat in similar manner to the effect of settling in tanks, or in natural lakes. Correlation to power law function may indicate filter operation: high correlation of $r > 0.90$ may indicate filter performance at working stage, while $r < 0.90$ may indicate that the filter is beyond working stage (at breakthrough stage).
- The parameter β may reflect on filter performance. High values of β may be typical to filter influent and effluent at initial filtration stages. Decrease in β values may indicate filtration cycle progress towards breakthrough. Lower values of β, with low PSD correlation to power law function, may indicate low filtration performances or breakthrough.
- It is difficult as yet to decide whether a filtrate particle is a particle that has penetrated throughout the entire bed depth or a detached particulate for particles smaller than 100 μm. Average particle size in the filtrate often increased with filtration time. It is probable that as the filtration run progresses, the ratio between original flowing through particles and detached particles is changing, hence affecting β values and PSD correlations.

PARTICLE MORPHOLOGY AND COMPOSITION

Particle shape as observed through the SEM measurements varies with the type of wastewater treatment. Thus, activated sludge treatment normally yields particles of well defined, typically oval shape (Figure 5.3). It seems also that tiny particles stick together and form a kind of "spongelike" medium. For a mixture of particles like this, it is more difficult to design a filtration process due to unpredictable behavior within the filter medium. Wastewater reservoir effluents contain a variety of particles but they are

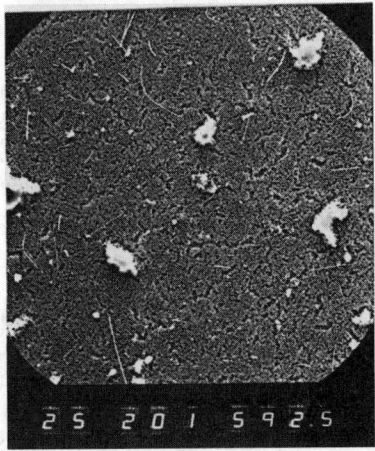

Figure 5.3. Activated sludge bioflocs in effluent.

typically characterized by gelatinous particle (film) shape. The latter are typified by their extremely high filter clogging potential.

The differences observed in shape corresponding to different origins of the wastes are reflected in the particle surface chemical composition as detected by the SEM X-ray analysis (Adin et al., 1989). Results point out that the particles from the open reservoir are compositionally dominated by Si whereas the particles from the activated sludge plant are dominated by Cl, Si, and Ca. At the same time, the particles from the aerated lagoons maturation ponds are mostly dominated by P, K and Ca which is typical to algal environment. Open effluent reservoir particle content may change, however, with the season (Dor et al., 1987).

ELECTROKINETIC POTENTIAL

Particles in wastewater effluents are mostly colloidal in nature and negatively charged, thus repelling each other and being stabilized. Positive ions are attracted to the negatively charged particle, forming what is known as the electrical double-layer model, which is presented and discussed in many papers and textbooks. A common, somewhat qualitative measure of the stabilizing energy is the electrokinetic surface potential, often referred to as ζ (zeta) potential (ZP).

Practically and roughly, three "stability zones" represented by ZP may be determined:

I	0 to −7 mv	Low potential	Low stability, tendency to coagulate
II	−10 to −15 mv	Medium potential	Medium repulsion, low to medium stability
III	≥ | −20 | mv	High potential	High repulsion, stable

ZP measurements of particles originated in various types of treatment plants effluents in Israel do not indicate any marked difference between the electrokinetic properties of particles from different sources. ZP of the particles is negative and varies in the range of -10 mV to -18 mV. Those values indicate that sometimes the repulsive forces among the particles are not as strong as repulsive forces in different types of natural surface water sources and a certain coagulation potential does exist in these systems (Adin et al., 1989).

BIOFILMS

Biological films and slimes, termed biofilms, derived from bacteria and algae, are among the major biological clogging factors in drip irrigation systems applying wastewater effluents. The individual cells in biofilms are surrounded by extracellular polymers, polysaccharides, and glycoproteins (Fletcher, 1979; Jones et al., 1969). Biofilms primarily utilize organic compounds to provide energy and carbon for cell growth and maintenance.

Such slimy gelatinous deposits of amorphous shape serve as triggers for serious blockage. Particles of a definable shape are found in a matrix of the gelatinous substance and form the primary sediment in the drippers. The primary sediment then acts as a trap for algae and zooplankton limbs that increase the volume of the sediment until they block the water pathway (Adin and Sacks, 1991). Thus, clogging of emitters using reservoir effluents is caused primarily by suspended solids in the water, but they do not necessarily initiate the clogging process.

Algae can be considered a special group of particles of a high indirect clogging potential within the size range of a few to 60 μm. One of the prominent findings, with respect to clogging mechanism by algae, is that this type of clogging develops only when initial deposits of minerals or a gelatinous material occurred previously. Algae particles are trapped and compressed into the primary sediments in the dripper. That happened when their concentrations were above 10^5 cells/ml for continuous period of a few weeks.

Kanaani et al. (1992) demonstrate that the adsorption coefficient for polysaccharide constituents of biofilms onto Kaolin clay increases with effluent ionic strength, while adding calcium ion, and when the pH is reduced. These phenomena are explained by conformational changes of the polysaccharide macromolecules, which result in respective changes in its electric charge. The phenomena are also in accord with results obtained by determination of the microelectrophoretic mobility of the clay and clay-polysaccharide particles. For instance, at reduced pH levels the enhanced adsorption was accompanied by decreasing of the negative reduced electric density of polysaccharide and clay-polysaccharide particles when the pH is reduced.

An adsorbed polysaccharide macromolecule may simultaneously attach to several particles to form a larger aggregate with a smaller charge density. This idea is supported by the observation that polysaccharide improves the alum flocculation process. Riddick (1961) has shown that the addition of a high molecular weight anionic polymer to alum flocculation system causes the flocs to build into large particles which settle rapidly. At this stage of flocculation, the long polymer chains become simultaneously attached to several floc particles, and this results in a mechanical as well as interlocking, and produces floc of tremendous size which settles rapidly.

The bridging effect of polysaccharide that causes the attachment of several clay particles to form floc of tremendous size may explain the role of polysaccharide in emitters and pipeline clogging. Thus a tremendous floc may trap algae or other aquatic substances present in the effluents, causing the formation of big biomass that eventually clogs the pipelines.

Clogging potential may be decreased by modifying the emitter internal design in order to decrease deposition and by chemical pretreatment with oxidants and flocculation-filtration.

COAGULATION AND FLOCCULATION

TERTIARY TREATMENT BY COAGULATION-FLOCCULATION

The physicochemical processes associated with coagulation are rapid mixing when chemical coagulants are quickly and uniformly dispersed in the water and particles are destabilized, and then slow mixing or flocculation when particles slowly aggregate and form a settleable or filterable mass. The four distinct mechanisms of coagulation include double layer compression, adsorption and charge neutralization, sweep coagulation, and interparticle bridging (Amirtharajah and O'Melia, 1991). Secondary effluents contain a wide variety of suspended and colloidal particles that cause color and turbidity. Chemical coagulation and flocculation are most important steps to remove colloidal particles and turbidity from wastewater and is quite effective in removing viruses. Chemical coagulation-flocculation processes are known to aggregate wastewater constituents within the size ranging from 0.1 μm to 10 μm (Levine et al., 1985).

Aluminum or iron salts may be added at various points in a wastewater treatment plant to enhance solids removal. The metal salt (coagulant or flocculant) destabilizes colloidal solids that would otherwise remain in suspension and thus can be used to improve effluent quality. In spite of the increased use of chemical coagulants, coagulation theory still fails in providing the use of the process in an optimal manner, particularly under transient conditions. This could lead to diminished effluent quality, increased chemical costs by routine overdosing, or both. An improved coagulation might alleviate its adverse effects on sludge digestibility. Coagulation has been the subject of much research, some of which have proposed coagulation in the context of water treatment. However, wastewater treatment differs from water treatment in several ways: particulate matter is present in substantially greater concentrations in wastewater; the average particle size is also greater. These factors are likely to affect both coagulant demand and flocculation behavior. The particulates to be removed include a much greater proportion of organic material than in the case of water treatment. The more hydrophilic surfaces of these particles may react differently to a coagulant.

Wastewater involves greater concentrations of soluble species that may affect coagulation in several ways. The mechanisms of coagulation by Al^{+3} salts may involve some chemical factors including the hydrolyzing and polymerizing tendencies of the Al^{+3} atom, the adsorptivity of such aluminum hydroxide species; the solubility of such species; the nature and extent of aluminum hydroxide precipitation including interaction with other colloidal surfaces and effects of other solutes of surface properties on the metal hydroxide species. In light of this apparent complexity it is not surprising that a number of different models have been proposed to explain the way aluminum salts can destabilize colloids.

The earliest explanation considered only double-layer compression. Others suggested that the hydrolysis products of aluminum play a more important role in particle destabilization. More current explanations of coagulation hypothesize the existence of two distinct mechanisms, which are:

- The adsorption of positively charged, polynuclear aluminum species to neutralize the charge of negatively charged particles, and
- The physical enmeshment of particulates in an aluminum hydroxide precipitate.

The classical coagulation agent Al^{+3} ions are usually obtained by dissociation of dissolved alum salt $[Al_2(SO_4)_3]$ containing 14 to 18 molecules of bound water. It may be added at various points in a water treatment scheme to enhance solid and colloids removal. The removal efficiency is a function of coagulant dosage, the pH and the mixing (dispersion) condition as well.

The mechanism(s) of coagulation by Al^{+3} salts may involve some chemical factors including the hydrolyzing and polymerizing tendencies of the Al^{+3} ion; the adsorbtivity of some hydroxide species, their solubility; the nature and extent of aluminum hydroxide precipitation, including interactions with other colloidal surfaces.

Organic substances that exhibit colloidal behavior range from organic macromolecular compounds such as humic substances, proteins, etc. up to viruses, bacteria and algae. Humic substances are often present in surface waters and it was shown that they constitute a significant fraction of residual organics in secondary effluents (Rebhun and Manka, 1971; Manka and Rebhun, 1982).

These studies have shown that coagulation of humic substances by alumino or iron salts involves chemical interaction with alumino or ferro hydrolysates and in formation of alumohydroxo-humic or ferric hydroxo-humic complexes-precipitates. These result in transforming humic substances from "soluble" form or "true color" to a precipitated form or "apparent color." Cationic polyelectrolytes interact with humic substances through charge neutralization and form strong association complexes in the form of precipitates.

There is a stoichiometric relationship between flocculants and humic substances to form the complexes-precipitates. These precipitates may be in a colloidal form and need an additional amount of flocculant for efficient settling or filtration.

Kavanaugh et al. (1977) reported that contact coagulation of secondary effluent with ferric salts and polymeric acids gave an indication of poor attachment of organo-alumino or ferric flocs in deep bed filtration and low capacity to breakthrough. Also, Rebhun et al. (1984) observation of alumino humic flocs in previous studies indicated their fragile nature and poorer settleability in comparison with the toughness of alumino-mineral turbidity flocs.

The generalized concepts of coagulation for drinking water can be used to determine the chemical dosages required for efficient flocculation and filtration of wastewater. The coagulation diagrams developed for water by Amirtharajah and co-worlers (Amirtharajah and Mills, 1982; Johnson and Amirtharajah, 1983; Amirtharajah, 1988) can be used as tools for predicting the optimum conditions for particle destabilization and effective filtration.

Ghosh et al. (1994) illustrated a methodology for using bench scale tests to optimize chemical coagulation for tertiary treatment of activated sludge secondary effluent in terms of overall quality measured by several parameters such as turbidity, BOD_5, TOC, and total-P. Alum was used as the test coagulant. The operational approach was to choose the lowest chemical dose that best achieved the effluent objective.

As previously mentioned, two different mechanisms for particle destabilization are considered, namely, charge-neutralization and sweep coagulation (Ghosh et al., 1994). At the lower and narrower 3–5 pH range the assumed mechanism of particle destabilization and coagulation is charge neutralization which occurs as a result of the interaction between the highly charged cationic aluminum hydrolysis species and the negatively charged colloidal particles present in the wastewater. A stoichiometric relationship exists between these positively charged species and the negatively charged particles and charge reversal of the colloids occurs by overdosing. This causes the settled water turbidity removals to go down after reaching a maximum value. On the other hand, for alum dosages greater than 20 mg/L and between final pH values of 6 and 9, rapid formation of amorphous, solid phase aluminum hydroxide takes place and in this case turbidity removal occurs by the physical entrapment of the colloidal particle within the mesh of aluminum hydroxide precipitate. As a result, instead of observing a maximum in turbidity removal, an asymptotic approach to a steady value in the settled turbidity is noticed beyond a certain alum dosage.

The occurrence of charge neutralization and then restabilization can be further corroborated by zeta potential data. The zeta potential of secondary effluent samples taken from R.M. Clayton wastewater treatment plant, Atlanta, were analyzed to be −30 mV. In the low pH zone of 3–5, it was observed that with the increase in the alum dosage, the zeta potential of the resultant suspension after flocculation first increased in the positive direction confirming the occurrence of charge neutralization and particle destabilization, and finally changed sign and acquired a positive value at higher dosages confirming the occurrence of charge reversal and restabilization. The zeta potential data, coupled with the settled turbidity variations are shown in Figure 5.4. By interpolating the data, it was observed that a zeta potential of zero corresponded to an alum dose of about 63 mg/L which produced a corresponding settled turbidity of about 0.6 NTU and this value is close to the minimum turbidity level of 0.5 NTU. The actual minimum turbidity corresponded to an alum dosage of 68 mg/L and zeta potential of about +5 mV.

These results are consistent with a study by Kirkpatrick and Asano (1986) involving tertiary treatment of secondary effluent at Castroville Wastewater Treatment Plant, USA. They observed optimum sweep coagulation at an alum dose of 50 mg/L and a pH value of about 7.0 in the charge-neutralization zone marked by pH 3 to 5 and alum dosages of 10 mg/L and more, the percent turbidity removals both after settling and filtration was slightly less than those at the sweep coagulation zone. The data in this region also indicated an upper boundary for turbidity removal which was attributed to charge reversal and restabilization due to overdosing. Since the turbidity removals after filtration and settling were found to be the maximum in the optimum sweep zone, in this region principal attention was subsequently focused to determine the variation in the concentrations of different parameters of the wastewater after conventional treatment.

Based on the removal data for various parameters (turbidity, TOC, BOD_5, and total-P), optimum design conditions for chemical pretreatment

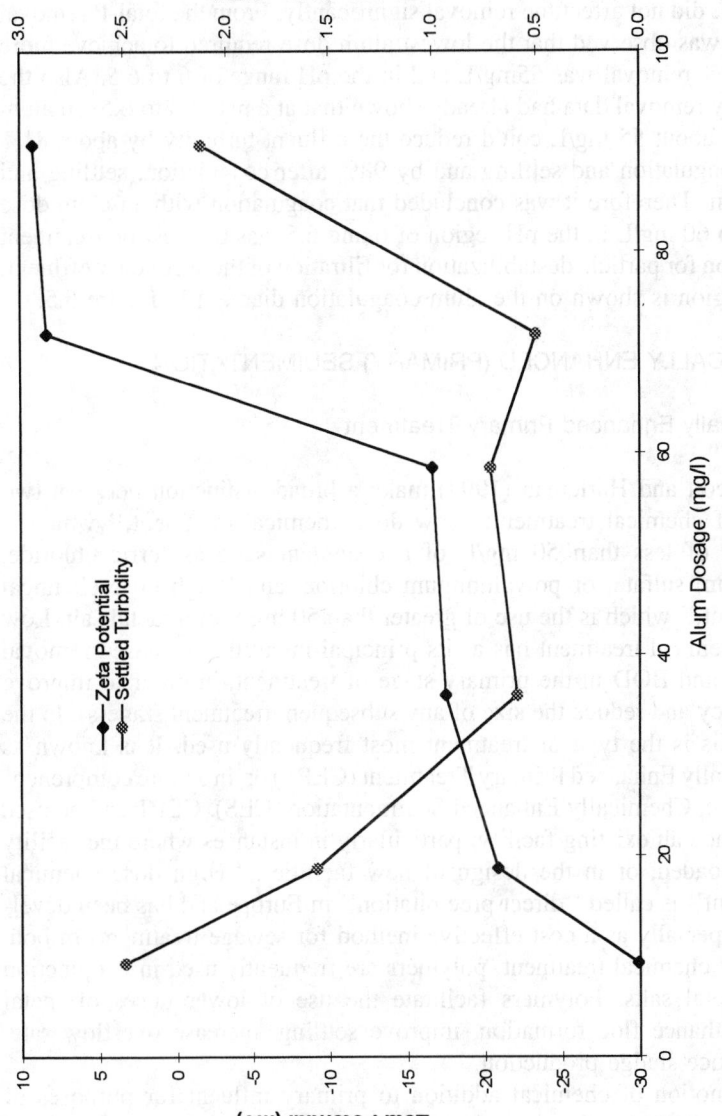

Figure 5.4. Variation in turbidity and zeta potential of the settled effluent as a function of alum dose in the charge neutralization zone (Initial pH = 6.93, Initial turbidity = 11.5 NTU) (after Ghosh et al., 1994).

for tertiary filtration of secondary effluent from the wastewater treatment plant was determined. The best TOC and BOD_5 removal occurred between pH 6 and 6.7 and in that pH range, an increase in the dosage from 36 to 70 mg/L did not affect the removal significantly. From the total-P removal data, it was observed that the lowest alum dose required to achieve more than 99% removal was 55mg/L and in the pH range of 6 to 6.5. Also the turbidity removal data had already shown that at a pH of 6 to 6.5, an alum dose of about 55 mg/L could reduce the effluent turbidity by about 81% after coagulation and settling and by 98% after coagulation, settling and filtration. Therefore it was concluded that coagulation with an alum dose of 55 to 60 mg/L in the pH region of 6 and 6.5 was the best pretreatment condition for particle destabilization for filtration of the secondary effluent. This region is shown on the alum coagulation diagram in Figure 5.5.

CHEMICALLY ENHANCED (PRIMARY) SEDIMENTATION

Chemically Enhanced Primary Treatment

Murcott and Harleman (1994) make a broad distinction between two types of chemical treatment: "low dose chemical treatment," which is the use of less than 50 mg/L of a coagulant such as ferric chloride, aluminum sulfate, or polyaluminum chloride, and "high dose chemical treatment," which is the use of greater than 50 mg/L of a metal salt. Low dose chemical treatment has as its principal intent the increased removal of TSS and BOD in the primary stage of treatment, in order to improve efficiency and reduce the size of any subsequent treatment stage(s). In the U.S., this is the type of treatment most frequently used. It is known as Chemically Enhanced Primary Treatment (CEPT) or, in a more comprehensive term, Chemically Enhanced Sedimentation (CES). CEPT can be used to upgrade an existing facility; particularly in instances where the facility is overloaded, or in the design of new facilities. "High dose chemical treatment" is called "direct precipitation" in Europe and has been developed especially as a cost effective method for sewage treatment. In both types of chemical treatment, polymers are frequently used in conjunction with metal salts. Polymers facilitate the use of lower doses of metal salts, enhance floc formation, improve settling, increase overflow rate, and reduce sludge production.

The notion of chemical addition to primary influent for purposes of relieving BOD loading on subsequent systems is not new, and has been utilized in Europe for decades (Odegaard, 1992). Chemical addition prior to primary sedimentation has received renewed attention in the U.S. as a probable secondary consequence of phosphate effluent regulations yielding the aftermath of concurrent organic loading reductions. Chemical addition

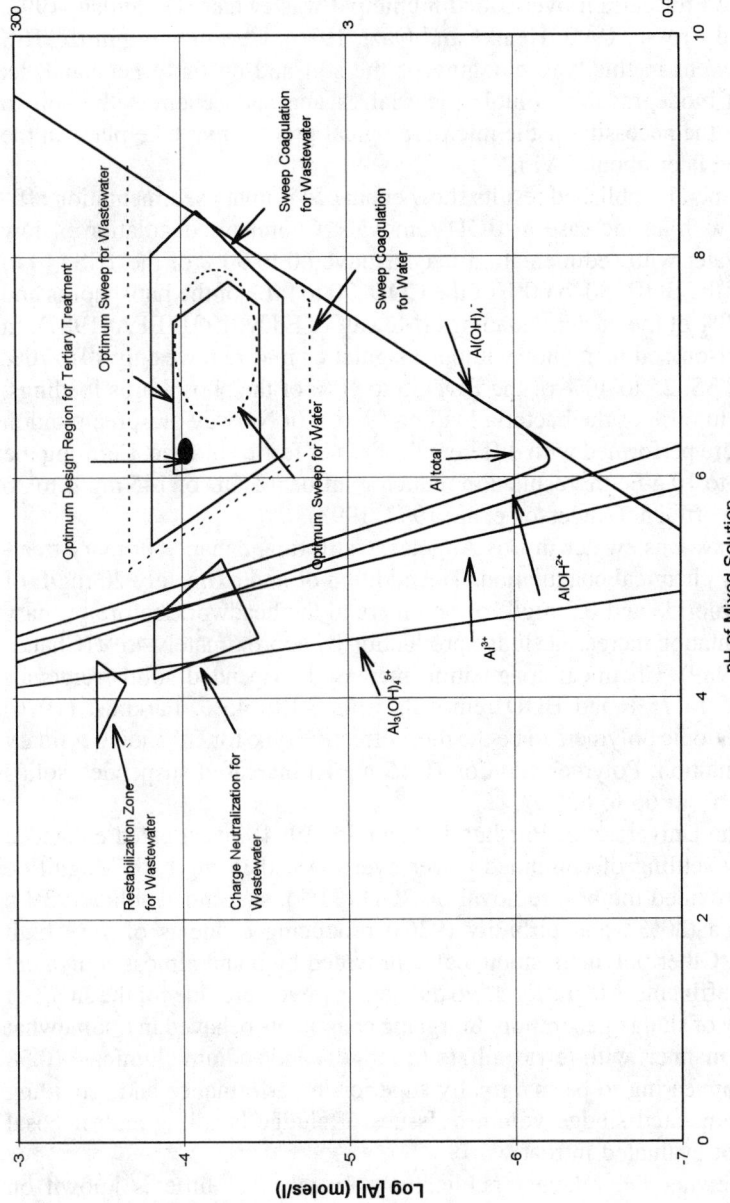

Figure 5.5. Alum coagulation diagram showing the optimum design condition for particle destabilization for tertiary filtration of secondary effluents (after Ghosh et al., 1994).

175

for enhanced sedimentation is employed for pulp and paper industry wastes also yielding BOD_5 reductions (Okey, 1983). This technique has also been evaluated for certain overloaded municipal wastewater (Harleman, 1992; Rad and Crosse, 1990; Heinke and Qazi, 1976). CES may minimize TSS and prevent in this way clogging of the soil and on the other hand, let enough biodegradable soluble material escape and penetrate the soil to provide the necessity of the microbiological process that take place in the soil (see later about SAT).

In general, published results show enhanced primary sedimentation efficiency with an increase in BOD removals. Chemical coagulation of raw wastewater with sedimentation may remove 60 to 90% of the TSS, 40 to 70% of the BOD, 30 to 60% of the COD, 70 to 90% of the phosphorus and 80 to 90% of the bacteria loadings (Metcalf & Eddy, 1991; EPA, 1987). In comparison, sedimentation without coagulation may remove only 40 to 70% of the TSS, 25 to 40% of the BOD, 5 to 10% of the phosphorus loadings, and 50 to 60% of the bacteria loading (ibid.). In Norway pre-precipitation tests were performed with different dosages of ferric chloride. Limiting the dosage to 10 g Fe/m^3 resulted in reducing influent BOD of 145 mg/L to 36 mg/L in effluent (Anderson et al., 1992, 1993).

All raw wastewater in Los Angeles, California, enhance primary treatment by chemical coagulation. The addition of approximately 20 mg/L of ferric chloride and 0.2 mg/L of polymers to the headwork before primary sedimentation increased sludge production by approximately 45% (Chaudhary, 1989). Chemical coagulation increased suspended solids removals from 65 to 75% and BOD removals from 30 to 45%. Parkhust (1976) found anionic polymers to be the most effective type for enhancing primary sedimentation. Polymer addition (0.15 mg/L) increased suspended solids capture from 66 to 83%.

At the University of Pittsburgh, Neufeld (1994) investigated enhanced primary settling of combined sewer overflows. 0.75 mg/L of Magnifloc 1596 provided the best removal of BOD (91%), suspended solids (93%), ammonia (80%), and turbidity (93%) producing effluents of very high quality. Other polymers studied also provided high and almost equivalent quality effluents. Magnifloc 1596 did not, however, provide for the smallest quantity of sludge generation. Inorganic coagulants behaved in a somewhat similar manner, with ferric sulfate (27 mg/L) and sodium aluminate (0.38 mg/L) appearing to be marginally superior in performance and providing small generated sludge volumes. Issues of sludge handling and disposal were not evaluated in that work.

Reviewing the relevant publications reveals that little is known on the linkage between primary CES with polyelectrolyte addition and the following secondary and tertiary treatment. It may be regarded in two ways: (a) CES may minimize organic loading, nutrients and suspended matter transported to the activated sludge unit; and (b) destabilized primary particles, escaping from the primary settling tank, may affect the bioflocculation process. While they might be smaller in number (and some of them perhaps a bit larger in size) than in conventional gravity sedimentation overflow, their effect on the microbial floc formation and settling characteristics may be greater, due to the partially adsorbed polymer chains carried on their surfaces. This, in turn, may influence the gravity floc separation in the secondary tank and hence, possibly, particle retention mechanisms within the filter media. The above phenomena should be studied in parallel with the effect of polymer type and dosage on primary sludge volume minimization.

It is believed that CEPT can reduce the size and cost of any biological process that follows. This results from a reduction of BOD_5 loading to the biological stage and from an increase in the oxidation rate of largely soluble BOD_5 remaining after CEPT (Murcott and Harleman, 1994). The use of biological aerated filters after CEPT may take the best advantage of CEPT's performance improvements.

The chemical pretreatment is expected to increase the soluble fraction in the wastewater, which may affect the biomass activity. COD and nitrogen compounds monitoring along the treatment train facilitate the understanding and determination of this effect.

The CES technological approach is for the wastewater treatment plant operator to add calibrated quantities of special organic polymeric coagulant into the raw sewage to promote organic substance removal in the form of sludge. Such sludge may be recovered via land application or disposed in the same manner as is current practice for each municipality. The thrust here, however, is to cause organic pollutants to be collected as a sludge thus minimizing the organic loading to the biological treatment unit. In this fashion, the municipality will maximize their utilization of existing plant settling capacity consistent with the minimization of organic loadings to biological systems (Neufeld, 1995).

Polymers to be used are selected on the basis of standard "jar tests" with the wastewater, concentrating on low anionic high molecular weight polymers as flocculant aids to alum/iron coagulants and on cationic polymers either as aid or as primary coagulants. The effect of CES on the bioflocculation process may be studied by also using the overall primary particle removal rate proposed by Wahlberg et al. (1994):

$$N_t = \alpha + be^{-t} \qquad (2)$$

where N_t = primary particle number concentration, α = equilibrium particle number concentration, b = difference between initial (i.e. unflocculated) and equilibrium particle number concentrations, and t = time.

The application of polymers as primary flocculants is interesting from the viewpoint of wastewater reutilization in agricultural irrigation, since there is no interest there in phosphorus removal (which serves as a nutrient to the plants). While a number of different substances can be employed, the expected quantity of sludge to be produced can vary widely.

CES as Pretreatment for Soil-Aquifer-Treatment (SAT)

Typical sewage treatment plants in high-income countries give primary and secondary treatment. Primary treatment is a mechanical process that removes floatables and sinkables. Secondary treatment is a biological process that removes most of the remaining organic material (dissolved and suspended) by bacteria in trickling filters or aeration tanks. This is the typical treatment prior to discharge of effluent into surface water. The treatment can be enhanced by further nutrient removal (nitrogen and phosphorus), disinfection, and high-rate filtration. When the effluent is to be used for irrigation or groundwater recharge, however, this expensive high-tech treatment to remove organic material is not necessary and simpler, primary-type treatment can suffice. As a matter of fact, the higher organic carbon level of primary effluent may enhance denitrification of nitrate in the soil, and degradation of refractory synthetic organics by bacteria in the soil or aquifer through co-metabolism and secondary utilization (Bouwer, 1974).

Denitrification is an anaerobic microbiological process whereby denitrifying bacteria use the oxygen of the nitrate to metabolize organic carbon (i.e., use organic substrate as electron acceptor), thereby producing free nitrogen gas and nitrogen oxides. Typically, about 1 mg organic carbon is required to denitrify 1 mg nitrate nitrogen (Bouwer, 1974). Strong sewage may contain 100–200 mg/L nitrogen, mostly as ammonium and organic nitrogen. This amounts to 1000 to 2000 kg/ha per m of effluent applied (typical summer season irrigation) to the field and is about ten times as high as the normal fertilizer requirement. After nitrification in the treatment plant and soil, it would take about 1000–2000 mg/L organic carbon to denitrify the remaining nitrogen. This amount of carbon can be present after primary treatment, but it is much less than that after secondary treatment. The still relatively high level of dissolved organic carbon in the effluent after primary treatment should enhance removal of nitrogen by supplying an adequate energy source for the denitrifying bacteria (Rice and Bouwer, 1984). For this reason, primary treatment is the preferred treatment where the effluent is used for irrigation or groundwater recharge and serious problems of excessive nitrate in crops or groundwater are to be avoided. Thus, treatment of sewage effluent must be adapted to its reuse, rather than blindly following treatment processes developed for disposal into streams or other surface water.

The additional organic carbon in primary effluent also promotes breakdown of synthetic organic compounds in the soil or aquifer. While the compounds themselves are resistant to biodegradation, bacteria will degrade them more if plenty of readily biodegradable substrate is also present, as in primary effluent. This phenomenon is called secondary utilization. A commonly recognized form of secondary utilization is co-metabolism, in which the refractory synthetic carbon cannot be used as an energy source for bacteria, but is still degraded or transformed by enzymes formed through metabolism of the readily degradable primary substrate (McCarty et al., 1985). For this reason too, effluent after primary treatment is better than after secondary treatment if the effluent is used for irrigation or groundwater recharge. Besides, primary treatment is a lot simpler and cheaper than secondary treatment.

Drip irrigation systems are dependent on virtually particle free water since even colloidal particles can clog dripper pathways (Adin and Sacks, 1991). On the other hand soluble organic and nutrients existing in the wastewater are welcome by the plants. This is another reason why the CES process is beneficial.

In summary, chemically enhanced primary treatment leaves an effluent with a low suspended solids content and a high dissolved organic carbon level. This is favorable for direct irrigation with subsurface drip irrigation, disinfection and subsequent other irrigation, and recharge of aquifers. The low suspended solids content prevents clogging of drip irrigation emitters, increases the effectiveness of disinfection by chlorine or UV irradiation, and reduces clogging of the soil surface when the effluent is used for surface irrigation or groundwater recharge. The high organic carbon content of the effluent enhances denitrification in the soil, thus preventing or reducing adverse effects of excessive nitrogen on crops and of nitrate contamination of underlying groundwater. Also, the high organic carbon level will increase biodegradation of synthetic organic compounds, which reduces potential uptake of these compounds by plants and residual levels in underlying groundwater, thus enhancing the potability of this water. Finally, the primary treatment is not only simple and doable for low-income areas and small towns, elimination of the secondary treatment, which until now is routinely used prior to discharge or reuse, will save a lot of money and will make it possible for large and small communities to safely use their sewage effluent for irrigation and other reuse purposes. The sludge problem has, of course, to be properly addressed.

ELECTROCHEMICAL COAGULATION

Electrochemical processes in water treatment which are gaining ground are the electroflotation and disinfection. The main advantage of the electrochemical process of disinfection is the production of disinfective chemicals in situ in the treatment device. Flotation is a gravity separation process in which air bubbles are attached to individual solid particles, thereby reducing their density so that they float to the surface of the liquid. In the electroflotation process gas bubbles are generated by electrolysis of the liquid. A 5–20V direct current is used at a current density of approx. 100 amp/m^2 of electrode. The particular attraction of flotation, as compared to sedimentation, is that particles rise generally at a much higher rate than their settling rate, so that the size of the unit for a given duty is approximately one third of that of a clarifier. Power needed to float one ton dry solids is 90–100 kw.

An emerging technology for flocculation or chemically enhanced sedimentation is the electrocoagulation process consisting in introduction in the raw wastewater by DC pulses aluminum cations (Al^{+3}) supplied by a special aluminum electrode (Adin and Vescan, 1996). The applied tension and current permit the diffusion of the metal ions in the solute. The disinfection effect of the electrostatic field created during the electrolysis process can be regarded as a secondary advantageous effect of the process (Vescan et al., 1986). However its intensity and its control should be studied.

A logical consequence from the point of view of the reactions kinetics (direct contact between the Al^{+3} ion and the particle without formatting the double-layer by dissolution) is that introduction of Al^{+3} ions (as ionic form resulting from an electrolysis process) in the solute will improve the coagulating behavior of the aluminum, and for the same effect the needed amount would be reduced together with the necessary mechanical stirring energy replaced with the dissolution and dispersion effect of the electrolysis process.

The classical coagulation agent Al^{+3} ions is usually obtained by dissociation of dissolved alum [$Al_2(SO_4)_3$] salt. The molecular weight of the alum is 666.7 and the aluminum from this represents only 8% (i.e., 54) as the active component of the flocculation process. In the electro-flocculation process, Al^{+3} ions are released in wastewater (as electrolyte) from the aluminum electrodes immersed, under the effect of a DC (direct current). This may save considerable amount in flocculant quantity, transportation and storage. In addition, the negative effects on particle attraction forces introduced by the sulfate anions in the conventional alum addition will be eliminated.

The electrocoagulation (electro-flocculation) technology consists of the introduction of aluminum cations supplied by conducting DC to an aluminum electrode, into wastewater. The applied voltage is less then 2.4 V, so that gas formation by electrolysis does not occur. The diffusion of the metal ions in the solute is produced primarily by the current, and could be enhanced by mechanical stirring.

A series of laboratory scale preliminary experiments were performed at the Hebrew University of Jerusalem. Sample of raw wastewater having initial turbidity of 400 NTU and TSS of 660 mg/L was conventionally treated with a dosage of 40 mg/L alum. Reductions to 65 NTU and TSS to 135 mg/L were obtained. By electroflocculation of the same sample 2.5 NTU and TSS of 8.8 mg/L were obtained. Another sample was treated with a dosage of 100 mg/L alum resulting in a turbidity of 4.5 NTU and a TSS of 20 mg/L, by electroflocculation the TSS was reduced to 12 mg/L and the NTU to 2.2. VSS of 490 mg/L was reduced to 105 mg/L after flocculation with alum; after electroflocculation the obtained VSS value was 6.8 mg/L. In those electroflocculation experiments the applied current was of 0.16 amp/cm^2.

With another series of experiments the needed time for treatment, the current (amp) and turbidity reduction were as follows: Initial turbidity of the sample was 80 NTU, the applied current was 4.0 amp; after 1 minute the turbidity was 20 NTU and after two minutes it was 5 NTU which remained the same for the next five minutes (93.7% reduction of the turbidity). The aluminum content of the resulting sludge and effluent was analytically determined. 13, 28 and 54 mg aluminum respectively were found in the sludge. The effluent aluminum content was 0.06 mg only!

The process is simple to construct and to operate and maintain. Additionally, it is applicable for small communities. The preliminary evaluation shows a reduction potential of the operating costs of more than 40–45%, versus the classical flocculation schemes and enhanced environmental protection by decreasing haloorganic precursors, decreasing heavy metals concentrations, and decreasing chlorine dosages for disinfection. The electrochemical treatment may also be conducted as a disinfection process.

If one works in a potential range below the decomposition potential of water (1.6–1.8 V) gas formation does not occur. If the gas formation is allowed after the coagulation-flocculation phase, the separation process along the gravitational settling will be completed by electroflotation. The separation of floating particles takes place with a much higher separation rate than that of settling. The process results in a very clean effluent with more than 90% turbidity and suspended solids removal.

It is supposed that a positive effect could be obtained also on the sludge dewaterability—by the correct dosing possibility of the coagulant and facilitate the sludge handling and disposal, aspects which would also contribute to increase the economical merit of the process.

FILTRATION FOR WATER REUSE

STRAINING

Different types of filters are currently being used to cope with the clogging problem. Most of them are based on straining mechanisms whose filter (strainer) pores are smaller than most of the particles to be strained. This differentiates them from granular filters, which operate on the principle of in-depth filtration. The more sophisticated strainers, by incorporating automatic cleaning mechanisms, relieve the farmer of time-consuming cleaning work. Most problems associated with such filters derive from incomplete back-flushing, which causes clogging phenomena that can only be corrected by individual filter servicing. Irrigation waters often contain considerable amounts of suspended matter (silt, algae, etc.) which affects the proper functioning of the flush mechanisms. This may result in too frequent back-flushing and also may lead to complete pore clogging.

The straining process is normally based on the principle that the pores of the medium are smaller than the particle diameters. SEM analysis demonstrates, however, that an enhanced screen filter clogging may be caused by particles adhering to the corners of the pores and peripheral wires or to the fewer, formerly settled, larger particles. It provides evidence to the claim that screen nominal pore size has nothing to do with the actual size of particles that are removed by it (Adin et al., 1989).

A mathematical expression of exponential growth in pressure losses as a function of volume filtered through a steel screen is known as Boucher's Law (Boucher, 1947), it assumes the form

$$H = H_0 e^{IV} \tag{3}$$

where H_0 = the pressure loss across a clean screen; V = the volume filtered; and I = a filterability index, related to the suspended material content of the water. I increases in value as water becomes more difficult to filter. Increase in head loss rate is caused by: (1) Increase in suspended material; (2) decrease in pore diameter in various screens; and (3) increase in filtration velocity.

This formula is the resultant of two simultaneous processes (Ives, 1960): (1) The deposition of material within the pores of the screen and the filter cake which obstructs the passage of the suspension; and (2) the deposition of material at a constant rate on the surface of the screen. An extension of the pores is made by the formation of capillaries and a porous filter cake is formed, the pores of which correspond to the pores upon the filter screen. The fouling rate of a filter depends to a considerable extent upon the distribution of the capillary diameters and entrapped particle dimensions.

It should be noted that Boucher's formula provides solely for the connection between the pressure drop through the screen itself and the flow rate (or volume filtered) without considering a specific filter configuration. In addition, it does not provide a definite idea about the behavior of the cake formed on the screens of the strainers. The filterability index, I, can be used as an index for the filter and cake resistance. With increasing I values, more frequent filter clogging can be expected. Thus the derivation of the filterability index from pilot experiments can be possibly used for predicting clogging rates in full-scale filter screens. It cannot, however, provide useful information as to the expected effluent quality and the filter cake behavior (Adin and Alon, 1986). An important cake parameter is its compressibility, which is particularly important when the rinsing of the filter is to be considered. The following formula can be used to define the cake's character in terms of compressibility (McCabe and Smith, 1971)

$$dP/dL = k_1\mu u \ [(1 - \epsilon)^2 \ (S_p)^2]/[\epsilon^3 \ (V_p)^3] \tag{4}$$

where dP/dL = the pressure gradient at cake thickness L; S_p = single particle surface area; V_p = single particle volume; ϵ = cake porosity; k_1 = constant; μ = hydrodynamic viscosity; and u = stream velocity. Further development of the above expression for an incompressible cake, yields

$$p = [k_1\mu u/\rho_p A] \ [(S_p)^2/(V_p)^2] \ [(1 - \epsilon)/\epsilon^3] \ m_c \tag{5}$$

where m_c = the total mass of the filter cake; and ρ_p = the specific gravity of a single particle, or

$$P = K \ m_c \tag{6}$$

Filter cakes adhering to this relation are incompressible.

A filter can usually serve a wide range of flow rates (Dickey, 1961). As the flow rates passing through a given filter increase, the work cycle (i.e. length of time until back-flushing) is shortened. Experimental and practical evidence indicates that increasing the flow rate in a granular medium may bring about a detachment process (Adin et al., 1979; Adin and Rebhun, 1977). This is caused by extended hydraulic shear forces on the particles in a granular medium. Similar forces act on the filter cake in straining, as described by Hosseini and Rushton (1979). Evidence of particle release from strainers in effluent irrigation lines was demonstrated by Oron et al. (1982). It has been postulated by Tien and Payatakes (1979) that the detachment phenomenon in strainers is frequent, and contributes to the relative inefficiency of this process, particularly in its initial stage. Also, colloidal particles in granular filters that do not penetrate into the effluent due to attachment processes succeed in doing so when straining is applied. Thus it can be concluded that it is impossible to establish a clear and universal optimum for the operation of screen filters.

Microscopic analysis and particle size measurements confirm the formation of larger particles in the filtrate (Figure 5.6). A detachment phenomenon does exist that prevails toward the end of the run. In this case, the build-up of particles on the cake competes with detachment of particles from the cake (sometimes labeled "channeling phenomenon"), thus causing the decrease in head loss rate. The detachment phenomenon may cause clogging in drip irrigation system components that are placed after the filters, and in the drippers themselves. Evidence for this was received from farmers and others who attempted to evaluate the reasons for clogging by field observations.

This type of behavior can be observed by microscopic examination of the aggregates in the effluent. They are compacted amorphous particles typically formed on a filter cake, as compared to the original particles in the suspension.

Although many researchers stress the advantage of granular filtration over straining in filtering colloidal and submicron particles (Adin, 1978; Rothwell, 1978), the possibility of retaining such particles cannot be excluded in the cake formed by straining. Adin and Alon (1986) demonstrate that the majority of particles in the suspension are smaller within one order of magnitude than the smallest screen pore, yet clogging of screens occurs. The clogging phenomenon is a result of both a bridging mechanism in the screen, and an attachment mechanism in the filter cake. The existence of particle bridging in strainers is also mentioned by Ives (1975) and Dickey (1979).

DEEP-BED FILTRATION

Filter Performance

Granular, deep bed filtration plays a major role in the protection of system components vulnerable to clogging in agricultural and industrial water reuse projects (Adin, 1991). It is commonly agreed that suspended solids provide the major cause of serious clogging problems in filters and low-rate applicators that serve in effluent transport schemes (Adin, 1978). Suspended solids cause rapid pressure drops and flow disturbances in screen filters commonly used for the protection of sensitive appurtenances. However, filter design in this field still lacks a sound scientific basis.

Filtration through porous media is a process which is based in principle on the capture of particles, rather than on removal of masses of solids. The main tool for evaluation of filter performance in wastewater filtration practice has been, so far, the removal ratio of TSS and sometimes of turbidity.

In conventional water treatment by granular filters, removal of suspended solids is usually improved by the use of a finer medium, deeper filter bed or a lower filtration rate while headloss buildup is increased by

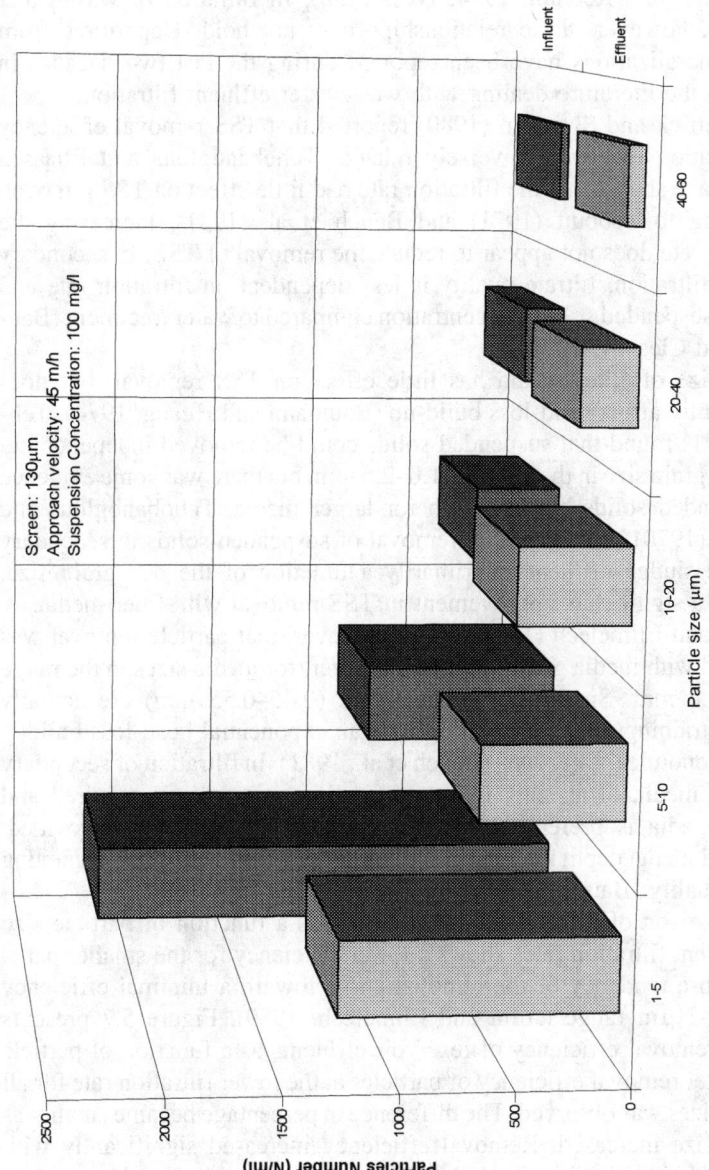

Figure 5.6. Particle size range in influent and effluent of a strainer.

Screen: 130μm
Approach velocity: 45 m/h
Suspension Concentration: 100 mg/l

a finer medium, deeper filter bed or a higher filtration rate (Ives and Sholji, 1965; Adin and Rebhun, 1974; Ives, 1980). In filtration of wastewater effluents, however, these relationships may not hold. Departures from these generalizations have been reported during the last two decades in works in the literature dealing with wastewater effluent filtration.

Fitzpatrick and Swanson (1980) reported that TSS removal efficiency and filtration rate were inversely related. Tchobanoglous and Eliassen (1970) have shown that the filtration rate had little effect on TSS removal. According to Tebbutt (1971) and Bench et al. (1981), increasing the filtration rate does not appear to reduce the removal of TSS. In secondary effluent filtration, filtrate quality is less dependent on filtration rate and influent suspended solids concentration compared to water treatment (Baumann and Cleasby, 1974).

The size of filter media has little effect on TSS removal, but does significantly affect head-loss build-up (Baumann and Huang, 1974). Tebbutt (1971) found that suspended solids could be removed independently of media grain size in the range of 1.0–2.5 mm, but there was some evidence of suspended solids breakthrough for larger media. Tchobanoglous and Eliassen (1970) concluded that removal of suspended solids in secondary activated sludge effluent is primarily a function of the bed grain size, showing a significant improvement in TSS removal with finer media.

Adin and Elimelech (1989) noted, however, that particle removal was improved with media grain size getting larger, for media sizes in the range of 0.7–1.2 mm. Small size media filters (0.45–0.55 mm) are actually surface straining devices, resulting with an exponential head-loss buildup and uneconomical filter runs (Bench et al., 1981). In filtration of secondary effluents media of at least 1.2 mm effective grain size is required and coarser media is preferred if appropriate backwash is to be provided. Increased media depth may not compensate for coarser media in achieving filtrate quality (Baumann and Cleasby, 1974).

Comparison of filter removal efficiency as a function of particle size for different filtration rates shows a lower efficiency for the smaller particles, with a tendency of the removal curve toward a minimal efficiency in the 1–2 μm range (Adin and Elimelech, 1989). Figure 5.7 presents particle removal efficiency of reservoir effluents as a function of particle size. Better removal efficiency of particles at the lower filtration rate for all particle sizes was observed. The difference in percentage became smaller as particle size increased. Removal efficiency increased significantly with increase in particle size up to 9–10 μm. At the higher filtration rate, an instantaneous phenomenon of more particles in the filtrate than in the effluent was observed in the size range of 1–3 μm. Particles larger than 10 μm were removed with a constant removal efficiency of about 80%, most probably by the layer formed on the medium surface.

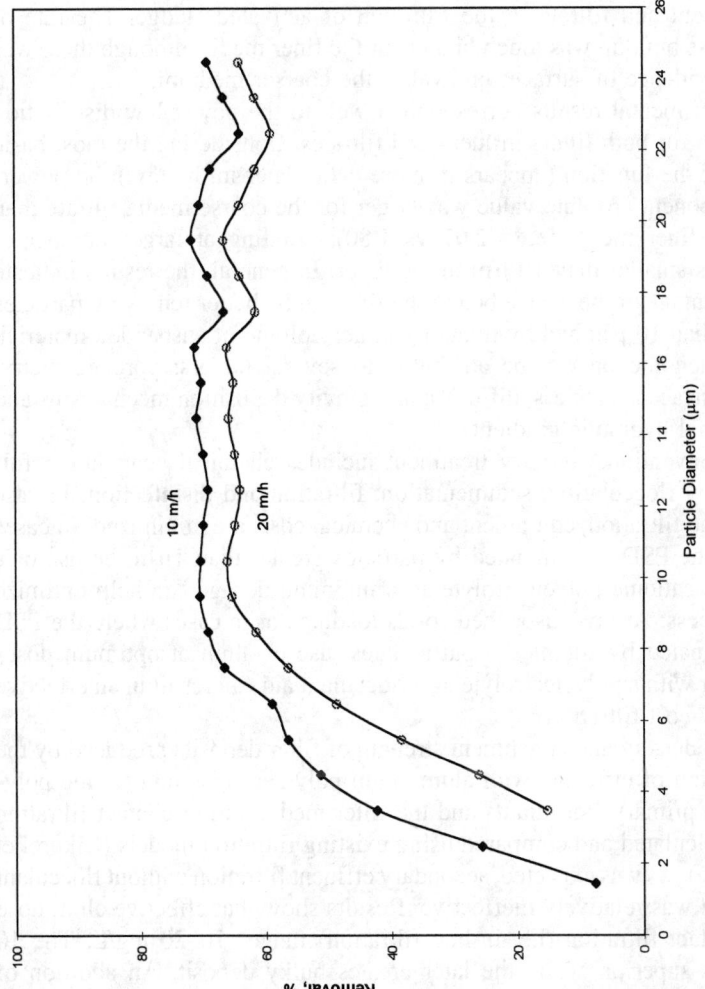

Figure 5.7. Removal ratio vs particle size for different filtration rates. Wastewater reservoir effluents, effective grain size = 0.7 mm, bed depth = 150 mm (after Adin and Elimelech, 1989).

The effect of media grain size on particulate removal for the reservoir effluent was also studied, with interesting results: larger grain sizes clearly showed better removal efficiency of very fine particles (1–10 μm). For particles larger than 10 μm (up to 60 μm), there was no difference in the rate of particle removal. This finding was confirmed by PSD measurements in influent and filtrate in the filtration of activated sludge. The rate of head loss buildup was much higher for the finer media although there was some evidence of surface removal in the coarser medium.

Experimental results corresponded well to the power-law distribution function for both filters influent and filtrates. Considering the most basic form of the function (appears in more detail later in the text) the power-law exponent absolute value was larger for the coarse media filtrate than for the finer media (e.g., 2.01 vs 1.80), pointing at larger number of particles smaller than 10 μm in the latter. In general, the results indicate an advantage of the coarse bed to the fine bed by better removing particles larger than 10 μm and by retaining larger volume of suspended material. This phenomenon can be attributed to several factors: pore geometry, grain surface roughness, diffusion and gravity deposition mechanisms and interstitial hydraulic gradients.

A conventional tertiary treatment includes chemical coagulation followed by flocculation, sedimentation, filtration and disinfection. In case of in-line filtration, equipment and chemical costs are minimized. In cases where the PSD is dominated by particles greater than 1μm the use of a suitable cationic polyelectrolyte at a minimum dosage can help optimize the process. At low suspended solids loadings or in cases where the PSD is dominated by submicron particulates, use of alum at optimum doses together with a polyelectrolyte as a flocculant aid can result in an effective and low cost filtration.

Size, density and attachment strength of filter deposits produced by the interaction of effluents with alum, alum-polymer aids and cationic polymers as primary flocculants and the filter media during contact filtration were calculated and compared using existing filtration models (Cikurel et al., 1996). As was expected, secondary effluent filtration without flocculant addition was relatively ineffective. Results show that effective alum dose for contact filtration (i.e. in-line filtration) ranges 10–20 mg/L. The 10 mg/L is superior, while the later creates bulky deposit. An addition of 0.05–0.1 mg/L of low anionic high molecular weight polymer strengthens the alum-particle bond. Grain size influence seems to be more pronounced at low approach velocities (5–10 m/hr). High cationic, medium to high molecular weight polymers performed equally or better than alum as primary flocculants. High molecular weight cationic polymer-effluent particle deposit has a higher attachment strength than that of the alum, resulting in a smaller detachment with increase of velocity or increase of filtration run length. Cationic, high molecular weight polymers are effective at doses as low as 0.5 mg/L, at different charge densities. High cationic, medium molecular weight polymers are effective at doses >5 mg/L. The pressure gradient increase and the filtration efficiency is proportional to the polymer dose in the 1–7 mg/L range. Compared to the high molecular weight polymer, the pressure build up is milder. High cationic, low molecular weight polymers are not effective at doses up to 7 mg/L. The pressure build up is slow following a slow ripening. In high particle loadings, the differences among the various treatments are less pronounced.

Mathematical Modeling

Particle Removal Theory

The way by which particles in a suspension are removed inside a granular filter bed is part of a complex physical and chemical process, consisting of three mechanisms: transport, attachment and detachment. The particle is to be transported from the carrying stream line to the vicinity of the media grain surface, where it may be attached (depending on its colloidal stability and the attachment forces). When pore clogging becomes pronounced, detachment forces come into effect.

Most of the filtration research has been so far concerned with drinking water. Recently there is a growing interest in studying wastewater filtration. Considerable research has been performed using artificial particles, such as latex, commonly uni-size, suspensions (Mackie, 1993). Some of the basic assumptions for current filtration theory are: (1) for the medium—grains are ideal, smooth spheres and are uniform by size; (2) for the particle—the particles suspended in wastewater are single spheres; (3) for the liquid—the liquid is a dilute suspension; (4) For the flow—the flow regime is laminar.

The facts in reality are different than the assumptions used as basis for the theoretical models. While the last two above assumptions may be correct, the first and particularly the second are incorrect: The particle in wastewater mostly appear in various shapes and textures. Particle size may range from several nanometers up to more than 100 micrometers, as can be seen from several studies (Ghosh, 1975; Adin and Alon, 1986, 1993; Adin and Elimelech, 1989; Darby et al., 1991; Moran et al., 1993).

Levine et al. (1985) and Wiesner and Mazouine (1989) proposed that treatment should be selected according to particle size distribution characteristics. Darby et al. (1991) showed that initially larger particles were removed more efficiently than small particles. The smallest particles were not effectively removed in the first few hours, but a significant increase in removal efficiency was later observed. Moran et al. (1993) studied particulate behavior focusing on the beginning and the end of a filtration cycle (ripening and breakthrough). Their results indicate that: (a) the ripening and breakthrough are not distinct stages but occur simultaneously for particle of different sizes; and (b) breakthrough consequently occurs by particle detachment.

Basic Equations, Adin & Rebhun Filtration Model

The filtration process can be described macroscopically by three sets of mathematical expressions—continuity (mass balance) equation, kinetic equation and head loss development expression (Adin and Rebhun, 1977). The continuity equation is given by:

$$v(\partial C/\partial x)_t + (\partial \sigma/\partial t)_x = 0 \tag{7}$$

where x and t are bed depth and time, C is the concentration of particulates in the water, v is the approach velocity, and σ is the specific deposit (mass of particulate per volume of media). A kinetic equation to be solved with the above expression is given by:

$$\partial \sigma/\partial t = K_a v C(F - \sigma) - K_d \sigma J \tag{8}$$

Here K_a is the attachment coefficient and K_d the detachment coefficient. F is the theoretical filter capacity (mass/volume) and J is the hydraulic gradient. Head loss through the granular media as a function of the specific deposit based on Shektman's expression is:

$$(J_0/J) = [1 - (\sigma/\epsilon_0\gamma)^{0.5}]^3 \tag{9}$$

J_0 is the initial hydraulic gradient, ϵ_0 is the clean bed porosity and γ is the solid concentration in the deposit, and is calculated (Rebhun et al., 1984) from

$$\gamma = F/\epsilon_0 \tag{10}$$

Furthermore, the density of the deposit including water can be expressed as:

$$\rho_d = 1 + \gamma(1 - 1/\rho_S) \tag{11}$$

where the specific gravity of water is considered 1. ρ_S is the density of the dry matter. From approximate densities of flocculant particles (Kavanaugh et al., 1977) and discrete kaolin particles (Tanaka, 1982), a mean value of 1.405 for ρ_s for comparative evaluation of various flocculants is assumed. This leads to:

$$\rho_d = 1 + 0.286\gamma \tag{12}$$

The change of suspension concentration with depth can be expressed for a single collector model as (Amirtharajah, 1988):

$$-(\partial C/\partial x)_t = [\eta \ (\sigma)(1 - \epsilon_0)C]/d_c \tag{13}$$

where $\eta \ (\sigma)$ is the single collector efficiency and is a function of the specific deposit and d_C is the grain diameter. η can be written as:

$$\eta = \eta_0\alpha \tag{14}$$

η_0 being the clean bed single collector efficiency factor, accounts for all transport properties and α is the collision efficiency factor which is a function of physicochemical parameters including particle attachment strength.

Ives (1978) proposed a filterability number (F.N.) to be used for design purposes:

$$\text{F.N.} = H(C/C_0)/(vt) \tag{15}$$

where C is average filtrate quality (NTU), C_0 is filter influent quality (NTU), H is head-loss (m), v is filtration rate (m/h), and t is filtration time (h). The lower the filterability number, the better is the filtration performance.

Using TSS, PSD and head-loss data, above mentioned factors were calculated for different treatments (Table 5.1).

Transport Efficiency Models

Expressions for single-collector efficiency for transport mechanisms of diffusion (η_D), sedimentation (η_G) and interception (η_I) are given by:

(a) Diffusion:

$$\eta_D = 4P^{-2/3} = 0.9[kT/(\mu d_p d_c U)]^{2/3} \tag{16}$$

(b) Sedimentation:

$$\eta_G = v_s/U = [(\rho_p - \rho)gd_p^2]/(18\mu U) \tag{17}$$

TABLE 5.1. Effect of Approach Velocity, Media Grain Size, Initial Particle Loading and Flocculant Type on Filtration Efficiency.

Approach Velocity (m/h)	Grain Size d_c (mm)	Total Filtr. Time (h)	Type of Treatment	C_{OTSS} (mg/L)	C_{OTurb} (NTU)	PSD (N_0/ml)	ΔH (cm)
10	1.1	7(4)	no flocculant	23.5	6.2	5500	22(12)
10	1.1	3	no flocculant	45.3	21	22450	14.5
10	1.1	3	10 mg/L Al	56.3	25.2	25640	50
10	1.1	3	20 mg/L Al	58.9	25.2	25640	66
10	1.1	3	10 Al + 0.1 A100	59.3	19.8	28560	81
10	1.1	3	0.5 mg/L C498	16	12		92.5
10	1.1	3.5	0.5 mg/L C442	46	27.7		135
10	1.1	3	0.5 mg/L C444	16	12		40
10	1.1	3.5	0.5 mg/L C444	46	27.7		138
10	1.1	3	0.5 mg/L C444	37	15.5	18100	71.5
10	1.1	3.5	0.5 mg/L C448	46	27.7		95
10	1.1	4	0.5 mg/L C567	20	12.3		14.5
10	1.1	4	7 mg/L C567	18.3	9	12070	17
10	1.1	6	0.5 mg/L C573	18	14.7		32
10	1.1	4	0.5 mg/L C581	20	12.2		14.5
10	1.1	3	7 mg/L C581	18.2	9	12070	95
15	1.1	4	no flocculant	28	10.5	22300	32
15	1.1	3	10 Al + 0.1 A100		13.5	31230	170
15	1.1	3	0.5 mg/L C444	20.4	12		45.5
15	1.1	3	2.0 mg/L C581	23	15.7		24
10	1.5	3	0.5 mg/L C444	39.7	19.6	47040	46
10	1.5	3	5.0 mg/L C581	39.7	19.6	47040	77
15	1.5	2.5	0.5 mg/L C444	28.5	16.5		69
15	1.5	2.5	5.0 mg/L C581	28.5	16.5		64

*Bed depth = 20 cm.

(c) Interception:

$$\eta_l = 1.5(d_p/d_c)^2 \tag{18}$$

where P = Peclet's Number, k = Bolzman's Coefficient, T = absolute temperature (°K), μ = absolute viscosity, d_p = particle diameter, d_c = collector (filter grain) diameter, U = flow velocity, v_s = particle gravity settling velocity, ρ_p = particle density, ρ = liquid density, and g = gravitational acceleration.

Total collector efficiency (η_0) is regarded as the sum of the above:

$$\eta_0 = \eta_D + \eta_G + \eta_l \tag{19}$$

The above equations show the dependency of the particle transport efficiency upon particle size. In diffusion, the relation is by power of 2/3 to the particle size. In sedimentation and interception, the proportion is to the second power. It is worth noting that:

TABLE 5.1. (continued).

Approach Velocity (m/h)	$(C/C_0)_{TSS}$	F (mg/cc)	Density (g/cc)	K_a (10^{-3})	K_d (10^{-3})	Ives Filtr. Number	λ_{max}
10	0.5	18.9	1.014	3	19.3	2.32	0.034
10	0.57	22	1.016	2	69.2	3.22	0.035
10	0.454	14	1.01	5.1	19.3	8.1	0.048
10	0.477	12.5	1.009	6.2	33.5	11.5	0.045
10	0.27	15	1.012	11.3	10.5	8.7	0.07
10	0.3	5.6	1.004	19.4	34	9.5	0.063
10	0.34	13.2	1.01	8.5	16.1	11.8	0.06
10	0.27	10.3	1.008	11.9	66	5	0.075
10	0.3	14.4	1.01	7.7	29.1	11.8	0.065
10	0.2	16.5	1.012	16.8	11	9.8	0.09
10	0.32	14.3	1.01	6.4	40	8.6	0.07
10	0.62	12.2	1.009	3	43.8	2.25	0.026
10	0.74	3.2	1.002	3.9	37.5	2.8	0.023
10	0.47	13.5	1.01	5.4	62	3	0.045
10	0.56	16.3	1.012	3.1	33.9	2.27	0.033
10	0.28	5.7	1.003	20	39.6	12	0.1
15	0.5	28.7	1.021	2.4	60	4.2	0.04
15				7.6		10.2	
15	0.32	16.4	1.012	7.1	20	5.1	0.063
15	0.62	19.4	1.014	1.9	65	2.9	0.04
10	0.26	10.7	1.008	19	14	4.5	0.08
10	0.38	7.4	1.005	19.9	30	9	0.06
15	0.4	13.1	1.01	6.1	57.2	8.7	0.07
15	0.64	6	1.004	4.8	80	10	0.05

(1) Removal efficiency of a single collector for diffusion mechanism is proportional to the Stephan-Boltzman constant and the temperature (in the power 2/3). Diverse proportion (in the power of 2/3) to the viscosity, the grain size and the velocity.

(2) Removal efficiency by the sedimentation mechanism is proportional to the relation between Stokes' settling velocity and filtration velocity. It also depends on the difference in water and particle density. There is an inverse proportion to the viscosity and to filtration velocity.

(3) Removal efficiency by geometric interception is proportional to the second power of particle size to the pore size ratio.

By calculation for high rate filtration at 20 m/h, 1.05 g/cm³ average particle density, temperature of 20°C, collector diameter of 1 mm, it can be assessed that the diffusion mechanisms has no effect on the transport of particles larger than 1 μm. It can also be demonstrated by such calculation, that particles between 1–50 μm in size follow sedimentation rules and from 50 μm and up the interception rules.

Checking the transport efficiency (coefficients of diffusion, sedimentation and interception) using standard conditions of filtration, the calculated values for particles larger than 1 μm are:

$$\eta_D = 10e^{-8} - 10e^{-9}$$
$$\eta_G = 10e^{-5} - 10e^{-3}$$
$$\eta_I = 10e^{-5} - 10e^{-2}$$

It can be concluded, that transport mechanism by diffusion is negligible for particles larger than $1\mu m$. In granular filtration the main transport mechanisms are sedimentation due to the gravity force and interception due to geometrical size (ratio of particle size to pore size).

PSD Introduction into Filtration Model

While dealing with effluent suspensions it seems to be an obvious conclusion that particle size should not be represented in the filtration models by only one representative diameter (Adin and Alon, 1993; Alon and Adin, 1994). Assuming a power-law PSD function, the total number N_t of p particles of sizes i to n entrapped in the bed is expressed by

$$N_t = A \sum_{p=i}^{n} X_p^{-a} - B \sum_{p=i}^{n} X_p^{-b} \tag{20}$$

where A,a and B,b are PSD function parameters of the filter influent and the filtrate respectively, and X represents variable particle size. N_t can be substituted, for example, into a mathematical model such as the Kozeny-based one developed by O'Melia and Ali (1978), which incorporates "mean" particle diameter. Applying some mathematical manipulations yields the hydraulic gradient

$$h/L = 36(k/g)(\mu/\rho)v(1 - \epsilon)^2\epsilon^{-3}(S_1/6)^2 d_c^{-2} T^2 \tag{21}$$

where

$$T = [(1 + \delta QS_2/N_c d_c^2)/(1 + R/N_c d_c^3)]^2 \tag{22}$$

$$Q = A \sum_{p=i}^{n} X_p^{2} - {}^a - B \sum_{p=i}^{n} X_p^{2} - {}^b \tag{23}$$

$$R = A \sum_{p=i}^{n} X_p^{3} - {}^a - B \sum_{p=i}^{n} X_p^{3} - {}^b \tag{24}$$

h is headloss, L is bed depth, k is Kozeny's constant, g is gravity constant, μ is water absolute viscosity, ρ is water density, v is filtration rate, ϵ is bed porosity, S_1, S_2 is media grain and deposit shape factors, respectively

(Vigneswaran, 1986, 1988), d_c is media grain diameter, δ is fraction of particles serving as collectors retained by a grain, N_c is number of grains per unit bed volume. Q and R represent differences between the filter influent and the filtrate PSD.

Secondary effluent from a municipal activated sludge plant was filtered at a rate of 8m/h through three columns containing 0.15 m pretreated, uniform, cleaned sand media. The sand grains, as previously stated, had geometric mean sizes of 0.767 mm, 0.917 mm and 1.3 mm. Filter influent and filtrate samples were taken 60 minutes after the start of each run. A sample of experimental results corresponding to the PSD function in a logarithmic form are depicted in Figure 5.8. Correlation coefficients lie in the 0.96–0.99 range.

For implementation purposes, the PSD for the filter influent and the PSD for the filtrate may be expressed, respectively, as:

$$Y_1 = AX^{-a} \tag{25}$$

$$Y_n = BX^{-b} \tag{26}$$

The degree of correlation between the distribution function coefficients of the filter influent (A,a) and those of the filtrate (B,b) was evaluated. The best correlation was found to be linear, as shown in Figures 5.9 and 5.10. The linear correlation between the coefficients A and B may be expressed as follows:

$$B = mA + n \tag{27}$$

and the correlation between a and b as:

$$b = pa + q \tag{28}$$

$m, n, p,$ and q are constants. Consequently, the filtrate PSD may be expressed as:

$$Y = (mA + n)X^{-(pa + q)} \tag{29}$$

Thus, the filtrate PSD can be predicted from determining the filter influent PSD provided that the filter media parameters are known. Furthermore,

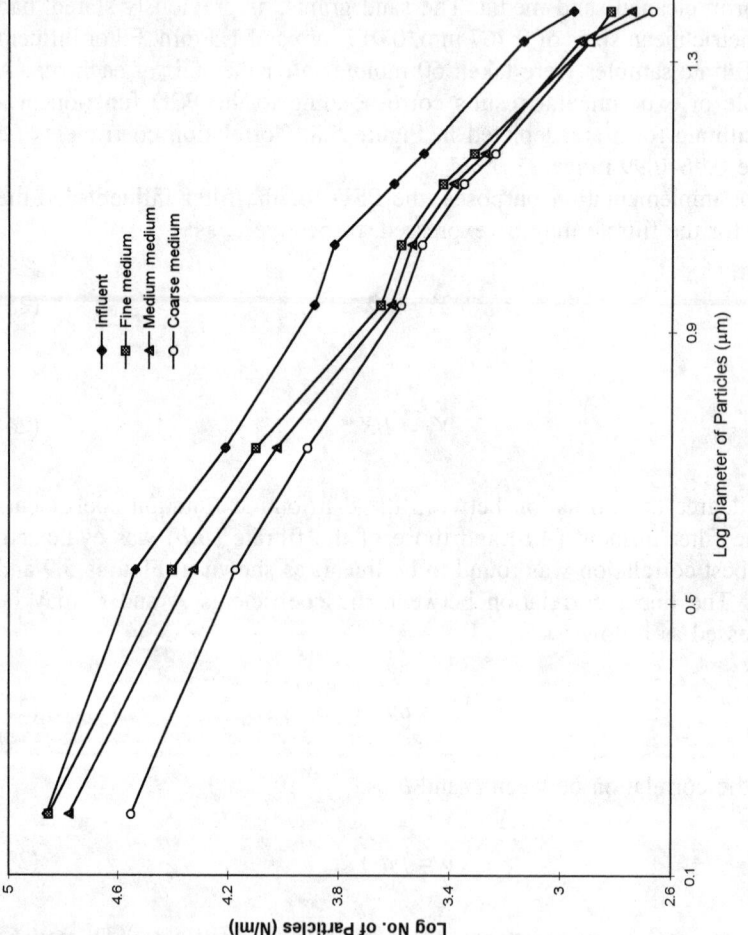

Figure 5.8. Particle size distribution before and after filtration (after Adin and Alon, 1993).

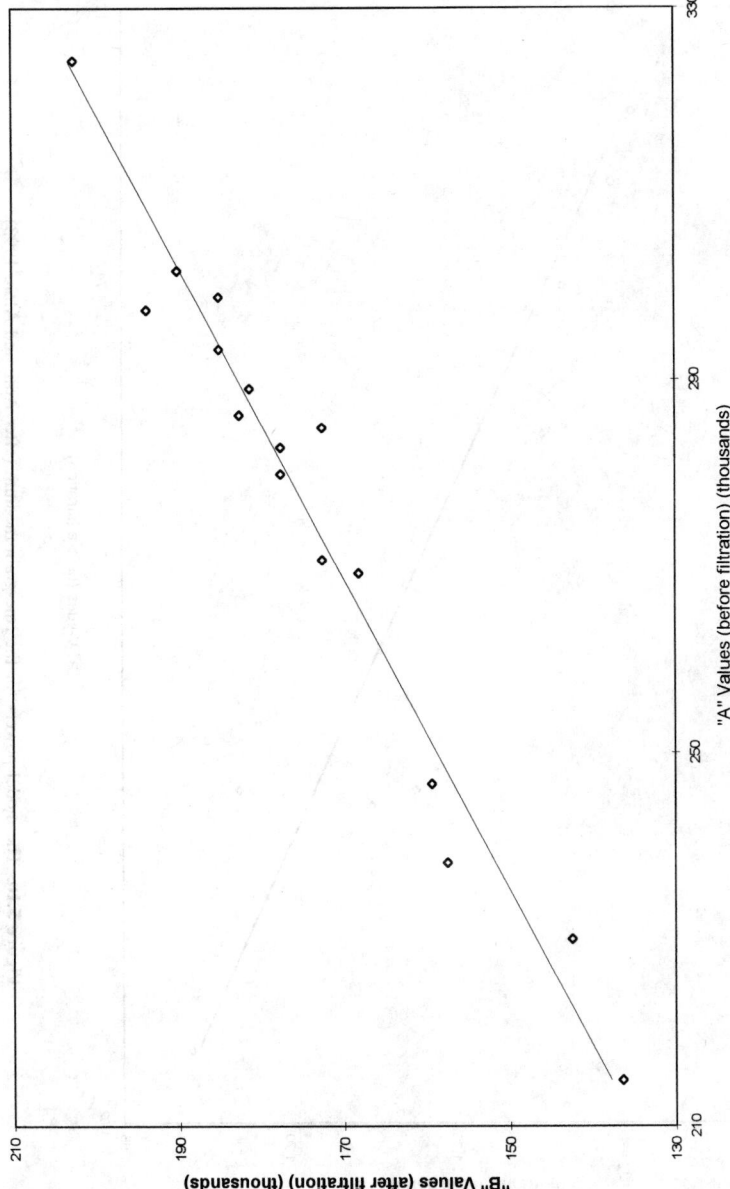

Figure 5.9. The model mantissae for the fine-grain medium (after Adin and Alon, 1993).

197

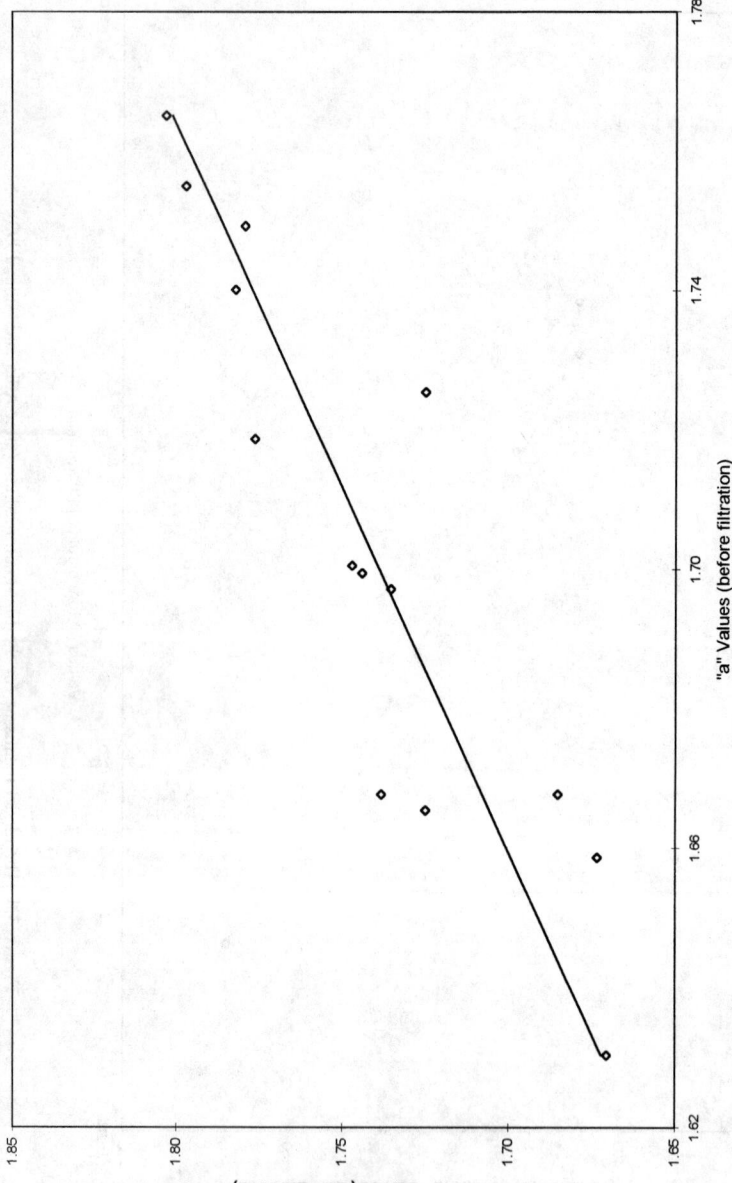

Figure 5.10. The model powers for the fine-grain medium (after Adin and Alon, 1993).

Equation (29) can be substituted in Equation (21), allowing prediction of head loss for a granular filter under specific influent and media conditions.

Optimization Model for Filter Bed

Design practice of chemically pretreated high-rate filters is commonly based on previous experience or on extensive pilot plant experimentation. The application of existing mathematical models simulating the filtration process to filter design has still many limitations. Furthermore, there are more than one combination of filter media and filter operation variables that could give similar filtration efficiency. The real question then is which combination is also the most cost effective. Adin and Hatukai (1994) present a methodology for obtaining design and operation parameters of granular filters which brings about filtered water at the lowest cost.

A model is proposed, composed of two successive stages: process optimization and economic optimization. In the first stage, the filtration system is divided into three subsystems, which are: filter bed, chemical pretreatment, and backwashing. The optimization of each subsystem is carried out while parameters of the other subsystems remain without change. Presently filtration rate is considered to be an independent variable.

Parameters of filter beds capable of producing the filtrate goal are provided by the process optimization model from data obtained by pilot plant runs. The parameters obtained provide the basic data for cost estimates further employed for economical optimization. At this stage, Equation (1) is proposed for the overall annualized capital cost of the filtration plant:

$$\text{TACC} = M_1 A_1^X + M_2 (AL)_2^X + M_3 (AV_B)_3^X \qquad (30)$$

in which the first right term relates the cost of filter house to filter area; the second term relates media cost to media volume: the third one relates backwashing facilities cost to flow rate of backwash. Values of the constants and exponents are determined by regression analysis of cost data provided by existing filtration plants. Costs affecting operation and maintenance, i.e., energy, wash water and wasted chemicals costs, are computed using separate equations.

SLOW SAND FILTRATION

Slow sand filtration (SSF) is known to be a most simple to operate, low cost, efficient and reliable technique for potable water treatment. The process enables to achieve substantial improvement in the physical, chemical and bacteriological characteristics of the water via one single process, while several separate steps are required to achieve similar quality by other technologies.

Slow sand filtration (SSF) is one of the earliest water treatment designs and is still in wide-spread use around the world. Thus, 20% of the drinking water in the U.K. is still slow sand filtered as is 80% of all London water (Ellis, 1991). In London SSF is the third stage in a four-stage treatment process consisting of long term storage, rapid sand filtration (or micro-straining), SSF, and disinfection. However, in spite of the fact that the use of SSF is wide spread, available information on the processes controlling SSF is very limited compared to that available for other treatment processes. Particularly limited is information available on SSF for the advanced treatment of effluents. Accordingly, a large portion of the information presented below will be based on work done with drinking water. A thorough investigation will be needed in order to define the design and operational parameters for the slow sand filtration of effluents.

SSF has lately become more attractive in the treatment of drinking water supplies due to its ability to remove cysts and oocysts such as *Giardia lamblia* (Bellamy et al., 1985; Schuler et al., 1991) and *Cryptosporidium* that may not be effectively removed by high-rate filters. These advantages often overshadow the fact that SSF implies the allocation of a relatively large area due to the slow rate of filtration.

The typical slow sand filter is basically a box containing uniform-size quartz sand, about one meter in-depth. The water floods the bed and percolates slowly through it (0.1–0.2 cu m/sq m bed area per hour) and is then collected via a drainage system for further application. The removal of contaminants takes place mainly in the upper few centimeters of the bed where a biologically active layer—referred to as *Schmutzdecke*—is formed. This layer is periodically removed and eventually washed and recycled.

Among the more important mechanisms for the removal of water impurities through slow sand filtration are: (a) enhanced biological activity on surfaces, (b) mechanical filtration, (c) adsorption, and (d) surface catalyzed degradation. These process features make SSF most attractive for advanced treatment of effluents.

Practically, SSF is not limited by hydrogeological considerations. The filtered water can be supplied for irrigation either following seasonal storage in open reservoirs or directly, placing the filter ahead of the irrigation system. The latter arrangement is especially important for the protection of drip irrigation systems from clogging (Adin and Alon, 1986; Adin and Elimelech, 1989; Adin and Sacks, 1991). The high quality effluent produced by slow sand filters can also be stored directly in the aquifer through recharge pumping wells, thus eliminating the need for soil-aquifer treatment and reduces considerably the area required for spreading fields. However, the difference in water quality which normally exists between surface water and secondary effluents does not allow direct application of SSF as used for clean-up of potable waters for the treatment of effluents.

Secondary effluents are characterized by: (a) high organic contents of differing nature; (b) suspended particulates of various origins; (c) higher concentrations of nutrients, including various nitrogenous compounds; (d) larger number and variety of pathogens—bacteria, virus, parasites; and (e) industrial contaminants, including heavy metals and toxic organics. Such characteristics may impose design and operational parameters very different from those used for potable water.

An important factor by which secondary effluents differ greatly from drinking water or from most surface waters is the effluents' high content of dissolved humic substances. The effects of dissolved humics on biodegradation have been reported in the literature without attributing them to specific mechanisms. Several investigators have shown that the biodegradation of various substances is markedly decreased in the presence of dissolved humics (Speital and DiGiano, 1987). This may also be due to the formation of substrate-humic complexes, the bioavailability of which might differ from that of the substrate alone. The larger and different-in-nature organic load may dictate a more intensive biological activity. This, in turn, may necessitate either a greater depth for biological activity, which may be achieved by a larger grain size, and/or preoxidation to facilitate biodegradation.

The nature of the solid phase (filter medium) will determine to a very large extent the efficiency of the process and its operational parameters. The predominating surface *Schmutzdecke* may not prevent the penetration of colloidal particulates into the depth of the filter bed; the extent of penetration, as yet undetermined, is a function of the size and shape distribution of the grains in the filter medium and their adsorptive capacity, and the extent of their destabilization by humic acids. Bacteria and viruses are also included in this category and their penetration may be similar to or associated with that of other solid particles. Particle penetration is a very important factor in determining filter design parameters.

SSF Basics for Design

As with rapid gravity sand filters, slow sand filters consist basically of three sections: underdrainage, gravel and filter media, of which only the filter media has any direct role to play in the purification process. Unlike rapid sand filters, the water reservoir above the sand bed in a slow filter plays an active role in the purification process as the settlement of suspended solids and the aerobic biodegradation of organic compounds can occur. However, the key parameters in SSF are the depth of the bed filter media and the effective size of the sand (Visscher, 1990). The effective size of the sand is usually 0.15–0.4 mm for potable water, and the depth of the sand bed between 0.6 and 1 m. The few studies of SSF of wastewater reported in the literature show an overall removal of 60–80% of suspended solids, and a removal of over 1 order of *E. coli* bacteria, with superiority to coarser sand (grain size approximately 0.6 mm), while fine size (approximately 0.3 mm) improves filtration only slightly, causing very short filtration runs (Ellis, 1987; Farook and Yousef, 1993; Farooq et al., 1993; Adin et al., 1995).

It is common to use an initial depth of 0.8–0.9 m to allow for a sufficient number of scrapings before resanding is needed. In order to improve the efficiency of purification it is better to increase the depth of the sand bed, rather than to reduce the grain size, because if the sand is too fine it may cause an excessive head loss. Improved purification efficiency can also be achieved by addition of fibers to the bed's surface layer, thus providing an additional matrix for the development of the *Schmutzdecke*.

The *Schmutzdecke*

Submerging a sand bed in a shallow layer of nutrient-rich water for prolonged periods of time leads to the formation of biologically active mats consisting of a mixture of photosynthetic microorganisms and hetero-trophic bacteria. These types of mats are in general referred to as *Schmutz-decke*. The development of a *Schmutzdecke* and the functioning of its mature form are expected to play a major role in the overall performance of slow filtration systems. In turn, operational measures are expected to have a strong effect on the biological properties of the sand bed.

Colonization of particulate matter by individual microorganisms forms the basis for the development of microbial mats. One may expect, even at the early stages of mat formation, that changing surface properties of filter particles may influence filter performance. Mature mats may contain micro-organisms capable of excreting copious amounts of exopolymers, in general exopolysaccharides (Katznelson, 1989; Uhlinger and White, 1983). Poly-saccharides have been found to posses specific adsorption and complexation affinities for inorganic and organic matter, such as, clays, bacteria and soluble hydrophobic compounds (Kanaani et al., 1992). Thus, the nature of this mass of cells and exudates determines the physiological properties of the *Schmutz-decke*. Its implications on microbiological/physiological processes, filter performance and water quality are addressed below.

Wastewater effluents, certainly those originating from oxidation/stabili-zation ponds, normally contain appreciable amounts of ammonium. This ammonium can be removed by the following processes: (a) stripping to the atmosphere by pH elevation, (b) incorporation into cellular material, and (c) transformation to N_2 through nitrification/denitrification. The latter is by far the desirable process, since it avoids the negative side-effects of air pollution or increasing the amount of suspended solids accompanying the other two processes. Both nitrification and denitrification are microbial processes that occur optimally in microbial mats or biofilms (Van Rijn and Rivera, 1990). The first process is strictly aerobic whereas the second is strictly anaerobic. However, even anoxic wastewaters are known to contain significant amounts of viable nitrifying bacteria. Depending on the operational conditions of the sand bed filter, the concurrent development of nitrification/denitrification may be expected as a result of alternating oxic/anoxic conditions resulting from algal activity in light/dark cycles (Post et al., 1985). Alternatively, operational measures such as regular changes in water level may induce the required alternation of oxic/anoxic conditions. A successful combination of nitrifying and denitrifying activity has been reported for lagoons with dissolved O_2-concentrations of >1 mg/L (Constable et al., 1989a, 1989b).

The development of microbial mats on a sand bed filter is a function of environmental conditions. The main factors determining the conditions are related to the quantity and quality of the water overlying the filter. Turbidity and water depth will together determine the amount of light available to the water column (Loogman et al., 1980) and whether light will reach the surface of the filter bed. Light conditions are extremely important in determining which type of microalgae develops in the overlying water (Post et al., 1985) and hence determine, through competition for available resources, which type of microalgae will dominate in the microbial mats. Light penetration in and mixing of the water column set the parameters for the DO-regime during the day and night (Post, 1991). The carbon flux in the microbial mat will be largely determined by the available light and the extent of anaerobiosis. Anoxic conditions will yield partial mineralization of organic matter and the release of organic compounds including organic nitrogen and sulfur compounds, which are less desirable from the point of water quality.

Effluent turbidity differs in magnitude and in nature surface water turbidity applied to SSF, making it an important water quality parameter.

Filtering Materials

The common medium used for SSF of drinking water is sand. Some other materials have also been employed for the filter medium in slow granular filters. Crushed coral has been used; volcanic ash has been employed in Ethiopia, and the use of burnt rice husks have been investigated in South East Asia (Ellis, 1991).

The nature of the solid phase will determine to a very large extent the relative importance of each of the above mentioned processes. Thus, the use of local types of sand should be examined (which is important also from economic viewpoint). Determination of sand grain size is a process of optimization, the larger the grain size the smaller is the head loss (i.e., clogging), however, hydraulic conductivity, as a rule, is inversely related to the surface area and activity.

Bacteria

It is widely accepted that slow sand filtration is a highly effective means of removing bacteria from water. Schuler et al. (1991) showed that slow sand filtration removed more than 90% of the total coliform bacteria. Huisman (1978) suggested that the total bacteria count in water is reduced by a factor of between 10^2 and 10^3 and that the factor for the removal of *E. coli* varies between 100 and 1000, with usually none appearing in the filtrate. Van Dijk and Ooman (1978) found that between 99 and 99.9% of pathogenic bacteria are removed during slow filtration. Bowles et al. (1983) studied the effect of temperature on the removal of coliform bacteria by slow sand filtration. Their studies have shown a removal of 98.25% during the summer months, whereas the removal of coliform dropped to 11.5% during the winter months (6°C) (Bowles et al., 1983). Paramasivan et al. (1981) recorded the effect of shading on bacterial removal in slow sand filters. They found that 98% of the filtrate samples were free from E. coli with a completely shaded filter, but only 68% being free with a filter open to the sky. These contradictory results require further study.

Most studies conducted on the removal of pathogenic agents by slow sand filtration were performed with waters of low turbidity and relatively low organic content that were treated for potable purposes. Very few studies were conducted on the treatment of secondary effluents by slow sand filtration and specifically the microbial quality of slow filtered effluents has not been sufficiently tested.

Mathematical Modeling of SSF

A considerable amount of work on rapid sand filters modeling has been developed; however, the interest in modeling of slow sand filtration has been minimal.

The continuity equation is normally given by Equation (7):

$$v(\partial C/\partial x)_t + (\partial \sigma/\partial t)_x = 0$$

where x and t are bed depth and time, C is the concentration of particulates in the water, v is the approach velocity, and σ is the specific deposit (mass of particulate per volume of media).

A common, conventional way to express filtration efficiency is in terms of the filter coefficient λ from Iwasaki-Ives equation (Ives, 1975):

$$-(\partial C/\partial x)_t = \lambda C \tag{31}$$

λ is a function of the specific deposit and presents the combined effects of the transport and attachment mechanisms that dominate particle removal.

Two recent mathematical models specifically aimed at SSF (for drinking water supply) has been proposed. Ojha and Graham (1991) propose that the removal of inert and biodegradable particles that contribute to the build up of the biomass are additive, arriving to the following mathematical expression:

$$\Delta\sigma = \{\beta viC_0[1\text{-exp}(- \lambda_0 x)] \ \Delta t\}/\Delta x + \tag{32}$$
$$\{\gamma\beta v(1 - i)C_0[1\text{-exp}(- \lambda_0 x)] \ \Delta t\}/\Delta x$$

where

i = inert material fraction in the influent
β = volume factor for the deposited material
γ = microbiological activity coefficient

The first term represents the removal of the inert material and the second, the removal due to the biological activity. $\gamma = 1$ during the filter ripening period. The above equation is too general to enable a profound study of the microbial mat.

Retamoza (1994) and Retamoza et al. (1994) claim that both rapid and slow filtration processes are similar with respect to the specific deposit in both biomass or inert material accumulation is greater in the upper layers and decreases with depth. Therefore, it seems reasonable to simulate slow sand filtration process by utilizing rapid filtration models. In their study on ground water treatment for iron and manganese removal, the following mathematical simulation model is proposed:

$$\Delta\sigma = v[(C_0 - C_e) \ \Delta t]/\Delta x + vY[(S_0 - S_e) \ \Delta t]/\Delta x = 0 \tag{33}$$

where

$$C_e = C_0 \exp(-\lambda x) \text{ and } S_e = S_0 \exp(-K_{rem} x) \tag{34}$$

where Y is the yield coefficient and K_{rem} is the substrate removal constant.

The model incorporates a material balance equation and two kinetic equations; one refers to substrate removal $(S_0 - S_e)$ and another includes the attachment term [see Equation (8)]. The proposed model was validated with experimental data for simulated groundwater, but has not been tried for natural groundwater nor for wastewater effluents. However, due to its "biological" approach it may serve as a starting point for the latter.

Observing slow sand filter functioning as two fixed bed reactors in series (Adin et al., 1995), it is expected that microbial mass formation in the upper thin layer will induce exponential assent in head loss (H) with time or water volume passing through a surface unit area (V). For such a case, Boucher's law (1947) for surface clogging of screens may apply [Equation (3)]:

$$H = H_0 e^{IV}$$

Hence,

$$H/H_0 = e^{Ivt} \tag{35}$$

H_0, in this case, is the initial head loss through the few upper centimeters of the bed which contribute the most to the filtration process. Since the flow regime under SSF conditions implies very low Reynolds numbers, i.e., laminar flow, Kozeny-Carman equation can be introduced. The value of I, a filter index, can be calculated from experimental data for comparing the various experimental conditions. The advantage of using vt at the exponent is that it is a way of normalization of the filtration velocity (v), i.e., comparison of filter performance at different filtration rates on a similar basis.

Equation (35) can be linearized by using it in a logarithmic form. The faster the microbial growth, the faster the clogging rate so the greater the I (the steeper the headloss curve).

OXIDATION-FILTRATION COMBINATIONS

A few studies have been reported concerning the use of oxidants for preventing clogging. According to these studies chlorine can be applied either by sequential pulses of relatively low chlorine concentrations (1–20 mg/L) aimed to prevent biofilm development, or by filling the drippers with high chlorine concentrations of 500–1000 mg/L for 24 h aimed to break down clogging aggregates that have already been formed in the drippers (English, 1985; The Water Commission of Israel, 1985).

Rav-Acha et al. (1995) tried to isolate the effects of chlorine and chlorine dioxide on several effluent constituents. Both oxidants affected very markedly the algal viability as expressed by its chlorophyll content and replication ability. Both chlorophyll content and replication ability were reduced by more than 90%, 20 min after the application of 10 mg/L Cl_2 or ClO_2, although the number of algae cells and their structure were not affected by the oxidants (i.e., viability reduction does not necessarily imply elimination of clogging potential).

Similar results to those described above were found by Kott (1973) with respect to the number of algae cells (he concluded that oxidants such as Cl_2 or ClO_2 have no effect on the number of algae), and by Susenik et al. (1987) with regard to algal viability.

The stronger effect of ClO_2 on algal viability, as compared to Cl_2, can be attributed to easier penetration of the non-ionic ClO_2 molecule into the algal cell, when compared with a molecule of chlorine which is present at pH 7–8 mainly in the form of hypochlorite ion (ClO^-). The aggregation of suspended particles caused by the oxidants, as expressed by the shifts in both particle size and particle volume distributions towards larger particles, was in accordance with the decrease in the absolute value of the suspended particles zeta potential, which indicated a destabilization effect due to a decrease in the electrostatic repulsion between the particles.

The observation that the same phenomena were found when algae was removed from the effluents (Figure 5.11), but not when humic materials were removed, indicates a direct link between the aggregation effect caused by the oxidants and the action of oxidants on humic material. Narkis and Rebhun (1975) have indicated that humic materials adsorbed on suspended particles increase the colloidal stability of suspended particles via steric hindrance, which increases the potential energy barrier for aggregation. Another reason for the destabilization caused by the action of oxidants on humics can be related to the pH of the effluents (7–8) at which humic materials are mostly in the ionized form. Thus, humic substances which are adsorbed on clay particles increase the electrostatic charge density on the colloid surfaces, thereby increasing stability. The oxidation of humic material by the oxidants, which has already been noticed by Grasso and Weber (1988) may, therefore, cause a colloidal destabilization by decreasing both the charge density on the colloid and its steric hindrance.

All the observations described above pointed a priori to two possible strategies for the prevention of the clogging phenomena in drip irrigation systems by the application of oxidants: (a) applying a small concentration (5–10 mg/L) of oxidant continuously into the line, (b) applying a higher dose (20 mg/L) to the effluent reservoir or a holding tank, waiting for settling (in order to enhance aggregation flocculation), and then feeding the emitter with the treated effluents. As it has already been shown that

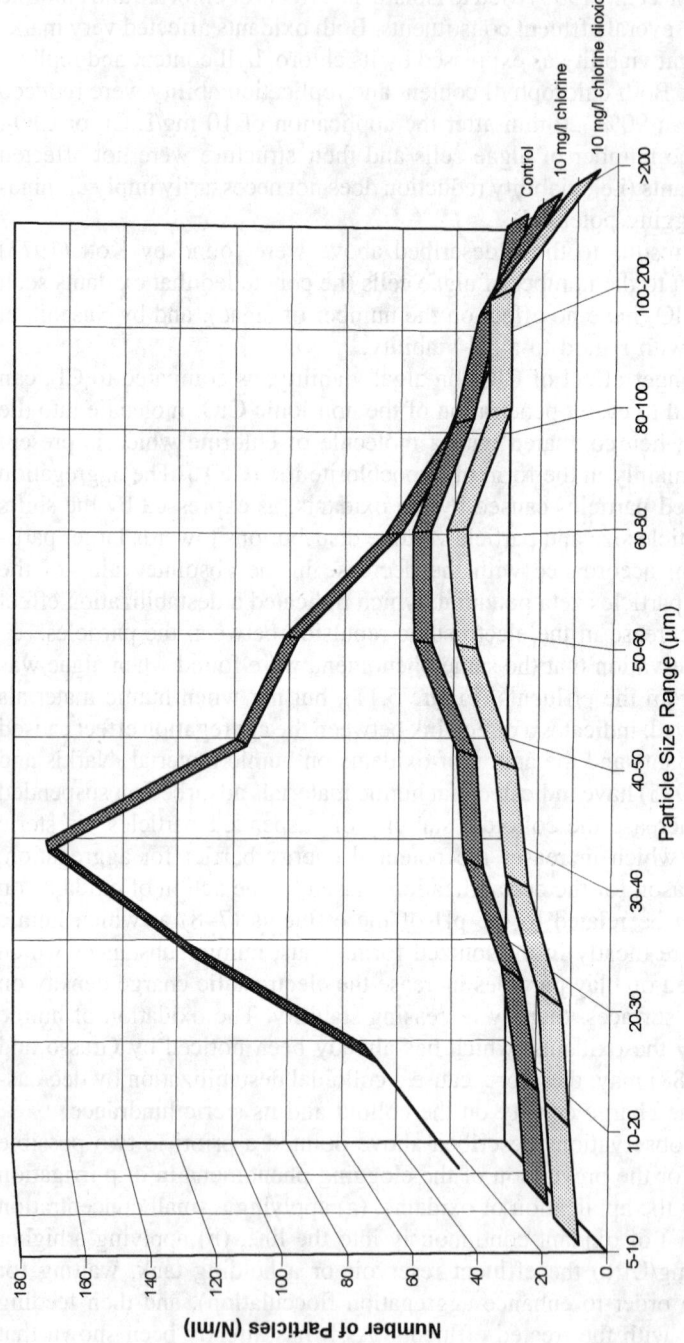

Figure 5.11. The effect of oxidants on particle size distribution in synthetic effluents.

208

the triggers for clogging are biofilms produced by bacteria (Adin and Sacks, 1991), in case the excretion of polysaccharides (the main building-stones of biofilms) and the processes leading to clogging took place inside the drippers or emitters the first strategy would be preferred, for it has been shown that a small concentration of oxidant can successfully inactivate bacteria. On the other hand, a batch treatment followed by settling would be required if the clogging agents were produced as early as in the reservoir, and would be transferred to the drippers by the effluent. In the latter case, a continuous application of a small concentration of chlorine would not be effective, as it was shown that chlorine is effective merely against free bacteria in the effluent, but not against bacteria which were protected by biofilm or incorporated in the sediment, which kept reproducing.

In batch treatment combined with settling, the chlorine acts not only as a disinfectant, but also as a coagulant, and as such it contributes to a significant reduction of the clogging potential. Such a process could have also increased the effectiveness of a filter, had it been introduced subsequently to the batch chlorination process. A filter would cause a decrease in the concentration of residual chlorine, and a second chlorination would be possibly required. On the other hand, a filtration process prior to batch chlorination could remove transported slimes, but the aggregation that follows chlorination might aggregate the clogging problem. Further studies would be necessary to optimize the coupling of these two treatment processes for clogging prevention.

MEMBRANE TREATMENT

Membrane treatment by ultrafiltration (UF) for essentially colloid removal and nanofiltration (NF) for colloid and mainly soluble organics removal have recently emerged as competitive solutions for drinking water treatment and seem to be highly promising in wastewater treatment for reuse purposes.

- Ultrafiltration technology was tested so far for removal of suspended and colloidal particles from surface water, no experience as to performance with wastewater exists.
- Nanofiltration (and low pressure reverse osmosis—LPRO) is a recently developed membrane process particularly suited for the removal of soluble organics. However the respective influence of suspended solids (including colloids) and organics fraction in fouling phenomena is not fully established.

The assumption that UF will be an excellent pretreatment to NF is supported by the evidence accumulated from the operation of RO plants using surface water (not municipal wastewater) that was pretreated by UF:

- The RO fluxes were increased by 20–30% when the surface water was pretreated by UF as compared to the pretreatment with $FeCl_3$ alone using sand filters.
- The frequency of cleaning the RO membranes was reduced from once every 3 weeks to once every 6 months.
- Since SDI (Silt Density Index) after UF is lower than 1 as compared to SDI between 3–5 when pretreated with more conventional processes, membrane life time will be increased from 1–2 years in the second case to 7–8 years in the first. Similar trends, but larger gaps can be expected using municipal wastewater.

With the availability of the new generations of low pressure RO membranes that have similar fluxes to NF with much better retention of organics, it will be wise to incorporate these RO into effluent reuse projects.

The UF technology was tested so far only for removal of suspended matter, colloidal particles and of all kinds of pathogens including viruses using as feed: drinking water after sand filtration, backflush water from the sand filters and also untreated surface water.

Mathematical Modeling of Membrane Performance

Ben Aim points at some recent studies which represent an important advance for a better understanding and precision of membrane operation (Bacchin et al., 1995; Liu et al., 1991; Liu, 1992). From the characteristics of the suspension to be filtered and of the membrane, it is possible to have a semi quantitative determination of the separation characteristics (filtrate flux).

Based on the phenomena responsible of the decrease of the permeability of the membrane, the following equations have been proposed for giving the filtrate flux in steady state conditions:

$$J = k \ln(C_m/C_0) + \lambda/(3\mu) \; d_p^2 \; \eta_w \qquad (36)$$

and

$$J_{crit} = D/\delta \ln (v_B/\delta) \qquad (37)$$

defining the flux J_{crit} below which no modification of the permeability of the membrane occurs.

j = filtrate flux
k = transfer coefficient
C_m, C_0 = concentrations of the species retained by the membrane (m—at the membrane, 0—bulk)
λ = hydrodynamic constant
μ = dynamic viscosity
d_p = diameter of the suspended particle
η_w = wall shear stress
D = diffusion coefficient
δ = thickness of the boundary layer
v_B = value of interaction potential barrier

These equations and mainly the values of λ and v_B derived from experimental results can be used to quantify the influence of pretreatment and permit extrapolation.

Layout Options

Considering deep-bed filtration-membrane combination, three strategies are possible (Ben Aim, 1997):

(1) Removing partially organics, which can be done by direct filtration producing an effluent adequate for irrigation but which could need disinfection for safety reason with the risk of organochlorinated compounds production

(2) Removing partially organics with simultaneous bacteria removal (flocculation + microfiltration or UF) producing waste adequate for irrigation purpose or some industrial application

(3) Removing organics as far as possible by NF (or LPRO) with or without pretreatment—the treated water would be adequate for irrigation purpose, for some specific industrial applications or, eventually with the necessary post treatment and mixed with other sources, as potable water.

Perry (1997) illustrates an example of a type of UF membrane that is produced by a company that manufactures PES (Polyether Sulfone) capillary membrane that can be supplied with different pore sizes, different capillary diameters (0.7–4 mm). It is based on a special hydrophilic polymer blend that combines good anti-fouling properties, excellent chemical stability (cleaning with up to 500 ppm of chlorine or peroxide, 2% NaOH). Because of these properties it is believed that this is a very powerful candidate for treating municipal wastes. The necessary diameter to cope with varying loads of suspended matter and the optimal pore size can be selected. Aggressive chemicals can be used to restore the membrane if fouled or plugged and if it will be necessary, the properties of such membrane for further optimization of the properties can be modified. Working in a "dead end" mode might save considerably on pumping energy. Usually the membranes are operated for 10–15 minutes in a dead end filtration mode and then backflushed for a period of 30 seconds. Periodically there is an injection into the backflushing liquid of a disinfecting chemical such as 200 ppm of hypochlorite or peroxide. The whole operation is controlled by a PLC and is done automatically.

Membrane technology has been hardly tested with municipal wastes. Furthermore, the reclamation of municipal wastewater for potable uses is becoming one of the hottest subjects in the water reclamation area.

REFERENCES

Adin, A. 1987. "Problems Associated with Particulate Matter in Water Reuse for Agricultural Irrigation and Their Prevention," *Water Science and Technology,* 18:185–195.

Adin, A. and Sacks, M. 1991. "Dripper Clogging Factors in Wastewater Irrigation," *J. Irrigation and Drainage Eng. ASCE,* 117(6):813–826.

Adin, A. 1979. Problems and Advanced Methods in Filtration for Low Rate Applicators. *Proceedings of the International Conference on Water Systems and Applications,* Israel Center of Waterworks Appliances, Israel. V-3, pp. 1–9.

Adin, A. and Alon, G. 1986. "Mechanisms and Process Parameters of Filter Screens," *J. Irrigation and Drainage Eng., ASCE,* 112 (4): 293–304.

Adin, A. and Alon, G. 1993. "The Role of Particle Characterization in Advanced Wastewater Treatment," *Water Science and Technology,* 27(10):131–139.

Adin, A. and Elimelech, M. 1989. "Particle Filtration in Wastewater Irrigation," *J. Irrigation and Drainage Eng. ASCE,* 115(3):474–487.

Adin, A. and Hatukai, S. 1994. "Combined Design, Operation and Economic Optimization of Contact Filtration," *Chemical Water and Wastewater Treatment: III.* Hahn, H. H., Klute, R. (eds.). Springer-Verlag. pp. 173–187.

Adin, A., Mingelgrin, U. and Kanarek, A. 1995. Slow Granular Filtration for Advanced Wastewater Treatment: Design, Performance and Operation. Annual Scientific Report. Granted by BMFT, Germany through NCRD. Jerusalem, Israel.

Adin, A. and Rebhun, M. 1974. "High Rate Contact-Flocculation Filtration with Polyelectrolytes," *J. American Water Works Association,* 66(2):109–117.

Adin, A. and Rebhun, M. 1977. "A Model to Predict Concentration and Head-Loss Profiles in Filtration," *J. American Water Works Association,* 8:444–451.

Adin, A., Rubinstein, L. and Zilberman, A. 1989. "Particle Characterization in Wastewater Effluents in Relation to Filtration and Irrigation," *Filtration and Separation,* 26(4):284–287.

Adin, A. and Vescan, N. 1996. Electro-Flocculation Process for Water and Wastewater Treatment. Yissum, R&D Company of the Hebrew University of Jerusalem. Patent No. 2256.

Alon, G. and Adin, A. 1994. "Mathematical Modeling of Particle Size Distribution in Secondary Effluent Filtration." *Water Environm. Res.,* 66(6):836–841.

Amirtharajah, A. and Mills, K. M. 1982. "Rapid Mix Design for Mechanisms of Alum Coagulation," *J. American Water Works Association,* 74(4):210–216.

Amirtharajah, A. 1988. "Some Theoretical and Conceptual Views of Filtration," *J. American Water Works Association,* 80(12):36–46.

Amirtharajah, A. and O'Melia, C. R. 1991. "Coagulation Processes: Destabilization, Mixing, and Flocculation," *Water Quality and Treatment,* 4th ed., AWWA, Pontius, F.W. (tech. ed.). McGraw-Hill, pp. 269–127.

Anderson, B., Nyberg, U., Aspegren, H., La Cour Jansen, J. and Odegaard, H. 1992. "Evaluation of Pre-Precipitating in a Wastewater Treatment System for Extended Nutrient Removal," *Chemical Water and Wastewater Treatment System: II.* Hahn, H.H., Klute, R. (eds.). Springer-Verlag. pp. 341–355.

Anderson, B., Nyberg, U., Aspegren, H., La Cour Jansen, J. and Odegaard, H. 1993. Nutrient Removal at a Wastewater Treatment Plant in Malmo. *IFAT 93,* Munich, Germany.

Bacchin, P., Aimar, P. and Sanchez, V. 1995. "Model for Colloidal Fouling of Membranes," *AIChET* 41(2)368–376.

Baumann, E. R. and Huang, J. Y. C. 1974. "Granular Filters for Tertiary Wastewater Treatment," *J. Water Pollution Control Federation,* 46(8):1958–1973.

Baumann, E. R. and Cleasby, J. L. 1974. *Wastewater Filtration: Design Considerations, Technology Transfer,* Series No. EPA-625/4-74-007a: US EPA.

Bellamy, W. D., Silverman, G. P., Hendricks, D. W. and Logsdon, G. S. 1985. "Removing Giardia Cysts with Slow Sand Filtration," *J. American Water Works Association,* 77(2):52–60.

Ben Aim, R. 1997. Personal communication.

Bench, B. I., Middlebrooks, E. J., George, D. B. and Reynolds, J. H. 1981. "Evaluation of Wastewater Filtration," *Water Quality Series,* No. UWRL/Q-81/01, Utah Water Research Laboratory.

Boucher, P. L. 1947. "A New Measure of The Filterability of Fluids With Applications to Water Engineering," *J. Inst. Civ. Engrs,* 27:415–423.

Bouwer, H. 1996. Personal communication.

Bouwer, H. 1974. "Nitrification-Denitrification in the Soil." Proc. Section 6, in *Nitrogen Removal from Municipal Wastewater.* Correspondence Conf. sponsored by Water Resources Research Center, 1973, Univ. of Mass., Amhurst.

Bowles, D. A., Drew, M. W. and Hirth, G. 1983. *The Application of Slow Sand Filtration Process to the Treatment of Small Town Water Supplies.* State Rivers and Water Supply. Australia: Commission of Victoria.

Bucks, D. A., Nakayama, F. S. and Warrick, A. W. 1982. "Principles, Practices and Potentialities of Trickle Irrigation." *Advances in Irrigation,* D. Hillel, ed., Vol. 1, San Diego, California: Academic Press, p. 219.

Chaudhary R. et al. 1989. Evaluation of Chemical Addition in the Primary Plant at Los Angeles Hyperion Treatment Plant, presented at *Water Pollut. Control Fed. Annu. Conf.,* San Francisco, Calif.

Cikurel, H., Rebhun, M., Amirtharajah, A. and Adin, A. 1996. "Wastewater Effluent Reuse by In-Line Flocculation Filtration Process," *Water Science and Technology,* 33:203–211.

Constable, J. D., Connor, M. A. and Scott, P. H. 1989. Fixed bed nitrification as a potential means of enhancing nitrogen removal rates in a sewage lagoon. *Proc. Australian Water & Wastewater Ass.* pp. 192–196.

Constable, J. D., Connor, M. A. and Scott, P. H. 1989a. The Operative Importance of Different Nitrogen Removal Mechanisms in 5 West Lagoon, Werribee Treatment Complex. *Proc. Australian Water & Wastewater Ass.* pp. 187–191.

Dansberg, S. and Bresler, E. 1985. *Drip Irrigation Manual,* International Irrigation Information Center (IIIC), Volcani Center for Agricultural Research, Beit-Dagan, Israel.

Darby, J. L., Lawler, D. F. and Wilshusen, T. P. 1991. "Depth Filtration of Wastewater: Particle Size and Ripening," *J. Water Pollut. Contr. Fed.,* 63(3):228–238.

EPA. 1987. *Design Manual for Phosphorus Removal,* EPA /625/1-87/001, Cincinnati, Ohio: US EPA.

Dickey, G. D. 1961. *Filtration,* Reinhold Publishing Co., New York, N.Y.

Dor, I., Schechter, H. and Bromley, H. J. 1987. "Limnology of Hypertrophic Reservoir Storing Wastewater Effluent for Agriculture at Kibbutz Naan, Israel," *Hydrobiologia,* 150:225–241.

Ellis, K. V. 1987. "Slow Sand Filtration as a Technique for the Tertiary Treatment of Municipal Sewages," *Water Research,* 21:403–410.

Ellis, K. V. 1991. "Slow Sand Filtration," CRC Critical Reviews in Environmental Control, 15(4):315–330.

English, S. 1985. Filtration and Water Treatment for Micro-Iirrigation. *Proceeding of the Third International Drip Trickle Irrigation Congress,* ASAE, pp. 50–57.

EPA. 1993. *Manual Combined Sewer Overflow Control.* US EPA Office of Research and Development, Washington, D.C.: EPA/625/R-93-007.

Farooq, Sh. and Yousef, A. K. 1993. "Slow Sand Filtration of Secondary Effluent," *J. Env. Eng.,* 19(4):615–630.

Farooq, Sh., Yousef, A. K., Al-Layla, R. I. and Ishaq, A. M. 1993. "Tertiary Treatment of Sewage Effluent via Pilot Scale Slow Sand Filtration," *Environmental Technology Letters,* 15:15–28.

Fitzpatrick, J. A. and Swanson, R. 1980. *Evaluation of Full-Scale Tertiary Wastewater Filters.* EPA/600/2-80-005:USEPA.

Fletcher, M. 1979. *The Attachment of Bacteria to Surfaces in Aquatic Environments. Adhesion of Microorganisms to Surfaces.* London, U.K.: Academic Press.

Ghosh, M., Amirtharajah, A. and Adin, A. 1994. "Particle Destabilization for Tertiary Treatment of Municipal Wastewater by Filtration," *Water Science and Technology,* 30:209–218.

Grasso, D. and Weber, W. J. 1988. "Ozone Induced Particle Destabilization," *J. American Water Works Association,* 80:73–81.

Habibian, M. T. and O'Melia, C. R. 1975. "Particles, Polymers and Performance in Filtration," *J. Environmental Engineering Division, ASCE,* Vol. 101, No. EE4, Proc. Paper 11512, Aug, 1975, pp. 567–583.

Harleman, D. F. 1992. Chemically Enhanced Primary Treatment for Municipal Wastewaters, presented at *Flocculations, Coagulants and Precipitants for Drinking Water and Wastewater Treatment. Intertech Conf.,* Herndon, VA.

Heinke, G. W. and Qazi, M. A. 1976. Upgrading Primary Clarifier Performance by Chemical Addition, *Proceedings of the Technology Transfer Seminar on High Quality Effluents,* sponsored by the Canada-Ontario Agreement on Great Lakes Quality, Toronto.

Hosseini, M. and Rushton, A. 1979. "Shear Effects in Cake Formation Mechanisms," *Filtration and Separation.* 16 (Sept.–Oct.): 456–460.

Hsiung, K. I. 1980. Chlorine Effect on Secondary Effluent Filtration. *J. Environmental Engineering Division, ASCE,* 106, EE3, Technical Note, Proc. Paper 15447, 649.

Huisman, L. 1978. "Developments of Village-Scale Slow Sand Filtration," *Prog. Water Tech.* 11:59.

Ives, K. J. 1960. "Filtration Through a Porous Septum: A Theoretical Consideration of Boucher's Law," *J. Institute of Water Engineers,* 17:333–338.

Ives, K. J. 1975. "Capture Mechanisms in Filtration," *The Scientific Basis of Filtration,* Ives, K.J. ed., Netherlands: Noordhoff-Leyden.

Ives, K. J. 1978. *Prog. Water Technology,* 10:128.

Ives, K. J. 1980. "Deep Bed Filtration: Theory and Practice," *Filtration and Separation,* 17:157–166.

Ives, K. J. and Sholji, I. 1965. "Research Variables Affecting Filtration," *J. Environmental Engineering Division, ASCE,* 91 (SA4, Proc. Paper 4436):1–18.

Johnson, P. N. and Amirtharajah, A. 1983. "Ferric Chloride and Alum as Single and Double Coagulants," *J. American Water Works Association,* 75(5):232.

Jones, H. C., Roth, I. L. and Sanders, W. M. 1969. "Electron Microscope Study of Slime Layer," *J. Bacteriol.,* 99(1):316–325.

Kaminsky, I. and Adin, A. 1996. Particle Size Distribution and Wastewater Filter Performance. *Proc. IAWQ/IWSA Joint Specialist Group on Particle Separation 4th Intern. Conf.,* October 28–30, 1996, Jerusalem, Israel.

Kanaani, Y. M., Adin, A. and Rav-Acha, Ch. 1992. "Biofilm Interactions in Water Reuse System: Adsorption of Polysaccharides to Kaolin," *Water Science and Technology,* 26(3/4):673–682.

Kasper, D. R. 1971. Theoretical Investigation of Flocculation. Ph.D. thesis, Caltech.

Katznelson, R. 1986. "Cyanobacterial Mats in Groundwater Recharge Basins of the Dan Region Wastewater Reclamation Project," *Environmental Quality and Ecosystem Stability,* Dubinsky, Z. and Steinberger, Y., eds., Ramat Gan, Israel: University Press.

Katznelson, R. 1989. "Clogging of Groundwater Recharge Basins By Cyanobacterial Mats. FEMS," *Microbiol. Ecol.,* 62:231–242.

Kavanaugh, M., Sigster, K., Weber, A. and Boller, M. 1977. "Contact Filtration For Phosphorous Removal," *J. Water Pollut. Control Fed.,* 49:2157.

Kavanaugh, M. C., Tate, C. H., Trussell, A. R. and Treweek, G. 1980. "Use of Particle Size Distribution Measurement for Selection and Control of Solid/Liquid Separation Processes. Particles in Water," *Advances in Chemistry Series,* M.C. Kavanaugh and Luckie, J.O. ed., Washington, D.C.: American Chem. Soc. p. 305.

Kirkpatrick, W. and Asano, T. 1986. "Evaluation of Tertiary Treatment Systems for Wastewater Reclamation And Reuse," *Water Science and Technology,* 18(10):83–95.

Klute, R. et al. 1990. "Particle Destabilization in Pipe Flow: The Effect of Aluminum Hydrolysis Species," *Z. Wasser-Abwasser-Forschung.*

La Mer, V. K. and Healy, T. W. 1963. "Adsorption-Flocculation Reactions of Macromolecules at the Solid-Liquid Interface," *Rev. Pure Appl. Chem.,* 13:1835–1851.

Kott, Y. 1973. "Hazards Associated with the Use of Chlorinated Oxidation Pond Effluents for Irrigation," *Water Research,* 7:853–862.

Lauglier, W. F. 1921. "Coagulation of Water with Alum by Prolonged Agitation," *Eng. News Res.,* 86:924.

Lawler, D. F., O'Melia, C. R. and Tobiason, J. E. 1980. "Integral Water Treatment Plant Design," in *Particles in Water,* Kavanaugh, M.C. and Luckie, J.O. (ed.), *Advances in Chemistry Series,* Washington, D.C.: American Chemical Society, pp. 353–328.

Levine, D. A., Tchobanoglous, G. and Asano, T. 1985. "Characterization of the Size

Distribution on Contaminants in Wastewater Treatment and Reuse Implications," *J. Water Pollut. Control Fed.*, 57(7):805–816.

Liu, M. G. 1992. Etude de la Texture des Membranes et de Leur Interaction avec des Suspensions au Cours D'une Filtration Dynamique. These de Dr.Sc. Université de Technologie de Compiegne.

Liu, M. G., Ben Aim, R., Mietton and Peuchor, M. 1991. Modeling of Deposit Formation on a Rotating Tubular Membrane. *Proceedings 2nd ICIM,* Montpellier. France.

Loogman, J. G., Post, A. F. and Mur, L. R. 1980. The influence of periodicity in light conditions, as determined by the trophic state of the water, on the growth of the green alga *Scenedesmus protuberans* and the Cyanobacterium *Oscillatoria agardhii.* In *Developments in Hydrobiology,* Barica, J. and Mur, L.R. (eds.), Junk Press, The Hague, pp. 79–82.

Mackie, R. I. 1993. Numerical Solution of the Filtration Equations for Polydispersed Suspensions. *Proceedings 6th World Filtration Congress,* Nagoya, Japan, pp. 244–247.

Manka, J. and Rebhun, M. 1982. "Organic Groups and Molecular Weight Distribution in Tertiary Effluent Sand Renovated Waters," *Water Research* 16:4:399–403.

McCabe, W. L. and Smith, J. C. 1971. *Unit Operations of Chemical Engineering,* McGraw-Hill Book Co., New York, N.Y. pp. 885–904.

McCarty, P. L., Rittmann, B. E. and Reinhard, M. 1985. "Processes Affecting The Movement and Fate of Trace Organics in the Subsurface Environment," *Artificial Recharge of Groundwater,* Asano, T. ed., Butterworth Publ., pp. 627–646.

Metcalf & Eddy, Inc., 1991. *Wastewater Engineering: Treatment, Disposal,* Reuse. 3rd Ed., NewYork: McGraw-Hill Inc.

Moran, C. D., Moran, M. C., Cushing, R. S. and Lawler, D. F. 1993. "Particle Behaviour in Deep-Bed Filtration: Part 1—Ripening and Breakthrough," *J. American Water Works Association,* 85(12):69–81.

Moran, C. D., Moran, M. C., Cushing, R. S. and Lawler, D. F. 1993. "Particle Behaviour in Deep-Bed Filtration: Part 2—Particle Detachment," *J. American Water Works Association,* 85(12):69–81.

Murcott, S. and Harleman, R. F. 1994. "Chemically Enhanced Primary Treatment," CEPT. M.I.T., *The Water Mirror,* pp. 3–5.

Nakayama, F. S., Bucks, D. A. and French, O. F. 1977. "Reclaiming Partially Clogged Trickle Emitters," *Transactions of the ASAE,* pp. 278–280.

Narkis, N. and Rebhun, M. 1975. "The Mechanism of Flocculation Processes in Presence of Humic Substances,." *J. American Water Works Association,* 67:101–108.

Neufeld, R. D. 1995. Personal communication.

Neufeld, R. D. 1994. Chemically Enhanced Settling of P Influents. University of Pittsburgh, Pittsburgh.

O'Mella, C. R. and Ali, W. 1978. "The Role of Retained Particles in Deep Bed Filtration," *Progress in WaterTech.,* 10:123.

Odegaard, H. 1992. "Norwegian Experiences with Chemical Treatment of Raw Waste-water," *Water Science and Technology,* 25(12):255–264.

Ojha, C. S. P., and Graham, N. J. D. 1991. Computer-Aided Modeling of Slow Sand Filtration: Preliminary Assessment. *Proc. International Slow Sand Filtration Workshop,* Durham, USA, pp. 1–24.

Okey, R. W. 1983. Enhanced Primary Treatment of Pulping Wastewaters. Presented at the *38th Annual Purdue Industrial Waste Conference,* W. Lafayette, IN.

Oron, G., Ben-Asher, J. and Demalach, Y. 1982. "Effluent in Trickle Irrigation of Cotton in Arid Zones," *J. Irrigation and Drainage Division, ASCE,* 108(No. IR2, June):115–126.

Packman, R. I. 1962. "The Coagulation Process," *J. Appl. Chem.,* 12:556–558.

Paramasivam, R., Mhaisalkar, V. A. and Berthouex, P. M. 1981. "Slow Sand Filter Design and Construction in Developing Countries," *J. American Water Works Association,* 73(4):178–185.

Parkhurst, J. D. et al. 1976. Wastewater Treatment for Ocean Disposal. Paper presented at *Am. Soc. Eng. Natl. Conf. Environ. Eng. Res. Dev. Design.,* Univ. Wash., Seattle.

Perry, M. 1997. Personal communication.

Post, A. F., Eygenraam, F. and Mur, L. R. 1985. "Influence of Light Period Length on Photosynthesis and Synchronous Growth of the Green Alga Scenedesmus Protuberans," *Br. Phycol. J.,* 20:391–397.

Rad, H., Crosse, J. 1990. Chemically Assisted Primary Treatment: A Viable Alternative to Upgrading Overloaded Treatments Plants—The Hyperion Experience. Presented at *66th Annual Water Pollution Control Federation Conf.,* October 1990, Washington.

Rav-Acha, Ch., Kummel, M., Salomon, I. and Adin, A. 1995. "The Effect of Chemical Oxidants on Effluent Constituents for Drip Irrigation," *Water Research,* 29(1):119–129.

Rebhun, M. and Manka, J. 1971. Classification of organics in secondary effluents. *Envi. Sci. and Tech.* 5:606–609.

Rebhun, M., Furer, Z., Adin, A. 1984. "Contact Flocculation-Filtration of Humic Substances," *Water Research,* 18(8):963–970.

Retamoza, J. G. 1994. Models of Slow Sand Filtration for Water Potabilization. Doctoral thesis, Faculty of Engineering, Nacional Autonomic University of Mexico (U.N.A.M.). 140 p. In Spanish.

Retamoza, J. G., Martinez, P. and Adin, A. (1994). Mathematical Models of Sand Filtration. *Proc. 24 Inter-American Congress of Sanitary and Environmental Engineering,* November 1994, Buenos Aires, Argentina.

Rice, R. C. and Bouwer, H. 1984. "Soil-Aquifer Treatment Using Primary Effluent.," *J. Water Pollut. Contr. Fed.,* 56(1):84–88.

Riddick, M. T. 1961. "Zeta Potential and its Application to Difficult Waters," *J. American Water Works Association,* 1007–1031.

Rothwell, E. 1978. "Fabric Filter Failures," *Filtration and Separation.* 15(Nov.–Dec.):586–593.

Schuler, P. F., Ghosh, M.M. and Gopalan, P. 1991. "Slow and Diatomaceous Earth Filtration of Cysts and Other Particulates," *Water Research,* 25:995–1005.

Sekulov, I. 1982. "A Modified Cross-Flow Filter Unit for the Advanced Wastewater Treatment," Preprints of the *Symposium on Water Filtration, European Federation of Chemical Engineering,* Antwerp, Belgium, (April 21–23):3.1–3.6.

Speitel, G. E. and Digiano, F. A. 1987. "The Bioregeneration of GAC Used to Treat Micropolutants," *J. American Water Works Association,* 79:64–73.

Spielman, L. A. and Fitzpatric, J. A. 1973. "Theory for Particle Collection Under Gravity Forces," *J. Coll. Int. Sci.,* 42.

Susenik, A., Telch, B., Wachs, A. W., Shelef, G. and Levanon, D. 1987. "Effect of Oxidants on Microalgal Flocculations," *Water Research,* 21:533–539.

Tambo, N. and Watanabo, Y. 1979. "Physical Characteristics of Flocs. I. The Floc Density Function and Aluminum Floc," *Water Research,* 13:409–419.

Tanaka, T. 1982. Kinetics of Deep-Bed Filtration. Ph.D. thesis, UCLA.

Tchobanoglous, G. and Eliassen, R. 1970. "Filtration of Treated Sewage Effluent," *J. Sanitary Engineering Division, ASCE,* 96, SA2, Proc. Paper 7210, 243.

Tebbutt, H.Y. 1971. "An Investigation into Tertiary Treatment by Rapid Sand Filtration," *Water Research,* 5(3):81–92.

Tien, C., and Payatakes, A. C. 1979. "Advances in Deep Bed Filtration," *J. American Institute of Chemical Engineering,* 4(Sept.):737–748.

The Water Commission of Israel, 1985. Report of the Committee for Irrigation in Small Volume. Israel.

Thomas D. C. 1964. "Turbulent Disruption of Flocs in Small Particles," *AIChE J.,* 10:517–523.

Uhlinger, D. J. and White, D. C. 1983. "Relationship between Physiological Status and Formation of Extracellular Polysaccharide Glycocalyx in Pseudomonas Atlantica," *Appl. Env. Microb.,* 45:64–70.

Van Dijk, J. C. and Ooman, J. H. C. M. 1978. *Slow Sand Filtration for Community Water Supply in Developing Countries: A Design and Construction Manual,* WHO International Reference Center for Community Water Supply, Hague, The Netherlands.

Van Rijn, J. and Rivera, G. 1990. "Aerobic and Anaerobic Biofiltration in an Aquaculture Unit—Nitrite Accumulation as a Result of Nitrification and Denitrification," *Aquacultural Eng.,* 9:217–234.

Vescan, N. et al. 1986. Electrochemical Purification and Decolorizing of the Effluent at Lupeni Textile Plant, IPCFS Report, Bucharest.

Vigneswaran, S. and Chang, J. S. 1986. "Mathematical Modeling of the Entire Cycle of Deep Bed Filtration," *Water Air and Soil Pollution,* 29:155–164.

Vigneswaran, S. and Tulachan, R. K. 1988. "Mathematical Modeling of Transient Behavior of Deep Bed Filtration," *Water Research,* 22:1093–1100.

Visscher, J. T. 1990. "Slow Sand Filtration: Design, Operation and Maintenance," *J. American Water Works Association,* 82(6):67–71.

Wiesner, M. R. and Mazounie, P. 1989. "Raw Water Characteristics and the Selection of Treatment Configurations for Particle Removal," *J. American Water Works Association,* 81(5):80–89.

Yao, K. M., Habibian, M. T. and O'Melia, C. R. 1971. "Water and Wastewater Filtration: Concepts and Applications," *Environmental Science and Technology,* 5(11):1105.

CHAPTER 6

The Role of Filtration for Wastewater Reuse

INTRODUCTION

BACKGROUND OF WATER REUSE PROGRAM

DURING the period of 1927 to 1962, the County Sanitation Districts of Los Angeles County (CSDLAC) had already operated several wastewater reclamation plants (WRPs) for incidental water reuses [1]. However, the first large-scale intentional reuse was initiated on August 20, 1962, when the 10 MGD (0.44 m³/s) Whittier Narrows Water Reclamation Plant (WNWRP) was completed. It produced reclaimed water for groundwater recharge with quality in full compliance with the California Regional Water Pollution Control Board No. 4 requirements [2].

With the successful water reuse of the WNWRP effluent, the CSDLAC launched an ambitious water reuse program by constructing and operating several more WRPs. The combined flow rate of the well oxidized and disinfected reclaimed water from these WRPs approached 100 mgd (4.38 m³/s) by 1974. Groundwater recharge, irrigation and wildlife refuge ponding were accounted for the major areas of water reuses.

During the period of 1965 to 1973, the Districts successfully conducted several pilot-plant activated carbon filtration studies at their Pomona Research Facility. As results of these studies, the CSDLAC was funded in 1974 by the California State Water Resources Control Board (CSWRCB)

Ching-lin Chen, Jih-Fen Kuo, and James F. Stahl, County Sanitation Districts of Los Angeles County, P.O. Box 4998, Whittier, CA 90607.

to construct an activated carbon filtration plant at the CSDLAC's Pomona WRP (PWRP). This carbon filtration plant was constructed with an intention of producing effluent to meet the discharge standards established by the Regional Water Quality Control Board of the Los Angeles Region [3]. It has successfully provided highly renovated water for a variety of reuses including recreational, industrial and agricultural reuses.

Virus monitoring, conducted by the CSDLAC during a pilot-plant activated carbon filtration study from 1972 to 1973, never found detectable quantities of viruses in the effluent of a two-stage activated carbon filtration system. However, the construction of such a system to provide effluent for non-restricted recreational reuses would constitute a variance from the Title 22 of the California Administrative Code for wastewater reclamation criteria [4]. These criteria require a full treatment system consisting of a series of individual coagulation/flocculation, sedimentation, filtration, and disinfection unit processes, as shown in Figure 6.1, for recreational reuses which may involve human contact. The intent of the full treatment is to limit the potential of human exposure to waterborne viruses. Since this so-called Title 22 standard full treatment system was considered to be very expensive for the intended water quality objectives, a study was thus initiated to determine the cost effectiveness of other tertiary treatment alternatives for achieving the same virus control objectives. The study, commonly referred to as the Pomona Virus Study, was jointly funded by the U.S. EPA, the CSWRCB, and the CSDLAC. It was conducted by the CSDLAC at their Pomona Research Facility.

POMONA VIRUS STUDY

Three alternative tertiary treatment systems were evaluated and compared with the Title 22 standard system during the Pomona Virus Study. These four pilot-plant systems are briefly described as follows:

(1) System A: high dose alum coagulation/flocculation $\rightarrow$ sedimentation $\rightarrow$ filtration $\rightarrow$ disinfection (Title 22 full treatment system)
(2) System B: low dose alum coagulation $\rightarrow$ filtration $\rightarrow$ disinfection
(3) System C: activated carbon filtration $\rightarrow$ disinfection $\rightarrow$ activated carbon filtration
(4) System D: low dose alum coagulation $\rightarrow$ filtration $\rightarrow$ disinfection (with nitrified secondary effluent)

Both chlorination and ozonation disinfection processes were applied to the effluents of these systems (except for System D which was chlorinated only) to compare their costs in inactivating the pathogens, based on the monitoring results of total coliform and virus. The flow diagrams for these

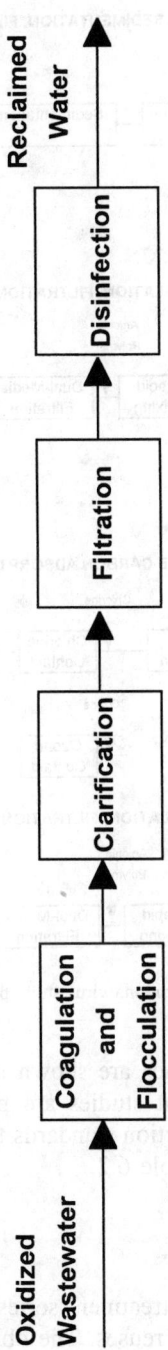

Figure 6.1 So-called Title 22 full treatment scheme.

SYSTEM A: COAGULATION, SEDIMENTATION, FILTRATION AND DISINFECTION

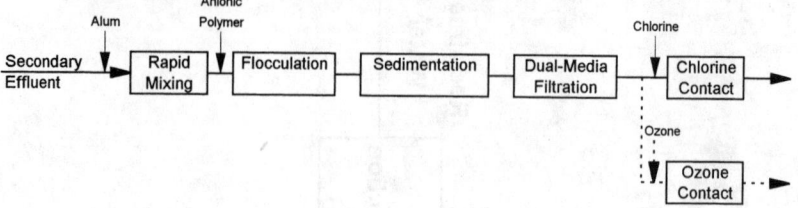

SYSTEM B: COAGULATION, FILTRATION AND DISINFECTION

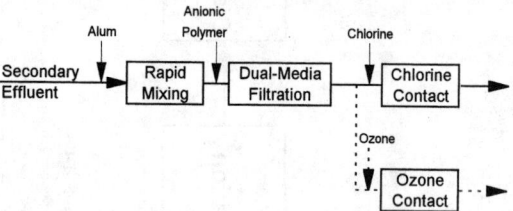

SYSTEM C: TWO-STAGE CARBON ADSORPTION AND DISINFECTION

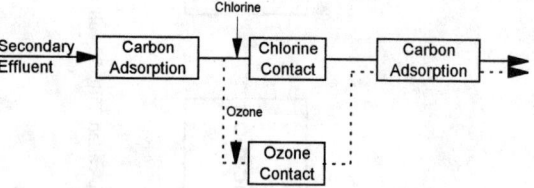

SYSTEM D: COAGULATION, FILTRATION AND DISINFECTION

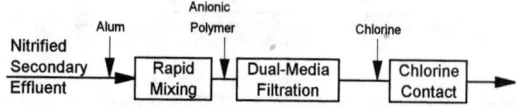

Figure 6.2 Pomona virus study pilot systems.

four tertiary treatment schemes are shown in Figure 6.2. The operating parameters for the pilot-plant studies are presented in Table 6.1. The effluent turbidity and disinfection standards for nonrestricted recreational reuses are summarized in Table 6.2.

System A

System A provided a full treatment series as required by the Title 22 for nonrestricted recreational reuses. The objective of this full treatment is to minimize the amount of particles which could shelter the pathogens

TABLE 6.1. Summary of Operating Parameters for Pomona Virus Study Pilot Systems.

	System A	System B	System D
Chemical dosage			
Alum, mg/L	155 (13.5 as Al)	5 (0.45 as Al)	
Polymer—Calgon WT-3000, mg/L	0.2	0.06	
Chlorine residual, mg/L	5 & 10 (combined)	5 & 10 (combined)	5 (free)
Ozone dosage, mg/L	10 & 50	10 & 50	
Flow rates			
System flow	40 gpm through coagulation, flocculation, and sedimentation tanks 25 gpm through filtration & chlorination (20 gpm for ozonation system)	25 gpm through filtration & chlorination (20 gpm for ozonation system)	
Sedimentation overflow rate, gpd/ft² (m/d)	800 (32)		
Filter loading rate, gpm/ft² (m/h)	5 (12)	5 (12)	
Backwash flow rate, gpm/ft² (m/h)	20 (48)	20 (48)	
Air scour rate, scfm/ft² (m/min)	4 (1.2)	4 (1.2)	
Detention times			
Rapid mixing tank, minutes	4		
Flocculation tank, minutes	62		
Sedimentation tank, minutes	126		
Chlorine contactor, minutes	120	120	
Ozone contactor, minutes	18	18	
Filter Backwash Schedule	daily	daily	

TABLE 6.1. (continued).

| | System C | | | |
| | Filtration (Activated Carbon) | | Disinfection | |
	First Stage	Second Stage	Chlorination	Ozonation
Flow rates				
System flow rate, gpm (L/sec)	100 (6.3)	50 (3.1)	50 (3.1)	50 (3.1)
Filter loading rate, gpm/ft² (m/h)	3.5 (8.4)	4.0 (9.6)		
Detention time				
Empty-bed detention time, minutes	10	10		
Contact time, minutes			120	18
Chemical dosages				
Chlorine residual, mg/L			5 & 10 (combined)	
Ozone dosage, mg/L				6
Filter backwash schedule	daily	weekly		

1 gpd/ft² = 0.040 m/d; 1 gpm = 0.063 L/sec; 1 gpm/ft² = 2.4 m/h; 1 scfm/ft² = 0.3 m/min.

TABLE 6.2. Title 22 Reclaimed Water Standards for Nonrestricted Recreational Reuse.

1. **Turbidity Standard:** The turbidity cannot exceed an average of two turbidity units and cannot exceed five turbidity units more than 5 percent of the time during any 24-hour period. 2. **Coliform Standard:** The seven-day median of coliform organisms does not exceed 2.2 per 100 milliliters and the number of coliform organisms does not exceed 23 per 100 milliliters in more than one sample within any 30-day period.

during the disinfection process and thus prevent them from effective inactivation. The treatments prior to filtration are supposed to reduce the concentration of the particles as low as possible before being polished further by the filtration system.

The turbidity of well-treated secondary effluents usually ranges from 2 to 5 nephelometric turbidity units (NTU). The amount of turbidity-causing submicron particles in these well-treated secondary effluents is relatively small. By using a coagulation/flocculation/sedimentation process to further remove these particles from the secondary effluent, a rather high dosage of coagulant followed by an addition of anionic polymer coagulant aid is required to form settleable chemical flocs. Alum, instead of iron salts, is commonly selected for the water reclamation because of its relatively more favorable pH range for floc formation and its rather minimal maintenance problem (Some esthetic and storage problems are usually associated with the use of iron salts). Adsorption and sweeping of the impurity particles by the settling chemical flocs are believed to be the primary mechanisms for the particle removal in this coagulation/flocculation/sedimentation process.

The clarified effluent is then fed to the filtration system for polishing the unsettled particles as well as the chemical flocs. Figure 6.3 shows the effect of alum dose on the overall results of the clarification and filtration systems. As shown in the figure, the addition of filter-aid made an insignificant improvement on the filter effluent turbidity, but it adversely increased the headloss of the filter operation. The headloss was measured at the end of the first 7.5 hours of a filter run. An optimum dosage of alum for the overall performance of System A was around 150 mg/L, as indicated by the minimum effluent turbidity of 0.2 turbidity unit (TU) in the figure.

System B

The principle behind the treatment scheme of System B is to enhance the filterability of the filter influent particles. A low dose of coagulant,

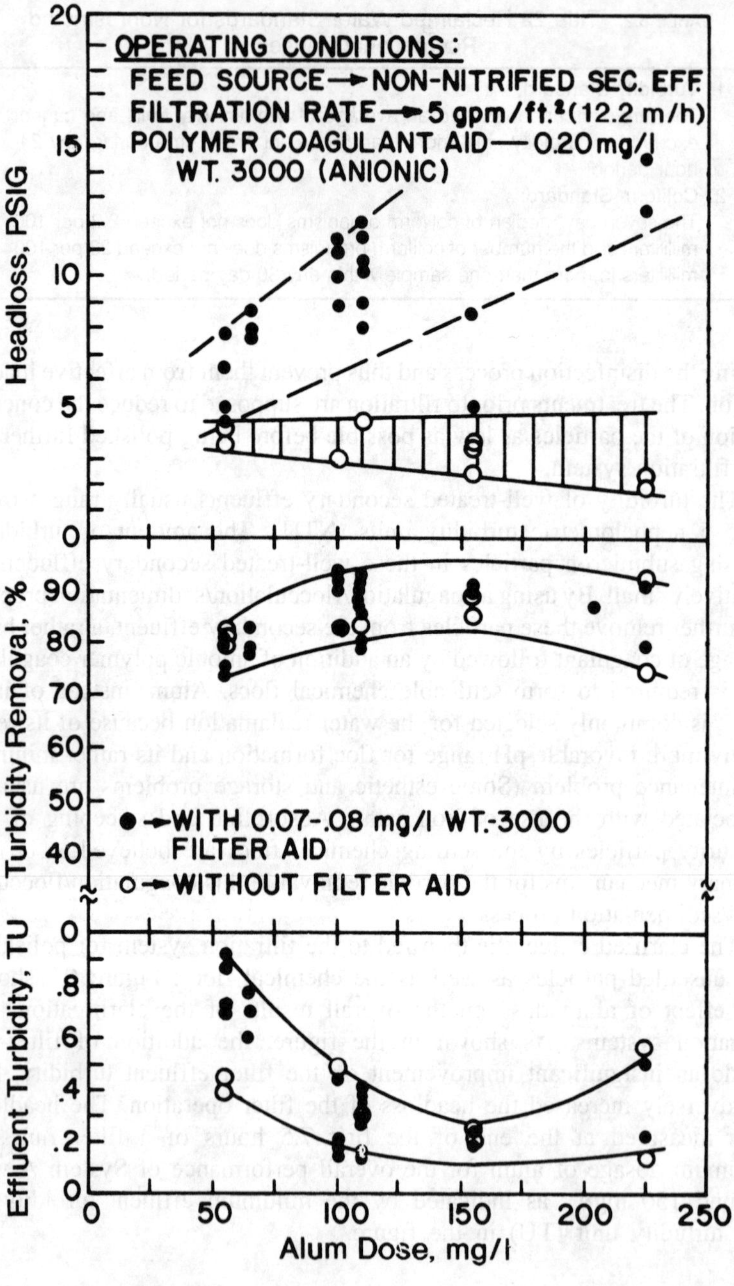
OPERATING CONDITIONS:
FEED SOURCE → NON-NITRIFIED SEC. EFF.
FILTRATION RATE → 5 gpm/ft.² (12.2 m/h)
POLYMER COAGULANT AID — 0.20 mg/l
WT. 3000 (ANIONIC)
Headloss, PSIG
20
15
10
5
0
Turbidity Removal, %
90
80
70
60
50
40
0
● → WITH 0.07-.08 mg/l WT.-3000
FILTER AID
○ → WITHOUT FILTER AID
Effluent Turbidity, TU
.8
.6
.4
.2
0
0
50
100
150
200
250
Alum Dose, mg/l

Figure 6.3 Effect of alum dose on dual media filter performance in system A.

alum or cationic polymer, is injected into the filter feed line just ahead of the filter. A static in-line mixer is usually installed between the coagulant injection point and the filter inlet to provide efficient mixing. The added coagulant destabilizes the colloidal particles by neutralizing or slightly reversing the negative surface charges of the particles. When alum is used as the coagulant, it is usually followed by an addition of about 0.06 mg/L of anionic polymer to enhance the coagulation/flocculation effects of the particles. Through the mechanisms of chemical linkage or bridging, the destabilized microflocs will gradually grow in size as they travel through the filter bed. This progressive size growth process of the flocculated particles will enable them to be effectively removed through the full depth of the filter media without blinding the surface of the filter bed nor causing premature particle break-through.

However, formation of large flocculated particles before reaching the filter bed should be avoided to prevent a rapid build-up of a cake layer on the surface of the filter bed. Such build-up will quickly cause a steep increase of headloss and result in a short filter run. The entire filter bed is to be fully utilized as a temporary storage of the flocculated particles for an optimal operation. Figure 6.4 shows the effect of alum dose on dual-media filter performance. Direct filtration with dual-media filter and low alum doses was unable to achieve a filter effluent turbidity of 0.4 TU or less. However, a high dose of alum (160 to 183 mg/L) with non-ionic polymer (1.2 to 2.1 mg/L) was shown to achieve 0.3 to 0.4 TU filter effluent turbidity in this study. Unfortunately, the filter run for this type of operation lasted only 20 to 75 minutes, too short to be economically feasible for field applications.

System C

The use of an activated carbon adsorption/filtration system at the head end of the tertiary treatment train was to remove a substantial amount of viruses in addition to the removal of bacteria and suspended solids [5].

The first stage granular activated carbon process was followed by either chlorination or ozonation for inactivating the bacteria and viruses to below the control limits. The disinfection byproducts were then removed by the second stage activated carbon process. Each stage of activated carbon process provided an empty bed contact time of approximately 10 minutes. The exhausted carbon beds were thermally regenerated at the Pomona Research Facility.

System D

System D basically had the same layout as System B. However, their feed waters were different. The feed water to System B was a non-nitrified

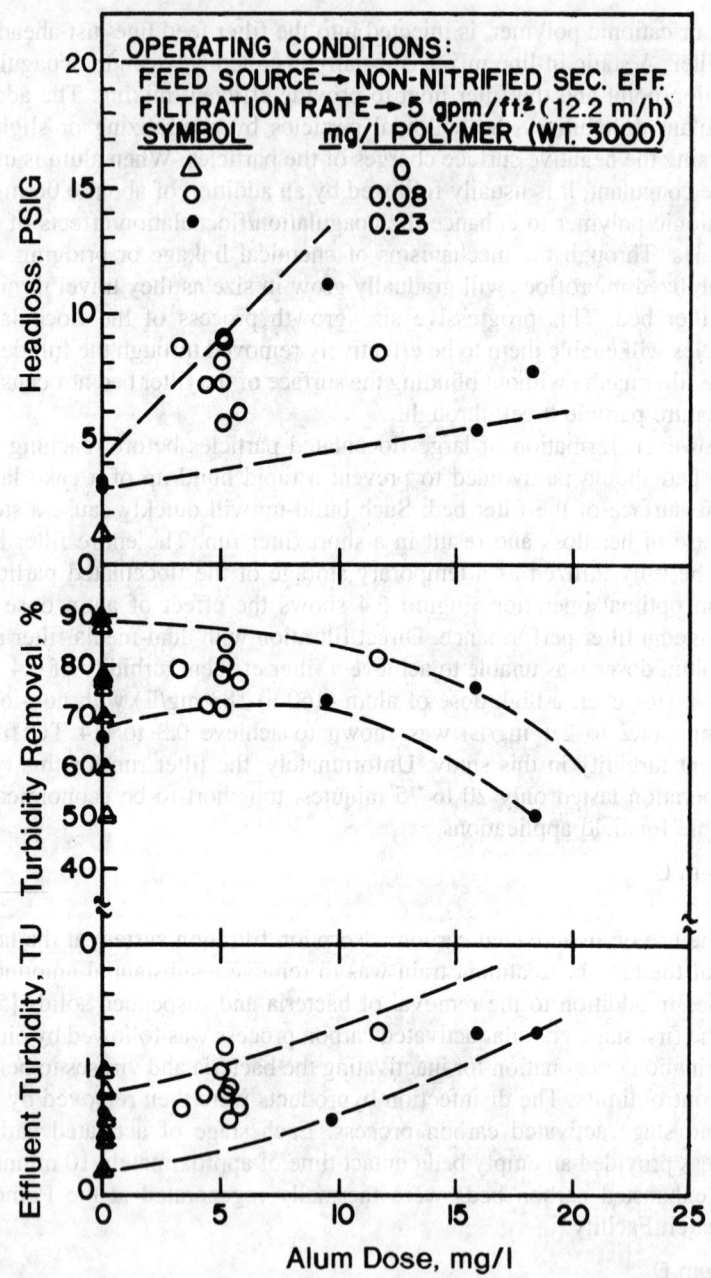

Figure 6.4 Effect of alum dose on dual media filter performance in system B.

secondary effluent from the PWRP, while the feed to System D was a nitrified secondary effluent from the old activated sludge plant of the PWRP, which was retrofitted for a two-sludge nitrification study.

Under the similar chlorine dosage level as required by System B, a breakpoint chlorination always occurred in System D. Therefore, the chlorination of System D was mostly achieved by the free chlorine, while System B was by the chloramines. Free chlorine was believed to provide more effective viruses inactivation than the chloramines under a shorter contact time condition. However, it was found that the chloramines with two hours of contact time would perform equally effective as the free chlorine for virus inactivation. Furthermore, free chlorine was found to form a substantially high amount of trihalomethanes in the chlorinated effluent [6]. Therefore, System D with breakpoint chlorination was not considered to be a desirable treatment alternative for providing virus risk-free reclaimed water for reuse.

Results of the Pomona Virus Study

The frequency distributions of filter influent and effluent turbidities for the four tertiary pilot-plant systems are shown in Figure 6.5. As indicated in the figure, the frequencies for Systems B, C, and D to achieve an effluent turbidity of 2 TU were about 90 percent, while for System A it was better than 99 percent. The full treatment System A did provide more consistent and better results prior to disinfection. However, the more economical System B with direct filtration could also provide an adequate turbidity control for effective disinfection. The 90 percent frequency for System B to achieve an effluent turbidity of 2 TU could be greatly improved if the filter influent turbidity could be maintained at 4 TU or less through proper pretreatment, which is normally done in the full scale plant. The secondary effluent supplied to the pilot plant usually contained slightly higher turbidity than the actual secondary effluent of the PWRP. This higher turbidity was contributed by the biosolids formed in the long transfer pipe between the full scale plant and the pilot plant facility.

Figures 6.6 and 6.7 show the impacts of four different treatment schemes on the cumulative virus removals by chlorination and ozonation, respectively. As indicated in the figures, different treatments did accomplish different levels of pre-disinfection virus removals. The highest level of virus removal was achieved by System A, which had the most extensive as well as expensive pretreatment. It was followed by System C with an activated carbon adsorption/filtration scheme. Systems B and D achieved the third and fourth levels of virus removals, respectively. In spite of the lower pre-chlorination virus removal, System B did adequately demonstrate its capability of achieving an overall virus removal of more than 5

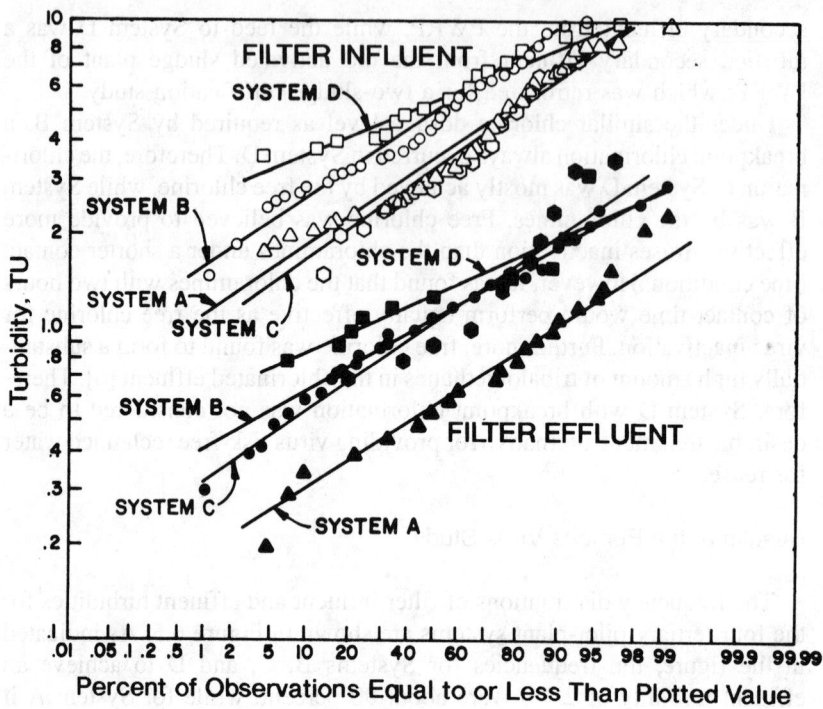

Figure 6.5 Frequency distributions of filter influent and effluent turbidities.

logs with chlorination or ozonation, similar to the overall removal by System A (The log removal refers to the negative logarithm of the ratio between the remaining seeded polio viruses after treatment to the influent polio viruses).

Based on the comparisons of turbidity and virus removals, as well as treatment cost analyses, the direct filtration (System B) was considered as the most cost-effective pre-disinfection treatment alternative to the so-called standard full treatment (System A) for producing an essentially pathogen-free disinfected effluent. The California State Department of Health Services (CSDOHS) had consequently recommended the following specific design and operational requirements be followed when one employing the alternatives evaluated in the Pomona Virus Study [7]:

(1) Secondary effluent turbidity of less than 10 TUs
(2) Coagulation ahead of the dual media filters
(3) Comparable filter (activated carbon or dual media) depths and loading rates
(4) Average turbidity in the filtered wastewater of less than or equal to 2.0 TUs

(5) High energy rapid mix of chlorine
(6) Theoretical chlorine contact time of 2 hours and a modal time between 90 to 100 minutes based on peak dry weather flow
(7) Chlorine contact chamber length to depth or width ratios of 40 : 1
(8) Median (2.2/100 ml) and maximum (23/100 ml) total coliform requirements

The CSDOHS's recommendations have led to the construction of tertiary filters in WRPs in the State of California to ensure that the use of reclaimed water would not impose any significant health risks. The Pomona Virus Study results are applicable to those cases where an essentially virus free effluent must be maintained. These include reuses of spray irrigation of food crops, landscape irrigation and nonrestricted recreational impoundment, as well as discharges to ephemeral or low flow streams where body contact may occur.

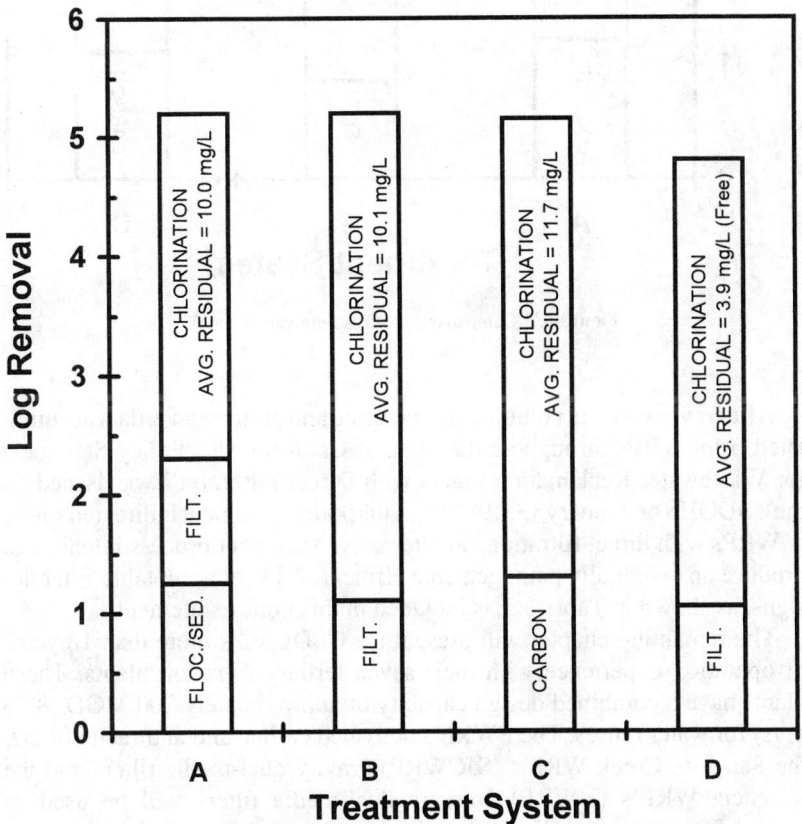

Figure 6.6 Comparison of virus removals (high chlorine residuals).

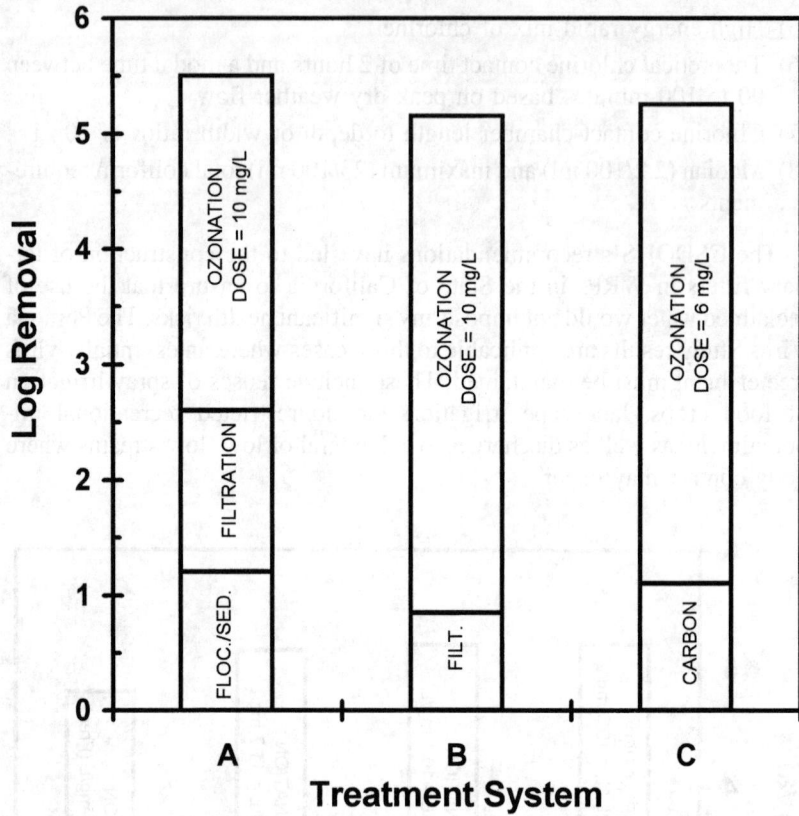

Figure 6.7 Comparison of virus removals (ozone).

After reviewing the voluminous operating and performance data accumulated from WRPs throughout the State of California, a "Policy Statement for Wastewater Reclamation Plants with Direct Filtration" was issued by the CSDOHS on January 13, 1988 [8]. This policy statement is directed solely at WRPs with direct filtration, an alternative treatment process intended to produce an essentially pathogen-free effluent. A list of acceptable filter designs, as shown in Table 6.3, is included in this policy statement.

The remaining chapter will present the CSDLAC's more than 18 years of operation experience with their seven tertiary filtration plants. These plants have a combined design capacity of approximately 200 MGD (8.76 m³/s) for water reuses. The PWRP's activated carbon and anthracite filters, the San Jose Creek WRP's (SJCWRP) gravity dual-media filters and the Valencia WRP's (VWRP) pressure dual-media filters will be used as typical filter types for discussions in this chapter.

TABLE 6.3. Specifications of Media for Direct Filtration Systems.

Type of Filter	Media Depth (inches)	Effective Size (mm)	Uniformity Coefficient
Dual media	Anthracite: 24	1.00–1.20	1.30–1.40
	Sand: 12	0.55–0.60	1.15–1.20
Mixed media	Anthracite: 18–24	1.00–1.20	1.60–1.65
	Sand: 9	0.40–0.45	1.30–1.50
	Garnet: 4–6	0.30–0.35	1.40–1.50
HydroClear	Sand: 10–12	0.45	1.50
Anthracite	Anthracite: 48	1.50	1.40
Parkson DynaSand	Sand: 40	1.30	1.50
Traveling Bridge	Sand: 11	0.55	1.50

1 inch = 2.54 cm.

PRINCIPAL MECHANISMS OF FILTRATION

FILTRATION MECHANISMS

Depending on the characteristics of filter media and influent particles, filter design configuration, influent water qualities, filter operation conditions, and application of coagulant and flocculant, particle removal by filtration may be attributed to the results of following mechanisms:

- straining: Particles larger than the water passages through the filter media are filtered out of the flow streams.
- adsorption: Particles smaller than the water passages in the filter are removed by adsorption through surface effect.
- sedimentation: Particles settle out in the localized quiescent spaces in the filter bed.
- impingement: Particles impinge on the surfaces of the media through inertial force.
- coagulation/flocculation: Particles modified by the added coagulant and flocculant coagulate with the filter media or form larger flocculated particles and are removed from the flow streams.

EFFECTS OF FILTER MEDIUM AND CONFIGURATION ON TURBIDITY REMOVAL

All the above-mentioned particle removal mechanisms can be enhanced by increasing the depth of filter bed or reducing the filtration rate; while removal through adsorption mechanism can be further enhanced by use of granular activated carbon as filter media.

Figure 6.8 shows that the granular activated carbon filter was more efficient in particle removal than the mono-medium anthracite filter with the same 6-ft (1.8-m) filter media depth. This effect was more pronounced when the filter influent turbidity was in the lower range. The lower turbidity water generally contains a higher proportion of finer particles as shown in Figure 6.9. The active sites of the granular activated carbons seemed to adsorb the submicron organic particles more effectively and resulted in a better performance than the inert anthracite media. It was also noticed that the activated carbon filter generally maintained a more consistent particle removal throughout the entire secondary effluent turbidity range. In addition, the activated carbon filter was found to perform less sensitive to the fluctuation of the filter influent turbidity than the dual-media or anthracite filters.

As shown in Figure 6.8, a better particle removal was achieved by the 6-ft (1.8-m) depth anthracite filter than the 3-ft (0.9-m) depth anthracite/ sand dual-media filter for the lower turbidity waters. This might be attributed to the greater depth of anthracite media in providing a better contact chance for removal of fine particles. When the influent turbidity was high, the particle removals by filters of different media and designs were very similar. Since the higher turbidity waters contained a smaller proportion of fine particles as indicated in Figure 6.9, all four tertiary filters would perform equally well in removing the bigger particles from these high turbidity water. All filter effluents were found to be essentially free from particles larger than 50 μm, as indicated in Table 6.4 of particle size distribution data.

The dual-media pressure filters at the VWRP seemed to perform slightly better than the dual-media gravity filters at the SJCWRP, as shown in Figure 6.8. This better performance might be partly contributed by the partial nitrification practiced at the VWRP, and partly by the relatively higher compaction of the filter media in the pressure filters. With a longer mean cell residence time (MCRT), about 8.5 days, for the partial nitrification, the floc strength would be stronger to resist shear force imposed by the flow stream [9]. Therefore, a better particle removal was achieved by the pressure dual-media filters at the VWRP.

FILTER DESIGN FEATURES

In addition to the inorganic dissolved and suspended solids, the secondary effluents usually contain bacteria, viruses, organic compounds and bacteria debris. The accumulation of these impurities in the filter bed tends to promote the bacterial growth and thus accelerates the headloss build-up through the filter bed. Therefore, this additional headloss build-up

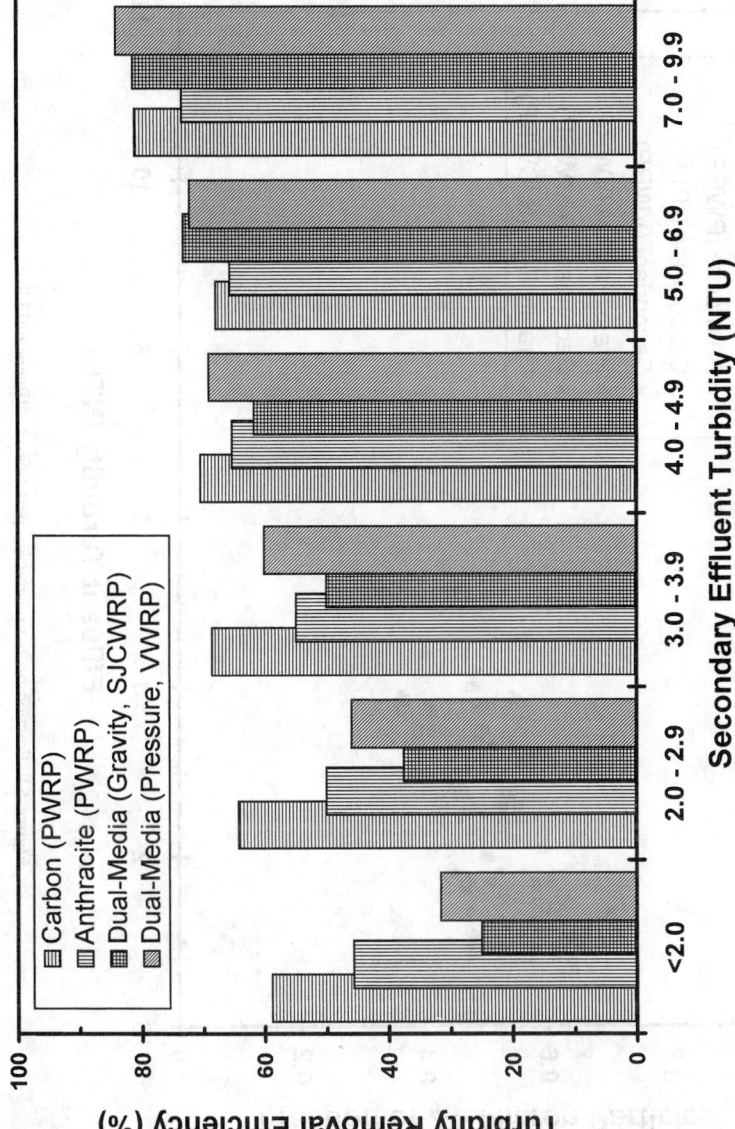

Figure 6.8 Impacts of filter media and secondary effluent turbidity on turbidity removal efficiency.

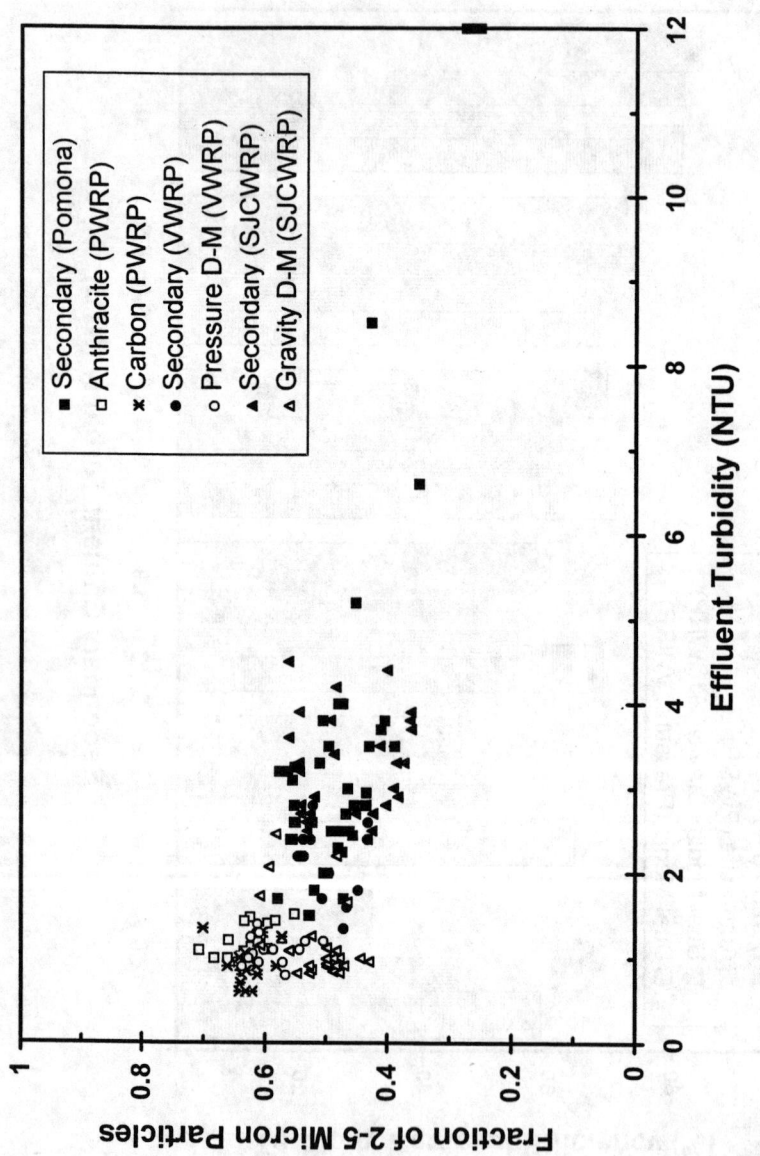

Figure 6.9 Fraction of 2- to 5-micron particles vs. effluent turbidity.

TABLE 6.4. Particle Size Distribution Data (particles/ml except as noted).

	PWRP			SJCWRP		VWRP	
	Secondary	Carbon (Gravity)	Anthracite (Gravity)	Secondary	Dual Media (Gravity)	Secondary	Dual Media (Pressure)
Particle size range, μm							
1.2–2	1018	325	339	979	529	933	236
2–5	2514	750	804	2056	945	1800	505
5–10	861	83	99	441	185	340	67
10–20	861	28	36	367	85	305	31
20–50	399	7	8	278	25	161	8
50–100	35.8	0.1	0.1	45.9	2.7	26.3	0.2
100–200	2.36	0.01	0.01	9.95	0.34	4.19	0.02
>200	0.02	0.00	0.00	0.19	0.01	0.04	0.00
Total Particle Count	5691	1193	1286	4176	1772	3570	847
Beta Value (β)	2.90	3.40	3.41	2.49	3.05	2.74	3.24
Turbidity, NTU	3.0	0.9	1.2	3.2	1.2	2.1	1.1
Particle surface area, cm²/mL	0.0321	0.0010	0.0011	0.0306	0.0030	0.0174	0.0009
Particle volume, μL/L	23.1	0.3	0.3	37.4	2.1	18.0	0.3
Number of data set	18	18	18	32	32	16	16

Note: Particle size distribution was performed with a Hiac/Royco model 8000 particle counter equipped with a laser-diode light obscuration sensor, model HRLD-400 (Hiac/Royco, Silver Spring, MD).

factor along with the following basic parameters should be carefully considered in designing an efficient filtration system for wastewater reclamation.

FILTER TYPES

Currently there are three different filter systems in operation at the CSDLAC's seven tertiary filtration plants. They are dual-media gravity filters, dual-media pressure filters, and mono-medium anthracite gravity filters. The CSDLAC had previously operated granular activated carbon gravity filters and mixed-media pressure filters in their PWRP and VWRP, respectively. These filter systems have been respectively replaced by mono-medium anthracite gravity filters and dual-media pressure filters for better operation efficiency and cost-effectiveness in meeting the effluent standards. Table 6.5 shows the typical filter design features for the tertiary filtration plants owned and operated by the CSDLAC.

The mixed-media pressure filters at the VWRP contained 3 in (7.6 cm) of 0.21 × 0.32 mm garnet at the bottom layer of the filter bed. Under normal plant conditions, these mixed-media pressure filters functioned satisfactorily. However, a rapid headloss build-up usually occurred whenever the filter influent quality was deteriorated as results of minor treatment plant upsets. Such high headloss seriously shortened the filter runs and practically failed the entire filter backwash program. Therefore, the mixed-media were replaced with the dual-media as specified in Table 6.5.

The granular activated carbon filters at the PWRP were originally designed to remove color, in addition to turbidity, for water reuse by local paper companies [10]. The activated carbon media were replaced with anthracite in 1992, because the color treatment is performed by the users at their plant sites and is no longer a treatment requirement for the PWRP.

In addition to the above five types of filters, the CSDLAC have conducted a seven-month of pilot-plant evaluation studies of a pulsed-bed (HydroClear) and a continuous-backwash (DynaSand) filter systems at their Joint Water Pollution Control Plant (JWPCP) in 1994 [11]. The objective of the study was to evaluate several tertiary treatment alternatives, including disinfection processes, for in-plant water reuses. Table 6.6 presents some of the major design parameters for these pilot-plant filters. Their schematic diagrams are also shown in Figures 6.10 and 6.11. These two types of filters performed equally well, as the 6-ft (1.8-m) anthracite filter, under parallel operation conditions for the secondary effluent from the pure-oxygen activated sludge plant.

TABLE 6.5. Design Features of CSDLAC's Typical Tertiary Filters.

	PWRP	PWRP	VWRP	SJCWRP
Filter type	Carbon/Gravity[a]	Anthracite/Gravity	Dual-media/Pressure	Dual-media/Gravity
Filter rate control	Influent weir	Influent weir	Master controller	Influent weir
Plant design flow rate, MGD (m³/sec)	15 (0.66)	15 (0.66)	7.5 (0.33)	100 (4.41)
Filter media	#8 × #30 Carbon	1 × 1.25 mm anthracite	1 × 1.25 mm anthracite 0.5 × 0.6 mm sand	1 × 1.25 mm anthracite 0.5 × 0.6 mm sand
Filter depth, inches (m)	72 (1.8)	72 (1.8)	24 (0.6) anthracite 12 (0.3) sand	24 (0.6) anthracite 12 (0.3) sand
Avg. filtration rate, gpm/ft² (m/h)	3.4 (8.2)	3.4 (8.2)	3.9 (9.4)	3.6 (8.6)
Backwash				
Surface wash rate, gpm/ft² (m/h)	3–5 (7.2–12)	None	3–5 (7.2–12)	None
Air scouring, scfm/ft² (m/min)	None	2–5 (0.6–1.5)	None	2–5 (0.6–1.5)
Max. back wash rate, gpm/ft² (m/h)	15 (36)	18 (43)	18 (43)	22 (53)
Typical filtration run, hr	12	18	24	48
Backwash water supply	Chlorine contact tank effluent for all filtration systems			
Chemical dosing capability				
Polymer (filter influent)	No	No	Yes	Yes
Alum (filter influent)	No	Yes	Yes	Yes
Chlorine (filter influent)	No	Yes	Yes	Yes
Chlorine (filter effluent)	Yes	Yes	Yes	Yes

[a]Converted to anthracite/gravity filters in 1992.

239

TABLE 6.6. Design Parameters for HydroClear and DynaSand Pilot Filters.

	HydroClear	DynaSand
Filter media	Sand	Sand
Media size (mm)	0.45 E.S.	0.9 E.S.
Uniformity Coefficient	1.5	1.8
Media depth (m)	0.25	2.03
Filter area (m²)	0.37	0.99
Flow direction	Downward	Upward
Backwash mode	Pulsed	Continuous

FLOW CONTROLS

The following three types of filtration rate control mechanisms are used in the CSDLAC's tertiary filtration plants: (1) flow splitting, constant rate; (2) flow splitting, declining rate; and (3) master controller, constant rate.

The flow splitting, constant rate control method is used for the tertiary filtration plants at the WNWRP and SJCWRP. The influent flow is split equally to all operating filters by means of a weir box at each filter. A constant water level is maintained in each filter by the action of an automatic effluent flow control valve. Level control is considered to be a simpler, less expensive and more reliable system than rate-of-flow and master controller methods. The filters can be operated in the declining-rate mode, if desired.

The rate controls for the tertiary filters at the PWRP for both activated carbon and anthracite media are similar to those of WNWRP and SJCWRP. The influent flow is also split equally to the operating filters. However, the water level in the filters is allowed to rise during filtration as the filter becomes clogged with solids. The filter effluent flow rate thus declines as the headloss increases.

The dual-media pressure filters at the VWRP are equipped with a master controller for flow rate regulation. The controller can be set to either split the flow equally to the operating filters or maintain a constant flow through two filters while the third filter handles the remaining plant flow.

Since the influents to all filters are normally of excellent quality with turbidity ranging from 2 to 4 NTU, the impacts of flow control modes on the filter effluent quality are insignificant. All these modes of flow control are found to be equally functional and efficient for the CSDLAC's tertiary filtration plants.

FILTER MEDIA CHARACTERISTICS

The bacterial growth on the wastewater filtration media may cause clogging and result in serious filter operation problems. These problems

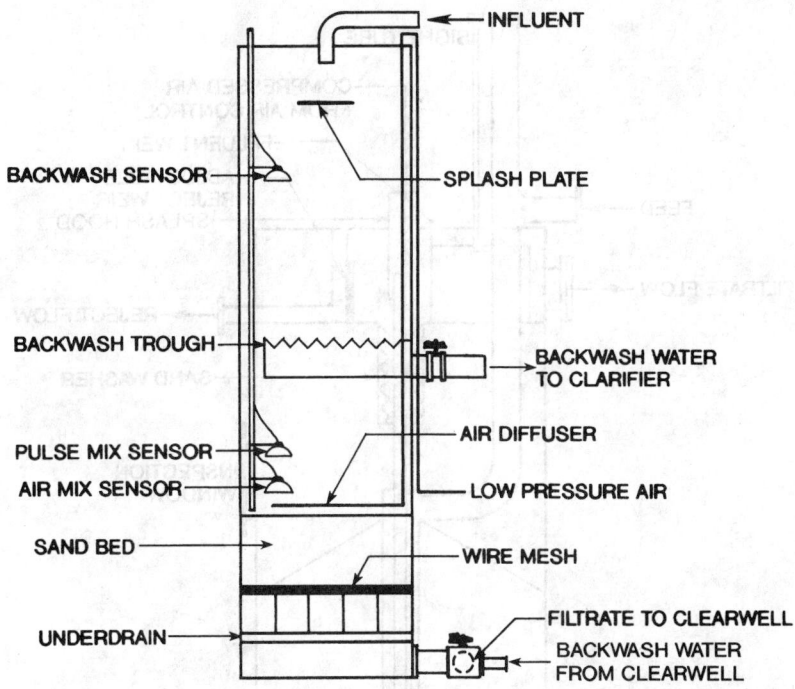

Figure 6.10 Schematic diagram of the HydroClear pilot filter.

may be minimized by selecting a media uniformity coefficient (U.C.) between 1.2 and 1.5 to obtain a more favorable porosity in the filter bed. This would allow a better distribution and passage for the liquid stream through the filter bed. Other factors, such as size and shape of filter media, total surface area of filter media per filter bed area, and ratio of bed depth to media size, are also considered important in filter media design and selection [12]. Currently, the anthracite and sand are the most common media used for wastewater filtration systems. The effective size (E.S.) normally falls between 1.0 and 1.2 mm with a bed depth of 24 in (0.6 m) for anthracite, and 0.5 and 0.6 mm with a depth of 12 in (0.3 m) for sand in a conventional dual-media filter. However, the U.C. and E.S. may vary substantially for other media and filter designs. Table 6.3 shows the specifications for the filter types and their media characteristics currently accepted by the State of California for WRPs with direct filtration.

FILTRATION RATES

The filtration rate is important because the required filter surface area would decrease if the rate can be increased. However, in order to maximize

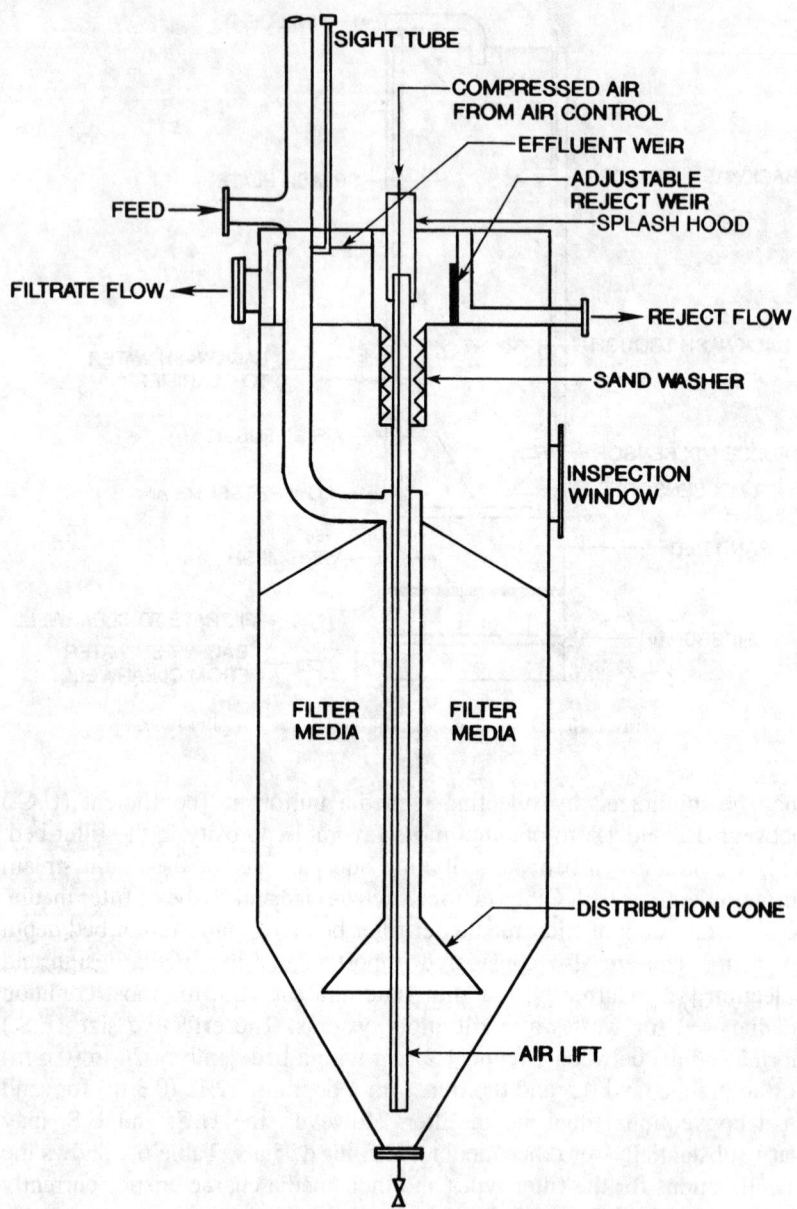

Figure 6.11 Schematic diagram of the DynaSand pilot filter.

the efficiencies of particle removal through a filtration system, it is necessary to limit the filtration rate to allow an adequate detention time for providing enough contact chance between the particles and the filter media. The limitation of the filtration rate can also avoid premature particle breakthrough. Figure 6.12 shows the impact of filtration rate on the particle concentration of a 6-ft (1.8-m) pilot-plant anthracite filter effluent. The effluent particle concentration increases with the filtration rate (The concentration at 3.5 gpm/ft^2 was used as the baseline). Under low influent turbidity condition, the impacts of filtration rates on the removals of turbidity and suspended solids are not as obvious. These two parameters are insignificantly affected by the filtration rate, varying within the range of 3 to 5 gpm/ft^2 (7.3 to 12.2 m/h), in the daily operation of the CSDLAC's tertiary filtration plants.

Depending on filter design configurations, wastewater filtration rates may vary substantially. However, a filtration rate of 3.5 gpm/ft^2 (8.5 m/h) is commonly practiced for a gravity dual-media or anthracite filter system, while a filtration rate of 5 gpm/ft^2 (12.2 m/h) is normally practiced for a pressure dual-media or mixed-media filter system. For filters other than the types listed in Table 6.3, their acceptable filtration rates are to be determined by pilot-plant studies as required by the State of California [8].

BACKWASH MECHANISMS

The characteristics of filter influent, filter designs, filter pretreatment methods, and efficiency of filter backwashing will cause the filter runs varying from a couple of hours to several days. A good practice is to maintain a filter run of 24 to 48 hours to achieve a 95% or higher water recovery. An extended filter run is often not desirable because of the potential of causing an excessive biological growth and accumulation of suspended solids in the filter bed.

A filtration system is normally provided with a surface wash mechanism and/or an air-scouring mechanism in its filter backwashing system. For good filter influent waters with low turbidity, an effective surface wash followed by a backwash is usually adequate to clean the filter bed. For filters with high turbidities and suspended solids in their influents, an air-scouring system is necessary to assure an adequate filter cleaning. The air-scouring mechanism is usually operated without simultaneously using backwash water in the beginning of the backwash cycle. The air scouring may continue to operate while backwash water is on later to thoroughly scour and wash away the well attached and deeply trapped suspended solids. Only a surface wash mechanism is provided in the CSDLAC's dual-media pressure filters, while an air-scouring system, with or without a surface wash mechanism, is installed in most of the CSDLAC's dual-media or anthracite gravity filters. Both fixed grid and rotary designs are

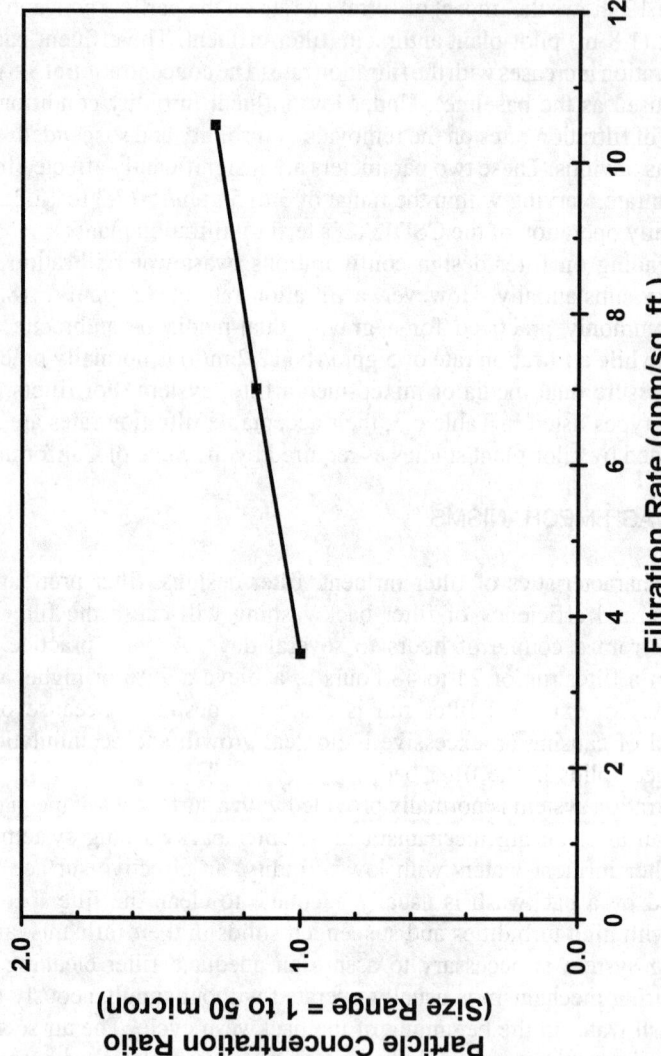

Figure 6.12 Impact of filtration rate on effluent particle concentration.

used among these tertiary filter surface wash systems. All surface wash systems are designed with orifices drilled 15 degrees downward from horizontal.

The backwash water rate is normally between 15 to 20 gpm/ft^2 (36.4 to 48.8 m/h). The water used per backwash cycle is usually about 5 times of the filter bed volume. The air-scouring rate is about 5 SCFM/ft^2 (1.5 m/min) of filter surface area. The total backwash wastewater volume is about 5% or less of the daily filter plant flow. The high flux volume of backwash water is temporarily collected in a wastewater storage tank or a specially designated final clarifier for wastewater recovery. The collected backwash water is then pumped to the influent distribution channel of the final clarifiers at a regulated flow rate. An appropriate dose of cationic polymer, 0.1 to 0.3 mg/L, may be added to the backwash water recovery stream to enhance settling of the particles, if necessary. The backwash water may also be pumped to the aeration tanks through the foam spray lines in some plants.

Since filter backwash is a very important step for the filter operation and performance, some alternative backwash mechanisms have been developed, such as pulse-bed, traveling-bridge, and continuous-backwash filter bed cleaning systems. These new nonconventional backwash systems are designed to extend the filter run, to minimize filtration interruption, or to achieve concurrent filtration and backwash operations without interrupting the service. Some of these new backwash systems may not require wastewater storage tanks for recovery.

PRETREATMENT REQUIREMENTS

NEED FOR INFLUENT TURBIDITY CONTROL

The tertiary filtration systems in CSDLAC are used to treat the well-oxidized secondary effluents from activated sludge plants. Most of these plants are equipped with fine bubble diffusers and are operated with step-feed mode of operation. Figure 6.13 shows the typical distribution of secondary effluent turbidity values for PWRP, SJCWRP and VWRP. As indicated in the figure, approximately 50% of the turbidity samples of each plant showed a secondary effluent turbidity range of 2 to 2.9 NTU. The occurrences for turbidity higher than 5 NTU were very infrequent. This is attributed to the control measures taken to prevent the secondary effluent turbidity from getting higher than 5 NTU. The maximum secondary effluent turbidity acceptable by each specific filter type to comply with the effluent turbidity standard of 2 NTU is found to be different. According to the long term average of the CSDLAC's filter performances,

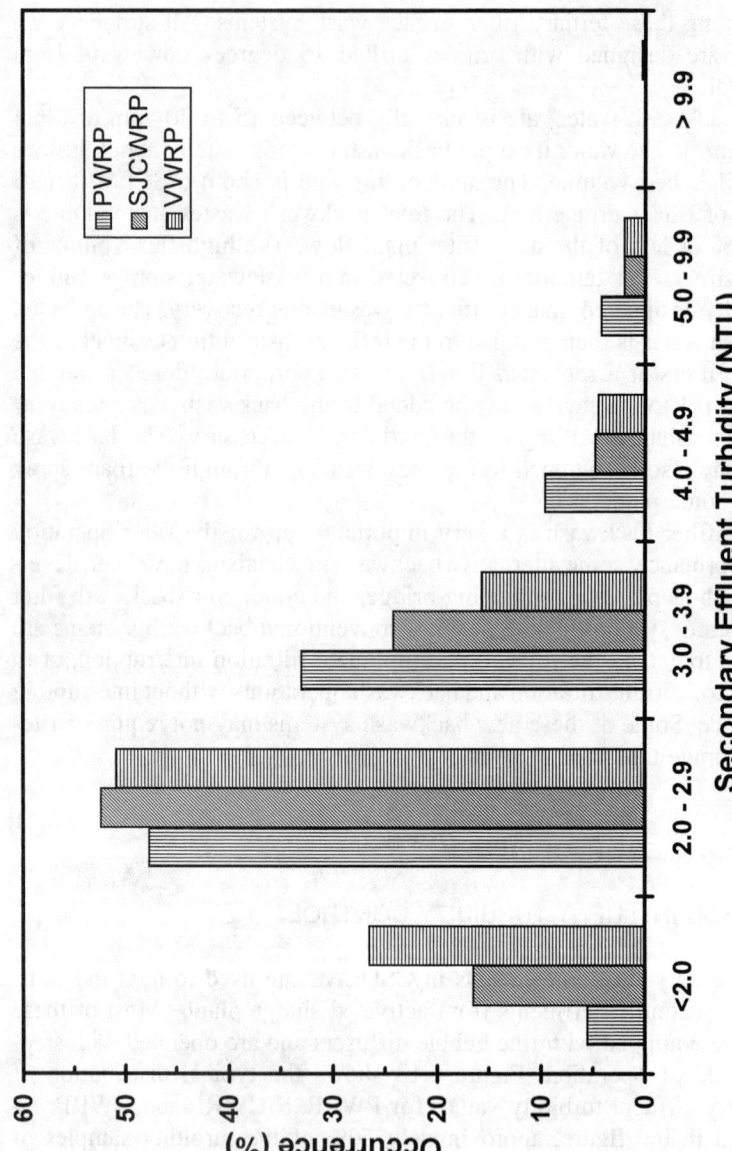

Figure 6.13 Occurrence of secondary effluent turbidity values in the specified ranges.

a limit of secondary effluent turbidity of 5 NTU is necessary for a typical dual-media gravity filter to achieve a filter effluent turbidity of 2 NTU, while a limit of 6 NTU seems to be acceptable for the dual-media pressure and anthracite gravity filters. Recent pilot studies indicated that the limit of secondary effluent turbidity for a pure-oxygen activated sludge plant could be slightly higher for tertiary filters to achieve the same 2 NTU goal [11].

The limitation of the filter capability requires that an appropriate and effective pretreatment be applied to the secondary effluent to assure the achievement of the filter effluent turbidity goal. The extent and method of pretreatment are dependent on the secondary effluent quality, type of filter, and the filter effluent turbidity limit. The general practices adopted by the CSDLAC in their pretreatment for tertiary filter operations are described in the following sections.

IMPROVEMENT OF PARTICLE FILTERABILITY

Most of the particles in the secondary effluents of activated sludge plants are very fragile biological microflocs, organic colloids, and cell debris. The surface characteristics of these particles need to be chemically modified to enhance their removals through the filters.

When the secondary effluent turbidity is less than 5 NTU, the particle concentration is generally low, the water can be fed into the filter system with low dosage of alum and followed by an appropriate amount of anionic polymer as coagulant-aid. These coagulants are added directly into the filter influent line and static in-line mixers are used to accomplish necessary mixings. The dosage of alum is normally limited to 15 mg/L, while the anionic polymer is about 0.05 mg/L. Higher dosage of alum may cause breakthrough of the alum flocs which will contribute to effluent turbidity. An excessive dosage of alum may also cause post-precipitation in the liquid stream after passing through the filter.

The addition of alum may partially neutralize the surface charges of the influent particles and the filter media to facilitate the attachment of these particles onto the media. The alum may also form positively charged aluminum hydroxide microflocs, which can effectively adsorb the submicron particles and then be removed altogether by the filter.

REDUCTION OF PARTICLE CONCENTRATION

When the secondary effluent turbidity exceeds 5 NTU, the particle concentration is usually too high to rely solely on a direct or contact filtration to meet effluent turbidity standard of 2 NTU. Therefore, an appropriate amount of cationic polymer (up to 2 mg/L) is usually added near the exit end of the aeration tanks. This will provide adequate mixing for coagulation and flocculation of the biological flocs to achieve an efficient settling in the final clarifiers. The objective of this pretreatment step is to control the secondary effluent turbidity to less than 5 NTU. This chemically improved secondary effluent can be further treated with the same low alum pretreatment described above, if necessary, to enhance the filterability of the remaining particles.

STABILIZATION OF BIOLOGICAL OXIDATION PROCESS

If the secondary effluent turbidity exceeds 10 NTU, the biological oxidation process is usually out of effective control. After applying the above cationic polymer treatment, if the secondary effluent turbidity is still higher than 5 NTU, a significant plant upset situation must be prevailing. This may be caused by an unlawful industrial waste discharge or a temporary plant operation problem. Therefore, an immediate adjustment of the plant operation parameters is necessary to prevent a drastic deterioration of the biological process from occurring. The adjustments may involve rate changes in aeration and returned activated sludge flow, modifications of step-feed pattern, and others as necessary. If these operation changes fail to quickly and effectively correct the plant performance, the secondary or filter effluent will be temporarily returned to the downstream sewer system until the secondary effluent turbidity can be reduced to 5 NTU or less. The wasted effluent will be treated by downstream treatment plants.

After the biological oxidation process is stabilized, the pretreatments mentioned above will be applied in accordance with the actual need to achieve a filtered effluent of 2 NTU or less.

BIOLOGICAL GROWTH CONTROL

Serious biological growth could occur within the filter bed for wastewater filtration. The extent of growth could be related to the influent water quality, type of filter media, filter backwash effectiveness, length of filter run, and degree of prechlorination. The growth causes an impact on the headloss of the filter operation. The filter run may be shortened from a few days to a few hours without prechlorination. The heavy growth may also cause the filter effluent turbidity to fluctuate due to slough-off of the biological films.

An occasional shock treatment of the heavy growth situation with chlorination may cause the biological films to slough off substantially and reduce the filter headloss. However, this kind of treatment will create a temporary high flux of suspended solids in the filter effluent and, consequently, the effluent turbidity standard may be violated.

A continuous prechlorination, with either a full or fractional dosage required for complete disinfection, is able to effectively control the biological growth in a dual-media or a mono-medium anthracite filter system. However, the prechlorination was found to cause a slight reduction of suspended solids removal by the filter. This effect may become so significant in a shallow-depth filter, such as HydroClear filter, that prechlorination is not appropriate for its operation. The prechlorination is not effective for an activated carbon filter neither, because the adsorption of chlorine or chloramines by the carbon in the top layer makes the disinfectant not available for the bulk of the filter media for growth control. A yearly thermal regeneration of the carbon media seemed to be adequate to maintain a satisfactory operation of the carbon filters for the purpose of turbidity control [17].

FILTER PERFORMANCE

TURBIDITY AND SUSPENDED SOLIDS REMOVAL

During the initial period of filter startup operations, the filter effluent turbidity standard of 2 NTU or less was only met at approximately 90% of the time. This level of compliance was very much consistent with the Pomona Virus Study results. Figure 6.5 indicates that the direct filtration with a dual-media filter (System B) could achieve a filter effluent turbidity of 2 or less for 90% of samples taken. No significant improvement in the degree of compliance was ever accomplished by simply manipulating the filter operation parameters. With the realization of this limited capability of filter performance, efforts were thus made to control the filter influent turbidity with proper pretreatments. The pretreatment strategies as discussed in the above sections have since been generally adopted and practiced by all the tertiary filtration plants of the CSDLAC to fully achieve the compliance of the effluent turbidity standard.

Table 6.7 compares the suspended solids and turbidity removals by the four typical tertiary filtration systems. The suspended solids removal is shown to be generally higher than the turbidity removal in all four types of tertiary filtration plants. According to the study of Levine et al. [13], particles smaller than 1.2 μm are not detectable by the standard suspended solids test. Therefore, the presence of sub-micron particles may significantly increase the turbidity value but not the suspended solids concentration.

Higher percentage removals of suspended solids and turbidity could be achieved by the filters with higher influent turbidities. Table 6.8 shows

TABLE 6.7. Comparison of Suspended Solids and Turbidity Removals.

| | PWRP | | | | SJCWRP | | VWRP | |
| | Carbon (Gravity) | | Anthracite (Gravity) | | Dual Media (Gravity) | | Dual Media (Pressure) | |
	Turbidity (NTU)	SS (mg/L)	Turbidity (NTU)	SS (mg/L)	Turbidity (NTU)	SS (mg/L)	Turbidity (NTU)	SS (mg/L)
Filter influent (median)	2.9	7	3.2	8	2.6	7	2.3	6
Filter effluent (median)	1.0	1	1.4	2	1.5	2	1.2	2
Percent removal efficiency (median)	66.7	83.3	55.2	75.0	41.0	71.4	46.4	75.0
No. of data	1908	1908	344	344	2189	2189	1730	1730

TABLE 6.8. Effect of Secondary Effluent Turbidity on SS and Turbidity Removals.

Secondary Effluent Turbidity (NTU)	PWRP				SJCWRP		VWRP	
	Carbon (Gravity)		Anthracite (Gravity)		Dual Media (Gravity)		Dual Media (Pressure)	
	Turbidity (%)	SS (%)	Turbidity (%)	SS (%)	Turbidity (%)	SS (%)	Turbidity (%)	SS (%)
<2.0	58.8	83.3	45.6	50.0	25.0	66.7	31.6	66.7
2.0–2.9	64.3	83.3	50.0	66.7	37.5	66.7	46.0	75.0
3.0–3.9	68.6	81.8	54.8	71.4	50.0	75.0	60.0	77.8
4.0–4.9	70.5	82.4	65.3	80.0	61.7	83.3	69.1	84.2
5.0–6.9	67.9	83.3	65.6	83.3	73.1	86.7	72.1	88.7
7.0–9.9	81.0	89.5	73.3	88.2	81.3	89.5	84.0	94.6
>9.9	93.5	95.0	89.4	93.3	86.0	62.5	89.3	92.3

the effect of secondary effluent turbidity on percentage removals of suspended solids and turbidity by the tertiary filtration plants.

CONSISTENCY OF EFFLUENT TURBIDITY

The filter effluent turbidity generally fluctuates with the changes of the corresponding influent turbidity. The degree of fluctuation seems to be greatly influenced by the type of filter media and filter depth. The activated carbon filter with 6-ft (1.8-m) depth produced the most consistent effluent turbidity. The mono-medium anthracite filter with a depth of 6 ft (1.8 m) was able to maintain a reasonably consistent effluent turbidity; while both gravity and pressure dual-media filters, having 3-ft (0.9-m) depth, seemed to be more responsive to the fluctuations of the influent turbidities.

Controlling the filter influent turbidity with the pretreatment methods is considered as the most reliable way to assure a consistent and acceptable filter effluent turbidity. Consistently good quality water will greatly enhance disinfection to minimize associated health risk for water reuse. With the effective pretreatments and excellent plant operations, the CSDLAC have been able to maintain an outstanding compliance with the effluent turbidity and disinfection standards.

IMPACT ON PARTICLE SIZE DISTRIBUTION

A particle size distribution study was conducted among several of the tertiary treatment plants of the CSDLAC. The particle counter used for the study had a size range of 1.2 to 400 μm. Samples for the particle size distribution analyses were conducted within 30 minutes of sampling, using a Hiac/Royco model 8000 particle counter equipped with a laser-diode light obscuration (light blocking) sensor model HRLD-400 (Hiac/Royco, Silver Spring, MD).

As indicated in Table 6.4, the total particle count within the size range of 1.2 μm to 200 μm for the three secondary effluents varied from 3,570 to 5,691 particles/mL, while for the four filter effluents it varied from 847 to 1,772 particles/mL. The individual particle removals by the activated carbon, anthracite, dual-media pressure, and dual-media gravity filters were 79.0%, 77.4%, 76.3% and 57.6%, respectively. The overall average particle removal by these filters was approximately 71.4%.

According to the results in Table 6.4, approximately 99.9% of the particles accountable by the particle counter were within the size range of 1.2 μm to 100 μm for the secondary effluents, while about 99.9% of the particles in the filter effluents were within the size range of 1.2 μm to 50 μm. There were a great number of particles having size less than 1.2 μm which were not accountable by the particle counter used in this study. These sub-micron particles might not affect the suspended solids concentration values. However, they could impact the turbidity readings.

Table 6.9 shows the average ratios of suspended solids concentration to the turbidity for different types of tertiary filtration systems at CSDLAC. The ratios for the four secondary effluents ranged from 2.3 to 2.97, while the ratios for the four final effluents ranged from 1.44 to 1.47. Assuming the specific gravities of the particles in the secondary and filtered final effluents were equal, then a smaller suspended solids concentration to turbidity ratio might imply a greater proportion of finer particles in the effluent. Therefore, the final effluents could contain a higher proportion of finer particles than the secondary effluents. It also indicates that the filters could preferentially remove larger particles as expected. The column marked as ''Final'' in Table 6.9 closely represents the filtered effluent results, since the impacts of both chlorination and dechlorination, which were the intermediate processes between the filter and the final effluents, on the suspended solids concentration and turbidity were found to be insignificant. The additional removals of suspended solids and turbidity by the chlorination/dechlorination processes were found to be only 1 to 2% [14].

The β values listed in Table 6.4 indicates that the sub-micron particles cause a more significant impact on the turbidity values of the filter effluents than those of the secondary effluents. As shown in the table, the β values of the secondary effluent ranged from 2.49 to 2.90, while the filter effluents from 3.05 to 3.41. When β is smaller than 3, the amount of transmitted light for turbidity measurement is primarily controlled by particles with sizes of 5 μm or larger. For β values larger than 3, the particles in the colloidal range of 0.01 μm to 2 μm become responsible [15]. This effect is also clearly demonstrated in Figure 6.9.

The β value is defined as the empirical constant for the following two-parameter power law distribution function [16]:

$$\frac{\mathrm{d}N}{\mathrm{d}l} = A \times (l)^{-\beta}$$

where
  N = the particle count per unit volume
  l = the arithmetic mean particle diameter
A and β = empirical constants

RECOVERY OF BACKWASH WATER

The waters used to backwash the filters are typically drawn from the filter effluent wet well. Some plants have the capability of drawing the

TABLE 6.9. Ratio's of Suspended Solids to Turbidities.

| | PWRP | | | | SJCWRP | | VWRP | |
| | Activated Carbon (Gravity) | | Anthracite (Gravity) | | Dual Media (Gravity) | | Dual Media (Pressure) | |
	Sec.	Final	Sec.	Final	Sec.	Final	Sec.	Final
No. of data	1908	1588	344	344	2189	2163	1730	1485
SS: Turbidity ratio	2.46	1.46	2.31	1.44	2.97	1.47	2.90	1.47

254

water from the chlorine contact tank, if the need is beyond the effluent flow of the remaining operating filters and the storage capacity of the effluent well. Since prechlorination is practiced by the CSDLAC for their tertiary filter systems, both sources of backwash waters are chlorinated with an average chlorine residual of 5 mg/L.

The amount of backwash wastewater produced per backwashing cycle for each filter is approximately five times of the filter bed volume. Depending on the length of filter run, the volumetric ratio of the backwash wastewater to the filtered effluent may range from 2% to 8%, with an average of 5%.

The turbidities of the backwash wastewaters normally start out with a high turbidity range of 150 to 300 NTU, the values are then rapidly reduced to the level of 3 NTU after approximately three filter volumes of wash waters have passed through the filters. The 3 NTU level maintains quite stable for the remaining two bed volumes of filter backwashing flow. The entire water backwashing period usually lasts about 8 minutes, while the overall backwashing cycle, including draining, air scouring or surface washing, and water backwashing, may take 15 to 30 minutes.

Basically the backwash wastewaters can be recovered without going back to the plant headwork for another cycle of full treatment. The CSDLAC apply the following means for recovering the backwash wastewaters:

(1) The backwash wastewater is pumped to a decant tank with addition of appropriate amounts of cationic polymer for coagulation/flocculation and sulfur dioxide for dechlorination. The effluent from the decant tank is mixed with the effluents from the secondary clarifiers for further treatment by filtration.

(2) The backwash wastewater is collected in an underground waste storage tank or a specifically designated secondary clarifier. A regulated constant flow pump is then used to transfer the backwash wastewater to the aeration tanks through the existing foam spray lines and nozzles. No flocculating agents are normally added in the stream being recovered. However, sulfur dioxide solution may be added to remove the chlorine residual to minimize toxicity to the microorganisms in the aeration process.

(3) The backwash wastewater is collected first in an underground storage tank. A pump is then used to lift the wastewater to the distribution channel of the secondary clarifiers at a regulated constant flow. Appropriate amounts of cationic polymer and sulfur dioxide are injected into the wastewater transfer line for enhancing settling and minimizing chlorine toxicity to the microorganisms in the returned activated sludge.

Depending on the size of the storage or decant tank in relation to the amount of backwash wastewater flow, a temporary small increase of secondary effluent turbidity may occur during the wastewater recovery period. A proper adjustment of the coagulant feed rate would control this situation without causing any violation of the stringent effluent discharge turbidity standard.

ORGANICS REMOVAL

The granular activated carbon filter can serve a dual-role as a tertiary filter and an adsorber of organics. However, the organics removal efficiency depends greatly on the frequency of carbon regeneration. The expensive cost of regeneration may render the carbon filter disadvantageous to be used only as a tertiary filter, in spite of its better performance than the other types of filters with inert media. Nevertheless, if organics removal is required for a tertiary treatment to meet higher water reuse requirements, then the carbon filter may be considered cost-effective for achieving both turbidity and organics removals.

A 10 MGD (0.44 m³/s) granular activated carbon plant was operated by the CSDLAC at their PWRP from 1977 to 1992 to provide color removal in addition to meeting the effluent discharge and general water reuse requirements. The activated carbon filtration was to meet the demand of the City of Pomona Water Department (CPWD) for purchasing this higher quality reclaimed water, with color of 10 units or less, to meet the requirements of their major industrial reusers [10]. The activated carbon filters were replaced by the anthracite filters in 1992, because color removal is no longer demanded by the CPWD [17]. Figure 6.14 shows the typical removal efficiencies for the total COD (TCOD), dissolved COD (DCOD), apparent color (AC), and true color (TC) by the carbon filters at the PWRP during the twelve-month period of May 1986 to April 1987. During this period, the carbon was on a twice-a-year regeneration program, one at the beginning of May, 1986 and the other approximately 6 months later in October, 1986. This regeneration schedule provided a minimum frequency needed to meet the color requirement of 10 units or less. On the other hand, a yearly regeneration cycle was found to be adequate for operating the carbon filters for the sole objective of effluent turbidity control.

In 1991, four anthracite filters were added to the carbon filtration plant at the PWRP to accommodate the increased plant flow needed for an expanded water reuse program. This addition offered an opportunity to study the carbon and the anthracite filters on a side-by-side basis. The anthracite was found to have little or no adsorption capacity for true color removal. However, the anthracite filter did reduce the apparent color of the secondary effluent. This reduction was attributed to the removal of

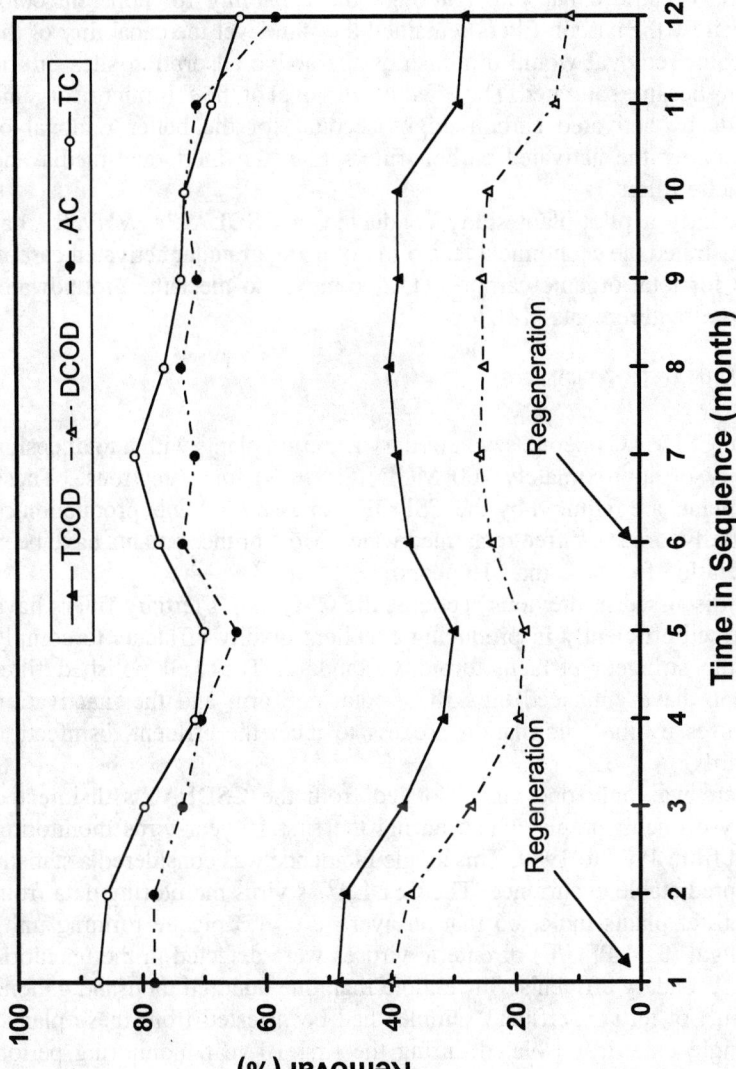

Figure 6.14 Typical organics removal efficiencies by the granular activated carbon filters at PWRP.

257

the fine particles which contributed to the apparent color values. The reduction was greater when apparent color in the secondary effluent was higher. The similar effects were exhibited by the carbon filters when the carbons became exhausted. Although the capability for apparent color removal by the carbon filters remained the same, yet the capability of the true color removal would diminish as the active adsorption sites on the carbons became saturated. The effective removal of the sub-micron organic colloids by activated carbon might account for the better removal of turbidity by the activated carbon filters than the inert dual-media and anthracite filters.

Recently a pilot-plant study conducted at CSDLAC's WNWRP has demonstrated the economical feasibility of using granular activated carbon filters for total organic carbon (TOC) removal to meet the groundwater recharge requirements [18].

VIRUS INACTIVATION

The CSDLAC operate seven tertiary filtration plants with a total design capacity of approximately 200 MGD (8.8 m³/s) for water reuse. These filter plants are required by the CSDOHS to assure reliable production of essentially virus risk-free reclaimed water. Most of these plants have been in operation for more than 16 years.

As discussed in previous sections, the CSDLAC's tertiary filters have performed efficiently in producing excellent quality effluents to comply with the stringent effluent turbidity standard. The well polished filter effluents have enhanced the kill of total coliform and the inactivation of viruses by the chlorination process to meet the effluent disinfection standards.

There was only one virus isolated from the CSDLAC's disinfected tertiary effluent samples taken during the first 10-year virus monitoring period from 1979 to 1989. This single incidence was considered a statistically predictable occurrence. The CSDLAC's virus monitoring data from these filter plants indicated that an average of 1.3 plaque forming units (PFU)/gal (0.34 PFU/L) of enteric viruses were detected in the unchlorinated secondary effluents. Since more than one hundred thousand gallons (380 m³) of tertiary effluent samples had been tested from these plants, the single one virus isolated during the first 10-year monitoring period was thus considered to be consistent with the results of the Pomona Virus Study, which concluded that a 5-log virus removal could be achieved by WRPs with direct filtration and chlorination processes [19]. The continuous virus monitoring program since 1989 has indicated negative virus results for all CSDLAC's tertiary effluents. The outstanding virus inactivation results accomplished by these tertiary treatments have confirmed the ade-

quacy of the existing California wastewater reclamation plant design criteria for producing virus risk-free effluents for water reuses. In addition to
effective virus inactivation, the total coliform disinfection standard has
also been fully complied with.

COST OF WASTEWATER FILTRATION

Cost analyses have been prepared to compare the total process costs
for operating the different types of tertiary filtration systems at the
CSDLAC for water reclamation and reuse.

CRITERIA FOR COST ANALYSIS

Design Criteria

The various design factors used for the process cost analysis are basically
similar to the factors shown on Table 6.5. A plant flow of 10 MGD (0.44
m³/s) is used as a typical plant size for this cost estimate. The chemical
and equipment costs for filtration pretreatment to accomplish the effluent
turbidity standard are included in the process cost estimates.

Unit Cost

The following unit costs for the various items are used in the process
cost estimates:

Electricity ($/kWh)	0.08
Compressed air, 100 cfm @10psi ($/hr)	0.35
Backwash water ($/1,000 gal)	0.13
Alum, as 7.8% Al_2O_3, ($/lb)	0.06
Polymer ($/lb)	1.00
Carbon ($/lb)	0.95
Anthracite ($/lb)	0.35
Sand ($/lb)	0.25
Labor ($/hr)	26.00
Carbon regeneration ($/lb)	0.50

Other Cost Factors

The following cost categories are based on appropriate percentages of
the total equipment (including the filter media) and construction costs:

Electrical	10%
Instrumentation	7%

TABLE 6.10. Cost Comparison for Four Types of Tertiary Filtration Plants.

	Dual Media (Pressure)	Dual Media (Gravity)	Anthracite (Gravity)	Carbon (Gravity)
Capital cost ($)				
Chemical feeding system	36,000	36,000	36,000	36,000
Pumping system	150,000	100,000	100,000	100,000
Filter shell/tank	1,000,000	1,200,000	1,200,000	1,200,000
Filter media	200,000	200,000	300,000	380,000
Electrical (10% of equipment cost)	138,600	153,600	163,600	168,000
Instrumentation (7% of equipment cost)	97,020	107,520	114,520	117,600
Contingencies (20% of equipment cost)	277,200	307,200	327,200	336,000
Engineering (15% of equipment cost)	207,900	230,400	245,400	252,000
Total capital cost ($)	2,106,720	2,334,720	2,486,720	2,553,600
Amortization at 20 yrs, 8% Interest ($/1000 gallons)	0.059	0.065	0.069	0.071
Operating and maintenance costs ($/1000 gallon)				
Chemical (alum and polymer)	0.012	0.012	0.012	
Power	0.025	0.015	0.015	0.025
Backwash water	0.011	0.005	0.015	0.022
Labor	0.012	0.010	0.012	0.011
Maintenance materials	0.009	0.010	0.011	0.011
Carbon make-up (7% loss)				0.006
Carbon regeneration (once per year @$0.50/lb)				0.044
Total O&M cost ($/1000 gallons)	0.069	0.052	0.065	0.120
Total treatment cost ($/1000 gallons)	0.128	0.117	0.134	0.191

| Contingencies | 20% |
| Engineering | 15% |

COST COMPARISONS

The results of the cost estimates for the four types of tertiary filtration systems are presented in Table 6.10. The total treatment costs for the activated carbon (gravity), anthracite (gravity), dual-media (pressure) and dual-media (gravity) are approximately $0.191, $0.134, $0.128 and $0.117 per 1,000 gallons (3.79 m^3) of filtered effluent, respectively.

The above decreasing order of the total treatment process costs is very much in the same trend of decreasing average filtered effluent turbidity removals. However, all four types of filter systems are capable of meeting the effluent turbidity standard. Therefore, the selection of any of these filter types for tertiary treatment is very much influenced by other considerations, such as land availability, need for organics removal, conformity with the existing filters, and so forth.

The O&M costs for the inert media tertiary filtration systems represent 10% to 15% of the total plant O&M costs (excluding sludge treatment and disposal costs). The actual percentage depends heavily on the aeration system designs, operation modes and treatment requirements of the water reclamation plants. The O&M costs of the activated carbon system represent approximately 25% of the total plant O&M costs.

The total process costs, including capital and O&M costs, for the HydroClear and the DynaSand filtration processes were reported to be in the same range of total process costs as the 6-ft (1.8-m) anthracite filter system for tertiary filtration [11]. This cost equivalency may allow these new emerging systems to be considered as alternative filter types for tertiary filtration plants.

REFERENCES

1 "Annual Status Report on Reclaimed Water Reuses" (1995). County Sanitation Districts of Los Angeles County, Whittier, California.
2 "A Plan for Water Reuse" (1963). County Sanitation Districts of Los Angeles County, Whittier, California.
3 "Pomona Virus Study—Final Report" (1977). County Sanitation Districts of Los Angeles County, Whittier, California.
4 "California Administrative Code, Title 22, Division 4, Environmental Health—Wastewater Reclamation Criteria" (1978). State of California, Department of Health Services, Berkeley, California.
5 Cookson, J. T. (1970). "Design of Activated Carbon Adsorption Beds," *J. Water Pollution Control Federation,* 42: 2124–2134.

6 "The Effects of Advanced Wastewater Treatment on Trace Organic Compounds," Publication No. 61, July 1978. California State Water Resources Control Board.

7 California State Department of Health Services' Internal Memorandum, dated April 28, 1978.

8 "Policy Statement for Wastewater Reclamation Plants with Direct Filtration," Department of Health Services, State of California, January 1988.

9 ASCE Task Committee on Design of Wastewater Filtration Facilities (1986). "Tertiary Filtration of Wastewaters," *J. Environ. Eng.*, 112(6): 1008–25.

10 Garrison, W. E., J. C. Gratteau, B. E. Hansen, and R. F. Luthy, Jr., (1978). "Gravity Carbon Filtration to Meet Reuse Requirements," *J. Environ. Eng. Div. Proc. ASCE*, 104: 1165.

11 Kuo, J. F., K. M. Dodd, C. L. Chen, R. W. Horvath, and J. F. Stahl (1995). "Evaluation of Tertiary Treatment Alternatives for Pure-Oxygen Activated Sludge Effluent," *Proceedings of the Water Environment Federation 68th Annual Conference & Exposition*, October 21–25, Miami Beach, Florida.

12 Kawamura, S. (1991). *Integrated Design of Water Treatment Facilities*. John Wiley & Sons, Inc., p. 197.

13 Levine, A. D., G. Tchobanoglous, and T. Asano (1985). "Characterization of the Size Distribution of Contaminants in Wastewater Treatment and Reuse Implications," *J. Water Pollution Control Federation*, 57: 805–816.

14 Kuo, J. F., C. L. Chen, J. F. Stahl, and R. W. Horvath (1994). "Evaluation of Four Different Tertiary Filtration Plants for Turbidity Control," *Water Environment Research*, 66: 879–886.

15 Kavanaugh, M. C., C. H. Tate, A. R. Trussell, R. D. Trussell, and G. Treweek (1980). "Use of Particle Size Distribution Measurements for Selection and Control of Solid/Liquid Separation Processes," *ACS Advances in Chemistry Series*, 189: 305.

16 Hargesheimer, E. E., C. M. Lewis, and C. M. Yentsch (1992). Evaluation of Particle Counting as a Measure of Treatment Performance. AWWARF, Denver, CO.

17 Kuo, J. F., J. F. Stahl, C. L. Chen and P. V. Bohlier (1994). "Gravity Activated Carbon Filtration for Water Reuse," *Proceedings AWWA/WEF Water Reuse Symposium*, Feb. 27–March 2, Dallas, Texas.

18 Argo, D., T. Lyon, J. Norman, and J. Helsley (1993). "GAC Treatment Studies to Optimize Groundwater Spreading at Whittier Narrows," *66th WEF Annual Conference*, October 3–7, 1993, Anaheim, California.

19 Yanko, W. A. (1993). "Analysis of 10 years of Virus Monitoring Data from Los Angeles County Treatment Plants Meeting California Wastewater Reclamation Criteria," *Water Environment Research*, 65: 221–226.

CHAPTER 7

Reverse Osmosis Membranes for Wastewater Reclamation

INTRODUCTION

THE RO (reverse osmosis) technology started as a scientific experiment at the University of Florida in the 1950s where Reid and Breton [1] were able to demonstrate desalination properties of a cellulose acetate film. After the development of the first asymmetric membrane material from cellulose acetate by Loeb and Sourirajan in 1950s [2,3], the subsequent progress included the development of better membrane chemistry, development and optimization of membrane module configurations and the improvement of the process and RO system design. Within the next four decades of continuous development, inventions and improvements, the RO process has been transferred from a scientific curiosity into a self supporting, rapidly growing industry. The scientific experiment of the 1950s, which produced a few drops of desalted water per hour, resulted today in a worldwide network of RO plants of combined desalting capacity of about 1.7 billion gallons per day (equivalent to the approximate capacity of 2 million acre-foot/year). Reverse osmosis technology is used today in large municipal plants to produce potable water quality from brackish and seawater sources, reclaim contaminated water sources and reduce water salinity for industrial applications. At the other end of the application spectrum reverse osmosis membrane elements are used in small under the sink units to produce a few gallons per day of drinking water. A wide variety of membrane material chemistry and membrane module configura-

Mark Wilf, Director, Technical Support, Hydranautics, 401 Jones Rd., Oceanside, CA 92054-1216.

TABLE 7.1. Evolution of Performance of Flat Sheet Brackish Membranes.

Year	69	75	82	82	90	95
Membrane type	CA asymm.	CA asymm.	CA asymm.	PA comp.	PA comp.	PA comp.
Specific flux, gfd	0.013	0.031	0.050	0.070	0.120	0.240
Salt rejection, %	96.0	98.0	98.0	98.5	99.5	99.0
Salt transport, cm/sec	1.5E-5	1.2E-5	1.9E-5	2.0E-5	5.8E-6	6.0E-6
NDP required for flux 15 gfd, psi	1150	483	300	214	125	62.5

Note: CA asymm.—cellulose acetate membrane material, asymmetric structure of membrane layer; PA comp.—polyamide membrane material, composite structure. Ultra-thin membrane barrier cast on a porous support.

tion have been developed over the years. Current commercial membrane modules are almost exclusively made of cellulose acetate or aromatic polyamide membrane materials in spiral wound and hollow fiber configurations. Significant efforts were made to reduce investment and operating cost of desalination systems. Better understanding of feed water quality requirements and the introduction of organic scale inhibitors resulted in the simplification of the feed water pretreatment process and the increase of feed to permeate conversion ratio. The development of high flux membrane elements and the incorporation of variable speed drivers and power recovery equipment into RO system design resulted in significant reductions of specific power consumption. One of the remaining unresolved problems is the effective control of biofouling in RO systems equipped with polyamide membrane elements. This is especially important in applications involving treatment of feed water from surface or waste water sources. The initial cellulose acetate membrane manufactured in late sixties had a specific flux of about 0.013 gfd per psi of net driving pressure and salt transport coefficient of 1.5E-05 cm/sec. The early RO membranes required a net driving pressure of over 1000 psi in order to produce a permeate flux rate of 15 gfd. The latest polyamide brackish water membranes have specific flux of 0.24 gfd/psi and salt transport coefficient below 1E-05. The corresponding net driving pressure required to produce a flux rate of 15 gfd is only 62 psi with higher salt rejection than the initial CA membranes. This improvement of specific permeate flux translates in over a twenty fold reduction of the specific power consumption of the RO process pumps. The evolution of membrane performance is summarized in Table 7.1 and Figure 7.1 and 7.2. A comprehensive review of early development work and theory of RO process is included in a book edited by Merten [4]. The

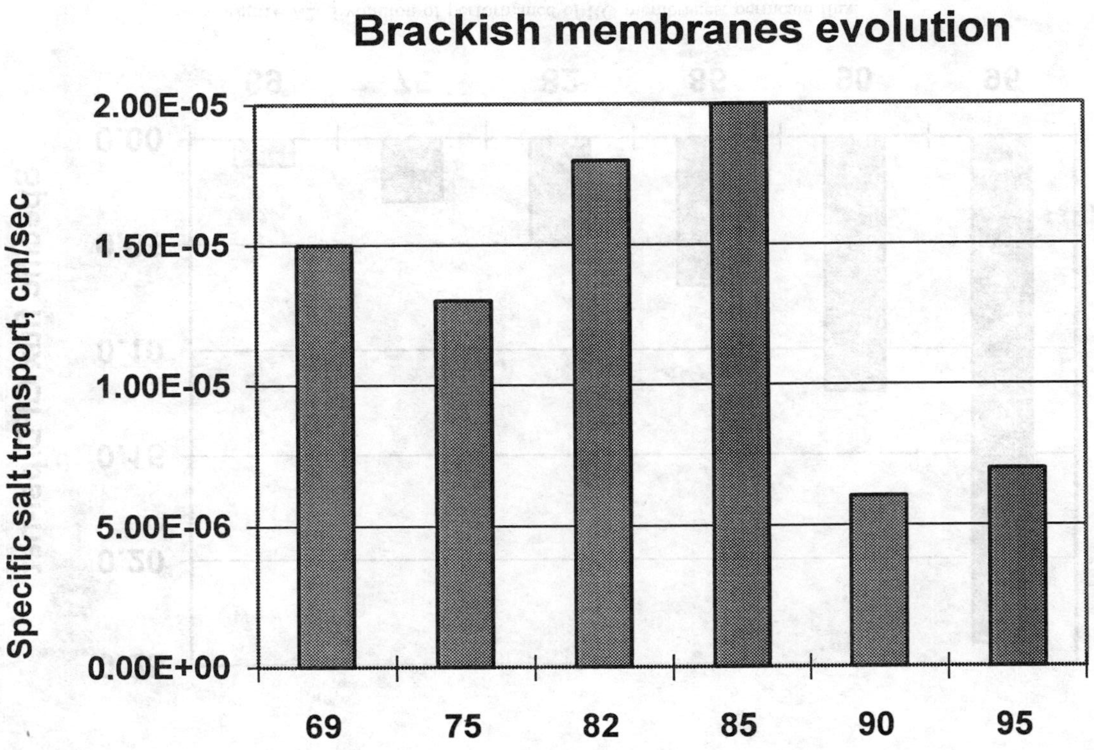

Brackish membranes evolution

Specific salt transport, cm/sec

Figure 7.1. Evolution of performance of RO membranes, salt transport.

265

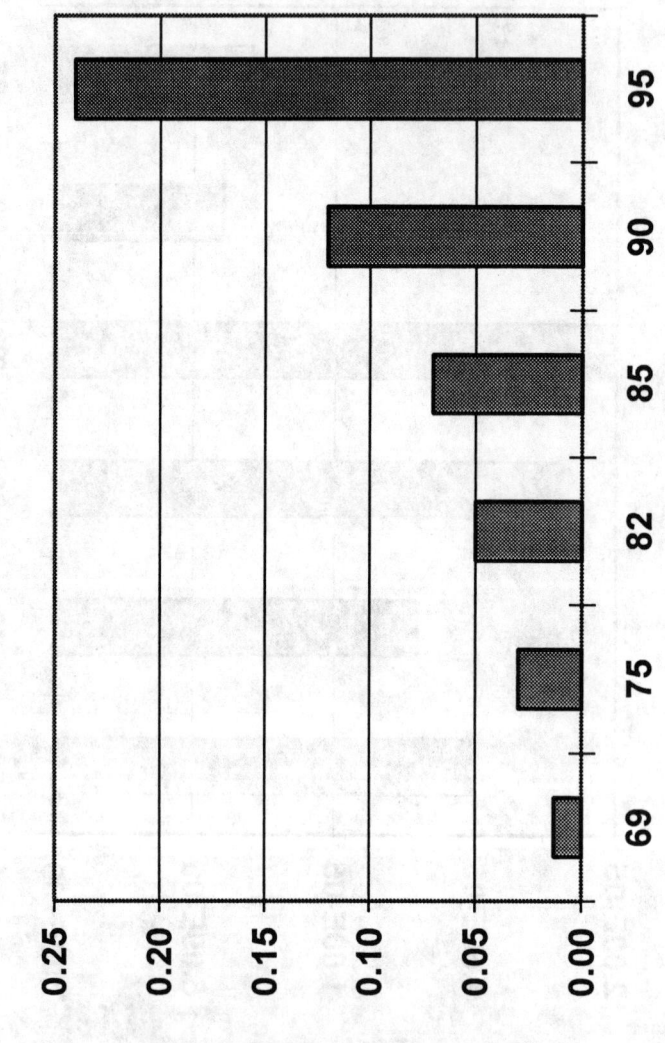

Figure 7.2. Evolution of performance of RO membranes, permeate flux.

early review of RO applications and RO systems design is included in publication edited by Buros [5].

INTRODUCTION TO REVERSE OSMOSIS

Osmosis is a natural process involving fluid flow across a semipermeable membrane barrier. It is selective in the sense that the solvent passes through the membrane at a faster rate than the dissolved solids. The difference of passage rate results in solvent solids separation. The direction of solvent flow is determined by its chemical potential, which is a function of pressure, temperature, and concentration of dissolved solids. Pure water in contact with both sides of an ideal semipermeable membrane at equal pressure and temperature has no net flow across the membrane because the chemical potential is equal on both sides. If a soluble salt is added on one side, the chemical potential of this salt solution is reduced. Osmotic flow from the pure water side across the membrane to the salt solution side will occur until the equilibrium of chemical potential is restored [Figure 7.3(a)]. Equilibrium occurs when the hydrostatic pressure differential resulting from the volume changes on both sides is equal to the osmotic pressure. This is a solution property independent of the membrane. Application of an external pressure to the salt solution side equal to the osmotic pressure will also cause equilibrium. Additional pressure will raise the chemical potential of the water in the salt solution and cause a solvent flow to the pure water side, because it now has a lower chemical potential. This phenomenon is called reverse osmosis [Figure 7.3(b)].

OSMOTIC PRESSURE

The osmotic pressure, P_{osm}, of a solution can be determined experimentally by measuring the concentration of dissolved salts in solution:

$$P_{osm} = 1.19 \ (T + 273) * \sum(m_i) \qquad (1)$$

where P_{osm} = osmotic pressure (in psi), T is the temperature (in °C), and $\sum(m_i)$ is the sum of molal concentration of all constituents in a solution. An approximation for P_{osm} may be made by assuming that 1000 ppm of total dissolved solids (TDS) equals about 11 psi (0.76 bar) of osmotic pressure. The mechanism of water and salt separation by reverse osmosis is not fully understood. Current scientific thinking suggests two transport models: porosity and diffusion. That is, transport of water through the membrane may be through physical pores present

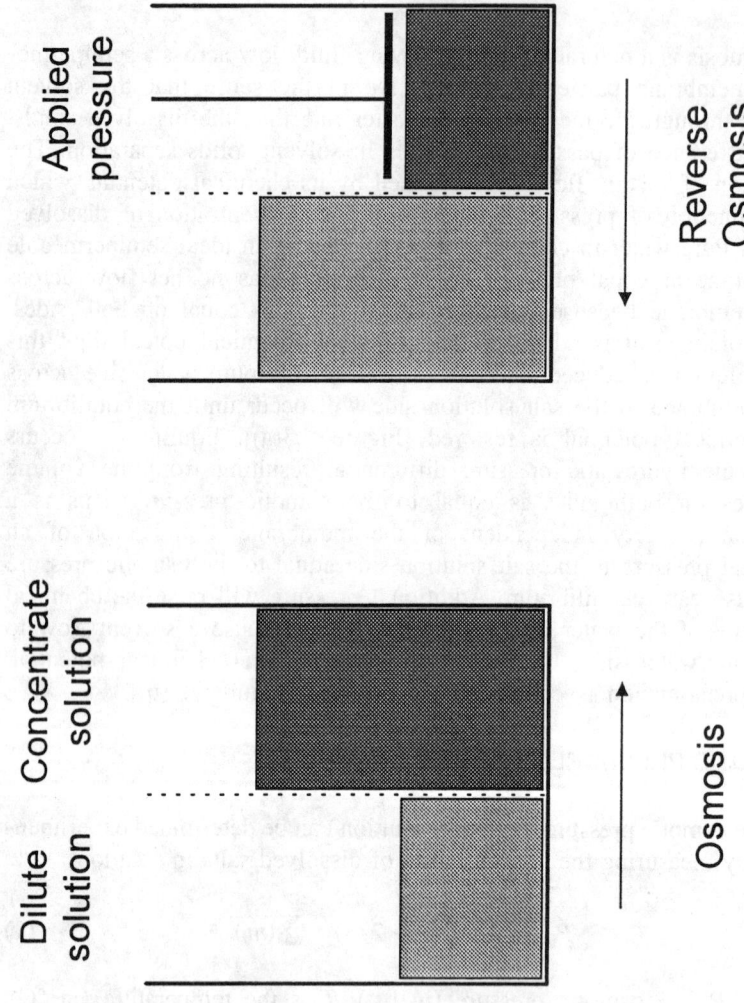

Figure 7.3. Osmosis and reverse osmosis process.

in the membrane (porosity), or by diffusion from one bonding site to another within the membrane. The theory suggests that the chemical nature of the membrane is such that it will absorb and pass water preferentially to dissolved salts at the solid/liquid interface. This may occur by weak chemical bonding of the water to the membrane surface or by dissolution of the water within the membrane structure. Either way, a salt concentration gradient is formed across the solid/liquid interface. The chemical and physical nature of the membrane determines its ability to allow for preferential transport of solvent (water) over solute (salt ions).

WATER TRANSPORT

The rate of water passage through a semipermeable membrane is defined in Equation (2):

$$Qw = (\Delta P - \Delta P_{osm}) * K_w * S/d \qquad (2)$$

where Q_w is the rate of water flow through the membrane, ΔP is the hydraulic pressure differential across the membrane, ΔP_{osm} is the osmotic pressure differential across the membrane, K_w is the membrane permeability coefficient for water, S is the membrane area, and d is the membrane thickness.

This equation is often simplified to:

$$Q_w = A * (NDP) \qquad (3)$$

where A represents a unique constant for each membrane material type, and NDP is the net driving pressure or net driving force for the mass transfer of water across the membrane.

SALT TRANSPORT

The rate of salt flow through the membrane is defined by Equation (4):

$$Q_s = \Delta C * K_s * S/d \qquad (4)$$

where Q_s is the flow rate of salt through the membrane, K_s is the membrane permeability coefficient for salt, ΔC is the salt concentration differential across the membrane, S is the membrane area, and d is the membrane thickness.

This equation is often simplified to

$$Q_s = B*(\Delta C) \tag{5}$$

where B represents a unique constant for each membrane type, and ΔC is the driving force for the mass transfer of salts.

Equations (4) and (5) show that for a given membrane:

(1) Rate of water flow through a membrane is proportional to net driving pressure differential (NDP) across the membrane.
(2) Rate of salt flow is proportional to the concentration differential across the membrane and is independent of applied pressure.

Salinity of the permeate, C_p, depends on the relative rates of water and salt transport through reverse osmosis membrane:

$$C_p = Q_s/Q_w \tag{6}$$

The fact that water and salt have different mass transfer rates through a given membrane creates the phenomena of salt rejection. No membrane is ideal in the sense that it absolutely rejects salts; rather the different transport rates create an apparent rejection. Equations (2), (4) and (5) explain important design considerations in RO systems. For example, an increase in operating pressure will increase water flow without changing salt flow, thus resulting in lower permeate salinity.

SALT PASSAGE

Salt passage is defined as the ratio of concentration of salt on the permeate side of the membrane relative to the average feed concentration. Mathematically, it is expressed in Equation (7):

$$SP = 100\% * (C_p/C_{fm}) \tag{7}$$

where SP is the salt passage (in %), C_p is the salt concentration in the permeate, and C_{fm} is the mean salt concentration in feed stream.

Applying the fundamental equations of water flow and salt flow illustrates some of the basic principles of RO membranes. For example, salt passage is an inverse function of pressure; that is, the salt passage increases as applied pressure decreases. This is because reduced pressure decreases permeate flow rate, and hence, dilution of salt (the salt flows at a constant rate through the membrane as its rate of flow is independent of pressure).

SALT REJECTION

Salt rejection is the opposite of salt passage, and is defined by Equation (8):

$$SR = 100\% - SP \tag{8}$$

where SR is the salt rejection (in %), and SP is the salt passage as defined in Equation (7).

PERMEATE RECOVERY RATE (CONVERSION)

Permeate recovery is another important parameter in the design and operation of RO systems. Recovery or conversion rate of feed water to product (permeate) is defined by Equation (9):

$$R = 100\% * (Q_p/Q_f) \tag{9}$$

where R is recovery rate (in %), Q_p is the product water flow rate, and Q_f is the feed water flow rate. The recovery rate affects salt passage and product flow. As the recovery rate increases, the salt concentration on the feed-brine side of the membrane increases, which causes an increase in salt flow rate across the membrane as indicated by Equation (5). Also, a higher salt concentration in the feed-brine solution increases the osmotic pressure, reducing the NDP and consequently reducing the product water flow rate according to Equation (2).

CONCENTRATION POLARIZATION

As water flows through the membrane and salts are rejected by the membrane, a boundary layer is formed near the membrane surface in which the salt concentration exceeds the salt concentration in the bulk solution. This increase of salt concentration is called concentration polarization. The effect of concentration polarization is to reduce actual product water flow rate and salt rejection versus theoretical estimates. The effects of concentration polarization are as follows:

(1) Greater osmotic pressure at the membrane surface than in the bulk feed solution, ΔP_{osm}, and reduced net driving pressure differential across the membrane ($\Delta P - \Delta P_{osm}$)
(2) Reduced water flow across membrane (Q_w)
(3) Increased salt flow across membrane (Q_s)
(4) Increased probability of exceeding solubility of sparingly soluble salts

at the membrane surface, and the distinct possibility of precipitation causing membrane scaling

The concentration polarization factor (CPF) can be defined as a ratio of salt concentration at the membrane surface (C_s) to bulk concentration (C_b):

$$CPF = C_s/C_b \tag{10}$$

An increase in permeate flux will increase the delivery rate of ions to the membrane surface and increase C_s. An increase of feed flow increases turbulence and reduces the thickness of the high concentration layer near the membrane surface. Therefore, the CPF is directly proportional to permeate flow (Q_p), and inversely proportional to average feed flow (Q_{favg}).

$$CPF = K_p * \exp(Q_p/Q_{favg}) \tag{11}$$

where K_p is a proportionality constant depending on system geometry.

Using the arithmetic average of feed and concentrate flow as average feed flow, the CPF can be expressed as a function of the permeate recovery rate a of membrane element (R_i).

$$CPF = K_p * \exp(2R_i/(2 - R_i)) \tag{12}$$

The value of the concentration polarization factor of 1.2, which is the recommended Hydranautics limit, corresponds to 18% permeate recovery for a 40″ long membrane element.

COMMERCIAL REVERSE OSMOSIS TECHNOLOGY

The semipermeable membrane for reverse osmosis applications consists of a film of polymeric material composed of a skin layer several thousands angstroms thick and spongy supporting layer approximately 0.1 mm thick cast on a fabric support. The commercial grade membrane must have high water permeability and a high degree of semipermeability; that is, the rate of water transport must be much higher than the rate of transport of dissolved ions. The membrane must be stable over a wide range of pH and temperature, and have good mechanical integrity. The stability of these properties over a period of time at field conditions defines the commercially useful membrane life, which is in the range of 3 to 5 years. There are two major groups of polymeric materials which can be used to produce satisfactory reverse osmosis membranes: cellulose acetate (CA) and polyamide (PA). Membrane manufacturing, operating conditions, and performance differ significantly for each group of polymeric material.

CELLULOSE ACETATE MEMBRANE

The original cellulose acetate membrane, developed in the late 1950s by Loeb and Sourirajan, was made from cellulose diacetate polymer [3]. Current CA membrane is usually made from a blend of cellulose diacetate and triacetate. The membrane is formed by casting a thin film acetone-based solution of cellulose acetate polymer with swelling additives onto a non-woven polyester fabric. Two additional steps, a cold bath followed by high temperature annealing, complete the casting process. During casting, the solvent is partially removed by evaporation. After the casting step, the membrane is immersed into a cold water bath which removes the remaining acetone and other leachable compounds. Following the cold bath step, the membrane is annealed in a hot water bath at a temperature of 60–90°C. The annealing step improves the semipermeability of the membrane with a decrease of water transport and a significant decrease of salt passage. After processing, the cellulose membrane has an asymmetric structure with a dense surface layer of about 1000–2000 angstrom (0.1–0.2 micron) which is responsible for the salt rejection property. The rest of the membrane film is spongy and porous and has high water permeability. Salt rejection and water flux of a cellulose acetate membrane can be controlled by variations in temperature and duration of the annealing step. Description of manufacturing process of cellulose acetate membranes and its properties can be found in number of publications [6–8].

COMPOSITE POLYAMIDE MEMBRANES

Composite polyamide membranes are manufactured in two distinct steps. First, a polysulfone support layer is cast onto a non-woven polyester fabric. The polysulfone layer is very porous and is not semipermeable; that is it does not have the ability to separate water from dissolved ions. In a second, separate manufacturing step, a semipermeable membrane skin is formed on the polysulfone substrate by interfacial polymerization of monomers containing amine and carboxylic acid chloride functional groups. This manufacturing procedure enables inde-pendent optimization of the distinct properties of the membrane support and salt rejecting skin. The resulting composite membrane is character-ized by higher specific water flux and lower salt passage than cellulose acetate membranes. Polyamide composite membranes are stable over a wider pH range than cellulose acetate membranes. However, polyamide membranes are susceptible to oxidative degradation by free chlorine, while cellulose acetate membranes can tolerate limited levels of exposure to free chlorine. Compared to a polyamide membrane, the surface of cellulose acetate membrane is smooth and has little surface charge. Because of the neutral surface and tolerance to free chlorine, cellulose acetate membranes will usually have a more stable performance than polyamide membranes in applications where the feed water has a high fouling potential, such as with municipal effluent and surface water supplies. The early composite membranes made of aliphatic polymers [9] were very sensitive to presence of oxidants and suffered from inadequate stability of performance in field conditions. The later generation of composite membranes based on aromatic polyamide, invented by Cadotte [10,11] have some tolerance to free chlorine and shows excellent performance stability with majority of feed water types. This type of membrane material is used today almost exclusively in commercial composite membrane elements. The structures of cellulose acetate and polyamide polymer are shown in Figures 7.4(a) and 7.4(b). Figures 7.5 and 7.6 contain ESM pictures of the surface and cross section of cellulose acetate membrane. Figures 7.7 and 7.8 show parallel pictures of a composite polyamide membrane.

MEMBRANE MODULE CONFIGURATIONS

The two major membrane module configurations used for reverse osmosis applications are hollow fiber and spiral wound. Two other configurations, tubular and plate and frame, have found good acceptance in the food and dairy industry and in some special applications. Modules of these configuration have been less frequently used in conventional reverse osmosis applications. New configurations of plate and frame module is being used for treatment of land fill leachate.

HOLLOW FINE FIBER (HFF) MEMBRANE ELEMENTS

The concept of hollow fiber configuration module has been introduced by Mahon [12] in the early 1960s. The HHF configuration utilizes semipermeable membrane in the form of hollow fibers which have been extruded from cellulosic or non-cellulosic material [13]. The fiber is asymmetric in structure and is as fine as a human hair, about 40–80 microns (0.0016–0.0030 inch) I.D. and 85–150 microns (0.0033–0.060 inch) O.D. Millions of these fibers are formed into a bundle and folded in half to a length of approximately 120 cm (4 ft). A perforated plastic tube, serving as a feed water distributor is inserted in the center and extends the full length of the bundle. The bundle is wrapped and both ends are epoxy sealed to form a sheet-like permeate

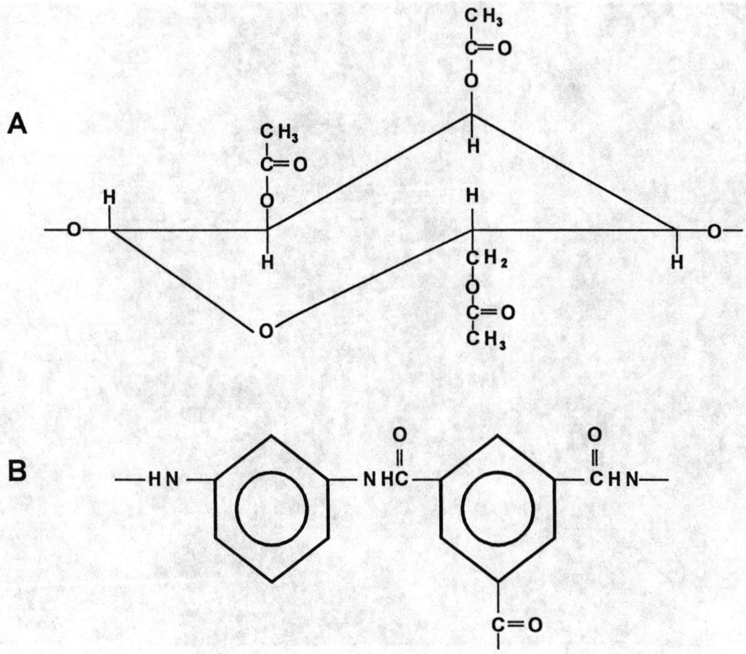

Chemical structure of cellulose triacetate (A) and polyamide (B) membrane material

Figure 7.4. Structure of cellulose acetate and polyamide polymer.

tube end and a terminal end which prevents the feed stream from bypassing to the brine outlet. The hollow fiber membrane bundle, 10 cm to 20 cm (4 to 8 inches) in diameter, is contained in a cylindrical housing or shell approximately 137 cm (54 inches) long and 15–30 cm (6–12 inches) in diameter. The assembly is called a permeator. The pressurized feed water enters the permeator feed end through the center distributor tube, passes through the tube wall, and flows radially around the fiber bundle toward the outer permeator pressure shell. Water permeates through the outside wall of the fibers into the hollow core or fiber bore, through the bore to the tube sheet or product end of the fiber bundle, and exits through the product connection on the feed end of the permeator. In a hollow fiber module, the permeate water flow per unit area of membrane is low, and therefore, the concentration polarization is not high at the membrane surface. The net result is that hollow fiber units operate in a non-turbulent or laminar flow regime. The HFF membrane must operate above a minimum reject flow to minimize concentration polarization and maintain even flow distribution through the fiber bundle.

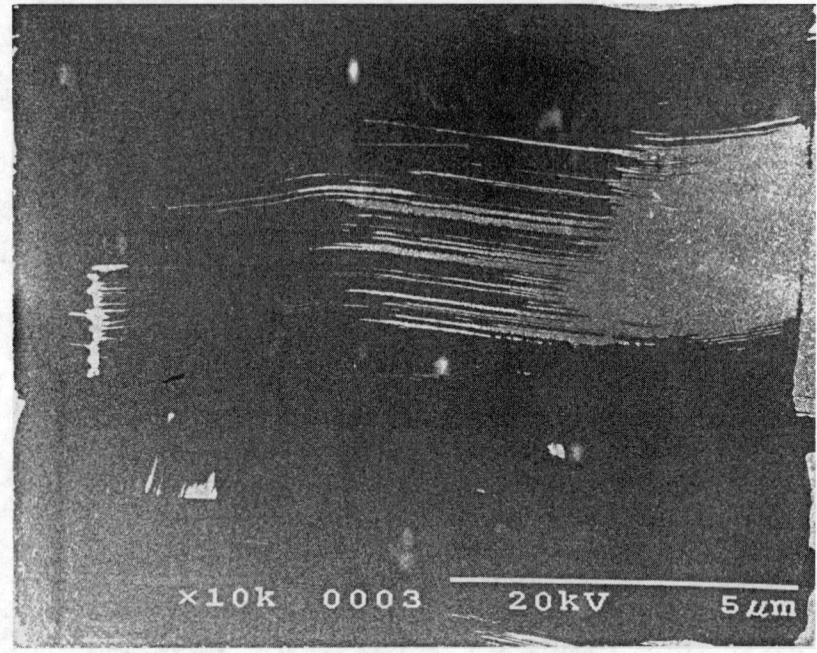

Figure 7.5. ESM picture of cellulose acetate membrane surface.

Typically, a single hollow fiber permeator can be operated at up to 50 percent recovery and meet the minimum reject flow required. The hollow fiber unit allows a large membrane area per unit volume of permeator which results in compact systems. Hollow fiber membrane modules are available for brackish and seawater applications. Membrane materials are cellulose acetate blends and aramid (a proprietary polyamide type material in an anisotropic form). Because of very close packed fibers and tortuous feed flow inside the module, hollow fiber modules require feed water of better quality (lower concentration of suspended solids) than the spiral wound module configuration. The hollow fiber modules are used mainly for desalting of seawater and treatment of good quality brackish water (well water). Due to fouling susceptibility of the conventional hollow fiber configuration, these type of modules are not used for the desalting of municipal wastewater.

SPIRAL WOUND MEMBRANE ELEMENTS

The concept of spiral wound membrane element device was introduced shortly after the invention of the hollow fiber configuration [14]. In a spiral wound configuration two flat sheets of membrane are separated with a permeate collector channel material to form a leaf. This assembly is sealed on three sides with the fourth side left open for permeate to exit. A feed/brine spacer material sheet is added to the leaf assembly. A number of these assemblies or leaves are wound around a central plastic permeate tube. This tube is perforated to collect the permeate from the multiple leaf assemblies. The typical industrial spiral wound membrane element is approximately 100 or 150 cm (40 or 60 inches) long and 10 or 20 cm (4 or 8) inches in diameter. The feed/brine flow through the element is a straight axial path from the feed end to the opposite brine end, running parallel to the membrane surface. The feed channel spacer induces turbulence and reduces concentration polarization. Manufacturers specify brine flow requirements to control concentration polarization by limiting recovery (or conversion) per element to 10–20 percent. Therefore, recovery (or conversion) is a function of the feed-brine path length. In order to operate at acceptable recoveries, spiral systems are usually staged with three to six membrane elements connected in series in a pressure tube. The brine stream from the first element becomes the feed to the following element, and so on for each element within the pressure tube. The brine stream from the last element exits the pressure tube to waste. The permeate from each element enters the permeate collector tube and exits the vessel as a

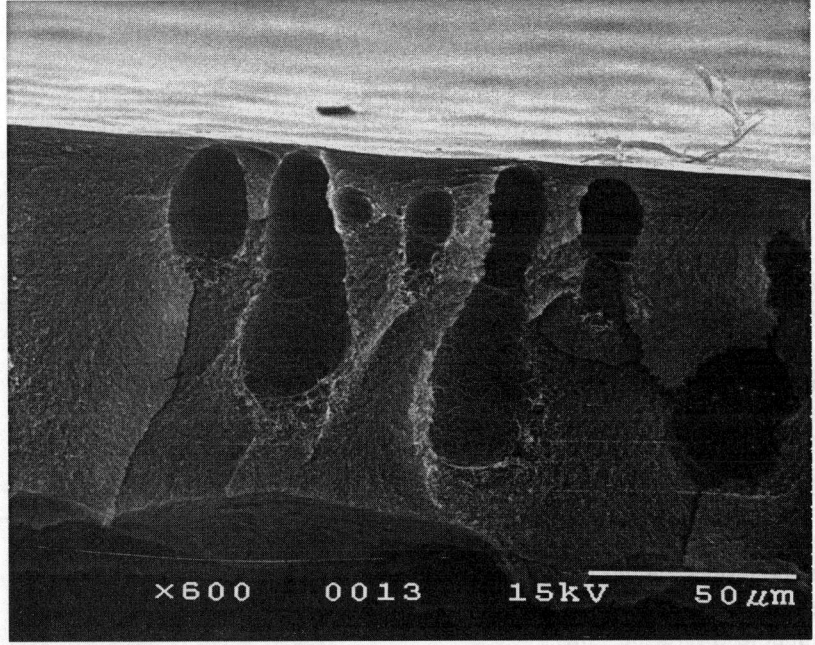

Figure 7.6. ESM picture of cross section of cellulose acetate membrane.

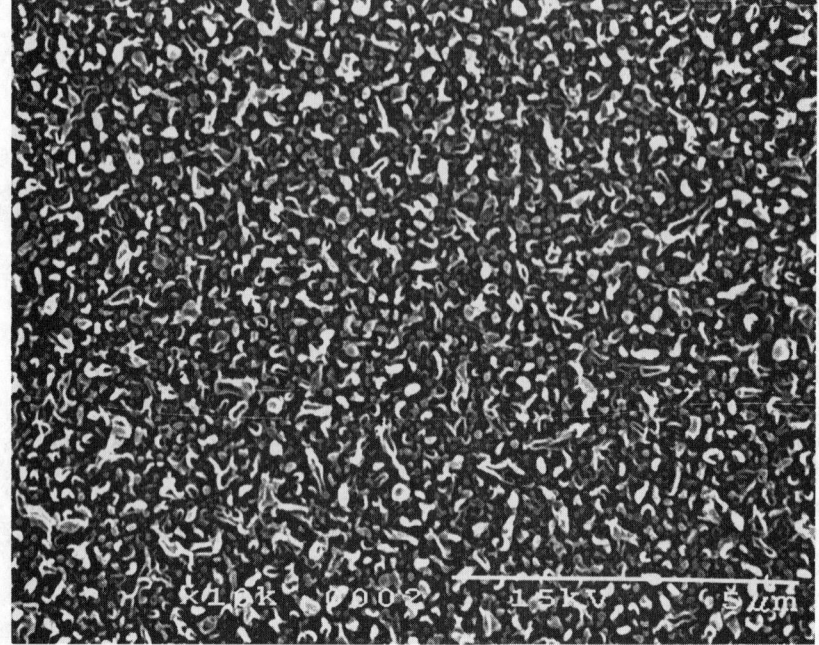

Figure 7.7. ESM picture of composite polyamide membrane surface.

common permeate stream. A single pressure vessel with four to six membrane elements connected in series can be operated at up to 50 percent recovery under normal design conditions.

The brine seal on the element feed end seal carrier prevents the feed/brine stream from bypassing the following element. Spiral wound elements are most commonly manufactured with flat sheet membrane of either a cellulose diacetate and triacetate (CA) blend or a thin film composite. A thin film composite membrane consists of a thin active layer of one polymer cast on a thicker supporting layer of a different polymer. The composite membranes usually exhibit higher rejection at lower operating pressures than the cellulose acetate blends. The composite membrane materials may be polyamide, polysulfone, polyurea, or other polymers. The spiral wound modules are used with all types of feed water: seawater, brackish well water and surface water. Elements of this configuration are also used to treat highly fouling municipal effluents. The structure of composite and hollow fiber membrane and the corresponding modules configurations are shown in Figures 7.9, 7.10, 7.11, and 7.12. Figure 7.13 shows spiral wound modules of various sizes.

PLATE AND FRAME MEMBRANE ELEMENTS

The plate and frame configuration has been introduced at the early stages of development of reverse osmosis technology [15] and later on was almost abandoned in favor of higher packing density spiral wound and hollow fiber configurations. Today the plate and frame modules are still used in applications where spiral wound and hollow fiber modules can not provide sufficient reliability or performance. One of such applications is reduction of volume of land fill leachate. The example of modern plate and frame configuration is shown in Figures 7.8 and 7.9. This design developed by Rochem, Germany, consists of a chamber of disks interleaved with membrane cushions. The treated feed water is fed into the tubular chamber where its flow is controlled as it passes through the discs and over the membranes. Clean water is progressively removed and the waste material concentrated. The flow regime provides turbulent flow and short feed flow path. Therefore the tendency for membrane to scale or foul is significantly reduced. Flow schematic of plate and frame configuration and corresponding module configuration is shown in Figures 7.14 and 7.15.

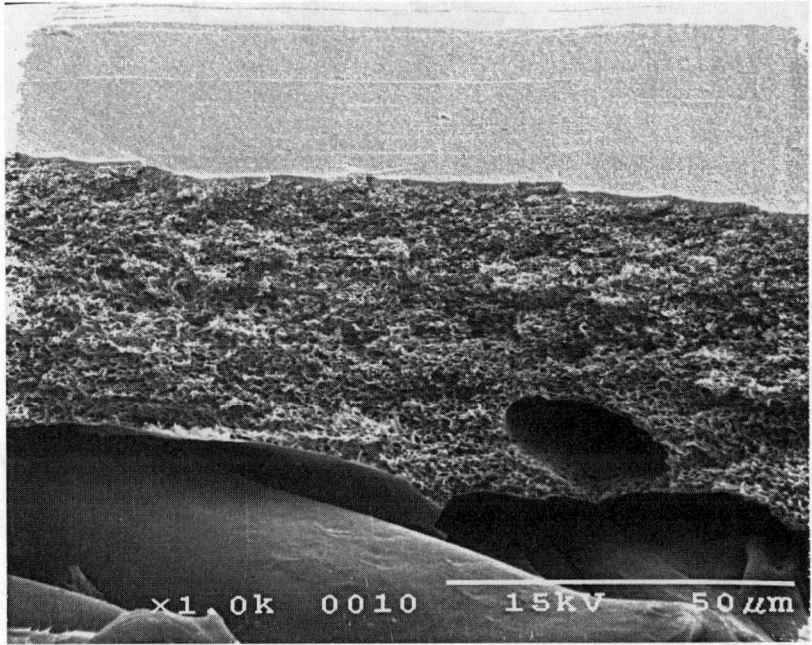

Figure 7.8. ESM picture of cross section of composite polyamide membrane.

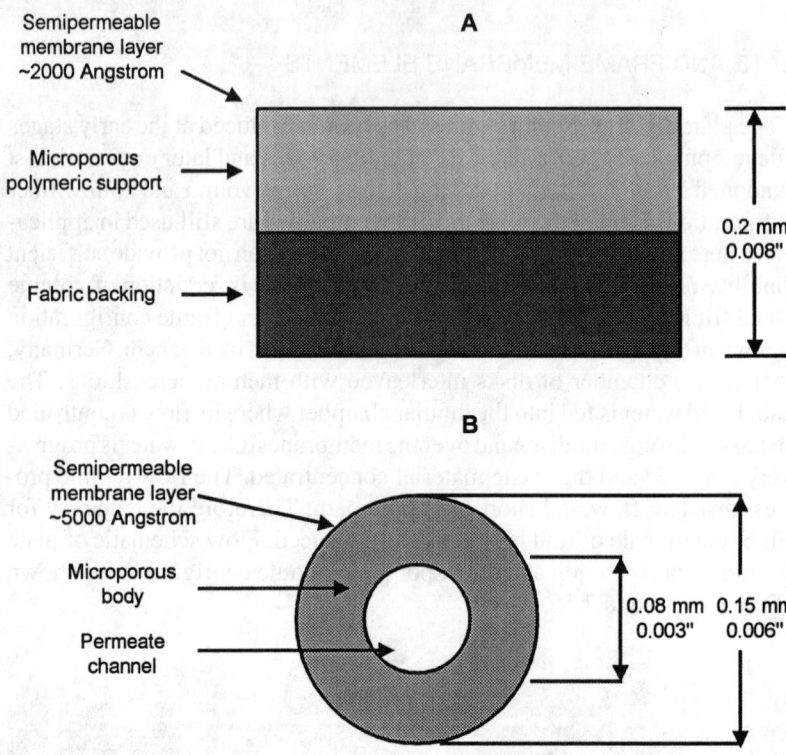

Cross section of flat (A) and hollow fiber (B) membranes

Figure 7.9. Diagram of flat sheet and hollow fiber membrane configuration.

RO SYSTEM DESIGN PROCESS

SYSTEM DESIGN REQUIREMENTS

Design of an RO system is based on the requirements of system performance and specifications of the feed water. System performance requirements may include the following parameters:

- permeate capacity
- permeate quality
- permeate recovery ratio
- membrane type
- average permeate flux rate
- size of RO train(s) or number of RO units

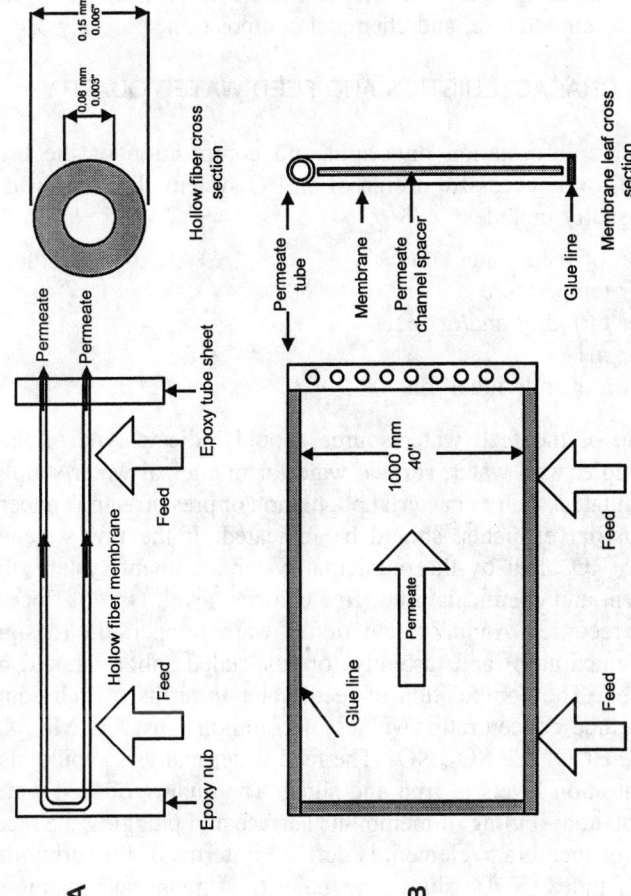

Figure 7.10. Diagram of operation of hollow fiber and spiral wound element configuration.

Hollow fiber (A) and Spiral wound (B) module configuration

A

0.15 mm
0.006"

0.08 mm
0.003"

Hollow fiber cross section

Permeate   Permeate

Feed

Epoxy tube sheet

Hollow fiber membrane

Feed

Feed

Epoxy nub

B

Permeate tube

Membrane

Permeate channel spacer

Glue line

Membrane leaf cross section

1000 mm
40"

Feed

Permeate

Feed

Glue line

Feed

Usually the required permeate quality is specified in terms of total dissolved solids (TDS). In some cases the concentration of selected individual ions in the permeate is specified as well. The ions, which are usually specified, are those which are regulated by the drinking water quality standards or by the requirements of the industrial process, for example, pressure boiler make-up water quality specifications. System specifications also should include design values of the feed water in terms of feed water type, feed water temperature, and chemical composition.

FEED WATER CHARACTERISTICS AND FEED WATER QUALITY

Comprehensive information on origin and composition of the feed water is critical for a successful design of an RO system. The feed water specification usually includes:

- description of feed water source
- feed water temperature
- feed water turbidity and/or SDI
- feed water pH
- composition of individual ions

The description of the feed water source should indicate type of feed water, for example: well water, surface water or municipal water supply. Known contaminates, such as bacterial population or presence and concentration of industrial effluents, should be indicated. If the raw water is treated prior to RO plant by the municipal water treatment system, the type of treatment and chemicals used (free chlorine level, type of flocculant) should be recorded. Annual fluctuation of water temperature (design, minimum and maximum) and turbidity (or suspended solids) should be specified as well. The composition of feed water in terms of individual ions should include concentration values of common ions: Ca, Mg, Na, K, Ba, Sr, NH_4, HCO_3, Cl, SO_4, NO_3. The feed water analysis should also include concentration levels of iron and silica. The quality of feed water in respect of potential fouling of membrane surface and plugging the feed/brine channels of membrane element is defined in terms of the turbididty and Silt Density Index (SDI) values. The majority of membrane manufactures defines the limits for feed water turbidity as 1.0 NTU and SDI at 4–5 range. The meaning of the above limits is that RO system should not be operated with feed water with such quality for any significant period of time. It is common experience that operation at the very limits of feed water quality will result in rapid membrane fouling. In order to maintain a stable membrane performance, in respect of permeate flux, the feed water quality indicators should not exceed 0.5 NTU and/or 2.5 SDI units. The turbidity is a simpler quality parameter to measure and it

Hollow fiber module assembly

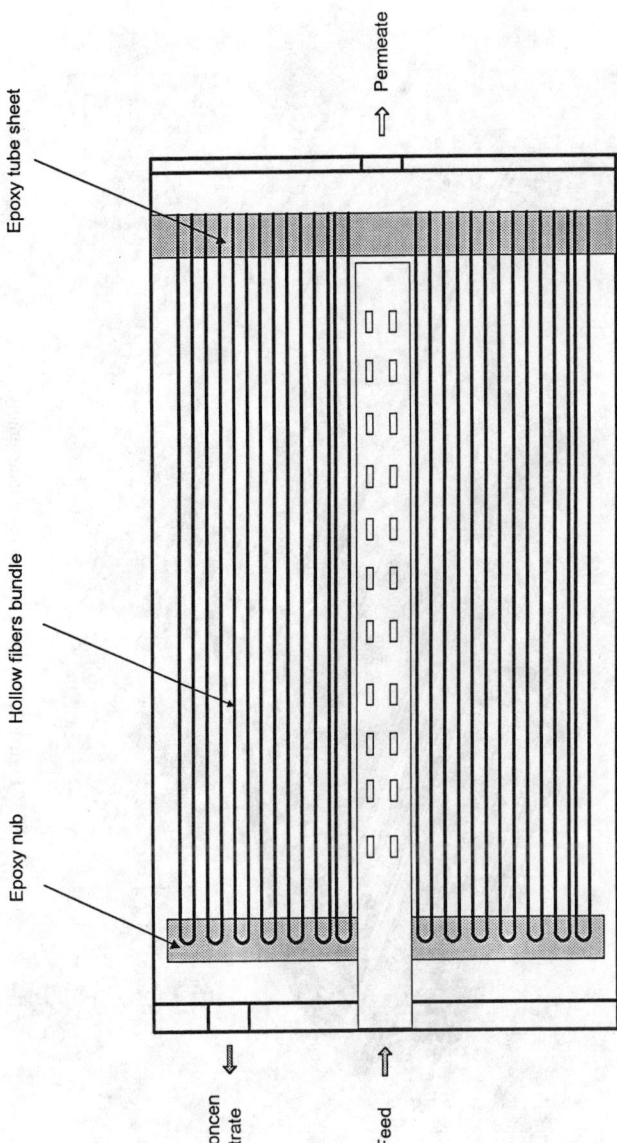

Permeate

Epoxy tube sheet

Hollow fibers bundle

Epoxy nub

Concen
trate

Feed

Figure 7.11. Hollow fiber module assembly.

283

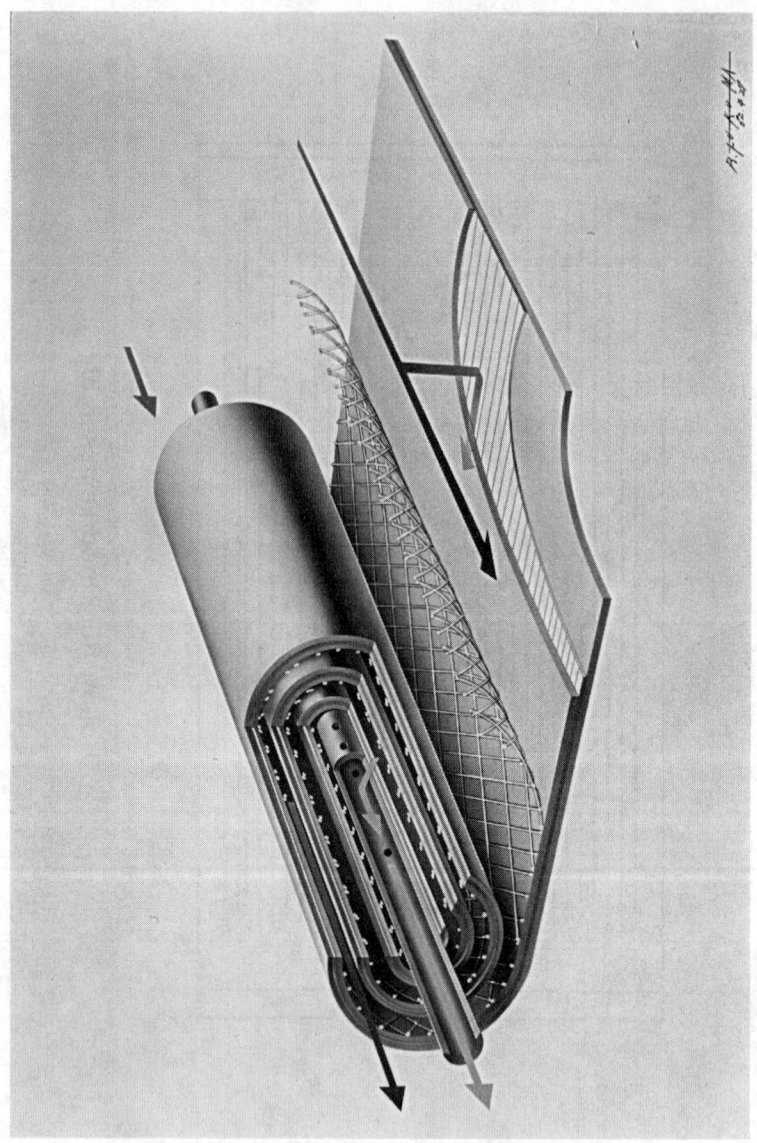

Figure 7.12. Spiral wound module assembly.

Figure 7.13. Spiral wound elements of various sizes: 8"–2" diameter.

can be monitored continuously, using automatic instrumentation. SDI is obtained in discrete measurements, each measurement taking approximately 20 to 25 minutes. The measurements are usually taken manually by operators, though automatic equipment is available commercially. It is of common opinion that the SDI is a more sensitive indicator of feed water quality then turbidity and provides a better prediction of membrane fouling rate. The procedure for SDI measurement is described in the Annual Book of ASTM Standards [16]. It is based on measurement of rate of declining flow at a constant pressure of a water tested through a porous filter membrane. The time measured for filtration of a constant volume (500 ml) at the beginning of the test (t_1) and after 15 minutes (t_2) is used to calculate the SDI according to the following equation:

$$SDI = 100\% * (1 - t_1/t_2)/15 \qquad (13)$$

If the filter plugs too fast for meaningful determination of the filtration

Disc Tube™ Flow Path Schematic

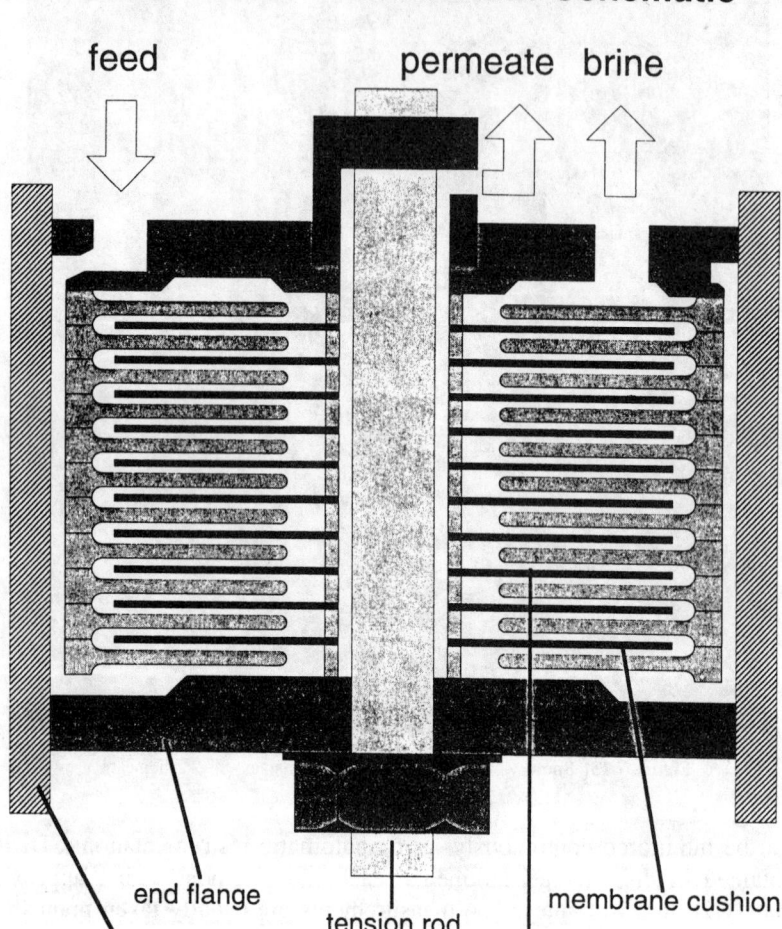

Figure 7.14. Plate and frame flow configuration (courtesy of ROCHEM).

time, the volume of filtrate being collected or time between measurements can be decreased. The SDI method is very sensitive to foulant concentration but it is not very accurate. No meaningful correlation could be established between SDI and turbidity. Attempts to improve accuracy of this method led to introduction of Modified Fouling Index [16]. The test for MFI is based on measurement of pressure increase required for maintaining of constant filtration rate through well defined membrane filter. The MFI results are more reproducible but the test is difficult to perform manually and automatic equipment is necessary [17] for a routine determination in the field conditions. SDI, MFI and turbidity are useful indicators of fouling potential of feed water of a relatively good quality. Usually, these parameters cannot be measured for municipal waste water, even after an extensive pretreatment, because the plugging rate of the filter is too fast. For the reclamation of municipal waste water the prediction of membrane performance are based entirely on results of pilot unit operation, using similar pretreatment and the same membrane type as will be used in the commercial plant.

MEMBRANE ELEMENTS SELECTION

Selection of membrane elements type for a given RO unit is related to feed water type, operating parameters and required permeate quality. It is also related to the intrinsic property of cellulose acetate and polyamide

THE DISC TUBE™ MODULE SYSTEM

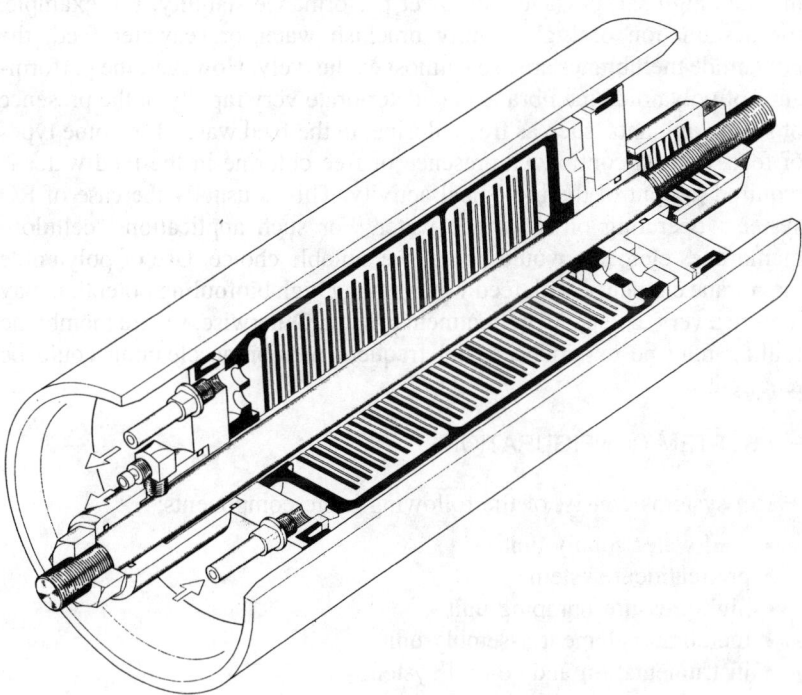

Figure 7.15. Plate and frame module assembly (courtesy of ROCHEM).

membrane material. Cellulose acetate membrane has a low affinity for organic foulants and can operate in the presence of low concentration of free chlorine, up to 1 ppm. However, cellulose acetate membrane material can undergo hydrolytic degradation at low and high pH. Hydrolytic degradation is accelerated at high feed water temperature. For stable performance the cellulose acetate membrane should be operated at feed pH range of 5 to 6 and feed temperature should be below 35°C. Cellulose acetate is characterized by low specific permeate flux and a relatively high salt passage. Therefore, systems equipped with cellulose acetate membranes have to operate at feed pressure higher by 50%–100% than the systems equipped with composite polyamide membranes. The permeate salinity produced by systems using cellulose acetate membrane is also higher than for composite membranes. The composite polyamide membrane is characterized by good stability of membrane material over a wide range of feed water pH and temperature. It has high specific permeate flux and high salt rejection. For those reasons, polyamide membrane is the preferred membrane type for the majority of applications. For applications which involve desalination of high salinity feed, cellulose acetate membrane may not be capable of reliably producing the required permeate quality due to intrinsic high salt passage and lower performance stability. For example, for desalination of high salinity brackish water or seawater feed, the polyamide membranes are used almost exclusively. However, the performance of polyamide membrane may deteriorate very rapidly in the presence of strong oxidants, such as free chlorine, in the feed water. For some types of feed water a continuous presence of free chlorine in the feed water is required to control the biological activity. This is usually the case of RO systems operating on a surface water. For such applications, cellulose membranes elements would be a more suitable choice. Use of polyamide membrane elements, with feed water having high biofouling potential, may require a very extensive pretreatment system. Otherwise, severe membrane fouling may be experienced and frequent membrane cleaning could be required.

RO SYSTEM CONFIGURATION

RO systems consist of the following basic components:

- feed water supply unit
- pretreatment system
- high pressure pumping unit
- membrane element assembly unit
- instrumentation and control system
- permeate treatment and storage unit
- cleaning unit

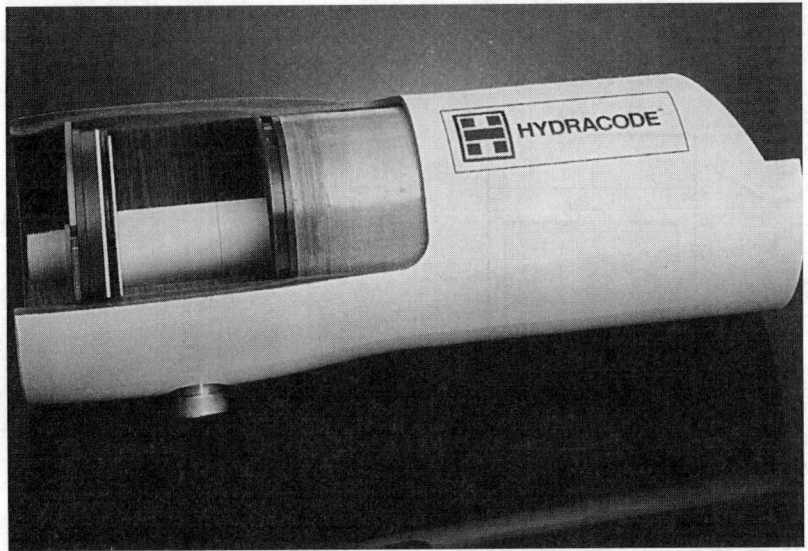

Figure 7.16. Cut away picture of spiral wound element installed in pressure vessel.

The membrane assembly unit (RO block) consists of a stand supporting the pressure vessels, interconnecting piping, and feed, permeate and concentrate manifolds. Membrane elements are installed in the pressure vessels. The pressure vessel has permeate ports on each end, located in center of the end plate, and feed and concentrate ports, located on the opposite ends of the vessel. Each pressure vessel may contain from one to seven membrane elements connected in series. The cut away picture of membrane element in pressure vessel is shown in Figure 7.16. Figure 7.17 shows schematic configuration of pressure vessel containing three membrane elements. As shown in Figure 7.17, the permeate tube of the first and the last element is connected to the end plates of the pressure vessel. Permeate tubes of elements in the pressure vessel are connected to each other using interconnectors. On one side of each membrane element there is a brine seal, which closes the passage between outside rim of the element and inside wall of the pressure vessel. This seal prevents feed water from bypassing the membrane module, and forces it to flow through the feed channels of the element. As feed water flows through each subsequent membrane element, part of the feed volume is removed as permeate. The salt concentration of the remaining feed water increases along the pressure vessel. Permeate tubes conduct the permeate from all connected elements. The collected permeate has the lowest salinity at the feed end of the pressure vessel, and increases gradually in the direction of the concentrate flow. A system is divided into groups of pressure vessels, called concentrate

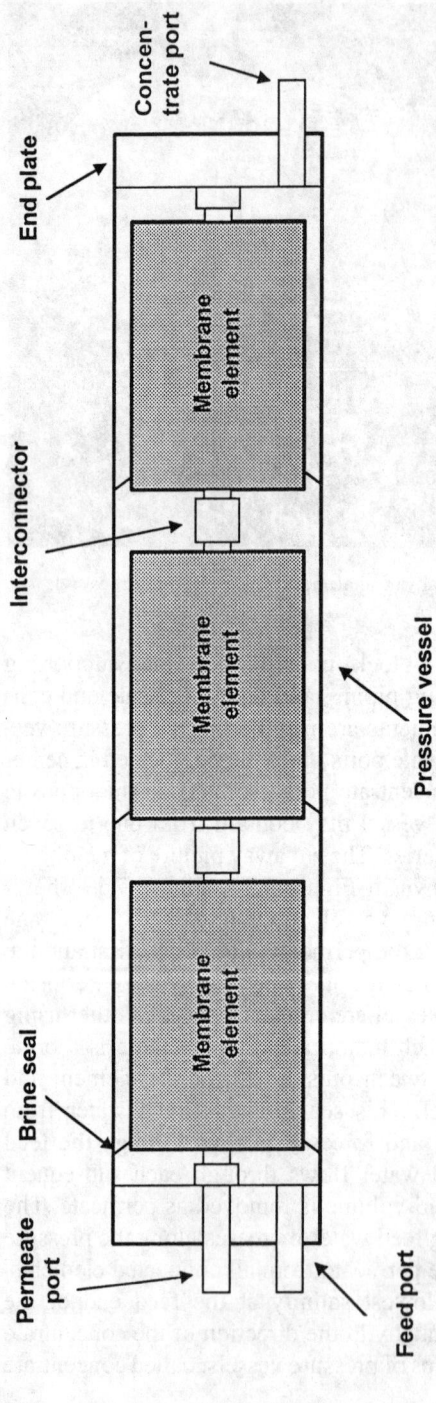

Pressure vessel with three membrane elements

Figure 7.17. Diagram of membrane elements installed in pressure vessel.

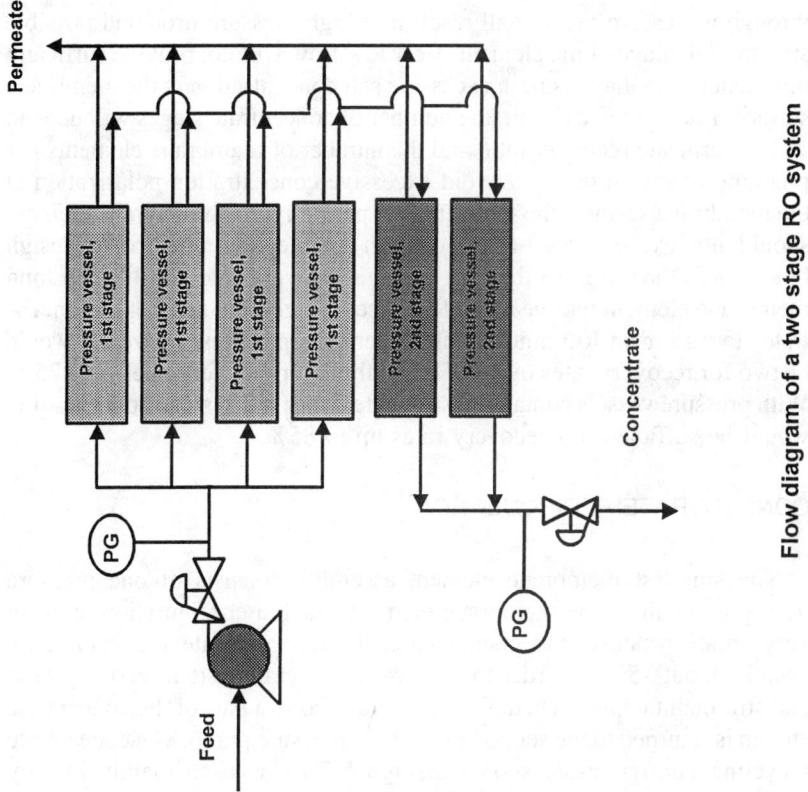

Figure 7.18. Flow diagram of pressure vessel array in a two stage RO unit.

stages. In each stage pressure vessels are connected in parallel, with respect to the direction of the feed/concentrate flow. The number of pressure vessels in each subsequent stage decreases in the direction of the feed flow, usually in the ratio of 2:1, as shown in Figure 7.18. Thus, one can visualize that the flow of feed water through the pressure vessels of a system resembles a pyramid structure: a high volume of feed water flows in at the base of pyramid, and a relatively small volume of concentrate leaves at the top. The decreasing number of parallel pressure vessels from stage to stage compensates for the decreasing volume of feed flow, which is continuously being partially converted to permeate. The permeate of all pressure vessels in each stage, is combined together into a common permeate manifold. The objective of the taper configuration of pressure vessels is to maintain a similar feed/concentrate flow rate per vessel through the length of the system and to maintain feed/concentrate flow within the limits specified for a given type of membrane element. Very high flow through a pressure vessel will result in a high pressure drop and possible structural damage of the element. Very low flow will not provide sufficient turbulence, and may result in excessive salt concentration at the membrane surface. For a given RO unit, the number of concentrate stages will depend on the permeate recovery ratio and the number of membrane elements per pressure vessel. In order to avoid excessive concentration polarization at the membrane surface, the recovery rate per individual membrane element should not exceed 18%. It is common engineering practice to design brackish RO systems so that the average recovery rate per 40 inch long membrane element will be about 9%. Accordingly, the number of concentrate stages for an RO unit having 6 elements per pressure vessel would be two for recovery rates over 60%, and three for recovery rates over 75%. With pressure vessels containing seven elements, a two stage configuration would be sufficient for recovery rates up to 85%.

CONCENTRATE RECIRCULATION

The simplest membrane element assembly consists of one pressure vessel, containing one membrane element. Such a configuration, used in very small systems, can operate at a limited permeate recovery ratio, usually about 15%. In order to increase the overall system recovery ratio and still maintain an acceptable concentrate flow, a part of the concentrate stream is returned to the suction of the high pressure pump. The concentrate recycling configuration, shown in Figure 7.19, is used mainly in very small RO units. An advantage of such a design is the compact size of the RO unit. The disadvantage of concentrate recirculation design is related to the need for a larger feed pump to handle higher feed flow. Accordingly, the power consumption is relatively higher than that required in a multistage configuration. Due to blending of the feed with the concentrate stream, the average feed salinity is increased. Therefore, both the feed pressure and the permeate salinity are higher as well.

CONCENTRATE STAGING

A commercial RO unit usually consists of single pump and a multistage array of pressure vessels. A simplified block diagram of a two stage RO unit is shown in Figure 7.20. The concentrate from the first stage becomes the feed to the second stage; this is what is meant by the term ''concentrate staging.'' The flows and pressures in the multistage unit are controlled with the feed and concentrate valves. The feed valve, after the high pressure pump, controls feed flow to the unit. The concentrate valve, at the outlet of RO block, controls the feed pressure.

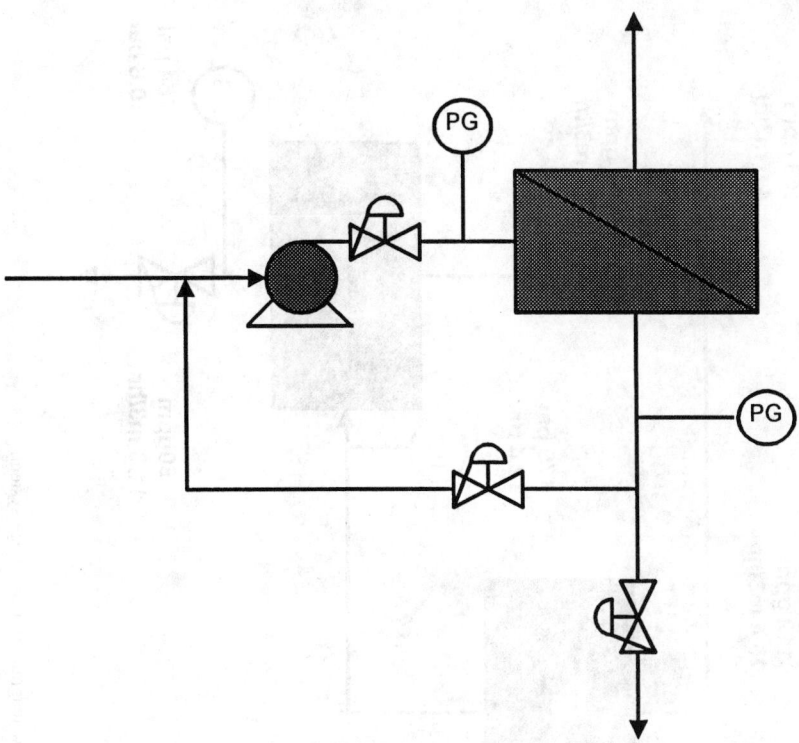

Figure 7.19. Flow diagram of RO unit with concentrate recirculation.

FLOW DISTRIBUTION

In some cases it is necessary to equilibrate permeate flow between stages (i.e. decrease permeate flow from the first stage and increase permeate flow from the last stage). This can be accomplished in one of two design configurations. One solution is to install a valve on the permeate line from the first stage (Figure 7.21). By throttling this valve, permeate back pressure will increase, reducing net driving pressure and reducing permeate flux from the first stage. The differential permeate flux is produced from the second stage by operating the RO unit at a higher feed pressure. The other solution is to install a booster pump on the concentrate line between the first and the second stage (Figure 7.22). The booster pump will increase feed pressure to the second stage resulting in higher permeate flow. The advantage of the permeate throttling design is simplicity of the RO unit and low capital cost. However, this design results in additional power losses due to permeate throttling and higher power consumption. The interstage pump design requires modification of the interstage manifold

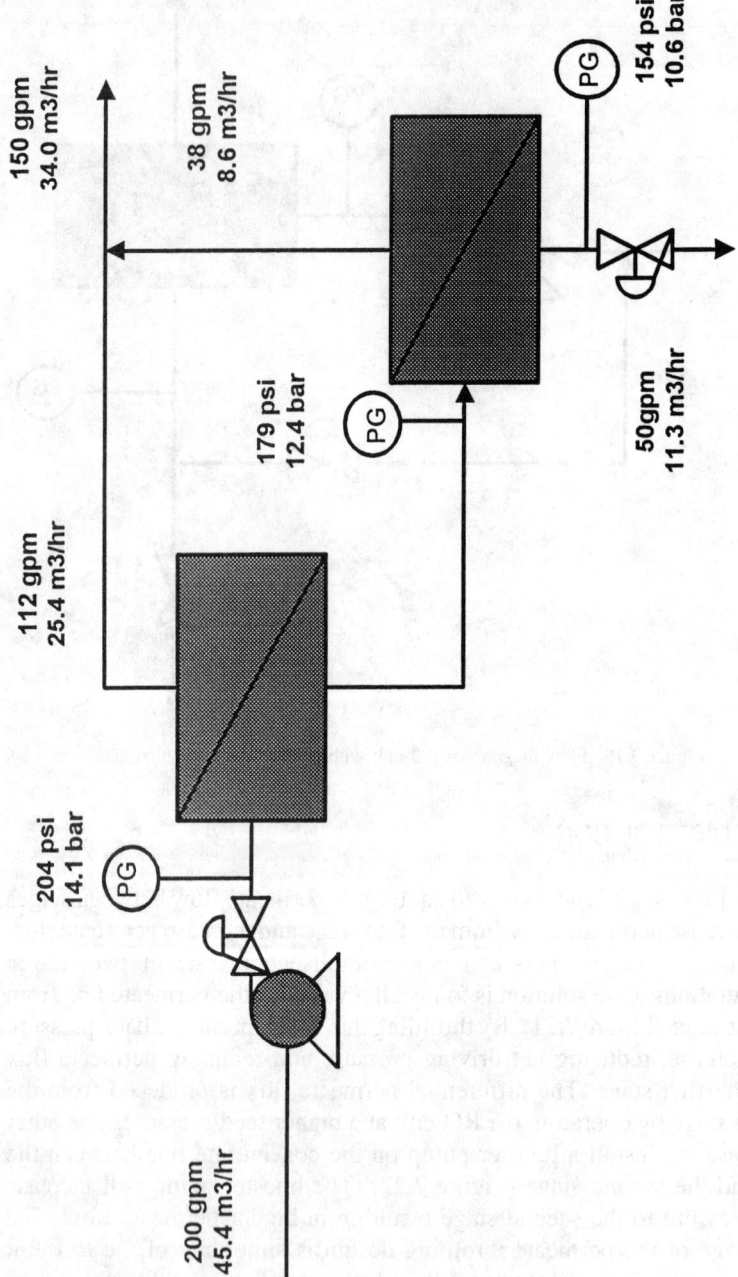

Figure 7.20. Flow diagram of a two stage system.

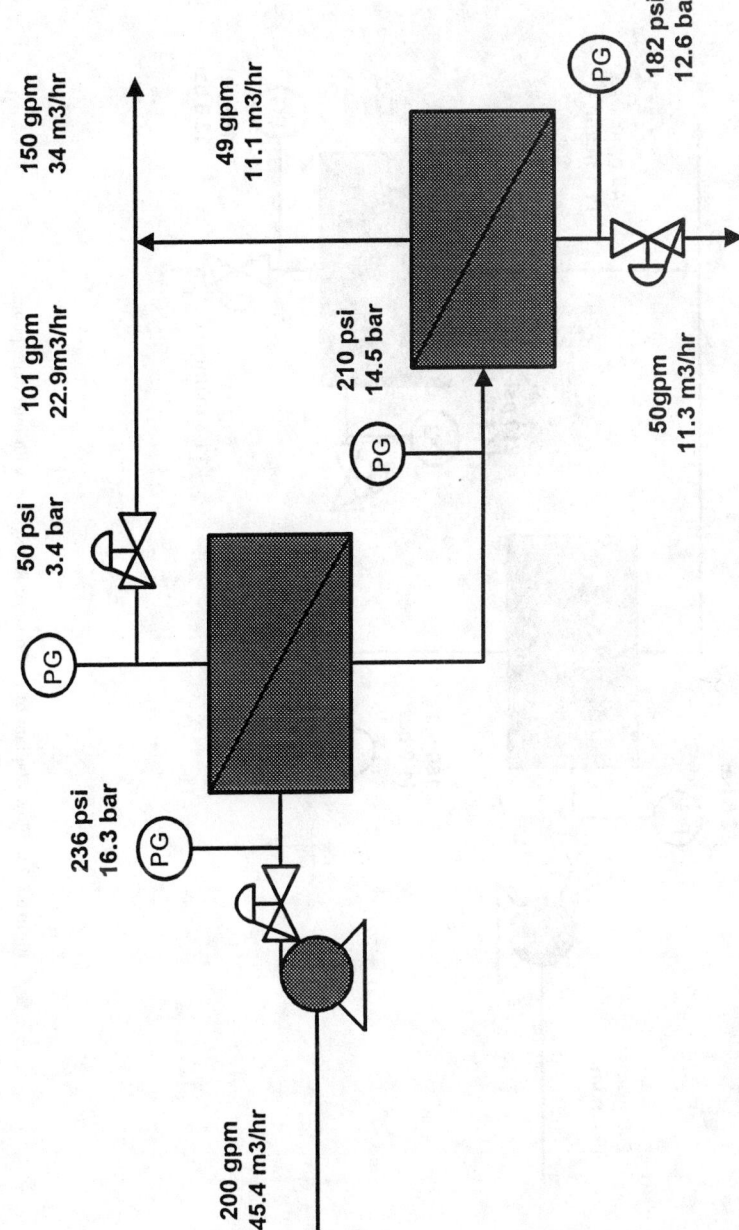

Figure 7.21. Flow diagram of a two stage RO system with permeate throttling.

200 gpm
45.4 m3/hr

236 psi
16.3 bar

50 psi
3.4 bar

101 gpm
22.9m3/hr

150 gpm
34 m3/hr

49 gpm
11.1 m3/hr

210 psi
14.5 bar

182 psi
12.6 bar

50gpm
11.3 m3/hr

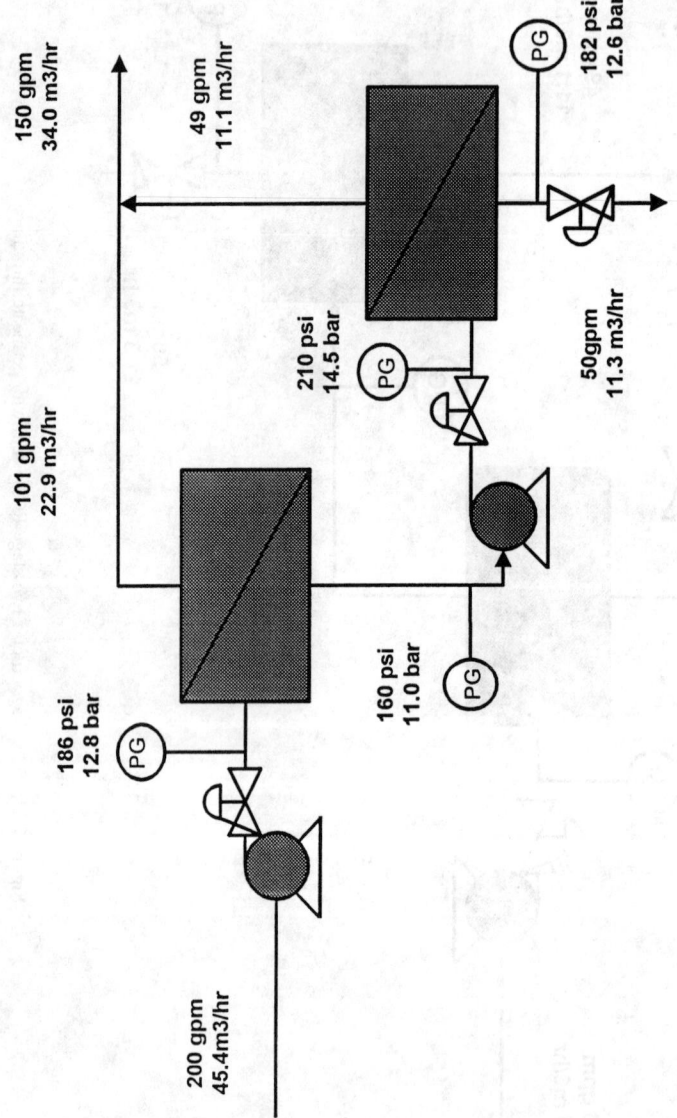

Figure 7.22. Flow diagram of a two stage RO system with interstage pump.

150 gpm
34.0 m3/hr

49 gpm
11.1 m3/hr

182 psi
12.6 bar

210 psi
14.5 bar

50gpm
11.3 m3/hr

101 gpm
22.9 m3/hr

160 psi
11.0 bar

186 psi
12.8 bar

200 gpm
45.4m3/hr

and an additional pumping unit. The investment cost is higher than in the first design, but the power consumption is lower.

PERMEATE STAGING

For some applications, the single pass RO system may not be capable of producing permeate water of a required salinity. Such conditions may be encountered in two types of RO applications:

- seawater RO systems, which operate on a very high salinity feed water, at high recovery ratio and/or at high feed water temperature
- brackish RO applications which require very low salinity permeate such supply of makeup water for pressure boilers or production of rinse water for microelectronics applications

To achieve an additional reduction in permeate salinity, the permeate water produced in the first pass is desalted again in a second RO system. This configuration is called a two pass design, or ''permeate staging.'' Depending on quality requirements, all or part of the first pass permeate volume is desalted again in the second pass system. The system configuration is known as a complete or partial two pass system depending on whether all of the permeate is fed to the second pass or not. The first pass permeate is a very clean water. It contains very low concentrations of suspended particles and dissolved salts; therefore, it does not require any significant pretreatment. The second pass system can operate at a relatively high average permeate flux and high recovery rate. The common design parameters for the second pass RO unit are average flux rate of 20 gfd and recovery rate of 85%–90%. In a two pass system, the permeate from the first pass flows through a storage tank or is fed directly to the suction of the second pass high pressure pump. There are a number of possible configurations of the two pass RO units. One configuration, which is a partial two pass system is shown in Figure 7.23, splits the permeate from the first pass into two streams. One stream is processed by the second pass unit, and is then combined with the unprocessed part of the permeate from the first pass. Provided that the partial second pass system can produce the required permeate quality, this configuration results in smaller capital and operating costs, as well as higher combined permeate recovery rate (utilization of the feed water), compared to a complete two pass system. It is a common procedure in two pass systems to return concentrate from the second pass unit to the suction of the high pressure pump of the first pass unit. The dissolved salts ion concentration in the concentrate from the second pass is usually lower than the concentration of the feed to the first pass unit. Therefore, blending feed water with the second pass

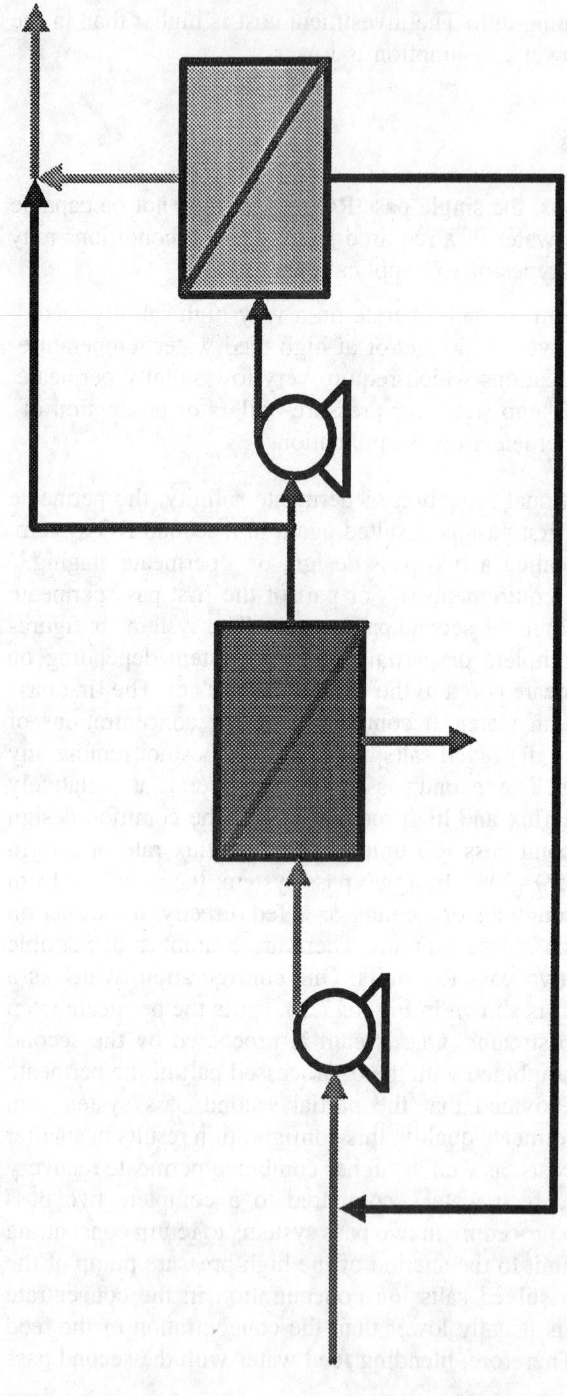

Figure 7.23. Flow diagram of a partial two pass system.

concentrate reduces slightly the salinity of the feed, and increases the overall utilization of the feed water.

SYSTEM DESIGN PARAMETERS

RECOVERY RATIO

Permeate recovery ratio, which is the rate of conversion of feed water to the permeate affects both the capital and operating cost of the RO desalting process. The volume of raw water required for a given permeate capacity of the RO system is directly determined by the design recovery ratio. Therefore, the size of the raw water supply system, capacity of the pretreatment system, size of the high pressure pump and feed manifold are all functions of the recovery ratio as well. The major part of power requirement of an RO process is related to operation of the high pressure feed pump. The power used by the high pressure feed pump is directly proportional to flow and pressure of the feed stream. With increased recovery ratio, the feed flow decreases, but the pressure requirement increase to some extent. With an increase in recovery ratio, both the concentrate salinity and the average feed salinity increase. Correspondingly, the feed pressure has to be increased to compensate for a higher average osmotic pressure and decreased available net driving pressure. The net result is decrease of power consumption with increasing recovery rate.

Figure 7.24 displays the relation between permeate recovery ratio, feed pressure, flow and resulting power consumption for an RO brackish system operating at recovery rates between 60% and 90%. The feed flow rate depends only on recovery ratio. The feed pressure is a complex function of recovery ratio, feed salinity, feed temperature and specific permeate flux of the membrane. The power requirement of the high pressure pump is proportional to flow and pressure. In the usual range of operating parameters, for an increase in recovery ratio, the decrease of the feed water flow has a greater effect on the power consumption then the parallel increase of the feed pressure requirement. The maximum permeate recovery ratio in brackish RO systems is mainly limited by the concentration of sparingly soluble salts in the feed water. With increasing recovery rates, concentrations of dissolved ions in the concentrate stream increase and may reach a level of spontaneous precipitation.

The precipitant layer, or scale, reduces water flow along the feed channels and diffusion across the membrane. In extreme cases, a complete blockage of membrane elements may occur. The salts which are of practical importance in determining the design recovery rate are: sulfates of calcium, barium and strontium, and the concentration of reactive silica. The concen-

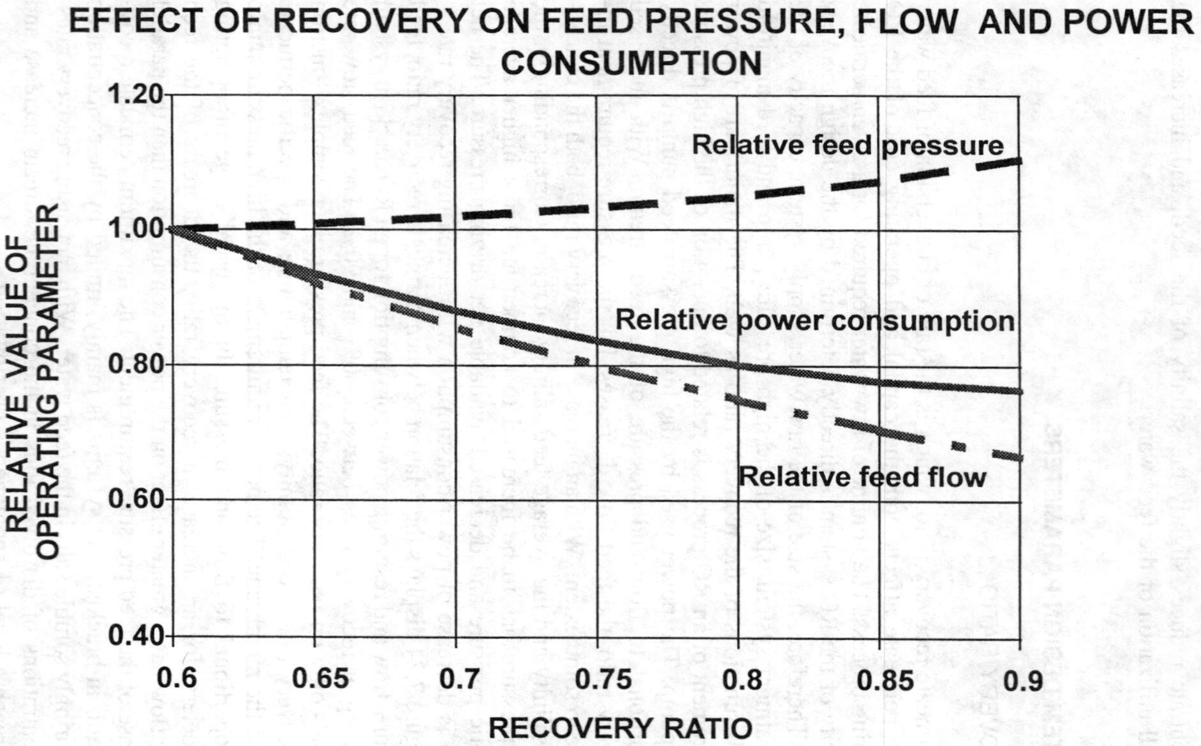

EFFECT OF RECOVERY ON FEED PRESSURE, FLOW AND POWER CONSUMPTION

Relative feed pressure

Relative power consumption

Relative feed flow

RELATIVE VALUE OF OPERATING PARAMETER

RECOVERY RATIO

Figure 7.24. Effect of recovery ratio on feed pressure, feed flow and power consumption.

tration limits of these constituents are listed in Table 7.2. The practical limit of recovery rate for brackish water systems is about 90% for low salinity (>1000 ppm TDS) feed, down to 60–65% for highly brackish (5000–7000 ppm TDS) water. In RO seawater systems, the upper limit of permeate recovery is determined by the maximum allowable feed pressure (usually based on pressure vessel limits) and permeate salinity specifications. The usual range of permeate recovery of commercial seawater systems is 35%–45%. The rationale for this limit is based on the following discussion of pressure requirements for seawater RO systems. Osmotic pressure of common seawater ranges from 380 psi to 480 psi (26–33 bar). At a recovery ratio of 45% the average concentration factor is about 1.33, which results in average osmotic pressure range of 505 psi to 635 psi (35–44 bar) depending on the salinity of the initial feed. This value has to be increased by about 60 psi (4 bar) to account for pressure drop over the system.

Since the majority of seawater systems are designed to operate at an average flux rate of about 8 GFD, the net driving pressure required is about 220 psi (17 bar). This value of NDP has to be increased by about 30% to compensate for expected flux decline due to fouling and membrane compaction over time. Therefore the design feed pressure should be in the range of 850 psi to 980 psi. For the purpose of these calculations feed temperature of 25°C has been assumed. To compensate for the decrease of permeate flux with decreasing temperature, the NDP has to be increased about 3% for each degree C under 25°C. While membrane elements can operate at significantly higher feed pressure, pressure vessels are rated at 1000–1200 psi (70–83 bar). Thus, to avoid damage to pressure vessels, the recovery rate should not be increased since this would result in higher feed pressure requirements. Another limiting factor of higher recovery rate in seawater systems is permeate salinity. With higher recovery, the average feed salinity increases, resulting in higher permeate salinity. Higher feed water temperature also increases the rate of salt passage to the permeate. For these reasons, at most locations, the recovery rate in seawater systems is limited to 45%. For locations where seawater has low relative salinity and feed water temperature does not exceed 25°C, slightly higher recovery rates are possible.

PERMEATE FLUX RATE

Permeate flux represents the water flow rate through a specified area of membrane surface. That is, part of the feed water crosses the membrane leaving on the feed side rejected dissolved constituents and water born particles. The value of the average flux rate (the flux rates averaged over the entire system) is a very important consideration in designing a system.

TABLE 7.2. Hydranautics Design Limits.

Water Type	SDI	Flux	% Flux Decline/Year
Average flux rates and expected % decrease in flux per year:			
Surface water	(SDI 2–4)	8–14 GFD	7.3–9.9
Well water	(SDI < 2)	14–18 GFD	4.4–7.3
RO Permeate	(SDI < 1)	20–30 GFD	2.3–4.4

Membrane Type	Abbreviation	% SP Increase/Year
Expected % salt passage/increase per year:		
Cellulosic membrane	CAB1, CAB2, CAB3	17–33
Composite membrane		
Brackish, low pressure	ESPA	3–17
Brackish, polyamide	CPA2, CPA3, CPA4	3–17
Seawater	SWC1, SWC2	3–17
Softening, polyvinyl deriv.	PVD1	3–17

Membrane Diameter (in)	Max (GPM)	Max (m³/hr)	Min (GPM)	Min (m³/hr)
Maximum feed flow and minimum concentrate flow rates per vessel:				
4	16	3.6	3	0.7
6	30	8.8	7	1.6
8	75	17.0	12	2.7
8.5	85	19.3	14	3.2

TABLE 7.2. (continued).

Salt	Saturation %
Saturation limits for sparingly soluble salts in the concentrate:	
$CaSO_4$	230
$SrSO_4$	800
$BaSO_4$	6000
SiO_2	100

Condition[a]	LSI Value
Limits of saturation indices:	
LSI and SDSI without scale inhibitor	≤ -0.2
LSI & SDSI with SHMP	≤ 0.5
LSI & SDSI with organic scale inhibitor	≤ 1.8

[a]Langelier and Stiff & Davis Saturation Indices.

303

Depending on the feed water source, and how "clean" the water is, if the flux rate chosen is too high, severe fouling from the retained constituents becomes more likely. The design average permeate flux rate of the RO system uniquely defines number of membrane elements and pressure vessels required for a given system capacity. Value of the average flux rate also affects the operating parameters. Higher permeate flux rate results in lower permeate salinity and requires higher feed pressure. Relations between permeate flux, permeate salinity and feed pressure are defined by Equations (2) to (6). As mentioned above, as water flows through the system, rejected dissolved constituents and water borne particles are retained in the concentrate stream. The concentration of the retained constituents near the membrane surface is higher than in the bulk of the feed stream. With increased concentration, formation of a fouling layer on the membrane surface may occur. This excess of concentration at the membrane surface depends on the flux rate and the concentration of the rejected constituents in the feed water. The constituents which are of special concern in respect of potential membrane fouling are particles, biofragments and dissolved organics. The recommended range of average flux rate for an RO system is defined according to the quality of the feed water. The quality is defined in terms of the Silt Density Index. Because in most cases prior to construction of the desalination system, SDI values are not available, feed water is qualified in terms of water source. From the conventional sources, one can expect that the surface water will have a high concentration of potential foulants and that their concentration may experience significant seasonal variations. Feed water from deep wells is usually of good and consistent quality. It can be assumed that RO permeate, used as a feed in two pass systems, does not have any significant fouling tendency. The corresponding design range of the average flux rate is 8–14 GFD (13–23 L/m²-hr) for surface water, 14–18 GFD (23–30 L/m²-hr) for well water and 20–25 GFD (33–42 L/m²-hr) for RO permeate. Feed water to seawater systems originates either from an open intake or shallow beach wells. The range of design flux rate for seawater systems is 7–8 GFD for feed water from an open intake and 8–10 for a beach well source.

DISSOLVED SALTS SATURATION LIMITS

The saturation ratios are defined as the percentage ratio of product of concentrations of salt forming ions to solubility product of a given salt. For example the saturation ratio (SR) for $CaSO_4$ would be calculated according to the following equation:

$$SR = 100* [Ca]*[SO_4]/K_{sp} \qquad (14)$$

The [Ca] and [SO$_4$] are molar concentrations of calcium and sulfate ion.

The solubility products are based on analytical determinations of solubility, corrected for ionic strength and the temperature. In addition to the actual concentration of sparingly soluble ions in the feed water, the scaling tendency depends on a number of system parameters which can not be easily quantified. These may include the condition of membrane surface, presence and sizes of particles which may serve as crystallization centers, kinetics of nucleation and crystal formation. Presence of dissolved organics in the feed water may inhibit crystal growth. For those reasons the saturation limits, listed in Table 7.2, are based on a positive experience, i.e., based on field results. A significant safety margin exists for operation within the recommended range of concentrations. The saturation limits listed in Table 7.2 apply to operation with presence of scale inhibitor in the feed water. Operation without scale inhibitor requires that the concentrations of all constituents in the concentrate stream be below their saturation values.

RECOMMENDED DESIGN GUIDELINES

In the design of RO systems, good engineering practice should be followed. The RO membrane elements, which are critical components of the system, are affected by the operation of all system components. Special attention should be devoted to provide the best possible feed water quality. System components and materials of construction should be selected to provide stable, long-term performance. Pretreatment systems should be designed to produce water quality compatible with the membrane element type used. Instrumentation and control equipment should be adequate to give a complete set of information on system performance and to alert an operator of operation conditions outside the design limits. Table 7.2 contains a summary of the recommended range of design operating parameters for RO systems equipped with Hydranautics membrane elements. In addition to operating parameters, which were discussed already, Table 7.2 includes the expected range of values for salt passage increase and flux decline. These parameters are incorporated in calculations of projected long term permeate salinity and feed pressure. These two parameters account for the change in membrane performance during long-term operation. The performance changes are a result of membrane fouling and cleaning, exposure to residual free chlorine (PA membranes) and hydrolysis (CA membrane). Except for an RO system operating on very clean water, membrane fouling to various degrees is eventually experienced. Depending how severe the fouling is, cleaning may not always be capable of restoring performance to the initial level. A good system design should include some provision for membrane performance deterioration with time.

PARAMETERS AFFECTING SYSTEM PERFORMANCE

FEED WATER SALINITY

Some RO systems experience fluctuation of feed water composition during operation. This may be due to seasonal fluctuation of feed water salinity supplies, or due to intermittent operation of a number of water sources of different salinity. As long as different feed water compositions will not require a change in the system recovery ratio, changing feed water composition will affect only the required feed pressure and permeate water salinity. Figure 7.25 shows the change in required feed pressure and projected permeate salinity as a function of feed salinity for a system operating at an average flux rate of 15 gfd and recovery rate of 85%. Calculations were made for two composite polyamide membrane types: ESPA and CPA2. The ESPA membrane has specific flux of 0.24 gfd/psi net, which is twice as high as specific flux of CPA2 membrane. It can be seen that for both membranes feed pressure and permeate salinity increase with feed salinity in a similar way. The rate of increase in permeate salinity is higher than the rate of increase in feed pressure. If the different feed water also contains concentrations of sparingly soluble salts higher than in the design feed water, then the recovery ratio may have to be reduced to avoid the possibility of precipitation of scale from the concentrate stream.

FEED PRESSURE

RO systems equipped with spiral wound membrane elements are designed to operate at a constant flux rate (i.e., to produce a constant permeate flow). Over operating time, the feed pressure is adjusted to compensate for fluctuation of feed water temperature, salinity and permeate flux decline due to fouling or compaction of the membrane. For the purpose of specifying the high pressure pump, it is usually assumed that specific flux of the membrane will decline by about 20% in three years. The pump has to be designed to provide feed pressure corresponding to the initial membrane performance and to compensate for expected flux decline. If the RO system is equipped with centrifugal pump, then the conventional approach is to use an oversized pump and regulate feed pressure by throttling (partially close the feed valve). Today an increasing number of RO system use electric motors with variable speed drives, which enable adjustment of flow and feed pressure of the pump over a wide range with very little loss in efficiency. The variable speed drive reduces unproductive pressure losses which were common in the past. Some RO systems use positive displacement pumps (piston or plunger pumps) as a high pressure process

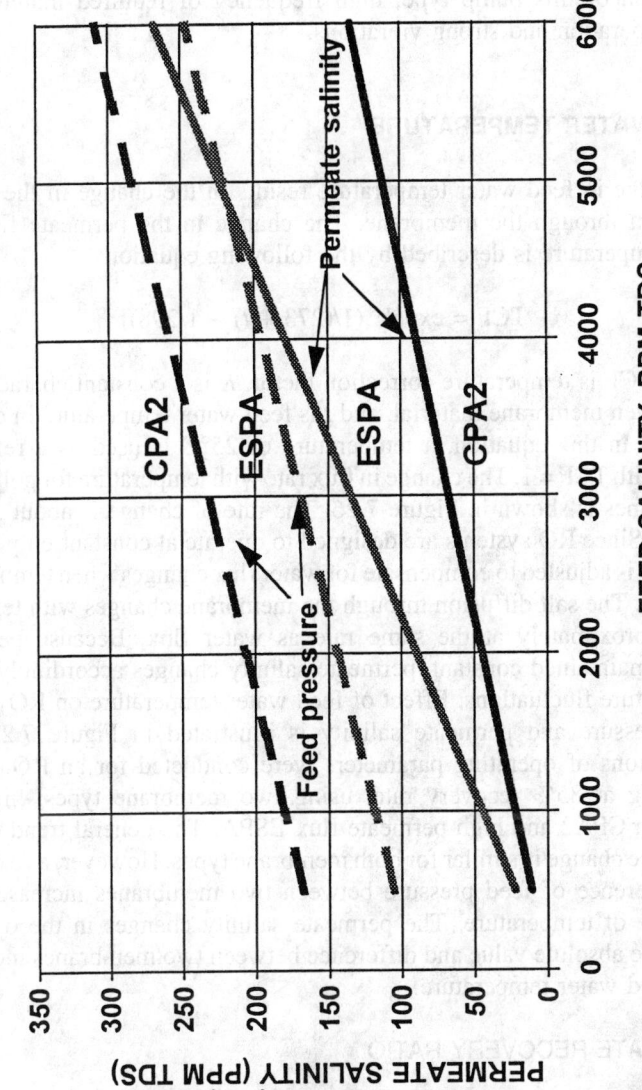

Figure 7.25. Effect of feed salinity on system performance.

pump. The positive displacements pumps enable regulation of feed pressure at constant pump output, with little change of pump efficiency. Positive displacement pumps are less common in RO systems due to capacity limitation of this pump type, high frequency of required maintenance, noisy operation and strong vibrations.

FEED WATER TEMPERATURE

Change in feed water temperature results in the change in the rate of diffusion through the membrane. The change in the permeate flux rate with temperature is described by the following equation:

$$TCF = \exp(K*(1/(273 + t) - 1/298)) \qquad (15)$$

where TCF is temperature correction factor, K is a constant characteristic for a given membrane material, and t is feed water temperature in degrees Celsius. In this equation, a temperature of 25°C is used as a reference point, with TCF = 1. The change in flux rate with temperature for polyamide membranes is shown in Figure 7.26. The rate of change is about 3% per degree. Since RO systems are designed to operate at constant output, feed pressure is adjusted to compensate for water flux changes when temperature changes. The salt diffusion through the membrane changes with temperature approximately at the same rate as water flux. Because permeate flux is maintained constant, permeate salinity changes accordingly to the temperature fluctuations. Effect of feed water temperature on RO system feed pressure and permeate salinity is illustrated in Figure 7.27. The calculations of operating parameters were conducted for an RO system operating at 85% recovery rate, using two membrane types; high salt rejection CPA2 and high permeate flux ESPA. The general trend of performance change is similar for both membrane types. However, as expected the difference of feed pressure between two membranes increases with decrease of temperature. The permeate salinity changes in the opposite way. The absolute value and difference between two membranes increases with feed water temperature.

PERMEATE RECOVERY RATIO

Recovery ratio affects system performance, i.e. permeate salinity and feed pressure, by determining the average feed salinity. The average feed salinity is calculated from feed salinity using the average concentration factor. For calculation of the average concentration factor (ACF), a logarithmic dependency on recovery ratio (R) is assumed:

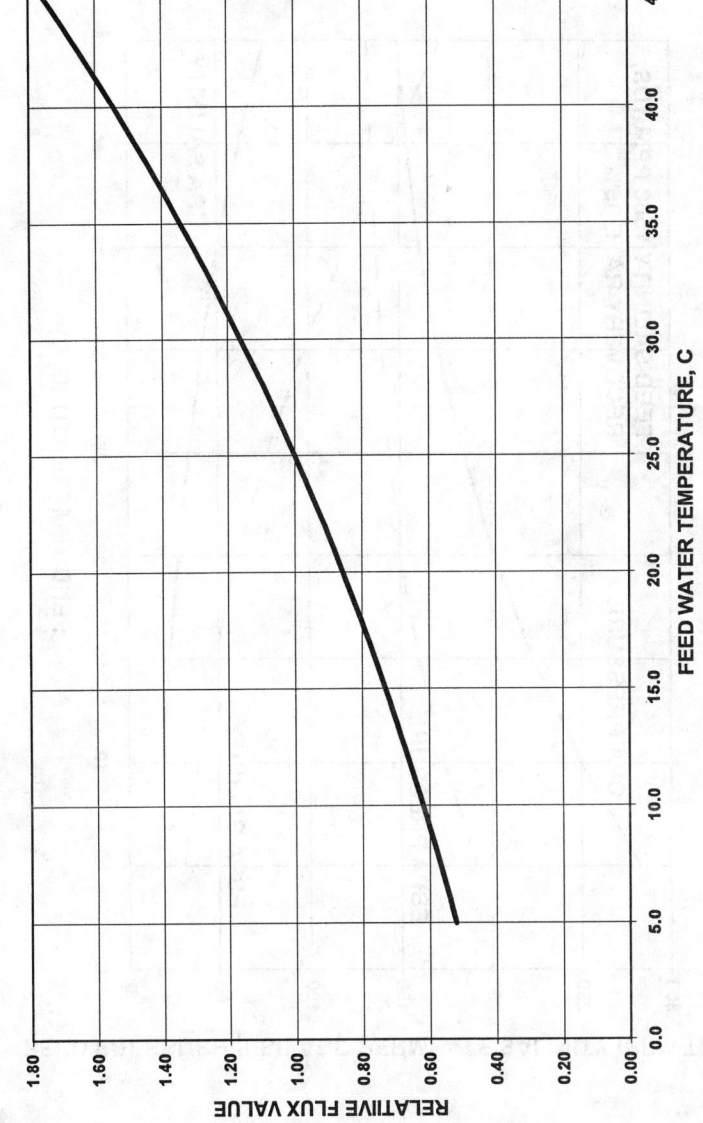

Figure 7.26. Effect of feed water temperature on specific permeate flux.

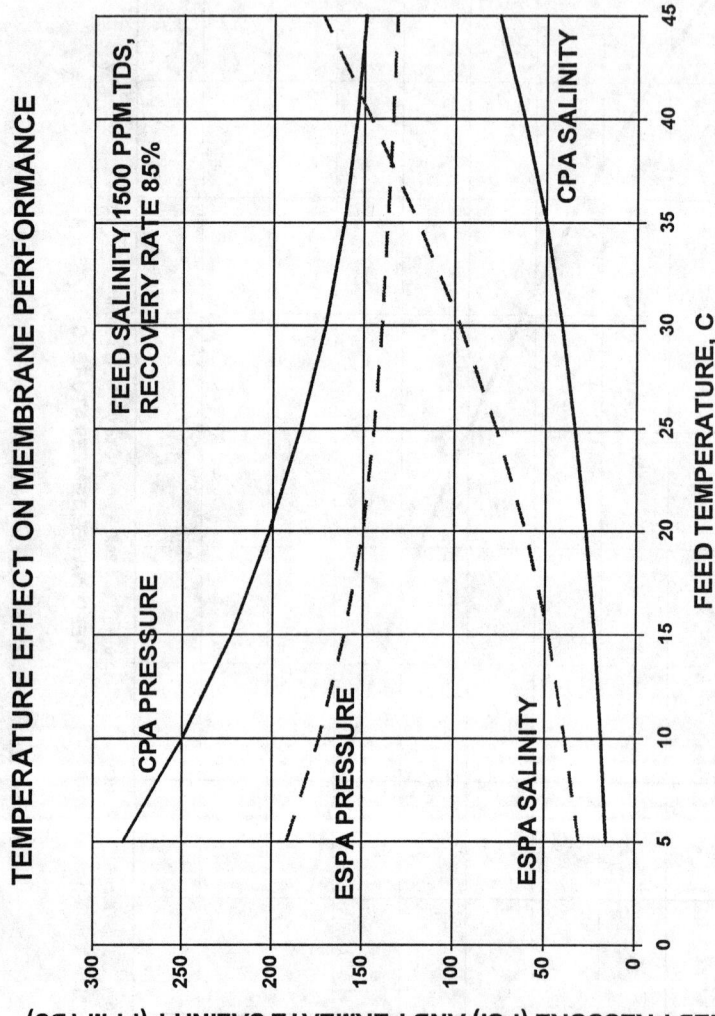

Figure 7.27. Effect of feed water temperature on feed pressure and permeate salinity.

310

$$ACF = \ln(1/(1 - R))/R \tag{16}$$

where "ln" represents the natural log.

Because recovery rate strongly affects process economics, there is a tendency to design operation of RO systems at the highest practical value.

MEMBRANE COMPACTION AND FOULING

During operation of RO systems, membrane material is exposed to high pressure of the feed water. Exposure of membranes to high pressure may result in an increase in the density of membrane material (called compaction), which will decrease the rate of diffusion of water and dissolved constituents through the membrane. As a result of compaction, higher pressure has to be applied to maintain the design permeate flow. In parallel, a lower rate of salt diffusion will result in lower permeate salinity. The effect of compaction is more significant in asymmetric cellulose membranes than in composite polyamide membranes. In seawater RO, where the feed pressure is much higher than in brackish applications, the compaction process will be more significant. Higher feed water temperature will also result in a higher compaction rate. Usually membrane compaction results in few percent flux decline, and has strongest effect during the initial operating period. Membrane fouling has a negative effect on membrane performance, and in extreme cases may result in non-reversible membrane degradation. The membrane fouling process is usually referred to as deposition of inorganic or organic substances on the membrane surface and/or blockage of feed channels. In the initial stages of membrane fouling, performance changes are similar to those caused by the compaction process. The fouling process is usually associated with an increase in pressure drop. An uncontrolled fouling process may lead to very severe performance degradation and even to complete destruction of membrane elements. The most effective way to control membrane fouling is to identify the origin of the fouling process and eliminate it by the modifying pretreatment process or operating conditions. Foulant deposits can be removed from the membrane surface by chemical cleaning. However, success of the cleaning procedure depends on the age of the foulant deposit, and on proper selection of the cleaning solution.

FEED WATER PRETREATMENT

MEMBRANE FOULING CONSIDERATIONS

The feed water, depending on its source, may contain various concentrations of suspended solids and dissolved matter. Suspended solids may consist of inorganic particles, colloids and biological debris such as micro-organisms and algae. Dissolved matter may consist of highly soluble salts, such as chlorides, and sparingly soluble salts, such as carbonates, sulfates, and silica. During the RO process, due to removal of permeate, the volume of feed water decreases along the RO unit, and the concentration of suspended particles and dissolved ions increases. Suspended particles may settle on the membrane surface, thus blocking feed channels and increasing friction losses (pressure drop) across the system. Sparingly soluble salts may precipitate from the concentrate stream, create scale on the membrane surface, and result in lower water permeability through the RO membranes (flux decline). This process of formation of a deposited layer on a membrane surface is called membrane fouling and results in performance decline of the RO system.

The objective of the feed water pretreatment process is to improve the quality of the feed water to the level which would result in reliable operation of the RO membranes. The quality of the feed water is defined in terms of concentration of suspended particles and saturation levels of the sparingly soluble salts. The common indicators of suspended particles used in the RO industry are turbidity and Silt Density Index (SDI). The maximum limits specified by majority of membrane manufacturers are: turbidity of 1 NTU and SDI of 4–5. Continuous operation of an RO system with feed water which has turbidity or SDI values near the limit levels may result in a significant membrane fouling. For long-term, reliable operation of the RO unit, the average values of turbidity and SDI in the feed water should not exceed 0.5 NTU and/or 2.5 SDI units, respectively. The indicators of saturation levels of sparingly soluble salts in the concentrate stream are the Langelier Saturation Index (LSI) and the saturation ratios. The LSI provides an indication of the calcium carbonate saturation. Negative values of LSI indicate that the water is aggressive and that it will have a tendency to dissolve calcium carbonate. Positive values of LSI indicate the possibility of calcium carbonate precipitation. The LSI was originally developed by Langelier for potable water of a low salinity. For high salinity water encountered in RO applications, the LSI is an approximate indicator only. The saturation ratio is the ratio of the product of the actual concentration of the ions in the concentrate stream to the theoretical solubility of the salts at given conditions of temperature and ionic strength. These ratios are applicable mainly to sparingly soluble sulfates of calcium, barium and strontium. Silica could be also a potential scale forming constituent. Other potential scale forming salts, such as calcium fluoride or phosphate which may be present in RO feed, seldom represent a problem.

PRETREATMENT SYSTEM CONFIGURATION

Depending on the raw water quality, the pretreatment process may consist of all or some of the following treatment steps:

- removal of large particles using a coarse strainer
- water disinfection with chlorine
- clarification with or without flocculation
- clarification and hardness reduction using lime treatment
- media filtration
- reduction of alkalinity by pH adjustment
- addition of scale inhibitor
- reduction of free chlorine using sodium bisulfite or activated carbon filters
- water sterilization using UV radiation
- final removal of suspended particles using cartridge filters

The initial removal of large particles from the feed water is accomplished using mesh strainers or traveling screens. Mesh strainers are used in well water supply systems to stop and remove sand particles which may be pumped from the well. Traveling screens are used mainly for surface water sources, which typically have large concentrations of biological debris. It is common practice to disinfect surface feed water in order to control biological activity. Biological activity in a well water is usually very low, and in majority of cases, well water does not require chlorination. In some cases, chlorination is used to oxidize iron and manganese in the well water before filtration. Well water containing hydrogen sulfide should not be chlorinated or exposed to air. In presence of an oxidant, the sulfide ion can oxidize to elemental sulfur which eventually may plug membrane elements.

Settling of surface water in a detention tank results in some reduction of suspended particles. Addition of flocculants, such as iron or aluminum salts, results in formation of corresponding hydroxides; these hydroxides neutralize surface charges of colloidal particles, aggregate, and adsorb to floating particles before settling at the lower part of the clarifier. To increase the size and strength of the flock, a long chain organic polymer can be added to the water to bind flock particles together. Use of lime results in increase of pH, formation of calcium carbonate and magnesium hydroxide particles. Lime clarification results in reduction of hardness and alkalinity, and the clarification of treated water. Well water usually contains low concentrations of suspended particles, due to the filtration effect of the aquifer.

The pretreatment of well water is usually limited to screening of sand, addition of scale inhibitor to the feed water, and cartridge filtration (Fig. 7.28). Surface water may contain various concentrations of suspended particles, which are either of inorganic or biological origin. Surface water usually requires disinfection to control biological activity and removal of suspended particles by media filtration. The efficiency of filtration process can be increased by adding filtration aids, such as flocculants and organic polymers. Some surface water may contain high concentrations of dissolved organics. Those can be removed by passing feed water through an activated carbon filter. Depending on composition of the water, acidification and addition scale inhibitor may be required.

The flow diagram of pretreatment system configuration for surface water source is shown in Fig. 7.29. This system consists of chlorination, in line flocculation with coagulant and polyelectrolyte. After passing through static mixer water is filtered in sand filter. Concentration of dissolved organics is reduced in activated carbon filter. The activate carbon filter also reduces free chlorine to chloride ion, protecting composite membranes against oxidation damage. Before cartridge filters acid and scale inhibitor are added to feed stream to prevent precipitation of sparingly soluble salts from the RO concentrate.

At the pH range encountered in the natural waters, the carbonate alkalinity is mainly associated with the presence of the bicarbonate ion, HCO_3^-. At sufficiently high concentrations of bicarbonate and calcium ions, a calcium carbonate scale can be formed according to the following reaction:

$$Ca^{++} + 2HCO_3^- = CaCO_3 + H_2CO_3 \qquad (17)$$

Formation of carbonate scale can be controlled by acidification and conversion of the bicarbonate ion to carbon dioxide and water:

$$HCO_3^- + H^+ = CO_2 \uparrow + H_2O \qquad (18)$$

The acidification is usually conducted with sulfuric or hydrochloric acid. The amount of required addition of acid to the feed stream is calculated to obtain a pH of the concentrate stream that would result in a negative value of LSI. The tendency of carbonate scale formation can be reduced by maintaining a low concentration, several ppm, of scale inhibitor in the feed/concentrate stream. In the past sodium hexametaphosphate was the scale inhibitor of choice. Today, more effective, modern scale inhibitors used in RO applications are organic polymers—mainly derivatives of acrylic acid. They retard scale formation by binding crystallization centers and delaying growth of the scale crystals. These scale inhibitors enable

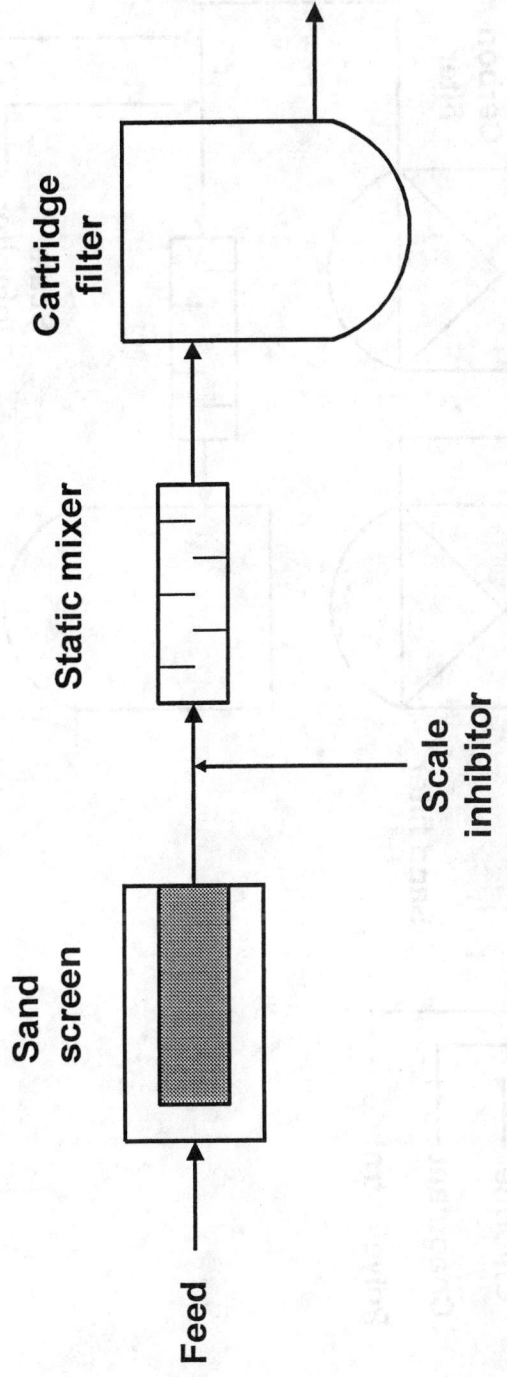

Figure 7.28. Diagram of pretreatment system operating on feed water from a well.

315

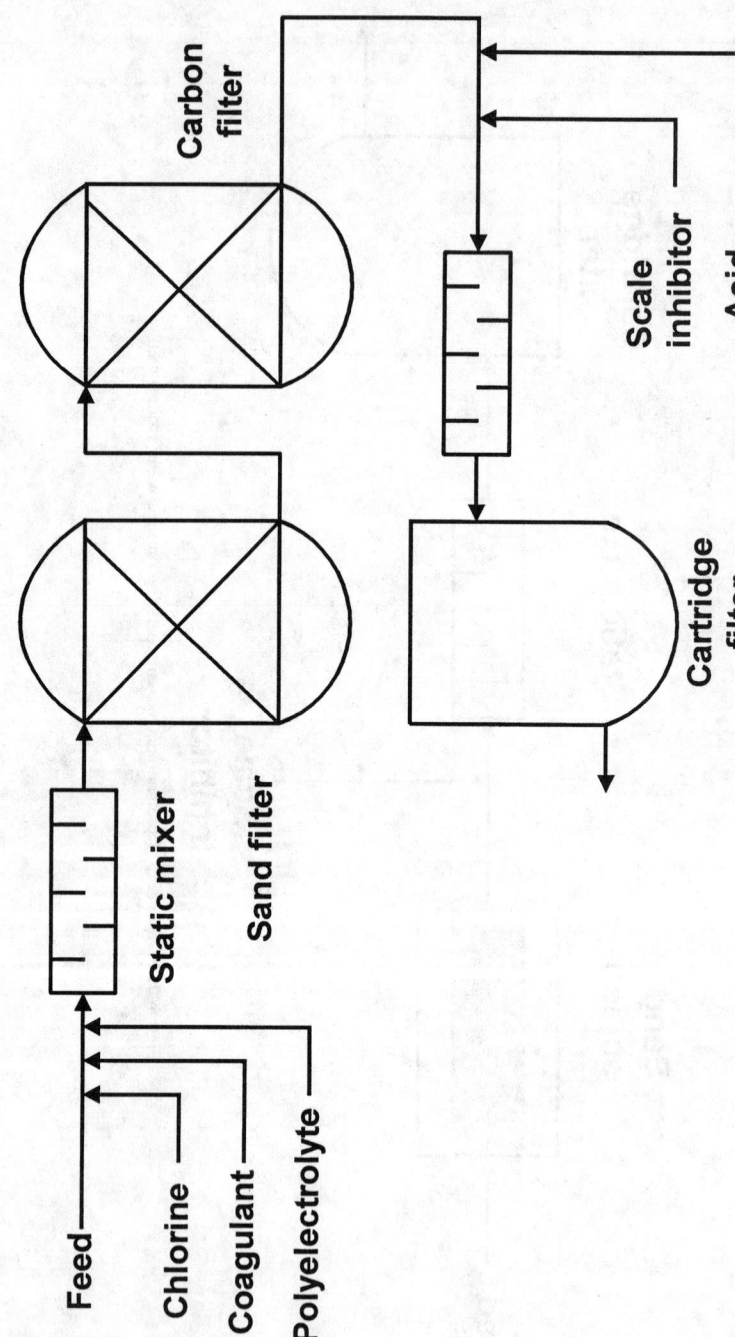

Figure 7.29. Diagram of pretreatment system operating on feed water from a surface source.

operation of an RO system at high concentrations of calcium and bicarbonate ions in the concentrate stream. These concentrations correspond to the LSI values of 1.5–2. The organic scale inhibitors also enable one to significantly exceed the saturation values of other scale forming salts, namely sulfates of calcium, barium and strontium. It was a common belief that the rate of formation of silica scale is not affected by the presence of scale inhibitors. The practical limit was a concentration of about 150 ppm of reactive silica, as SiO_2, in the concentrate. Presently, some suppliers of scale inhibitors claim availability of special scale inhibitors that can prevent formation of silica scale from solutions with a silica concentration up to 250 ppm.

The practical saturation limits for common scale forming salts, encountered in RO applications, are listed in Table 7.3. The effectiveness of organic scale inhibitors can be adversely affected by the presence of the oxidized iron in the feed water. Free chlorine (and chloramine) is commonly used as a disinfectant in potable water systems and for control of biological activity in the feed water originating from surface sources. Cellulose acetate membranes can operate with free chlorine in the feed water up to the concentration of about 1 ppm. The common level of chlorine concentration is 0.3–0.5 ppm. The stability of polyamide barrier of composite membranes is adversely affected by free chlorine. Long term exposure of polyamide membranes to free chlorine results in an increase of salt passage and water flux. In RO systems equipped with polyamide membrane, free chlorine present in the feed water has to be reduced to chloride ion before entering the RO membranes. This is usually accomplished by dosing sodium bisulfite to the feed water, at the inlet to the cartridge filters. The stoichiometric requirement is 1.4 ppm of sodium bisulfite for 1.0 ppm of free chlorine. In practice the sodium bisulfite dosing rate is about twice as high as required by the stoichiometry of the chemical reaction.

In some RO systems activated carbon filters are used to reduce free chlorine. An activated carbon filter, being a static device, provides higher reliability than the sodium bisulfite dosing pump. However, the activated carbon has to be maintained in its active state by periodic regeneration. Also, high surface area carbon particles may be a breeding ground for bacteria in the carbon layers close to the filter outlet. In some RO systems equipped with polyamide membranes UV radiation is applied to sterilize feed water. This method of disinfection is very effective and does not represent a problem from the point of stability of polyamide membrane polymer. The major disadvantage of UV equipment is high cost and lack of biocidal effect down stream of the UV sterilizer. Cartridge filters, almost universally used in all RO systems prior to the high pressure pump, serve as the final barrier to water born particles. The nominal rating commonly used in RO applications is in the range of 5–15 microns. Some systems use cartridges with micron ratings as low as 1 micron. There seems to be little benefit from lower micron rated filters as such filters require a high replacement rate with relatively small improvement in the final feed water quality. Recently, new pretreatment equipment has been introduced to the RO market. It consists of backwashable capillary microfiltration and ultrafiltration membrane modules. This new equipment can operate reliably at very high recovery rates and low feed pressure. The new capillary systems can provide better feed water quality than a number of conventional filtration steps operating in series. The cost of this new equipment is still very high compared to the cost of an RO unit.

WASTEWATER RECLAMATION USING RO TECHNOLOGY

INTRODUCTION

Application of reverse osmosis technology for municipal wastewater reclamation traces back to the early stages of commercialization of the RO process. It was realized soon that application of membrane technology to treatment of municipal effluent represents a unique challenge due to a very high fouling potential of the treated stream. Therefore, development of effective pretreatment process and demonstration of performance stability was the main objective of the early works [18–21]. As early as the 1960s the City of San Diego conducted first testing and demonstration program of application RO technology for augmentation of water supply by reclamation of wastewater. This attempt was not successful due to severe membrane fouling [22]. Since then, a number of field tests have conducted at different sites, which enabled development of process parameters and system components of the future commercial plants [23–26]. Application of reverse osmosis technology for wastewater reclamation in Southern California traces back to the early stages of commercialization of the RO process. The first large reverse osmosis plant, which processes wastewater is a part of what is known as "Water Factory 21" in Orange County, California, commenced operation in the late 1970s. This RO system has 5 MGD of product capacity and reduces salinity of municipal waste water after tertiary treatment. Product water after blending is injected into local aquifers to prevent seawater intrusion [27,28]. The next large RO system for water reclamation, the Arlington Desalter, located in Riverside County, California, commenced operation in 1990. This system processes agricultural drainage water of about 1000 ppm TDS salinity, which contains high concentration of NO_3 (100 ppm) and SiO_2 (40 ppm). The plant produces 6 MGD of low salinity water by blending 4 MGD of RO permeate with 2 MGD of ground water. The blending ratio is determined by the limit of nitrate ion concentration in the blend water, which has to be below 40 ppm [29]. Today, a large number of new membrane projects of water reclamation are under design or extensive pilot testing. In the majority of them, the new advanced, membrane pretreatment methods are evaluated. The objective is to improve stability of performance of RO membranes and improve process economics. More information on the Water Factory 21 are found in Chapter 23 by Mills et al.

CONVENTIONAL PRETREATMENT

The municipal effluent after secondary treatment contain high concentration of colloidal particles, suspended solids and dissolved organics. The municipal treatment process usually includes biological treatment (activated sludge clarification) which results in high level of biological activity in the effluent. Prior to RO this water has to be treated to reduce concentration of colloidal and solid particles and arrest biological activity. A typical configuration of conventional pretreatment is shown in Figure 7.30, which outlines the tertiary pretreatment process applied currently at Water Factory 21. The current pretreatment process is a result of evolution, improvements and simplification of the original design [22]. The pretreatment consists of flocculation, lime clarification, recarbonation with CO_2 and settling and slow gravity filtration. The biological activity is controlled applying chlorination. Lime clarification is a very effective process in improving feed water quality, but is expensive, requires large area and produces sludge, which can be difficult to dispose. In some smaller systems the lime clarification and gravity filtration is replaced by in line flocculation followed by two stage pressure filtration and cartridge filtration. On the average, this simplified pretreatment produces effluent of lower quality than after the lime clarification process, but the equipment is significantly smaller and simpler to operate. The feed water after a conventional pretreatment has a high fouling potential. It is not uncommon that RO membranes would experience 25%–30% per year average flux decline, even with frequent membrane cleaning.

ADVANCED PRETREATMENT

Use of membranes as a definite barrier in the RO pretreatment process has been proposed in the past [22]. Ultrafiltration (UF) and microfiltration (MF) membranes have the ability to produce feed water of significantly better quality than the conventional pretreatment process based on lime clarification, followed by media and cartridge filtration. However, the conventional, spiral wound configuration of ultrafiltration membrane ele-

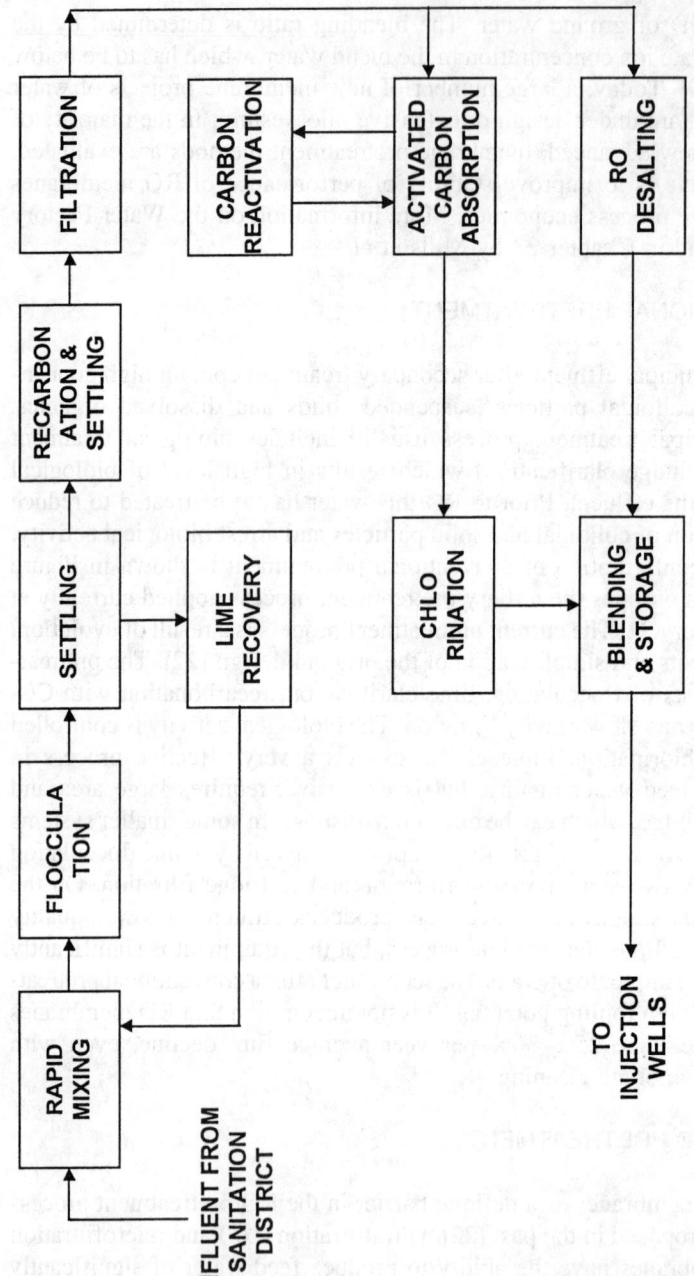

Figure 7.30. Flow diagram of water treatment process at Water Factory 21 plant.

ments was not suitable for treatment of highly fouling waste water. The UF elements could not operate at high flux rates without severe fouling of membrane surfaces and plugging of feed channels. High cross flow feed velocities, required to reduce concentration polarization, resulted in a high power consumption. Membrane cleaning, frequently required, was cumbersome and not very effective in restoring permeate flux. New micro-filtration and ultrafiltration technology offered recently [29] is based on a fat capillary membrane configuration. The capillary bore is of 0.7–0.9 mm diameter. Outside diameter of the capillary is in the range of 1.3–1.9 mm. Membrane material consists of polypropylene, sulfonated polyether sulfone or cellulose acetate polymer. In some capillary elements design configuration the feed-permeate flow direction is outside-in (i.e., feed water flushes the outside of the capillary fiber, water permeates through the wall and is collected as a permeate inside the fiber). Other element configurations have inside-out flow direction.

There are two common novel properties of the new commercial capillary equipment:

(1) Frequent, short duration, automatically sequenced flushing (or back-flushing in some models) of the capillary fibers, which cleans the membrane surface and enables to maintain a stable permeate flux rates with a little off-line time

(2) Ability to operate at a very low feed cross flow velocity, or even in a direct filtration flow (dead end) mode

The off-line time due to pulse cleaning is very short, comparable to off line time of conventional filters due to filter backwashing. The frequent pulse cleaning of the capillary device results in a stable permeate flux rates. The feed water pressure is in the range of 1 to 2 bar. Operation at low feed pressure and low rate of feed cross flow or in a direct filtration mode results in high recovery rates and very low power consumption, of about 0.4 kWhr/kgallon (0.1 kWhr/m^3) of filtrate. The membrane type is either microfiltration (nominal pore size 0.2 micron) or ultrafiltration (molecular weight cut off 100,000–200,000 Dalton). The dimensions of capillary ultrafiltration modules are in the range of 40″–52″ (100–130 cm) long and 8″–13″ (20–32 cm) in diameter. In actual field operation, a single module can produce 8,000–40,000 gallons (30–150 m^3/day) of filtrate.

This new capillary technology has been developed initially for treatment of a potable water, which originated from surface sources. Compared to the conventional water treatment technology, the new process offers modular design, high output capacity from a small foot print, no need for continuous handling and dosing of chemicals, and limited labor requirements. The major advantage, however, is inherent to membrane technology: the existence of a membrane barrier between feed and permeate which enables a several log reduction of colloidal particles and pathogens. The above technology has been extensively tested and a large number of systems, mainly utilizing microfiltration membranes, are already in operation.

Following the successful applications in potable water applications, the capillary technology has been proposed and tested as a potential pretreatment for RO systems operating on a highly fouling water. One of the first targets was RO processing of municipal effluents. The objectives were to replace the expensive and cumbersome conventional tertiary effluent treatment and to increase the current level of the design flux rate of the RO system. The field tests have been conducted for over three years now [30]. Results are promising and large commercial installations are under consideration.

Another area of application is pretreatment in RO systems operating on feed water from surface sources. The objectives here are similar: replacement of the conventional pretreatment and increase of the design permeate flux rate from the RO membranes. In addition, it is expected that adding a second membrane barrier (in addition to the RO membrane) will reduce the presence of pathogens in the permeate to a level that will eliminate the need for continuous disinfection of potable water with strong oxidants. The field tests confirmed the technical and economic feasibility of such design. Large capacity systems, combining capillary UF pretreatment with RO technology are currently being designed. The cost of the capillary membrane pretreatment is estimated to be similar to the cost of an extensive conventional pretreatment which is usually required for RO reclamation of the municipal waste water. Use of capillary technology will simplify the pretreatment system and reduce the use of chemicals. The new capillary technology is capable to reliably produce RO feed water with a very low concentration of colloidal particles and bacteria. Present offering of commercial products for treatment of municipal effluent is included in Table 7.3. A comprehensive description of field tests of the new technology conducted at the Orange County Water Factory 21 test facility has been published recently [30].

CASE STUDIES

RO SYSTEM AT THE WATER FACTORY 21

The objective of operation of Water Factory 21, water treatment facility, is to produce water for underground injection to prevent seawater intrusion into the fresh water aquifer. The flow diagram of the current treatment process is shown in Figure 7.30. The current pretreatment process has evolved from the initial design [27,28]. The pretreatment process has been

TABLE 7.3. Offering of Commercial Capillary MF-UF Products for Treatment of Municipal Effluents.

Manufacturer	Memcor	Zenon	X-Flow	Aquasource
Membrane material	PP	Proprietary	SPES	CA
Type	MF	MF	UF	UF
Configuration	Capillary	Capillary	Capillary	Capillary
Nominal pore size/MWCO	0.2 micron	0.2 micron	150,000 Dalton	100,000 Dalton
Fiber I.D.	0.3 mm	0.9 mm	0.8 mm	0.94 mm
Fiber O.D.	0.7 mm	1.9 mm	1.3 mm	1.3 mm
Membrane area per module	22 m² / 233 ft²	14 m² / 150 ft²	25 m² / 265 ft²	55.4 m² / 596 ft²
Flow direction	out-in	out-in	in-out	in-out
Feed pressure range	0.5–2 bar	< 0.5 bar vacuum	0.5–2 bar	0.5–2 bar
Operation type	Dead end	Immersed fibers	Dead end	Partial recirculation
Capillary backflush	Compressed air backflush every 20 min.	Permeate backflush and air scrubbing	Permeate backflush every 20 min.	Permeate backflush every 20 min.
Typical module capacity	27 m³/d / 7000 gpd	13 m³/d / 3600 gpd	60 m³/d / 16000 gpd	130 m³/d / 36000 gpd

Membrane material: CA—cellulose acetate, SEPS—sulfonated polyethersulfone, PP—polypropylene.

optimized and some pretreatment steps have been totally removed. The use of activated carbon filters, to treat RO feed has been discontinued, after it was found that carbon filter effluent contains large amount of carbon fines which block feed channels of the RO elements. The degasifiers, initially used to remove ammonia from feed water, were bypassed after some period of operation. The degasification process results in significant cooling of the feed water and reduction of RO system capacity. It was also established that ammonia present in feed water is being converted to chloramines during chlorination and is beneficial in controlling biological activity.

PROCESS DESCRIPTION

The influent received from the Orange County Sanitation District passes the treatment stages of lime clarification, recarbonation, chlorination and media filtration. After media filtration, one-third of the stream is directed to the RO system. The rest of the media filtration effluent passes through granulated, activated carbon bed, and after chlorination is blended with the RO permeate. Figure 7.31 shows the top view of the clarifier tanks, where calcium carbonate flock together with colloidal and suspended particles is settled and removed as a sludge. Figure 7.32 shows the recarbonation tanks where feed water pH is adjusted using CO_2 gas to prevent post precipitation of calcium carbonate. The CO_2 is produced on site by thermal decomposition of calcium carbonate sludge formed in the lime clarifier. Figure 7.33 shows the top view of the gravity multimedia filters. The objective of the gravity filters is to retain suspended particulate matter carried over from the clarifier.

The early pilot study of different membrane module configurations has indicated that spiral wound and tubular configuration modules were least affected by fouling. Spiral wound configuration technology was selected for the above system because of better economics (investment cost and power consumption). In spite of continuous testing program of commercially available membrane elements conducted on site, cellulose acetate membrane elements in a spiral wound configuration seem to be the membrane of choice for treatment of municipal wastewater. This type of membrane element has been continuously specified for all subsequent membrane replacements at Water Factory 21.

The RO unit consists of six banks of membrane elements arranged in three pass array: 24:12:6 pressure vessels, each containing four elements. The membrane elements are 8″ diameter, 60″ long, 510 ft^2 nominal membrane area. Figure 7.34 shows a front view of the RO banks operating at the Water Factory 21 plant. The RO unit operates at 85% permeate recovery. Feed water at Water Factory 21, even after tertiary treatment has a very high fouling potential. Foulants existing in the RO feed at Water Factory 21 consist mainly of dissolved organics and biological fragments (algae and bacteria). The common indicator of RO feed water quality, the Silt Density Index, is not even measurable in the local feed water. TOC and COD are in the range of 10–20 ppm, in addition to a high concentration of biological debris. Due to high concentration of ammonia in the raw water, chlorination produces chloramine in a concentration of approximately 10 ppm in the RO feed. In spite of this adverse feed water quality, the membrane performance is relatively stable, the key factor being sustaining desired water flux by effective membrane cleaning procedures. Initially, the RO system was designed with 35 pressure vessels per unit, which resulted in water flux of 12.4 gfd. At this flux rate, an irreversible flux decline was experienced. To maintain rated product flow, seven pressure vessels were added to each unit, which resulted in reducing the average flux rate to about 10 gfd.

In 1987, when Hydranautics was awarded a membrane replacement contract from a competitive biding process, the RO system at Water Factory

Figure 7.31. Top view of clarifiers at Water Factory 21 plant (courtesy of OCWD).

Figure 7.32. Top view of recarbonation tanks at Water Factory 21 plant (courtesy of OCWD).

Figure 7.33. Top view of gravity media filters at Water Factory 21 plant (courtesy of OCWD).

Figure 7.34. RO trains at Water Factory 21 plant (courtesy of OCWD).

21 had already a 10-year history of successful operation. Hydranautics' scope of responsibility was limited to supply of membrane elements at specified performance, supply of new pressure vessels, review of RO feed pretreatment process, recommendations regarding cleaning procedure which would be effective and compatible with Hydranautics' membrane elements, and guarantee of membrane performance. The membrane elements supplied by Hydranautics were 8060-HSY-CAB2 membrane elements, 8″ diameter, 60″ long, made of blended cellulose acetate membrane material in a spiral wound configuration. The nominal performance of this element model at 400 psig (2.76 MPa) applied pressure for a 2 g/L sodium chloride test solution is 11,000 gpd product flow, 98% salt rejection. At the early stages of the membrane replacement process, some problems have been encountered which were related to partial incompatibility of materials of construction of the membrane elements (mainly o-rings and seals) with the harsh environment (possibly the high chloramine concentration in the Feed water). These early problems have been solved in cooperation with staff of the Orange County Water District. Currently all six RO blocks at the Water Factory 21 are equipped with Hydranautics elements.

MEMBRANE PERFORMANCE

Figures 7.35, 7.36, 7.37 and 7.38 contain performance of subunit 2C, for the period of April '89 until February '91. This unit still operates with the original load of membrane elements. The rejection rate is very stable and a salinity reduction of 94%–96% is obtained consistently (Figure 7.36). The product flow was maintained at the level of 580 GPM (Figure 7.38). Membrane fouling and its affect on productivity was initially compensated by increasing feed pressure to a maximum agreed upon value of 350 psig (2.41 MPa) (Figure 7.37). After 275 days, this maximum feed pressure was reached for the first time during the first period of low feed temperature; the winter of 1989, the required feed pressure dropped during the warmer months corresponding to 350 days through 550 days of operation, as indicated in Figure 7.38; product flow had declined between 10% to 14%. As a result of subsequent cleaning, permeate flow could be usually restored to rated system flow by cleaning membrane elements with a cleaning solution containing sodium tripolyphosphate, EDTA and anionic detergent. Membrane cleaning was conducted about every three to four weeks. The objective of application of reverse osmosis technology at the Factory 21 Plant is to reduce concentration of harmful constituents in the water. This is necessary because product water after underground injection mixes to some extend with potable water aquifer. Table 7.4 contains representative composition of the effluent obtained from the Orange County Sanitation District, RO permeate and Water Factory 21 blended

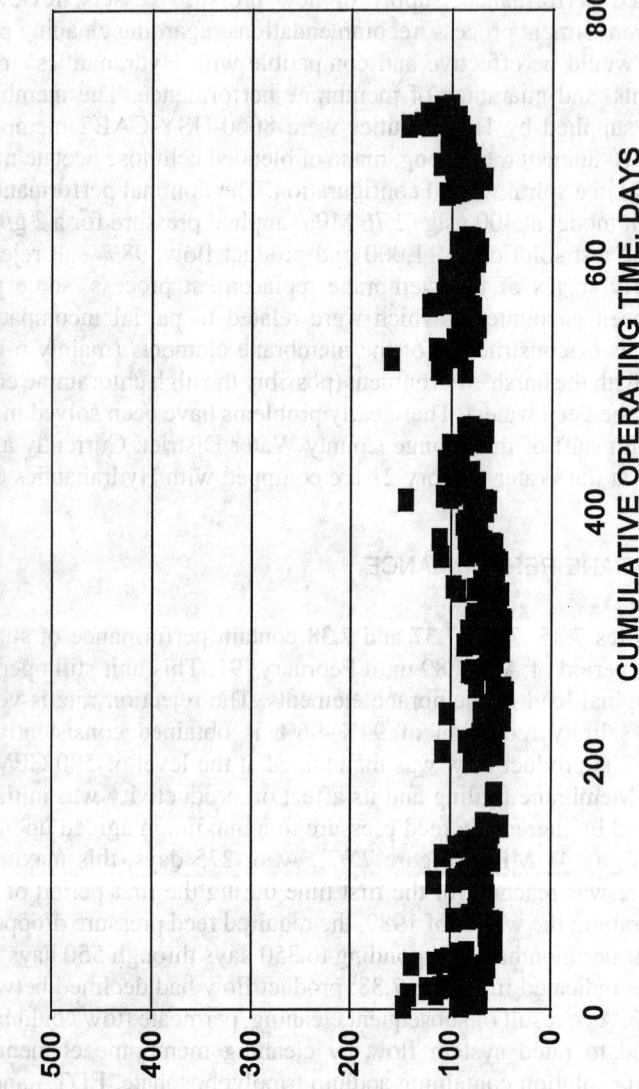

Figure 7.35. Performance of Hydranautics CAB membranes at Water Factory 21, permeate conductivity vs. time.

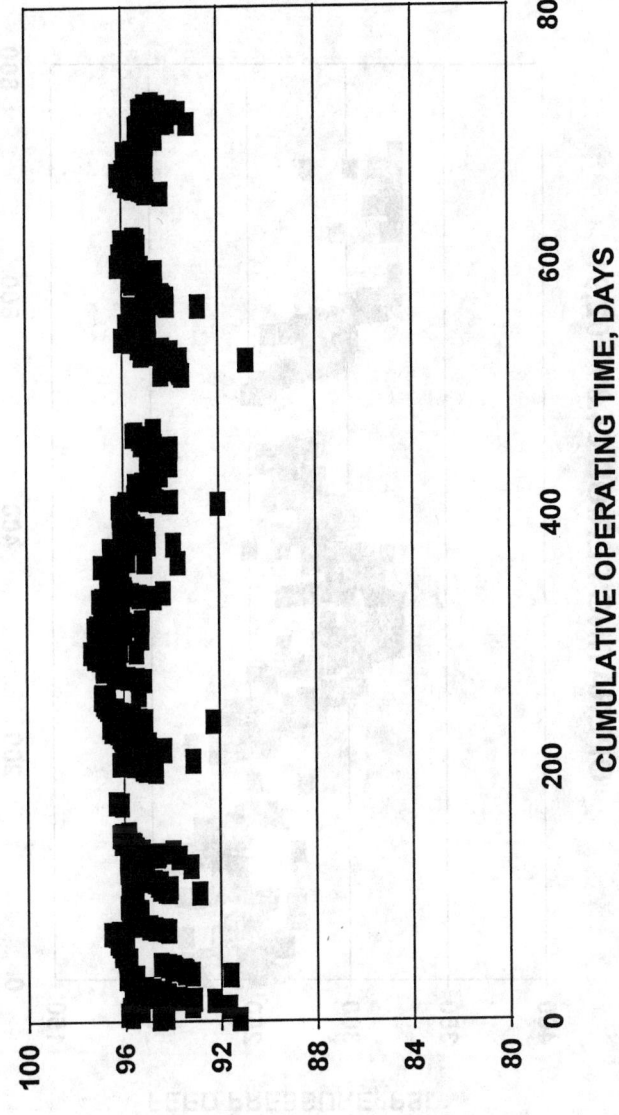

Figure 7.36. Performance of Hydranautics CAB membranes at Water Factory 21, salinity rejection vs. time.

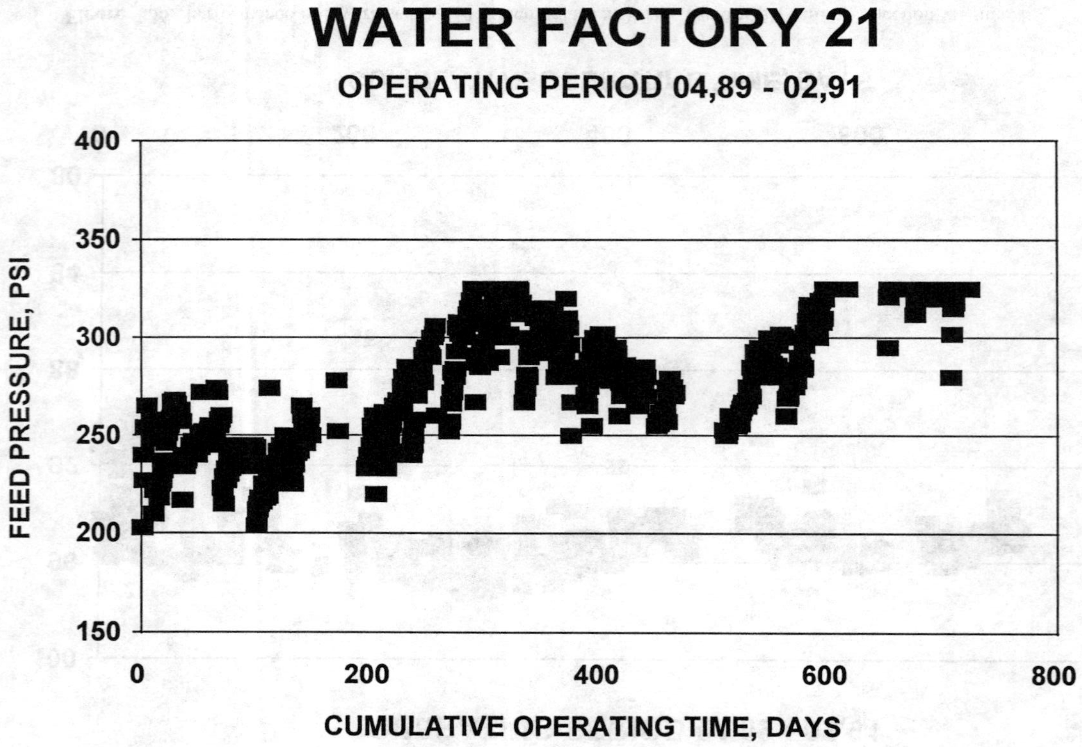

Figure 7.37. Performance of Hydranautics CAB membranes at Water Factory 21, feed pressure vs. time.

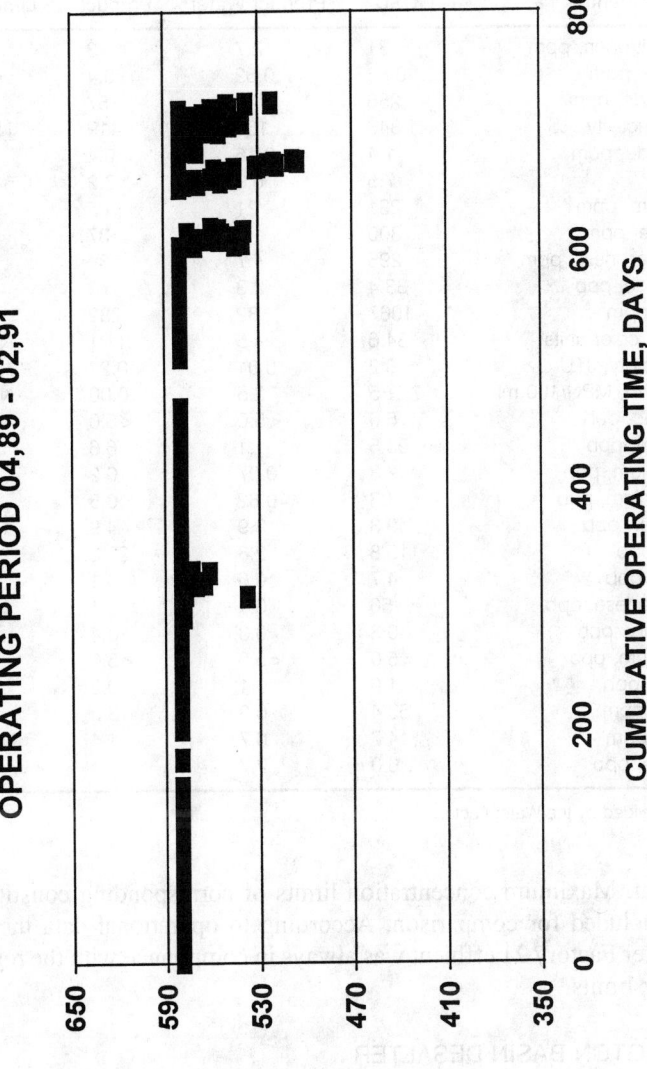

Figure 7.38. Performance of Hydranautics CAB membranes at Water Factory 21, permeate conductivity vs. time.

333

TABLE 7.4. Representative Results of Water Quality at Water Factory 21.

Constituent	Influent OCSD	From RO Product Water	WF-21 Blended Product	Regulatory Limits
Tot. nitrogen, ppm	31	2.7	3	10
Boron, ppm	0.85	0.52	0.4	0.5
Chloride, ppm	256	29	57	120
Conductivity, μS	1,848	150	419	None
Fluoride, ppm	1.4	0.16	0.4	1.0
pH	7.5	6.9	7.2	6.5–8.5
Sodium, ppm	231	21	65	115
Sulfate, ppm	300	1.4	37	125
Tot. hardness, ppm	298	4.7	34	180
Cyanide, ppb	33.4	2.3	6.3	
TDS, ppm	1067	82	232	500
Color, color units	34.6	<5	11.1	None
Turbididty, JTU	6.2	<0.01	0.27	None
Coliform, MPN/100 ml	7.2E5	2.5	0.00	None
Arsenic, ppb	<5.0	<5.0	<5.0	50
Barium, ppb	93.5	1.1	6.6	1000
Cadmium, ppb	9.3	0.07	0.2	100
Chromium, ppb	33	0.82	0.5	50
Copper, ppb	49.3	3.9	4.9	1000
Iron, ppb	113.8	2.8	22.2	300
Lead, ppb	4.7	0.6	0.1	50
Manganese, ppb	56	0.1	2.1	50
Mercury, ppb	0.3	0.3	0.4	2
Selenium, ppb	<5.0	<5.0	<5.0	10
Silver, ppb	1.6	0.1	0.2	50
COD, ppm	53.4	0.8	3.5	30
TOC, ppm	14.7	0.7	1.4	None
THM's, ppb	6.0	2.7		100

Data provided by the Water Factory 21.

effluent. Maximum concentration limits of corresponding constituents is also included for comparison. According to operational data the quality of Water Factory 21 effluent was always in compliance with the regulatory quality limits.

ARLINGTON BASIN DESALTER

The Southern California Arlington Basin contains approximately 300,000 acre feet of water. The groundwater in this basin has degraded by agricultural leachate from historic citrus grove farming operations. The agricultural drainage has increased salt concentration to a level that this groundwater is no longer usable for domestic purposes. Because of poor groundwater quality, all pumping has been discontinued in the area. This resulted in impaired groundwater seeping to the surface and draining into the Santa Ana River and other adjacent groundwater basins, thereby degrading downstream water supplies.

The Arlington Basin Desalting project initiated, as a solution to this problem, serves the following purposes:

(1) Reduce salts entering into the Santa Ana River.

(2) Provide a clean water supply.

(3) Restore the groundwater in the Arlington Basin to a usable condition.

(4) Restore the condition of the Arlington Basin for future water storage.

The Arlington reverse osmosis plant was designed by Camp Dresser & McKee of Ontario, California. Hydranautics furnished, installed and commissioned the RO system. Hydranautics also was responsible for operator training. The RO system was designed to provide 6 MGD of blended product water containing less than 500 ppm TDS by mixing 4 MGD of degassifed permeate from the three RO trains (each rated at 1.33 MGD) with 2 MGD of groundwater treated by granular activated carbon followed by aeration stripping. The RO system operates at a permeate recovery rate of 77%. The RO plant is designed to allow for expansion up to 8 MGD, by addition of a fourth RO train. The Plant data is summarized in Table 7.5.

PROCESS DESCRIPTION

The schematic process flow diagram of the Arlington Desalter is given in Figure 7.39. Feed water from the five local brackish wells is pumped to the plant site where it is split into two streams. Out of the total raw water flow of 7 MGD, provision exists for passing 2 MGD through Granular Activated Carbon (GAC) filters, to remove dissolved organic compounds, mainly dibromochloropropane (DBCP). At present, due to lower than expected concentration of DBCP in the groundwater, the GAC filters are bypassed and the blend stream is only treated by aeration stripping. The remaining flow, 5 MGD, is used as feed for the RO system. The RO feed water is treated by dosing of scale inhibitor and sulfuric acid to a pH of 6.9 and is filtered through 5 micron cartridge filters. After the filtration feed water is pressurized to approximately 210 psi with Afton vertical turbine pumps, the pressurized feed enters three parallel RO trains operating at 77% permeate recovery.

Each train contains 44 pressure vessels, 8″ diameter, in a two pass 33:11 array. The pressure vessels each contain six Hydranautics spiral wound, composite, membrane elements, model 8040-LSY-CPA2. The average water flux rate of the membranes is 13.8 gfd. The first pass

TABLE 7.5. **Arlington Desalter Plant Data.**

Start up:	July 1990
Commissioning:	September 1990
Plant capacity	6 MGD
RO permeate	4 MGD
GAC effluent (blend water)	2 MGD
Number of RO trains	3
Array	33:11
Number of elements per train	264
Element type	8040-LSY-CPA2
Permeate recovery	77%
Feed water type	Well water
	1200 ppm TDS
	90–100 ppm NO_3
	44 ppm SiO_2
Product water quality	< 500 ppm TDS
Requirement, after blending	40 ppm NO_3
Pretreatment	pH adjustment, scale inhibitor cartridge filtration
Post-treatment	Degasifier
Design feed pressure	160–205 psi
High pressure pump	
Power consumption (1)	1.3 KWHR/1,000 gallons final effluent
Wire to water efficiency	82.5%

(1) Pump equipped with Power Recovery Turbine.

averages 15 gfd and the second pass averages 11 gfd. Permeate flow from the RO trains is combined with the blend stream at the ratio 2:1. The design blend ratio was based on the projected concentration of nitrate in the wells and in the permeate water, with a target concentration corresponding to California drinking water standard of not more than 45 ppm of nitrate in the total plant effluent. This blended effluent is of potable water quality and flows to the storm water channel and eventually recharges the groundwater basin. The concentrate stream from each RO train passes through an energy recovery turbine, which is a reverse running pump mechanically coupled with the high pressure pump. The combined concentrate from the plant is conveyed to the Orange County Sanitation District through the Santa Ana Regional Interceptor (SARI) line. After mixing with municipal sewage, and primary and secondary treatment of the Sanitation District effluent is split for further treatment by OCWD Water Factory 21, or direct disposal to the ocean.

The flow diagram for the Arlington Basin Desalter Plant is shown in Figure 7.32. The RO unit incorporates innovative design features to assure efficient and stable plant operation. One of the features is the control of system output by permeate throttling. The objective of using permeate

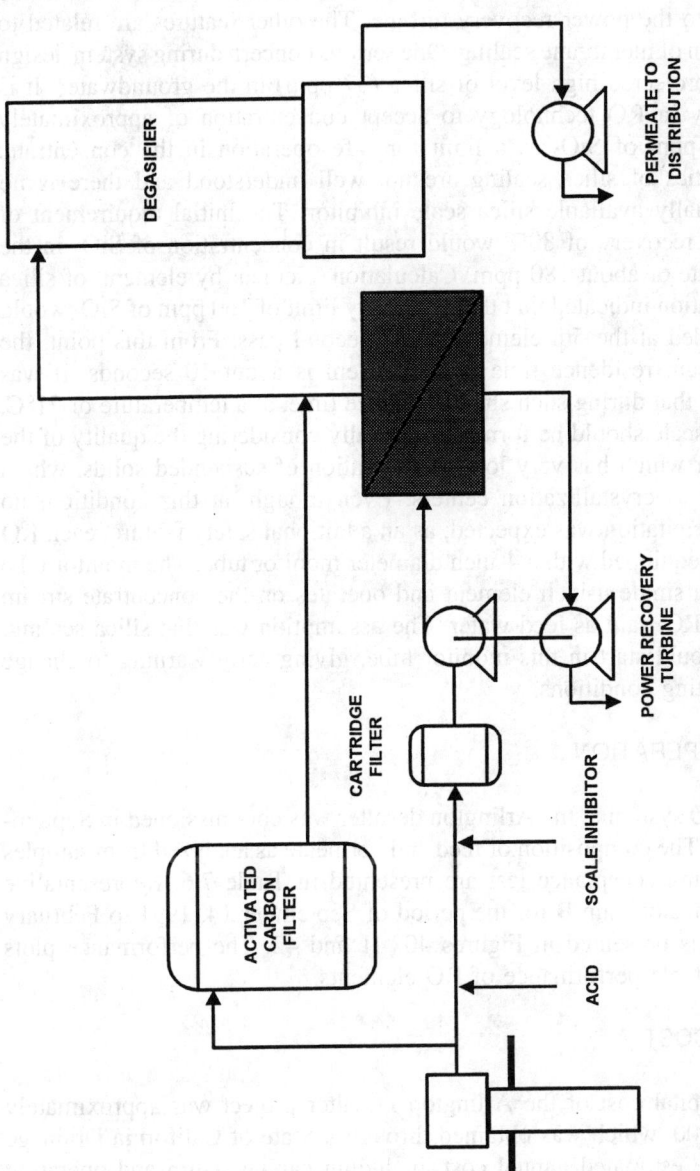

Figure 7.39. Flow diagram of water treatment process at Arlington Desalter.

Flow diagram of the Arlington RO plant

PERMEATE TO
DISTRIBUTION

DEGASIFIER

POWER RECOVERY
TURBINE

CARTRIDGE
FILTER

SCALE INHIBITOR

ACID

ACTIVATED
CARBON FILTER

throttling rather than conventional feed throttling to control productivity is to eliminate loss of recoverable energy and maintain constant concentrate pressure to the power recovery turbine. The other features are related to prevention of membrane scaling. One serious concern during system design was the projected high level of silica (37 ppm) in the groundwater. It is customary in RO technology to accept concentration of approximately 120–160 ppm of SiO_2 as a limit for safe operation in the concentrate. The kinetics of silica scaling are not well understood and there is no commercially available silica scale inhibitor. The initial requirement of permeate recovery of 80% would result in concentration of SiO_2 in the concentrate of about 180 ppm. Calculation, element by element, of silica concentration indicated that the customary limit of 160 ppm of SiO_2 would be exceeded at the 5th element of the second pass. From this point, the brine stream residence time in the system is about 10 seconds. It was estimated that during such short residence time, at a temperature of 21°C, no silica scale should be formed, especially considering the quality of the raw water which has very low concentration of suspended solids, which could act as crystallization centers. Even though, at this condition, no silica precipitation was expected, as an additional safety feature, each RO train was equipped with a 4-inch diameter monitor tube. The monitor tube contains a single 4-inch element and operates on the concentrate stream from the RO train as feed water. The assumption was that silica scaling, if any, would start in this monitor tube, giving early warning to change the operating conditions.

PLANT OPERATION

The RO system at the Arlington desalter was commissioned in September 1990. The composition of feed and permeate as analyzed from samples taken during acceptance test are presented in Table 7.6. Representative performance of train B for the period of September 14, 1991 to February 27, 1991 is presented in Figures 40, 41 and 42. The performance plots indicate stable performance of RO elements.

WATER COST

The capital cost of the Arlington Desalter project was approximately $14,000,000, which was obtained through a State of California Drainage Loan. The estimated annual cost, including capital return and operating cost, is $2,000,000. Assuming annual production of 1980 million gallons of blended water, the specific water cost is $0.27/m³ or $1.4 per 1,000 gallons. The cost of reclaiming municipal wastewater, based on the process used in Water Factory 21 is estimated to be about $0.32/m³ or $1.21 per

TABLE 7.6. Water Analysis of Water Samples Taken During Acceptance Test at the Arlington Plant.

Constituents	RO Feed Water (g/m³)	RO Permeate (g/m³)	Salt Passage (%)	Blended Water (g/m³)
Calcium	140	0.3	0.22	0.33
Magnesium	42	0.01	0.24	0.15
Sodium	168	7.1	4.21	66
Potassium	3.8	0.2	5.22	1.5
Bicarbonate	367	8.4	2.72	140
Chloride	162	2.3	1.44	60
Sulfate	243	0.6	0.21	89
Nitrate	93	8.3	8.96	37
Fluoride	0.41	0.1	24	0.2
Silica	40	0.1	0.25	15
TDS	1260	28	2.21	460
Temperature, deg.C	21			

1,000 gallons. The cost of water reclamation compares favorably with other alternatives of augmentation of Southern California water supply. About one-third of the water demand of Southern California is supplied from local wells. The rest is imported from Northern California through the aqueduct systems. Both water sources are limited and they are affected by the annual precipitation level. The rate of Metropolitan Water District, which governs the water distribution in Southern California, for imported treated water is $0.19/m³ or $0.72 per acre foot. The additional water supply for Southern California can be produced by desalination programs. Due to limited local availability of brackish water, only seawater desalination can provide a significant volume of new water. The cost of water from a large scale seawater desalination system is estimated to be in the range of $0.82/m³ or $3.1 per 1,000 gallons. Considering the fact that most of the water used in Southern California is used once and discharged to the ocean, and the cost of alternative water supplies using advanced treatment technology combined with membrane desalting is not significantly higher than the cost of imported water, water reclamation looks like a very economically attractive alternative provided public acceptance can be established for this procedure.

CONCLUSIONS

Reverse osmosis process is today a mature water treatment technology. It is applied commercially in water treatment plants off all sizes, whenever a separation of water dissolved species is required. Stable system perform-

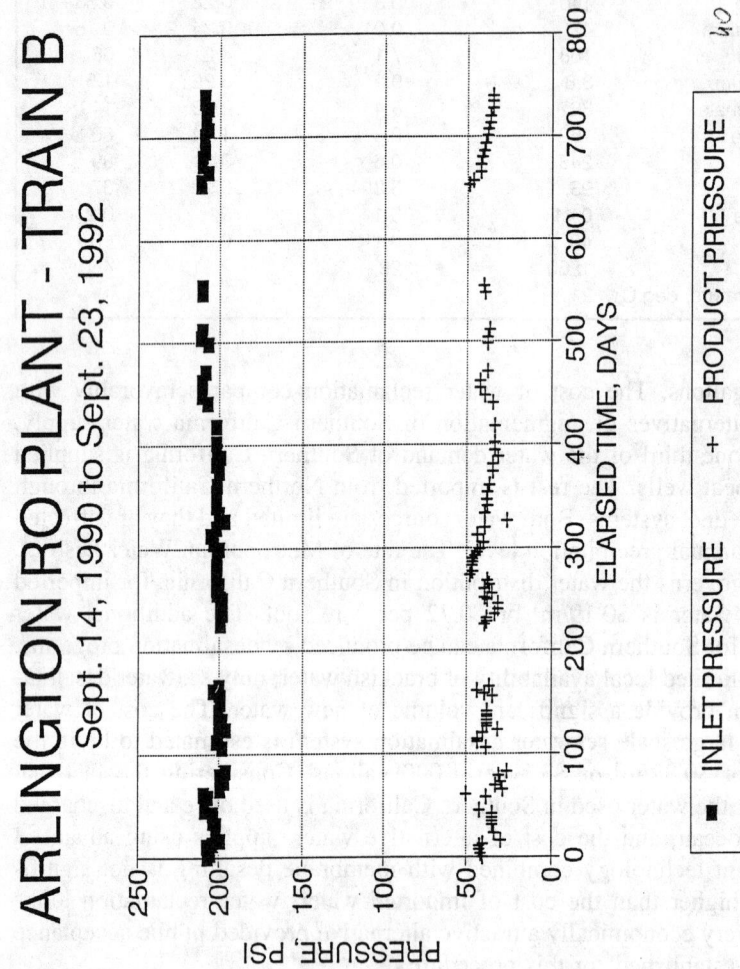

Figure 7.40. Performance of Hydranautics CPA2 membranes at Arlington Desalter, pressure vs. time.

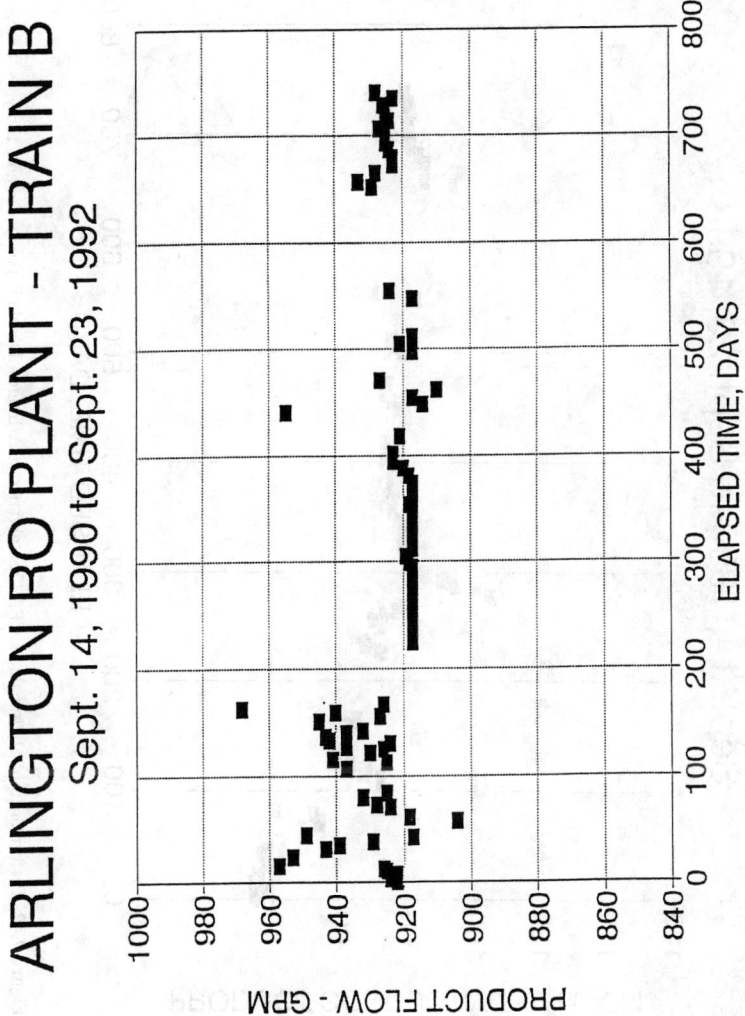

Figure 7.41. Performance of Hydranautics CPA2 membranes at Arlington Desalter, product flow vs. time.

341

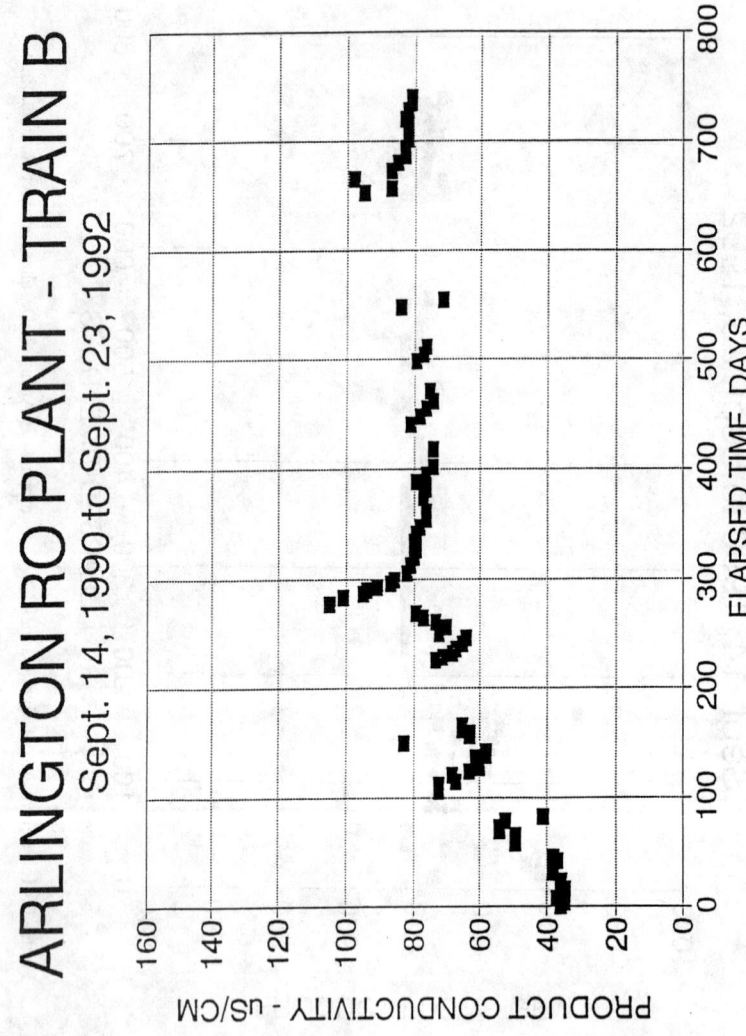

Figure 7.42. Performance of Hydranautics CPA2 membranes at Arlington Desalter, product conductivity vs. time.

342

ance and long term membrane element life is achievable in well design and operated RO systems.

Combination of advanced water treatment and membrane desalination technology can be used effectively to treat municipal effluent and agricultural drainage water to reduce contaminants level to potable water quality.

The cost of reclaimed water, produced by applying currently available advanced treatment technology, is only slightly higher than the cost of water imported to Southern California. It is significantly lower than an alternative of water augmentation by desalting seawater.

REFERENCES

1   C. E. Ried and E. J. Berton, *J. Apply. Polymer Sci. 1,* 133 (1959).

2   Sea water Demineralization by the Surface Skiming Process, T. Yuster, S. Sourirajan and K. Bernstein, UCLA Department of Engineering Report 58-26 (1958).

3   S. Loeb and S. Sourirajan, *Adv. Chem. Ser. 38,* 117 (1962).

4   *Desalination by Reverse Osmosis,* U. Merten, Editor, The M. I. T. Press, Cambridge, Mass., 1966.

5   *The USAID Desalination Manual,* O. K. Buros at. al., IDEA 1981.

6   *Reverse Osmosis and Synthetic Membranes, Theory—Technology—Engineering,* S. Sourirajan Editor, National Research Council Canada Publications, Ottawa 1977.

7   *Membrane Processes,* R. Rautenbach and R. Albrecht, John Willey and Sons, New York 1989.

8   *Reverse Osmosis Technology Application for High-Purity Water* Production, S. Parekh Editor, Marcel Dekker Inc., New York 1988.

9   *Spiral Wound Thin Film Composite Membrane Systems for Brackish and Seawater Desalination by Reverse Osmosis,* R. L. Riley, C. E. Milstead, A. L. Lloyd, M. W. Seroy and M. Tagami, Fluid System Division, UOP, Inc. (1977).

10 *Material Science of Synthetic Membranes,* J. E. Cadote, D. R. Lloyd Editor, Amer. Chem. Soc. Symposium Series, Washington DC, pp. 273–294, (1985).

11 *Development of FT-30 Membranes in Spiral Wound Modules,* R. J. Petersen, J. E. Cadotte and J. M. Buettner, Report to Office of Water Research and Technology, October 1982.

12 H. I. Mahon, *Proc. of Desalination Research Conference,* Nat. Acad. Sci., Publ. 942, 345 (1963).

13 DuPont Hollow Fibers Membranes, V. P. Caracciolo, N. W. Rosenblatt and V. J. Tomsic, *Reverse Osmosis and Synthetic Membranes,* S. Sourirajan ed., National Research Council of Canada Publications, Ottawa (1977).

14 *Design Study of Reverse Osmosis Pilot Plant,* D. T. Bray and H. F. Menzel, Office of Saline Water, Research and Development Progress Report No. 176 (1966).

15 Preliminary Economic Study U.C.L.A. Reverse Osmosis Process for Brackish Water Desalination, J. W. McCutchan, S. Loeb, P. A. Buckingham and A. W. Ayers, UCLA Engineering Department Report 63-62, (1963).

16 Standard Test Method for Silt Density Index (SDI) of Water, ASTM procedure D4189-82

17   An Automated Modified Fouling Index Device for Monitoring and Controlling Pre-
     treatment Process, G. B. Tanny, M. Sayar, J. Ricklis and M. Wilf, *Proceedings of
     the 15th National Symposium on Desalination,* Ashkelon, Israel (June 1983).

18   Water Renovation of Municipal Effluents by Reverse Osmosis, J. E. Curver, J. E.
     Beckman and E. Bevage, Report EPA 670/2-75-009, Prepared for National Environ-
     mental Research Center, General Atomic Company, March 1975.

19   Control Of Fouling of Reverse Osmosis Membranes when Operating on Polluted
     Surface Water, J. E. Beckman, E. Bevage, J. E. Curver, I. Nusbaum, and S. S. Kremen,
     Office of Saline Water Report CA-10488, Gulf Environmental System, February 1971.

20   Application of Hyperfiltration to Treatment of Municipal Sewage Effluents, K. A.
     Kraus, Eater Pollution Research Series: 17030E OHO1/70, January 1970.

21   Evaluation of Membrane Processes and Their Role in Wastewater Reclamation, Final
     Report of Contract for U.S. Department of Interior OWRT, David Argo and Martin
     Rigby, November 30, 1981.

22   Municipal Wastewater Reclamation and Reverse Osmosis, Richard G. Sudak, William
     Dunivin and Martin G. Rigby, *Proceedings of the National Water Supply Improvement
     Association 1990 Biennial Conference,* Florida, August 1990, p. 225.

23   Renovation of Municipal Waste Water by Reverse Osmosis, J. M. Smith, A. N. Masse
     and R. P. Mile, Water Pollution Research series: 17040-05/70, 1970.

24   Study and Experiments in Waste Water Reclamation by Reverse Osmosis, I. Nusbaum,
     J. H. Sleigh and S. S. Kremen, Water Pollution Research Series 17040-05/70 (1970).

25   Reverse Osmosis of Treated and Untreated Secondary Sewage Effluent, D. F. Boen
     and G. L. Johnson, Report to the Environmental Protection Agency, EPA-6702-74-
     74-007, September 1974.

26   Demineralization of Sand Filtered Secondary Effluent by Spiral Wound Reverse
     Osmosis Processes, C. Chen and R. P. Miele, Report to the Environmental Protection
     Agency, EPA-600/2-77-169, September 1977.

27   Evaluation of Membrane Processes and Their Role in Wastewater Reclamation, D.
     G. Argo et al., Volume I, Report to U.S. Department of Interior, Office of Water
     Research and Technology, November 1979.

28   Evaluation of Membrane Processes and Their Role in Wastewater Reclamation, D.
     G. Argo et. al., Volume III, Report to U.S. Department of Interior, Office of Water
     Research and Technology, November 1981.

29   Membrane Process in Water Reuse, J. Lozier and R. Bergman, *Proceedings of Water
     Reuse Symposium,* Denver, Colorado, 1994.

30   Pilot Testing of Microfiltration and Ultrafiltration Upstream of Reverse Osmosis
     During Reclamation of Municipal Wastewater, G. L. Leslie, W. R. Dunivin, P.
     Gabillet, S. R. Conklin, W. R. Mills and R. G. Sudak, *Proceedings of ADA 1966
     Biennial Conference,* Monterey, California, 1996.

CHAPTER 8

Ultraviolet (UV) Disinfection for Wastewater Reuse

T HE germicidal properties of the radiation emitted from ultraviolet (UV) light sources have been used in a wide variety of applications. First used in the early 1900's on high quality water supplies, the use of UV light as a wastewater disinfectant has evolved during the last 10 years with the development of new lamps, ballasts, and ancillary equipment. With the proper dosage, UV radiation has proven to be an effective bacteriocide and virocide for wastewater, while not contributing to the formation of any toxic compounds. The purpose of this chapter is to provide: (1) a brief introduction to disinfection with UV radiation, (2) a review of current UV technologies, (3) a description of the methods used to determine the required number of UV lamps, (4) key design considerations for UV disinfection facilities, (5) a description of the operation of UV systems, and (6) a review of compliance monitoring.

DISINFECTION WITH UV RADIATION

In considering the application of UV radiation for wastewater disinfection, it is appropriate to review the source of UV radiation, the mechanisms of UV disinfection, the definition of UV dose, the effect of wastewater characteristics, the effectiveness of UV disinfection, and the environmental impacts of UV disinfection.

Frank J. Loge, Jeannie L. Darby, and George Tchobanoglous, Department of Civil and Environmental Engineering, University of California at Davis, Davis, CA 95616; David Schwartzel, Trojan Technologies Inc., 3020 Gore Rd., London, Ontario, Canada N5V 4T7.

SOURCE OF UV RADIATION

To produce UV energy special UV lamps that contain mercury vapor are charged by striking an electric arc. The energy generated by the excitation of the mercury vapor contained in the lamp results in the emission of UV light. In general, UV disinfection systems are categorized based on the internal operating parameters of the UV lamp. Available UV technology is reviewed in greater detail in a later section.

MECHANISMS OF DISINFECTION

The principal mechanisms that have been put forth to explain the action of disinfectants are (1) damage to the cell wall, (2) alteration of cell permeability, (3) alteration of the colloidal nature of the protoplasm, (4) inhibition of enzyme activity, and (5) damage to the DNA and RNA in a cell. Damage or destruction of the cell wall will result in cell lysis and cell death. Alteration of the selective permeability of the cell membrane will allow vital nutrients, such as nitrogen and phosphorus, to escape. Heat, radiation, and highly acid or alkaline agents alter the colloidal nature of the protoplasm. Heat will coagulate the cell protein and acids or bases will denature proteins, producing a lethal effect. Oxidizing agents, such as chlorine, can alter the chemical arrangement of enzymes and inactivate the enzymes. If the DNA and RNA in a cell are damaged (e.g., through the formation of double bonds due to UV radiation) the organism will be unable to reproduce. The mode of action of UV radiation is summarized in Table 8.1. Information on chlorine and ozone is also included for comparison.

DOSE RELATIONSHIPS

The dose of disinfectant can be defined as follows for physical agents [Equation (1)] and chemical agents [Equation (2)], respectively.

$$\text{Dose}_{\text{physical agent}} = I \times t \tag{1}$$

$$\text{Dose}_{\text{chemical agent}} = C \times t \tag{2}$$

where

I = intensity of physical agent (e.g., UV radiation)
t = contact time
C = concentration of chemical agent (e.g., chlorine)

TABLE 8.1. Mechanisms of Disinfection Using UV, Chlorine, and Ozone.

UV	Chlorine	Ozone
1. Photochemical damage to RNA and DNA (e.g., formation of double bonds) within the cells of an organism.	1. Oxidation	1. Direct oxidation/destruction of cell wall with leakage of cellular constituents outside of cell
2. The nucleic acids in microorganisms are the most important absorbers of the energy of light in the wavelength range of 240–280 nm.	2. Reactions with available chlorine	2. Reactions with radical byproducts of ozone decomposition[a]
3. Because DNA and RNA carry genetic information for reproduction, damage of these substances can effectively inactivate the cell.	3. Protein precipitation	3. Damage to the constituents of the nucleic acids (Purines and pyrimidines)
	4. Modification of cell wall permeability	4. Breakage of carbon-nitrogen bonds leading to depolymerization
	5. Hydrolysis and mechanical disruption.	

Adapted from Crites and Tchobanoglous (1988).

In UV disinfection the contact time is in seconds (7 to 24 s) as compared to minutes for chlorine (90 to 120 min) disinfection. Thus, it is apparent that the contact time is a critical factor in UV disinfection.

Currently there are three approaches utilized to estimate UV dose: chemical actinometry, biological assays, and mathematical models. Some field UV systems use intensity probes to provide an indicator of intensity at a single location, but the readings from these cannot be used to determine the average UV intensity in a reactor. Chemical actinometry and biological assays provide a direct measure of UV dose, whereas mathematical methods provide an estimate of UV intensity, which when coupled with an estimate of average exposure time can be used to calculate UV dose. Biological assays and mathematical methods are used most frequently.

Biological Assays

In the bioassay method, the relationship between UV dose and inactivation of a selected organism is developed under highly controlled conditions. Typically the selected organism is placed in solution and exposed to a collimated beam of UV light of known intensity (measured with a calibrated radiometer) for a measured amount of time. A typical collimated beam apparatus is illustrated in Figure 8.1. The UV dose is then the product of the average UV intensity, corrected for the depth of sample, and exposure time. The dose can be varied by varying the exposure time. An example of a log-survival versus dose curve developed for MS2 coliphage is illustrated in Figure 8.2. By seeding a flow-through UV system with the selected organism and measuring its response (the log survival), the bioassay results can then be used to back calculate an estimate of the UV dose occurring in the system. Published reports of bioassays used in determining UV dose in bench or pilot-scale UV systems include Qualls and Johnson (1983, 1989), Harris et al. (1987), and Suidan and Severin (1986).

Mathematical Methods

The most common mathematical method used to determine average UV intensity in a UV disinfection system is the point source summation (PSS) method. In the PSS method, the average UV intensity for a particular reactor geometry and lamp configuration is determined as a function of the UV transmittance of the water being disinfected. The cylindrical UV lamps within a reactor are treated as a finite series of point sources radiating in all directions. The light from every point source is assumed to be spread over spheres. The intensity at any point in the reactor is calculated by summing the intensities received at that point from all point sources in

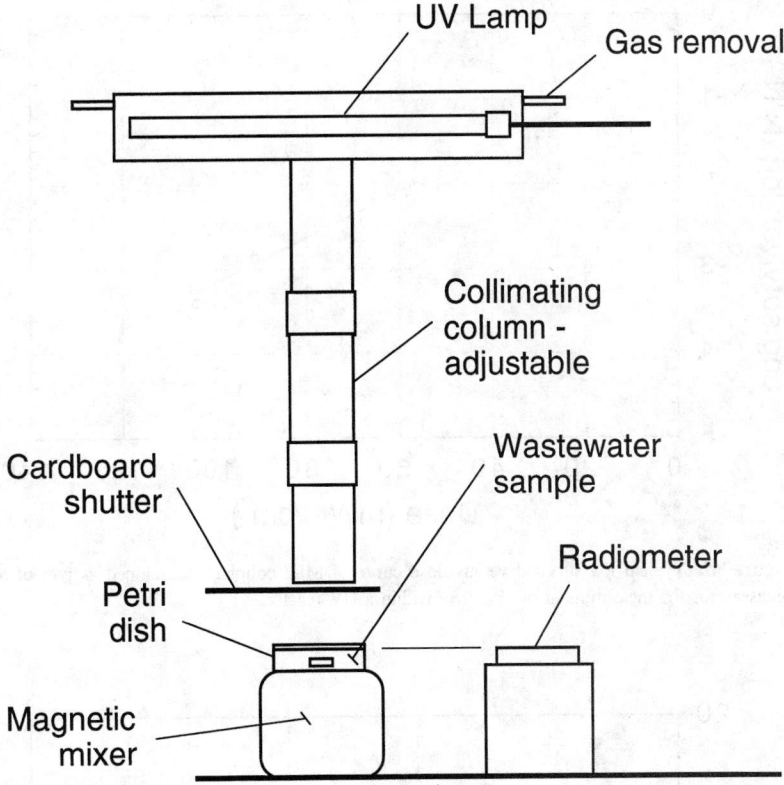

Figure 8.1. Schematic of a typical collimated beam apparatus.

the system. Important parameters in the PSS method include lamp and reactor geometry, transmittance of the quartz sleeves, and transmittance of the water flowing through the system. Dissipation and absorption mechanisms are assumed to attenuate the light intensity with increasing distance from the point source (Qualls and Johnson, 1983, 1985; Suidan and Severin, 1986; USEPA, 1986; Qualls et al., 1989).

To determine the UV dose using the computed average intensity, the average exposure time must also be known. In a UV disinfection system approximating plug-flow conditions, the average exposure time in one bank of the UV system can be estimated as the net bank volume over the arc length of the UV lamps divided by the flowrate through the system. In a UV system dominated by non plug-flow behavior, the residence time distribution (RTD) must be taken into account. An example of a PSS curve developed for a two by two lamp array spaced on three inch centers is illustrated in Figure 8.3.

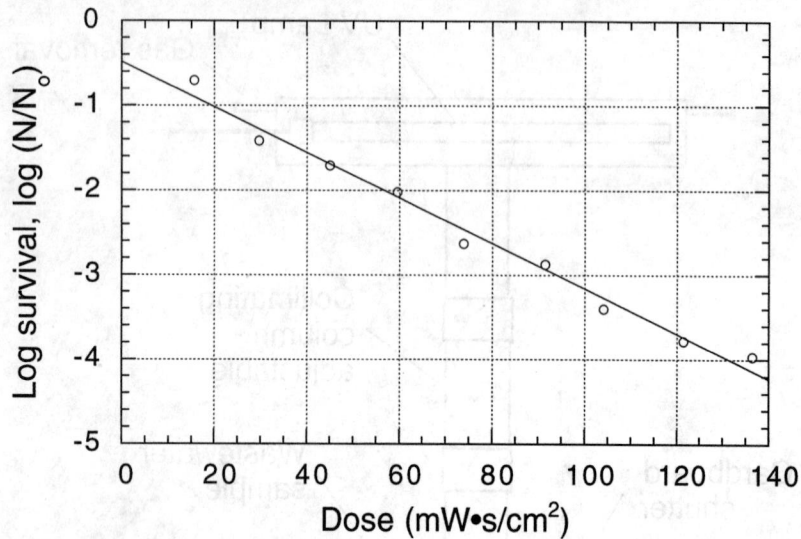

Figure 8.2. Typical log survival versus dose curve of MS2 coliphage developed as part of a bioassay for the measurement of UV dose within a UV reactor.

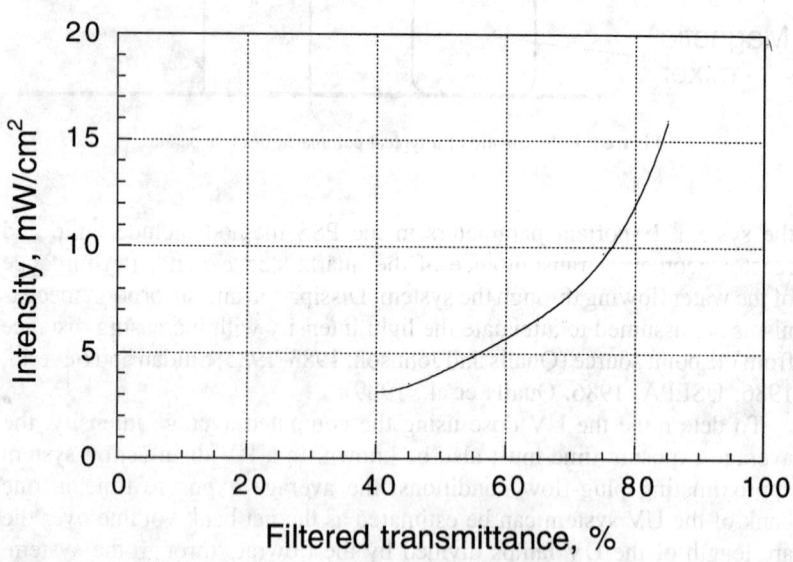

Figure 8.3. Average intensity within a 2 by 2 lamp array calculated using the point source summation method (based on a rated UV output of 26.7 W, 23 mm OD quartz sleeve, and 75 mm centerline lamp spacing).

EFFECT OF WASTEWATER CHARACTERISTICS

Because the only UV radiation that is effective in destroying bacteria is that which reaches the bacteria, the water must be relatively free from turbidity that would absorb the ultraviolet energy and shield the bacteria. Wastewater characteristics that have been found to influence UV disinfection performance include the type and concentration of particulate and dissolved material as well as the nature and degree of particle-association of targeted organisms. The impact of wastewater characteristics on UV, chlorine, and ozone disinfection are summarized in Table 8.2. The impact of total suspended solids (TSS) on the effectiveness of UV disinfection is illustrated in Figure 8.4. The relative significance of the wastewater characteristics will vary from plant to plant. The variability of these characteristics and their relative influence within a single plant must also be considered. As a consequence, it is generally recognized that it is unlikely that a purely deterministic approach can be developed to predict the overall influence of the many parameters.

GERMICIDAL EFFECTIVENESS OF UV RADIATION

The principal infectious agents that occur in wastewater are classified into three broad groups: bacteria, parasites (principally protozoa and helminths), and viruses. Diseases caused by waterborne bacteria include typhoid, cholera, paratyphoid, and bacillary dysentery. Among the parasites, the protozoan organisms *Cryptosporidium parvum, Cyclospora, Giardia lamblia* are of greatest concern. The most important helminthic parasites that may be found in wastewater are intestinal worms, including the stomach worm *Ascaris lumbricoides,* the tapeworms *Taenia saginata* and *Taenia solium,* the whipworm *Trichuris trichiura,* the hookworms *Ancylostoma duodenale* and *Necator americanus,* and the threadworm *Strongyloides stercoralis.* Diseases caused by waterborne viruses include poliomyelitis and infectious hepatitis. Because filtration is used in most reuse applications to meet the turbidity requirement, the greatest concern is with bacteria and viruses. The relative effectiveness of UV radiation is summarized in Table 8.3.

ENVIRONMENTAL IMPACTS OF UV DISINFECTION

Because UV light is not a chemical agent, no toxic residuals are produced. However, certain chemical compounds may be altered by the UV radiation. Based on the evidence to date, it appears that the compounds formed are harmless or are broken down into more innocuous forms.

TABLE 8.2. Comparison of Impact of Wastewater Characteristics on UV, Chlorine, and Ozone Disinfection.

Wastewater Characteristic	UV Disinfection	Chlorine Disinfection	Ozone Disinfection
Ammonia	No or minor effect	Combines with chlorine to form chloramines	No or minor effect, can react at high pH
BOD, COD, etc.	No or minor effect, unless humic materials comprise a large portion of the BOD	Organic compounds that comprise the BOD and COD can exert a chlorine demand. The degree of interference depends on their functional groups and their chemical structure.	Organic compounds that comprise the BOD and COD can exert an ozone demand. The degree of interference depends on their functional groups and their chemical structure.
Hardness	Effects solubility of metals that may absorb UV light. Can lead to the precipitation of carbonates on quartz tubes	No or minor effect	No or minor effect
Humic materials	Strong adsorbers of UV light	Reduces effectiveness of chlorine	Affects the rate of ozone decomposition and the ozone demand
Iron	Strong adsorber of UV light	No or minor effect	No or minor effect
Nitrite	No or minor effect	Oxidized by chlorine	Oxidized by ozone
Nitrate	No or minor effect	No or minor effect	Can reduce effectiveness of ozone
pH	Can effect solubility of metals and carbonates	Effects distribution between hypochlorous acid and hypochlorite ion	Effects the rate of ozone decomposition
TSS	UV absorption and shielding of embedded bacteria	Shielding of embedded bacteria	Increase ozone demand and shield embedded bacteria

Adapted from Crites and Tchobanoglous (1988).

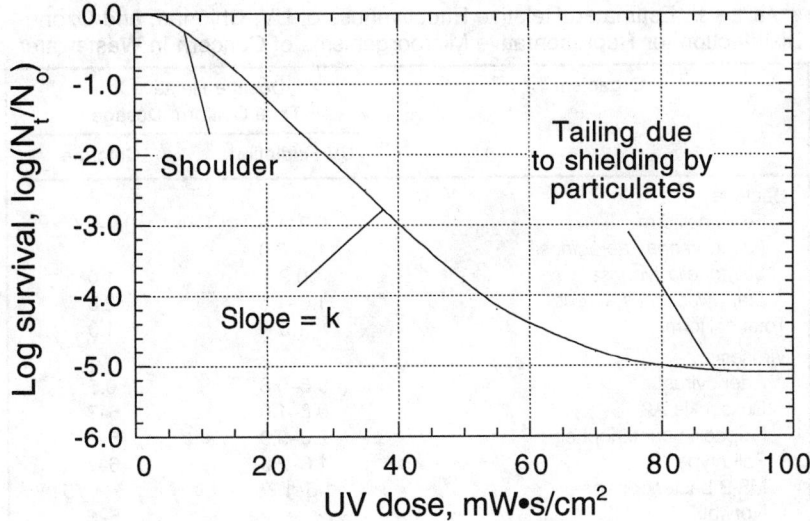

Figure 8.4. Log survival versus dose curve illustrating the tailing region generally associated with particulate material in wastewater.

At present, disinfection with UV light is considered to have no direct adverse environmental impacts.

REVIEW OF CURRENT UV DISINFECTION TECHNOLOGIES

The classification of UV disinfection technologies follows two general categories based on the internal operating parameters of the UV lamp itself; namely low and medium pressure lamp systems. In both cases mercury-argon lamps are used to generate the UV-C region wavelengths.

LOW PRESSURE UV SYSTEMS

The majority of UV disinfection systems now in operation are based on the use of low pressure low intensity UV lamps, although low pressure high intensity lamps are available. The lamps are a slimline design and operate optimally at a lamp wall temperature of 40°C and an internal fill pressure of 7×10^{-3} torr. Approximately 88 percent of the lamp output is at 253.7 nm, making it an efficient choice for disinfection processes. Although low intensity lamps produce small amounts of UV at the 185 nm and 365 nm wavelengths, the lamp quartz is doped to prevent the emission of the 185 nm wavelength that produces ozone.

The low intensity G64T5L lamp design is approximately 60 years old and

TABLE 8.3. Estimated Relative Effectiveness of UV, Chlorine, and Ozone Disinfection for Representative Microorganisms of Concern in Wastewater.

Organism	Dosage Relative to Total Coliform Dosage	
	UV Radiation	Chlorine
Bacteria		
Fecal coliform	0.9–1.0	0.9–1.0
Pseudomonas aeruginosa	1.5–2.0	
Salmonella typhosa	0.9	1.0
Staphylococcus aureus	1.0–1.5	2.5
Total coliform	1.0	1.0
Viruses		
Adenovirus	0.6–0.8	0.5
Coxsackie A2	0.8–1.0	6–7
F specific bacteriophage	2.0–3.0	5–6
Polio type 1	1.8–2.0	6–7
MS-2 bacteriophage	1.5–1.75	1.5–1.75
Norwalk		5–6
Protozoa		
Cryptosporidium parvum oocysts	1.0–2.0	1.0–2.5
Giardia lamblia cysts	2.0–3.0	2.0–3.0

Adapted from Crites and Tchobanoglous (1998).

has been optimized over this time period with respect to the lamp geometry and lamp physics. Manufacturers of low pressure UV systems utilize the same G64T5L lamp but use their own unique electronic ballast architecture to drive the lamps. Competition among low pressure UV manufacturers generally centers on the type of ballast used in the UV system.

With both low and medium pressure designs, the UV output is proportional to lamp input power, with lamp current the general indicator of relative UV system performance. In a low pressure lamp, there is an excess of liquid mercury in the lamp and the mercury vapor pressure is controlled by the coolest part of the lamp wall. If the lamp wall does not remain at its optimum temperature of 40°C, some of the mercury in the lamp condenses back to its liquid state thereby decreasing the number of mercury atoms available to release photons of UV and hence UV output declines. In most cases, the effluent has a cooling effect on the lamp wall and additional power must be provided by the ballast to compensate for lamp wall heat loss. The quartz sleeve buffers the effluent temperature extremes to which the UV lamps are exposed thereby maintaining uniform UV lamp output.

Electronic ballast efficiency as determined by electrical conversion efficiency (e.g., 60 Hz to 17 kHz) is uniform, within 5 percent, across the industry and reflects the current state of the art in ballast design. The physics-related design parameters of the standard low intensity lamp are listed in Table 8.4, along with the range of performance to be anticipated based on the choice of ballast or input power to the lamp.

MEDIUM PRESSURE UV SYSTEMS

Medium pressure high intensity UV disinfection systems were first introduced about ten years ago, well after the first low pressure UV disinfection system. The medium pressure high intensity UV lamp operates at a higher mercury vapor pressure (10^2 to 10^4 torr) and at a higher lamp wall temperature (600 to 900°C), therefore the UV output of these lamps is unaffected by effluent temperature. The medium pressure high intensity UV lamp operates at temperatures at which all the mercury is vaporized; therefore the UV output can be modulated across a range of power settings without significantly changing the spectral distribution of the lamp. A typical spectral distribution for a low and a medium pressure lamp is illustrated in Figure 8.5.

While the medium pressure lamp is less efficient at generating the 253.7 nm and other germicidal wavelengths (only 27–44 percent of the total energy of a medium pressure lamp is in the germicidal wavelength range) it does generate approximately 50 to 80 times higher germicidal UV output than its low intensity counterpart. This higher output is advantageous in poor quality effluent applications, whether primary, secondary, or combined sewer overflows (CSOs). It may also be advantageous with better quality effluents such as those used for unrestricted reuse because the number of lamps per system can be

TABLE 8.4. Typical Low Intensity G64T5L UV Lamp Parameters and Performance Range.

Parameter	Units	Standard Lamp Design	Performance Range
Nominal length[a]	in.	64	N/A
Arc length	in.	58	N/A
Lamp wattage	W, Watts	65	N/A
Lamp input current	A	4.25×10^{-1}	3.0–5.25×10^{-1}
UV output per lamp @ 253.7 nm[b]	W	26.7	15.4–32.0
Ozone generation		0	0
Lamp life: guaranteed (hours)[c]	hr	8,760	4,000–13,000

[a] A 36 in lamp is also available although it is seldom used in designs.
[b] Range of UV output is a function of lamp current and water temperature.
[c] Lamp manufacturer guarantees 8760 hours, lamp life in field depends on lamp current and predetermined end of lamp life UV output intensity. The lower the operating current the sooner the lamp will reach end of lamp life, typically defined as 65 to 70 percent of new lamp intensity.
N/A = Not applicable.

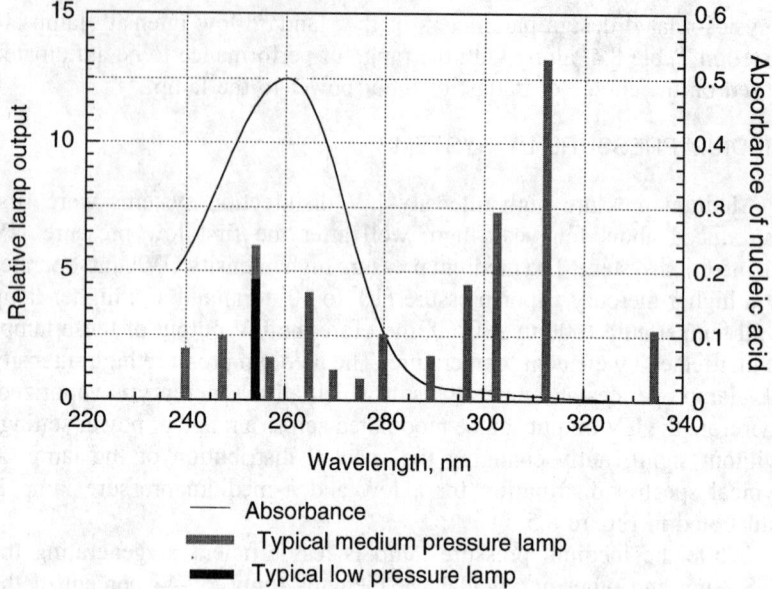

Figure 8.5. Spectral distribution of a typical low and medium pressure UV lamp and the absorption spectrum of nucleic acid.

significantly reduced. In tertiary applications, up to a 20 fold reduction in the number of medium pressure lamps required for a particular UV disinfection system design over the corresponding low pressure lamp alternative may be achieved.

There are a substantial number of medium pressure lamps currently on the market. The lamp selected by any one UV system manufacturer is chosen on the basis of an integrated system design where lamp selection and reactor design are interdependent. Medium pressure lamps can also be modulated from 15 to 100 percent of maximum output with no significant changes in the spectral distribution. Characteristics of a typical medium pressure lamp are summarized in Table 8.5.

REACTOR LAMP CONFIGURATION

UV lamps can be oriented in the flow in one of two orientations: horizontal and parallel to the flow, or vertical and perpendicular to the flow. The critical issues in UV system design are promotion of "plug-flow" conditions and minimization of head loss. Both issues are optimized most easily using lamps that are oriented horizontal and parallel to the flow.

Horizontal and Parallel to the Flow Orientation

The majority of UV disinfection systems have lamps that are horizontal and parallel to the flow for two principal reasons: contact time per lamp is greatest when the flow travels parallel to the length of the UV arc, and headloss is least when lamps are parallel to the flow, for a given flowrate. This orientation affords the best UV disinfection efficiency while minimizing the chance for short-circuiting of the UV reactor.

Vertical and Perpendicular to the Flow Orientation

Manufacturers sometimes orient the UV lamp vertically and perpendicular to the flow, mainly to facilitate the lamp change operation. Lamps only (not lamps and sleeves) can often be removed from the UV reactor in this configuration by simply unscrewing an end cap and sliding the lamp out of the sleeve. While lamp removal becomes a simple operation, other maintenance activities are complicated by this orientation. For example, to remove/replace quartz sleeves, the UV reactor must either be removed from the flow stream (due to the tendency of the lamp sleeves to break against other sleeves in the flow) or taken off-line (because removal of the sleeve would put a "hole" in the side of the UV reactor).

UV LAMPS AND CONTACT METHODS

There are two general methods of contacting the waste to be disinfected with the UV radiation: (1) through the use of quartz sleeves and (2) with teflon contactors.

TABLE 8.5. **Range of Medium Pressure UV Lamp Design Parameters and Performance.**

Parameter	Units	Lamp Design and Performance
Nominal length	in.	64
Arc Length	in.	58
Lamp wattage	W, Watts	1,000–3,000
Lamp input current	A	3–11
Germicidal output per lamp nm[a]	W/in.	10–30
Ozone generation	0	0
Lamp life: guaranteed		
100 percent output	hr	3,000
30 percent output	hr	5,000

[a]Germicidal output depends on the lamp and power modulation to the lamp.

Quartz Sleeves

Most UV systems employ a UV lamp encased in a larger diameter fused quartz circular sleeve. The quartz sleeve is made of 99.99 percent SiO_2 and has a low coefficient of thermal expansion and a high resistance to thermal shock, making it ideal for use in wastewater applications. Wall thickness varies from approximately 1.5 to 2.0 mm for larger diameter sleeves.

Fused quartz is one of the best known transmitters of UV radiation available. The cut-off for fused quartz in the UV region is between 155 and 175 nm and is dependent on the fusion process used in the manufacture of the material. Some UV systems are characterized as "non-ozone producing" by doping the fused quartz sleeves with 0.01 to 0.02 percent titanium dioxide which prevents the formation of ozone generated when the 185 nm wavelength strikes oxygen.

Fused quartz transmits approximately 89 percent of the UV energy that the lamps generate at the 253.7 nm wavelength through its wall and into the effluent being disinfected. This 89 percent average UV transmittance value for the quartz sleeves also includes surface reflection losses. Most types of fused quartz will occasionally turn brown or discolor slightly due to the apparent chemical reduction of the silicon dioxide.

Teflon Contactors

A limited number of UV systems use teflon tubes to separate the wastewater from the UV lamps. In the case of Teflon tubes, the tubes are normally used to contain the water or wastewater and the UV lamps are mounted around them so that UV light penetrates the Teflon tube into the water. In another design the Teflon is wrapped around the UV lamp. The UV lamps operate in air with or without reflectors and are air-cooled to maintain the optimum 40°C lamp wall temperature. Practically speaking, this type of system is limited to small flowrates.

The UV transmittance of the tube is approximately 70 to 90 percent when new depending on wall thickness, diminishing to as low as 30 percent under significant fouling conditions. Cleaning the Teflon tubes is problematic because the UV reactor assembly must be taken off line and disassembled for cleaning. The cleaning process appears to reduce the ability of the Teflon tubes to transmit UV light over time resulting in an ever decreasing UV transmittance for the system. Teflon tubes can also become heavily coated with dust due to electrostatic precipitation because of the highly charged atmosphere that exists between the lamp and the tube.

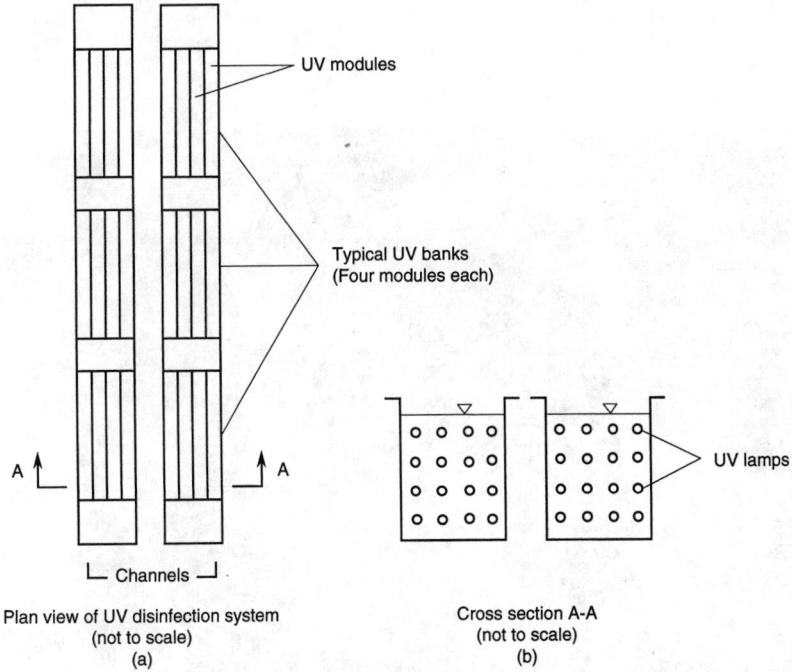

UV modules

Typical UV banks
(Four modules each)

UV lamps

A

A

Channels

Plan view of UV disinfection system
(not to scale)
(a)

Cross section A-A
(not to scale)
(b)

Figure 8.6. Plan and cross-sectional views of a typical UV disinfection system (shown are 2 channels, 3 banks per channel, 4 modules per bank, and 4 lamps per module).

COMPONENTS OF A UV DISINFECTION SYSTEM

The principal components of a UV disinfection system and the typical terminology used in design are illustrated in Figure 8.6(a). Typically, the design flow rate is divided equally among a number of channels. Each channel typically contains two or more banks of UV lamps in series, and each bank is comprised of a specified number of modules (or racks of UV lamps). Each module contains a specified number of lamps; typically manufacturers of UV systems use 2, 4, 8, 12 or 16 lamps per module. A spacing of 3 in. between the centers of UV lamps is currently the most frequently used lamp configuration by UV manufacturers [Figure 8.6(b)]. A typical example of low and medium pressure UV installations are shown in Figures 8.7 and 8.8, respectively.

METHODS USED TO DETERMINE THE NUMBER OF LAMPS REQUIRED FOR DISINFECTION

The design of a UV disinfection system requires two general steps: (1)

Figure 8.7. Low pressure low intensity UV installation for disinfection of water extracted from a rapid infiltration and extraction system in Southern California (Courtesy of Fischer Porter, Inc.).

Figure 8.8. Example of medium pressure high intensity UV system with lamps removed from the UV reactor (courtesy of Trojan Technologies, Inc.).

determination of the number of lamps required for disinfection and (2) determination of an optimal process configuration (e.g., the number of lamps per module, modules per bank, banks per channel, and the overall number of channels). Following is a detailed discussion of the approaches that can be used to determine the number of lamps that will be required. Factors that affect the minimum number of UV lamps necessary for disinfection are (1) wastewater quality and its variability and (2) the nature of the discharge permit itself and the confidence desired in meeting that permit. The characteristics of wastewater that have been found to influence UV disinfection performance were reviewed previously (see Table 8.2). Wastewater discharge permits vary in both the stated coliform discharge limit (e.g., 2.2, 23, 240, 1000 coliform per 100 mL) and in the method used to calculate the limit (e.g., 7- or 30-day running median or value not be exceeded more than once in a 30-day period).

FOUR APPROACHES FOR DETERMINING THE REQUIRED NUMBER OF LAMPS

In what follows, two non-probabilistic and two probabilistic methods are considered for determining the number of UV lamps required for disinfection at a WWTP. The first non-probabilistic method is described to provide an overview of the current design approaches used by most UV manufacturers. The second non-probabilistic method is part of the California Wastewater Reclamation Criteria (NWRI, 1993). Two probabilistic approaches are described because each approach is appropriate to use in certain types of design situations. The first probabilistic approach is appropriate for use at WWTPs with an extensive existing wastewater quality data set. The second probabilistic approach is appropriate for use at WWTPs that do not have an extensive water quality data set.

Non-Probabilistic Design Approach 1

In the first non-probabilistic approach, used by many UV manufacturers, the results obtained from a bench-scale collimated beam study are used to determine the minimum UV dose necessary to meet the most stringent permit requirements of a WWTP. The probabilistic nature of permit requirements is ignored. The variability in wastewater characteristics is accounted for indirectly through the application of various safety or adjustment factors. The actual steps used in this design approach are outlined below.

(1) Plot the log of the effluent coliform concentration, obtained from a collimated beam study, as a function of UV dose (see Figure 8.9).

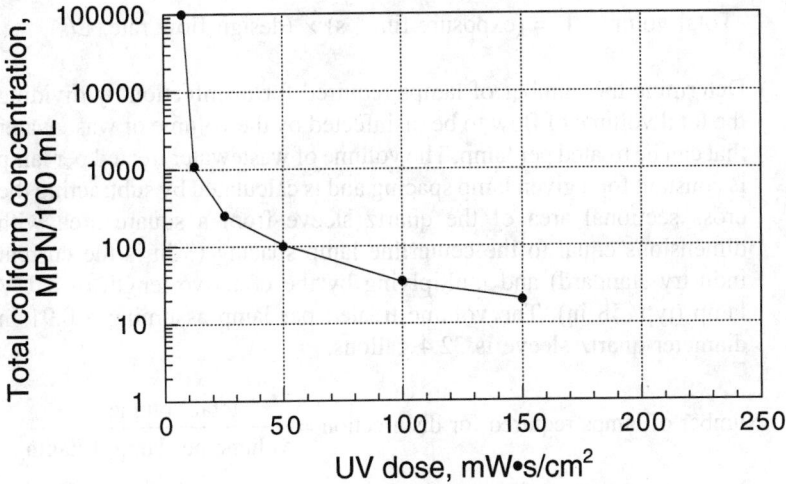

Figure 8.9. Total coliform concentration after exposure to a collimated beam of UV light (data, from a single wastewater sample, used in Non-Probabilistic Design Approach 1).

(2) Determine the UV dose that corresponds to the lowest discharge limit stated in the WWTP permit. For example, as shown in Figure 8.9 a dose of approximately 140 mW·s/cm² is necessary to meet a permitted effluent coliform concentration of 23 MPN/100 mL.

(3) Adjust the required UV dose to account for lamp aging and fouling anticipated in the full scale facility. Increase the UV dose by 30 percent to account for lamp aging and an additional 30 percent to account for lamp fouling. Additional safety or adjustment factors can be applied to further increase the UV dose to account for the uncertainties associated with variable wastewater characteristics. UV manufacturers with extensive experience in system design typically have a vast and varied array of data sources available to aid in determining appropriate adjustment factors for various design situations. This same data base is typically not available to municipalities and their consultants.

(4) Use the PSS method (or equivalent) to determine the average UV intensity in the UV system (see Figure 8.3).

(5) Given the average intensity in step 4, determine the exposure time necessary to provide the required UV dose:

$$\text{Exposure time, s} = \frac{\text{UV dose, mW·s/cm}^2}{\text{average intensity, mW/cm}^2}$$

(6) Based on the design flow rate to be treated, determine the volume of flow that must be retained for the required exposure time:

Total volume, L = (exposure time, s) × (design flow rate, L/s)

(7) Determine the number of lamps required for disinfection by dividing the total volume of flow to be disinfected by the volume of wastewater that can be treated per lamp. The volume of wastewater treated per lamp is constant for a given lamp spacing and is calculated by subtracting the cross sectional area of the quartz sleeve from a square area with dimensions equal to the centerline lamp spacing (3 in is the current industry standard) and multiplying by the effective length of a UV lamp (typ. 58 in). The volume treated per lamp assuming a 0.91 in diameter quartz sleeve is 32.4 gallons.

$$\text{Number of lamps required for disinfection} = \frac{\text{total volume, L}}{\text{volume per lamp, L/lamp}}$$

Non-Probabilistic Design Approach 2

In the second non-probabilistic approach, a UV disinfection system is designed to meet a highly restrictive standard under the worst possible operating conditions. Typically, the standard and the worst possible operating conditions are delineated by a governmental agency responsible for oversight of wastewater reclamation regulations. Because this second approach involves no pilot scale testing, it is most appropriate for use with high quality wastewater with limited variability, such as would be available following effluent filtration. The actual steps used in this design approach discussed below are based on the California Wastewater Reclamation Criteria (NWRI, 1993):

(1) A minimum UV design dose of 140 mW·s/cm² at the maximum weekly flow and 100 mW·s/cm² at peak flow (maximum day)

(2) End of lamp life factor = 70% of nominal (new) UV lamp output (unless a higher end of lamp life is demonstrated)

(3) Dose calculations must incorporate a UV transmittance loss through the quartz sleeve.

(4) Fouling factor = 30% (incorporates quartz sleeve UV transmission loss) (In the future, the magnitude of the fouling factor may be reduced if the proposed system uses mechanical/chemical cleaning or another chemical-based cleaning system that has been demonstrated to maintain a higher transmission of UV energy under effluent quality similar to that being considered.)

(5) Minimum allowable wastewater transmittance = 55% (If continuous transmittance data have been collected for a minimum period of six

months including wet weather periods, a higher transmittance value may be allowed.)

(6) UV intensity calculation method = point source summation (PSS) (Or equivalent method, verifiable by bioassay. Collimated beam data coupled with field pilot data on the same effluent quality may also be used to verify intensity model.)

(7) UV dose is to be achieved with a minimum of three UV disinfection units in series delivering an EPA UV dose equivalent to 1/3 the total required dose (e.g., 100,000 or 140,000 $\mu W \cdot s/cm^2$ design dose) from each disinfection unit in series for open channel gravity flow systems with lamps parallel to the flow.

(8) A reduced dose for design may be used if it can be demonstrated to the satisfaction of the Department of Health Services that the plant has produced a consistent water quality for an extended period of time, including seasonal variations in effluent quality, through pilot scale or collimated beam studies.

The number of lamps required for disinfection are determined using steps 3 through 7 as discussed previously in Non-Probabilistic Design Approach 1.

Probabilistic Design Approach 1

In this approach, described in detail elsewhere (Darby et al., 1995; Loge et al., 1996), the results of a pilot scale field test are used to calibrate an empirical UV disinfection model. Typical pilot scale UV units used to conduct field studies are illustrated in Figure 8.10. The model is then used in conjunction with a statistical technique (Monte Carlo analysis) to account for the variability in wastewater characteristics and the probabilistic nature of permit requirements to determine the number of lamps required for disinfection. The steps involved in this approach are as follows:

(1) First the wastewater quality and disinfection performance data collected from the pilot UV reactor are used to calibrate the following empirical equation:

$$N = A \; (SS)^a (N_o)^b \; (UFT)^c \; (I)^n \; (t)^n \tag{3}$$

where

N = coliform density after exposure to UV light, MPN/100 mL
SS = suspended solids, mg/L
UFT = unfiltered transmittance at 253.7 nm, %
N_o = influent coliform density, MPN/100 mL
I = average intensity of UV light, mW/cm^2 (from PSS method, Figure 8.3)

Figure 8.10. Typical examples of pilot plants used to conduct UV disinfection studies: (a) linear channel arrangement (Fisher Porter, Inc.).

Figure 8.10 (continued). Typical examples of pilot plants used to conduct UV disinfection studies: (b) stacked channel arrangement (Trojan Technologies, Inc.).

t = exposure time, s (assuming approximate plug flow conditions)

A,a,b,c,n = empirical coefficients (site specific)

The advantage of a flexible empirical model is that statistical techniques can be used to determine the significance of variables thought to impact the performance of a UV disinfection system, rather than relying on human intuition in prejudging the variables of importance.

(2) A wastewater quality data set characterizing the significant model variables is collected from the WWTP to characterize the variability in wastewater quality.

(3) A log transformed version of the calibrated model is used to predict the effluent coliform concentration after exposure to UV light for each of the days in the water quality data set.

$$\text{Log } N = \text{Log } A + a \text{ Log (SS)} + b \log(N_o) + \\ n \text{ Log } (l) + n \text{ Log } (t) + \epsilon \quad (4)$$

The error term, ϵ, is a normally distributed random number with mean of zero and standard deviation equal to the adjusted root mean square error (RMSE′) of the calibrated model (see Darby et al., 1995 or Loge et al., 1996 for the equation used to calculate the RMSE′). The error term accounts for the uncertainty associated with each empirical model prediction.

(4) The set of predicted coliform concentrations are revised to reflect the different methods of calculating the permit requirements at the WWTP (e.g., 7- or 30-day running median). The percent of time each permit requirement is not violated is then plotted as a function of UV hydraulic loading on log-probability paper (see Figure 8.11). The maximum allowable UV hydraulic loading is defined as the UV loading at which there is an acceptable likelihood the specified permit requirement will not be violated.

(5) The UV hydraulic loading can be adjusted to reflect reductions in lamp output associated with lamp aging and fouling anticipated in the full scale UV reactor, as described previously for the non-probabilistic approach (see step 3). The design flow rate is then divided by the UV hydraulic loading to obtain the number of lamps required for disinfection, given a desired level of confidence in meeting the permit conditions.

Probabilistic Design Approach 2

Long-term existing water quality data were used in Probabilistic Design Approach 1. The second probabilistic design approach described below

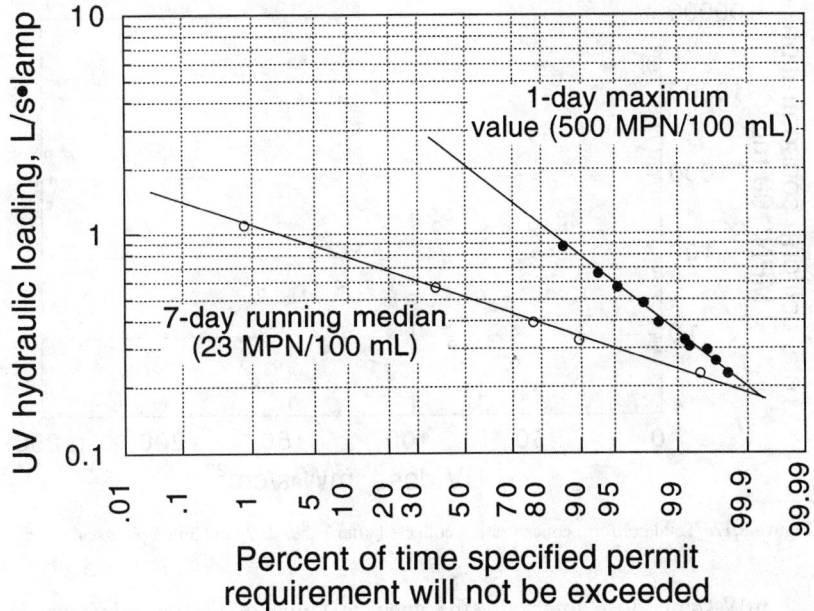

Figure 8.11. Confidence of not violating the specified permit requirements as a function of UV hydraulic loading (data used in Probabilistic Design Approach 1).

is most appropriate for situations where the wastewater quality at a WWTP is not well characterized either because data have not been collected for an extended period of time or because the plant is new. In this approach, the results of an extensive pilot scale field test are used to determine the number of lamps necessary to meet the various permit requirements of a WWTP. The impact of variable wastewater characteristics on disinfection performance is incorporated in this design approach through the use of confidence intervals placed on a non-parametric curve fit through the disinfection performance data. The actual steps used in this approach are outlined below:

(1) Arrange the coliform concentrations measured after each bank of the pilot UV reactor to reflect the different methods of calculating the permit requirements (e.g., 7- or 30-day median). No modification is necessary for a permit requirement based on a 1-day maximum (value not to be exceeded any given day). Plot the log of the effluent values as a function of UV dose for each of the data sets (see Figure 8.12).

(2) Use the method of moving averages, a non-parametric statistical technique, to fit a curve through the disinfection performance data as illustrated in Figure 8.13. The curve is generated by calculating the geometric mean of the total coliform concentrations within 25

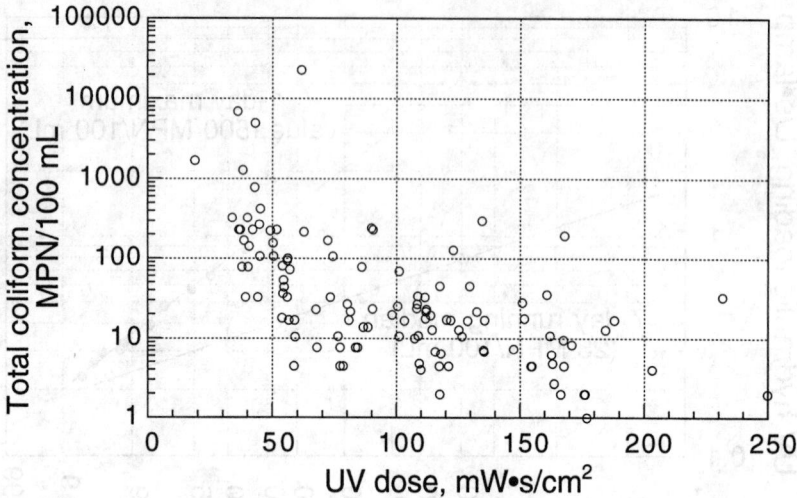

Figure 8.12. Total coliform concentration collected after banks 1, 2, and 3 in a pilot scale study.

mW·s/cm² dose intervals (the specific value of the dose interval is dependent on the data set collected from the WWTP). A curve is then fit through the disinfection performance data based on the geometric mean of the coliform concentration and the median UV dose within each interval as illustrated in Figure 8.13.

(3) Plot the residuals of the non-parametric curve, calculated as the log of each observed value minus the log of the predicted value (e.g., from the curve fit in step 2), on probability paper as shown in Figure 8.14.

(4) Calculate the upper confidence intervals for the non-parametric curve in Figure 8.10 using the following procedure. For a desired level of confidence (e.g., 70, 80, 90, 95, or 99 percent), select the corresponding residual from Figure 8.11 (see step 3). Add the value of the residual to the non-parametric curve to obtain the upper confidence interval. Upper confidence intervals of 70, 80, 90, 95, and 99 percent are illustrated in Figure 8.13 for the non-parametric curve fit through the disinfection performance data.

(5) From the figure developed in step 4, select the UV dose corresponding to a desired level of confidence the permit requirement will not be violated.

(6) Repeat steps 2–5 for each of the revised data sets in step 1 (e.g., for each data set revised to reflect the different permit requirements). The highest value of UV dose required for the different permit requirements is used to determine the number of lamps required for disinfection. Using the highest value of UV dose, an identical procedure as outlined

in steps 3 through 7 in the Non-Probabilistic Design Approach 1 is used to determine the number of lamps required for disinfection with the exception that the average filtered transmittance occurring during the analysis period is used.

COMPARISON OF THE FOUR APPROACHES FOR LAMP DETERMINATION

Principal factors influencing the number of lamps required for disinfection, regardless of the design approach, include the wastewater characteristics, variability in the wastewater characteristics, and the permit requirements of a particular WWTP. All the design approaches except Non-Probabilistic Design Approach 2 account in some fashion for the impact of wastewater characteristics on disinfection performance through either collimated beam or pilot studies conducted at the particular WWTP.

In the first non-probabilistic design approach, the results of a collimated beam study conducted on a single wastewater sample are used to determine the number of lamps required to meet the most stringent discharge limit of a WWTP. The probabilistic nature of permit requirements is ignored. In this approach, variability in wastewater characteristics is either neglected or accounted for through the application of various adjustment or safety factors. Depending on the wastewater characteristics of the single sample tested, the actual variability that

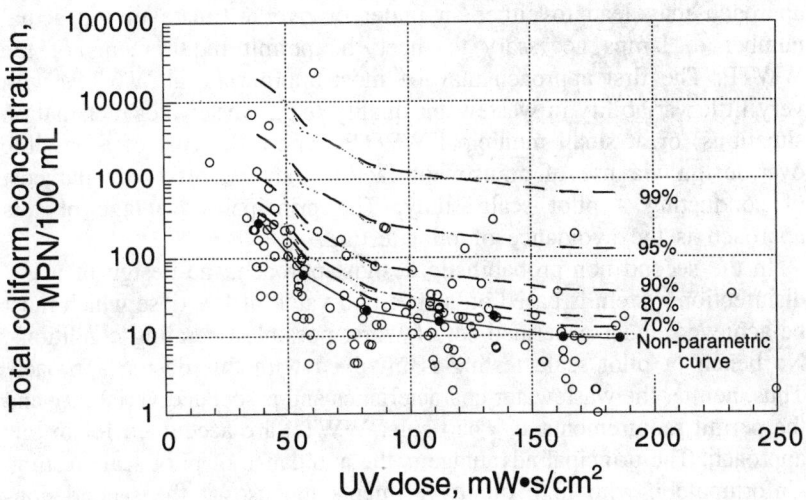

Figure 8.13. Non-parametric curve and associated upper confidence intervals fit through the disinfection performance data collected during the pilot scale study (data used in Probabilistic Design Approach 2).

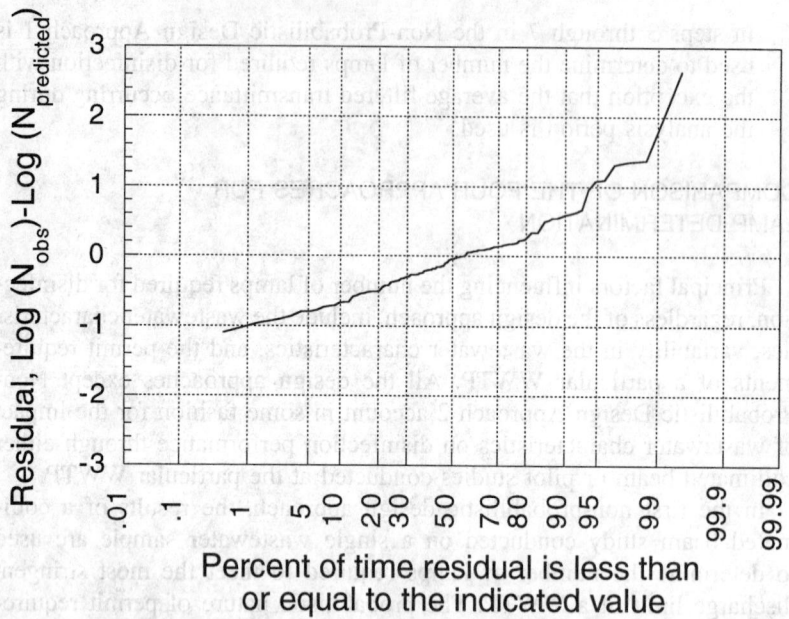

Figure 8.14. Residuals of the non-parametric curve fit through the disinfection performance data.

occurs, and the resources (e.g., data base, experience, judgment) available to the designer to adjust the design to account for the variability, this approach may lead to either an under or over estimate of the actual number of lamps necessary to meet the permit requirements of the WWTP. The first approach may be most appropriate at WWTPs with very little variability in wastewater quality (e.g., wastewater reclamation situations) or at small municipal WWTPs where the cost of a possible over-design via use of adjustment factors may be offset by the cost of conducting a pilot scale study. The principal advantage of this approach is the avoidance of pilot testing.

In the second non-probabilistic design approach, the design of a UV disinfection system is based on a specified value of UV dose which must be achieved under a defined set of worst possible operating conditions. No bench or pilot scale testing is involved with this design approach. Thus, neither the wastewater characteristics and associated variability and the permit requirements at a particular WWTP are accounted for in this approach. The principal advantage is the avoidance of pilot scale testing. Unfortunately, with high quality effluents the use of the second non-probabilistic design approach results in a very conservative design. Because of the potential cost savings, pilot scale testing is usually conducted to reduce the UV dose.

The probabilistic approaches require different amounts of pilot testing. In both, the impact of wastewater quality and variability and the probabilistic nature of permit requirements on the number of lamps required for disinfection are explicitly considered. In the first probabilistic approach, the use of an empirical UV disinfection model to predict the performance of a UV disinfection system allows a rational method to quantify the probabilistic nature of wastewater variability and permit requirement while minimizing pilot scale testing. The use of an existing wastewater quality data set minimizes the need for extensive pilot testing. This approach is appropriate to use at WWTPs that have adequate records of wastewater quality data.

In the second probabilistic approach, extensive pilot scale testing is required to characterize the impact of unknown variability of wastewater characteristics on disinfection performance. This approach is appropriate to use in the design of a full scale UV disinfection facility at a WWTP that does not have an existing wastewater quality data set.

DESIGN CONSIDERATIONS FOR UV DISINFECTION FACILITIES

Considerations in the design of a UV disinfection facility include the selection of an optimal system configuration, hydraulic constraints, and system reliability.

SELECTION OF AN OPTIMAL SYSTEM CONFIGURATION

Once the number of lamps required for disinfection are determined using one of the four approaches outlined above, the lamps must be put into a particular process configuration. A process configuration consists of the number of lamps per module, modules per bank, banks per channel, and the overall number of channels. The primary consideration in this step is to generate a process configuration that minimizes the difference between the number of lamps required for disinfection and the actual number of lamps in the system. The actual number of lamps in a system is likely to be different than the number of lamps required for disinfection for two reasons. First, there is typically a slight increase in the total number of lamps that results from adding integer numbers of components together to develop an acceptable system configuration. Second, the system configuration may need to be enlarged to reduce the headloss to an acceptable value. A detailed discussion of the steps involved in selecting an optimal process configuration is contained elsewhere (Darby et al., 1995; Loge et al., 1996).

HYDRAULIC DESIGN—LOW PRESSURE SYSTEMS

Hydraulic design consideration for low pressure UV systems relate primarily to the approach channel, the headloss along the UV channel, and the outlet control facilities.

Approach Channel

Approach conditions for low pressure lamp systems are recommended to be two times the channel depth or at least 4 ft upstream of the first UV bank. The intent is to minimize the potential for non-uniform velocity profiles. A deepened inlet floor just upstream of the UV channel can be used to limit the formation of non-uniform velocity profiles. This dropped section imparts a stilling effect to the velocity of the effluent entering the channel. Each UV channel generally has a stilling plate located at the front of the channel, otherwise known as a perforated plate or orifice plate, that acts to create a uniform headloss and hence velocity distribution across its frame. This controlled, uniform headloss is used in multi-channel UV systems to control the distribution of effluent equally into the available channels and to equalize the velocities across the cross-section of the channel under all operating conditions.

For multi-channel UV systems the inlet structures to each channel should contain an isolation gate which can be used to hydraulically isolate each channel during maintenance or low flow conditions when the UV banks are not required for disinfection. A distance of at least 2 to 4 ft should be left between UV banks to allow for maintenance and UV bank removal during the cleaning process.

Headloss along UV Channel

As the water to be disinfected passes through a UV bank there is a loss of head due to turbulence and friction, which is a function of the velocity of the flow through the UV channel. In an open channel, with three UV banks in series, the headloss is usually limited to about 0.5 in. per bank. If the headloss is greater, there is the potential for short circuiting which may result in a lower average UV dose. The headloss through a bank will depend on the flow that can be treated per UV lamp (see Non-Probabilistic Design Approach No. 1, step 7). For a very high quality wastewater, the flow rate that can be treated may be so great the headloss through a UV bank may be excessive. The situation can be corrected by increasing the number of channels or by adjusting the lamp spacing. Increasing the number of channels increases the total number of lamps required and, in turn, the UV dose that can be delivered. Increasing the lamp spacing reduces the average intensity within each UV bank, and thereby average UV dose. To account for the reduced UV intensity, a greater number of lamps will be required. Both approaches should be assessed to determine the most cost effective configuration. Although alternative lamp spacings are not commonly used at this time, it is anticipated that this approach will be used more commonly in the future, particularly with the high quality effluents typical of reclaimed wastewater.

Outlet Design

Outlet design for low pressure systems is closely tied to the effluent level control in the channel. The types of effluent level controllers include a weir, automatic level controller (flap gate), or modulating weir gate. The selected effluent level controller must keep the uppermost lamp in the last UV bank submerged for all flow conditions. A distance of at least two times the channel depth or 4 ft minimum should be left behind the last UV bank in the channel particularly where a weir or automatic level controller is used to prevent short-circuiting of the reactor or stagnant flow areas upstream of the level control device.

The channel floor should continue at least 2 ft past the level controller to provide a stable backwater pressure on the flap gate when an automatic level controller is used. Mud valves upstream of the level control device are recommended so that the channels can remain dry at all times when out of service and so that accumulated debris upstream of the level controller can be flushed from the channel floor. A floor drain is critical where post aeration or post UV processes are in use to inhibit the growth of bacteria where UV dose is unavailable for further disinfection.

HYDRAULIC DESIGN—MEDIUM PRESSURE SYSTEMS

As with the low pressure UV systems, hydraulic design considerations for medium pressure UV systems are also related to the approach channel, the headloss through the UV reactor, and outlet control facilities.

Approach Channel

The length of approach for a medium pressure system may not be important if the flow through the UV reactor is turbulent, approaching 5 times the upstream channel velocity of the UV reactor. The particular UV reactor design is critical to deciding whether a significant approach length is warranted. Some medium pressure systems with lamps oriented perpendicular to the flow use rapid transition sections between the UV reactor and the adjoining piping. Under these conditions the flow regime upstream of the UV reactor is variable based on flowrate and a cursory analysis would suggest that a longer approach section (8 ft minimum) upstream of the UV lamps is warranted to ensure the best opportunity for uniform lateral mixing across the width of the UV reactor. Every UV reactor design must be evaluated for its own unique hydraulic and disinfection performance across a range of flow conditions. A side-by-side comparison with a known low pressure UV system can be used to assess the performance of different UV reactor designs.

Headloss Through UV Reactor

In medium pressure UV systems the UV lamps are placed in a reactor with fixed dimensions. Thus, the headloss through the UV reactor will depend on the velocity of flow through the unit. The headloss will be the largest with a high quality wastewater. Headlosses in the range from 4 to 10 in have been observed in the field with medium pressure UV reactors. Accordingly, the approach and outlet facilities must be designed to accommodate such variations.

Outlet Design

The principal decision regarding outlet design concerns the choice of the downstream effluent level controller. Precise downstream level control is essential to ensure that the UV lamps remain covered at all times during all flow regimes. This is particularly important for medium pressure systems because the lamps operate at temperatures in excess of 600°C and effluent cooling is required at all times. All systems must be equipped with safety interlocks that shut down the UV modules if they are moved out of position during operation, or if the water level falls exposing the lamps to the air.

RELIABILITY DESIGN

Special attention must be devoted to the reliability of any proposed UV disinfection system, especially in wastewater reclamation applications. The provisions of standby power, electrical safety, and seismic design (in some locales) are of critical importance.

Backup Power

Backup power and power supply reliability should be provided wherever the disinfection performance is required under all operating conditions. To assure a continuous supply of power to the UV disinfection facilities if one of the power supply lines fail, a dual feed power distribution system should be considered. Another feature that could enhance the system reliability at minimal cost increase would be to divide the disinfection system components of the same type (i.e., UV disinfection unit) between two or more power distribution panel boards or switch boards.

For water reuse applications it would be prudent as a minimum standard, to provide enough standby capacity to power 3/4 of the available UV banks for open channel systems, recognizing that the fourth UV bank is generally used to replace a UV bank that is off-line. Similarly, other UV system layouts would require that all UV disinfection units except the redundant units have standby power provided.

Electrical Safety Design

All UV systems should be provided with ground fault interrupt circuitry on each UV module, or at the load side of the power supply for smaller UV systems. In addition it is recommended that UV systems have a manual, fused disconnect on the power supply to each bank of lamps, and that for medium pressure lamps an individual circuit breaker be provided for each UV module to isolate it electrically for maintenance purposes. Medium pressure UV lamps should also have a mercury switch or similar device that ensures that the UV lamps cannot turn on when moved from their horizontal operating condition. Where possible the electrical circuitry should be reduced to as many parallel feeds as is reasonable to minimize the working voltages of the equipment.

Seismic Design

The UV system should be designed in accord with the seismic design requirements (e.g., Section 2312 of the Uniform Building Code for "special occupancy structures") in the appropriate seismic zone. These same seismic design standards should apply to structures in which UV replacement lamps are stored on site. The seismic requirements generally reduce to providing larger diameter anchor bolts to account for the heightened shear forces during seismic activity.

Seismic design considerations are particularly important for UV systems because of the fragile components (especially lamps and quartz sleeves) used in the systems. The seismic safety design of the UV disinfection system should be at least equivalent to the design of the reclamation facilities prior to disinfection. This provision will assure that whenever the plant is capable of producing effluent, the UV system will provide adequate disinfection at all times.

OPERATION OF UV SYSTEM

Important considerations in the operation of a UV disinfection system include system monitoring and alarms, lamp cleaning, lamp aging, and sequencing of UV lamps.

MONITORING AND ALARMS

The ability to monitor operating parameters continuously is of fundamental importance to the operation and ultimate performance of a UV disinfection system.

Continuous Monitoring

Continuous monitoring should be provided for the following parameters:

Wastewater

(1) Flow rate
(2) UV transmittance
(3) Turbidity
(4) Liquid level in UV disinfection channels

UV Disinfection System

(1) Status of each UV disinfection unit, on/off
(2) Status of each UV lamp, on/off
(3) UV intensity measured by at least one probe per UV disinfection unit
(4) Lamp age in hours
(5) Applied UV dose
(6) Status of UV control system communications link
(7) Cleaning system status
(8) Time to next cleaning cycle
(9) UV bank power level

Continuous Monitoring of UV Dose

The average UV dose delivered by each UV disinfection unit (or equivalent multiple UV disinfection units) is usually delivered in increments ranging from 50 percent (e.g., two banks in series) down to 20 percent (e.g., 5 banks in series) and should be monitored continuously. The average UV intensity within each UV disinfection unit is to be computed using a proven UV dose calculation methodology. Flowrate, UV transmittance, lamp power setting, and lamp age are needed to determine the average UV dose.

The average UV dose is then computed by multiplying the average exposure time, determined from the flowrate, by the average UV intensity. The UV dose should be increased automatically to compensate for high turbidity readings or low UV dose, or both, by implementing (1) an extra UV disinfection unit in the UV channel, (2) an additional UV channel of UV disinfection units, or (3) by increasing the amount of power to the medium pressure lamps and hence the UV output per disinfection unit. The alarm system should be activated if the average UV dose falls below a predetermined set-point for more than three minutes.

Alarms and Alarm Records

All alarms generated by the UV system controller should be logged and displayed at the operator interface. Each alarm will be time and date stamped when it occurs. The UV system should be designed to record automatically each alarm condition.

CLEANING OF UV LAMPS

Lamp cleaning of horizontal low pressure lamp based systems is accomplished by one of two methods. The most elegant methodology involves the removal of an entire UV bank via overhead monorail and placement on a water spray-down pad and then into an air agitated acid bath. The conventional method for smaller UV systems is a manual process whereby 3 modules at a time are placed by hand into a portable cleaning tank where a small compressor agitates a phosphoric acid solution (pH = 2.0) to remove organic sleeve deposits. Vertical lamp low pressure systems use an integral air scour at the base of each UV module to extend the time between cleanings. Vertical modules must also be removed via overhead jib crane, or monorail to an external acid cleaning basin for air agitated cleaning.

There are two types of cleaning systems for medium pressure lamps: (1) a separate mechanical wiper and isolated chemical cleaning process where the channel is evacuated of effluent and filled with acid to chemically clean the lamps or (2) an integral mechanical/chemical sleeve wiper collar that cleans the lamps on demand, without the need to take the channel off-line for chemical cleaning (see Figure 8.15).

LAMP AGING

UV lamps are manufactured with a finite amount of thermionic emissive material on the filaments. This material, when heated by the electrical

Figure 8.15. Wiping mechanism used to mechanically and chemically clean medium pressure UV lamps (courtesy of Trojan Technologies, Inc.).

current, liberates electrons that travel from filament to filament under the lamp voltage potential striking mercury atoms and liberating photons of UV energy. Over time the electron pool is depleted to the point where the UV output of the lamp is insufficient for the energy being put into the lamp. For low pressure systems using electronic ballasts, it is standard practice to assume that after approximately 13,000 hours of use, assuming less than 3 on/off cycles per day, the lamp UV output will have declined to about 65 percent of the output of a new lamp after 100 hours burn in time. In cases where the input current to the lamp is less than 425 mA, the lamp starting intensity is lower which can potentially reduce the lamp life to 3000 to 4000 hours.

Medium pressure lamps arrive at their 80 percent, theoretical end of lamp life in shorter time because the higher operating currents deplete the electrons at a faster rate. The higher the lamp current the shorter the lamp life. Typical industry figures for longevity of the medium pressure lamp are as follows assuming no more than 3 on/off cycles per 24 hour period.

- high power setting 100% input power = 3,000 hours lamp life
- medium power setting 60% input power = 5,000 hours lamp life
- low power setting 30% input power = 8,000 hours lamp life

COMPLIANCE MONITORING

Compliance monitoring for UV disinfection systems will include both grab samples, a variety of continuous on-line measurements, and continuous monitoring of the average UV dose. Routine monitoring based on representative daily grab samples should include the following: total or fecal coliform bacteria and total suspended solids. The samples for both coliform bacteria and suspended solids should be collected daily at a time when wastewater characteristics are most demanding on the treatment facilities and disinfection facilities. A sampling program that includes UV transmission testing will help to establish seasonal effluent variations with the net result that the UV dose delivery system can be fine tuned during these periods to optimize disinfection performance as well as minimizing power consumption.

REFERENCES

Crites, R. and G. Tchobanoglous (1988) *Small and Decentralized Wastewater Management Systems.* McGraw-Hill, New York, NY.

Darby, J. L., M. Heath, J. Jacangelo, F. Loge, P. Swain, and G. Tchobanoglous, G. (1995) *Comparative Efficiencies of UV Irradiation to Chlorination: Guidance for Achieving*

Optimal UV Performance. Project 91-WWD-1, Water Environment Research Foundation, Alexandria, VA.

Harris, G. D. et al. (1987) Potassium Ferrioxalate as Chemical Actinometer in Ultraviolet Reactors. *J. Environ. Eng.,* 113:612.

Loge, F. J., R. W. Emerick, M. Heath, J. Jacangelo, G. Tchobanoglous, and J. Darby (1996) Ultraviolet Disinfection of Secondary Wastewater Effluents: Prediction of Performance and Design. *Water Environ. Res.,* 68:900–916.

National Water Research Institute (NWRI, 1993). UV Disinfection Guidelines for Wastewater Reclamation in California and UV Disinfection Research Needs Identification. Prepared for the California Department of Health Services. Sacramento, CA.

Qualls, R. G., and Johnson, J. D. (1983) Bioassay and Dose Measurement in UV Disinfection. *Appl. and Environ. Microbiol.,* 45:872.

Qualls, R. G., and Johnson, J. D. (1985) Modeling and Efficiency of Ultraviolet Disinfection Systems. *Water Res.,* 19:1039.

Qualls, R. G. et al. (1989) Evaluation of the Efficiency of Ultraviolet Disinfection Systems. *Water Res.,* 23:317.

Suidan, M. T., and Severin, B. F. (1986) Light Intensity Models for Annular UV Disinfection Reactors. *AIChE J.,* 32: 1902.

U.S. Environmental Protection Agency (1986) *Design Manual: Municipal Wastewater Disinfection.* EPA/625/1-86/021, Cincinnati, Ohio.

CHAPTER 9

Reclaimed Water Distribution and Storage

INTRODUCTION

THIS chapter provides an overview of the various factors which go into developing a reclaimed water distribution system. From the first steps in determining the available markets and potential demands, through system layout, operational and seasonal storage issues, and finally institutional and regulatory challenges. In addition to providing guidelines for the distribution of the reclaimed water to the different users, this chapter also presents the recommendations for the user's systems.

BACKGROUND

A discussion of reclaimed water distribution and storage begins with determining current and potential uses of reclaimed water which exist within a service area. The expected demand for reclaimed water that each user represents allows a purveyor to determine how many users can be accommodated from the reclaimed water supply. Layout of possible networks to connect all of the targeted users to the reclaimed water source follows along with establishing appropriate peaking factors for each type of use and the resultant needs for operational storage, seasonal storage, and, if appropriate, emergency and flood control containment storage.

Thomas R. Holliman, Director, Engineering and Planning/Chief Engineer, Long Beach Water Department, 1800 E. Wardlow Rd., Long Beach, CA 90807-4994.

After analyzing several alternative networks, final adjustments are made to either add or delete users depending on the balance established through modeling efforts.

With a network plan and with size and locations established for the various types of storage needed, design of the distribution system and storage facilities can be performed. Finally, with the distribution and storage systems established, the task of developing site specific plans for using the reclaimed water on-site can be prepared, and facilities can be constructed and operated.

This chapter provides the basis for fully developing a reclaimed water distribution and storage system from initial planning through developing on-site reclaimed water facilities. It will include a detailed discussion of the various uses of reclaimed water and their specific quality requirements, provide an overview of the criteria for sizing system storage, include recommendations for reclaimed water design standard, and discuss the necessary provisions needed for on-site use of reclaimed water.

PLANNING ISSUES

The first phase in developing a reclaimed water distribution system is the development of a system "plan." This involves defining customers, demands, alternative system configurations, and the size and location of key facilities such as pump stations and storage reservoirs. This phase also includes an analysis of the obstacles to implementing the plan which might arise from institutional or regulatory constraints.

USES

The first step in developing a reclaimed water distribution system is defining the reclaimed water demand base. This demand base should include all existing reclaimed water users, as well as all potable water users that could be converted to reclaimed water both now and in the future.

Customer Base and Demands

Existing irrigation water customers and demands serve as a partial basis for establishing current potential reclaimed water use and for estimating future demands. Identifying potential new customers and forecasting their demands necessitate a market survey. The market survey involves identification and classification of potential reclaimed users and assessment of their projected demands.

The most common uses for reclaimed water are: irrigation, both land-scape and agriculture, impoundments, industrial process water and cooling towers, construction uses, laundries, carpet dying operations, plating processes, washdown water at sewage treatment plants, and groundwater recharge.

Irrigation Systems

Irrigation systems provide reclaimed water for various non-potable purposes, including parks, school yards and playing fields, highway right-of-way, landscaped slopes, landscaped areas surrounding public buildings and facilities, and agricultural irrigation. In some areas reclaimed water is being used to irrigate single-family residential landscaping and provide car washing. Common areas in multi-family complexes such as condominiums and apartments are being irrigated with reclaimed water. Irrigation of landscaped areas surrounding commercial, office and industrial developments is also common in many areas. However, large irrigation areas such as golf courses, cemeteries, highway right-of-way, and parks still represent the largest users of urban reclaimed water.

Industrial Uses

Industrial process water for cooling purposes represents a tremendous use of reclaimed water in areas such as Florida. In Southern California, several large refineries are replacing potable and well water with reclaimed water for industrial cooling, freeing up millions of gallons per day for potable water use. Other industrial/commercial uses for reclaimed water include fleet vehicle washing facilities and mixing water for pesticides, herbicides, and liquid fertilizers. Another use is the production of redi-mix concrete products and other types of on-site process water such as rinse water for carpet dying operations. Commercial uses could also include ornamental landscape uses and decorative water features such as fountains, reflecting pools, water falls, and on a limited basis, toilet and urinal flushing in commercial and industrial buildings.

Industrial reuse represents a significant potential market for reclaimed water in the U.S. and other developed countries. Although industrial uses accounted for only about 8% of the total U.S. water demands in 1985, in some states, industrial demands accounted for as much as 43% of a state's total water demands. Reclaimed water is ideal for many industries where processes do not require water of potable quality or that must treat the potable extensively before use. Also, industries are often located near populated areas where centralized wastewater treatment facilities already generate an available source of reclaimed water. Perhaps the largest potential market for industrial reuse is cooling water which is discussed in later sections.

Construction Uses

The primary use of reclaimed water in general construction is for dust control and concrete production on construction projects. In addition, reclaimed water has been used in general earth grading operations and in trench backfill compaction. Reclaimed water has also been used to provide temporary fire protection during construction.

Groundwater Recharge

One of the largest uses for reclaimed water may ultimately be for recharging groundwater basins or as a water supply to augment surface water supplies. Several significant projects are under way to determine the suitability of reclaimed water to augment surface water supplies in Colorado, California, Arizona, and Nevada. If successful, this type of use would present a tremendous opportunity to fully use all available reclaimed water in areas that have limited traditional uses. Groundwater recharge involves pumping reclaimed water into groundwater basins to provide either pressure barriers to saltwater intrusion, as in the case in Southern California, or spreading the reclaimed water in ponds where it percolates into the groundwater table over time. Thousands of acre-ft of reclaimed water are annually spread into holding ponds to provide recharge water to groundwater basins in Los Angeles and Orange County, California.

MARKET WATER QUALITY REQUIREMENTS

In assessing the potential for reclaimed water use, the distribution and storage planning should recognize the unique needs of each segment of the market. Each user has specific water quality parameters which should be met in order for them to use reclaimed water successfully. This differs from potable water systems where health departments closely monitor the quality of potable water supplies and where essentially there is no difference between the water quality demands of each segment of the market. The differences in water quality between potable water and reclaimed water are caused primarily due to the wastewater origin of reclaimed water. The following sections outline the general water quality requirements for the major potential users of reclaimed water. It should be noted that many potable water customers provide their own on-site treatment in order to use the water and these on-site treatment requirements should be considered when converting customers to reclaimed water.

LANDSCAPE IRRIGATION REQUIREMENTS

Of all of the current and potential uses for reclaimed water, landscape irrigation has always constituted a large share of the market. Turf irrigation has been successfully implemented in many areas due to the ability of most turf grasses to tolerate the elevated levels of the some of the constituents normally found in reclaimed water such as sodium, chloride, ammonia, phosphate, and nitrates. However, because this market also includes ornamental nurseries and flower and shrub irrigation, care must be taken in planning for this type of use to make sure that the plants' specific needs are compatible with the specific reclaimed water available for use in the service area.

In golf course irrigation, the higher salinity of reclaimed water has sometimes had adverse effects on the "greens." This has generally been linked to the specific species of grass being used on the greens. In California, there have been several cases of green "browning," where the grass dies prematurely or turns yellowish brown, where reclaimed water was being used. In those cases, the combination of the grass type, typically Poa Annua, and site specific irrigation practices has led to problems. Irrigation of greens with reclaimed water should include periodic deep watering to avoid the browning problem. If the greens are especially sensitive to salinity, the use of potable water should be considered. Greens and tees are already under stress due to being cut so short and can be significantly impacted by construction, and the ability to drain excess water effectively.

Salinity has been the chief concern among landscape irrigators; conversely, the elevated levels of nitrates found in the reclaimed water have provided benefits by reducing the amount of nitrogen base fertilizers required. In some estimates, the nitrates and other fertilizer chemicals in reclaimed water provide a user savings of as much as $30.00/acre-ft, depending on the area irrigated and the normal amount of fertilizers required with potable water.

Table 9.1 lists the general water quality criteria for landscape irrigation.

Agricultural Irrigation Requirements

Benefits to agricultural users of reclaimed water include: reliability of supply, reduced fertilizer costs, compatibility with wastewater treatment policies/legislation, water conservation, and reduced cost of water.

Standards for crop irrigation (Figure 9.1) with reclaimed water are based on public health. However, agronomic factors should also be considered. For example, the nitrogen concentration of the reclaimed water may be too high for general crop irrigation because it could produce undesirable

TABLE 9.1. General Water Quality Criteria for Landscape Irrigation (concentrations are maximum limits).

Parameter	Concentration (mg/L, except as noted)	Parameter	Concentration (mg/L, except as noted)
Aluminum	5.0	Manganese	0.20
Arsenic	0.10	Molybdenum	0.01
Bacteria, fecal coliform, MPN/100 ml	2.2		0.20
Beryllium	0.10	Nickel	
Biological oxygen demand	20.00	Nitrogen	5.00
			6.0–9.0
Cadmium	0.01	pH, units	
Chromium	0.10	Phenols	50.0
Cobalt	0.05	Oil and Grease	Nil
Copper	0.20	Selenium	0.02
Fluoride	2.00	Sodium absorption ration	8–18
Iron	5.00	Suspended solids	15.0
Lead	5.00	Sulfate	200–400
Lithium (general)	2.50	Vanadium	0.10
Lithium (citrus)	0.075	Zinc	2.00
Chloride		TDS	

Source: Reference [1].

effects such as too much vegetative growth and not enough fruit or seed production, delayed maturity and harvestablity (cotton), reduced flavor and poor texture (fruits and vegetables), reduced sugar content (beets or cane), or reduced starch content (of potatoes). Nitrogen in irrigation water reportedly causes no problems if its concentration is below 5 mg/L, increasing problems if it is between 5 and 30 mg/L, and severe problems if it exceeds 30 mg/L. Because secondary wastewater effluent normally contains 20 to 40 mg/L nitrogen, agronomic problems can be expected if certain crops are irrigated with wastewater effluent. Crops like hay and other forage can tolerate and benefit from large amounts of nitrogen. Other quality parameters (Table 9.2) that should be considered in using reclaimed water for irrigation are the total salt content (no problems below 500 mg/L, increasing problems between 500 and 2,000 mg/L, and severe problems above 2,000 mg/L), the sodium adsorption ration (SAR) (no problems below 6, increasing problems between 6 and 9, and severe problems above 9), and the concentration of potentially phytotoxic ions (e.g., sodium, fluoride, boron, and heavy metals) [2].

Many plants are quite sensitive to chloride and boron. Salts can cause plant stunting, leaf burn, leaf drop, and stem die back and may lower frost resistance, and can cause reduced crop yields in avocados and citrus crops.

It is difficult to prevent boron excess, which can also cause leaf burn. When reclaimed water use is planned, farmers may wish to consider plants that have a high tolerance for salts and boron.

Industrial Use Requirements

One of the largest potential markets for reclaimed water use is to supplement or replace the potable water demands of industrial process and cooling water. In many instances, the water quality requirements are less than drinking water quality, and adequate continuous supplies from the reclaimed water system year round is the critical concern.

For this chapter, industrial processing water demands include boiler feed, washing, transport of materials, and rinse water. Cooling water, the most predominant reuse application, presently accounts for 99% of the total volume of industrial reuse water [2].

The major factors that influence the potential for industrial process and cooling water reuse include water quality, volume, and economics. In planning for industrial reuse, several key water quality concerns should be addressed. The cost of treating the reclaimed water to deal with constituents of concern should be included in determining the feasibility of using reclaimed water as a substitute process water.

Reclaimed water for industrial reuse may be derived from in-plant recycling of industrial wastewater and/or municipal water reclamation

Figure 9.1 Citrus grove irrigated with reclaimed water.

TABLE 9.2. Water Quality Criteria for Agricultural Irrigation—Unrestricted Irrigation for Waters Used Continuously on All Soils (concentrations are maximum limits).

Parameter	Concentration (mg/L, except bacteria)	Parameter	Concentration (mg/L, except bacteria)
Aluminum	5.0	Lead	5.00
Arsenic	0.10	Lithium (general)	2.50
Bacteria, fecal coliform, MPN/100 ml	1000		0.075
		Lithium (citrus)	
Beryllium	0.10	Manganese	0.20
Cadmium	0.010	Molybdenum	0.010
Chromium	0.10	Nickel	0.20
Cobalt	0.050	Selenium	0.02
Copper	0.20	Vanadium	0.10
Fluoride	1.00	Zinc	2.00
Iron	5.00	TDS	
Chloride		Boron	

Source: Reference [1].

facilities. Recycling within an industrial plant is usually an integral part of the industrial process and should be developed on a case-by-case basis. Industries from steel mills, breweries, electronics, and many others treat and recycle their own wastewater (Figure 9.2) either to conserve water or to meet stringent regulatory standards for effluent discharges. This chapter doesn't review in-plant recycling. However, information and guidelines for these types of uses are readily available from the applicable industrial association or, in some cases, the local regulatory authorities.

Dissolved chlorides and sulfates contribute to the corrosive potential of industrial water and add total dissolved solids that may limit the use of reclaimed water. Dissolved gases, such as oxygen, carbon dioxide, hydrogen sulfide, and ammonia, are all significant water quality considerations. Depending on their concentration, they may contribute to metallic corrosion or add impurities to the industrial product.

Organic contaminants generally are undesirable in reclaimed water used for industrial processes. The organic substances present in municipal wastewater and certain industrial effluents may limit the use of the reclaimed water as well.

Cooling Applications

Industrial uses for reclaimed water include: evaporative cooling water, boiler-feed water, process water, irrigation of plant grounds, and general washing and cleaning applications. Of these uses, cooling water is currently the predominant industrial reuse application. In most industries, cooling creates the single largest demand for water within a plant. Worldwide, the majority of industrial plants using reclaimed water for cooling (Figure 9.3) are utility power stations (see Table 9.3).

Alkalinity of the reclaimed water, as determined by its bicarbonate, carbonate, and hydroxyl content, is of concern. Excessive alkalinity concentrations in boiler feed water may contribute to foaming and other forms of carryover, resulting in deposits in super-heater, re-heater, and turbine units. Bicarbonate alkalinity in feed water breaks down under the influence boiler heat to release carbon dioxide, a major source of localized corrosion in steam using equipment and condensate return systems. The presence of alkalinity in cooling tower makeup water accelerates precipitation of calcium carbonate hardness, leading to deposit formation in heat exchangers and tower basins [2].

Recreational Applications

Reclaimed water fits naturally into recreational applications such as the development and maintenance of recreational lakes and scenic parks. This type of use can include ''non-body contact'' activities, such as boating or sport fishing, or primary body contact activities, such as swimming and skiing (Figure 9.4). The acceptance of using reclaimed water for recreational purposes will vary regionally as alternative water supplies dwindle

Figure 9.2 Cement processing plant.

Figure 9.3 Industrial cooling unit.

and as increased population pressures overcrowd existing water-related recreational activities [2].

The standards for contact recreational water quality are higher than for non-contact reuse applications. In fact, it is common, and it is highly recommended, for effluents intended for body contact recreational uses to receive tertiary treatment.

For primary contact recreation, the water should be aesthetically enjoyable, clear, and not have objectionable odor. The latter would include both the residual smell normally associated with reclaimed water and higher-than-normal levels of disinfection odors, as is sometimes the case with chlorine. The water shouldn't contain substances that are toxic in any way or, if ingested, are irritating to the eyes or skin. The water must not contain pathogenic organisms. Although not a specific characteristic of reclaimed water, the temperature of the water should be suitable for swimming and fishing, and the pH range should between 6.5 to 8.3 to avoid irritation to the eyes.

Reclaimed water with a BOD and suspended solids values of less than 20 mg/L should provide satisfactory water quality for recreational use. However, phosphorus and nitrogen should be controlled. In several instances, lakes filled and maintained with reclaimed water have experienced higher-than-acceptable levels of algae growth caused by nutrients present in treated wastewater. Maintaining sufficiently low levels of nitrogen should minimize this problem [7].

TABLE 9.3. General Cooling Water Quality Criteria.

Parameter[a]	Recommended Limit [4]	Recommended Limit [5]	Comments
Cl	500	100–500	
TDS	500	500–1650	
Hardness	650	50–130	
Alkalinity	350	20	
pH	—	6.9–9.0	
COD	75	75	Preferably 6.8–7.2
TSS	100	25–100	Preferably < 10
Turbidity	—	50	Preferably < 10
BOD	—	25	Preferably < 10
Organics	1	2	Preferably < 5
NH₄	—	4	2 is good
PO₄	—	1	Preferably < 1
SiO₂	50	—	1 is good
Al	0.1	0.1	
Fe	0.5	0.5	
Mn	0.5	0.5	
Ca	50	50	
Mg	—	0.5	
HCO₃	24	24	
SO₄	200	200	
Ammonia			

[a]All values in mg/L except pH.
Source: Reference [6].

PUBLIC PERCEPTION, LEGAL ASPECTS, AND INSTITUTIONAL CONSTRAINTS

In addition to the specific water quality requirements for each type of use there are significant social, legal, and institutional aspects related to using reclaimed water in both new applications and in retrofit systems.

The first obstacle is that of public acceptance. Because reclaimed water is developed from reclaimed wastewater, a psychological objection to close contact frequently occurs. Many industries such as laundries and fabric mills could benefit from the use of reclaimed water for nonpotable uses such as rinse and wash water, but the potential for adverse public opinion regarding the finished products make many industries reluctant to use reclaimed water even if there are significant cost savings which could be passed on to the customer. Recent polls have indicated a fairly widespread public acceptance of the use of reclaimed water for irrigation purposes and in industrial applications such as cooling which don't require public contact with the water or products manufactured with reclaimed

Figure 9.4 Landscaping and water feature using reclaimed water.

water. The recent emphasis on conservation and recycling in other industries such as plastics and paper has helped develop a recycling and conservation ethic in the public which has been beneficial in expanding the acceptance of reclaimed water.

Legal obstacles focus primarily on aspects of liability and water rights. In a society that is quick to seek legal remedy for real and/or perceived damages, agencies considering the use of reclaimed water should pay close attention to this area. Explicit disclosure of the use of reclaimed water and the potential health effects has helped many agencies head off lawsuits based on fears rather than substance. As of this writing, there have been no reported cases of reclaimed water related damages either to property or to personal health. Water rights issues have become a concern in several areas where there are competing interests for the water or competing suppliers. In some jurisdictions where reclaimed water has been historically discharged to streams, the ownership of the water has been brought into question when the reclaimed water agency wants to begin using the water for irrigation uses. Such cases have led to disputes over the rights of the reclaimed water producer to suspend discharges which now support thriving ecosystems.

Finally, institutional barriers should be evaluated including competing uses for the reclaimed water, financing considerations, health regulations, local support or opposition, restriction on growth which might affect future sales, and the effects of internal agency polices and governance.

The following case study illustrates the extensive institutional, regulatory, and public relations constraints that had to be overcome in order to initiate the use of reclaimed water for toilet and urinal flushing in high-rise office buildings in Irvine, California.

Case Study [8]: Use of Reclaimed Water for Flushing Toilets and Urinals, and Floor Drain Trap Priming in the Restroom Facilities at Jamboree Tower 2C, 3 Park Plaza, Irvine, CA

Introduction

The Irvine Ranch Water District (IRWD) is a full service water and sewer agency serving approximately 120 square miles of southern Orange County, California since 1964. IRWD has gained international recognition for its work in the field of reclaimed water, with over 20 years of experience in producing and distributing high quality reclaimed water to its customers.

In 1987, IRWD begin investigating the feasibility of using reclaimed water in commercial buildings for uses which do not require potable water quality. After working closely with the Health Department, the City of Irvine, and local builders and developers, IRWD succeeded in facilitating the construction of six dual plumbed high-rise office buildings. On March 27, 1991, the first health department approved building to use reclaimed water from a municipal reclaimed water system for interior uses in the United States was put into service. Referred to as Jamboree Tower 2C, this high-rise is the subject of this case study.

IRWD RECLAIMED WATER SYSTEM

Since the late 1960s, IRWD has maintained a separate irrigation/reclamation distribution system (dual system) which provides reclaimed and untreated water for irrigation uses. During 1992–1993, the dual system served 12,670 acre-feet of reclaimed water and was operated under the conditions of the IRWD Primary Producer/User Permit issued by the California Regional Water Quality Control Board. Water for the irrigation system is provided as a combination of tertiary treated wastewater (''reclaimed water'') which is produced by IRWD's Michelson Water Reclamation Plant (MWRP), and raw water from IRWD's Irvine Lake.

The MWRP produces high quality reclaimed water through in-line coagulation and dual media filtration followed by disinfection. All of the treatment processes meet the requirements of the State of California Department of Health Services (DOHS) ''Wastewater Reclamation Criteria'' (California Administrative Code, Title 22, Division 4, Environmental Health, 1987) referred through the remainder of this paper as ''Title 22.''

Dual System Water Quality

The reclaimed water referred to in this case study is Type I, or Class A reclaimed water. The reclaimed water contains a coliform organism count of not more than 2.2 per 100 milliliters, as determined from the bacteriological results of the last 7 days for which analyses were completed. The number of coliform organisms will not exceed 23 per 100 milliliters in any sample, as sampled at the MWRP. The water has passed through MWRP's primary, secondary, and tertiary treatment processes and received an adequate level of treatment.

In addition to the reclaimed water produced by the MWRP, untreated water from Irvine Lake could be introduced into the IRWD irrigation/reclaimed water distribution system from time to time. This water is of a quality that could be treated and used for potable purposes.

All water used in this project originated directly from the MWRP, or came from one of several storage reservoirs. The use of multiple sources for the IRWD irrigation water distribution system provided a level of reliability consistent with that of the potable system and IRWD standards. This reliability was a key concern of the developers in accepting this program.

Dual System Safety

A major reason why IRWD considered the use of reclaimed water in non-residential buildings to be an appropriate use was the high level of attention paid to the prevention of cross-connections. IRWD had two full-time employees assigned to the exclusive task of investigating and identifying potential and actual cross-connection occurrences. To compliment these efforts, IRWD had a fully equipped water quality laboratory to monitor the quality of the effluent leaving the treatment facilities, and a large operations and maintenance staff to maintain the necessary treatment works. In addition, the five person IRWD On-site Water Systems Group provided on-site construction inspection and checked these facilities on a routine basis for cross-connections during and after construction.

Reclaimed Water Use in Non-residential Buildings—Program Overview

In selecting high-rise buildings for study, IRWD considered several factors.

ECONOMICALLY FEASIBLE

In 1987, IRWD completed a study of the economic feasibility of expanding reclaimed water use into high-rise buildings. This study determined that it was economically feasible to expand reclaimed water into high-rise buildings which were approximately six stories or taller for flushing water for toilets and urinals, and water for priming floor drain traps.

LARGE POINT LOADS

IRWD master plans its domestic water system to provide a base allocation of 2,400 gallon/acre/day to industrial and commercial sites. This average allocation is usually adequate to meet the water demands of standard office commercial or light industrial developments. A multi-story high-rise, however, normally in the six to sixteen story range within IRWD, often exceeds this base allocation. Although IRWD has implemented a high volume surcharge for these large point demands, the surcharge doesn't mitigate the effect of the higher point demand on the distribution systems which are normally installed several years before the actual buildings are built. High-rises converted to reclaimed water use were expected to use approximately 80% less domestic water and therefore lessen the effect on the distribution system.

CENTRAL UTILITY CORE

High-rise buildings of the type IRWD has evaluated are designated Type I buildings in the Uniform Building Code and are designed with a central utility core. The elevators, electrical, bathrooms, etc., are all located in the center core of the building. When the floors are completed, utilities are extended radially out from the central core. This design feature allows all of the reclaimed water fixtures to be fed from a common riser, and all of the domestic water fixtures (sinks, drinking fountains) to be fed from a separate common riser. By placing the risers on opposite sides of the core, the separation of the two risers was maximized, and should aid in reducing cross-connections.

PRIMARY OCCUPIED BY ADULTS

High-rise buildings are typically commercial office buildings with an employee population made up of adults. Posting signs, and providing advisory information is normally sufficient to inform this sector of the "public" about the precautions which should be observed when dealing with reclaimed water (i.e., do not drink). In addition, the restrooms of most commercial high-rises are restricted to employees only, making the risk of the "general public" coming in contact with the reclaimed water even lower.

CONTROLLED ACCESS TO PLUMBING

Cross-connections between domestic and reclaimed water would most probably occur where both systems are in close proximity and not clearly delineated. Because the reclaimed water piping will almost always be behind finished walls, such as tile or painted sheet rock, it is expected that an accidental or careless cross-connection will not take place.

NORMALLY HAS A DESIGNATED MAINTENANCE STAFF

Most of the high-rises in the IRWD service area are leased to tenants on a floor by floor basis or as an entire building. As a result, the building owners routinely retain a professional maintenance company with staff and adequate supervision to handle building repairs and minor modifications. Since the tenants must have all major modifications approved by the lease holder and the City of Irvine, plumbing changes can be more easily monitored and inspected by IRWD on-site inspectors.

MISCELLANEOUS

The use of reclaimed water in commercial buildings involved the development of several regulations and procedures. The most significant of these were joint IRWD/City of Irvine building code plan checking guidelines for the installation of the dual systems. These installation and testing guidelines laid out the features of the dual systems, the marking techniques required, the necessary warning and identification signs, and annual cross connection control test procedures.

In addition to the joint IRWD/City of Irvine installation and testing guidelines, the Department of Health Services (DOHS) outlined 22 conditions which had to be met before the DOHS would approve reclaimed water use in high-rise office buildings. The most significant requirements were the development of a project-specific Engineer's Report, a comprehensive management plan in accordance with the requirements of the health department, and project-specific annual testing requirements.

Project Qualifications

One of the larger development areas within IRWD is known as the Irvine Business Complex (IBC). This area consists primarily of local commercial developments, industrial buildings, and high density residential apartment buildings. This area and Jamboree Tower 2C lies on the western boundary of the reclaimed water distribution system and was constructed by The Irvine Company (TIC). Jamboree Tower 2C met all of the program level qualifications listed in Section II and had several other project specific qualifications which helped to make this project successful. These additional qualifications were:

(1) The developer, TIC, had extensive experience using reclaimed water for landscape irrigation of residential and non-residential developments through irrigation of parks and common landscape areas.

(2) TIC was a major builder in the IRWD service area and developed a 29-year working relationship with IRWD staff.

(3) The project facilities were not available to the general public, but rather to a focused group of users who could be identified and communicated with effectively.

(4) TIC had the capability and willingness to implement an effective tenant education program.

(5) TIC owned and controlled the facilities and employed the operation and maintenance staffs which would service the facilities built under this project.

Project Description—Jamboree Tower 2C

PHYSICAL FEATURES

Jamboree Tower 2C, is a twenty (20) story, Type I, high-rise office tower (Figure 9.5). It is one of three sister towers constructed at Jamboree Road and the San Diego Freeway in Irvine, CA. The restroom facilities consist of a men's and women's restroom for each of the nineteen floors. The men's restrooms typically contain three sinks, two urinals, one toilet, and one floor drain. The women's restrooms typically contain two sinks, three toilets, and one floor drain. Only the urinals, toilets, and floor drain traps are plumbed to use reclaimed water. The sinks and all other water appurtenances are connected to the potable water system.

Reclaimed water to the building is provided through a 6" pipeline connected to the IRWD 8" main line in Union Avenue. The reclaimed meter assembly is located in the rear of the building, adjacent to the road leading to the building's parking structure. All of the internal reclaimed water piping is identified to clearly differentiate it from potable water piping. This identification involves wrapping the pipe with purple mylar warning tape. With the exception of the portions of the piping protruding through the walls and connecting directly into the fixtures, no other access to the piping is available from the bathrooms. All of the internal reclaimed water systems were installed in accordance with the joint IRWD and City of Irvine guidelines. In addition to the purple mylar warning tape, an approved reduced pressure principle backflow prevention device was in-

Figure 9.5 Jamboree Tower 2C.

stalled on the domestic water line leading to the building. Within each restroom and the utility room, signs were posted identifying the use of reclaimed water. The equipment room signs consisted of a highly visible color (white) on contrasting background (purple) with a letter size of 1/2" high and 1/2" wide. In each reclaimed water control valve access door, two signs were installed indicating that the system was filled with reclaimed water. The sign on the inside of the access door was a decal type sign, while the sign which hung in the opening was a plastic sign. The plastic signs in the access door opening were equipped with locking valve seals to detect tampering. Finally, after all of the systems were checked and no cross-connections had been determined, all of the system valves were locked in the open position with locking valve seals. These seals will alert IRWD inspectors in the future if the valves have been operated without notification of IRWD and will be evidence of an unauthorized access to the reclaimed water plumbing.

The design features of the reclaimed water portion of the dual system are listed below:

(1) Risers and headers were constructed of Type L copper with the domestic and reclaimed water risers on opposite sides of the utility core.

(2) All piping was wrapped with purple marking tape.

(3) All valves were locking ball, lever valves and sealed with IRWD seals after initial startup.

(4) All control valve access panels were equipped with warning signs supplied by IRWD and sealed after initial start-up (Figure 9.6).

(5) All valves, pumps, and appurtenances were painted purple.

(6) All restrooms and equipment rooms were equipped with information signs notifying the users of the presence of reclaimed water.

(7) IRWD used existing full-time staff to perform routine monitoring and annual cross connection testing.

(8) The entire program was and continues to be managed under the terms of a comprehensive management plan developed by IRWD and approved for this program by the regulatory agencies.

COMPREHENSIVE MANAGEMENT PLAN

In order to obtain DOHS approval for this project, the DOHS required that IRWD develop a comprehensive management plan for the entire reclaimed water use in non-residential building program. The management plan details the engineering review, construction coordination, and annual cross connection testing provisions of the program. Because this program is still in the developmental stage, it is expected that some of the features

Figure 9.6 Reclaimed water signs in valve door opening at Jamboree Tower 2C.

of the plan will change over time. The primary focus of the management plan is to demonstrate to the DOHS that IRWD is adequately monitoring these uses and providing a systematic means of tracking and reporting on such items as annual cross connection testing, operational problems, access violations, system modifications, etc.

COST OF THE DUAL SYSTEM

In the early stages of reclaimed water in the non-residential building program, many of the developers were concerned about the increased cost of the dual systems. To mitigate those concerns and to encourage them to accept the idea of a dual plumbed building, IRWD agreed to pay the incremental cost of the dual system. The cost of dual plumbing one of the two high-rise buildings is shown in Table CS-1.

TABLE CS-1.   **Cost of Dual Plumbing System Koll-Transamerica No. 3 (1991 dollars).**

Total cost of building	$14,091,000
Total cost of plumbing systems	$358,000
Incremental cost for dual plumbing	$32,000
Percentage increase in total building cost for dual plumbing	0.23%
Percentage increase in total plumbing system cost for dual plumbing	8.94%

TABLE CS-2. Cost of Dual Plumbing System
Jamboree Tower 2C (1991 dollars).

Total est. cost of building	$28,240,000
Total est. cost of plumbing systems	$717,470
Est. incremental cost for dual plumbing	$64,130
High volume water use surcharge fee deferred because of dual plumbing	$35,110

Subsequent to the completion of two IRWD subsidized buildings, IRWD adopted changes to its rules and regulations making the use of reclaimed water for flushing toilets and urinals in high-rise buildings where reclaimed water was available or would be available in the near future a requirement. IRWD further rescinded the high volume use surcharge for dual plumbed buildings. This project was constructed under the revised rules and regulations and therefore no separate accounting of incremental plumbing costs was developed. In order to estimate the average cost of the dual plumbing system in Jamboree Tower 2C, average unit prices from the two IRWD subsidized buildings was used in developing Table CS-2. The actual high volume use fees that were deferred for the project are shown in Table CS-2. Based on this information, the developer's capital cost that must be recovered through water cost reductions would equal approximately $29,000.

WATER AND COST SAVINGS

Water Savings

In commercial buildings, IRWD estimated that approximately 70–90% of the water used inside the building was being used for toilet and urinal flushing. Further, it was estimated that 80% of the total water used for this type of development, including toilet, urinal flushing, and landscape irrigation could use reclaimed water. The following shows the actual water use for Jamboree Tower 2C for a sample month:

Total Water Use at Jamboree Tower 2C (data from March 1990).

Percentage of domestic water used	25%
Percentage of reclaimed water used	75%

Data substantiates IRWD's projections that approximately 75% of the water used in a high-rise development could be serviced from the reclaimed water system. It should also be pointed out that over 80% of domestic water used was directed toward the operating of two large cascade type cooling towers. IRWD has installed a sub-meter on the feeder line to the

TABLE CS-3. **Water Charges Comparison for Jamboree Tower 2C (1991 dollars).**

Actual charges:	
1991, 1992, and 1993	$6,238.74
Total actual 3-year charges	$6,238.74
Projected charges at conservation rates:	
1991, 1992, and 1993	$28,998.65
Total projected 3-year charges	$28,998.65
Total estimated 3-year savings	$22,759.91

two cooling towers and is monitoring actual flow rates. It is expected that as data is gathered and coordination with the DOHS and the local building department brings approval, the cooling towers can also be switched to reclaimed water bringing the total domestic water use down to approximately 6% of the total demand.

Cost Savings

In response to the current drought in California, IRWD has adopted an ascending block rate structure to encourage water conservation. Under the water conservation program, users have been required to reduce water consumption by 20% to avoid paying higher water rates. In determining the time frame for recovering the $29,000 residual capital investment for dual plumbing, actual water charges for the months of April, May, and June of 1991 were compared with what Jamboree Tower 2C would have paid had they used only domestic water, and not reduced consumption by 20%. Table CS-3 compares the actual charges to projected charges. Based on this calculation, it is expected that the residual capital cost of $29,000 would be recovered in water cost savings in approximately four years.

STARTUP AND OPERATION ISSUES

This project has been in operation for several years. During this period, two startup and operational issues were raised. They were odor and color concerns, and user acceptance (aesthetics).

Odor and Color

During the first few months of operation, California experienced higher than normal rainfall in March 1991, dubbed ''Miracle March'' by the California media. It brought with it the problem of potentially overflowing IRWD reclaimed water storage reservoirs. Because of the drought, IRWD had maintained most of the storage reservoirs at near capacity in anticipation of the long hot summer ahead. When the rainfall began, Jamboree Tower 2C was receiving reclaimed water directly from the MWRP, with water color ranging from 15 to 25 color units. The reclaimed water had a slight chlorine odor since the water had a residual chlorine level of 8 to 10 ppm. When the heavy rains continued, the treatment plant was taken off-line to avoid overflowing the reservoirs. As a result, reclaimed water for Jamboree Tower 2C was supplied from the Rattlesnake Reservoir, which at that time was supplying unfiltered, unchlorinated, lake water (the filters and chlorinators were down for maintenance and repair). During this period the color units climbed to a high of 114, and the water had a strong musty odor. The Jamboree Tower 2C maintenance staff received over 100 complaints from the building occupants during the one week that the treatment plant was off-line.

IRWD met with the maintenance staff and has worked with them to install a bag filter system. Bench tests showed that the bag filter was capable of reducing the color units down to approximately 5 microns. This color level is consistent with IRWD's domestic water supply. IRWD is continuing to study the need for system-wide color upgrading as more and more non-residential buildings come on-line as well as monitor the long term performance of the bag filter system.

With the resumption of reclaimed water service from MWRP, the musty odor associated with the lake water was eliminated. User complaints dropped to virtually none as soon as the treatment plant was put back on-line and the lake water was flushed from the building's plumbing system with fresh reclaimed water. IRWD has completed repairs to the filters and chlorination facilities at the Rattlesnake Reservoir and review their status when considering taking the treatment plant off-line for repairs or routine maintenance.

IRWD now notifies the maintenance staff of Jamboree Tower 2C whenever the treatment plant is taken off-line so they can be prepared for the possibility of reclaimed water quality changes. This will enable them to explain the potential temporary color and odor changes to the building occupants if and when they receive complaints.

In addition, staff is coordinating with customers to help extend the life of the filters by providing a turbidimeter and a three way control valve. An in-line turbidimeter will provide a 4–20 mAmp output signal which will be relayed to the dual current switch. The dual current switch has two set points or positions (on or off) that correspond to specific turbidity levels, A and B. If the turbidity level is greater than A, the three way control valve will be instructed by the switch via a 4–20 mAmp signal to divert flow through a first port, A; and if the turbidity level is less than A, the three way control valve will be instructed via a 4–20 mAmp signal from the switch to divert flow through a second port, B.

Filter replacement is approximately $3,000.00 each set. To extend the life of this equipment, two filter sets per year are required. If we project to double the life of the filter, the costs are approximately $9,100 to purchase, and the payoff would be 3 years (all figures in 1991 dollars).

A.	Turbidimeter (with NEMA enclosure)	$2,500
B.	Three way control valve (with electric actuator)	$5,000
C.	Dual current switch (with relay)	$ 400
D.	Labor and materials	$1,200
	Total	$9,100

Aesthetics

The second concern was public acceptance and overcoming misconceptions about reclaimed water quality and potential health effects. To overcome these concerns, IRWD focused on educating the maintenance staff for the buildings, meeting with management teams, and distributed printed information on the IRWD system and reclaimed water in general. This was followed up with presentations made by District staff to the building occupants and maintaining close communication ties.

Conclusion

The Irvine Ranch Water District is presently using reclaimed water for toilets and urinals at two 20-story high-rise buildings, the IRWD administration facility as well as restroom facilities at a local park. The District believes these projects are a positive step towards conserving water and represent another significant step towards a water efficient future.

At present, there are five additional high-rise buildings dual plumbed for reclaimed water use in restrooms. It is expected that in the near future, IRWD will provide reclaimed water service.

In the past, only exterior uses, such as landscaping, golf courses, parks, and ornamental ponds were considered appropriate regulated uses for reclaimed water. With droughts in California and the heightened interest in water reclamation and water recycling, more agencies are examining internal uses as a means of expanding the use of reclaimed water and preserving our precious domestic water resources.

Future uses will include additional high-rise buildings, single family residential lots and cooling towers for commercial buildings. With these additional uses it is realized that the District's three open reservoirs could negatively effect our system adding color and odor to our reclaimed water. Staff therefore continues to monitor and prepare for system wide changes.

The District continues to review systems operation changes and in the future will look at filtration systems at our open reservoirs.

In winter months during the last three years, we experienced on rare occasions color and odor in high-rise buildings when the treatment plant is off-line. The District has assisted this customer with on-site filtration using bag cartridge filters to help remove contaminants from the water.

The District acknowledges this is only a short term solution for point of use with a long term solution being examined.

The District believes using reclaimed water for these projects has been a very successful means of conserving water as we lead the way for others to help preserve our natural resources [8].

DEMANDS/STORAGE

In order to properly plan the reclaimed water distribution system, criteria to size the system must be derived from known and potential demands in the service area. Once the demands are derived and the variability of that demand over time is established, adequate storage can be provided in the system to provide reliable service to all of the customers. This section discusses methods for deriving current demands and projecting future demands and defines recommendations for various types of system storage.

Nature and Location of Demands

Urban public water supplies are treated to satisfy the requirements for potable use. However, potable use represents only about 28 percent of the total daily residential use of treated potable water. The remaining 72 percent represents opportunities for substituting reclaimed water for potable water for uses where potable water is not required and, in many cases, not appropriate. Because potential users for reclaimed water are scattered throughout the system in existing cities it is sometimes difficult to link them all economically [3].

Reclaimed water demand is not generally uniform across a service area. This situation is further complicated by extremely variable demand patterns. Planning a distribution system should include identifying the location of each type of demand, the expected level of demand, and the short term and long term viability of the demand. In service areas undergoing tremendous development or re-development, short term reclaimed water customers might be replaced in the future with customers who cannot or will not use reclaimed water, and the reverse is also true.

Quantity

Several methods used to estimate the major types of projected demands are described in the following sections.

Irrigation Demands

It is not surprising that landscape irrigation currently accounts for the largest urban use of reclaimed water in the United States. This is particularly true of urban areas with substantial residential areas and a complete mix of landscaped areas ranging from golf courses to office parks to shopping malls. In a "typical" American city, 70 percent of the landscaped acreage surrounds residential properties, primarily single-family homes [9]. The urban areas also have schools, parks, and recreational facilities which require regular irrigation. Studies also show that more than half of the potable water produced in urban areas is used by single-family homes.

Agricultural irrigation, representing 40 percent of the total water demand nationwide, represents another significant opportunity for water reuse, particularly in areas where agricultural sites are near urban areas and can easily be integrated with urban reuse applications [3]. Methods of estimating irrigation demands include potable water irrigation meter records, application rates times a known or estimated irrigated area, and a percentage of potable water use.

Irrigation application rates can also be derived by estimating reclaimed water demands. A review of reclaimed water master plans for other Southern California areas showed that turf irrigation application rates used for planning purposes varies between 2.5 acre-ft per year/acre for coastal areas to 4.0 acre-ft per year/acre for inland areas.

- Irvine Ranch Water District—4.0 AFY/acre
- Eastern Municipal Water District—4.0 AFY/acre
- City of Ventura—3.0 AFY/acre
- San Diego Clean Water Program—2.5 AFY/acre
- Eastlake Development, Chula Vista—3.8 AFY/acre
- Otay Water District—4.0 AFY/acre

Based on existing irrigation rates and the experience of a sample of communities, an application rate of 2.5 AFY/acre should be used to estimate reclaimed water demands for turf irrigation, where applicable using this second method. However if the weather patterns in the study area are very high such as in desert areas this should be increased accordingly [10].

Using an application rate of 2.5 AFY/acre, it was possible to estimate the reclaimed water demand for a potential landscape user by approximating the user's irrigated area. Potential irrigation areas were estimated based on either (1) results of the market survey, or (2) land use category or customer type. Table 9.4 compares different sources estimating the percentage of each land use category that could be irrigated.

TABLE 9.4. Comparison of Irrigation Percentages for Various Land Use Categories.

Land Use Category	Percent of Land Irrigated for Indicated Source			
	IRWD [11]	San Diego [14]	Eastlake Development, Chula Vista [15]	Recommended Average for LBWD
Gold courses	100	100	100	100
Parks	100	100	100	100
Institutional	20	10	50	20
Commercial	15	5	30	10
Industrial	15	5	30	5
Multifamily residential	15	12	30	10
Street/highway greenbelts	100	100	100	100

Source: Reference [10].

Industrial Process Water Demands

For discussion purposes industrial processing water demands include boiler fed, washing, transport of materials, and rise waters. The major factors that influence the potential for industrial process and cooling water reuse include water quality, volume, and economics. Process water is industry specific and water quality requirements range from potable quality for segments of the food processing industry to extremely pure water for the electronics industry to low quality for leather tanning. The quality requirements are further complicated by the fact that each step of the process generally requires water of different quality. Although this makes a specific industry's water requirements somewhat more difficult to identify, it often leads to enhanced opportunities for both planned direct reuse of wastewater effluents as well as water recycling from one process to another. The actual volume of water that will be reused depends on the geographical location of the discharges and potential users, the availability at compatible rates of flow, and seasonal variations [2].

Recreational Demand Projections

Recreational demands will be determined by the extent the public will accept reclaimed water in place of potable water. In previous applications public acceptance for non-contact recreational uses has been well documented. Common practices include substituting reclaimed water for potable water as make-up supply to lakes and ponds in golf courses, as well as for fishing and boating.

While non-contact recreational use is fairly well accepted, full contact recreational uses such as swimming pools, or even swimming in reclaimed water filled lakes, has not yet received widespread acceptance in the community within the health agencies. In deriving potential recreational uses and demands this distinction between contact and non-contact is essential in establishing both current and future demands.

Groundwater Recharge Demand Projections

Estimating the potential for converting existing injection systems to reclaimed water or reclaimed blended waters follows the same guidelines as recreational use. Local and state health regulations will severely impact the ease of migrating toward this type of use of reclaimed water from traditional uses such as golf courses and parks. Currently, there is only one instance known of health department approval for full reclaimed water injection: Water Factory 21, operated by the Orange County Water District, Fountain Valley, CA. However, even in this case, extensive additional treatment beyond tertiary is required for injection. This facility currently treats reclaimed water to levels beyond tertiary and then blends the water with potable on a 50-50 basis before injection. Conservative estimates would be to consider 100% of current groundwater injection as a potential market for tertiary treated reclaimed water. Even though the reclaimed water is blended at a 50-50 rate, the advanced treated reclaimed water requires larger quantities of tertiary treated reclaimed water to accommodate the reject stream from the treatment processes.

In addition to direct injection, many agencies spread reclaimed water in gravel filled ponds to seep into the groundwater table. Currently, the County Sanitation Districts of Los Angeles County spread 50,000 acre-ft/year of tertiary treated reclaimed water in holding ponds which seep into the groundwater table. Plans are underway in Los Angeles and Orange County to significantly increase the amount of spreading. In determining the potential of this use of reclaimed water in other areas, a study of the infiltration rates possible through spreading should be conducted. This study would focus on the infiltration rate of the soil, ease of access to the groundwater table from the spreading area, and the detention time in the groundwater table. The first two factors would determine the recharge rate and therefore the rate at which the aquifer could take the reclaimed water. The last would determine the safe infiltration rate to allow the detention to mitigate the potential of virus contamination.

Demand Variations Over Time

After determining the potential demand for reclaimed water it must be recognized that those demands will vary over the day, week, month, and year depending on the nature of that demand or industry.

Existing irrigation customers can be analyzed by daily water usage, historical monthly use, and annual demands for each existing service. Average annual demands are typically used to represent potential reclaimed water usage. However, the distribution systems are based on average daily and peak hour demands, while storage requirements look at average daily, monthly, and seasonal demand patterns.

Seasonal demands for irrigation water varies significantly from month to month. Seasonal peaking factors can be derived from actual use records for current customers. These peaking factors are calculated as the ratio of each month's actual demands to the overall average monthly demand in order to evaluate and predict seasonal trends. An example of the variations in seasonal demands for Long Beach, CA are shown in Table 9.5 [10].

TABLE 9.5.   Seasonal Demands for Existing Reclaimed
Water Customers in Long Beach, California.

Month	Total Demand (AFY)[a]	Seasonal Peaking Factor[b]
January	116.39	0.35
February	118.77	0.36
March	109.73	0.33
April	279.88	0.84
May	373.06	1.12
June	412.07	1.24
July	695.25	2.09
August	551.74	1.66
September	492.32	1.48
October	429.93	1.29
November	200.76	0.60
December	211.82	0.64
Total annual	3,991.72	
Average monthly	332.64	

[a] Actual use records from July 1989–June 1990.
[b] Ratio of actual monthly demand: average monthly demand, Long Beach Water Department.

Maximum Day Demands

The maximum day demand is assumed to occur during the maximum month seasonal demand period. For many communities this maximum day would occur during July when the monthly demand is 2.09 times Average Demand (AAD). For the maximum day during July, an even higher peak demand would be expected. A peaking factor of 2.6 times AAD was used to estimate the maximum day demand (Table 9.6) [10].

Peaking Factors

Fluctuations in demand need to be established for each user or type of user and are usually analyzed through the use of "peaking factors." These peaking factors automatically adjust the demands to reflect variations in use pattern and are integral part of the computer modeling process. Peaking factors are also extremely important in determining pipeline sizing, pump station capacity, and storage requirements. Because of the significant impact on the final distribution and storage system, it is extremely important for the planner to select appropriate peaking factors and to calibrate those factors against actual use records for the potential reclaimed water customers.

Diurnal Peaking Factors

Irrigation uses are usually confined to evening and nighttime periods, with a duration of 8–10 hours because watering outside those hours presents an inconvenience to the public or the business. Assigning these nighttime uses an 8-hour application period, a diurnal peaking factor of 3.0 results.

Other irrigation uses, such as freeway medians and nurseries, generally occur during the day. Similarly, dual plumbing uses in office buildings would, for the most part, occur during the day and early evening when these buildings are occupied. Assuming a 16-hour application period, a diurnal peaking factor of 2.0 results for these daytime uses.

Most industrial process water uses, on the other hand, would exhibit relatively constant round-the-clock demands. Groundwater recharge would also require steady flows.

The reclaimed water system can experience two separate peak hour demand conditions within a 24-hour period. The first peak demand occurs during the nighttime landscape irrigation as discussed above. A second peak demand for toilet flushing, process water, cooling towers, etc. can occur during the day similar to the domestic water peak hour demand.

Peak Hour Demand

The peak hourly distribution mains rate of use, which is a critical consideration in sizing the delivery pumps and distribution mains, may best be determined by observing and studying local urban practices and considering time of day and rates of use by large users to be served by the system. A wide range of design peak factors have been used in designing urban reuse systems from a low of 2.00 in Sea Pines, South Carolina, to a high of 4.00 in Apopka, Florida.

For the irrigation peak demand condition, a peaking factor of 6.8 is applied to all demands to simulate the nighttime irrigation demands. For the second condition, it is assumed that 70% of the domestic water demand for the commercial and industrial uses would be supplied by the reclaimed

TABLE 9.6.   **Summary of Proposed Landscape Reclaimed Water Demand Forecasting Criteria.**

Average Annual Demands (AAD), AFY/acre	
Turf irrigation (parks, golf courses, greenbelts, intensive landscaping)	2.5
Peaking Factors (times AAD)	
Maximum month	
Maximum day	2.09
Peak hour (nighttime)	2.6
Peak hour (daytime)	6.8
	5.2

Source: Reference [10].

water system. These demands are then peaked using the domestic water peak hour demand factor of 2.35. The Irvine Ranch Water District, Irvine, CA, uses a peak hourly use factor of 6.8 when reclaimed water is used for landscape irrigation and a peak factor of 2.0 for agricultural and golf course irrigation. The higher peak factor for landscape irrigation is due to the restricted use of reclaimed water between 9 P.M. and 6 A.M. This restriction does not apply to agricultural or golf course irrigation if they have on-site storage [11].

Seasonal Storage

Reclaimed water supply is fairly constant throughout the year with effluent flows from the wastewater treatment plants generally showing little seasonal variation. Demands for reclaimed water do, however, vary seasonally, with higher than average demands during summer months and lower than average demands during winter months.

The need to provide adequate seasonal storage facilities to hold large amounts of reclaimed water could require construction of a dam and large reservoir. When considering the size of the reservoirs to meet irrigation requirements, open reservoirs may prove to be the most economical alternative. In a highly developed area, siting such a large reservoir would be extremely difficult and the cost to construct such a storage facility with all the associated pumping, pipelines and land purchase expenses, make the seasonal storage option cost prohibitive, even if all significant adverse environmental impacts caused by the reservoir could be acceptably mitigated.

In general, reclaimed water quality criteria difficulties related to long term storage will fall into three categories.

- regulatory constraints—Many states specify water quality requirements for various uses. For example, the growth of algae may result in a suspended solids level in excess of a regulatory limit. Algae growth and suspended solids from open reservoirs have been recognized as sources of particles which may clog the sprinkler system. Most sprinkler system control valves and sprinkler heads can readily pass particles which go through a 30-mesh screen. This corresponds to a screen opening of 0.0233 inch or 600 microns. It is recommended that all irrigation water that enters the distribution system from open reservoirs be filtered through a process similar in performance to the filters used at the reclamation plant or, as a minimum, screened through a micro strainer with a 200-mesh screen. The use of a very fine strainer or filter will maximize the suspended solids' removal at central reservoir sites and reduce any special maintenance of the local sprinkler systems.

- aesthetic—where excessive algae growth may result in a product that is not aesthetically suitable for the intended use. Difficulties may include degradation in both appearance and increased odor.
- functional—where quality degradation may result in operational difficulties in downstream irrigation systems

Sizing Seasonal Storage

Using monthly supply and demand factors, the required seasonal storage can be obtained from the cumulative supply and demand. If the expected annual average demands of a reclaimed water system are approximately equal to the average annual available supply, storage is required to hold water for peak demand months. The amount of seasonal storage needed is the total of the monthly differences between supply and demand, plus allowances for the amount of water necessary to maintain the integrity of the seasonal storage reservoir if it is a storage "lake." There may be operational reasons to keep a fixed amount of water in the seasonal storage lakes throughout the year.

In many cases, a system will seek to provide reclaimed water to a diverse customer base. Each water use will have a distinctive seasonal demand pattern and thereby impact the need for storage.

Operational Storage

The purpose of operational storage is to provide a reliable supply of water during periods of down time at the treatment plant, meet peak daily fluctuations in water demands, and allow for optimum plant operation (Figure 9.7). The size of the storage facilities depends on the degree of fluctuation and availability of supplemental supplies. Frequently, the reservoir is constructed to save cost by reducing peak period pumping charges. If supplementary sources are used to meet peak demands, smaller operational storage facilities may be used to control supplies into the distribution system.

Operational storage is generally accommodated by covered tanks or open ponds. Covered storage in ground (Figure 9.8) or elevated tanks is used for unrestricted urban reuse where aesthetic considerations are important. Ponds are less costly, in most cases, but require more land per gallon stored. Where property costs are high or sufficient property is not available, ponds may not be feasible. Open ponds also result in water quality degradation from biological growth, and a chlorine residual is difficult to maintain, as discussed in the section on seasonal storage. Ponds

Figure 9.7 Reclaimed water seasonal storage lake.

are appropriate for on-site applications such as agricultural irrigation and golf courses. Ground or elevated tanks may be used. When located within the system, operational storage is generally elevated.

Sizing Operational Storage

Sufficient storage to accommodate diurnal flow variation is essential in the operation of a reclaimed water system. The volume of storage required can be determined from the daily reclaimed water demand and supply curves. Reclaimed water is normally produced 24 hours per day in accordance with the diurnal flow at the water reclamation plant and may flow to ground storage to be pumped into the system or into a clear well for high-lift pumping to elevated storage facilities. Covered storage is preferred to preclude biological growth and maintain a chlorine residual.

Operational storage facilities should be sized to hold at least one and one-half to two times the average summer day demand volume. Operational storage is necessary in reuse systems with varying daily demands. Primarily, operational storage is provided for irrigation users with demands that occur during late evening and early morning hours. The required usable volume of operational storage is the difference between the total peak hour demand and the total maximum day demand under the worst case, either daytime or nighttime conditions. Because most irrigation demands occur at night, nighttime peak hour and maximum day demands are generally used to calculate required usable storage.

To keep from completely draining the reservoirs and to allow for the situation when peak hour nighttime demands occur when the reservoirs are not full, operational storage is typically designed for at least 5 percent reserve. Under this criteria a maximum day demand would draw the reservoir down to an almost empty state.

Storm Water Storage

In some areas, a requirement may be placed on storage facilities, particularly open reservoirs, that no discharge be allowed during storm events. This requirement usually stems from the effect of introducing reclaimed water to streams, lakes, and other water bodies when it is mixed with storm runoff. If this requirement is placed on the storage reservoir, an analysis of the potential volume of water represented by the storm events should be calculated from past rainfall records. Since many open reservoirs have significant drainage areas, the impact of a storm water retention requirement could have a dramatic impact on the size of the storage and may make the open reservoir uneconomical. If this requirement is placed on the reservoir, several alternatives should be evaluated: changing the configuration of the reservoir to a closed tank type reservoir or installing a floating cover, or providing adequately sized diversion structures to

Figure 9.8 Reclaimed water covered operational storage reservoir.

allow the storm water to be diverted around the reservoir and avoid co-mingling with the water. Another alternative might be to seek regulatory relief from the requirement. However, this process is usually very involved and could add a significant amount of time to the development of the reclaimed water project.

Emergency Storage and Potable Water Back-up

The reclaimed water distribution system should have supplementary sources to meet its demand in case of a plant upset or other main supply interruption. Each system's required emergency storage capacity will be different, depending on the reliability of treatment processes, peak summer-time demands, availability of other sources, the proposed reliability of the system, and the ability to recover to normal conditions.

Seasonal or operational storage facilities may be able to meet emergency storage requirements, depending on their storage capacities. It is recommended that the reclaimed water system have some amount of emergency storage. If a system is lacking necessary emergency storage capacity, it should have at least one reliable supply source to meet its demand. This is usually provided through a potable water back-up supply, through an air gap, or other non-potable supplies such as raw water lakes and streams, or groundwater through wells.

If the reclaimed water system is expected to also provide fire flow capacity, then an amount equal to meet the standard four hour fire with a 20 psi residual should be included in the emergency storage calculations. Care should be taken however in providing additional emergency or standby capacity in the storage reservoirs so that the reservoirs will still be able to fully cycle all of the water in the reservoirs and mitigate the potential of water quality problems associated with stagnant water, or loss of chlorine residual.

SYSTEM LAYOUT AND MODELING

Having established the location, quantity, and demand characteristics of each use, preliminary networks are developed to link each user to the source of reclaimed water. Pipelines, pump stations, control valves, and various storage facilities are incorporated into the networks, which should then be analyzed by means of computer simulation programs to optimize the system and to develop alternative networks for economic analysis, and final alternative selection. The following sections detail the methods of selecting the various hydraulic components in the distribution plan.

Hydraulic Components

The design of an urban distribution system is similar in many respects to that of a municipality's potable water distribution system, and the use of materials of equal quality for construction is recommended. System integrity should be assured, and delivery of reclaimed water should be without interruption and on demand. No special measures are required to pump, deliver, and use the water. Also, no modifications other than identification of equipment or materials are required because reclaimed water is being used. However, for service lines in urban settings, different materials may be desirable for more certain identification.

The design of distribution facilities is based on topographical conditions as well as reclaimed water demand requirements. If topography has wide variations, multi-level systems may be needed. Distribution mains should be sized to provide the peak daily demands at a pressure adequate for the user being served. Pressure requirements for a dual-distribution system vary depending on the type of user being served. Pressures for irrigation systems can be as low as 10 psi if additional booster pumps are provided at the point of delivery or as high as 100 to 150 psi.

For reclaimed water systems that include fire protection as part of its service, fire flow plus the maximum daily demand should be considered when sizing the distribution system. This scenario is not as critical in sizing the delivery pumps since it will likely result in less pumping capacity, but is critical in sizing the distribution mains due to the fact that fire flow could be required at any point in the system resulting in high localized flows.

Maximum/Minimum Pressures

A maximum system operating pressure of about 120 psi and a minimum pressure of about 60 psi are recommended for the reclaimed water system. These pressures however should be set so that they are close to the existing potable water pressure for retro-fits areas. This range also provides sufficient pressure for most types of spray irrigation systems. For future expansion, it is important to consider the effects of pressure differentials between current potable pressure and planned reclaimed water pressure. Since most of the larger users will be landscaping and prefer higher pressures, conversion of some of the current potable users which operate below 80 psi to the higher pressure may cause problems such as excessive overspray. In addition, potable water irrigation systems constructed for 80 psi or less operating pressures cannot safely handle higher pressure reclaimed water without retrofitting the systems and using higher pressure pipe, fittings, and valves. All of these will add additional cost to the retrofit projects and may cause some to become unfeasible.

Pressure Zones

Generally an urban reuse system have both ''high-pressure'' and ''low-pressure'' users. The high-pressure users receive water directly from the system at pressures suitable for the particular type of reuse. Examples include residential and landscape irrigation, industrial process and cooling water, car washes, fire protection, and toilet flushing in residential, commercial, and industrial buildings. The low-pressure users receive reclaimed water into an on-site storage pond to be repumped into their reuse system. Typical low-pressure users are golf courses, parks, and condominium developments which utilize reclaimed water for irrigation. Other low pressure uses include delivery of reclaimed water to landscape or recreational impoundments.

Typically, urban dual-distribution systems operate at a minimum pressure of 50 psi, which will satisfy the pressure requirements for irrigation of larger landscaped areas such as multi-family complexes and offices and commercial and industrial parks. Based on requirements of typical residential irrigation equipment, a minimum delivery pressure of 30 psi is used for the satisfactory operation of in-ground residential irrigation systems. A minimum pressure of 50 psi should also satisfy the requirements of car washes, residential toilet flushing, construction dust control, and some industrial users [3].

For users who operate at higher pressures than other users on the system, additional on-site pumping will be required. For example, golf course irrigation systems typically operate at higher pressure (100–200 psi) and, if directly connected to the reclaimed water system, will likely require a booster pump station. Additionally, some industrial users may operate at higher pressures as well as high rise buildings utilizing reclaimed water for toilet and urinal flushing [3].

Pressure requirements should be based on system design and practice. In any case, minimum pressure at the user's meter should be maintained at the peak demand hour. It is desirable that a pressure differential of 10 psi or greater be maintained with the potable water supply having the higher pressure; however this is often not physically possible or economically feasible, but should be considered as a target goal [3].

Flow Velocities

Reclaimed water pipelines are sized to generally keep maximum velocities below ten feet per second (fps) and head loss per thousand feet below five feet. A Hazen-Williams C-factor of 130 should be used for hydraulic analyses. A minimum velocity of 1 foot per second under average daily flows, and a minimum 2 feet per second under peak hour demand, is recommended to minimize stagnant water in the reclaimed water system [3].

Extended Period Computer Simulations

The design of a reuse transmission system is usually accomplished through the use of computer modeling, with portions of each of the sub-area distribution systems representing demand nodes in the model. The demand of each node is determined from the irrigated acreage tributary to the node, the irrigation rate, and the daily irrigation time period. Demands for uses other than irrigation such as fire protection, toilet flushing, and industrial uses should also be added to the appropriate node.

Because of significant variations in demands throughout the day, month, and year the use of computer network simulators with an extended simulation mode are recommended. Several current software packages, such as CYBERNET and EPANET, offer extended period simulations. Through these models, the systems can be studied throughout all of the various demand profiles. In addition these models can help predict the degradation of disinfection residuals in the system. The loss of adequate disinfection in the system has been identified as a major cause of water quality complaints. The extended period simulations will also allow the reservoirs to go through complete filling and emptying cycles to make sure they are adequately circulating the reclaimed water through the reservoirs.

Growth and Future Expansion

In many reclaimed water systems the initial distribution systems are constructed to meet the demands of large users. As the system matures, smaller users and added and new markets are explored for potential users. Because this pattern is common, planning for future growth in both the quantity of reclaimed water user and the types and locations of future customers should be factored into reclaimed water plans.

Identifying Ultimate Demand

Once the marketing and cost-benefits analyses have been made and the decision taken to develop a reuse system, attention should be given to ensure that the reclaimed water components of the system are designed to allow for future expansion to potential customers. When economic market, or demand conditions make such expansion feasible, the system should be able to extend service by simple extension of the distribution network.

Clustering or concentration of users will result in lower absolute costs than a delivery system to dispersed users. To expand in an economically responsible manner, initially a primary skeletal system is designed to serve large institutional users who are clustered and closest to the treatment plant. A second phase expands the system to more scattered and smaller users which receive reclaimed water from the central arteries of the reclaimed water system. Initial customers are often institutional (e.g. schools, golf courses, urban green space, and commercial); however, the lines should be sized to make allowance for future service to residential customers.

To ensure that expansion can occur to the projected future markets, the initial system design should model sizing of pipes to satisfy future customers within any given zone within the service area. At points in the system where future network of connections is envisioned turnouts should be installed. Pump stations and other major facilities involved in conveyance should be designed to allow for planned expansion. Space should be provided for additional pumps, or the capacities of the pumps may be expanded by changes to impellers and motor size. Increasing a pipe diameter by one size is economically justified since over half the initial cost of installed pipeline is for excavation, backfill, and pavement.

A potable system is designed to provide round-the-clock, "on demand" service. Some non-potable systems allow for unrestricted use while others place limits on the hours when service is available. A decision on how the system will be operated will significantly affect system design. Restricted hours for irrigation (i.e. to evening hours) may shift peak demand and require greater pumping capacity than if the water was used over an entire day or may necessitate a programmed irrigation cycle to reduce peak demand.

When the system is unregulated, on-demand operational storage needs may become significant. Storage requirements may be greatly reduced by designing the system for totally scheduled irrigation. Also, energy costs, which are the most significant component of O&M, may be reduced by taking advantage of time-of-use or off-peak power pricing.

New Users/ Markets

In cases where the parts of the system are being upgraded and some of the discarded potable water lines are transferred over to the reclaimed water system, care should be taken to prevent any cross connection. As each section is completed, the new system should be shut down and drained and each water user checked to ensure that there are no connections. In existing developments where an in-place irrigation system is being converted to carry reclaimed water, the new installation should be inspected and tested with tracers or some other method to ensure separation of the potable from the reclaimed water supply.

DISTRIBUTION SYSTEM DESIGN ELEMENTS

The design considerations for reclaimed water distribution systems are in most respects the same as those for potable water systems. The major differences are that the reclaimed water distribution system must make provisions for more stringent safeguards to protect the public health, and the same level of reliability found in the potable water systems may not be required. A consistent program of safeguards should be in effect in all phases of project development, from planning, design, construction, and installation, and operations and maintenance.

The major concern which guides design, construction, and operation of a reclaimed water distribution system is the prevention of cross-connections. A cross connection is an unprotected or potential connection between a potable water system used to supply water for drinking purposes and any source containing reclaimed water. Another major concern is to prevent improper use or inadvertent use of reclaimed water as potable water. To protect the public health from the outset, a dual distribution system should be accompanied by health codes, procedures for approval (and disconnection) of service, regulations governing design and constructions specifications, inspections, and operation and maintenance staffing.

PIPELINES

Design and construction of reclaimed water pipelines is essentially the same as that for potable water pipelines except for those differences noted in this section. The primary focus is on adjustments to accommodate differences in water quality, the need to protect public health, and design features which reflect the differences in demand patterns for reclaimed water.

Sizes

Common sizes for reclaimed water are divided into off-site and on-site pipelines. Off-site reclaimed water facilities typically consist of transmission or distribution mains in public rights of way. Typically these facilities are the upstream side of the water meter. On-site reclaimed water facilities typically consist of facilities which will be owned, operated, and maintained by the customer and are downstream of the water meter. On-site reclaimed water facilities are typically 4″ and smaller.

Material

Reclaimed water pipelines can be constructed of a variety of different materials. However, the recent introduction of color coded C-900 PVC

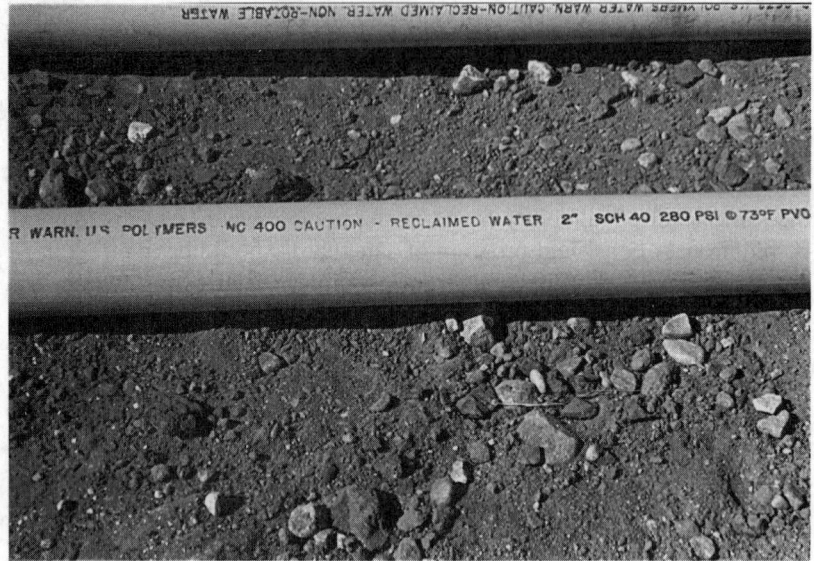

Figure 9.9 Reclaimed water warning tags quick connect outlet.

pipe either Class 150 or 200, plastic pipelines has made plastic pipe almost an industry standard. Many potable water pipelines are constructed of asbestos cement, ductile iron, or steel. PVC pipeline materials have the ability to be fabricated in different colors so that they are quickly distinguishable from potable water pipelines. This is particularly important in a dual distribution system [12]. If asbestos cement, ductile iron, or steel lines are used they should be adequately marked with color coded paint, tape, or polyethylene wrapping (Figure 9.9).

Location/Alignment/Depth/Clearances from Other Utilities

Historically, a ten foot horizontal interval and a one foot vertical distance has been used as the standard offset to be maintained between potable (or sewer) lines and reclaimed water lines running parallel to each other. When these distances cannot be maintained special construction requirements are placed on the pipeline by the local health agencies, although a minimum lateral distance of four feet is generally mandatory. This arrangement allows for the installation of reclaimed water lines between water and force mains that are separated by ten feet. In addition to the required clearances, best design practice recommends that the potable water be placed above the reclaimed water if possible [12].

The top of distribution mains 4-inch and smaller, should be a minimum of 42 inches below the finished street grade. Transmission mains, 6-inch and larger, should be a minimum of 48 inches below the finished street grades. Off-site reclaimed water facilities should typically be located in a standard alignment which is different from that set for the potable water mains. In the Irvine Ranch Water District service area the reclaimed water pipelines are placed either four (4) feet or eight (8) feet from the curb face on the opposite side of the street from the potable water mains [12].

Looping

In most potable water systems reliability is a key factor in design. When potable and fire protection systems are combined, and they are in many areas, the reliability to fight fires dictates multiple supply points to each node in the distribution network. The "best practices" approach is to provide at least two points of connection to each fire protection hydrant or outlet. This also provides for looping pipelines throughout the distribution grid and often results in smaller main sizes. This is because the fire flow is divided between the two pipelines feeding the fire hydrant and adequate flow velocities, even at low flow periods, to maintain water quality.

In reclaimed water systems fire protection is not normally included in the demands placed on the reclaimed water supply. As a result there is a tendency to eliminate the need for looping the distribution system. Many reclaimed water systems are constructed as backbone transmission mains from the reclaimed water source to major users such as golf courses and parks. Other users are added to the system like branches on a tree and represent a series of dead-end connections. The false perception that looping is not required for reclaimed water systems overlooks some of the basic characteristics of reclaimed water.

First, due to its normally higher organic loading, reclaimed water tends to use up the available disinfectant faster and to yield faster biological regrowth in the pipelines. Allowing the reclaimed water to move very slowly, if at all, during non-irrigation hours, permits bacteria in the pipelines to grow, causing fouling and slime build-up. Ultimately, biological growth in the pipelines could result in clogged irrigation facilities, unacceptable color and odor, and increased customer complaints.

Second, looping allows for some level of redundancy in the system which customers who were formerly using potable water have come to expect. Because of the looping in potable systems, interruptions in irrigation water are minimized. Without looping, main breaks would leave all downstream users without water until repairs are made. This could lead again to increased customer complaints and difficulties in expanding reclaimed water use to other potable customers in the system.

Finally, as the uses for reclaimed water expand into industrial and

Figure 9.10 PVC pipe with reclaimed water markings.

commercial uses, the reliability of the water supply becomes critically important to the customer. Outages, even for short periods, could have significant implications if the reclaimed water is a primary need for the manufacturing facilities. For example, in a carpet dying operation, the loss of the reclaimed water would mean stopping all production/operation and would cause the business to suffer economic impacts.

It is therefore recommended that where economically feasible some level of looping be provided, especially in the 12-inch and larger transmission mains and laterals.

Color Coding and Signs

All components and appurtenances of the reclaimed water system (pipes, pumps, outlets, valve boxes, etc.) should be clearly and consistently identified throughout the system. Identification should be through a specified color coding and marking system The methods most commonly used are: unique colorings, labeling, and markings [12].

In California, reclaimed water piping and appurtenances are painted purple and stamped or marked "CAUTION NON-POTABLE WATER—DO NOT DRINK" or "CAUTION: RECLAIMED WATER—DO NOT DRINK" (Figure 9.10). The pipe may also be wrapped in purple polyethylene vinyl wrap. The City of St. Petersburg, Florida, uses brown coloring to distinguish reclaimed water piping [12].

Another identification method is the marking of pipe with colored marking tape or adhesive vinyl tape. Color coded identification (caution) tape differentiating the reclaimed water piping from other utility lines should be consistent throughout the service area. All valve covers on reclaimed water distribution and transmission mains should be of non-interchangeable shape with potable water covers and with a recognizable inscription cast on the top surface. They should also be consistently color-coded to differentiate them from potable water or wastewater facilities.

STORAGE RESERVOIRS, OPEN AND CLOSED

The design of reclaimed water storage reservoirs, both open and closed, is based on the same hydraulics and structural calculations as potable water storage reservoirs. Steel, reinforced concrete, and plastic lined ponds are materials used with equal frequency for both types of storage facilities. The differences between the reclaimed water and potable water storage reservoirs are primarily based on differences in demand profiles, water quality, frequency of cleaning, and monitoring requirements.

Depth to Surface Ratio

In both closed and open reservoir configuration, depth to surface ratios are important in maintaining water quality. Because of the higher levels of nutrients typically found in reclaimed water, shallow open reservoirs are problematic because they tend to be plagued by high levels of algae production. In addition to creating aesthetic problems the algae increases the loading on the sand filters and could possibly cause clogging of irrigation emitters in the distribution system.

In addition, if the reservoir's depth to surface ratio is such that the surface area is not large enough, the water may stratify and cause anaerobic bacteria to create undesirable sludge build up in the bottom of the reservoir. The sludge may also cause turbidity problems and, because of the release of hydrogen sulfide, cause odor problems as well.

Depth to surface ratios for closed reservoirs should allow for proper turnover of the reservoir volume, allowing for adequate surface contact for oxygen transfer.

Inlet/Outlet Designs

Reclaimed water reservoirs should be designed with inlet and outlets placed at opposite ends so that proper flow through can be achieved. Care should be taken to ensure that no "dead spaces" are created in the reservoir. Although a single inlet/outlet design is common in many reservoirs, this configuration is not recommended for reclaimed water. In addition, many potable water reservoirs are designed with the inlet/outlet situated six inches or more above the bottom of the reservoir floor to avoid drawing sediments into the distribution system. In reclaimed water reservoirs it is important not to create this "dead area" where anaerobic bacteria can thrive.

Reservoir Maintenance

Reclaimed water reservoirs, both closed and open, require more frequent cleaning than their potable water counterparts. Potable water from filtration plants may be stored for ten to fifteen years before tank cleaning is necessary. Reclaimed water reservoirs, on the other hand, may require cleaning as frequently as every three to five years. This increased frequency is caused by the build up of sludge on the bottom.

Because the sludge is usually removed through pumping and flushing, it is very important to provide additional access hatches in the closed reservoirs for access. In addition, high pressure hose connections need to be provided near the storage tanks, as well as connection to the sewer to accommodate the flows from the flushing operations. In open reservoirs, facilities for dredging the material from the bottom need to be included in the design as well as pre-arranging for the disposal of the sludge material.

Backflow Prevention Devices

Backflow prevention devices should be considered on reclaimed water services where the customer boosts the distribution system pressure through on-site pumps, or if chemicals are injected into the reclaimed water by the customer. If the reclaimed water system is being used for fire protection, the appropriate backflow device to prevent over-pressurization of the reclaimed water system from fire engine booster pumps should be considered.

Blow-offs

Even with sufficient chlorination, residual organics and bacteria may grow at dead spots in the system, leading to odor and clogging problems. Blow-off valves and periodic maintenance of the system can significantly allay the problem.

Either an in-line or end-of-line type blow-off or drain assembly should be installed for removing water or sediment from the pipe. The line tap for the assembly should be no closer than 18-inches to a valve, couplings, joint, or fitting unless it is at the end of the line. If there are restrictions on discharge or runoff, the regulatory agencies should be consulted to find an acceptable alternative.

System Monitoring

Even when a system is unmetered, accurate flow recording is essential to manage the system's growth. Flow data are needed to confirm total system use and spatial distribution of water supplied. Such data allow for efficient management of the reclaimed water pump stations and formulations of policies to guide system growth. Meters placed at the treatment facility may record total flow, and flow monitoring devices may be placed along the system, particularly in high consumption areas.

Reclaimed water systems are typically "customer driven" because customer demand patterns are a significant consideration in the operation of the distribution system. Thus, consideration should be given to providing remote sensing points at the meters of large customers or non-traditional users such as laundries and industrial applications. Monitoring the usage of the largest customers allows their impact on the system to be assessed, as well as their demand profiles. This information can then be used to develop a maintenance program which will allow the largest users, and those that generate the majority of the reclaimed water sales revenue, to be properly served. It also allows the reclaimed water supplier to validate daily and seasonal demand patterns and validate time of use and peaking factor assumptions.

Finally the ability to monitor chlorine residuals in the system would be beneficial to the operators by providing a glimpse into the potential for biological regrowth. By monitoring disinfection levels, operators can identify hot spots and maximize the use of the storage reservoirs. This will avoid compromising water quality by holding the chorine residual in the holding reservoirs within proper operating limits. Several large reclaimed water suppliers have noted a significant correlation between the drop off of chlorine residual in the distribution system and increased levels of bio-fouling of irrigation equipment and odor and turbidity complaints.

System Maintenance

Maintenance requirements for the hydraulic components of the reclaimed water distribution system are the same as for potable. The reclaimed water lines should be treated like potable lines. As the system matures, any disruption of service due to operational failures is intolerable to the users. Hence, from the outset, such items as isolation valves, which allow for repair on parts of the system without affecting a large area, should be designed into the reclaimed water system. Flushing the line after construction should be mandatory to prevent sediment from accumulating and hardening and becoming a serious future maintenance problem.

Training and Public Outreach

A frequently underestimated component of reclaimed water facilities design is the development and design of a proper training program and a public outreach program. A comprehensive training program for the water system operators will help them understand the differences between operating a reclaimed water system and operating a potable water system. Water quality, safety, cross-connection control, notification, differing demand patterns, and the specific needs of large customers, should be addressed. In addition, training programs for on-site reclaimed water customer personnel should be developed to help the customers use the reclaimed water properly and avoid operational problems. Public education should also be provided to help the general public better understand this type of water. Public acceptance of reclaimed water is not only essential both to the short and long term use of reclaimed water in public areas such as parks and golf courses, but also important to industries considering the use of the water for various processes.

ON-SITE RECLAIMED WATER SYSTEMS

In order to complete a discussion of reclaimed water distribution and storage, specific issues relative to the use of the reclaimed water on-site should be addressed. On-site reclaimed water systems primarily deliver irrigation water throughout the use site through a network of pipes, valves, pumps, and sprinklers. Reclaimed water irrigation systems are designed the same way as a potable system, but the inherent differences in water quality prompt the need for some modifications to the standard potable water design. This section will address those modifications and provide additional guidelines for the operation and maintenance of on-site reclaimed water systems.

Distribution System Modifications

The following sections describe the modifications that are recommended to standard irrigation system design, construction, and operation and maintenance.

Pipeline Identification

New on-site pipelines should be identified as reclaimed water pipes by using a color code differentiating them from potable water piping. In California, purple has become the standard color coding for reclaimed water facilities.

Purple colored pipe stamped "CAUTION: RECLAIMED WATER—DO NOT DRINK" or "CAUTION: RECLAIMED WATER—DO NOT DRINK", on opposite sides of the pipe, repeated every three feet should be used. As an alternative, warning tape should be installed on pressure and/or non-pressure lines. The tape should run continuously the entire length of the pipe and should be at least 3-inches in width. It is recommended that the identification tape be locator type marking tape to help find the non-metallic pipelines in the future [12].

When potable water is being supplied to an area also being supplied with reclaimed water, the potable water main should also be identified. A warning tape with words "CAUTION—DRINKING WATER LINE" should be fastened directly to the top of the potable water pipe and run continuously the entire length of the pipe. This tape should be at least 3-inches in width. The color code should differentiate potable water from reclaimed water [12].

Pipeline Clearances

A 10-foot horizontal separation of the reclaimed water pipe should be maintained at all times between a potable water pipe and/or a parallel sanitary sewer system. If a 10-foot separation is not possible, special construction methods should be considered. Common trench construction should not be permitted. In any event, a minimum of 4 foot horizontal separation should be maintained.

Vertical separation should be maintained so that the potable water pipe is installed a minimum of one foot above the reclaimed water pipe which, in turn, should be installed a minimum of one foot above a sanitary sewer system. If a one foot separation is not possible, the approval for special construction requirements should be obtained from the regulatory agencies [12].

System Appurtenances

All system appurtenances should be selected with the use of reclaimed water in mind and the differences in water quality. It is particularly important to make sure that reclaimed water appurtenances cannot be interchanged with potable water devices and equipment. Gauges, test devices, and other tools which come in contact with the water should be designed and reserved for the reclaimed water system.

Pipeline Identification

Hose bibbs should not be allowed on reclaimed water irrigation systems. Quick couplers should be used if hose connections are necessary. Fittings should prevent inter-connection between the potable and reclaimed water systems. Hoses used with the reclaimed water system should not be usable with potable water systems. Signs should be used to identify the reclaimed water quick coupling. When potable quick couplers are within 60 feet of the reclaimed water system, both should be equipped with appropriate signs [12].

Strainers

Strainers should be installed on all reclaimed water meters and service connections to prevent material from clogging on-site irrigation systems. Wye strainers are not recommended for below ground (in vaults) installations. Basket strainers are suitable for above or below ground (in vaults) installations. Filter strainers are normally used above ground on drip irrigation systems.

Backflow Protection

In California, if a premises is supplied with both potable water and reclaimed water, then backflow protection with an approved air gap (AG) must be provided at the potable water service connection. A reduced pressure principle device (RPPD) backflow prevention device may be provided only when approved by the health department and potable water supplier [13].

Backflow prevention devices are not normally used on reclaimed water systems. However, an agency should maintain the water quality in a reclaimed water distribution system. A backflow prevention device may therefore be needed at a specific meter where on-site exposures would impact the quality of the reclaimed water supply. If temporary potable water connections to the reclaimed water system are required, the connections should be protected in the same manner as a permanent connection. Exceptions may be necessary under special circumstances, but, in any case, should not be allowed unless approved by the potable water supplier and regulatory agencies.

Drinking Fountains/Public Facilities

Potable water drinking fountains and other public facilities should be placed out of the irrigation area in which reclaimed water is used, or otherwise be protected.

Operation and Maintenance Guidelines

Construction Water

Water trucks, hoses, drop tanks, etc., should be identified as containing reclaimed water, not suitable for drinking. Reclaimed water used for construction purposes may be used for soil compaction during grading operations, dust control, and consolidation and compaction of backfill in reclaimed water, sanitary sewer, storm drain, gas, and electric pipeline trenches. Reclaimed water may be suitable for water jetting and consolidation or compaction of backfill in potable water pipeline trenches with health agency approval.

Runoff

Conditions which directly or indirectly cause a runoff outside of the approved use area should be prohibited. This provides a benefit in terms of water conservation and also minimizes public contact with the reclaimed water.

Ponding

Conditions which directly or indirectly cause ponding outside of or within the approved use area should be prohibited.

Overspray

Conditions which directly or indirectly permit windblown spray or overspray to pass outside of the approved use area should be prohibited.

Irrigation Application Rates and Practice

An irrigation system designed with reclaimed water should specify type of sprinkler, placement of sprinkler, type of soil, type of plants, slope, landscape, etc., to be used so as to prevent runoff, ponding, and overspray.

Reclaimed water should be applied at a rate that does not exceed the infiltration rate of the soil. The irrigation system should not be allowed to operate for a time longer than the landscape's water requirement. If runoff or ponding occurs before the landscape's water requirement is met, the automatic controls should be reprogrammed with additional watering cycles to meet the requirements and prevent runoff.

To the extent possible, the irrigation system should operate during periods of minimal public use of the approved area. Such periods of operation should remain within any general period of reclaimed water irrigation operation specified by the water purveyor or regulatory agency.

Equipment and Facilities

Any equipment or facilities such as tanks, temporary piping or valves, and portable pumps which have been used with reclaimed water should be cleaned and disinfected before removal from the approved use area for use at another job site. This disinfection and cleaning should ensure the protection of the public health in the event of any subsequent use as approved by the water purveyor or regulatory agency, and the disinfection process should be performed in his or her presence.

Warning Labels and Signs

Agency warning labels should be installed on designated facilities such as, but not limited to, controller panels and wash down or blow-off hydrants on water trucks, and temporary construction services. The labels should indicate that the system contains reclaimed water that is unsafe to drink (Figure 9.11).

Figure 9.11 Reclaimed water outlet tags.

Figure 9.12 Reclaimed water sign at Community Park.

Where reclaimed water is used for recreational impoundments and or irrigation, warning signs should be installed to notify that the water in the impoundment is unsafe to drink. A detailed plan should be prepared showing placement and spacing of the proposed signs. Warning signs and labels should be in both English and other language(s) common to the particular area. The signs should include the international system sign for "DO NOT DRINK" (Figure 9.12) [12].

SUMMARY

The planning, design, construction, and operation of reclaimed water systems is in almost all respects identical to potable water systems. The source of reclaimed water and the associated institutional, regulatory, aesthetic, and water quality demands are the major factors which must be considered when developing a reclaimed water system.

The principal differences in the systems lie in the requirements for physical separation, identification, and limitations on its use in lieu of potable water supplies. Operationally the system must be designed and operated to maintain good water quality by making sure storage facilities turn over in a timely manner and the unique demand profiles of the users is adequately addressed in sizing all of the facilities.

REFERENCES

1 National Academy of Science and National Academy of Engineering. *Water Quality Criteria, 1972.* EPA-R3-73-033, NTIS PB 236-199 (1972).

2 *Water Reuse Manual of Practice SM-3,* Water Pollution Confederation, 1983.

3 *Guidelines for Water Reuse,* Environmental Protection Agency, 1981.

4 National Research Council. 1973. *Water Quality Criteria: A Report of the Committee on Water Quality Criteria.* Prepared for the U.S. Environmental Protection Agency, EPA Report R3-73-033, Washington, D.C.

5 Goldstein, D. J.; I. Wei; and R. E. Hicks. 1979. Reuse of Municipal Wastewater as Make-up to Circulating Cooling Systems. In: *Proceedings of the Water Reuse Symposium, Vol. 1,* AWWA Research Foundation, Denver, CO.

6 Donovan, J. F.; J. E. Bates; and C. H. Powell. 1980. *Guidelines for Water Reuse.* Prepared for the U.S. Environmental Protection Agency by Camp, Dresser, and McKee Inc. Boston, MA.

7 Lake, L. M.; and Perrine, R. L., *Institutional Barriers to Wastewater Reuse in Southern California.* U.S. Office of Water Res. and Technol., OWRT/RU-79/4.

8 Young, R. E.; Holliman, T. R.; and Parsons, J.; "Using Reclaimed Water For Bathroom Flushing—A Case Study", *1994 Water Resources Symposium Proceedings,* American Water Works Association/Water Environment Federation, 1994.

9 University of California, Division of Agriculture and Natural Resources, 1985.

10 Long Beach Water Department, "Reclaimed Water Master Plan", 1992. Prepared by Black and Veatch.

11 Irvine Ranch Water District. 1991. Water Resource Master Plan. Irvine, California.

12 "Guidelines for the Distribution of Non-Potable Water", AWWA California/Nevada Section, 1992.

13 *Manual of Cross Connection Control,* 9th Edition, Foundation for Cross-Connection Control and Hydraulic Research, University of Southern California, 1993.

14 City of San Diego, Clean Water Program "Water Reclamation and Reuse Master Plan," James M. Montgomery, Consulting Engineers, March and May 1989.

15 "Preliminary Reclaimed Water Use Study for Eastlake Development Co.", NBS/Lowry, February 1990.

CHAPTER 10

Microbial Considerations in Wastewater Reclamation and Reuse

INTRODUCTION

FROM an acute disease standpoint, the contaminants of greatest concern when considering the reuse of treated wastewater are the enteric microorganisms. Microorganisms were responsible for 90% of the reported waterborne disease outbreaks in the United States between 1971 and 1994 (Table 10.1) (Craun, 1991; Herwaldt et al., 1992; Moore et al., 1994; Kramer et al., 1996). More than 560,000 people were affected in the 589 documented outbreaks. The reported illnesses included infectious hepatitis, cryptosporidiosis, giardiasis, typhoid fever, and gastroenteritis.

In this chapter, the numbers and types of different microorganisms that may be present in domestic wastewater, as well as their removal by various wastewater treatment processes are described. The survival of enteric microorganisms on food crops that have been irrigated with reclaimed water, as well as disease outbreaks associated with consumption of contaminated foods are discussed. In addition, the fate and transport of microorganisms in reclaimed water that has been applied to land for purposes such as artificial recharge of ground water are reviewed. Finally, methods that have been used to quantify the potential risks associated with exposure to pathogens in reclaimed water will be discussed and illustrated.

Marylynn V. Yates, Department of Soil and Environmental Sciences, University of California, Riverside, CA 92521; Charles P. Gerba, Department of Soil, Water, and Environmental Science, University of Arizona, Tucson, AZ 85721.

TABLE 10.1.   Causative Agents of Waterborne Disease
in the United States, 1971–1994.

	Outbreaks		Illnesses	
Causative agent	No.	% of Total	No.	% of Total
Gastroenteritis, unknown cause	321	49.1%	81229	14.4%
Giardia	119	18.2%	27039	4.8%
Chemical poisoning	65	9.9%	4233	0.7%
Shigella	43	6.6%	9219	1.6%
Viral gastroenteritis	27	4.1%	12699	2.2%
Hepatitis A virus	26	4.0%	772	0.1%
Campylobacter	15	2.3%	5456	1.0%
Salmonella typhimurium	13	2.0%	2995	0.5%
Cryptosporidium	10	1.5%	419939	74.3%
Salmonella typhi	5	0.8%	282	< 0.1%
Yersinia	2	0.3%	103	< 0.1%
Toxigenic E. coli	2	0.3%	1243	0.2%
Vibrio cholera	2	0.3%	28	< 0.1%
Chronic gastroenteritis	1	0.2%	72	< 0.1%
Dermatitis	1	0.2%	31	< 0.1%
Amebiasis	1	0.2%	4	< 0.1%
Cyclospora	1	0.2%	21	< 0.1%
Total	654	100.0%	565365	100.0%

Source: Craun, 1991; Herwaldt et al., 1992; Moore et al., 1994; Kramer et al., 1996.

TYPES AND OCCURRENCE OF PATHOGENS IN WASTEWATER

There are hundreds of different types of microorganisms that may be present in domestic wastewater. The pathogenic microorganisms include bacteria, viruses, parasites, fungi, and algal toxins. The major source of pathogenic microorganisms in domestic wastewater is the fecal material of infected individuals; however, urine may also be a source of certain pathogenic viruses (Hurst, 1989). The numbers and types of pathogens found in wastewater will vary both spatially and temporally, depending on the disease incidence in the population producing the wastewater, season, water use, economic status of the population, and quality of the potable water (Rose and Carnahan, 1992).

BACTERIA

Bacteria are microscopic (generally 0.1 to 10 micrometers in size), single-celled organisms. Given the necessary nutrients (e.g., carbon, nitrogen, oxygen) and appropriate environmental conditions, bacteria are capable of growth and reproduction. The bacteria of most concern in domestic wastewater are the enteric bacteria, those that infect the gastrointestinal tract of man, and are shed in the fecal material. Enteric bacteria are adapted to the conditions of the gastrointestinal tract: high organic carbon and other nutrients, as well as a relatively high temperature (37°C). When these organisms are introduced to the wastewater, water or soil environment, the conditions are generally very different from those in the gastrointestinal tract. As a result, the enteric bacteria are not always capable of competing with the indigenous bacteria for the scarce nutrients available. Thus, their ability to reproduce, and even survive, in the environment tends to be limited.

Human fecal material typically contains up to 10^{12} bacteria per gram, the majority of which are non-pathogenic. However, an infected individual may excrete high numbers of pathogenic bacteria in her/his feces. For example, an individual infected with *Shigella* may excrete up to 10^9 organisms per gram of feces (Bitton, 1994). These pathogens are transmitted by direct contact with an infected individual, by consumption of contaminated water, and by consumption of contaminated food. A list of pathogenic bacteria found in domestic wastewater, the diseases they cause, and the reported concentrations in wastewater is given in Table 10.2.

VIRUSES

In contrast to the bacteria, viruses typically are not found in the feces of humans. They are present only in the feces of individuals who have been infected either intentionally (e.g., poliovirus vaccination) or inadvertently through contact with an infected individual, or contaminated water or food. Viruses can be excreted in very high numbers in feces. For example, the concentration of rotaviruses may be as high as 10^{12} particles per gram feces (Flewett, 1982). The duration of excretion of viruses varies. Rotavirus excretion usually lasts for 1 to 3 weeks; however two months of excretion has been observed in some individuals (Kapikian and Chanock, 1990). The excretion of enteroviruses (i.e., poliovirus, echoviruses, and coxsackieviruses) may persist for 16 weeks (Melnick and Rennick, 1980).

There are more than 140 types of enteric viruses that can contaminate wastewater. A list of pathogenic human enteric viruses is shown in Table 10.3. The results of infection caused by enteric viruses range widely—from inapparent, undetectable infections to a variety of diseases including gastroenteritis, respiratory illness, hepatitis, paralysis, encephalitis, and conjunctivitis.

PARASITES

A third group of pathogenic microorganisms that can be found in domestic wastewater is the parasites. The parasites that are pathogenic to

TABLE 10.2. Bacterial Pathogens in Wastewater.

Organism	Disease	Numbers in Wastewater (per 100 ml)	Reference
Salmonella	Typhoid, paratyphoid, salmonellosis	2.3-8000	Feachem et al., 1983
Shigella	Bacillary dysentery	1-1000	Feachem et al., 1983
Enteropathogenic E. coli	Gastroenteritis		
Yersinia	Gastroenteritis		
Campylobacter	Gastroenteritis		
Vibrio	Cholera	3700	Holler, 1988
Leptospira	Leptospirosis	10-10000	Kott & Betzer, 1972
Legionella	Acute respiratory illness		
Mycobacterium	Tuberculosis		

TABLE 10.3. Viral Pathogens in Wastewater.

Organism	Disease	Number in Wastewater (per liter)	Reference
Adenovirus	Respiratory illness, conjunctivitis, vomiting, diarrhea		
Astrovirus	Vomiting, diarrhea		
Calicivirus	Vomiting, diarrhea		
Coronavirus	Vomiting, diarrhea		
Enterovirus:			
Poliovirus	Paralysis, meningitis, fever	182–492,000	Irving, 1982
Coxsackie A	Meningitis, fever, herpangina, respiratory illness		
Coxsackie B	Myocarditis, congenital heart anomalies, rash, fever, meningitis, respiratory illness, pleurodynia		
Echovirus	Meningitis, encephalitis, respiratory illness, rash, diarrhea, fever		
Enterovirus 68–71	Meningitis, encephalitis, respiratory illness, acute hemorrhagic conjunctivitis, fever		
Hepatitis A virus	Infectious hepatitis		
Hepatitis E virus	Hepatitis		
Norwalk virus	Epidemic vomiting and diarrhea		
Reovirus	Not clearly established		
Rotavirus	Diarrhea, vomiting	400–85000	Gerba et al., 1984
Small round viruses	Diarrhea, vomiting		

441

humans generally can be classified in two groups: the protozoa and the helminths. Protozoa are single-celled organisms whose life cycles include a vegetative (trophozoite) as well as a resting (cyst) stage. The resting stage of the organism is generally relatively resistant to inactivation during conventional wastewater treatment processes. Most of the intestinal protozoa are transmitted by fecally contaminated water, food, or other materials.

The helminths are a group of multi-celled parasitic worms, which includes the nematodes (roundworms), the trematodes (flukes), and the cestodes (tapeworms).

The concentrations of parasites in the fecal material of infected individuals can also be quite high. For example, the reported concentrations of *Giardia* and *Cryptosporidium* are 10^6 to 10^7 per gram feces (Jakubowski, 1984; Robertson et al., 1992). Excretion of *Giardia* may persist for up to six months (Pickering et al., 1984). A list of the parasites that can be found in domestic wastewater, the diseases they cause, and the reported concentrations in wastewater are given in Table 10.4.

REMOVAL OF PATHOGENS BY TREATMENT PROCESSES

There have been a number of reviews on the removal of pathogenic microorganisms by activated sludge and other wastewater treatment processes (Feachem et al, 1983; Leong et al, 1983). This information suggests that significant removals, especially of enteric bacterial pathogens, can be achieved by these processes (Table 10.5). However, disinfection and/or advanced tertiary treatment are necessary for many reuse applications to ensure pathogen reduction. Current issues with pathogen reduction are treatment plant reliability, removal of new and emerging enteric pathogens of concern, and the ability of new technologies to effect pathogen reduction. While the ability of conventional treatment processes to remove pathogens has been demonstrated in pilot and full scale systems, the question of how often this can be achieved over time given the variation in quality raw wastewater and dynamics of plant processes remains. Wide variation in pathogen removal could result in significant numbers of pathogens passing through a treatment plant for various periods of time. The issue of reliability is of major importance if the reclaimed water is to be used for recreational or potable reuse where short term exposures to high levels of pathogens could result in significant risk to the exposed population (Tanaka et al., 1993). Recent advances in methods for the detection of pathogens and recognition of the importance of water and food in the transmission of emerging enteric pathogens has created a need for information on the ability of treatment processes to remove these pathogens.

TABLE 10.4. Parasitic Pathogens in Wastewater.

Organism	Disease	Numbers in Wastewater (per liter)	Reference
Protozoa:			
Entamoeba histolytica	Amoebic dysentery	4	Kott & Kott, 1967
Giardia lamblia	Diarrhea, malabsorption	125–100,000	Fox & Fitzgerald, 1979; Jakubowski & Ericksen, 1979; Rose & Carnahan, 1992
Balantidium coli	Mild diarrhea, colonic ulceration	28–52	Kott & Kott, 1967; Wang & Dunlop, 1954
Cryptosporidium	Diarrhea	0.3–122	Rose & Carnahan, 1992
Toxoplasma	Toxoplasmosis		
Cyclospora			
Microsporidium			
Helminths:			
Ascaris (roundworm)	Ascariasis	5–111	Feachem et al., 1983; Rose and Carnahan, 1992
Ancylostoma (hookworm)	Anemia	6–188	Feachem et al., 1983
Necator (hookworm)	Anemia		
Taenia (tapeworm)	Taeniasis		
Trichuris (whipworm)	Diarrhea, abdominal pain	10–41	Liebmann, 1965; Rowan and Gram, 1959
Toxocara (roundworm)	Fever, abdominal pain		
Strongyloides (threadworm)	Diarrhea, abdominal pain, nausea		

TABLE 10.5. Pathogen Removal in Treated Wastewater.

	Enteric Viruses	Salmonella	Giardia	Cryptosporidium
Concentration in raw wastewater (no. L^{-1})	100,000–1,000,000	5,000–80,000	9,000–200,000	1–3960
Removal during:				
Primary treatment[a]				
% removal	50–98.3	95.5–99.8	27–64	0.7
No. remaining L^{-1}	1,700–500,000	160–3,360	72,000–146,000	
Secondary treatment[b]				
% removal	53–99.92	98.65–99.996	45–96.7	
No. remaining L^{-1}	80–470,000	3–1,075	6,480–109,500	
Tertiary treatment[c]				
% removal	99.983–99.9999998	99.99–99.999995	98.5–99.99995	2–7[d]
No. remaining L^{-1}	0.007–170	0.000004–7	0.099–2,951	

[a]Primary sedimentation and disinfection.
[b]Primary sedimentation, trickling filter/activated sludge, and disinfection.
[c]Primary sedimentation, trickling filter/activated sludge, disinfection, coagulation, filtration, and disinfection.
[d]Filtration only.
Source: Yates, 1994; Robertson et al., 1994; Enriquez et al., 1995; Madore et al., 1987.

ACTIVATED SLUDGE

Compared to other biological treatment methods (i.e., trickling filters) activated sludge is relatively efficient in reducing the numbers of pathogens in raw wastewater. Both sedimentation and aeration play a role in pathogen reduction. Primary sedimentation is more effective in the removal of the larger pathogens such as helminth eggs, but solid-associated bacteria and even viruses are also removed. During aeration, pathogens are inactivated by antagonistic microorganisms, and environmental factors such as temperature. The greatest removal probably occurs due to the adsorption or entrapment of the organisms within the biological floc which forms. The ability of the activated sludge process to remove viruses is related to the effectiveness of solids removal. This is because viruses tend to be solid-associated and are removed along with the floc. Activated sludge typically removes 90% of the enteric bacteria and 80 to 90–99% of the enteroviruses and rotaviruses (Rao et al., 1986). Ninety percent of *Giardia* and *Cryptosporidium* can also be removed (Rose and Carnahan, 1992; Casson et al., 1990), and are largely concentrated in the sludge. Because of their large size, helminth eggs are effectively removed by sedimentation and are rarely found in wastewater effluent in the United States, although they may be detectable in the sludge. However, while the removal of the enteric pathogens may seem large, it is important to remember that initial concentrations are also large (i.e., the concentration of all enteric viruses in one liter of raw wastewater may be as high as 100,000 per liter in some parts of the world).

OXIDATION PONDS

Given sufficient retention times, oxidation ponds can cause significant reductions in the concentrations of enteric pathogens, especially helminth eggs. For this reason, they have been promoted widely in the developing world as a low cost method of pathogen reduction for wastewater reuse for irrigation (WHO, 1989). However, a major drawback of ponds is the potential for short circuiting because of thermal gradients, even in multi-pond systems designed for long retention times (e.g., 90 days). Even though the amount of short-circuiting may be small, detectable levels of pathogens can often be found in the effluent from oxidation ponds.

Inactivation and/or removal of pathogens in oxidation ponds is controlled by a number of factors including temperature, sunlight, pH, infection by bacteriophages, predation by other microorganisms, and adsorption to or entrapment in settlable solids. Indicator bacteria and pathogenic bacteria may be reduced by 90–99% or more depending upon retention times. The die-off of indicator bacteria in waste stabilization ponds has been studied extensively and several models have been proposed largely based on retention time, solar radiation and temperature (Marais, 1974; Mill et al., 1992; Sarikaya and Saarci, 1987). The greater the amount of sunlight and temperature the more rapid the rate of die-off of enteric pathogens. Enteric viruses are also affected by these same factors, but to a lesser degree. Because these factors vary greatly by location and season, the removal is somewhat site specific. For example, in one study it was found that it took five days for a 99% inactivation of poliovirus type 1 in a model oxidation pond, while in the winter 25 days was required (Funderburg et al., 1978). In a study of waste stabilization ponds in Africa, it was found that a retention time of at least 37.3 days would be required to ensure at least a 99.9% reduction of *Giardia* and *Cryptosporidium* (Grimason et al., 1992). Helminth eggs can be removed below detection with a ten day detention time (Schwartzbrod et al., 1989).

TERTIARY TREATMENT PROCESSES

Tertiary treatment processes involving physical/chemical processes can be effective in further reducing the concentration of pathogens and enhancing the effectiveness of disinfection processes by the removal of soluble and particulate organic matter (Table 10.6). Filtration is probably the most common tertiary treatment process. Mixed media filtration is most effective in the reduction of protozoan parasites. Usually greater removal of *Giardia* cysts than *Cryptosporidium* oocysts occurs because of the larger size of the cysts (Rose and Carnahan, 1992; Gerba and Rose, 1996). Removal of enteroviruses and indicator bacteria is usually 90% or less. Addition of coagulant can increase the removal of poliovirus to 99% (USEPA, 1992).

Coagulation, particularly with lime, can result in significant reductions of pathogens. The high pH conditions (pH 11–12) which can be achieved with lime can result in significant inactivation of enteric viruses. To achieve removals of 90% or greater, the pH should be maintained above 11 for at least an hour (Leong, 1983). Inactivation of the viruses occurs by denaturation of the viral protein coat. The use of iron and aluminum salts for coagulation can also result in 90% or greater reductions in enteric viruses. The degree of effectiveness of these processes, as in other solids-separating processes, is highly dependent on the hydraulic design and operation of the coagulation and flocculation. The degree of removal observed in bench scale tests may not approach those seen in full-scale plants where the process is more dynamic.

The removal of enteric viruses by granular activated carbon has been found to be highly variable and not very effective. Viruses are believed to be adsorbed to the activated carbon, but it appears that sites available for adsorption are quickly exhausted (Gerba et al., 1975).

TABLE 10.6. Average Removal of Pathogens and Indicator Microorganisms in a Wastewater Treatment Plant, St. Petersburg, FL.

Microorganism	Raw to Secondary		Secondary to Postfiltration		Postfiltration to Postdisinfection		Postdisinfection to Poststorage		Raw to Poststorage	
	%	log$_{10}$	%	log$_{10}$	%	log$_{10}$	%	log$_{10}$	%	log$_{10}$
Total coliforms (cfu/100 ml)	98.8	1.75	69.3	0.51	99.99	4.23	75.4	0.61	99.999992	7.1
Fecal coliforms (cfu/100 ml)	99.1	2.06	10.5	0.05	99.998	4.95	56.8	0.36	99.999996	7.42
Coliphage 15597 (pfu/ml)	82.1	0.75	99.98	3.81	90.5	1.03	90.3	1.03	99.99996	6.61
Enterovirus (pfu/100 l)	98	1.71	84	0.81	96.5	1.45	90.9	1.04	99.99997	5.01
Giardia (cysts/100 l)	93	1.19	99	2	78	0.65	49.5	0.3	99.999	4.13
Cryptosporidium (oocysts/100 l)	92.8	1.14	97.9	1.68	61.1	0.41	8.5	0.04	99.95	3.26

cfu = colony-forming units; pfu = plaque-forming units.
Source: Modified from Rose and Carnahan, 1992

447

Reverse osmosis and ultrafiltration are also believed to result in significant reduction in enteric pathogens, although few studies have been done on full-scale facilities. Removals of enteric viruses in excess of 99.9% can be achieved (Leong et al., 1983). Although the pore size of the membranes used in this process are smaller than even viruses, the smallest waterborne pathogen, they should not be considered absolute barriers. In an evaluation of an on-site wastewater recycling system, it was found that while poliovirus was reduced by 99.9998% by the ultrafiltration of the wastewater, coliphage MS2 (of similar size and shape as the poliovirus) was reduced only 99.93% (Naranjo et al., 1993). This may be due to differences in the two viruses in the degree of attachment to solids in the wastewater. While poliovirus readily adsorbs to solids, MS2 coliphage does not (Hurst et al., 1980). Nanofiltration was also observed to achieve a 99% reduction of MS2 coliphage (Yahya, 1993). It is possible that viruses may find a few openings in the membranes through which to pass or they may pass through the seals. Because such "leakage" occurs at a very low level, it is not observed by measurement of other water quality parameters used to judge performance of the membranes (e.g., conductivity).

ARTIFICIAL WETLANDS

There has been an increasing interest in the use of natural systems for the treatment of municipal wastewater as a form of tertiary treatment (Kadlec and Knight, 1996). Artificial or constructed wetlands have a higher degree of biological activity than most ecosystems; thus transformation of pollutants into harmless by-products or essential nutrients for plant growth can take place at a rate which is useful for application to the treatment of municipal wastewater. Most artificial wetlands in the United States use reeds or bulrushes, although floating aquatic plants such as water hyacinths and duckweed have been used. To reduce potential problems with flying insects, subsurface flow wetlands have also been built. In these types of wetlands all of the flow of the wastewater is below the surface of a gravel bed in which plants tolerant to water saturated soils are grown. Most of the existing information on the performance of these wetlands is on coliform and fecal coliform bacteria. Kadlec and Knight (1996) have summarized the existing literature on this topic. They point out that natural sources of indicators in treatment wetlands never reach zero since wetlands are open to wildlife. Reductions in fecal coliforms are generally greater than 99%, but there is a great deal of variation depending on the season, type of wetland, numbers and type of wildlife and retention time within the wetland. Volume-based and area-based bacterial die-off models have been used to estimate bacterial die-off in surface flow wetlands (Kadlec and Knight, 1996).

Virus removal has been studied in several types of wetlands. Gersberg et al. (1989) found that coliphage MS2 removal was closely related to the removal of suspended solids in a surface flow wetland. In a study of a mixed species surface flow wetland with a detention time of approximately four days, *Cryptosporidium* was reduced by 53 percent, *Giardia* by 58 percent, and enteric viruses by 98 percent (Karpiscak et al., 1996). In the same wetlands, fecal coliform reduction averaged 98 percent. In the same study greater removals of the protozoan parasites (*Giardia* 98% and *Cryptosporidium* 87%) than fecal coliform bacteria (57%) were seen in a duckweed pond of a similar retention time. The differences may reflect greater opportunity for settling of the large parasites in the duckweed pond or input of the parasites from animals in the mixed species wetland. Additional research is needed to understand the influence of wetland design on its potential for pathogen reduction.

DISINFECTION

Disinfection is the most important barrier in the treatment process for control of pathogenic microorganisms. Several important factors control the effectiveness of disinfection against microorganisms. These include the type and strain of organism, temperature, pH, amount and type of suspended matter, and dissolved inorganic and organic compounds. The inactivation of microorganisms ideally follows first-order reactions kinetics and can be expressed as:

$$N_t/N_0 = e^{-kt}$$

where N_0 = number of microorganisms at time 0; N_t = number of microorganisms at time t; k = decay constant (time^{-1}); and t = time. However, studies have shown that deviation from first-order kinetics often occurs, especially in field studies (Hoff and Akin, 1986) where tailing off of the inactivation curve is observed. This may be due to sub-populations which may be more resistant due to attachment to protective solids or genetic differences in the microorganism's resistance to the disinfectant.

A common way of expressing disinfectant effectiveness is c·t where C = the disinfectant concentration and t the time required to inactivate a certain percentage of the microorganism under a specific set of conditions (pH and temperature). For most disinfectants the effectiveness increases with temperature (i.e., c·t decreases). Chlorine disinfection effectiveness decreases with an increase in pH above neutral, while chlorine dioxide becomes more effective. The effectiveness of various disinfectants against waterborne pathogens from laboratory experiments is shown in Table 10.7.

Protozoan parasites are more resistant to most common disinfectants

TABLE 10.7. Summary of c·t (mg/L × min) for 99% Inactivation of
Enteric Microorganisms by Disinfectants at 5°C.

Microorganism	Chlorine[a]	Chloramine[b]	Chlorine Dioxide	Ozone
E. coli	0.04	95–130	0.48	0.02
Poliovirus	1.7	1,420	0.2–6.7	0.1–0.2
G. lamblia	50–250	430–580	—	0.5–0.6
G. muris	250	1,400	10.7–15.5	1.8–2.0
Cryptosporidium	7,200	—	—	3

[a]pH 6.0.
[b]Preformed.
Source: Adapted from Hoff, 1986.

than viruses and bacteria. *Cryptosporidium* is currently the most resistant of all waterborne pathogens known to common water disinfectants. For this reason numerous waterborne outbreaks documented in recent years have involved disinfected water supplies. Because of its extreme resistance, disinfection alone can not be relied upon for *Cryptosporidium's* removal from wastewater. Filtration with coagulation is essential for effective removal of protozoan cysts.

Enteric viruses are more resistant to inactivation by chlorine, ozone, and ultraviolet irradiation than bacteria. Enteric viruses can usually be detected after chlorination of activated sludge effluents at levels and contact times normally practiced (Tyrrell et al., 1995). With extended contact times, free chlorine can reduce enteric viruses to undetectable levels (USEPA, 1992). Ultraviolet light may also be useful for inactivation of enteric bacteria and enteroviruses in filtered secondary effluent, however higher levels may be needed to inactivate other enteric viruses such as adenoviruses (Meng and Gerba, 1996). Inactivation of protozoan parasites would require doses in excess of 100,000 W/cm^2/sec to be effective (Rice and Hoff, 1981; Campbell et al., 1995).

SURVIVAL AND FATE OF PATHOGENS ON CROPS

Transmission of foodborne illness by enteric pathogens due to irrigation with reclaimed water has been well established for more than 100 years. This is the reason why irrigation of food crops, especially those eaten raw, is usually forbidden. Reclaimed water intended for crops eaten raw should only be irrigated with reclaimed water that meets the same standards for reclaimed water intended for potable reuse. Lower water quality standards have been applied to reclaimed water used for orchard crops such as citrus and other fruits. However, recent outbreaks associated with apples and raspberries caused by the parasites *Cryptosporidium* (Milard et al., 1994) and *Cyclospora* (MMWR, 1996) suggest that reclaimed water of a high microbial quality should be used for these crops as well.

Most foodborne illness associated with enteric microorganisms occurs during the mishandling of food, typically when an ill food handler does not practice proper sanitation (i.e., hand washing). Outbreaks due to contamination of the crops in the field are far more difficult to trace, because the contamination of the crops would likely be more random and the food could be dispersed over a larger geographic area. In spite of this, outbreaks of hepatitis A virus (Niu et al., 1992) and *Cryptosporidium* have been traced to crops likely contaminated in the field. Recent outbreaks of hepatitis A virus from imported lettuce (Rosenblum et al., 1990) and *Cyclospora* likely from imported raspberries (MMWR, 1996) illustrate the potential for the transmission of enteric pathogens by foods. If the crop is used for animal feed or grazing, there is a potential for the transmission of enteric pathogens, such as *Salmonella,* enteropathogenic *E. coli,* and *Cryptosporidium,* that infect both man and the animal.

Association of illness with non-food crops is likely only a potential health concern with persons involved in the production and processing of the crop. In the case of persons involved in landscape irrigation there may be exposure to plant surfaces wetted by reclaimed water. This exposure also extends to persons in potential contact with these plants, such as golfers on reclaimed water irrigated golf courses or children on playgrounds.

The most important factor determining the survival of pathogens on crops are temperature, moisture, and exposure to sunlight. The lower the temperature, the longer the survival of an enteric pathogen in any environment. Although temperatures in the field during the growing season in temperate climates usually exceed 20°C during the day, many crops are stored post-harvest at temperatures in the range of 4 to 8°C. In this temperature range, little pathogen inactivation is expected to occur. Since desiccation causes the rapid inactivation of viruses and protozoan parasites, moisture on the surface of the crop will affect survival. The ultraviolet light in sunlight will inactivate microorganisms by inducing damage to the nucleic acid. However, leaves and other plant surfaces will act to reduce temperatures, exposure to direct sunlight, and evaporation. Also, periods of rainfall will increase the humidity of the air, reducing evaporation. Clouds associated with these events will reduce exposure to direct sunlight. Because of these factors, enteric pathogens can be expected to eventually die-off on the surface of crops. As a general rule, it would appear that in the field most enteric pathogens survive for a shorter period of time on crops than in the soil (Feachem et al., 1983). However, crops can become contaminated by soil anytime during the growing and harvesting of the crop.

FOOD CROPS

Foodborne outbreaks implicating fresh produce have been associated with hepatitis A and E, rotavirus, Norwalk virus, echoviruses (Hui et al., 1994) and small round viruses (Djuretic et al., 1996). Although most of these outbreaks were believed to have been caused by contamination during food handling, they demonstrate the potential for transmission of viral disease by crops. Only the survival of enteroviruses and adenovirus on food crops has been studied. No published information could be found on the survival of hepatitis A virus, which is likely to persist longer because of its greater resistance to thermal inactivation and ultraviolet irradiation present in sunlight. From a review of the literature, it would appear that enteric viruses are unlikely to survive more than 2 weeks during a summer growing season and 6 weeks during the fall or spring growing seasons. However, under post-harvest storage conditions, enteric viruses may persist for many weeks (Badawy et al., 1985). Larkin et al. (1976) reported that poliovirus survived on lettuce or radishes for 14 to 36 days after irrigation with water that had been artificially contaminated with poliovirus. However, the virus concentration decreased 99 percent during the first 4 to 5 days. Survival time was longer for plants grown in the fall than those grown during the summer. In a study conducted in Australia, poliovirus could not be detected after 13 days on spinach, but less than 4 days on tomatoes and lettuce in the fields with maximum daytime temperatures of 14 to 40°C (Ward and Irving, 1987). However, on storage post harvest in a refrigerator, the virus had decreased only 90% on celery after 76 days.

Outbreaks caused by enteric bacterial pathogens in developed countries have been associated, in recent years, with vegetables imported from developing countries or from areas where lower water quality is the suspected source (Beuchat, 1996). Unlike enteric viruses and parasites, some enteric bacteria may grow on contaminated fruits and vegetables. For example, *Shigella* has been reported to grow in cut cubes of watermelon (Beuchat, 1996). Certain strains of *Salmonella* can grow on the surfaces of stored tomatoes and cut pieces of watermelon and cantaloupes (Beauchat, 1996). Enteric bacteria are often reported as having a shorter survival time than enteric bacteria. Of the enteric bacteria, *Salmonella* appears to survive the longest on food crops in the field. Foodborne outbreaks of enterotoxigenic *E. coli* and *Campylobacter* have been associated with contaminated fruits and vegetables (Beauchet, 1996). Like *Salmonella* and *Shigella, E. coli* can grow in cut fruits. Information on the survival of these organisms on field crops is not currently available.

The fact that protozoan cysts and oocysts can survive long enough to get into the food supply is evident by the isolation of *Cryptosporidium, Giardia lamblia,* and *Entamoeba histolytica* on market vegetables in Mexico City (Kowal, 1982) and Costa Rica (Monge and Chinchilla, 1995). *E. histolytica* has been observed to survive less than three days on tomatoes (Rudolfs et al., 1950). However, information on the survival of the other protozoan parasites on vegetables is currently not available. Because they are rapidly inactivated by desiccation, they are unlikely to survive long on dry surfaces of fruits and vegetables. Because of desiccation and exposure to sunlight, helminth eggs deposited on plant surfaces die off more rapidly than in the soil. Thus Rudolfs et al. (1950) in their review concluded that Ascaris eggs, the longest lived of the helminth eggs, sprayed on tomatoes and lettuce, to be completely degenerated after 27–35 days.

NON-FOOD CROPS

Rotavirus and enteroviruses appear to be inactivated very rapidly on plant materials under the conditions found in the arid southwestern United States. Badawy et al. (1990) found that poliovirus type one was inactivated at a rate of 0.06 $\log_{10}$/hr and rotavirus SA-11 at 0.04 $\log_{10}$/hr during the winter (4–10°C). The rates of inactivation during the summer were 0.37 and 0.2 $\log_{10}$/hr. They concluded that to achieve virus inactivation on turf grasses after the application of wastewater effluents, 8 to 10 hours would be needed during the summer and 16 to 24 hours would be needed during the winter. *Salmonella* and other enteric bacteria can survive for several weeks on grass if sufficient organic matter and moisture is available. Transmission of *Salmonella* from untreated or improperly treated human wastes to domestic animals demonstrates this potential (Haugaard, 1984). Helminth eggs such as *Ascaris* are believed to survive for 30 to 60 days, although they may survive for many months in the soil itself (Feachem, 1983).

REMOVAL AND FATE OF PATHOGENS DURING LAND TREATMENT

FACTORS AFFECTING FATE AND TRANSPORT

The potential for pathogens in reclaimed water to contaminate the underlying ground water is dependent on a number of factors including the physical characteristics of the site (e.g., soil texture), the hydraulic conditions (e.g., wastewater application rate, wetting/drying cycles), the environmental conditions (e.g., rainfall, temperature) at the site, and the characteristics of the specific pathogens present in the reclaimed water. The factors that influence the fate and transport of pathogens in the subsurface have been the subject of a number of reviews (Bitton and Harvey,

1992; Gerba and Goyal, 1985; Yates and Yates, 1988; Vaughn and Landry, 1983).

Temperature

Temperature is probably the most important factor influencing virus inactivation in the environment (Bitton, 1980). In a study using groundwater samples collected throughout the U.S., none of the measured water characteristics including pH, nitrate, ammonia, sulfate, iron, hardness, turbidity, and total dissolved solids, except temperature were significantly correlated ($p < 0.01$) with the inactivation rate of the viruses (polioviruses, echoviruses, and MS2 bacteriophages) (Yates et al., 1985). However, linear regression analysis of virus inactivation rate as a function of temperature produced a correlation coefficient, r, of 0.88. The coefficient of determination, r^2, was 0.775, indicating that 77.5% of the variation in virus inactivation rates may be explained by temperature effects. This finding has also been observed by other investigators (Jansons et al., 1989; Yahya et al., 1993) studying the inactivation rates of enteric viruses and bacteriophages in ground water.

Temperature also affects the persistence of viruses in soils. Lefler and Kott (1974) found that it took 42 days for 99% inactivation of poliovirus in sand at 20–25°C, whereas more than 175 days were required at 1–8°C. Poliovirus was found to persist for more than 180 days in saturated sand and sandy loam soils at 4°C, whereas no viruses could be recovered from the soils incubated at 37°C after 12 days (Yeager and O'Brien, 1979a). Hurst et al. (1980a) studied the survival of poliovirus at three temperatures: 1, 23, and 37°C in a loamy sand. They also found that the inactivation rate was significantly correlated ($p < 0.01$) with incubation temperature, noting faster inactivation rates at the higher temperatures.

The exact mechanism whereby temperature inactivates viruses in soils has not been determined, but several theories have been proposed. The inactivation may be due to denaturation of the viral capsid (Bitton, 1980), however, it has been shown that the RNA released during thermal denaturation remains infective as it is more resistant to heat inactivation (Larkin and Fassolitis, 1967). Dimmock (1967) suggested that different mechanisms may be at work, depending on the temperature. At low temperatures (less than 44°C), the rate of virus inactivation corresponds with the inactivation rate of the viral RNA. However, at high temperatures (greater than 44°C), the rate of virus inactivation exceeds that of the inactivation rate of the viral RNA, and is associated with structural changes in the viral capsid. Kapuscinski and Mitchell (1980) suggested that temperature itself may not be responsible for virus inactivation, but may merely control whether other inactivation mechanisms can occur.

Microbial Activity

There are conflicting reports regarding the role of microorganisms in virus inactivation; many of these have been reviewed (Sattar, 1981; Kapuscinski and Mitchell, 1980). Sobsey et al. (1980) found that the inactivation rates of poliovirus and reovirus in eight different soil suspensions were almost always greater under non-sterile conditions compared to sterile conditions. Hurst et al. (1980a) reported that virus inactivation rates were similar in non-sterile anaerobic soils and sterile soils at 1, 23, and 37°C. Under aerobic conditions, poliovirus and echovirus inactivation rates were more rapid in non-sterile soil preparations than sterile at 23 and 37°C. At very low temperatures, 1°C, inactivation rates were similar in both sterile and non-sterile soils.

Further studies by Hurst (1988) investigated the influence of the oxygen status of the soil and conditions of sterility on the survival of poliovirus 1 in soil. These studies showed that the virus inactivation rate was significantly affected by the presence of aerobic microorganisms. Two-to-three-fold increases in virus inactivation rates were found in the presence of aerobic microorganisms. This observation was made at all three incubation temperatures, 1, 23, and 37°C. Anaerobic microorganisms did not have any significant effects on the virus survival in this study.

The influence of ground-water bacteria on the persistence of viruses has also been investigated. In a study using over 30 ground-water samples, the presence or absence of the indigenous bacteria was not significantly correlated with the inactivation rate of the virus (Yates et al., 1990). The bacteria had an inconsistent effect on virus persistence: in some sterile samples, viruses persisted longer than in the non-sterile counterparts, while in other samples, the opposite effect was observed.

Similar results were obtained in a study by Sobsey et al. (1986) on the survival of hepatitis A virus in several soil-ground water mixtures. In no case did the virus persist for a longer period of time in the presence of the microorganisms. However, in some of the mixtures (i.e., kaolinite and Cecil clay loam) there appeared to be no effect of bacteria on the length of time the virus remained infective.

Several mechanisms by which microorganisms may influence virus inactivation have been suggested (Melnick and Gerba, 1980). Proteolytic enzymes released by certain bacteria were capable of destroying the capsid of coxsackievirus A9, but polioviruses were unaffected (Cliver and Herrmann, 1972). Some microorganisms produce substances which render viruses more susceptible to inactivation by photodynamic processes (Fukada et al., 1968) or enzymes (Salo and Cliver, 1978). Other compounds produced by microorganisms such as humic acids, tannins, phenols, and ascorbic acid may act as oxidizing or reducing agents which lead to virus inactivation (Melnick and Gerba, 1980). Studies by Lipson and Stotzky (1985) on the adsorption of reovirus to clay particles suggested that microorganisms may enhance virus inactivation in two ways. One is by the production of a soluble metabolic product that is virucidal; the other is by interfering with virus adsorption to soil particles.

Moisture Content

The moisture content of the soil also influences virus persistence in soils. Although some investigators have observed no difference in inactivation rates in dried vs. saturated sand (Lefler and Kott, 1974), the majority of the reports have indicated a difference. Bagdasaryan (1964) observed that several enteroviruses, including poliovirus 1, coxsackievirus B3, and echoviruses 7 and 9, could survive for 60 to 90 days in soil with 10% moisture as compared with only 15 to 25 days in air-dried soils. In another study, 99 percent inactivation of poliovirus occurred in one week as the soil moisture content was reduced from 13 to 0.6%; however, 7 to 8 and 10 to 11 weeks were required for the same amount of inactivation in soils with 25 and 15% moisture content, respectively (Sagik et al., 1978).

Hurst et al. (1980a) found that the moisture content affected the survival of poliovirus in a loamy sand. The inactivation rate increased as the moisture content was increased from 5 to 15%, then decreased as more liquid was added. It was noted that the fastest inactivation rate occurred near the saturation moisture content of the soil (15 to 25%). The slowest inactivation rates were observed at the lowest moisture contents, 5 and 10%.

In another study, it was observed that virus inactivation rates were greater in more rapidly drying soils. In a field study of virus inactivation during rapid infiltration of wastewater, it was shown that allowing the soils to periodically dry and become aerated enhances virus inactivation (Hurst et al., 1980b). Using radiolabeled viruses, Yeager and O'Brien (1979a) showed that the viruses are inactivated during the drying process, rather than becoming irreversibly bound to the soil particles.

Other studies by Yeager and O'Brien (1979b) suggested that the mechanisms responsible for poliovirus inactivation are different in moist and drying soils. In moist soils, under both sterile and non-sterile conditions, the viral RNA was recovered in a degraded form, suggesting that the nucleic acid was degraded prior to its release from the capsid. In dried, non-sterile soils, complete loss of viral infectivity was observed, and the RNA was recovered in degraded form. In contrast, the nucleic acid was recovered intact from sterile, dry soils.

The moisture content also influences the movement of viruses through soil. In a column of loamy sand, poliovirus penetrated 40 cm under unsaturated flow conditions. However, virus could be recovered at a depth of 160 cm during saturated flow conditions (Lance and Gerba, 1984). Under unsaturated conditions, some of the soil pores are filled with gas rather than water, leaving a film of water around the particles. Thus, viruses can get closer to the soil than in saturated conditions when the pores are completely filled with water. This increases the opportunity for virus adsorption to soil under unsaturated conditions (Gerba and Bitton, 1984).

The results of laboratory studies support this hypothesis. Bitton et al. (1984) conducted experiments of virus transport using unsaturated soil cores. They postulated that the very limited virus movement observed was due to the fact that under unsaturated conditions, water flow is through micropores (rather than macropores), which promotes virus adsorption to the soil surface. Jorgensen (1985) detected no penetration of coxsackievirus B3 or adenovirus 1 below 3.5 cm in an unsaturated soil column after 28 weeks of operation. Both virus types were recovered from the top soil layers at the end of the experiment, indicating that the lack of recovery in the deeper soil layers was not due to inactivation over time. These results were contrasted with studies using coxsackievirus B3 and the same soil columns operated under saturated flow conditions. Coxsackievirus B3 seeded into wastewater sludge was able to penetrate a 30-cm-long column of sandy loam soil. Powelson et al. (1990) studied the transport and survival of MS2 bacteriophages in saturated and unsaturated columns of Vint fine loamy sand. These investigators found that the concentration of viruses in the effluent of the saturated column reached the influent concentration after less than two pore volumes. In contrast, in the unsaturated columns, the effluent concentration never exceeded 5% of the input concentration. In addition, these investigators found that only 39% of the input viruses could be recovered from the unsaturated soil column, while no reduction in virus number was seen in the saturated column. In another study by Powelson and Gerba (1994), the removal of poliovirus 1, PRD1, and MS2 was three times greater in unsaturated columns containing coarse alluvial sand compared with parallel saturated columns. Significant inactivation of viruses during transport through unsaturated soils has also been observed by Poletika et al. (1995).

pH

The effect of pH on virus survival in soil has not been extensively studied. It has been suggested that pH indirectly influences virus survival by controlling adsorption onto soil particles which, in turn, affects virus survival (Gerba and Bitton, 1984). The direct effects of pH on virus persistence has been studied by a few investigators. Sobsey (1983) reported the results of a study using simian adenovirus SV-11 in which it was shown that inactivation was more rapid at pH 5.0 and 6.0 than at pH 4.0 and 7.0.

The stability of five enteroviruses including hepatitis A virus, coxsackievirus A9, coxsackievirus B1, poliovirus 1, and echovirus 9, was studied under conditions of extreme acidity at room temperature (Scholz et al., 1989). Of the viruses studied, hepatitis A virus was the most resistant to inactivation. In order to eliminate the possibility that attachment to cellular material was responsible for the high resistance to inactivation, the studies were repeated with highly purified hepatitis A virus. The purified virus was found to be even more resistant to inactivation, remaining infectious for 8 hours at pH 1.

Salo and Cliver (1978) found that in the aqueous environment, pH affects different viruses in different ways. While poliovirus 1 was inactivated more rapidly at pH 3 and 9 than the near-neutral values of 5 and 7, coxsackievirus A9 was inactivated more rapidly at pH 5 than the extremes of 3 and 9. Murphy et al. (1958) observed that mouse encephalomyelitis virus survival was longer in neutral soils than in soils with the pH adjusted to 3.7 or 8.5. Hurst et al. (1980a) studied the survival of poliovirus 1 and two bacteriophages, MS2 and T2, in nine soils with pH values ranging from 4.5 to 8.2. They found that virus inactivation was significantly correlated ($p < 0.05$) with soil saturation pH, with longer survival at the lower pH values.

The exact mechanism whereby pH causes virus inactivation has not been fully elucidated, but one investigator (Prage et al., 1970) found that the inactivation of adenovirus 2 was accompanied by an increase in sensitivity to DNase and the loss of structural proteins from the virus capsid. Results obtained by Salo and Cliver (1978) also suggested that virus inactivation involves alterations in the virus capsid. These investigators found that the RNA of the inactivated virus particles became sensitive to ribonuclease at all pH levels tested (pH 3–9), and at pH 5 and 7 the RNA was hydrolyzed in the absence of ribonuclease.

The effects of pH on the movement of viruses through soil, or, more specifically, the retardation of virus movement by adsorption to soil particles has been extensively studied. Goyal and Gerba (1979) measured the effects of seven soil properties on the adsorption of 15 viruses to nine soils. They found that, in general, soils with a saturation pH of less than 5 were good virus adsorbers. The adsorption of five of the viruses (four echoviruses and phage phiX174) showed a significant negative correlation ($p < 0.05$) with soil pH. Subsequent analysis of these data using factor analysis (Gerba et al., 1981) divided the viruses into two general groups based on their adsorption behavior. For viruses in group I (two coxsackie B4 viruses, four echo 1 viruses, phiX174, and MS2), pH was an important factor affecting virus adsorption to soil. However, for viruses in group II (polio 1, echo 7, coxsackie B3, and phages T4 and T2) none of the studied soil characteristics, including pH, was correlated with adsorption to soil.

Burge and Enkiri (1978) found that the adsorption of bacteriophage phiX174 to five soils showed a significant negative correlation with soil pH. The adsorption of poliovirus 1 and reovirus 3 to eight soils suspended in wastewater at pH values from 3.5 to 7.5 was generally greater at the lower pH values (Sobsey et al., 1980). The adsorption of encephalomyelitis virus to montmorillonite clay was greatest at pH 5.5 (Schaub and Sagik, 1975). At pH 9.5, the adsorption was less, but the poorest adsorption was observed at pH 3.5. In an extensive study of poliovirus adsorption by 34 minerals and soils, no significant correlation was found between substrate pH and virus adsorption (Moore et al., 1981). The investigators reported that although adsorption by most of the neutral and acidic materials was strong, the variation in percentage of virus bound in the alkaline materials was so great that no significant correlation could be detected.

Loveland et al. (1996) studied the effects of pH on the reversibility of virus attachment to mineral surfaces. Their experiments on the attachment of PRD1 to quartz and ferric oxyhydroxide-coated quartz indicated that attachment is controlled by the isoelectric point of the mineral surface. Attachment of PRD1 was observed at pH values 2.5 to 3.5 units above the isoelectric point of the mineral surface. Below this pH, the attachment of PRD1 was found to be complete and irreversible, while above this pH, the adsorption was minimal and reversible.

The mechanism(s) whereby pH affects virus adsorption to soil particles has been explained in terms of the electrochemical nature of the virus and soil surfaces (Gerba and Bitton, 1984; Sobsey, 1983; Gerba, 1984). The outer surface of the enteric viruses is made of protein, therefore, the surface charge is influenced by the ionization of the carboxyl and amino groups in the capsid. The isoelectric point (pI) of many enteric viruses is below 7 (Gerba, 1984), thus, at neutral pH, most viruses are negatively charged. Most soils are also negatively charged at neutral pH, and virus adsorption is not favored due to the mutual repulsion. If the pH of the environment is decreased, the surface charge of the virus will become positive (or less negative) due to increased ionization of the amino groups and decreased ionization of the carboxyl groups. While soils also become less negatively charged at lower pH values, soil pI values are generally lower than those of viruses, thus they may still have a net negative charge at acidic pH levels. This results in an electrostatic attraction between the virus particle and the soil, which leads to adsorption. In reality, the effect of pH on virus adsorption is not so clear-cut. There are many complicating factors which can interfere with the mechanism discussed above. One is that a given virus may have more than one isoelectric point: poliovirus 1 (Brunhilde strain) has isoelectric points at pH 4.5 and 7.0 (Gerba, 1984). The factors responsible for passage from one form to another are unknown at this time. Other soil factors such as cations and humic and fulvic acids may also influence the net surface charge of viruses.

Salt Species and Concentration

The presence of certain chemicals may render a virus more or less susceptible to inactivation. Burnet and McKie (1930) found that bacteriophage inactivation at 60 C was partially prevented in the presence of 0.002 to 0.01 M $CaCl_2$, or $BaCl_2$. However, when the concentration was increased to 0.15 M or greater, thermal inactivation was increased. Enhanced stabilization of poliovirus at temperatures ranging from 4°C to 50°C in the presence of high concentrations (1 M) of $MgCl_2$ has been reported (Wallis and Melnick, 1961). Echoviruses have also been found to possess this property (Dulbecco and Ginsberg, 1980). Cords et al. (1975) found that many type A coxsackieviruses were inactivated more rapidly in low ionic strength media than in high ionic strength media.

Thurman and Gerba (1987) studied the effects of modifying soil with several metals on the survival of MS2 and poliovirus. The addition of aluminum metal, magnesium oxide, and magnesium peroxide had a significant negative effect on the inactivation rate of MS2 when compared with unmodified control soils. The addition of unrefined substances such as zeolite, bauxite, limonite, and glauconite (which contain multivalent cations and oxides of iron and aluminum) did not have a significant negative effect on the inactivation rate of MS2. Rather, they seemed to have a protective effect, as indicated by the lower inactivation rate as compared with the unamended control soil. Exposure of poliovirus to the aluminum-amended soil also resulted in significantly higher rates of inactivation as compared with controls. Further experiments by these investigators indicated that contact between the aluminum and virus was necessary for inactivation of the viruses. They postulated that a combination of electrodynamic van der Waals interactions and electrostatic double layer interactions promoted virus adsorption to the surface of the aluminum, where inactivation of the virus could then take place.

The types and concentrations of salts in the environment have a profound influence on the extent of virus transport in the subsurface. In general, it has been found that transport is retarded in the presence of increasing concentrations of ionic salts and increasing cation valences due to increased virus adsorption (Sobsey, 1983). Taylor et al. (1981) studied the effects of pH and electrolytes on the adsorption of poliovirus 2 to five soils, ranging from a sand to a montmorillonite clay. These investigators found that both the type of electrolyte and its concentration affected poliovirus 2 adsorption to the soils. A "critical" pH region was observed for each soil in which there was a rapid transition from weak to strong uptake. When the pH was at or above the critical region, the virus adsorption increased with electrolyte concentration. Sobsey et al. (1980) found that viruses could be made to adsorb to some relatively poor adsorbents upon the addition of divalent cations, especially Mg^{2+}.

The role of the soil cation exchange capacity (CEC) in virus adsorption has also been investigated. Burge and Enkiri (1978) found that the CEC of four of five soils was correlated significantly ($p < 0.05$) with virus adsorption. In contrast, Goyal and Gerba (1979) did not find a significant correlation between soil CEC and adsorption of 15 different viruses. Additionally, no correlation was found between virus adsorption and total phosphorus or total and exchangeable iron, calcium and magnesium.

Experiments to determine the effects of water quality on virus retention in soil columns were conducted by Sobsey et al. (1986) using poliovirus 1, echovirus 1, and hepatitis A virus. Little difference in the amount of virus adsorption was noted when the columns were dosed with either ground water or primary wastewater effluent. Dosing with simulated rain water resulted in considerable elution of all three viruses from the Corolla sand columns at 5°C. Elution of the three viruses from columns of FM loamy sand was also observed, although the effects were more pronounced for echovirus 1 than for the other two viruses.

A decrease in the salt concentration or ionic strength of the soil water, such as would occur during a rainfall event, can cause desorption of viruses from soil particles (Gerba, 1983). This has been demonstrated in the laboratory by applying distilled water to simulate rainfall to soil columns containing viruses (Gerba and Lance, 1978; Lance et al., 1976; Landry et al., 1979; Landry et al., 1980). While some virus particles were readsorbed farther down the column, others were detected in the column effluent. This phenomenon has also been observed at a field site. Wellings et al. (1975) detected viruses in wells which had previously been virus-free at a land application site in Florida after a period of heavy rainfall.

Virus Adsorption to Soil

The adsorption of viruses to soils and other surfaces may prolong or reduce survival, depending upon the properties of the sorbent. The survival of poliovirus and reovirus was not always prolonged in soil suspensions as compared with soil-free controls in a study using eight different soil materials (Sobsey et al., 1980). Murray and Laband (1979) found that poliovirus adsorption onto some inorganic substances, such as CuO, results in decreased infectivity of the virus. These investigators suggested that van der Waal's interactions between the virus and the particle surface induced spontaneous disassembly of the virus.

More commonly, however, adsorption to soils has been found to prolong virus survival. Jorgensen and Lund (1985) compared the survival of viruses in ground water to survival in soil-ground water mixtures. Coxsackievirus B3, adenovirus 1, and echovirus 7 were seeded into ground water and ground water-soil mixtures and incubated at 4 to 7°C. Their studies showed that the viruses survived for longer periods of time in the soil-ground water mixtures than in the ground water alone. Their results indicated that the majority of the viable virus particles were contained in the soil fraction of the mixture, rather than in the ground water. A similar study on the survival of hepatitis A virus in several soils showed that, in all cases but one, virus survival was prolonged in the presence of soil as compared with freely suspended in ground water (Sobsey et al., 1986). These studies suggest that the protective effects of soil are the result of adsorption to the solid surface.

The mechanisms whereby adsorption to a solid surface prolongs or reduces virus survival have not been elucidated. However, Gerba and Schaiberger (1975) have suggested several possibilities, including interference with the action of virucides, increased stability of the viral protein capsid, prevention of aggregate formation, and adsorption of enzymes and other inactivating substances.

The transport of viruses through soil is retarded by association with the soil particulates. As discussed previously, however, an adsorbed virus is not necessarily permanently immobilized. A change in ionic strength or salt concentration of the surrounding medium, such as would be induced by a rainfall event, can cause virus desorption, allowing further migration in the soil. Virus adsorption acts to slow the apparent rate of virus movement in the soil.

Virus Aggregation

The formation of aggregates influences virus survival in natural waters. It has been suggested that this is because virus particles within the aggregates are highly resistant to destruction by environmental factors (Bitton, 1980). It has been shown that aggregation renders virus particles more resistant to inactivation by chemical disinfectants such as bromine (Young and Sharp, 1977), free chlorine, and monochloramine (Sobsey et al., 1991). Although there are no definitive reports on the effects of aggregation on virus survival in soil, the results of studies in aqueous media would suggest that aggregates would survive longer in soils than would monodispersed viruses (Sobsey, 1983).

Soil Properties

The influence of soil type on virus survival is probably related to the degree of adsorption. Hurst et al. (1980a) suggested that the correlation between pH and virus survival was probably mediated through its influence on virus adsorption. A positive correlation between virus survival and soil exchangeable aluminum and a negative correlation with resin-extractable phosphorus were also attributed to influencing virus adsorption to the soil.

Sobsey et al. (1986) compared the rate of inactivation of hepatitis A virus in five different soil types including a clay soil, a clay loam, a loamy sand, a sand, and an organic muck. The survival was greatest in the clay soils, in which at least 8 weeks were required to inactivate 99% of the infectious viruses.

The soil properties have a profound influence on virus movement in the subsurface. In general, virus migration is greater in coarse-textured and karstic terrain than in fine-textured soils. The migration of bacteriophage in chalk aquifers has been studied by Skilton and Wheeler (1988, 1989). These investigators injected different bacteriophage into piezometers that intersected the water table and then collected samples at different sites downgradient to determine the extent of movement. Very high velocities were observed at one site, which is due to the fact that the majority of the water flow at that site is through fissures, fractures, solution openings, and cavities. All three phage types were detected 355 m from the injection site approximately 5 hours after introduction. It is noteworthy that viable phage were being recovered more than 150 days after they were injected into the aquifer.

Virus Type

As is obvious from the previous discussions, different viruses vary in their susceptibility to inactivation in the subsurface environment. Sobsey et al. (1980) found that the rates of inactivation of poliovirus and reovirus in eight soils were different. Hurst et al. (1980a) also showed that the inactivation rates of seven different viruses varied even when incubated under the same conditions. In a comparative study of the survival of poliovirus, echovirus and coliphage MS2 in several different ground-water samples, no significant difference ($p < 0.01$) was found overall (Yates et al., 1985). There were, however, differences in the inactivation rates of the viruses in individual water samples. In a recent report on the survival of hepatitis A virus (HAV), it was shown that this virus generally survived longer than poliovirus and echovirus at 25°C (Sobsey et al., 1986).

The survival of poliovirus 1 and f2 coliphage in dry sand was compared by Lefler and Kott (1974). Poliovirus survived for at least 77 days at room temperature. The f2 phage survived considerably longer, possibly for as long as one year under dry conditions. Based on the previous discussions, it is possible that the survival times of both viruses would be greatly increased if water were added to the soil.

The survival of human rotavirus strain Wa was compared to that of poliovirus 1, which is commonly used as a model for virus behavior in the environment (Pancorbo et al., 1987). Inactivation experiments were conducted using five water samples (lake water, creek water, ground water, nonchlorinated secondary effluent, and tap water) at 20°C. The inactivation rates for the two viruses were significantly different from one another in all water samples except the tap water. This led to the conclusion that poliovirus is not a good model for the behavior of human rotavirus in water. The authors proposed that human rotavirus strain Wa can be used as a model for the persistence of other rotaviruses in water as it survived for longer than has been reported for simian rotavirus SA11 in similar environmental conditions.

In addition to affecting the survival, the properties of the specific virus also affect how the virus is transported in the subsurface. The experiments by Goyal and Gerba (1979) which studied the adsorption of 15 different viruses to soils found a high degree of variability, not only among different virus types but among different strains of the same virus type. Using factor analysis, they were able to place the viruses in three groups based on adsorption characteristics: good sorbers, poor sorbers, and f2, which did not seem to fit in the other two groups (Gerba et al., 1981). Sobsey et al. (1986) found that the movement of HAV in soil was intermediate between poliovirus and echovirus. Differences in the adsorption and elution behavior of several enteroviruses were also observed by Landry et al. (1979).

The differences in the transport of two bacteriophages, MS2 and phiX174, were studied in Ottawa sand columns (Jin et al., 1997). These investigators found that a significant amount of the phiX174 was reversibly retained by the soil, while no retention of the MS2 was observed. The differences in the behavior of the two viruses was attributed to differences in their isoelectric points: the pH_{iep} of MS2 is 3.9, while that of phiX174 is 6.6.

Field tracer experiments with bacteriophages MS2 and PRD1 were conducted in a weathered and fracture-rich clay till by McKay et al. (1993). These investigators attributed a tenfold difference in the breakthrough concentrations of the viruses to differences in size. They hypothesized that the smaller size of MS2 (25 nm) allowed it to diffuse into the fractured matrix to a greater extent than the larger (62 nm) PRD1 bacteriophages.

In contrast, in column studies of poliovirus and echovirus movement through soil columns, Lance et al. (1982) found that the viruses behaved very similar to one another, suggesting that poliovirus would be an appropriate model of virus movement in soil.

Organic Matter

The influence of organic matter on virus survival has not been firmly defined. In some studies, it has been found that proteinaceous materials present in wastewater may have a protective effect on viruses, however, in others no effect has been observed (Bitton, 1980). Lefler and Kott (1974) found that poliovirus survived longer in sand columns saturated with oxidation pond effluent than distilled water. They suggested that the prolonged survival was probably due to the protective effect of protein-aceous materials in the effluent. Hurst et al. (1980a) did not find a signifi-cant correlation between virus survival and percent soil organic matter.

Humic and fulvic acids in soils may cause loss of virus infectivity and prevent adsorption (Bixby and O'Brien, 1979). The infectivity could be partially restored by treatment with 3% beef extract, pH 9. Sobsey (1983) has observed a similar phenomenon: poliovirus infectivity was reduced in the presence of humic acid. At least some of the infectivity could be restored when the samples were filtered, probably by disruption of the humic acid-virus complexes.

Dissolved organic matter has generally been found to decrease virus adsorption by competing for adsorption sites on soil particles. In their study of poliovirus adsorption by 34 minerals and soils, Moore et al. (1981) found that organic matter showed a strong negative correlation ($p < 0.001$) with virus adsorption. Bitton et al. (1976) suggested that the organic material in secondary effluent interfered with virus adsorption to a sandy cypress dome soil. The effects of natural humic material and wastewater sludge-derived organic matter on the transport of MS2 bacterio-phages in unsaturated soil was studied by Powelson et al. (1991). The transport of MS2 was found to be higher in a loamy fine sand column that had been treated with the organic material than in a parallel column that had not been treated. In a series of experiments, Bales et al. (1991, 1993) studied the effects of hydrophobic organic material on the attachment of bacteriophages MS2 and PRD1 and poliovirus 1 to silica beads. These investigators found that even very small amounts of hydrophobic organic material (≥0.001%) can retard the transport of viruses in porous media.

Several investigators have found that organic material can act not only as a competitor for virus adsorption sites, but also as an eluting agent, that is, it can cause sorbed viruses to desorb from the soil (Gerba, 1984). Pieper et al. (1997) studied the effects of wastewater-derived organic matter on the transport of bacteriophage PRD1 in a sand and gravel aquifer on Cape Cod, MA. They found that the removal of the virus was greater in an uncontaminated area (83%) than in an area that had been contaminated by secondary wastewater effluent (42%). They concluded that the differ-ence in removal was due to the higher content of organic matter in the contaminated zone (2.0–4.4 mg/L vs. 0.4–1.0 mg/L dissolved organic carbon), which blocked the adsorption of the PRD1. The attachment of the viruses to the soil surfaces was reversible, as demonstrated by the recovery of viruses upon addition of linear alkylbenzene sulfonates (LAS) to the system. In the contaminated zone, 87% of the attached PRD1 bacteriophages were remobilized. Jin et al. (1997) also found that addition of a solution of 1.5% beef extract, which is high in organic matter, could remobilize phiX174 bacteriophages from Ottawa sand.

Widespread use has been made of this property of organic matter in the area of removing viruses from filters in order to detect them in environmental samples. Mucks and other soils with high organic matter content are poor adsorbers of viruses (1979) and may not be suitable for wastewater application sites (Sobsey, 1983).

In contrast to the other studies, Powelson et al. (1993) found that the type of effluent (primary, secondary, and tertiary) had no significant effect on the transport of three viruses (poliovirus 1, MS2 and PRD1) through columns of an alluvial coarse sand.

Hydraulic Conditions

The rate at which water or effluent is applied to the soil affects the amount of virus removal or adsorption to the soil particles. Several investigators have shown that the amount of virus removal increases as the application rate decreases. Lance and Gerba (1984) found that increasing the application rate from 0.6 to 1.2 m day^{-1} caused an increase in the number of virus particles in the column effluent. Increasing the application rate above 1.2 m day^{-1} up to 12 m day^{-1} did not, however, cause any further increase in virus movement through the columns. Vaughn et al. (1981) also found that application rate had a large influence on the movement of poliovirus through a coarse sand-fine gravel soil. They speculated that the formation of a surface mat of wastewater solids may have been responsible for the greater virus removals at the lower application rates.

FIELD STUDIES OF VIRUS TRANSPORT AT LAND APPLICATION SITES

A number of studies of pathogen transport have been conducted at wastewater application sites during the last 25 years. The focus of these studies has been on the transport of enteric viruses, as they are thought to have the greatest potential to be transported due to their relatively small size. The results of these studies are summarized in Table 10.8. The majority of the field studies were conducted in the late 1970s and early 1980s; these studies were reviewed by Gerba and Goyal (1985). Recently, however, there has been a resurgence in studies in the United States that have monitored the transport of viruses through soil at land application sites. This increased activity is likely due to the need to augment fresh water resources in the increasingly urban areas, such as those in the arid southwestern United States.

In Colton, California, land application and extraction of wastewater was studied as an alternative to conventional tertiary treatment (filtration and disinfection) of secondary effluent (Foreman et al., 1993). At this site, unchlorinated secondary effluent was applied to infiltration basins consisting of coarse sands with clean water infiltration rates of approximately 15 m day^{-1}. Virus levels in the effluent were as high as 316 pfu per 200 L. After infiltration through 24.4 m of soil, one sample of the extracted water was found to contain infective human enteroviruses.

Jansons et al. (1989) compared the behavior of indigenous enteroviruses and seeded polioviruses at a site in Western Australia where wastewater was used to artificially recharge the ground water. In this study, vaccine strain polioviruses, at a concentration of 4000 most probable number of cytopathic units per liter, were added to the infiltrating wastewater. A 1.6-log (40-fold) reduction in virus numbers was found after transport through 0.5 m of sandy soil. The viruses were isolated on a single occasion at a depth of 1.5 m; however no viruses were detected at the 3-m sampling depth although monitoring was conducted for 7 days. In contrast, several types of enteroviruses indigenous to the wastewater were isolated at the 3-m sampling depth. In addition, indigenous enteroviruses were isolated from a 9-m-deep well located 14 m from the recharge basin.

The removal of seeded bacteriophages during artificial recharge with wastewater has also been studied by Powelson et al. (1993). At this site near Tucson, Arizona, two experimental recharge basins, 3.65 m × 3.65 m, were constructed for the seeding studies. Two bacteriophages, MS2 and PRD1, were added to the infiltrating wastewater at a constant concentration of 10^5 pfu ml^{-1}. These researchers found that between 36 and 99.7% of the viruses were removed after transport through 4.3 m alluvial sand and gravel. In addition, significantly greater removal of the PRD1 was observed when suspended in secondary effluent compared to tertiary effluent. This may have been due to adsorption of the virus particles to particulate material in the secondary effluent, which were then filtered out by passage through the soil pores.

Yanko (1994) studied the removal of indigenous wastewater coliphages in the Montebello Forebay recharge area. In this area, tertiary effluent is infiltrated into alluvial sand and gravels for the purpose of ground water

TABLE 10.8. Transport of Viruses at Wastewater Application Sites.

Site Location	Virus Types	Wastewater Type	Soil Type	Maximum Transport Distance (m)		Reference
				Vertical	Horizontal	
Canning Vale, Australia	Echovirus 11	Secondary effluent	Bassendean sand	9	14	Jansons et al., 1989
	Echoviruses 14, 24, 29, 30; Coxsackievirus B4; Adenovirus 3			3		
	Poliovirus 1			2		
	Echovirus 6; Poliovirus 3; Reovirus			1		
	Poliovirus 2; Coxsackievirus B5			0.5		
Tucson, AZ	Bacteriophages MS2 and PRD1	Secondary effluent	Coarse alluvial sand and gravel	4.6		Powelson et al., 1993
Gainesville, FL	Coxsackievirus B4; Poliovirus 1, 2	Tertiary effluent	Sand	6.1	7	Wellings et al., 1975
Tucson, AZ	Bacteriophage PRD1	Chlorinated secondary effluent	Coarse alluvial sand and gravel	6.1	46	Gerba et al., 1991
Kerrville, TX	Enteroviruses	Secondary effluent	Loams to clays	1.4		Duboise et al., 1976
East Meadow, NY	Echovirus 12	Secondary effluent	Coarse sand and fine gravel	*11.3	3	Vaughn et al., 1978
Holbrook, NY	Echovirus 6, 21, 24, 25	Tertiary effluent	Coarse sand and fine gravel	6.1	45.7	Vaughn et al., 1978

TABLE 10.8. (continued).

Site Location	Virus Types	Wastewater Type	Soil Type	Maximum Transport Distance (m)		Reference
				Vertical	Horizontal	
Fort Devens, MA	Bacteriophage f2	Secondary effluent	Silty sand and gravels	28.9	183	Schaub and Sorber, 1977
Phoenix, AZ	Coxsackievirus B3	Secondary effluent	Fine loamy sand over coarse sand and gravel	18.3	3	Gerba, 1982
Colton, CA	Enterovirus	Secondary effluent	Coarse sand	24.4		Foreman et al., 1993
Lubbock, TX	Coxsackievirus B3	Secondary effluent	Loam	1.4		Goyal et al., 1984
San Angelo, TX	Enterovirus	Secondary effluent	Clay loam	27.5		Goyal et al., 1984
Muskegon, MI	Enterovirus	Chlorinated aerated effluent	Rubicon sand	10		Goyal et al., 1984
12 Pines, NY	Poliovirus 2	Tertiary effluent	Coarse sand and fine gravel	6.4		Vaughn et al., 1981
North Massapequa, NY	Echovirus 11, 23; Coxsackievirus A16	Stormwater	Coarse sand and fine gravel	9.1		Vaughn and Landry, 1977
Vineland, NJ	Poliovirus; Echovirus; Coxsackievirus B3	Primary effluent	Coarse gravel and sand	16.8	250	Koerner and Haws, 1979
Montebello Forebay, CA	Male-specific coliphages	Tertiary effluent	Alluvial sand		7.6	Yanko, 1994

recharge. A 3-log removal within 15 feet of the surface was documented at this site. However, questions have been raised regarding the appropriateness of indigenous coliphages as surrogates for human pathogenic viruses in studies of transport through soils. Further studies by this researcher revealed that male-specific phages in the reclaimed water were not transported beyond 7.6 m of the recharge basin.

The recent advances in molecular techniques have enhanced the ability to monitor the transport of a variety of pathogens in the subsurface. Ongoing studies in southern California and in Arizona are using polymerase chain reaction and specific nucleic acid probes to determine whether viruses such as hepatitis A virus and rotaviruses, in addition to the traditionally-studied enteroviruses, are present in ground waters impacted by artificial recharge with treated wastewaters.

ASSESSING MICROBIAL RISKS

From a public health perspective, the central issues associated with the artificial recharge of ground water relate to pathogenic microorganisms and potentially carcinogenic chemicals. The chemicals of greatest concern are the disinfectants (D) used to reduce pathogen concentrations in wastewater and the disinfection by-products (DBPs) formed by the reaction of disinfection with organic compounds in the water. The question becomes one of balancing the risks from pathogenic microorganisms against those from the D-DBPs. In order to develop an optimum strategy for the protection of public health, the relative risks associated with various concentrations of pathogenic microorganisms and D-DBPs in reclaimed wastewater must be quantified.

A conceptual framework for assessing the risks associated with pathogenic microorganisms in drinking water has been proposed by Sobsey et al. (1993). An integral part of the framework is to conduct a quantitative risk assessment. There are four steps in a formal risk assessment: (1) hazard identification, (2) dose-response determination, (3) exposure assessment, and (4) risk characterization. The information required for each of the steps in the risk assessment process will be discussed in the following sections. Gaps in the data required for an accurate assessment of the risks associated with the presence of microorganisms in reclaimed water will also be identified.

HAZARD IDENTIFICATION

Hundreds of different types of pathogenic microorganisms (i.e., bacteria, viruses, and parasites) can be found in municipal wastewater, as discussed previously. The number and types of pathogenic microorganisms present in wastewater varies by location and over time at a given location. A variety of factors influence the pathogen content of wastewater including the incidence of disease in the population producing the wastewater, the season of year, the economic status of the population, water use patterns, and the quality of the potable water supply (Rose and Carnahan, 1992). Diseases caused by waterborne microorganisms range from mild gastroenteritis to severe illnesses such as infectious hepatitis, cholera, typhoid, and meningitis, as discussed earlier.

The microorganisms of concern when using wastewater to artificially recharge ground water can be identified using data from waterborne disease outbreaks. The most commonly identified causative agents in waterborne outbreaks in the United States between 1971 and 1994 were *Giardia,* chemicals, and *Shigella* sp. (Table 10.1). *Giardia lamblia* caused over 18% of the illness associated with waterborne disease outbreaks. Enteric viruses (viral gastroenteritis and hepatitis A) were identified as the causative agents of disease in 8.1% of the outbreaks during this period.

In the 1980's use of untreated or inadequately treated ground water was responsible for 44% of the outbreaks that occurred in the United States (Craun, 1991). When considering outbreaks that have occurred due to the consumption of contaminated, untreated ground water, from 1971 to 1985, wastewater was most often identified as the contamination source. In ground-water systems, etiologic (disease-causing) agents were identified in only 38% of the outbreaks, with *Shigella* sp. and hepatitis A virus being the most commonly identified pathogens (Craun, 1990).

At the present time, we cannot completely identify the hazard because of our inability to identify causative agents in approximately one-half of the waterborne disease outbreaks in this country. In over one-half of the outbreaks, no etiologic agent could be identified, and the illness was simply listed as gastroenteritis of unknown etiology. However, retrospective serological studies of outbreaks of acute nonbacterial gastroenteritis from 1976 through 1980 indicated that 42% of these outbreaks (i.e., the 62% for which no etiologic agent was identified) were caused by the Norwalk virus (Kaplan et al., 1982). Thus, it has been suggested that the Norwalk virus is responsible for approximately 23% of all reported waterborne outbreaks in the United States (Keswick et al., 1985).

The difficulty in the isolation of many enteric viruses from clinical and environmental samples probably accounts for the limited number of viruses identified as causes of waterborne disease. For example, there are no standardized, routine procedures available for isolating and identifying hepatitis A and E viruses in environmental (i.e., soil and water) samples. The ability to analyze samples for *Cryptosporidium* is restricted to only a few laboratories. There are currently no methods available for culturing the Norwalk virus in the laboratory. As methods for the detection of enteric viruses and parasites have improved, so has the percentage of waterborne disease identified as having a viral or parasitic etiology.

In addition to the hazards of acute microbial disease that result in waterborne disease outbreaks, the level of endemic microbial disease associated with drinking water must be identified. In the single epidemiological study that has been conducted to determine the contribution of drinking water to endemic gastrointestinal illness, it was found that approximately one-third could be the result of consuming treated drinking water that met all water quality standards and contained no pathogens detectable by current technologies (Payment, 1991).

It has been recognized recently that exposure to microbial pathogens in drinking water may also lead to chronic health problems such as diabetes (Gerba and Rose, 1993). This association must be further investigated so that the microbial hazard can be accurately identified.

DOSE-RESPONSE DETERMINATION

When determining the dose-response relationship for microorganisms, three different responses to microbial exposure are possible: infection that remains subclinical, infection that results in clinical illness, and infection that leads to illness and subsequent death. The response used in the determination will vary based on the purpose of the risk assessment. For example, in balancing the risks between pathogenic microorganisms and D-DBPs in water, it might be desirable to use mortality as the endpoint for both cases. However, other endpoints such as number of life-years lost or "quality adjusted" life years lost may be used (Putnam and Graham, 1993). In the latter two cases, illness and mortality could be used in the determination.

Dose-response data on the ability of a microorganism to cause infection are generally obtained by exposing a group of animal or human volunteers to different doses of the microorganisms of interest. In the case of enteric viruses, human volunteers must be used because animals are not infected by the viruses of interest. Infection is determined directly by detecting the microorganisms in the fecal material or other bodily fluids, or indirectly by detecting an antibody response to the microorganism. The infective dose of several enteric microorganisms is shown in Table 10.9.

These experimentally obtained dose-response data have been analyzed to determine whether mathematical models can be used to describe them. Several models are available for describing dose-response relationships. Two models, the simple exponential and the modified exponential (beta-poisson), have generally been used to model the response of humans to enteric pathogens (Haas, 1983; Regli et al., 1991). The experimental data

TABLE 10.9. Probability of Infection for Enteric Microorganisms
(Rose and Gerba, 1991).

Microorganism	Probability of Infection from Exposure to One Organism
Campylobacter	7×10^{-3}
Salmonella	2.3×10^{-3}
Salmonella typhi	3.8×10^{-5}
Shigella	1.0×10^{-3}
Vibrio cholera classical	7×10^{-6}
Vibrio cholera El Tor	1.5×10^{-5}
Poliovirus 1	1.49×10^{-2}
Poliovirus 3	3.1×10^{-2}
Echovirus 12	1.7×10^{-2}
Rotavirus	3.1×10^{-1}
Entamoeba coli	9.1×10^{-2}
Entamoeba histolytica	2.8×10^{-1}
Giardia lamblia	1.98×10^{-2}

are fit to a particular model, enabling one to determine the value of the constants in the model. Once the values of the constants are known, the probability of infection from ingestion of any number of microorganisms can be calculated.

When morbidity or mortality are the endpoints of interest, data are generally obtained from medical and hospital records. The relationships between infection and clinical illness for several enteric viruses are shown in Table 10.10. Mortality rates for enteric viruses are given in Table 10.11.

Dose response data are generally obtained from studies of relatively small groups of healthy volunteers, thus they represent average or possibly best-case situations. Certain populations may be more at risk from exposure to a given dose of pathogens than others. For example, very young and very old individuals have a higher risk of severe illness and even death from exposure to pathogens than do other population groups. Individuals with suppressed immune systems may also be more susceptible to infection, illness and death than healthy individuals.

Another issue that has not been adequately addressed in the dose-response determination or the exposure assessment is that of aggregated (i.e., a group of several microorganisms stuck to one another) or solids-associated microorganisms. Most dose-response studies are performed using laboratory grown and purified, monodispersed pathogens. However, in the environment, most enteric viruses occur as aggregates or associated with cellular debris (Sobsey et al., 1991). During sample analysis, a virus aggregate comprised of several tens to hundreds of virus particles may be counted as only one infectious unit. Thus, the exposure to pathogens may

TABLE 10.10. Ratio of Clinical Illness to Subclinical Infections with Enteric Viruses (Gerba and Rose, 1993).

Virus	Frequency of Clinical Illness
Polio 1	0.1–1
Coxsackie	
A16	50
B2	11–50
B3	29–96
B4	30–70
B5	5–40
Echo	
Overall	50
9	15–60
18	rare–20
20	33
25	30
30	50
Hepatitis A (adults)	75
Rota	
Adults	56–60
Children	28
Astro (adults)	12.5

be higher in drinking water than is reflected by analysis of a contaminated water sample.

EXPOSURE ASSESSMENT

Assessing the exposure of an individual to pathogens in reclaimed water or ground water that has been impacted by reclaimed water is the most difficult and uncertain aspect of the risk assessment process. Exposure to pathogens in reclaimed water may occur by direct ingestion of the reclaimed water if recharge is by infiltration, or by ingestion of potable ground water that has been contaminated by the recharge process. The level of exposure as well as the pathogens of concern will depend on which route of exposure is of concern. For example, exposure to bacteria, viruses, and parasites is likely if reclaimed water in an infiltration basin is directly ingested. On the other hand, bacteria and parasites are generally removed to a greater extent than enteric viruses during infiltration through soil; thus viruses will be of greater concern when exposure is to impacted ground water.

The number and types of pathogens in reclaimed water will depend on the level of treatment the water has received, as discussed previously. As stated earlier, the concentrations of different pathogens in raw wastewater will vary from community to community, and vary over time within a given community. Reported concentrations of microorganisms in raw wastewater range from zero to several hundred thousand *Giardia* cysts per liter, one million virus particles per liter, and several tens of thousands of bacterial pathogens per liter.

In order to determine the exposure of an individual to pathogens in reclaimed water, the number of each of the pathogens of concern ingested must be known. For direct ingestion of reclaimed water, the only report on this topic used an exposure of 100 ml (Rose and Carnahan, 1992). The concentrations of the pathogens were the concentrations remaining after treatment of the water. In the case of ingestion of ground water that has been impacted by reclaimed water, an ingestion of 2 L of water per day is usually assumed. The concentration of microorganisms in the ground water is more difficult to determine, as the removal of the various microorganisms during infiltration through the soil and aquifer must be determined.

Another aspect of the exposure assessment which is relevant to microbial disease is that of secondary exposure. In other words, disease can be spread to contacts of the individuals who were directly exposed to the contaminated water. In some cases, secondary spread can result in a significant proportion of the illnesses (see Table 10.12).

RISK CHARACTERIZATION

Using the data from Tables 10.9–10.11, and assuming an exposure of 2 L per day of water containing a known concentration of pathogens,

TABLE 10.11. **Mortality Rates for Enteric Microorganisms (from Gerba and Rose, 1993).**

Microorganism	Mortality Rate (%)
Polio 1	0.90
Coxsackie	
A2	0.50
A4	0.50
A9	0.26
A16	0.12
Coxsackie B	0.59–0.94
Echo	
6	0.29
9	0.27
Hepatitis A	0.6
Rota	
Total	0.01
Hospitalized	0.12
Norwalk	0.0001
Adeno	0.01

TABLE 10.12. Secondary Attack Rates of Enteric Viruses.

Virus	Secondary Attack Rate (%)
Poliovirus	90
Hepatitis A virus	78
Coxsackievirus	76
Echovirus	40
Norwalk virus	30

Source: Data from Gerba and Rose, 1993.

the probability of infection, illness, and death from exposure to a given microorganism can be calculated. Regli et al. (1991) compared the simple exponential and modified exponential (beta-poisson) with experimental dose-response data. They found that the beta-poisson model fit the echovirus 12, poliovirus III, and rotavirus exposure data best; while the exponential model fit the poliovirus 1 data best. The annual risk of infection, disease, and mortality from exposure to three different concentrations of hepatitis A virus and rotavirus in drinking water is shown in Table 10.13.

A study in Florida examined the removal rates of enteric viruses, indicator bacteria and viruses, *Giardia, Cryptosporidium,* and helminths in a full-scale operating wastewater treatment plant (see Table 10.6) (Rose and Carnahan, 1992). The measured pathogen concentrations in the final effluent were used to calculate the risks associated with ingestion of 100 ml of the water. The results of these calculations are shown in Table 10.14.

Asano et al. (1992) used the beta-poisson model to calculate the risk associated with exposure to viruses in ground-water recharge operations. They assumed that the nearest domestic well to a groundwater recharge site could draw water that contains 50% reclaimed wastewater which has been underground for six months after percolating through 3 m of unsaturated soil. They also assumed that an individual consumes two liters of water per day for 70 years. Results of their calculations for poliovirus 1, echovirus 12, and poliovirus 3 are presented in Table 10.15. The results of this analysis have been questioned because of the choice of dose-response models used by these investigators. The equation used to calculate the fraction of viruses remaining after infiltration through soil must also be carefully examined to determine its applicability to a variety of soil types, pathogens, and hydrogeologic and environmental conditions.

Tanaka et al. (1998) calculated the risks associated with exposure to reclaimed water produced by several treatment plants in California. In this risk assessment, measured concentrations of enteroviruses in unchlorinated secondary effluent were used as the starting point. Full tertiary treatment (consisting of coagulation, flocculation, sedimentation, filtration, and disinfection) was assumed to reduce the virus concentration by 5.2 logs from

TABLE 10.13. Annual Risk to Adults from Exposure to Hepatitis A Virus and Rotavirus in Water.

| | Concentration (pfu/100 L) | Daily Exposure (pfu/2 L) | Infection (per 100) | Annual Risk | |
				Illness (per 100)	Mortality (per 1 million)
Rotavirus	1	0.02	98	49	49
	0.1	0.002	34	17	17
	0.01	0.0002	4	2	2
Hepatitis A virus	1	0.02	98	74	4420
	0.1	0.002	34	25	1530
	0.01	0.0002	4	3	184

TABLE 10.14. Probability of Infection from Accidental Ingestion of 100 ml Reclaimed Water.

Concentration in Treated Water per 100 L	Exposure per 100 ml	Estimated Risk of Infection to Exposed Population		
		Rotavirus Model	Echovirus Model	Giardia Model
Viruses:				
0.01 pfu	1.0×10^{-5}	6.2×10^{-6}	2.0×10^{-3}	
0.13 pfu	1.3×10^{-4}	6.0×10^{-5}	2.7×10^{-7}	
Giardia:				
0.49 cysts	4.9×10^{-4}			9.8×10^{-6}
0.89 cysts	8.9×10^{-4}			1.88×10^{-5}
1.67 cysts	1.77×10^{-3}			3.3×10^{-5}
3.3 cysts	3.3×10^{-3}			6.6×10^{-5}
Cryptosporidium:				
0.75 oocysts	7.5×10^{-4}			1.5×10^{-5}
5.35 oocysts	5.35×10^{-3}			1.1×10^{-4}

Source: Rose and Carnahan, 1992.

478

TABLE 10.15. Risk of Contracting Infection from Exposure to Ground Water Recharged with Reclaimed Water.

		Risk		
	Concentration (pfu/100 l)	Daily (per trillion)	Annual (per billion)	Lifetime (per billion)
Echovirus 12	1	1.5	0.5	37
	111	162	59	4130
Poliovirus 1	1	0.1	0.05	3.4
	111	15	5.4	378
Poliovirus 3	1	0.6	0.2	14
	111	62	23	1590

Source: Asano et al., 1992.

the secondary effluent concentration. Direct chlorination of the secondary effluent was assumed to reduce the virus concentration by 3.9 logs. These researchers assumed that the individual was exposed to 2 L water containing 50% reclaimed water per day. They also assumed that the rate of virus inactivation and/or removal during transport through 3 m unsaturated soil and 6 months retention in the ground water was 0.1 per day. The annual risks of virus infection from exposure to ground water recharged with various types of reclaimed water ranged from 8 per 100 million for chlorinated secondary effluent to 5.8 per 100 billion for tertiary effluent.

REFERENCES

Asano, T., L. Y. C. Leong, M. G. Rigby, and R. H. Sakaji. 1992. Evaluation of the California wastewater reclamation criteria using enteric virus monitoring data. *Wat. Sci. Tech.* 26:1513–1524.

Badawy, A. S., C. P. Gerba, and L. M. Kelley. 1985. Survival of rotavirus SA-11 on vegetables. *Food Microbiol.* 2:199–205.

Badawy, A. S., J. B. Rose, and C. P. Gerba. 1990. Comparative survival of enteric viruses and coliphage on sewage irrigated grass. *J. Environ. Sci. Health* A25:937–952.

Bagdasaryan, G. "Survival of Viruses of the Enterovirus Group (Poliomyelitis, Echo, Coxsackie) in Soil and on Vegetables," *J. Hyg., Epidemiol., Microbiol., and Immunol.* 8:497 (1964).

Bales, R. C., S. R. Hinkle, T. W. Kroeger, K. Stocking, and C. P. Gerba. 1991. Bacteriophage adsorption during transport through porous media: chemical perturbations and reversibility. *Environ. Sci. Technol.* 25:2088–2095.

Bales, R. C., S. Li, K. M. McGuire, M. T. Yahya, and C. P. Gerba. 1993. MS-2 and poliovirus transport in porous media: hydrophobic effects and chemical perturbations. *Wat. Resour. Res.* 29:957–963.

Beuchat, L. R. 1996. Pathogenic microorganisms associated with fresh produce. *J. of Food Protection.* 59:204–216.

Bitton, G. 1980. *Introduction to Environmental Virology.* New York: John Wiley & Sons.

Bitton, G. 1994. *Wastewater microbiology.* Wiley-Liss, New York.

Bitton, G. and R. W. Harvey. 1992. Transport of pathogens through soils and aquifers. In: *Environmental microbiology,* R. Mitchell, ed., Wiley-Liss, Inc., New York, pp. 103–124.

Bitton, G., N. Masterson, and G. E. Gifford. "Effect of a Secondary Treated Effluent on the Movement of Viruses Through a Cypress Dome Soil," *J. Environ. Qual.* 5:370 (1976).

Bitton, G., O. C. Pancorbo, and S. R. Farrah. "Virus Transport and Survival After Land Application of Sewage Sludge," *Appl. Environ. Microbiol.* 47:905–909 (1984).

Bixby, R. L. and D. J. O'Brien. "Influence of Fulvic Acid on Bacteriophage Adsorption and Complexation in Soil," *Appl. Environ. Microbiol.* 38:840 (1979).

Burge, W. D. and N. K. Enkiri. "Virus Adsorption by Five Soils," *J. Environ. Qual.* 7:73 (1978).

Burnet, F. M. and M. McKie. "Balanced Salt Action as Manifested in Bacteriophage Phenomena," *Austral. J. Exp. Biol. Med. Sci.* 7:183 (1930).

Campbell, A. T., L. J. Robertson, M. R. Snowball, and H. V. Smith. 1995. Inactivation of oocysts of *Cryptosporidium parvum* by ultraviolet irradiation. *Water Res.* 29:2583–2586.

Casson, L. W., C. A. Sorber, J. L. Sykora, P. D. Cavaghan, M. A. Shapiro, and W. Jakubowski. 1990. *Giardia* in wastewater—Effect of treatment. *J. Water Pollut. Control Fed.* 62:670–675.

Cliver, D. O. and J. E. Herrmann. "Proteolytic and Microbial Inactivation of Enteroviruses," *Water Res.* 6:797 (1972).

Cords, C. E., C. G. James, and L. C. McLaren. "Alteration of Capsid Proteins of Coxsackievirus A13 by Low Ionic Conditions," *J. Virol.* 15:244 (1975).

Craun, G. F. 1990. *Methods for investigation and prevention of waterborne disease outbreaks,* U.S. Environmental Protection Agency, Office of Research and Development, Report No. EPA-600/1-90/005a, September, 1990, 22 pp.

Craun, G. F. 1991. Causes of waterborne outbreaks in the United States, *Water Sci. Technol.,* 24:17–20.

Dimmock, N. J. "Differences Between the Thermal Inactivation of Picornaviruses at 'High' and 'Low' Temperatures," *Virol.* 31:338 (1967).

Djuretic, T., P. G. Wall, M. J. Ryan, H. S. Evans, G. K. Adak, and J. M. Cowden. 1996. General outbreaks of infectious intestinal diseases in England and Wales 1992 to 1994. *Communicable Disease Report,* Review No. 4. 6:R57–R68.

Duboise, S. M., B. E. Moore, and P. B. Sagik. 1976. Poliovirus survival and movement in a sandy forest soil. *Appl. Environ. Microbiol.* 31:536–543.

Dulbecco, R. and H. S. Ginsberg. *Virology* (Philadelphia, PA: Harper and Row, 1980), p. 1112.

Enriquez, V., J. B. Rose, C. E. Enriquez, and C. P. Gerba. 1995. Occurrence of *Cryptosporidium* and *Giardia* in secondary and tertiary wastewater effluents. *In: Protozoan Parasites and Water.* W. B. Betts, D. Casemore, C. Fricker, H. Smith, and J. Watkins, Eds. pp. 84–86. Royal Society of Chemistry, Cambridge, UK.

Feachem, R. G., D. J. Bradley, H. Garelick, and D. D. Mara. 1983. *Sanitation and disease: health aspects of excreta and wastewater management.* John Wiley & Sons, New York.

Flewett, T. H. 1982. Clinical features of rotavirus infections. In: *Virus infections of the gastrointestinal tract.* D. A. J. Tyrell and A. Z. Kapikian, eds. Marcel Dekker, New York.

Foreman, T. L., G. Nuss, J. Bloomquist, and G. Magnuson. 1993. Results of a one-year rapid infiltration/extraction (RIX) demonstration project for tertiary filtration. *Proc. Water Environment Federation 66th Annual Conference,* pp. 21–36.

Fox, J. C. and P. R. Fitzgerald. 1979. The presence of *Giardia lamblia* cysts in sewage and sewage sludges from the Chicago area. In: *Waterborne transmission of giardiasis.* EPA-600/9-79-001.

Fukada, T., M. Hoshino, H. Endo, M. Mutai, and M. Shirota. "Photodynamic Antiviral Substance Extracted from *Chlorella* Cells," *Appl. Environ. Microbiol.* 16:1809 (1968).

Funderburg, S. W., B. E. Moore, C. A. Sorber, and B. P. Sagik. 1978. Survival of poliovirus in model wastewater holding pond. *Prog. Water Technol.* 10:619–629.

Gerba, C. P. 1982. Unpublished data.

Gerba, C. P. "Virus Survival and Transport in Groundwater," *Dev. Indust. Microbiol.* 24:247 (1983).

Gerba, C. P. "Applied and Theoretical Aspects of Virus Adsorption to Surfaces," *Adv. Appl. Microbiol.* 30:133 (1984).

Gerba, C. P. and G. Bitton. "Microbial Pollutants: Their Survival and Transport Pattern to Groundwater," in *Groundwater Pollution Microbiology,* G. Bitton and C. P. Gerba, Eds. (New York: John Wiley & Sons, 1984), Chap. 4.

Gerba, C. P. and S. M. Goyal. 1985. Pathogen removal from wastewater during groundwater recharge. In: *Artificial recharge of groundwater,* T. Asano, ed. Butterworth Publishers, Boston, pp. 283–317.

Gerba, C. P., S. M. Goyal, I. Cech, and G. F. Bogdan. "Quantitative Assessment of the Adsorptive Behavior of Viruses to Soils," *Environ. Sci. Technol.* 15:940 (1981).

Gerba, C. P. and J. C. Lance. "Poliovirus Removal from Primary and Secondary Sewage Effluent by Soil Filtration," *Appl. Environ. Microbiol.* 36:247 (1978).

Gerba, C. P., D. K. Powelson, M. T. Yahya, L. G. Wilson, and G. L. Amy. 1991. Fate of viruses in treated sewage effluent during soil aquifer treatment designed for wastewater reclamation and reuse. *Wat. Sci. Tech.* 24:95–102.

Gerba, C. P., M. D. Sobsey, C. Wallis, and J. L. Melnick. 1975. Factors influencing the adsorption of poliovirus onto activated carbon in wastewater. *Environ. Sci. Technol.* 9:727–731.

Gerba, C. P. and J. B. Rose. 1993. Estimating viral disease risk from drinking water. In: *Comparative environmental risk assessment.* C. R. Cothern, ed. Lewis Publishers, Boca Raton, FL. pp. 117–135.

Gerba, C. P., and J. B. Rose. 1996. Quantitative microbial risk assessment for reclaimed wastewater. *Proceedings Water Tech.* 1996. pp. 254–260. Australian Water and Wastewater Assoc. Inc., Artarmon, NSW.

Gerba, C. P. and G. E. Schaiberger. "Effect of Particulates on Virus Survival in Seawater," *J. Water Poll. Contr. Fed.* 47:93 (1975).

Gerba, C. P., S. N. Singh, and J. B. Rose. 1984. Waterborne gastroenteritis and viral hepatitis. *CRC Crit. Rev. Environ. Contr.* 15:213–268.

Gersburg, R. M., R. A. Gerhart, and M. Ives. 1989. Pathogen removal in constructed wetlands. In: *Constructed Wetlands for Wastewater Treatment: Municipal, Industrial, and Agricultural.* D. A. Hammer, Ed. pp. 431–445. Lewis Publishers, Chelsea, MI.

Goyal, S. M. and C. P. Gerba. "Comparative Adsorption of Human Enteroviruses, Simian Rotavirus, and Selected Bacteriophages to Soils," *Appl. Environ. Microbiol.* 38:241 (1979).

Goyal, S. M., B. H. Keswick, and C. P. Gerba. 1984. Viruses in groundwater beneath sewage irrigated cropland. *Wat. Res.* 18:299–302.

Grimason, A. M., H. V. Smith, W. N. Thitai, P. G. Smith, M. H. Jackson, and R. W. A.

Girdwood. 1992. Occurrence and removal of *Cryptosporidium* spp. oocysts and *Giardia* cysts in Kenyan waste stabilization ponds. *Water Sci. Technol.* 27:97–104.

Haas, C. N. 1983. Wastewater disinfection and infectious disease risks. *Crit. Rev. Environ. Contr.* 17:1–20.

Herwaldt, B. L., G. F. Craun, S. L. Stokes, and D. D. Juranek. 1992. Outbreaks of waterborne disease in the United States: 1989–1990. *J. Amer. Water Works Assoc.,* 84:129–135.

Hoff, J. C. 1986. *Inactivation of Microbial Agents by Chemical Disinfectants.* EPA-600/2-86/067. U.S. EPA, Water Engineering Research Laboratory, Cincinnati, OH.

Hoff, J. C., and E. W. Akin. 1986. Microbial resistance to disinfectants: Mechanisms and significance. *Environ. Health Perspect.* 69:7–13.

Holler, C. 1988. Quantitative and qualitative investigations of *Campylobacter* in a sewage treatment plant. *Zbl. Bakt. Hyg. B.* 185:326–339.

Hui, Y. H., J. R. Gorham, K. D. Murell, and D. O. Cliver. 1994. *Foodborne disease handbook. Vol. 2,* Marcel-Dekker, Inc., NY.

Hurst, C. J. "Influence of Aerobic Microorganisms upon Virus Survival in Soil," *Can. J. Microbiol.* 34:696–699 (1988).

Hurst, C. J. 1989. Fate of viruses during wastewater sludge treatment processes. *CRC Crit. Rev. Environ. Contr.* 18:317–343.

Hurst, C. J., C. P. Gerba, and I. Cech. 1980a. Effects of environmental variables and soil characteristics on virus survival in soil. *Appl. Environ. Microbiol.* 40:1067–1079.

Hurst, C. J., C. P. Gerba, J. C. Lance, and R. C. Rice. "Survival of Enteroviruses in Rapid-Infiltration Basins During the Land Application of Wastewater," *Appl. Environ. Microbiol.* 40:192 (1980b).

Irving, L. G. 1982. Viruses in wastewater effluents. In: *Viruses and disinfection of water and wastewater.* M. Butler, A. Medlen, and R. Morris, eds. University of Surrey, United Kingdom.

Jakubowski, W. and T. H. Ericksen. 1979. Methods for detection of *Giardia* cysts in water supplies. In: *Waterborne transmission of giardiasis.* EPA-600/9-79-001.

Jakubowski, W. 1984. Detection of *Giardia* cysts in drinking water: state of the art. In: *Giardia and Giardiasis, biology, pathogenesis, and epidemiology.* Plenum Press, New York.

Jansons, J., L. W. Edmonds, B. Speight, and M. R. Bucens. 1989. Movement of viruses after artificial recharge. *Wat. Res.* 23:293–299.

Jansons, J., L. W. Edmonds, B. Speight, and M. R. Bucens. 1989. Survival of viruses in groundwater. *Wat. Res.* 23:301–306.

Jin, Y., M. V. Yates, S. S. Thompson, and W. A. Jury. 1997. Sorption of viruses during flow through saturated sand columns. *Environ. Sci. Tech.* 31:548–555.

Jorgensen, P. H. "Examination of the Penetration of Enteric Viruses in Soils Under Simulated Conditions in the Laboratory," *Wat. Sci. Tech.,* 17:197–199 (1985).

Jorgensen, P. H. and E. Lund. "Detection and Stability of Enteric Viruses in Sludge, Soil, and Groundwater," *Wat. Sci. Tech.* 17:185–195 (1985).

Kadlec, R. H., and R. L. Knight. 1995. *Treatment Wetlands.* Lewis Publishers, Boca Raton, FL.

Kapikian, A. Z. and R. M. Chanock. Norwalk group of viruses. In: *Virology.* B. N. Fields, D. M. Knipe, R. M. Chanock, M. S. Hirsh, J. L. Melnick, T. P. Monath, and B. Roizman, eds. pp. 671–684. Raven Press, New York.

Kaplan, J. E., G. W. Gary, R. C. Baron, W. Singh, L. B. Schonberger, R. Feldman, and

H. Greenberg. 1982. Epidemiology of Norwalk gastroenteritis and the role of Norwalk virus in outbreaks of acute nonbacterial gastroenteritis. *Ann. Intern. Med.* 96:756–761.

Kapuscinski, R. B. and R. Mitchell. "Processes Controlling Virus Inactivation in Coastal Waters," *Water Res.* 14:363 (1980).

Karpiscak, M. M., C. P. Gerba, P. M. Watt, K. E. Foster, and J. A. Falabi. 1996. Multispecies plant systems for wastewater quality improvements and habitat enhancement. *Water Sci. Technol.* 33:231–236.

Keswick, B. H., T. K. Satterwhite, P. C. Johnson, H. L. DuPont, S. L. Secor, J. A. Bitsura, G. W. Gary, and J. C. Hoff. 1985. Inactivation of Norwalk virus in drinking water by chlorine. *Appl. Environ. Microbiol.* 50:261–264.

Koerner, E. L. and D. A. Haws, *Long-term effects of land application of domestic wastewater: Vineland, New Jersey, rapid infiltration site,* U.S. Environmental Protection Agency pub. no. EPA-600/2-79-072, 1979.

Kott, H. and Y. Kott. 1967. Detection and viability of *Entamoeba histolytica* cysts in sewage effluents. *Sewage Works J.* 140:177–180.

Kott, Y. and N. Betzer. 1972. The fate of *Vibrio cholerae* (El Tor) in oxidation pond effluents. *Israel J. Med. Sci.* 8:1912–1922.

Kowal, N. E. 1982. *Health effects of land treatment: Microbiological report.* EPA-600/1-82-007. USEPA, Cincinnati, OH.

Kramer, M. H., B. L. Herwaldt, G. F. Craun, R. L. Calderon, and D. D. Juranek. 1996. Waterborne disease: 1993 and 1994. *J. Amer. Wat. Works Assoc.* 88:66–80.

Lance, J. C. and C. P. Gerba. "Virus Movement in Soil During Saturated and Unsaturated Flow," *Appl. Environ. Microbiol.* 47:335 (1984).

Lance, J. C., C. P. Gerba, and J. L. Melnick, Virus movement in soil columns flooded with secondary sewage effluent, *Appl. Environ. Microbiol.,* 32:520 (1976).

Lance, J. C., C. P. Gerba, and D.-S. Wang. "Comparative Movement of Different Enteroviruses in Soil Columns," *J. Environ. Qual.* 11:347 (1982).

Landry, E. F., J. M. Vaughn, and W. F. Penello. "Poliovirus Retention in 75-cm Soil Cores After Sewage and Rainwater Application," *Appl. Environ. Microbiol.* 40:1032 (1980).

Landry, E. F., J. M. Vaughn, M. Z. Thomas, and C. A. Beckwith. "Adsorption of Enteroviruses to Soil Cores and Their Subsequent Elution by Artificial Rainwater," *Appl. Environ. Microbiol.* 38:680 (1979).

Larkin, E. P., J. T. Tierney, and R. Sullivan. 1976. Persistence of virus on sewage irrigated vegetables. *J. Environ. Engr. Div., ASCE* 102:29–35.

Larkin, E. P. and A. C. Fassolitis. "Viral Heat Resistance and Infectious Nucleic Acid," *Appl. Environ. Microbiol.* 38:650 (1967).

Lefler, E. and Y. Kott. "Virus Retention and Survival in Sand," in *Virus Survival in Water and Wastewater Systems,* J. F. Malina and B. P. Sagik, Eds. (Austin, TX: Center for Research in Water Resources, 1974), p. 84.

Leong, L. Y. C. 1983. Removal and inactivation of viruses by treatment processes for potable water and wastewater—A review. *Water Sci. Tech.* 15:91–114.

Liebmann, H. 1965. Investigations about helminth ova in sewage and wastes in Japan. *Berl. Muench. Tieraerztl. Wochenschr.* 78:106–114.

Lipson, S. M. and G. Stotzky. "Effect of Bacteria on the Inactivation and Adsorption on Clay Minerals of Reovirus," *Can. J. Microbiol.* 31:730–735 (1985).

Loveland, J. P., J. N. Ryan, G. L. Amy, and R. W. Harvey. 1996. The reversibility of virus attachment to mineral surfaces. *Colloids and Surfaces* 107:205–221.

MacDonald, R. J., and A. Ernst. 1987. Disinfection efficiency and problems associated with maturation ponds. *Wat. Sci. Tech.* 19:557–567.

Madore, M. S., J. B. Rose, C. P. Gerba, M. J. Arrowood, and C. R. Sterling. 1987. Occurrence of *Cryptosporidium* in sewage effluents and selected surface waters. *J. Parasitology.* 73:702–705.

Marais, G. V. R. 1974. Fecal bacterial kinetics in stabilization ponds. *J. Sanit. Eng. Div. Am. Soc. Civ. Eng.* 100:119–139.

McKay, L. D., J. A. Cherry, R. C. Bales, M. T. Yahya, and C. P. Gerba. 1993. A field example of bacteriophage as tracers of fracture flow. *Environ. Sci. Tech.* 27:1075–1079.

Melnick, J. L. and C. P. Gerba. "The Ecology of Enteroviruses in Natural Waters," *CRC Crit. Rev. Environ. Contr.* 10:65 (1980).

Melnick, J. L. and V. Rennick. 1980. Infectivity of enterovirus as found in human stools. *J. Med. Virol.* 5:205–220.

Meng, S. M., and C. P. Gerba. 1996. Comparative inactivation of enteric adenoviruses, poliovirus and coliphages by ultraviolet irradiation. *Water Res.* 30:2665–2668.

Milard, P. S., K. F. Gensheimer, D. G. Addiss, D. M. Sosin, G. A. Beckett, A. Houck-Jankoski, and A. Hudson. 1994. An outbreak of cryptosporidiosis from fresh-pressed apple juice. *J. Amer. Med. Assoc.* 272:1592–1600.

Mill, S. W., G. P. Alabaster, D. D. Mara, H. W. Pearson, and W. N. Thitai. 1992. Efficiency of faecal bacterial removal in waste stabilization ponds in Kenya. *Water Sci. Technol.* 26:1739–1748.

Monge, R., and M. Chinchilla. 1995. Presence of *Cryptosporidium* oocysts in fresh vegetables. *J. of Food Prot.* 59:202–203.

Moore, A. C., B. L. Herwaldt, G. F. Craun, R. L. Calderon, A. K. Highsmith, and D. D. Juranek. 1994. Waterborne disease in the United States, 1991 and 1992. *J. Amer. Wat. Works Assoc.* 86:87–98.

Moore, R. S., D. H. Taylor, L. S. Sturman, M. M. Reddy, and G. W. Fuhs. "Poliovirus Adsorption by 34 Minerals and Soils," *Appl. Environ. Microbiol.* 42:963 (1981).

Morbidity and Mortality Weekly Report. 1996. Outbreaks of Cyclospora cayatanensis infection—United States, 1996. 45:549–550.

Murray, J. P. and S. J. Laband. "Degradation of Poliovirus by Adsorption onto Inorganic Surfaces," *Appl. Environ. Microbiol.* 37:480 (1979).

Murphy, W. J., Jr., O. R. Eylaz, E. L. Schmidt, and J. T. Sylverton. "Adsorption and Translocation of Mammalian Viruses by Plants. I. Survival of Mouse Encephalomyelitis and Poliomyelitis Viruses in Soil and Plant Root Environment," *Virol.* 6:612 (1958).

Naranjo, J. E., C. P. Gerba, S. M. Bradford, and J. Irwin. 1993. Virus removal by an on-site wastewater treatment and recycling system. *Water Sci. Technol.* 27:441–444.

Nasser, A. M., L. Zev, and B. Fattal. 1994. Removal of indicator microorganisms and enteroviruses from wastewater in a single cell stabilization ponds. *Inter. J. of Environ. Health Res.* 4:7–16.

Niu, M. T., L. B. Polish, B. H. Robertson, B. K. Khanna, B. A. Woodruff, C. N. Shapiro, M. A. Miller, J. D. Smith, J. K. Kedrose, M. J. Alter, and H. S. Margolis. 1992. Multistate outbreak of hepatitis A associated with frozen strawberries. *J. of Infectious Diseases.* 166:518–524.

Pancorbo, O. C., B. G. Evanshen, W. F. Campbell, S. Lambert, S. K. Curtis, and T. W. Woolley. "Infectivity and Antigenicity Reduction Rates of Human Rotavirus Strain Wa in Fresh Waters," *Appl. Environ. Microbiol.* 53:1803–1811 (1987).

Payment, P., L. Richardson, M. Edwardes, E. Franco, and J. Siemiatycki. 1991. A prospective epidemiological study of drinking water related gastrointestinal illness. *Wat. Sci. Technol.* 24:27–28.

Pickering, L. K., H. W. Kim, and C. D. Brandt. 1987. Occurrence of *Giardia lamblia* in children in day care centers. *J. Pediatr.* 104:522–526.

Pieper, A. P., J. N. Ryan, R. W. Harvey, G. L. Amy, T. H. Illangasekare, and D. W. Metge. 1997. Transport and recovery of bacteriophage PRD1 in a sand and gravel aquifer: effect of sewage-derived organic matter. *Environ. Sci. Technol.* 31:1163–1170.

Poletika, N. N., W. A. Jury, and M. V. Yates. Transport of bromide, simazine, and MS-2 coliphage in a lysimeter containing undisturbed, unsaturated soil. *Wat. Resour. Res.* 31:801–810.

Powelson, D. K. and C. P. Gerba. 1994. Virus removal from sewage effluents during saturated and unsaturated flow through soil columns. *Wat. Res.* 28:2175–2181.

Powelson, D. K., C. P. Gerba, and M. T. Yahya. 1993. Virus transport and removal in wastewater during aquifer recharge. *Wat. Res.* 27:583–590.

Powelson, D. K., J. R. Simpson, and C. P. Gerba. 1990. Virus transport and survival in saturated and unsaturated flow through soil columns. *J. Environ. Qual.* 19:396–401.

Powelson, D. K., J. R. Simpson, and C. P. Gerba. 1991. Effects of organic matter on virus transport in unsaturated flow. *Appl. Environ. Microbiol.* 57:2192–2196.

Prage, L., V. Patterson, S. Hoglund, K. Lonberg-Holm, and L. Philipson. "Structural Proteins of Adenoviruses. IV. Sequential Degradation of the Adenovirus Type 2 Virion," *Virol.* 42:341 (1970).

Putnam, S. W. and J. D. Graham. 1993. Chemicals versus microbials in drinking water: a decision sciences perspective. *J. Amer. Water Works Assoc.* 85:57–61.

Rao, V. C., T. G. Metcalf, and J. L. Melnick. 1986. Removal of pathogens during wastewater treatment. pp. 531–554. In: *Biotechnology. Vol. 8,* H. J. Rehm and G. Reed, Eds., VCH, Berlin.

Regli, S., J. B. Rose, C. N. Haas, and C. P. Gerba. 1991. Modeling the risk from *Giardia* and viruses in drinking water. *J. Amer. Water Works Assoc.* 83:76–84.

Rice, E. W., and J. C. Hoff. 1981. Inactivation of *Giardia lamblia* cysts by ultraviolet irradiation. *Appl. Environ. Microbiol.* 42:546–547.

Robertson, L. J., A. T. Campbell, and H. V. Smith. 1992. Survival of *Cryptosporidium parvum* oocysts under various environmental pressures. *Appl. Environ. Microbiol.* 58:3494–3500.

Robertson, L. J., H. V. Smith, and C. A. Paton. 1995. Occurrence of *Giardia* cysts and *Cryptosporidium* oocysts in sewage influent and six sewage treatment plants in Scotland and prevalence of cryptosporidiosis and giardiasis diagnosed in the communities by those plants. In: *Protozoan Parasites and Water.* W. B. Betts, D. Casemore, C. Fricker, H. Smith, and J. Watkins, Eds. pp. 47–49. Royal Society of Chemistry, Cambridge, UK.

Rose, J. B. and R. P. Carnahan. 1992. Pathogen removal by full scale wastewater treatment. Report to Department of Environmental Regulation, State of Florida.

Rose, J. B. and C. P. Gerba. 1991. Use of risk assessment for development of microbial standards. *Wat. Sci. Technol.* 23:29–34.

Rosenblum, L. S., I. R. Mirkin, D. T. Allen, S. Safford, and S. C. Hadler. 1990. A multifocal outbreak of hepatitis A traced to commercially distributed lettuce. *AJPH.* 80:1975–1079.

Rowan, W. B. and A. L. Gram. 1959. Quantitative recovery of helminth eggs from relatively large samples of feces and sewage. *J. Parasitol.* 45:615–623.

Rudolfs, W., L. I. Frank, and R. A. Ragotskie. 1950. Literature review on the occurrence and survival of enteric pathogenic and relative organisms in soil, water, sewage, and sludges, and on vegetation. *Sewage Indust. Wastes* 22:1261–1281.

Sagik, B. P., B. E. Moore, and G. A. Sorber. "Infectious Disease Potential of Land Application of Wastewater," in *State of Knowledge of Land Treatment of Wastewater,* H. L. McKim, Ed. (Hanover, NH: U.S. Army Corps of Engineers Cold Regions Research and Engineering Laboratory, 1978), p. 35.

Salo, R. J. and D. O. Cliver. "Inactivation of Enteroviruses by Ascorbic Acid and Sodium Bisulfite," *Appl. Environ. Microbiol.* 36:68 (1978).

Saqqar, M. M. and M. B. Pescod. 1992. Modelling coliform reduction in wastewater stabilization ponds. *Water Sci. Technol.* 26:1667–1677.

Sarikaya, H. Z. and A. M. Saarci, 1987. Bacterial die-off in waste stabilization ponds. *J. Environ. Eng. Div. ASCE* 113:366–382.

Sattar, S. A. "Virus Survival in Receiving Waters," in *Viruses and Wastewater Treatment,* M. Goddard and M. Butler, Eds. (New York: Pergamon Press, 1981) p. 91.

Schaub, S. A. and B. P. Sagik. "Association of Enteroviruses with Natural and Artificially Introduced Colloidal Solids in Water and Infectivity of Solids-Associated Virions," *Appl. Microbiol.* 30:212 (1975).

Schaub, S. A. and C. A. Sorber, Virus and bacteria removal from wastewater by rapid infiltration through soil, *Appl. Environ. Microbiol.,* 33, 609, 1977.

Scheuerman, P. R., G. Bitton, A. R. Overman, and G. E. Gifford. "Transport of Viruses Through Organic Soils and Sediments," *J. Environ. Engr. Div., Amer. Soc. Civ. Engr.* 105:629 (1979).

Scholz, E., U. Heinricy, and B. Flehmig. "Acid Stability of Hepatitis A Virus," *J. Gen. Virol.* 70:2481–2485 (1989).

Schwartzbrod, J., J. L. Stien, K. Bouhoum, and B. Baleux. 1989. Impact of wastewater treatment on helminth eggs. *Water Sci. Technol.* 21:295–297.

Skilton, H. and D. Wheeler. "Bacteriophage tracer experiments in groundwater," *J. Appl. Bacteriol.* 65:387–395 (1988).

Skilton, H. and D. Wheeler. "The Application of Bacteriophage as Tracers of Chalk Aquifer Systems," *J. Appl. Bacteriol.* 66:549–557 (1989).

Sobsey, M. D. "Transport and Fate of Viruses in Soils," in *Microbial Health Considerations of Soil Disposal of Domestic Wastewaters,* U.S. Environmental Protection Agency, Cincinnati, Ohio, 1983.

Sobsey, M. D., C. H. Dean, M. E. Knuckles, and R. A. Wagner. "Interactions and Survival of Enteric Viruses in Soil Materials," *Appl. Environ. Microbiol.* 40:92 (1980).

Sobsey, M. D., A. P. Dufour, C. P. Gerba, M. W. LeChevallier, and P. Payment. 1993. Using a conceptual framework for assessing risks to health from microbes in drinking water. *J. Amer. Wat. Works Assoc.* 85:44–52.

Sobsey, M. D., T. Fujii, and R. M. Hall. 1991. Inactivation of cell-associated and dispersed hepatitis A virus in water. *J. Amer. Wat. Works Assoc.* 83:64–67.

Sobsey, M. D., P. A. Shields, F. H. Hauchman, R. L. Hazard, and L. W. Caton. "Survival and Transport of Hepatitis A Virus in Soils, Groundwater, and Wastewater," *Water Sci. and Technol.* 18:97–106 (1986).

Tanaka, H., T. Asano, E. D. Schroeder, and G. Tchobanoglous. 1998. Estimating the safety of wastewater reclamation and reuse using enteric virus monitoring data. *Water Environment Research,* Vol 70, No. 1, pp. 39–51, January/February 1998.

Taylor, D. H., R. S. Moore, and L. S. Sturman. "Influence of pH and Electrolyte Composition on Adsorption of Poliovirus by Soils and Minerals," *Appl. Environ. Microbiol.* 42:976 (1981).

Thurman, R. B. and C. P. Gerba. "Protecting Groundwater From Viral contamination by Soil Modification," *J. Environ. Sci. Hlth.* A22:369–388 (1987).

Tyrrell, S. A., S. R. Rippey, and W. D. Watkins. 1995. Inactivation of bacterial and viral indicators in secondary sewage effluents, using chlorine and ozone. *Water Res.* 29:2483–2490.

U.S. EPA. 1992. *United States Environmental Protection Agency. Guidelines for water reuse.* EPA/625/R-92/004. Washington, D.C.

Vaughn, J. M. and E. F. Landry. 1983. Viruses in soils and groundwater. In: *Viral pollution of the environment.* G. Berg, ed. CRC Press, Boca Raton, FL., pp. 163–210.

Vaughn, J. M. and E. F. Landry, Data Report: An Assessment of the Occurrence of Human Viruses in Long Island Aquatic Systems, Department of Energy and Environment, Brookhaven National Laboratory, Upton, New York, 1977.

Vaughn, J. M., E. F. Landry, L. J. Baranosky, C. A. Beckwith, M. C. Dahl, and N. C. Delihas, Survey of human virus occurrence in wastewater-recharged groundwater on Long Island, *Appl. Environ. Microbiol.,* 36:47 (1978).

Vaughn, J. M., E. F. Landry, C. A. Beckwith, and M. Z. Thomas, Virus removal during groundwater recharge: effects of infiltration rate on adsorption of poliovirus to soil, *Appl. Environ. Microbiol.,* 41:139 (1981).

Wallis, C. and J. L. Melnick. "Stabilization of Poliovirus by Cations," *Tex. Rep. Biol. Med.* 19:683 (1961).

Wang, W.-L. L. and S. G. Dunlop. 1954. Animal parasites in sewage and irrigation water. *Sewage. Ind. Wastes* 26:1020–1025.

Ward, B. K. and L. G. Irving. 1987. Virus survival on vegetables spray-irrigated with wastewater. *Wat. Res.* 21:57–63.

Wellings, F. M., A. C. Lewis, C. W. Mountain, and L. V. Pierce. "Demonstration of Viruses in Groundwater After Effluent Discharge onto Soil," *Appl. Microbiol.* 29:751 (1975).

World Health Organization. 1989. *Guidelines for the safe use of wastewater and excreta in agriculture and aquaculture.* Geneva, Switzerland.

Yahya, M. T., C. B. Cluff, and C. P. Gerba. 1993. Virus removal by slow sand filtration and nanofiltration. *Water Sci. Technol.* 27:445–448.

Yahya, M. T., L. Galsomies, C. P. Gerba, and R. C. Bales. 1993. Survival of bacteriophages MS-2 and PRD-1 in ground water. *Wat. Sci. Tech.* 27:409–412.

Yanko, W. 1994. Soil removal of male-specific coliphage during groundwater recharge with reclaimed water in Montebello Forebay recharge areas. Progress report prepared by County Sanitation Districts of Los Angeles County, Whittier, CA.

Yates, M. V. 1994. Monitoring concerns and procedures for human health effects. In: *Wastewater Reuse for Golf Course Irrigation.* pp. 143–171. CRC Press, Boca Raton, FL.

Yates, M. V., L. D. Stetzenbach, C. P. Gerba, and L. M. Kelley. 1990. The effect of indigenous bacteria on virus survival in ground water. *J. Environ. Sci. Eng.* A25:81–100.

Yates, M. V., C. P. Gerba, and L. M. Kelley. "Virus Persistence in Groundwater," *Appl. Environ. Microbiol.* 49:778 (1985).

Yates, M. V. and S. R. Yates. 1988. Modeling microbial fate in the subsurface environment. *CRC Crit. Rev. Environ. Contr.* 17:307–344.

Yeager, J. G. and R. T. O'Brien. "Enterovirus Inactivation in Soil," *Appl. Environ. Microbiol.* 38:694 (1979a).

Yeager, J. G. and R. T. O'Brien. "Structural Changes Associated with Poliovirus Inactivation in Soil," *Appl. Environ. Microbiol.* 38:702 (1979b).

Young, D. C. and D. G. Sharp. "Poliovirus Aggregates and Their Survival in Water," *Appl. Environ. Microbiol.* 33:168 (1977).

CHAPTER 11

Infectious Disease Concerns in Wastewater Reuse

INTRODUCTION

THE treatment of domestic wastewater for reuse results in two distinct products, reclaimed water and biosolids (sludge). Because the source of these products is sanitary waste they have the potential of being contaminated by a variety of pathogenic microorganisms. The use of these products in a manner that exposes humans, either through such activities as irrigation, indirect potable reuse or the application of sludges to land, has the potential for the transmission of disease and is of concern to public health authorities. In order to understand the public health implications of reclamation and reuse it is essential that one has some understanding of the disease microorganisms involved and the variables associated with infectious disease transmission. The control of these risks requires waste treatment processes that are effective in pathogen reduction as well as methods of monitoring for the sanitary quality of the reuse products.

INFECTIOUS DISEASE CONSIDERATIONS

In order to transmit infectious disease the infectious agent must be present and in numbers adequate for the infection of an exposed and

Robert C. Cooper, Professor Emeritus, School of Public Health, University of California at Berkeley, Berkely, CA 94720, and Vice President, Bio Vir Laboratories, Inc., 685 Stone Rd., Benicia, CA 94510; Adam W. Olivieri, Vice President, Eisenberg, Olivieri & Associates, 1410 Jackson St., Oakland, CA 94612.

489

TABLE 11.1. **Examples of Pathogens Associated with Wastewater and Solids.**

Class of Pathogen	Example	Disease
Bacteria	*Salmonella typhi*	Typhoid fever
	Salmonella sp.	Gastroenteritis
	Shigella sp.	Shigellosis/bacillary dysentery
	Vibrio sp.	cholera
	Escherichia coli	Gastroenteritis
	Campylobacter jejuni	Gastroenteritis
Virus	Hepatitis A	Infectious hepatitis
	Poliovirus	Poliomyelitis
	Echovirus	Enteric infection
	Coxsackievirus	Enteric infection
	Adenovirus	Conjunctivitis
	Rotavirus	Gastroenteritis
	Norwalk virus	Gastroenteritis
Parasites	*Entamoeba hystolytica*	Amoebic dysentery
	Giardia lamblia	Giardiasis
	Crysptosporidium parvum	Cryptosporidiosis
	Tanaea sp.	Tape worm infection
	Ascaris lumbercoides	Round worm infection

susceptible individual. As stated above, domestic sewage and associated solids can be contaminated with any microbial agent that can enter the sewer. Understandably, the most prevalent agents are those whose source is the feces of infected individuals. These agents include bacterial, viral, protozoan and metazoan pathogens. Examples of microbial pathogens that might be found in untreated sewage are listed in Table 11.1. The number and types of pathogens present in untreated wastewater is a function of the infectious disease prevalence in the community from which the waste is derived. In communities with high sanitary standards the occurrence of typhoid fever or cholera is extremely rare and the occurrence of the agents of these disease in wastewater would be equally as rare. The occurrence of salmonellosis, shigellosis, campylobacteriosis and *E. coli* associated enteritis is much more common in most communities and, therefore, the bacteria associated with these diseases are much more likely to be present in wastewater.

There are well over 100 strains of enteric virus that can be endemic in a community and which, in most cases, are shed dependent on the season of the year and the age distribution of the population. Poliovirus is common in wastewater because of live vaccine immunization programs. Infectious hepatitis is one of the more severe of the viral disease agents apt to be present in wastewater. An estimated 23,112 cases were reported to have occurred in the United States in 1992, of which 8 percent are thought to be associated with food or water [1].

Norwalk agent and rotavirus infections are considered to be quite common in most populations [2]. The Centers for Communicable Disease and Prevention determined that between 1976 and 1980, 42 percent of 74 non-bacterial gastroenteritis outbreaks, associated with a variety of sources including water were attributable to Norwalk agent. They concluded that although there are a great variety of enteric viruses present in stool only a few serotypes cause most outbreaks of gastroenteritis [3].

As in the case of pathogenic bacteria and viruses, there are a variety of intestinal parasites that might be found in raw sewage and associated sludge. Some examples are *Taenia* sp. (Tape worm), *Ascaris lumbercoides* (round worm) and the protozoa *Giardia lamblia* and *Cryptosporidium parvum*. The cysts and oocysts of the protozoa and the ova of the helminths are of greatest public health concern since they can remain viable for extended periods outside of their human host. At the present time there is considerable concern with waterborne outbreaks of cryptosporidiosis the most notorious being that which occurred in Milwaukee, Wisconsin, and in which an estimated 400,000 cases were reported [4]. Recently other protozoan parasites have been reported to be associated with waterborne gastroenteritis including members of the genus *Cyclosporum* and *Microsporidium*. The magnitude of the public health significance of these latter protozoa remains to be determined.

Typical concentrations of these pathogens in raw sewage and sludge are shown in Table 11.2. Wastewater and sludge treatment processes currently in use will reduce the numbers of these pathogenic microbes to various levels depending upon the degree of treatment applied; the more intimate the human contact with the effluent the more rigorous the treatment

TABLE 11.2. **Examples of Concentration of Microbial Pathogens in Raw Wastewater and Sludge.**

Microbial Agent	Raw Wastewater No./L	Sludge No./gm
Salmonella	4×10^3 MPN[a]	2×10^3 MPN
Enteric virus	3×10^4 PFU[b]	1×10^3 PFU
Giardia	2×10^2 Cysts	1×10^2 Cysts
Cryptosporidium	2×10^2 Oocysts	ND[c]
Helminths	8×10^2 Ova	3×10^1 Ova

[a]Most probable number.
[b]Plaque forming unit.
[c]No data.
Source: Data from EPA [5,6] and Mamala Bay Study [7].

TABLE 11.3. Estimate of Percent Removal of Selected Microbial Pathogens Using Conventional Treatment Processes.

Microbial Agent	Primary Treatment	Secondary Treatment	Digested Sludge
Salmonella	50	99	99
Enteric virus	70	99	15
Giardia cysts	50	75	30
Helminth ova	90	99.99	30

Adapted from EPA [5], Feachem et al. [8].

required. Typical levels of removal through undisinfected secondary are shown in Table 11.3. Primary treatment does not make a significant reduction in the numbers of bacteria, viruses and protozoan parasites, while the larger helminth ova reduction is about 90 percent. It should be kept in mind that since the various agents can be present in large numbers, such as 10,000 PFU[1] of virus per liter, a reduction of 70 to 90 percent leaves a substantial remainder of from 1,000 to 3,000 PFU per liter. In most instances wastewater reuse will require further treatment to reduce the number of microbial agents to low levels. Tertiary treatment usually includes coagulation with or with out sedimentation followed by sand filtration and disinfection, commonly chlorination. These processes normally reduce the numbers of microbes to non detectable levels. Enteric viruses, for example, can be reliably reduced by up to five orders of magnitude [9]. In the author's experience we have seen low levels of *Giardia lamblia* cysts in very large samples (5 to 16 cysts per 800 L) of chlorinated tertiary effluent; the viability of these cysts remained unconfirmed. Advanced treatment is used to further refine the chemical quality of tertiary effluent using such technology as ion exchange, activated carbon, reverse osmosis or micro and ultra filtration. Such processes will reduce the level of pathogens to even lower levels and, where monitored, have consistently been below detection limits. In the case of sludge the biosolids can be treated in a variety of ways to reduce the number of microbial and metazoan pathogens. These processes include sludge digestion followed by such further treatment as composting, heat drying, lime stabilization or direct land application with restricted human exposure.

It is clear that wastewater treatment processes are available which can reduce the concentration of pathogens to levels that are presently considered to be "safe." The important public health issue is the ability of these

[1] PFU: A plaque is a localized focus (clearing) of virus infected tissue culture cells. Enumeration of these foci is a quantitative method for determining virus concentration and reported as plaque forming units (PFU).

processes to *reliably* produce a reclaimed water or biosolids of a quality consistent with their design criteria.

An important element in infectious disease risk considerations is the concentration of the microbial agent required to elicit a response in those exposed. This response can be seen as (1) infection without illness, (2) infection resulting in clinical disease, or (3) no response. Rational microbial water quality standard development, particularly with the advent of exquisitely sensitive detection methods, will require as much dose response information as will be practical to obtain. At the present time the amount of this type of data available is quite limited and somewhat disjointed. Salmonella dose response information can be used as an example for most bacterial pathogens. Blaser and Newman [10] reviewed human salmonellosis and the associated infective dose. From epidemiological evidence and extrapolation based upon length of incubation period they concluded that the minimum disease producing dose in a human population is 1,000 salmonella per exposure. Byron [11] estimates the doses required for a 25 to 75 percent response (Illness) ranges from 100 to 1000 bacteria for *Shigella* to as high as 1,000,000 to 10,000,000 bacteria for pathogenic *Escherichia coli.*

There are limited reports on dose response relationships between human populations and enteric viruses. It has been estimated that the number of viruses required for infection in an individual can be as low as one tissue culture infective dose-50 percent $(TCID_{50})^2$ as is the case for infants exposed to poliovirus. Ward et al. [12] fed rotavirus to human adult volunteers and found a dose as low as 10 focus forming units (FFU).[3] These authors indicate that the infective dose was between 1 and 919 PFU or $TCID_{50}$ per dose. Because the infectious hepatitis virus has not been isolated feeding studies have been done using infectious fecal material; the results of which are not very quantitative. Virus dose response studies have, for the most part, used infection as the endpoint. It has been estimated that one out of 100 poliovirus infections and one out of 1,000 coxsackie virus infections result in clinical illness [13]. In adults approximately 25 percent infected with hepatitis A virus are asymptomatic and the proportion is much larger in children [14].

There is a developing literature on the dose response relationships between human subjects and pathogenic protozoans [15–17]. For example in human feeding studies in which a *Giardia* dose of from 10 to 1,000,000 cyst were ingested it was noted that as few as 10 cysts produced infection

---

[2] $TCID_{50}$: That dilution of a virus culture that will infect 50% of the tissue culture units challenged.
[3] Focus forming units (FFU) are analogous to plague forming units except the cells infected with virus (foci) are located using indirect methods such as fluorescent antibody staining.

(Giardia in stool) in the two subjects challenged. One hundred percent of those challenged with 100 or more cysts became infected. The illness to infection ratio has been reported to be between 20 and 67 percent. In the case of a recent *Cryptosporidium* feeding study human subjects ingested between 30 to 1,000,000 oocysts. Doses from 30 to 500 oocysts resulted in from 20 to 85 percent of the subjects becoming infected. Any dose of 1,000 or greater resulted in 100 percent infection. There is little data on the illness to infection ratio; however, it has been observed that 50 to 100 percent of infected children excreting oocysts were asymptomatic while in another instance the ratio was only 4 percent. In the case of helminths a single ova can be infectious for humans. The degree of infection is often associated with repeated exposure because, in most cases, the worms do not go through an entire life cycle in the human host [18].

The final consideration to be made in evaluating the health risk potential of using recycled water or biosolids is the manner in which the population is exposed. The important route of infection is by ingestion of contaminated material. The potential to ingest significant amounts of material is a function of the intimacy of the contact. For example, potable reuse via ground water recharge or planned indirect reuse would result in the most direct ingestion of the product. Other examples of exposure, in decreasing potential for ingestion, would be: the consumption of food grown in areas that are irrigated with reclaimed water or fertilized with biosolids; swimming in reclaimed water; aerosols generated by sprinkle irrigation using product water; non-swimming recreation such as boating and fishing; and, occupational exposures to wastewater such as experienced by treatment plant operators. There is a segment of the population in which the results of an exposure could be quite serious. Immune compromised individuals are at the greatest risk from exposure to the various agents that might be present in reclaimed products. These individuals are not necessarily more susceptible to infection than the general population but their infection can more readily result in serious illness. A recent outbreak of cryptosporidiosis primarily among AIDs patients and immunocompromised children was reportedly associated with a high quality drinking water supply [19]. These are special cases that might require some control strategies that go beyond the maintenance of acceptable water quality standards.

ENVIRONMENTAL MONITORING FOR PATHOGENS

The number and variety of microbial agents that might be present in domestic wastewater is considerable. The routine monitoring for all the possibilities is either impossible or impractical. The real time required to complete most of the analyses precludes their utility as a water quality control feedback tool. The solution to this problem has been the use of indicator microorganisms that would be present when potential pathogen containing materials were present. Three important criteria for the ideal indicator organism are (1) that it be present only when fecal contamination is present, (2) that it exhibits the same or greater survival characteristics as the "target" pathogen for which it is a surrogate and does not reproduce outside of the host, and (3) that its presence can be readily monitored in a timely fashion. At the present time no one indicator meets all of these ideal criteria. The following indicators; coliforms, fecal streptococci and *Clostridium perfringens* are, in that order, the three most common indicator bacteria presently in use.

Historically the coliform group of organisms has been used to indicate the presence of bacteria of fecal origin and therefore the possible presence of pathogenic microbes. Soon after the inception of the coliform as an indicator of fecal contamination it became clear that a number of bacterial members of this "total coliform" population were not of fecal origin. This situation resulted in the development of the "fecal coliform" test which is designed to enumerate thermotolerant ($45 \pm 0.5°C$) coliforms which are most apt to be of fecal origin. Even among the fecal coliform there are certain thermotolerant bacteria, such as *Klebsiella* sp., which are of questionable sanitary significance. Tests have been devised which will differentiate between *E. coli* and other "fecal" coliforms in environmental samples. Standards for drinking water quality have traditionally been based upon the total coliform count which are quite conservative since they allow no more than low numbers (<1 per 100 ml) of coliform regardless of their source. The USEPA suggest that the fecal coliform should be the standard indicator bacteria for recycled water and biosolids. Some regulatory agencies, to be conservative, require total coliform standards for recycled water. If there is some question as to the sanitary significance of the fecal coliform being measured the *E. coli* test might be applied.

The coliforms do not meet the ideal indicator criteria in a number of ways and certainly can not be used as the sole indicator of the sanitary quality of reuse products. Most importantly many of the viruses and most of the parasite cysts and ova are more resistant to removal by certain waste treatment processes, particularly disinfection, as well as demonstrating slower decay rates when exposed to the external environment.

There are some instances in which coliform regrowth might occur. This is seen most often in dechlorinated effluent. These regrowth coliforms are of questionable sanitary significance but can cause administrative problems for dischargers and regulators. Coliform regrowth has been observed in biosolids composting which, again, have questionable sanitary significance but cause the biosolids product to be out of compliance with coliform standards. In this case the sanitary significance of the regrowth coliforms is determined by monitoring for *Salmonella* [20].

Despite the negative aspects of using coliforms as a pathogen surrogate they are useful as an indicator of bacterial pathogens. The use of these indicator bacteria has been, and will continue to be, of great service to the public health community; however, coliform results must be interpreted in light of their known limitations.

Another group of intestinal bacteria, the fecal streptococci, are also used as indicators of fecal pollution. These streptococci all belong to Lansfield's group D (a serological typing scheme) and are in the feces of all warm-blooded animals. It should be noted that there is a subset of the fecal streptococci known as the enterococcus group which include those fecal streptococci that have the ability to grow: (1) at both 10 and 45°C, (2) in medium containing 6.5 percent sodium chloride, and (3) in medium at pH 9.6. This subgroup has been reported to be a more useful indicator of the sanitary quality of recreational waters than are the fecal coliforms [21,22].

There is a need to be able to determine whether the source of fecal contamination is from humans or animals. This distinction can not be made using coliforms. In the 1950s and 1960s there was considerable interest in using the ratio between fecal coliforms and fecal streptococci (FC/FS) as an indication of the source of contamination [23]. It was known that the feces of animals tended to have considerably more fecal streptococci than coliform in their scats and the reverse was true with human feces. If the FC/FS ratio was less than one the source was animal; a ratio greater than 4 was human; and, in between was inconclusive. It has been shown that the utility of this ratio is questionable because the decay rates of the various bacteria involved, whether in natural waters or in the presence of disinfectant, vary so as to change this ratio resulting in misleading conclusions [24]. At this time the use of fecal streptococci as a water quality indicator is limited to monitoring recreational waters.

Spore forming anaerobic bacilli are common to feces and *Clostridium perfringens* has been used, particularly in Britain, as an indicator of present and past fecal pollution of water. Historically American microbiologists have been unenthusiastic about the utility of this bacterium as an indicator of fecal pollution [25]. More recently there has been a renewed interest in using this bacterium as a surrogate for microbial pathogens in water and wastewater [26]. This revived interest has come about because of improved methods of detection and because of the desire to have a surrogate that is more resistant to environmental factors, including disinfection, than are the various pathogens that might be present. This latter need would be met by the resistant endospores formed by these bacteria. The presence of vegetative cells would indicate recent pollution while the presence of spores (the amount of growth recorded after heating to 60°C) would indicate past pollution. Their usefulness in monitoring recycled water is questioned by Grabow [27] who feel that although they are resistant to disinfection their pre-disinfection numbers, after other treatment, are reduced to low levels such that detection becomes difficult. This latter objection may be overcome with newer membrane filter detection methods. At this time there is little data available on their application to either water or biosolids reclamation.

As stated previously there are a great variety of animal viruses that may be present in wastewater and sludge. Monitoring for the presence of all of these viruses is impossible and, as bacterial indicators fall short, a surrogate virus indicator would be most useful. Any one of the animal viruses that might be selected to act as an indicator would not meet two of the criteria for an acceptable surrogate. The animal viruses are shed intermittently and therefore the chosen indicator may well be absent when other viruses are present and their low numbers in wastewater makes their isolation and enumeration tedious, expensive and of a duration that limits timely detection of contamination.

Coliphage have often been suggested as a possible surrogate for animal viruses. They are always present in raw sewage and in reasonably large numbers. Their detection is much simpler and less costly than for animal viruses and results can be attained within 24 hours. An important concern has been the question of their physical correspondence and resistance characteristics as compared with animal viruses. The F-specific RNA coliphages have a number of attributes that make them candidate enterovirus-surrogates in water and wastewater treatment process monitoring. They resemble human enteric viruses in physical characteristics, are present in sewage in numbers ranging from 100 to more than 1,000 per ml of raw sewage, demonstrate equal or greater resistance to environmental factors, including disinfection as do the human viruses and are easy to detect [28,29].

The pathogens that are the most resistant to disinfection and which present the greatest survival capacity when exposed to the environment are the protozoan cysts and helminth ova; therefore, the absence of bacterial or viral indicators in treated water or biosolids does not necessarily demonstrate the absence of enteric parasites. Protozoan cysts would be of most concern in recycled water with *Cryptosporidium* oocysts being among the most resistant to treatment and natural decay. At the present time there is no surrogate for protozoan parasites that would meet the criteria for an acceptable indicator. Perhaps, for the present, the best surrogate for this class of parasite would be the oocysts of *Cryptosporidium*. The major stumbling block in this application is the lack of an accurate and precise method of detection and the difficulty of determining viability of any oocyst that might be present. In biosolids a number of helminth ova may be present in the raw sludge the most resistant of which are the ova of the round worm, *Ascaris lumbercoides*. A method has been developed for the detection of viable *Ascaris* ova in biosolids that is reported to result in a recover efficiency of 80 percent. The method is cumbersome but at the present there seems to be no viable parasitological substitute procedure [30,31]. One drawback to this monitoring procedure is the time required, up to one month, to complete the analysis.

There is great promise in the application of molecular biological techniques to the identification and enumeration of specific pathogenic agents. There is much interest in the use of the polymer chain reaction (PCR) phenomenon which can be used to magnify small amounts of a specific nucleic acid fraction. The application of polymer chain reaction (PCR) methodology has the potential of detecting very low amounts of specific pathogen nucleic acid and by inference the presence of pathogenic microbes. While the application of these sensitive detection methods could result in more definitive monitoring, there remain questions as to the viability of the microbes detected; as to the public health significance of the presence of very low numbers of these agents in water and biosolids that are applied to land; and, as to which pathogen(s) should act as the monitoring target. With the potential of the increased detection sensitivity that these methods may provide one may find it difficult to produce materials that have ''no'' pathogens present. Thus to set a zero tolerance for these agents would be analogous to the situation encountered with chemical contaminants in which the analytical methods are becoming so exquisitely sensitive that ''zero'' concentration is impossible. The application of these sophisticated methods to environmental microbiology are still in an embryonic state and the gestation period may be a long one. Until that time dischargers and regulators will continue to rely on conventional methods. Nevertheless, these sophisticated analyses will continue to be performed and regulators will be required to interpret the results.

STANDARDS

RECYCLED WATER

Thus far it has been established that: domestic wastewater can contain a variety of pathogenic agents; the various reclamation processes of wastewater and sludge will reduce the concentration of these agents in the recycled product; and methods have been devised, using indicator organisms, and in some cases specific pathogens, to monitor for the presence or potential presence of these agents in the recycled material. The next step is the establishment of standards for acceptable levels of these infectious agents or their surrogates in the recycled product, and criteria for the type and reliability of the processes used to produce the product. Ideally these microbial standards should be based upon

TABLE 11.4. Examples of Microbial Quality Standards Used by Various States.

Exposure Route	Total Coliform per 100 ml n^a	Total Coliform Range of Values	Fecal Coliform per 100 ml n	Fecal Coliform Range of Values	Enteric Viruses per 40 L n	Enteric Viruses Range of Values
Spray Irrigation[b]	4	2.0–100	3	2.2–200	1	1[e]
Surface Irrigation[b]	2	100	9	10–1,000	0	—
Parks and Playgrounds[c]	8	2.2–100	3	10–100	1	125
Golf Courses and Open Space[d]	6	2.2–1,000	5	0–100	0	—

[a]Number of states involved out of the 13 selected.
[b]Includes food crop irrigation.
[c]Includes playgrounds.
[d]Includes cemeteries.
[e]Arizona is the only state that has a virus standard.
Adapted from Water Pollution Control Federation, 1989 [32].

known relationships between indicator numbers and the number of pathogens that might be present, and dose of agent(s) that would not adversely affect the exposed population. Unfortunately, for the most part, such relationships remain undefined and certainly would never be constant because of varying disease prevalence in the community and the demographics of the exposed population. Rather the standard values that have been established are based more upon experience and philosophy than on science. For reasons already described most standards for the presence of pathogens are based upon indicator organisms and in particular upon the coliform bacteria.

Some examples of microbial water quality standards that are applied in the United states are summarized in Table 11.4. As can be seen they are all based upon either total of fecal coliforms (the former being a more conservative value), and in one State a very conservative virus standard is in place. Crook [33] points out that, as regards water recycling, some states only require that water quality standards be followed; others require only a specified treatment process; while, in others both process and water quality criteria are imposed. In practice the standards should be based upon the process used and the expected microbiological quality of the product; the former is relied upon to reduce the infectious disease potential to a level that, from experience, is acceptable, while the latter is used to determine the reliability of the process. Using California regulations for spray irrigation as an example the treatment requirements include secondary treatment of the wastewater followed by coagulation, sedimentation, filtration and disinfection (chlorination) to produce a water that contains a median total coliform value of 2.2 per 100 ml (or less) not to exceed 23 in more than one sample in 30 days. Samples are collected daily and the median value is based on the last 7 days of sampling [34]. Through a number of studies conducted in California it has been shown that the above treatment process can reduce the number of enteric viruses by 4 to 5 orders of magnitude [9,35]. Thus if any substitute reclamation process is proposed it must be shown that it can reliably produce an effluent that meets the coliform requirement and can reduce the influent virus content by five orders. Although there is no standard for numbers of viruses in the effluent the process requirement indirectly sets a standard. A similar requirement based upon $\log_{10}$ reduction can also be applied to protozoan cysts and oocysts. The National Primary Drinking Water Regulations [36] refers to a 3 $\log_{10}$ reduction in numbers of these cysts and the same criteria might be applied to recycled water treatment processes.

On the other hand the World Health Organization (WHO) feels that such restrictive standards as those promulgated by California were not justified by the available epidemiological evidence [37]. Based upon the conclusion that the major infectious disease health risk in many developing countries are helminth diseases, then treatment processes should be used that are effective in the removal of helminth ova. A series of stabilization ponds is suggested to be the most effective means to reduce helminth ova to one or less per liter and a fecal coliform number of 1,000 MPN per 100 ml. The WHO feels that since helminth infections are more persistent, have a long latent period, low infective dose and weak immune response they are of the greatest public health concern. WHO states, " Where a significant amount of transmission occurs by other routes, as it often does with many fecal-oral infections (Categories I and II[4]) a small amount of transmission due to wastewater reuse may be of relatively minor importance" [37].

Potable reuse provides the most intimate contact of a population with recycled water. There are enough qualms, both technical and psychological, about direct potable reuse[5] to proscribe its application. However, at the present time, there is a great deal of interest in engineered indirect potable reuse[6] in which highly treated recycled water will be considered as a raw water to be further treated prior to distribution (see Chapter 12 for details of such an application in San Diego, California). In the United States, and elsewhere in

---

[4] Categories I and II include bacterial, viral and parasitical infections which are infectious upon excretion and whose transmission is directly affected by personal hygiene.

[5] Direct potable reuse is the direct distribution of a reclaimed water to the consumer commonly defined as "flange to flange."

[6] Engineered indirect potable reuse implies the intentional use of the reclaimed water as a source of raw water supplement for a drinking water supply as opposed to non-intentional indirect potable reuse such as might occur during groundwater recharge activities.

the world, there seems to be a dichotomy in the application of standards applied to recycled water for engineered indirect potable reuse. Chemical and biological standards are in place for finished drinking water supplies whose source waters can be of poorer quality than that normally reached using recycled water treatment processes, but are considered to be from natural sources. Thus, if a drinking water that comes from a ''natural'' source meets the standards it is considered safe for the consumer; if the water is from an engineered reclamation process and meets these same standards the water is not acceptable. Somehow the ''natural'' nature of the source water involves the intervention of the goddess of health, Hygeia, and the standards apply. Recycled water does not enjoy Hygeia's blessing and the same standards are not applicable. In fact recycled water that will be used as a source of *raw* water will, most probably, be held to a *finished* drinking water standards prior to further treatment. In California it is proposed that the ''further treatment'' will include mixing with indigenous water in a lake or reservoir prior to standard drinking water treatment; thus, the ''natural'' element is included as an additional barrier. Ideally, drinking water quality standards should be applicable to all waters regardless of source; however, for the present, such an ideal seems to be beyond our grasp. Prescribed treatment process trains must also be in place for indirect potable reuse, be the best available technology, and be of such variety and reliability as to afford multiple barriers to the passage of infectious agents. For example these multiple barriers could include those described previously for spray irrigation followed by activated carbon, membrane filtration (reverse osmosis for example), disinfection, dilution by mixing with indigenous water and storage time.

BIOSOLIDS

Microbiological quality standards have been developed by the USEPA for biosolids to be disposed on the land and are commonly known as the ''503'' rule [30,38]. These quality standards do not stand alone but are in concert with treatment processes or with contact restrictions. Two classes of sludge are defined, Class A and B. The former can be applied with no restrictions and the latter, because of a lesser treatment requirement, primarily sludge digestion, has restrictions placed on human and animal contact with the application site. The pathogen and treatment requirements for Class A biosolids are summarized in Table 11.5. In all cases Class A biosolids must meet either a fecal coliform level of <1,000 MPN per gram of dry solids or a Salmonella level of <3 Salmonella per 4 grams of dry solids. There are six process-monitoring alternatives that can be used to produce and classify a sludge as class A. The requirement for further pathogen testing beyond the fecal coliform or Salmonella test depend upon the treatment process used. Alternative processes 1, 2, 5 and 6 have been

TABLE 11.5.   503 Class "A" Biosolids Pathogen Requirements.[a]

Alternative	Test	Pathogen Limits
All[b]	Fecal Coliform or *Salmonella* sp.	<1,000 MPN/gram total solids <3 MPN/4 grams total solids
#1	Time and temperature	
#2	Alkaline treatment/temperature	pH >12 for at least 72 hours. Temperature >52°C for at least 12 hours. Then air dry sludge to ≥50% total solids.
#3	Enteric virus and helminth ova *prior* to pathogen treatment.	<1 PFU/4 grams total solids for virus and <1 viable ova/4 grams total solids for helminth ova
	Or If enteric virus and/or helminth ova prior to pathogen treatment are ≥1 PFU or viable ova respectively test *after* treatment.	<1 PFU/4 grams total solids for virus and <1 viable ova/4 grams total solids for helminth ova and document process operating parameters
#4	Enteric virus and helminth ova after treatment and ready to distribute, etc.	<1 PFU/4 grams total solids for virus and <1 viable ova/4 grams total solids for helminth ova
#5	Treat to Process to Further Reduce Pathogens (PFRP) requirements.	
#6	Treat *equivalent* to Process to Further Reduce Pathogens (PFRP) requirements.	

[a]This table represents a simplified version of the Class "A" alternatives and does not contain all details, requirements, or exceptions.
[b]For Class "A" biosolids salmonella and fecal coliforms densities are measured at the time the biosolid is used or disposed, sold or given away, etc.

shown to effectively remove virus and helminth ova; thus, assuming the processes are run correctly, no further microbial monitoring is required. Processes to further reduce pathogens (PFRP in options 5 and 6) are specifically spelled out and include composting, alternate heat and drying regimes and irradiation. Option 6 allows for other methods which have been shown to the regulator to be equivalent to an accepted PFRP. Options 3 and 4 require monitoring for enteric viruses and helminth ova for those processes that do not meet the process requirements of 1, 2, 5 or 6. In option 3 the demonstration that the process used can meet the bacterial, viral and helminth concentration requirements allows for the microbial monitoring to be replaced by strict monitoring of the process to assure that no variations have occurred. In option 4, which applies to sludges of unknown treatment or past history or those that have been treated by a questionable process, one cannot substitute effective operating parameters for microbiological monitoring.

Class B sludge standing can be achieved under two alternative options. The first option requires that the solids have a fecal coliform density no greater than 2 million per dry gram as determined by a defined sampling program while the second alternative requires the use of a treatment process that has been classified as one that significantly reduces Pathogens (PSRP) and requires no microbiological monitoring since these processes are known to be able to reach or exceed the coliform standard listed in Class B option 1. Examples of PSRP would include composting, anaerobic and aerobic digestion and lime stabilization all with specific operating parameters. Even though the pathogen level in these sludges have been "significantly" reduced it is assumed that some level of these agents might persist, therefore application site restriction are applied. These restrictions are all time related based upon projected decay rates of target pathogens. The length of time varies with the intimacy of the contact, for example; food crops below ground (not less than 20 months at the time of harvest); cattle grazing (one year before cattle are allowed to graze); and, public access (one year prior to use). The standards set by the EPA for pathogens are heavily governed by the use of specified treatment processes rather than by actual pathogen, or pathogen surrogate, monitoring. In most cases the microbiological testing depends upon the fecal coliform test which acts as an indicator of the reliability of the given certified process. More extensive microbial monitoring is applied when the treatment process used is of uncertain efficacy.

Vector attraction is an added public health dimension, primarily aesthetic, to the addition of biosolids to the land. Because of the putrescible nature of sludge there is the potential for the production of odors and the attraction of flies and other vermin causing a nuisance as well as a possible health hazard. Vector attraction can be managed by treatments that; reduce the volatile solids content of the sludge; reduce oxygen uptake rates (a measure of stability); aerobically compost the biosolids; maintain alkaline conditions; or, applications in which the solids are covered such as injection below ground. Compliance with the vector attraction reduction requirements should be demonstrated separately from compliance with pathogen reduction requirements. For more details on the USEPA vector attraction reduction requirements see reference [30].

EPIDEMIOLOGY

Reports on the occurrence of disease transmission associated with water reuse are related to the use of untreated sewage or effluents of questionable quality. To date there have been no recorded incidents of infectious disease transmission associated with engineered wastewater reuse or with the application of treated biosolids to land [37,39,40]. A review of the extensive literature concerning the health of sewage plant operators and maintenance personnel, who are exposed to wastewater in all stages of treatment, indicates that the occurrence of infectious disease associated with this exposure is rarely reported [41]. It would be inappropriate to extrapolate these findings to the general public health since the exposed working population is, most probably, composed of healthy and relatively young individuals which may not reflect a cross section of the general population; however, it does imply a low infectious disease risk associated with such exposure. It appears, then, that the risk of contracting infectious disease via contact with recycled water and biosolids is of such a small magnitude that traditional epidemiological methods are not sensitive enough to ''tease out'' cases that might be associated with recycled water or biosolids from the background incidence of these ailments in the community. Current exposures to environmental chemicals and pathogens are often quite low, and empirical studies can no longer produce information that is specific enough to be the sole means of assessing these risks [42]. Because of the insensitivity of epidemiological methods, risk estimators and risk managers are increasingly turning to methodologies that rely on indirect measures of risk by using analytical models for the estimation of the intensity of human exposure and the probability of an adverse human response from this exposure.

INFECTIOUS DISEASE RISK ASSESSMENT

Microbial risk assessment involves evaluating the likelihood that an adverse health effect may occur when humans are exposed to one or more potential pathogens. Most microbial risk assessments to date have used a framework originally developed for chemicals, defined by four major steps: hazard identification, dose-response identification, exposure assessment, and risk characterization. The problem with this framework is that it does not explicitly acknowledge the differences between the health effects due to chemical and microbial exposure, such as secondary spread, infection without disease and immunity.

One framework that is consistent with current views on infectious disease process is based on three components: the microbial agent, the human host, and the environment in which the infection process is mediated. Since it is the interaction of these components that produce human disease, risk is dependent on defining the following:

(1) The exposed population including demographic profile and susceptibility status

(2) The microbial characteristics including such properties as viability and virulence

(3) The environmental setting in which the exposure occurs

After the population, disease agent, and environment are defined then the corresponding health effects are described, including such issues as symptoms, severity, and the potential for secondary infection. This alternative framework, based on host/microbe interaction, makes explicit the mechanistic aspects of the infectious disease process and provides a structure from which data are gathered. Some of the data required to characterize risk may be available in the literature. It may be necessary to obtain other information, such as site specific data, through surveillance or experimentation. This data collection process is analogous to the analysis phase in the ecological risk assessment framework.

The risk characterization step uses the data obtained in the analysis phase to evaluate risk. This step includes the development of risk assessment methods that incorporate uncertainty and variability. The following is a discussion of one quantitative approach to risk characterization that (1) takes advantage of available infectious disease and environmental data defined above, and (2) explicitly acknowledges the uncertainty and variability inherent in infectious disease processes.

Because environmental risk assessment is subject to a variety of uncertainties, the process is often cast in probabilistic terms. Moreover, field data are frequently unavailable to quantify some elements of the process, and mathematical modeling is used to bridge these data gaps. The principal advantage of mathematical modeling in risk assessment applications is that it makes assumptions explicit, including structural mechanisms relating human exposure to pathogens and the public health outcome and quantitative assumptions such as the dose-response relationship. A mathematical model organizes data and assumptions in a framework leading to quantitative predictions and can be an indispensable tool for decision making. However, the model itself brings no new data or information to the process. Thus the biological significance of a model's output is completely dependent on the appropriateness and accuracy of the assumptions used to build the model.

Attempts to provide a quantitative framework for the assessment of human health risks associated with the ingestion of waterborne pathogens have generally focused on the static models that calculate the probability of individual infection or disease as a result of a single exposure event [43–46]. These models, all of the same generic form, are based on dose-response data which is used to fit a standard distribution function such as either an exponential or beta function.

A somewhat different approach to risk analysis of waterborne disease was carried out for the U.S. Army in which a specific population group was considered and the analysis was carried beyond the risk of infection to an individual by estimating the probability distribution of the number of infected/diseased people in the exposed population [47]. A key feature of the Army model was its probabilistic treatment of exposure data. In the Army model, each member of the population received a different dose, resulting in a distribution of risk over the population.

Most model structures do not provide ways to incorporate epidemiological data such as incubation period, immune status, duration of disease, and the rate of symptomatic development, or exposure data such as processes affecting the pathogen concentration. These are all factors important in the disease transmission process and knowledge of which is required to track variables such as the number of susceptible, infected, diseased, and immune within a population group.

A MICROBIAL RISK MODEL

To take advantage of available infectious disease and dose-response data, Eisenberg et al. [48] took a population perspective in the development of a mathematical model that characterizes the human disease risk of waterborne pathogen exposure. An existing dose-response model [43,44] was imbedded into an epidemiological framework, relying on a large base of literature describing the use of dynamic population models in the study of epidemics [49]. These models emphasize the importance of how the susceptible, infected, or immune status of an individual, within a defined population group, vary over time. An additional state variable was added to account for dynamics of pathogen concentration at the site of exposure.

The model is composed of five state variables, one output variable and up to 15 parameters associated with pathogen concentration and rates of "state" changes. A simplified diagram of the model is presented in Figure 11.1. Four of the state variables represent the human population, which is divided into four epidemiological groups:

- X—susceptible individuals
- Y—infectious/asymptomatic individuals
- Z—non-infectious/asymptomatic individuals
- D—infectious/symptomatic individuals

Individuals in state X are susceptible to infection. For the remaining groups, the terms infectious or non-infectious define whether or not individ-

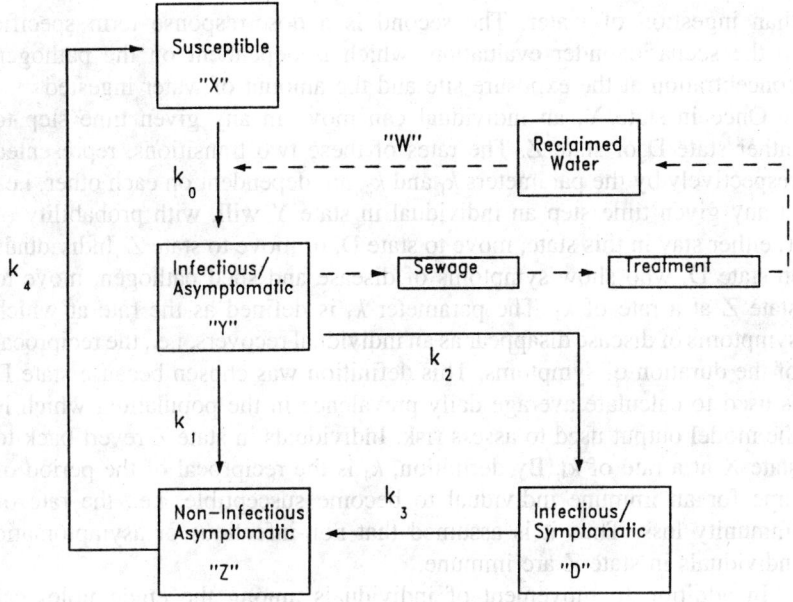

k_0 = Baseline transmission rate plus infection rate due to ingestion of "W".
k_1 = Fraction in state of "Y" who move to state "Z".
k_2 = Fraction in state of "Y" who move to state "D".
k_3 = Fraction in state of "D" who move to state "Z".
k_4 = Fraction in state of "Z" who move to state "X".
W = Concentration of pathogens in recycled water.

———————▶   Movement of Individuals
– – – – ▶   Movement of Pathogens

Figure 11.1. Simplified diagram of microbial disease risk model.

uals shed pathogens in their stool, and the terms symptomatic and asymptomatic define whether or not an individual exhibits symptoms of disease. The state variables X, Y, Z and D keep track of the population levels in each group. A fifth state variable, W, traces the concentration of pathogen in the water to which the population is exposed. The movement of individuals from one state to another and the concentration of pathogen are governed by a set of five differential equations.

The rate at which members of the population move from state X to state Y is governed by two factors. One is the background rate of infection, which accounts for non-outbreak transmission due to exposure routes other than ingestion of water. The second is a dose-response term specific to the scenario under evaluation, which is dependent on the pathogen concentration at the exposure site and the amount of water ingested.

Once in state Y, an individual can move in any given time step to either state D or state Z. The rates of these two transitions, represented respectively by the parameters k_1 and k_2, are dependent on each other, i.e., at any given time step an individual in state Y will, with probability of 1, either stay in this state, move to state D, or move to state Z. Individuals in state D, who show symptoms of disease and shed pathogen, move to state Z at a rate of k_3. The parameter k_3 is defined as the rate at which symptoms of disease disappear as an individual recovers, i.e., the reciprocal of the duration of symptoms. This definition was chosen because state D is used to calculate average daily prevalence in the population, which is the model output used to assess risk. Individuals in state Z revert back to state X at a rate of k_4. By definition, k_4 is the reciprocal of the period of time for an immune individual to become susceptible, i.e., the rate of immunity loss. Thus it is assumed that non-infectious or asymptomatic individuals in state Z are immune.

In addition to movement of individuals among the epidemiological states, the model also describes the concentration of the waterborne pathogen at the exposure site. The specifics of this portion of the model depends on the scenario of interest. For example, for a recreational swimming scenario the pathogen may arrive at the exposure site in two ways. First, individuals in state Y directly shed pathogen into the water used for swimming and second, pathogens from wastewater outfalls and non-point sources may migrate to the recreational waters.

Uncertainty Analysis

A methodological problem with such a complex model is that the high levels of uncertainty and variability inherent in environmental processes preclude use of traditional parameter estimation techniques. In general, biological systems have a high degree of variability due to both genetic and other differences between individuals and environmental factors that are explicitly modeled. Standard analytical tools, such as curve fitting techniques and sensitivity analysis, become less useful when data such as that produced from surveillance of infectious diseases is so variable and uncertain. Traditionally, a sensitivity analysis procedure involves selecting a point in the parameter space and perturbing the parameter values about this point. Unfortunately, in many biological models there is sufficient uncertainty in parameter values to make the selection of any particular parameter set about which to conduct the sensitivity analysis a questionable procedure. This is particularly true with infectious disease data.

Regional sensitivity analysis (RSA) was developed in the context of a class of problems in environmental systems analysis in which simulation models were employed but in which field data on the state of the specific system being studied was very limited [50,51]. The approach involves describing the uncertainty or variability in each model parameter by a statistical distribution function. The structure of the model and this set of parameter distributions now define an ensemble of models, one of which is considered to be the model of the "real system." By defining the essential characteristics of the system under study from a biological perspective it is possible to develop a classification algorithm which allows one to apply a variety of powerful multivariate statistical procedures to the sensitivity analysis task. The strength of the approach is that it overtly acknowledges both the uncertainly and variability in parameter values in a structured fashion that avoids the pitfalls of worst-case analysis. The RSA procedure has now been applied to a variety of problems. Some of these are: the control of DO in a river receiving treated effluent [52]; Control of systems with uncertain parameters [53]; Mechanisms of benzene carcinogenesis [54]; and Microbial risk assessment [48].

A Case Study

There is continual interest in the use of reclaimed water in California and other regions of the United States and the world. The extended California drought dramatized the need for reuse in urban areas and it is becoming clear that, even without a drought, population pressures are such that reclaimed water will be an important water supply adjunct. With this interest in the use of reclaimed water, the effectiveness of the water treatment process, including disinfection of pathogenic agents, has become a central public health issue. Regulators in charge of monitoring the prevalence of infectious diseases within the human population are in constant need of a way to estimate the risk of waterborne disease transmission. This need has become more acute with the increased use of reclaimed water.

Irvine Ranch Water District sponsored a project to explore new approaches to microbial risk assessment applicable to assessing the population risks of ingestion of reclaimed water [55,56]. The epidemiological model form was used to investigate the risk of contracting giardiasis while swimming in an impoundment filled with reclaimed water. The literature on the ecology and epidemiology of *Giardia lamblia* is fairly extensive including reports which provide data on the non-outbreak-background incidence of giardiasis in a community setting which allowed a "calibration" of a number of the biological parameters of the model [62]. Regional sensitivity analysis was used to establish a set of parameters that reproduced non-outbreak conditions. Using these "calibrated parameters" computer simulations were performed to calculate the disease risk for defined ranges of parameter uncertainty and variability.

Seven wastewater reclamation alternatives and corresponding levels of wastewater treatment were evaluated: (a) Non-Restricted (Swimming) Recreational Impoundment, (b) Golf Course Irrigation, (c) Restricted Recreational Impoundment, (d) Park Irrigation, (e) Industrial Cooling Tower, (f) Irrigation of Food Crops, and (g) Groundwater Recharge. In California, the uses selected require disinfected tertiary, disinfected secondary-2.2 (2.2 MPN/100 ml) or disinfected secondary-23 reclaimed (23 MPN/100 ml) water. Disinfected tertiary is the highest level of conventional treatment and is defined as an "essentially" virus free water suitable for all uses except food preparation and potable reuse. In addition to secondary treatment, disinfected tertiary (2.2 MPN/100 ml) requires contact/direct filtration with coagulation (if secondary effluent turbidity is greater than 5 NTU) plus 90 minute modal chlorine contact time with minimum 5 mg/L residual. Disinfected secondary-2.2 does not require filtration. Disinfected secondary-23 reclaimed water represents typical secondary treatment plant effluent widely used for restricted access golf course irrigation and landscape impoundments. The assumed removal of *Giardia lamblia* in a disinfected secondary treatment plant ranged from 0.3 to 1.5 logs of removal, while the removal in disinfected tertiary plant ranged from 1.8 to 6.3 logs of removal [57].

Recent literature and reports were reviewed to develop estimates of the amount of reclaimed water that could potentially be ingested at the above water reclamation sites. The results of the review are summarized for each alternative below. For the following scenarios, exposure is assumed to be 100% oral ingestion. Indirect ingestion through spray inhalation of irrigation or cooling tower aerosols, other exposure pathways, and factors such as microorganism attenuation are not addressed.

(1) **Non-Restricted (Swimming) Recreational Impoundment**—This water reclamation alternative is a recreational impoundment filled with disinfected tertiary reclaimed water. Human exposure occurs via incidental ingestion of water during swimming, with estimates ranging from 10 to 130 ml/day of swimming. Estimates of swimming days ranged from 1 to 40 per year, the rate for water ingestion ranged from 30 to 50 ml/hour of swimming, and the national average swimming time for children was reported as 2.6 hours/day and for adults was 0.5 hours per day [58–60].

(2) **Golf Course Irrigation**—This water reclamation alternative is a golf course irrigated with disinfected tertiary, secondary-2.2 or secondary-23 reclaimed water, depending on the degree of public

access. It was assumed that 1 ml/day for 2 days per week year round is ingested by golfers handling and cleaning golf balls [60,61]. For exposure to joggers/golfers or children playing via incidental ingestion of spray, we used the swimming ingestion rate of 50 ml/hour for 10 minutes/day (i.e., ~8 ml/day) and 61 days per year exposure (irrigation every third day for six months per year) [59].

(3) **Restricted Recreational Impoundment**—This water reclamation alternative is a restricted recreational impoundment filled with disinfected secondary-2.2 reclaimed water. Fishing and boating are allowed in this impoundment, but not swimming. Exposure occurs via incidental ingestion of spray or incidental ingestion following handling of wet objects such as a fishing line or boat tiller. A highly conservative spray exposure estimate that has been applied is the swimming exposure rate assumption of 50 ml/hour [59]. For handling of wet objects, 1 ml/day was assumed to be ingested, similar to golfers handling and cleaning golf balls at a golf course irrigated with reclaimed water, as described above. Both exposure categories were assumed to occur 4 hours/day and 14 days/year.

(4) **Park Irrigation**—This water reclamation alternative is a public park irrigated with disinfected tertiary reclaimed water. Exposure occurs to adult joggers via accidental contact with spray and to children via contact with wet grass while playing. For joggers or children playing exposed to spray, the above assumptions for golf course irrigation could be applied. For children playing on wet grass, a 1 ml/day exposure rate, analogous to handling and cleaning golf balls, and 14 days per year exposure was assumed.

(5) **Industrial Cooling Tower**—This water reclamation alternative is a cooling tower using disinfected tertiary, secondary-2.2 or secondary-23 reclaimed water. Exposure occurs via incidental ingestion of reclaimed water during routine maintenance of the cooling tower's heat exchangers and piping. Industrial or commercial cooling or air conditioning with cooling towers, evaporative condensers, or spraying that creates a mist requires disinfected tertiary reclaimed water. For exposure, we assumed the swimming ingestion rate of 50 ml/hour for 4 hours/day (i.e., 200 ml/day) and 6 days/year. This is an extremely conservative assumption, since direct immersion is highly unlikely, unless a pressurized pipeline were to burst. A more likely exposure route is ingestion of residue on hands, analogous to the golf ball handling scenario (1 ml/day). Indirect ingestion due to inhalation of mist downwind of cooling towers was not evaluated.

(6) **Irrigation of Food Crops**—This water reclamation alternative is the irrigation of food crops with disinfected tertiary reclaimed water.

Exposure occurs via daily direct consumption of food crops with 10 ml reclaimed water assumed left on the portion of the crop eaten raw [60].

(7) **Groundwater Recharge**—This water reclamation alternative is a groundwater basin recharged with 50% by volume disinfected tertiary reclaimed water. Exposure occurs when an individual consumes well water at an assumed rate of 2 L per day, equivalent to 1 L of reclaimed water (given a minimum 50% dilution).

For each reclaimed water alternative, nine sets of 6,000 simulations were performed. The nine sets consisted of the background scenario, the non-restricted (swimming) recreational impoundment with water from a pathogen-free source scenario and each of the seven water reclamation alternative scenarios. The mean and standard deviation for average daily prevalence were computed for each set and are summarized in Figure 11.2.

Review of Figure 11.2 reveals that even in the presence of substantial uncertainty, exposures in restricted recreational impoundments and from industrial cooling towers result in public health risks indistinguishable from background levels. It should be noted that the prevalences calculated are case-specific and, while the information found in the literature relating to the values of the biological parameters can be used in more general analyses, the exposure-related parameters must be selected for specific sites, populations and water reuse applications.

In order to assess which parameters were important determinants of high prevalence for each water reclamation alternative, the Kolmogorov-Smirnov (K-S) statistical test was performed. A summary of the most important parameters for each alternative is presented below.

Water Reuse Alternative	Important Parameter
Swimming Impoundment	Days of exposure
Golf Course Irrigation	Treatment efficiency
Restricted Recreational Impoundment	Duration of symptoms
Park Irrigation	Treatment efficiency
Industrial Cooling Tower	Treatment efficiency
Irrigation of Food Crops	Treatment efficiency
Groundwater Recharge	Concentration of pathogens
	Treatment efficiency
	Duration of symptoms

Examination of the above summary reveals that treatment efficiency was, overall, the most often identified important parameter for the water reclamation alternatives analyzed in this study. Fortunately, this parameter is one which is controllable and further investigation can be conducted to help resolve the uncertainty and variability. The second-most important parameter was the duration of symptoms.

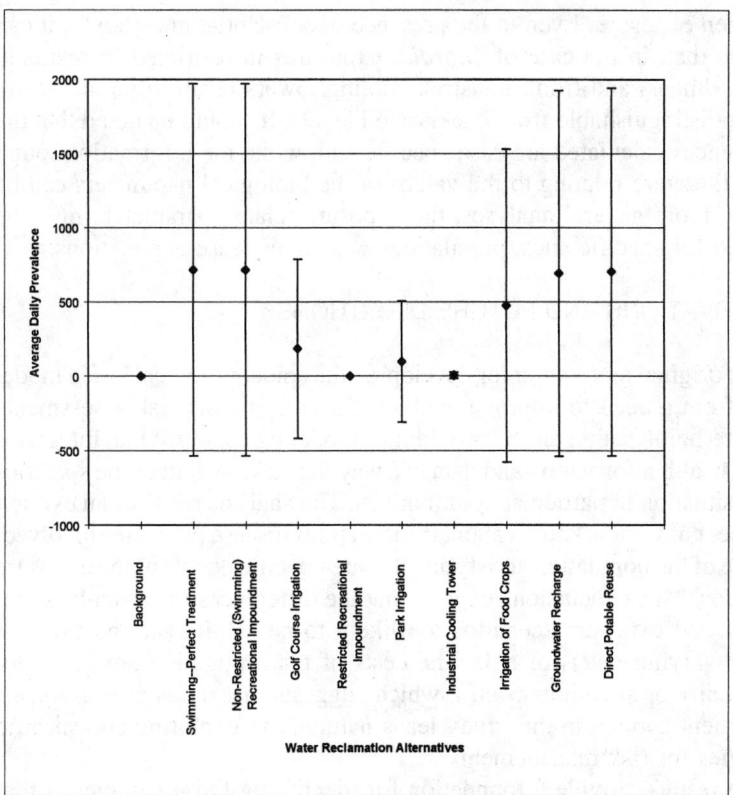

The figure above shows the results for *Giardia*. 6,000 simulations were performed for each water reclamation alternative.
For each alternative, the average daily prevalence is shown with the standard deviation indicated with error bars.

Figure 11.2. *Giardia* average daily prevalence.

The simulations forecast significantly higher prevalence of giardiasis in a community with a public swimming impoundment than in a community without such an impoundment. Filling the impoundment with tertiary treated wastewater did not increase mean prevalence levels above those obtained when the impoundment was filled with water from a pathogen-free source. However, the predicted disease prevalence is highly variable under both scenarios due to the uncertainty and variability of the model's parameter values and this may have masked differences in predicted public health impact. The results of the recreation impoundment simulations for a hypothetical closed community point towards two findings: (1) Shedding by swimmers may be an important source of pathogens in swimming areas, and (2) the levels of uncertainty and variability inherent in the system preclude a conclusive assessment of the importance of shedding by swimmers relative to treated wastewater in producing a given level of pathogen exposure. Even in the presence of substantial uncertainty, it can be seen that, in the case of *Giardia,* exposures in restricted recreational impoundments and from industrial cooling towers result in public health risks indistinguishable from background levels. It should be noted that the prevalences calculated are case-specific and, while the information found in the literature relating to the values of the biological parameters can be used in more general analyses, the exposure-related parameters must be selected for specific sites, populations and water reuse applications.

OBSERVATIONS AND FUTURE DIRECTIONS

The original motivation for developing the epidemiological based model arose from a need to improve methods for carrying out risk assessments for waterborne pathogens by providing a modeling approach that integrates public health information and data in a way that acknowledges the specifics of the situation in particular communities. The analysis requires an explicit statement of what is known about the infectious disease processes involved, the size of the population at risk, and the site-specific details of the exposure scenarios. This conclusion suggests that the differences in particular community and exposure scenarios are likely to be significant and produce widely varying levels of risk. The costs of risk mitigation are generally community or situation-specific which suggests that the approach to risk assessment applied in this study leads naturally to exploring site-specific strategies for risk management.

The results provide a foundation for identifying those parameters that tend to influence the risk assessment results as well as a general sense of the relative relationship and significance, in terms of public health risk, among various alternative uses (i.e., exposure scenarios) and background levels of risk. This information, while not community or situation-specific, can be useful in generally evaluating the relative degree of public health risk posed by different water-related exposure scenarios (i.e., water reclamation projects, recreational exposure issues, and drinking water exposure). From the risk management perspective, although risk assessment modeling using generic or site-specific input data can be valuable in providing general information pertaining to risk, the results should not be considered definitive. Risk assessment modeling should instead be considered one of many tools providing information to policy makers in the decision making process.

The benefits of this modeling approach are three fold. First, the structure of the model allows for a more realistic analysis of waterborne pathogen dynamics. Second, model simulations can be used to assess the comparative risks of various waterborne pathogens, the difference between exposure scenarios on various demographic populations and the potential effectiveness of different management strategies. Third, the analysis can directly identify dominant sources of uncertainty in the model.

While the above application of the model was successful in showing the feasibility and benefits of a new and effective approach to assessing the public health risks of waterborne pathogens, the risk assessment methodology remains in an early stage of development. More work is particularly needed in studying the dynamic properties of these models in the risk assessment application. It is clear that expanding the scope of the model to incorporate seasonal patterns of exposure and the susceptibility status of different groups within the population will be fruitful directions to pursue. In addition, the uncertainty analysis needs to be refined to include nonlinear multivariate techniques.

SUMMARY AND CONCLUSIONS

From a public health perspective one must assume that all domestic wastewater will contain any of a variety and concentration of infectious disease agents that are endemic to the population from which the sewage arises. This requires that reclamation processes be employed that will *effectively* and *reliably* reduce the concentration of these pathogens to acceptable levels. One must be aware of all possible pathogens that might be involved and recognize new developments of which we were previously not aware; for example, the occurrence of cryptosporidiosis in human populations, a relatively recent addition to the list of diseases that may be waterborne.

Wastewater recycling processes do exist that effectively and reliably reduce the concentration of pathogens to acceptable levels. The existent "acceptable" level has been derived from experience gained from those water recycling activities, throughout the world, from which no adverse health effects have been observed in the populations exposed. Generally, for human contact, these processes include primary and secondary aerobic treatment followed by tertiary treatment which mimics present day drinking water treatment processes. Variations in the treatment trains employed in these traditional processes must be shown to reduce pathogens to levels that are equal to, or greater than, those reductions achieved by the "traditional" methods.

Treatment processes must not only be capable of reducing pathogens to acceptable levels but must do so reliably. Ideally recycling plant effluent should be closely monitored to ascertain that pathogen concentration is kept within acceptable limits. To monitor for every pathogen that might be expected is not practical, and in many cases impossible; therefore, indicator microorganisms need to be employed. The indicator(s) chosen should be representative of the target pathogenic agent both in terms of their concentration and their ability to survive the treatment process. Coliforms have been the traditional indicator of choice; however, they are proving to be an inadequate surrogate for those pathogens which show greater resistance to disinfection and are less readily removed by physical processes such as filtration. At the present time more reliance is placed upon the pre-determined ability of a process to reduce all types of pathogens. There is a need for the development of better microbial indicators along with reliable analytical methods for their determination and enumeration in environmental matrices. A companion to the development of improved indicator methods will be the evolution of rational water quality standards using the new surrogates.

Historically the public health acceptance of microbial water quality levels was developed around the epidemiological evidence that drinking water treatment had a significant impact on the transmission of diseases such as cholera and typhoid fever. When these improvements were being made, at the turn of the 19th century, the risk of contracting these diseases was high, relative to today's incidence, and the lessening of these risks through water treatment was obvious. In contrast the risk involved among those exposed to recycled water that meet present day standards are recognized to be such that traditional epidemiological methods are not sensitive enough to detect or quantitate. If rational standards are to be developed in the future the industry will have to rely more upon risk models that will produce realistic predictions and be useful in making risk management decisions by both the regulator and water producer.

Even with the recognition of "new" diseases and the need for better indicators for monitoring, our experience with the use of recycled water, that meets current standards, has been very positive. The movement towards more intimate human contact with recycled water, such as indirect potable reuse, may require stricter quality and performance standards and sharper estimates of the risks involved with the various treatment alternatives that may be proposed.

REFERENCES

1 Centers for Disease Control and Prevention. 1994. *Hepatitis Surveillance Report No. 65.* Atlanta, GA. pp. 1–36.

2 Cukor, G. and Blacklow, N. R. 1984. "Human Viral Gastroenteritis," *Microbiological Reviews,* 48(2):157–179.

3 Kaplan, J. E, Gary, G. W., Baron, R. C., Singh, N., Schornberger, L. B., Feldman, R. and Greenberg, H. B. 1982. "Epidemiology of Norwalk Gastroenteritis and the Role of Norwalk Virus in Acute Nonbacterial Gastroenteritis," *Ann. Intern. Med.* 96:756–761.

4   MacKenzie, W. R., Hoxie, N. J., Procter, M. E., Gradus, M. S., Blair, K. A., Peterson, D. E., Kazimerczak, J. J., Addiss, D. G., Fox, K. R., Rose, J. B. and Davis, J. P. 1994. "A Massive Outbreak in Milwaukee of Cryptosporidium Infection Transmitted Through the Public Water Supply", *The New England Jour. Med.* 331(3):161–167.

5   EPA. 1991. *Preliminary Risk Assessment for Parasites in Municipal Sewage Sludge Applied to Land.* EPA/600/6-91//001. March 1991. 96 pp.

6   EPA. 1992. *Manual of Guidelines for Water Reuse.* EPA/625/R-92/004. 245 pp.

7   Cooper, R. C., Olivieri, A. W., Konnan, J., Eisenberg , J. and Seto, E. Y. W. 1995. "Mamala Bay Study Commission Report; Project MB-10 Infectious Disease Public Health Risk Assessment". Eisenberg Olivieri Associates, Inc. Oakland, CA. August 1995. 111 pp.

8   Feachem, R. G., Bradley, D. J., Garelick, H. And Mara, D. D. 1983. *Sanitation and Disease: Health Aspects of Excreta and Wastewater Management.* New York, NY. John Wiley and Sons, 501 pp.

9   Sheikh, B., Cooper, R. C., Cort, R. and Jaques, R. S. 1996. "Tertiary-Treated Reclaimed Water for Irrigation of Raw-Eaten Vegetables," in *Wastewater Reclamation and Reuse,* T. Asano, Ed., Technomic Publishing Co., Inc., Lancaster, PA, Chapter 17.

10 Blaser, M. J. and Newman, L. S. 1982. A Review of Human Salmonellosis I: Infective Dose. *Rev. of Inf. Dis.* 4(6):1096–1106.

11 Bryan, F. L. 1977. Diseases Transmitted in Food Contaminated with Wastewater. *J. Food Protecton* 40:45–56.

12 Ward, R. L., Bernstein, D. I., Young, E. C., Sherwood, J. R., Douglas, R. K. and Schiff, G. M. 1986. "Human Rotavirus Studies in Volunteers: Determination of Infectious Dose and Serological Response to Infection," *J. Inf. Dis.* 154(5):871–880.

13 Gamble, D. R. 1979. "Viruses in Drinking Water: Reconsideration of Evidence for Postulated Healthy Hazard and Proposals for Virological Standards of Purity," *Lancet.* 8113:425–428.

14 Purcell, R. H., Hoofnagelo, J. H., Ticehurst, J. and Gervin, J. L. 1989. "Hepatitis Viruses," *Diagnostic Procedures for Viral, Rickettsial and Chlamydial Infections,* Schmidt, N. J. and Emmons, R. W., Washington, D.C., American Public Health Association.

15 Rendoff, R. C. and Holt, C. J. 1954. "The Experimental Transmission of Human Intestinal Protozoan Parasites, IV. Attempts to Transmit *Entamoeba coli* and *Giardia lamblia* Cysts by Water", *Am. J. Hyg.* 60:327–338.

16 Rendoff, R. C. 1954. "The Experimental Transmission of Human Intestinal Protozoan Parasites, II. *Giardia lamblia* Cysts Given in Capsules". *Am. J. Hyg.* 59:209–220.

17 Dupont, H. L., Chappell, C. L., Sterling, C. R., Okhuysen, P. C., Rose, J. B. and Jakubowski, W. 1995. "The Infectivity of *Cryptosporidum parvum* in Healthy Volunteers", *N. Engl. J. Med.* 332:855–859.

18 EPA. 1992. *Technical Support Document for Reduction of Pathogens and Vector Attraction in Sewage Sludge.* EPA 822/R-93-004, pp. 2-24–2-26.

19 Goldstein, S. T., Juranek, D. D., Ravenholt, O., et al. 1996. "Cryptosporidiosis: An Outbreak Associated With Drinking Water Despite State-of-the-Art Water Treatment", *Ann. Internal Medicine.* 124:459–468.

20 Yanko, W. A., Walker, A. S., Jackson, J. L., Libao, L. L. and Garcia, A. L. 1995. "Enumerating Salmonella in Biosolids for Compliance with Pathogen Regulations." *Water Environment Research* 67(3):364–370.

21 Cabelli, V. J. 1983. *Health Effects Criteria for Marine Waters.* EPA 600/1-80-031. USEPA, Cincinnati, Ohio, 98 pp.

22 Dufour, A. P. 1984. *Health Effects Criteria for Fresh Recreational Waters*. EPA-600/1-84-004, USEPA, Cincinnati, Ohio, 87 pp.

23 Geldreich, E. E. 1966. *Sanitary Significance of Feacal Coliforms in the Environment*. Water Pollution Control Research Series Publication No. WP-20-3. U.S. Dept. of the Interior. pp. 93–109.

24 American Public Health Association. 1995. *Standard Methods for the Examination of Water and Wastewater*. 19th. Ed. Wash. D.C. pp 9-70–9-71.

25 Gainey, P. L. and Lord, T. H. 1952. *Microbiology of Water,* New York: Prentice-Hall Inc. 430 pp.

26 Bisson, J. W. and Cabelli, V. J. 1980. *"Clostridium perfringens* as a Pollution Indicator". *JWPCF*. 55(2):241–248.

27 Grabow, W. O. K. 1990. "Microbiology of Drinking Water Treatment: Reclaimed Wastewater." *Drinking Water Microbiology*. G. A. McFetters, Ed. New York: Springer-Verlag. pp. 185–203.

28 Havelaar, A. H., van Olphen, M. And Drost, Y. C. 1993. "F-Specific RNA Bacteriophage are Adequate Model Organisms for Enteric Viruses in Fresh Water." *Appl. Env. Microbiol.* 59(9):2956–2962.

29 Hsu, F., Shieh, C., Van Duin, J., Beekwilder, M. J. and Sobsey, M. D. 1995. "Genotyping Male Specific RNA Coliphages by Hybridization with Oligonucleotide Probes." *Appl. Env. Microbiol.* 61(11):3960–3966.

30 EPA. 1992. *Environmental Regulations and Technology: Control of Pathogens and Vector Attraction in Sewage Sludge*. EPA/625/R-92/013, Dec. 1992. 151 pp.

31 Reimers, R. S., Little, M. D., Englande, A. J., Leftwich, D. B., Bowman, D. D. and Wilkerson, R. F. 1981. *Parasites in Southern Sludges and Disinfection by Standard Sludge Treatment*. Tulane University, New Orleans, LA. NTIS Publication No. PB 82-102344, Springfield, VA. 171 pp.

32 Water Pollution Control Federation. 1989. *Water Reuse: Manual of Practice SM-3*. Alexandria, VA. Water Pollution Cont. Fed., 241 pp.

33 Crook, J. 1990. "Water reclamation." *Encyclopedia of Physical Sciences and Technology, 1990 Year book.* R. A. Myers, Ed. NY. Academic Press. pp. 158–186.

34 State of California. 1978. Wastewater Reclamation Criteria. California Administrative Code, Title 22, Div.4., Chapter 16. California State Department of Health Services, Sanitary Engineering Section, Berkeley, California. 16 pp.

35 Sanitation District of Los Angeles County. 1977. *Pomona Virus Study: Final Report for the California State Water Resources Control Board and The USEPA,* 225 pp.

36 EPA. 1989. Drinking Water: National Primary Drinking Water Regulations. Part II. 40 CFR parts 141 and 142. Wash. D.C. *Federal Register* 54(124):27486.

37 World Health Organization. 1989. *Health Guidelines for the Use of Wastewater in Agriculture and Aquaculture*. Report of a WHO Scientific Group. Geneva, Switzerland. WHO Technical Report Series. 778 pp.

38 EPA. 1993. Standards for the Use or Disposal of Sewage Sludge; Final Rule. 40CFR Part 257. *Federal Register*. Washington, D.C. 58(32):9248.

39 National Research Council. 1996. *The Use of Treated Municipal Wastewater Effluents and Sludge in the Production of Crops for Human Consumption*. Washington, D.C. National Academy Press. 178 pp.

40 National Research Council. 1994. *Ground Water Recharge Using Waters of Impaired Quality*. Washington, D.C. National Academy Press. 279 pp.

41 Cooper, R. C. 1991. *Disease Risk Among Sewage Plant Operators: A Review*. Sanitary

Engineering and Environmental Health Research Laboratory, Report 91-1. Berkeley, CA. University of California. 18 pp.

42 Taubes, G., 1995. "Epidemiology Faces Its Limits," *Science.* 269(5221):164–169.

43 Regli, S., Rose, J. B., Haas, C. N. and Gerba, C. P. 1991. "Modeling the Risk from *Giardia* and Viruses in Drinking Water," *Journal of the American Water Works Association.* 83(11):76–84.

44 Haas, C. N. 1983. "Estimation of Risk Due to Low Doses of Microorganisms: A Comparison of Alternative Methodologies," *American Journal of Epidemiology.* 55(4):573–582.

45 Fuhs, G. W. 1975. "A Probabilistic Model of Bathing Beach Safety," *The Science of the Total Environment,* Amsterdam: Elsevier Scientific Pub. Co. pp. 165–175.

46 Dudely, R. H., Hekimain, K. K. and Mechalas, B. J. 1976. "A Scientific Basis for Determining Recreational Water Quality Criteria," *Journal of the Water Pollution Control Federation.* 48(12):2761–2777.

47 Cooper, R. C., Olivieri, A. W., Danielson, R. E., Badger, P. G., Spear, R. C. and Selvin, S. 1986. "Infectious Organisms of Military Concern Associated With Consumption: Assessment of Health Risks And Recommendations for Establishing Related Standards." *Evaluation of Military Field-Water Quality, Volume 5.* Livermore, CA. Lawrence Livermore National Laboratory. 250 pp.

48 Eisenberg, J. N., Seto, E. W. Y., Olivieri, A. W. and Spear, R. C. 1996. "Quantifying Water Pathogen Risk in an Epidemiological Framework," *Risk Analysis.* 16(4):549–563.

49 Anderson, R. M., and May, R. 1991. *Dynamics and Control of Human Infectious Diseases,* New York. Oxford University Press. 757 pp.

50 Hornberger, G. M., and Spear, R. C. 1980. "Eutrophication in Peel Inlet: 1. The Problem Defining Behavior and a Mathematical Model for the Phosphorous Scenario," *Water Research* 14:29–42.

51 Humphries, R. B., Hornberger, G. M., Spear, R. C. and McComb, A. J. 1984. "Eutrophication in Peel Inlet III. A Model for Nitrogen Scenario and a Retrospective Look at the Preliminary Analyses," *Water Research* 18:389–395.

52 Spear, R. C., and Hornberger, G. M. 1983. "Control of DO level in a River Under Uncertainty," *Water Resources Research.* 19:1266–1270.

53 Tsai, K. C., and Auslander, D. M. 1988. "A Statistical Methodology for the Design of Robust Process Controllers," *Trans. ASME, Dyn. Sys. Meas. and Control.* 110(2):126–133.

54 Bois, F. Y., Smith, M. T, and Spear, R. C. 1991. "Mechanisms of Benzene Carcinogenesis: Application of a Physiological Model of Benzene Pharmacokinetics and Metabolism," *Toxicology Letters.* 56:283–297.

55 Eisenberg, J. N., Konnan, J., Seto, E. Y. W., et. al. 1995. "Microbial Risk Assessment of Reclaimed Water," Prepared by Eisenberg, Olivieri and Associates for the Irvine Ranch Water District and the National Water Research Institute, Irvine, CA.

56 Olivieri, A. W., Eisenberg, J. N., Seto, E. Y. W., Spear, R. C. and Thompson, K. 1997. "Microbial Risk Assessment Using an Epidemiological Based Model: A Case Study of the Public Health Risk of Giardiasis via Exposure to Reclaimed Water," *Water Environment Federation Specialty Conference on Beneficial Reuse of Water and Biosolids,* Marbella, Spain.

57 United States Golf Association (USGA). 1994. *Wastewater Reuse for Golf Course Irrigation,* Lewis Publishers, Chelsea, MI.

58   United States Environmental Protection Agency (USEPA). 1988. *Superfund Exposure Assessment Manual,* Office of Remedial Exposure, Washington, D.C.

59   Parametrix, Inc. 1993. *Metro Effluent Reuse Baseline Risk Assessment, Volume II: Human Health,* Kirkland, Washington.

60   Asano, T., Leong, L. Y. C., Rigby, M. G. and Sakaji, R. H. 1992. "Evaluation of California Wastewater Reclamation Criteria Using Enteric Virus Monitoring Data," *Water Science and Technology* 26:1513–1524.

61   Asano, T. and Sakaji, R. H. 1990. "Virus Risk Analysis in Wastewater Reclamation and Reuse," in *Chemical Water and Wastewater Treatment,* H. H. Hahn and R. Klute, Eds., pp. 483–496.

62   Birkhead, G. And Voigt, R. L. 1988. "Epidemiologic Surveillance for Endemic *Giardia lamblia* Infection in Vermont: The Role of Waterborne and Person-to-Person Transmission", *Amer. Jour. of Epidemiology,* 129(5):762–768.

CHAPTER 12

City of San Diego Health Effects Study on Potable Water Reuse

W ATER reclamation is becoming an increasingly common component of water resource planning. In the past, the driving motivation for water reuse was to provide a means of avoiding effluent disposal into surface waters. With increased demand brought on by drought and increasing population, reclaimed wastewater is now considered an important water resource. Thus, nonpotable and potable use of reclaimed water can provide communities with the option of maximizing and extending the use of limited water resources.

Public health jurisdictions have been reluctant to accept the concept of potable reuse (indirect or direct) because of concerns that the proposed water supply, reclaimed water in this instance, could contain infectious and toxic materials; because of the uncertainty as to the quality requirements for a safe drinking water regardless of the source; and, because public health authorities have worked for decades to provide "pure and safe" drinking water the deliberate reuse of reclaimed water runs counter to traditional patterns of water use.

The public health implications of potable reuse have been examined for a number of wastewater reclamation projects. In all cases, no adverse health effects have been observed in the exposed population. These projects, for the most part, were all retrospective epidemiologic studies, and may not have been sensitive enough to detect small effects in a large population [1]. In the absence of definitive epidemiological evidence it is necessary to define the ability of water reclamation processes to reliably

Adam W. Olivieri and Don M. Eisenberg, Eisenberg, Olivieri & Associates, Western Consortium for Public Health, 1410 Jackson St., Oakland, CA 94612; Robert C. Cooper, Professor Emeritus, School of Public Health, University of California at Berkeley, Berkeley, CA 94720, and Vice President, BioVir Laboratories, Inc., 685 Stone Rd., Benicia, CA 94510.

521

reduce the concentration of toxic chemicals and infectious agents that might be present. Such information brings more certainty to the risk assessment and risk management process. These issues have been addressed in three studies: the San Diego Total Resource Recovery Project— Health Effects Study [2]; the Denver Potable Reuse Demonstration Project [3]; and the Tampa Water Resource Recovery Plan [4]. In this chapter, the results of the San Diego Total Resource Recovery Project are presented.

The City of San Diego California imports virtually all of its water supply from other parts of the state. New sources of imported water are not readily available, and the City is actively investigating advanced water treatment technologies for utilization of potential resources that may already exist locally. Municipal wastewater, presently discharged to the Pacific Ocean, is among the potential local water resources that are being investigated. The City has received Federal funding under the Clean Water Act to demonstrate a unique total resource wastewater recovery system in which one of the resources would be a raw water source for potable use.

The City of San Diego and the California Water Resource Control Board have recognized that the assessment of potential human health risk is a critical factor in evaluating the acceptability of the proposed reuse, and that such evaluation is closely related to testing technical feasibility of the proposed system. The City of San Diego's Health Effects Study (HES) represents the product of a substantial research effort to evaluate the potential health risk associated with reclaimed water relative to an existing raw water supply to the City [2].

HEALTH EFFECTS STUDY (HES) APPROACH AND METHODOLOGY

A report by a panel of experts on the Health Aspects of Wastewater Reclamation for Groundwater Recharge [5] identified seven research recommendations relevant to the potable reuse of wastewater. These included: (1) characterization of contaminants, (2) treatment processes for organics removal, (3) disinfection techniques and viruses, (4) toxicological risk, (5) toxicological risks, (6) epidemiological studies, and (7) monitoring techniques and strategies.

The National Research Council Panel on Quality Criteria for Water Reuse [6] provided specific guidance on health effects criteria for use in the evaluation of water recycling for human consumption. The Panel identified the following investigations: (1) that the prime health concerns of wastewater reuse are the long-term health outcomes of ingesting chemical contaminants found in recycled water; (2) that the health risk of using recycled water as a potable water supply should be compared against similar risk by conventional water supplies; and (3) that an intense toxicity program be instituted.

The concept and approach for the City of San Diego's HES emerged from the review of ongoing reclamation efforts, and the considerable thought by the expert panel and the NRC in formulating their recommendations. The primary objective of the HES was to determine if an advanced wastewater treatment system could reliably reduce contaminants of public health concern to levels such that the health risks posed by any proposed potable use of the treated water are no greater than those associated with the present water supply [7].

HES APPROACH

Since the early 1970s, the City of San Diego has been experimenting with an innovative aquatic system for the secondary treatment of wastewater. In 1977, a Technical Advisory Committee (TAC) was established by the California State Water Resource Control Board (SWRCB) to investigate the feasibility of a Total Resource Recovery Program. Under the TAC's direction, the City's initial research was merged with the Total Resource Recovery Program. The Total Resource Recovery Program is an experimental program being conducted by the City of San Diego and the TAC to study the technical and public health issues associated with, and to develop a strategy for, wastewater reclamation that includes evaluating the feasibility of recycling a portion of the City's wastewater.

Overall guidance on the project was provided by the TAC. A separate Health Advisory Committee (HAC) was also established to address the Public Health issues. The make up of the HAC is shown in Table 12.1. The main goal of the HAC was to assess the health risks associated with the use of a treated wastewater as a raw water supply. The key HAC tasks were:

- to determine the specific kind of health data required to address the health issues adequately
- to aid in planning the means and methods to be used in collecting the required data
- to systematize and interpret the data thus collected
- to make a health risk evaluation
- to make a final recommendation concerning the use of the product for human consumption

To address these tasks the HAC produced a work plan, reviewed the status and results of the investigations, modified the program as appropriate and produced annual reports [8–12] and the Health Effects Study Report for the Mission Valley pilot plant in September 1992 [2].

TABLE 12.1. **Members of the Health Advisory Committee (HAC) for the Health Effects Study.**

Name	Expertise
Robert C. Cooper, Ph.D.	Chairman, Microbiology
A. Ralph Barr, Sc.D.	Vectors
Herschel Griffin, M.D.	Epidemiology
Donald Heyneman, M.D.	Parasitology
Dennis Hsieh, Ph.D.	Toxicology
Edwin Lennette, M.D., Ph.D.	Virology
Martyn Smith, Ph.D.	Toxicology
George Tchobanoglous, Ph.D.	Engineering
Edward Wei, Ph.D.	Toxicology
Ex Officio Members	**Affiliation**
Harvey Collins, Ph.D.	CA Department of Health Services
Robert Hultquist, M.S	CA Department of Health Services
Paul Gagliardo, P.E.	City of San Diego
Donald Ramras, M.D.	San Diego County Health Officer
Gary Hoffmann, P.E.	CA State Water Resource Control Board
Cecil Martin, M.P.H.	CA State Water Resource Control Board

The work was carried out by the Western Consortium and a group of subcontractors, each with responsibility for work-plan elements in their respective areas of expertise. These subcontractors and their subject areas are listed in Table 12.2.

HES METHODOLOGY

Water quality factors of particular significance in situations where potable reuse is contemplated include: microbiological (bacteria, virus, and parasites), total minerals, heavy metals, and stable organics. Thus the HES included elements for: identifying, characterizing, and quantifying infectious disease agents and potentially-toxic chemicals, and for screening mutagenicity and bioaccumulation of the chemicals present, in both the advanced treated wastewater and the City's untreated raw water supply. In this study the two water sources that were compared were the untreated raw reservoir water from the City's Miramar drinking water treatment plant and the effluent from the City's advanced water treatment plant (AWT).

The study included a reliability analysis of the mechanical systems, characterized the variability of the water reclamation treatment process effluent quality, and determined the probabilities of effluent characteristics meeting, or exceeding, specified concentrations. An epidemiology evaluation was performed to collect baseline health information on residents of the San Diego County who used the current water supply. Lastly, the study design included a quantitative risk assessment based on the results of the chemical and biological monitoring conducted on both the raw water supply and the water supplied by the AWT plant.

The initial HES investigations were conducted on the pilot wastewater treatment system (Aqua II) located in Mission Valley. The California State Water Resources Control Board (SWRCB) and the Department of Health Services (DHS) required that the results of the HES conducted at the Aqua II Pilot Plant be confirmed on a full-scale plant. Therefore, a limited additional HES investigation was conducted on the City's San Pasqual treatment facility (Aqua III), located in the San Pasqual Valley, to corroborate the pilot plant HES findings.

The overall findings from the Aqua II and Aqua III HES investigations are presented below. All the data and results of the Aqua III HES are not currently available, however, where data are available from the Aqua III HES they are presented. A detailed discussion of the results for the Aqua II and Aqua III Health Effects Studies is contained in two separate HES reports [2,35].

DESCRIPTION OF WASTEWATER TREATMENT SYSTEM

The initial HES investigations were based on a pilot wastewater treatment system (Aqua II) located in Mission Valley, San Diego, California.

TABLE 12.2. List of Subcontractors and Principal Investigators for the Health Effects Study.

Subcontractor	Element	Principal Investigator
SDSU	Epidemiology	Dr. Craig Molgaard
CPHF (MDL)	Enteric Pathogens[a]	Dr. Ted Midura
CPHF (VRDL)	Enteric Viruses[a]	Dr. John Riggs
UCSF, TDL	Parasites[a]	Dr. Robert Goldsmith
UCB	Chemical Database	Dr. Edward Wei
UCLA	Chemical Screening Bioassay	Dr. John Froines
SDSU	Biomonitoring	Dr. Ann de Peyster
SDSU	QA/QC Biological[b]	Dr. Ann de Peyster
UCLA & WCPH	Chemical Risk Assessment	Dr. John Froines, Dr. Adam Olivieri

[a]All Aqua II pathogen analyses conducted by Dr. Riggs at BioVir Laboratories, Benicia, CA.
[b]Aqua II QA/QC conducted by Dr. Rick Danielson at BioVir Laboratories, Benicia, CA.
Notes:
  SDSU San Diego State University.
  CPHF California Public Health Foundation.
  MDL Microbial Diseases Laboratory, California Department of Health Services.
  VRDL Virus and Rickettsial Disease Lab, California Department of Health Services.
  UCSF University of California at San Francisco.
  TDL Tropical Diseases Laboratory.
  UCB University of California at Berkeley.
  UCLA University of California at Los Angeles.
  WCPH Western Consortium for Public Health.
  QA/QC Quality Assurance and Quality Control.

The pilot treatment system was a unique and innovative system, utilizing water hyacinths for secondary treatment, followed by solids contact clarification with ferric chloride plus gravity dual media filters providing tertiary treatment and ultraviolet disinfection, cartridge filtration, reverse osmosis, air stripping, and granular activated carbon providing an advanced water treatment (AWT). The capacity of the water hyacinth ponds was approximately 156,000 gal/d (four ponds in operation) and the capacity of the tertiary and AWT facility was approximately 50,000 gal/d. A more detailed discussion of the evolution of the aquatic portion of the wastewater treatment pilot plant is contained in a separate technical paper [13]. A discussion of the overall performance of the pilot plant system is presented in latter sections of this chapter as it relates to public health issues.

The San Diego Total Resource Recovery Program relocated to the San Pasqual Valley as a result of the limited demand for reclaimed water and land development pressures within Mission Valley. The San Pasqual Facility (Aqua III) is similar to the Mission Valley plant design but is on a larger scale having nominal flow rates of 1 Mgal/d through the water hyacinth secondary treatment facilities, 1 Mgal/d through the chemical coagulation and filtration process, and 0.5 Mgal/d through the reverse osmosis, activated carbon and air stripping facilities. The SWRCB and the DHS required that the results of the HES conducted at Aqua II be confirmed using a full-scale plant. This portion of the study is ongoing at the time of this writing (May 1996).

INFECTIOUS DISEASE AGENTS

One major element of the HES was to collect data sufficient to make an informed judgment as to the public health risk of infectious disease from pathogens that might be present in reclaimed water. The waters involved were monitored for the presence of: microbiological indicators of water quality; representative pathogenic bacterial genera; enteric viruses, and parasites. A major effort of the HES was directed toward monitoring of the raw wastewater (RWW), the AWT plant effluents from Aqua II and Aqua III, and the raw water used for drinking in the San Diego area (Miramar) so that the sanitary quality of each of the three waters could be compared. Monitoring at Aqua II and Aqua III was conducted over a period of three years and for about 18 months, respectively.

MICROBIOLOGICAL INDICATORS OF WATER QUALITY

During the HES at Aqua II, samples were collected from the AWT plant effluent, Miramar, and RWW from June 1987 through June 1990.

TABLE 12.3. Comparison of Microbiological Indicator
Data Raw Wastewater (RWW).

Indicator Organism	Sampling Period	Log Mean	Log SD	Log Min	Log Max	N
Total Coliform	Aqua III	7.48	0.41	6.3	8.48	79
	Aqua II	7.37	0.31	6.11	8.20	212
Fecal Coliform	Aqua III	7.01	0.36	6.30	8.20	79
	Aqua II	7.05	0.32	6.32	8.20	212
Fecal Strep	Aqua III	6.69	0.41	5.60	8.20	77
	Aqua II	6.48	0.34	5.6	7.38	196

Duplicate excluded from analysis.
Data was log transformed, and then statistics were computed.
Values less than 2 MPN/100 mL were computed at 2 MPN/100 ml.
N = number of samples.

At Aqua III, samples were collected from the plant effluent, Miramar raw water, and the RWW over a period from March 1994 through August 1995. Sampling occurred two to three times a week for all the varieties of indicator bacteria. All samples were collected as grab samples. For the Miramar samples, sodium thiosulfate was added to the bottles to neutralize the effects of residual chlorine (chlorine was sometimes added to the Miramar water, prior to treatment, to control the formation of biofilms in the transit lines). The total coliform and fecal coliform method used was the standard, multiple tube MPN test as described in *Standard Methods for the Examination of Water and Wastewater,* 16th ed. (henceforth referred to as *Standard Methods*) [14]. A five tube MPN method for fecal strepto-cocci was also conducted as per *Standard Methods.*

A summary of the RWW data for the number of indicator bacteria found in the three water sources for the Aqua II and Aqua III HES sampling periods is presented in Tables 12.3 and 12.4. As anticipated, the RWW samples yielded the highest total and fecal coliform and fecal streptococci

TABLE 12.4. Comparison of Microbiological Indicator Data Ratio of Samples
Greater Than or Equal to 2MPN/100 ml to Total Number of Samples.

Water Source	Sampling Period	Total Coliform	Fecal Coliform	Fecal Strep
AWT	Aqua III	7/105 (0.07)	2/105 (0.02)	2/105 (0.02)
Effluent	Aqua II	131/413 (0.32)	2/413 (0.004)	2/411 (0.005)
Miramar	Aqua III	176/193 (0.91)	133/193 (0.69)	152/192 (0.79)
	Aqua II	389/425 (0.92)	216/425 (0.51)	272/425 (0.64)

Values in parentheses are the calculated ratios of samples with values greater than or equal to 2 MPN/100 ml to the total number of samples analyzed.

counts of the three waters, and were in the range expected for raw domestic wastewater. The Aqua II and Aqua III AWT effluents and Miramar waters contained low coliform and enterococci values at one time or another. The majority of coliform and streptococci values found in the Aqua II and III AWT water and the Miramar raw water, as shown in Table 12.4, are less than 2.2 per 100 ml. In the AWT product water the highest observed value for total coliform (TC), fecal coliform (FC) and fecal streptococci (FS) was an MPN per 100 ml of 14, 4 and <2.2, respectively. The values seen in the Miramar raw water were, in the same order, 5,000, 316, and 1,500 MPN per 100 ml.

Based on a statistical comparison (Wilcoxon Rank Sum Test) of the total and fecal coliform and streptococci data, the probability of observing an Aqua II or Aqua III value greater than that in the Miramar is too low to measure.

The Aqua II and Aqua III AWT effluent had consistently fewer indicator coliform and fecal streptococci bacteria than did the untreated City drinking water supply. Total coliforms were detected in only 19 percent of the samples taken from the Aqua II effluent over a three year period, and none of these samples contained more than 80 total coliform MPN/100 ml. Total coliforms were detected in only 7 percent of the samples taken from the Aqua III effluent over a one and one-half year period, and none of these samples contained more than 14 total coliform MPN/100 ml. Conversely, total coliforms were detected in the untreated City drinking water approximately 92 percent of the time during both sampling periods. As a raw water source, the microbiological water quality of that currently employed by the City of San Diego (Miramar) is very similar to raw drinking water sources observed in the United States. In surface waters employed as raw drinking water sources, total coliforms range from 2 to 1000 MPN/100 ml. Fecal coliform levels observed in Miramar water were at the lower end of the reported nationwide concentrations, but were about average in regard to the streptococci. On the other hand, the Aqua II and Aqua III product waters appear to be of a much higher microbiological quality than Miramar or for surface waters nationwide concentration. The Aqua II and Aqua III product waters tend to be most comparable to concentrations reported in groundwater for all indicators.

There were only 2 occasions each (0.4 percent of the total samples at Aqua II and 2 percent at Aqua III) where fecal coliforms were detected in the Aqua II and Aqua III waters. At Aqua II, both samples were at the detection limit of 2.2 MPN/100 ml. At the Aqua III facility, a plant operational event could be associated with each event. Fecal coliforms were sixty times as likely to be detected in Miramar water as in the Aqua II and III effluent. Streptococci were never detected in the Aqua II AWT effluent and were detected on two occasions in the Aqua III AWT plant effluent. On both occasions the fecal streptococci concentration was at the detection limit and on one occasion was related to a plant operational activity. Fecal streptococci were detected in almost 60 percent of the Miramar samples during the first sampling period, and 79 percent of the time during the second.

Overall, it was observed that the Aqua II and Aqua III treatment plants could reduce the number of measurable indicators by greater than 99.9999 percent. It should be noted that the results of the HES reflect the water quality of the Aqua II and Aqua III treatment plant effluents without disinfection.

At Aqua II, about 20 percent of all plant events resulted in detectable total coliform concentrations, while at Aqua III, less than 2 percent of all plant events resulted in detectable total coliform concentrations. Because plant failures are random and the sampling schedules are fixed, it can be speculated that there may have been more occasions where the indicator levels may have been detected (greater than 2.2 MPN/100 ml). However, with over 400 samples taken over a three year period at Aqua II and over 100 samples taken over an 18 month period at Aqua III, it seems likely that substantial breakthroughs were a rare event. In any case, the concentrations of total and fecal coliform and fecal streptococci in the AWT effluent were far below those observed at Miramar.

ENTERIC VIRUSES

The potential presence of viruses in wastewater and reclaimed water are of concern to public health officials and other decision makers. For this reason, a great deal of effort was expended during the health effects study to examine for the presence and numbers of enteric viruses in the input and output of the Aqua II and Aqua III treatment plants. At Aqua II, the raw wastewater (RWW) influent to the plant and the plant product water were monitored for in situ virus from June 1987 through April 1990. At Aqua III, in situ virus monitoring occurred between February 1994 and August 1995. Virus seeding studies were also conducted at both plants to gain some better quantification of the virus removing capability of the reclamation process. For purposes of comparison, in situ virus sampling was performed on Miramar raw water, a major water source for the San Diego area, in parallel with the sampling programs at the Aqua II and Aqua III plants.

It was essential that continuity be maintained, where feasible and practical, between the Aqua II and Aqua III sampling periods to provide comparable data. Therefore, with few exceptions, the sampling and analysis procedures followed were essentially the same. During the Aqua II sampling period, the virus assay work was conducted by the California Department of Health Services Viral and Rickettsial Laboratory (VRDL). During the Aqua III sampling period the assay work was conducted by BioVir Laboratories, Inc., Benicia, CA as VRDL was not able to provide the previous analysis conducted during the Aqua II sampling period.

During the Aqua II sampling period, 53,576 gallons of Aqua II effluent and 43,028 gallons of Miramar water were concentrated for viruses. Two Miramar in situ virus samples, out of 67, were found to have virus present in July and November 1989. In these cases, the virus isolated was Polio 2 at concentrations of 0.11 and 0.15 pfu/L, respectively. In both instances, virus spiking studies for recovery efficiencies, using Polio 2, were conducted by laboratory personnel using the concentrator a few days prior to the in situ sampling.

As part of the Aqua III virus monitoring program, measurement of male-specific RNA (F + RNA) coliphage were included as part of the sampling routine. The advantages of using these surrogate viruses are their presence in raw wastewater in large numbers, relative to enteric animal viruses, regardless of season and their resistance to environmental factors that are equal to, or greater, than that of enteric animal viruses. A portion of the enteric animal virus concentrates (usually 50 mL) from the Aqua III AWT and Miramar samples was examined for F + RNA coliphage by plaque assay. *Escherichia coli* F + Amp was the bacterial host used. The concentration of F + RNA coliphage was determined in raw wastewater without prior sample concentration.

In the Aqua II effluent, of 49 samples for in situ virus testing, three were found to be positive. One sample was taken in May 1988 and the other two occurred in February of 1989. The virus concentration was 0.04, 0.05 and 0.003 pfu/L, respectively. In all cases, the virus present was identified as Polio 2, the same virus serotype used to evaluate virus recover efficiency and treatment plant virus removal efficiency. In the case of the virus isolations in the Aqua II effluent the latter two virus isolations occurred shortly after the Polio 2 virus seeding exercises were performed. The first virus isolation, in May 1988, occurred two months after a virus seeding episode and, thus in this instance, laboratory contamination is a somewhat less creditable explanation. During these time periods there were no indications of Aqua II plant malfunction or any untoward water quality changes at Miramar. Because of the singular nature of the virus type involved in every case and because of the use of Polio 2 virus in seeding experiments the most likely source of the virus was laboratory contamination or seed virus carry over.

The raw wastewater was initially sampled for virus by taking one gallon grab samples. The median virus concentration in these samples was none detected in 1000 ml with a high value of 405 pfu/L. Only 23 percent of the samples contained detectable virus. This low concentration of virus

TABLE 12.5. Aqua III Sampling Program Results for Phage and Enteric Viruses (pfu/L) in AWT, Miramar, and RWW Water Sources.

Analyte	Source	Median	n	Min	Max
Phage	AWT	<0.05	31	<0.01	<0.07
(F + RNA)[a]	MIR	<0.06	34	<0.007	0.72
	RWW	1.8 × 106	35	1.5 × 105	3.3 × 107
Enteric	AWT	<0.001	32	<0.001	<0.009
Virus	MIR	<0.001	36	<0.0005	<0.004
	RWW	22	36	<1	1067

[a]F + RNA = male specific coliphage.

was an unexpected result and the sampling regime was changed to using a composite sample. In Aqua II, the percentage of composite samples containing detectable viruses was 56; thus, slightly less than half the samples were virus negative. The median virus concentration was 9 pfu/L, ranging from nondetected to 17,300 pfu/L. Obviously, the composite sampling method was superior to the initial grab sample method. However, the number of samples that had no detectable viruses present per liter of wastewater was surprising. A perusal of the VRDL reports reveals that on many occasions the number of virus plaques were found to increase upon dilution of the sample. This observed increase might be the result of some inhibitory effect associated with the undiluted wastewater. This inhibitory effect might also be the cause of the relatively large percentage of negative virus results.

During the Aqua III monitoring period, 14,916 gallons of effluent and 17,144 gallons of Miramar raw water were concentrated for viruses. As shown in Table 12.5, of 32 Aqua III effluent samples for in situ virus testing none were found to be positive. Of 31 samples for Phage (F + RNA), none were found to be positive. In the Miramar water, of 36 samples for in situ virus testing, none were found to be positive. Of 34 samples for Phage, 33 of 34 were negative. One sample was positive at 0.72 pfu/L. The average detection level for enteric viruses was 0.001 pfu/L and for F + RNA coliphage was 0.05 pfu/L.

The median virus concentration in the RWW was 22 pfu/L, ranging from nondetected to 1067 pfu/L. The percentage of composite samples containing detectable viruses was only 20 percent due to the inhibitory nature of the raw wastewater to the assay procedure. Thirty-five RWW samples, collected and analyzed for coliphage, had a median concentration of 1.8E + 06 pfu/L, ranging from a minimum value of 1.5E + 05 to a high maximum value of 3.33E + 07 pfu/L. Of 31 Aqua III AWT samples, no phage were enumerated and one Miramar sample contained 0.72 pfu/L of phage. The results of the bacteriophage monitoring at the Aqua III plant corroborate the *in situ* enteric virus monitoring results and that, based on bacteriophage, from 5 to 7 logs of virus removal is occurring through the entire Aqua III plant, without disinfection. Based on these results, it can be concluded that there is no difference in the virus concentration found between the Aqua II and III product waters and between both treatment plant product waters and the Miramar raw water.

ENTERIC BACTERIAL PATHOGENS

An important public health concern associated with wastewater reclamation and reuse is the potential for transmission of infectious disease. With this concern in mind, an intensive monitoring was developed for specific bacterial enteric pathogens.

During the Aqua II sampling period, each of the three water sources: San Diego raw wastewater; Aqua II reclaimed water; and the associated community raw drinking water (Miramar) were examined for the presence of the bacterial pathogens: *Salmonella* spp., *Shigella* spp., and *Campylobacter* spp. During the Aqua III sampling period, each of the three water sources were examined only for the presence of *Salmonella* spp. The same protocols utilized for Aqua II were used for Aqua III.

The available methods for isolation of *Campylobacter* in RWW proved to be inadequate. Because of the large number of background bacteria present in raw wastewater, the selective media used in the analysis was always overgrown. Under these conditions *Campylobacter* was not detectable. The testing for *Campylobacter* was carried out for only three months because of the bacterial background interference problem.

The same condition that interfered with the isolation of *Campylobacter* from the RWW water source was encountered in isolating *Shigella*. Because the concentration of *Shigella* was expected to be low, high dilutions of the sample were avoided. At these low dilutions, the standard selective/differential medium used was overgrown with background microorganisms, none identifiable as *Shigella*. After nine months of sampling for *Shigella* with no isolations from any water type, sampling for *Shigella* was discontinued.

Aqua II monitoring results for two years of monitoring for *Salmonella* are summarized in Table 12.6. From the results presented, it can be seen that of the three waters investigated, only the raw wastewater (84 samples) was found to contain detectable levels of *Salmonella*. The median value for in situ *Salmonella* in RWW was 13/L. There was a great deal of variation from sample to sample, with a range from less than 3 to 128,000/L. No *Salmonella* were detected in either the Aqua II AWT effluent or Miramar water (55 and 51 samples, respectively).

The results of one year of monitoring for *Salmonella* during the Aqua

TABLE 12.6. Summary of Observed *Salmonella* spp. in Aqua II AWT, Miramar, and RWW Water Sources.

Water Source	Number of Samples	Volume Filtered[a] (L)	Salmonella Concentration[a] (MPN/L)
RWW	84	0.76	12.5[b]
Miramar	55	10.10	< 3[c]
AWT	51	10.10	< 3[c]

[a]Median values.
[b]Range of values 0 > 28,000 MPN/L method.
[c]Lowest level of detection using MPN/L method.

III sampling period are summarized in Table 12.7. From the results presented, it can be seen that of the three waters investigated, only the raw wastewater (11 out of 32 samples) was found to consistently contain detectable levels of *Salmonella*. The median value for in situ *Salmonella* was 22 MPN/L. None were detected in the 29 samples of Aqua III product water. *Salmonella* were detected in two out of 32 Miramar water samples, at concentrations equal to the detection limit of 0.22 MPN/L.

ENTERIC PARASITES

During the Aqua II sampling period, a large portion of the parasite monitoring effort was dedicated to methods development. A detailed discussion of this effort is contained in the HES Final Project Report [2,18]. No *Giardia* spp. were observed in 55 Aqua II AWT effluent samples or 59 Miramar water samples. *Giardia* spp. were observed in 49 RWW samples at an average of 160 cysts per liter.

The Aqua III monitoring program concentrated on evaluating the three waters for two pathogenic parasites: *Giardia lamblia* cysts and *Cryptosporidium* oocysts. Results of the one year Aqua III monitoring program are summarized in Table 12.8. Approximately 14,333 gallons of Aqua III AWT effluent and 17,694 gallons of Miramar water were concentrated for parasite analysis, and 36 one liter composite samples of RWW were examined. *Giardia lamblia* were detected with regularity in the RWW,

TABLE 12.7. Summary of Observed *Salmonella* spp. in Aqua III AWT, Miramar and RWW Water Sources.

Analyte	Source	Median	n	Min	Max
Salmonella spp. (MPN/L)	AWT	< 0.22	29	< 0.22	< 1.6
	MIR	< 0.22	32	< 0.22	22
	RWW	22	32	< 22	92

TABLE 12.8.   Summary of *Giardia lamblia* and *Cryptosporidium* in Aqua III
AWT, Miramar and RWW Water Sources.

Analyte	Source	Median	n	Min	Max
Giardia lambia	AWT	< 0.002	29	< 0.001	< 0.01
(cysts/L)	MIR	< 0.002	34	< 0.001	< 0.006
	RWW	325	34	< 0.004	3200
Cryptosporidium	AWT	< 0.002	29	< 0.001	0.02[a]
(oocysts/L)	MIR	< 0.002	34	< 0.001	< 0.006
	RWW	2[b]	36	< 0.004	25

[a]Two samples were positive after plant spike.
[b]86% (31/36) of the samples were below detection limits.

with a geometric mean of 325 cysts/L. *Cryptosporidium* spp. were only rarely detected (5 out of 36 samples, 13 percent of the time) with a mean concentration of 2 oocysts/L. During routine monitoring, no *Giardia lamblia* and *Cryptosporidium* oocysts were found in the Aqua III and Miramar waters at a detection limit of <0.001 cysts or oocysts/L.

PLANT VIRUS SEEDING STUDY

To evaluate the overall ability of the Aqua II and the Aqua III processes to remove measurable human enteric viruses, a series of dosing experiments were carried out. The first step in the virus dosing experiment was designed to predict the flow characteristics of the plants. At the Aqua II plant, lithium chloride was added, as a tracer, to the influent. The effluent of the ponds and the AWT system were monitored for lithium concentration to determine the dilution rate and when the maximum concentration of the added lithium would appear in the effluents. Lithium chloride was also used at the Aqua III plant as a tracer, however, it was added to the effluent of the water hyacinth ponds and used to predict the flow characteristics and optimum sampling times of only the tertiary and advanced water treatment processes.

In both plants, a known dose of attenuated vaccine strain of Poliovirus 2 was used as the virus seed. Using results from the lithium chloride dosing experiments it was possible to predict the optimum time to sample for the virus and what concentration would be expected if dilution alone was affecting virus concentration. Any difference between the dilution effect and the number of viruses recovered would be an indication of the percent virus removal through the treatment process. At Aqua II, one 1,000 gal sample was taken of the pond effluent and another from the final plant effluent at 21 and 25 hours after spiking, respectively. Samples were concentrated and shipped to the laboratory for analysis. Four separate plant seeding tests were conducted. At Aqua III, approximately 280 to 420 L

TABLE 12.9. Virus Removal Efficiency of the Aqua II Pond and AWT Systems.

Run Number	Virus in Pond Effluent		Virus Removal in Ponds (%)	Virus in AWT Effluent (pfu/L)	Virus Removal[d] (%)
	Expected[a] (pfu/L)	Observed (pfu/L)			
1	4.7×10^4	1.9×10^2	99.6	ND[b]	>99.9999
2	9.6×10^4	1.2×10^4	87.5	ND[c]	>99.99999
3	1.0×10^5	4.0×10^3	96.1	ND	>99.99999
4	5.6×10^4	3.7×10^3	93.5	ND	>99.99999

[a]The number expected if no virus removal occurred (based on lithium tracer study).
[b]ND = none detected in 3800 L.
[c]The occurrence of 0.2 pfu/L in this sample was due to cross contamination.
[d]Calculation based upon 3800 L sample.

of advanced treated water were sampled during each of three separate spiking events, between three and five hours after release of the seed.

Results of the Aqua II treatment plant virus dosing experiments are presented in Table 12.9. It was observed that the Aqua II ponds removed 88 to 99 percent of the added Poliovirus Type 2, while the entire treatment plant reduced the seeded virus concentration by more than 6 to 8 logs. Results of the Aqua III treatment plant virus dosing experiments are presented in Table 12.10. None of the seeded viruses were recovered in the final effluent of the Aqua III plant. This observation indicates that the combined tertiary and advanced water treatment processes at the Aqua III plant removed 10 logs or 99.99999999 percent of the added Poliovirus Type 2.

The four virus seeding studies performed on the entire Aqua II plant indicated a great virus removal capability. None of the seeded viruses were recovered in the final effluent which indicates a removal efficiency as great as eight logs (99.999999 percent). The three virus seeding studies performed on the tertiary and advanced water treatment processes of the Aqua III plant also indicated a great virus removal capability. None of the seeded viruses were recovered in the final effluent which indicates a

TABLE 12.10. Virus Removal Efficiency of the Combined Aqua III Tertiary and Advanced Water Treatment Plant.

Run Number	Expected Concentration (pfu/L)	Observed Concentration (pfu/L)	Log_{10} Removal
1	5.7×10^7	<0.002	>10.5
2	8.51×10^7	>0.003	>10.4
3	6.42×10^7	<0.003	>10.3

removal efficiency as great as 10 logs (99.99999999 percent). This removal rate is considerably greater than that reported in either the Pomona virus study [16] or the Monterey (MWRSA) study [17] which used filtration and chlorination of secondary effluent, but not reverse osmosis (RO). At both the Aqua II and Aqua III, RO was used, but there was no disinfection. Virus removals at both plants is both effective and reliable.

The results of the male specific coliphage monitoring at Aqua III corroborate the in situ enteric virus monitoring results and indicates that at least, based on coliphage, from 5 to 7 logs of virus removal is occurring through the entire Aqua III plant, without final disinfection.

CHEMICAL AGENTS

The investigation of chemical agents was performed as four separate sub-tasks that included: Chemical Screening and Monitoring; Ames Assay Testing for Mutagenicity; Chemical Analyses of Resin Extracts; and Fish Biomonitoring. At Aqua II, the HES chemical investigations were carried out over a three-year period with year one as the chemical screening phase, and years two and three as the chemical monitoring phase. The objective of the screening phase was twofold; (1) to determine which chemicals are present and their concentrations in the final Aqua II advanced water treatment water, the Miramar water, and the Raw Wastewater (RWW) and (2) to identify which of these chemicals are known to be of concern to human health. Estimates of the unidentifiable fraction of the organics were also made. The Aqua III investigations were limited to monitoring for a one year period.

CHEMICAL MONITORING RESULTS

The data presented in this section are from samples collected at the Miramar water treatment plant, the influent to the Aqua II treatment facility (RWW), and the final effluent from Aqua II. Data collected during the monitoring phase of the Aqua II HES investigations were measured over a two-year period, May 1988 through March 1990 [2,19]. Routine analysis during the monitoring phase included the following:

- analysis for sixty-three semi-volatile organic compounds using liquid-liquid extraction and Gas Chromatographic Mass Spectrometric analysis (GCMS)
- analysis for thirty-five volatile organic compounds by purge and trap GCMS, including the trihalomethanes and benzene
- determination of twenty-seven inorganic elements by inductively

coupled argon plasma—atomic emission spectroscopy (ICP-AES) and atomic absorption spectroscopy (AA), and
- total organic halogen and total organic carbon

Pesticide/Polychlorinated Biphenyl (PCB) analysis of water samples were performed four times during the second year of the monitoring phase, and chlorinated dibenzodioxin/dibenzofuran analysis was conducted once. Sampling and analyses were performed routinely in duplicate and at two sites. As part of the quality control program, blank samples were analyzed to monitor the purity of reagents and to monitor the potential carryover from one sample to the next. Essentially the same sampling and analysis protocols were utilized during the investigations at Aqua III as those discussed above for the Aqua II. At the time of this writing, not all of the monitoring results were available for Aqua III; however, data where available are presented.

Metal Concentrations

Inductively coupled plasma atomic emission spectrometry (ICP-AES) was used to measure all metals except for arsenic, mercury and selenium in which atomic adsorption-mass spectrometry (AA-MS) was used.

Inductively Coupled Plasma Atomic Emission Spectrometry Metals (ICP-AES)

The (ICP-AES) Method Detection Limits (MDLs) during the Aqua II monitoring period ranged from 0.001–1.25 mg/L. The elements that occurred in measurable quantities in both waters included boron, calcium, magnesium, manganese, molybdenum, phosphorus, sodium, and strontium. Boron occurred regularly in Aqua II water in amounts higher than those found in Miramar water. Phosphorus in Aqua II water was found at lower concentrations than in Miramar. Concentrations of each element at the beginning of the monitoring phase were relatively low but increased at a regular rate until approximately day 430 (mid-July 1989). At that time the amounts observed decreased radically and remained at low levels for the duration of the monitoring phase. This occurrence of increasing concentrations was consistent with the degradation of the original reverse osmosis membranes and after replacement of the membranes the original performance was again achieved.

During the Aqua III monitoring period, the Method Detection Limits (MDLs) ranged from 0.001–0.13 μg/L. The average concentrations of most metals were at their method detection limits. The elements that occurred in measurable quantities in both waters included barium, boron, calcium, lithium, magnesium, manganese, molybdenum, phosphorous, silicon, sodium, and strontium.

Overall, the Miramar water consistently showed higher metal concentrations than Aqua III water. The concentrations of cobalt, copper, lead, nickel and silver ranged from 0.001 to 0.003 mg/L higher on average than the concentrations measured in Miramar water. These differences may not be of any significance as the concentrations of these metals were all less than the method quantitation limits.

Findings from a comparison of the Aqua II and Aqua III monitoring results are as follows:

- the concentrations of most metals in both Aqua II and Aqua III waters were at their method detection limits
- of the elements that were quantifiable, Aqua III water had lower metal concentrations than Aqua II water, and Miramar waters from both sampling periods, and
- calcium, magnesium, and sodium concentrations were highest in Miramar water during the Aqua III monitoring period

AA-MS and ICP-MS Metals

Arsenic and selenium were measured during the Aqua II monitoring period using atomic adsorption mass spectrometry (AA-MS). During the Aqua III monitoring period, arsenic, mercury and selenium were measured by inductively coupled plasma mass spectrometry (ICP-MS). These constituents were selected for measurement, using analytical techniques that provided lower detection limits, because of the potential public health significance. The AA-MS detection limits for arsenic and selenium were 2.0 and 2.0 μg/L, respectively, while the ICP-MS detection limits were 0.2, 0.01, and 0.1 μg/L, respectively. Arsenic and selenium concentrations in the Aqua II final effluent averaged 0.3 and 0.1 μg/L, respectively. In Miramar water, arsenic and selenium averaged 1.7 and 0.7 μg/L, respectively.

Based on preliminary data, the concentration of arsenic and selenium in the Aqua III effluent averages 0.19 and 0.62 μg/L, respectively. Miramar water concentrations of arsenic and selenium average 2.6 and 3.1 μg/L, respectively. Mercury concentrations in the Aqua III effluent averaged 0.61 μg/L and in Miramar water averaged 0.57 μg/L. During both the Aqua II and Aqua III monitoring periods, the Miramar raw water consistently had higher arsenic and selenium concentrations than either the Aqua II or III waters. Mercury concentrations for Aqua III and Miramar water are similar. Arsenic concentrations in Aqua II and III product water are comparable, whereas selenium shows higher concentrations in Aqua III.

Purgeable and Extractable Organics

The analytes evaluated at Aqua II and Aqua III included the analysis for (1) thirty-five volatile organic compounds, (2) sixty-three semi-volatile organic compounds, including the trihalomethanes and benzene, (3) pesticides/PCBs, and (4) Chlorinated Dibenzodioxins/Dibenzofurans.

During the Aqua II monitoring period for purgeable compounds, a comparison of the three water sources (Aqua II AWT, Miramar and blank waters) revealed that three of the analytes were determined to be above the MDL in Aqua II water and were also found at increased levels in blanks and sometimes in Miramar water. These compounds included: benzene, 2-butanone, chloroform, chloromethane, methylene chloride and toluene. Further review indicated that three analytes (2-butanone, chloroform, and methylene chloride) were found above the MQL. Analytes detected above the MQL suggests the possibility of sample contamination or increased instrumental background.

Chloromethane and bromomethane were observed in all three water sources during the final months of the Aqua II monitoring phase. Based on analytical equipment maintenance records and corresponding analytical results, changing the gas chromatograph trap in November 1989 may have been the source of these compounds.

Toluene was detected throughout the project in higher amounts in Aqua II water than in Miramar. A temporary increase from a MQL of 0.05 μg/L to an average concentration of 0.5 μg/L in toluene concentration in Aqua II AWT water was noted about midway through the monitoring phase. A similar increase in 4-methyl-2-pentanone concentration was observed in the Aqua II water during approximately the same period.

Trihalomethanes, by-products of upstream chlorination en route from Lake Skinner, were among the compounds detected in the Miramar water. Concentrations of these compounds varied considerably during the course of the project coinciding with chlorination activities. Chloroform was found in Aqua II water at low levels. Based on tests conducted to identify the THM formation potential in both waters it was found that the formation potential of THM in Miramar water was 10 times that of the AWT water. The only extractable organic compounds measured above detection limits in either Aqua II or Miramar water were the phthalates: Bis(2-ethylhexyl)phthalate, diethyl phthalate and di-*n*-octyl phthalate.

Pesticide/PCB analysis was performed four times during the second year (1989–1990) of the monitoring phase. Heptachlor epoxide was measured in the Miramar water near the detection limit. Dioxin/furans were not detected in either Aqua II or Miramar water during the monitoring phase. Detection limits ranged from 2.4–20.7 pg/L (parts per quadrillion).

During the Aqua III monitoring period for purgeable compounds, a comparison of the three water sources (Aqua III AWT, Miramar and blank waters) reveals that five of the analytes were determined to be above the MDL in Aqua II water and were also found at increased levels in blanks and sometimes in Miramar water. These compounds included: benzene, chloroform, chloromethane, methylene chloride and tetrachloroethylene. Further review indicated that methylene chloride was found above the MQL, suggesting the possibility of sample contamination or increased instrumental background.

Trihalomethanes were again detected in Miramar samples. Chloroform was found in Aqua III water at levels above those detected in Aqua II AWT samples. The increase in chloroform concentration can be attributed to chlorination prior to the filtration units. The extractable organic compounds measured above detection limits in the Aqua III or Miramar water were the phthalates and pyrene. The concentration of pyrene in blank samples was comparable to those found in the water samples indicating the possibility of laboratory contamination.

Total Organic Carbon (TOC) and Total Organic Halogen (TOX)

Based on a limited number of samples collected during the Aqua II monitoring phase, the organic carbon concentration in the RWW was estimated at approximately 35 mg/L. The average total organic carbon concentration found in AWT water was 1.37 mg/L, while TOC concentration averaged 9.83 mg/L in Miramar water. The method detection limit for this analysis was 1.0 mg/L.

Total organic halogen averaged 20.2 μg/L in the Aqua II AWT water and 43.3 μg/L in Miramar water during the Aqua II monitoring and blanks averaged 12.7 μg/L total organic halogen. During the Aqua III monitoring phase the organic carbon concentration in the RWW averaged 86 mg/L. The average total organic carbon concentration found in AWT water was 0.27 mg/L (determined by probit analysis), while TOC concentration averaged 5.56 mg/L in Miramar water. The method detection limit for this analysis was 0.5 mg/L.

CHEMICAL SPIKING STUDY

A spiking study was performed to estimate the efficiency of removal of selected organic and inorganic chemicals by the Aqua II facility. The purpose of the study was to determine the capability of the facility for handling large pulses of contaminants that could potentially enter the plant. The study consisted of adding selected chemicals to the primary wastewater as it entered the water hyacinth ponds. Samples were then collected from the water hyacinth pond effluent and the AWT plant effluent. The chemicals added were consid-

TABLE 12.11. Aqua II Chemical Removal Efficiency by Water Hyacinth Ponds and The AWT Facility-Spike Study Results.

Compound	Spike (μg/L)	Pond Effluent (μg/L)	Removal by pond (%)	AWT (μg/L)	Removal by AWT (%)	Removal Overall (%)
Chromium	182	74	59.3	< 1	> 98.6	> 99.5
Tetrachloroethylene	900	29	96.8	< .03	> 99.9	> 99.997
Tetrachloroethane	224	53	76.3	< .1	> 99.8	> 99.996
Trichlorobenzene	813	130	84.0	< .3	> 99.8	> 99.996
Tetrachlorobenzene	140	2	98.6	< .03	> 98.5	> 99.98
Lindane	70	12	82.9	> .03	> 99.7	> 99.96

ered representative of industrial chemicals that could appear in any wastewater stream and included chromium, tetrachloroethylene, tetrachlorobenzene, trichloroethan, tetrachlorobenzene, and lindane.

Peak concentrations of organic and inorganic spike components were found in pond effluent after 20–30 hours and in Aqua II AWT effluent four hours later. As shown in Table 12.11, the ponds removed 60–99 percent of the inorganic and organic spike components and the AWT process removed from 98.5 to 99.9 percent of the compounds in the pond effluent. Overall, the water hyacinth ponds and the advanced water treatment process used at the Aqua II plant were observed to remove greater than 99.5 to greater than 99.997 percent of the added chemicals.

COMPARISON OF HES MIRAMAR AND AQUA II WATER WITH U.S. SUPPLIES

A literature review was performed to compare Miramar and Aqua II raw waters with other raw water sources sampled throughout the United States [2]. The raw water supplies were compared based on metal and organic constituents present in the waters. Results of the comparison of arithmetic mean values are shown in Tables 12.12 and 12.13. As shown in Table 12.12, metal concentrations detected in AWT and Miramar waters ranged from 2×10^{-5} mg/L to 2.5×10^{-5} mg/L and were in the lower mean values of or below the U.S. average range for metal concentrations in raw water supplies.

The majority of organic constituents detected in AWT and Miramar waters were either in the lower portion of the U.S. average range or below for organics in raw water supplies (Table 12.13). Trihalomethanes ranged from 0.004 to 3.79 μg/L in AWT and Miramar waters. Volatile organic compounds ranged from 0.004 to 11.62 μg/L. Methylene chloride was the highest volatile organic compound detected in Miramar and AWT waters (9.55 and 11.62 μg/L, respectively). However, as mentioned pre-

TABLE 12.12.   Comparison of Aqua II AWT and Miramar Water Quality with
U.S. Raw Water Supplies—Metals (mg/L)

Constituent	MCL	AWT[a]	Miramar[a]	U.S. Groundwaters[b]	U.S. Surface Waters[b]
Arsenic	0.005	0.0003[c]	0.0017[c]	0.001–0.100	0.001–0.020
Cadmium	0.010	2E-05[c]	4E-05[c]	0.001–0.058	0.001–0.440
Chromium	0.050	0.008	0.003	0.0008–0.050	0.0002–0.650
Copper	NA	0.005	0.009	0.001–0.100	0.0004–0.200
Lead	0.050	0.0004[c]	0.0005[c]	0.002–0.100	0.0002–1.560
Nickel	NA	0.008	0.003	0.001–0.130	0.0013–0.500
Selenium	0.01	0.0001[c]	0.0007[c]	0.001–0.061	0.0001–0.047
Zinc	NA	0.025	0.017	0.0027–1.3767	0.001–8.600

[a]AWT and Miramar values are arithmetic mean concentrations. Data generated by University of California at Los Angeles during the chemical monitoring phase of the HES project.
[b]U.S. groundwater and surface water values are a range of arithmetic mean concentrations.
[c]Arsenic, cadmium, lead and selenium were analyzed using the AA method.
MCL: Maximum contaminant level; NA: not applicable.

viously, methylene chloride is a common laboratory contaminant and extremely difficult to detect with accuracy in any laboratory. The remaining volatile organic compounds were all detected at less than 1 μg/L.

Extractable organics ranged from 0.023 to 8.05 μg/L in AWT and Miramar waters. As shown in Table 12.13, limited data were available for extractable organic compounds in U.S. ground and surface waters. Some of the values reported are maximum rather than average values while others are only for a small region of the United States. As in the case of volatile organic compounds, the majority of the extractable organics detected in AWT and Miramar waters were in the lower portion of the U.S. average range for raw water supplies.

BIOASSAY

One important concern associated with the use of reclaimed water for human consumption is the potential for adverse health effects that may be associated with low concentrations of residual chemicals in the reclaimed water. The potential adverse health effects of most concern are the long-term chronic effects, such as cancer or genetic damage. To estimate the potential genetic toxicity associated with reclaimed water, a series of short-term bioassays were employed to compare effluent water from San Diego's Aqua II treatment facility with the raw Miramar water.

The low levels of chemicals in both AWT and Miramar raw water, and the expected sensitivity of the bioassay procedures necessitated the concentration of these chemical constituents from the water samples. During the Aqua II investigations the concentration procedure involved passing large volumes of water through XAD resin columns to absorb organic material from the water. The organic materials were then eluted from the column using an acetone/hexane solvent and concentrated into a small volume (referred to as the total concentrate) which was used for both bioassay and chemical analysis.

Four separate types of bioassay systems were employed to assess genetic toxicity and potential cancer-causing effects of the Aqua II and the Miramar waters. A summary of the results from each type of testing is presented in this section.

TABLE 12.13. **Comparison of Aqua II AWT and Miramar Water Quality with U.S. Raw Water Supplies—Organics (μg/L)**

Constituent	MCL	AWT[a]	Miramar[a]	U.S. Groundwaters[b]	U.S. Surface Waters[b]
Trihalomethanes					
Bromoform	100[c]	0.004	0.972	0.100–43.333	0.133–27.000
BDCM	100[c]	0.008	3.79	0.100–11.500	0.100–57.000
Chloroform	100[c]	0.130	2.78	0.267–19.975	0.267–198.000
DBCM	100[c]	0.017	3.15	0.100–19.733	0.100–92.000
Volatile organics					
Benzene	1.0	0.048	0.054	0.200–11.000	0.233–0.467[e]
1,1-Dichloroethane	6.0	0.004	0.004	0.280–17.700	7.2[d]
Methylene Chloride	NA	11.62	9.55	0.100–72.000	0.367–1.067[e]
Tetrachloroethene	5.0	0.006	0.009	0.100–106.000	0.100–2.033[e]
Trichlorofluoromethane	150	0.857	0.400	0.300–4.200	Not Available
Toluene	NA	0.099	0.042	0.100–8.000	0.200–21.000
Xylene	1750	0.015	0.003	0.200–7.580	0.300–400[f]
Extractable Organics					
Benzo-a-pyrene	NA	0.023	0.037	0.001–0.01	0.001–0.0100
Butylbenzyl phthalate	NA	0.166	0.152	38.000[g]	0.034–0.059[h]
Dibutyl phthalate	NA	0.994	0.912	470[g]	0.79–0.090[i]
Di-*n*-octyl phthalate	NA	1.395	0.705	2.4[j]	0.024–0.477[k]
Bis(2ethylhexyl) phthalate	4.0	8.05	4.65	0.400–12.356[k]	7.467–61.067

[a]AWT/Miramar: AWT and Miramar values are arithmetic mean concentrations. Data generated by University of California at Los Angeles during the chemical monitoring phase of the HES project.
[b]U.S. groundwater and surface water values are a range of arithmetic mean concentrations.
[c]Total Trihalomethane Limit.
[d]Only one value found. Florida urban stormwater runoff.
[e]California surface waters only.
[f]Colorado surface water only.
[g]Maximum value reported in New York.
[h]Values reported for the Mississippi River in Ill. and Tn.
[i]Values reported for the Mississippi River in Ill., Minn. and Tn.
[j]Average values near landfill sites in Norman, Ok.
[k]California groundwater only.
BDCM: Bromodichloromethane.
DBCM: Dibromochloromethane.
MCL: Maximum contaminant level.
NA: Not applicable.

Ames Assay

The Ames assay (*Salmonella*/microsome reversion assay) is used to measure the induction of hereditable genetic alterations (mutations) in special strains of *Salmonella* bacteria (*Salmonella* tester strains TA98 and TA100) following exposure to chemical or physical agents. During the Aqua II sampling period (February 1988 and June 1990) forty-eight water samples were collected, extracted and tested. Twenty-five samples from the Miramar source and twenty-three samples from the Aqua II source were collected and the organic extract, fractions, and sub-fractions were tested, at various doses, for mutagenicity using the Ames assay. A large number of the tested samples from both the Aqua II and the Miramar sources exhibited weak but statistically significant activity. These effects were seen using both strains with and without metabolic activation. Data from the individual assays were combined and a comparison of the slopes of the dose-response curves for the Aqua II and the Miramar water extracts was made. Based on the slope of the dose-response curves the extracts from the Miramar water exhibited more mutagenic activity than the extracts from the Aqua II AWT water.

During the Aqua III sampling period (September 1994 and July 1995) twelve water samples were collected from both the Miramar and Aqua III plant effluent. The samples were extracted and tested in the Ames assay. The Aqua III bioassay investigations were limited to the Ames assay since the most interesting and useful results of the aqua II investigations stemmed from the Ames assay. In addition to testing the total concentrate, many of the fractions and a few of the sub-fractions were also tested. The organic extract, fractions and sub-fractions were tested at various doses for mutagenicity. Preliminary data indicate that positive Ames dose/response curves have been achieved at 200 L/equivalents from Miramar water along, while 800 L/equivalents from the Aqua III water did not show positive dose/response curves. Increasing the L/equivalents to 1200 for the Aqua III water resulted in positive dose/response results. A preliminary comparison of the Miramar and Aqua III Ames data suggests that the Aqua III effluent water exhibits lower mutagenic response. A preliminary comparison between the Aqua II and Aqua III effluent Ames data also suggests that the Aqua III water exhibits a lower mutagenic response.

Micronucleus Test

The micronucleus test is a short-term assay to assess genetic damage occurring in the bone marrow following the in vivo exposure of mice to chemical agents. Eight samples collected between the summer of 1988 and the winter of 1990 from both the AWT and the Miramar sources were tested using the mouse micronucleus assay. Water samples were run through resin columns representing between 4000 to 6000 liters of water.

Water samples were tested at increasing dose levels throughout a two year period, resulting in acute daily exposures to the animals exceeding, by conservative estimate 100 times their expected lifetime dose via drinking water. The total administered doses contained the organic extracts from the equivalent of 300 to 900 liters of water.

Statistically significant increases in the frequency of micronuclei were not observed in the majority (95 percent) of the samples from either the Aqua II or the Miramar sources. At the higher near-lethal doses (approximately 600 liter equivalents of water), a possible trend towards increased micronuclei frequency with both Aqua II and Miramar samples was noted. An approximately three-fold increase in the frequency of micronuclei was observed in the two high doses of the AWT concentrated samples from Winter 1989. The increased frequency of micronuclei was not confirmed by follow-up experiments.

6-Thioguanine Resistance Assay

The 6-Thioguanine Resistance assay measures the mutagenic inactivation of the Hypoxanthine-Guanine Phosphoribosyl Transferase (HGPRT) gene in a cell line established from the ovary of a Chinese hamster. One sample collected during the fall of 1988 from both the AWT and the Miramar sources was tested repeatedly. In addition, four Miramar fractions collected during Winter 1989 and one Miramar fraction collected during Winter of 1990 were also tested. None of the tested samples induced a significant increase of 6-thioguanine resistant cells.

Cellular Transformation Assay

C3H10T1/2 cells, a cell line exhibiting a unique appearance and growth characteristics, is non-tumorigenic when injected into immunosuppressed mice. One sample collected during the fall of 1988 from both the AWT and the Miramar sources was tested several times. Although some tests indicated a possible increase in transformed cells following treatment with the Miramar water, a consistent reproducible and statistically significant increase in the number of transformed cells was not observed for either the AWT or the Miramar water extracts.

In summary, the genotoxic effects detected in this study were observed primarily in the Ames test, but some indications of potential mutagenic activity were also observed in the other bioassays. Based on the Ames test data, the tested Miramar extracts exhibited somewhat higher mutagenic activity than the comparable Aqua II and III extracts. The reason for this difference in observed mutagenicity is unknown, but may reflect differences in composition of the original source waters including chlorination of Miramar source water prior to entering the Miramar water treatment plant. From these data, based on short-term bioassay results of organic extracts of the Aqua II, III and Miramar waters, it can be concluded that water from the Aqua II and III facilities are unlikely to be more genotoxic or mutagenic than Miramar raw water.

CHEMICAL ANALYSES OF RESIN EXTRACTS

As described previously, during the Aqua II sampling period, synthetic resins were used to concentrate trace organics from Aqua II, Miramar, and reagent (blank) water. A total of 71 resin extracts were analyzed by gas chromatography-mass spectrometry to characterize the component mixture. The resin extracts were derived from water collected each month, February through June 1990, from the Aqua II and Miramar sites. Prior to the resin columns, the samples were filtered through a teflon wool pre-filter. The size of the extracted water samples ranged from 70 to 200 L equivalents. The resin extract for each sample was fractionated by High Performance Liquid Chromatography (HPLC) into four fractions (B, C, D, E). The C fraction was fractionated a second time to give the C_1, C_2, and C_3 subfractions. Therefore a total of 6 fractions (B, C_1, C_2, C_3, D, E) were analyzed by GCMS for each water sample collected. Additionally, a fractionated method blank and the B, C, D and E fractions of the extracted pre-filter for the Miramar water (February sample) were also analyzed.

Results of the GCMS analysis of resin accumulated organic residues from Aqua II and Miramar waters are discussed in detail in the Aqua II HES final report. Analysis of resin accumulated organic residues from Aqua II and Miramar waters revealed the presence of many classes of organic compounds in low to trace amounts in the waters. Several herbicides were found in four of the five Miramar water samples. Only one compound that was predicted to be an active mutagen and reported positive in the Ames test, 2-hexenal, was found in the monthly water samples. The compound 2-hexenal was found in small amounts in the B fraction of both Aqua II and of Miramar samples. Of the compounds found, 26 percent were detected in both Aqua II and Miramar samples.

FISH BIOMONITORING

To complement information gained from other HES activities, fish biomonitoring experiments were conducted to provide information on trace contaminants contained in the effluent from the Aqua II plant and the untreated raw city water that accumulate in tissue but do not show activity in genetic toxicity screening bioassays [20,21]. Juvenile fathead minnows (*Pimephales promelas*) were exposed to Aqua II water, Miramar water, or charcoal filtered San Diego tap water (acclimation or control water) in flow-through aquaria. Three 28-day bioaccumulation experiments were completed using water at each site under standardized conditions following Environmental Protection Agency (EPA) [22] and American Society for Testing and Materials (ASTM) [23] bioconcentration guidelines. In addition, swim speed and optomotor response tests were evaluated for inclusion in the biomonitoring component. After experience with both tests, the swim speed test was selected for inclusion in all biomonitoring experiments.

A total of 120 fish tissue samples and 60 aquarium water samples were collected and analyzed for 69 base/neutral/acid extractable organics, 27 pesticides/ PCBs, and 26 inorganics by investigators at the University of California at Los Angeles (UCLA). Quadruplicate samples of fish tissue and either duplicate or triplicate samples of water were collected and analyzed for each sampling time (Days 0, 7, 14, and 28 of the 28-day experiments). The concentrations of most analytes in these samples were below the detection limit. Statistical comparisons were only made if at least 75 percent of the samples exceeded the detection limit for a given analyte. Chemical analysis of fish tissue and water in the 28-day bioconcentration experiments revealed no statistically significant differences in chemical composition between the fish exposed to Aqua II water and those exposed to Miramar water. The fish also displayed no significant difference in survival, growth, or swim speed after up to 28 days of continuous whole body exposure to test waters. This finding was the same for all three tests conducted on each of the three waters (i.e., Aqua II, Miramar, and control waters).

PLANT RELIABILITY

Treatment plant reliability can be defined as the probability that a system can meet established performance criteria consistently over extended periods of time. In the case of a water reclamation plant that produces an effluent for potable reuse, reliability can be defined as the likelihood of the plant achieving an effluent that matches, or is superior to, predetermined standards. If predetermined standards are not available, reliability can be defined as the likelihood of achieving a consistent effluent of acceptable quality, consistently over extended periods without mechanical failures or excessive variability in performance. For the purposes of this study, overall plant reliability was determined by considering both normal effluent variability, comparison with drinking water standards and with performance of other similar plants, and the treatment plant operational reliability.

Normal effluent quality variability (i.e., inherent reliability) can be evaluated by estimating the probability that the plant effluent will achieve or exceed predetermined effluent standards. When predetermined effluent standards are not available, specific observed concentrations can be used to describe expected mean values and variability. This characterization will also allow comparison of predicted effluent concentrations with effluent standards defined in the future. Inherent reliability only encompasses the variability associated with effluent quality related to the in-plant treatment processes, and assumes that the specified plant is operated consistently and maintained properly. Operational reliability, on the other hand, is defined as the probability that the plant will be non-functional at any given time. This definition of reliability requires evaluating plant operational reliability as a function of mechanical, design, process, or human factors. This analysis is separate from the normal effluent quality variability.

The reliability analysis can be used to quantify the reliability of a plant and to identify weak points in the treatment process requiring modification or correction. By identifying and improving the weak points of the treatment process, the reliability of the treatment system increases. Often the corrections or modifications are simple. Even a well maintained, well operated plant is not perfectly reliable. Variations in influent flow and quality create variation in effluent characteristics, which affect the reliability of the system. Design values and other factors such as power outages, equipment failure, and operational (human) error may impact plant reliability and have been incorporated into the reliability analysis.

INHERENT RELIABILITY

A wastewater treatment system can be viewed as a "black box"; wastewater flows in and treated water flows out, or the treatment system can be viewed as a composite of many processes; each with a definite means of altering the system output. Both perspectives were used to evaluate the inherent variability of the Aqua II and Aqua III facilities. Assessing inherent effluent variability involves the basic statistics associated with frequency analysis including mean values, standard deviations, variances, and coefficients of variation. Frequency distributions, using a log-normal distribution, were plotted for various plant parameters to provide a more precise description of projected performance. The percent removals of measured effluent parameters by each unit process were calculated to provide further information on the ability of the plant to achieve a consistent level of treatment.

Plant Performance

The plant process performance was reviewed to characterize the normal effluent quality and variation through each of the key treatment plant units. The plant performance analysis involved reviewing and analyzing, statistically, influent and effluent data for individual plant unit processes and for the entire plant. The analysis covered Aqua II pilot plant operating data for approximately two and a half years (July 1987 through December 1989) and Aqua III full scale plant operating data for one year (October 1994 through September 1995).

Sampling and Analysis

Water samples were collected before and after each of the individual unit processes and from the final Aqua II and Aqua III plant effluent. Sampling locations include plant influent (RAW), rotary disk filter (RDF), aquaculture pond effluent/package plant influent (PPI), package plant effluent (PPE), reverse osmosis influent (ROI), reverse osmosis effluent (ROE), aeration tower effluent (ATE), and final plant effluent/carbon tower effluent (CTE). The water samples were analyzed for physical constituents, metals, nitrogen compounds, anions, and coliform.

Sampling frequency varied for each parameter and location: samples for physical constituents were collected daily or weekly; for nitrogen compounds, silica, and coliform were collected weekly; and for anions, organic compounds, and metals were collected monthly.

Analytical Results

Several methods for evaluating inherent plant reliability, including: time series analysis, probability plots, and statistical analysis of process performance were applied. Log normal probability plots were generated for most constituents for each sampling locations, except where the majority of the sample results were below detection limit. No probability plots were generated for total and fecal coliform and organic compounds as the values for these analytes in the City's data were below analytical detection limits. All of the plots are available in separate documents [24,25]. Most of the water quality data generated in the Aqua II and III studies fit a log normal cumulative frequency distribution. To illustrate the results of this analysis, the log normal probability plots and TOC in the Aqua II and Aqua III plants are included as examples (Figures 12.1 and 12.2).

Whisker-box plots provide a graphic overview of the unit process per-

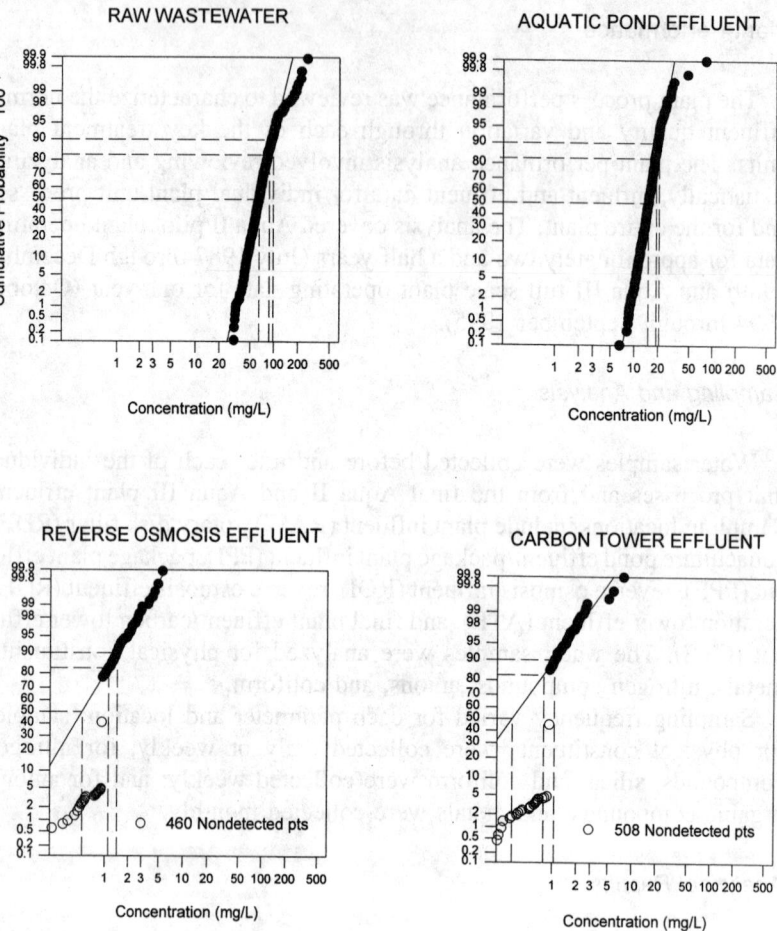

Figure 12.1 Aqua II lognormal probability plots for TOC.

formances and variabilities for removal of constituents along the entire treatment process. To illustrate the utility of these plots, Figures 12.3 and 12.4 were generated for CBOD, COD, TOC, turbidity, total solids, and total suspended solids. The ''box'' on these plots represents the middle 50 percent of the data (the 25th to the 75th percentile), while the ''whiskers'' extend to the extreme values. The line through the box represents the median value of the data. The trends exhibited by these constituents are representative of the treatment effectiveness and variability through the treatment train for certain classes of contaminants. Review of the example figures illustrates the treatment effectiveness and variability through the treatment processes of the Aqua II and III facilities. Further, the significant reduction in overall variability of water quality indicators achieved by the treatment facilities is also illustrated.

Unit Process Performance

To evaluate unit process performance, the percent removals of the analytes was calculated. The Aqua II and III treatment plants contain four major unit processes:

- primary—rotary drum and rotary disk filters
- secondary—water hyacinth ponds

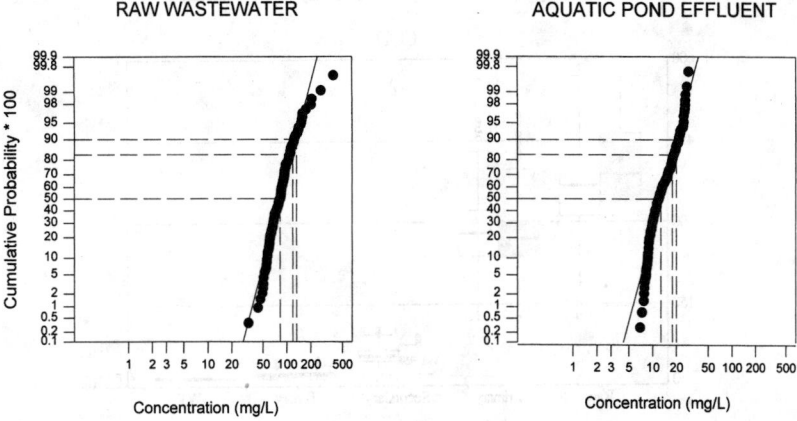

Figure 12.2 Aqua III lognormal probability plots for TOC.

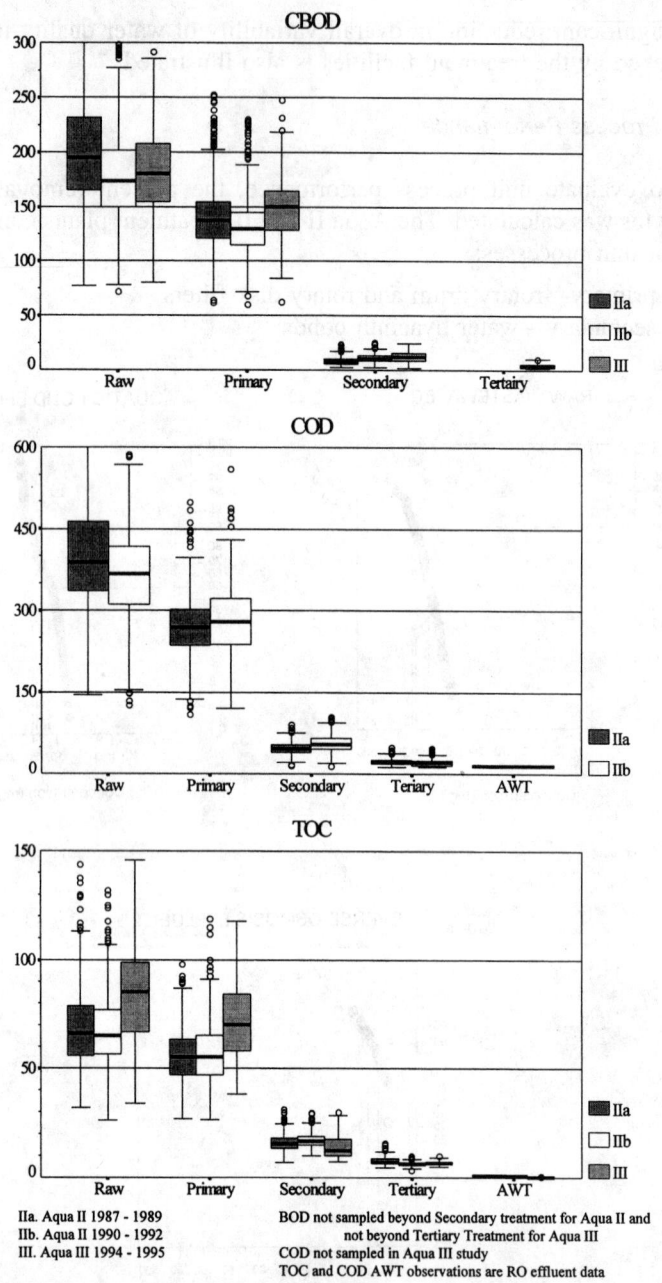

Figure 12.3 Aqua II and III—unit process CBOD, COD, and TOC removals. All units are mg/L.

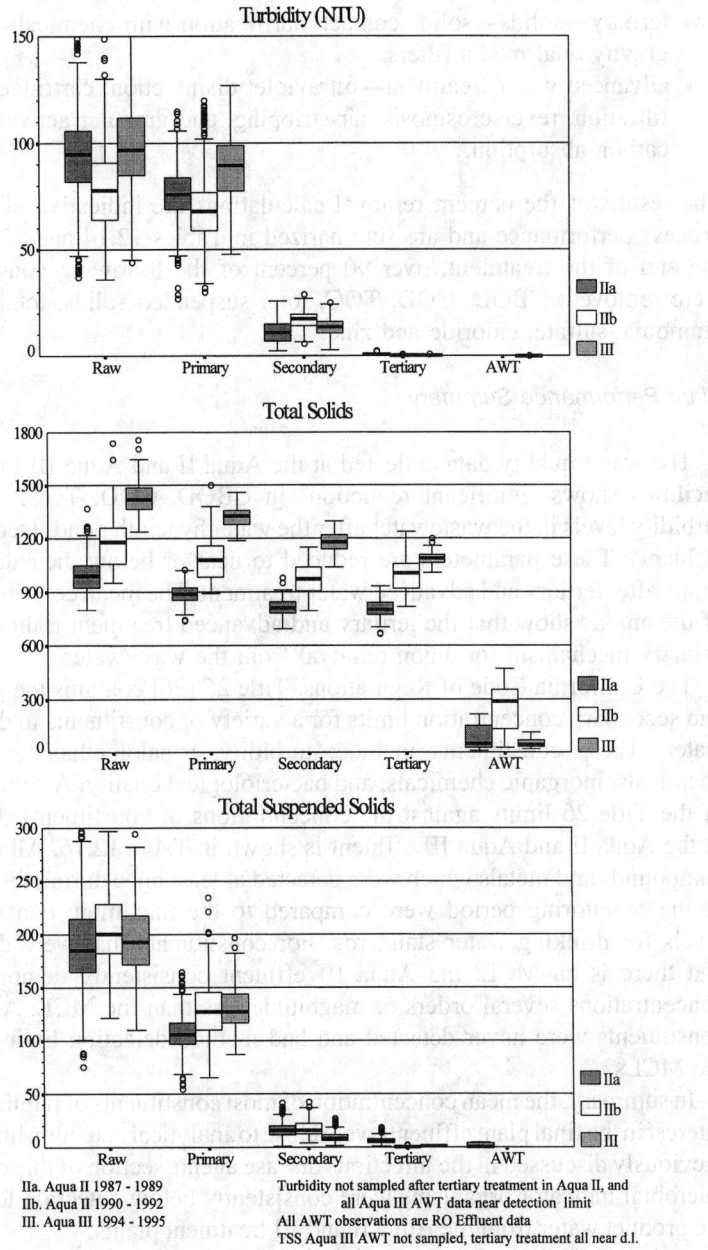

Figure 12.4 Aqua II and III—unit process turbidity, total solids, and total suspended solids removals. Units are mg/L unless otherwise noted.

IIa. Aqua II 1987 - 1989
IIb. Aqua II 1990 - 1992
III. Aqua III 1994 - 1995

Turbidity not sampled after tertiary treatment in Aqua II, and all Aqua III AWT data near detection limit
All AWT observations are RO Effluent data
TSS Aqua III AWT not sampled, tertiary treatment all near d.l.

- tertiary—solids—solids contact clarification with chemicals plus gravity dual media filters
- advanced water treatment—ultraviolet disinfection, cartridge filtration, reverse osmosis, air stripping, and granular activated carbon absorption.

The results of the percent removal calculations are indicative of overall process performance and are summarized in Tables 12.14 and 12.15. By the end of the treatment, over 90 percent of the following constituents were removed: CBOD, COD, TOC, total suspended solids, total solids, ammonia, sulfate, chloride and zinc.

Plant Performance Summary

The water quality data collected at the Aqua II and Aqua III treatment facilities shows significant reductions in CBOD, COD, TOC, SS, and turbidity levels in the wastewater after the water hyacinth ponds (secondary effluent). These parameters are reduced to near or below their detection limits after tertiary and advanced water treatment. The mean concentrations of the anions show that the tertiary and advanced treatment train was the primary mechanism for anion removal from the wastewater.

The California Code of Regulations, Title 26 [26] contains the primary and secondary concentration limits for a variety of constituents in drinking water. These constituents include: turbidity, trihalomethanes, organic chemicals, inorganic chemicals, and bacteriological quality. A comparison of the Title 26 limits against the concentrations of constituents detected in the Aqua II and Aqua III effluent is shown in Table 12.16. All organic compounds and metals which were detected at least once during the course of the monitoring period were compared to the maximum contaminant levels for drinking water standards. For constituents that were detected and there is an MCL, the Aqua III effluent consistently demonstrated concentrations several orders of magnitude less than the MCL. All other constituents were never detected and had method detection limits below the MCLs.

In summary, the mean concentration of most constituents of public health interest in the final plant effluents were close to analytical detection limits. As previously discussed in the infectious disease agents section of this chapter, microbial indicator organisms were consistently below detection levels in the product water from the Aqua II and III treatment plants.

Comparison with Inherent Reliability of Other AWT Plants

In evaluating the inherent reliability of the Aqua II and III plants

TABLE 12.14. Aqua II Performance Summary (all units are mg/L unless otherwise noted).

Constituent	RAW Conc.	Primary Conc.	Effluent % Rem	Secondary Conc.	Effluent % Rem	Tertiary Conc.	Effluent % Rem	AWT Conc.	Effluent % Rem	Total Plant % Rem
Physical 87–89										
CBOD[a]	212	142	33	9	63	NA		NA		96
TSS	209	108	48	17	44	7	5	2.7	2	99
COD	427	270	37	48	52	24	6	15	2	96
TOC	71	56	21	16	56	8	11	1.1	10	98
TS	1008	898	11	815	8	804	1	81	72	92
Turb (NTU)	96	76	21	11	68	0.7	11	NA		99+
Physical 90–92										
CBOD[a]	179	132	26	11	68	NA		NA		94
TSS	211	129	39	18	53	1.4	8	1.3	0	99+
COD	371	282	24	57	61	22	9	15	2	96
TOC	68	57	16	17	59	6.5	15	1	8	99
TS	1180	1083	8	976	9	997	0	254	61	78
Turb (NTU)	69	79	0	17	75	0.4	24	NA		99+
Nutrients 87–89										
Ammonia—N	24.8	23.3	6	9.8	54	9	3	1.1	32	96
Nitrate—N	0.1	0.1	17	2.9	0	3.6	0	0.6	0	0
Phosphate—P	14.1	12.6	11	12.7	0	2.2	74	1.6	4	89
Sulfate	177	176	1	172	2	169	2	3.1	94	98
Chloride	195	188	4	185	2	192	0	16	87	92

TABLE 12.14. (continued).

Constituent	RAW Conc.	Primary Conc.	Effluent % Rem	Secondary Conc.	Effluent % Rem	Tertiary Conc.	Effluent % Rem	AWT Conc.	Effluent % Rem	Total Plant % Rem
Metals 87–89										
Arsenic	0.0025	0.0026	0	0.0025	0	0.0017	32	0.0016	4	36
Cadmium	0.0028	0.0024	14	0.0018	21	0.0015	11	0.0010	18	64
Chromium	0.017	0.014	15	0.008	35	0.004	26	0.002	13	89
Copper	0.103	0.094	9	0.029	63	0.023	6	0.017	6	83
Lead	0.029	0.030	0	0.017	41	0.010	26	0.003	22	89
Manganese	0.097	0.094	3	0.060	35	0.118	0	0.015	46	85
Mercury	0.0012	0.0010	17	0.0011	0	0.0010	0	0.0010	0	17
Nickel	0.021	0.019	13	0.016	13	0.019	0	0.004	54	79
Selenium	0.005	0.005	4	0.003	45	0.004	0	0.003	0	31
Silver	0.008	0.006	27	0.007	0	0.003	33	0.005	0	42
Zinc	0.109	0.095	13	0.024	65	0.021	3	0.008	12	92
Boron	0.26	0.25	5	0.26	0	0.26	0	0.23	8	13
Calcium	67.7	70.5	0	64.4	5	65.3	0	3.6	90	95
Iron	0.80	0.68	14	0.50	23	0.13	46	0.04	12	95
Magnesium	29.8	29.7	0	27.1	9	29.9	0	1.6	86	95
Sodium	127	121	5	131	0	136	0	11.3	86	91

[a]Raw and Primary effluent values are BOD not CBOD.

Notes:

Percent removals are the percentage of the total constituent removal due to the individual treatment. Treatment removals sum to equal the total plant removal.

Negative removals are reported as zero.

NA: not available.

Physical and nutrient concentration values are arithmetic means. Any nondetected observations were assumed to be present at the corresponding detection limit.

Meal concentration values are geometric means determined through probit analysis.

TABLE 12.15. Aqua III Performance Summary (10/94–9/95) (all units are mg/L unless otherwise noted).

Constituent	RAW Conc.	Primary Conc.	Effluent % Rem	Secondary Conc.	Effluent % Rem	Tertiary Conc.	Effluent % Rem	AWT Conc.	Effluent % Rem	Total Plant % Rem
Physical										
CBOD[a]	185	149	19	13	74	4.3	5	NA		98
TSS	219	131	40	9.8	55	1.3	4	NA		99+
TOC	91	72	21	14	64	7.1	8	0.6	7	99+
TS	1452	1322	9	1183	10	1090	6	43	72	97
Turb (NTU)	100	88	12	14	74	0.5	14	0.27	0	99+
Nutrients										
Ammonia—N	22	21	5	9.5	52	9.3	1	0.8	39	96
Nitrate—N	0.1	0.1	0	1.4	0	1.7	0	0.7	0	0
Phosphate—P	6.1	5.1	16	3.4	28	0.1	54	0.1	0	98
Sulfate	312	283	9	309	0	368	0	0.1	91	99+
Chloride	240	232	3	238	0	284	0	15	90	94
TKN	31.5	30.6	3	13.9	53	14.2	0	0.9	41	97

TABLE 12.15. (continued).

Constituent	RAW Conc.	Primary Conc.	Effluent % Rem	Secondary Conc.	Effluent % Rem	Tertiary Conc.	Effluent % Rem	AWT Conc.	Effluent % Rem	Total Plant % Rem
Metals										
Arsenic	0.0032	0.0031	3	0.0025	19	0.0015	30	0.0003	40	92
Cadmium	0.0006	0.0005	17	0.0012	0	0.0001	67	0.0001	0	83
Chromium	0.003	0.004	0	0.002	32	0.001	24	0.001	28	83
Copper	0.063	0.070	0	0.043	33	0.009	52	0.011	0	83
Lead	0.008	0.008	0	0.008	0	0.001	93	0.001	0	91
Manganese	0.065	0.062	4	0.039	37	0.002	57	0.002	0	97
Mercury	0.0003	0.0002	33	0.0001	33	0.0001	0	0.0001	0	67
Nickel	0.007	0.010	0	0.004	33	0.004	11	0.001	45	89
Selenium	0.003	0.003	0	0.002	16	0.002	0	0.001	64	80
Silver	0.002	0.003	0	0.001	75	0.001	0	0.001	0	75
Zinc	0.081	0.076	6	0.024	64	0.002	27	0.002	0	97
Boron	0.35	0.38	0	0.42	0	0.31	13	0.29	3	17
Calcium	74.4	72.2	3	66.7	7	70.1	0	1.0	88	99
Iron	0.60	0.53	11	0.18	59	0.05	22	0.04	2	94
Magnesium	38.5	38.1	1	39.3	0	6.4	82	1.5	13	96
Sodium	198	192	3	198	0	211	0	11.9	91	94

[a]Raw and primary effluent results are BOD not CBOD.

Notes:

Percent removals are the percentage of the total constituent removal due to the individual treatment. Treatment removals sum to equal the total plant removal.

Negative removals are reported as zero.

NA: not available.

Physical and Nutrient concentration values are arithmetic means. Any nondetected observations were assumed to be present at the corresponding detection limit.

Metal concentration values are geometric means determined through probit analysis.

TABLE 12.16. Comparison of Aqua III AWT Effluent Detected Constituents with CA MCLs (all units are μg/L).

Constituent	MDL	MQL	# > MDL	Total #	Method 1 Mean[a]	Method 2 G.M.[b]	MCL[c]
Purgeables							
Benzene	0.02	0.1	8	26	0.024	0.009	1.00
Bromochloromethane	0.16	0.8	1	26	0.085		
Bromodichloromethane[e]	0.04	0.2	17	26	0.500	0.138	100[e]
Bromoform[e]	0.06	0.3	1	26	0.035		100[e]
Bromomethane	0.05	0.25	1	26	0.027		
2-Butanone	0.07	0.35	1	26	0.036		
Carbon disulfide	0.04	0.2	10	26	0.100	0.040	
Chloroform[e]	0.03	0.15	19	26	0.624	0.390	100[e]
Chloromethane	0.04	0.2	18	26	0.156	0.062	
Dibromochloromethane[e]	0.05	0.25	5	26	0.042	0.030	100[e]
1,3-Dichlorobenzene	0.04	0.2	1	26	0.033		
1,4-Dichlorobenzene	0.04	0.2	1	26	0.034		5.0
1,1-Dichloroethane	0.05	0.25	1	26	0.026		5.0
trans-1,2-Dichloroethene	0.05	0.25	1	26	0.088		10.0
Dichlorodifluoromethane	0.18	0.9	2	26	0.104	0.056	
2,2-Dichloropropane	0.02	0.1	1	26	0.011		
1,1-Dichloropropene	0.04	0.2	1	26	0.048		
Ethyl Benzene	0.01	0.05	1	26	0.008	0.007	680
2-Hexanone	0.06	0.3	2	26	0.035		
Methylene chloride	0.05	0.25	17	26	7.908	0.912	
4-Methyl-2-pentanone	0.05	0.25	1	26	0.032		
Tetrachloroethylene	0.01	0.05	2	26	0.037	1.3E-06	5.0
Toluene	0.06	0.3	3	26	0.045	0.003	150
Trichloroethene	0.07	0.35	1	26	0.040		5.0
Trichlorofluoromethane	0.07	0.35	3	26	0.049	0.039	150
m,p-Xylene	0.10	0.50	1	26	0.055		1750
o-Xylene	0.01	0.05	1	26	0.009		

559

TABLE 12.16. (continued).

Constituent	MDL	MQL	# > MDL	Total #	Method 1 Mean[a]	Method 2 G.M.[b]	MCL[c]
Extractables							
Bis(2-ethylhexyl) phthalate	0.048	0.24	25	25	7.903	4.519	4.0
Butyl benzyl phthalate	0.30	1.5	13	25	0.411	0.293	
Chrysene	0.068	0.34	2	25	0.038	0.039	
1,3-Dichlorobenzene	0.084	0.42	2	25	0.048	0.069	
Diethyl phthalate	0.06	0.3	15	25	0.079	0.058	
Di-n-butyl phthalate	0.06	0.3	25	25	2.481	1.826	
2,4-Dinitrotoluene	1.163	5.815	7	25	2.208	3.984	
Di-n-octyl phthalate	0.064	0.32	21	25	0.433	0.273	
Phenanthrene	0.064	0.32	1	25	0.037		
Pyrene	0.069	0.345	8	25	0.124	0.183	
Metals							
Copper	0.2		15	15		10.8	
Lead	0.5		9	15		0.71	15[d]
Manganese	0.5		15	15		1.80	
Nickel	0.5		3	15		0.70	100
Zinc	1		8	15		2.30	
Boron	5		13	13		293	
Iron	30		6	12		37	
Potassium	500		12	14		700	
Sodium (m g/L)	10		10	14		11.9	

[a] Arithmetic Mean computed using all values <MDL at one half of the MDL.
[b] Geometric Mean determined through probit analysis. Computed in cases where there were 2 or more detected observations.
[c] Source: Barclays California Code of Regulations, 1/20/95.
[d] Federal Standard.
[e] THMs regulated at a total of 100 µg/L.
Source: Organic Compounds—UCLA labs; Metals—City of San Diego Labs.

an effort was made to identify comparable facilities, and compare the effectiveness and variability of treatment at similar facilities. Reports and publications from two other major advanced wastewater treatment reclamation studies were reviewed for information regarding the quality and variability of influent and effluent, and the effectiveness of removal of contaminants. The Denver Potable Water Reuse Project, and the Water Factory 21 Advanced Treatment Wastewater Reclamation Project were compared with the San Diego Aqua II and III plants [2,3,27]. Each of these projects included evaluation of effluent quality with respect to the potential for potable reuse of wastewater after treatment in an advanced wastewater treatment (AWT) plant.

Metals data and selected physical data are summarized in Table 12.17. for each of the above facilities to provide a comparison of the effectiveness and variability of treatment achieved at various facilities. The 98th percentile is presented, as *calculated* by multiplying the geometric mean times the square of the spread factor. This calculation results in a value for the concentration that is exceeded less than 2.5 percent of the time for the data set. The calculated 98th percentile value is shown for each of the selected constituents, for the influent and effluent. For the selected compounds shown in the tables, with the possible exception of copper, it can be concluded that approximately equivalent effluent concentrations are achieved from the four advanced water treatment plants. For the general parameters (TOC, TDS, nitrate) the performance of the Aqua II and III plants and the Denver plants were approximately equivalent, while the Water Factory 21 facility produced an effluent with somewhat higher TOC and TDS concentrations. In general, influent values were similar, although not identical.

MECHANICAL RELIABILITY: CRITICAL COMPONENT ANALYSIS

The Critical Component Analysis approach was selected to evaluate mechanical component reliability of the Aqua II and Aqua III plants. This method has been developed and used by Environmental Protection Agency (EPA) to determine the in-service reliability, maintainability, and operational availability of selected critical wastewater treatment components [28]. The objective of the critical component analysis is to determine the mechanical components in the wastewater treatment plant that would have the most immediate impact on effluent quality should failure occur.

Performance statistics calculated on data gathered from the critical components were divided into the following main categories:

- reliability statistics: (1) overall mean time between failures (overall MTBF) and (2) lower limit 90 percent confidence level for MTBF point estimates

TABLE 12.17 Comparison of Secondary and AWT Effluent Concentrations at Aqua II, Aqua III, Water Factory 21, and Denver Treatment Facilities (all units are mg/L).

Constituent	Source	Aqua III		Aqua II		Water Factory 21		Denver	
		Mean	98th %ile	Mean	98th %ile	Mean	98th %ile	Mean	98th %ile
Arsenic	Sec. Eff.	0.0025	0.0030	0.0025	0.015	0.005	NA	<0.001	NA
	AWT Eff.	<0.0005	NA	0.0020	0.030	0.005	NA	<0.001	NA
Boron	Sec. Eff.	0.42	0.61	0.26	0.54	0.85	1.17	0.41	0.59
	AWT Eff.	0.29	0.50	0.23	0.49	0.52	0.78	6.23	6.33
Cadmium	Sec. Eff.	0.0012	NA	0.0018	0.0093	0.0093	0.054	<0.001	NA
	AWT Eff.	<0.0002	NA	0.0010	0.0016	0.00007	0.00075	<0.001	NA
Calcium	Sec. Eff.	66.7	96.0	64.4	114.0	85.9	116.5	52.1	75.0
	AWT Eff.	<2.0	NA	3.6	58	0.01	0.85	0.8	2.6
Chromium	Sec. Eff.	<0.004	NA	0.008	0.061	0.033	0.093	0.002	0.0088
	AWT Eff.	<0.001	NA	0.002	0.014	0.008	0.025	<0.001	NA
Copper	Sec. Eff.	0.0425	0.109	0.029	0.080	0.049	0.147	0.023	0.066
	AWT Eff.	0.0108	0.052	0.017	0.058	0.004	0.019	0.009	0.036
Iron	Sec. Eff.	0.18	0.46	0.50	1.93	0.11	0.33	0.25	0.423
	AWT Eff.	0.037	0.148	0.039	0.29	0.003	0.0079	<0.001	NA
Lead	Sec. Eff.	0.008	0.0157	0.017	0.277	0.0048	0.021	<0.001	NA
	AWT Eff.	0.0007	0.013	0.003	0.019	0.00065	0.0029	<0.001	NA
Magnesium	Sec. Eff.	39.3	47.6	27.1	132.9	24.5	36.8	12.6	15.2
	AWT Eff.	<3.0	NA	1.6	20.6	0.02	1.3	<0.2	NA
Manganese	Sec. Eff.	0.0385	0.099	0.06	0.17	0.056	0.13	0.1	0.14
	AWT Eff.	0.0018	0.020	0.015	0.081	0.0013	0.027	<0.01	NA
Mercury	Sec. Eff.	<0.0002	NA	0.0010	0.0016	0.00033	0.0032	0.0001	0.00036
	AWT Eff.	<0.0002	NA	0.001	0.001	0.00028	0.0023	<0.00005	NA
Nickel	Sec. Eff.	0.0044	0.0063	0.016	0.14	0.077	0.23	0.007	0.037
	AWT Eff.	0.0007	0.0047	0.004	0.02	0.00015	0.003	<0.001	NA

TABLE 12.17 (continued).

Constituent	Source	Aqua III		Aqua II		Water Factory 21		Denver	
		Mean	98th %ile	Mean	98th %ile	Mean	98th %ile	Mean	98th %ile
Potassium	Sec. Eff.	10.9	15.7	4.8	15.1	14.3	26.1	12.7	18.3
	AWT Eff.	0.7	1.0	0.91	7.6	1.1	3.6	0.6	1.2
Selenium	Sec. Eff.	0.0021	0.0026	0.0025	0.053	0.005	NA	<0.001	NA
	AWT Eff.	<0.001	NA	0.0030	0.01	0.005	NA	<0.001	NA
Silver	Sec. Eff.	<0.001	NA	0.007	0.03	1.58	6.2	0.001	0.003
	AWT Eff.	<0.001	NA	0.004	0.01	0.00016	0.0017	<.001	NA
Sodium	Sec. Eff.	198	240	131	260	232	277	117	142
	AWT Eff.	11.9	17.1	11.3	139.3	20.9	50.7	4	10.2
Zinc	Sec. Eff.	0.024	0.047	0.024	0.062	0.086	0.20	0.38	0.86
	AWT Eff.	0.0023	0.011	0.008	0.032	0.10	NA	0.005	0.04
TOC	Sec. Eff.	13.5	26.6	15.8	26.0	16.7	30.9	16.4	23.5
	AWT Eff.	0.27[a]	1.50	0.33	2.33	1.09	7.66	0.2	1.0
TDS	Sec. Eff.	1180[b]	1428	796.5	939.2	1066.6	1249.6	681	703
	AWT Eff.	42.0[c]	60.5[c]	15.0	71.9	82.4	257.5	17.0	55.1
Nitrate	Sec. Eff.	0.8	9.2	2.18	6.47	0.25	1.35	0.1	3.6
	AWT Eff.	0.6	3.5	1.23	14.69	0.07	0.68	0.1	0.63

[a] Reverse osmosis effluent data.
[b] Computed from (TS-TSS).
[c] Total solids data.

563

- maintainability and maintenance statistics: (1) mean time to repair (MTTR): mean time the unit was out of service and (2) corrective maintenance person hours per unit per year
- availability statistics: (1) inherent availability (AVI): fraction of calendar time the unit was operating and (2) operating availability (AVO): fraction of time the unit can be expected to be operational excluding preventative maintenance

The statistics calculated from the data serve as the best estimate of the mechanical reliability, maintainability of maintenance, and subcomponent failure distribution parameters for the critical components of the Aqua II and Aqua III facilities.

Mechanical Reliability Analysis

Performance statistics were calculated from maintenance data collected over a two and a half year period (June 26, 1987 to January 6, 1990) at Aqua II and for approximately one year period (October 1994 through September 1995) at Aqua III. Performance statistics were calculated from maintenance data collected at Aqua III over the one year period between October 1994 and September 1995.

A brief description and summary of the mechanical reliability analysis is presented below. A detailed description of the methodology, raw data, and analyses are presented in the Aqua II and Aqua III HES plant reliability reports [24,25].

The first step in the analysis was to identify the "critical" components. The critical components are the pieces of equipment identified as most important to maintaining continuous, uninterrupted operation of the treatment plant. Equipment and maintenance data were collected from three main sources: data collection forms, summary of entries in weekly plant operations binder, and discussion with treatment plant water superintendent.

A summary of the results of the statistical analysis completed on the nine process units for Aqua II and for the eleven process units for Aqua III is shown in Table 12.18. Results of the mechanical reliability analysis indicate that, over a two and one-half year period at Aqua II and for a one year period at Aqua III, the critical equipment were operational nearly 100 percent of the time. In addition, the mechanical reliability analysis confirms statistically what is known practically about the Aqua II and Aqua III Plants: observed equipment failures did not significantly effect water quality or cause any significant "down time" in the facilities.

Further, the reliability statistics for the two facilities are very comparable. Mean time between failures are generally slightly higher for Aqua III

TABLE 12.18. Mechanical Reliability Summary Statistics for Aqua II and Aqua III.

Treatment Unit	MTBF[c]		90% C.L. MTBF		AVO[d]		AVI[e]	
	Aqua II[a]	Aqua III[b]	Aqua II	Aqua III	Aqua II	Aqua III	Aqua II	Aqua III
Headworks	0.35	0.79	0.28	0.57	0.992	0.9953	0.9956	0.99977
Primary	0.92	0.82	0.67	0.65	0.9977	0.99668	0.9994	0.99806
Primary Odor Control System		0.68		0.45		0.9985		0.99889
Secondary	1.29	2.12	1.09	1.75	0.9975	0.97575	0.9998	0.9963
Package Plant	0.38	2.24	0.33	1.78	0.9989	0.99944	0.9997	0.99949
Ultraviolet Disinfection	5.36	0.58	3.58	0.25	0.9988	0.99906	0.9999	0.99843
Reverse Osmosis	0.39	1.22	0.35	0.99	0.997	0.99004	0.9993	0.99034
Aeration Tower	0.14	1.16	0.12	0.50	0.9989	0.7835	0.9996	0.99951
Carbon Tower	1.30	1.86	0.89	1.02	0.9999	0.99626	0.9999	0.9999
Product Water	0.18	0.56	0.15	0.45	0.9996	0.9771	0.9997	0.99641
Solids Processing		0.96		0.41		0.99943		0.99905

[a] Aqua II data 6/20/87 through 1/6/90.
[b] Aqua III data 10/9/94 through 9/30/95.
[c] MTBF = Point estimate for the mean number of years between failures.
[d] AVO = Fraction of operating time that components in the unit were functioning.
[e] AVI = Fraction of time that the unit can be expected to be operational excluding preventative maintenance downtime.

than for Aqua II, with the exception of the ultraviolet disinfection unit. Although, based on available data, the MTBF is lower, it is not expected that failures would occur more frequently than once every 20 days. AVO and AVI are high for all units in both facilities, with the exception of the aeration tower AVO in Aqua III. The lower AVO in Aqua III results from plant staff voluntarily leaving a non critical plant component, a transfer pump, out of service.

CHEMICAL RISK ASSESSMENT

Chemical risk assessment was carried out to compare the potential health risks of the existing drinking water supply (based on untreated potable water) with the water supplied by Aqua II and/or Aqua III (based on water quality without final disinfection and no additional water treatment). This risk assessment incorporated the general methodology prescribed by the Environmental Protection Agency (EPA) and

- summarized and evaluated existing data
- addressed exposure pathways
- assessed potential adverse effects of constituents detected in the waters
- characterized the potential health risks associated with exposure to each constituent
- compared the potential health risks associated with each water source sampled

DATA EVALUATION AND CHEMICALS OF POTENTIAL CONCERN

Data collected during the chemical monitoring phase of the Aqua II and Aqua III HES were evaluated to determine statistically significant concentrations of the chemicals detected and to determine which chemicals to include in the risk assessment. To determine chemicals of potential concern, average values were compared with the average values for the travel blanks (i.e., quality assurance samples used as a check on sample contamination). In many cases, the long term average values for the travel blanks were similar in order of magnitude to the long term average values of the Aqua II and Aqua III product water and the Miramar water samples. Therefore, the long term average concentrations (mean difference) were calculated by subtracting the blank averages from the sample averages.

To determine if significant amounts of contaminant are present above background levels, a statistical significance test of the difference between the mean of the sample and the mean of the blanks was performed. As a

TABLE 12.19. Comparison of Compounds of Potential Concerns in Aqua II AWT and Raw Miramar Waters.

Compound	AWT (µg/L)	Miramar (µg/L)	MCL (µg/L)
Organics			
Benzyl butylphthalate	< 0.300[d]	0.153	NA
Benzoic Acid	< 4.011[d]	0.114	NA
Bis (2-ethylhexyl) phthalate	< 0.048–7.584[c]	< 0.045[d]	4
Bromodichloromethane	< 0.04[d]	3.726	100[a]
Bromoform	< 0.06[d]	0.957	100[a]
Chloroform	< 0.03[d]	2.194	100[a]
Dibromochloromethane	< 0.05[d]	3.153	100[a]
Toluene	0.101	0.042	100[b]
Trichlorofluromethane	< 0.07[d]	0.389	NA
Metals			
Aluminum	63	206	1000
Arsenic	< 2[d]	1.542	50
Barium	< 24[d]	77	1000
Boron	195	< 25[d]	NA
Calcium	5082	45740	NA
Iron	13	43	NA
Magnesium	1749	19010	NA
Manganese	9	9	NA
Phophorous	738	1293	NA
Potassium	< 1255[d]	2878	NA
Sodium	23540	69040	NA
Silicon	909	4386	NA
Strontium	46	67.40	NA

[a]Limit applies to total trihalomethane level.
[b]Drinking Water Action Level recommended by the California Department of Health Services.
[c]BEHP was detected within the listed range. For conservatism, 7.584 was used in the risk assessment.
[d]Not detected above method detection limit; not evaluated in risk assessment.
NA = not applicable.

conservative measure, the significance level was set at the 95 percent confidence interval [i.e., 1.96 multiplied by the standard error (S.E.)]. The compound was considered present at significant levels if it was detected (i.e., values above the MDL) and the lower confidence limit of the differences between the means is greater than zero [i.e., mean difference −(1.96) (S.E.)].

Based on the above statistical analysis of the sampling data 22 compounds were detected in significant quantities in the Aqua II water and 24 were identified in the Aqua III water and justify further review. A listing of the compounds of potential concern for the Aqua II, Aqua III and Miramar waters is given in Tables 12.19 and 12.20. The drinking water standards available for the compounds detected in the Miramar and AWT waters are also shown in these tables. Review of the Aqua II and

TABLE 12.20. Comparison of Compounds of Potential Concern in Aqua III AWT and Raw Miramar Waters.

Compound	AWT (μg/L)	Miramar (μg/L)	MCL (μg/L)
Organics			
Bromodichloromethane	0.48	0.41	100[a]
Chloroform	0.51	0.35	100[a]
Chloromethane	0.11	< 0.05[b]	NA
Dibromochloromethane	< 0.05[b]	0.34	5
Trichloroethene	< 0.07[b]	0.08	100[a]
Bis(2-ethylhexyl) phthalate	3.40[b]	6.46	4
Metals			
Aluminum[c]	17.9	81.9	1000
Arsenic (ICP-MS)	< 0.2[b]	2.48	50
Barium	4.3[b]	118	1000
Boron	193	198	NA
Calcium[c]	1540[b]	67142	NA
Iron[c]	9.2[b]	34.9	NA
Lithium[c]	2.45	51.6	NA
Magnesium[c]	555[b]	27490	NA
Manganese	4.4[b]	9.7	NA
Molybdenum	< 2[b]	15.5	NA
Phosphorus[c]	31.6	143	NA
Potassium[c]	474	4035	NA
Selenium (ICP-MS)	0.44	2.92	50
Silicon[c]	458	4201	NA
Sodium[c]	13668	95143	NA
Strontium	22.6[b]	998	NA
Titanium[c]	0.33	2.69	NA
Vanadium[c]	< 1[b]	0.67	NA

[a] Limit applies to total trihalomethane level.
[b] Not detected in significant quantity above laboratory blank level. Not evaluated in risk assessment.
[c] Health risk data not available in IRIS; not evaluated in the risk assessment.
ICP-AES Method used unless otherwise noted.
NA = not applicable.

Miramar water data in Table 12.19 indicates that the key difference between the two waters is that Miramar contains both arsenic and Trihalomethanes (THMs) while the AWT water does not. Review of the Aqua III data shown in Table 12.20 indicates that the Aqua III water contains low concentrations of THMs and the Miramar water contains slightly higher average concentrations than previously measured. As discussed previously, the THMs in the Aqua III water can be attributed to chlorination prior to the filtration process. Finally, it is interesting to note that the concentration of phalates in the Aqua III water is at the low end of the range found in the Aqua II water.

RISK ASSESSMENT RESULTS

The risk assessment was performed for all compounds of potential concern whose reference doses (RfD) and/or unit risk values were available in the Integrated Risk Information System (IRIS) database, a toxicological database maintained by EPA. The results of the non-carcinogenic and carcinogenic risk assessments are presented below.

Non-carcinogenic Risk Assessment

Reference doses (RfD) are established based on reported results from human epidemiological data, animal studies, and other available toxicological information. In general, the RfD is an estimate (with uncertainty spanning perhaps an order of magnitude) of a daily exposure to the human population (including sensitive subgroups) that is likely to be without an appreciable risk of deleterious effects during a lifetime. A summary of the reported RfD values for compounds of potential concern detected in Miramar and Aqua II AWT waters are shown in Table 12.21 and in Table 12.22 for the Aqua III water.

The hazard index method has been used for assessing the overall potential for noncarcinogenic effects posed by the chemicals in the test waters [29]. In this approach, it is assumed that multiple sub-threshold exposures could result in an adverse effect and that the magnitude of the adverse effect will be proportional to the sum of the ratios of the sub-threshold exposures to acceptable exposures.

Hazard indices (HI) calculated for Aqua II AWT and Miramar waters are reported in Table 12.21. The preliminary results for the Aqua III AWT water are shown in Table 12.22. As shown, the hazard indices in the Miramar and Aqua II AWT waters during the Aqua II HES monitoring period are essentially the same, both are well below one, and would not be anticipated to result in any significant public health risk. The preliminary HI results from the Aqua III HES monitoring period indicate that the Aqua II and Aqua III AWT waters are similar and well below one. The preliminary Miramar HI results are still well below one but have increased because several additional analytes were identified. Overall, based on the HI results for both the Aqua II and Aqua III HES monitoring periods, a significant public health risk would not be anticipated.

Carcinogenic Risk Assessment

In assessing the carcinogenic potential of a chemical, the EPA classifies chemicals into one of the following groups:

- Group A—Human Carcinogen

TABLE 12.21. Non-carcinogenic Risk Assessment for Miramar Raw and Aqua II AWT Water.

Compound	Concentration, (µg/L)	Dose[a] [(mg/kg)/day]	Rf Dose [(mg/kg)/day]	Hazard Index
Miramar Raw Water				
Organics				
Bis (2-ethylhexyl) phthalate	7.58	2.17E-4	2E-2	1.08E-2
Toluene	0.101	2.8E-6	2E-1	1.4E-5
Metals				
Boron	195	5.57E-3	9E-2	6.3E-2
Manganese	8	2.3E-4	1E-1	2.3E-3
Total hazardous index				0.077
Aqua II AWT Water				
Organics				
Benzylbutyl phthalate	0.153	4.37E-06	2E-01	2.19E-05
Benzoic Acid	0.114	3.26E-06	4	8.14E-07
Bromoform	0.957	2.73E-05	2E-02	1.37E-03
Bromodichloromethane	3.726	1.07E-04	2E-02	5.3E-03
Chloroform	2.194	6.27E-05	1E-02	6.27E-03
Dibromochloromethane	3.153	9.01E-05	2E-02	4.51E-03
Toluene	0.042	1.2E-06	2E-01	6E-06
Trichlorofluoromethane	0.389	1.1E-05	3E-01	3.7E-05
Metals				
Barium	77	2.2E-03	7E-02	3.14E-02
Manganese	9	2.57E-04	1E-01	2.57E-03
Total Hazard Index				0.051

aBased on consumption of 2 L/day.

TABLE 12.22. Non-carcinogenic Risk Assessment for Miramar Raw and Aqua III AWT Water.

Compound	Concentration (μg/L)	Dose[a] [(mg/kg)/day]	Rf Dose [(mg/kg)/day]	Hazard Index
	Miramar Raw Water			
Organics				
Bromodichloromethane	0.41	1.2E-05	2E-02	6E-04
Chloroform	0.35	1.0E-05	1E-02	1E-03
Dibromochloromethane (DBCM)[b]	0.34	9.7E-06	2E-02	5E-04
Bis (2-ethylhexyl) phthalate	6.46	1.8E-04	2E-02	9E-03
Metals[c]				
Barium	118.22	3.4E-03	7E-02	5E-02
Boron	197.77	5.7E-03	9E-02	6E-02
Manganese	9.67	2.8E-04	1E-01	2E-03
Molybdenum	15.52	4.4E-04	5E-03	9E-02
Selenium (ICP-MS)	2.92	8.4E-05	5E-03	2E-02
Strontium	997.94	2.9E-02	6E-01	5E-02
Total Hazard Index				3E-01
	Aqua III AWT Water			
Organics				
Bromodichloromethane	0.48	1.4E-05	2E-02	7E-04
Chloroform	0.51	1.5E-05	1E-02	1E-03
Chloromethane[d]	0.11	3.1E-06	6E-02	5E-05
Metals				
Boron	193.48	5.5E-03	9E-02	6E-02
Selenium (ICP-MS)	0.44	1.3E-05	5E-03	3E-03
Total Hazard Index				7E-02

[a]Based on consumption of 2 L/day.
[b]The Rf Dose for DBCM is currently under review. Therefore, the 1992 Rf Dose was used.
[c]The non-carcinogenic effects of arsenic were under review in 1992 and were not included in the Aqua II HES. Therefore only the carcinogenic effects of arsenic were evaluated in Aqua III HES. Inclusion of arsenic, with an RfD of 3E-04, yields a Hazard Index of 2E-01 for arsenic and a total Hazard Index of 5E-01.
[d]The Rf Dose for chloromethane is currently under review. Therefore, the 1992 Rf Dose was used.

571

- Group B—Probable Human Carcinogen: B[1] indicates that limited human data are available; B[2] indicates sufficient evidence on animals and inadequate or no evidence in humans.
- Group C—Possible Human Carcinogen
- Group D—Not Classifiable as to Human Carcinogenicity
- Group E—Evidence of Noncarcinogenicity for Humans

Quantitative carcinogenic risk assessments are performed for chemicals in Groups A and B, and on a case-by-case basis for chemicals in Group C. Carcinogenicity is generally assumed to be a nonthreshold phenomenon meaning that any exposure is assumed, in the absence of sufficient negative information, to represent some finite level of risk. The data used in these risk estimates usually come from lifetime exposure studies in animals. For contaminants in water, concentrations are correlated with carcinogenic risk estimates by employing a cancer potency (unit risk) value together with the assumption for lifetime exposure via ingestion of 2 L of water per day. The unit risk value is derived typically from a linearized multistage model with a 95 percent upper confidence limit providing a low dose estimate. The unit risk values are published in several sources including: Integrated Risk Information System (IRIS) [30], and EPA Health Effects Assessment Summary Tables [31].

An important issue surrounding the cancer potency factors is whether these factors should be characterized as a distribution rather than the standard upper-confidence-limit value representing the 95th percentile of cancer potency as used by EPA's Carcinogen Assessment group. To inves-

TABLE 12.23.   Aqua II Carcinogenic Risk Assessment Results

Compound	50th Percentile Median	Arithmetic Mean	95th Percentile
Miramar water			
Bromoform	1.6E-8	5.1E-8	2.0E-7
Chloroform	2.1E-8	1.0E-7	3.8E-7
Dibromochloromethane	5.5E-7	1.8E-6	7.0E-5
Bromodichloromethane	1.0E-6	3.8E-6	1.4E-5
All trihalomethanes	1.6E-6	5.8E-6	2.2E-5
Arsenic	7.9E-5	7.9E-5	9.6E-5
Total	8.1E-5	8.5E-5	1.2E-4
AWT Water			
BEHP	2.0E-7	8.2E-7	3.2E-6
Total	2.0E-7	8.2E-7	3.2E-6

[a]This simulation is based on ingestion of water at 2L/day and does not include adjustments for inhalation or dermal exposure. The concentrations and cancer potency factors are modeled as distributions except arsenic.
[b]Values reported are those from the cumulative distribution for all pollutants in the water source.

TABLE 12.24. Aqua III Carcinogenic Risk Assessment Results[a]
(1992 Cancer Potency Factors).

Compound	50th Percentile Median	Arithmetic Mean	95th Percentile
Miramar water			
Bromodichloromethane	1.0E-07	4.7E-07	1.7E-06
Chloroform	3.8E-09	1.5E-08	6.1E-08
Trichloroethene	1.4E-09	6.0E-09	2.3E-08
BEHP	1.7E-07	6.4E-07	2.5E-06
Arsenic	1.2E-04	1.2E-04	1.3E-04
Total	1.2E-04	1.3E-04	1.3E-04
AWT Water			
Bromodichloromethane	1.3E-07	4.9E-07	1.8E-06
Chloroform	6.1E-09	2.2E-08	8.9E-08
Chloromethane	1.6E-09	6.1E-09	2.5E-08
Total	1.6E-07	5.2E-07	1.8E-06

[a]This simulation is based on ingestion of water at 2L/day and does not include adjustments for inhalation or dermal exposure. The concentrations and cancer potency factors are modeled as distributions except arsenic.
[b]Values reported are those from the cumulative distribution for all pollutants in the water source.

tigate cancer potency factors, a statistical analysis procedure proposed by Hattis et.al. was employed [32]. These studies indicate that the median value of the cancer potency distribution is approximately 0.072 times the EPA upper confidence limit (where the EPA value was derived from animal data), and that the distribution is reasonably represented by a lognormal distribution with a standard deviation of 4.93 (geometric standard deviation of 0.6931).

The median, the mean, and the 95th percentiles of the calculated risk distributions for each chemical are shown separately in Table 12.23, for some aggregates of chemicals, and for the total risk in the Aqua II AWT and Miramar waters. The preliminary risk assessment results for the Aqua III AWT water are shown in Table 12.24. The median represents the middle value of each distribution of 5000 values, which has a 50 percent chance of either being larger or smaller than the true risk. The mean value is the arithmetic value of all 5000 values. The key advantage for creating a distribution of potency factors from the EPA UCL factor is the ability to characterize the distribution of risk.

CONCLUSIONS

Based on the Aqua II sampling period results and the assumptions inherent in the risk assessment, the overall mean estimate of lifetime risk from 2 L/day consumption of Miramar water is about 3 cancers per 10,000 people. About 98–99 percent of this risk is derived from the presence of inorganic arsenic; trihalomethanes from chlorination represent the remaining assessed carcinogenic risk. The concentration of arsenic in Miramar water is well within applicable federal and state standards, approximately 2 percent of the existing drinking water standards. Without inclusion of arsenic risk values, the mean estimate of lifetime risk is less than one cancer per 100,000, based on THMs. The concentration of THMs in the Miramar water is 1/10th of the existing drinking water standard. Tests conducted to identify the THM formation potential in both waters indicated that Miramar water had the potential to form THM 10 times the amount of the Aqua II water.

By contrast, the water from the Aqua II system that was sampled does not have enough arsenic to be reliably detected. The risk calculations indicate that the Aqua II water poses a mean risk that is almost two orders of magnitude lower than the Miramar water, 8.5×10^{-5} versus 8.2×10^{-7} (Table 12.23, arithmetic mean values). In general, the cancer risk posed by the AWT water is less than one cancer per million people. The cancer risk estimates for the Aqua II water were all based on Bis(2-ethylhexyl) phthalate (BEHP), which may be a laboratory artifact and is possibly not present in AWT water.

The preliminary risk assessment results for the Aqua III AWT water are similar to those estimates made for the Aqua II HES investigations. A major difference is that the Aqua III AWT risk assessment results are based on THMs where the Aqua II results were based on BEHP.

HEALTH ADVISORY COMMITTEE CONCLUSION

Based on the above research, the overall conclusion reached by the Health Advisory Committee [2] was that: *"The health risk associated with the use of the Aqua II AWT water as a raw water supply is less than or equal to that of the existing City raw water as represented by the water entering the Miramar water treatment plant."* Preliminary results of the Aqua III HES confirm the observations of the Aqua II HES.

LONG-TERM HEALTH EFFECTS MONITORING PROGRAM RECOMMENDATIONS

Objectives of the HES were to determine if an advanced wastewater treatment (AWT) system could reliably reduce contaminants of public health concern to levels such that the health risks posed by an assumed potable use of the treated water were no greater than those associated with the present water supply. In this regard, the HES was designed to identify possible health risks of direct reuse of reclaimed wastewater. Based on the results of the Aqua II and Aqua III facility HES investigations, it has been found that the levels of treatment at the AWT facilities produce product water that meets all existing potable drinking water standards. Based on statistical and mechanical reliability analysis results, it was concluded that compliance with State and Federal drinking water standards is reliably maintained at the Aqua II and Aqua III facility on a long-term basis, with a low probability of exceeding the standards.

The positive performance and operational results obtained from the Aqua II pilot treatment facility and the Aqua III full-scale plant along with the results of the health effects study investigations provided the impetus along with the technical and public health support for the City of San Diego to actively pursue indirect potable reuse of AWT treated effluent via blending in a drinking water reservoir. At the time of this writing, feasibility studies have been completed, and conditional approvals have been obtained from health authorities. Any decision to actually implement such a project will be subject to additional unit process testing for plant design, public acceptance, cost evaluations and development, and implementation of a long-term health effects monitoring program. Specific components of a long-term health effects monitoring program should include:

- epidemiology
- AWT Plant Performance and Reliability Monitoring
- delivered water quality monitoring

These monitoring elements should be designed to provide direct monitoring of the population (epidemiology study) and direct monitoring of key distribution points (AWT water quality and delivered water quality). Recommended monitoring for future potable reuse of AWT water is described below.

EPIDEMIOLOGY

A four year epidemiology study was performed as a component of the San Diego Total Resource Recovery Program—Aqua II HES. The objective of this study was to provide baseline health information on the residents of San Diego County while the current water supply is in use. The epidemiological evaluation focused on establishing the normal occurrence of selected morbidities and mortalities for the San Diego area. This baseline information can then be used for comparison should repurified wastewater be used in the future. A detailed discussion of the results of this work is contained in the following references [2,33].

The epidemiological evaluation was based on selecting biological conditions that offer the most potential as environmental warning systems for environmental contamination. Therefore, the health of the residents was evaluated by characterizing the reproductive health and analyzing vital statistics of San Diego County using the following sources of information:

- reproductive outcomes; a survey of women of reproductive age (15 through 44), conducted by the San Diego State University Foundation, used to characterize the reproductive health of San Diego County residents
- published vital statistics for San Diego County residents from 1980 through 1989, used to calculate the normal occurrence rates for deaths and selected illnesses
- neural tube birth defects survey of data from 1980 through 1989 from the California Birth Cohort Perinatal Files, used to estimate the prevalence rates, at birth, of certain neural tube defects

Experience with these three elements of the baseline epidemiology study would indicate that, at a minimum, the neural tube defects component should be continued to ensure production of suitable end points (rates and confidence intervals by outcome) and to allow eventual time series analysis.

AWT PLANT PERFORMANCE AND RELIABILITY MONITORING

Monitoring of the overall plant reliability and of AWT effluent quality provides confirmation that the plant is meeting established performance criteria. During the Aqua II and Aqua III health effects studies the AWT treatment systems were monitored and evaluated to determine if the level of treatment consistently achieved a water quality protective of human health and that the mechanical system components operated reliably. Plant reliability studies and AWT effluent monitoring performed during the HES investigations provided the bases and methodology for assessing AWT plant performance and reliability. An AWT plant performance and reliability monitoring program should be included in the routine monitoring requirements in any full-scale implementation of indirect potable reuse.

DELIVERED WATER QUALITY MONITORING

It is anticipated that future potable reuse projects will involve blending the AWT water with existing water supplies prior to distribution to the public. It is important to characterize the quality of the water delivered for potable consumption as part of documenting the level of exposure to chemicals via drinking water. Therefore, it is recommended that a monitoring program similar in format to the AWT plant performance monitoring program be included in any full-scale implementation of indirect potable reuse.

INDEPENDENT THIRD PARTY REVIEW

The Health Advisory Committee (HAC) was established to provide independent third party review on the development, implementation and analysis of results for the Aqua II and Aqua III Health Effects Studies. The need for an independent third party, similar to the HAC, to develop and review the results of long-term health effects monitoring was recognized as a necessary element of any indirect potable reuse program by the State of California in a recently published document entitled "A Proposed Framework for Regulating the Indirect Potable Reuse of Advanced Treated Reclaimed Water by Surface Water Augmentation in California, 1996" [34]. It is recommended that independent third party review of monitoring plans and results be included in any full-scale implementation of indirect potable reuse.

ACKNOWLEDGEMENTS

A number of City, SWRCB, Health Advisory Committee (HAC), and Technical Advisory Committee (TAC) members have been associated with the TRR project since its inception. City participants included Mr. Paul Gagliardo (Project Manager), Mr. Jack Swerlein (Chief Plant Operator), Mr. Abel Hernandez (Senior Chief Operator), Mr. John Chaffin (Water Production Superintendent), and Mr. Greg Elliott (Chemistry Section). The operations, maintenance, laboratory, and entomology section staff are recognized for their contribution to the success of Aqua II and Aqua III. WCPH project staff included Ms. Lori Pettegrew, Mr. Jeff Soller and on-site liaison Mr. Dave Ross.

REFERENCES

1 National Research Council 1994, *Ground Water Recharge Using Water of Impaired Quality,* National Academy Press, Washington, D.C., pp. 1–275.

2 Cooper, R. C., Olivieri, A. W., Eisenberg, D., Pettegrew, L., Danielson, R. E., Fairchild, W. A., and Sanchez, L. A. 1992, City of San Diego Total Resource Recovery Project: Health Effects Study, Western Consortium for Public Health, Berkeley, CA.

3 Lauer, W. C., Johns, F. J., Wolfe, G. W., Meyers, B. A., Condie, L. W., and Borazelleca, J. F. 1990, Comprehensive Health Effects Testing Program for Denver's Potable Water Reuse Demonstration Project, *J. Toxicol. Environ. Health* 30:305–321.

4 Rowney, A. C., Wright, D. L., and MacIntyre, D. F. 1987, Large-scale Groundwater Recharge Via Infiltration Basins in Orange County, *Proceedings of the Florida Water Reuse Symposium IV,* AWWA Research Foundation, Denver, CO.

5 State of California 1975, *A State-of-the-Art Review of Health Aspects of Wastewater*

Reclamation for Groundwater Recharge, State Water Resources Control Board, Department of Water Resources, Department of Health, Sacramento, CA.

6   National Research Council 1982, *Quality Criteria for Water Reuse,* National Academy Press, Washington, D.C.

7   Eisenberg, D. M., Olivieri, A. W., Cooper, R. C. 1987, Evaluation of Potential Health Risk of Direct Potable Reuse of Reclaimed Wastewater, *Proceedings of the ASCE, Environmental Engineering Speciality Conference,* pp. 346–352.

8   Cooper, R. C., Olivieri, A. W., and Eisenberg, D. M. 1984, Workplan for the Evaluation of Potential Health Risks Associated with the San Diego Total Resource Recovery Program, Western Consortium for Public Health, Berkeley, CA.

9   Cooper, R. C., Olivieri, A. W., and Eisenberg, D. M. 1986, San Diego Health Effects Study First Year Project Status Report, Western Consortium for Public Health, Berkeley, CA.

10   Cooper, R. C., Olivieri, A. W., and Eisenberg, D. M. 1988, San Diego Health Effects Study Second Year Project Status Report, Western Consortium for Public Health, Berkeley, CA.

11   Cooper, R. C., Olivieri, A. W., and Eisenberg, D. M. 1989, San Diego Health Effect Study Third Year Project Status Report, Western Consortium for Public Health, Berkeley, CA.

12   Cooper, R. C., Olivieri, A. W., and Eisenberg, D. M. 1990, San Diego Health Effect Study Fourth Year Project Status Report, Western Consortium for Public Health, Berkeley, CA.

13   Tchobanoglous, G., Maitski, F., Thompson, K., Chadwick, T. H. 1987, "Evolution and Performance of City of San Diego Pilot Scale Aquatic Wastewater Treatment System Using Water Hyacinths," *Research Journal WPCF,* 61(11/12):1623–1633.

14   American Public Health Association (APHA), American Water Works Association (AWWA), and Water Pollution Control Federation (WPCF) 1985, *Standard Methods for the Examination of Water and Wastewater,* 16th Ed., Washington, D.C.

15   Water Environment Federation (WEF), APHA and AWWA 1992, *Standard Methods for the Examination of Water and Wastewater,* 18th Ed., Washington, D.C.

16   Los Angeles County Sanitation District 1977, Pomona Virus Study Final Report, Whittier, CA.

17   Sheikh, B. 1987, Monterey Wastewater Reclamation Study for Agriculture, Engineering-Science, Berkeley, CA.

18   Cooper, R. C., Olivieri, A. W., Eisenberg, D. M., Danielson, R. E., Donohue, R., and Pettegrew, L. A. 1991, Bacterial Pathogens and Parasites, Final Report, Western Consortium for Public Health, Berkeley, CA.

19   Froines, J. R. 1991, Evaluation of Potential Health Risks Presented by Chemical Agents Associated with the San Diego Total Resource Recovery Program, School of Public Health, University of California, Los Angeles, CA.

20   de Peyster, A., Rudnicki, R., Slymen, D. J., Kimball, A. 1991, Biomonitoring Program—Final Report, San Diego State University, San Diego, CA.

21   de Peyster, A., Donohue, R., Slymen, D. J., Froines, J. R., and Olivieri, A. W. 1993, "Aquatic Biomonitoring of Reclaimed Water for Potable Use—The San Diego Health Effects Study," *Journal of Toxicology and Environmental Health,* 39(1):121–141.

22   Environmental Protection Agency 1985, Fish Bioconcentration Test Guidelines, *Federal Register,* 50 188:39347.

23 American Society for Testing and Materials 1975, *Standard Practice for Conducting Bioconcentration Tests with Fishes and Saltwater Bivalve Molluscs,* Philadelphia, PA.

24 Cooper, R. C., Olivieri, A. W., Eisenberg, D., Danielson, R. E., Pettegrew, L. A., Fairchild, W. A., Sanchez, L. A. 1992, San Diego Aqua II Pilot Plant Reliability Study, Western Consortium for Public Health, Berkeley, CA.

25 Soller, J., Cooper, R. C., Eisenberg, D., Olivieri, A. W., Pettegrew, L. A. 1996, Aqua III Reliability Report, Western Consortium for Public Health, Berkeley, CA.

26 California Department of Health Services, Domestic Water Quality and Monitoring, Title 26 of the California Code of Regulations, Division 4, Chapter 15.

27 McCarty, P. L., Reinhard, M., Goodman, N. L., Graydon, J. W., Hopkins, G. D., Mortelmans, K. E., Argo, D. G., and English, J. N. 1982, Advanced Treatment for Wastewater Reclamation at Water Factory 21, Dept. of Civil Engineering, Stanford University, Stanford, CA.

28 Environmental Protection Agency 1982, *Evaluation and Documentation of Mechanical Reliability of Conventional Wastewater Treatment Plant Components,* Washington, D.C.

29 United States Environmental Protection Agency 1989, *Risk Assessment Guidance for Superfund, Volume I, Human Health Evaluation Manual (Part A),* Office of Emergency and Remedial Response, Washington, D.C. (EPA/540/1-89/002).

30 Integrated Risk Information System (IRIS), Electronic On-line Database of Summary Health Risk Assessment and Regulatory Information on Chemical Substances, U.S. EPA.

31 Health Effect Assessment Summary Tables (HEAST), Annual FY 1991 Office of Solid Waste and Emergency Response, Publication #9200.6-303, (91-1), U.S. EPA.

32 Hattis, D. and Wasson, J. 1987, A Pharmacokinetic/Mechanism-Based Analysis of the Carcinogenic Risk of Butadiene, National Technical Information Service No. NTIS/PB88-202817, M.I.T. Center for Technology, Policy and Industrial Development, CTPID 87-3.

33 Molgaard, C. A., Golbeck, A. L., Elder, J. P. 1990, The Epidemiology Component of the San Diego Total Resource Recovery Program—Final Report, San Diego State University, San Diego, CA.

34 California Potable Reuse Committee 1996, A Prepared Framework for Regulating the Indirect Potable Reuse of Advanced Treated Reclaimed Water by Surface Water Augmentation in California, California Department of Health Services and California Department of Water Resources.

35 Cooper, R. C., Olivieri, A. W., Eisenberg, D., Pettegrew, L., Soller, J., Danielson, R. E. 1996, City of San Diego Total Resource Recovery Project: Health Effects Study—San Pasqual, Western Consortium for Public Health, Berkeley, CA.

CHAPTER 13

Evaluating Methods of Establishing Human Health-Related Chemical Guidelines for Cropland Application of Municipal Wastewater

INTRODUCTION

ARCHAEOLOGICAL evidence has traced the advent of domestic wastewater conveyance and disposal practices to antiquity (Asano and Levine, 1996). The idea of establishing a community-wide systematic collection, treatment, and disposal of wastewater, however, did not emerge until late in the 19th century. From the very beginning, land spreading was an integral component of municipal wastewater treatment and disposal scheme, and many large urban centers around the world applied wastewater they generated on nearby land. Noted examples were the "sewerage farms" in Paris, France; Berlin, Germany; and Melbourne, Australia (Jewell and Seabrook, 1979). Early propositions of wastewater spreading considered soil as a "treatment" medium and stressed benefits of wastewater as a source of plant nutrients, in contrast to the then customary "direct" discharge of untreated sewage into a surface water body. In land application, the need for securing a reliable discharge outlet for wastewater, however, was not always compatible with the seasonal water and nutritional requirements of plants. Most early systems were plagued by both hydraulic and pollutant overloading that public nuisance and pollution were rampant (Jewell and Seabrook,

Andrew C. Chang and Albert L. Page, Department of Soil and Environmental Sciences, University of California at Riverside, Riverside, CA 92521; Takashi Asano, Department of Civil and Environmental Engineering, University of California at Davis, Davis, CA 95616; Ivanildo Hespanhol, Department of Hydraulic and Sanitary Engineering, Escola Politécnica da Universidade de São Paulo, P.O. Box 61548, 05508-900 São Paulo, S.P. Brazil

1979). The sanitary engineering profession turned to other options for wastewater disposal (Chase, 1964a).

During 1950–60, interests in applying wastewater on land resurfaced in the western hemisphere as wastewater treatment technology advanced and quality of treated effluents steadfastly improved (Chase, 1964b). Land application became a cost-effective alternative of discharging effluent into surface water bodies. It was especially attractive in the semi-arid and arid regions of the world where treated effluents could supplement growing demands for water (Stone et al., 1952). The pursuing investigations concentrated on hydraulics of dissipating water and on pathogen survival and transport in soils (Orlob and Butler, 1955; Gotaas et al., 1955). This time around, the circumstances had changed. Unlike earlier practices, land treatment systems were successful because hydraulic and pollutant assimilation capacities of soils were better understood. Many technical difficulties resulting from pollutant overloading and, consequently, public nuisance were remedied by properly treating the wastewater prior to spreading. When appropriately designed and operated, the soil mantle served as an effective treatment system in attenuating pollutants carried in wastewater (McGauhey et al., 1966). Long term environmental impacts of reclaimed wastewater irrigation on soils and plants appeared to be minimal (Benham-Blair & Associates, Inc. and Engineering Enterprise, Inc., 1979; Renolds et al., 1979).

There had been more than 3000 land application sites in the United States alone (Jewell and Seabrook, 1979; U.S. Environmental Protection Agency, 1981 and 1992b). In developing countries, land application has always been the primary means of disposing municipal wastewater as well as meeting irrigation needs (Bahri and Houmane, 1988; Kansel and Singh, 1983; Zhang and Liu, 1989; Abdel-Reheem et al., 1986; Ul-Haque et al., 1986; Sanai and Shayegen, 1980). In China, at least 2×10^6 hectares of crop land were irrigated with untreated or partially treated wastewater from cities (Wang, 1984). The combined flow of untreated domestic sewage and industrial wastewater from Mexico City, a metropolis of 15 million inhabitants, was discharged into a river and used to irrigate more than 85,000 hectares of cropland downstream (Villalobos et al., 1981; Gutierrez-Ruiz et al., 1994). In this case, the only wastewater treatment afforded before land spreading was provided by self-purification processes of the stream channel and storage reservoirs. Heaton (1981) reported wastewater reclamation and reuse activities in Israel, Saudi Arabia, Iran, South Africa, The Netherlands, and Germany. Judging from the locations, land application is practiced essentially on every continent, except perhaps Antarctica. As wastewater is commonly used in crop production around the world, universally applicable water quality guidelines would standardize the practice and define expectations in its performance throughout the world (Marecos do Monte, 1996).

Historically, pathogen elimination has been the primary concern of wastewater treatment. Wastewater, even after treatment, is always a potential reservoir for human pathogens that could spread via a fecal-oral route (Rudolfs et al., 1950; Benarde, 1973; Burge and Marsh, 1978; Frankenberger, 1985). Using wastewater for crop irrigation short-circuits the transmission route by bringing potentially contaminated water into direct contact with human food supply and thus increase the probability of spreading epidemic diseases. Studies showed that, with proper treatment of the wastewater and proper control of its use, disinfected wastewater could be safely used in crop and landscape irrigation (Uiga and Crites, 1980). Results of an epidemiological investigation involving the populace in and around a 1,500 hectare wastewater application site in Texas demonstrated that spray irrigation of wastewater produced no statistically significant increase in incidences of disease (George et al., 1986).

Because of its potential public health harms, the use of wastewater for crop irrigation has always been regulated (Crook and Surampalli, 1996). In the U.S., 43 of the 50 states have regulations controlling use of reclaimed wastewater for crop irrigation (U.S. Environmental Protection Agency, 1992a). The requirements are based primarily on defining the extent of treatment a wastewater must receive together with numerical limits on bacteriological quality, turbidity, and suspended solids content of the water. The wastewater *Reclamation Criteria* adopted by California may be used as an example.[1] This regulation does not set any numerical limit, except for the bacteriological quality of water. Instead, it defines the minimum treatment level required for a given type of water use and, if necessary, limits access of general public and grazing animals to application sites. If the reclaimed wastewater meets prescribed requirements, it is assumed that the chemical constituents in the treated effluents will not be harmful to human health or cause water pollution. This assumption, however, may not be valid if industrial waste discharge is not restricted. Unlike pathogens which contaminate crops through direct contact, chemical pollutants are incorporated into the plants tissue through root absorption and can not be effectively removed by washing and/or cooking.

Several comprehensive guidelines and manuals for using reclaimed wastewater in crop irrigation have been published (Pettygrove and Asano, 1985; Pescod, 1992; U.S. Environmental Protection Agency, 1992a). None of them, however, addressed potential human health issues associated with toxic chemical pollutants introduced via wastewater irrigation. The World Health Organization developed microbiological standards for proper use of treated wastewater in agriculture and called to the attention that potential

[1] California Administrative Code of Regulations, Title 22 Social Welfare, Division 4 Environmental Health, Chapter 3 Reclamation Criteria.

public health consequences associated with toxic chemical pollutants in wastewater had not yet been fully addressed when food crops were irrigated with municipal wastewater (World Health Organization, 1989c).

TOXIC CHEMICALS IN MUNICIPAL WASTEWATER—HUMAN HEALTH IMPLICATIONS

Municipal wastewater is a source of chemical pollutants that may affect human health. According to information compiled by a study group organized by National Research Council, over 65,000 chemicals are being used routinely in manufacturing, in agricultural production, and in daily living (National Academy of Science, 1984). A fraction of the used chemicals inadvertently finds their way into municipal wastewater collection systems. Because they enter the wastewater collection systems from defused sources, most of them are present in municipal wastewater only some of the time and often in variable concentrations (Blakeslee, 1973; Chen et al., 1974; Feiler, 1979; Klein et al., 1974; Levins, et al., 1979; Minear et al., 1981). There is, however, not an expedient method to monitor routinely the presence of potentially hazardous pollutants in wastewater (World Health Organization, 1975). Surveys of POTWs in the United States showed that the concentrations of priority pollutants in influent wastewater often exceed the allowable concentration limits (U.S. Environmental Protection Agency, 1982; Bhattacharya et al., 1990). Results from a pilot-scale system treating wastewater spiked with selected toxic chemicals indicated that up to 90% of the added chemicals were removed from the secondary effluent. During the course of treatment, some of the spiked compounds such as di- and tri-chlorobenzenes, hexachlorobutadiene, dibutyl phthalate, butyl benzyl phthalate, bis(2-ethylhexyl) phthalate, naphthalene, lindane, dieldrin were found in the sludge fraction even following anaerobic digestion (Bhattacharya et al., 1990). Conventional wastewater treatment systems, however, were not designed specifically to remove potentially toxic chemicals. Even effluents processed with the most advanced wastewater treatment technologies contained traces of pollutants (McCarty and Reinhard, 1980).

Results of several field studies indicated that it was safe to use secondary or tertiary treated wastewater effluents for food crop production as pathogens were not found and trace element concentrations in harvested food were not elevated (Pound et al., 1978; Benham-Blair and Associates, Inc. and Engineering Enterprises, Inc., 1979; Renolds et al, 1979; Hinesly et al., 1978; George et al., 1986; Sheikh et al., 1990). In China, human health problems related to long-term irrigation with wastewater heavily polluted by industrial waste discharges had been documented. A report estimated that 8.4% of the 2.1×10^6 hectares of wastewater-irrigated farm land in

China were seriously polluted and almost 50% of the total acreage were noted for pollutant accumulation in soils (Anonymous, 1992b). Yuan and Zhao (1987) investigated adverse health effects resulting from irrigating farm land with a wastewater containing petroleum refinery waste and reported that, among the 13,621 human subjects examined, the incidence of hepatomegaly were 10.61%, 1.92% and 0.42% of the subjects residing in the heavily polluted area (>50 years of wastewater irrigation), the lightly polluted area (22 years of wastewater irrigation), and the control area (no wastewater irrigation), respectively. The soils and the rice produced on these soils contained higher levels of mineral oil, phenol, and benzo(a)pyridine than their respective controls and drinking water was also contaminated by the same chemicals.

The prevalence of water-borne diseases of chemical etiology is not easy to account. For example, it took 30 years to discover that fluoride ion in drinking water was the cause of dental fluorosis. Some diseases may result from exposure to a chemical and exhibit distinctive symptoms, such as exposure to methyl mercury is the cause of Minamata disease. More often, chemical agents are merely one factor in a complex multiple causal relationship. For example, Itai-Itai disease is a disorder in which Cd toxicity is only one of the causal factors (Friberg et al., 1974). Literally, hundreds of chemical toxicants are appearing in the priority pollutant lists of regulatory agencies in the United States and the European Community (Anthony and Breimhurst, 1981; Rogers, 1987). Many designated chemicals exhibit carcinogenic, mutagenic, or teratogenic effects. But they are present in environmental media in low concentrations and have long latency periods between exposures to the chemicals and expression of the symptoms. As a result, it is difficult to establish a baseline exposure level and to distinguish exposures received from multiple pathways. As statistically significant epidemiological evidence is difficult, if not impossible, to establish, dose-response relationships are derived from data obtained from animal bioassays or other surrogate tests.

The human health issues involving toxic chemicals during the land application of wastewater are cases in point. Toxic chemicals are omnipresent in the wastewater. There is, however, no unequivocal epidemiological evidence to demonstrate any harm caused by even one potentially toxic chemical present in irrigation-bound wastewater. It is a challenge to develop criteria that are not overly restrictive to beneficial use of wastewater and yet protect humans from potential harm that could be caused by hundreds of toxic chemicals that may or may not be present in a wastewater.

EXISTING IRRIGATION WATER QUALITY CRITERIA

Table 13.1 outlines irrigation water quality criteria developed by seven

TABLE 13.1. Irrigation Water Quality Criteria of Selected Nations.

Parameter	Unit	Canada	USA	Taiwan	Hungary	People Republic of China			Saudia Arabia	Tunisia
		All Soil 1991	All Soil 1973	All Soil 1978	All Soil 1991	Rice Paddy Not Dated	Dryland Not Dated	Vegetable Not Dated	All Soil Not Dated	All Soil Not Dated
pH				6.0-9.0	6.5-8.5	5.5-8.5	5.5-8.5	5.5-8.5	6.0-8.4	6.5-8.5
Total dissolved solids	mg/L	500-3500				1000-2000	1000-2000	1000-2000		
Electric conductivity	dS/m			0.75						0.7
Suspended solids	mg/L			100		150	200	100	10	30
Chloride	mg/L			175	700	250		250	280	2000
Sulfate	mg/L			200						
Total Kjeldahl nitrogen	mg/L			1	80	12	30	30		
BOD	mg/L					150	80	10		30
COD	mg/L					200	300	150		90
Temperature	degree C			35		35	35	35		
Al	µg/L			5000	5000				5000	
As	µg/L	100		1000	200	50	100	50	100	100
Ba	µg/L				4000					
B (total)	µg/L	500-600		750	700	1000-3000	1000-3000	1000-3000	500	3000
Cd	µg/L	10		10	20	5	5	5	10	10
Cr (total)	µg/L	100		100	1000	100	100	100	100	100
Co	µg/L	50		50					50	100
Cu	µg/L	200-1000		200	2000	1000	1000	1000	400	500
F (total)	µg/L	1000		1000	1000	2000-3000	2000-3000	2000-3000	2000	3000
Fe	µg/L			100	100				5000	5000
Pb	µg/L	200		2500	1000	100	100	100	100	1000
Li	µg/L			2500					70	
Mn	µg/L	200		2000	5000				200	500
Hg	µg/L			5		1	1	1	1	1

TABLE 13.1. (continued).

Parameter	Unit	Canada All Soil 1991	USA All Soil 1973	Taiwan All Soil 1978	Hungary All Soil 1991	People Republic of China Rice Paddy Not Dated	Dryland Not Dated	Vegetable Not Dated	Saudia Arabia All Soil Not Dated	Tunisia All Soil Not Dated
Mo	µg/L	10–50	10	10					100	
Ni	µg/L	200	200	500	1000				20	200
Se	µg/L	20–50	20	20		20	20	20	20	50
Ag	µg/L				100					
V	µg/L	100	100	10000	5000					
Zn	µg/L	1000–5000	2000	2000	5000	2000	2000	2000	4000	2000
Cyanide (total)	µg/L				10000	500	500	500	50	
Surfactant (ABS)	µg/L			5000	50000	5000	3000	5000		
Oil and grease	µg/L			5000	8000				absent	
Benzene	µg/L				2500	2500	2500	2500		
Tar	µg/L				30000					
Petroleum	µg/L				500	1000	500	500		
Methanol	µg/L				100					
Trichloroacetylaldehyde	µg/L					1000	500	500		
Propionaldehyde	µg/L					500	500	500		
Phenol	µg/L								2000	

nations, including Taiwan. Five of the seven guidelines were developed specifically for wastewater irrigation (e.g. Taiwan, Hungary, People's Republic of China, Saudi Arabia, and Tunisia). In these criteria, numerical limits for selected chemical constituents are set to protect crops from being injured (Pratt, 1972; National Academy of Science, 1973; Ayers and Westcot, 1985). Some of the elements (such as Cd, Pb, and Hg) are potentially toxic, the maximum contaminant levels (MCLs) of drinking water are adopted (Pratt, 1972).

In general, numerical limits are limited to selected inorganic chemical elements and their concentrations are considerably higher than those of the same substances typically found in natural water and treated wastewater effluents. In some cases (for example, concentrations of As, Cr, Cu, and Zn), even the concentrations of pollutants in untreated municipal waste-water are not likely to exceed these numerical limits. There are also similarities among these criteria in terms of the constituents being included and their numerical limits. Most notably, some of the chemical constituents (e.g. As, B, Cd, Cr, Cu, Pb, and Zn) appeared in all of the lists and have identical concentrations in several sets of criteria. Chronologically, these seven irrigation water quality criteria evolved over 20 years. Scientific knowledge concerning pollutant behavior in wastewater and in soil has greatly improved during this period. Conceptually, irrigation water quality criteria have not changed significantly. While they are adequate to protect the plants from injuries, the pollutants may accumulate in the soils.

For wastewater irrigation, the criteria often include numerical limits for suspended solids, biological oxygen demand (BOD), chemical oxygen demand (COD) and organic chemicals (oil and grease, trichloroethylene, petroleum hydrocarbon, and detergent residues). These parameters are included to prevent potential adverse effects on operation and maintenance (BOD, suspended solids, detergent residues, and oil and grease), ground-water pollution, (TCE, petroleum hydrocarbon, and benzene), or plant growth (petroleum hydrocarbon). The numerical limits for these parameters in many cases are identical in several criteria, but in other cases they varied by one to two orders of magnitude. It is, however, not clear whether they reflected a true difference in opinion or simply the location-specific soil and plant growing conditions. The inclusion of some constituents (petroleum hydrocarbons, detergent residues, benzene, etc.) in the waste-water irrigation undoubtedly indicated the prevalence of these pollutants in the wastewater stream.

The pollutant inputs could be substantial, even when it is irrigated with a water meeting recommended numerical limits in irrigation water quality criteria (Table 13.1). Existing irrigation water quality guidelines do not necessarily provide protection of human health.

CONCEPTUAL FRAMEWORK OF GUIDELINE DEVELOPMENT

Guidelines may be viewed as industry-wide technical advisories (or specifications) in the form numerical limits and/or narrative statements which may be used to accomplish a specific goal. In the case of a human health-related chemical guideline for reclaimed wastewater irrigation, its goal is to protect consumers from potential harmful effects of toxic chemicals present in irrigation water. For a guideline to be recognized and accepted, numerical limits and narrative statements must be scientifically defensible, technically adaptable effective, and reasonable. Although they lack the enforcement power of regulations, well conceived guidelines often have far reaching implications in standardizing practices and defining performance expectations. Before its development, objectives of an intended guideline, however, must be clearly defined. Approaches that may be employed to accomplish the goals can then be identified. For each adopted approach, there will be a set of scientifically-derived numerical limits. Conceivably, not all of the numerical limits is technically achievable and cost effective. A strategy must then be employed to formulate a sensible and practical guideline by adjusting the numerical limits. Narrative statements often are introduced at this stage to qualify alterations made on numerical limits. Dependent on approaches taken, the final products may be quite different in format and substance. In principle, they all accomplish the preset goals.

OBJECTIVE

Reclaimed wastewater irrigation is an important option for water quality management. Practiced in a proper manner, it provides cost effective wastewater disposal for a community, prevents environmental pollution, and conserves water resource. Toxic chemicals however may be present in the wastewater. The objective of this guideline, like many other water quality management guidelines, will be on protecting human health. More specifically, the purpose of this guideline is to safeguard individuals from unsuspected harm of consuming harvested crops that are irrigated with municipal wastewater.

APPROACHES

Two approaches may be used to accomplish this goal: (1) prevent accumulation of pollutants in soils or (2) take maximum advantage of the capacity of the soil to assimilate, attenuate, and detoxify pollutants (Chang et al., 1995).

Preventing Pollutant Accumulation in Wastewater-Receiving Soils

The soil is an irreplaceable natural resource of mankind. If the soil is free from pollutants, it can be used for a variety of purposes without any undue restriction. If pollutants introduced via land application of wastewater are allowed to accumulate in the soil, its potential use, over the long term, may become limited. Because soil also modulates the water flows of an hydrologic cycle, pollutants accumulate in the soil may also be transported to contaminate surface and/or groundwater. Under this approach, land application of wastewater should not result in a net increase of the pollutant level in the soil. If this requirement is met, the soil's ability to support any future land use is insured and the food chain transfer of potentially hazardous pollutants may be stopped or kept to a minimum. Guidelines based on this approach accomplish their goals by preventing pollutants from accumulating in wastewater-receiving soils. Human health and safety, therefore, are protected if the soil is contaminated.

The concept fundamental to this type of guidelines is conceptually in agreement with principles of ecology. An advantage of using this approach to develop guidelines is that detailed knowledge of exposure pathways and the dose-response relationships are not needed. The numerical limits for pollutants may be calculated by simple mass balances by matching pollutant input from wastewater application with pollutant outputs and the results are universally applicable. The net cost of wastewater treatment and disposal, however, will be high because more advanced technologies must be employed in treating wastewater, or a larger land area is required to dissipate the same amount of wastes.

Maximizing Soil's Capacity to Assimilate and Detoxify Pollutants

There are many options for disposing of wastewater but none is as universally applicable and, from the overall environmental quality point of view, more desirable than land application. Biochemical cycling of nutrients in the soil will render pollutants in applied wastewater innocuous over time. Therefore, during land application of wastewater, the capacity of soils to attenuate pollutants should be utilized fully. If land applications operations are managed properly, the agronomic benefits of wastewater are realized and accumulation of pollutants in the soil does not reach levels that can be harmful to the exposed individuals. Based on this approach, the land application guidelines set only the maximum permissible pollutant loading limits and/or maximum permissible pollutant concentrations of the soil and provide the users of this disposal option flexibility of developing safe and site-specific land application operations.

Taking this approach, the pollutant level in the soil will be permitted to rise, eventually to a maximum tolerable level. In wastewater irrigation, there are multiple exposure pathways through which pollutants may be transported to humans. Any one of the pollutants or exposure pathways may limit the input of pollutants. Thorough knowledge of the exposure pathways and knowledge on parameters defining rate constants and amount of pollutant transfer in each pathway are necessary to determine the maximum permissible pollutant loading limits. In order for the guidelines to be comprehensive, every possible toxicant contained in the wastewater must be tracked through these pathways, and regulated.

RISK-BASED ASSESSMENT OF HUMAN HEALTH HAZARDS

Risk assessment is a complex, multifaceted undertaking that identifies the sources and pathways of chemical exposure and quantifies the risks resulting from such events. Regulatory agencies and international bodies concerned about public health, food safety, and environmental protection are interested in assessing the exposure of the general human population or subgroups to hazardous chemicals present in the environment. Results of risk assessment are often used as bases for taking regulatory action or for issuing advisory guidelines. During land application of wastewater, potentially toxic pollutants are introduced into soils, the starting points of the human food chain and other exposure pathways. Through the food chain transfer, the potentially toxic pollutants may affect the health and well-being of consumers. Pollutants accumulated in the soil, as the result of land application, may subsequently contaminate surface and ground water, resulting in additional exposures.

''Risk'' is conceptually a random variable related to chance or peril. Applied to environmental protection and management, the risk assessment is an objective procedure used to determine the likelihood of harm resulting from exposure to pollutants emitted through an actual episode or a hypothetical event. A comprehensive risk assessment usually involves four essential elements, namely, hazard identification, exposure analysis and estimation, dose-response evaluation, and risk characterization (National Academy of Sciences, 1983; U.S. Environmental Protection Agency, 1984). In a routine exercise, the chemicals to be assessed are identified and selected first via an objective screening process; the routes of exposure (transfer of the toxicant from the source to the receptor) and the dose received through each route are determined; the extent of harmful effect is evaluated through a quantitative dose-response for each selected chemical; and finally, human health implications resulting from this level of exposure are determined.

Essentially, the same procedural elements are needed in establishing human health related numerical limits for toxic chemicals in land application of wastewater. In this case, the maximum permissible exposure of the receptor (exposed human population) is first determined based on an acceptable level of "risk"; a hypothetical exposure scenario for each pollutant transport route is then constructed; and maximum contaminant levels through wastewater application is calculated according to the mathematical relationships that define the transfer of the toxicant through exposure pathways. The concept of risk assessment is not new or foreign to the environmental sanitation profession. The numerical limits for toxic chemicals in the drinking water guidelines are developed in this manner (National Academy of Sciences, 1977; World Health Organization. 1984ab). In the case of land application of wastewater, the evaluation will be more complicated because risk of exposure to pollutants deposited in soil will involve assessing the fate and transport of toxic chemicals through a multitude of exposure routes. Nevertheless, information obtained from other human health risk assessments may be used to aid the development of this guideline.

The risk assessment, in this case, is employed to establish upper limits for inputs of toxic chemicals that prevent adverse human health effects when wastewater is used in crop irrigation. Uncertainties and errors, however, may be introduced at every step of the assessment process because of data deficiencies, incorrect assumptions, and inappropriate computational models. In addition, results of a risk assessment are probabilistic; the actuality of such events to take place oftentimes cannot be verified experimentally. Uncertainties and errors in risk assessment, unfortunately, could lead to erroneous conclusions, unwarranted advisories, and unjustified restrictions for land application practices. To insure that risk assessors are aware of the uncertainties and to minimize errors in selecting data, guidelines which formalize protocols of risk assessment are necessary to promote consistency among various risk and exposure assessment activities (U.S. Environmental Protection Agency, 1986a,b).

To determine the numerical limits for pollutants in land application of wastewater, the following steps should be followed:

Step	Description of Tasks
1. Hazard identification	Use epidemiological and environmental toxicological data and information on chemical composition of wastewater and sewage sludge to identify potentially toxic pollutants that should be considered.
2. Dose and response analysis	Determine the maximum permissible pollutant intake of the exposed population in the form of acceptable daily intake (ADI).

3. Exposure
   pathway and
   scenario
   analysis

Identify environmental exposure pathways and exposure scenarios through which the population may be exposed to the pollutants. Quantitatively determine pollutant transfer coefficients for each step of the exposure pathways.

4. Pollutant
   loading
   computation

Based on the ADI information, exposure pathways, and exposure scenario, determine the amount of pollutant permitted in the soil.

HAZARD IDENTIFICATION—SELECTING TOXIC CHEMICALS

Hazard identification is the procedure by which potentially toxic chemicals in wastewater may be identified, and the need for being included in the guideline is assessed. Data from epidemiological and environmental toxicological investigations usually are the starting point of the selection process. If a chemical is potentially toxic, as indicated by epidemiological or toxicological data, its occurrence in wastewater, fate during treatment, concentrations in effluents should be examined to insure it will not be a health hazard when the final effluents are applied on land.

For a specific episode, the hazard identification exercise usually involves a limited number of chemicals. For selecting toxic chemicals in wastewater, the hazard identification process is quite different because a collection system receives wastewater discharges from numerous sources whose flow and chemical composition cannot be predicted. Tens of thousands of chemicals are being used worldwide as manufacturing stock and consumer products (Pimental and Coonrod, 1987). The adverse environmental and health effects of only a very small number of chemicals, however, are known (National Academy of Sciences, 1984). During the course of manufacturing, industrial processing, and product consumption, these chemicals may be released inadvertently or discharged deliberately into the municipal wastewater collection system. The toxic release inventory data collected by the U.S. Environmental Protection Agency indicated that, in the U.S., approximately one billion kilograms of toxic chemicals were sent to municipal wastewater treatment plants for processing and disposal in 1987. Conventional wastewater treatment systems are designed to remove suspended solids, BOD and coliform group bacteria from the influent wastewater stream, but the fate and decomposition of potentially toxic chemicals during the course of wastewater treatment cannot be predicted accurately. Conservative constituents, such as the toxic metals, are expected to either remain in the treated effluent or to be concentrated into the sludge fraction.

Based on surveys conducted in many parts of the world, certain chemical constituents, such as the heavy metals, appear to be ubiquitous and can be found in almost any municipal wastewater stream; others, especially the organic chemicals, are present only in some wastewater or are present only sometimes (World Health Organization, 1975). Table 13.2 illustrates the occurrence of selected organic chemicals in wastewater in one such survey. Logically, chemicals appears in wastewater frequently and in high concentrations in relation to their toxicity should be targeted for action. The frequency of the chemicals' presence in one wastewater stream, however, cannot be a guide for their presence or absence in another wastewater stream. If the wastewater from the entire world must be included in the hazard identification process for land application of wastewater, it is imperative to assume that any potentially toxic chemical may be found in the wastewater and treated effluents. There is no logical justification to exclude any, because there is not a comprehensive worldwide survey of toxic chemicals in wastewater of which a decision may be based on.

We examined the potentially harmful chemicals which have been identified through protocols similar to a hazard identification exercise by selected public agencies and international bodies (Table 13.3). For example, 236 inorganic and organic chemicals were found in the *Drinking Water Regulations and Health Advisories* of the U.S. Environmental Protection Agency (1991a). It identified these chemicals (30 inorganic and 206 organic) for potential regulatory action. A few substances are included in this list because of their possible effects on performance of water purification processes or on consumer water preferences. An overwhelming majority

TABLE 13.2. Occurrence to Toxic Organic Chemicals in Wastewater from 23 Treatment Plants in Canada.

Compound	% Occurrence at Concentration		
	> 1 mg L^{-1}	> 10 mg L^{-1}	> 100 mg L^{-1}
Chloroform	92	13	
Benzene	71	25	
Di-*n*-butyl phthalate	46	17	4
Anthracene	25	4	4
1, 3 Dichloropropane	17	13	4
1, 4 Dichlorophenol	13		
Pentachlorophenol	13		
Pyrene	8		
Bis (2-chloroethyl) ether	4		
2, 4, 6 Trichlorophenol	4		
Acenephythylene	4		
Hexachlorobenzene	4		
Benzo(a)pyrene	4		
Bis (2-chloroethyl) methane	4		

Source: Data derived from Black et al., 1984.

TABLE 13.3. Potentially Toxic Chemicals Identified by Selected Regulatory Agencies and International Bodies.

Source	Organic	Inorganic
Drinking Water Regulation and Health Advisories (U.S. Environmental Protection Agency, 1996)	206	30
Identification and Listing of Hazardous Wastes (U.S. Code of Federal Regulations, Title 40, Part 261)	189	15
Guidelines for Drinking Water Quality (World Health Organization, 1984)	40	16
Environmental Carcinogens, Selected Methods of Analysis (WHO/International Agency for Research on Cancer, 1979–86)	61	6
EEC Directive on Pollution Caused by Certain Dangerous Substances Discharged into Aquatic Environment of the Community (EEC, 1976)	127	22
Recommended Public Health Related Groundwater Standards (Wisconsin Division of Health, 1986)	16	1
Pollutants Selected for Environmental Profile/Hazard Indices (U.S. Environmental Protection Agency, 1993)	33	15
Standards for the Use and Disposal of Sewage Sludge (U.S. Code of Federal Regulations, Title 40, Part 503)	0	10

of them are listed because of their potential adverse effects on human health. Another document from the same agency, *Identification and Listing of Hazardous Wastes*[2] contained 204 chemicals. Many chemicals appeared on both lists. Since 1976, 129 substances and substance groups, which are toxic, persistent, and bioaccumulative, have been identified by the European Community, and the member nations are required to eliminate environmental pollution caused by these chemicals (EEC, 1976). Other national and international entities have produced similar lists.

While the number of chemicals (or group of chemicals) on these lists varied from 10 to several hundred, many substances appeared in almost all of the lists (Table 13.3). The discrepancy in numbers is related to the date when the list was drafted and to the purpose of listing. In general, (1) more substances are found in lists drafted more recently (e.g. 1990s) in contrast to those prepared earlier (e.g. 1970s) and (2) regulations containing numerical limits usually list fewer chemicals than advisory guidelines. The increase in the number of candidate toxic chemicals reflects improvements in the toxicological data bases and in analytical methods for detecting chemicals. Despite the hazard potentials, the dose-response relations and/or the food chain transfer for many toxic chemicals cannot be defined accurately because of a lack of data. For example, in establishing the land

[2] U.S. Code of Federal Regulations, Title 40, Part 261 Identification and Listing of Hazardous Chemicals.

TABLE 13.4. **Frequently Regulated Toxic Chemicals.**[a]

Inorganic Substances	Organic Compound	
Arsenic	Aldrin	Hexachloroethane
Barium	Benzene	Pyrene
Beryllium	Benzo(a)pyrene	Lindane
Cadmium	Carbon Tetrachloride	Methoxychlor
Chromium	Chlorodane	Pentachlorophenol
Cyanide	Chlorobenzene	PCBs
Fluoride	Chloroform[b]	Tetrachloroethane
Lead	Dichloroethanes	Tetrachloroethlene
Mercury	Dichlorophenols	Tolueune
Nickel	2,4-D	Toxaphene
Selenium	Dieldrin	Trichloroethane
Silver	Heptachlor	Trichlorophenol
	Hexachlorobenzene	

[a]Frequently appeared substances in lists in Table 13.3.
[b]Include trihalomethanes (THMs).

application regulations for sludge, the U.S. Environmental Protection Agency (1992b) screened approximately 200 pollutants and selected approximately 32 for further evaluation. When the rule was finalized, numerical limits were set only on 10 elements.

There appears to be little disagreement among various water quality regulatory authorities worldwide about which inorganic substances are potentially hazardous. The entry of these substances into the human food chain often is restricted by regulations or recommended by guidelines. The selection of organic chemicals, however, varies considerably among regulatory authorities. It is impractical to include chemicals of everybody's choice and to establish maximum permissible levels for hundreds of organic chemicals that could sometimes be present in the wastewater only in minute quantities. Without a better approach for reducing the numbers, we subsequently compared the existing lists (Table 13.3) and selected substances that appeared on at least 4 of the lists (Table 13.4). The toxic chemicals outlined in Table 13.3 represent the regulated or potentially regulated chemicals for several different concerns, such as drinking water and hazardous waste management. More frequent appearances on these lists indicate a chemical's potentially serious toxicological effects and ubiquitous nature of their presence in the environment. The set of compounds in Table 13.4 (26 organic and 12 inorganic) will be included in this universally applicable guidelines for land application of wastewater. Many organic chemicals in this list are industrial solvents and are expected to be removed/degraded during wastewater treatment. They probably do not need to be considered in land application. Since partially treated and untreated wastewater are frequently used for irrigation in many parts of the world, they must be included in the assessment and the potential impact of these chemicals on human health needs to be evaluated.

DOSE-RESPONSE ANALYSIS—QUANTIFYING ACCEPTABLE DAILY INTAKE FOR TOXIC CHEMICALS

While the hazard identification portion of the risk assessment can establish causality, it does not provide quantitative information on adverse effect due to exposures. The dose-response relationship quantitatively defines the responses resulted from exposures to various dose levels of a chemical. This relationship is specific for a chemical and may be derived from data obtained in human epidemiological investigations, lifetime feeding studies of experimental animals, and toxicity assays of mammalian or bacterial cells. Depending on the nature of the data, the resulting dose-response relationships may vary considerably (Cohrsen and Covello, 1989).

Epidemiological studies have their origin in investigating causes of epidemic diseases of the 19th century. They are designed to establish a direct cause and effect relationship between a causative factor and the outbreak of a disease via an exposure route. Although epidemiological data provide the most realistic dose-response relationships, studies are conducted only for hazards to which exposures have already occurred. Inherently, data are available only for a limited number of chemicals. Generally, epidemiological data are unable to detect small increase in risk, particularly in case of diseases that take many years to develop. The long latency period of certain health effects, such as cancers, further reduces the data quality as difficulties in tracking the exposed population over many decades, sorting out confounding exposures from other sources, and finding appropriate experimental controls become overwhelming.

Despite their flaws, toxicological studies based on animal bioassays often are the basis to quantify dose-response relationships. Uncertainties caused by extrapolations from animals to humans, high dose to low dose, and short term acute exposure to long term sub-chronic exposures in establishing the dose-response relationship have been reviewed in *Drinking Water and Health* (National Academy of Sciences, 1977). Many of the technical issues remain unresolved. It is, nevertheless, beyond the scope of this study to resolve the uncertainties of dose-response analysis. Through the dose-response analysis performed by toxicologists, the acceptable daily intakes (ADI) of toxic chemicals have been determined (Lu and Sielken, 1991). We chose to use the values published in the *Drinking Water Regulations and Health Advisories* (U.S. Environmental Agency, 1991a) as the primary source of information and supplemented it with data used in other

TABLE 13.5. Exposure Assessment Pathways Used by U.S. Environmental Protection Agency in Establishing Numerical Limits for Land Application of Sewage Sludge (U.S. Environmental Protection Agency, 1992b).

Pathway	Description of Pollutant Transport
1	Sewage sludge-soil-plant-human
2	Sewage sludge-soil-plant-home gardener
3	Sewage sludge-soil-child
4	Sewage sludge-soil-plant-animal-human
5	Sewage sludge-soil-animal-human
6	Sewage sludge-soil-plant-animal
7	Sewage sludge-soil-animal
8	Sewage sludge-soil-plant
9	Sewage sludge-soil-soil biota
10	Sewage sludge-soil-soil biota-predator of soil biota
11	Sewage sludge-soil-airborne dust-human
12	Sewage sludge-soil-surface water-fish-human
13	Sewage sludge-soil-air-human
14	Sewage sludge-soil-groundwater-human

risk assessment exercises (U.S. Environmental Protection Agency, 1992b; National Academy of Sciences, 1977). *Drinking Water Regulations and Health Advisories* may also be accessed via internet website http://www. epa.gov/docs/ostwater/tools/dwstds0.html. The Integrated Risk Information System (IRIS)[3] of the U.S. Environmental Protection Agency (a collection of computer-based files that contains descriptive and quantitative information on the reference dose and ADI of many chemicals) provides the most current estimates and is accessible via electronic mail networks (U.S. Environmental Protection Agency, 1991b).

EXPOSURE ANALYSIS

The U.S. Environmental Protection Agency (1993) employed 14 pollutant transport pathways, completed with a well-defined exposure scenario for each pathway to establish numerical limits for land application section of the *Standards for the Use and Disposal of Sewage Sludge*[4] (Table 13.5). Its technical support document on which the standards are based is by far the most comprehensive risk-based analysis on land application of sewage sludge ever undertaken (U.S. Environmental Protection Agency, 1992b). The pathways identified in this document described almost all of the possible environmental transport

[3] IRIS is a collection of computer-based data files that contain descriptive and quantitative information on reference dose and ADI of many toxic chemicals.
[4] U.S. Code of Federal Regulations, Title 40, Part 503 Standards for the Use and Disposal of Sewage Sludge.

routes for pollutants released through land application of wastes with receptors of pollutants ranging from child and adult humans, livestock animals, plants, soil biota, to soil biota predators. They may be used as a reference in selecting exposure pathways for wastewater irrigation. The quasi-equilibrium approach was employed to track the transfer of pollutants along the pathway. The underlying rationale is as follows: (1) if maximum permissible pollutant input of the receptor is defined and pollutant transfer coefficients between two stages of pollutant transfer are known, the maximum permissible pollutant loading may be back calculated via the exposure pathway; (2) to be protective for all exposure scenarios, MCL for a pollutant is the lowest maximum permissible loading derived from all pathways considered.

Comprehensive coverage of the exposure routes, however, does not imply completeness or accuracy of the analysis. The plausibility of the outcome is dependent on how realistic the exposure scenarios and transfer coefficients are. Many pollutant transport routes involve multiple stages and, at each stage, there may be more than one receptor of the pollutant. The pollutant transfer coefficient for each receptor, however, may not be the same. There is a great deal of uncertainty, as a slight change in exposure scenario will significantly affect the outcome.

Nine of the 14 pathways affect human health and safety. The mathematical expressions for establishing the pollutant loading limits using these 9 pathways are summarized as follows (U.S. Environmental Protection Agency, 1992b).

PATHWAYS 1 AND 2: SLUDGE-SOIL-PLANT-HUMAN

Pathway 1 describes the food chain transfer of toxic pollutant from sewage sludge through food plants grown on sludge-treated soils to human beings. Under this pathway, the reference application rate (maximum permissible application rate under the defined exposure scenario) of a pollutant (RP) is calculated as:

$$RP = RIA / \Sigma (UC_i \times DC_i \times FC_i) \qquad (1)$$

where RIA = adjusted reference intake (μg/day) derived from RfD and q^* which is adjusted for body weight and relative effectiveness, RfD = the reference dose of the chemical under consideration (mg/kg/day), q^* = human cancer potency (mg/kg/day) at a given risk level, UC_i = uptake response slope for food group i (pollutant concentration of plant in μg/g DW per unit pollutant loading of soil (kg/ha), DC_i = daily dietary consumption of the food group i (g/day), and FC_i = fraction of food group i assumed to originate from sludge-treated soil.

To determine permissible pollutant loading under this pathway, the daily dietary intake of each food group (DC_i) and the pollutant uptake slope for each food group (UC_i) must be known. In addition, the fraction of the pollutant RfD (ADI is estimated from RfD by introducing a conservative bias, depending on the reliability of the data) that may come from this exposure scenario (FC_i) must also be known.

Pathway 2 may be considered as a special case for pathway 1 where percentages of the daily dietary intake affected pollutant inputs are different for these two cases.

PATHWAY 3: SLUDGE-SOIL-CHILD

Pathway 2 describes the pollutant's toxicity to children ingesting sludge-treated soil. Several steps are needed to arrive at the appropriate pollutant loading rate. For non-threshold chemicals, the adjusted reference intake, RIA (μg/day) is determined as

$$RIA = \{[(RL \times BW)/(q^* \times RE)] - TBI\} \times 10^3 \qquad (2)$$

where RL = cancer risk level (e.g. 10^{-5}, 10^{-6} ...), BW = human body weight (kg), RE = relative effectiveness of ingestion exposure, and TBI = total background intake rate of pollutant (mg/day), from all other sources.

For threshold chemicals,

$$RIA = \{[(RfD \times BW)/RE] - TBI\} \times 10^3 \qquad (3)$$

The reference pollutant concentration in the soil, RLC (μg/g DW), may be calculated by

$$RLC = RIA/(I_s \times DA) \qquad (4)$$

where I_s = soil ingestion rate (g/day DW) and DA = exposure duration adjustment factor. Finally, the reference annual application rate of pollutant, RP_a (kg/ha) is determined by

$$RP_a = RLC \times MS \times 10^{-3} \times e^{-kT} \times [1 + D \times e^{-k} + D^2 \qquad (5)$$
$$\times e^{-2k} \ldots + D^{n-1} \times e^{(1-n)k}]^{-1}$$

where MS = mass of soil in top 15 cm, 2×10^3 mT/ha, e = base of natural logarithm, 2.718, k = loss rate constant for the pollutant, T = waiting period (year), $D = (MS - AR_a)/MS$, Ar_a = annual application rate (mT DW/ha), and n = number of years involved.

PATHWAY 4: SLUDGE-SOIL-PLANT-ANIMAL-HUMAN

Pathway 4 describes the pollutant's toxicity to humans ingesting animal products which come from animals consuming crops grown on sludge-treated soil. The RIA of non-threshold chemicals and threshold chemicals are first determined according to Equations (2) and (3), respectively. The reference animal feed concentration, RFC (μg/g DW), is determined as

$$RFC = RIA/(UA_i \times DA_i \times FA_i) \qquad (6)$$

where UA_i = uptake response slope of pollutant in animal tissue food group i (pollutant concentration in μg/g DW per unit concentration of pollutant in feed, μg/g), DA_i = daily dietary consumption of the animal tissue food group i (g DW/day), and FA_i = fraction of food group i assumed to be derived from sludge-amended soil or feedstuff.

Based on the RFC, a cumulative pollutant loading rate, RP_c (kg/ha) for inorganic conservative constituent is calculated as

$$RP_c = RFC/UC \qquad (7)$$

where UC = linear response slope of forage crop (pollutant concentration in crop μg/g DW per unit of pollutant input kg/ha). For degradable organic chemicals, a reference soil concentration, RLC (μg/g DW), is calculated before determining the annual pollutant application rate.

$$RLC = (RFC/UC) + BS \qquad (8)$$

where BS = background soil concentration of pollutant (μg/g DW). The final reference annual pollutant application rate, RP_a (kg/ha/yr), is determined according to Equation (5).

PATHWAY 5: SLUDGE-SOIL-ANIMAL-HUMAN

This pathway concerns human toxicity from the ingestion of meat products derived from animals that accidentally ingested sludge-treated soil. The adjusted reference intake, RIA (μg/day), and reference feed concentration of pollutant, RFC (μg/g DW), are calculated according to Equations (2), (3), and (6). The reference application rate of pollutant, RP (μg/g DW), is determined as

$$RP = (RLC - BS) \times MS \times 10^{-3} \qquad (9)$$

and

$$RLC = (RFC/FL) + BS \qquad (10)$$

where FL = fraction of diet that is adhering soil/sludge (g soil DW/g diet DW)

PATHWAY 11: SLUDGE-SOIL-AIRBORNE PARTICULATE-HUMAN

Pathway 11 involves human exposure to pollutants in sludge-treated soil through particulate inhalation resuspended by tillage operations. The mathematical formula to be used to calculate the maximum pollutant application rate, RP (kg/ha), is

$$RP = (NIOSH/TDA) \times MS \qquad (11)$$

Where NIOSH = Occupational health and safety standard for the pollutant set by the National Institute of Occupational Health and Safety or the Occupation Safety and Health Act (mg/m^3), TDA = total dust standard, 10 mg soil/m^3 air, MS = mass of soil in top 15 cm, 2×10^3 mT/ha.

PATHWAY 12: SLUDGE-SOIL-SURFACE RUNOFF-SURFACE WATER-HUMAN

This pathway concerns the transfer of pollutant from sludge-treated soil via surface runoff and surface water body to human (drinking water) or aquatic life (fish). The exposure scenario assumes that the sludge-treated area is located in a large size watershed, employs the Universal Soil Loss Equation (Wischemeier, 1977) to calculate the sediment load, routes the surface runoff through a buffer zone where the sediment and thus the pollutant load are modulated, and determines the pollutant concentration in the surface water body by merely simple dilution. Data required to undertake the computation include the following:

- area of land over which sludge is applied
- sludge application rate
- time period over which application is proposed
- first order decay rates for organic chemicals
- bulk density of soil/sludge mixture
- depth of incorporation
- recharge rate
- concentration of pollutant in leachate of sludge
- pollutant concentration in sludge (DW)
- length of buffer zone

- width of buffer zone
- mean annual flow for the receiving river
- average annual precipitation

In addition, slope, soil type, and management practice for the site must be known to select the appropriate input parameters for the Universal Soil Loss Equation.

PATHWAY 13: SLUDGE-SOIL-ATMOSPHERE-HUMAN

This pathway examines the inhalation hazard to residents downwind from a land application site and adopts an analytical model that evaluates pollutant vapor generation at a hazardous waste site for the computation (ESE, 1985). Namely, the rate of pollutant vaporization is calculated by a mass balance which partitions the chemicals into those degraded, leached, vaporized, and stored. The concentration of pollutant in the air is arrived at by distributing the vaporized pollutant into a defined volume of the atmosphere. Finally, the inhalation is determined as the volume of air breathed multiplied by pollutant concentration in the air.

PATHWAY 14: SLUDGE-SOIL-VADOSE ZONE-
GROUNDWATER-HUMAN

In this pathway, the hazard of drinking contaminated ground water is evaluated after the pollutant has leached through the vadose zone. The computer algorithm CHAIN (Van Genuchten, 1985) is used to solve the one-dimension convective-dispersive transport equation for the unsaturated zone. The pollutant concentration at the entrance point of the aquifer is compared to a reference water concentration which is derived from the ADI (estimated from RfD and q^*, if ADI is not available) for the pollutant under consideration. In the case of inorganic minerals, MINTEQ is used to account for precipitation and dissolution of mineral species.

Both computer codes require inputs of site- and chemical-specific parameters which must be defined specifically for the exposure scenario. It is difficult to construct an exposure scenario that can be representative of all possible situations. Simulating the movement of a pollutant pulse through the vadose zone with an analytical solution of one-dimensional convective-dispersive transport equation casts further questions about the credibility of the final result.

These 9 exposure routes are applicable to all land application situations worldwide, but it becomes obvious immediately that constructing globally relevant exposure scenarios from these exposure routes would be very difficult, if not impossible. Due to the inherent data requirements and difficulties in defining representative exposure scenarios, the risk-based methodology is more appropriate for site- and case-specific situations. Despite the deficiencies in methodology, one may always obtain a set of risk-based numerical limits for pollutants by arbitrarily defining the exposure scenarios. But, how realistic are these numerical limits? To what extent are the guidelines able to protect human health and safety? These two questions would be difficult to answer.

NUMERICAL LIMITS BASED ON PREVENTING POLLUTANT ACCUMULATION IN WASTEWATER RECEIVING SOILS

If one assumes that the irrigation water application does not exceed 1.2 m/yr (approximately the amount of water required to produce a crop in an arid climate), the maximum pollutant inputs to soils which is irrigated with water meeting the requirements of selected irrigation water quality guidelines may be calculated and compared with typical metal concentrations of soils and typical annual metal removal by plants (Table 13.6). It appears that amounts of pollutant extracted by the plants would be small, relative to inputs from irrigation. In crop production systems, plant uptake is the most significant pathway for removing metals from soils. If the pollutant loading (input) must be balanced by the pollutant removal by plants (output), the numerical limits for pollutants in wastewater used in crop irrigation must be 100–1000 times smaller than those currently listed in Table 13.1.

Based on the normal elemental concentrations of plant tissue, the plant removal of potentially toxic inorganic elements may be estimated (Adriano, 1986; Alloway, 1991). Assuming pollutant inputs via wastewater irrigation must equal pollutant output of via plant removal and volume of irrigation water is 1.2 m yr^{-1}, the MCL of pollutants may be calculated (Table 13.6).

In untreated and primary treated wastewater, the pollutant concentrations often exceed those calculated by equating input to output (Table 13.7). When the wastewater has undergone complete secondary treatment or advanced treatment, concentrations of potentially toxic pollutants in water are reduced by 2–3 orders of magnitude. If industrial waste discharge pretreatment programs are used to control pollutant input from influent wastewater and the treatment is properly designed and operated, it is possible for reclaimed wastewater from such installations to meet the calculated MCL in Table 13.8. Under the same circumstances, concentration of potentially toxic organic pollutants in reclaimed wastewater are also expected to be low.

Besides imposing stringent numerical limits on concentrations of toxic chemicals in wastewater, guidelines employing this principle may also be drafted by:

TABLE 13.6. Permissible Toxic Element Concentration of Wastewater Based on Preventing Pollutant Concentration in Soil to Increase.

Element	Concentration (mg kg^{-1})[a]	Removal by Plant (kg ha^{-1} yr^{-1})[b]	Irrigation Water (m^3 ha^{-1} yr^{-1})[c]	MCL in Wastewater (mg L^{-1})[d]
As	0.02–7	$(0.4-140) \times 10^{-3}$	1.2×10^4	$(0.33-117) \times 10^{-4}$
Ba	1–150	$(20-3000) \times 10^{-3}$	1.2×10^4	$(17-2500) \times 10^{-4}$
Be	0.0006–0.037	$(0.12-0.74) \times 10^{-3}$	1.2×10^4	$(0.01-0.62) \times 10^{-4}$
Cd	0.1–2.4	$(2-48) \times 10^{-3}$	1.2×10^4	$(1.7-40) \times 10^{-4}$
Cr	0.03–14	$(0.6-280) \times 10^{-3}$	1.2×10^4	$(0.5-233) \times 10^{-4}$
Cyanide	e	e	e	e
F	0.2–24	$(4-480) \times 10^{-3}$	1.2×10^4	$(3.3-400) \times 10^{-4}$
Pb	0.2–20	$(4-400) \times 10^{-3}$	1.2×10^4	$(3.3-333) \times 10^{-4}$
Hg	0.005–0.17	$(0.1-3.4) \times 10^{-3}$	1.2×10^4	$(0.8-3) \times 10^{-4}$
Ni	0.02–5	$(0.4-100) \times 10^{-3}$	1.2×10^4	$(0.3-83) \times 10^{-4}$
Se	0.001–2	$(0.02-40) \times 10^{-3}$	1.2×10^4	$(0.02-33) \times 10^{-4}$
Ag	0.1–0.8	$(2-16) \times 10^{-3}$	1.2×10^4	$(1.7-13) \times 10^{-4}$

[a] Normal concentration range for plant tissue.
[b] Assuming annual biomass production of 20×10^3 kg ha^{-1}.
[c] Typical amount of irrigation water required for crop production in arid climate.
[d] Calculated by dividing values in column 2 by value in column 3.
[e] Plant absorption of cyanide unknown.

TABLE 13.7. Selected Mean Trace Element Concentrations of Treated Urban Wastewater in Portugal (Marecos do Monte et al., 1996).

	Treatment Effluent		
Parameter	Primary	Secondary	Facultative Ponds
As (mg L^{-1})	5.8	5.9	
Cd (mg L^{-1})	2.3×10^{-3}	2.1×10^{-3}	$<5.0 \times 10^{-3}$
Co (mg L^{-1})	$<30 \times 10^{-3}$	$<30 \times 10^{-3}$	$<5 \times 10^{-3}$
Cu (mg L^{-1})	148×10^{-3}	126×10^{-3}	6×10^{-3}
Fe (mg L^{-1})	400×10^{-3}	312×10^{-3}	105×10^{-3}
Mn (mg L^{-1})	60×10^{-3}	58×10^{-3}	13×10^{-3}
Ni (mg L^{-1})	68×10^{-3}	73×10^{-3}	$<5 \times 10^{-3}$
Pb (mg L^{-1})	22×10^{-3}	22×10^{-3}	18×10^{-3}
Zn (mg L^{-1})	140×10^{-3}	129×10^{-3}	571×10^{-3}

- requiring the wastewater to undergo rigorous treatment to remove pollutants and meet preset performance standards prior to land application
- setting pretreatment requirements to prevent point source pollutants from entering wastewater collection and treatment systems
- setting very stringent pollutant loading limits for soils
- employing a combination of the previous three approaches

For reclaimed water meeting these performance requirements, human health related numerical limits are not needed when the reclaimed wastewater is used in crop irrigation. Guidelines based on stringent pollutant loading limits and/or rigorous industrial wastes pretreatment requirements and performance standards, as outlined here, tend to discourage the POTWs from considering the land application option.

TABLE 13.8. Concentration of Trace Metals in Effluent of Two Advanced Wastewater Treatment Facilities in San Diego, California (AWT and Miramar) in Comparison with Those in the U.S. Groundwater and Surface Water (U.S. Environmental Protection Agency, 1992a).

Constituent	AWT	Miramar	U.S. Groundwater	U.S. Surface Water
As (mg L^{-1})	0.3×10^{-3}	1.7×10^{-3}	0.001–0.100	0.001–0.020
Cd (mg L^{-1})	2×10^{-5}	8×10^{-5}	0.001–0.058	0.001–0.12
Cr (mg L^{-1})	8×10^{-3}	3×10^{-3}	0.0008–0.050	0.0002–0.650
Cu (mg L^{-1})	5×10^{-3}	9×10^{-3}	0.001–0.100	0.0004–0.200
Pb (mg L^{-1})	0.4×10^{-3}	0.5×10^{-3}	0.002–0.100	0.0002–1.56
Ni (mg L^{-1})	8×10^{-3}	3×10^{-3}	0.001–0.130	0.0013–0.500
Se (mg L^{-1})	0.1×10^{-3}	0.7×10^{-3}	0.001–0.061	0.0001–0.047
Zn (mg L^{-1})	25×10^{-3}	17×10^{-3}	0.0027–1.3767	0.001–8.60

NUMERICAL LIMITS BASED ON MAXIMIZING POLLUTANT ATTENUATION CAPACITY OF SOIL

Soil has capacity to attenuate pollutants from anthropogenic sources. This capacity may render pollutants innocuous or serve as a "buffer" in preventing pollutant in the soil from being transferred to crop plants. To determine the maximum contaminant level a soil may bear, one must start with the maximum tolerable pollutant level for human (RfD) and back calculate through the exposure pathways which accounted for the attenuation capacity of soils to arrive at a corresponding concentration in soil or in wastewater as illustrated in previous sections.

SELECTION OF AN EXPOSURE ANALYSIS METHOD

The input data required to define exposure scenarios outlined in the previous section was immense. Simplifications were necessary because it is unrealistic that an exposure scenario with so many input parameters can be representative of situations worldwide. A working group organized by the World Health Organization Regional Office for Europe to evaluate the human health risk caused by chemicals in sewage sludge applied to land has considered pollutant transfer on essentially the same exposure pathways (Dean and Suess, 1985). They concluded that food chain transfer is the most significant exposure route and Cd appeared to be the most serious contaminant. Other metals in sludge are unlikely to cause adverse health effects because of low plant absorption. Organic pollutants are not absorbed significantly by plants through their roots but may be ingested with soil by some grazing animals. Nevertheless, overall exposure to organic pollutants through the land application of sludge is minor and unlikely to result in any harmful effect.

Perwak et al. (1986) reviewed methods for assessing environmental pathways of food contamination. They concluded that very little quantitative information was available to evaluate pollutant transfer from soil to plants, a key process for food chain transfer of pollutants. As a first-cut approximation, however, plant absorption of pollutant in the vapor phase and through root extraction might be approximated by the Henry's law constant (H) or octanol water partition coefficient (K_{ow}) of the chemical.

$$C_p = (C_a \times R \times T)/H \tag{13}$$

and

$$\text{Log } (RCF - 0.82) = 0.77 \times \log K_{ow} - 1.52 \tag{14}$$

$$TSCF = 0.784 \times e^{-[(\log K_{ow}-1.78)**2]/2.44} \tag{15}$$

where

C_p = concentration of pollutant in plant (μg g^{-1})
C_a = concentration of pollutant in air (μg m^{-3})
H = Henry's law constant (m^3 atm. Mol^{-1})
R = gas constant (8.2 × 10^{-5} m^3 atm. Mol^{-1} °K^{-1})
T = temperature (°K)
RCF = root concentration factor (concentration in root/concentration in external solution)
TSCF = transpiration stream concentration factor (concentration in transpiration stream/concentration in external solution) used to describe the uptake of pollutant by shoots

McKone and Ryan (1989) described the fruit and vegetable and grain contamination by pollutants in the soil as

$$F_{sv} = (I_v \times f_v \times K_{sp}) \times 0.25 \tag{16}$$

$$F_{sg} = (I_g \times f_g \times K_{sp}) \times 0.25 \tag{17}$$

where

F_{sv} = pathway exposure factor from soil to vegetable and fruit (kg/kg · day)
F_{sg} = pathway exposure factor from soil to grain (kg/kg · day)
I_v = human intake of fruit and vegetable (kg/kg · day fresh mass)
I_g = human intake of grain (kg/kg · day fresh mass)
f_v = fraction of target population's fruit and vegetables that were grown in contaminated source
f_g = fraction of the target population's grain that comes from the contaminated source
K_{sp} = plant/soil partition factor (mg/kg DW of pollutant in plant tissue per mg/kg DW of pollutant in soil)

The numeral 0.25 is a fresh weight to dry weight conversion factor. The daily average ingestion exposure E_{vg} (mg/day), may then be determined by

$$E_{vg} = (F_{sv} + F_{sg}) \times C_s \tag{18}$$

where C_s = pollutant concentration in soil (mg/kg)

The plant soil partition coefficients (K_{sp}) for organic chemicals have been related to the log K_{ow} of the chemicals (Travis and Arms, 1988).

$$\log B_v = 1.588 - 0.578 \times \log K_{ow} \tag{19}$$

where B_v, the vegetation bioconcentration factor, is defined as the ratio of the concentration in above ground parts (mg kg^{-1} of plant DW) to the concentration in soil (mg kg^{-1} of soil). It can be used to estimate the K_{sp} of organic pollutants. Ryan et al. (1988) developed methods to estimate the plant uptake of non-ionic organic chemicals from soil also using K_{ow}.

The K_{sp} of inorganic chemicals can be estimated with experimental data in the published literature. The maximum permissible pollutant concentration in the soil may be determined using Equation (18), if the fraction of ADI attributable to vegetables, fruits, and grains (E_{vg}) is known. Although these are not exact computations, an order-of-magnitude estimate of the outcome is possible. Because they do not rely on so many site- and event-specific parameters for the estimation, they can be the alternative computation method to the multiple exposure route and risk-based assessment methods previously discussed.

Pollutant intake via food consumption is by far the most significant environmental exposure route for pollutants to reach human subjects (McKone and Ryan, 1989). The pollutant ingested due to consumption of cereal, vegetable, root/tuber and fruit customarily accounts for about 75% of the total dietary intake of pollutant (Galal-Gorchev, 1991). Any increase of pollutant concentration in cereal, vegetable, and root/tuber would therefore have far greater implications for dietary intake of pollutants than other exposure routes. Available data indicate that dietary intake of pollutants among the worldwide population is, with few exceptions, significantly below the Provisional Tolerable Weekly Intake (PTWI) published by the Joint FAO/WHO Expert Committee on Food Additives (Ryan et al., 1982; Galal-Gorchev, 1991; World Health Organization, 1989b). But the situation for each pollutant is not the same. There also are considerable variations among nations. On the average, one may assume that the dietary intake of pollutants does not exceed 50% of the PTWI. Except for individuals who are exposed through occupational activities or residents who live near a pollution point source, pollutant intakes through the atmosphere (inhalation route) and drinking water (ingestion route) are minor. For this reason, we chose only to focus on the food chain pathway of wastewater-soil-plant-human to evaluate transfer of pollutants.

The food consumption patterns that must be used along with the ADI to determine the permissible pollutant concentration in food vary considerably from region to region (Ryan et al., 1982; World Health Organization, 1989c; Galal-Gorchev, 1991). Even if the total dietary intake can be normalized, there are still significant differences in regional preferences for food groups whose K_{sp} are not always known. For the purposes of estimating pesticide residue intake at the international level, the Food and

TABLE 13.9.   Regional Food Consumption Pattern and Global Diet.

Food Group[a]	Middle Eastern	Far Eastern	African	Latin American	European	Global Diet
Cereals	432	452	320	254	226	405
Roots/tubers	62	108	321	159	242	288
Pulses	21	27	18	21	9.3	23
Sugar/honey	95	50	43	104	105	58
Nuts/oil seeds	4.3	18	15	19	12	18
Vegetable oil/fat	38	15	24	26	48	38
Stimulants	8	1.5	0.5	5.3	14	8.1
Spices	2.3	2.0	1.6	0.3	0.3	1.4
Vegetables	193	168	74	124	298	194
Fish/seafood	13	32	32	40	46	39
Eggs	14	13	3.6	12	38	38
Fruits	220	94	85	288	213	235
Milk and products	132	33	42	168	338	55
Meat/offal	72	47	32	78	221	111
Animal oil and fat	0.5	—	0.5	7.0	2.0	3.2
Total	1311	1062	1012	1311	1822	1520

[a]All categories expressed in grams person^{-1} day^{-1}.

Agriculture Organization recommends that average food consumption data listed in FAO Food Balance Sheets (the global diet) be used (Table 13.9). The food consumption described in the global diet does not resemble any specific diet, but it is a generic aggregation of diets from various regions. In the global diet, grain/cereal, vegetables, root/tuber, and fruit account for 76% of the total daily food consumption. The potential for wastewater-borne pollutants to be transferred to humans through other food groups (dairy and animal products, oil/fat/shortening, sugar/honey, etc.) is relatively small because amounts consumed are low and pollutant concentrations in these food groups are low.

EXPOSURE SCENARIO

Guidelines are established to protect the exposure of an adult whose dietary intakes of grain/cereal, vegetable, root/tuber, and fruit are entirely from plants grown on wastewater-soils. The conditions used in the computation are summarized as follows:

(1) Based on the global diet, the daily intake of food is 0.405, 0.212, 0.288, and 0.235 kg fresh weight for grain/cereal, vegetable, root/tuber, and fruit, respectively. Fresh weight to dry weight conversion factors are 0.90, 0.05, 0.20, and 0.05 for grain/cereal, vegetable, root/tuber, and fruit, respectively. Other portions of the global diet are assumed not to be affected by pollutants in the land application of

wastes and their contribution to the ADI is considered a part of the background exposure.

(2) Normally, only a fraction of the exposed individual's diet will be affected by the land application of wastewater. For example, the U.S. Environmental Protection Agency (1992b) assumes that 2.5% of an urban dweller's diet and 45% of a home gardener's diet may be affected by application of sewage sludge. Globally, there is little information to justify any specific level. In the computation, we assume that 100% of the exposed individual's intake of grain/cereal, vegetable, root/tuber, and fruit were derived from wastewater irrigated land. It was a conservative bias and represents an exposure scenario for the inhabitants who consumed entirely the locally-produced food of a land application region.

(3) The pollutant partition factor between plant (food) and soil, K_{sp} (mg of pollutant per kg of food vs mg of pollutant per kg of soil) for inorganic chemical was determined with data in Kabata-Pendias and Pendias (1984) which was compiled from geochemical studies world-wide. The K_{sp} for grain/cereal, vegetable, root/tuber, and fruit are determined separately. The K_{sp}s were derived from the averages of tabulated data from many unrelated sources worldwide, and represent at best, a first cut estimation. As more realistic data for As, Cd, Pb, Hg, and Ni were available, their K_{sp} are adjusted according to information in the U.S. Environmental Protection Agency (1992b).

The K_{sp} of non-ionic organic pollutants are estimated using Equation (18). The K_{ow} needed for the estimation are obtained from Ryan et al. (1988), Travis and Arms (1988) and Howard (1989, 1990). Because the equation does not distinguish food groups, the K_{sp} for four food groups are assumed to be the same.

(4) ADI (mg per kg body weight per day) of the pollutants are obtained from *Drinking Water Regulation and Health Advisories* (U.S. Environmental Protection Agency, 1991b) and *Technical Support Document for Land Application of Sewage Sludge* (U.S. Environmental Protection Agency, 1992b). It denotes ''the daily intake which, during a lifetime, appears to be without appreciable risk, on the basis of all the facts known at the time'' (World Health Organization, 1989b). For the inorganic pollutants and non carcinogens, the ADI is the reference pollutant dose (RfD) and corresponds to the no-observed-adverse-effect pollutant level. The ADI for carcinogens is the estimated pollutant dose with 10^{-6} risk of causing cancer (International Agency for research in cancer, 1978–86). The body weight (BW) of an adult is estimated at 60 kg.

(5) The background exposures for these pollutants varied from country to country (Peterson, 1990; Galal-Gorchev, 1991). Depending on the

background exposure of each pollutant, the portion of the ADI that might be assigned to pollutant exposure due to land application of wastes might be determined. For the purpose of this computation, we assumed that pollutant exposure (background + land application-induced) from consumption of grain/cereal, vegetable, root/tuber and fruit is limited to 50% of the ADI for all the pollutants.

(6) The maximum soil pollutant concentration, C_s (mg of pollutant per kg of soil), was computed as:

$$C_s = [\text{ADI}(\text{mg kg}^{-1} \text{ day}^{-1}) \times 60 \text{ (kg)} \times 0.5]/\Sigma[I_i \times f_i \times K_{spi} \times \text{FD}_i] \quad (20)$$

where
  I = the food consumption (kg day^{-1})
  f = the fraction of food affected by land application (no unit)
  K_{sp} = is the plant/soil partition factor [(mg/kg)/(mg/kg)]
  FD = is the fresh weight to dry weight conversion factor
   i = the index representing grain/cereal, vegetable, root/tuber and
      fruit

In principle, the maximum pollutant mass loading might be derived from the maximum permissible pollutant concentration of soil, once the background concentration of pollutant and the depth and bulk density of soils were accounted for. However, the background pollutant concentrations and antecedent conditions of soils varied considerably. The maximum permissible pollutant concentration of the soil, in our opinion, was a better reference point for a global guideline. The pollutant mass loading might be determined accordingly for specific conditions.

Parameters used in the computation and the final results of the computation (last column of the computation table) were summarized in Table 13.10. The computed maximum concentration for organic pollutants might be adjusted upward in relation to the degradation half-life of each pollutant using mathematical expressions similar to Equation (6). Results in Table 13.10 were derived from data with a great deal of generalization. For example, the maximum pollutant concentration of the soil would increase approximately 40 and 2 fold, if the percentages of the diet affected by the wastewater and sewage sludge application are 2.5% and 45%, respectively, as assumed by the U.S. Environmental Protection Agency (1992b), instead of the 100% used in our computation. The soil and plant partition coefficient (K_{sp}) for the pollutants may not be representative of the global situation, because the values we used were generalizations of arbitrary

TABLE 13.10. Calculated Pollutant Numerical Limits of Wastewater-Treated Soils.

Element	Food Consumption (kg day^{-1})				Fraction Contaminated	Fresh/Dry Weight Conversion			
	Grain	Vegetable	Root/Tuber	Fruit		Grain	Vegetable	Root/Tuber	Fruit
As	0.405	0.212	0.288	0.235	1	0.9	0.05	0.2	0.05
Ba	0.405	0.212	0.288	0.235	1	0.9	0.05	0.2	0.05
Be	0.405	0.212	0.288	0.235	1	0.9	0.05	0.2	0.05
Cd	0.405	0.212	0.288	0.235	1	0.9	0.05	0.2	0.05
Cr	0.405	0.212	0.288	0.235	1	0.9	0.05	0.2	0.05
F	0.405	0.212	0.288	0.235	1	0.9	0.05	0.2	0.05
Pb	0.405	0.212	0.288	0.235	1	0.9	0.05	0.2	0.05
Hg	0.405	0.212	0.288	0.235	1	0.9	0.05	0.2	0.05
Ni	0.405	0.212	0.288	0.235	1	0.9	0.05	0.2	0.05
Se	0.405	0.212	0.288	0.235	1	0.9	0.05	0.2	0.05
Ag	0.405	0.212	0.288	0.235	1	0.9	0.05	0.2	0.05

Chemical	Food Consumption (kg day^{-1})				Fraction Contaminated	Fresh/Dry Weight Conversion			
	Grain	Vegetable	Root/Tuber	Fruit		Grain	Vegetable	Root/Tuber	Fruit
Aldrin	0.405	0.212	0.288	0.235	1	0.9	0.05	0.2	0.05
Benzene	0.405	0.212	0.288	0.235	1	0.9	0.05	0.2	0.05
Benzo(a)pyrene	0.405	0.212	0.288	0.235	1	0.9	0.05	0.2	0.05
Chlorodane	0.405	0.212	0.288	0.235	1	0.9	0.05	0.2	0.05
Chlorobenzene	0.405	0.212	0.288	0.235	1	0.9	0.05	0.2	0.05
Chloroform	0.405	0.212	0.288	0.235	1	0.9	0.05	0.2	0.05
Dichlorophenol	0.405	0.212	0.288	0.235	1	0.9	0.05	0.2	0.05
2,4-D	0.405	0.212	0.288	0.235	1	0.9	0.05	0.2	0.05
Dieldrin	0.405	0.212	0.288	0.235	1	0.9	0.05	0.2	0.05
Heptachlor	0.405	0.212	0.288	0.235	1	0.9	0.05	0.2	0.05
Hexachlorobenzene	0.405	0.212	0.288	0.235	1	0.9	0.05	0.2	0.05
Hexachloroethane	0.405	0.212	0.288	0.235	1	0.9	0.05	0.2	0.05
Pyrene	0.405	0.212	0.288	0.235	1	0.9	0.05	0.2	0.05
Lindane	0.405	0.212	0.288	0.235	1	0.9	0.05	0.2	0.05
Methoxychlor	0.405	0.212	0.288	0.235	1	0.9	0.05	0.2	0.05
Pentachlorophenol	0.405	0.212	0.288	0.235	1	0.9	0.05	0.2	0.05
Tetrachloroethane	0.405	0.212	0.288	0.235	1	0.9	0.05	0.2	0.05
Tetrachloroethylene	0.405	0.212	0.288	0.235	1	0.9	0.05	0.2	0.05
Toluene	0.405	0.212	0.288	0.235	1	0.9	0.05	0.2	0.05
Toxaphene	0.405	0.212	0.288	0.235	1	0.9	0.05	0.2	0.05
Trichloroethane	0.405	0.212	0.288	0.235	1	0.9	0.05	0.2	0.05
DDT	0.405	0.212	0.288	0.235	1	0.9	0.05	0.2	0.05
PCBs	0.405	0.212	0.288	0.235	1	0.9	0.05	0.2	0.05
2,4,5-T	0.405	0.212	0.288	0.235	1	0.9	0.05	0.2	0.05
2,3,7,8 TCDD	0.405	0.212	0.288	0.235	1	0.9	0.05	0.2	0.05

TABLE 13.10. (continued).

Element	Plant/Soil Partition				Pollution Partition Factor				ADI (mg/kg BW/day)	ADI × BW × 0.5 (mg/day)	Soil Concentration (mg/kg)
	Grain	Vegetable	Root/Tuber	Fruit	Grain	Vegetable	Root/Tuber	Fruit			
As	0.004	0.037	0.004	0.003	0.001458	0.000749	0.000324	6.08E-04	0.0008	0.024	9.26
Ba	0.005	0.01	0.02	0.02	0.000101	0.000106	0.000288	0.000235	0.07	2.1	2875.73
Be	0.14	0.02	0.14	0.14	0.002835	0.000212	0.002016	0.001645	0.005	0.15	22.36
Cd	0.036	0.223	0.008	0.09	0.000729	0.002364	0.000115	0.001058	0.001	0.03	7.03
Cr	0.0014	0.0007	0.0003	0.006	2.84E-04	7.42E-05	4.03E-02	7.05E-05	0.005	0.15	3182.01
F	0.006	0.09	0.005	0.02	0.000122	0.000954	0.000072	0.000235	0.12	3.6	2603.98
Pb	0.002	0.00156	0.00002	0.00014	4.05E-04	1.65E-04	2.88E-07	1.65E-05	0.0003	0.009	152.62
Hg	0.085	0.009	0.002	0.009	0.001721	9.54E-04	2.88E-04	0.000106	0.0003	0.009	4.61
Ni	0.007	0.032	0.01	0.007	0.000142	0.000339	0.000144	8.02E-01	0.02	0.6	848.42
Se	0.002	0.015	0.042	0.021	4.05E-04	0.000159	0.000605	0.000247	0.005	0.15	142.71
Ag	1	1	1	1	0.02025	0.0106	0.0144	0.01175	0.005	0.15	2.63

Chemical	Plant/Soil Partition				Pollutant Partition Factor				ADI (mg/kg BW/day)	ADI × BW × 0.5 (mg/day)	Soil Concentration (mg/kg)
	Grain	Vegetable	Root/Tuber	Fruit	Grain	Vegetable	Root/Tuber	Fruit			
Aldrin	0.083	0.083	0.083	0.083	0.001681	0.00088	0.001195	0.000975	0.00003	0.0009	0.19
Benzene	2.27	2.27	2.27	2.27	0.045968	0.024062	0.032688	0.02673	0.00014	0.0042	0.03
Benzo(a)pyrene	0.012	0.012	0.012	0.012	0.000243	0.000127	0.000173	0.000141	0.00006	0.0018	2.63
Chlorodane	0.127	0.127	0.127	0.127	0.002572	0.001346	0.001829	0.001492	0.00006	0.0018	0.25
Chlorobenzene	0.884	0.884	0.884	0.884	0.017901	0.00937	0.01273	0.010387	0.02	0.6	11.91
Chloroform	2.814	2.814	2.814	2.814	0.056984	0.029828	0.040522	0.033065	0.01	0.3	1.87
Dichlorophenol	0.479	0.479	0.479	0.479	0.0097	0.005077	0.006898	0.005628	0.003	0.09	3.3
2,4-D	0.441	0.441	0.441	0.441	0.00893	0.004675	0.00635	0.005182	0.01	0.3	3.3
Dieldrin	0.816	0.816	0.816	0.816	0.016524	0.00865	0.01175	0.009588	0.00005	0.0015	11.93
Heptachlor	0.216	0.216	0.216	0.216	0.004374	0.00229	0.00311	0.002538	0.0005	0.015	0.03
Hexachlorobenzene	0.01	0.01	0.01	0.01	0.000203	0.000106	0.000311	0.000188	0.0008	0.024	1.22
Hexachloroethane	0.239	0.239	0.239	0.239	0.00484	0.002533	0.003442	0.002808	0.001	0.03	42.11
Pyrene	0.033	0.033	0.033	0.033	0.000668	0.00035	0.000475	0.000388	0.03	0.9	2.2
Lindane	0.274	0.274	0.274	0.274	0.005549	0.002904	0.0003946	0.00322	0.0003	0.009	478.47
Methoxychlor	0.111	0.111	0.111	0.111	0.002248	0.001177	0.001598	0.001304	0.005	0.15	0.58
Pentachlorophenol	0.049	0.049	0.049	0.049	0.000992	0.000519	0.000706	0.000576	0.03	0.9	23.71
Tetrachloroethane	1.283	1.283	1.283	1.283	0.025981	0.0136	0.018475	0.015075	0.01	0.3	322.23
Tetrachloroethylene	0.42	0.42	0.42	0.42	0.008505	0.004452	0.006048	0.004935	0.2	6	4.1
Toluene	1.079	1.079	1.079	1.079	0.02185	0.011437	0.015538	0.012678	0.1	3	250.63
Toxaphene	0.231	0.231	0.231	0.231	0.04678	0.002449	0.003326	0.002714	0.004	0.12	48.78
Trichloroethane	2.156	2.156	2.156	2.156	0.043659	0.022854	0.031046	0.025333	0.035	1.05	9.11
DDT	0.014	0.014	0.014	0.014	0.000284	0.000148	0.000202	0.000165			8.54
2,4,5-T	0.167	0.167	0.167	0.167	0.003382	0.00177	0.002405	0.001962	0.01	0.3	31.52
PCBs	0.443	0.443	0.443	0.443	0.008971	0.004696	0.006379	0.005205	0.000001	0.00003	
2,3,7,8 TCDD	0.011	0.011	0.011	0.011	0.000223	0.000117	0.000158	0.000129	0.0007		0

data from around the world. Any change in these values again will further affect the outcomes of the computations. Depending on the extent of background exposure one receives, the fraction of ADI assigned to the food chain exposure route could also vary. We feel, however, further refinement may be counterproductive, as technical information is not available to define the statistical distribution of these variables at the present time and the sensitive analysis, which commonly accompanies this type of assessment, cannot be carried out. Because of the uncertainties surrounding the selection of data and the difficulty of defining a scenario that is truly global, we feel these numerical limits are by no means absolute. They should not be taken at their face values and should only be viewed as the first attempt in formulating such a guideline. If the methodology is deemed acceptable, these values provide estimates of what the numerical limits may look like and permit us to identify what types of technical information are necessary to refine the estimates.

Based on this analysis, it is clear that some constituents (As, Cd, Hg, aldrin, chlorodane, dieldrin, lindane, 2,4,5-T), if present, are more likely to limit the land application of wastewater because the permissible levels of these substances in soil (C_s) are rather low, due to low ADI or high K_{sp}. Others (Ba, Cr, F, etc.) probably do not need to be regulated because their C_s are so large that the upper thresholds are not likely to be exceeded under normal use of the wastewater in agriculture. Further investigations, however, are necessary to confirm the initial findings. For example, the C_s for chloroform is 1.87 mg kg^{-1} of soil because K_{sp} for this compound (2.814) is computed from an empirical equation [Equation (19)]. We are unable to locate any information to substantiate the actual plant/soil partition factor for chloroform. We are also concerned that the computed upper thresholds for soil F and Se may be too high. Some forms of Se in the soil are readily available to plants and chronic Se intoxication of humans has been reported (Yang et al., 1983). These inconsistencies in the final results point out the weakness of our evaluation—a paucity of available data for (1) defining precisely the exposure scenario, (2) obtaining accurate soil/plant partition coefficients for the pollutants, and (3) determining the extent of background exposure.

While we believe the technical data base for determining the numerical limits can be refined, the inherent variability in diets, environmental exposure to pollutants, and soil and plant partition coefficients will always be problematic. There will never be universally acceptable global numerical limits. Should there be one guideline for the entire world? Or should the pollutants be controlled and their numerical limits be set on a site- and case-specific basis with a standardized evaluation procedure?

FORMULATION OF GUIDELINES

In our opinion, the guiding principle of water pollution control as well as wastewater reuse must be source reduction for pollutant discharge. If pollutants are prevented from entering the wastewater collection systems, the treated effluents will be essentially free of pollutants. Infrequent and small amounts of pollutants discharged from ordinary households do not constitute a serious threat to quality of treated effluent for crop irrigation. Under this circumstance, even the untreated wastewater may be used without producing any harmful effect. If discharges of toxic chemical pollutants from industrial sources are not controlled, the land application of reclaimed wastewater must be carefully monitored. Because conventional treatment systems are unable to consistently remove toxic chemicals from influent wastewater. A guideline which sets the upper threshold for pollutant loading is necessary.

In a community which has an effective industrial wastewater pretreatment program, the toxic pollutant input to the wastewater collection system is minimized. The pathogen density and toxic chemical concentrations may be effectively removed by appropriate wastewater treatment and reclaimed wastewater does not contain significant quantity of pollutants. Therefore, its use as irrigation water does not need to be restricted, as the toxic chemical pollutant input via irrigation probably will be equal to or less than the loss through plant uptake. Therefore, only pretreatment requirements and performance standards for wastewater treatment are needed in a human health-related chemical guideline for reclaimed wastewater irrigation.

For untreated wastewater, partially treated wastewater, and wastewater from collection systems that do not have rigorous industrial waste discharge pretreatment requirements numerical limits are necessary. Because the toxic chemical pollutant input via wastewater application may be significant. The numerical limits, in the form of maximum permissible pollutant concentration in soil are presented in Table 13.11.

These values may be used as preliminary estimates of the upper limits, before a better approach can be devised to estimate pollutant exposures from land application of wastes or more representative technical data become available to refine the computation.

For many volatile organic pollutants, the likelihood that they will ever accumulate in the soil to the computed threshold concentration is slim. These chemicals are common in industrial processing and may appear in the wastewater in significant quantities, if industrial waste pretreatment programs are not implemented. Not to acknowledge them in the guideline would imply their absolute safety.

In our opinion, the maximum pollutant concentration of soil is a better

TABLE 13.11. Maximum Pollutant Concentration in Soil.

Inorganic Elements:	
Constituent	Concentration in soil (mg/kg DW)
Arsenic	9
Barium	2900
Beryllium	20
Cadmium	7
Chromium	3200
Fluorine	2600
Lead	150
Mercury	5
Nickel	850
Selenium	140
Silver	3
Organic Compounds:	
Compound	Concentration in Soil (mg/kg DW)
Aldrin	0.2
Benzene	0.03
Benzo(a)pyrene	3
Chlorodane	0.3
Chlorobenzene	ND
Chloroform	2
Dichlorophenols	ND
2,4-D	10
DDT	ND
Dieldrin	0.03
Heptachlor	1
Hexachlorobenzene	40
Hexachloroethane	2
Pyrene	480
Lindane	0.6
Methoxychlor	20
Pentachlorophenol	320
PCBs	30
Tetrachloroethane	4
Tetrachloroethylene	250
Toluene	50
Toxaphene	9
2,4,5-T	ND
2,3,7,8 TCDD	30

ND = Numerical limit not determined because of insufficient data

global reference than the pollutant mass loading because it defines a condition in the soil beyond which the food chain transfer of pollutants may become unacceptable regardless of the location or the source of pollutants. It does not need to account for pollutant attenuation in the soil. By defining the maximum permissible pollutant concentrations in soils, it also precludes soils already contaminated from being used for wastewater applications.

SUMMARY

(1) Land application has been a popular option for disposing of municipal wastewater worldwide for centuries. While most of the operations appear to be successful, reports from China (based on reviews of technical literature printed in Chinese) suggest that large-scale irrigation of crops with mostly untreated municipal and industrial waste containing wastewater could be harmful to crops and cause injuries to humans because of poorly controlled toxic and hazardous waste discharges.

(2) Two general approaches may be used to establish human health-related numerical limits for toxic chemicals in land application of wastewater. The two approaches, as explained below, differ in their underlying principles and result in the promulgation of distinctly different application limits, but conceptually both are equally valid.

- **preventing pollutant accumulation in wastewater receiving soil:** The first approach equates pollutant input to pollutant output, ascribing minimal value to the pollutant-attenuating properties of soil. Conceptually, no net accumulation of pollutants is permitted in the receiving soil; therefore, numerical limits are set to prevent the pollutant concentration of the soil from rising during the course of land application, which maintains the soil's original ecological and chemical integrity. When these requirements are met, the sustainability of the soil to maintain any future land uses is guaranteed, and the transfer of pollutants up the food chain is kept to a minimum.

- **maximizing soil's capacity to assimilate, attenuate, and detoxify pollutants:** The primary principle underlying the second approach in setting numerical limits is that the capacity of soil to attenuate pollutants should be utilized fully. The agronomic benefits of applying wastewater may be realized under this scenario because, when land application operations are managed properly, accumulation of pollutants in soil can be controlled so that they will not reach levels harmful to human health. Land application guidelines based on this approach set the maximum permissible pollutant loading and provide users the flexibility to develop suitable management practices for using wastewater and sewage sludge. Under this scenario, pollutant levels in the soil, however, will rise eventually to levels considerably higher than the background levels, and future land uses may be restricted. The technical data needed to define the pollutant transfer parameters of the exposure pathways are not always available.

(3) For the purposes of maximizing the beneficial uses of wastewater and sewage sludge in agriculture, the most comprehensive method of deriving the numerical limits for pollutant input is to establish first the acceptable daily human intake (ADI) for a particular pollutant and then quantitatively back track the pollutant transport through various environmental exposure routes and arrive at an acceptable pollutant concentration for the receiving soil. To assess the risk of human exposure to pollutants released via land application of wastes, we have determined that at least seven (7) exposure pathways are involved, and each pathway requires a mathematical model. More importantly, each exposure pathway requires an exposure scenario to define the model parameters and to select input data for the computation. As globally representative exposure scenarios are almost impossible to define, this method has little practical utility. It is more suitable to use site- and case-specific evaluations in which the exposure scenarios and environmental parameters needed for the mathematical models can be defined more precisely.

(4) U.S. Environmental Protection Agency (USEPA) employed 14 exposure pathways for the development of its regulations on sewage sludge disposal. Instead of considering all exposure routes, we concentrated on the food chain transfer of toxic chemical pollutants via the wastewater-soil-plant-human route and considered only the pollutant intake from food consumption of grain, vegetable, root/tuber, and fruit. According to technical data in the literature, food chain transfer is the primary route of human exposure to environmental pollutants. Based on the global diet, daily intake of grain/cereal, vegetable, root/tuber, and fruit account for about 75% of daily adult food consumption. We used an exposure scenario that identifies residents who live inside a land application area and whose daily consumption of grain/cereal, vegetable, root/tuber, and fruit are entirely products from wastewater-affected soils. We assumed that they are, for the purposes of this guideline, the most exposed population and their daily intake of pollutants from consumption of grain/cereal, vegetable, root/tuber, and fruit

is limited to 50% of the ADI. The numerical limits for pollutants, therefore, should be developed to protect these exposed people from potential harmful effects.

(5) The pollutant concentration of soil is a more suitable global reference point than the pollutant mass loading rate for assessing potential negative impacts of pollutants in soil, primarily because crop uptake of pollutants is a function of pollutant concentration in soil and because soil properties and environmental conditions are variable around the world. The same pollutant mass loading to soils with different background concentrations may result in different soil pollutant concentrations.

(6) We have demonstrated that numerical limits for toxic chemical pollutants could be developed to safeguard human health from possible exposures to pollutants released during the land application of wastewater. Because this was the first attempt to evaluate the feasibility of developing globally-applicable chemical guidelines, we focused primarily on examining the methodology and adequacy of available data. The computations were used to illustrate the feasibility of the method. These numerical limits should be used only as a basis for further discussion of the issues. While we believe that the technical data base for determining the numerical limits can be refined, the inherent variability in diets, environmental exposure to pollutants, and soil and plant partition coefficients will always be problematic.

CONCLUSIONS

(1) In communities that have a comprehensive and effective industrial waste pretreatment program, secondary and tertiary treated wastewater effluents can be applied on cropland without any restriction, if the amount used is in accordance with the water requirement of crops.

(2) Due to the fact that untreated and partially treated wastewater and treated effluents of wastewater that received unregulated industrial wastewater discharge contain significant amounts of chemical pollutants, their land application must be restricted because of the potential for food chain transfer of pollutants. For these irrigation water, the best agronomic management practices must be followed and the pollutant concentration of the soil should not exceed an upper threshold may be derived from the methodology outlined in this chapter.

ACKNOWLEDGEMENT

The manuscript was rewritten from a report we prepared for World Health Organization. The report was distributed by Urban Environmental

Health, Division of Operational Support in Environmental Health, World Health Organization, 1221 Geneva 27, Switzerland as WHO/EOS/95.20.

REFERENCES

Abdel-Reheem, M. A., R. L. Faltas, R. M. El-Awady, and W. E. Ahmed. 1986. Changes in Trace Elements in Sandy Soil Irrigated with Sewage Water. *Bull. Faculty Agri., Cairo Univ.* 37:969–980.

Adriano, D. C. 1986. *Trace Elements in Terrestrial Environment.* Springer-Verlag, Berlin, Germany.

Alloway, B. J. 1990. *Heavy Metals in Soils.* Wiley Science, New York. 339 pp.

Anonymous. 1978. Irrigation Water Quality Standards, Province of Taiwan. No. 59931, Bureau of Hydraulics, Commission of Reconstruction, Province of Taiwan, Republic of China.

Anonymous. 1991. Decree by the Ministry of Agriculture; Environment and Regional Development; Transport, Telecommunications, and Water Management; and Social Welfare to Regulate Disposal of Effluent and Sludge on Farmland. Hungary.

Anonymous. 1992a. Standards for Irrigation Water Quality. GB 5084–92, National Standard People's Republic of China.

Anonymous. 1992b. Recommendations for Improving the Environmental Management, Monitoring, and Research of Wastewater Irrigated Farm Land in China. A Correspondence Circulated to Officials of the Chinese Agricultural Ecology and Environmental Protection Association (October 28, 1992), (In Chinese).

Anthony, R. M. and L. H. Breimhurst. 1981. Determining maximum influent concentrations of priority pollutants for treatment plants. *J. Water Pollut. Control Fed.* 53:1457–1468.

Asano, T. and A. D. Levine. 1996. Wastewater Reclamation and Reuses: Past, Present and Future. *Water Science and Technology.* 33(10–11):1–14.

Ayers, R. S. and D. W. Westcot. 1985. Water Quality for Agriculture. Irrigation and Drainage Paper No. 29. Food and Agriculture Organization of the United Nations, Rome, Italy.

Bahri, A. and B. Houmane. 1988. Effect of Treated Wastewaters and Sewage Applications on The Characteristics of A Sandy Soil of Tunisia. *Sci. de Sol* 25:267–278, (In French).

Benarde, M. A. 1973. Land Disposal and Sewage Effluent: Appraisal of Health Effects of Pathogenic Organisms. *J. Amer. Water Works Assoc.* 65:432–440.

Benham-Blair & Associates, Inc. and Engineering Enterprise, Inc. 1979. *Long-Term Effects of Application of Domestic Wastewater: Dickinson, North Dakota, Slow Rate Irrigation Site.* EPA-600/2-79-144, U.S. Environmental Protection Agency, Robert S. Kerr Environmental Research Laboratory, Ada, Oklahoma.

Bhattacharya, S. K., R. V. R. Anfara, D. F. Bishop, Jr., R. A. Dobbs, and B. M. Austern. 1990. *Removal and Fate of RARC and CERCLA Toxic Organic Pollutants in Wastewater Treatment.* EPA/600/2-89/026, U.S. Environmental Protection Agency, Risk Reduction Research Laboratory, Cincinnati, Ohio.

Black, S. A., D. N. Graveland, W. Nicholaichuk, D. W. Smith, R. S. Tobin, M. D. Webber, and T. R. Bridle. 1984. *Manual for Land Application of Treated Municipal Wastewater and Sludge.* EPS 6-EP-84-1, Environment Canada. 216 pp.

Blakeslee, P. A. 1973. Monitoring Considerations for Municipal Wastewater Effluent and Sludge Application to Land. pp. 138–198. In: *Recycling Municipal Sludges and Effluents on Land.* National Assoc.. State University and Land Grant Colleges, Washington, D.C.

Burge, W. D. and P. B. Marsh. 1978. Infectious Disease Hazards of Landspreading Sewage Wastes. *J. Environ. Qual.* 7:1–9.

Chang, A. C., A. L. Page, and T. Asano. 1995. *Developing Human Health-related Chemical Guidelines for Reclaimed Wastewater and Sewage Sludge Applications in Agriculture.* WHO/EOS/95.20, World Health Organization, Geneva. 114 pp.

Chase, E. S. 1964a. Nine Decades of Sanitary Engineering, Part II—The Awakening. *Water Work and Waste Engineering* 34(6):48–49, 79.

Chase, E. S. 1964b. Nine Decades of Sanitary Engineering, Part III—Back to The Land. *Water Work and Waste Engineering* 34(7):49–50, 78.

Chen, K. Y., C. S. Young, T. K. Jan, and N. Rohatgi. 1974. Trace Metals in Wastewater Effluents. *J. Wastewater Pollut. Contr. Fed.* 45:2663–2675.

Cohrsen, J. J. and V. T. Covello. 1989. *Risk Analysis: A Guide to Principles and Methods for Analyzing Health and Environmental Risks.* Council on Environmental Quality, Executive Office of The President of the United States. 407 pp.

Crook, J. and R. Y. Surampalli. 1996. Water Reclamation and Reuse Criteria in the U.S. *Water Science and Technology.* 33(10–11):451–462.

Dean, R. B. and M. J. Suess. 1985. The Risk to Health of Chemicals in Sewage Sludge Applied to Land. *Waste Management and Research* 3:251–278.

ESE (Environmental Science and Engineering). 1985. *Exposure to Airborne Contaminants Released from Land Disposal Facilities—A Proposed Methodology.* U.S. Environmental Protection Agency, Washington, D.C.

Environment Canada. 1987. *Canadian Water Quality Guidelines.* Water Quality Branch, Environment Canada, Ottawa, Ontario.

EEC (European Economic Council). 1976. *Dangerous Substances Directive* (76/464/EEC). European Economic Council, Brussels, Belgium.

Feiler, H. 1979. *Fate of Priority Pollutants in Publicly Owned Treatment Works.* EPA 440/1-79-300. U.S. Environmental Protection Agency, Washington, D. C.

Frankenberger, Jr., W. T. 1985. Chapter 14. Fate of Wastewater Constituents in Soil and Groundwater: Pathogens. pp. 14-1 to 14-25 In: G. S. Pettygrove and T. Asano (eds.) *Irrigation with Reclaimed Municipal Wastewater.* California State Water Resources Control Board Report Number 84-1 wr. Lewis Publishers, Chelsea, Michigan.

Friberg, L., M. Piscator, G. F. Nordberg, and T. Kjellstrom. 1974. *Cadmium in the Environment.* CRC Press, Cleveland, Ohio, USA.

Galal-Gorchev. H. 1991. Dietary Intake of Pesticide Residues, Cadmium, Mercury, and Lead. *Food Additives and Contaminants* 8:793–806.

George, D. B., N. L. Altman, D. E. Camann, B. J. Claborn, P. J. Graham, M. N. Guntzel, H. J. Harding, R. B. Harris, A. H. Holguin, K. T. Kimball, N. A. Klein, D. B. Leftwich, R. L. Mason, B. E. Moore, R. L. Northrup, C. B. Popescu, R. H. Ramsey, C. A. Sorber, and R. M. Sweazy. 1986. *The Lubbock Land Treatment System Research and Demonstration Project.* EPA/600/2-86/027, U.S. Environmental Protection Agency, Robert S. Kerr Environmental Research Laboratory, Ada, Oklahoma.

Gotaas, H. B., P. H. McGauhey, and W. J. Kaufman. 1955. Studies in Water Reclamation. Sanitary Engineering Research Laboratory, University of California, Berkeley, Technical Report No. 13. 65 pp.

Gutierrez Ruiz, M. E., Ch. Siebe, and I. Sommer. 1994. Environmental Aspect of land Application of Wastewater from Mexico City Metropolitan Area: A Bibliographical Review and Analysis of Implications. Transactions, *15th World Congress of Soil Science* 3a:445–466.

Heaton, R. D. 1981. Worldwide Aspect of Wastewater Reclamation and Reuse. pp. 43–74 In: F. M. D'Itri, J. Aguirre-Martinez, and M. Athle-Lambarri (eds.) *Municipal Wastewater in Agriculture*. Academic Press, New York.

Hinesly, T. D., R. E. Thomas, and R. G. Stevens. 1978. *Environmental Changes from Long-Term Land Application of Municipal Effluents*. EPA 430/9-78-003, U.S. Environmental Protection Agency, Office of Water, Washington, D.C. 32 pp.

Howard, P. H. 1989. *Handbook of Environmental Fate and Exposure Data for Organic Chemicals, Vol. 1 Large Production and Priority Pollutants*. Lewis Publishers, Chelsea, Michigan. 574 pp.

Howard, P. H. 1990. *Handbook of Environmental Fate and Exposure Data for Organic Chemicals, Vol. 2 Solvents*. Lewis Publishers, Chelsea, Michigan. 546 pp.

International Agency for Research on Cancer (IARC). 1978–86. Environmental Carcinogens Selected Methods of Analysis.

> *Volume 1. Analysis of Volatile Nitrosamines in Food.* IARC Scientific Publication No. 18 (1979).
> *Volume 2. Methods for the Measurement of Vinyl Chloride in Poly (vinyl chloride), Air, Water, and Foodstuffs.* IARC Scientific Publication No. 22 (1979).
> *Volume 3. Analysis of Polycyclic Aromatic Hydrocarbons in Environmental Samples* (1979).
> *Volume 4. Some Aromatic Amines and Azo Dyes in the General and Industrial Environment.* IARC Scientific Publication No. 40 (1981).
> *Volume 5. Some Mycotoxins.* IARC Scientific Publication No. 44 (1982)
> *Volume 6. N-Nitroso Compounds.* IARC Scientific Publication No. 45(1983)
> *Volume 7. Some Volatile Halogenated Hydrocarbons.* IARC Scientific Publication No. 68.(1985).
> *Volume 8. Some Metals: As, Be, Cd, Cr, Ni, Pb, Se, Zn.* IARC Scientific Publication No. 71 (1986)

International Agency for Research on Cancer, Lyon, France.

Jewell, W. J, and. B. L. Seabrook. 1979. *A History of Land Application as a Treatment Alternative: Technical Report.* EPA 430/9-79-012 U.S. Environmental Protection Agency, Office of Water Program Operation, Washington, D.C.

Kabata-Pendias, A. and H. Pendias. 1984. *Trace Elements in Soils and Plants*. CRC Press, Boca Raton, Florida. 315 pp.

Kansel, B. D. and J. Singh. 1983. Influence of the Municipal Waste Water and Soil Properties on the Accumulation of Heavy Metals in Plants. pp. 413–416 in *Proceedings of International Conferences of Heavy Metals in the Environment,* Heidelberg, Germany. CEP Consultants, Edinburgh, UK.

Klein, L. A., M. Lang, N. Nash, and S. L. Kirscher. 1974. Sources of Metals in New York City Wastewater. *J. Water Pollut. Contr. Fed.* 46:1563–65.

Levins, P., J. Adams, P. Brenner, S. Coons, G. Harris, C. Jones, K. Thurn, and A. Wechsler. 1979. *Sources of Toxic Pollutants Found in Influents to Sewage Treatment Plants: VI. Integrated Interpretation.* EPA 68-01-3857. U.S. Environmental Protection Agency, Washington, D.C.

Lu, F. C. and R. L. Sielken, Jr. 1991. Assessment of Safety/risk of Chemicals: Inception and Evolution of the ADI and Dose-response Modeling Procedures. *Toxicology Letters* 59:5–40.

Marecos do Monte, M. H. F., A. N. Angelakis, and T. Asano. 1996. Necessity and Basis for the Establishment of European Guidelines for Reclaimed Wastewater Reuse for Irrigation in the Mediterranean Region. *Water Science and Technology* 33(10):303–316.

McCarty, P. L. and M. Reinhard. 1980. Trace Organics Removed by Advanced Wastewater Treatment. *J. Water Pollut. Contr. Fed.* 52:1907–1922.

McGauhey, P. H., R. B. Krone, and J. H. Winneberger. 1966. *Soil Mantle as a Wastewater Treatment System—Review of Literature.* Sanitary Engineering Research Laboratory, University of California, Berkeley Technical Report No. 67. 120 pp.

McKone, T. E. and P. B. Ryan. 1989. Human Exposure to Chemicals Through Food Chains: An Uncertainty Analysis. *Environ. Sci. Tech.* 23:1154–1163.

Minear, R. A., R. O. Ball, and R. Church. 1981. *Data Bases for Influent Heavy Metals in Publicly Owned Treatment Works.* EPA-660/2-81-220. U.S. Environmental Protection Agency, Cincinnati, OH.

National Academy of Sciences and National Academy of Engineering. 1973. *Water Quality Criteria.* EPA-R3-73-033. U.S. Environmental Protection Agency, Washington, D.C.

National Academy of Sciences. 1977. *Drinking Water and Health.* Safe Drinking Water Committee, Advisory Center on Toxicology, Assembly of Life Science, National Research Council. National Academy of Sciences, Washington, D.C. 939 pp.

National Academy of Sciences. 1983. *Risk Assessment in the Federal Government: Managing the Process.* National Academy Press, Washington, D.C.

National Academy of Sciences. 1984. *Toxicity Testing: Strategies to Determine Needs and Priorities.* National Academy Press, Washington, D.C.

Orlob, G. T. and R. G. Butler. 1955. *An Investigation of Sewage Spreading on Five California Soils.* Sanitary Engineering Research Laboratory, University of California, Berkeley, Technical Report No. 23. 71 pp.

Perwak, J. H., J. H. Ong, and R. Whelan. 1986. *Methods for Assessing Exposure to Chemical Substances, Vol. 8 Methods for Assessing Environmental Pathways of Food Contamination.* EPA 560/5-85/008, U.S. Environmental Protection Agency, Washington, D.C.

Pescod, M. B. 1992. Wastewater Treatment and Use in Agriculture. FAO Irrigation and Drainage Paper 47. Food and Agriculture Organization of the United Nations, Rome. 125 pp.

Peterson, P. J. 1990. Global Environmental Issues. pp. 10–15. In: Tan Jiun'an, P. J. Peterson, Li Ribang, and Wang Wuyi (eds.). *Environmental Life Elements and Health.* Science Press, Beijing, People's Republic of China.

Pettygrove, G. S. and T. Asano (eds.). 1985. Irrigation with Reclaimed Municipal Wastewater, A Guidance Manual. Report No. 84-1 wr. California State Water Resources Control Board, Sacramento, California, Lewis Publishers, Chelsea, MI.

Pimental, G. C. and J. A. Coonrod. 1987. *Opportunities in Chemistry: Today and Tomorrow.* National Academy Press, Washington, D.C., 244 pp.

Pound, C. E., R. W. Crites, and J. V. Olsen. 1978. *Long-Term Effects of Land Application of Domestic Wastewater, Hollister, California, Rapid Infiltration Site.* EPA-600/2-78-084. U.S. Environmental Protection Agency, Washington, D.C.

Pratt, P. F. 1972. Quality Criteria for Trace Elements in Irrigation Waters. University of California, Agricultural Experiment Station and Citrus Research Center, Riverside, California.

Renolds, J. H., L. R. Anderson, R. W. Miller, W. F. Campbell, and M. O. Braun. 1979. *Long-Term Effects of Land Application of Domestic Wastewater: Toole, Utah, Slow-*

Rate Site, Volume I: Field Investigation. EPA-600/2-79-171a, U.S. Environmental Protection Agency, Robert S. Kerr Environmental Research Laboratory, Ada, Oklahoma. 283 pp.

Rogers, H. R. 1987. Occurrence and Fate of Synthetic Organic Compounds in Sewage and Sewage Sludge A Review. Organic Contaminants in Sewage Sludge (EC 9322 SLD). WRc Environment, Medmenham Laboratory, Henley Road, Medmenham, P.O. Box 16, Marlow, Becks, SL7 2HD UK. 83 pp.

Rudolfs, W., L. L. Frank, and R. A. Ragotzkie. 1950. Literature Review on the Occurrence and Survival of Enteric Pathogenic and Relative Organisms in Soil, Water, Sewage, and Sludges, and on Vegetation. *Sewage and Industrial Wastes* 22:1261–1281.

Ryan, J. A., H. R. Pahran, and J. B. Lucas. 1982. Controlling Cadmium in Human Food Chain: A Review and Rationale Based on Health Effects. *Environ. Research* 28:251–302.

Ryan, J. A., R. M. Bell, J. M. Davidson, and G. A. O'Connor. 1988. Plant Uptake of Nonionic Organic Chemicals from Soils. *Chemosphere* 12:2299–2323.

Sanai, M. and J. Shayegan. 1980. Field Experiments on Application of Treated Municipal Waste Waters to Vegetated Lands. *Water Pollut. Control* 79:126–135.

Sheikh, B., R. P. Cort, W. R. Kirkpatrick, R. S. Jaques, and T. Asano. 1990. Monterey Wastewater Reclamation Study for Agriculture. *Research J. Water Pollut. Control Fed.* 62:216–226.

Stone, Jr., R. V., H. B. Gotaas, and W. W. Bacon. 1952. Economic and Technical Status of Water Reclamation from Sewage and Industrial Wastes. *J. Amer. Water Works Assoc.* 44:503–517.

Travis, C. C., and A. D. Arms. 1988. Bioconcentration of Organics in Beef, Milk, and Vegetation. *Environ. Sci. Tech.* 22:271–274.

Uiga, A. and R. W. Crites. 1980. Relative Health Risks of Activated Sludge Treatment and Slow-rate Land Treatment. *J. Water Pollut. Contr. Fed.* 52:2865–2874.

Ul-Haque, I., M. Saleem, and S. Naheed. 1986. Uptake of Some Metal Contaminants by *Coriandrum sativum* Irrigated with Raw Sewage Effluents. *Biologia* 32:1–9.

U.S. Environmental Protection Agency. 1981. *Process Design Manual for Land Treatment of Municipal Wastewater.* EPA 625/1-81-013. Center for Environmental Research Information, U.S. Environmental Protection Agency, Cincinnati, OH.

U.S. Environmental Protection Agency. 1982. *Fate of Priority Pollutants in Publicly Owned Treatment Works, Volume 1.* EPA 400/1-823.3, U.S. Environmental Protection Agency, Washington, D.C.

U.S. Environmental Protection Agency. 1984. *Risk Assessment and Management: Framework for Decision Making.* EPA 600/9-85-002, U.S. Environmental Protection Agency, Washington, D.C.

U.S. Environmental Protection Agency. 1986a. *Guidelines for Carcinogenic Risk Assessment.* 51:33992–34003.

U.S. Environmental Protection Agency. 1986b. Guidelines for Estimating Exposures. *Federal Register* 51:34042–34054.

U.S. Environmental Protection Agency. 1991a. *Drinking Water Regulations and Health Advisories.* Office of Water, U.S. Environmental Protection Agency, Washington, D.C.

U.S. Environmental Protection Agency. 1991b. Access EPA. EPA/IMSD-91-100, *U.S. Environmental Protection Agency, Administration and Resources,* Washington, D.C.

U.S. Environmental Protection Agency. 1992a. *Guidelines for Water Reuse.* EPA/625/R-92/004, U.S. Environmental Protection Agency, Washington, D.C. 247 pp.

U.S. Environmental Protection Agency 1992b. Technical Support Document for Land

Application of Sewage Sludge. Prepared for Office of Water, U.S. Environmental Protection Agency by Eastern Research Group, 110 Hartwell Avenue, Lexington, MA 02173. November 23, 1992.

U.S. Environmental Protection Agency. 1993. Standards for the Use or Disposal of Sewage Sludge. *Federal Register* 58 (32):9248–9415.

Van Genuchten, M. Th. 1985. Convective-dispersive Transport of Solutes Involved in Sequential First-order Decay Reaction. *Comput. Geosci.* 11:129–147.

Villalobos, G. G., G. M. Gamez and F. F. Herrera. 1981. Program for the Reuse of Wastewater in Mexico. pp. 105–144. In: *Municipal Wastewater in Agriculture* (F. M. D'Itri, J. Aguirre-Martinez, and M. Athle-Lambarri; eds). Academic Press, New York.

Wang, H. K. 1984. Sewage Irrigation in China. *Intl. J. Dev. Tech.* 2:291–301.

Wischemeier, W. H. 1977. Soil Erodibility by Rainfall and Runoff. pp. 45–56 In: T. J. Toy ed. *Erosion: Research Techniques, Erodibility and Sediment Delivery.* Geo. Abstracts Ltd., Norwich, England.

Wisconsin Division of Health. 1986. Summary of Scientific Support Document for NR 140.10 Recommended Public Health Related Groundwater Standards—1986. Wisconsin Division of Health, Madison, Wisconsin, USA.

World Health Organization. 1975. Health Effects Relating to Direct and Indirect Re-Use of Waste Water for Human Consumption, Report of an International Working Meeting Held at Amsterdam, The Netherlands, January 13–16, 1975, World Health Organization Technical Paper No. 7. 164 pp.

World Health Organization. 1984a. *Guidelines for Drinking Water Quality, Volume 1. Recommendations.* World Health Organization, Geneva, Switzerland.

World Health Organization. 1984b. *Guidelines for Drinking Water Quality, Volume 2. Health Criteria and Other Supporting Information.* World Health Organization, Geneva, Switzerland.

World Health Organization. 1989a. *Guidelines for Predicting Dietary Intake of Pesticide Residues.* Prepared by the Joint UNEP/FAO/WHO Food Contamination Monitoring Programme in Collaboration with the Codex Committee on Pesticide Residues. World Health Organization, Geneva, Switzerland.

World Health Organization. 1989b. *Guidelines for the Study of Dietary Intake of Chemical Contaminants.* WHO Offset Publication No. 87, World Health Organization, Geneva, Switzerland. 102 pp.

World Health Organization. 1989c. *Health Guideline for the Use of Wastewater in Agriculture and Aquaculture.* World Health Organization Technical Report Series 778, World Health Organization, Geneva, Switzerland.

Yang, G. S., S. Wang, R. Zhou, and R. Sun. 1983. Endemic Selenium Intoxication of Humans in China. *Amer. J. Clin. Nutrition* 37:872–880.

Yaun, Y. and J. Zhao. 1987. On the Liver Toxicity Caused by Polluted Well Water in Petroleum Waste Water Irrigated Area. pp. 22–24 In: *Proceedings of the Third National Symposium on Environmental Geochemistry and Health.* December, 1987, Canton, China. Committee of Environmental Geology and Geochemistry, Chinese Society of Mineralogy, Petrology, and Geochemistry.

Zhang, L. and Z. Liu. 1989. A Methodological Research on Environmental Impact Assessment of Sewage Irrigation Region. *China Environ. Sci.* 9:298–303 (in Chinese).

CHAPTER 14

Water Reclamation and Reuse Criteria

INTRODUCTION

W ATER reuse is well established in water-short regions of the United States and is receiving increasing consideration in other parts of the country where traditional water supply sources are being stretched to their limits. Generally, water reclamation and reuse regulations or guidelines are first developed in response to a need to adequately regulate water reuse activities that are occurring or expected to occur in the near future. Regulations and guidelines differ in that regulations are enforceable by law, while guidelines are not legally enforceable and compliance is voluntary. Guidelines are sometimes included in regulations by reference and thus become enforceable requirements.

Water reclamation and reuse criteria are principally directed at health and environmental protection and typically address wastewater treatment, reclaimed water quality, treatment reliability, distribution systems, and use area controls. There are no federal regulations governing water reclamation and reuse in the U.S.; hence, the regulatory burden rests with the individual states. The first water reclamation and reuse standards were adopted by the State of California in 1918 and addressed the use of reclaimed water for agricultural irrigation. California continually revised its water reuse standards since that time to address additional reclaimed water applications, advances in wastewater treatment technology, and increased knowledge in the areas of microbiology and public health protection. As water reuse became recognized as an integral component in water

James Crook, Black & Veatch, 230 Congress Street, Suite 802, Boston, MA 02110.

resources management in other parts of the country, several states have followed California's lead and developed water reclamation and reuse regulations.

The criteria vary among the states that have developed regulations, and some states have no regulations or guidelines. Some states have regulations or guidelines directed at land treatment or land application for further treatment or as a means of wastewater disposal rather than regulations oriented to the intentional beneficial use of reclaimed water, even though the effluent may be used for irrigation of agricultural sites, golf courses, or public access lands. No states have regulations that cover all potential uses of reclaimed water and few states have criteria that address potable reuse. The U.S. Environmental Protection Agency (EPA) published guidelines in 1992 that are intended to provide guidance to states that have not developed their own criteria or guidelines. The World Health Organization (WHO) has published recommended guidelines for wastewater use in agriculture and aquaculture, which are presented for comparative purposes.

STATUS OF REUSE

Although comprehensive statistical information on water reuse in the U.S. has not been compiled, it has been estimated that the use of reclaimed water approached 5.7×10^6 m^3/d (1.5 bgd) in 1990 [23]. The most recent statewide survey of reclaimed water use in California was performed in 1987, at which time an average of 0.91×10^6 m^3/d (240 mgd) of municipal wastewater was reclaimed, representing about 12 percent of the total wastewater produced in the state. Sixty-three percent of the reclaimed water was used for agricultural irrigation, 14 percent for groundwater recharge, 13 percent for landscape irrigation, and the remaining 10 percent for wildlife habitat, recreational impoundments, and industrial or other applications [13]. In Florida, 1.1×10^6 m^3/d (290 mgd), or about 30 percent of the state's municipal wastewater, was reused in 1992. Thirty-eight percent of the reclaimed water was used for landscape irrigation, 30 percent for agricultural irrigation, 14 percent for groundwater recharge, 8 percent for wetlands creation, restoration, or enhancement, 6 percent for commercial and industrial uses, and the remaining 4 percent for other applications [36]. In 1995, approximately 1.4×10^6 m^3/d (360 mgd) of reclaimed water from about 390 wastewater treatment facilities was reused in Florida [113]. Twenty-three percent of all domestic wastewater produced in the state in 1995 was reused for beneficial purposes, and it is projected that 3.3×10^6 m^3/d (860 mgd) will be reused by the year 2020 [113].

TABLE 14.1. Uses of Reclaimed Water.

Category of Use	Specific Types of Use
Landscape irrigation	Parks, playgrounds, cemeteries, golf courses, roadway rights-of-way, school grounds, greenbelts, residential lawns
Agricultural irrigation	Food crops, fodder crops, fiber crops, seed crops, nurseries, sod farms, silviculture, frost protection
Nonpotable urban uses (other than irrigation)	Toilet and urinal flushing, fire protection, air conditioner chiller water, vehicle washing, street cleaning, decorative fountains and other water features
Industrial uses	Cooling, boiler feed, stack scrubbing, process water
Impoundments	Ornamental, recreational
Environmental uses	Stream augmentation, marshes, wetlands, fisheries
Groundwater recharge	Recharge aquifers, salt water intrusion control, ground subsidence control
Potable water supply augmentation	Groundwater recharge, surface water augmentation
Miscellaneous	Aquaculture, snow-making, soil compaction, dust control, equipment washdown, livestock watering

Historically, the largest-volume uses of reclaimed water were those that do not require high quality water, e.g., pasture or nonfood crop irrigation, and were often perceived as a method of wastewater disposal. Reclaimed water is now valued as a resource and, in recent years, the trend has shifted toward higher level uses such as urban irrigation, toilet and urinal flushing, industrial uses, and indirect potable reuse. Table 14.1 lists many of the currently-practiced types of water reuse.

WATER QUALITY CONCERNS

The acceptability of reclaimed water for any particular use is dependent on the physical, chemical, and microbiological quality of the water. Industrial wastes discharged to municipal sewerage systems can introduce chemical constituents that may adversely affect biological wastewater treatment processes and subsequent reclaimed water quality. Assurance of treatment reliability is an obvious, yet sometimes overlooked, quality control measure. Distribution system design and operation are important to assure that the reclaimed water is not degraded prior to use and not subject to misuse. Open storage may result in water quality degradation by microorganisms, algae, or particulate matter, and may cause objectionable odor or color in the reclaimed water.

Water reuse criteria and guidelines include provisions intended to assure a reasonable degree of public health protection. Therefore, a discussion of microbial and chemical constituents of concern is warranted to provide an appreciation of the need to adopt and enforce criteria for the production, distribution, and use of reclaimed water.

MICROORGANISMS

The potential transmission of infectious disease by pathogenic agents is the most common concern associated with nonpotable reuse of treated municipal wastewater. Sanitary engineering and preventive medical practices have combined to reach a point where waterborne disease outbreaks of epidemic proportions have, to a great extent, been controlled. However, the potential for disease transmission through the water route has not been eliminated. With a few exceptions, the disease organisms of epidemic history are still present in today's sewage, and the status is more one of severance of the transmission chain than a total eradication of the disease agent. While there is no reliable epidemiological evidence that the use of appropriately-treated wastewater for any of the applications identified in Table 14.1 has caused a disease outbreak in the U.S., the potential spread of infectious diseases through water reuse remains a public health concern.

The principal infectious agents that may be found in raw municipal wastewater can be classified into three broad groups: bacteria; parasites (protozoa and helminths); and viruses. Table 14.2 lists many of the infectious agents potentially present in raw municipal wastewater, and some of the important waterborne pathogens are discussed below.

Bacteria

Bacteria are microscopic organisms ranging from approximately 0.2 to 10 μm in length. They are distributed ubiquitously in nature and have a wide variety of nutritional requirements. Many types of harmless bacteria colonize the human intestinal tract and are routinely shed in the feces. Pathogenic bacteria are present in the feces of infected individuals; thus, municipal wastewater can contain a wide variety and concentration range of bacteria.

One of the most common pathogens found in municipal wastewater is the genus *Salmonella*. The *Salmonella* group contains a wide variety of species that can cause disease in humans and animals. The most severe form of salmonellosis is typhoid fever, caused by *Salmonella typhi*. A less common genus of bacteria is *Shigella*, which produces an intestinal disease

TABLE 14.2. **Infectious Agents Potentially
Present in Raw Municipal Wastewater.**

Pathogen	Disease
Bacteria	
Campylobacter jejuni	Gastroenteritis
Escherichia coli (enteropathogenic)	Gastroenteritis
Legionella pneumophila	Legionnaire's disease
Leptospira (spp.)	Leptospirosis
Salmonella typhi	Typhoid fever
Salmonella (2400 serotypes)	Salmonellosis
Shigella (4 spp.)	Shigellosis (dysentery)
Vibrio cholerae	Cholera
Yersinia enterocolitica	Yersiniosis
Protozoa	
Balantidium coli	Balantisiasis (dysentery)
Cryptosporidium parvum	Cryptosporidiosis, diarrhea, fever
Entamoeba histolytica	Amebiasis (amebic dysentery)
Giardia lamblia	Giardiasis
Helminths	
Ancylostoma duodenale (hookworm)	Ancylostomiasis
Ascaris lumbricoides (roundworm)	Ascariasis
Echinococcus granulosis (tapeworm)	Hydatidosis
Enterobius vermicularis (pinworm)	Enterobiasis
Necator americanus (roundworm)	Necatoriasis
Schistosoma (spp.)	Schistosomiasis
Strongyloides stercoralis (threadworm)	Strongyloidiasis
Taenia (spp.) (tapeworm)	Taeniasis, cysticercosis
Trichuris trichiura (whipworm)	Trichuriasis
Viruses	
Adenovirus (51 types)	Respiratory disease, eye infections
Astrovirus (5 types)	Gastroenteritis
Calicivirus (2 types)	Gastroenteritis
Coronavirus	Gastroenteritis
Enteroviruses (72 types) (polio, echo, coxsackie, new enteroviruses)	Gastroenteritis, heart anomalies, meningitis, others
Hepatitis A virus	Infectious hepatitis
Norwalk agent	Diarrhea, vomiting, fever
Parvovirus (3 types)	Gastroenteritis
Reovirus (3 types)	Not clearly established
Rotavirus (4 types)	Gastroenteritis

Source: Adapted from References [44,75].

known as bacillary dysentery or shigellosis. Waterborne outbreaks of shigellosis have been reported from recreational swimming and where wastewater has contaminated wells used for drinking water [57].

Other bacteria isolated from raw wastewater include *Vibrio, Mycobacterium, Clostridium, Leptospira* and *Yersinia* species. *Vibrio cholerae* is the disease agent for cholera, which is not common in the United States but is still prevalent in other parts of the world. Humans are the only known hosts, and the most frequent mode of transmission is through water. *Mycobacterium tuberculosis* has been found in municipal wastewater [40], and outbreaks have been reported among persons swimming in sewage-contaminated water [11].

Waterborne gastroenteritis of unknown cause is frequently reported, with the suspected agent being bacterial. One potential source of this disease is certain gram-negative bacteria normally considered to be non-pathogenic. These include enteropathogenic *Escherichia coli* and certain strains of *pseudomonas,* which may affect the newborn. Waterborne enterotoxigenic *E. coli* have been implicated in gastrointestinal disease outbreaks [58].

Campylobacter jejuni has been identified as the cause of a form of bacterial diarrhea in humans. While it has been well established that this organism causes disease in animals, it has also been implicated as the etiologic agent in human waterborne disease outbreaks [20].

Coliform bacteria are commonly used as indicators of fecal contamination and the potential presence of pathogenic species. While coliforms generally respond similarly to environmental conditions and treatment processes as many bacterial pathogens, coliform bacteria determinations by themselves do not adequately predict the presence or concentration of pathogenic viruses or protozoa. Concerns for newly emerging pathogenic organisms which may arise from nonhuman reservoirs, e.g., *Giardia lamblia* or *Cryptosporidium parvum,* have led to questioning the use of indicators that arise primarily from human fecal inputs. *Giardia* cysts and *Cryptosporidium* oocysts are not as readily inactivated by chlorine as bacterial surrogates now in use.

Protozoa

In general, protozoan parasite cysts and oocysts are larger than bacteria and range in size from 2 μm to over 60 μm. While parasitic cysts are present in the feces of infected individuals who exhibit disease symptoms, they also can be excreted by carriers with inapparent infections. As with viruses, cysts do not reproduce in the environment. They are, however, capable of surviving in the open environment for extended time periods, e.g., up to 7 years in soil under ideal conditions [77].

Several pathogenic protozoan parasites have been detected in municipal wastewater. One of the most important of the parasites is the protozoan *Entamoeba histolytica,* which is responsible for amoebic dysentery and amoebic hepatitis. The diseases are found worldwide, but in the U.S., *Entamoeba histolytica* has not been an important disease agent since the 1950s.

Waterborne disease outbreaks around the world have been linked to the protozoans *Giardia lamblia* and *Cryptosporidium parvum* in drinking water and recreational water, and, although no giardiasis or cryptosporidiosis cases related to water reuse projects have been reported, giardiasis and cryptosporidiosis are emerging as major waterborne diseases. Infection is caused by ingestion of *Giardia* cysts or *Cryptosporidium* oocysts. The cysts and oocysts are present in most wastewaters and are more difficult to inactivate by chlorination than are bacteria and viruses. Cryptosporidiosis can be fatal to immuno-compromised individuals.

Helminths

The most important helminthic parasites that may be found in wastewater are intestinal worms, including the stomach worm *Ascaris lumbricoides,* the tapeworms *Taenia saginata* and *Taenia solium,* the whipworm *Trichuris trichiura,* the hookworms *Ancylostoma duodenale* and *Necator americanus,* and the threadworm *Strongyloides stercoralis.* The infective stage of some helminths is either the adult organism or larvae, while the eggs or ova of other helminths constitute the infective stage. The free-living nematode larvae stages are not pathogenic to human beings. The eggs and larvae, which range in size from about 10 μm to more than 100 μm, are resistant to environmental stresses and may survive usual wastewater disinfection procedures, although eggs can be removed by commonly used wastewater treatment processes such as sedimentation, filtration, or stabilization ponds.

Viruses

Viruses are obligate intracellular parasites able to multiply only within a host cell, where they are assembled as complex macromolecules utilizing the cell's biochemical system. Viruses occur in various shapes and range in size from 0.01 to 0.3 μm in cross-section. They are composed of a nucleic acid core surrounded by an outer coat of protein [8]. Bacteriophages are viruses that infect bacteria as the host; they have not been implicated in human infections. More than 100 different types of enteric viruses capable of producing infection or disease are excreted by humans. Enteric viruses multiply in the intestinal tract and are released in the fecal matter of infected persons.

The most important human enteric viruses are the enteroviruses (polio, echo, and coxsackie), Norwalk virus, rotaviruses, reoviruses, caliciviruses, adenoviruses, and hepatitis A virus [44,106]. The reoviruses and adenoviruses, which are known to cause respiratory illness, gastroenteritis, and eye infections, have been isolated from wastewater. Of the viruses that cause diarrheal disease, only the Norwalk virus and rotavirus have been shown to be major waterborne pathogens [71]. There is no evidence that the human immunodeficiency virus (HIV), the pathogen that causes the acquired immunodeficiency syndrome (AIDS), can be transmitted via a waterborne route [39,70].

It has been reported that viruses and other pathogens in wastewater used for crop irrigation do not readily penetrate fruits or vegetables unless the skin is broken [9]. In one study where soil was inoculated with poliovirus, viruses were detected in the leaves of plants only when the plant roots were damaged or cut [80]. Although absorption of viruses by plant roots and subsequent acropetal translocation has been reported [55], it probably does not occur with sufficient regularity to be a mechanism for the transmission of viruses. Therefore, the likelihood that pathogens would be translocated through trees or vines to the edible part of crops is extremely low.

PRESENCE AND SURVIVAL OF PATHOGENS

The occurrence and concentration of pathogenic microorganisms in raw municipal wastewater depend on a number of factors, and it is not possible to predict with any degree of assurance what the general characteristics of a particular wastewater will be with respect to infectious agents. These factors include the sources contributing to the wastewater, the general health of the contributing population, the existence of disease carriers in the population, and the ability of infectious agents to survive outside their hosts under a variety of environmental conditions.

The occurrence of virus in municipal wastewater fluctuates widely. Virus concentrations are generally highest during the summer and early autumn months. Viruses as a group are generally more resistant to environmental stresses than many of the bacteria, although some viruses persist for only a short time in municipal wastewater. As a group, parasitic cysts maintain their viability for longer time periods in the open environment than either bacteria or viruses. The infectious doses of selected pathogens and the concentration ranges of some microorganisms in untreated municipal wastewater are presented in Tables 14.3 and 14.4, respectively.

Under favorable conditions, pathogens can survive for long periods of time on crops or in water or soil. While various pathogens exhibit a wide range of survival characteristics, environmental factors that affect survival include moisture content (desiccation generally adversely affects survival), soil organic matter content (presence of organic matter aids survival), temperature (longer survival at low temperatures), humidity (longer survival at high humidity), pH (bacteria survive longer in alkaline soils than

TABLE 14.3. Infectious Doses of Selected Pathogens.

Organisms	Infectious Dose[a]
Escherichia coli (enteropathogenic)	$10^6–10^{10}$
Clostridium perfringens	$1–10^{10}$
Salmonella typhi	$10^4–10^7$
Vibrio cholerae	$10^3–10^7$
Shigella flexneri 2A	180
Entamoeba histolytica	20
Shigella dysentariae	10
Giardia lamblia	< 10
Cryptosporidium parvum	1–10
Ascaris lumbricoides	1–10
Enteric virus	1–10

[a]Some of the data for bacteria are given as ID_{50}, which is the dose that infects 50 percent of the people given that dose. People given lower doses also could become infected.
Source: Adapted from References [29,30].

in acid soils), amount of rainfall, amount of sunlight (solar radiation is detrimental to survival), protection provided by foliage, and competitive microbial fauna and flora. Survival times for any particular microorganism exhibit wide fluctuations under differing conditions. Typical ranges of survival times for some common pathogens on crops and in water and soil are presented in Table 14.5.

At low temperatures some microorganisms can survive in the underground for months or years. Viruses have been isolated in groundwater after various migration distances by a number of investigators examining

TABLE 14.4. Microorganism Concentrations in Raw Wastewater.

Organisms	Concentration (number/100 mL)
Total coliforms	$10^7–10^{10}$
Clostridium perfringens	$10^3–10^5$
Enterococci	$10^4–10^5$
Fecal coliforms	$10^4–10^9$
Fecal streptococci	$10^4–10^6$
Pseudomonas aeruginosa	$10^3–10^4$
Protozoan cysts	$10^3–10^5$
Shigella	$1–10^3$
Salmonella	$10^2–10^4$
Helminth ova	$10–10^3$
Enteric virus	$10^2–10^4$
Giardia lamblia cysts	$10–10^4$
Entamoeba histolytica cysts	1–10
Cryptosporidium parvum oocysts	$10–10^3$

TABLE 14.5. Typical Pathogen Survival Times at 20–30°C.

Pathogen	Survival Time (days)		
	Fresh Water and Sewage	Crops	Soil
Viruses[a]			
Enteroviruses[b]	<120 but usually <50	<60 but usually <15	<100 but usually <20
Bacteria			
Fecal coliforms[a]	<60 but usually <30	<30 but usually <15	<70 but usually <20
Salmonella spp.[a]	<60 but usually <30	<30 but usually <15	<70 but usually <20
Shigella spp.[a]	<30 but usually <10	<10 but usually <5	
Vibrio cholerae[c]	<30 but usually <10	<5 but usually <2	<20 but usually <10
Protozoa			
E. histolytica cysts	<30 but usually <15	<10 but usually <2	<20 but usually <10
Helminths			
A. lumbricoides eggs	Many months	<60 but usually <30	Many months

[a]In seawater, viral survival is less, and bacterial survival is very much less, than in fresh water.
[b]Includes polio, echo, and coxsackie viruses.
[c]V. cholerae survival in aqueous environments is a subject of current uncertainty.
Source: Adapted from Reference [30].

a variety of recharge operations. Horizontal migration distances ranged from 3 m (10 ft) to more than 400 m (1,300 ft) [37]. Depending on soil conditions, bacteria and larger organisms associated with wastewater can be effectively removed by soil after percolation through as little as 8 cm (3 in) of the soil mantle.

DISEASE INCIDENCE RELATED TO WATER REUSE

Epidemiological investigations directed at wastewater-contaminated drinking water supplies, the use of raw or minimally-treated wastewater for food crop irrigation, health effects on farm workers who routinely come in contact with poorly treated wastewater used for irrigation, and the health effects related to aerosols or windblown spray emanating from spray irrigation sites using undisinfected wastewater have all provided evidence of infectious disease transmission from such practices [28,51,77,81]. The majority of documented disease outbreaks have been the result of contamination by bacteria or parasites. Several incidences of typhoid fever were reported in the early 1900s, and a major outbreak of cholera in Jerusalem in 1970 was reportedly caused by food crop irrigation with undisinfected wastewater [81]. Ascaris and hookworm infections have been attributed to the irrigation of vegetables and salad crops with untreated wastewater in Germany and India [77,81]. Human infection with the adult stage of the beef tapeworm *Taenia saginata,* due to ingestion of the cyst form, has resulted from irrigation of grazing land with raw and settled sewage. Evidence of tapeworm transmission via infected cattle and sheep in Europe, Australia, and elsewhere has been well-documented [77].

Excluding the use of raw sewage or primary effluent on sewage farms in the late 19th century, there have not been any confirmed cases of infectious disease resulting from reclaimed water use in the U.S. In developing countries, on the other hand, the irrigation of market crops with poorly treated wastewater is a major source of enteric disease [81].

Although pathogen-free water is not needed for all reclaimed water applications, the general practice is to provide water of a quality appropriate for the highest use of reclaimed water in a community. Since regulatory agencies may require essentially pathogen-free reclaimed water for the highest level nonpotable uses, e.g., residential landscape irrigation, toilet flushing, and the irrigation of parks, in most cases all reclaimed water distributed throughout a community meets that requirement. Wastewater treated to that level would not present unreasonable risks of infectious disease from infrequent, inadvertent ingestion [5,73,110].

CHEMICAL CONSTITUENTS

The chemical constituents potentially present in municipal wastewater generally are not a major health concern for urban uses of reclaimed water but may affect the acceptability of the water for uses such as food crop irrigation, industrial applications, and indirect potable reuse. With the exception of the possible inhalation of volatile organic compounds (VOCs) from indoor exposure, there are minimal health concerns associated with chemical constituents where reclaimed water is not intended to be consumed or used for food crop irrigation. Chemical constituents may be of concern when reclaimed water percolates into potable groundwater aquifers as a result of irrigation, groundwater recharge, or other uses. Some inorganic and organic constituents and their potential significance in water reclamation and reuse are discussed below.

- biodegradable organics: Organics provide food for microorganisms, adversely affect disinfection processes, make water unsuitable for some industrial or other uses, consume oxygen, and may cause acute or chronic health effects if reclaimed water is used for potable purposes. Regulatory agencies typically use biochemical oxygen demand (BOD) as a gross measure of biodegradable organic constituents.
- stable organics: Some organic constituents tend to resist conventional methods of wastewater treatment and are toxic in the environment. If not eliminated or reduced to very low levels in reclaimed water, their presence may limit the suitability of reclaimed water for some types of irrigation or other uses, particularly potable reuse. Total organic carbon (TOC) is the most common monitoring parameter for gross measurement of organic content in reclaimed water used for potable purposes.
- nutrients: Nitrogen, phosphorus, and potassium are essential nutrients for plant growth, and their presence normally enhances the value of the water for irrigation. When discharged to the aquatic environment, nitrogen and phosphorus can lead to the growth of undesirable aquatic life. When applied at excessive levels on land, the nitrate form of nitrogen will readily leach through the soil and may cause groundwater concentrations to exceed drinking water standards.
- heavy metals: Some heavy metals accumulate in the environment and are toxic to plants and animals. Their presence in reclaimed water may limit the suitability of the water for irrigation or other purposes. Heavy metals in reclaimed water that has received at least secondary treatment are generally within acceptable levels for

most uses; however, certain industrial wastewaters that may be discharged to a municipal wastewater collection system can contribute significant amounts of heavy metals, particularly if industrial wastewater pretreatment programs are not enforced.

• residual chlorine: Residual chlorine is toxic to many aquatic organisms and must be removed from the reclaimed water prior to discharge to aquatic environments. The reaction of chlorine with organics in water creates a wide range of byproducts, some of which may be harmful to health when ingested over the long term. Inadvertent, infrequent ingestion of highly-treated reclaimed water intended for nonpotable uses would present minimal health risks.

• suspended solids: Organic constituents, heavy metals, etc. are adsorbed on particulates. Suspended matter can shield microorganisms from disinfectants, react with disinfectants such as chlorine or ozone and thus lessen disinfection effectiveness, and lead to sludge deposits and anaerobic conditions if discharged to the aquatic environment. It is essential to remove suspended solids where ultraviolet radiation is used as the disinfection process to destroy or inactivate bacteria and viruses to low or undetectable levels in reclaimed water.

The concentrations of inorganic constituents in reclaimed water depend mainly on the nature of the water supply, source of wastewater, and degree of treatment provided. Residential use of water typically adds about 300 mg/L of dissolved inorganic solids, although the amount added can range from approximately 150 mg/L to more than 500 mg/L [54]. The presence of total dissolved solids, nitrogen, phosphorus, heavy metals, and other inorganic constituents may affect the suitability of reclaimed water for different reuse applications. Wastewater treatment can readily reduce many trace elements to below recommended maximum levels for irrigation and other uses with existing conventional technology.

The presence of toxic chemicals and microbial pathogens in wastewater creates the potential for adverse health effects where there is contact, inhalation, or ingestion of chemical or microbiological constituents of health concern. Effects of physical parameters, e.g., pH, color, temperature, and particulate matter, and chemical constituents, e.g., chlorides, sodium, and heavy metals, are well known, and recommended limits have been established for many constituents [56,101,106,107]. However, the effect of organic constituents in reclaimed water used for crop irrigation may warrant attention if industrial wastes contribute a significant fraction to the wastewater.

Health effects related to the presence of organic constituents are of primary concern with regard to potable reuse. Both organic and inorganic constituents need to be considered where reclaimed water is utilized for food crop irrigation, where reclaimed water from irrigation or other beneficial uses reaches potable groundwater supplies, or where organics may bioaccumulate in the food chain, e.g., in fish-rearing ponds.

WATER QUALITY CONSIDERATIONS FOR REUSE APPLICATIONS

IRRIGATION

Trace elements in reclaimed water normally occur in low concentrations that are not hazardous, but some are toxic at elevated concentrations. The mechanism of potential food contamination from irrigation with reclaimed water include: physical contamination, where evaporation and repeated application may result in a buildup of contaminants on crops; uptake through the roots from the applied water or the soil; and foliar uptake. Some constituents are known to accumulate in particular crops, thus presenting potential health hazards to both grazing animals and/or humans. The concentrations of heavy metals and other trace elements in reclaimed water generally are much less than those in biosolids from wastewater treatment plants, which also may be applied to agricultural land. The chemical composition of reclaimed water that has received secondary or higher levels of treatment, although highly variable, normally meets existing guidelines for irrigation water [61].

The elements of greatest concern at elevated levels are cadmium, copper, molybdenum, nickel, and zinc. Cadmium, copper, and molybdenum can be harmful to animals at concentrations too low to affect plants. Cadmium is of particular concern because it can accumulate in the food chain. It does not adversely affect ruminants in the small amounts they ingest. Most milk and beef products are unaffected by livestock ingestion of cadmium because it is stored in the liver and kidneys of the animal rather than the fat or muscle tissues. Copper is not toxic to monogastric animals but may be toxic to ruminants; however, their tolerance to copper increases as available molybdenum increases [101]. Molybdenum can also be toxic when available in the absence of copper. Nickel and zinc are a lesser concern than plants at lower concentrations than the levels harmful to animals and humans. Zinc and nickel toxicities decrease as pH increases.

Crop uptake of certain pesticides has been studied [56,66], and uptake of polychlorinated biphenyls by root crops has been demonstrated under field conditions [48]. Uptake of organic compounds is affected by: solubility, size, concentration, and polarity of the organic molecules; organic content, pH, and microbial activity of the soil; and climate. It has been postulated that most trace organic compounds are too large to pass through the semipermeable membrane of plant roots [101].

Many of the concerns associated with agricultural irrigation water are directed at effects to the crops and soil, and guidelines for evaluating irrigation water quality are available [6,56,107]. As with other reclaimed water applications, water reuse criteria typically focus on public health implications of using the water, and nonhealth-associated water quality criteria usually are not included in reuse regulations.

Wastewater containing pathogens can contaminate crops directly by contact during irrigation or indirectly as a result of soil contact. Crops can also be contaminated by blowing dust or by workers, birds, and insects that convey organisms from irrigation water or soil to the edible portion of the crop. Where there is minimal health risk based on water quality, type of crop irrigated, and degree of contact between the crop and reclaimed water, most state regulations are liberal and allow the use of reclaimed water that has received a relatively low level of treatment, e.g., secondary treatment and disinfection to achieve a fecal coliform concentration that does not exceed, on average, 200/100 mL.

As previously stated, many pathogens can survive for extended periods on plants and in soil, and simply providing extensive time periods between irrigation and crop harvest, or providing commercial storage before public sale cannot be relied upon to destroy all pathogens. Consequently, in the case of food crop irrigation with reclaimed water, there are three options to minimize risks of disease transmission: (1) eliminate pathogens from reclaimed water before irrigation; (2) process the crop to destroy pathogens before sale to the public or others; or (3) prevent direct contact between the reclaimed water and the edible portion of the crop to minimize risks of disease transmission.

Spray irrigation of food crops that grow above the ground surface requires more stringent requirements than surface irrigation because of the direct contact between the reclaimed water and the crops. Organisms contaminating food crops remain viable on the food surface unless they succumb to desiccation, exposure to sunlight, starvation, or action of other organisms or chemical agents. The reliability and completeness of pathogen inactivation by these mechanisms are questionable. Therefore, reclaimed water that is essentially free of measurable levels of pathogens is typically required for the spray irrigation of all crops that are eaten or sold raw. The surface irrigation of root crops, such as carrots, beets, and onions also results in direct contact between the crop and reclaimed water; hence, irrigation of those and similar root crops should be subject to the same requirements.

Exceptions to stringent reclaimed water quality requirements may be made for the irrigation of food crops that undergo sufficient physical or chemical commercial processing to destroy pathogens before they are sold for human consumption. Because of opportunities for transmission of infectious organisms created by handling crops that may be contaminated, some states prohibit selling or distributing the crops before processing. This provision assures that the transmission chain is severed and that contaminated raw foods are not brought into food preparation environments.

Landscape irrigation involves the irrigation of golf courses, parks, cemeteries, school grounds, freeway medians, residential lawns, and similar areas. The concern for pathogenic microorganisms is somewhat different than for agricultural irrigation in that landscape irrigation frequently takes place in urban areas where control over the use of the reclaimed water is more critical. Depending on the area being irrigated, its location relative to populated areas, and the extent of public access or use of the grounds, the water quality requirements and operational controls placed on the system may differ. Both agricultural and landscape irrigation with reclaimed water are well-accepted and widely practiced in the U.S.

DUAL SYSTEMS

Increasing use of reclaimed water in urban areas has resulted in the development of several large dual water systems that distribute two grades of water to the same service area—one potable water and the other nonpotable reclaimed water. Some regulatory agencies prohibit hose bibbs on reclaimed water distribution systems to reduce the potential for misuse of the water. At use areas that receive both potable and reclaimed water, backflow prevention devices are usually required on the potable water supply line to each site. This reduces the potential of contaminating the entire potable drinking water system in the event of an inadvertent cross-connection at a use area. Detailed information on the planning, design, construction, operation, and management of dual water systems is available in several publications [13,24,104,106].

Where reclaimed water is used in an urban setting, most states having criteria or guidelines require a high degree of treatment and disinfection. Where there is likely to be public contact with the reclaimed water, tertiary treatment to produce finished water that is essentially pathogen-free is typically required [33,83,85]. The EPA manual, *Guidelines for Water Reuse* [104], recommends similar treatment and quality for reclaimed water use in urban areas.

Reclaimed water used inside buildings for toilet and urinal flushing or for fire protection presents cross-connection control concerns. Although such uses do not result in frequent human contact with the water, regulatory agencies usually require that the reclaimed water be essentially pathogen-free to reduce health hazards upon inadvertent cross-connection to potable water systems [33,68,92].

While the need to maintain a chlorine residual in reclaimed water distribution systems to prevent odors, slimes, and bacterial regrowth was recognized early in the development of dual water systems [63], only recently have regulatory agencies begun to require such residuals. In Washington, for example, criteria require maintenance of a chlorine residual in distribution systems carrying reclaimed water [93]. The EPA *Guidelines for Water Reuse* recommends that a chlorine residual of at least 0.5 mg/L be maintained in reclaimed water distribution lines.

Reclaimed water distributed via a dual water system throughout a community for irrigation and other uses where significant portions of the population will be exposed to the water should be microbiologically safe such that inadvertent contact or ingestion does not constitute a health hazard. Chemical constituents in tertiary treated reclaimed water are generally not a problem for most types of nonpotable urban reuse, but, if necessary, can be removed by advanced wastewater treatment unit processes such as granular activated carbon adsorption or reverse osmosis.

INDUSTRIAL REUSE

Due to the myriad of industrial processes that use water and site-specific conditions, regulatory agencies generally prescribe water reuse requirements on an individual case basis, except for some common uses such as cooling water. Pathogenic microorganisms in reclaimed water used in cooling towers present potential hazards to workers and to the public in the vicinity of cooling towers from aerosols and windblown spray. In practice, however, biocides are usually added to all cooling waters onsite to prevent slimes and otherwise inhibit microbiological activity, which has the secondary effect of eliminating or greatly diminishing the potential health hazard associated with aerosols or windblown spray. Aerosols produced in the workplace or from cooling towers also may present hazards from the inhalation of VOCs, and although little definitive research has been done in this area, there has been no indication that VOCs have created health problems at any existing water reuse site. Closed-loop cooling systems using reclaimed water present minimal health concerns unless there is inadvertent or intentional misuse of the water.

Legionella pneumophila, the bacterial agent that causes Legionnaire's Disease, is known to proliferate in air conditioning cooling water systems under certain conditions. There is no indication that reclaimed water is more likely to contain *Legionella* bacteria than waters of non-sewage origin. All cooling water systems should be operated and maintained to reduce the *Legionella* threat, regardless of the origin of the source water.

The suitability of reclaimed water for use in industrial processes depends on the particular use. Reclaimed water is used in the manufacture of a wide variety of paper products, ranging from kraft pulp newsprint to high quality paper for stationery and wrappings. Regulatory agencies are likely to either prohibit the use of reclaimed water in the manufacture of paper products used as food wrap or beverage containers or require the water to be pathogen-free and not contain any health-hazardous contaminants that could leach into consumable products.

RECREATIONAL AND ENVIRONMENTAL USES

Impoundments may serve a variety of functions from aesthetic, non-contact uses, to boating, fishing, and swimming. As with other uses of reclaimed water, the level of treatment required will vary with the intended use of the water. Required treatment levels increase as the potential for human contact increases. The appearance of the reclaimed water is important when it is used for impoundments, and treatment for nutrient removal may be required. Without nutrient control there is a potential for algae blooms, resulting in odors, an unsightly appearance, and eutrophic conditions. Reclaimed water used for recreational impoundments where fishing and boating are allowed should not contain high levels of pathogenic microorganisms or heavy metals that accumulate in fish to levels that present health hazards to the consumers of the fish. For use in nonrestricted recreational impoundments where full-body contact with the water is allowed, the water should be microbiologically safe, colorless, and non-irritating to eyes or skin.

Stream augmentation is different than surface water discharge in that augmentation seeks to accomplish a beneficial end, whereas surface discharge is a disposal alternative. As with impoundments, the water quality requirements for stream augmentation are based on the designated use of the stream and maintenance of required water quality standards. In addition, there may be an emphasis on creating a product that improves existing stream quality to sustain or enhance aquatic life. To achieve aesthetic goals, both nutrient removal and high-level disinfection are often needed. Dechlorination may be required to protect aquatic wildlife where chlorine is used as the wastewater disinfectant.

In most cases, the primary intent in applying reclaimed water to wetlands is to provide additional treatment of effluent prior to discharge, although wetlands are sometimes created solely for environmental enhancement. In such cases, secondary treatment is usually acceptable as influent to the wetland system.

GROUNDWATER RECHARGE

The two principal means of recharging groundwater basins with reclaimed water are surface spreading and injection. The purposes of groundwater recharge using reclaimed water can include establishing saltwater intrusion barriers in coastal aquifers, providing soil-aquifer treatment (SAT) for future reuse, providing storage of reclaimed water, controlling or preventing ground subsidence, and augmenting potable or nonpotable aquifers. In most surface spreading projects, the wastewater receives at least secondary treatment and disinfection. Where the surface spreading of reclaimed water is used to augment potable groundwater supplies, tertiary treatment, i.e., secondary treatment followed by filtration and disinfection, or advanced wastewater treatment processes may be needed—and, in some cases, are required by regulatory agencies—to assure that the recharged water does not contain pathogens or health-significant levels of chemical constituents.

It is not uncommon to impose design and operational requirements on surface spreading operations. For example, a cycling regime of filling, draining, and drying spreading basins aids in maintaining an aerobic zone and may be required. Use of algaecides, herbicides, and pesticides at spreading areas may be regulated to prevent groundwater contamination. If SAT is an expected benefit, controls or limits may be placed on the percolation rate and depth of the vadose zone. Depending on the purpose of the recharge, regulatory controls also may be placed on the amount of reclaimed water that can be spread, the retention time in the underground prior to withdrawal, and the distance to the first extraction well that receives reclaimed water.

Injection involves pumping reclaimed water directly into the groundwater zone, which is usually a confined aquifer. Injection requires water of higher quality than surface spreading to prevent clogging because of the absence of soil matrix treatment afforded by surface spreading, and, more importantly, to have the injection water meet drinking water standards or match or exceed the quality of the groundwater into which it is injected. Treatment processes beyond secondary treatment that may be used—or required—prior to injection include chemical coagulation/clarification, filtration, air stripping, ion exchange, granular activated carbon, reverse osmosis or other membrane processes, and disinfection. With various subsets of these processes in appropriate combinations, it is possible to satisfy the full range of water quality requirements for injection. If the intent is to augment potable groundwater supplies, constraints are likely to be placed on the amount of reclaimed water that can be injected, the retention time in the underground prior to withdrawal, and the distance to the extraction well that first receives reclaimed water. Some states prohibit injection of treated wastewater to potable aquifers under any circumstances.

EVOLUTION OF WATER REUSE REGULATIONS

The early history of public health in the environmental field was one of efforts to provide safe water supply and safe disposal of wastewater. With the latter, the first efforts were directed at eliminating indiscriminate discharges of raw sewage to the environment and at providing wastewater treatment. These efforts progressed to providing higher levels of treatment; in particular, biological oxidation to restore receiving waters to aerobic conditions and disinfection of effluents to protect against health hazards from public contact with recreational waters and to reduce contamination of potable water supplies.

Standards for acceptable performance evolved from these practices—standards that represented good practice, that could be attained by well designed and operated treatment plants, and that were validated by indications that the resulting conditions were no longer producing epidemic disease. Hence, standards evolved as part of the process to cleanup major public health hazards associated with domestic water supply and municipal wastewater disposal. The evolution of regulations pertaining to water reuse followed the same pattern.

Although agricultural irrigation with low quality wastewater was practiced in some areas of the U.S. in the late 1800s, there were no significant regulations or restrictions on the practice until the early part of the 20th century. As urban areas began to encroach on sewage farms and as the scientific basis of disease became more widely understood, concern about the possible health risks associated with irrigation using wastewater grew among public health officials. This led to the establishment of guidelines and regulations to control the use of wastewater for agricultural irrigation, which was the first reclaimed water application to be regulated.

Water reuse began to increase in both number of projects and types of reclaimed water applications as wastewater treatment and disinfection processes and microbiological analytical techniques became more sophisticated during the first half of the 20th century. Similarly, water reuse standards evolved during this period to regulate the use of reclaimed water, which was mainly used for irrigation. During this period, water resources were generally adequate to meet all potable and nonpotable needs, and decisions to use reclaimed water were often based on opportunity, convenience, and economics. For the most part, projects were implemented when water reuse constituted the most economical method for disposing of sewage, because natural freshwater supplies could be obtained in the necessary amounts at reasonable cost.

Burgeoning population growth in the second half of the 20th century began to strain available fresh water resources and increased water demands in certain areas to the point where no more readily available natural freshwater was available, and development of additional supplies was costly. It was in these areas that reclaimed water first came to be viewed as a beneficial resource, particularly in the western states and most notably in California. The few state reuse regulations that existed in the 1960s and 1970s reflected the state-of-the-art of the times and a conservative approach taken by public health officials, addressing only nonpotable uses of reclaimed water. During this time period, regulations addressed a limited number of reuse applications, such as irrigation and impoundments.

As the need for additional water became apparent, additional types of reclaimed water applications were proposed. An era of intensive research and demonstration studies beginning in the late 1960s provided valuable information to regulatory agencies that came under increasing pressure to adopt rational water reuse regulations. This enabled states to develop more comprehensive water reclamation and reuse criteria that addressed a wide variety of uses, including indirect potable reuse. Most states, including California and Florida, developed their respective regulations independent of each other and required that pertinent research and demonstration studies be conducted in their particular state. While this resulted in a duplication of effort in some cases, it did produce a large database of information that has been useful to other states embarking on development of water reuse criteria. In the 1990s, several states have adopted or are developing or revising criteria, and it is common practice to base reuse criteria on those of other states that have comprehensive criteria and background information to support them. EPA's *Guidelines for Water Reuse* [104], which was published by EPA in 1992 and includes suggested criteria, also is used as a resource by states that have limited or no regulations or guidelines.

We are at a point in time where there no longer is a need for each individual state to require studies for nonpotable uses of reclaimed water that only serve to verify operating experience or studies previously performed in other states. Research, demonstration studies, experience at operational water reclamation and reuse facilities, and advances in treatment technology, water quality monitoring, and risk assessment in the last 20 years have provided a scientific basis for rational water reclamation and reuse criteria. Existing water reclamation and reuse criteria are not intended to—nor do they—provide for zero risk. They are intended to ensure that the use of reclaimed water for any approved beneficial use does not impose "undue" risks to public health.

FACTORS AFFECTING CRITERIA DEVELOPMENT

Water quality criteria are based on a variety of considerations, including the following:

- public health protection: Reclaimed water should be safe for the intended use. Most existing water reuse regulations are directed principally at public health protection. For nonpotable uses of reclaimed water, most state regulations address only microbiological and environmental concerns. Currently, no state reclaimed water microbiological quality criteria are based on rigorous risk assessment determinations.

- use requirements: Many industrial uses and some other applications have specific physical and chemical water quality requirements that are not related to health considerations. The physical, chemical, and/or microbiological quality may limit user or regulatory acceptability of reclaimed water for specific uses. Water quality requirements not associated with public health or environmental protection are seldom included in water reuse criteria by regulatory agencies.

- irrigation effects: The effect of individual constituents or parameters on crops or other vegetation, soil, and groundwater or other receiving water should be evaluated for potential reclaimed water irrigation applications. Numerous guidelines with suggested or recommended water quality limits are available. User water quality concerns often fall outside the scope of regulatory responsibility.

- environmental considerations: The natural flora and fauna in and around reclaimed water use areas and receiving waters should not be adversely impacted by the reclaimed water.

- aesthetics: For high level uses, e.g., urban irrigation and toilet flushing, the reclaimed water should be no different in appearance than potable water, i.e., clear, colorless, and odorless. For recreational impoundments, reclaimed water should not promote algal growth.

- political realities: Regulatory decisions regarding water reclamation and reuse are influenced by public policy, technical feasibility, and economics. Although regulatory agencies take into account the costs that regulations impose on reclaimed water producers and users, they are prone to set standards thought to be safe and, quite properly, do not lower health or environmental standards for the sole purpose of making projects cost-effective.

While there have been no documented disease outbreaks resulting from the use of reclaimed water in the U.S., adverse health consequences associated with the reuse of raw or improperly treated wastewater in other countries are well documented [30,51,77,81]. As a result, water reuse standards and guidelines for nonpotable reuse are directed principally at public health protection and generally are based on the control of pathogenic organisms. Health risks associated with both pathogenic microorganisms and chemical constituents need to be addressed where reclaimed water is used for indirect potable water supply augmentation.

Making reclaimed water suitable and safe for reuse applications is achieved by eliminating or reducing the concentrations of microbial and chemical constituents of concern through wastewater treatment and/ or by limiting public or worker exposure to the water via design and operational controls. Factors that affect the quality of reclaimed water include source water quality, wastewater treatment processes and treatment effectiveness, treatment reliability, and distribution system design and operation. Most states require implementation of industrial source control programs to limit the input of chemical constituents that may adversely affect biological treatment processes and subsequent acceptability of the water for specific uses. Assurance of treatment reliability is an obvious, yet sometimes overlooked, quality control measure. Distribution system design and operation is important to assure that the reclaimed water is not degraded prior to use and not subject to misuse. For example, open storage may result in water quality degradation by microorganisms, algae, or particulate matter, and may cause objectionable odor or color in the reclaimed water.

States that have water reuse regulations or guidelines typically set standards for reclaimed water quality and specify minimum treatment requirements. The most common parameters for which water quality limits are imposed are BOD, turbidity or total suspended solids (TSS), total or fecal coliform bacteria, nitrogen, and chlorine contact time and residual.

REGULATORY ISSUES

There are a number of specific issues related to water reuse criteria. Controversial topics associated with water reclamation and reuse in the U.S. and elsewhere include the following:

- inclusion of treatment unit process requirements in regulations in lieu of relying solely on reclaimed water quality requirements
- determination and use of the "best" indicator organism
- selection of reclaimed water quality parameters to be monitored, parameter limits, sampling frequency, and monitoring compliance point
- health hazards associated with aerosols or windblown spray from spray irrigation sites and industrial cooling towers
- the value of epidemiological investigations
- the use of risk assessment models to determine health risk and as a tool to determine appropriate water quality requirements
- the safety and acceptability of potable reuse

TREATMENT REQUIREMENTS

Use of water quality criteria alone, particularly those involving surrogate parameters, do not adequately characterize reclaimed water quality. Numerous studies and operational data clearly demonstrate that there is little correlation between the concentration of indicator organisms and pathogenic organisms, particularly viruses and parasites, when evaluated in the absence of wastewater treatment processes. Similarly, several studies and water reclamation plant operational data indicate that, to provide assurance that reclaimed water will be produced that is essentially free of measurable levels of pathogens, it is necessary to prescribe both treatment unit processes and water quality limits. As an added benefit, treatment reliability is enhanced with this multiple barrier approach.

A combination of treatment and quality requirements known to produce reclaimed water of acceptable quality obviates the need to monitor the finished water for certain constituents, e.g., some health-significant chemical constituents or pathogenic microorganisms such as viruses. Expensive and time-consuming monitoring for some pathogenic organisms, including viruses, may be eliminated without compromising health protection. As an example, for uses where the California Department of Health Services (DHS) has determined that reclaimed water should be essentially free of pathogenic organisms, DHS specifies treatment processes (oxidation, filtration, and disinfection), operational requirements (filtration rates, chlorine contact time, etc.), and water quality parameters (turbidity and coliform organisms) that, in combination have been demonstrated to result in the production of water of the desired quality [85].

It is sometimes argued that inclusion of treatment process requirements in regulations may result in economic unfeasibility, treatment process requirements would stifle the development and implementation of innovative treatment techniques, and, thus, that selection of treatment processes to meet established water quality limits should be left to project proponents. While regulatory agencies do consider economic and technical feasibility during regulation development, their primary responsibility is public health and environmental protection. Thus, regulatory agencies do not, nor should they, compromise health and welfare in the interest of making reuse projects more cost-effective. In addition, states with comprehensive regulations allow for alternative methods of treatment provided that they are demonstrated—in the opinion of the regulatory agency—to be as effective as those specified in the regulations.

INDICATOR ORGANISMS

In the U.S. and many other industrialized countries, either total or fecal coliform organisms are the preferred indicator organisms for reclaimed water. The total coliform analysis includes enumeration of organisms of both fecal and nonfecal origin, while the fecal coliform analysis is specific for coliform organisms of fecal origin. Therefore, fecal coliforms are better indicators of fecal contamination than total coliforms. Other indicator organisms, e.g., enterococci and bacteriophages, have been proposed but for various reasons are not recommended or required in any existing reuse regulations or guidelines.

Regulatory decisions regarding the selection of which coliform group to use are somewhat subjective. Where low levels of coliform organisms are required to indicate the absence of pathogenic bacteria, there is no consensus among microbiologists that the total coliform analysis is superior to the fecal coliform analysis. The use of total coliforms provides an added safety factor that appeals to regulatory agencies that adhere to a conservative approach to water reuse.

An argument has been made that nonpotable reclaimed water should be acceptable if it meets a geometric mean fecal coliform level of 200/100 mL, a limit previously recommended by the U.S. Environmental Protection Agency for recreational (bathing) waters [98]. The added gastro-intestinal illness rate at this fecal coliform level has been calculated to be 8 illnesses per 1,000 swimmers at freshwater beaches and 19 illnesses per 1,000 swimmers at marine beaches [103]. Public acceptability of this illness rate is inferred, perhaps incorrectly. There is no indication that the public has knowledge of the illness rate in such situations, and the public may consider 8 illnesses/1,000 swimmers to be an excessive rate. Some regulatory agencies consider illness rates of this magnitude to present an excessive, unreasonable, and unacceptable risk. In addition, the source and, thus, the significance, of fecal coliform organisms in recreational water, irrigation water, or other waters of mostly nonsewage origin may be different than those in municipal wastewater, e.g., animal versus human origin, which may impact the use of a specific coliform level as an indicator of human pathogens in the different waters. Similarly, use of the fecal coliform level of 200/100 mL as a disinfection standard can be questioned regarding its ability to provide an assured level of protection from viral pathogens.

WATER QUALITY MONITORING

Water quality monitoring is often the most prominent issue during development of reuse standards or guidelines. Decisions involving monitoring include: selection of water quality parameters; numerical limits; sampling frequency; and the monitoring compliance point. It would be impractical to monitor reclaimed water for all of the toxic chemicals and pathogenic organisms of concern and surrogate parameters are universally accepted. In addition to the previously described issue of whether to use total or fecal coliform organisms, important issues include the need to monitor for viruses and the appropriate parameter for measurement of particulates.

There is agreement that: low levels of viruses can initiate infection or disease; raw sewage contains a myriad of pathogenic viruses; some viruses are more resistant to disinfection than coliform organisms; there is little direct correlation between coliform level and virus concentration; and virus monitoring can be a valuable tool in gaining public acceptance of reuse. However, there is no consensus among public health experts regarding the health significance of low levels of viruses in reclaimed water, and as previously stated, a significant body of information exists indicating that viruses are removed or inactivated to low or immeasurable levels via appropriate wastewater treatment, including filtration and disinfection [22,26,76].

The identification and enumeration of viruses in wastewater are hampered by relatively low virus recovery rates, the complexity and high cost of laboratory procedures, and the relatively small number of laboratories with capability to perform the analyses. The laboratory culturing procedure to determine the presence or absence of viruses in a water sample takes about 14 days, and another 14 days are required to identify the viruses. While recent advances in recombinant DNA technology have provided new tools to rapidly detect viruses in water, currently used methods do not allow quantification of viruses or determination of infectivity of the virus particles. Improvements will have to be made before these methods can be put into general use for detecting viruses and other pathogens in environmental samples.

The removal of suspended matter is related to health protection. It is known that many pathogens are particulate-associated and that particulate matter can shield both bacteria and viruses from disinfectants. Also, organic matter consumes chlorine, resulting in less of the disinfectant being available for disinfection. There is general agreement that particulate matter should be reduced to low levels, e.g., 2 nephelometric turbidity units (NTU) or 5 mg/L TSS, prior to disinfection to insure reliable destruction or inactivation of microbial pathogens.

Suspended solids measurements are typically performed daily on a 24-hour composite sample and produce an average value for the sampling period. A common argument in support of monitoring for suspended solids is that the required sampling frequency for most other important parameters is daily on either grab or composite samples, and, therefore, more frequent monitoring for particulate matter is unjustified. It is clear that continuously monitored turbidity is superior to daily suspended solids measurements as an aid to treatment performance. Reliable instrumentation is available for continuous on-line measurement of turbidity, and turbidity monitoring has found wide application as a water quality parameter at water reclamation facilities. Low turbidity or suspended solids values by themselves do not indicate that reclaimed water is devoid of microorganisms. As such, turbidity and suspended solids measurements are not used as an indicator of microbiological quality but rather as a quality criterion for wastewater prior to disinfection.

Reclaimed water quality standards, particularly microbiological limits, vary considerably among U.S. states and among countries, although several states and countries with extensive reuse experience have comparable, conservatively-based criteria or guidelines. Arguments for less restrictive standards are most often predicated upon a lack of documented health hazards rather than upon any certainty that hazards are small or nonexistent. In the absence of a common interpretation of scientific and technical data, selection of water quality limits will continue to be somewhat subjective and inconsistent among the states.

The location of the monitoring point for regulatory compliance has been an issue in some states. One viewpoint is that the reclaimed water should meet all water quality requirements at the point of use. Arguments in favor of this position generally center around the possible regrowth of microorganisms between the treatment plant and the point of reuse and algal growth in storage ponds that may create aesthetic problems and clog sprinkler heads. However, for uses where human contact with reclaimed water via contact, inhalation, or ingestion is likely, restrictive coliform requirements insure that pathogenic bacteria are destroyed during disinfection and any bacterial regrowth would only be that of nonpathogenic coliform organisms. Viruses require living cells to invade and replicate themselves and do not increase in concentration in the open environment.

Many regulatory agencies subscribe to the rationale that any degradation that may occur during storage and distribution would be no different than that which would occur with the use of other water. This is not meant to imply that subsequent water quality control should be ignored. For example, maintenance of a chlorine residual will reduce slime growths in distribution systems, help eliminate musty odors, and provide an added disinfection safety factor.

AEROSOLS

Pathogen levels in aerosols caused by spraying of wastewater is a function of their concentration in the applied wastewater and the aerosolization efficiency of the spray process. During spray irrigation, the amount of water that is aerosolized can vary from less than 0.1 percent to almost

2 percent, with a mean aerosolization efficiency of 1 percent or less [7,16,46,47]. Infection or disease may be contracted directly by inhalation or indirectly by aerosols deposited on surfaces such as food, vegetation, and clothes. The infective dose of some pathogens is lower for respiratory tract infections than for infections via the gastrointestinal tract; thus, for some pathogens, inhalation may be a more likely route for disease transmission than either contact or ingestion [41,82].

In general, bacteria and viruses in aerosols remain viable and travel farther with increased wind velocity, increased relative humidity, lower temperature, and lower solar radiation. Other important factors include the initial concentration of pathogens in the wastewater and droplet size. Aerosols can be transmitted for several hundred meters under optimum conditions. Some types of pathogenic organisms, e.g., enteroviruses and *Salmonella,* appear to survive the wastewater aerosolization process much better than the indicator organisms [96]. Bacteria and viruses have been found in aerosols emitted by spray irrigation systems using untreated and poorly treated wastewater [14,15,16,96].

Studies [1,2] indicate cooling tower aerosols can include substantial numbers of bacteria. Some state regulatory agencies require that reclaimed water used in cooling towers be essentially free of measurable levels of pathogens. Typical requirements include secondary treatment followed by filtration and a high level of disinfection to produce reclaimed water containing very low levels of coliform organisms.

Studies in the U.S. directed at residents in communities subjected to aerosols from sewage treatment plants have not detected any definitive correlation between exposure to aerosols and disease [17,28,49]. There have not been any documented disease outbreaks resulting from spray irrigation with disinfected reclaimed water, and studies indicate that the health risk associated with aerosols from spray irrigation sites using disinfected reclaimed water is low [100]. The general practice is to limit exposure to aerosols produced from reclaimed water that is not highly disinfected through design or operational controls. Design features include: setback distances, which are sometimes called buffer zones; windbreaks, such as trees or walls, around irrigated areas; low pressure irrigation systems and/or spray nozzles with large orifices to reduce the formation of fine mist; low-profile sprinklers; and surface or subsurface methods of irrigation. Some operational measures are: spray only during periods of low wind velocity; do not spray when wind is blowing toward sensitive areas subject to aerosol drift or windblown spray; and irrigate at off-hours when the public or employees would not be in areas subject to aerosols or spray.

Windblown spray of reclaimed water droplets may present a greater potential health hazard than that from aerosols. Most states with reuse regulations or guidelines include buffer zone or setback distances from spray areas to property lines, buildings, or public access areas. The intent is to prevent direct or indirect human contact with the reclaimed water. Although predictive models have been developed to estimate microorganism concentrations in aerosols or larger water droplets resulting from spray irrigation of wastewater [102], setback distances are somewhat arbitrarily determined by regulatory agencies based on experience and engineering judgment.

EPIDEMIOLOGICAL STUDIES

As stated previously in this chapter, epidemiological investigations and other health effects studies have provided evidence of disease transmission associated with wastewater-contaminated drinking water supplies, use of raw or minimally treated wastewater for food crop irrigation, and aerosols or windblown spray emanating from spray irrigation sites using undisinfected wastewater. The situation is different where wastewater is treated to relatively high levels and includes disinfection.

Most health experts agree that epidemiological studies of the exposed population at nonpotable water reuse sites are of limited value due to the following: (1) the mobility of the population; (2) the small size of the study population; (3) the difficulty in determining the actual level of exposure of each individual; (4) the low illness rate—if any—resulting from the reuse practice; (5) insufficient sensitivity of current epidemiological techniques to detect low-level disease transmission; and (6) other confounding factors. It is particularly difficult to detect low-level transmission of viral disease, because many enteric viruses cause such a broad spectrum of disease syndromes that scattered cases of acute illness would probably be too varied in symptomology to be attributed to a single etiologic agent. Thus, while the few epidemiological studies that have been conducted on nonpotable reuse operations in the U.S. have not indicated adverse health responses and lead some to believe that stringent water quality standards are not justified, most regulatory agencies have chosen not to use such studies as a basis for determining water quality standards.

In contrast to epidemiological studies directed at nonpotable reuse, observation of health effects associated with large, stable populations consuming reclaimed water via indirect potable reuse is one way—albeit imperfect—to estimate whether an effect on human health has occurred by long-term ingestion of reclaimed water. However, because the minimal observed latency period is about 15 years between first exposure and disease for human cancers, it is unlikely that epidemiological studies will detect cancer incidence and mortality if they are performed before 15 or more years have elapsed since initiation of a potable reuse project. In general, *in vivo* toxicological studies provide a better indication of possible human health effects associated with complex chemical mixtures in reclaimed water.

RISK ASSESSMENT

Because of the insensitivity of epidemiological studies to provide a direct empirical assessment of microbial health risk due to low level exposure to pathogens, methodologies have increasingly relied on indirect measures of risk by using analytical models for estimation of the intensity of human exposure and the probability of human response from the exposure. Microbial risk assessment involves evaluating the likelihood that an adverse health effect may occur from human exposure to one or more potential pathogens. Most microbial risk assessments in the past have used a framework originally developed for chemicals that is defined by four major steps: hazard identification; dose-response identification; exposure assessment; and risk characterization. However, this framework does not explicitly acknowledge the differences between the health effects due to chemical and microbial exposure, such as secondary spread and immunity.

Risk analyses require several assumptions to be made, e.g., minimum infective dose of selected pathogens, concentration of pathogens in reclaimed water, quantity of reclaimed water or pathogens ingested, inhaled, or otherwise contacted by humans, and probability of infection based on infectivity models. Mathematical models that have been used to represent dose-response data include the log-normal, exponential, beta, and logistic models.

Risk analysis has been used as a tool in assessing relative health risks for microorganisms in drinking water [19,38,64,69,74], seawater [65], and reclaimed water [4,27,72,73,95]. Most of the models calculated the probability of individual infection or disease as a result of a single exposure. One of the more sophisticated models calculates a distribution of risk over the population by utilizing epidemiological data such as incubation period, immune status, duration of disease, rate of symptomatic development, and exposure data such as processes affecting pathogen concentration [27].

Operation and management practices, such as treatment reliability features and use area controls, play an important role in reducing estimated health risks, and risk analysis models for microorganisms are not used as the sole basis for regulatory decisions affecting water reuse. At the present time, no wastewater disinfection or reclaimed water standards or guidelines in the U.S. are based on risk assessment using microorganism infectivity models. However, microbial risk assessment methodology is a useful tool in assessing relative health risks associated with water reuse and undoubtedly will play a role in future criteria development as epidemiological-based models are improved and refined.

INDIRECT POTABLE REUSE

Direct potable reuse, where reclaimed water is piped directly into a potable water distribution system, is not practiced anywhere in the U.S. Indirect potable reuse, where treated wastewater is discharged into a water course, a raw water reservoir, or an underground aquifer and withdrawn downstream or downgradient at a later time for treatment and subsequent distribution as drinking water, is being practiced, if often inadvertently. A major consideration in the planned use of reclaimed water for indirect potable reuse via groundwater recharge or surface water augmentation is the possible presence of chemical and microbiological agents in the source water that may be hazardous to human health. Public health issues notwithstanding, planned indirect potable reuse is receiving increasing attention.

While it may be technically possible to produce reclaimed water of almost any desired quality, some health authorities and others have been reluctant to allow or support the planned augmentation of groundwater or surface water supplies with reclaimed water and generally subscribe to the maxim of using natural waters derived from the most protected source as raw water supplies. This principle has guided the selection of potable water supplies for almost 150 years and was affirmed in EPA's 1976 Primary Drinking Water Regulations: "Priority should be given to selection of the purest source. Polluted sources should not be used unless other sources are economically unavailable" [99].

Indirect potable reuse, on its face, is less desirable than using a higher quality source water for drinking, and reclaimed water is inherently suspect as a source water supply since untreated wastewater contains potentially harmful contaminants, including pathogens, heavy metals, and organic compounds. Ultimately, the water must meet all physical, chemical, radiological, and microbiological drinking water standards. Since drinking water standards are not intended to apply to highly-contaminated source waters such as municipal wastewater, the product water may have to meet additional water quality criteria for known or unknown potentially harmful constituents, particularly organic compounds. Public health concern over the use of reclaimed water centers on water quality, treatment reliability, and the difficulty of identifying and estimating human exposures to the potentially toxic chemicals and microorganisms that may be present. To some extent the assessment of possible health risks can rely on the vast body of knowledge that has been developed for water supplies using conventional source waters.

Studies have been made of the chemical and microbiological characteristics of reclaimed water, although they are limited in number and scope. Several studies have indicated that reclaimed water can meet drinking water standards. Such findings lead some experts to conclude that reclaimed water is acceptable as a drinking water source. Other experts disagree, stating, for example, that: disinfection of reclaimed water may create different and often unidentified disinfection byproducts (DBPs) than those found in conventional water supplies; only about 15 percent by weight of the organic compounds in drinking water have been identified and the health effects of only a few of the individual identified compounds have been determined; the health effects of mixtures of two or more of the hundreds of compounds in any reclaimed water used for potable purposes are not easily characterized; and throughout the whole process, there is increased reliance on technology and management [60].

The assessment of health risks associated with indirect potable reuse is not definitive due to limited chemical and toxicological data and inherent limitations in the available toxicological and epidemiological methods. The results of epidemiological studies directed at drinking water have generally been inconclusive, although they provide evidence for maintaining a hypothesis there may be a health risk [59]. Recognizing the limitations of epidemiological studies because of the many confounding variables, health-related studies do provide a basis for concern where water that may contain significant levels of organic constituents is subsequently chlorinated and distributed for potable use. The limited data and extrapolation methodologies used in toxicological assessments provide a source of limitations and uncertainties in the overall risk characterization.

Quality standards have been established for many inorganic constituents and treatment and analytical technology has demonstrated our capability to identify, quantify, and control these substances. Similarly, available technology is capable of eliminating pathogenic agents from contaminated waters. On the basis of available information, there is no indication that health risks from using highly-treated reclaimed water for potable purposes are greater than those from using existing water supplies or that the concentrations of regulated chemicals or microorganisms in the product water are higher than those established in drinking water standards established by EPA. However, unanswered questions remain with organic constituents, due mainly to their potential large number and unresolved health risk potential resulting from long-term exposure to extremely low concentrations.

A multiple barrier system using demonstrated treatment technologies is essential to assure that reclaimed water used to augment drinking water supplies is at least as safe and reliable as other alternative supplies. Existing treatment technology is able to produce reclaimed water that meets all current drinking water standards. However, in consideration of the source water, meeting drinking water standards does not necessarily mean that the water is "safe." Intensive water quality monitoring and contingency plans for response to system failures should be a part of a conservative regulatory approach. Monitoring programs for reclamation projects should be adequate to verify the performance of treatment processes and to detect potentially harmful contaminants.

From a regulatory standpoint, few states have addressed the challenge of developing criteria for potable reuse. California and Florida are in the forefront of developing discrete criteria relating to planned indirect potable reuse of reclaimed water. California has prepared draft criteria for groundwater recharge, and Florida has adopted criteria for both groundwater recharge and surface water augmentation. Some of the other states rely on EPA's Underground Injection Control regulations to protect potable groundwater basins, while some states prohibit potable reuse altogether. There are no federal regulations that specifically address potable reuse.

WATER RECLAMATION AND REUSE CRITERIA

There are no federal regulations governing water reclamation and reuse in the U.S.; hence, the regulatory burden rests with the individual states. This has resulted in differing standards among states that have developed reuse criteria. The *Guidelines for Water Reuse* manual [104] indicates that, as of 1992, 18 states had some form of water reuse regulations, 18 states had guidelines, and 14 states had neither regulations nor guidelines (see Table 14.6). These numbers are misleading. On closer examination of the data, some states have regulations or guidelines that pertain to land application or land disposal of wastewater, rather than beneficial reuse of reclaimed water; inclusion of criteria from those states in Table 14.6 as being water reuse regulations or guidelines is, at best, questionable.

No states have regulations that cover all potential uses of reclaimed water, but several states have extensive regulations or guidelines that prescribe requirements for a wide range of end uses of the reclaimed water. Other states have regulations or guidelines that focus on land treatment of wastewater effluent, emphasizing additional treatment or effluent disposal rather than beneficial reuse, even though the effluent may be used for irrigation of agricultural sites, golf courses, or public access lands. The absence of state criteria for specific reuse applications does not necessarily prohibit those applications; many states evaluate specific types of water reuse on a case-by-case basis. The standards in states having the most reuse experience tend to be conservative, as are criteria in several other

TABLE 14.6. Summary of State Water Reuse Regulations and Guidelines.

State	Regulations or Guidelines	No Regulations or Guidelines	Urban Reuse Unrestricted	Urban Reuse Restricted	Agricultural Reuse Food Crops	Agricultural Reuse Nonfood Crops	Recreational Reuse Unrestricted	Recreational Reuse Restricted	Environmental Reuse	Industrial Reuse
AK		•								
AL										
AR	•					•				
AZ	•		•	•	•	•	•	•	•	•
CA	•		•	•	•	•	•	•		•
CO	•		•		•	•	•	•		
CT		•								
DE	•					•				
FL	•		•	•	•	•				
GA	•				•	•				
HI	•		•	•	•	•				
IA		•								
ID	•			•		•				
IL	•				•	•				
IN		•								
KS	•				•	•				
KY										
LA		•								
MA	•		•		•	•				
MD	•			•		•				
ME		•								
MI		•								
MN		•								
MO		•								
MS	•					•				

TABLE 14.6. (Continued).

State	Regulations or Guidelines	No Regulations or Guidelines	Urban Reuse		Agricultural Reuse		Recreational Reuse		Environmental Reuse	Industrial Reuse
			Unrestricted	Restricted	Food Crops	Nonfood Crops	Unrestricted	Restricted		
MT	•		•	•	•	•				
NC	•					•				
ND	•					•				
NE		•				•				
NH		•				•				
NJ	•			•		•				
NM	•		•	•	•	•				
NV	•		•		•	•	•	•		•
NY	•					•				
OH	•					•				
OK	•					•				
OR	•		•	•		•	•	•		•
PA		•								
RI		•				•				
SC	•		•			•				
SD	•		•			•			•	
TN	•		•			•				
TX	•		•	•	•	•		•		•
UT	•			•		•				•
VA		•								
VT		•				•				
WA	•		•		•	•				
WI	•					•				
WV	•					•				
WY	•		•	•	•	•				

technologically-advanced countries where reuse is prevalent, such as Israel [79], Japan [52], and South Africa [105].

Several states with active reuse programs, e.g., Arizona, California, Florida, and Texas, have comprehensive regulations and prescribe requirements according to the end use of the water. While water reclamation and reuse standards in those states are similar in many respects, there are differences among them, such as: California uses total coliform as the indicator organism, while the other three states use fecal coliform; Florida is the only one of the four states that requires monitoring for total suspended solids to determine particulate levels—the other states use turbidity; California and Florida prescribe treatment processes in addition to water quality limits, while Arizona and Texas do not specify treatment processes, although Arizona is considering the inclusion of treatment process requirements in the state's reuse regulations [68]; and Arizona and California permit the use of reclaimed water for spray irrigation of any food crop eaten raw, while Florida and Texas are more restrictive in the types of food crop irrigation allowed.

An exhaustive review of each state's water reclamation and reuse criteria is beyond the scope of this chapter. Arizona's criteria are currently being revised and will not be discussed in detail; however, regulations relating to indirect potable reuse via groundwater recharge are not being revised at this time and are summarized below. As with Arizona, the Texas regulations addressing water reuse are undergoing revision and are not presented here. California and Florida have developed the most comprehensive reuse criteria, and criteria adopted by these two states have served as models for regulations developed by some other states. California and Florida criteria are summarized below.

ARIZONA

The state of Arizona water reclamation and reuse regulations permit reclaimed water to be used for a variety of nonpotable applications, including all types of landscape and agricultural crop irrigation, livestock watering, recreational impoundments—including those with full-body contact, industrial uses, and wetland formation. The water reuse regulations do not address indirect potable reuse and specifically state that the use of reclaimed water for direct human consumption is prohibited [83].

Nonpotable Reuse

Arizona's water reuse criteria include limits for viruses and parasites for high-level nonpotable reclaimed water uses [72]. For example, the current (as of 1997) standards for the irrigation of food crops to be consumed raw require that the fecal coliform level not exceed a geometric mean of 2.2/100 mL, the turbidity not exceed 1 NTU, enteric viruses not exceed 1 pfu/40 L, and that there be no detectable *Entamoeba histolytica, Giardia lamblia,* or *Ascaris lumbricoides.* Routine monitoring for enteric viruses or parasites is not required unless the Arizona Department of Environmental Quality (ADEQ) finds or has reason to believe that such contaminants are present in excess of the allowable limits. The Arizona regulations are currently under revision, and indications are that the virus and parasite monitoring requirements will be dropped from the regulations [68]. The latest draft revisions to the Arizona reuse criteria relating to nonpotable reuse applications are based on, and are similar to, criteria recommended in water reuse guidelines published by EPA [104], which are discussed in a subsequent section of this chapter.

Indirect Potable Reuse

In Arizona, regulations addressing groundwater recharge with treated wastewater are independent from the state's reuse criteria. The use of reclaimed water for groundwater recharge in Arizona is regulated under statutes and administrative rules administered by ADEQ and the Arizona Department of Water Resources (ADWR). Several different permits are required by these agencies prior to implementation of a groundwater recharge project. In general, ADEQ regulates groundwater quality and ADWR manages groundwater supply.

A groundwater recharge project is considered to be a discharging facility which is regulated under the Aquifer Protection Permit program that is administered by ADEQ. The owner or operator of a groundwater recharge project utilizing reclaimed water is required to obtain an individual Aquifer Protection Permit before any reclaimed water can be recharged [3]. A wastewater treatment plant that provides reclaimed water for groundwater recharge also is required to obtain an individual Aquifer Protection Permit. A single permit may be issued if the same applicant applies for both permits and the permits are sought for facilities located in a contiguous geographical area.

In order to obtain an individual Aquifer Protection Permit for a project to recharge groundwater using reclaimed water, it must be demonstrated that the project will not cause or contribute to a violation of an aquifer water quality standard at the point or points of compliance that are established for the project. If aquifer water quality standards are already being violated in the receiving aquifer, the permit applicant must demonstrate that the groundwater recharge project will not further degrade aquifer water quality.

All aquifers in Arizona currently are classified for drinking water protected use, and the state has adopted National Primary Drinking Water

Maximum Contaminant Levels (MCLs) as aquifer water quality standards. These standards apply to all groundwater in saturated formations which yield more than 20 liters/day (5 gpd) of water. For all practical purposes, all groundwater in Arizona is protected by drinking water standards. Any groundwater recharge project involving injection of reclaimed water into an aquifer is required to demonstrate compliance with aquifer water quality standards at the point of injection. Thus, reclaimed water must be treated to meet drinking water standards before it can be injected into an aquifer.

A groundwater recharge project that uses reclaimed water is required to obtain an underground storage facility permit from ADWR. In order to obtain an underground storage facility permit, it must be demonstrated that: (1) the applicant has the technical and financial capability to construct and operate the groundwater recharge project; (2) storage of the maximum amount of reclaimed water that could be in storage at any one time is hydrologically feasible; (3) the storage of reclaimed water will not cause unreasonable harm to land or to other water users within the area of impact of the maximum amount of reclaimed water that could be in storage at any one time over the duration of the permit; (4) the applicant has applied for and received any required floodplain use permit from the county flood control district; and (5) the applicant has applied for and received an Aquifer Protection Permit from ADEQ to recharge reclaimed water. Underground storage facility permits include provisions which prescribe the design capacity of the groundwater recharge project, the maximum annual amount of reclaimed water that may be stored, and monitoring requirements. ADWR may require the permit holder to monitor the operation of the groundwater recharge project and the impact of reclaimed water storage on land and other water users within the area of impact of the groundwater recharge project.

ADWR requires a water storage permit to actually store reclaimed water in an underground storage facility. The water storage permit allows the permit holder to store a specific amount of reclaimed water at a specific underground storage facility. In issuing a water storage permit, ADWR must determine that the applicant has the legal right to use the reclaimed water for groundwater recharge, the underground storage of reclaimed water will occur at a permitted underground storage facility, and the applicant has applied for and received an Aquifer Protection Permit to store the reclaimed water.

Before recovering any of the reclaimed water that has been stored underground, the person seeking to recover the water must apply to ADWR for a recovery well permit. If the recovery well permit is for a new well, ADWR must determine that the proposed recovery of the stored water will not unreasonably increase damage to surrounding land or other water users resulting from the concentration of wells. If the recovery well permit is for an existing well, the applicant must demonstrate that it has a right to use the existing well. A recovery well permit includes provisions that specify the maximum pumping capacity of the recovery well.

CALIFORNIA

The State of California has a long history of water reclamation and reuse and developed the first water reuse regulations in the U.S. in 1918. These have been modified and expanded through the years. In California, the Department of Health Services (DHS) has the authority and responsibility under California law to establish health-related standards for water reclamation and reuse. The Water Code [91] provides for the nine California Regional Water Quality Control Boards (RWQCBs) to establish water quality standards, to prescribe and enforce waste discharge requirements, and, in consultation with DHS, to prescribe and enforce reclamation requirements. Thus, DHS's reclamation criteria are enforced by the regional boards, and each water reclamation project must have a permit from the appropriate RWQCB conforming to DHS criteria. As is the case in many states, local health agencies have independent authority and may, if they deem necessary, impose requirements more stringent than those specified by the California DHS or RWQCB.

Nonpotable Reuse

The state's current Wastewater Reclamation Criteria [85], which are in the process of being revised, were adopted by DHS in 1978. The reclamation criteria include water quality standards, treatment process requirements, operational requirements, and treatment reliability requirements. The treatment and quality criteria summarized in Table 14.7 include the latest (as of 1997) proposed revisions to the Wastewater Reclamation Criteria. Coliform samples must be collected at least daily and compliance is based on a running seven-day median number. Turbidity must be monitored continuously. While the regulations include disinfection design, operational, and reliability requirements based on the use of chlorine as the disinfectant, other methods of disinfection, such as ultraviolet radiation or ozone, are not precluded if demonstrated to the satisfaction of DHS to provide an equivalent degree of disinfection and reliability.

For uses where direct or indirect human contact with reclaimed water is likely, the proposed criteria state that the turbidity in the reclaimed water after filtration shall not exceed a daily average of 2 NTU, shall not exceed 5 NTU more than 5 percent of the time, and shall not exceed 10 NTU at any time. The turbidity standard is predicated on studies conducted

TABLE 14.7. California Treatment and Quality Criteria[a] for Nonpotable Uses of Reclaimed Water.

Type of Use	Total Coliform Limits[b]	Treatment Required
Irrigation of fodder, fiber, and seed crops, orchards and vineyards,[c] and processed food crops; flushing sanitary sewers	None required	Secondary
Irrigation of pasture for milking animals, landscape areas,[d] ornamental nursery stock, and sod farms; landscape impoundments; industrial or commercial cooling water where no mist is created; nonstructural fire fighting; industrial boiler feed; soil compaction; dust control; cleaning roads, sidewalks, and outdoor areas	23/100 mL	Secondary and disinfection
Surface irrigation of food crops; restricted landscape impoundments	2.2/100 mL	Secondary and disinfection
Irrigation of food crops[e] and open access landscape areas;[f] nonrestricted recreational impoundments; toilet and urinal flushing; industrial process water; decorative fountains; commercial laundries; snow-making; structural fire fighting; industrial or commercial cooling where mist is created	2.2/100 mL	Secondary, coagulation,[g] filtration,[h] and disinfection

[a] Includes proposed revisions.
[b] Based on running 7-day median.
[c] No contact between reclaimed water and edible portion of crop.
[d] Cemeteries, freeway landscaping, restricted access golf courses, and other controlled access irrigation areas.
[e] Contact between reclaimed water and edible portion of crop; includes edible root crops.
[f] Parks, playgrounds, schoolyards, residential landscaping, unrestricted access golf courses, and other uncontrolled access irrigation areas.
[g] Not required if the turbidity of the influent to the filters does not exceed 5 NTU more than 5 percent of the time.
[h] The turbidity of filtered effluent cannot exceed a daily average of 2 nephelometric turbidity units (NTU).
Source: Reference [92].

several years ago to determine the virus removal capability of tertiary treatment processes. A fundamental decision was made that the standard to be applied was to be the absence of measurable levels of enterviruses, based on the assumptions that very low levels of virus can initiate infection and wastewater treatment processes controlling enteroviruses would produce reclaimed water free of any human pathogen, including parasites such as *Giardia lamblia* and *Cryptosporidium parvum,* and would be safe for the intended uses [21]. At that time, the required treatment processes included oxidation, chemical coagulation and clarification, filtration, and disinfection to achieve a total coliform level not exceeding 2.2/100 mL.

The required treatment train for high level uses has been relaxed in the last decade, to the point where chemical clarification is no longer required and chemical addition is not required under certain conditions. For high level reclaimed water uses, such as irrigation of food crops eaten raw and unrestricted access landscape irrigation areas, the proposed criteria require secondary treatment, chemical coagulation if the influent to the filters exceeds 5 NTU more than 5 percent of the time, filtration, and disinfection such that the concentration of total coliform bacteria does not exceed a 7-day median of 2.2/100 mL, does not exceed 23/100 mL in more than one sample in any 30-day period, and does not exceed 240/100 mL in any sample. It is currently proposed to require either: (1) a chlorine disinfection process that provides a residual chlorine concentration times modal contact time (CT) value of at least 450 mg-min/L at all times with a modal contact time of at least 90 minutes; or (2) a disinfection process that, when combined with the filtration process, has been demonstrated to reduce the concentration of MS-2 bacteriophage or virus by 5 logs, i.e., to 1/100,000 of the concentration in the filter influent.

In consideration of the likelihood of ingesting reclaimed water while swimming in nonrestricted recreational impoundments and the paucity of information regarding parasite removal where a discrete chemical clarification process is omitted, the proposed regulations require that reclaimed water used for nonrestricted recreational impoundments be monitored for enteric viruses, *Giardia lamblia,* and *Cryptosporidium parvum* during the first two years of operation if the treatment chain does not include a sedimentation unit process between the coagulation and filtration processes.

The Wastewater Reclamation Criteria also include requirements for treatment reliability. The reliability requirements address standby power supplies, alarm systems, multiple or standby treatment process units, emergency storage or disposal of inadequately treated wastewater, elimination of treatment process bypassing, monitoring devices and automatic controllers, and flexibility of design. For example, where filtration is a required unit process, one of the following features must be provided: an alarm to indicate failure of the filtration process and multiple filter units capable of treating the entire flow with one unit not in operation; an alarm, short-term (at least 24 hours) retention or disposal provisions, and standby replacement equipment; an alarm and long-term (at least 20 days) storage or disposal provisions; automatically actuated long-term storage or disposal provisions; or an alarm and standby filtration unit process.

The proposed revisions to the reclamation criteria include the following reclaimed water use area setback distance requirements: no irrigation or impoundment of undisinfected reclaimed water within 50 m (150 ft) of any domestic water supply well; no irrigation of disinfected secondary-treated reclaimed water within 30 m (100 ft) of any domestic water supply well; no irrigation with tertiary-treated (secondary treatment, filtration, and disinfection) reclaimed water within 15 m (50 ft) of any domestic water supply well unless special conditions are met, and no impoundment of tertiary-treated reclaimed water within 30 m (100 ft) of any domestic water supply well; and only tertiary-treated reclaimed water can be sprayed within 30 m (100 ft) of a residence or places where more than incidental exposure is likely. Other use area controls include the following: runoff must be confined to the reclaimed water use area; drinking water fountains must be protected against contact with reclaimed water; signs, e.g., RE-CLAIMED WATER—DO NOT DRINK, at sites using reclaimed water; prohibition of hose bibbs on reclaimed water irrigation systems; prohibition of direct connections between public water supply systems and reclaimed water systems; and backflow prevention devices on public water systems at sites receiving both potable and reclaimed water [92].

The state's proposed revisions to its cross-connection control regulations [90] require that water systems serving residences through a dual water system that uses reclaimed water for landscape irrigation must, as a minimum, be protected by a double check valve assembly bacfkflow preventer, as does the public water system in buildings using reclaimed water in a separate piping system within buildings for fire protection. A reduced pressure principle prevention device is required as a minimum to protect the potable system at sites other than those mentioned above.

Indirect Potable Reuse

The Wastewater Reclamation Criteria include general requirements for groundwater recharge of domestic water supply aquifers by surface spreading. The regulations state that reclaimed water used for groundwater recharge of domestic water supply aquifers by surface spreading "shall be at all times of a quality that fully protects public health" and that DHS recommendations "will be based on all relevant aspects of each project, including the following factors: treatment provided; effluent quality and quantity; spreading area operations; soil characteristics; hydrogeology; residence time; and distance to withdrawal" [85]. Until more definitive criteria are adopted, proposals to recharge groundwater by either surface spreading or injection will be evaluated on a case-by-case basis, although draft groundwater recharge criteria described below will guide DHS decisions.

The use of reclaimed water to recharge groundwater basins that are sources of domestic water supply has been under study by the State of California since the early 1970s. Reports [25,53,84,87,88] by nationally recognized experts in water quality and public health have been prepared to provide information needed to assess health issues and establish criteria for groundwater recharge with reclaimed water.

While aspects of its regulatory development process has been protracted, California has developed a comprehensive approach to groundwater recharge with treated municipal wastewater. Currently-proposed regulations [89] have gone through several iterations and, when finalized, will replace the general regulations for groundwater recharge in the Wastewater Reclamation Criteria. The proposed regulations address both surface spreading and injection projects and are focused on indirect potable reuse of the recovered water. The criteria are designed to assure a groundwater supply that meets all drinking water requirements and requires no treatment prior to distribution. The proposed criteria also address source control, water quality standards, recharge methods, operational controls, distance to withdrawal, time in the underground, and monitoring wells. A summary of the proposed treatment process and site requirements is presented in Table 14.8. The criteria are intended to apply only to planned groundwater recharge projects using reclaimed water, i.e., any water reclamation project designed for the purpose of recharging groundwater suitable for use as a drinking water source [42].

The proposed regulations prescribe stringent microbiological and chemical constituent limits. The concentration of general mineral, inorganic chemicals (except nitrogen compounds), and organic chemicals in reclaimed water prior to recharge must not exceed the maximum contaminant levels established in the state's drinking water regulations. The total nitrogen concentration of the reclaimed water cannot exceed 10 mg/L as nitrogen unless it is demonstrated that the standard can be consistently met in the water prior to reaching the groundwater level.

The proposed regulations specify total organic carbon (TOC) as a surrogate for unregulated organics that may be of concern. Although TOC is not a measure of specific organic compounds, it is considered to be a suitable measure of gross organics content of reclaimed water for the purpose of determining organics removal efficiency in practice. Based principally on information and recommendations contained in a report [87] prepared by an expert panel commissioned by DHS and other state agencies in California, DHS concluded that extracted groundwater should contain no more than 1 mg/L TOC of wastewater origin. The requirements shown in Table 14.8 are intended, in part, to ensure that the TOC concentration of wastewater origin is limited to 1 mg/L in public water supply wells. Reductions in TOC concentrations are less restrictive for surface spreading projects than for injection projects because of additional TOC removal that has been demonstrated to occur in the unsaturated zone with surface spreading projects [62].

For surface spreading operations, criteria pertaining to percolation rates and depth-to-groundwater are included to ensure removal of pathogens and trace organic constituents. Maintenance of an unsaturated vadose zone

TABLE 14.8. Proposed California Groundwater Recharge Criteria.

Treatment and Recharge Site Requirements	Project Category[a]			
	I	II	III	IV
Required treatment				
Secondary	X[b]	X	X	X
Filtration	X	X		X
Disinfection	X	X	X	X
Organics removal	X			X
Maximum allowable reclaimed water in extracted well water (%)	50	20	20	50
Depth to groundwater at initial percolation rate of:				
<0.5 cm/min (<0.2 in/min)	3 m (10 ft)	3 m (10 ft)	6 m (20 ft)	NA[c]
<0.8 cm/min (<0.3 in/min)	6 m (20 ft)	6 m (20 ft)	15 m (50 ft)	NA[c]
Minimum retention time underground (months)	6	6	12	12
Horizontal separation[d]	150 m (500 ft)	150 m (500 ft)	300 m (1000 ft)	600 m (2000 ft)

[a]Categories I, II, and III are for surface spreading projects. Category IV is for injection projects.
[b]X means that the treatment process is required.
[c]Not applicable.
[d]From edge of recharge operation to the nearest potable water supply well.
Source: Reference [89].

670

will allow development of aerobic biological systems that can retain or degrade organic chemicals and remove microorganisms from the water. The required vadose zone depth varies from 3 m (10 ft) to 15 m (50 ft) depending on site-specific conditions and treatment provided. Studies [79] have shown that the initial percolation capacity in a soil should not exceed 0.8 cm/min (0.3 in/min) to assure at least minimal benefit from the soil column. Where the initial percolation capacity is less than 0.5 cm/min (0.2 in/min) additional credit for soil column treatment is recognized and the required vadose zone percolation distance is reduced. The maximum percolation capacities are initial percolation test results, done before the recharge operation starts, not equilibrium infiltration rates. DHS deemed it impractical to establish a mature or equilibrium infiltration rate in regulation due to the difficulty in determining compliance [42].

The criteria for minimum underground retention are to assure further inactivation of enteric virus. The retention times required are typical of those in projects determined by DHS to be safe [42]. The values vary by category to compensate for virus reduction differences in the other project features. The retention time underground must be determined annually and at the first (in time) domestic water supply well to receive reclaimed water.

DHS has determined that the composition of the water at the point of extraction shall not exceed either 20 percent or 50 percent water of reclaimed water origin, depending on site-specific conditions, type of recharge, and treatment provided. The dilution requirement must be met at all extraction wells. The reclaimed water contribution has to be determined annually and at the domestic water supply well which receives the highest percentage of reclaimed water.

Any project sponsor proposing a groundwater recharge project using reclaimed water must submit an engineering report on the proposed project. The report will consist of a thorough investigation and evaluation of the groundwater recharge project, impacts on existing and potential uses of the impacted groundwater basin, and proposed means for achieving compliance with the criteria. The report must include a description of the methods of determination and results for initial percolation capacity, maximum reclaimed water contribution, and minimum retention time underground. Methods have not been specified in the criteria, and groundwater flow models can be used to demonstrate conformance to these criteria.

Criteria for indirect potable reuse via surface water augmentation have not been developed in California and projects will be considered on an individual case basis. Any intentional augmentation of drinking water sources with reclaimed water in California requires two state permits. A waste discharge or reclamation permit is required from a California Regional Water Quality Control Board, and a public drinking water system using an impacted source is required to obtain an amended water supply permit from DHS to address changes to the source water.

State-Mandated Reuse

In California, laws and regulations exist that mandate water reuse under certain conditions. The California Water Code states that the use of potable domestic water for nonpotable uses, including, but not limited to, cemeteries, golf courses, highway landscaped areas, and industrial and irrigation uses, is a waste or an unreasonable use of the water if reclaimed water is available which meets certain conditions [91]. The conditions are: (1) the source of reclaimed water is of adequate quality for these uses and is available for these uses; (2) the reclaimed water may be furnished for these uses at a reasonable cost to the user; (3) after concurrence with the State Department of Health Services, the use of reclaimed water from the proposed source will not be detrimental to public health; and (4) the use of reclaimed water for these uses will not adversely affect downstream water rights, will not degrade water quality, and is determined not to be injurious to plant life, fish and wildlife.

The Water Code mandates that no person or public agency shall use water from any source or quality suitable for potable domestic use for nonpotable uses if suitable reclaimed water is available and meets the conditions stated above. Other sections of the code allow for mandating reclaimed water use for irrigation of residential landscaping, industrial cooling applications, and toilet and urinal flushing in nonresidential buildings [91]. Some local jurisdictions in the state have taken action to require the use of reclaimed water in certain situations. For example, the Irvine Ranch Water District enacted an ordinance in 1990 requiring all new buildings higher than 17 m (55 ft) to install a dual distribution system for toilet flushing in areas where reclaimed water is available [47].

FLORIDA

Until the late 1970s, the primary driving force behind implementation of reuse projects in Florida was effluent disposal. This was reflected in Florida's first reuse-related regulations, which were directed at land application of municipal wastewater [34]. As the need for additional water supplies became apparent, treated wastewater began to be viewed as a beneficial resource, and the state embarked on a program to encourage water reuse and develop regulations that would provide appropriate public health and environmental protection. Chapter 17-610, Florida Administrative Code, *Reuse of Reclaimed Water and Land Application,* was adopted by the Florida Department of Environmental Regulation (DER) in 1989 and revised in 1990. DER merged with the Florida Department of Natural Resources in 1993 to form the Florida Department of Environmental Protection (DEP). Chapter 17-610, F.A.C. was transformed to Chapter 62-610, F.A.C. in 1993, and the regulations were again revised in 1996 [33].

Nonpotable Reuse

The Florida DEP coordinates the state's water reuse program, administers the domestic wastewater permitting program, and has the primary responsibility for administering water quality programs. There are five water management districts in the state, each of which has jurisdiction over water quantity issues in its particular district, including the consumptive use of water permitting program. The water management districts have authority to incorporate water reuse requirements into consumptive use permits. Another agency that plays an important role in the state's reuse program is the Florida Public Service Commission. This commission regulates rates for investor-owned utilities in about one-half of the state's counties. Utilities regulated by the commission are able to distribute water reuse facility costs among water, wastewater, and reclaimed water customers.

Wastewater treatment and reclaimed water quality criteria are summarized in Table 14.9. Daily monitoring is required for fecal coliform organisms, carbonaceous BOD (CBOD), and TSS. Class I reliability, as defined by the U.S. EPA [97], is required at water reclamation facilities where filtration and high-level disinfection are provided. Exceptions to the prescribed standards are allowed under certain conditions. For example, for uses where filtration of the wastewater is needed, chemical feed facilities are required to be installed ahead of the filtration process, although the chemical feed facilities may be idle if reclaimed water limits are met without chemical addition.

Filtration and high-level disinfection (no detectable fecal coli/100 mL in at least 75 percent of the samples analyzed over a 30-day period, maximum of 25 fecal coli/100 mL in any sample, TSS limit of 5.0 mg/L, and a minimum total chlorine residual of 1.0 mg/L after at least 15 minutes contact at peak hour flow) are required for reclaimed water applications at all new or expanded projects where the regulations specify that level of treatment. The rules acknowledge that higher chlorine residuals or longer contact times may be needed to meet the design and operational requirements for high-level disinfection and include CT requirements based on fecal coliform levels in the wastewater at a point immediately prior to disinfection. For wastewater containing 1,000 fecal coliform organisms/100 mL prior to disinfection, a minimum CT value of 25 is required, while a CT value of at least 40 is required where the fecal

TABLE 14.9. Florida Treatment and Quality Criteria for Reclaimed Water.

Type of Use	Water Quality Limits	Treatment Required
Restricted public access area irrigation,[a] industrial uses[b]	200 fecal coli/100 mL 20 mg/L TSS[g] 20 mg/L CBOD	Secondary and disinfection
Public access area irrigation,[c] food crop irrigation,[d] toilet flushing,[e] fire protection, aesthetic purposes, dust control, commercial laundries, vehicle washing, other uses[f]	No detectable fecal coli/100 mL[h] 5.0 mg/L TSS 20 mg/L CBOD	Secondary, filtration, and disinfection
Rapid infiltration basins, absorption fields	200 fecal coli/100 mL 20 mg/L TSS 20 mg/L CBOD 12 mg/L NO₃ (as N)	Secondary and disinfection
Rapid infiltration basins in unfavorable geohydrologic conditions	No detectable fecal coli/100 mL[h] 5.0 mg/L TSS Primary & secondary drinking water standards	Secondary, filtration, and disinfection
Injection to groundwater	No detectable fecal coli/100 mL[h] 5.0 mg/L TSS Primary & secondary drinking water standards	Secondary, filtration, and disinfection

TABLE 14.9. (continued).

Type of Use	Water Quality Limits	Treatment Required
Sod farms, forests, pasture land, areas used to grow trees and fodder, fiber, and seed crops, or similar areas.	No detectable fecal coli/100 mL[h] 5.0 mg/L TSS 5 mg/L TOC 0.2 mg/L TOX[i] Primary and secondary drinking water standards	Secondary, filtration, disinfection, and activated carbon adsorption
Injection to formations of Floridan or Biscayne aquifers having TDS <500 mg/L	No detectable fecal coli/100 mL[h] 5 mg/L TSS 20 mg/L CBOD 10 mg/L NO$_3$ (as N) Primary and secondary drinking water standards	Secondary, filtration, and disinfection
Discharge to Class I surface waters (used for potable supply)		

[a] Sod farms, forests, pasture land, areas used to grow trees and fodder, fiber, and seed crops, or similar areas.
[b] Contact between reclaimed water and food or beverage products prohibited.
[c] Residential lawns, golf courses, cemeteries, parks, landscaped areas, highway medians, or similar areas.
[d] Direct contact between reclaimed water and tobacco and citrus is allowed, as is direct contact between reclaimed water and edible crops that are peeled, skinned, cooked, or thermally processed before consumption; also allowed for all edible crops where irrigation methods preclude direct contact between reclaimed water and crops.
[e] Only allowed where residents do not have access to plumbing system. Not allowed in single-family residences.
[f] Flushing of sanitary sewers and reclaimed water lines, mixing of cement, manufacture of ice for ice rinks, and cleaning roads, sidewalks, and outdoor areas.
[g] TSS in reclaimed water used for subsurface irrigation systems cannot exceed 10 mg/L.
[h] No detectable fecal coliform organisms/100 mL in at least 75 percent of the samples, with no single sample to exceed 25 fecal coliform organisms/100 mL.
[i] TOX = total organic halogen.
Source: References [31, 33].

675

coliform level is between 1,000/100 mL and 10,000/100 mL prior to disinfection. If the fecal coliform level is greater than 10,000/100 mL prior to disinfection, a minimum CT value of 120 is required.

Direct contact between reclaimed water and tobacco and citrus is allowed for agricultural irrigation projects meeting all food crop irrigation requirements stated in the regulations, as is direct contact between reclaimed water and edible crops that are peeled, skinned, cooked, or thermally processed before consumption. Irrigation also is allowed where irrigation methods preclude direct contact between reclaimed water and crops. Direct application of reclaimed water to crops that are not peeled, skinned, cooked, or thermally processed before human consumption may be approved based on demonstration plant data which indicate that, in the opinion of DEP, public health protection will not be compromised.

In addition to the wastewater treatment and water quality requirements presented in Table 14.9, Florida's regulations contain design and use area requirements including the following: minimum system size of 380 m³/d (0.1 mgd) for any irrigation system; minimum system storage of at least 3 days if no backup system is provided; prohibition of cross-connections with potable systems; backflow prevention devices on potable water lines entering property served by reclaimed water systems; use area controls, including groundwater monitoring, surface runoff control, public notification, and setback distances; and ordinances or user agreements to document controls on individual users. The minimum system size requirement applies only to certain high level uses of reclaimed water, e.g., public access area irrigation, food crop irrigation, aesthetic purposes, and dust control, and is not required if reclaimed water will be used only for toilet flushing or fire protection. By eliminating the 380 m³/d (0.1 mgd) minimum size requirement, small onsite treatment facilities that may be housed in the basement of, or adjacent to, a building (other than a single-family residence) and provide reclaimed water for toilet and urinal flushing and/or fire protection within the building are not excluded from consideration.

Where a wastewater treatment plant permittee uses reclaimed water or disposes of effluent using property owned by another party, a binding agreement between the involved parties is required to ensure that all construction, operation, maintenance, and monitoring requirements are met. Such binding agreements are required for all reuse or disposal sites not owned by the permittee. The permittee has primary responsibility for ensuring compliance with all regulatory requirements.

Reclaimed water distribution requirements include separation distances between reclaimed water lines and either potable water lines or sewage collection lines, signage warning the public and employees that the water is not intended for drinking, and color-coded (Pantone Purple 522C) reclaimed water lines and appurtenances. Florida requires a minimum 23-

m (75-ft) setback from areas irrigated with highly disinfected reclaimed water containing no detectable fecal coliforms/100 mL to potable water supply wells, but no setbacks to surface water or developed areas are required when this quality of reclaimed water is used. Other setback distances depend on reclaimed water quality and range up to 150 m (500 ft) at some areas receiving reclaimed water that has a fecal coliform limit of 200/100 mL.

The Florida DEP requires establishment of an operating protocol for all projects where high levels of treatment (secondary treatment and filtration) and disinfection (no detectable fecal coliform organisms/100 mL in the reclaimed water) are specified. An operating protocol is a document which describes how a wastewater treatment facility is to be operated to assure that only reclaimed water meeting applicable standards is released to a reuse system. The operating protocol must address: (1) the criteria used to make continuous determinations of the acceptability of the reclaimed water being produced; (2) the steps and procedures to be followed when substandard water is produced; (3) the steps and procedures to be followed when the water reclamation plant returns to normal operation and acceptable quality reclaimed water is again produced; and (4) procedures to be followed during periods when an operator is not present at the treatment facility and when the operator returns following unattended periods.

In addition to monitoring for suspended solids and fecal coliform organisms, on-line monitoring for turbidity and chlorine residual is required to be included in the operating protocol. Limits for turbidity and chlorine residual are established in the operating protocol based on water quality conditions needed to assure that the fecal coliform and TSS limits can be met consistently. Thus, while the regulations do not directly impose turbidity and chlorine residual limits, operational limits are determined on a case-by-case basis by each facility's operating protocol.

Indirect Potable Reuse

The regulations include sections addressing rapid-rate infiltration basin systems and absorption field systems, both of which may result in groundwater recharge. Because nearly all groundwater in Florida is classified as G-II, which is defined as groundwater containing 10,000 mg/L or less of total dissolved solids (TDS) and is designated as a potable supply source, any land application system located over a G-II groundwater could function as an indirect potable reuse system. For absorption fields, a TSS limitation of 10 mg/L is imposed to protect against formation plugging. If more than 50 percent of the wastewater applied to the systems is collected after percolation, the systems are considered to be effluent disposal systems and not beneficial reuse. Loading to these systems is limited to 23 cm/d (9 in/d), and wetting and drying cycles must be used. For systems having higher loading rates or a more direct connection to an aquifer than normally encountered, the reclaimed water must receive secondary treatment, filtration, and high-level disinfection, and must meet primary and secondary drinking water standards. These reclaimed water treatment and quality are similar to those in the California regulations for surface spreading of reclaimed water.

The Florida regulations also include criteria directed at planned indirect potable reuse by injection into water supply aquifers and augmentation of surface supplies. The injection regulations pertain to G-I, G-II, and F-I groundwaters, all of which are classified as potable aquifers. Secondary treatment, filtration, and high-level disinfection are required, as is Class I reliability. Reclaimed water must meet G-II groundwater standards prior to injection. G-II groundwater standards are, for the most part, primary and secondary drinking water standards [112]. For injection into formations of the Floridan and Biscayne Aquifers where the TDS does not exceed 500 mg/L, the regulations are more restrictive and specify that reclaimed water meet drinking water standards, require activated carbon adsorption as an organics removal process, limit the average TOC and total organic halogen (TOX) concentrations to 5.0 mg/L and 0.2 mg/L, respectively, in the product water, and require a full-scale, two-year testing program.

Chapter 62-610 , entitled *Reclaimed Water and Land Application,* cross-references Chapter 62-600, F.A.C., entitled *Domestic Wastewater Facilities,* which addresses domestic wastewater treatment facilities, treatment requirements, management facilities, and planning. It includes requirements for discharges to surface waters used for drinking water supplies and for injection into aquifers which are or may be used for potable supply [31]. Under this chapter, reclaimed water may be used to augment supplies in Class I surface waters. Class I waters are those used as potable water supply sources. Reclaimed water must meet primary and secondary drinking water standards and a total nitrogen limit of 10 mg/L (as N). The treatment facility must provide Class I reliability and high-level disinfection. Outfalls for discharge of reclaimed water cannot be located within 150 m (500 ft) of a potable water intake.

Ongoing rulemaking (as of 1997) is refining indirect potable reuse requirements as well as requirements for aquifer storage and recovery and use of reclaimed water for salinity barrier systems. For projects involving injection into formations of the Floridian and Biscayne Aquifers where the TDS does not exceed 500 mg/L, the full-scale testing requirements may be replaced with streamlined pilot testing requirements, and the average and maximum TOC limits may be reduced to 3 mg/L and 5 mg/L, respectively [112]. TOC and TOX limitations and other requirements currently applicable only to high quality (TDS <500 mg/L) portions of the Floridan and Biscayne Aquifers probably will be extended to a wider range of injection applications during the ongoing rulemaking process.

State-Mandated Reuse

The Florida Water Policy [32] establishes a mandatory reuse program which, in contrast to California's mandated reuse regulations, is actively enforced. The policy requires that the state's water management districts identify water resource caution areas that have water supply problems that have become critical or are anticipated to become critical within the next 20 years. State legislation requires preparation of water reuse feasibility studies for treatment facilities located within the water resource caution areas, and a "reasonable" amount of reclaimed water use from municipal wastewater treatment facilities is required within the designated water resource caution areas unless reuse is not economically, environmentally, or technically feasible. Water reuse also may be required outside of designated water resource caution areas if reclaimed water is readily available, reuse is economically, environmentally, and technologically feasible, and rules governing the imposition of requirements for reuse have been adopted in those areas by the water management district having jurisdiction. In addition, if reuse is found to be feasible, disposal by surface water discharge or deep well injection will be limited to backups for reuse systems.

Florida's antidegradation policy applies to new or expanded surface water discharges and requires demonstration that the proposed discharge is in the public interest, which, in part, is determined by evaluation of water reuse feasibility. If reuse is determined to be feasible, it will be preferred over a new or expanded surface water discharge.

In response to a need to provide guidance to applicants, Florida developed *Guidelines for Preparation of Reuse Feasibility Studies for Applicants Having Responsibility for Wastewater Management* [35]. This document describes comprehensive requirements for the preparation of reuse feasibility studies. The requirements include evaluation of alternatives, including at least a "no action" alternative and a public access/urban reuse system alternative. The feasibility evaluation must include economic considerations, reuse benefits, and technical feasibility. Under certain conditions, a municipality, government entity, or utility that has developed and is implementing a reuse master plan can have the master plan accepted in lieu of a reuse feasibility study report.

U.S. EPA WATER REUSE GUIDELINES

The U.S. Environmental Protection Agency, in conjunction with the U.S. Agency for International Development, published *Guidelines for Water Reuse* in 1992 [104]. The primary purpose of the document is to provide guidelines, with supporting information, for utilities and regulatory agencies in the U.S., particularly in states where standards do not exist or are being revised or expanded. EPA has determined that guidelines will encourage reuse in areas where it is not now allowed or practiced and may eliminate some of the inconsistencies that characterize current regulations. It is EPA's view that national water reclamation and reuse standards are not necessary and that comprehensive guidelines, coupled with flexible state regulations, will foster increased consideration and implementation of reuse projects [67].

The guidelines address all important aspects of water reuse and include recommended treatment processes, reclaimed water quality limits, monitoring frequencies, setback distances, and other controls for various water reuse applications. The guidelines address water reclamation and reuse for nonpotable applications as well as indirect potable reuse by groundwater recharge and augmentation of surface water sources of supply. The treatment processes and reclaimed water quality limits recommended in the guidelines for various reclaimed water applications are given in Table 14.10.

Both reclaimed water quality limits and wastewater treatment unit processes are recommended for the following reasons: (1) water quality criteria involving surrogate parameters alone do not adequately characterize reclaimed water quality; (2) a combination of treatment and quality requirements known to produce reclaimed water of acceptable quality obviate the need to monitor the finished water for certain constituents; (3) expensive, time-consuming, and in some cases, questionable monitoring for pathogenic microorganisms is eliminated without compromising health protection; and (4) treatment reliability is enhanced.

In the U.S., total and fecal coliforms are the most commonly used indicator organisms in reclaimed water. The total coliform analysis includes organisms of both fecal and nonfecal origin, while the fecal coliform analysis is specific for coliform organisms of fecal origin. Therefore, fecal coliforms are better indicators of fecal contamination than total coliforms, and the authors of the guidelines, upon the recommendation of noted microbiologists, chose to use fecal coliform as the indicator organism. The guidelines state that either the membrane filter technique or the multiple-tube fermentation technique may be used to quantify the coliform levels in the reclaimed water.

The guidelines suggest that, regardless of the type of reclaimed water use, some level of disinfection should be provided to avoid adverse health

TABLE 14.10. EPA Suggested Guidelines for Reuse of Municipal Wastewater.

Types of Reuse	Treatment	Reclaimed Water Quality[a]	Reclaimed Water Monitoring	Setback Distances[b]	Comments
Urban Reuse All types of landscape irrigation (e.g., golf courses, parks, cemeteries)— also vehicle washing, toilet flushing, use in fire protection systems and commercial air conditioners, and other uses with similar access or exposure to the water.	• Secondary[c] • Filtration[d] • Disinfection[e]	• pH = 6–9 • ≤10 mg/L BOD[f] • ≤2 NTU[g] • No detectable fecal coli/100 mL[h,i] • ≥1 mg/L Cl$_2$ residual[j]	• pH—weekly • BOD—weekly • Turbidity—continuous • Coliform—daily • Cl$_2$ residual—continuous	• 50 ft (15 m) to potable water supply wells	• Consult recommended agricultural (crop) limits for metals. • A lower level of treatment, e.g., secondary treatment and disinfection to achieve ≤14 fecal coli/100 mL, may be appropriate at controlled-access irrigation sites where design and operational measures significantly reduce the potential of public contact with reclaimed water. • Chemical (coagulant and/or polymer) addition prior to filtration may be necessary to meet water quality recommendations. • The reclaimed water should not contain measurable levels of pathogens.[k] • Reclaimed water should be clear, odorless, and contain no substances that are toxic upon ingestion. • A higher chlorine residual and/or a longer contact time may be necessary to assure that viruses and parasites are inactivated or destroyed. • A chlorine residual of 0.5 mg/L or greater in the distribution system is recommended to reduce odors, slime, and bacterial regrowth. • Provide treatment reliability.

TABLE 14.10. (continued).

Types of Reuse	Treatment	Reclaimed Water Quality[a]	Reclaimed Water Monitoring	Setback Distances[b]	Comments
Restricted Access Area Irrigation					
Sod farms, silviculture sites, and other areas where public access is prohibited, restricted, or infrequent	• Secondary[c] • Disinfection[e]	• pH = 6–9 • ≤30 mg/L BOD[f] • ≤30 mg/L SS • ≤200 fecal coli/100 mL[h,i,m] • ≤1 mg/L Cl₂ residual[l]	• pH—weekly • BOD—weekly • SS—daily • Coliform—daily • Cl₂ residual—continuous	• 300 ft (90 m) to potable water supply wells • 100 ft (30 m) to areas accessible to the public (if spray irrigation)	• Consult recommended agricultural (crop) limits for metals. • If spray irrigation, SS less than 30 mg/L may be necessary to avoid clogging of sprinkler heads. • Provide treatment reliability.
Agricultural Reuse—Food Crops Not Commercially Processed[o]					
Surface or spray irrigation of any food crop, including crops eaten raw	• Secondary[c] • Filtration[d] • Disinfection[e]	• pH = 6–9 • ≤10 mg/L BOD[f] • ≤2 NTU[g] • No detectable fecal coli/100 mL[h,i] • ≥1 mg/L Cl₂ residual[l]	• pH—weekly • BOD—weekly • Turbidity—continuous • Coliform—daily • Cl₂ residual—continuous	• 50 ft (15 m) to potable water supply wells	• Consult recommended agricultural (crop) limits for metals. • Chemical (coagulant and/or polymer) addition prior to filtration may be necessary to meet water quality recommendations. • The reclaimed water should not contain measurable levels of pathogens.[k] • A higher chlorine residual and/or a longer contact time may be necessary to assure that viruses and parasites are inactivated or destroyed. • High nutrient levels may adversely affect some crops during certain growth stages. • Provide treatment reliability.

TABLE 14.10. (continued).

Types of Reuse	Treatment	Reclaimed Water Quality[a]	Reclaimed Water Monitoring	Setback Distances[b]	Comments
Agricultural Reuse—Food Crops Commercially Processed[n]; Surface Irrigation of Orchards and Vineyards					
	• Secondary[c] • Disinfection[e]	• pH = 6–9 • ≤30 mg/L BOD[f] • ≤30 mg/L SS • ≤200 fecal coli/100 mL[h,m] • ≥1 mg/L Cl$_2$ residual[l]	• pH—weekly • BOD—weekly • SS—daily • Coliform—daily • Cl$_2$ residual—continuous	• 300 ft (90 m) to potable water supply wells • 100 ft (30 m) to areas accessible to the public	• Consult recommended agricultural (crop) limits for metals. • If spray irrigation, SS less than 30 mg/L may be necessary to avoid clogging of sprinkler heads. • High nutrient levels may adversely affect some crops during certain growth stages. • Provide treatment reliability.
Agricultural Reuse—Non-Food Crops					
Pasture for milking animals; fodder, fiber, and seed crops	• Secondary[c] • Disinfection[e]	• pH = 6–9 • ≤30 mg/L BOD[f] • ≤30 mg/L SS • ≤200 fecal coli/100 mL[h,m] • ≥1 mg/L Cl$_2$ residual[l]	• pH—weekly • BOD—weekly • SS—daily • Coliform—daily • Cl$_2$ residual—continuous	• 300 ft (90 m) to potable water supply wells • 100 ft (30 m) to areas accessible to the public (if spray irrigation)	• Consult recommended agricultural (crop) limits for metals. • If spray irrigation, SS less than 30 mg/L may be necessary to avoid clogging of sprinkler heads. • High nutrient levels may adversely affect some crops during certain growth stages. • Milking animals should be prohibited from grazing for 15 days after irrigation ceases. A higher level of disinfection, e.g., to achieve ≤14 fecal coli/100 mL, should be provided if this waiting period is not adhered to. • Provide treatment reliability.

TABLE 14.10. (continued).

Types of Reuse	Treatment	Reclaimed Water Quality[a]	Reclaimed Water Monitoring	Setback Distances[b]	Comments
Recreational Impoundments					
Incidental contact (e.g., fishing and boating) and full body contact with reclaimed water allowed	• Secondary[c] • Filtration[d] • Disinfection[e]	• pH = 6–9 • ≤10 mg/L BOD[f] • ≤2 NTU[g] • No detectable fecal coli/100 mL[h,j] • ≥1 mg/L Cl$_2$ residual[l]	• pH—weekly • BOD—weekly • Turbidity—continuous • Coliform—daily • Cl$_2$ residual—continuous	• 500 ft (150 m) to potable water supply wells if bottom not sealed	• Dechlorination may be necessary to protect aquatic species of flora and fauna. • Reclaimed water should be non-irritating to skin and eyes. • Reclaimed water should be clear, odorless, and contain no substances that are toxic upon ingestion. • Nutrient removal may be necessary to avoid algae growth in impoundments. • Chemical (coagulant and/or polymer) addition prior to filtration may be necessary to meet water quality recommendations. • The reclaimed water should not contain measurable levels of pathogens.[k] • A higher chlorine residual and/or a longer contact time may be necessary to assure that viruses and parasites are inactivated or destroyed. • Fish caught in impoundments can be consumed. • Provide treatment reliability.
Landscape Impoundments					
Aesthetic impoundments where public contact with reclaimed water is not allowed	• Secondary[c] • Disinfection[e]	• ≤30 mg/L BOD[f] • ≤30 mg/L SS • ≤200 fecal coli/100 mL[l,m,n] • ≥1 mg/L Cl$_2$ residual[l]	• pH—weekly • SS—daily • Coliform—daily • Cl$_2$ residual—continuous	• 500 ft (150 m) to potable water supply wells if bottom not sealed	• Nutrient removal processes may be necessary to avoid algae growth in impoundments. • Dechlorination may be necessary to protect aquatic species of flora and fauna. • Provide treatment reliability.

TABLE 14.10. (continued).

Types of Reuse	Treatment	Reclaimed Water Quality[a]	Reclaimed Water Monitoring	Setback Distances[b]	Comments
Construction Uses					
Soil compaction, dust control, washing aggregate, making concrete	• Secondary[c] • Disinfection[e]	• ≤30 mg/L BOD[f] • ≤30 mg/L SS • ≤200 fecal coli/100 mL[l,m,n] • ≥1 mg/L Cl_2 residual[j]	• BOD—weekly • SS—daily • Coliform—daily • Cl_2 residual—continuous		• Worker contact with reclaimed water should be minimized. • A higher level of disinfection, e.g., to achieve ≤14 fecal coli/100 mL, should be provided where frequent worker contact with reclaimed water is likely. • Provide treatment reliability.
Industrial Reuse					
Once-through cooling	• Secondary[c]	• pH = 6–9 • ≤30 mg/L BOD[f] • ≤30 mg/L SS • ≤200 fecal coli/100 mL[h,l,m] • ≥1 mg/L Cl_2 residual[j]	• pH—weekly • BOD—weekly • SS—weekly • Coliform—daily • Cl_2 residual—continuous	• 300 ft (90 m) to areas accessible to the public	• Windblown spray should not reach areas accessible to users or the public.
Recirculating cooling towers	• Secondary[c] • Disinfection[e] (chemical coagulation and filtration[d] may be needed)	• Variable, depends on recirculation ratio		• 300 ft (90 m) to areas accessible to the public. May be reduced if high level of disinfection is provided	• Windblown spray should not reach areas accessible to the public. • Consult recommended water quality limits for make-up water. • Additional treatment by user is usually provided to prevent scaling, corrosion, biological growths, fouling and foaming. • Provide treatment reliability.

685

TABLE 14.10. (continued).

Types of Reuse	Treatment	Reclaimed Water Quality[a]	Reclaimed Water Monitoring	Setback Distances[b]	Comments
Environmental Reuse Wetlands, marshes, wildlife habitat, stream augmentation	• Variable • Secondary[c] and disinfection[e] (min.)	Variable, but not to exceed: • ≤30 mg/L BOD[f] • ≤30 mg/L SS • ≤200 fecal coli/100 mL[h,l,m]	• BOD—weekly • SS—daily • Coliform—daily • Cl_2 residual—continuous		• Dechlorination may be necessary to protect aquatic species of flora and fauna. • Possible effects on groundwater should be evaluated. • Receiving water quality requirements may necessitate additional treatment. • The temperature of the reclaimed water should not adversely affect ecosystem. • Provide treatment reliability.
Groundwater Recharge By spreading or injection into nonpotable aquifers	• Site-specific and use dependent • Primary (min.) for spreading • Secondary[c] (min.) for injection	• Site-specific and use dependent	• Depends on treatment and use	• Site-specific	• Facility should be designed to ensure that no reclaimed water reaches potable water supply aquifers. • For injection projects, filtration and disinfection may be needed to prevent clogging. • Provide treatment reliability.

TABLE 14.10. (continued).

Types of Reuse	Treatment	Reclaimed Water Quality[a]	Reclaimed Water Monitoring	Setback Distances[b]	Comments
Indirect Potable Reuse					
Groundwater recharge by spreading into potable aquifers	• Site-specific • Secondary[c] and disinfection[e] (min.) • May also need filtration[d] and/or advanced wastewater treatment[o]	• Site-specific • Meet drinking water standards after percolation through vadose zone	Includes, but not limited to, the following: • pH—daily • Coliform—daily • Cl₂ residual—continuous • Drinking water standards—quarterly • Other[i]—depends on constituent	• 2000 ft (600 m) to extraction wells. May vary depending on treatment provided and site-specific conditions	• The depth to groundwater (i.e., thickness of the vadose zone) should be at least 6 feet (2 m) at the maximum groundwater mounding point. • The reclaimed water should be retained underground for at least 1 year prior to withdrawal. • Recommended treatment is site-specific and depends on factors such as type of soil, percolation rate, thickness of vadose zone, native groundwater quality, and dilution. • Monitoring wells are necessary to detect the influence of the recharge operation on the groundwater. • The reclaimed water should not contain measurable levels of pathogens after percolation through the vadose zone.[k] • Provide treatment reliability.

TABLE 14.10. (continued).

Types of Reuse	Treatment	Reclaimed Water Quality[a]	Reclaimed Water Monitoring	Setback Distances[b]	Comments
Indirect Potable Reuse (continued)					
Groundwater recharge by injection into potable aquifers	• Secondary[c] • Filtration[d] • Disinfection[e] • Advanced wastewater treatment[o]	Includes, but not limited to, the following: • pH = 6.5–8.5 • ≤2 NTU[g] • No detectable fecal coli/100 mL[h,i] • ≥1 mg/L Cl$_2$ residual[j] • Meet drinking water standards	Includes, but not limited to, the following: • pH—daily • Turbidity—continuous • Coliform—daily • Cl$_2$ residual—continuous • Drinking water standards—quarterly • Other[p]—depends on constituent	• 2000 ft (600 m) to extraction wells. May vary depending on site-specific conditions	• The reclaimed water should be retained underground for at least 1 year prior to withdrawal. • Monitoring wells are necessary to detect the influence of the recharge operation on the groundwater. • Recommended quality limits should be met at the point of injection. • The reclaimed water should not contain measurable levels of pathogens at the point of injection. • A higher chlorine residual and/or a longer contact time may be necessary to assure virus inactivation. • Provide treatment reliability.
Augmentation of surface supplies	• Secondary[c] • Filtration[d] • Disinfection[e] • Advanced wastewater treatment[o]	Includes, but not limited to, the following: • pH = 6.5–8.5 • ≤2 NTU[g] • No detectable fecal coli/100 mL[h,i] • ≥1 mg/L Cl$_2$ residual[j] • Meet drinking water standards	Includes, but not limited to, the following: • pH—daily • Turbidity—continuous • Coliform—daily • Cl$_2$ residual—continuous • Drinking water standards—quarterly • Other[p]—depends on constituent	• Site-specific	• Recommended level of treatment is site-specific and depends on factors such as receiving water quality, time and distance to point of withdrawal, dilution and subsequent treatment prior to distribution for potable uses. • The reclaimed water should not contain measurable levels of pathogens.[k] • A higher chlorine residual and/or a longer contact time may be necessary to assure virus inactivation. • Provide treatment reliability.

TABLE 14.10. (continued).

a Unless otherwise noted, recommended quality limits apply to reclaimed water at the point of discharge from the treatment facility.

b Setbacks are recommended to protect potable water supply sources from contamination and to protect humans from unreasonable health risks due to exposure to reclaimed water.

c Secondary treatment processes include activated sludge processes, trickling filters, rotating biological contactors, and many stabilization pond systems. Secondary treatment should produce effluent in which both the BOD and SS do not exceed 30 mg/L.

d Filtration means the passing of wastewater through natural undisturbed soils or filter media such as sand and/or anthracite.

e Disinfection means the destruction, inactivation, or removal of pathogenic microorganisms by chemical, physical, or biological means. Disinfection may be accomplished by chlorination, ozonation, other chemical disinfectants, UV radiation, membrane processes, or other processes.

f As determined from the 5-day BOD test.

g The recommended turbidity limit should be met prior to disinfection. The average turbidity should be based on a 24-hour time period. The turbidity should not exceed 5 NTU at any time. If SS is used in lieu of turbidity, the average SS should not exceed 5 mg/L.

h Unless otherwise noted, recommended coliform limits are median values determined from the bacteriological results of the last 7 days for which analyses have been completed. Either the membrane filter or fermentation tube technique may be used.

i The number of fecal coliform organisms should not exceed 14/100 mL in any sample.

j Total chlorine residual after a minimum contact time of 30 minutes.

k It is advisable to fully characterize the microbiological quality of the reclaimed water prior to implementation of a reuse program.

l The number of fecal coliform organisms should not exceed 800/100 mL in any sample.

m Some stabilization pond systems may be able to meet this coliform limit without disinfection.

n Commercially processed food crops are those that, prior to sale to the public or others, have undergone chemical or physical processing sufficient to destroy pathogens.

o Advanced wastewater treatment processes include chemical clarification, carbon adsorption, reverse osmosis and other membrane processes, air stripping, ultrafiltration, and ion exchange.

p Monitoring should include inorganic and organic compounds, or classes of compounds, that are known or suspected to be toxic, carcinogenic, teratogenic, or mutagenic and are not included in the drinking water standards.

Source: Adapted from Reference [104].

689

consequences from inadvertent contact or accidental or intentional misuse of a water reuse system. For nonpotable uses of reclaimed water, only two different levels of treatment and disinfection are recommended. Reclaimed water used for applications where no direct or indirect public or worker contact with the water is expected should receive at least secondary treatment and be disinfected to achieve a fecal coliform concentration not exceeding 200/100 mL for the following reasons: most bacterial pathogens will be destroyed or reduced to low or insignificant levels in the water; the concentration of viable viruses and parasites will reduced somewhat; disinfection of secondary effluent to a fecal coliform level of 200/100 mL is readily achievable at minimal cost; and significant health-related benefits associated with disinfection to lower, but not pathogen-free, levels are not obvious.

For uses where direct or indirect contact with reclaimed water is likely or expected, and for dual water systems where there is a potential for cross-connections with potable water lines, disinfection to produce reclaimed water having no detectable fecal coliform organisms/100 mL is recommended. This more restrictive disinfection level is intended to be used in conjunction with tertiary treatment and other water quality limits, such as 2 NTU in the wastewater prior to disinfection. This combination of treatment and water quality has been shown to be capable of producing reclaimed water that is essentially free of measurable levels of pathogens.

The guidelines include limits for fecal coliform organisms but do not include parasite or virus limits. Parasites have not been shown to be a problem at reuse operations in the U.S. at the treatment levels and reclaimed water limits recommended in the guidelines, although there has been considerable interest in recent years regarding the occurrence and significance of *Giardia lamblia* and *Cryptosporidium parvum* in reclaimed water. Where filtration and a high level of disinfection are recommended in Table 14.10 to produce reclaimed water that is essentially pathogen-free, it may be necessary to provide chemical addition prior to filtration to assure complete removal of parasites.

While viruses are a concern in reclaimed water, virus limits are not recommended in the guidelines for the following reasons: (1) a significant body of information exists indicating that viruses are inactivated or removed to low or immeasurable levels via appropriate wastewater treatment [22,26,76]; (2) the type and concentration of viruses in wastewater are difficult to determine accurately because of low virus recovery rates; (3) there are a limited number of facilities having the personnel and equipment necessary to perform the analyses; (4) the laboratory analyses can take as long as 4 weeks to complete; (5) there is no consensus among public health experts regarding the health significance of low levels of viruses in reclaimed water; and (6) there have not been any documented cases of viral disease resulting from the reuse of wastewater in the U.S. While recombinant DNA technology provides new tools to rapidly detect viruses in water, e.g., nucleic acid probes and polymerase chain reaction technology, methods currently in use are not able to quantify viruses or differentiate between infective and non-infective virus particles [111].

Unplanned or incidental indirect potable reuse occurs in many states in the U.S., while planned or intentional indirect potable reuse via groundwater recharge or augmentation of surface supplies is less widely practiced. Whereas the water quality requirements for nonpotable water uses are tractable and not likely to change significantly in the future, the number of water quality constituents to be monitored in drinking water (and, hence, reclaimed water intended for potable reuse) will increase and quality requirements are likely to become more restrictive. Consequently, it would not be prudent to suggest a complete list of reclaimed water quality limits for all constituents of concern. In addition to some specific wastewater treatment and reclaimed water quality recommendations, the guidelines provide some general recommendations to indicate the extensive treatment and water quality requirements that are likely to be imposed where indirect potable reuse is contemplated. The guidelines do not advocate direct potable reuse and do not include recommendations for such use.

The guidelines published by EPA include recommended water quality limits other than those specified in Table 14.10. The guidelines document includes suggested chemical constituent limits for most of the uses presented in Table 14.10. For example, for urban uses of reclaimed water, the guidelines recommend that the product water be nontoxic upon ingestion. This is recommended to protect against inadvertent and infrequent ingestion; it is not meant to imply that wastewater meeting the requirements for urban reuse is acceptable as a source of potable water. Other recommendations addressing urban use of reclaimed water include the following: clear, colorless, and odorless product water; a setback distance of 15 m (50 ft) from irrigated areas to potable water supply wells; maintenance of a chlorine residual of at least 0.5 mg/L in the distribution system; treatment reliability and emergency storage or disposal of inadequately-treated water; and cross-connection control via reduced pressure principle backflow prevention devices on potable water service lines at areas receiving reclaimed water and color-coded or taped reclaimed water lines and appurtenances. Similar design and operational recommendations are included in the guidelines for the other reclaimed water applications presented in Table 14.10.

It is explicitly stated in the *Guidelines for Water Reuse* that the recommended treatment unit processes and water quality limits presented in the guidelines "are not intended to be used as definitive water reclamation and reuse criteria. They are intended to provide reasonable guidance for water reuse opportunities, particularly in states that have not developed their own criteria or guidelines'' [104].

WORLD HEALTH ORGANIZATION GUIDELINES

Guidance in establishing water reclamation and reuse regulations also is provided by the World Health Organization (WHO). In 1971, WHO sponsored a meeting of experts on water reuse, culminating in a report recommending health criteria and treatment processes for various reclaimed water applications [108]. The applications ranged from irrigation of crops not intended for human consumption, for which the criteria were freedom from gross solids and significant removal of parasite eggs, to indirect potable reuse for which secondary treatment followed by filtration, nitrification, denitrification, chemical clarification, carbon adsorption, ion exchange or membranes, and disinfection were recommended. For nonpotable urban reuse and contact recreation, secondary treatment followed by sand filtration and disinfection were recommended. However, the health criteria differed in that for the urban reuse only a general requirement for effective bacteria removal and some removal of viruses was specified, while for contact recreation a bacterial standard of not more than 100 coliform/100 mL in 80 percent of samples and the absence of skin-irritating chemicals were specified.

In 1985, a meeting of scientists and epidemiologists was held in Engelberg, Switzerland, to discuss the health risks associated with the use of wastewater and excreta for agriculture and aquaculture. The meeting did not consider other uses of reclaimed water. The meeting was sponsored by WHO, the World Bank, United Nations Development Programme, and the International Reference Centre for Wastes Disposal. Health-related and other research made available since publication of the 1973 WHO guidelines were reviewed, and a revised approach to the nature of health risks associated with agriculture and aquaculture was developed. A model was developed of the relative health risks from the use of untreated excreta and wastewater in agriculture or aquaculture. The meeting concluded that the health risks resulting from irrigation with well treated wastewater were minimal and that current standards and guidelines are overly conservative and unduly restrict appropriate project development, thereby encouraging unregulated use of wastewater [46].

The Engelberg Report developed tentative microbial quality guidelines for reclaimed water used for irrigation. It was recommended that the number of intestinal nematodes should not exceed a geometric mean of one viable egg/L for all irrigation and that for the irrigation of edible crops, sports fields, and public parks, the number of fecal coliform organisms should not exceed 1,000/100 mL. The participants reasoned that, if those limits are met, other pathogens such as trematode eggs and protozoan cysts also are reduced to undetectable levels. The participants recognized, in addition, that social and behavioral patterns are of fundamental importance in the design and implementation of reuse projects. The meeting recommended that WHO initiate revision of the 1973 guidelines.

A WHO Scientific Group on Health Aspects of the Use of Treated Wastewater for Agriculture and Aquaculture met in Geneva in 1987, and their report has been published by WHO as *Health Guidelines for the Use of Wastewater for Agriculture and Aquaculture* [109]. These WHO guidelines reaffirm the recommendations of the Engelberg Report. The recommended microbiological quality guidelines for irrigation are summarized in Table 14.11. The guidelines are based on the conclusion that the main health risks are associated with helminthic diseases and, therefore, a high degree of helminth removal is necessary for the safe use of wastewater in agriculture and aquaculture. The intestinal nematodes covered serve as indicator organisms for all of the large settleable pathogens. The guidelines indicate that other pathogens apparently become non-viable in pond systems with long retention times, implying that all helminth eggs and protozoan cysts will be removed to the same extent. The helminth egg guidelines are intended to provide a design standard, not a standard requiring routine testing of the effluent.

The scientific group concluded that no bacterial guidelines are necessary in cases where the only exposed populations are farm workers, due to a lack of evidence indicating a health risk from bacteria. The recommended bacterial guideline of a geometric mean fecal coliform level of 1,000/100 mL was based on the evaluation of epidemiological studies and was considered by the scientific group to be technically feasible in developing countries. Most of the epidemiological investigations studied the application of untreated or poorly-treated wastewater for irrigation of food crops in developing countries. They mainly focused on disease incidence related to parasites and paid little attention to bacteria and viruses [78]. The scientific group indicated that the potential health risks associated with the use of reclaimed water for lawn and park irrigation may present greater potential health risks than those associated with the irrigation of vegetables to be eaten raw and, hence, recommended a fecal coliform limit of 200/100 mL for such urban irrigation.

The WHO guidelines recognize that there are limited health effects data for reclaimed water used for aquaculture and do not recommend definitive bacteriological quality standards for this use. However, tentative bacterial guidelines in the guidelines recommend a geometric mean of 1,000 fecal coliforms/100 mL, which is intended to insure that invasion

TABLE 14.11 World Health Organization Recommended Microbiological Guidelines for Wastewater Use in Agriculture.[a]

Category	Reuse Conditions	Exposed Group	Intestinal nematodes[b] (arithmetic mean no. of eggs per liter)[c]	Fecal Coliforms (geometric mean no. per 100 mL)[c]	Wastewater Treatment Expected to Achieve the Required Microbiological Quality
A	Irrigation of crops likely to be eaten uncooked, sports fields, public parks[d]	Workers, consumers, public	≤1	1,000[d]	A series of stabilization ponds designed to achieve the microbiological quality indicated, or equivalent treatment
B	Irrigation of cereal crops, industrial crops, fodder crops, pasture and trees[e]	Workers	≤1	No standards recommended	Retention in stabilization ponds for 8–10 days or equivalent helminth and fecal coliform removal
C	Localized irrigation of crops in Category B if exposure of workers and the public does not occur	None	Not applicable	Not applicable	Pretreatment as required by the irrigation technology, but not less than primary sedimentation

[a] In specific cases, local epidemiological, sociocultural, and environmental factors should be taken into account, and the guidelines modified accordingly.
[b] *Ascaris* and *Trichuris* species and hookworms.
[c] During the irrigation period.
[d] A more stringent guideline (200 fecal coliforms per 100 mL) is appropriate for public lawns, such as hotel lawns, with which the public may come into direct contact.
[e] In the case of fruit trees, irrigation should cease 2 weeks before fruit is picked, and no fruit should be picked off the ground. Sprinkler irrigation should not be used.
Source: Reference [109].

of fish muscle is prevented. The same fecal coliform standard is recommended for pond water in which aquatic vegetables (macrophytes) are grown. Since pathogens may accumulate in the digestive tract and intraperitoneal fluid of fish and pose a risk through cross-contamination of fish flesh or other edible parts—and subsequently to consumers if hygiene standards in fish preparation are inadequate—a recommended public health measure is to ensure maintenance of high standards of hygiene during fish handling and gutting. A total absence of viable trematode eggs, which is achievable by properly-designed and operated stabilization pond systems, is recommended as the appropriate helminth quality guideline for aquaculture use of reclaimed water (see Appendix III for detail).

The 1989 WHO guidelines identify waste stabilization ponds as the method of choice in meeting these guidelines in warm climates where land is available at reasonable cost. Based on helminth removal, the guidelines recommend a pond retention time of eight to ten days, with at least twice that time required in warm climates to reduce fecal coliforms to the guideline level of 1,000/100 mL. Experience at some existing full-scale and demonstration stabilization pond systems indicates that the desired reductions of helminths and fecal coliform organisms may be difficult to achieve in practice [18,43].

The Scientific Group that developed the WHO guidelines criticized the California *Wastewater Reclamation Criteria* as being too stringent, not based on epidemiological evidence, unattainable, and not appropriate for developing countries. The California standards, which are discussed elsewhere in this chapter, are intended to "establish acceptable levels of constituents of reclaimed water and to prescribe means for assurance of reliability in the production of reclaimed water to ensure that the use of reclaimed water does not impose undue risks to health" [85]. They are not based on epidemiological studies for reasons previously mentioned in this chapter. The California criteria are based on the control of all wastewater-associated pathogens of concern in the U.S., including parasites, bacteria, and viruses. They were developed for use in that state, where they have been shown to be readily attainable at more than 250 reclamation facilities. It is not surprising that the California standards and those of other states and industrialized countries are not achievable in developing countries due to economic, technological, and institutional differences between developed and developing countries.

The WHO guidelines are significantly less restrictive than regulations or guidelines of many industrialized countries. The intentions of international organizations such as the World Bank and United Nations Development Programme, who sponsored early work in this area, were to introduce at least some treatment of wastewater prior to crop irrigation, particularly in developing countries. This concept is understandable and commendable, and the WHO guidelines satisfy that intent. The WHO guidelines are appropriate as an interim measure in some countries until there is an ability to produce higher quality reclaimed water. It is unlikely that the WHO guidelines will replace existing criteria in most industrialized countries.

SUMMARY

Well-established and commonly-used treatment technology can reduce in concentration, or eliminate entirely, the myriad of toxic chemicals and pathogenic microbial organisms that may be present in municipal wastewater. Water reuse criteria have evolved through the years to the point where reclaimed water is now acceptable for almost any nonpotable water use. Reclaimed water for indirect potable reuse also is acceptable, and practiced, in some states and is receiving increasing attention in water-short areas of the U.S.

Several states in the U.S. have independently developed water reclamation and reuse criteria or guidelines and, as a consequence, there is considerable variation among the different state regulations. Only a few states, such as California and Florida, have comprehensive regulations addressing most facets of water reuse. The U.S. Environmental Protection Agency published *Guidelines for Water Reuse* in 1992 in an effort to provide guidance to utilities, regulatory agencies, and others in states where standards do not exist or are being revised or expanded. Research is needed to resolve information shortfalls and controversial issues related to reuse criteria. Research needs include microbial risk assessment modeling, identification of more representative indicators of the presence and concentration of pathogenic microorganisms, real-time on-line monitoring of parameters of concern, evaluation of alternatives to chlorine for disinfection, and continuing study of the concentration and health significance of organic chemical constituents in reclaimed water used for potable purposes.

REFERENCES

1   Adams, A. P., M. Garbett, H. B. Rees, and B. G. Lewis. 1978. Bacterial Aerosols from Cooling Towers. *Jour. WPCF,* 50(10):2362–2369.

2   Adams, A. P., M. Garbett, H. B. Rees, and B. G. Lewis. 1980. Bacterial Aerosols Produced from a Cooling Tower Using Wastewater Effluent as Makeup Water. *Jour. WPCF,* 52(3):498–501.

3   Arizona Department of Water Resources. 1995. *Environmental Quality Act.* Arizona Revised Statutes, Section 49-241. Arizona Department of Water Resources, Phoenix, Arizona.

4   Asano, T., and Sakaji, R. H. 1990. Virus Risk Analysis in Wastewater Reclamation and Reuse. In: *Chemical Water and Wastewater Treatment,* pp. 483–496, H. H. Hahn and R. Klute (eds.), Springer-Verlag, Berlin, Germany.

5   Asano, T., Y. C. Leong, M. G. Rigby, and R. H. Sakaji. 1992. Evaluation of the California Wastewater Reclamation Criteria using Enteric Virus Monitoring Data. *Wat. Sci. Tech.,* 26(7/8):1513–1524.

6   Ayers, R. S. and D. W. Westcot. 1985. Water Quality for Agriculture. FAO Irrigation and Drainage Paper 21, Rev. 1, United Nations Food and Agriculture Organization, Rome, Italy.

7   Bausum, H. T., S. A. Schaub, R. E. Bates, H. L. McKim, P. W. Schumacher, and B. E. Brockett. 1983. Microbiological Aerosols From a Field-Source Wastewater Irrigation System. *Jour. WPCF,* 55(1):65–75.

8   Bitton, G. 1980. *Introduction to Environmental Virology.* John Wiley & Sons, New York, New York.

9   Bryan, F. L. 1974. Diseases Transmitted by Foods Contaminated by Wastewater. In: *Wastewater Use in the Production of Food and Fiber,* pp. 16–45. EPA-660/2-74-041, U.S. Environmental Protection Agency, Washington, D.C.

10 California Department of Health. 1974. Turbidity Standard for Filtered Wastewater. Report prepared by the State of California Department of Health, Water Sanitation Section, Berkeley, California.

11 California Department of Health and R. C. Cooper. 1975. Wastewater Contaminants and Their Effect on Public Health. In: *A "State-of-the-Art" Review of Health Aspects of Wastewater Reclamation for Groundwater Recharge,* pp. 39–95. State of California Department of Water Resources, Sacramento, California.

12 California-Nevada Section AWWA. 1992. *Guidelines for Distribution of Nonpotable Water.* Published by California-Nevada Section, American Water Works Association, San Bernardino, California.

13 California State Water Resources Control Board. 1990. *California Municipal Wastewater Reclamation in 1987.* California State Water Resources Control Board, Office of Water Recycling, Sacramento, California.

14 Camann, D. E. and B. E. Moore. 1988. Viral Infections Based on Clinical Sampling at a Spray Irrigation Site. In: *Proceedings of Water Reuse Symposium IV,* pp. 847–863, August 2–7, 1987, Denver, Colorado. Published by the AWWA Research Foundation, Denver, Colorado.

15 Camann, D. E. and M. N. Guentzel. 1985. The Distribution of Bacterial Infections in the Lubbock Infection Surveillance Study of Wastewater Spray Irrigation. In: *Proceedings of the Water Reuse Symposium III,* pp. 1470–1495, August 2–7, 1984, Denver, Colorado. Published by the AWWA Research Foundation, Denver, Colorado.

16 Camann, D. E., B. E. Moore, H. J. Harding and C. A. Sorber. 1988. Microorganism Levels in Air Near Spray Irrigation of Municipal Wastewater: the Lubbock Infection Surveillance Study. *Jour. WPCF,* 60(11):1960–1970.

17 Camann, D. E., D. E. Johnson, H. J. Harding, and C. A. Sorber. 1980. Wastewater Aerosol and School Attendance Monitoring at an Advanced Wastewater Treatment Facility: Durham Plant, Tigard, Oregon. In: *Wastewater Aerosols and Disease,* pp. 160–179, H. Pahren and W. Jakubowski (eds.), EPA-600/9-80-028, U.S. Environmental Protection Agency, Cincinnati, Ohio.

18 Camp Dresser & McKee International. 1993. As-Samra Wastewater Stabilization Ponds Emergency Short-Term Improvement System. Design report prepared for the Hashemite Kingdom of Jordan, Ministry of Water and Irrigation, by Camp Dresser & McKee International, Cambridge, Massachusetts.

19 Cooper, R. C., A. W. Olivieri, R. E. Danielson, P. G. Badger, R. C. Spear, and S. Selvin. 1986. *Evaluation of Military Field-Water Quality: Volume 5: Infectious Organisms of Military Concern Associated with Consumption: Assessment of Health*

Risks and Recommendations for Establishing Related Standards. Report No. UCRL-21008 Vol. 5, Environmental Sciences Division, Lawrence Livermore National Laboratory, University of California, Livermore, California.

20 Craun, G. F. 1988. Surface Water Supplies and Health. *Jour. AWWA,* 80(2):40–52.

21 Crook, J. 1985. Water Reuse in California. *Jour. AWWA,* 77(7):60–71.

22 Crook, J. 1989. Viruses in Reclaimed Water. In: *Proceedings of the 63rd Annual Technical Conference,* pp. 231–237, sponsored by the Florida Section American Water Works Association, Florida Pollution Control Association, and Florida Water & Pollution Control Operators Association, November 12–15, St. Petersburg Beach, Florida.

23 Crook, J. 1992. Water Reclamation. In: *Encyclopedia of Physical Science and Technology, Vol. 17,* pp. 559–589, Academic Press, San Diego, California.

24 Crook, J., D. A. Okun, and A. B. Pincince. 1994. *Water Reuse.* Project 92-WRE-1, Published by Water Environment Research Foundation, Alexandria, Virginia.

25 Culp/Wesner/Culp. 1979. *Water Reuse and Recycling Volume 2, Evaluation of Treatment Technology.* U.S. Department of the Interior, Office of Water Research and Technology, OWRT/RU-79/2.

26 Engineering-Science. 1987. Monterey Wastewater Reclamation Study for Agriculture: Final Report. Prepared for the Monterey Regional Water Pollution Control Agency by Engineering-Science, Berkeley, California.

27 EOA, Inc. 1995. Microbial Risk Assessment for Reclaimed Water. Report prepared for Irvine Ranch Water District by EOA, Inc., Oakland, California.

28 Fannin, K. F., K. W. Cochran, D. E. Lamphiear and A. S. Monto. 1980. Acute Illness Differences with Regard to Distance from the Tecumseh, Michigan Wastewater Treatment Plant. In: *Wastewater Aerosols and Disease,* pp. 117–135, H. Pahren and W. Jakubowski (eds.), EPA-600/9-80-028, U.S. Environmental Protection Agency, Health Effects Research Laboratory, Cincinnati, Ohio.

29 Feachem, R. G., D. J. Bradley, H. Garelick, and D. D. Mara. 1981. *Health Aspects of Excreta and Sullage Management: A State-of-the Art Review.* The World Bank, Washington, D.C.

30 Feachem, R. G., H. Bradley, H. Garelick, and D. D. Mara. 1983. *Sanitation and Disease—Health Aspects of Excreta and Wastewater Management.* Published for the World Bank by John Wiley & Sons, Chichester, England.

31 Florida Department of Environmental Protection. 1993. Domestic Wastewater Facilities. Chapter 62-600, Florida Administrative Code. Florida Department of Environmental Protection, Tallahassee, Florida.

32 Florida Department of Environmental Protection. 1995. State Water Policy. Chapter 62-40, Florida Administrative Code, Florida Department of Environmental Protection, Tallahassee, Florida.

33 Florida Department of Environmental Protection. 1996. Reuse of Reclaimed Water and Land Application. Chapter 62-610, Florida Administrative Code. Florida Department of Environmental Protection, Tallahassee, Florida.

34 Florida Department of Environmental Regulation. 1983. *Land Application of Domestic Wastewater.* Florida Department of Environmental Regulation, Tallahassee, Florida.

35 Florida Department of Environmental Regulation. 1991. *Guidelines for Preparation of Reuse Feasibility Studies for Applicants having Responsibility for Wastewater Management.* Florida Department of Environmental Regulation, Tallahassee, Florida.

36 Florida Department of Environmental Regulation. 1992. *1992 Reuse Inventory.* Florida Department of Environmental Regulation, Tallahassee, Florida.

37 Gerba, C. P., and S. M. Goyal. 1985. Pathogen Removal from Wastewater during Groundwater Recharge. In: *Artificial Recharge of Groundwater,* pp. 283–317, T. Asano (ed.), Butterworth Publishers, Boston, Massachusetts.

38 Gerba, C. P. and C. N. Haas. 1988. Assessment of Risks Associated with Enteric Viruses in Contaminated Drinking Waters. In: *Chemical and Biological Characterization of Sludges, Sediments, Dredge Spoils, and Drilling Muds,* pp. 489–494, ASTM STP 976, J. J. Lichtenberg, J. A. Winter, C. I. Weber, and L. Fradkin (eds.), American Society for Testing and Materials, Philadelphia, Pennsylvania.

39 Gover, N. 1993. HIV in Wastewater Not a Threat. *Water Environ. & Tech.,* 5(12):23.

40 Greenberg, A. E. and E. Kupka. 1957. Tuberculosis Transmission by Wastewater— A Review. *Sew. & Ind. Wastes,* 29(5):524–537.

41 Hoadley, A. W. and S. M. Goyal. 1976. Public Health Implications of the Application of Wastewater to Land. In: *Land Treatment and Disposal of Municipal and Industrial Wastewater,* p. 1092, R. L. Sanks and T. Asano (eds.), Ann Arbor Science Publishers, Inc., Ann Arbor, Michigan.

42 Hultquist, R. H. 1995. Augmentation of Ground and Surface Drinking Water Sources with Reclaimed Water in California. Paper presented at *AWWA Annual Conference,* Workshop on Augmenting Potable Water Supplies with Reclaimed Water. June 18, 1995, Fountain Valley, California.

43 Huntington, R. and J. Crook. 1993. Technological and Environmental Health Aspects of Wastewater Reuse for Irrigation in Egypt and Israel. WASH Field Report No. 418. Report prepared for the U.S. Agency for International Development, Near East Bureau, Washington, D.C.

44 Hurst, C. J., W. H. Benton, and R. E. Stetler. 1989. Detecting Viruses in Water. *Jour. AWWA,* 81(9):71–80.

45 Idelovitch, E., R. Terkeltoub, and M. Michall. 1980. The Role of Groundwater Recharge in Wastewater Reuse: Israel's Dan Region Project. *Jour. AWWA,* 72(7): 391–400.

46 International Reference Centre for Waste Disposal. 1985. Health Aspects of Wastewater and Excreta Use in Agriculture and Aquaculture: The Engelberg Report. *IRCWD News,* No. 23, Dubendorf, Switzerland.

47 Irvine Ranch Water District. 1990. Engineer's Report: Use of Reclaimed Water for Flushing Toilets and Urinals, and Floor Drain Trap Priming in the Restroom Facilities at Jamboree Tower. Irvine Ranch Water District, Irvine, California.

48 Iwata, Y. and F. A. Gunther. 1976. Translocation of the Polychlorinated Biphenyl Oroclor 1254 from Soil into Carrots under Field Conditions. *Arc. of Environ. Contam. & Tox.,* 4(1):44–59.

49 Johnson, D. E., D. E. Camann, D. T. Kimball, R. J. Prevost and R. E. Thomas. 1980. Health Effects from Wastewater Aerosols at a New Activated Sludge Plant: John Egan Plant, Schaumburg, Illinois. In: *Wastewater Aerosols and Disease,* pp. 136– 159, H. Pahren and W. Jakubowski (eds.), EPA-600/9-80-028, U.S. Environmental Protection Agency, Cincinnati, Ohio.

50 Johnson, D. E., D. E. Camann, J. W. Register, R. E. Thomas, C. A. Sorber, M. N. Guentzel, J. M. Taylor, and W. J. Harding. 1980. *The Evaluation of Microbiological Aerosols Associated With the Application of Wastewater to Land: Pleasanton, CA.* EPA-600/1-80-015, U.S. Environmental Protection Agency, Cincinnati, Ohio.

51 Lund, E. 1980. Health Problems Associated with the Re-Use of Sewage: I. Bacteria, II. Viruses, III. Protozoa and Helminths. Working papers prepared for WHO Seminar on Health Aspects of Treated Sewage Re-Use, 1–5 June 1980, Algiers, Algeria.

52   Maeda, M., K. Nakada, K. Kawamoto, and M. Ikeda. 1995. Area-Wide Use of Reclaimed Water in Tokyo, Japan. In: *Proceedings of the Second International Symposium on Wastewater Reclamation and Reuse: Symposium Preprint Book 1*, pp. 55–62, A. Angelakis, T. Asano, E. Diamadopoulos, and G. Tchobanoglous (eds.), October 17–20, 1995, Iraklio, Crete, Greece.

53   McCarty, P. L., M. Reinhard, N. L. Goodman, J. W. Graydon, G. P. Hopkins, K. E. Mortelmans, and D. G. Argo. 1982. Advanced Treatment for Wastewater Reclamation at Water Factory 21. Technical Report 267, Department of Civil Engineering, Stanford University, Stanford, California.

54   Metcalf & Eddy, Inc. 1991. *Wastewater Engineering: Treatment, Disposal, Reuse.* McGraw-Hill, Inc., New York, N.Y.

55   Murphy, W. H. and J. T. Syverton. 1958. Absorption and Translocation of Mammalian Viruses by Plants. II. Recovery and Distribution of Viruses in Plants. *Virology*, 6(3):623–636.

56   National Academy of Sciences-National Academy of Engineering. 1973. *Water Quality Criteria 1972*. EPA/R3/73/033. Prepared by the Committee on Water Quality Criteria, National Academy of Sciences-National Academy of Engineering, for the U.S. Environmental Protection Agency, Washington, D.C.

57   National Communicable Disease Center. 1969. *Shigella* Surveillance Second Quarter. National Communicable Disease Center, Report 20, Atlanta, Georgia.

58   National Communicable Disease Center. 1975. *Morbidity and Mortality, Weekly Report*. National Communicable Disease Center, 24(31):261.

59   National Research Council. 1980. *Drinking Water and Health, Vol. 2*. pp. 252–253, National Academy Press, Washington, D.C.

60   National Research Council. 1994. *Ground Water Recharge Using Waters of Impaired Quality*. National Academy Press, Washington, D.C.

61   National Research Council. 1996. *Use of Reclaimed Water and Sludge in Food Crop Irrigation*. National Academy Press, Washington, D.C.

62   Nellor, M. H., R. B. Baird, J. R. Smyth. 1984. Health Effects Study—Final Report. County Sanitation Districts of Los Angeles County, Whittier, California.

63   Okun, D. A. 1979. Criteria for Reuse of Wastewater for Nonpotable Urban Water Supply Systems in California. Report prepared for the California Department of Health Services, Sanitary Engineering Section, Berkeley, California.

64   Olivieri, A. W., R. C. Cooper, R. C. Spear, R. E. Danielson, D. E. Block, and P. G. Badger. 1986. Risk Assessment of Waterborne Infectious Agents. In: *ENVIRONMENT 86: Proceedings of the International Conference on Development and Application of Computer Techniques to Environmental Studies*, P. Zannetti (ed.), Los Angeles, California.

65   Olivieri, A. W., R. C. Cooper, and R. E. Danielson. 1989. Risk of Waterborne Infectious Illness Associated with Diving in the Point Loma Kelp Beds, San Diego, CA. In: *Proceedings of the ASCE 1989 Environmental Engineering Specialty Conference*, pp. 70–79, J. F. Malina (ed.), July 10–12, 1989, Austin, Texas. Published by the American Society of Civil Engineers, New York, New York.

66   Palazzo, A. J. 1976. The Effects of Wastewater Applications on the Growth and Chemical Composition of Forages. Report 76-9, U.S. Army Corps of Engineers, Cold Regions Research and Engineering Laboratory, Hanover, New Hampshire.

67   Passarew, L., Watkins, E., Bastian, R., and Hoyge, S. 1989. Should EPA Take a More Active Role in Wastewater Reuse? Paper presented at the 62nd Annual WPCF Conference, Preconference Workshop: New Directions in Reuse for the 1990's, October 16–19, 1989, San Francisco, California.

68 Pawlowski, S. 1992. Rules for the Reuse of Reclaimed Water. Paper presented at the Salt River Project Water Reuse Symposium, November 2, 1992, Tempe, Arizona.

69 Regli, S., J. B. Rose, C. N. Haas, and C. P. Gerba. 1991. Modeling the Risk from *Giardia* and Viruses in Drinking Water. *Jour. Amer. Water Works Assoc.,* 83(11):76–84.

70 Riggs, J. L. 1989. AIDS Transmission in Drinking Water: No Threat. *Jour. AWWA,* 81(9):69–70.

71 Rose, J. B. 1986. Microbial Aspects of Wastewater Reuse for Irrigation. *CRC Critical Reviews in Environ. Control,* 16(3):231–256.

72 Rose, J. B. and R. P. Carnahan. 1992. Pathogen Removal by Full-Scale Wastewater Treatment. Report prepared for the Florida Department of Environmental Regulation, Tallahassee, Florida.

73 Rose, J. B. and C. P. Gerba. 1991. Assessing Potential Health Risks from Viruses and Parasites in Reclaimed Water in Arizona and Florida, U.S.A. *Wat. Sci. Tech.,* 23:2091–2098.

74 Rose, J. B., C. N. Haas, and S. Regli. 1991. Risk Assessment and Control of Waterborne Giardiasis. *Amer. Jour. Pub. Health,* 81(6):709–713.

75 Sagik, B. P., B. E. Moore, and C. A. Sorber. 1978. Infectious Disease Potential of Land Application of Wastewater. In: *State of Knowledge in Land Treatment of Wastewater, Volume 1,* pp. 35–46. Proceedings of an International Symposium, U.S. Army Corps of Engineers, Cold Regions Research and Engineering Laboratory, Hanover, New Hampshire.

76 Sanitation Districts of Los Angeles County. 1977. Pomona Virus Study: Final Report. California State Water Resources Control Board, Sacramento, California.

77 Sepp, E. 1971. *The Use of Sewage for Irrigation: A Literature Review.* California Department of Public Health, Bureau of Sanitary Engineering, Berkeley, California.

78 Shelef, G. 1991. Water Reclamation and Water Resources Management. *Wat. Sci. Tech.,* 24(9):251–265.

79 Shelef, G. and Y. Azov. 1995. The Coming of Intensive Wastewater Reuse in the Mediterranean Region. In: *Proceedings of the Second International Symposium on Wastewater Reclamation and Reuse: Symposium Preprint Book 1,* pp. 137–146, A. Angelakis, T. Asano, E. Diamadopoulos, and G. Tchobanoglous (eds.), October 17–20, 1995, Iraklio, Crete, Greece.

80 Shuval, H. I. 1978. Land Treatment of Wastewater in Israel. In: *State of Knowledge in Land Treatment of Wastewater, Volume 1. Proceedings of an International Symposium,* pp. 429–436, U.S. Army Corps of Engineers, Cold Regions Research and Engineering Laboratory, Hanover, New Hampshire.

81 Shuval, H. I., A. Adin, B. Fattal, E. Rawitz, and P. Yekutiel. 1986. Wastewater Irrigation in Developing Countries—Health Effects and Technical Solutions. World Bank Technical Paper Number 51, The World Bank, Washington, D.C.

82 Sobsey, M. 1978. Public Health Aspects of Human Enteric Viruses in Cooling Waters. Report to NUS Corporation, Pittsburgh, PA.

83 State of Arizona. 1991. Regulations for the Reuse of Wastewater. Arizona Administrative Code, Chapter 9, Article 7, Arizona Department of Environmental Quality, Phoenix, Arizona.

84 State of California. 1975. A "State-of-the-Art" Review of Health Aspects of Wastewater Reclamation for Groundwater Recharge. Report prepared by the State of California Water Resources Control Board, Department of Water Resources, and Department of Health. Published by the State of California Department of Water Resources, Sacramento, California.

85   State of California. 1978. Wastewater Reclamation Criteria. California Administrative Code, Title 22, Division 4, California Department of Health Services, Sanitary Engineering Section, Berkeley, California.

86   State of California. 1979. Minimum Guidelines for the Control of Individual Wastewater Treatment and Disposal Systems. California Regional Water Quality Control Board, San Francisco Bay Region, Oakland, California.

87   State of California. 1987. *Report of the Scientific Advisory Panel on Groundwater Recharge with Reclaimed Water.* G. Robeck (ed.). Prepared for the State of California Water Resources Control Board, Department of Water Resources, and Department of Health Services. Published by the State of California Department of Water Resources, Sacramento, California.

88   State of California. 1987. Wastewater Disinfection for Health Protection. State of California Department of Health Services, Sanitary Engineering Branch, Sacramento, California.

89   State of California. 1993. Draft Proposed Groundwater Recharge Regulation. Prepared by the State of California Department of Health Services, Division of Drinking Water and Environmental Management, Sacramento, California.

90   State of California. 1994a. Draft Regulations Relating to Cross Connection Control. California Department of Health Services, Drinking Water Program, Sacramento, California.

91   State of California. 1994b. *Porter-Cologne Water Quality Control Act.* California Water Code, Division 7. Compiled by the State Water Resources Control Board, Sacramento, California.

92   State of California. 1997. Draft Water Recycling Criteria. California Department of Health Services, Drinking Water Program, Sacramento, California.

93   State of Washington. 1993. Water Reclamation and Reuse Interim Standards. State of Washington, Department of Health, Olympia, Washington.

94   Strauss, S. D. and P. R. Puckorius. 1984. Cooling Water Treatment for Control of Scaling, Fouling, Corrosion. *Power,* June 1984, 1–24.

95   Tanaka, H., T. Asano, E. D. Schroeder, and G. Tchobanoglous. 1993. Estimating the Reliability of Wastewater Reclamation and Reuse Using Enteric Virus Monitoring Data. In: *Proceedings of the 66th WEF Annual Conference and Exposition,* pp. 105–118, October 3–7, 1993, Anaheim, California. Published by the Water Environment Federation, Alexandria, Virginia.

96   Teltsch, B., S. Kidmi, L. Bonnet, Y. Borenzstajn-Roten, and E. Katzenelson. 1980. Isolation and Identification of Pathogenic Microorganisms at Wastewater-Irrigated Fields: Ratios in Air and Wastewater. *Applied Environ. Microbiol.,* 39:1184–1195.

97   U.S. Environmental Protection Agency. 1974. *Design Criteria for Mechanical, Electric, and Fluid System and Component Reliability. MCD-05.* EPA-430-99-74-001, U.S. Environmental Protection Agency, Office of Water Program Operations, Washington, DC. 1974.

98   U.S. Environmental Protection Agency. 1976. *Quality Criteria for Water.* EPA-R3-73-033, A Report of the National Academy of Science-National Academy of Engineering Committee on Water Quality Criteria. U.S. Environmental Protection Agency, Washington, D.C.

99   U.S. Environmental Protection Agency. 1976. *National Interim Primary Drinking Water Regulations.* EPA-570/9-76-003. U.S. Environmental Protection Agency, Washington, D.C.

100 U.S. Environmental Protection Agency. 1980. *Wastewater Aerosols and Disease. Proceedings of a Symposium,* H. Pahren and W. Jakubowski (eds.), September 19–21, 1979, EPA-600/9-80-028, U.S. Environmental Protection Agency, Health Effects Research Laboratory, Cincinnati, Ohio.

101 U.S. Environmental Protection Agency. 1981. *Process Design Manual: Land Treatment of Municipal Wastewater.* EPA/625/1-81-013, U.S. Environmental Protection Agency, Center for Environmental Research Information, Cincinnati, Ohio.

102 U.S. Environmental Protection Agency. 1982. *Estimating Microorganism Densities in Aerosols from Spray Irrigation of Wastewater.* EPA-600/9-82-003. U.S. Environmental Protection Agency, Health Effects Research Laboratory, Cincinnati, Ohio.

103 U.S. Environmental Protection Agency. 1986. *Ambient Water Quality Criteria for Bacteria—1986.* EPA A440/584-002, U.S. Environmental Protection Agency, Office of Water Regulations and Standards, Washington, D.C.

104 U.S. Environmental Protection Agency. 1992. *Guidelines for Water Reuse.* EPA/625/R-92/004, U.S. Environmental Protection Agency, Center for Environmental Research Information, Cincinnati, Ohio.

105 van Leeuwen, N. H. 1988. *Reuse of Wastewater—A Literature Survey.* Project No. 670 2723 5, National Institute for Water Research, Council for Scientific and Industrial Research, Pretoria, South Africa.

106 Water Pollution Control Federation. 1989. *Water Reuse: Manual of Practice SM-3.* 2nd ed. Water Pollution Control Federation, Alexandria, Virginia.

107 Westcot, D. W. and R. S. Ayers. 1985. Irrigation Water Quality. In: *Irrigation with Reclaimed Municipal Wastewater—A Guidance Manual.* G. S. Pettygrove and T. Asano (eds.), pp. 3-1–3-37. Prepared by the California State Water Resources Control Board. Published by Lewis Publishers, Inc., Chelsea, Michigan.

108 World Health Organization. 1973. Reuse of Effluents: Methods of Wastewater Treatment and Health Safeguards. Report of a WHO Meeting of Experts, Technical Report Series No. 17, World Health Organization, Geneva, Switzerland.

109 World Health Organization. 1989. Health Guidelines for the Use of Wastewater in Agriculture and Aquaculture. Report of a WHO Scientific Group, Technical Report Series 778, World Health Organization, Geneva, Switzerland.

110 Yanko, W. A. 1993. Analysis of 10 Years of Virus Monitoring Data from Los Angeles County Treatment Plants Meeting California Wastewater Reclamation Criteria. *Water Environ. Research,* 65(3):221–226.

111 Yates, M. V. 1994. Monitoring Concerns and Procedures for Human Health Effects. In: *Wastewater Reuse for Golf Course Irrigation,* pp. 143–171, Lewis Publishers, Chelsea, Michigan.

112 York, D. W. and E. A. Potts. 1996. The Evolution of Florida's Reuse Program. In: *1996 Water Reuse Conference Proceedings,* pp. 653–667, February 25–28, 1996, San Diego, California. Published by the American Water Works Association, Denver, Colorado.

113 Young, H. W. and D. W. York. 1996. Reuse of Reclaimed Water in Southwest Florida. In: *Integrated Water Resource Management: A Commitment to the Future. Proceedings of the 71st Annual Water Resources Conference,* pp. 136–153, sponsored by the Florida Section American Water Works Association, Florida Water Environment Association, and Florida Water & Pollution Control Operators Association, May 5–8, 1996, Ft. Myers, Florida.

CHAPTER 15

Microbial Risk Assessment and Its Role in the Development of Wastewater Reclamation Policy

INTRODUCTION

WASTEWATER reclamation holds promise as an important water resource as the desire to develop arid regions continues to place increasing demands on finite water resources. Wastewater reclamation and reuse can easily fill a portion of the demand for which potable water is not needed, i.e., landscape irrigation, crop irrigation, golf course irrigation, or industrial cooling water. However, wastewater reclamation proponents envision a future in which reclaimed wastewater may provide a fraction or all of a potable water supply.

Recognizing the origin of this resource and understanding the specific public health concerns continues to fuel debates over the development of public policy regarding wastewater reuse. The shortcomings in analytical methods, inability to develop specific operational criteria, and limited water quality monitoring requirements contribute to the uncertainty and unknowns surrounding public health issues such as exposure to pathogens. The unknowns and uncertainties continue to raise doubt about the efficacy of using reclaimed wastewater when human exposure is high.

The views expressed in this chapter are those of the authors and do not necessarily reflect the policies and views of the California Department of Health Services.

Richard H. Sakaji, California Department of Health Services, Division of Drinking Water and Environmental Management, 2151 Berkeley Way, Berkeley, CA 94704; Naoyuki Funamizu, Department of Sanitary & Environmental Engineering, Hokkaido University, Kita-13, Nishi-8, Kita-ku, Sapporo, 060 Japan.

The debate surrounding the consumption of reclaimed wastewater finds risk managers (regulators or public policy makers) pondering the question of what types of water quality standards might be set in order to provide the proper level of safety associated with the use of reclaimed wastewater (Pinholster 1995). In the quest to provide a rationale basis for establishing public policy, *risk assessment* protocols have been developed in an attempt to evaluate the risk[1] associated with wastewater reclamation practices. Such protocols not only need to answer the broad question of what is safe, but also must provide some assurance that such a decision has not been based on poor assumptions or false negative results.

This chapter is an attempt to illustrate how microbial risk assessment can be made more robust by using a stochastic rather than deterministic approach to assessing risk. Using such an approach in the development of wastewater reclamation policy can provide assurance that adequate public health protection is provided by a means that is consistent with the needs of the population at risk. Generally, there is a need to examine risk assessment using a single model to avoid variations between risk assessment methodologies. Dissimilarities in the risk assessment models may result in practices having similar levels of risk but may lead to dissimilar levels of protection, i.e., inconsistent and nonuniform wastewater reclamation regulations. Although the ideal microbial risk assessment should provide a quantitative measure of the absolute risk associated with an activity, it must be recognized that whenever models are used in risk assessment, the strength of the result will not exceed the least accurate data or the weakest assumption used. It may be more appropriate to compare risks between wastewater reclamation applications as being relative and not absolute. One should not be lulled into the false assumption that the use of models in risk assessment will precisely and accurately predict the risk of exposure. However, this chapter attempts to identify why microbial risk assessment can and should be used judiciously in the development of public policy.

As this chapter will attempt to illustrate, the current state of microbial risk assessment is good, but there is room for improvement. As our knowledge and understanding of waterborne pathogens and disease transmission improves, our ability to provide confidence and assurance regarding the microbiological safe use of reclaimed wastewater will increase.

WASTEWATER RECLAMATION AND PUBLIC POLICY

Currently wastewater reclamation practices are not regulated by the United States Environmental Protection Agency (USEPA). Though it is

---

[1] A measure of the probability or loss.

common to turn toward federal regulation to ensure some degree of consistency in the development of public policy between the states, regulation of wastewater reclamation activities are left to the discretion of the individual states. The degree of regulation and the types of regulated reclaimed water uses between the states is not consistent. Crook et al. (1992) provide an excellent summary of wastewater reclamation regulations from the fifty states. Generally, while there is some variation in the stringency of the wastewater reclamation requirements, the general trend is to make reclaimed wastewater quality standards a function of the intended use of the reclaimed wastewater, i.e., the greater the possibility of human exposure, the more stringent the requirements.

From a risk management perspective, water reuse policies have historically relied on a qualitative evaluation of treatment techniques and water quality standards to provide a reliable and consistent means of minimizing public exposure to pathogens. Large quantities of reclaimed wastewater have been used for food, fodder, and fiber irrigation in which public exposure is minimal with few documented problems. However, as the use of reclaimed water for indirect potable reuse, i.e., groundwater recharge or for recreational lakes (in which body contact can occur) becomes more popular, there is an increasing need to provide a more quantitative means of evaluating the risk of pathogen exposure associated with these activities.

Good policy development requires a broad fundamental understanding of the shortcomings and highlights of the statistics and research results being used to set acceptable risk levels. As Milloy et al. (1994) point out, the process of risk assessment has evolved over time as a means of providing regulators (risk managers) with a technical basis for implementing their statutory responsibility, that is, establishing regulations that protect human health and safety. Scientific hypotheses and technical opinions are often incorporated into policy (not necessarily into regulations) in order to bridge knowledge gaps and uncertainties in the risk assessment process. Most people in the business of risk management understand the shortcomings of microbial risk assessment. Yet, while risk managers convert such understanding into public policy, risk communication must provide a framework for communicating this same uncertainty to the lay public in order to enable a more integrated approach to water resource management.

In order to move forward, our understanding of reclamation and the impacts it can have on the quality of water supplies must be understood. Concerned about the use of reclaimed wastewater for recharging groundwater basins used for potable water supply, the State of California assembled a Scientific Advisory Panel (SAP) to review recharge practices (State of California 1987). The SAP identified organic compounds and pathogens as the primary public health concerns associated with groundwater recharge practices. The panel felt comfortable with continuing the existing groundwater recharge project because the available technology was capable of inactivating or removing known human pathogens from reclaimed wastewater to below detectable limits. They further concluded that the risks associated with the reclamation projects of the time were small and not dissimilar to hypothesized risks for surface waters commonly used for drinking water supplies. While this finding was useful, problems arose when the risk managers attempted to convert this finding into public policy. Without a formal quantitative risk assessment, the estimate of risk and the evaluation of the reclamation practice was entirely qualitative in nature. Regulators were not left with any means of evaluating the risk of infection from pathogens in reclaimed wastewater nor did policy makers have the information needed to establish a clear set of guidelines for groundwater recharge. While current research is seeking to resolve many of the unknowns and uncertainties, new wastewater reclamation technologies and uses are being proposed to regulators who do not have the means to properly evaluate proposed projects that are not comparable to the project reviewed by the SAP.

ELEMENTS OF MICROBIAL RISK ASSESSMENT

Regardless of whether the risk assessment covers organic chemicals, radionuclides, heavy metals, inorganic chemicals, or pathogens, risk assessment protocols contain four primary elements listed by the National Research Council (NRC) (1983). These four elements are: identification of the hazard, an assessment of the extent of exposure and routes of exposure, the response of humans to exposure to the pathogens, and a description of the risk. Although these primary risk assessment components remain the same, regardless of whether the risk assessment is for a chemical or microbiological threat to public health, the risk assessment elements and the manner in which each of these elements is assembled and used will vary depending on the needs of a specific risk assessment.

For microbial risk assessment these four primary elements are defined generally as follows:

(1) Hazard identification: general identification of classes and species of waterborne pathogens that can cause infection or disease

(2) Dose-response: establishes a relationship between the dose of a microbial agent administered and the outcome following population exposure

(3) Exposure assessment: estimates the intensity, frequency, and duration

of human exposure to a microbial agent. It describes the magnitude, duration, schedule, uncertainties, and route of exposure to the human population. These estimates are the direct result of combining environmental occurrence data, and dose-response information, in an exposure scenario that can be used to determine the risk of an adverse outcome.

(4) Risk characterization: attempts to describe the magnitude of the risk, how the risk might relate to other hazards, and a qualitative description of the uncertainty associated with the risk estimate

Too often risk managers move from risk assessment directly to risk management without recognizing the individual importance and limitations of the primary elements. As one of the cornerstones for public policy making, it is important to understand the role that each of the primary elements of risk assessment play in the development of public policy. "The committee believes that uncertainty analysis is the only way to combat the false sense of security, which is caused by a refusal to acknowledge and (attempt to) quantify the uncertainty in risk prediction." (NRC 1994).

The four previously mentioned primary elements of a microbial risk assessment are assembled in a linear series of steps, as shown in Figure 15.1, which produces an estimate of the risk associated with a specific endpoint for microbial risk assessment in reclaimed wastewater (infection). Each of these elements is detailed below in an attempt to explain how the elements can be melded into a risk assessment process. Although some overlap exists between the elements, generally each element represents a distinct step in the risk assessment process.

HAZARD IDENTIFICATION

Waterborne Disease

For a long time, public health protection in drinking water has relied on the primary sanitary engineering principle of using the highest quality water for drinking or potable (McGaughey 1968). Throughout the early 1950s the annual number of reported waterborne diseases in the United States has steadily declined. This trend may have been due to the increased use of chlorine and filters on potable water supplies. However, since the early 1970s there has been a disturbing increase in the number of waterborne disease outbreaks in the United States as detailed in a series of articles by Craun (1986a, 1986b, 1988, 1990).

Based on summary statistics, which are thought to underestimate the true incidence waterborne disease, this disturbing trend may be an artifact of increased awareness, increased vigilance, changing terminology, and

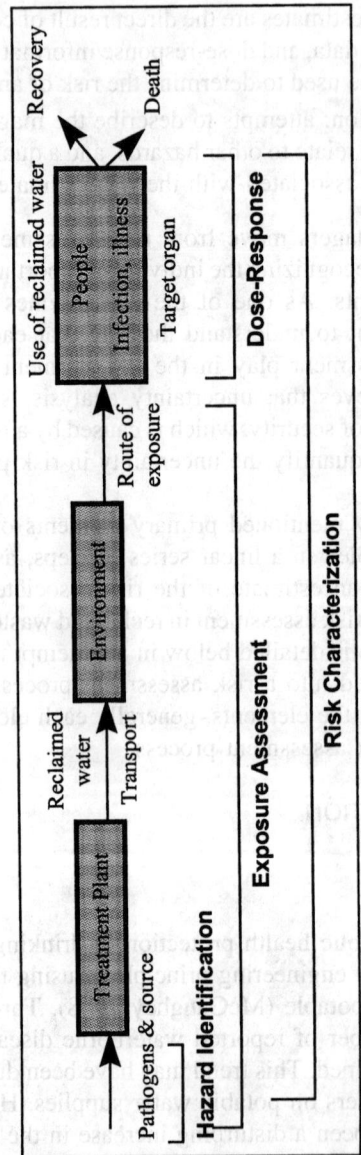

Figure 15.1 Four primary elements in microbial risk assessment for wastewater reclamation (adapted from NRC 1994).

improved reporting. While the term *waterborne disease* covers a broad spectrum of concerns without identifying specific etiologic agents or their sources, it should be recognized that exposure to waterborne microbial pathogens can occur through recreational activities not associated with wastewater reclamation, e.g., swimming in rivers, lakes, streams, public swimming pools, or wave pools. Also included in these statistics are illnesses associated with acute exposure to chemicals present in water, such as nitrates which can result in methemoglobinemia. Still, recent waterborne disease outbreaks [e.g., the 1993 cryptosporidiosis outbreak in Milwaukee (Fox and Lytle 1996)] and subsequent epidemiological investigations provide evidence of the continued threat posed by microbial contaminants in source waters.

While the risk of infection from waterborne pathogens is not well known, the list of known waterborne human pathogens reported originally by Hutzler and Boyle (1982) (modified as presented in Table 15.1) continues to expand as the biochemical tools for identifying and isolating human pathogens improves. Improved microbiological techniques continue to raise major interpretive issues for public health and water agency officials today. Although improved analytical methodology can detect a variety of pathogens beyond the standard coliform tests, questions regarding viability and infectivity of detected microorganisms remain unanswered. In addition, many analytical methods are limited to only identifying pathogens to the genus level, without being able to identify the species responsible for the disease. For example, the protozoan parasite, *Giardia* as identified by current analytical techniques (direct fluoresence anitbody staining) does not differentiate between the species *G. muris* and *G. lamblia,* only one of which (*G. lamblia*) is thought to be pathogenic to humans. Yet, even with these limitations, the water industry, in concert with regulatory agencies continue to collect data on the environmental occurrence of human and emerging pathogens.

The term *emerging pathogens*[2] has been coined to identify pathogens that are not really new or emerging, nor have these agents necessarily been isolated and identified from environmental samples as etiologic agents of human disease. These pathogens go by the names *Giardia lamblia, Cryptosporidium parvum, Cyclospora, Blastocystis hominis,* and *Mycobacterium avium* complex. Recent concerns include the possibility that the transmission of these infectious agents can be waterborne and may be of critical public health importance to sensitized subgroups (such as AIDS

---

[2] These pathogens are not "new" nor are they really "emerging." It has just been that, until recently, drinking water had not been considered a primary route of human exposure. Due to the development of new analytical methods and case-sensitive populations, these pathogens and their associated illnesses have been associated with water sources.

TABLE 15.1. **Waterborne Pathogens and Diseases.**

	Pathogen	Disease
Bacterial	*Legionella* spp.	Legionellosis
	Shigella spp.	Shigellosis
	Salmonella spp.	Salmonellosis
	Vibrio cholerae	Cholera
	Vibrio non-cholera	
	Mycobacterium leprae	Leprosy
	Brucella tularensis	Tularaemia
	Salmonella spp.	Typhoid, paratyphoid
	Escherichia coli	
Viral	Dengue virus (flavivirus)	Dengue fever
	Yellow fever virus	Yellow fever
	Poliovirus	Poliomyelitis
	Arboviruses	Arboviral diseases
	Hepatitis A virus	Hepatitis A
	Coxsackie virus A	Herpangina, aseptic meningitis
	Coxsackie virus B	Paralysis fever, pleurodynia
	Snow Mt. Agent	Gastroenteritis
	Norwalk like virus	Gastroenteritis
	Norwalk agent (Calcivirus)	Gastroenteritis
	Astrovirus	Gastroenteritis
	Calcivirus	Gastroenteritis
	Rotavirus	Infantile gastroenteritis
	Reovirus	Respiratory disease
	Adenovirus	Gastroenteritis
	Non-A and Non-B Hepatitis	Acute conjunctivitis, diarrhea
	Echovirus	Hepatitis
Protozoa	*Entamoeba histolytica*	Amebiasis
	Balantidium coli	Balantidiasis
	Cryptosporidium parvum	Cryptosporidiosis
	Giardia lamblia	Giardiasis
	Plasmodium spp.	Malaria
	Trypanosomium spp.	Trypanosomiasis
Helminths	Hook worms	
	Tape worms	
	Whip worms	
	Ascari lumbricoides	Ascariasis
	Schistosoma spp.	Schistosomiasis
Spirochaetes	*Leptospira* spp.	Leptospirosis
	Treponema pertenue	Yaws
	Trichophyton concentricum	Ringworm

Source: Adapted from Hutzler and Boyle, (1982) and NRC (1993).

patients with weakened immune systems or cancer patients undergoing chemotherapy). While many of these pathogens may not represent a serious threat of infection to persons with normal immune systems, they do pose a threat to individuals with compromised or suppressed immune systems. Serious health complications, even death, can arise from an infection of these sensitized individuals, e.g., dehydration caused by the loss of bodily fluids due to severe diarrhea and possibly death. Under the Safe Drinking Water Act amendments of 1996, the EPA has been tasked with the responsibility of identifying and evaluating the risk of infection to these subpopulations.

Unfortunately, until the Milwaukee (Fox and Lytle 1996) and Las Vegas (Roefer et. al. 1996) outbreaks of cryptosporidiosis, most major drinking water systems believed and were confident in the microbiological safety of their potable water supplies. However, these outbreaks served to resensitize the general public and the drinking water industry to the threat posed by microbial pathogens while reopening old questions on source water quality, treatment reliability, treatment performance, process operation, process control, and public health.

Epidemiology

Epidemiology is the study of exposure factors and the occurrence of disease in the human populations (Mausner and Kramer 1985). However, since environmental epidemiology attempts to draw a correlation between a risk factor and a health effect, the fact that such studies are based on populations exposed to other environmental factors, makes it impossible to rule out other factors that might contribute to the health effect. Therefore, drawing a direct cause and effect relationship is not possible.

Among the different types of epidemiological studies, ecological studies are the easiest to conduct, but there are other types of studies, cross-sectional studies, case-control studies, and cohort studies from which to choose. The type of study undertaken depends on whether one is looking from the point of disease to a potential cause backwards (retrospective) or are starting at some point collecting data as time moves forward (prospective).

In retrospective studies, information may be limited by the number of people exposed to the causative agent or limited by the ability to define a control group that was not influenced by the causative agent. In prospective studies such information is easier to obtain because the study populations are established at the beginning of the study, but identifying enough people to participate over long periods of time, due to emigration and immigration from the study area remains a problem. A baseline for at least one long-term prospective epidemiological study on the effects of reclaimed wastewater have been established by the City of San Diego, as part of its water

TABLE 15.2   **Morbidity and Mortality Statistics
for Selected Pathogens (NRC 1993).**

Pathogen	Morbidity (%)	Mortality (%)
Poliovirus I	1	0.01
Rotavirus	56	0.01
Hepatitis A virus		
Children	5	not known
Adults	75	0.6

reuse study (Western Consortium for Public Health 1992). However, it may be several years before useful information can be gathered from this research effort and time is the one physical variable that cannot be accelerated or circumvented in these studies. One of the previous criticisms of the Los Angeles County Sanitation Districts (LACSD) Health Effects study (Nellor, Baird, and Smyth 1984) was that the study period did not cover an adequate timeframe. A Science Advisory Panel (State of California 1987), reviewing the LACSD's Health Effects Study, noted that bladder and kidney cancers took twenty years or more to develop and that the study period was not sufficient on which to judge any impacts of wastewater reclamation on groundwater recharge with respect to bladder and kidney cancers.

Although the sensitivity of disease statistics may not be sufficient to establish a cause and effect relationship at low numbers, the morbidity and mortality statistics provide important information on the health status and the incidence of disease and death in a population (Table 15.2). It is recognized that the ability of epidemiological techniques to establish or detect a causal relationship between a pathogen source and disease at a 1:1,000,000 risk level requires an incredibly large sample population. However, information on the prevalence of disease might aid in identifying segments of the population that are more vulnerable to the disease. Acute problems observed by disease surveillance programs can be used to illustrate paradoxes posed by the results of retrospective studies. Since the incidence of waterborne disease is believed to be underreported, due in part to poor analytical tools (the reliance on physical and clinical diagnostics to call for special testing), risk mangers must turn to alternative sources of information, such as pathogen dose-response models to predict rates of infection during exposure to low concentrations.

In 1993 the city of Las Vegas experienced an outbreak of cryptosporidiosis in immunosuppressed populations that was detected by a disease surveillance program in Clark County (Roefer et al. 1996). A subsequent study by epidemiologists from the United States Environmental Protection Agency (USEPA) found a causal relationship between the water supply and those that had contracted the disease (Goldstein et al. 1996). A review of the water treatment plant records and water quality samples collected after the outbreak, produced no physical evidence to link the water supply to the outbreak. To some this investigation raises serious questions about the adequacy of water quality standards, e.g., turbidity and coliforms, because waterborne disease outbreaks can be associated with water supply, without physical evidence of treatment failure. Identified outbreaks, like Las Vegas, raise questions about the strength and adequacy of the multiple treatment barriers used to treat drinking water. To others, the Las Vegas findings raise questions about limitations associated with the investigative techniques.

Studies such as Las Vegas raise legitimate concerns that public health programs and the development of public health policy may not be providing adequate protection of subpopulations, i.e., those sensitized populations that may inadvertently become sentinel populations for the general population. While disease surveillance programs should be looking in more sensitive populations, public health officials are faced with the dilemma about providing special safeguards to subpopulations that are more susceptible to infection by microbial pathogens. Future drinking water regulations will be required to address protection of these populations, but it has yet to be seen how the development of such policies may impact future policy decisions regarding microbial water quality of reclaimed water.

DOSE-RESPONSE

Once the etiologic agent has been identified and human exposure to the pathogen is known to cause infection and disease, the next step is to determine how exposure to different concentrations of the pathogens might elicit a response in humans, i.e., infection or disease. Current pathogen dose-response data is limited. Yet limited dose-response information is often used to extrapolate risk estimates into regions for which no data exists to supply public policymakers with some estimate of the risk of infection due to exposure to reclaimed wastewater. Such estimates may demonstrate that the risk of infection is lower than endemic levels of infection and may provide a rationale means for establishing public policy. Such an approach may raise some serious ethical questions and concerns about approaches to microbial risk assessment that are appropriate for the healthy general population, but may not provide adequate protection to sensitive subpopulations.

Pathogen infectivity can vary due to a range of factors. In part, such variability may be due to an organism's response to changing environmental conditions in order to ensure survival and propagation. Both the host and parasite are living organisms and as such, do not respond to environmental factors in an absolute manner, i.e. the same way or to the same extent in all cases. The infectivity of a pathogen may be preconditioned to specific environmental factors that may change when entering the host. Upon entering the host, for a short period of time, the pathogen must adapt to a new environment (incubation). During this period the pathogen may be undergoing biochemical transformations that will allow the pathogen to utilize the host's resources which can result in infection or colonization of the host (viruses require the host's cellular DNA or RNA for reproduction). While the pathogen is adapting to the new environment, the host's immune system may be attempting to fend off the infection by producing antibodies that will stop the infection. If the pathogen is successful in colonization and reproduction, the host may manifest symptoms of the disease. Unsuccessful attempts to colonize and reproduce may mean the host may not show signs of illness, but may only show evidence of infection through biochemical tests designed to find the antibodies produced to fend off the pathogen.

Infection is often assumed to be synonymous with disease. However, disease or illness usually means that an infection was severe enough to produce symptoms of the disease in a person, e.g., vomiting, diarrhea, fever, etc., but infection can be asymptomatic. Infection only indicates that a person has been exposed to the pathogenic agent, but the body's immune system has fought off the pathogen such that the person does not exhibit any symptoms of illness.

Given this oversimplified version of pathogen-host interactions, it should be somewhat clear that dose-response relationships between the host and the pathogen depend on many factors. Feeding studies, using healthy individual volunteers cited in Haas (1983), have been used to develop a range of infection based dose-response models. Yet the observed responses are not uniform and the dose-response information for a given concentration of microorganisms shows a range of responses. This illustrates the problem of using single point dose-response estimates to predict the risk of infection. Biological populations often exhibit a range of responses (variation) and attempts to characterize the populations as single points ignore the variation in response.

Modeling Dose-Response

Current knowledge about the etiology of disease remains limited. Because of limited analytical or diagnostic tools, microbial risk assessments attempt to cover a range of situations often extrapolating into regions of low pathogen concentration by using models. Such an approach assumes that small pathogen concentrations may produce an adverse response. In this case, a threshold, below which there is no observable response, has not been established. Ideally, a dose-response model should be able to differentiate between infection and illness, because although the terms may be used interchangeably by some, the terms have very distinct meanings. The current models used for microbial risk assessment only predict the risk of infection, not the risk of contracting the disease.

The following models used to determine the probability of infection are outlined in detail in Haas (1983) and used by Haas (1983) and Regli et al. (1993) in their respective microbial risk assessment discussions for swimming or bathing in a wastewater effluent and for pathogens in drinking water. There are three models that Regli et al. (1993) considered for drinking water microbial risk assessment. These are the log-normal, exponential, and beta-Poisson models. Each dose-response model attempts to establish and provide the best correlation between the model and the dose-response data in an effort to better explain human dose response to pathogen exposure by making specific assumptions and hypotheses about the interaction between the pathogen and host. With a sufficiently strong correlation between the pathogen model and the dose-response data, the extrapolation of risk estimates may be possible, at pathogen concentrations below those used in the development of the risk model. Given a specific number of microorganisms, each dose-response model allows the calculation of a risk or probability of infection.

In the *log-normal* model it is assumed that each individual in a population has a minimum infective dose and that this minimum infective dose varies between individuals. It is further assumed that the distribution of the minimum infective dose, among the individuals in the population follows a log-normal distribution. The fraction of the population that are predicted to respond to a given dose is given by the following equations:

$$P^* = \frac{1}{\sqrt{2\pi}} \int_{-\infty}^{Z} \exp\left(-\frac{z^2}{2}\right) dz \qquad (1)$$

$$Z = \frac{\ln(D - \mu)}{\sigma} \qquad (2)$$

where

D = the number of pathogens (dose)
μ = average logarithm
σ = log standard deviation
P^* = probability that a single individual exposed to a single dose, N will become infected

Both μ and σ are developed from the observed dose-response data, e.g., from human feeding studies.

The *exponential* and *beta-Poisson* models are described as single-hit models. In the exponential model, the calculation to determine the probability of infection is the product of two independent events. First, the probability that a single dose will include j organisms and second, the probability that k organisms will survive to reach a site at which infection can be initiated. When the host-pathogen interaction is assumed to be constant, then one pathogen equals one infection. If the distribution of microorganisms in the environment follows a Poisson distribution and the probability of survival to the target organ can be derived from binomial theory, then the probability of infection is given by:

$$P^* = 1 - \exp(-rD) \tag{3}$$

where

D = number of pathogens (dose)
r = fraction of pathogens that survives to produce an infection

Unfortunately, this derivation assumes the probability of survival is constant, r. While this single hit, constant survival model fits the data from in vitro studies, it does not fit all dose-response data.

In reality, human response to a given dose probably follows some sort of distribution as no individual in a population can be singled out as being representative of the whole group. Each person in the population will react to pathogen exposure given the susceptibility of the individual to the disease, i.e., the state of the person's immune system, their immunological condition at the time of exposure, and their genetic makeup. It is more likely that a range or distribution of responses from the general population better accounts for reaction to pathogen exposure.

If r, the fraction of organisms that survives to produce an infection, is replaced by a probability distribution, such as a beta distribution (Furumoto and Mickey 1967) then the following beta-Poisson model results.

$$P^* = 1 - \left(1 + \left(\frac{D}{\beta}\right)\right)^{-\alpha} \tag{4}$$

where

D = number of organisms (dose)
α, β = parameters to fit the dose-response curve

Note the difference in the dose-response curves (Figure 15.2) which compares the exponential dose-response model to the beta-Poisson model for poliovirus III using data taken from Haas. The models give very

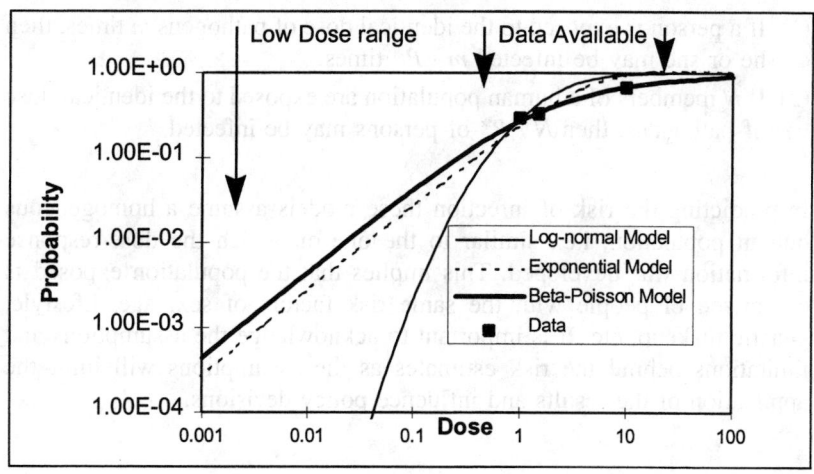

Figure 15.2 Comparison of three dose-response models for poliovirus III (Haas 1983).

different risk estimates for this one pathogen and may in the low dose region of the curve result in markedly different risk estimates. However, in this case, the beta-Poisson model best fits the dose-response data. As one might expect given the different survival and infection mechanisms of the various pathogens, a given model will not fit the dose-response data for every pathogen.

The risk assessor is now faced with several important decisions that will effect the results of the risk models:

(1) Selecting the dose-response model that best fits the dose-response information for a given pathogen

(2) Identifying which pathogen poses the greatest public health threat

(3) Deciding how the probability of infection should be extrapolated into the lower dose response region of the model

This means that before continuing with the microbial risk assessment, one needs to identify the specific pathogen that will produce the greatest risk of infection or the risk assessment must be conducted to cover a range of pathogens covering a range of conditions. If the approach is to cover a range of pathogens, then a confidence bound must also be selected as the basis for setting public policy. These conditions will more clearly define the scenarios under which risk is measured as an attempt is made to evaluate the risk of pathogen exposure in a "real world" setting.

The previous three dose-response models base the risk of infection, P^*, on a single exposure. However, P^* can be interpreted in at least two different ways:

(1) If a person is exposed to the identical dose of pathogens m times, then he or she may be infected $m \cdot P^*$ times.

(2) If N members of a human population are exposed to the identical dose of pathogens, then $N \cdot P^*$ of persons may be infected.

In predicting the risk of infection these models assume a homogeneous human population, i.e., similar to the one on which the dose-response information was developed. This implies that the population exposed is comprised of people with the same risk factors of sex, age, lifestyle, genetic makeup, etc. It is important to acknowledge the assumptions and limitations behind the risk estimates as the assumptions will limit the application of the results and influence policy decisions.

EXPOSURE ASSESSMENT

Up to this point the risk assessor has identified the pathogen(s) of concern and the pathogen dose response. Much more information and many more steps are required to complete the characterization of the risk of infection from using reclaimed wastewater. The exposure assessment includes the following steps that impact the survival, thereby allowing transport of the pathogen through the environment and to the host.

- pathogen concentration in wastewater
- pathogen removal and survival in treatment
- release to the environment
- survival in the environment
- route of human exposure
- survival in human
- survival and exposure to target organ
- exposure frequency
- infection/infectivity
- disease or outcome

The overall probability of infection from exposure to reclaimed wastewater is a function of the frequency of occurrence in wastewater, survival through the treatment process, survival in reclaimed wastewater (i.e., the environment), human exposure, and dose-response. Expressed mathematically, the overall probability of infection, P_i, depends on a series of independent events represented by Equation (5):

$$P_i = P_w \cdot P_t \cdot P_e \cdot P_d \tag{5}$$

where

P_w = probability of pathogen presence in the source water
P_t = probability of pathogen surviving the treatment process train
P_e = probability of pathogen surviving after exposure to the environment
P_d = probability of human infection from a given dose (calculated from
    dose-response models)

There are several steps or barriers that decrease the likelihood of pathogen survival, but exposure assessments are only as good as the understanding of the disease etiology, i.e., how the pathogen comes into contact with the human population and interacts with the host to produce disease. The less that is known about the etiology of the disease, the weaker the model and the less certain one can be about the outcome. Thus, the need for tools like dose-response models and epidemiology to define the route(s) of exposure is clear, but understanding their limitations illustrates the need to know how pathogens can survive environmental exposures to reach and infect human hosts. If each of the exposure steps could be identified, then the estimates of pathogen survival would accurately estimate the overall threat posed by the pathogen.

The form of Equation (5) is very similar to the approach taken by Haas (1983) in examining the risk of contracting a viral disease by exposure to recreational waters. In this paper Haas viewed the probability of contracting disease as a sequence of two independent events. First, that a single exposure will result in the ingestion of organisms and second, that once ingested, the organism will cause disease. In an analogous manner, the risk of infection due to exposure to reclaimed wastewater is the product of several sequential independent events.

For example, using Equation (5) and assuming the probability that a pathogen is present in reclaimed wastewater was 1 (always present), the probability of surviving the treatment process is 0.001, the odds of survival in the environment is 0.01, and the probability of infection is 0.0001, then, the probability of acquiring an infection due to exposure to a pathogen in reclaimed wastewater is small, $1 \cdot 10^{-3} \cdot 10^{-2} \cdot 10^{-4}$ or 10^{-9}. Using these point estimates to illustrate how to estimate the overall probability of infection shows the probability of survival to be slight. However, in reality, none of these point estimates accurately reflects the variability inherent in the risk predicted at each step in this chain of events. For example, one can ask how frequently the treatment processes perform as anticipated, i.e., on average or 95 percent of the time? Do all the point risk estimates have the same probability of occurrence? This leads to other policy questions regarding the frequency with which the probability at which event occurs and the subsequent question of what is the proper level (magnitude) of acceptable risk.

Equation (5) is a simple representation of the steps that might comprise an exposure assessment. Such an equation produces a single-point estimate (deterministic) of the probability of infection, i.e., the equation only calculates one probability value based on a single value for each element of the equation. Most risk assessments are deterministic in nature. They depend on the input of a single number and are less apt to account for temporal and spatial variations in the occurrence data and certainly will not reflect the variability in treatment performance or differing human dose-response. By their very nature, deterministic models mask the uncertainty in their answers by the assumptions described in the prose used in the descriptive prose. Regulatory agencies routinely, as a matter of policy use deterministic models (Kizer 1985; Lam et al. 1994) for the assessment of risk associated with exposure to chemicals. Similar modeling efforts have been applied to pathogen exposure in drinking water (Regli et al. 1991) and in wastewater reclamation (Asano et al. 1992). As most risk assessments look at risk relative to the general population, one must remember that populations are made of individuals and that genetic variations influence an individual's ability to fight off disease.

Up to this point the exposure assessment has only been discussed in general terms. Let us now consider each of the elements of Equation (5), the frequency, occurrence, survival, and source of the pathogen. In addition, let us consider some mechanism for the pathogen to reach the host, whether it be through person-to-person contact, aerosols, or direct contact with reclaimed wastewater. Ideally, the exposure assessment should be able to enumerate the number of pathogens at each step so that the probability of survival to the host organism can be determined in estimating the overall risk of infection.

Environmental Occurrence

The first element of Equation (5) involves reviewing the occurrence data for the pathogens found in wastewater. Several human pathogens present in wastewater have already been identified (Table 15.1). However, data to accurately estimate the relative concentration and frequency of occurrence for the individual pathogens for specific wastewater streams is not available. However, it does appear that assuming viruses or other pathogens are always present in raw wastewater is not a bad assumption.

Unlike chemicals there is nothing to suggest that pathogens are homogeneously distributed in wastewater. In fact, surveys conducted on raw and treated wastewater streams suggest that the distribution of pathogens is patchy and may exhibit temporal and spatial variations as pathogens may be clumped or associated with solids. Studies (Riggs and Spath 1981; Western Consortium for Public Health 1992) have characterized the occur-

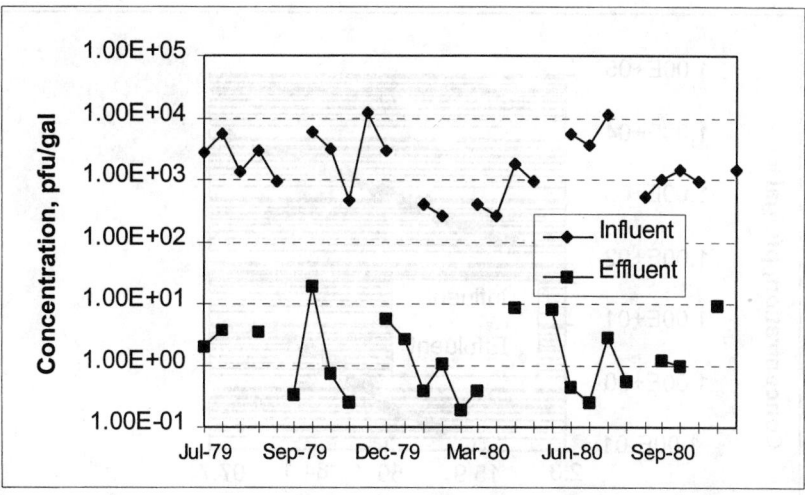

Figure 15.3 Time course of virus concentration in raw and treated wastewater based on data from the Livermore wastewater treatment plant (Riggs and Spath 1981).

rence of pathogens in raw wastewater and in treated wastewater streams. These studies seem to verify the fact that the concentration of viruses can vary in time and space.

The cost and analytical techniques limit the resolution with which one can establish the cumulative frequency distribution of pathogens. The spatial and temporal patterns of virus concentrations can be characterized by semi-log plots as shown in Figure 15.3 or in log-normal frequency distributions as shown in Figure 15.4.[3] While there is some question regarding the homogeneous distribution of microbial pathogens from a wastewater sample, there is reason to be concerned about the spatial and temporal (seasonal) distribution patterns of microbiological pathogens, like viruses.

Virus samples collected at the Livermore Wastewater Treatment Plant from June 1979 to October 1980 showed a seasonal pattern in the raw wastewater (Figure 15.3) (Riggs and Spath 1981). The peak plaque concentrations [using buffalo green monkey (BGM) cell cultures] were observed in the period of September 1979 to January 1980 with a second peak occurring between June 1980 to August 1980.

---

[3] Not all environmental data may be normally distributed, but it may be transformed mathematically to a form that is normally distributed. For example, environmental data may be log transformed so that the distribution of the data appears to follow a Gaussian distribution, especially when occurrence data covers a range spanning several orders of magnitude. Similarly, when discrete events are rare, the standard binomial distribution remains skewed and special techniques must be used to analyze the data.

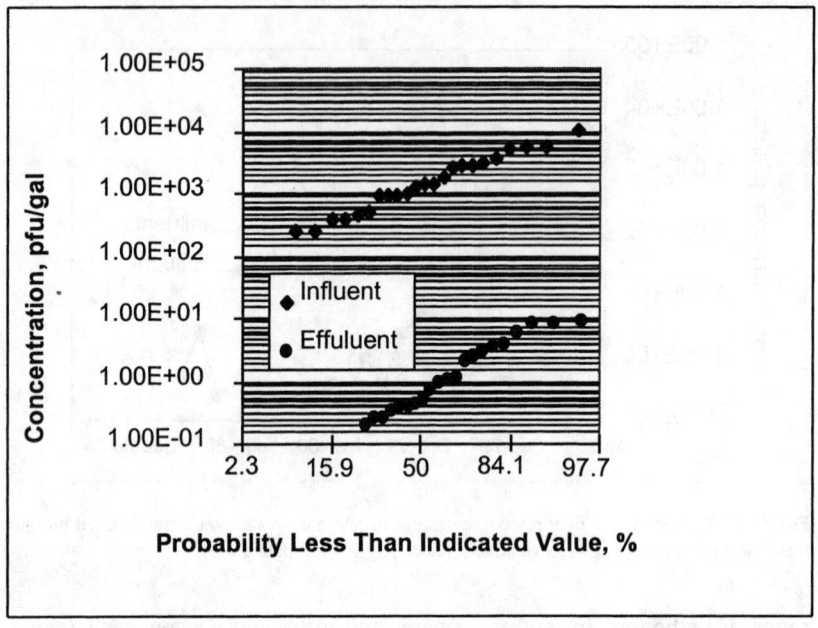

Figure 15.4 Plots of cumulative distribution function of virus concentration in the influent and effluent of the Livermore wastewater treatment plant.

Similar monthly average virus concentrations were observed in the City of San Diego's extensive pathogen monitoring program (Western Consortium for Public Health 1992) using a standard method for enumerating with the BGM cell line (Figure 15.5). Over a 2-1/2 year period virus peaks were observed in January, June, September, and October. The temporal distribution patterns are similar to the results reported by Riggs and Spath (1981) for the Livermore wastewater.

In addition to determining the frequency and occurrence in wastewater effluent streams or survival through treatment, there are the problems of trying to determine how many samples are required to characterize microbial water quality with confidence and the frequency with which these samples must be collected. Many statistical tests used to distinguish trends or significant differences in occurrence are predicated on a normal or Gaussian distribution of the occurrence data. As sample size increases, the distributions tend to follow a more normal distribution (central limiting theorem) thereby allowing the use of standard statistical tests. Alternatively the data can be transformed mathematically to achieve a normal distribution. However, the level of confidence in the data is a function of the number of samples collected, hence the adage, ''the more samples the better.''

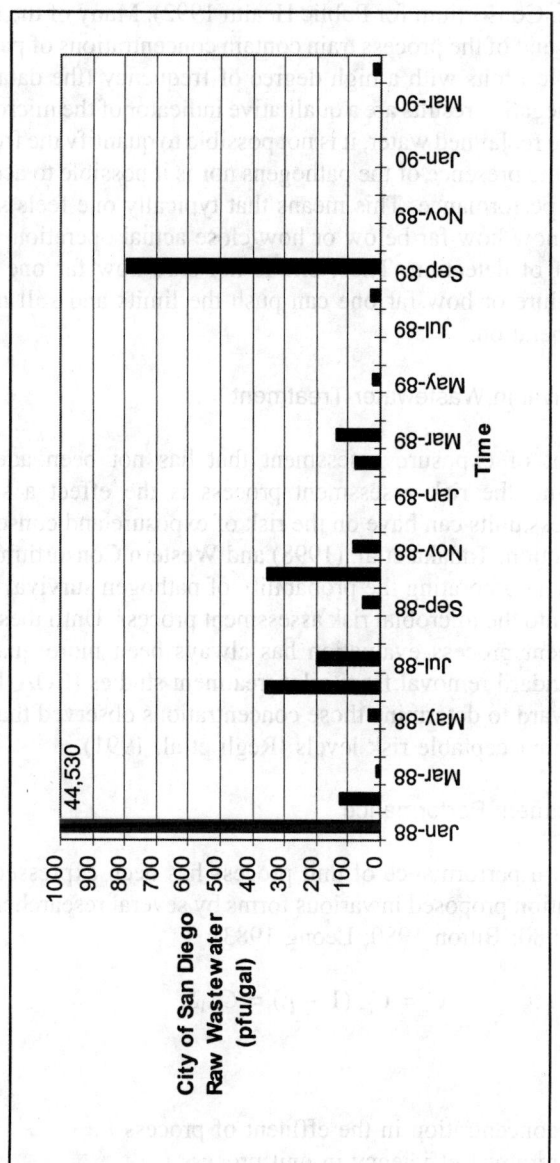

Figure 15.5 Virus occurrence data from the San Diego Study (Western Consortium for Public Health 1992).

One of the complications with trying to enumerate the frequency distribution of viruses is illustrated by work conducted for the City of San Diego (Western Consortium for Public Health 1992). Many of the samples collected at the end of the process train contain concentrations of pathogens below detectable limits with a high degree of frequency (the data is censored). While negative results are a qualitative indicator of the microbiological safety of the reclaimed water, it is not possible to quantify the frequency distribution of the presence of the pathogens nor is it possible to accurately assess process performance. This means that typically one feels safe, but one does not know how far below or how close actual operation is to the analytical limit of detection. Thus, one is not sure how far one is from the edge of failure or how far one can push the limits and still maintain a successful operation.

Microbial Survival in Wastewater Treatment

One element of exposure assessment that has not been adequately incorporated into the risk assessment process is the effect a series of treatment process units can have on the risk of exposure and consequently the risk of infection. Tanaka et al. (1998) and Western Consortium (1992) have looked at incorporating the probability of pathogen survival through process trains into the microbial risk assessment process. Until these recent studies, treatment process evaluation has always been more qualitative, assuming a standard removal from pilot treatment studies (EOA 1995) or working backward to determine those concentrations observed that would correspond to unacceptable risk levels (Regli et al. 1991).

Modeling Treatment Performance

Virus removal performance of unit process has been expressed by the following equation proposed in various forms by several researchers (Haas 1983; Gerba 1980; Bitton 1980; Leong 1983):

$$C_i = C_{i-1}(1 - r_i) = C_{i-1}S_i \qquad (6)$$

where

C_i = pathogen concentration in the effluent of process i
r_i = pathogen removal efficiency in unit process i
S_i = pathogen survival rate in unit process i

Overall virus removal performance of the treatment plant has been expressed by:

$$C_{\text{Re}} = C_0 \prod_i (1 - r_i) = C_0(1 - r_{\text{tot}}) = C_0 \prod_i S_i = C_0 S_{\text{tot}} \qquad (7)$$

where

C_{Re} = pathogen concentration in the effluent of a plant
C_0 = pathogen concentration in the influent of a plant
r_{tot} = overall efficiency of a plant
S_{tot} = overall survival rate

These removal efficiencies represent the log removal efficiency, R_i. This log removal efficiency is defined as the absolute value of the logarithm of a pathogen concentration after treatment divided by the pathogen concentration before treatment.

$$R_i = \left| \log(1 - r_i) \right| = \left| \log(S_i) \right| \qquad (8)$$

Some typical values of log removal efficiency are summarized in Table 15.3. It should be noted that the log removal efficiency fluctuates even at the same treatment plant. Figure 15.6 shows the cumulative distribution of the log removal efficiency calculated with concentration data of influent and effluent from the Livermore wastewater treatment plant (Figure 15.3).

Pilot studies have shown that plant removal efficiencies can be similar, therefore, the assumption of a constant removal efficiency for virus inactivation has been used. Treatment at the Livermore plant also appeared to be effective in reducing the peak virus concentration as the peaks were attenuated by three to four orders of magnitude in the filtered effluent. This occurred before the addition (Figure 15.3) of a disinfectant, but no sampling on the chlorinated effluent was conducted at either plant. The log removals[4] for the Livermore wastewater treatment plant (Figure 15.6) show a greater than 4-log removal of viruses for this plant, but the virus concentrations were still apparent in the effluent prior to chlorination (Figure 15.3). These results are consistent with the reported treatment efficiencies (two to three log removal) listed in Table 15.3 for secondary treatment.

In the absence of plant performance data, seeding studies are conducted on pilot- or full-scale wastewater treatment plants. Although seeding studies assume that viruses behave like fluid particles and that tracer studies of viruses are sufficient to characterize unit process performance, there is nothing to suggest that pathogens behave like water or tracer molecules as they pass through unit treatment processes. Yet the adsorptive interaction

---

[4] Engineers frequently view process performance on the basis of removal although a log removal concept, which is related to pathogen survival, can be used.

TABLE 15.3 Virus Removal Efficiency by Treatment Unit Process.

Treatment	Percent Removal
Primary	0–90%[a]
	0–65%[b]
	6.6%[c]
Secondary	
Trickling filter	0–90%[a]
	0–90%[b]
	54%[c]
Activated sludge	90–99%[a]
	0–99.4%[b]
	94%[c]
Stabilization ponds	99–99.99%[a]
	0–96%[b]
Tertiary	
Excess lime precipitation	90–99.99%[b]
	98.8%[c]
Alum precipitation (after lime)	90%[b]
Activated carbon adsorption	0–50%[b]
	90%[c]
Alum coagulation	95%[c]
Ferric coagulation	99.5%[c]
Coagulation (Fe + Al)	40–99%[b]
Coagulation (polyelectrolytes)	90–99.99%
Sand filtration	10–50%[b]
	73%[c]
Reverse osmosis	≥99%[c]
Disinfection	
Free chlorine	99.9%[c]
Ozone	99.9%[c]
Combined chlorine	90%[c]

[a] Shuval et al. (1986).
[b] Gerba (1980).
[c] Leong (1983).

between the pathogens and surfaces (particulates, filter media, or inert parts of the treatment process) in the unit processes have not been adequately characterized. While the loss of pathogens during seeding studies may be attributed to removal, such losses may also be attributed to problems associated with the analytical techniques, i.e., poor recoveries and poor precision. While these methods may be the best available at the time, they are deficiencies that must be understood and the associated error must be incorporated into the risk assessment as an aid to risk managers setting public policy.

In addition, it must be understood that conducting bench- or pilot-scale studies on a limited number of pathogens may not provide a sufficient

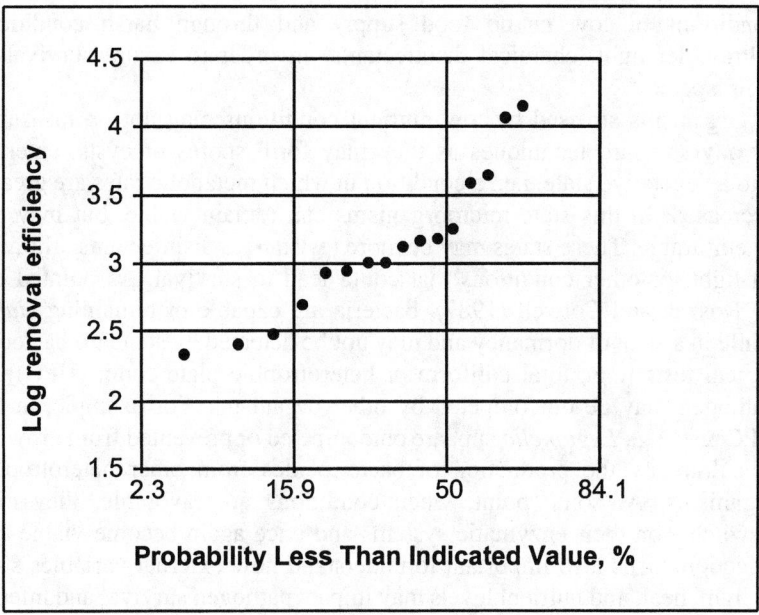

Figure 15.6 Plots of cumulative distribution functions of log removal rates based on data from the Livermore wastewater treatment plant.

range of organisms with which to model survival in a real wastewater treatment plant with nonseeded organisms. Due to their different genetic makeup or due to survival strategies (organism ecology) pathogen response to treatment can vary. For example, *Legionella* spp. is more resistant to chlorine than *E. coli* and in the environment *Legionella* spp. are known to be engulfed by amoebae which can shield or protect the organisms from disinfectants while providing nutrition for growth and reproduction. In addition to this form of facilitated transport, normally free swimming individual organisms can be hidden or cloistered in biofilms or particulates and shielded from disinfectants and environmental factors (e.g., desiccation).

PATHOGEN SURVIVAL IN THE ENVIRONMENT

The subject of microbial ecology covers material that is germane to understanding why wastewater treatment processes can fail and how pathogens can survive under harsh environmental conditions. It is important to recognize that microorganisms are life forms that respond to changing environmental conditions in order to survive and propagate. Pathogens must be able to survive in the environment under adverse conditions of low or no food supply and through harsh conditions (ultraviolet light, chemical disinfectants) in order to ensure survival of their species.

Organisms stressed by low nutrient conditions may not be measured by any standard techniques as they may form spores or cysts, entering into a vegetative state, i.e., a condition in which metabolic rates are greatly decreased. In this state microorganisms can remain viable, but may not be culturable. These states may be more resistant to disinfectants, ultraviolet light, or other conditions that could lead to survival. As pointed out by Roszak and Colwell (1987), bacteria are capable of remaining *viable* while in a state of dormancy and may not be detected by standard bacteriological tests [e.g., total coliform or heterotrophic plate count (HPC)]. A pathogen may be outcompeted by other organisms. For example, in the HPC test, e.g., *Legionella* spp. are outcompeted or prevented from growing in culture by the production of bacteriocides from other heterotrophic organisms. At some point, when conditions are favorable, they may "switch" on their enzymatic systems and once again become viable and infectious. It is also important to understand how external variables such as light, heat, and nutrient levels may impact pathogen survival and infectivity. With so many variables of concern, engineers turn to empirical modeling in an attempt to predict how well a pathogen may survive in the environment.

The use of models to simulate pathogen survival in the environment is not new. Pathogen die-off models were used for locating deep water outfalls in order to protect bathing and recreational areas. Predicting the survivability of microorganisms in wastewater was one means of determining the hazard associated with discharges located near beaches. In these models, *E. coli* or coliforms were used as indicators of pollution attributed to warm blooded mammals and first-order "die-off" equations were used to predict the probability that these organisms could survive to reach the beach. Similar approaches have been taken to model the survival of pathogens in wastewater reclamation practices.

Environmental data from Engineering Science (1987) was used by Asano et al. (1992) as a means of predicting the inactivation of pathogens in reclaimed wastewater. Although a pathogen has passed through the treatment process and has not been removed or inactivated, the integrity of the pathogen could be compromised, i.e., the pathogen could be severely wounded, making it more vulnerable to environmental factors that would lead to its demise. Environmental factors such as dessication or UV may have a significant impact on pathogen survival. While conveyance facilities may provide additional opportunities for pathogen regrowth, proper use-area controls may serve to mitigate the problem. Many studies have been conducted to determine the impact soils have on virus transport and survival in the environment. Similarly, studies on pathogen survival in agricultural applications have also been conducted (Engineering Science 1987).

Agricultural and Soil Studies

Physical factors in the environment reduce the likelihood of pathogen transport by introducing barriers that reduce the potential for the pathogen to reach its target by inactivating or removing the pathogen. Environmental factors also provide additional control points that can be used to reduce exposure to pathogens. Research done on agricultural applications, spray irrigation of food crops, flood irrigation of food and fodder crops, spray irrigation of golf courses, and groundwater recharge, all show the importance of use-area controls in providing additional barrier(s) to pathogens. By restricting access to fields irrigated with reclaimed wastewater for specific periods of time or by requiring residence times underground, some recognition and credit is given to environmental factors that play a role in pathogen die-off, decay, or removal.

Long before a person can be exposed to reclaimed wastewater there are other factors that can act as barriers to pathogen survival. For example, environmental factors such as sunlight (UV), moisture, or desiccation can affect the survival of viruses after being sprayed onto crops, especially those consumed raw. Haas (1983) and Asano et al. (1992) used the following first-order reaction formula for expressing loss of virus viability in the environment.

$$C(t) = C(0) \exp(-\lambda t) \qquad (9)$$

where

$C(0)$ = initial virus concentration
$C(t)$ = virus concentration at time, t
λ = first-order decay factor
t = time

Studies conducted by Engineering Science (1987) showed that poliovirus can survive on vegetables under the controlled conditions of an environmental chamber varied for the different crops. Figure 15.7 illustrates the varying survival rates of poliovirus on different vegetables. The difference in the curves was attributed to humidity and leaf structure (shape and packing).

Soil studies done by Engineering Science (1987) also show inactivation after application to soils in environmental chambers and in the field to be

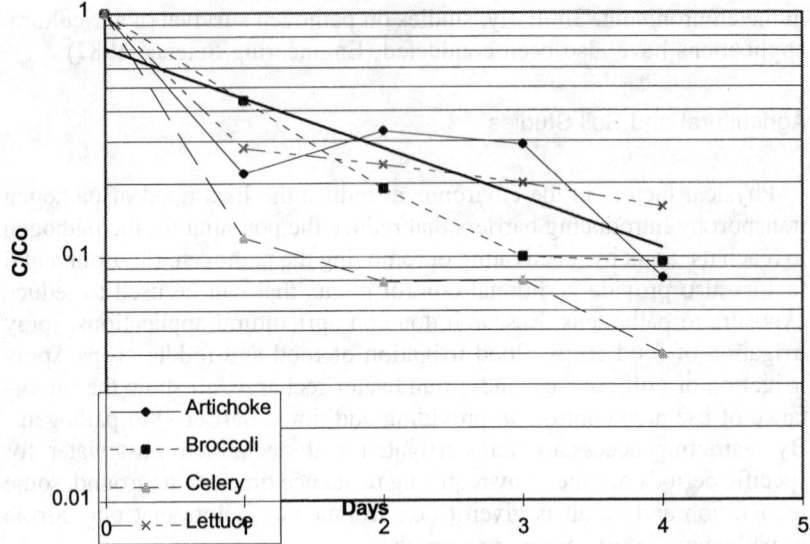

Figure 15.7 Virus inactivation on agricultural crops (Engineering Science 1987).

similar. T_{99}-values[5] for soil studies conducted in the environmental chamber were 5.4 days compared to T_{99}-values from the field studies which ranged from 4.8 to 5.2 days. Though these studies were conducted in an environmental chamber, several soil column studies reviewed and cited by Asano et al. (1992) show that soils can reduce viruses.

So far, treatment and environmental factors have been identified as barriers that can remove or inactivate pathogens in the reclaimed wastewater. If, after having survived each of these steps the pathogen remains alive and infectious, then the exposure assessment uses the dose-response models to predict the probability of infection should the pathogen reach the host organism.

EXPOSURE SCENARIOS

Until now only single elements of the exposure assessment have been identified. The next task is to assemble these components into a scenario that can be used to describe the fate of a pathogen in wastewater from the source to the point of ingestion. Recognizing that risk estimates are comprised of a series of events, as illustrated by Equation (5), a typical set of four exposure scenarios (consisting of exposure and infection) for

---

[5] The time it takes to reduce the initial concentration of pathogens to 1% of the original concentration.

reclaimed wastewater have been assembled and summarized in Table 15.4. The four scenarios identify the reclaimed wastewater use, the group at risk, the frequency at which the exposure will occur, the quantity of reclaimed wastewater ingested, and additional environmental factors that might attenuate the effects of exposure. Using Equation (5) as a template, the risk estimates summarized in Table 15.5 for selected pathogens were generated. As might be anticipated from the previous discussion on dose-response models, there are variations between the risk estimates for the different organisms.

Although the risk estimate may vary by an order of magnitude, a risk manager must decide if the differences are significant enough to warrant a separate risk assessment for each pathogen or if a specific pathogen could be selected as a limiting case. Without some estimate of uncertainty on the risk estimate, the task of determining if the differences in the risk estimates between the pathogens is significant remains a difficult qualitative task. If the risk is deemed to be significant, the next step, i.e., risk characterization is ready to be taken.

RISK CHARACTERIZATION

Characterization of risk is the final step in risk assessment. The goal of risk characterization is to provide an understanding of the type and magnitude of adverse effect that a particular pathogen could have under particular circumstances (NRC 1994). The results of this characterization are then communicated to the risk manager with an overall assessment of the risk of infection.

The elements of risk characterization at a minimum should include:

(1) Generation of a quantitative estimate of risk
(2) Description of uncertainty
(3) Presentation of the risk estimate
(4) Communication of the results of the risk analysis to the risk manager

The risk of infection values presented in Table 15.5, from the exposure scenarios described previously, represent a quantitative estimate of the risk of infection. These are singular values (deterministic models) with no estimate of the uncertainty associated with each of the estimates. Risk managers should understand that these elements are capable of producing values that represent an absolute risk only when the models can predict with absolute certainty the fate of the pathogen in each step of the risk assessment, otherwise the risk assessment models only provide a relative measure of risk. Although not absolute, a measure of relative risk may

TABLE 15.4. Exposure Scenarios for Microbial Risk Assessment for Reclaimed Municipal Wastewater (Asano et.al. 1992).

Reclaimed Wastewater Application	Risk Group	Exposure Frequency	Quantity of Reclaimed Wastewater Ingested in a Single Exposure	Environmental Factors Mitigating Exposure
Scenario I Golf course irrigation	Golfer	Twice per week	1 mL	24 hours of drying prior to allowing players on the course
Scenario II Crop irrigation	Consumer	Every day	10 mL	Stop spray irrigation two weeks before harvest
Scenario III Recreational impoundment	Swimmer	40 days per year	100 mL	No reduction
Scenario IV Groundwater recharge	Consumer	Every day	2000 mL[a]	Assumes the presence of a 10 ft. vadose zone and 6 months of retention in the aquifer

[a] This water is assumed to be diluted by 50% so the actual volume of reclaimed wastewater consumed is 1000 mL.

734

TABLE 15.5. Viral Infection Risk for Chlorinated Tertiary Effluent Containing 111 Viral Units per 100 L (Asano et. al. 1992).

	Landscape Irrigation for Golf Course	Spray Irrigation for Food Crops	Unrestricted Recreational Impoundments	Groundwater Recharge
Lifetime				
Echovirus 12	3.0E-02	3.1E-04	9.5E-01	4.1E-06
Poliovirus 1	1.0E-03	1.0E-05	1.0E-01	3.8E-07
Poliovirus 3	5.3E-01	7.9E-03	1.0	1.60E-06
Annual				
Echovirus 12	1.0E-03	4.5E-06	7.4E-02	5.9E-08
Poliovirus 1	3.5E-05	1.5E-07	2.6E-03	5.4E-09
Poliovirus 3	2.5E-02	1.1E-04	8.4E-01	2.4E-08

be suitable for the purpose of risk management. The relative risk estimates are even more valuable if the uncertainty associated with the risk estimate is known. Including uncertainty measurements as part of the risk assessment will aid risk managers in determining if the difference in risk estimates, which can differ by order of magnitude, are of real significance.

Dealing with Uncertainty in a World of Absolutes

Although a single risk value may result from the exposure assessment, the exposure assessment contains elements of uncertainty. As there are many steps in the exposure assessment, each with its own associated errors, there is a concern with the accumulation of error.

Uncertainty can be defined as a lack of precise knowledge as to what the truth, whether qualitative or quantitative, really is. There is uncertainty in the estimated risk number because of uncertainties in the data and models used at each step in the risk assessment process. Bogen (1990) provided a summary detailing some sources of uncertainty. The NRC (1983) recognized that the primary difficulty in decision-making based on risk assessments is the pervasive use of assumptions.

NRC (1994) suggested two types of uncertainty in risk assessment: parameter uncertainty and model uncertainty. The following are summaries of the descriptions on model and parameter uncertainty from the NRC report:

(1) *Parameter uncertainty:* Estimated values of parameters have uncertainties stemming from the following sources:

- measurement errors
- random errors in analytic devices
- systematic biases using generic or surrogate data instead of analyzing the desired parameter directly
- random sampling error
- non-representativeness

(2) *Model uncertainty:* These uncertainties arise because of unknowns in the scientific theory that require the use of reasonable assumptions.

- errors in model structures
- errors introduced by oversimplified representations of reality
- usage of surrogate variables for ones that cannot be measured
- failing to account for correlations that cause seemingly unrelated events and
- the extent of aggregation used in the model

Propagation of Uncertainty

In exposure assessment, the final estimate of the risk of infection is based on the dose of pathogens received in one exposure event. To reach this risk estimate, one must pass through several steps. At each step, the uncertainty in the estimated value changes as additional error from each measurement accumulates and is carried from one estimate to another.

Deterministic and Stochastic Analysis of Exposure Assessment

There are two mathematical procedures which can be used to propagate uncertainty in exposure assessment: one is a procedure using the single "best guess" value and its standard deviation as a measure of uncertainty, the other is a way of estimating probability density function by Monte Carlo simulation. The former is commonly referred to as a deterministic analysis, whereas the latter is known as a stochastic analysis. A comparison of the generalized deterministic and stochastic schemes is depicted in Figure 15.8.

Depending on how the steps of the exposure assessment are linked or how the probability of infection is calculated, the errors accumulate in different ways. For example, when the probability of infection is the product of two numbers, the total error in the final number is the square root of the sum of the squares of the errors associated with each number.

As an example, let us estimate the virus concentration, C_{Re}, in the effluent of water treatment plant using the treatment model:

$$C_{Re} = C_0 \cdot S_{tot} = f(C_0, S_{tot}) \tag{10}$$

At this juncture in the analysis, we can see that Equation (10) has two inputs, C_0 and S_{tot}, which are variables (i.e., are not constant) and one output, C_{Re}. Analyzing the propagation of uncertainty in the estimation step, we should consider two issues: the uncertainty associated with each input and its contribution to the uncertainty in the output. In this example, σ_{C_0} and $\sigma_{S_{tot}}$ are measures of uncertainty associated with the respective variables (inputs), leaving the uncertainty of the output $(\sigma_{C_{re}})$ to be estimated.

One approach to the uncertainty analysis is generally known as a first-order approximation or Gaussian approximation (Morgan and Henrion 1990). It is common to express uncertainty by the standard deviation, σ. The standard deviation provides a range of values that the true population mean is thought to lie between. Expressed as a confidence interval in a deterministic model, a multiple of the standard deviation provides a certain

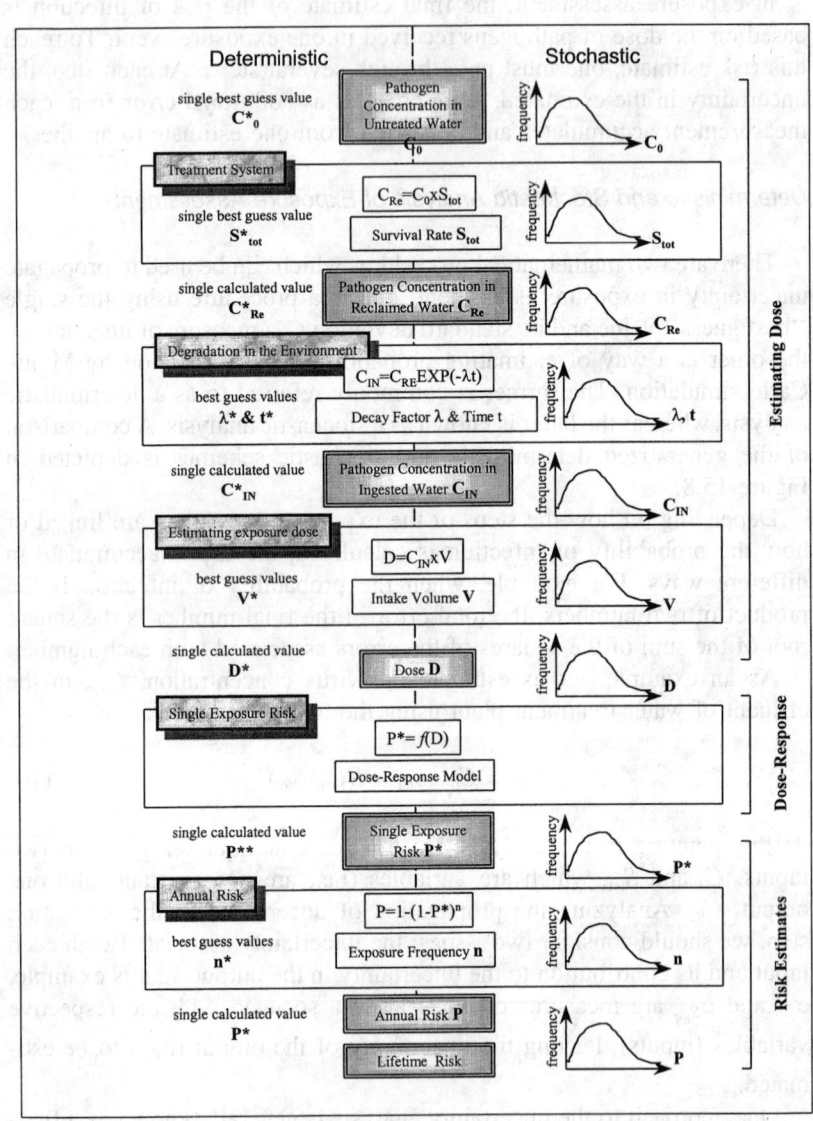

Figure 15.8 Deterministic and stochastic analysis in microbial risk assessment.

degree, say a 95% confidence level (~2r) that the true population mean lies within the confidence interval.

A method to determine how much the uncertainty in the input variables contributes to the uncertainty in the output is a measure of uncertainty importance. The simplest measure of which is termed "sensitivity" (Morgan and Henrion 1990). Sensitivity is defined as the output rate of change taken with respect to the input change. In this case, two partial derivatives of output C_{Re} with respect to inputs, C_0 and S_{tot} correspond to the sensitivities:

$$\left[\frac{\partial f}{\partial C_0}\right]_{(C_0^*, S_{tot}^*)}, \qquad \left[\frac{\partial f}{\partial S_{tot}}\right]_{(C_0^*, S_{tot}^*)} \tag{11}$$

where C_0^* and S_{tot}^* = initial "best guess" values for the inputs.

Since the variance of output, $\sigma^2_{C_{Re}}$ is estimated as the sum of the squares of the contributions from each input, then $\sigma^2_{C_{Re}}$ is given by the Gaussian approximation as:

$$\sigma^2_{C_{Re}} \approx \left(\left[\frac{\partial f}{\partial C_0}\right]^2_{(C_0^*, S_{tot}^*)} \cdot \sigma^2_{C_0}\right) + \left(\left[\frac{\partial f}{\partial S_{tot}}\right]^2_{(C_0^*, S_{tot}^*)} \cdot \sigma^2_{S_{tot}}\right) \tag{12}$$

Since the overall uncertainty is a function of the sum of the squares of the individual uncertainties (the probability of infection is the product of a series of events), the uncertainty in the measured risk will never be less than the least certain element of the risk assessment. Another way of stating this is the risk assessment will only be as strong as its weakest link.

Although there may be a temptation to use multiparameter models or models within a risk assessment to cover unknowns and uncertainties, risk assessments containing models developed using empirical relationships (when an understanding of how processes interrelate is not well known) will contain elements of uncertainty associated with the model predictions. While the use of multiparameter models may be appropriate, the limitations of such an approach need to be kept in mind. As noted in multiple parameter epidemiological studies, the greater the number of parameters used or examined, the greater the likelihood of observing a false positive association (Feinstein et al. 1981).

Whenever first principles can be used in model development, they should be. As in the development of any model, the less empiricism the greater the degree of applicability to a wider variety of situations with less uncertainty associated with the outcome.

A fallback position is to use a deterministic approach to estimate pathogen removal. Approaches like those of Regli et al. (1991) attempt to deal with the errors by establishing a confidence interval around a dose-response estimate by using maximum likelihood estimates (MLE) in order to determine the upper and lower concentration that corresponds to a given probability of infection. Unfortunately, the analysis may then fall back to a conservative approach of using a singular value at the end of a confidence interval when dealing with uncertainty or an unknown. These uncertainties are understood by those doing microbial risk assessments and the impacts of these statistical uncertainties is not lost on these researchers because they state that the equations used in their modeling exercises should be replaced by distributions whenever such information is available. However, the same message regarding uncertainty needs to be accurately delivered to the risk manager in all subsequent risk communications.

For a risk manager, the less empiricism the better. One has less confidence in empirically based risk estimates when the models used are asked to extrapolate or predict risks in situations for which no data or information exists. Strictly speaking, the application of models, especially empirical ones, should be limited to the conditions under which the model was developed. As the understanding of the mechanisms of disease, the ecology of the pathogens, and the effectiveness of treatment processes improves, the degree of empiricism needed will decrease and the accuracy and precision of the microbial risk assessment will improve.

Deterministic Models: In the deterministic analysis, such as the approach used by Regli et al. (1991), the "best" value is estimated from observed data in each step of exposure assessment process as shown in the left half of Figure 15.8. Single nominal inputs (C_0', S') that can be the mean, median or a value that covers a range of potential outcomes (e.g., upper or lower bound set for example by the 95th confidence level from the distribution) are selected.

If we can assume the distribution of data (either C_0' or S') is log-normal, a simple method of estimating the mean, μ, and variance, σ^2, is given by the following set of equations (Gilbert 1987):

$$\mu^* = \exp\left(\bar{y} + \frac{s_y^2}{2}\right) \tag{13}$$

and

$$\sigma^{*2} = \mu^{*2}[\exp(s_y^2) - 1] \tag{14}$$

where

$$\bar{y} = \frac{1}{n} \sum_{i=1}^{n} \ln x_i \tag{15}$$

$$s_y^2 = \frac{1}{n} \sum_{i=1}^{n} (\ln x_i - \bar{y}) \tag{16}$$

The upper one-sided $100(1 - \alpha)\%$ and the lower one-sided $100\alpha\%$ confidence limits for the mean are obtained by the following formulas (Land 1971, 1975):

$$UL_{1-\alpha} = \exp\left(\bar{y} + 0.5s_y^2 + \frac{s_y H_{1-\alpha}}{\sqrt{n-1}}\right) \tag{17}$$

$$LL_{\alpha} = \exp\left(\bar{y} + 0.5s_y^2 + \frac{s_y H_{\alpha}}{\sqrt{n-1}}\right) \tag{18}$$

where $H_{1-\alpha}$ and H_{α} are obtained from the table provided by Land (1975).

However, often the uncertainty associated with the development of pathogen occurrence data make the steps outlined in Equations (13)–(18) more complicated. Pathogen occurrence data that might be used to characterize, C_0', can be limited by the interpretation of the analytical results. When viruses are below detectable limits this should not be taken to mean that the reclaimed wastewater is free of viruses or other pathogens. Pathogens may be present at concentrations below which they can be detected by present analytical methods or not detected because the pathogen could not be cultured or grown on the specific media selected for the test.

The physical limitations of the analytical methodology, especially for viruses and protozoan pathogens, result in poor, low, or highly variable recoveries due to the accumulation of losses through multiple handling steps. Antibody staining techniques such as those used for *Giardia* and *Cryptosporidium* are not species specific and will result in the detection of nonpathogenic organisms. The antibody staining procedures for these protozoans and some bacteria are not capable of differentiating viable from nonviable organisms (nor are the assays capable of enumerating infectious organisms).

Historically, virus assays have shown poor recovery due to clumping of the viruses or other losses due to adsorption onto surfaces during the many steps of the assay. Although the tissue assays count "plaques" that are supposed to be the result of a single viron, there is no guarantee that the plaque truly is the result of a single viron. In addition, the poor recoveries from spiking studies and high variability between laboratories continue to hinder the interpretation of results.

In addition to poor precision and inaccurate results, a high degree of variability (that may be caused by a nonhomogeneous distribution of the pathogens) between the laboratory recovery rates has been noted. The City of San Diego study (Western Consortium for Public Health 1992) showed a range of recovery from 4 to 179%. Recoveries greater than 100% were attributed to the breakup of virus "clumps." Poor recovery of spikes, typically about 10%, that can infrequently exceed 100% recovery, are not unusual for this method.[6] Since the number of samples required to reach a statistically significant number of samples is directly proportional to the variance, the number of samples required to reach a statistical confidence in the data will be high when the variance is high.

In some situations, virus concentration may be very near zero, and the true value may be less than the method limit of detection. Data sets containing these types of data are "censored on the left" because data within the data set are below the limit of detection.

To handle data sets containing data censored on the left, the procedure proposed by Cohen (1959; 1961) can be used to calculate the maximum likelihood estimators of the mean and variance from censored data sets. Cohen's procedure is as follows:

(1) Step 1: computation of measurements below the limit of detection

$$h = \frac{n - k}{n} \tag{19}$$

where

n = total number of data
k = number out of n that are above the detection limit

(2) Step 2: computation of the sample mean and variance from data above the detection limit.

---

[6] Similar problems exist with the other microbiological assays, such as those for *Giardia* and *Cryptosporidium*. There has been much written about the inaccuracy and imprecision of these tests caused by the analytical methods which contain a large measure of uncertainty due to poor recoveries, heterogeneous distribution of pathogens, seasonal distribution, or other inherent errors.

$$\bar{y}_u = \frac{1}{k} \sum_{i=1}^{k} \ln x_i \tag{20}$$

and

$$s_u^2 = \frac{1}{k} \sum_{i=1}^{k} (\ln x_i - \bar{y}_u)^2 \tag{21}$$

(3) Step 3: estimate the mean and variance of the log-transformed data:

$$\bar{y} = \bar{y}_u - \lambda^*(\bar{y}_u - y_0) \tag{22}$$

$$s_y^2 = s_u^2 + \lambda^*(\bar{y}_u - y_0)^2 \tag{23}$$

where

$y_0 = \ln$ (limit concentration of detected)
$\lambda^* =$ parameter [values in Cohen (1961)]

(4) Step 4: estimate the maximum likelihood the mean and the variance of the log-normal distribution by Equations (17) and (18).

Redundancy versus Reliability

Using the previous procedures to estimate the distribution of process performance provides some insight as to why multiple barriers are required to ensure pathogen control. The quality of water produced by a treatment unit will vary as a function of time, source water quality, and operating conditions, i.e., the quality of process output is not constant. Due to this variability, treatment process unit performance can be characterized by a distribution. In order to ensure adequate performance to protect public health, it is important that overall process performance be robust enough to compensate for minor excursions in unit process performance.

In part, because the source of variance in process performance can be operational or analytical, a series of unit treatment processes has always provided multiple independent barriers to ensure the production of reclaimed water of consistent quality. On the other hand, process reliability should not be confused with redundancy, i.e., the presence of "superfluous" equipment that does not increase the log removal capacity of the unit process. Redundant equipment, such as an extra sedimentation basin or filter, allows treatment units to be taken "off-line" for routine maintenance without affecting the quality and quantity of recycled water.

Whereas redundancy may require "additional" treatment units, a series of independent unit processes improves the reliability of the treatment process train as the overall odds of failure (defined as achieving a specific log removal) for the treatment train diminishes with the addition of each independent treatment step. California's wastewater reclamation regulations specify a series of independent unit processes to ensure the reliable production of reclaimed wastewater that is "essentially pathogen free."

According to the American Society of Civil Engineers and the American Water Works Association (1990) reliability refers to the inherent dependability of a piece of equipment, a unit process, or the overall treatment process in meeting the design objective. Reliability is a means of characterizing routine process performance by determining the frequency with which a water quality standard, i.e., coliform or turbidity, is met. For example, the unit processes listed in Table 15.3 are associated with a range of virus removal/inactivation efficiencies. What is missing is measure of the frequency with which these unit processes meet their removal efficiencies so that the probability distribution of each unit process with respect to pathogen removal is known, e.g., what percent of the time can filtration meet a 2-log virus removal?

Redundancy can be viewed as a subset of reliability because initial design allows for future expansion by having duplicate process trains or additional unit processes to handle increased flow capacity. Until flow through the plant matches or exceeds design capacity, some degree of redundancy is built into the treatment process train. At some point in the future, as flows increase, the redundancy in the treatment units will disappear and the process will only be left with an inherent degree of reliability. If multiple units are designed to handle peak or high flow periods, then the units are not truly redundant, i.e., present all the time as a backup in case another unit fails. By extending the microbial risk assessment to cover the reliability of the treatment processes allows one to demonstrate the level of protection provided by the series of independent treatment processes.

Limited plant studies characterizing unit process performance have been conducted. In California both pilot- and full-scale studies have been conducted to determine the fate and survival of pathogens in the wastewater treatment process (Engineering Science 1987; Western Consortium for Public Health 1992; EOA 1995). Most of these studies have been reported in chapters throughout this book. Understanding, evaluating, and characterizing process performance based on the frequency of occurrence, i.e., how frequently a unit process can attain a specific degree of process performance will lead to risk assessments based on stochastic models. Stochastic models improve the ability of risk assessors to illustrate to risk managers, the uncertainties associated with the risk assessment in quantitative terms.

There are two key questions with regard to public policy development that needs to be asked at this juncture. These questions reflect the way water quality standards are set, i.e., by maximum contaminant level (MCL) or by specifying a minimum degree of treatment. The questions that concern us at this point are:

(1) What is the goal of the treatment process, i.e., what degree of pathogen removal or inactivation are we trying to achieve?
(2) What is the overall log removal capability of the process?

For example, the goal of trying to achieve a 10-log virus removal/inactivation 99.9999 percent of the time differs from simply trying to achieve a 10-log virus removal/inactivation by summing the log removals of the unit processes. Deciding on the approach to policy development remains outside the scope of this chapter, but the question requires careful consideration of the risk assessment results and a critical review of the risk assessment protocols by the risk manager making the final decision.

The approach to the first question might consist of setting the log removal requirement and then, by examining the individual components of the treatment process train, determine if the individual units can meet the log removal goals. If not, then it would be appropriate to place unit processes in series with the objective of meeting the log removal requirement. If any single unit process can meet the log removal requirement, then placing two unit processes in series can increase the percentage of time the overall process can meet the log removal requirement (reliability).

The danger of summing log removal credits to individual processes to determine the overall log removal credit is that the log removal credits for the individual unit processes may not all be achieved with the same frequency. In fact overall, the probability of meeting the overall or sum of the log removals will be no greater than the least reliable unit process.

For example,[7] given a two-step process in which each unit process can succeed (s_i) or fail (f_i), the only four possible outcomes are s_1s_2, s_1f_2, f_1s_2, and f_1f_2. The probability of any one of the four outcomes is the product of the probability of success or failure of unit 1 and the success or failure of unit 2. Adding log removals to determine the overall log removal of a series of unit processes means that one is assuming only one outcome (s_1s_2) will be successful, as all other outcomes will fail to meet the overall log removal requirement. Alternatively, if the goal is to meet a specific log removal, then three possible outcomes will fit the criteria, s_1s_2, s_1f_2, and f_1s_2 and the probability of success will be the sum of the probabilities

---

[7] The foundation for the following discussion can be found in Nash (1993).

of the individual outcomes. Only the outcome f_1f_2 fails to meet the minimum log removal requirement. As multiple outcomes are being evaluated in response to the second question, a stochastic approach to risk assessment becomes favored.

It is becoming common practice to sum log removals across unit treatment processes. Such an approach is not incorrect, however, it does not account for the variation in log removals for the individual unit processes. For example, a barrier which meets a 2-log removal of viruses 50% of the time (on average), will only meet the 2-log or better removal half of the time. This means that during the balance of the time, a lower log removal will be achieved. In order to ensure an adequate level of removal when the process is removing less than 2-logs, another barrier is put in place. In this example, assume the second barrier also achieves a 2-log removal of viruses 50% of the time. Together this series of two unit processes can achieve a 4-log removal of viruses about 25% of the time (s_1s_2). However, achieving a 2-log removal or greater, by the series of processes, may be accomplished more than 75% $(s_1s_2 + s_1f_2 + f_1s_2)$ of the time, as 2-log removal can be met in any of the three outcomes $(s_1s_2, s_1f_2,$ and $f_1s_2)$ because each unit acts as an independent barrier. Instead of a single unit process meeting 2-log removal 50% of the time, the two processes should now meet 2-log removal 75% of the time. Having two unit processes each of which is capable of achieving a 2-log removal, increases the reliability of the overall process, although the probability of achieving a total 4-log removal is less than 50%.

What happens in this example if the probability of achieving 2-log removal in unit 2 is changed to 25%? The corresponding probability of failure will be 75% (hold the probability of 2-log removal in unit 1 to 50%). The overall probability of achieving a 2-log removal will be 62.5% $(s_1s_2 + s_1f_2 + f_1s_2)$. With the probability of achieving 4-log removal dropping to 12.5% (s_1s_2), which is less than the least reliable unit process (unit 2 at 25%). The overall probability of achieving 2-log removal is greater than for a single unit process (improved reliability), but the overall probability of not meeting a 2-log removal also increased to 37.5%.

From this example, it is clear that a very important advantage to a multiple barrier is the greater reliability of achieving the overall goal (2-log removal). This example also serves to suggest that barriers in a treatment train can be evaluated by a stochastic process. However, approaching risk assessment using a stochastic analysis requires adequate knowledge of process performance. This means having sufficient water quality data to characterize process cycles and changes in water quality.

STOCHASTIC MODELING

One approach to analysis of uncertainty propagation is estimation of the probability density function of the output by using the probability density functions of each of the input variables. As shown in Figure 15.3, fluctuation of virus concentration in the influent follows a log-normal distribution. The virus survival rate in treatment plants also follows a log-normal distribution (Figure 15.5). Using the log-normal distributions of the input data and the virus survival rate yields the probability density distribution of the virus concentration in the effluent by Monte Carlo simulation (Dagpunar 1988).

In stochastic analysis, we set two probability density functions of the concentration in untreated water, C_0, and survival rate, S, in order to estimate pathogen concentration in reclaimed water C_{Re} by Monte Carlo simulation (right-hand column of Figure 15.8). The simulation in Figure 15.8 yields the probability density function of C_{Re}, which can be used as the input distribution to the next Monte Carlo simulation step, virus degradation in the environment. This kind of simulation is applied to the intake, single exposure, and annual risk computation, which results in the overall probability density distribution of risk value as the final result of the assessment. The advantage of the stochastic analysis is that the distribution of the final risk value is known (expressed as a log removal) and the degree of uncertainty (frequency of occurrence) associated with that final value is also known.

Figure 15.9 illustrates the results of this Monte Carlo simulation using data from Figures 15.3 and 15.6. The influent virus concentrations from Figure 15.3 follow a log-normal distribution ($\mu = 7.313$ (ln (1500)) and $\sigma = 1.226$). Using the geometric mean or the influent virus concentration distribution, the different effects of holding S constant (deterministic model) and varying S (stochastic model) can be illustrated. If the survival rate ($\mu = -8.06$ (ln (3.16×10^{-4})) and $\sigma = 1.336$), which were estimated from data in Figure 15.6, follows a log-normal distribution, the two distributions can be combined. The results from holding S constant (ln (S^*) = $\mu = -8.06$), are also plotted for comparison. When S is held constant, the slope of the influent and effluent lines is similar. If S is allowed to vary, the resulting distribution is a steeper line illustrating the effect of using the distributions rather than a constant value. Although the average (50th percentile) effluent virus concentrations for the variable and constant distributions are similar, the difference between the slopes of the distribution lines is significant. The steeper distribution from the variable S simulation shows a wider distribution or greater variability in the effluent virus concentration than would be expected from a deterministic evaluation. The distribution line of the variable S simulation yields good agreement with effluent data plotted in Figure 15.3. This example shows that by using the probability distribututions of the two variables, the Monte Carlo simulation will provide a better estimate of the probability distribution of the

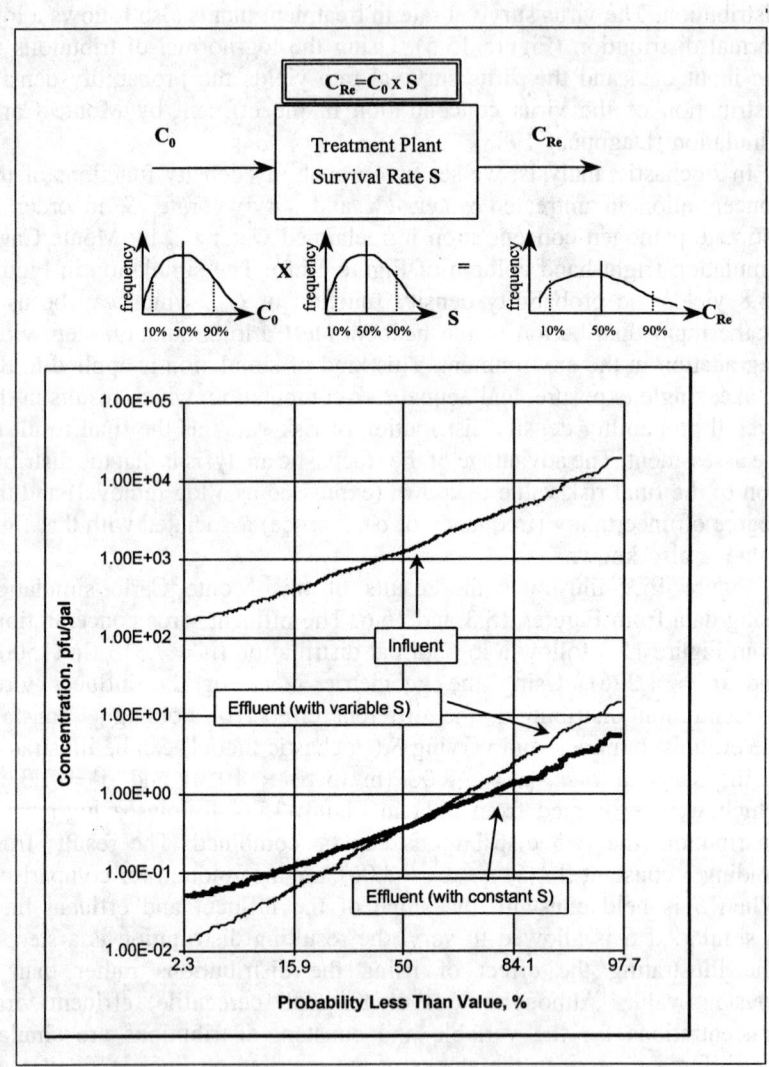

Figure 15.9 Propagation of uncertainty using Monte Carlo simulation.

output variable by incorporating the uncertainties associated with the input variables into the distribution of the output.

Because of their current shortcomings, microbial risk assessments, like chemical risk assessments, should be used in relative terms. However, to discount microbial risk assessments simply because they do not provide an absolute measure of the risk of infection would be inappropriate. As a risk management tool, Microbial risk assessments provide insight into the fate of pathogens in the environment, how pathogens survive, and potential means of pathogen control.

USE OF RISK ESTIMATES IN COST-BENEFIT ANALYSIS

A risk assessment can provide a statement of risk, but the risk manager still needs to decide what constitutes an adequate level of protection. Aside from quantifying the role uncertainty plays in determining the level of protection provided in the final policy document, another way to use a risk value in setting public policy is to balance costs of additional regulation (treatment and monitoring) with the risk of infecting a human population, i.e., in a cost-benefit analysis.

Rather than relying solely on the benefits of averting a theoretical illness, a cost-benefit analysis views societal costs of illness and lost productivity as providing an economic gain. However, selecting the appropriate measures and endpoints may not be the easiest choice a risk manager has to make. Meeting the objective of minimizing exposure to the risk of infection may require a delicate balancing act in which microbial risk assessment is used to weigh the benefits of changing wastewater reclamation policies. Though the risk manager may not fully appreciate the complexity of risk assessment and may doubt the statistical validity of the microbial risk assessment, it is important to recognize that risk assessment is just one tool that aids the risk manager in establishing realistic water quality objectives.

The concept of using risk assessment to achieve a balance in the setting of public policy (standards) is not foreign. The nuclear power industry and product testing safety routinely include risk assessment as part of a cost-benefit analysis. With an adequate microbial risk assessment, the cost-benefit analysis can help establish some sort of consistency in setting better public policy by providing a break-even point between public health protection and economics.

Figure 15.10 is an example that illustrates, in general terms, how costs and benefits might be assessed to balance risks and costs in the establishment of regulations. If the health risk is plotted against treatment, the risk decreases as treatment is added to the process train (solid line in Figure

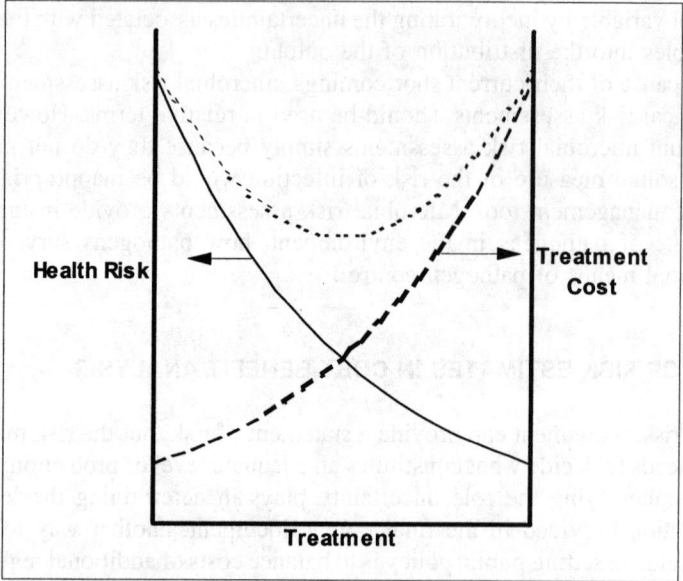

Figure 15.10 Idealized cost-benefit analysis.

15.10). If the cost of treatment is added to the plot, the cost will increase exponentially as new, more sophisticated processes are added to the treatment process train (dashed line in Figure 15.10). By combining the health risks and treatment cost graphs, there will be some point that produces the optimal cost to benefit ratio (lowest point of the dotted line in Figure 15.10).

Regulators are attempting to regulate to lower risk levels that are commensurate with lower pathogen concentrations. This is akin to attempting to reverse the second law of thermodynamics, i.e., trying to obtain order from disorder. As one tries to remove or inactivate to lower pathogen concentrations, the cost of treatment increases exponentially. Ideally, the risk manager must weigh societal costs and benefits against the cost of increased regulation.

It is interesting, but not surprising, to note that cost-benefit estimates reported by Baram (1981) cover a wide range in the dollar value associated with a human life. For example, the dollar value of human life as determined by Consumer Product Safety Commission ranges from $200,000 to $1,000,000. In 1985, Shodell (1985) reported that the USEPA dollar values on human life ranged from $400,000 to $7 million per life. A similar study by Tengs et al. (1995) reported, after examining several life-saving interventions, using published regulatory impact analyses, that there was still a significant disparity among the regulations in the cost of saving one year of life. In the interventions examined by Tengs et al. (1995),[8] the overall median intervention resulted in "cost effectiveness" of $42,000 per life-year saved with a wide distribution in the number of dollars per life-year. For chlorination, filtration, and sedimentation of drinking water, the cost effectiveness was $4,200/life-year saved. However, changing the water quality regulation for the carcinogen, trichloroethylene, from 2.7 to 11 μg/L in drinking water, resulted in the cost effectiveness being estimated at $34,000,000/life-year.

While the uncertainty with the risk estimate can play a role in the disparity of the dollar amount of the cost-to-benefit ratio, it is important to remember that the comparison of risks should be done on an equivalent basis. Disparities between cost estimates in a cost-benefit analysis can also result from the failure to use a common basis on which to compare risks, i.e., an annual risk versus risk accumulated over a lifetime. If benefits are clearly identified such as preventing a one in 10,000 *annual* risk of infection and comparison of cost/benefit ratios such benefits should not be compared to a cost/benefit ratio based on a benefit that may be based on a 1 in 100,000 *lifetime* risk because the basis, lifetime versus annual risk, cover different time spans. For example, the *annual* risk of infection in drinking water is 10^{-4}, but carcinogens are regulated on a *lifetime* estimate of risk that might be on the order of 10^{-5}. However, these risks are not equivalent time frames so how does one compare the risks? One means is to extend the annual risk estimate to a lifetime risk estimate.

When a person in a homogeneous population is exposed to identical pathogenic microorganisms n times in a year, the probability of infection, P_a, is given by:

$P_a = 1 -$ (the probability of not acquiring an infection after n exposures)

$$P_a = 1 - \prod_{i=1}^{n} (1 - P_i^*) \qquad (24)$$

where

$P_i^* =$ the probability of infection after a single exposure

If all exposures are at the same dose, then the relationship $P_i^* = P^*$ holds and P_a is written:

$$P_a = 1 - (1 - P^*)^n \qquad (25)$$

[8] Dollar values are based on 1993 dollars using a general consumer price index. The benefit is based on average number of years of life saved by averting a premature death.

where

P_a = annual risk of infection
P^* = risk of infection after a single exposure

Once again, the simple approach of Equation (24) can be used if the doses at each exposure event are the same. If, however, the doses during each exposure are different, then each exposure carries with it a different probability of infection. Consequently, each exposure becomes an independent event and the probability of infection can be calculated using an equation similar to Equation (5) for a series of independent events to calculate the annual risk of infection.

If the annual exposures are the same, then the lifetime risk of contracting an infection can be extrapolated simply by using Equation (26) below.

$$P_L = 1 - (1 - P_a)^y \qquad (26)$$

where

P_a = annual probability of infection
P_L = lifetime probability of infection
y = years

For example, using Equation (26) the 10^{-4} annual risk of microbial infection becomes equal to a 0.00698 lifetime risk ($y = 70$ years) showing the difference between the annual risk of a microbial infection and the lifetime risk of contracting a theoretical cancer (10^{-5}) may be almost three orders of magnitude.

Though the risk estimates may now be compared on a similar basis, cost-benefit analysis in general, raises several other questions. There may be specific community concerns not addressed by a cost-benefit analysis conducted at a national or state level, i.e., the national or state perspective does not adequately account for local concerns or economics. This raises more of an ethical dilemma for policy makers. How should the individual needs of the communities be incorporated into a public policy cost-benefit analysis without conducting a project by project analysis?

CONCLUSION

This chapter may raise more questions than provide succinct answers, which in part may be a reflection of the limited quantity of information available on which to base a microbial risk assessment. As greater pressure is placed on our water resources and the uses for reclaimed wastewater, public policy makers may be faced with need to accept greater risks. Water quality standards must be updated continually to reflect current understanding in disease occurrence and transmission. It becomes incumbent on risk managers to use microbial risk assessment and risk assessment models in a uniform and consistent manner to ensure that water quality standards reflect the current state of the art. Yet it is dangerous for public health officials to approach the development of public policy without the tools to make informed decisions. Risk managers require some degree of certainty to set rationale public policy, yet public policy must still be set in the absence of certainty. Consequently the risk manager is faced with many dilemmas formed by this circular process.

One should not forget that public policy based on a probability assessment developed in the absence of sound technical data may provide a false sense of security. Such a problem produces an interesting paradox for the policy manager, one does not know how close to allowing a failure the formulated public policy might be or if the policy exceeds the desired level of public health protection by several orders of magnitude. A risk manager's job can be likened to a blindfolded man standing near the edge of a cliff. The man knows that he is on a cliff, but he does not know whether his next step will take him over the edge or whether he will still be on solid ground.

Aside from the need to balance competing concerns, the risk manager is faced with the task of communicating the decision to the effected public, i.e., communicating the risk and the sources of uncertainty to a public that wants some assurance that a practice imposed on them will be "risk free." The risk manager must realize that there is no such thing as a "risk-free" practice, however, even with the uncertainty associated with the risk assessment the message that needs to be communicated clearly is that the practices being undertaken are reasonable, prudent, and essentially risk free. The practices will conserve precious water resources while protecting public health.

Although public policy may evolve at a slower rate than technical information, reducing research and debate on microbial risk assessment would not serve the best interests of public health. The lack of information and the absence of debate would only serve to delay, curtail, or encourage the overly conservative revision of wastewater reclamation criteria.

Last and certainly not least, public confidence and trust in wastewater reclamation projects must be cultivated through public education. A risk manager must build trust in the public policy and must be capable of communicating and instilling that trust, with confidence, to the public. In order to accomplish such a task, a risk manager must always keep in mind the shortcomings in the elements of the microbial risk assessment process, from the development of analytical methods to the application of dose-response information. In doing so, the primary objective of public policy should always be to protect the welfare and well being of the public. Risk managers need to strike a proper balance between the broad needs of a society and the protection of public health.

In the immortal words of Robert Frost:

> The woods are lovely, dark, and deep
>
> But I have promises to keep,
>
> And miles to go before I sleep.
>
> (Robert Frost, "Stopping by Woods on a Snowy Evening")

REFERENCES

American Society of Civil Engineers; American Water Works Association, 1990. *Water Treatment Plant Design,* Second Edition, McGraw-Hill Publishing Co. NY, NY.

Asano, T., et al., 1992. "Evaluation of the California Wastewater Reclamation Criteria using Enteric Virus Monitoring Data," *Wat. Sci. Tech.* 26 (7–8), pp. 1513–1524.

Baram, M. S., 1981. "The Use of Cost-Benefit Analysis in Regulatory decision-Making is Proving Harmful to Public Health," in *Management of Assessed Risk for Carcinogens,* W. J. Nicholson (ed), the New York Academy of Sciences, New York, NY.

Bitton, G., 1980. *Introduction to Environmental Virology,* John-Wiley & Sons, New York.

Bogen, K. T., 1990. *Uncertainty in Environmental Health Risk Assessment,* New York: Garland Publishing.

Cohen, A. C., Jr. 1959. Simplified estimators for the normal distribution when samples are single censored or truncated, *Technometrics,* vol. 1, pp. 217–237.

Cohen, A. C., Jr. 1961. Tables for maximum likelihood estimates: Singly truncated and singly censored samples, *Technometrics,* vol. 3, pp. 535–541.

Craun, G. F., 1986a. "Statistics of Waterborne Disease Outbreaks in the U.S. (1920–1980)," in *Waterborne Diseases in the United States,* G. F. Craun (ed). CRC Press, Inc., Boca Raton, FL.

Craun, G. F., 1986b. "Recent Statistics of Waterborne Disease Outbreaks (1981–1983)," in *Waterborne Diseases in the United States,* G.F. Craun (ed). CRC Press, Inc., Boca Raton, FL.

Craun, G. F., 1988. "Surface Water Supplies and Health," *JAWWA,* 80 (2), pp. 40–52.

Craun, G. F., 1990. *Waterborne Disease Outbreaks,* EPA/600/1-90/005b, Health Effects Research Laboratory, Office of Research and Development, USEPA, Cincinnati, OH.

Crook, J. et al., 1992. *Guidelines for Water Reuse,* United States Environmental Protection Agency, EPA/625/R-92/004.

Dagpunar, J., 1988. *Principles of Random Variate Generation,* Clarendon Press, Oxford.

Engineering Science, 1987. "Final Report Monterey Wastewater Reclamation Study for Agriculture," Engineering Science, Berkeley, CA.

Feinstein, A. R. et al., 1981. "Coffee and Pancreatic Cancer," *JAMA,* 246 (9), pp. 957–961.

Fox, K. R. and D. A. Lytle, 1996. "Milwaukee's Crytpo Outbreak: Investigation and Recommendations," *JAWWA,* 88(9) pp. 87–94.

Furumoto, W. A. and R. Mickey, 1967. "A mathematical model for the infectivity-dilution curve of tobacco mosaic virus: theoretical considerations," *Virology,* 32, pp. 216–23.

Gerba, C. P., 1980. "Virus Survival in Wastewater Treatment," in *Viruses and Wastewater Treatment,* Goddard, M. and M. Butler (Eds.) Pergamon Press, Oxford.

Gilbert, R. O., 1987. *Statistical Methods for Environmental Pollution Monitoring,* Van Nostrand Reinhold, New York.

Goldstein, S. T., et al., 1996. "Cryptosporidiosis: An Outbreak Associated with Drinking Water Despite State-of-the-Art Water Treatment," *Annals of Internal Medicine,* 124 (5), pp. 459–468.

Haas, C. N., 1983. "Effect of Effluent Disinfection on Risks of Viral Disease Transmission via Recreational Water Exposure," *Journal WPCF,* 55 (8), pp. 1111–1116.

Haas, C. N., 1983. "Estimation of Risk due to Low Doses of Microorganisms: A Comparison of Alternative Methodologies," *American Journal of Epidemiology,* 118, (4), pp. 573–582.

Hutzler, N. and W. C. Boyle, 1982. "Risk Assessment in Water Reuse," in *Water Reuse,* Middlebrook, E. J. (Ed.) Ann Arbor Science Pub. Inc., Ann Arbor, MI, pp. 293–346.

Kizer, K., 1985. "Guidelines for chemical Carcinogen Risk Assessments and Their Scientific Rationale," State of California Health and Welfare Agency, Department of Health Services, Sacramento, CA. McGaughey, P. H., 1968. *Engineering Management of Water Quality,* McGraw-Hill Book Co., New York, NY.

Lam, R. H. F., J. P. Brown, and A. M. Fan, 1994. "Chemicals in California Drinking Water: Source of Contamination, Risk Assessment, and Drinking Water Standards," in *Water Contamination and Health,* R. G. M. Wang (Ed.), Marcel Dekker, Inc., New York, NY.

Land, C. E., 1971. Confidence intervals for linear functions of the normal mean and variance, *Annals of Mathematical Statics,* vol. 42, pp. 1187–1205.

Land, C. E., 1975. Tables of confidence limits for linear functions of the normal means and variance, in *Selected Tables in Mathematical Statistics, vol. III.* American Mathematical Society, Providence, R.I., pp. 385–419.

Mausner, J. S. and S. Kramer, 1985. *Epidemiology—An Introductory Text,* W.B. Saunders Co. Philadelphia, PA.

Milloy, S. J., P. S. Aycock, and J. E. Johnston, 1994. "Choices in Risk Assessment," Regulatory Impact Analysis Project, Inc., Washington, D.C.

Morgan, M. G. and M. Henrion, 1990. *Uncertainty: A guide to dealing with uncertainty in quantitative risk and policy analysis,* Cambridge University Press, Cambridge.

Nash, F. R., 1993. *Estimating Device Reliability: Assessment of Credibility,* Kluwer Academic Publishers, Boston, MA.

National Research Council (NRC), 1983. *Risk Assessment in the Federal Government,* National Academy Press, Washington, D.C.

National Research Council, 1993. *Managing Wastewater in Coastal Urban Areas,* National Academy Press, Washington, D.C.

National Research Council, 1994. *Science and Judgment in Risk Assessment,* National Academy Press, Washington, D.C.

Nellor, M. H., R. B. Baird, and J. R. Smyth, 1984. Health Effects Study Final Report, County Sanitation Districts of Los Angeles County, Whittier, CA.

Pinholster, G., 1995. "Drinking Recycled Wastewater," *Environmental Science & Technology,* 29 (4), pp. 174A–179A.

Regli, S. et al., 1991. "Modeling the Risk From Giardia and Viruses in Drinking Water," *JAWWA,* 83 (11), pp. 76–84.

Riggs, J. L. and D. P. Spath, 1981. "Health Risks from Viruses in Reclaimed Wastewaters," Health Effects Research Laboratory Report, Office of Research and Development, U.S. Environmental Protection Agency, Cincinnati, OH.

Roefer, P. A., J. T. Monscvitz, and D. J. Rexing, 1996. "The Las Vegas Cryptospordiosis outbreak." *JAWWA,* 88(9), pp. 95–106.

Roszak, D. B. and R. R. Colwell, 1987. "Survival Strategies of Bacteria in the Natural Environment," *Microbiol. Rev.,* 51, pp. 365–379.

Shodell, M., 1985. "Risky Business," *Science* 85, 6(8), pp. 43–7.

Shuval, H. I. et al., 1986. "Wastewater Irrigation in Developing Countries, Health Effects and Technical Solutions," World Bank Technical Paper No. 51, The International Bank for Reconstruction.

State of California, 1987. "Report of the Scientific Advisory Panel on Groundwater Recharge with Reclaimed Wastewater," prepared for the State of California, State Water Resources Control Board, Department of Water Resources, and Department of Health Services.

Tanaka, H. et al., 1998. "Estimating the Reliability of Wastewater Reclamation and Reuse using Enteric Virus Monitoring Data," Water Environment Research, Vol. 70, No. 1, pp. 39–51, January/February 1998.

Tengs, T. O. et al., 1995. "Five-Hundred Life-Saving Interventions and Their Cost-Effectiveness," *Risk Analysis,* 15 (3), pp. 369–390.

Western Consortium for Public Health, 1992. "City of San Diego Total Resource Recovery Project Health Effects Study Final Summary Report," Western Consortium for Public Health, Berkeley, CA.

CHAPTER 16

Water Resources Management and Wastewater Reuse for Agriculture in Israel

INTRODUCTION

WATER RESOURCES IN ARID REGIONS

A RID and semi-arid regions such as Israel are characterized by small amounts of precipitation and low availability of natural water sources. Frequently, the limited precipitation and related water shortages are associated with geographical and seasonal distribution. The sharp geographical variations in precipitation and consequently in availability of water is typical of dry regions. Special ventures have to be undertaken in order to supply water at adequate quality for all needs.

It can be observed that along a distance of about 550 km in the state of Israel precipitation decreases from an approximate annual precipitation of 800 mm in the north (Zone A) to less than 50 mm in the south (zone D) as shown in Figure 16.1. This steep precipitation gradient is a well known phenomenon of many dry regions in the Mediterranean Basin and, similarly, in some regions along the western coast of the American continent (e.g. California, USA), which are located in similar latitudes. Variations in water supply, which is based on seasonal precipitation and subsequently on natural and artificial aquifer recharge, can to some extent be compensated for by construction of storage facilities. The reservoirs (either open-surface or subsurface) provide water for all usage primarily during the water shortage months. The shortage of water has therefore resulted

Gideon Oron, Ben-Gurion University of the Negev, The Institute for Desert Research, Kiryat Sde-Boker, Israel 84990.

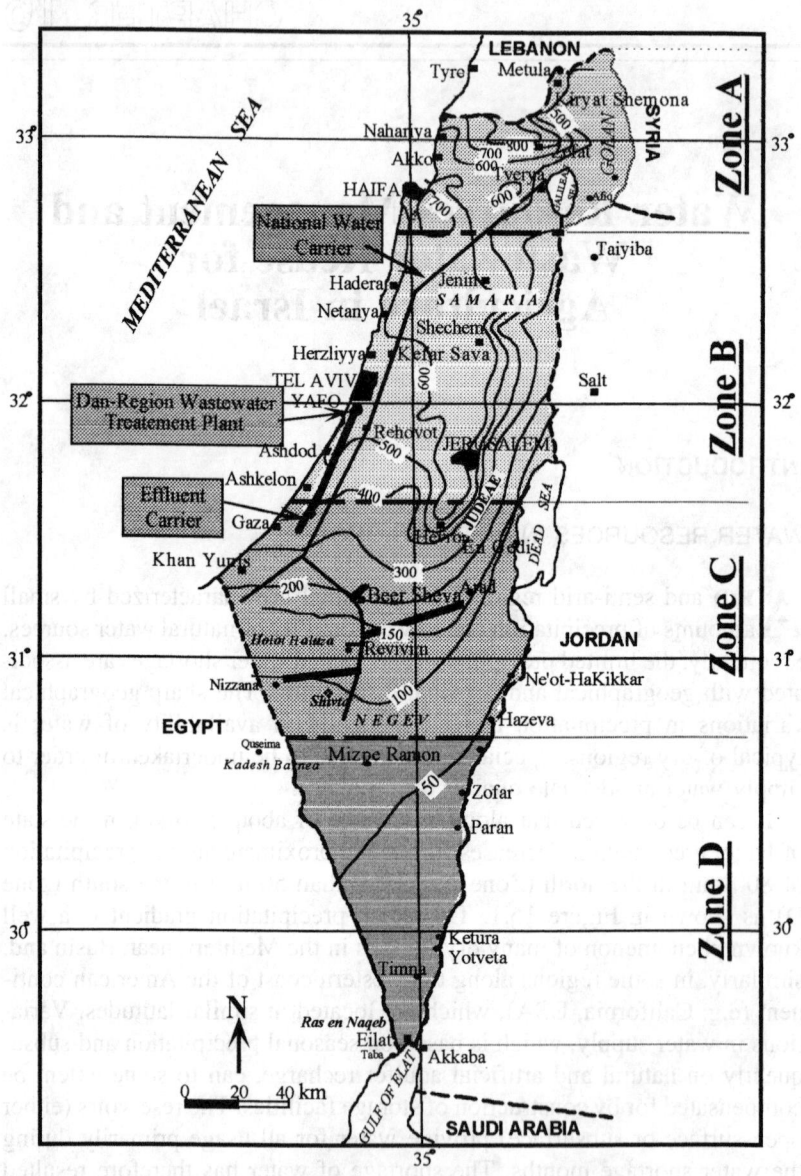

Figure 16.1 Annual mean variation of precipitation along the state of Israel.

TABLE 16.1. Water Sources and Approximate Annual Water Potential in Israel ($\times 10^6$ m^3) (Nativ and Issar, 1988).

Water Source	Coastal Aquifer	Mountain Aquifer	Sea of Galilee	Local Aquifers	Flood Water	Total Volume
Available Volume	283	330	455	700	32	1800

in intensive development of alternative solutions, including utilization of non-conventional sources.

WATER SUPPLY AND CONSUMPTION IN ISRAEL

Main water supply in Israel is based on groundwater pumping from the two main aquifers (Table 16.1). The coastal aquifer consists of a conglomerate of small sandstone soil aquifers, begins in the north (around 30 km north of the City of Akko) and extends along the sea shore (Mediterranean) into the Sinai Peninsula. Mean in-land width of the coastal aquifer is around 10 km and water is pumped from a depth ranging from 50 m to 100 m. Total water potential for annual use is assessed at 283×10^6 m^3/year. The water pumped from the coastal aquifer can be used for drinking with some complementary treatment phases. However, the water quality of the coastal aquifer water is very much affected by on-surface agriculture activity and urbanization processes. The main influences are leaching of nitrogen and other compounds into the water and reduced natural recharge processes.

The mountain aquifer is located along the main ridge of mountains in Samaria (Shomron) via the Jerusalem mountains and the Judaean Desert. The aquifer consists of limestone and dolomite. High quality water for unrestricted application is pumped from the mountain aquifer, from various depths of up to 400 m. Total pumping potential is assessed at 330×10^6 m^3/year. These two main aquifers and several local aquifers distributed over the entire country are insufficient to satisfy all water requirements, neither annually nor on a shorter time scale basis (Figure 16.2). Additional amounts of water are obtained from runoff (mainly the Jordan and Yarmuk rivers) and local distributed aquifers consisting of basalts, chalks and young alluvial deposits.

Most of available water is consumed for agricultural purposes (Figure 16.2). However, due to limited availability of water, this picture will soon change and agriculture will have to rely more on marginal water sources, primarily effluent. Human consumption is in the range of 70 to 95 m^3/cap per year. The domestic consumption is continuously increasing along

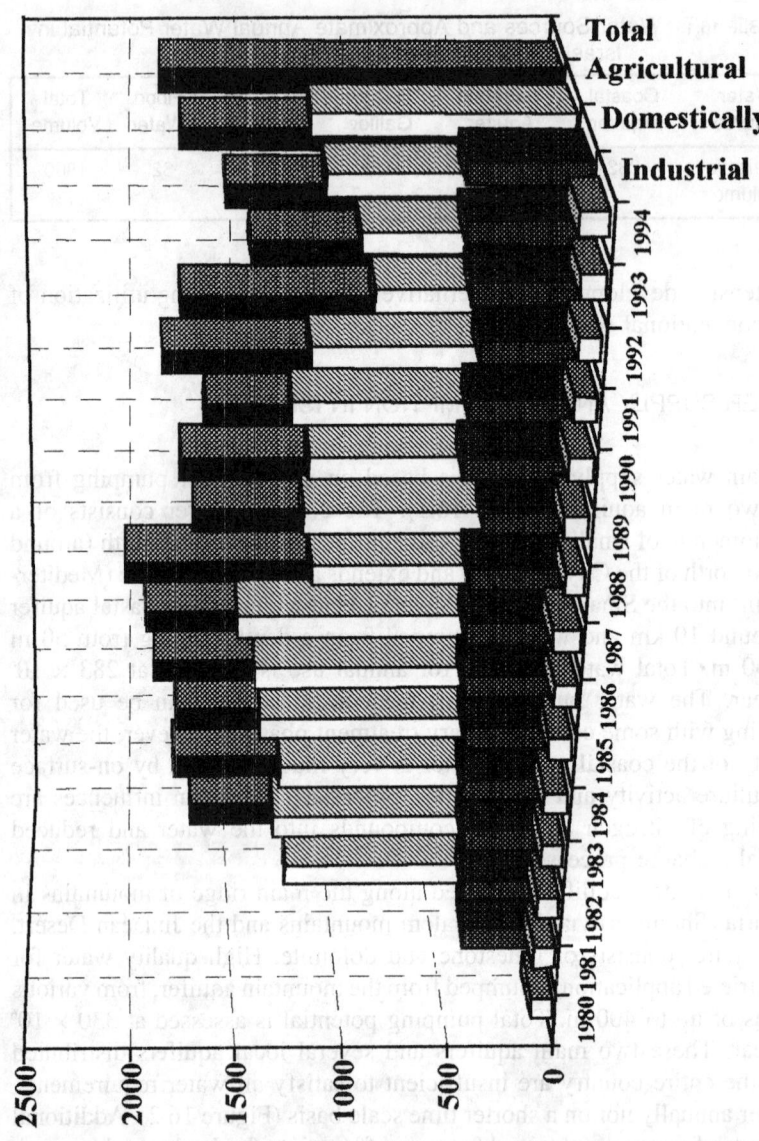

Figure 16.2 Annual water demand during last years in Israel.

with elevated living standards and increased population. Around 85% of the domestic water consumed is discharged to sewers and most of it is treated.

One way of reducing the gap between water supply and demand is to import water from external sources to the main consumption sites. For this reason, water is transported from the Sea of Galilee (Kinnerret Lake) via the National Water Carrier (NWC) to the southern parts of the country as shown in Figure 16.1. Consequently, a realistic total annual water potential in Israel is between 1.8×10^9 m^3/year and 2.0×10^9 m^3/year (Table 16.1). The available amounts are mainly subject to the recharge situations depending on winter precipitation and water demand, mainly for irrigation, subject to environmental conditions.

MARGINAL WATERS IN ARID ZONES

In addition to the above sources, water scarcity in arid zones can be overcome by gradual development of local marginal water sources. Additional water sources development is subject to regional and periodical needs, economic and environmental considerations, and future prospects (Brimberg et al., 1994). In order to reduce the dependence of water supply on external sources such as the NWC and alleviate the problems associated with over-pumpage, it has become necessary to develop the nonconventional and not yet fully exploited water sources existing primarily in the Negev Desert. These additional water sources have definite characteristics.

SALINE GROUND WATER (SGW)

Saline groundwater can be found in the deep fossil aquifers (approximately 1000 m) below the Negev Desert and Sinai Peninsula (Issar and Adar, 1992). The high variable expenses for pumping and the high salinity [the Electrical Conductivity (EC) varies between 4 and 8 dS/m] of the water generate extra utilization difficulties. The main advantages stem from the huge available volumes stored in the Negev Desert (billions of cubic meters) and the possibility of utilizing the water directly, or after further desalination, for a broader pattern of purposes. Advanced application techniques are frequently implemented to use the SGW for irrigation (Pasternak and DeMalach, 1987; Oron et al., 1995).

RUNOFF WATER (ROW)

Runoff water is generated during intermittent rainfall events during the wet (winter) season. Due to low soil surface permeability, the flood water can be diverted to special facilities and stored for future needs mainly for the summer irrigation season. Relatively high capital investments are required for the collection, storage and distribution of the water to the consumption sites. The stochastic nature of runoff water supply raises additional reliability difficulties to the efficient use of this water. Despite the low stability, the high-quality of the water, and the potential to retain large volumes are advantageous in regions with scarce conventional waters.

TREATED WASTEWATER (TWW)

Treated wastewater and, primarily, domestic treated sewage can be reused for a large pattern of possibilities, primarily for agricultural irrigation (Asano et al., 1992). The major drawbacks to TWW use are the high capital investment in the treatment facilities and equipment, the dual piping system required to distribute it separately from potable water, effluent quality control and additional required precautions to minimize health and environmental risks. Treatment level, as related to the purpose of reuse, is also of concern. The nutrients contained in the TWW are, however, beneficial for agricultural use (Oron et al., 1991).

WASTEWATER TREATMENT AND REUSE

WASTEWATER TREATMENT

As in most developed countries, wastewater treatment is compulsory in Israel. Domestic wastewater is primarily treated in stabilization pond systems. Stabilization ponds are mainly popular in small communities. In large urban areas, the wastewater is treated by more advanced methods. The advanced systems include aerated lagoons, activated sludge, anaerobic phases and, lately, also Sequential Batch Reactors (SBR). Industrial wastewater is mostly treated separately at the factory stage to adequate levels and is subsequently discharged to the main sewerage systems.

There are in Israel close to 400 treatment plants of various sizes, of which about 120 are under strict control of water and environmental authorities (Eitan, 1994; Shelef and Azov, 1995). Many of the larger plants treat the sewage of a cluster of several municipalities. Large plants treat also the wastewater of several small isolated communities which are located relatively adjacent to each other. Many of the small treatment facilities are based on a settling stage which is subsequently followed by oxidation and maturation stages and extended storage. The larger communities can afford to add aeration to the existing oxidation pond system.

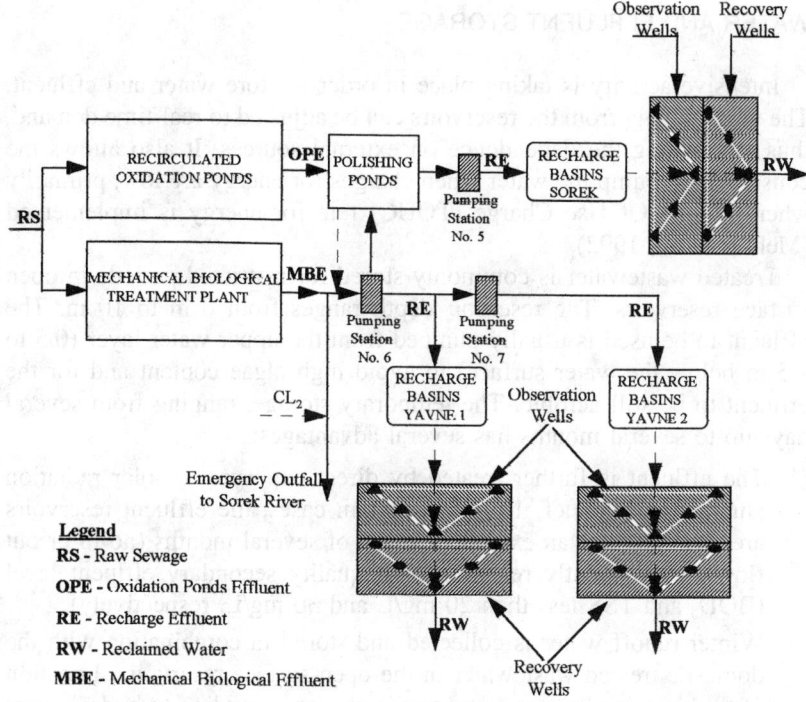

Figure 16.3 Schematic layout of the treatment plant of Greater Tel-Aviv (Dan Region, Israel).

In the large urban centers, the wastewater is treated by trickling filters, and different versions of activated sludge. In the greater Tel-Aviv area (Dan Region), the wastewater is treated by a combination of activated sludge, nitrification and denitrification. Annual raw sewage discharge is around 130×10^6 m³/year, treating the wastewater of a population of about 1.4×10^6 inhabitants (Kanarek and Michail, 1996). The effluent is subsequently injected into the aquifer (Soil Aquifer Treatment, SAT) and then pumped for irrigation (Figure 16.3). However, there are unique problems related to industrial wastewater and agricultural wastes as noted above. Industrial wastewater is commonly treated at the plant level and only then discharged to main sewerage system.

Much agricultural waste is produced by dairy farms. There are around 330 dairy farms in Israel. The number of milking cows is in the general range of 200 to 400 head per farm. A reasonable total agricultural waste flow rate is in the range of 150 to 200 L/day per milking cow. Treatment of agricultural wastes is maintained via a settling stage and subsequently in septic tanks and storage. The agricultural biosolids are primarily used for fertilization.

WATER AND EFFLUENT STORAGE

Intensive activity is taking place in order to store water and effluent. The water supply from the reservoirs can be adjusted to real-time demand, thus eliminating the dependence on external sources. It also allows the consumers to pump the water when charges for energy are low, primarily when a Time Of Use Charge (TOUC) rate for energy is implemented (Mehrez et al., 1992).

Treated wastewater is commonly stored for restricted periods in open surface reservoirs. The reservoir depth ranges from 6 m to 10 m. The effluent to be used is usually pumped from the upper water layer (0.5 to 1.5 m below the water surface) to avoid high algae content and for the effluent to be still aerobic. The temporary storage, ranging from several days up to several months has several advantages:

(1) The effluent is further treated by direct exposure to solar radiation (Juanico and Shelef, 1994). In certain cases, the effluent reservoirs are "closed" for an extended period of several months (no in or out flow), consequently reaching high quality secondary effluent level (BOD_5 and TSS less than 20 mg/L and 30 mg/L, respectively).

(2) Winter runoff water is collected and stored in combination with the domestic treated wastewater in the open-surface reservoirs. Location of the combined water reservoirs is chosen according to hydrological conditions and basin characteristics. Also, in these storage facilities, polishing phases can be expected (Oron et al., 1980). There are over 225 combined reservoirs for runoff water and effluent in Israel (Eitan, 1994). Total capacity is above 85×10^6 m^3.

In addition, there are storage facilities for tap water supply only. These are primarily relatively small reservoirs ("floating" reservoirs) which are part of the national water supply system for human consumption. These operational reservoirs are treated very carefully, including lining, with continuous tight quality water control.

WASTEWATER REUSE FOR AGRICULTURAL IRRIGATION

In regions with limited natural water sources, treated wastewater, primarily urban sewage, can be utilized for agriculture, industry, recreation and recharge of aquifers (Asano and Milles, 1990; Bouwer, 1991; Asano et al., 1992). Predominantly, effluent application for irrigation simultaneously solves water shortage and wastewater disposal problems.

The use of water in series is a necessity, mostly in regions with limited conventional and high quality water sources. Treated wastewater has been utilized for irrigation of a variety of field crops and orchards with serious efforts to expand the crop pattern to include processing fruits and edible vegetables (Smith, 1982; Oron et al., 1986; Burau et al., 1987; Asano and Pettygrove, 1987; Oron and DeMalach, 1987; Itoyama et al., 1990). Reuse of effluent is becoming practical worldwide and is expanding rapidly, due to the advantages noted. Effluent is reused in numerous countries throughout the world (Ghobrial, 1993: Siebe and Cifuentes, 1995: Salgot et al., 1996; Asano et al., 1996: Ju-Si, 1996). Commonly, prior to distribution, the wastewater is sufficiently treated to minimize environmental and health risks. However, traditional application methods include sprinkler irrigation, surface flooding and ridge furrow irrigation at a rate of 500 to 1500 mm per year, subject to the region and crops cultivated (Telsch and Katzsenelson, 1978; Oron et al., 1990). Spray application methods are frequently associated with aerosol distribution and direct contact of the contaminated effluent with the plants and the fruits. Open-surface irrigation is inefficient in terms of water use and is frequently associated with direct contact of the effluent with the plants, primarily under vegetable irrigation. Therefore, application of effluent can sometimes be associated with outbreaks of infectious and epidemiological diseases (Kott et al., 1978; Fattal et al., 1987; Blumenthal, 1989; Miller, 1994).

REUSE CRITERIA

Environmental and health problems were raised in the past when effluent was reused (Gilbert et al., 1976; Pescod and Alka, 1988). These aspects were given attention at the beginning of the century in the USA and followed by other countries throughout the world. The consequence was setting criteria for reuse by various states and international bodies, such as the World Health Organization (WHO, 1989) and others (ISQW, 1981; Nellor et al., 1985; CIHEAM, 1988; Pescod and Alka, 1988; Asano et al., 1992; US EPA, 1992). These criteria appeared mainly as recommendations with a restricted legislative kernel. The wide variety of criteria, mainly the ones issued by the WHO (1989) has increased the controversy between health authorities, managers, designers and users of treated wastewater.

One of the major drawbacks of nearly all reuse criteria is the focus on the treatment level and the purpose of use, while the disposal method is ignored. This approach imposes the burden on local authorities in the various countries to control the effluent quality and allows only a small degree of influence to the intelligence, experience and needs of the effluent consumers. However, advanced application technology can allow health and environmental risks to be reduced, even when low quality effluent is reused.

EFFLUENT REUSE AND DISPOSAL SYSTEMS

The intense use of effluent for irrigation has stimulated public awareness of environmental and groundwater pollution (Goyal et al., 1984; IRCWD, 1985). Public awareness raised mainly due to the fact that spray and open-surface irrigation are associated with aerosol distribution and direct contact with the edible products, respectively.

During the last decade, along with advanced innovative technology, drip methods have taken over in many crops. Drip irrigation is now a popular technology due to efficient use of water, high productivity and convenience of operation. Drip systems are common in orchards, greenhouses, vegetable and field crops such as corn, cotton, sunflower and jojoba. Micro and drip systems were originally introduced for irrigation of high cash crops in areas with limited conventional water sources as a water saving measure (Dawood and Hamad, 1985). The concept of water saving might seem to contradict the idea of obtaining maximum economic yields from irrigated crops. However, drip irrigation is associated with many additional advantages, alongside the related economic benefits. A reasonable combination of effluent reuse and disposal technology can minimize health and environmental risks, coupled with satisfying the demand for water.

EFFLUENT FILTRATION AND DISINFECTION

Filtration of the applied effluent is frequently required to prevent emitter clogging (Adin and Elimelech, 1989). During the filtration process, the suspended matter is split into small particles which can flow freely through the filter and subsequently through the emitters' passages. In order to minimize both filter and emitter clogging, the recommended range of filter opening is 80 to 120 "mesh" (177 and 125 micron, respectively). A higher filter "mesh" is frequently associated with frequent filter clogging and the need for perpetual flushing (manual or automatic).

Three prominent types of filters can be considered:

(1) Surface filtration in which metal or plastic screens are installed within the filter.
(2) Deep bed filtration consisting of a deep soil or gravel layer through which the effluent flows. The deep filtration resembles flow in porous media, hence it can be described adequately using the appropriate flow equations. The preferred filtration velocity is in the range of 10 m/hr to 20 m/hr.
(3) Disc (ring) filtration which essentially combines surface and deep filtration principles. The compressed discs provide in-depth three di-

mensional filtration. Releasing the internal pressure in the filter loosens the tightened discs. Separation of the discs and instituting a reverse direction flow allows a backflush stage and the discarding of the accumulated solids. The backflushing stage lasts only several seconds thus minimizing water losses. The head control in most drip irrigation systems consists of a battery of filters allowing an intermittent gradual backflushing and remediation cycle.

The backflushing stage in each kind of filter is subject to the difference in pressure head generated during flow. Usually when the head loss exceeds 2 m the backflush stage is automatically initiated.

Disinfection is a complementary phase which is commonly required in order to remove the contained pathogens. Various disinfectants can be used, primarily subject to efficiency of pathogen removal, environmental and economic considerations. Chlorine is still the most popular constituent. Other disinfectants include chlorine dioxide, bromine, ozone, sodium hypochlorite and ultraviolet (UV) radiation. Chlorine is attractive due to simplicity of implementation and potential of killing most pathogens excluding viruses and parasites. The chlorine is injected at the outlet of the reservoirs (or at the main inlet of the effluent supply system) and oxidizes the contained suspended matter (organic matter and algae), thus reducing the risk of filter and emitter clogging. Chlorine can be injected in a liquid form or as gas which is the least expensive form. Conventional chlorine dosage is in the range of 5 to 20 mg/L and related contact time until the effluent reaches the inlet to the drip irrigation system is in the range of 10 to 20 minutes. Residual chlorine at the drip lateral inlet system should be more than 0.5 mg/L.

BIOLOGICAL FILTERING PROPERTIES OF THE SOIL MEDIA

A viable modification to the conflict between completely safe effluent reuse and closing the gap between water demand and supply, is to dispose and utilize the treated wastewater efficiently and with minimal pollution risk. This approach can be implemented either by applying the effluent under conventional Drip Irrigation (DI) or primarily under Subsurface Drip Irrigation (SDI) systems (Bresler, 1977; Chase, 1985; Tollefson, 1985; Oron et al., 1992). Under SDI application, the soil encompassing the trickling tube, which is located between 30 to 70 cm below the soil surface, acts as a living biological filter (McCarthy and Zachara, 1989: Oron et al., 1992; Phene and Ruskin, 1995). The effluent which is discharged into the porous media flows in the direction of many tiny pumps in the form of the plants' roots, which absorb the wastewater and the contained nutrients. Since the effluent is consumed by the plants very close to the point sources (emitter outlet) it does not reach the soil surface (Bar-Yosef et al., 1989: Phene and Ruskin, 1995). The layer above the emitters therefore functions as a soil filter, which is a compulsory treatment stage in some reuse criteria for unrestricted irrigation (ISQW, 1981). The dry soil surface has minimal direct contact with the effluent and therefore diminishes any kind of pollution hazard.

The efficiency of effluent disposal and reuse depends on the combined effects of the filtering properties of the soil and irrigation procedure. The filtering capacity of the soil depends on soil size and distribution of particles, moisture content, organic matter content, chemical and electrical absorption characteristics. The effectiveness of the filtering process under SDI depends also on the emitter depth and irrigation regime. These will be essentially expressed by two main parameters: (a) hydraulic load (flowrate per unit area, L/hr per m^2) of the wastewater; (b) the organic load (commonly BOD_5/day or TSS/day per unit area) which depends on the soil properties and the interaction with the effluent. The combined hydraulic and organic loading will allow to assess contaminant migration and removal in the porous media during the application process (McCaulou et al., 1995). Similarly, the irrigation procedure is controlled by water requirements, wetting and drying cycle and the density of the water distributing outlets given by the emitter spacing and depth.

It is therefore imperative to consider the synergistic properties of the soil media together with the efficiency of treated wastewater disposal for irrigation. These properties, in combination with the loading, influence the migration and fate of pathogens and ultimately the irrigation efficiency and quality (Oron et al., 1990).

The fact that the soil surface remains dry under SDI has many additional agronomic, environmental and agrotechnical advantages: (a) reduced evaporation losses; (b) reduced generation of runoff and high water saving; (c) better control of weeds due to their low germination rate; (d) saving of herbicides which are otherwise regularly required to control weed development, and thus reducing pollution hazard; (e) improved traffic and maneuver conditions for the agricultural machinery equipment; and, (f) minimal environmental pollution due to restricted flow towards the groundwater direction and prevention of aerosol generation. Under DI and when the soil infiltration rate is relatively low (e.g. Loess soils) the effluent will be distributed on the soil surface with minimal contact with the plant foliage. Under these conditions, the exposure to direct solar radiation and high temperature enhance the die-off of pathogenic contamination agents.

In a series of case studies and on-going experiments, it has been confirmed that SDI can be considered as a complementary treatment phase.

Effluent disposal under SDI combines efficient water application for agricultural irrigation with minimal environmental and health aspects.

CASE STUDIES

BACKGROUND

The hypothesis that drip irrigation, and mainly, subsurface systems can be implemented for secondary wastewater disposal even for processing agricultural products and vegetables eaten raw has been examined recently in a series of field studies in Israel (Oron et al., 1990). Many commercial fields in Israel are currently irrigated with domestic effluent. Experiments with field crops were conducted at the commercial site of the Revivim and Mashabay-Sade Farm (RMF), located near the City of Beer-Sheva. Additional experiments are still in progress, applying the secondary effluent of the City of Arad on the Arad Heights and with the effluent of the Cities of Gedera and Yavnna in the fields of Kibbutz Chafets-Chaim. Effluent is also reused in several settlements in the Jezrael Valley with the treated wastewater of the City of Haifa and in the Rift Valley, applying the effluent of the City of Eilat. Secondary effluent from the Greater Dan Region treatment facility adjacent to the City of Tel-Aviv (around 130×10^6 m^3 per year) is transported via a branched pipe system for reuse in the southern Negev Desert (approximately 100 km long) after a Soil Aquifer Treatment (SAT) stage (Kanarek and Michail, 1996).

The crops on the RMF include corn, wheat, alfalfa, cotton, ryegrass, and in the Arad Heights vineyard, almond trees, sunflowers, wheat, and various field crops for seed production. A similar crop pattern is practiced in the fields of Kibbutz Chafets-Chaim. At all three sites, the domestic wastewater is treated in a stabilization pond system (Table 16.2). The effluent application methods include sprinkler irrigation, DI and currently also SDI. Secondary treated domestic wastewater and tap water is applied in the various experimental sites for comparison purposes.

The commercial alfalfa field on the RMF is irrigated with the effluent from the treatment plant (a facultative pond system) of the City of Beer-Sheva. Additional data was taken from an adjacent sprinkle irrigated alfalfa field which is located very close (several meters) to the SDI alfalfa field. Comparable data for corn irrigated with reclaimed wastewater were obtained on the RMF and Kibbutz Chafets-Chaim. On the Arad Heights, a commercial vineyard is being irrigated with a subsurface and drip irrigation system applying secondary wastewater and potable water (control treatment). Each experimental treatment is conducted under several replications

TABLE 16.2. Quality Range of the Secondary Effluent Applied for Irrigation on RMF, Arad Heights and Chafets-Chaim (mg/L).

Parameter	Revivim Mashabay-Sade Farm	Arad Heights	Chafets-Chaim
CODt	190–426	435–575	70–100
CODf	90–205	164–336	—[a]
BODt	48–65	141–180	11–20
BODf	10–15	35–120	—
TSS	90–135	160–244	12–48
Ammonia	28–49	50–85	15–22
PO_4	22–42	15–32	10–22
K	22–28	25–35	26–41
Na	161–252	227–386	139–225
Ca	78–83	43–65	—
Mg	36–48	16–37	—
EC, dS/m	1.7–1.8	1.4–2.1	1.6–2.3

[a] Not monitored.

in an area of approximately 1000 m². The residual constituents content in the applied effluent is adequate to provide all the plants' nutrient requirements. It allows artificial fertilization needs to be reduced. The effluent quality meets local criteria for reuse for field crop irrigation (Table 16.2; ISQW, 1981).

THE SOIL AND PLANTS

Sweet corn (*Zea mays* L., Saccharatum) was grown under conventional practice, but with a subsurface drip system. The corn was planted at row spacing of 0.96 m and one lateral drip tube served each plant row. Compensating emitters, with a flow rate of 2.3 L/h were spaced 100 cm apart on the laterals in all drip treatments. In the subsurface treatment, the drip tubes were installed 25 to 45 cm below the soil surface. Prior to effluent application via the drip systems, all treatments were sprinkle irrigated for germination. Nutrients were injected only in the fresh-water (control) treatment, at a dose similar to the nutrient content in the applied effluent. The plants' response in this study was based on yield sampling of each treatment. Yield samples were taken from a sampling area of about 10 m² in all replications.

The vineyard orchard on the Arad Heights was planted in the summer of 1991 at a plant spacing of 1.5 in the row and 3 m between them. Each row is irrigated by one drip lateral. Most of the orchard is irrigated with emitters with a flow-rate of 3.5 L/hr installed every 75 cm. Similar emitters were used in the experimental plots for SDI, installed at 25 and 40 cm below the soil surface. In addition, 2.3 L/hr emitters installed 50 cm apart, were used in the experimental plots. These emitters were also installed at two depths: 25 and 40 cm. The vineyard orchard is irrigated once a week and the total amount applied is around 4000 m^3/ha.

CONTAMINATION ANALYSIS

Soil, plant and fruit samples were collected from the experimental plots and the commercial fields. The samples were taken under sterilized conditions and transported to the laboratory for conventional and contamination constituents analysis (APHA, 1995). The contamination analysis included monitoring of fecal coliform concentrations in the soil, on the leaves of the plants, on the kernels of the cobs (the husk was taken off) and in the grapes. Phosphate Buffer Saline (PBS) solution was used for the bacteria analysis. Additional virus analysis followed conventional methods.

RESULTS

APPEARANCE AND YIELDS OF PLANTS

Under most experimental circumstances, the corn plants' behavior and appearance in the subsurface trickle treatments was similar to drip and sprinkler irrigation. The cob wet yield obtained under subsurface and drip irrigation was similar (Table 16.3; Figure 16.4). The slight differences can be attributed to the proper timing of changing from sprinkler irrigation for germination to subsurface application. The percentage of dry matter is in the range of 37 to 48 percent. The grape yield for the first experimental year is given in Figure 16.5. The grape yield is probably affected by the new drip system installation.

SOIL AND PLANT CONTAMINATION

Fecal coliforms in the applied effluent are approximately 10^5 per 100 ml.

TABLE 16.3. Cob Corn Yields (t/ha) under Drip Irrigation and Two Lateral Spacing on RMF (Oron et al., 1990).

Parameter	Wet Yield		Wet Yield	
Lateral spacing, m	0.96	1.92	0.96	1.92
Subsurface, effluent	25.2	22.1	5.9	5.5
Onsurface, effluent	26.1	21.6	6.0	4.3
Onsurface, tap-water	27.0	25.1	6.7	5.9

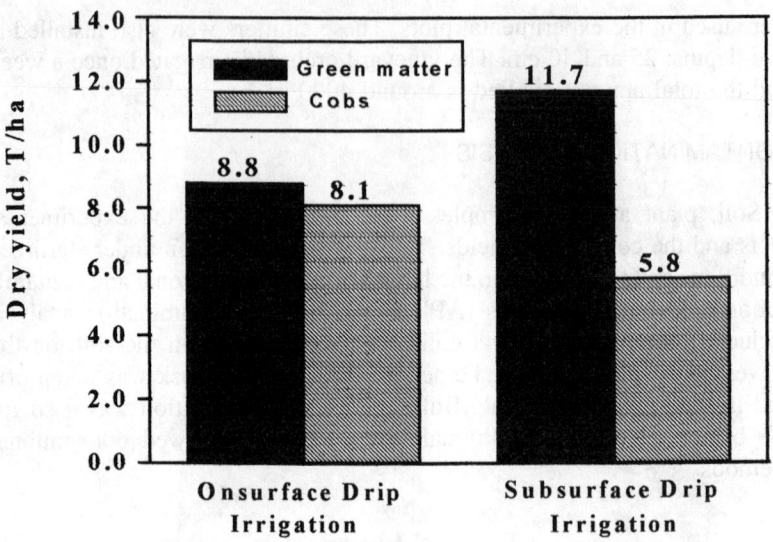

Figure 16.4 The corn yield in Kibbutz Chafets-Chaim, 1995.

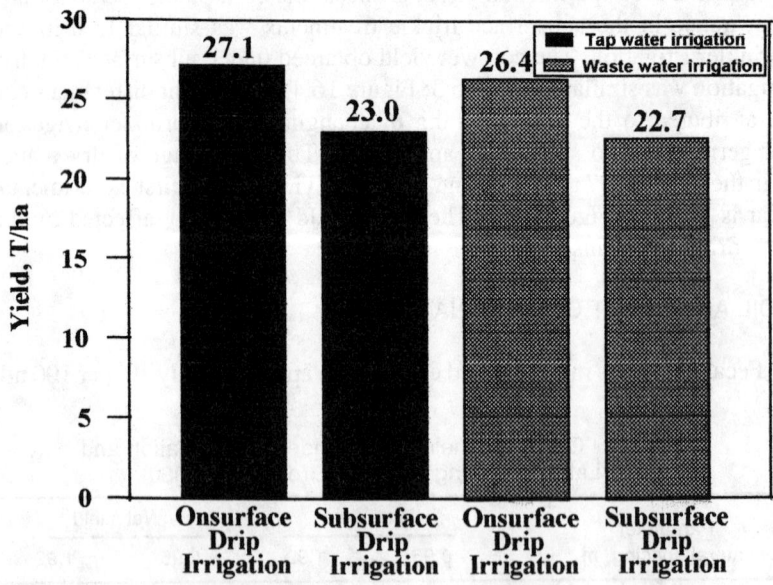

Figure 16.5 The grape yield in the vineyard orchard in Arad 1995 (emitters with 2.3 L/hr were installed at a depth of 40 cm).

TABLE 16.4. Soil Samples Contaminated by Fecal Coliforms in the Corn Field in Chafets-Chaim, 1995.

Item — Emitters Location	23 August, 1995		28 September, 1995	
	Onsurface Drip Irrigation	Subsurface Drip Irrigation	Onsurface Drip Irrigation	Subsurface Drip Irrigation
Sample size	23		24	
Number of contaminated samples at soil-surface	—	—	4	—
Number of contaminated samples at a depth of 30 cm	1	—	—	2
Number of contaminated samples at a depth of 60 cm	2	2	—	3

This level is typical for most stabilization ponds. Although the sampling was conducted one to two days after the end of the irrigation shift, bacteria were detected in the various system components. Maximal bacteria concentrations [expressed in Colony Forming Units (CFU)] were found under sprinkler irrigation at the soil surface, the leaves and cobs of the plants (Table 16.4). Essentially as anticipated, minimal bacteria concentrations were detected at the soil surface under subsurface effluent application. Maximal bacteria concentrations at a soil depth of 30 cm were detected for subsurface application. Similar results for soil contamination were found for the vineyard orchard (Table 16.5). These findings agree with previous results by Gerba et al. (1975), showing that moisture content is one of the major factors hindering die-off of microorganisms.

Maximal bacteria concentrations on the kernel cobs were found in the sprinkler plots. The limited bacteria concentrations that were detected on the cobs under subsurface and fresh water application, are probably due to drift effects of aerosols from adjacent sprinkle irrigated fields. Ineffectual bacteria concentrations on the cobs were detected at the DI plots irrigated with effluent. This can be attributed to rapid decay processes, caused by direct exposure to solar radiation. The concentration of fecal coliforms in the samples taken from the local market (control samples) were in a similar range. The field results from these experiments indicate that SDI with effluent is a promising technology to simultane-

TABLE 16.5. Fecal (CFU/gr dry soil) in the Arad Vineyard Irrigated with Domestic Effluent, 1994.

Irrigation Characteristics		Fecal Coliforms		
Emitter Discharge, L/hr	Emitter Depth, cm	At the Soil Surface	20 cm Deep	40 cm Deep
3.5	40	—[a]	—	n[b]
2.3	40	—	—	2
3.5	25	2.8×10^1	4.7×10^1	5.3×10^1
2.3	25	—	—	8.7×10^1

[a] Negligible, not detectable.
[b] Not monitored.

ously satisfy water shortage problems and reuse of treated wastewater for irrigation.

SUMMARY AND CONCLUSIONS

In a series of field experiments it was verified that drip technology, primarily SDI, can be combined with effluent disposal and meets, simultaneously, water shortage problems and environmental contamination control standards. Proper effluent filtration at the head control component will prevent emitter clogging. The dryness of the soil surface under subsurface effluent disposal minimizes any contamination risk. The limited fruit contamination indicates that delayed harvesting after last irrigation can be an additional acceptable safety barrier.

ACKNOWLEDGEMENT

The partial financial support for this study by AVICENE fund (European Community), research project, No. 93AVI076, by the Chief Scientist of the Ministry of Agriculture of the State of Israel (Research project number 857-0279-95) and by the Chief Scientist of the Ministry of Environment of the State of Israel (Research project number 120-5) is gratefully acknowledged.

REFERENCES

Adin, A., Elimelech, M. 1989. "Particle filtration for wastewater irrigation". *Journal of Irrigation and Drainage Engineering, ASCE,* 115(3): 474–487.

APHA, *Standard methods for the examination of water and wastewater.* 19th edition, American Public Health Association, Washington, D.C., 1995.

Asano, T. and Pettygrove, G. S. 1987 "Using reclaimed municipal wastewater for irrigation". *California Agriculture,* 42(3&4): 15–18.

Asano, T. and Mills, R. A. 1990. "Planning and analysis for water reuse projects". *Journal of the American Water Works Association,* January: 38–47.

Asano, T., Richard, D., Crites, R. W. and Tchobanoglous, G. 1992. "Evolution of tertiary treatment requirements in California". *Water Environmental and Technology,* 4(2): 37–41.

Asano, T., Maeda, M., and Takaki, M. 1996. "Wastewater reclamation and reuse in Japan: overview and implementation examples". *Water Science Technology,* 34(11): 216–226.

Bar-Yosef, B., Sagiv, B., and Markovitch, T. 1989. "Sweet corn response to surface and subsurface trickle phosphorus fertigation". *Agronomy Journal,* 81: 443–447.

Blumenthal, U. J., Strauss, M., Mara, D. D. and Cairncross, S. 1989. "Generalized model of the effect of different control measures in reduction health risks from wastewater reuse". *Water Science Technology,* 21: 567–577.

Bouwer, H. 1991. "Ground water recharge with sewage effluent". *Water Science Technology,* 23: 2099–2108.

Bresler, E. 1977. "Trickle-drip irrigation: principles and application to soil-water management". *Adv. Agronomy,* 29: 343–393.

Brimberg, J., Mehrez, A. and Oron, G. 1994. "Economic development of groundwater in arid zones with applications to the Negev Desert, Israel". *Management Science,* 40(3): 353–363.

Burau, H. G., Sheikh, B., Cort, R. P., Cooper, H. C. and Ririe, D. 1987. "Reclaimed water for irrigation of vegetables eaten raw". *California Agriculture,* 41(7&8): 4–7.

Chase, R. G. 1985. "P application through a sub-surface trickle system". *ASAE Proceedings of the IIIrd International Drip/Trickle Irrigation Congress,* 18–21 November, 1985, Fresno, California, Vol. II: pp. 393–400.

CIHEAM. 1988. Reuse of low quality water for irrigation in Mediterranean countries. *Proceedings of the Cairo/Aswan Seminar,* January 16–21, p. 204.

Dawood, S. A. and Hamad, S. N. 1985. A comparison of on-farm irrigation system performance. *Proceedings of the IIIrd International Drip/Trickle Irrigation Congress, ASAE,* II, 1985, pp. 540–545.

Eitan, G. 1994. *Wastewater collection and treatment in Israel—a survey* (in Hebrew). The State of Israel, Ministry of Agriculture, The Water Commission Department, p. 456.

Fattal, B., Margalith, M., Shuval, H. I., Wax, Y., and Morag, A. 1987. "Viral antibodies in agricultural populations exposed to aerosols from wastewater irrigation during a viral disease outbreak". *American Journal of Epidemiology,* 125(5): 899–906.

Gerba, C. P., Wallis, C. and Melnick, J. L. 1975. "Fate wastewater bacteria and viruses in soil". *Journal of Irrigation and Drainage Div., ASCE,* 101(IR3): 157–174.

Ghobrial, F. H. 1993. "Performance assessment of three wastewater treatment plants producing effluents for irrigation". *Water Science Technology,* 9(1): 139–146.

Gilbert, R. G., Gerba, C. P., Rice, R. C., Bouwer, H., Wallis, C. and Melnick, J. L., 1976. "Virus and bacteria removal from wastewater by land treatment." *Applied and Microbiology,* 32(3): 333–338.

Goyal, S. M., Keswick, B. H. and Gerba, C. P. 1984. "Viruses in groundwater beneath sewage irrigated cropland". *Water Research,* 18(3): 299–302.

Haas, C. N. and Rose, J. B. 1996. "Distribution of *Cryptosporidium* oocysts in a water supply". *Water Research,* 30(10): 2251–2254.

International Reference Center for Wastes Disposal. 1985. Health aspects of wastewater and excrete use in agriculture and aquaculture. The Engelberg Report, Dubendorf, Switzerland, IRCWD News, 23, pp. 11–18.

ISQW. 1981. Israel standards for quality of wastewater effluent to be reused for irrigation of agricultural crops. State of Israel, Israel Public Health Law No. 4263, paragraph 65, Aug., 1981.

Issar, A., Adar, E. 1992. "Integrated use of marginal water resources in arid and semi-arid zones." NATO ARW, Kluwer Scientific Publishers. In: H. J. W. Verplancke et al., (Ed.), *Water Saving Techniques for Plant Growth*, 229–239.

Itoyama, T., Yokose, H., Yoshida, S. and Kuwahara, M. 1990. "Evaluation of land application using secondary effluent in a forest slope: estimation of drained water quality and discussion of the effects upon soil or plants and behavior of bacteria". *Water Research*, 24(3): 275–288.

Juanico, M. and Shelef, G. 1994. "Design, operation, and performance of stabilization reservoirs for wastewater irrigation in Israel". *Water Research*, 28(1), 175–186.

Ju-Si, W. 1996. Wastewater treatment and reuse—a case study in a petrochemical industrial area. Presented at the *18th IAWQ Biennial International Conference*, 23–28 June, 1996, book 6: 261–267.

Kanarek, A. and Michail, M. 1996. Groundwater recharge with municipal effluent, Dan-Region Reclamation Project, Israel. Paper presented at the *18th biennial IAWQ International Conference on Water Quality*, 23–26 June, 1996, Singapore pp. 228–234.

Kott, Y., Ben-Ari, H., and Vinokur, L. 1978. "Coliphages survival as viral indicator in various wastewater quality effluents". *Progr. Water Technology*, 10: 337–346.

McCaulou, D. R., Bales, R. C. and Arnold, R. G. 1995. "Effect of temperature-controlled motility on transport of bacteria and microspheres through saturated sediment". *Water Resources Research*, 31(2): 271–280.

Mehrez, A., Percia, C. and Oron, G. 1992. "Optimal operation of a multisource and multiquality regional water system". *Water Resources Research*, 28(5): 1199–1206.

Miller, K. J. 1994. "Protecting consumers from Cryptosporidiosis". *Journal AWWA*, December: 108–110.

Nativ, R. and Issar, A. 1988. Problems of an over-developed water-system—the Israeli case. *Water Quality Bulletin (Groundwater Management—Part I)*, 13(4): 126–131, 146.

Nellor, M. H., Baired, R. B. and Smyth, J. R. 1985. "Health effects of indirect potable water reuse". *Journal American Water Works Association*, January: pp. 88–96.

Oron, G., Shelef, G. and Zur, B. 1980. "Storm water and reclaimed effluent in trickle irrigation". *Journal of the Irrigation and Drainage Division, ASCE*, 106(IR4): 299–310.

Oron, G., DeMalach, Y. and Bearman, J. E. 1986. "Trickle irrigation of wheat applying renovated waste water". *Water Resources Bulletin (AWWA)*, 22(3): 439–446.

Oron, G. and Enthoven, G. 1987. "Stochastic considerations in optimal design of a microcatchment layout of runoff water harvesting". *Water Resources Research*, 23(7): 1131–1138.

Oron, G., DeMalach, Y., Hoffman, Z. and Cibotaru, R. 1991. "Subsurface microirrigation with effluent". *Journal of Irrigation and Drainage Engineering, ASCE*, 117(1): 25–36.

Oron, G., DeMalach, Y., Hoffman, Z., Manor, Y. 1992. "Effect of effluent quality and application method on agricultural productivity and environmental control". *Water Science Technology*, 26(7&8): 1593–1601.

Pasternak, D. and DeMalach, Y. 1987. Saline water irrigation in the Negev Desert. Paper Presented at the *Regional Conference on Agriculture Food Production in the Middle East*, Athens, Greece, Jan. 21–26, 24p.

Pescod, M. B., and Alka, U. 1988. "Guidelines for wastewater reuse in agriculture". In: Pescod, M. B. and Arar, A. (Ed.), *Treatment and use of sewage effluent for irrigation.* Butterworth Scientific Ltd., 1988, Chapter 3: 21–37.

Phene, C. and Ruskin, R. 1995. "Potential of subsurface drip irrigation for management of nitrate in wastewater". In: Lamm, F. R. (Ed.), *Microirrigation for Changing World: Conservation Resources/Preserving the Environment. Proceeding of the 5th International Microirrigation Congress (ASAE),* 2–6 April, 1995, Orlando, Florida, USA: 155–167.

Salgot, M., Brissaud, F., and Campos, C. 1996. "Disinfection of secondary effluents by infiltration-percolation". *Water Science Technology,* 33(10–11): 271–276.

Shelef G., and Y. Azov (1995). "The coming era of intensive wastewater reuse in the Mediterranean Region". In: Angelakis, A., Asano, T., Diamadopoulos, E. and Tchobanoglous, G. (Editors): *Wastewater Reclamation and Reuse. Proceedings of the 2nd IAWQ Reuse Conference,* Iraklio, Crete, Greece, 17–20 October, 1995, pp. 137–146.

Siebe, C. and Cifuentes, E. 1995. "Environmental impact of wastewater irrigation in central Mexico: and overview". *International Journal of Environmental Health Research* 5: 161–173.

Smith, M. A. 1982. *Retention of bacteria, viruses and heavy metals on crops irrigated with reclaimed water.* Australian Water Resources Council, Canberra, p. 308.

Telsch, B., and Katzsenelson, E. 1978. "Airborne enteric bacteria and viruses from spray irrigation with wastewater". *Applied and Environmental Microbiology,* 35(2): 290–296.

The Great Man-Made River Project. 1989. Socialist People's Libyan Arab Jamahiriya, Libya, p. 25 (and additional descriptive plates).

Tollefson, S. 1985. "Sub-surface drip irrigation of cotton and small grains". *ASAE Proceedings of IIIrd International Drip/Trickle Irrigation Congress,* 18–21 November, 1985, Fresno, California, Vol. I, pp. 887–895.

U.S. Environmental Protection Agency (US EPA). 1992. *Guidelines for water reuse* (manual). US EPA, Washington, DC, EPA/625/R-92/004, September 1992, p. 247.

WHO. 1989. World Health Organization, Health guidelines for the use of wastewater in agriculture and aquaculture. Report of a WHO scientific group, WHO technical report, series 778, Geneva.

CHAPTER 17

Tertiary-Treated Reclaimed Water for Irrigation of Raw-Eaten Vegetables

INTRODUCTION

SCARCITY of water supplies has become recognized as a real economic threat to many communities in recent years, and is worsening apace with population growth, particularly in the less industrialized regions of the world. The more developed regions are not immune either. The official California Water Plan [1] projects a deficit of 235 m³/s (six million acre-ft per year) by the year 2020. California received a harsh sampling of the impending water shortfalls during the prolonged drought of 1988–1993. To meet the deficit, many strategies are being proposed. One of the most feasible at this time, with the promise of filling up to half of the gap, is water reclamation. But, at what level of treatment is reclaimed water good enough for reuse? More specifically, how extensively should wastewater be purified before it can be placed in irrigation systems for the food crops eaten raw?

Today, raw sewage is used in many parts of the world—practically on all continents—for irrigation of all types of crops including those consumed without processing or cooking. The result is both endemic and periodic epidemics of a variety of enteric diseases. Foreigners visiting these regions are warned not to drink the water or eat uncooked vegetables. International funding agencies are trying to improve sanitation in these countries by at-

Bahman Sheikh and Robin Cort, Parsons Engineering Science, Inc., 2101 Webster St., Suite 700, Oakland, CA 94612; Robert C. Cooper, School of Public Health, University of California at Berkeley, Berkeley, CA 94720 and BioVir Laboratories, Inc., 685 Stone Rd., Benicia, CA 94510; Robert S. Jaques, Monterey Regional Water Pollution Control Agency, #5 Harris Court, Bldg. D, Monterey, CA 93940.

tempting to provide advanced levels of wastewater treatment, at great costs. Nonetheless, the pace of population growth, and scarcity of adequate irrigation water place increasing pressures on local farmers to use whatever sources of water are "available," regardless of their quality. By contrast, California has maintained a successfully strict balance between its need for additional water supplies and its goal of protecting the public health.

Local environmental health officials and the farming community were not ready to give total approval to a project in Monterey County proposing to use tertiary reclaimed water for food crop irrigation. Out of their concerns grew the concept of a pilot project to investigate the safety and feasibility of using reclaimed water for irrigation of vegetable crops.

Water quality requirements established by agronomists for irrigation of crops are very different from those designed specially for the protection of public health. Both the public health criteria and agronomic concerns were high priority objectives of the Monterey Wastewater Reclamation Study for Agriculture (MWRSA). Two compelling reasons motivate public health concerns about reclaimed water used in the irrigation of food crops: (a) protection of individuals consuming the produce and handling it in the kitchen environment; and, (b) protection of farm workers coming in contact with potentially contaminated soil and crops. These concerns are of particular importance in the case of vegetables eaten raw, such as lettuce, carrots, celery, green onions, broccoli, and cauliflower. These are human food crops intimately manipulated and consumed with no intermediate processing in the home and in the restaurant.

OVERVIEW OF THE STUDY

PARTICIPANTS

The objectives of the Monterey Wastewater Reclamation Study for Agriculture (MWRSA) were set through extensive dialog among the stakeholders. Participants in the numerous sessions held to arrive at consensus included:

- several growers in the Castroville area of Monterey County
- local environmental health officer
- agricultural extension service representatives
- University of California professors of soils, agronomy and public health
- US Environmental Protection Agency
- State Water Resources Control Board
- Regional Water Quality Control Board—Central Coastal Region
- local water resources and flood control agency

These and several other entities formed a task force that oversaw MWRSA from inception to completion. The task force reviewed planning reports and annual reports and made mid-course corrections as necessary. The benefit of hands-on involvement of the task force was that it sought buy-in by the end users and other stake-holders.

OBJECTIVES

The ultimate objective of MWRSA was to demonstrate the overall feasibility of water reclamation for food crop irrigation in northern Monterey County. Specifically, the primary objective of MWRSA was to provide quantitative, unbiased, and authoritative information and data regarding the following concerns:
- safety of consumers and farm workers
- survival of viruses in water, on plants and in the soil
- survival of bacteria on crops and soils
- aerosol transmission of viruses and bacteria
- heavy metals in edible and residual plant materials
- impact of reclaimed water on soil (metals, salts, permeability)
- growth, yield, quality of crops
- marketability of the crops grown with recycled water
- economics of irrigation with reclaimed water

Secondary objectives were:
- evaluate wastewater treatment effectiveness
- provide design criteria for the full-scale regional plant
- provide criteria for full-scale use of reclaimed water in vegetable fields
- provide operational experience

MATERIALS AND METHODS

The MWRSA project spanned eleven years from planning and design (1975 to 1980) to field experiments (1980 to 1985) and final project reporting in 1986. Demonstration fields and test plots were located near the town of Castroville on the central coast of California, as shown in Figure 17.1. A portion of the material in this Chapter—excluding the virus data presented herein—was published in 1990 [2].

BASELINE STUDIES

A series of studies were conducted to select the most appropriate field site, to characterize its soil condition prior to the use of recycled water,

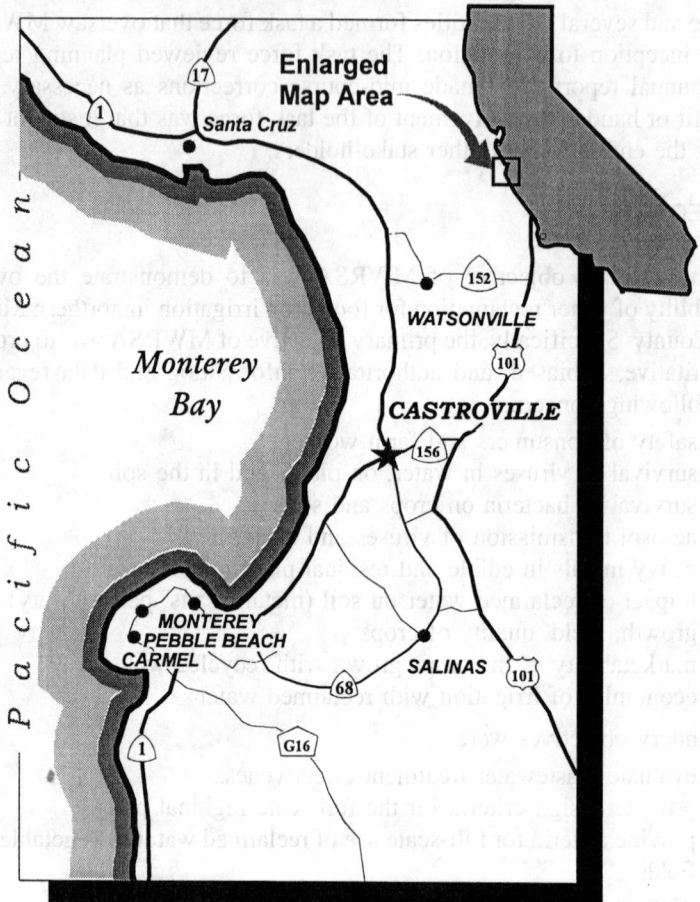

Figure 17.1 Project location map.

and to prepare design criteria for the pilot plant process trains used in the study. The results of these studies are reported in the annual and final project reports [3–9] published in limited circulation over the period from 1980 to 1987.

PILOT PLANT DESCRIPTION

The pilot plant was constructed as an adjunct to an existing secondary (activated sludge) plant located in Castroville, California. The unchlorinated effluent from this plant became the influent of the pilot plant.

The pilot plant consisted of two parallel processes: one that included coagulation, flocculation, sedimentation, filtration and chlorination; and, the second that consisted of the addition of coagulant and polymer followed by coagulation, flocculation, filtration and chlorination. Initially, both process trains also included dechlorination. However, dechlorination was later discontinued when it was found to be unnecessary. The former process complied with the requirement in effect at that time by the California Department of Health Services (DOHS) under Title 22 of the California Code of Administration [10] and is therefore referred to as the T22 process throughout this Chapter. The other process train is referred to as the filtered effluent (FE or FE-F in the case of filtered effluent receiving special flocculation opportunity) process. Virus removal from the two process streams was consistently monitored as a primary and direct comparison of the potential equivalency of the two treatment trains.

REPLICATED FIELD PLOTS

In order to fulfill the objective of MWRSA, it became necessary to design an experiment at the end of which rigorous statistical analysis of the results could be performed. This is essential whenever biological systems are subjected to independent variables, because so many other variables are simultaneously at work, randomly affecting the behavior of the system. Statistical methods can separate "noise" generated by extraneous variables from cause-and-effect attributable to specific variables. The so-called "split-plot" design was selected in consultation with the University of California, Davis agronomists and soil scientists. This choice enabled the use of Analysis of Variance (ANOVA) for analysis of data generated [11].

The experiment had two independent variables: irrigation water type and fertilizer application rate. Three water types (well water for control, filtered effluent and coagulated, settled, filtered effluent) and four fertilizer application rates (full, 2/3, 1/3, and zero) comprised the experimental variables. These variates were replicated four times in a total of 48 subplots, randomly assigned to the field by drawing lots before the start of the experiment. A succession of various vegetables (varieties of lettuce, broccoli, cauliflower, celery) were grown in these plots over the five-year period from 1980 to 1985. The entire experiment was replicated in another 48 subplots growing artichokes over the same five year-period. Each subplot measured 14 m (45 ft) in length and 9 to 12 m (30 to 40 ft) in width. A layout of the experimental plots is

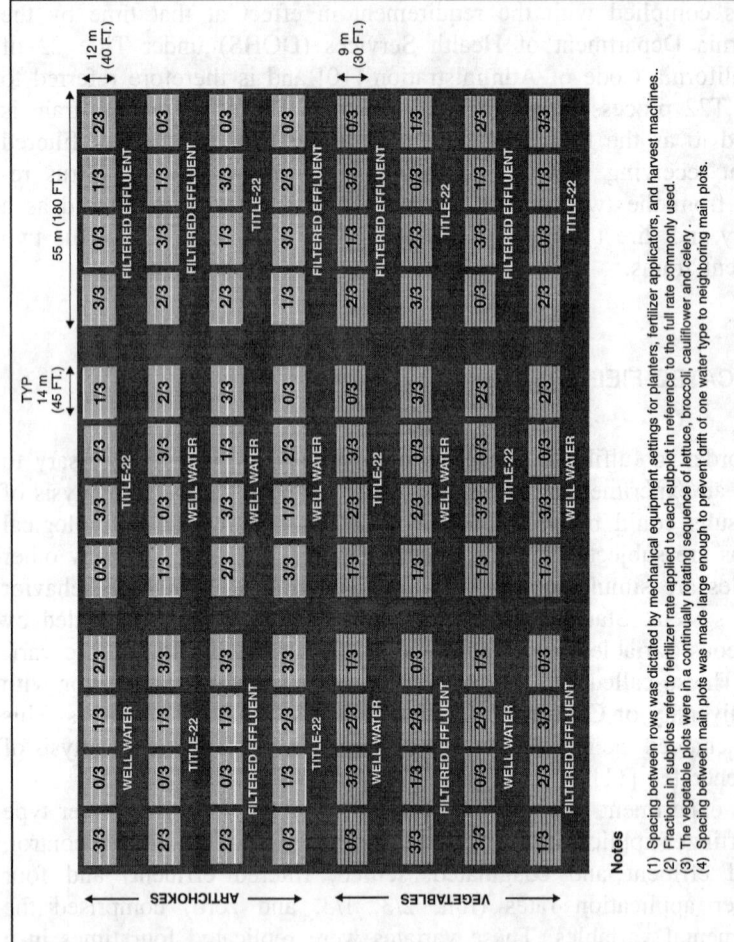

Figure 17.2 Layout of MWRSA experimental plots with randomly assigned irrigation water type and fertilization rate.

Notes

(1) Spacing between rows was dictated by mechanical equipment settings for planters, fertilizer applicators, and harvest machines..

(2) Fractions in subplots refer to fertilizer rate applied to each subplot in reference to the full rate commonly used.

(3) The vegetable plots were in a continually rotating sequence of lettuce, broccoli, cauliflower and celery .

(4) Spacing between main plots was made large enough to prevent drift of one water type to neighboring main plots.

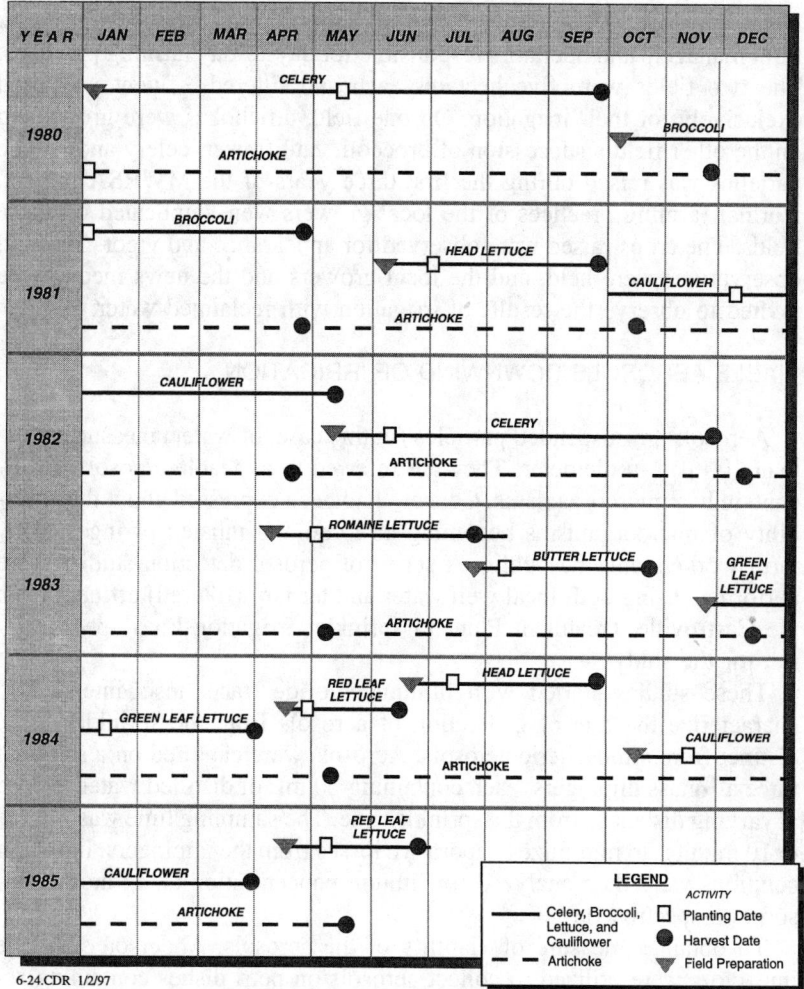

Figure 17.3 Crop rotation schedule for the five-year MWRSA pilot study.

presented in Figure 17.2. Crop rotation schedule for the duration of the experiment is shown on Figure 17.3.

DEMONSTRATION FIELDS

In addition to the replicated plots providing material for sampling, analysis and statistical evaluations, two relatively large demonstration fields also were utilized adjacent to the experimental plots. The purpose of these fields was to determine field-scale feasibility of using reclaimed water. This demonstration was of special importance to the local growers, farm managers and operators responsible for day-to-day farming practices. The two fields were five hectares each and filtered effluent was used exclusively for their irrigation. On one field, artichokes were grown and on the other field a succession of broccoli, cauliflower, celery and lettuce varieties was raised during the first three years of the MWRSA project. Normal farming practices of the local growers were duplicated on these fields. The crops raised were observed for appearance and vigor. Six field observations were held, and the local growers and the news media were invited to observe the results of irrigation with reclaimed water.

VIABLE AEROSOLS DOWNWIND OF IRRIGATION

Aerosols are suspended particles (in this case, of water) measuring less than 100 μm in diameter. They are referred to as viable aerosols if they contain live microorganisms. Concern had been expressed about the possibility of microorganisms becoming airborne and inhaled or ingested by farm workers and the public. A series of aerosol detection studies were performed using both local well water and tertiary (filtered) effluent from the Castroville Treatment Plant in sprinkler irrigation lines specifically run for the study.

These studies started with lithium chloride tracer tests intended to characterize the rate of generation of aerosols from the sprinkler lines, distinct from atmospheric aerosols. Aerosols were captured on a series of three all-glass impingers, each containing 20 mL of distilled water, placed at varying distances from the sprinkler line. The sampling time was limited to 10 minutes to minimize evaporative losses from the impingers. Impinger solutions were then analyzed for lithium concentration using atomic absorption spectroscopy.

To obtain a measure of viability of the aerosols, Anderson six-stage impactors were utilized to collect aerosols on petri dishes containing selected biological culturing media. These petri dishes were then incubated at 35°C for 24 hours, and colonies were counted and identified. Aerosols in six particle size ranges were captured at distances varying from 15 m upwind of the sprinkler line (for background measurements) to 30 and 60 m downwind. All sampling was performed at a height of 1.5 m, about the height of average adult human respiration. The three samplers were operated simultaneously by three field workers, suctioning air at the rate of 28 L per minute. Wind speed and direction, temperature, relative humidity, cloud cover, and time of day were recorded for each sampling event. Two types of media were used: standard plate count as an indicator of total aerosol bacteria, and EMB-Levine as an indicator of enteric bacteria. Both day-time and night-time samplings were performed to be representative of actual irrigation conditions. At the same time, samples of irrigation water were also taken to determine its bacteriological content.

VIRUS ASSAY METHOD

Viral plaque assays were performed using Buffalo Green Monkey Kidney Cells (BGM). The growth medium was 45 percent Hanks Minimum Essential Medium (MEM), 5 percent L-15 medium and 10 percent fetal bovine serum containing 0.01 percent L-glutamine, potassium penicillin (100 units/mL), streptomycin sulfate (0.1 mg/mL) and enough 7.5 percent $NaHCO_3$ to produce a pH of 7.2 to 7.4. Cells were incubated at 37°C until confluent. Each cell culture was inoculated with 0.2 to 0.5 mL of prepared sample and incubated at 37°C for one hour to allow virus adsorption. Each bottle then received overlay medium consisting of MEM with Hanks Balanced Salts containing two percent gammaglobulin-free bovine serum, 0.1 percent milk (Difco), 0.01 percent $MgCl$, potassium penicillin (100 units), streptomycin sulfate (0.1 mg/mL, 0.75 percent $NaHCO_3$), 1.5 percent agar, and 0.01 percent neutral red. These overlays were incubated for three to five days at 37°C prior to counting the resultant plaques.

Recovery of in situ Virus from Water

Because the number of enteric viruses present in any of the irrigation waters was expected to be low, it was necessary to concentrate the viruses. Large volumes of water were passed through fiberglass filters and the viruses present in the water adsorbed to the filters. The water was pretreated by adjusting 3800 L of the irrigation water to pH 3.5 using hydrochloric acid and adding aluminum chloride to a concentration of 0.0005M. This adjusted sample was then pumped through four 25-cm tall fiberglass-epoxy filters (Filterite Corp.) each with an effective porosity of 0.45 μm. The four filters were arranged in parallel and the total flow rate through the filters was 30 to 38 L/min. The adsorbed viruses were subsequently desorbed from the filters by passing through 3 L of 3.5 percent beef extract buffered to pH 10 with 0.5M glycine buffer. The 3 L of eluate was then adjusted to pH 3.5, resulting in a precipitate to which viruses would be adsorbed. This precipitate was collected using a continuous flow centrifuge at 12,000 RCF. The resulting pellet was then redissolved in up to 25 mL of 0.15M sodium biphosphate at pH 9.2, the pH neutralized, and the sample immediately assayed or stored at −70°C. Virus recovery efficiency using this method ranged from 25 to 70 percent which is to be expected with these concentration methods.

The volume of pilot plant influent sample (unchlorinated secondary effluent) was six liters. Each sample was adjusted to pH 3.5 and aluminum chloride added to 0.0005M, and filtered through a 142 mm Cox filter "sandwich." The "sandwich" was composed of an AP20 (Millipore Filter Corp.), a Cox 5 micrometer and a Cox 0.45 micrometer (Cox Research Corp.) filter. The adsorbed viruses were then eluted with 100 mL of pH 11.5 glycine buffer, quickly adjusted to pH 7.0, and the resultant material assayed for virus.

Virus Seeding Studies

During the course of the virus studies it became apparent that in situ virus concentration in the pilot plant influent water was very low; thus in order to estimate the virus removal efficiency of the two pilot plant processes virus seeding studies were initiated. The test virus employed was a vaccine strain poliovirus (Polio 1 LSC). This is an attenuated or "deactivated" form of the virulent poliovirus and poses no threat to human health. Since the volume of flow into the pilot plant was too large for continuous virus seeding it was necessary to inoculate slugs of virus. The high-dose challenge seed of a vaccine strain of poliovirus (1 LSC) was added to the plant influent along with a fluorescent tracer dye, pontacyl pink B, in order to determine the dilution factor associated with the final virus sample. For example if only 0.01 percent of the initial dye concentration were present in the virus sample one could assume that at least 99.99 percent of the observed reduction in virus was due to dilution. Dye measurements were made using a Fluorometer equipped with appropriate filters. A series of tests were conducted that determined that the dye did not interfere with the assay of poliovirus nor did the doses of chlorine used in the test processes have a significant affect on dye concentration. Chlorine residual in all virus samples was neutralized with sodium thiosulfate.

Recovery of Viruses from Soil and Vegetables

Soil and vegetable samples were collected from the experimental plots within 24 hours of the end of an irrigation set. Two plants and associated soil samples were selected at random from the outside rows of non-fertilized test plots and composited. In each instance, the final sample amounted to at least 50 g of material. Viruses were eluted from both plant and soil surfaces using three percent, pH 9 beef extract. The resultant eluate was then precipitated by adjusting the pH to 3.5 and the precipitate collected by centrifugation at 12,000 RCF.[1] The resultant pellet was redissolved in 0.15 M sodium biphosphate at pH 9.2, neutralized and assayed

---

[1] RCF = relative centrifugal field, a measure of the effective acceleration applied by the centrifuge.

for virus. Virus recovery efficiency using these methods ranged from 19 to 37 percent from soil and 36 to 83 percent from vegetables.

ENVIRONMENTAL CHAMBER STUDIES

To study the survival characteristics of viruses on plants and soil, experiments were conducted in an environmental chamber which simulated weather conditions in Castroville, California. Plants and soil were inoculated with the vaccine strain poliovirus, exposed to chamber conditions and the decay in virus concentration was measured over time. The environmental chamber located at the University of California Sanitary Engineering and Environmental Health Research Laboratory was a 6 × 2.4 m double wall insulated structure. Temperature and humidity control were provided by an integrated system which includes air circulation through the chamber and an air conditioning system. Lighting was provided with an enclosed light box equipped with a mixture of F40 BL and F40 CL fluorescent lamps. The former lights supplied the ultraviolet spectrum of sunlight and the latter the wave lengths required for plant growth. The chamber was operated at a temperature of 15.6°C (60°F) and at three different relative humidities of 60, 70 and 80 percent respectively. Known amounts of virus were sprayed on to tared test plants such that the increase in weight of the inoculated plants reflected the amount of inoculum and thus the zero time virus concentration. The exposed plants included: artichokes, black-seeded Simpson lettuce, celery, broccoli, and head lettuce. Each plant stem or root remained immersed in water throughout the exposure period.

The survival of animal virus in Castroville soil under controlled in situ conditions was studied. The investigation involved the use of Castroville soil seeded with poliovirus and exposed to ambient conditions at the University of California Engineering Field Station in Richmond, California. This latter site was chosen because: (a) the logistics of sampling soil and performing virus analysis frequently over a 20 day period made performing the study at Castroville impractical; and (b) the Richmond site is adjacent to the east shore of San Francisco Bay and has a climate very similar to that found at Castroville with foggy cool nights and mornings and sunny afternoons.

The methods employed for the virus survival in soil study were as follows: Castroville soil was collected from the demonstration site and homogenized in a soil mixing mill. Soil columns were prepared using twelve-inch long, three-inch diameter schedule 40 PVC pipes each divided into four-inch sections. The sections were glued together with silicone cement. For each of the two runs conducted 25 columns were prepared. Covered empty columns were placed symmetrically in a five-foot square excavated plot and the plot back filled with indigenous soil such that the top of each column was one quarter inch above the soil surface.

Each column was then filled with 1640g of homogenized Castroville soil and compacted by spraying the plot with 250 L (67 gal) of local water (75 mm or 3 inches of water). After the test columns were "irrigated" a 350 mL suspension of poliovirus in Castroville dechlorinated filtered effluent was dripped onto the surface of each column over a 2.5 hour period. This was accomplished by filling sterile plastic bags with the appropriate volume of virus seed and allowing the contents of each of the 24 bags to drip onto the surface of each of 24 soil columns. At each sample date two columns were randomly selected and removed from the soil plot. The exception to this was at time zero when four columns were selected to establish base line virus concentration. In each case the column was divided into its top, middle and bottom section. The soil from each section was then analyzed for the number of viruses present. Each selected pipe column was divided into the three sections and the soil from each carefully mixed. Fifty g of the mixed section was taken for virus assay. The 50 g of soil was suspended in 3 percent beef extract at pH 10 in 0.25 M glycine buffer for 10 min. and extracted at 200 RPM on a gyrorototory shaker and then centrifuged for 10 min. at 2000 RPM. Twenty mL of the supernate, adjusted to pH 7 was assayed for virus using previously described methods.

Virus survival on plants under in situ conditions was also studied. The method adopted required that a growing plant (or appropriate portion of same) be sprayed with a known volume of virus suspension in as uniform a manner as practicable. The average amount of virus retained on from four to eight plants immediately after spraying was used as the baseline for computing virus decay. The number of viruses present on the exposed plants at the end of any given time period was determined using elution and assay methods previously described. The test periods were one week for artichokes, 17 days for butter lettuce and 4 days for romaine lettuce. A test plot section of plants, irrigated with well water, was set aside for these tests. On each occasion, from 12 to 15 plants were sprayed with virus suspension. The virus concentration was determined as PFU per entire plant or artichoke rather than per weight or surface area. In this way plant growth, particularly in the case of butter lettuce, would not introduce an unknown dilution factor over the course of the test. Plants were selected at random during the experiment, and the number of viruses remaining were determined.

SAMPLE COLLECTION, PRESERVATION, HANDLING AND ANALYSIS

Sample Collection—Irrigation Water

Throughout the five years of field studies, samples of the three irrigation waters were taken at each irrigation event. Samples of the three water types were continuously composited during each irrigation. Depending on the irrigation schedule, samples were collected over a three-to-five-day period. The composite samples were then divided into subsamples for metal and chemical analysis. Grab samples of irrigation water were collected in sterilized bottles for bacteriological analysis and in unsterilized clean bottles for biochemical oxygen demand (BOD) analysis. These grab samples were stored in an ice chest and sent to the laboratory within 24 hours. Irrigation waters were sampled 91 times over the five years for 33 artichoke and 58 vegetable irrigations.

During furrow irrigation of row crops, tailwater samples were collected from runoff. Eight tailwater samples were collected during the five years of sampling. The remainder of the 58 irrigations were performed with sprinklers.

Analysis of Water Samples—Bacteriological Analyses

All bacteriological samples were analyzed for total and fecal coliform and parasites. Samples that were positive for fecal coliform were also analyzed for shigellae and salmonellae. Coliform analyses were performed using multiple-tube fermentation techniques. All positive tubes from the presumptive test were confirmed, and 10 percent of all confirmed tests were completed. Results were reported as most probable number (MPN). The formalin-ether sedimentation technique was used to recover helminth eggs and protozoan cysts.

Analysis of Water Samples—Metal Analyses

During Years One through Three, samples for metals were preserved with nitric acid and analyzed by atomic absorption spectroscopy for cadmium, zinc, iron, manganese, copper, nickel, cobalt, chromium, and lead. Because levels of cadmium, copper, nickel, cobalt, and lead were very low, a method was developed to concentrate metals. Metal samples in Years Four and Five were composited and analyzed using this new technique, which is detailed in the soil section under soil metal analyses.

Analysis of Water Samples—Chemical Analyses

Aqueous samples were analyzed for pH, electrical conductivity, carbonate, bicarbonate, nitrogen as nitrate, nitrogen as ammonia, BOD, and methylene-blue-active substances (MBAS) within 24 hours of receiving the sample. Waters were also analyzed for hardness, total phosphorus, sulfate, chloride, boron, total dissolved solids (TDS) and, for tailwater samples only, total Kjeldahl nitrogen. Calcium, magnesium, sodium, and potassium were analyzed by atomic absorption spectroscopy. To provide a measurement of the concentration of sodium in the waters relative to calcium and magnesium, sodium adsorption ratios (SAR) were calculated. Adjusted SAR (ASAR) values, which consider the tendency of calcium to precipitate or dissolve, were also calculated.

Soil Sample Collection

During the first three years of field studies, surface soil samples were taken for bacteriological analyses in the fall and spring. Samples were taken with a trowel from the surface 15 cm (6 in.) of soil within two days after irrigation. Five subsamples of 20 g of soil were taken from different locations within each plot and composited to produce a 100-g sample. Locations for subsamples were chosen randomly from the inner rows of each plot. Composited soil samples were cooled and shipped to the laboratory within 24 hours. Bacteriological sampling was discontinued after Year Three.

Throughout all five years of MWRSA, soil profile samples were collected and analyzed for a variety of metal, chemical, and physical parameters. Soil sampling was performed by the California Department of Water Resources. Soils were analyzed annually for metals and organic matter content. During the first two years, biannual sampling was conducted for cation exchange capacity, boron levels, and chemical parameters such as pH and salt content. After the first two years, sampling frequency was reduced to once each year. Biannual samples were taken at the end of the irrigation season (mid-October to mid-December) and again after artichoke cutback (mid-May). Annual samples were taken at the end of the irrigation season. A baseline soil profile sample was taken in December 1979, before the beginning of the MWRSA field operations, and analyzed for the full complement of metal, chemical, and physical parameters.

At each sampling event, soil samples were taken with a soil auger at depths of 30 cm (1 ft), 100 cm (3 ft), and 200 cm (6 ft). Soil samples gathered at the 30-cm and 100-cm depths were each a composite of five subsamples taken from within each of the 96 artichoke and vegetable plots. The 200-cm sample was taken only at the center of the plot.

During the first three years of MWRSA, a portion of the annual soil samples was used for permeability analyses performed in the laboratory. In Year Four, it was decided that measurement of field infiltration rates would provide a more realistic quantification of permeability. During Years Four and Five, field infiltration rates were measured three times in both the artichoke and vegetable fields.

Soil Analyses—Bacteriological Analyses

All soil samples collected for bacteriological quantifications were analyzed for total and fecal coliform and parasites. Any sample that was identified as positive for fecal coliform was also analyzed for shigellae and salmonellae. Soil samples were weighed aseptically into sterile containers containing an appropriate diluent, placed in a horizontal position in a reciprocating shaker for ten minutes, and then assayed as for liquid samples.

Soil Analyses—Metal Analyses

Soils were analyzed for nine trace metals: cadmium, zinc, iron, manganese, copper, nickel, cobalt, chromium, and lead. All soils were first air dried and ground in a Wiley mill. The ground soil was then extracted with DTPA (diethylenetriaminepentaacetic acid), which extracts those metals that are in a plant-available form. Samples were then analyzed by atomic absorption spectroscopy.

Using this technique, it was found that levels of a number of metals were at or below detection limits. With the help of Dr. Richard Burau (University of California, Davis), an extraction protocol was developed to concentrate metals. The extraction used organic chelators that bind metals in the saturation paste extract. The chelators were then digested and the metals suspended in nitric acid. Each extraction contained a set of internal standards to correct for any difference in the specific metal-binding efficiency of the chelator. Samples were again quantified using an atomic spectrophotometer. This extraction technique was used to analyze levels of cadmium, copper, nickel, cobalt, and lead in baseline and Year-Five soils. The chelation procedure was also used in the analysis of those metals in irrigation waters in Years Four and Five.

Soil Analyses—Chemical Analyses

After air drying, soils were blended with distilled water until saturated. This saturation paste was allowed to stand to equilibrate for at least four hours, and it was mixed before determining the pH. The paste was then filtered and the aqueous extract analyzed for electrical conductivity, calcium, magnesium, sodium, potassium, carbonate, bicarbonate, and chloride.

After appropriate extraction, analyses for total Kjeldahl nitrogen, nitrogen as nitrate, nitrogen as ammonia, and boron were performed by autoanalyzer. Acetate-soluble sulfate and phosphorus as phosphate were analyzed by colorimeter.

Soil Analyses—Physical Characteristics

Organic matter content of soils was determined by igniting oven-dried samples for two hours in a muffle furnace at 550°C (1,020°F) and measuring the percentage of the original sample burned off as volatile carbon compounds. Cation exchange capacity was quantified by measuring the amount of sodium that would absorb into the soil when shaken in a sodium acetate solution. Laboratory permeability analyses were performed by packing sieved soil into plexiglass columns, maintaining a constant head of water on the columns, and measuring the drainage rate. Field infiltration rates were measured using standard double-ring infiltrometers consisting of two concentric 30-cm (12-in.) tall cylinders, driven 15 cm (6 in.) into the soil, in the center of each plot. Water was poured into both rings, and the water level changes in the inner ring were recorded for four hours.

Plant Tissue—Sample Collection

At each major harvest, samples of plant tissue were collected for analysis. Edible and residual tissues were sampled for bacteriological and metals assays. Any portion of the plant that was left in the field after harvest was considered to be residual tissue.

Plant Tissues—Bacteriological Sampling

Samples of both edible and residual tissues were collected using aseptic gloves, aseptic bags, and alcohol sterilized knives. Fresh gloves were used for each plot, and knives were cleaned with alcohol before each sample. All bacteriological samples were kept on ice and shipped to the laboratory for analysis within 24 hours.

Plant Tissues—Metal Sampling

Edible portions of the crop were collected for metals analyses at each major harvest. Samples for analysis were composited from 8 to 20 plants, depending on the size of the harvestable portion of the crop. Crop residues were also sampled for analysis. The oldest leaves from 10 to 12 plants were gathered for boron assay.

Plant Tissues—Nutrient Sampling

Nutrient samples were taken from petioles of the most recently matured leaf. Samples were composited from 10 to 20 plants at each major harvest. Starting in Year Two, nutrient samples were also collected at each fertilization.

Plant Tissues—Neighboring and Random Field Sampling

In addition to sampling the experimental plots, at each artichoke harvest samples of edible tissue were also taken from the neighboring artichoke fields for bacteriological and metal assays. Because only artichokes were grown in the fields neighboring the project site, no neighboring field sample was collected at vegetable harvests. Artichokes were sampled at 15, 30, 60, 150, and 300 m (50, 100, 200, 500, and 1,000 ft).

A random selection of fields in the area was also sampled for bacteriological and metal analyses at each artichoke and row crop harvest. Samples of edible tissue were collected at four locations within a randomly selected field located at least 1 km (0.6 mi) away from the project site.

Plant Tissues—Bacteriological Analyses

Plant tissues were subjected to the same bacteriological analysis as were water and soils samples. For total coliform, fecal coliform, salmonellae and shigellae assays, tissue was weighed aseptically into sterile bottles containing phosphate-buffered diluent and 0.1 percent surfactant. The bottles were placed horizontally on a reciprocating shaker for ten minutes and then analyzed as for aqueous samples. For parasite analysis, tissues were weighed into a high-speed blender jar with physiological saline diluent, and they were blended before being assayed as for liquid samples.

Plant Tissues—Metal Analyses

Air-dried samples were ground and block digested. Edible tissue was analyzed for cadmium, zinc, copper, iron, magnesium, nickel, cobalt, chromium, and lead. Residual tissue was analyzed for cadmium, zinc, and boron. Extraction for boron was by the Association of Official Analytical Chemists (AOAC) method [12].

Plant Tissues—Nutrient Analyses

Samples were dried and ground before being extracted and analyzed for nitrogen as nitrate, phosphorus as phosphate, and potassium. Methods were the same as for aqueous samples.

ANALYSIS OF VARIANCE

Analysis of variance (ANOVA) was the primary statistical technique used to determine if significant differences existed between the characteristics of the soils and plants receiving different water types and the fertilization treatments. ANOVA provides a statistical measure of the likelihood that differences in sample means of the measured parameters (soil and plant, heavy metals, and chemicals) are attributable to the different water and fertilizer treatments or that apparent differences simply reflect natural random variation or errors in sampling. The hypotheses tested are that there are no differences in the measured parameters due to (a) water types, (b) fertilization rates, and (c) interactions[2] between water types and fertilization rates. ANOVA estimates the probability of no significant difference at a generally accepted (but arbitrarily defined) error rate of either 5 percent or 1 percent. Statistical significance at a 5 or 1 percent level indicates that there is a 95 or 99 percent chance, respectively, that differences noted in water types or fertilizer rates are not merely due to chance variation or sampling error. A more detailed discussion of the ANOVA procedure and the underlying assumptions are given in standard texts on field plot technique [11].

It should be noted that the occurrence of an apparently statistically significant result does not necessarily imply biological or agricultural significance. Statistical significance at the 5 percent level indicates that there is one chance in twenty that the observed difference is due to chance variation. For example, out of 1,000 analyses, 50 would appear to show statistically significant differences, even if no true differences existed. With the large number of analyses performed over the five years of MWRSA, a number of "significant" statistical analyses are undoubtedly spurious. The biological or agricultural significance of these results should be interpreted only in light of recurring results and trends observed over the years. Statistical analyses are a tool to guide the interpretation of numeric data.

YIELD AND QUALITY DETERMINATION

Artichokes were harvested about every two weeks during the period from early September until the cutback in May. The sample harvest was taken from the six central plants of the two middle rows of the artichoke plots. Because of the variation in time of maturity, broccoli and cauliflower generally required two or three cuttings spaced about a week apart. The yield for celery and lettuce was determined in a single harvest, which took from one to three days, depending on the size of the crop. The sample harvest was taken from the central 8 m (26 ft) of the four middle rows of the vegetable plots.

Crops were monitored to detect qualitative differences attributable to the different irrigation waters. An experienced agriculturalist made periodic field inspections of the experimental artichoke plots. Crop evaluations were made without knowing the type of irrigation water used. Control

---

[2] In this instance, an interaction means that the effect of water types on a parameter is different under different fertilization rates.

standards of quality, as enforced by the County Agricultural Commissioner's Office for the State Department of Food and Agriculture, were used in making these judgments.

Quality inspections for vegetable crops were performed at the same time as shelf life determinations. A portion of the sample harvest for vegetables from each plot was packed into three boxes and placed in cold storage where they were later inspected for shelf life characteristics. A representative from the County Farm Advisors Office inspected these crops, without knowing the type of irrigation water used, at intervals of approximately 7, 14, and 21 days following the harvest. Criteria used for judging celery were color, spoilage, and pithiness. Broccoli was judged for color, odor, decay, compactness, and general appearance. Lettuce was examined for tip burn, russet spotting, slime, injury spot (bruising), and firmness. Cauliflower criteria were color, shape, brown spotting, decay, riciness, stem color, and hollow stem.

GROUNDWATER MONITORING

Four groundwater monitoring wells (piezometers) were installed at a depth of approximately 2 m in the MWRSA demonstration fields at the inception of Phase III in 1980. The sites for these wells were selected because those fields were wholly irrigated with filtered effluent. at that time. When the applied irrigation water was changed to well water in the demonstration fields at the end of Year Two, these peizometers no longer provided a means of sampling irrigation leachates from the effluent. However, quarterly sampling of wells 1 through 4 continued in an effort to observe any change in hydrochemistry and to monitor the water levels across a greater area of the project site. The installation of 24 new piezometers in the artichoke experimental subplots irrigated with different water types took place at the end of 1983. Four of these new monitoring wells (wells 6, 7, 8, and 9) were chosen to provide quarterly sampling for constituents, including all major and minor cations and anions. The remaining 20 piezometers were sampled for nitrate, because it is the most mobile ion likely to affect the shallow groundwater quality. Monthly water level measurements were taken in all wells in Year Five, except at times when access to the site was not feasible because of rain.

AGROCLIMATIC MONITORING

Throughout the five-year field study, climatic parameters relevant to crop development were measured and recorded continually, analyzed periodically, and reported annually. In the first two years, a special weather station, erected to specifications for the project at the Castroville Treatment

Plant, provided the needed data. In the third year, an automated, remote-controlled station, part of the California Irrigation Management Information System (CIMIS) network was placed at project site, at the southeastern corner of the experimental plots. Both weather stations operated for several months, providing evidence of continuity and credibility before the first station was dismantled.

During the course of field work, the climatic data were used for irrigation scheduling and planning of other cultural practices, although other input (e.g., local farm manager's judgment or equipment/personnel availability) also played significant roles in frequently altering plans. The weather summaries are not repeated in this report, but they are available in the annual reports [3–7]. It is believed that over the five-year period of field operations, the normal range of climatic variations to be expected in the Salinas Valley was experienced, including extremes of wet and dry years and hot and cool periods.

MARKET STUDY

A market survey was commissioned as part of MWRSA to determine the attitude of buyers, wholesalers and shippers of vegetables vis-à-vis irrigation with reclaimed water. A specialized firm (The Marketing Arm of San Francisco, California) was selected to conduct the survey design, interviews and interpretation of the results. Buyers of vegetables from several geographical areas in the United States were selected for detailed interviews and asked to respond to a series of questions. The design of the survey instrument and the polling techniques followed those of standard survey research aimed at obtaining representative samples of the attitudes of the industry as a whole.

RESULTS

PILOT TREATMENT PLANT PERFORMANCE

Comparison of Filtered Effluent and Title-22 Treatment Processes

The performance of the filtered effluent (FE and FE-F) and Title-22 (T-22) tertiary treatment processes during the five-year MWRSA field operations in Phase III and subsequent Phase IV pilot treatment plant operation was evaluated primarily in terms of levels of total suspended solids (TSS), turbidity, coliform bacteria, and viruses. A flocculation chamber was added to the process train, and this expanded process is referred to as filtered effluent with flocculation (FE-F). A test series to determine the optimum operating parameters for the FE-F process was conducted. During the test series, rapid mixer speed, flocculation detention time and energy, and alum/polymer dosage were systematically varied. Based on these preliminary test series results the following operating parameters were selected for the subsequent operation of the pilot treatment facilities:

Rapid mixer tip speed and energy (G)	2 m/sec (360'/min) and 150 sec^{-1}
Flocculation theoretical detention time and energy (G)	500 sec and 35 sec^{-1}
Chemical dose	5 mg/L alum and 0.06 mg/L polymer

Table 17.1 summarizes the log normal mean TSS and turbidity levels of the secondary effluent (SE, tertiary plant influent), filtered effluent (both FE and FE-F), and Title-22 effluent during the six-year pilot tertiary reclamation facilities operation. The log normal mean turbidity of both the filtered effluent (FE and FE-F) and the Title-22 effluent was well below the DOHS standard of 2 NTU, except for the FE flow system during Year One. Both processes achieved 100 percent compliance with this standard during the intensive test period. The log normal mean turbidities of 0.7 NTU for the FE-F flowstream and 0.5 NTU for the T-22 flowstream during Phase IV indicate that both flowstreams are capable of producing excellent turbidity removal. Good average overall treatment plant removals of TSS were achieved by both flowstreams, with 99.57 percent removal using the FE-F process (92 percent removal of suspended solids present in the secondary effluent by FE-F facilities), and 99.64 percent removal using the T-22 process (93 percent removal of suspended solids present in the secondary effluent by T-22 facilities) during Phase IV.

During MWRSA Years One through Five, T-22 effluent TSS and turbidity levels were lower than those for filtered effluent (FE and FE-F) by a ratio of about 2:1. To determine if more closely controlled and motivated operations would enable this ratio to be reduced, a sixth year of dedicated treatment plant testing was added to the study. During this sixth year of intensive plant performance studies, the optimized FE process and increased operator attention to the FE-F flowstream reduced this ratio to 1.2:1 for suspended solids and 1.4:1 for turbidity.

Compliance with the DOHS coliform standard of 2.2 MPN/100 mL was achieved for months at a time in later years and most of the time during the nine-month-long period of intense operation in Phase IV. Both tertiary processes achieved compliance, with the Title-22 process being significantly more reliable. Compliance with the DOHS requirement that no more than one coliform sample exceed

TABLE 17.1. Longnormal Mean Concentrations of BOD$_5$, Total Suspended Solids, and Turbidity in Treatment Plant Effluents (in mg/L unless otherwise noted).

Parameters[a]	Process Optimization		Year 5		Year 4		Year 3		Year 2		Year 1	
	Number of Samples	Mean[b]	Number of Samples	Mean	Number of Samples	Mean	Number of Samples	Mean	Number of Samples	Mean	Number of Samples	Mean
BOD$_5$												
SE	115	12	74	13.4	54	11	60	8	54	8	18	22
Total suspended solids												
SE	157	14.3	302		282	11.2	228	10.2	220	8.7	192	12
FE	—	—	—	—	131	1.9	202	1.5	216	2.2	188	4.4
FE-F	155	1.2	286	1.6	132	1.5	—	—	—	—	—	—
FC	153	5.8	275	4.4	263	5.7	220	4.9	217	4.3	191	6.1
T-22	153	1.0	273	0.8	258	1.3	220	1.0	214	1.2	190	1.9
Turbidity[c]												
SE	155	3.7	288	3.8	217	3.2	212	3.6	218	2.9	—	—
FE	—	—	—	—	102	1.4	209	1.1	213	1.4	178	2.4
FE-F	152	0.7	282	1.1	103	1.0	—	—	—	—	—	—
T-22	149	0.5	262	0.6	195	0.9	205	0.6	211	0.5	183	0.6

[a] SE = secondary effluent; FE = filtered effluent without flocculator (September 1980–September 1983); FE-F = filtered effluent with flocculator (October 1983–April 1986); FC = flocculator-clarifier effluent; and T-22 = Title-22 effluent.

[b] Means are 50 percentile values from probability distribution analyses. Data are fitted to the Pearson Type III lognormal distribution.

[c] Nephelometric turbidity units (NTU).

23 MPN/100 mL within a 30 day period was achieved consistently in the T-22 flowstream. This criterion was violated only once in Phase IV, by the FE-F system.

It is difficult to predict the chlorine dose requirements due to the great variation in physical and chemical characteristics of the wastewater at the point of disinfection. Chlorine doses of 11 mg/L for T-22 and 15 mg/L for FE/FE-F were selected as the average doses required to achieve the desired target residuals: 3.5 mg/L for T-22 and 7.5 mg/L for FE/FE-F. These doses and residuals were chosen to achieve at least a 5 log virus removal rate based on extensive virus seeding data obtained at MRWSA during Phase IV. Each process train achieved 100 percent compliance with the DOHS standards for several consecutive months at a time with adequate chlorine doses.

During the intensive plant performance test period, nearly 100 percent FE coliform compliance was achieved. In addition, virus removal was also essentially 100 percent. To establish at what dose both bacteria and/or virus would begin to break through, the chlorine dose was gradually lowered, and chlorine residual management watched very closely. Both average dose and dose range were substantially lowered. As expected breakthroughs began. This phenomenon coupled with winter storms which caused periodic plant upsets, resulted in the increases in bacteria and somewhat reduced virus log removal (see the following virus seeding discussion). The end of this test period included an increased attention to chlorine residual control as well as fewer wet weather storm events. The chlorine dose was varied from hour to hour and a slightly higher residual was maintained. Coliform compliance rebounded back to 100 percent. To comply with both the DOHS coliform standards and a five log virus removal criterion, the FE-F flowstream requires a higher chlorine dose than the T-22 flowstream.

Additionally, during the final year of MWRSA the ratio of chlorine dose to ammonia nitrogen concentration (Cl_2:NH_3-N) was compared to the general results as well as to specific daily bacteria reduction. This comparison is important because of the chlorine demand that ammonia in reclaimed water imposes, giving rise to production of chloramine, itself a disinfectant. Throughout the nine-month test series, the average monthly ratio varied from 1.5 to 13 with the average being about 8. Table 17.2 shows the monthly average values.

It was noted that for March and April, the free chlorine residual dropped to less than 1.0, a phenomenon consistent with the low Cl_2:NH_3-N ratios. This corroborates other experience indicating that ammonia in well treated secondary effluents may not only stabilize the ability to control total chlorine residual, but as well, increase longevity of the disinfectant (chloramine) residual.

TABLE 17.2. **Monthly Average Values of the Concentration Ratio Cl_2:NH_3-N.**

Month	Average NH_3-N (mg/L)	Average Cl_2 Dose (mg/L)	Cl_2:NH_3-N Ratio	(range)
August	2.6	28	11	(6.5–22)
September	3.0	38	13	(6.8–50)
October	3.9	22	5.6	(3.2–70)
November	3.0	25	8.3	(3.0–30)
December	3.5	19	5.4	(3.0–28)
January	1.7	20	11.6	(3.0–45)
February	1.7	19	11.3	(3.0–40)
March	4.1	14	3.4	(1.5–16)
April	9.1	13	1.5	(1.4–7)

Note: Dose is calculated from chlorinator setting and flow rate. The pH of the secondary effluent held very steady at 7.4 throughout the period.

FINDINGS OF AEROSOL STUDIES

Concentrations of viable aerosols from filtered effluent measured during daytime and nighttime sampling with both types of biological culture media are presented in Table 17.3.

Using standard plate count media, the concentrations of viable aerosols in all samples were in the same range as background levels measured upwind during both daytime and nighttime samplings. Viable aerosols sampled with EMB-Levine media exhibited much lower counts than that with plate count agar. In the selective media, low background counts were observed during the day, compared with ten times higher background levels observed at night. The significance of these data is that coliform counts are basically the same at both the upwind and downwind stations

TABLE 17.3. **Concentration of Viable Aerosols from Sprinkler Irrigation Using Tertiary Reclaimed Water[a] (mean number of viable particles per m^3 air sampled).**

Distance from Sprinkler Line	Standard Plate Count Media[b] Day[d]	Standard Plate Count Media[b] Night[e]	EMB-Levine Medium[c] Day[d]	EMB-Levine Medium[c] Night[e]
15 m, upwind	594	777	17	170
30 m, downwind	365	981	165	67
60 m, downwind	479	369	51	67

[a] Collected with Anderson six-stage impactors.
[b] Nonselective media for gram-positive and gram-negative bacteria.
[c] Selective differential medium for lactose fermenting gram-negative bacteria (coliforms).
[d] Temperature = 17°C, wind speed = 0.4 m/s.
[e] Temperature = 11°C, wind speed = 0.8 m/s.

and that the types of coliform present were not the type typically found in human-origin wastewater. The enterobacteria identified on the EMB-Levine plates ranged from 41 to 67 percent *Enterobacter agglomerans* and from three to four percent *Citrobacter* spp. The remainder of the bacteria were unidentifiable using Enterotubes and Analytical Profile Index-Enteric screen identification methods. Subsequently reported studies by others have corroborated these findings and established the safety of aerosols from an FE spray [13].

NATURAL VIRUS IN SECONDARY AND TERTIARY EFFLUENTS

Both the influent and the effluent of the two pilot plant process streams were monitored routinely over a five-year period for the presence of in situ animal viruses. The natural virus assays amounted to 181 samples (114 of which were effluent samples). The sample volumes averaged 5.6 L for the influent and 3,086 L for the effluent samples. The influent to the two pilot processes (Castroville unchlorinated secondary effluent) contained measurable viruses in 53 of the 67 samples taken. The concentration of virus ranged from 1 to 734 plaque-forming units per liter (PFU/L), with a median 3 PFU/L. The mean concentration was 28 PFU/L, but the distribution of concentrations was highly skewed toward the lower end of the range, with 90 percent of the samples containing less than the mean.

During the five-year period, no in situ viruses were recovered from the chlorinated effluent of either process. A total of 114 samples with a volume of 186,025 and 159,402 L were concentrated and assayed for viruses from the T-22 and FE effluents, respectively.

SEEDED VIRUS ATTENUATION THROUGH TREATMENT PROCESS

Because of the low level of in situ virus present in the influent, it was necessary to perform seeding studies so that the virus removal efficiency of each process could be estimated. The test poliovirus was introduced into the process streams along with tracer dye. Two preliminary tests were conducted to determine the effect of the tracer dye on the virus assay system; and, the effect of chlorination on apparent dye concentration. The results of these two preliminary tests are shown in Tables 17.4 and 17.5.

The results of these tests indicate that the doses of chlorine used during the course of these studies (>10 mg/L) did not have an effect upon the observed dye concentration even at exposure times up to 21 hours. Likewise, the effect of relatively high concentrations of dye on low levels of virus (65 to 165 PFU/mL) had no effect on the virus assay system. Thus the tracer dye was suitable for this purpose.

TABLE 17.4. Effect of Chlorine (10 mg/L Residual) on Apparent Pontacyl Pink Dye Concentration in Distilled Water and Pilot Plant Influent.[a]

Elapsed Time (Minutes)	Distilled Water Fluorometer Reading	Elapsed Time (Minutes)	Plant Influent Fluorometer Reading
0	36	0	37
10	34	15	36
26	33	29	35
40	32	43	36
72	32	60	35
85	32	88	33
1,260	32	1,260	32

[a]Unchlorinated effluent of activated sludge secondary treatment process comprised influent to the two pilot tertiary plant process trains.

Post seeding virus samples were taken when the dye was at peak concentration in the effluent so that the sample would have the highest possible virus concentration. Virus samples were taken from the post-chlorination effluent of each process and the chlorine residual immediately neutralized with sodium thiosulfate.

Thirteen virus seeding experiments were conducted in which both the T22 and FE processes were simultaneously challenged. The results of these matched pair experiments are shown in Table 17.6. On examination of all the virus seeding results it is clear that there is variation in the pilot plant operation and that the statistical distribution of the virus removal data may not be normal. Thus to determine if there is a difference in virus removal between the two processes (T-22 and FE) the non-parametric Wilcoxon Signed Rank Test was applied in which the differences among a set of matched pairs of observations are investigated. In this instance there are 13 matched pairs of data (see Table 17.6) that were appropriate for statistical analysis. The results of the analysis indicates that there was

TABLE 17.5. Effect of Pontacyl Pink Dye on Poliovirus Recovery.

Dye Concentration (mg/L)	PFU/0.2 mL of Test Solution[a]				
	Exposure Time in Minutes				
	0	10	30	60	120
0	19.3	26.0	20.0	23.5	19.0
75	24.0	22.5	19.5	21.0	20.5
150	21.5	26.5	26.5	13.0	13.5
300	21.0	21.0	20.5	20.5	17.5
600	25.0	16.0	15.5	19.5	16.0

[a]Zero time data based upon 12 replicate samples; all other data based upon duplicates.

TABLE 17.6. Samples from Paired Runs Used in the Wilcoxon Signed Rank Test to Compare Virus Removing Effectiveness of "Title-22" And Filtered Effluent Pilot Processes.

Sample Date	Log Virus Removal by Each Process	
	"Title-22"	Filtered Effluent
29 Jul 81	7.3	6.6
6 Aug 81	8.3	5.7
14 Nov 84	3.5	3.1
27 Feb 85	3.2	6.4
20 Mar 85	6.2	6.3
24 Apr 85	2.8	3.1
01 May 85	4.2	3.3
26 Feb 86	5.6	3.0
02 Apr 86	6.1	6.0
09 Apr 86	4.5	4.5
16 Apr 86	4.9	5.2
23 Apr 86	6.7	6.1
30 Apr 86	5.2	5.0
Median	5.2	5.2

no difference in the virus removal efficiency of either process (alpha > 0.01.) The median virus removal capability of each process was greater than five logs.

A number of experiments were conducted to examine the effect of various combinations of alum and polymer (Dow Anionic 825) dosages in the FE process on virus removal. The results of these tests indicated that there was little if any statistically significant effect of coagulant addition on virus concentration in the non-chlorinated FE effluent.

From March 1984 to April 1986 seed virus recovery measurements were made on the prechlorinated post filter effluent from the T-22 and FE processes. The results of these determinations are summarized in Table 17.7 as percent poliovirus removal.

Examination of the results indicates two relatively distinct subsets of data; one from 14 Mar 84 to 01 May 85 and the other from 11 Sep 1985 onward. In the first subset the virus removal efficiency of the FE direct filtering process was similar in magnitude to that observed during the coagulant addition studies, averaging 61 ± 29.5 percent, while the T-22 process gave an average removal of 98.3 ± 3.9 percent. These results indicate a statistical difference between the two processes (alpha 0.05 using the Wilcoxon Signed Rank Test) and also indicate the wide variation in efficiency associated with the FE process, compared to the T-22 process. In the second subset of filtration data (11 September 85 onward) the results of the seeded virus removal tests were much more uniform. In this instance,

TABLE 17.7. Percent Poliovirus Removal from "Title-22" and Filtered Non-chlorinated Post Filter Effluents.

| | Process Stream | | | Process Stream | |
| | | Filtered | | | Filtered |
Test Date	"Title-22"	Effluent	Test Date	"Title-22"	Effluent
14 Mar 84	NM[a]	11.3	11 Sep 85	99.9	99.7
14 Mar 84	NM	0.0	18 Sep 85	99.6	99.0
02 May 84	NM	75.8	16 Oct 85	98.7	—
02 May 84	NM	65.0	23 Oct 85	99.0	—
23 May 84	NM	57.6	11 Dec 85	98.7	99.4
23 May 84	NM	55.3	15 Jan 86	98.4	96.0
19 Aug 84	99.8	95.6	22 Jan 86	98.7	97.5
29 Aug 84	99.2	48.3	05 Feb 86	99.5	98.7
14 Nov 84	99.9	99.6	26 Feb 86	99.0	98.7
05 Dec 84	(**)[b]	59.2	03 Mar 86	96.8	98.0
16 Jan 85	99.8	96.3	19 Mar 86	99.4	98.0
27 Feb 85	99.8	99.6	02 Apr 86	99.6	99.6
20 Mar 85	99.9	75.9	09 Apr 86	95.0	96.8
27 Mar 85	99.8	68.6	16 Apr 86	99.0	98.4
24 Apr 85	99.5	57.0	23 Apr 86	97.5	98.4
01 May 85	87.4	20.6	30 Apr 86	99.0	99.0

[a]NM = not measured.
[b]Pilot plant malfunctioned on the T-22 flowstream.

there was no difference in the virus removal efficiency of either process. The average removal for the T-22 stream was 98.6 ± 1.2 percent and for the FE it was 98.4 ± 1.0 percent.

The major difference between the first and second periods—in the plant operation—was that during the latter period substantial additional effort was made to have the filters freshly backwashed and the plant operating smoothly. From these data it can be concluded that the T-22 process (before chlorination), on average and with small variation, removes >98 percent of the seeded virus during both routine and optimized operating conditions. The FE direct filtration process is equivalent to the T-22 process when the plant is carefully operated but under less rigorous conditions it can be very inconsistent. Thus, from the point of view of process reliability, the T-22 treatment train (prior to disinfection) is superior to the FE system. It is also obvious that the chlorination step is absolutely necessary to achieve the 5 log virus reduction observed in the final effluent of both processes.

DECAY OF VIRUSES ON VEGETABLES AND SOILS

Natural Virus

Crops and soils irrigated with reclaimed water from both the T-22 and

TABLE 17.8. Time to Reach 99 Percent Die-off of Seeded Virus on Various Vegetables, in the Laboratory.

Vegetable	T-99
Artichoke	8.6
Broccoli	7.8
Celery	8.4
Iceberg Lettuce	15.1

the FE process were monitored for the presence of in situ virus from July 1980 to April 1983. Crops were irrigated by both sprinkler and furrow irrigation methods. No viruses were recovered from any of 34 soil samples or 32 samples of vegetables (20 samples of artichoke, 6 samples of celery, 3 samples of lettuce, 2 samples of cauliflower, and 1 sample of broccoli). These results would be expected since no viruses were detected in the irrigation water (reclaimed water).

Seeded Virus

Although no in situ viruses were recovered from irrigated plants and soil, it was decided that an estimate should be made of the ability of virus to survive in the environment. Viruses were seeded onto plants and soil and survival measurements were made in both environmental chamber, and in the field.

Vegetables

Time decay in the environmental chamber on four types of vegetables was determined as the geometric mean of six replicates. Number of days necessary to reach 99 percent die-off of seeded viruses (T-99) in the laboratory (using simulated field conditions of 70 percent relative humidity and 60°C) is shown in Table 17.8.

Plotting decay curves showed that the rate of virus die-off was similar to the rate of moisture loss, and both seem affected by the geometry of the plant. Decay curves were linear in artichokes, which dry slowly, and at a fairly constant rate. Other vegetables showed a rapid moisture loss and die-off of virus during the first 24 hours, with a more gradual die-off in the remaining days of exposure. The number of days required for 99 percent die-off in the field are shown in Table 17.9.

Twice the number of plants was used for establishing baseline virus concentration. Die-off rates were log linear with time on all plants. Rates of die-off were greater in the field than in the environmental chamber, possibly because conditions were harsher (i.e., there was greater variability in temperature and humidity). Differences in the T-99 for lettuce were

TABLE 17.9. Time to Reach 99 Percent Die-off of Seeded Virus
on Various Vegetables, in the Field.

Vegetable	T-99	Number of Runs	Number of Plants
Artichokes	5.4	4	2
Romaine Lettuce	5.9	1	4
Butter Lettuce	7.8	1	4

also likely due to the different geometry of iceberg lettuce and the leafy romaine and butter lettuce varieties.

Soils

The survival of virus in Castroville soil was determined both under environmental chamber conditions and under field conditions. In the environmental chamber study, 100 g of fresh Castroville soil was inoculated with a known amount of virus and the decline in virus number was measured over time. Three runs were performed, each at differing relative humidities, all at a temperature of 70°C. The T-99 values for the decay of virus are presented in Table 17.10.

The survival of animal virus was studied using Castroville soil exposed to field conditions at the University of California Sanitary Engineering and Environmental Health Research Laboratory at Richmond, California. Two test runs were made during the months of June and July inoculating 1.6×10^6 and 1.9×10^7 PFU per column, respectively. Temperature in the various soil sections varied between 36°C and 17°C. The relative humidity during these time periods averaged between a low of 59 percent and a high of 78 percent. The T-99s were 5.2 and 4.8 days for runs one and two, respectively. Thus, the rate of virus removal under chamber and field conditions was similar. No viruses were recovered from any soil section after 12 to 14 days of exposure. None were recovered below a depth of 20 cm.

COMPARISON OF PLOTS IRRIGATED WITH FILTERED EFFLUENT, T-22 AND WELL WATER

The primary statistical test used to determine if significant differences

TABLE 17.10. Time to Reach 99 Percent Die-off of
Seeded Virus on Soil, in the Laboratory.

Relative Humidity	T-99
60 percent	5.4
70 percent	9.7
80 percent	20.9

could be detected between the characteristics of the soils and plants receiving different water types was ANOVA (analysis of variance). The chemical characteristics of the irrigation waters are shown in Table 17.11.

Irrigation Waters

The two effluents had higher concentrations of specific ions and total dissolved salts relative to the well water. Agriculturalists are particularly interested in the levels of salts and nutrients in the irrigation water. Well water had the lowest levels of nitrogen, phosphorus and potassium. In nearly all cases, the FE water had the highest concentration of nutrients; with substantially more nitrogen and phosphorus than the T-22 water. Salinity (total dissolved solids) and sodicity (sodium concentration relative to calcium and magnesium) were significantly higher in the two reclaimed waters than in well water. Heavy metals concentrations in the irrigation waters are shown in Table 17.12. Levels of metals in irrigation waters were often below detection limits in all three water types. Therefore, the range of concentrations is also shown on table 17.12.

All three water types, including the well water, periodically exhibited high total coliform levels, due to regrowth of coliform bacteria in the storage tanks used for periodic storage for all three water types. Both the FE and T-22 processes were capable of producing reclaimed water that meets the most stringent requirements set by the California Water Reclamation Criteria. Fecal coliform levels—measured on samples before storage— in all three water types were at or below 2.2 MPN/100 mL most of the time. No salmonellae, shigellae, Ascaris lumbricoides, Entamoeba histolytica, or other parasites were ever detected in any of the irrigation waters.

Suitability of the reclaimed waters for irrigation of food crops was evaluated based on major agronomic criteria: total salts, sodium, sodium adsorption ratio (SAR), chloride, boron. The SAR[3] of the reclaimed waters and their corresponding salinity, taken together, consistently settled in a range of favorable quality for irrigation, as demonstrated in Figure 17.4. Using these criteria and the results obtained during the course of the study it was concluded that even though salts and sodium in the reclaimed water were in the "increasing potential problem" range of the spectrum, no detectable effects on soil, plant appearance and vigor, yield or quality were observed. It is reasoned, therefore, that with appropriate irrigation scheduling and provision of a suitable leaching fraction the reclaimed waters are suitable for irrigation of food crops from an agronomic point of view.

[3] Adjusted SAR refers to a calculated value taking into account the influence of pH and bicarbonate in the irrigation water and their effect in precipitation of calcium out of solution and resultant increase in effective SAR.

TABLE 17.11. Chemical Characteristics of Irrigation Waters Used in the Experimental Fields (mg/L unless otherwise noted).

Parameter	Well Water		Title-22 Water		Filtered Effluent	
	Range	Median	Range	Median	Range	Median
pH[a]	6.9–8.1	7.8	6.6–8.0	7.2	6.8–7.9	7.3
Electrical conductivity[b]	400–1344	700	517–2452	1256	484–2650	1400
Calcium	18–71	48	17–61.1	52	21–66.8	53
Magnesium	12.6–36	18.8	16.2–40	20.9	13.2–57	22
Sodium	29.5–75.3	60	77.5–415	166	82.5–526	192
Potassium	1.6–5.2	2.8	5.4–26.3	15.2	13–31.2	18
Carbonate, as CaCO₃	0.0–0.0	0.0	0.0–0.0	0.0	0.0–0.0	0.0
Bicarbonate, as CaCO₃	136–316	167	56.1–248	159	129–337	199.5
Hardness, as CaCO₃	154–246	202.5	187–416	217.5	171–435	226.5
Nitrate as N	0.085–0.64	0.44	0.18–61.55	8.0	0.08–20.6	6.5
Ammonia as N	ND[e]–1.04	—	0.02–30.8	1.2	0.02–32.7	4.3
Total phosphorus	ND–0.6	0.02	0.2–6.11	2.7	3.8–14.6	8.0
Chloride	52.2–140	104.4	145.7–784.1	221.1	145.7–620	249.5
Sulfate	6.4–55	16.1	30–256	107	55–216.7	84.8
Boron	ND–9	0.08	ND–0.81	0.36	0.11–0.9	0.4
Total dissolved solids	244–570	413	643–1547	778	611–1621	842
Biochemical oxygen demand (BOD)	ND–33	1.35	ND–102	13.9	ND–315	19
Adjusted SAR[c]	1.5–4.2	3.1	3.1–18.7	8.0	3.9–24.5	9.9
MBAS[d]	—	—	0.095–0.25	0.136	0.05–0.585	0.15

[a] Standard pH units.
[b] Decisiemens per meter (ds/m).
[c] Adjusted sodium adsorption ration (see text footnote 3): adj. SAR = $\dfrac{Na}{\sqrt{\dfrac{Ca + Mg}{2}}}[1 + (8.4 - pH_c)]$

[d] Methylene-blue active substance (MBAS).
[e] ND = Chemical concentration below detection limit. Detection limits are as follows: $NH_3\text{-}N$ = 0.02 mg/L; B = 0.02 mg/L; P = 0.01 mg/L; BOD = 1 mg/L; and MBAS = 0.05 mg/L.

TABLE 17.12. Heavy Metal Concentrations in Irrigation Waters (mg/Kg soil).

Parameter	Well Water		Title-22 Water		Filtered Effluent		Irrigation Water Criteria[a] (continuous)	Drinking Water Regulations
	Range	Median	Range	Median	Range	Median		
Cadmium	ND-0.1	ND	ND-0.1	ND	ND-0.1	ND	0.010	0.010[b]
Zinc	ND-0.6	0.02	0.07-6.2	0.33	ND-2.08	0.195	2.0	5[c]
Iron	ND-0.66	0.1	ND-2.3	0.05	ND-0.25	0.06	5.0	0.3[c]
Manganese	ND-0.07	ND[d]	ND-0.11	0.05	ND-0.11	0.05	0.20	0.05[c]
Copper	ND-0.05	0.02	ND-0.05	ND	ND-0.04	ND	0.20	1[c]
Nickel	0.001-0.200	0.04	0.002-0.180	0.04	0.004-0.200	0.04	0.20	—
Cobalt	ND-0.057	ND	0.001-0.062	0.002	ND-0.115	0.05	0.050	—
Chromium	ND-0.055	ND	ND	ND	ND	ND	0.10	0.05[b]
Lead	ND	ND	ND	ND	0.001-0.700	0.023	5.0	0.05[b]

[a] Water Quality Criteria 1972; Ecological Research Series, U.S. EPA.
[b] Primary Drinking Water Regulations (metals that post a potential adverse health effect), U.S. EPA.
[c] Secondary Drinking Water Regulations (metals that post an aesthetic problem), U.S. EPA.
[d] ND = Metal concentration below detection limit. Detection limits were improved during the 5-year period and were as follows (in mg/L):

Element	Year 1	Years 2 and 3	Years 4 and 5
Cd	0.1	0.001	0.001
Zn	0.5	0.02	0.02
Fe	0.03	0.03	0.03
Mn	0.05	0.05	0.05
Cu	0.02	0.02	0.001
Ni	0.2	0.05	0.001
Co	0.1	0.05	0.001
Cr	0.2	0.04	0.04
Pb	0.2	0.05	0.001

Soils

Statistically significant differences between soils irrigated with the three water types were found in several chemical parameters. In nearly all cases the relative values of the concentrations found in the soils followed the same relative relationships found in the irrigation waters. Concentrations of a typical parameter were generally highest in the FE-irrigated soil samples, and lowest in the well water-irrigated samples. In several cases, increased fertilizer application rates were found to affect concentrations of various soil chemical parameters, similar to the effect of irrigation with reclaimed water having high nutrient concentration. The following parameters increased significantly with higher fertilizer rates for more than half of the sampling events: electrical conductivity of the saturation extract (a measure of total salts), calcium, magnesium, potassium, nitrate, and sulfate. Reclaimed water irrigation also produced higher soil concentrations of these same parameters, except potassium. Reclaimed water also increased the SAR and chloride levels in the soil extract. Although many statistically significant differences were found, none of the data indicated that the soils irrigated with any of the three types of water were being affected adversely. Irrigating with reclaimed water did not have any harmful effects on the soil.

Soil Heavy Metals

Heavy metals concentration in the soil correlated most significantly with fertilizer application rates. By contrast, water types had no measurable effect. None of the nine heavy metals studied (cadmium, zinc, iron, manganese, copper, nickel, cobalt, chromium, or lead) manifested any consistently significant differences in concentration among plots irrigated with the different water types. Furthermore, except for copper, no increasing trends in metal concentrations with time over the five years were observed.

The gradual increase for copper occurred equally for all water types. At the end of the five years, copper concentrations were still below the average for California soils. As shown in Table 17.12, the concentrations of these metals in all three types of irrigation water were so low (mostly below detection level) that calculation of the theoretical accumulation rate would predict no significant accumulation over the five-year period. For much longer term periods, the same calculation would indicate potential accumulation in soil only of iron and zinc, were they not essential plant and animal micronutrients, uptaken and removed with each harvest. Iron was generally measured at higher concentrations in the well water than in either reclaimed water. Zinc, on the other hand, was higher in both

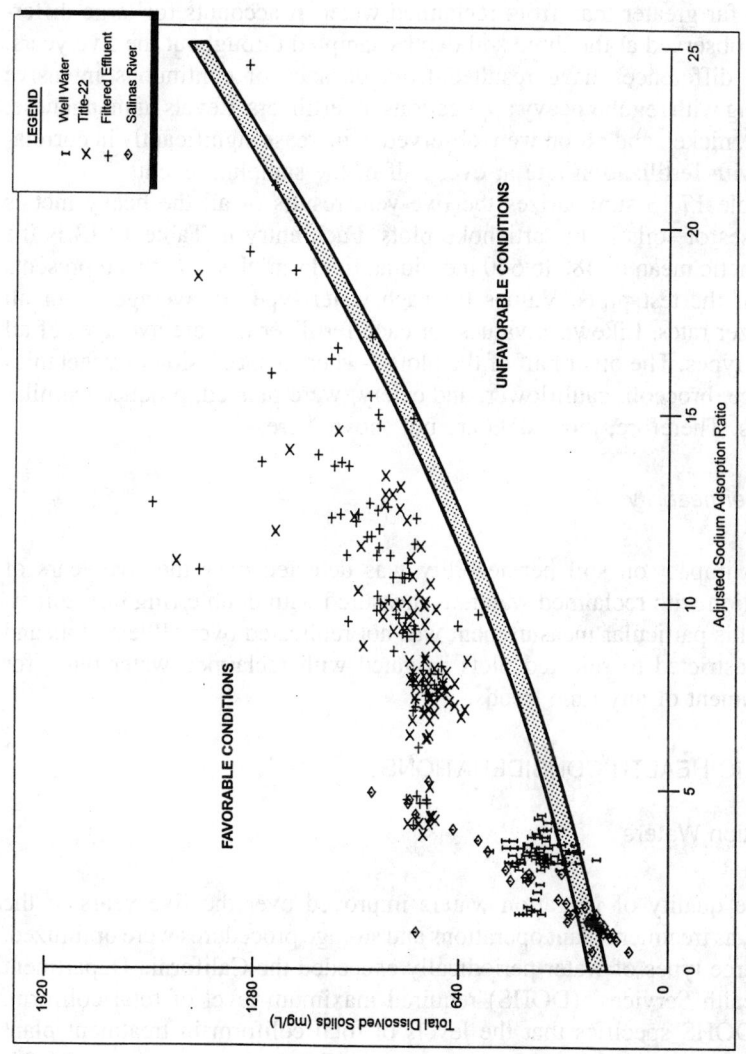

Figure 17.4 Relationship between adjusted SAR and salinity for four different irrigation waters.

types of reclaimed water than in well water. The actual concentrations were on the order of 0.1 mg/L in the two reclaimed waters. At these levels, uptake by plants is faster than accumulation from irrigation input.

Input of zinc and other heavy metals from commercial fertilizer impurities is far greater than from reclaimed water. It accounts for large differences observed at the three soil depths sampled throughout the five years. These differences have resulted from decades of continuous intensive farming with regular heavy applications of fertilizers. Levels of manganese, cobalt, nickel, and boron were observed to increase significantly in correlation with fertilization rate at over half of the sampling events.

Table 17.13 summarizes the five-year results of all the heavy metals analyses of soils in the artichoke plots. Each entry in Table 17.13 is the arithmetic mean of 480 to 640 individual field samples, and this represents half of the test plots. Values for each water type are averaged over all fertilizer rates. Likewise, values for each fertilizer rate are averages of all water types. The other half of the plots—where a succession of vegetables (lettuce, broccoli, cauliflower, and celery) were planted, produced similar results. Therefore, those data are not shown here.

Soil Permeability

No impact on soil permeability was detected over the five-years of irrigation with reclaimed water, as measured with double-ring infiltrometers. This particular measurement was not replicated over all test plots and was restricted to selected plots irrigated with reclaimed water only, for assessment of any time trends.

PUBLIC HEALTH CONSIDERATIONS

Irrigation Waters

The quality of irrigation waters improved over the five years of the study, as treatment plant operations and storage procedures were optimized. All three types of water periodically exceeded the California Department of Health Services' (DOHS) required maximum level of total coliform. The DOHS specifies that the levels of total coliform in treatment plant effluent used for irrigation not exceed a 7-day moving average of 2.2 MPN/100 mL with a maximum allowable level of 23 MPN/100 mL not to be exceeded more than once in 30 days. This standard, however, is applied to effluent measured at the treatment plant, and is not strictly applicable to irrigation waters sampled in the field. During the course of the study, it became evident that irrigation waters were exceeding these

TABLE 17.13. Average Concentration of Heavy Metals in Soil Profile of Artichoke Plots (mg/kg soil).

Heavy Metal	Soil Depth, cm	Water Type[a]			Fertilizer Rate[b]				
		WW	T-22	FE	0	33%	66%	100%	Average
Cadmium	30	0.46	0.44	0.45	0.43	0.44	0.46	0.45	0.45
	100	0.16	0.16	0.13	0.16	0.16	0.14	0.13	0.14
	200	0.13	0.08	0.10	0.12	0.10	0.11	0.09	0.10
Zinc	30	1.41	1.33	1.53	1.37	1.46	1.42	1.45	1.42
	100	0.33	0.36	0.34	0.35	0.35	0.35	0.33	0.34
	200	0.50	0.40	0.47	0.49	0.48	0.44	0.42	0.45
Iron	30	49.68	41.48	45.73	39.66	43.81	47.73	51.31	45.63
	100	8.19	8.13	6.07	7.59	7.82	7.44	7.00	7.47
	200	12.59	7.79	7.65	9.61	9.72	8.98	9.08	9.35
Manganese	30	23.30	20.21	24.53	16.01	18.64	24.18	31.86	22.67
	100	4.94	4.71	4.19	4.63	4.89	4.45	4.48	4.61
	200	6.26	4.63	5.04	5.87	5.13	5.33	4.90	5.31
Copper	30	2.06	2.02	2.09	2.05	1.97	2.11	2.11	2.06
	100	1.57	1.84	1.40	1.60	1.73	1.43	1.65	1.60
	200	1.79	1.25	1.51	1.68	1.55	1.41	1.41	1.52
Nickel	30	6.88	6.38	6.81	6.19	6.51	6.92	7.15	6.69
	100	0.91	0.93	0.69	0.92	0.85	0.83	0.78	0.84
	200	0.63	0.39	0.47	0.58	0.47	0.50	0.42	0.49
Cobalt	30	0.16	0.16	0.18	0.13	0.14	0.18	0.22	0.17
	100	0.09	0.09	0.09	0.09	0.09	0.10	0.09	0.09
	200	0.09	0.09	0.09	0.09	0.09	0.09	0.09	0.09
Chromium	30	0.15	0.13	0.15	0.14	0.14	0.14	0.15	0.14
	100	0.10	0.10	0.10	0.09	0.10	0.11	0.10	0.10
	200	0.10	0.09	0.10	0.09	0.10	0.09	0.09	0.10
Lead	30	0.98	0.92	0.96	0.97	0.94	0.93	0.98	0.95
	100	0.64	0.71	0.54	0.66	0.65	0.56	0.63	0.63
	200	0.70	0.49	0.60	0.66	0.62	0.56	0.54	0.59

[a]Averaged for all fertilizer rates.
[b]Averaged for all water types.

recommended levels even when total coliform levels measured at the treatment plant were below detection limits.

These high levels were generated by coliform regrowth in the redwood storage tanks. Regrowth problems were substantially reduced when dechlorination of the effluent was stopped.[4] Total coliform levels were generally highest in FE irrigation waters. There are no DOHS standards for fecal coliform. Levels in all three water types were at or below 2 MPN/100 mL most of the time. FE exceeded this level more often than the other two

---

[4] Dechlorination of the reclaimed water was found to be unnecessary as concern for possible leaf tip burn on plant materials diminished and was eventually put to rest.

water types. No salmonellae, shigellae, *Ascaris lumbricoides, Entamoeba histolytica,* or other miscellaneous parasites were ever detected in any of the irrigation waters.

Soils

The levels of total and fecal coliform in soils irrigated with all three types of water were generally comparable. No consistent significant difference attributable to water type was observed. No salmonellae, shigellae, *Ascaris lumbricoides, Entamoeba histolytica,* or other miscellaneous parasites were ever detected in soil samples.

Plant Tissue

Neither edible nor residual plant tissues showed any significant difference due to water type in levels of total or fecal coliform. No salmonellae, shigellae, *Ascaris lumbricoides, Entamoeba histolytica,* or other parasites were detected in edible or residual tissues of artichokes, broccoli, cauliflower, or lettuce. In Year One, parasites such as *Entamoeba histolytica, Ascaris lumbricoides,* and *Taenia* were found in both edible and residual celery tissue. Parasites were not limited to those crops irrigated with effluents; they were also found in tissues of crops irrigated with well water.

Sampling of neighboring fields detected no relationship between bacteriological levels and the distance from project site. The aerosol transmission of bacteria was thus deemed unlikely.

GROUNDWATER IMPACT

Groundwater quality data were collected over the five years of the MWRSA project to ascertain changes in shallow groundwater quality. If no significant change was observed in samples collected from shallow monitoring wells [depth of 5 m (15 ft)] as a result of applied irrigation water, then it could be assumed that impacts to the groundwater at greater depths would also be insignificant. Groundwater for municipal and agricultural purposes in this area is generally extracted from the "400-ft aquifer" (approximately 120 m deep).

An examination of the data indicates that no discernible relationship existed between the shallow groundwater quality and the type of applied irrigation water. Other trends commonly associated with shallow groundwater quality in agricultural areas were observed such as downgradient increases in TDS and seasonal effects.

The three most common types of pollutants associated with agricultural irrigation are nitrates, TDS, and pesticides. Nitrates are the residual of fertilizer application. Historically, nitrates are percolated to groundwater not necessarily through over-fertilization, but through over-irrigation. High levels of nitrates applied to soil will eventually be taken up by plants unless moved out of the root zone by excessive irrigation. Elevated TDS levels generally result from poor irrigation water, the leaching of ions from the unsaturated zone, and over-irrigation or ponding of water. Although pesticides have been applied at the project site, they have not been monitored in the groundwater because their use is widespread in the Castroville area.

The most common types of pollutants associated with treated effluent water application are nitrate and heavy metals. Nitrate levels in the two effluents were significantly higher than levels in well water. However, an examination of the filtered effluent water quality data shows no appreciably higher concentrations of metals compared with the Title-22 or well water quality. None of the three types of applied water (filtered effluent, Title-22, or well water) exceed any of the recommended maximum concentrations of trace elements in irrigation water adopted by the SWRCB.

Nitrate appears to be the only constituent potentially indicative of application of effluent. Soluble nitrate concentrations in the perched groundwater zone beneath the project site are best recorded in the artichoke experimental plots where 24 monitoring wells were installed in 1983–1984. Monitoring wells 1, 2, 6, 11, and 20 show consistently higher (10–90 mg/L) than ambient (0–5 mg/L) levels of nitrate in the groundwater. In addition, monitoring wells 9, 14, 21, and 25 have shown concentrations in excess of 10 mg/L nitrate in at least three of the seven sampling events.

The highest concentrations of nitrate are associated with the July, August, and September 1984 samplings, suggesting a direct relationship between nitrogen application and groundwater nitrate levels. Fertilizers were applied at rates of 56, 135, 135, and 135 kg/ha (50, 120, 120, and 120 lb/acre) of nitrogen on the artichoke experimental plots in July, September, October, and November, respectively. Water percolating through the soil will leach nitrate derived from nitrogen fertilizers. The greater the amount of percolating water, the greater the amount of nitrate that may be leached from the root zone. There appears to be no discernible correlation between the wells with high nitrate and a particular applied water type or fertilizer rate; subplots irrigated with well water, Title-22, and filtered effluent all showed high concentrations. The anomalously high nitrate values do not correlate with subplots fertilized at a particular rate; subplot 5L with well 6 installed in it had no fertilizers applied and yet it also had high dissolved nitrate. In addition, there is no relationship between water type and high nitrate values; wells 1 and 2, located in the demonstration fields irrigated with well water in the final two years of the study also show high nitrate concentrations.

In conclusion, an examination of all water quality data collected at the MWRSA site suggests that the groundwater quality trends are associated with trends generally applicable in irrigated areas such as increased TDS and nitrate. There is no apparent evidence of a unique contribution by filtered effluent application to the shallow groundwater quality over the five years of data reported.

ORGANIC COMPOUNDS IN RECLAIMED WATER

Individual organic compounds present in natural surface waters and wastewater effluents number in the thousands, although normally in trace concentrations detectable only at the part-per-billion level [14]. Toxicological characteristics of these compounds depend on their concentration in the water. The MRWPCA conducts an annual sampling and analysis program on its major treatment plants' effluents.

Volatile Organics

During the 1985 sampling, grab samples from the six MRWPCA treatment plants' effluents were taken and blended in proportion to their respective daily flows. Very low levels of six volatile organic priority pollutants (methylene chloride, chloroform, dichloroethene, tetrachlorothene, toluene, and ethylbenzene) and three nonpriority volatile organic pollutants (acetone, 2-butanone, and xylene) were detected in the blended waste streams. The sources of these pollutants are from the disposal of paints, paint thinners, cleaning and degreasing agents, perfumes, inks, dry cleaning solvents, dyes, and various other household products by residential and commercial users. Commercial users known to discharge these pollutants belong to the dry cleaning, industrial laundry, printing, machining, and autoshop business activities. Control of the discharge of these pollutants is being enforced through the issuance of industrial waste discharge permits to affected users, frequent on-site inspections, and monitoring of typical users belonging to each activity. Hence, levels of these pollutants are expected to remain at acceptably low levels (i.e., below the established action levels).

Semivolatile Organics

In the same sampling event, very low levels of four semivolatile organic priority pollutants [phenol, bis (2-ethylhexyl) phthalate, di-*n*-butyl phthalate, and diethyl phthalate] and three nonpriority semivolatile organic pollutants (4-methylphenol, 2 methylnapthalene, and benzyl alcohol) were detected. The presence of phenol and the phthalate esters: bis (2-ethylhexyl) phthalate, diethyl phthalate, and di-*n*-butyl phthalate is most likely the result of the washing and rinsing of plastic materials from both commercial and residential sources. Source control activities for those pollutants are aimed at the plastic forming businesses. The presence of 4-methyl phenol is most likely from the discharge of disinfectants, varnish, and raw materials for photographic developer by residential and commercial users. The presence of 2-methylnaphthalene is most likely from the discharge of metal-cutting fluids, various lubricants, and emulsion breakers by residential and commercial users. The source of benzyl alcohol is unknown at this time. Source control activities for these pollutants are being directed at the machine and autoshop businesses. Hence, levels of these pollutants are expected to remain at acceptably low levels, below established action levels.

Many mechanisms (including stripping in the spray process, adsorption on soil clay particles, decay, and decomposition) contribute to the further attenuation of any organic compound that may still be present after tertiary treatment. Because of these extremely low concentrations, non-potable use of the waters, and the existence of highly effective barriers in the irrigation process, a study of specific organic compounds was not included in MWRSA.

CROP YIELD, QUALITY AND SHELF LIFE OF PRODUCE

Yield of Crops

Statistically significant yield differences due to water type were observed in the celery and broccoli crops. In both cases, yields were significantly higher in the crops irrigated with reclaimed waters. The FE-irrigated crop yields were usually slightly higher than the T-22-irrigated crop yields. Yields of lettuce and celery showed an interaction of water type and fertilization rate; reclaimed water irrigation improved yields in unfertilized plots but had little or no effect on yield of plots receiving fertilizer. Artichoke yields were similar for all three water types. Yields of all crops leveled off at or below 66 percent the standard local fertilization rate. Use of the full (100 percent) local fertilization rate did not further improve crop yields. Thus, it is surmised that a reduction of up to 30 percent in fertilizer application rate may be possible for all of the crops—when reclaimed water is used for irrigation. Figure 17.5 depicts yield of lettuce by water type and fertilizer rate.

Quality and Shelf Life

Field inspection of crops showed no leaf damage from residual chlorine in the reclaimed waters. No differences were observed in the appearance

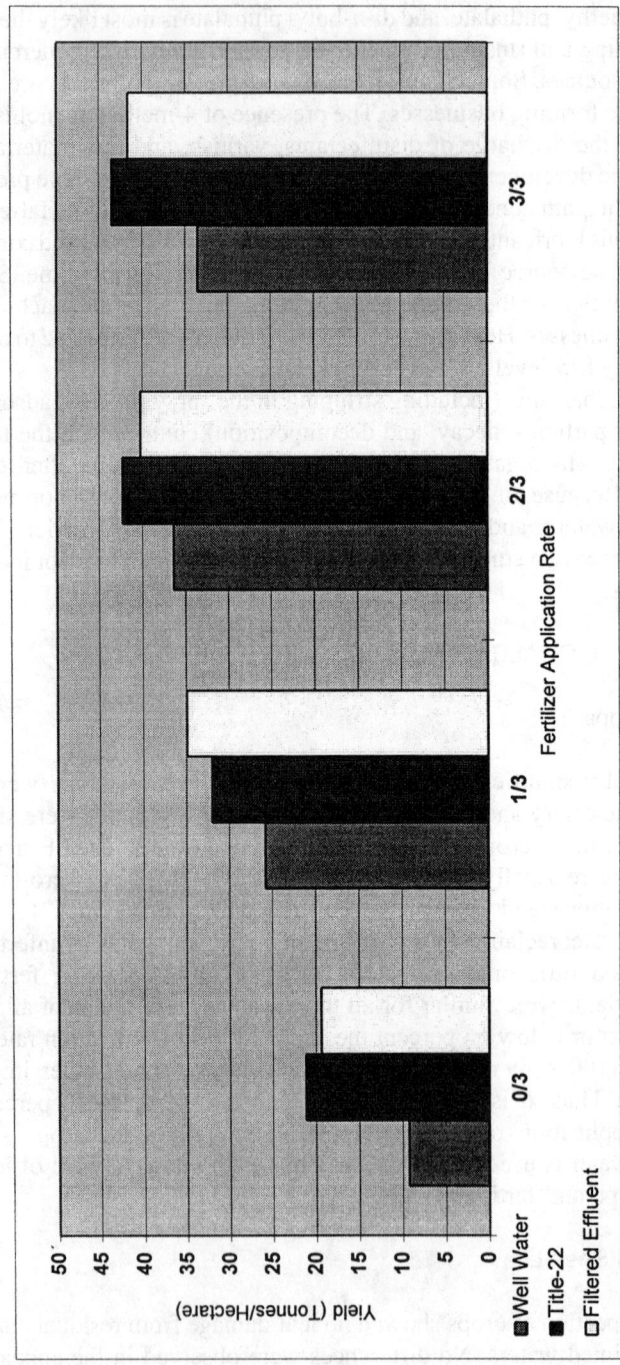

Figure 17.5 Mean lettuce yield as a function of fertilizer rate and water type.

or vigor of plants irrigated with different water types. Samples of each crop harvested were boxed and placed in cold storage warehouse for varying time periods—up to four weeks following each harvest. The produce was examined at weekly intervals during the storage period by the local farmers, an agricultural extension service representative, and project staff. The inspectors examined the boxed samples of vegetables for signs of pithiness, flaccidity, black heart, outside tissue breakdown, discoloration, decay, and any other indications of spoilage. In general, all of the produce was of excellent quality at these inspections and showed no unexpected deterioration over time. The quality and shelf life of all produce grown with the two reclaimed waters was as good as, and in some cases superior to, the produce grown with well water.

MARKET ACCEPTANCE

A special market acceptance study was conducted by The Marketing Arm, a firm from San Francisco, California with expertise in survey research in the field of agricultural commodities. One key issue of interest to the wholesale buyers and shippers of produce was whether or not produce grown with reclaimed water would be required to be labeled to identify the most recent origin of the water used for its irrigation. No regulatory or rational basis was found for labeling produce grown with disinfected tertiary reclaimed water—or any other kind of water used to irrigate crops imported from abroad, for example.

Further, the survey found that buyers, shippers, and other intermediaries would readily accept such produce as long as the regulatory agencies approved the use of reclaimed water. The intermediaries actually preferred that the produce not be labeled, as labeling adds an additional step in the handling and distribution of the produce. The marketing firm also investigated the potential for negative rumors and innuendoes, either spontaneously or intentionally generated, that could adversely affect produce sales.

It was concluded that the risk to growers from such rumors were extremely low, because rumors regarding the characteristics of produce are rare and unlikely to be generated—either by competitors or by the media. Unfounded rumors and negative PR can be contained and dispelled by dissemination of factual information.

DISCUSSION, MAJOR FINDINGS AND CONCLUSIONS

The Monterey Wastewater Reclamation Study for Agriculture (MWRSA) was a long-term field investigation of the feasibility of irrigating artichokes and raw-eaten vegetables such as lettuce, broccoli, celery and cauliflower with disinfected tertiary reclaimed water. The results obtained from the five-year field tests indicate that the use of disinfected tertiary reclaimed water for food crop irrigation is safe and acceptable. No drawbacks were observed in terms of soil or groundwater quality degradation. Conventional farming practices were followed, and there was no need to alter those practices in a significant way. Marketability of the produce did not appear to pose any obstacles.

No naturally occurring virus was detected in any of the monthly samples of irrigation waters taken over the five-year study period. Seeded virus studies showed that both tertiary treatment processes, given a 90-minute chlorine contact time, were able to remove more than five logs of virus. Seeding of virus on the surfaces of vegetables demonstrated rapid die-off under ambient environmental conditions. No significant differences in the chemical and bacteriological characteristics of plant tissues among irrigation water types were found.

The irrigation water quality characteristics of the reclaimed water were comparable to the locally used well water, although total salts and sodium adsorption ratio (SAR) were somewhat higher. Nonetheless, these and other parameters of concern for irrigation remained within safe and acceptable ranges. No impact on soil permeability was detected over the five years of irrigation with reclaimed water. Irrigation with tertiary reclaimed water produced normal yields (sometimes exceeding the yield of the controls) of crops with no difference in quality and shelf life compared to controls.

For accessibility and practical applications, the specific findings and general conclusions of MWRSA are bundled and summarized below:

(1) In situ enteric viruses were never detected in either of the two tertiary reclaimed waters tested—coagulated, settled, filtered effluent or direct filtered effluent. Viruses were present in the influent to the tertiary process (secondary effluent).

(2) Treatment plant virus seeding studies indicated that either process train (including chlorination with 90-minute modal contact time) was capable of reducing the numbers of viruses by five or more orders of magnitude—a demonstration of equivalency of the two process streams.

(3) The above findings corroborated those of a previous study conducted by the Sanitation Districts of Los Angeles County in Pomona, California. This led to the acceptance of direct filtration as an equivalent substitute for the full process train specified in California Code of Regulations Title-22—and eventually as the standard for tertiary reclaimed water.

(4) Since no in situ viruses were found in the reclaimed water, there would be no exposure to agricultural workers who must tend and harvest the crops irrigated with reclaimed water.

(5) In the unlikely event that some viruses reach the field, they would survive for only a short time. Virus seeding studies conducted both in an environmental chamber and in the field indicated that virus die-off, both on plants and in the soil, would allow workers a significant margin of protection, particularly since they must wait several days after the last irrigation to re-enter the fields.

(6) The absence of viruses in reclaimed water affords consumers primary protection. The die-off of virus over time afford an extra barrier to consumers eating produce irrigated with reclaimed water. This is the effective protection against viruses that might be on crops routinely imported from countries with far less rigid controls on use of irrigation waters of highly impaired quality, including raw sewage.

(7) MWRSA, and the aforementioned study in Pomona, provided the scientific basis for establishment of the five-$\log_{10}$ virus removal standard by the California Department of Health Services.

(8) Aerosols generated from sprinkler irrigation with this disinfected tertiary reclaimed water did not contain microorganisms of wastewater origin.

(9) Medical monitoring of project personnel who came in direct contact with reclaimed water over a five-year period did not indicate any adverse health effects attributable to the water.

(10) Crop protection criteria (yield, quality, appearance, shelf life, soil permeability) can be met with the use of tertiary reclaimed water as long as irrigation water quality characteristics fall within acceptable ranges for the specific crop, climatic and soil conditions.

(11) Normal agricultural practices need not be altered when reclaimed water is used.

(12) Macronutrients and micronutrients in reclaimed water contribute significantly to improved yield and quality of vegetables whose green leaves (not mature flowers and fruits) are harvested for sale. Nutrients in water had no observable effect on artichoke yield and quality.

(13) Available nutrients in reclaimed water can result in reduced commercial fertilizer application, with reduced potential for groundwater pollution with nitrogen.

(14) The majority of the wholesale buyers of raw-eaten produce surveyed indicated that they would accept vegetables irrigated with reclaimed water as long as regulatory agencies approved the practice and as long as labeling of the produce was not required. Such labeling would be impractical because of the perishable nature of produce and its rapid blending in the market with imported produce. Some major growers and food processors recently have raised new concerns about potential consumer rejection of their products in the market.

ACKNOWLEDGMENTS

This chapter is dedicated to the memory of Ken DeMent who was Agency Manager for the Monterey Regional Water Pollution Control Agency (MRWPCA) during most of the performance period of MWRSA. The authors are grateful to the many people and organizations contributing to the various facets of this multi-dimensional study, collectively known as the MWRSA project. Special thanks to Monterey County Supervisor Marc del Piero (now Board Member of State Water Resources Control Board), MWRSA Task Force Chairman Walter Wong, MRWPCA Board Member Granville Perkins, MWRSA pilot treatment plant design engineer Bill Kirkpatrick, University of California Soil Chemistry Professor Richard Burau, MWRSA field supervisors Sam Earnshaw and Jo Ann Baumgartner, Monterey County Agricultural Extension Service vegetable crops irrigation specialist Dr. David Ririe, University of California virology laboratory technician Dave Straube, Monterey County vegetable growers Silvio Bernardi and Ed Boutonnet, Artichoke Growers and Shippers Association representative Hugo Tottino, Castroville treatment plant senior operator Ed Anderson, laboratory technician Jon Popper, and State Water Resources Control Board representative on the MWRSA Task Force Dr. Takashi Asano. Members of the California Department of Health Services, particularly Dr. James Crook and Dr. David Spath contributed valuable advice and oversight.

Funding for the MWRSA Project was provided principally by the United States Environmental Protection Agency. Other funding was contributed by the California State Water Resources Control Board, the State Department of Water Resources, and the Monterey Regional Water Pollution Control Agency. Engineering-Science (now Parsons Engineering Science) was the prime contractor and the University of California Environmental Engineering Laboratory in Richmond performed all virus studies before, during and after the five-year field trials.

REFERENCES

1   State of California Department of Water Resources, October 1940. *California Water Plan Update Bulletin 160-93,* Sacramento, California, Volume 1, pp. 342–343.

2   Sheikh, B., Cort, R., Kirkpatrick, W., Jaques, R., Asano, T., May/June 1990. "Monterey Wastewater Reclamation Study for Agriculture," *Research Journal WPCF,* Volume 62 (3): 216–226.

3   Engineering-Science, June 1980. Monterey Wastewater Reclamation Study for Agriculture—Phase II Final Report, Monterey Regional Water Pollution Control Agency, Monterey, California.

4   Engineering-Science, July 1981. Monterey Wastewater Reclamation Study for Agri-

culture—Year One Annual Report, Monterey Regional Water Pollution Control Agency, Monterey, California.

5 Engineering-Science, July 1982. Monterey Wastewater Reclamation Study for Agriculture—Year Two Annual Report, Monterey Regional Water Pollution Control Agency, Monterey, California.

6 Engineering-Science, July 1983. Monterey Wastewater Reclamation Study for Agriculture—Year Three Annual Report, Monterey Regional Water Pollution Control Agency, Monterey, California.

7 Engineering-Science, July 1984. Monterey Wastewater Reclamation Study for Agriculture—Year Four Annual Report, Monterey Regional Water Pollution Control Agency, Monterey, California.

8 Engineering-Science, July 1985. Monterey Wastewater Reclamation Study for Agriculture—Year Five Annual Report, Monterey Regional Water Pollution Control Agency, Monterey, California.

9 Engineering-Science, April 1987. Monterey Wastewater Reclamation Study for Agriculture—Final Report, Monterey Regional Water Pollution Control Agency, Monterey, California.

10 State of California, 1978. California Administrative Code of Regulations, Title 22 Division 4: Environmental Health, Wastewater Reclamation Criteria.

11 Peterson, R. G., December 1979. *Introduction to Statistics and Experimental Design,* Technical Manual No. 7, ICARDA.

12 The Council of Soil Testing and Plant Analysis, 1980. *Handbook on Reference Methods for Soil Testing,* University of Georgia, Athens, Georgia, pp. 84–87.

13 Pahren, H., December 1980. "Assessment of Health Effects," Panel Discussion. *Proceedings of a Symposium on Wastewater Aerosols and Disease,* September 19–21, 1979, sponsored by Health Effects Research Laboratory, U.S. Environmental Protection Agency, Cincinnati, Ohio.

14 Pettygrove, G. Stuart and Takashi Asano (ed.), 1985. *Irrigation with Reclaimed Municipal Wastewater—A Guidance Manual.* Lewis Publications, Inc., Chelsea, Michigan.

CHAPTER 18

Agricultural Irrigation with Treated Wastewater in Portugal

INTRODUCTION

AGRICULTURAL DEVELOPMENT IN PORTUGAL

IN spite of its glorious navigator past, Portugal has always been mainly a rural country through its eight centuries of existence as an independent nation. This rural vocation, the difficulties of communication with other nations due to its location at the western end of Europe, and certainly some government policies are the main reasons for the poor development of Portuguese agriculture prior to the third quarter of the 20th century when the country initiated a new cycle, restoring democracy after 48 years of dictatorship. Eight years later Portugal formally joined the European Communities (EC), which had an enormous impact on the economy and society in general, which has meant a serious challenge for Portuguese agriculture, considered the weakest in the European Union (EU) as proved by the comparison of some indicators shown in Table 18.1.

The most relevant differences between Portuguese agriculture and the mean EU agriculture concern: the used agricultural area (UAA) per farm, which in Portugal is less than 50% of the mean EU value; the active population employed in agriculture, which in Portugal is three times that of the mean European value; and the proportion of Portuguese agriculture in the GNP which is higher than in the other Member States.

Maria Helena F. Marecos do Monte, Ministero, Laboratório Nacional de Engenharia Civil, Avenida do Brasil, 101, 1799 Lisboa Codex, Portugal.

TABLE 18.1. Comparison of General Indicators of the Portuguese and the European Union Agriculture (Ministry of Agriculture, 1993).

Indicator	EU 12[a]	Portugal
Surface (km^2)	2 259 873[b]	92 071
Population (10^3 inhabitants)	327 063[b]	10 355
Standard of living (GNP/inhab)	18 622[b]	10 373
UAA (10^3 ha)	127 499	4 532
Active population in agriculture, silviculture and fisheries (%)	6.6[b]	17.8
Used Agriculture Area/farm (ha)	13.3[c]	5.2
Percent of agriculture in GNP (GAV/GNP)	3.1[d]	5.5
Percent of imports of food and agricultural produces out of total imports	12.1[c]	26.8
Percent of exports	8.5[c]	10.5

[a] Mean values of the 12 EC countries.
[b] Data of 1990.
[c] Data of 1987.
[d] Data of 1989.
GNP—Gross National Product.
GAV—Gross Added Value.

Causes of Portuguese Agricultural Underdevelopment

Two main categories of factors combine for the current situation: natural resources (soil capacity, available water, climate) and structural factors, such as the small size of the majority of farms, the little specialisation of crop production (too many crops grown on small farms, frequently combined with cattle growth).

In spite of the fact that only 26% of the soil is suitable for farming 46.3% is used for agricultural purposes (Ministério da Agricultura, 1991), with subsequent low crop yields and adverse environmental impacts, especially soil erosion. Most of the soil capacity and use (59%) is suitable for forestry. Figure 18.1 illustrates the difference between soil capacity and utilisation.

Climate and soil are the natural factors that most powerfully affect agriculture. The climate of the Portuguese mainland[1] can be classified as Mediterranean with Atlantic influence in the northern and central littoral and continental influence in the northern, central and southern interiors. Summers are dry and hot and temperature is mild in winter. Rain falls during the cold season and is unevenly distributed over the country. Drought periods frequently arise, seriously affecting more than half the

[1] Excluding the Azores and Madeira archipelagos.

country. Shortage of water for irrigation is an important cause of the low development of agriculture in Portugal, as discussed later.

Consequences of Portuguese Agricultural Underdevelopment

The most important consequence of the low level of Portuguese agriculture is the chronic high deficit of the national food balance (Portugal imports more than 50% of food products), which has been increasing during the last decade. The main imported agricultural products include leather, furs and cotton followed by food products, such as cereals, oil-bearing crops, sugar and meat.

Being a member of the European Union Portugal would not need to increase its agricultural production in order to reduce the deficit of the food balance since the EU has an excess of food and Portugal could import all the required products from other Member-States. However, it is not a good strategy for a state to depend on others to such an extent and in such a vital area as food. Therefore, Portugal must develop its agriculture in order to reduce its deficit of food balance, in spite of the current objectives of the Common Agricultural Policy of the EU which are limitation of crop production, development of forests and protection of the rural environment. The reduction of the import of some important agricultural products, such as maize and sunflower, could be achieved if their production were to significantly increase in the country. Production increases of such crops could be assured by irrigation, as shown by maize, which is grown all

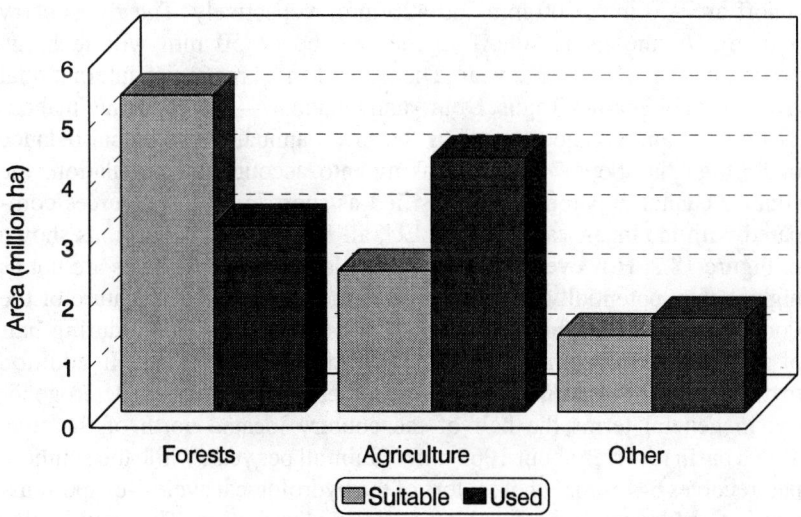

Figure 18.1 Soil capacity and use.

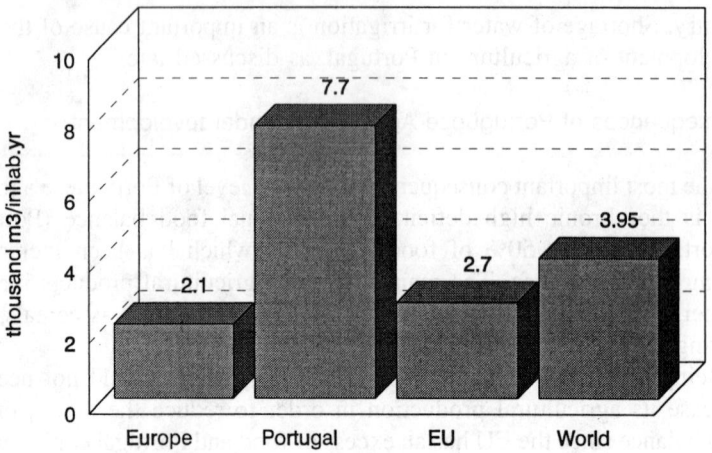

Figure 18.2 Comparison of water resources.

over the country, with higher yields in the regions where water for irrigation is readily available (the northwest and west coastal areas).

WATER RESOURCES AND WATER NEEDS FOR AGRICULTURE

Available Water Resources

In Portugal the mean annual values of rainfall, evapotranspiration and runoff are 920 mm, 500 mm and 420 mm, respectively. The rivers carry 370 mm of the mean runoff to the sea, hence 50 mm will recharge groundwater in an average year. The flow of the three major international rivers born in Spain—Tagus, Douro and Guadiana—used to bring in about 270 mm in an average year. Thus the mean annual hydrological balance in Portugal is about 740 mm. Taking into account the population, the country cannot apparently be classified as short in water resources compared with the mean values of the EU, all Europe or the world, as shown in Figure 18.2. However, the actually available water resources are not as high as they potentially could be, due to the Mediterranean feature of the Portuguese climate: about 66% of the annual rainfall occurs during half of the year, and in some cases about 30% falls in one month. In addition to the uneven time distribution of rainfall there is a clear spatial heterogeneity: in general terms the half of the country located north of the river Tagus basin receives about 1065 mm of rainfall per year, while the southern part receives 641 mm. Other factors of the hydrological cycle—evapotranspiration, infiltration runoff—show a similar distribution. The result is that under a natural regime 57.5% of the country mainland suffers a water deficit. These areas are located mainly in the southern part of the country and in northeast. Storage in reservoirs is the only way of making available the potential water resources. In many cases this is too expensive a solution. The forecasts for the ratio "available water/water need" in the year 2000 and 2020 is 0.99 and 1.42, respectively (Henriques, 1985), which points to a water shortage by the end of the century. The situation estimated for the year 2020 appears to be slightly better, because the growth rate of water needs is higher before the year 2000 than expected later due to the economic growth observed in the present decade. However, the situation is not as good as it appears, because 80% of the annual water needs concentrate in the dry season (May to October). Figure 18.3 presents the spatial distribution of the ratio "available water/water need" at present, in the years 2000 and 2020. The southern and eastern regions will be more affected than the northern and Tagus basin regions.

Water Needs in Agriculture

Basically, the water needed for agriculture is the amount required for irrigation since the water needed for livestock is a very small part (0.6%) of the total. The estimated mean annual water needs for irrigation during the dry season in mainland Portugal in the future are given in Table 18.2 (Henriques, 1985).

Should water be available, the increasing rate of water needed for irrigation will possibly be higher in the southern part of the country where irrigation projects will develop. Although the volume of water for agriculture tends to increase, its percentage will decrease in comparison with other water uses, such as industry, energy production, and drinking water supply, due to the development of industry and the use of more

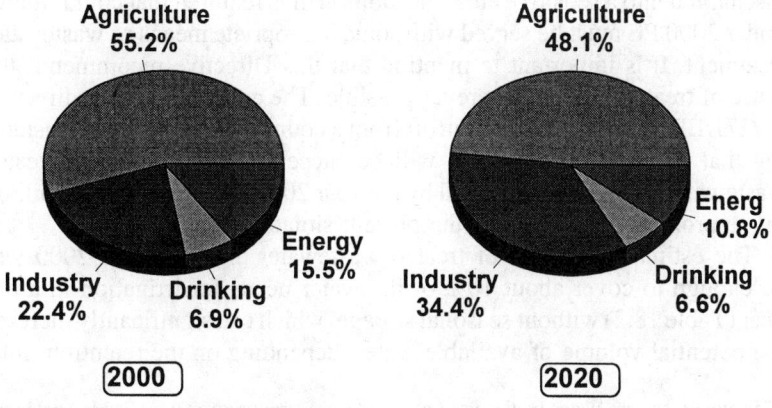

Figure 18.3 Estimation of water needs in the future.

TABLE 18.2. **Water Needs for Irrigation (hm³).**

Year	2000	2020
Wet	5145	6556
Mean	5932	7486
Dry	6717	8413

efficient irrigation technologies (allowing for less water losses in irrigation). Figure 18.3 compares the estimates of water needs in the near future (year 2000) and in the year 2020 (Henriques, 1985).

TREATED WASTEWATER IN PORTUGAL: PRESENT AND FUTURE

Treated wastewater is potentially a resource that may be used for several purposes, especially irrigation, because it is not so difficult to match the quality of treated wastewater with the required quality of water for irrigation. The annual volume of treated wastewater in Portugal is still much lower than expected by the year 2000. Very little wastewater treatment existed before 1974. Since then a major effort has been undertaken: in 1992 (Pires and Silva, 1992) the percentage of population of Portugal mainland served with drinking water systems, wastewater collection and wastewater treatment is 77%, 55.3% and 37.6%, respectively. This is the worst situation in the EU, which has to improve dramatically before the year 2005 according to the requirements imposed by the Directive of the Council of the European Communities regarding urban wastewater treatment (directive 91/271/EEC). Basically, Directive 91/271/EEC states that every community above 2000 PE must be served with secondary wastewater treatment or tertiary treatment where the effluent is to be discharged into a sensitive area;[2] in addition, it is required that communities under 2000 PE must be served with some appropriate means of wastewater treatment. It is important to mention that this Directive recommends the reuse of treated effluent wherever possible. The compliance with directive 91/271/EEC requires a strong effort from a country like Portugal. Considering that the effort to be made will be successful the volume of treated wastewater available in Portugal by the year 2000 will be about 580 million m³, approximately double of the present situation (Table 18.3).

The estimated volume of treated wastewater discharged in 2000 will be enough to cover about 10% of the water needs for irrigation in a dry year (Table 18.2) without seasonal storage, which can significantly increase the potential volume of available water, depending on the retention time.

---

[2] Sensitive areas are water bodies used as sources for drinking water, valuable ecological waters, and waters at risk of eutrophication.

TABLE 18.3. Annual Volume of Treated Wastewater.

Year	Population[a] (10^3 PE)	Drinking Water Consumption (L/PE/d)	Volume of Wastewater ($10^6 m^3$/year)			
			Total	Collected	Treated	Untreated
1977	8 911	140	181	122	4	118
1987	9 324	162	276	187	21	166
1990	9 866	185	533	295	111	422
2000	9 270	240	650	584	584	66

[a]Resident in mainland Portugal.

833

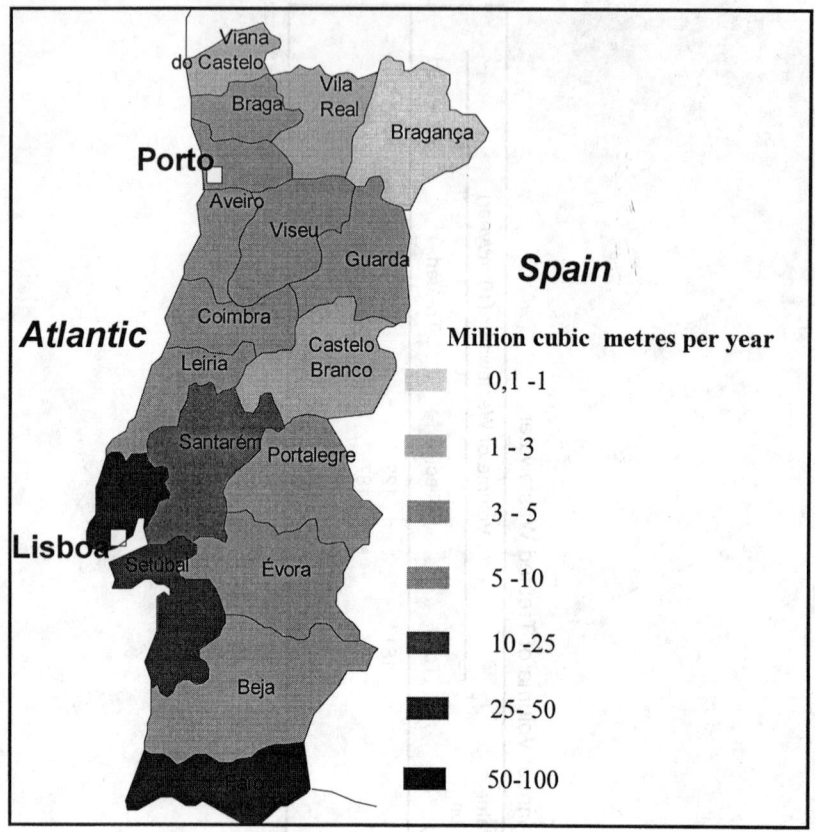

Figure 18.4 Treated wastewater in the year 2000.

The availability of treated wastewater will reflect the way the population is distributed in the country. In Portugal, most of the population lives in the western half of the mainland, as a migratory movement from the interior to the coast has been observed principally in the eighties. Another important characteristic of the distribution of the Portuguese population is that 48.6% lives in communities smaller than 2000 people, located principally in the northern region, which makes the reuse of treated wastewater more difficult, due to the small available volume (unless storage is provided) and the numerous mountains in the northern region, that make the cost of water distribution very high. Figure 18.4 illustrates the estimated availability of treated wastewater for reuse in the year 2000 according to a not very optimistic view that excludes the effluent discharged by plants treating less than 2000 PE and does not consider seasonal storage.

Figure 18.4 shows that the use of treated wastewater for irrigation may significantly contribute towards the agricultural development in the districts of Faro, Lisboa, Setúbal, Évora and Santarém. Major benefits will possibly be experimented in the districts of Faro (where water is badly needed for landscape irrigation and golf courses) and Évora (traditionally suffering of drought). It is very difficult to estimate the area that could be irrigated with treated wastewater, because this depends on a number of factors which have to be evaluated case by case, such as available flow, storage, distance to farms, local orography, etc. However, a rough estimate would range between 35 000 and 100 000 ha, depending on storage retention time. The interest of the use of treated wastewater for irrigation in Portugal is clear if we take in account that the current irrigated area in the country is 900 000 ha and this is about 22% of the total area used for agricultural.

Secondary treatment is most commonly used in Portugal. Usually, secondary effluents are not disinfected. Figure 18.5 shows the most frequent treatment processes in Portuguese WWTP. The tendency seems to be towards increasing the number of waste stabilization pond (WSP) systems, especially in the southern area of the country and a slight increase of fixed film reactors. The increase of WSP systems favours the subsequent use of the effluent for irrigation because of its high microbiological quality which makes it suitable for unrestricted irrigation, provided that the WSP system is properly designed.

BRIEF ASSESSMENT OF EXISTING GUIDELINES REGARDING THE USE OF RECLAIMED WASTEWATER FOR IRRIGATION

Reclaimed wastewater contains chemical and microbiological components which may pose a potential public health risk. There are no guidelines

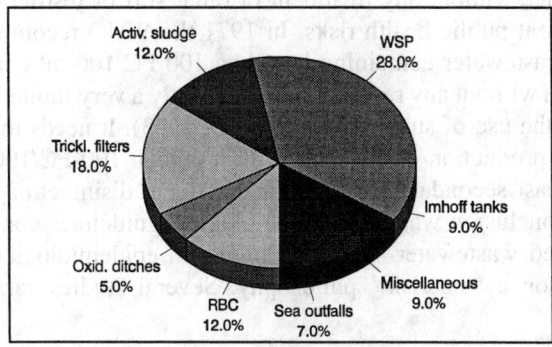

Figure 18.5 Type of treatment processes.

concerning the prevention of health risks due to the chemical characteristics of reclaimed wastewater used for irrigation. It is generally assumed that only the soil-plant biosystem is significantly affected by the chemical characteristics of irrigation water by means of the absorption of phytotoxic substances through roots and leaves, although it is known that some elements tend to accumulate within parts of plants, sometimes reaching levels dangerous to animal health. On the other hand there are several guidelines on the microbiological quality of reclaimed wastewater used for irrigation. The State of California was the pioneer and its wastewater reuse guidelines, first issued in 1918, have been adopted with more or less modifications in several North-American states as well as in other countries. *A Manual: Guidelines for Water Reuse* was published by the USEPA and the USAID in 1992 (EPA, 1992).

A survey of the existing regulations in the USA and other countries, such as South Africa, Israel, Kuwait, Tunisia, and France shows that regulations can be more or less comprehensive, varying from a list of limits of concentration for certain parameters to the specification of the required wastewater treatment and sometimes the inclusion of detailed monitoring procedures. The most common case is that regulations combine guidelines for the reclaimed wastewater quality with specifications on irrigation methods and crop restriction according to the water quality and hygiene recommendations.

Most of the existing regulations can be considered quite strict, in the sense that they require a high grade water quality, namely concerning the microbiological characteristics, only achievable by means of expensive technology requiring sophisticated operation. Such strict regulations based on a "no risk" philosophy could not be adopted by many countries badly needing to reuse wastewater for irrigation. The only options left in such countries were both undesirable: either no wastewater reuse for irrigation or wastewater reuse without any respect by unbearable regulations, meaning in practice without any treatment or other sort of restriction, with all the consequent public health risks. In 1971 the WHO recommended that reclaimed wastewater containing less than 100 FC/100 ml could be used for irrigation without any restriction because only a very limited risk would come from the use of such effluent (WHO, 1973). It needs to be pointed out that the production of effluents with less than 100 FC/100 ml would require at least secondary treatment and effluent disinfection.[3] A second important conclusion was that microbiological guidelines concerning the use of treated wastewater should be based on epidemiological evidence rather than on a "no risk" philosophy. Several studies carried out by

[3] At the time lagoon systems were not yet considered as capable of producing effluents with very low content of pathogens.

some international organizations (World Bank, UNEP, UNDP, FAO, IRCWD, and IDRC) identified some contradiction in the requirements of quality for water for irrigation and for other uses like bathing and lead to the conclusion that there was no justification for the stringency of most of the existing regulations on wastewater reuse (IRCWD, 1985). Moreover, it was found that the major health risk was not due to pathogenic bacteria or viruses but to helminths. These views were adopted by the WHO in 1989 when it published *"Health guidelines for the use of wastewater in agriculture and aquaculture"*. However, the WHO guidelines resulted in some heated discussions, with specialists being divided in what can be called two "schools": the less stringent epidemiological evidence school (WHO) and the nil risk school, represented by the US and several Israeli researchers.

THE ESTABLISHMENT OF PORTUGUESE GUIDELINES: NEED FOR EXPERIMENTAL DATA

The establishment of guidelines for the use of reclaimed wastewater for irrigation in Portugal must consider the existing guidelines in other countries, principally in Europe (France and Germany) on one hand, and on the other to take into account the experience from other countries with similar level of development, namely in the Mediterranean region. The guidelines must cover two main areas: (a) the agronomic aspects, related to the maximization of crop yields and soil and groundwater preservation; and (b) the sanitary aspects, related to public health protection. The document should have the following general characteristics: (a) simplicity, to prevent unnecessary discouragement of wastewater reuse; (b) robustness and reliability, to assure public health protection, good crop production, and to prevent adverse impact in the environment. Finally, the document must be flexible enough to allow for further improvement steps in the light of new scientific and technologic developments and acquired experience.

Guidelines can be defined in great detail or in a broad manner, such as a simple list of chemical and microbiological parameters whose values should range within a certain interval. This simple approach, however, presents a rather serious shortcoming: in most of the southern European countries, where wastewater reuse for irrigation is more necessary, the monitoring of effluent quality from wastewater treatment plants is not completely assured, due to several factors, including lack of trained staff, available laboratories and also financial reasons. A second simple approach would consist of specifications for minimum treatment level required for certain irrigation purposes. For instance, not less than primary treatment should be authorised in any circumstances; only disinfected effluent would be allowed to irrigate raw eaten vegetables and sport fields. This methodology is not satisfactory, however, because the operation of treatment plants is frequently unreliable in many areas and the effluent quality is unlike to correspond to the presumed level. A combination of both approaches, consisting of a reduction of the number of parameters in the list with some specification concerning the minimum treatment requested for certain irrigation purposes seems to be a better methodology. The selection of parameters and the requests for treatment specifications are now the question. As mentioned previously the list of parameters for the monitoring of reclaimed wastewater quality for irrigation should include both chemical parameters with major agronomic impact and microbiological parameters affecting public health. However, these impacts depend on the agriculture practices, such as irrigation methods and relevant crops in the region, for example. Therefore, the guidelines should be the interception of three factors: specification of limits for some chemical and microbiological quality parameters in combination with treatment specifications and as a function of the field practices (crops and irrigation methods). Experimental research can provide extremely useful information based on local conditions such as climate, quality of available treated wastewater, economically interesting crops, field practices. The obtention of such field data was the justification for the research project led by Laboratório Nacional de Engenharia Civil, which is described in the next section.

EXPERIMENTAL RESEARCH

OBJECTIVES OF THE STUDY

The main objectives of the study were to assess and compare the effects of irrigation of various types of treated wastewater versus the same crops irrigated with potable water and given commercial fertilizers (the crops irrigated with treated wastewater were fertilized by the nutrients contained in the effluent only). The assessed effects were crop quality, yield and contamination, soil contamination and changes in soil chemical characteristics.

The study also included a demonstration purpose: a videotape was produced in order to inform several publics about wastewater reuse for irrigation, namely farmers, agricultural, public health and municipal engineers.

CHARACTERIZATION OF THE EXPERIMENTAL INSTALLATIONS

Treated Wastewater Used for Irrigation in the Experiments

Treated wastewater used in the experiments corresponded to the most usual type of effluents from WWTP in Portugal. An assessment of the existing WWTP indicated that the majority provides secondary treatment, the first plant including tertiary treatment is expected to be commissioned in 1998. Activated sludge is the most used process (25% of total plants), followed by trickling filters (10% of total). Wastewater treatment ponds show a rapid increase in number, especially in the region of the Alentejo, where reuse for irrigation is badly needed. Secondary effluents from activated sludge and trickling filters have similar microbiological characteristics, hence it was decided to study the effects of one of these type of effluents. Due to logistic convenience final effluent of a high-rate trickling filter plant was selected for experiments. The availability of primary effluent at the same plant was the reason for the inclusion of this type of effluent in the tests. Effluent from a two-cell wastewater treatment pond system (consisting of an aerated lagoon followed by a facultative pond) was also selected for experimentation, due to the increasing importance of this type of plants in the country and to the specific characteristics of the effluent in comparison with activated and trickling filters effluents.

The controls of this study research were the crops irrigated with clean water (potable water free of pathogens) and given commercial fertilizers.

Experimental Facilities Description

Experiments were carried out in two sites: primary and secondary effluents were tested at the WWTP of Évora (150 km south-east of Lisbon) and the facultative pond effluent was taken from the WWTP of Santo André (135 km south of Lisbon). Figures 18.6 and 18.7 represent the plants' layouts as well as the experimental plots location.

Climate

Characterisation of the climate was obtained from the climatological stations nearest to the sites. The characteristics of most concern were those with greatest influence on agricultural production and the establishment of an effective irrigation schedule: air temperature, evapotranspiration rates and precipitation. Tables 18.4 and 18.5 present the mean values of meteorological data from May to October (months of experiments) recorded by INMG[4] at Évora and at Santo André for the last 45 and 30 years respectively.

According to Thorntwaite climatic criteria, local climate can be classified as semi-arid, with nil water excess and megatermic.

---

[4] Instituto Nacional de Meteorologia e Geofísica.

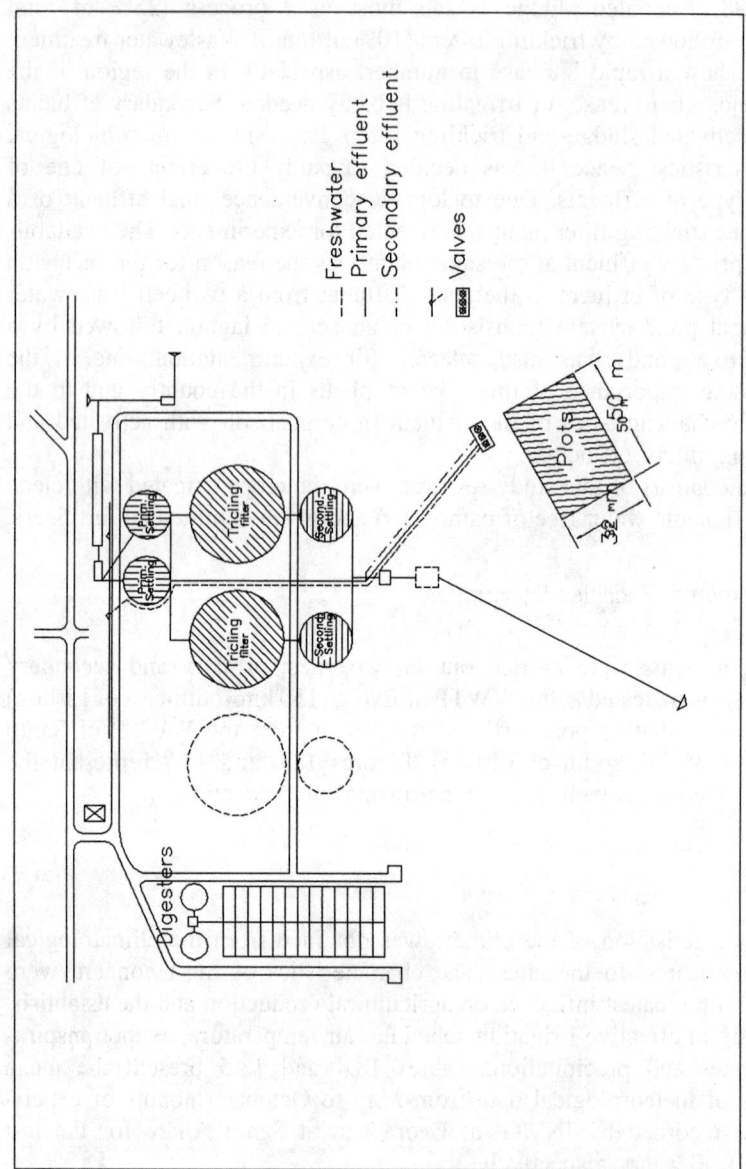

Figure 18.6 Location of the experimental plots in the experiments with primary and secondary effluents.

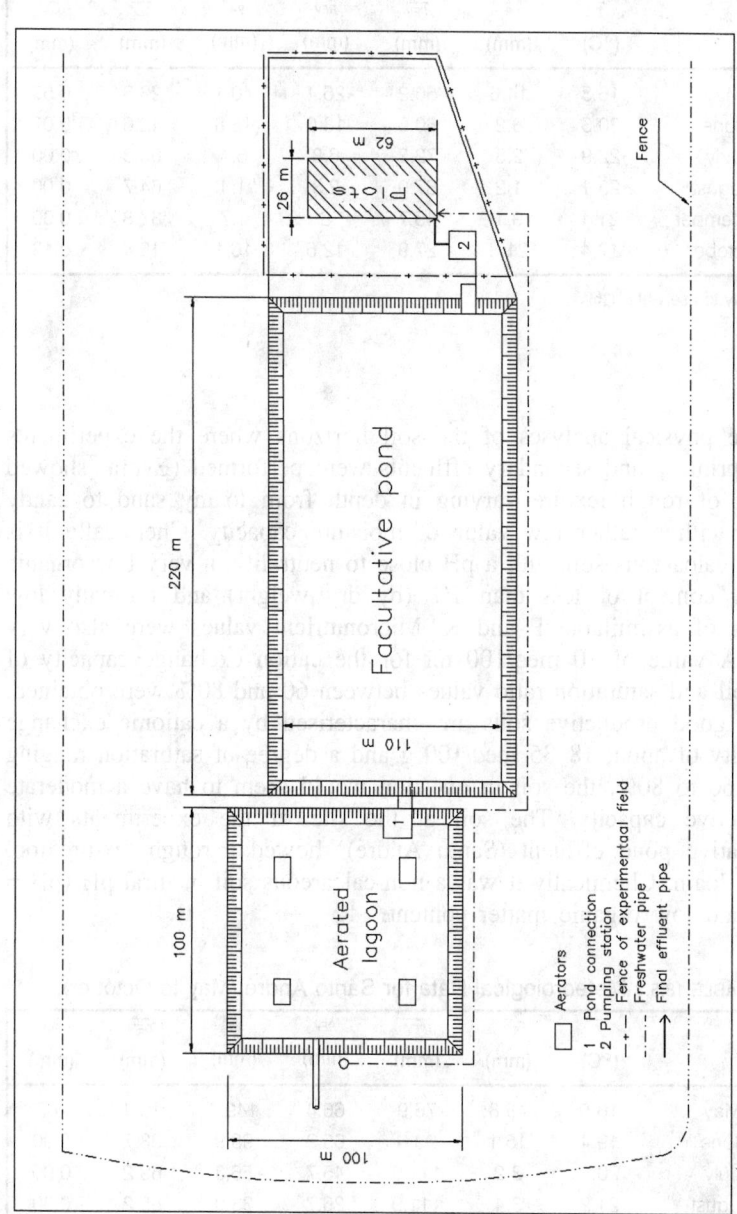

Figure 18.7 Location of the experimental plots in the experiment with facultative pond effluent.

Aerators

1 Pond connection
2 Pumping station
Fence of experimental field
Freshwater pipe
Final effluent pipe

Fence

Plots

62 m

26 m

Facultative pond

220 m

110 m

Aerated lagoon

100 m

100 m

TABLE 18.4. Meteorological Data for Évora (May to October).

	$\bar{T}$ (°C)	$\bar{R}$ (mm)	$\overline{PEV}$ (mm)	REV (mm)	SW^a (mm)	$\overline{DEF}$ (mm)	$\overline{SUP}$ (mm)
May	16.3	14.6	50.2	26.4	40.1	23.8	0.50
June	20.3	8.2	60.6	18.0	18.8	42.6	0.07
July	22.9	2.3	72.2	6.8	5.4	66.3	0.00
August	23.1	1.2	67.0	2.3	1.4	64.7	0.00
September	21.4	9.1	45.7	8	1.7	37.8	0.00
October	17.4	21.8	27.9	12.5	16.1	15.4	0.13

[a] Capacity to use water (%).

Soil

The physical analyses of the soil horizons where the experiments with primary and secondary effluents were performed (Évora) showed a soil of rough texture varying in depth from loamy sand to sandy loam, with a rather low value of moisture capacity. Chemically it is a non-calcareous soil with a pH close to neutrality, a very low organic matter content of less than 1% (by dry weight) and normally low values of assimilable P and K. Micronutrient values were also very low. A value of 10 meq/100 ml for the cation exchange capacity of the soil and saturation ratio values between 60 and 80% were obtained. Since good productive soils are characterised by a cationic exchange capacity of about 18–35 meq/100 g and a degree of saturation ranging from 60 to 80%, the soil in question would seem to have a moderate productive capacity. The soil at the site of the experiments with facultative pond effluent (Santo André) showed a rough texture too, sandy loam. Chemically it was a non-calcareous soil, neutral pH (pH = 7.3) and low organic matter content.

TABLE 18.5. Meteorological Data for Santo André (May to October).

	$\bar{T}$ (°C)	$\bar{R}$ (mm)	$\overline{PEV}$ (mm)	REV (mm)	SW^a (mm)	$\overline{DEF}$ (mm)	$\overline{SUP}$ (mm)
May	16.5	46.8	76.9	66.4	149.7	10.4	9.7
June	19.4	15.1	99.6	65.9	98.9	33.7	0.00
July	20.8	3.2	114.9	45.7	56.3	69.2	0.00
August	21.2	3.4	111.9	26.7	33.0	85.2	0.00
September	20.4	22.7	92.7	31.1	24.6	61.6	0.00
October	17.4	21.8	27.9	12.5	16.1	15.4	0.13

[a] Capacity to use water (%).

Crops Tested

The inclusion of several crops in the same experiment provides a broader view of the effects of irrigation with wastewater. The use of replicate subplots for the different types of wastewater used will increase the number of degrees of freedom and thus the statistical validity of the results obtained. The criteria used to select the crops to be tested were that they should be of economic importance to the region, that the consuming parts should not likely to be contaminated, that they should be fairly tolerant to salinity (as it was anticipated that treated wastewater would present high dissolved solids content) and finally that their number would not extend the experimental plots to an impractical number of replications. A cereal (grain-maize), a forage (sorghum) and an oil-bearing crop (sunflower) were the selected crops.

Field Layout

The field layout was governed by the objectives of the study and the need to avoid the risk of cross contamination between plots irrigated with different wastewater. The main objectives of the study can be shortly described as the comparison of the effects on crops of the irrigation with different effluents and with clean water. The layout was designed with large plots irrigated with the same type water, each one subdivided in three subplots, one per each crop.

In the experiments with three types of water (primary and secondary effluents, and freshwater) four replicate plots for irrigation water were used; and six replicates were considered in the experiment where only two irrigation waters were used (facultative pond effluent and freshwater). A total of 36 subplots were obtained in both experiments. Crops were distributed at random with the only limitation of the three types of crops irrigated by the same water, as shown in Figures 18.8 and 18.9. The experimental plots were located adjacent to the wastewater treatment plants and covered a total area of 1600 m^2 and 1664 m^2 at Évora and Santo André, respectively.

SETTING-UP AND RUNNING THE EXPERIMENTS

Analysis of Soil and Irrigation Waters before Experiments

After ploughing and levelling the soil, a composite sample, representative of the overall experimental area, was collected. The results of the preliminary analysis of this sample were used to determine the fertilizer type and amounts to be applied to the control subplots, that is, those that

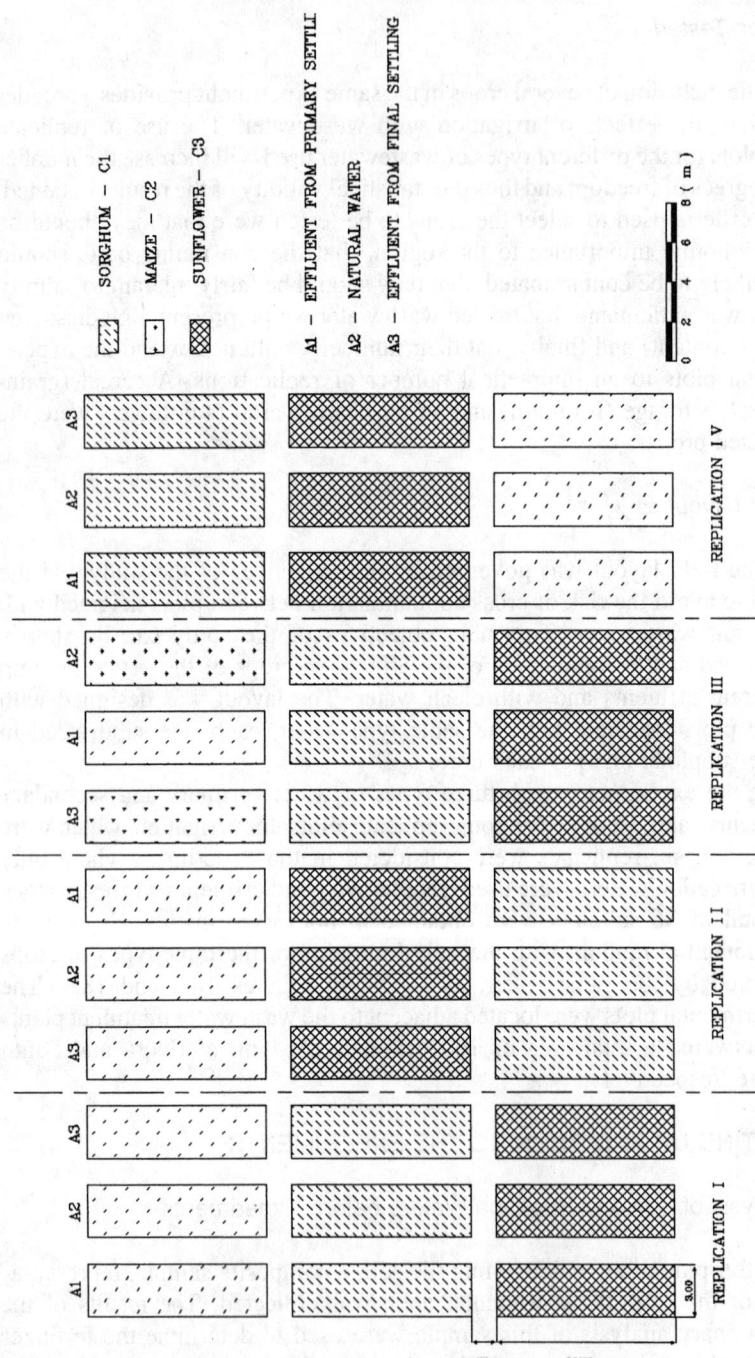

Figure 18.8 Layout of experiments with primary and secondary effluents.

SORGHUM – C1

MAIZE – C2

SUNFLOWER – C3

A1 – EFFLUENT FROM PRIMARY SETTLI

A2 – NATURAL WATER

A3 – EFFLUENT FROM FINAL SETTLING

0 2 4 6 8 (m)

REPLICATION IV

REPLICATION III

REPLICATION II

REPLICATION I

A3 A2 A1

A2 A1 A3

A1 A2 A3

A3 A2 A3

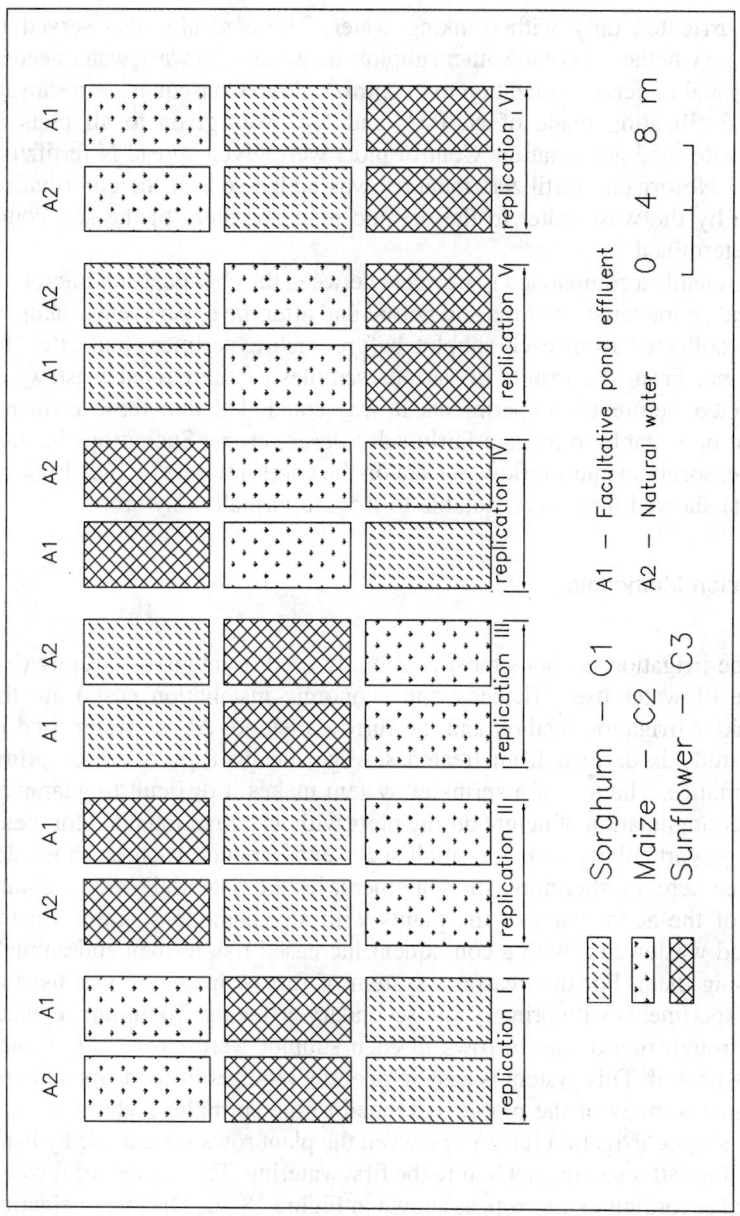

Figure 18.9 Layout of experiments with facultative pond effluent.

Sorghum — C1   A1 — Facultative pond effluent
Maize — C2   A2 — Natural water
Sunflower — C3

0   4   8 m

845

were irrigated only with drinking water. These results also served to indicate whether or not the other subplots irrigated with wastewater needed additional mineral fertilization to maximize yield. Consequently, a subsurface fertilization made of N, P_2O_5 and K_2O was given to all plots to stimulate seed germination. Control plots were given spread N fertilizers twice. No organic fertilizer (manure) was added so that the contribution made by the wastewater to the organic matter content of the soil could be determined.

To enable a comparison to be made between the chemical and microbiological characteristics of the soil before and after the experiments, samples were collected from each subplot before each experiment and after the last one. From the results of various samples of the treated wastewater collected before the experiments it was concluded that these effluents might be suitable to irrigate fairly salt-tolerant crops. Such crops include maize, sorghum and sunflower. Data on the control waters (actual drinking water) showed they were suitable to irrigate virtually any crop.

Irrigation Methodology

The irrigation methods most commonly used in intensive farming (because of water use efficiency and economic installation costs) are the sprinkler irrigation method and the furrow method. As the water used in this study is derived from treated sewage, health aspects are of prime importance. The use of a sprinkler system makes it difficult to guarantee non-contamination of neighbouring plots with water not intended for them. This is particularly true for small-scale experiments, such as those described here. Furthermore, spraying increases the potential for contamination of the aerial parts of the plants with any pathogens present in the treated wastewater, with a consequent increased risk to man and animals utilising them. For this reason a system of furrow irrigation was used in the experiments with primary and secondary effluents. To minimize water loss trough runoff, the furrows in each subplot were closed off at each end with soil. This system is easy to operate, requires little manpower and eliminates many of the problems related to the sprinkler method.

V-shaped irrigation furrows between the plant rows were made by hand with hoes after sowing but before the first watering. The number of furrows varied according to the crop, as shown in Figure 18.10. The three irrigation waters were taken in buried PVC pipes from the primary and secondary settling tanks and from a hydrant to a brick block near the plots. Three valves in the brick block were used to control the flow through three flexible plastic pipes. Three different colours were used for the pipes in order to avoid mistakes in the irrigation.

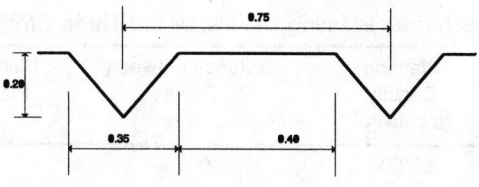

Maize and sunflower

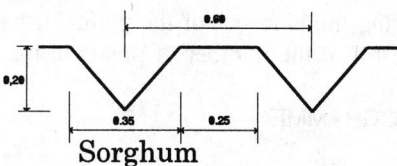

Sorghum

Figure 18.10 Furrow cross-section.

The water irrigation rate was calculated based upon the soil characteristics (available moisture, density, etc.) and the nutrient amount carried with the effluents. It was found that 21 irrigation sessions of 30 mm each was necessary, making a global irrigation rate of about 6 000 m³/ha/yr.

In the experiments with facultative pond effluent the irrigation water was applied by localized (drip) irrigation. The main objective of this choice was to assess the clogging potential of emitters when used with an effluent containing a high amount of suspended solids (mostly microalgae, in this case). The effluent was made to pass through a filter of 150 mesh and pumped to the irrigation system. This was made of pipes with inserted emitters at regular distance varying according to the type of crop: distance between emitters was 0.5 m for maize and sunflower and just 0.35 m in the case of sorghum. Emitters could be removed for cleaning purposes.

Similar calculations were made for the experiments. The volume of applied water was controlled by irrigation time. Daily irrigation sessions of two hours each were carried out, the frequency was double during August.

Cultivation Techniques

As the crops are to be watered and the soil is moderately productive, intermediate sowing densities were selected and proved to be satisfactory. Table 18.6 presents the planting details for the investigated crops.

To ensure that the required planting densities were achieved, the seeds

TABLE 18.6. Planting Details for the Three Crops.

Crop	Planting Density (plants/ha)	Distance between Rows (m)	Distance between Plants in a Row (m)
Maize	67 000	0.75	0.20
Sorghum	167 000	0.60	0.10
Sunflower	53 300	0.75	0.20

were sown at three times this density. Although this strategy has the drawback of requiring thinning out of the plants, it has the advantage of allowing better control of the number of plants in the subplots.

MONITORING PROGRAMME

The preparation and planning of a suitable monitoring programme for a study like this is very important. It requires as much attention as the setting up of the experimental field plots, since the success of the programme will depend on the correct type of data being collected to allow meaningful interpretation of the water-soil-plant interactions, which are the basis of the experimentation. Monitoring of the study has to include physicochemical and microbiological parameters of the treated wastewater, soil and crops and also measurements of crop productivity.

Physicochemical and Microbiological Analysis of Irrigation Waters

The irrigation waters used included three types of treated wastewater and drinking water from two cities. Only occasional grab samples of drinking water were taken, because its quality should be relatively constant and data should be available from the routine monitoring of quality carried out by the municipal supplier. Treated wastewater quality is likely to vary significantly and composite samples are to be collected to be representative. Perfectly representative samples were collected in the experiments with primary and secondary effluents, as they were made of subsamples collected at short intervals during the actual irrigation sessions. Automatic samplers were used for this purpose. Weekly samples were collected in the case of the facultative pond effluent, although irrigation sessions were carried out daily. However, the representativeness of data is considered good, since the high retention time of wastewater in the pond system provides a much lower quality variation of the effluent. Facultative pond effluent composite samples were taken from a tap after the filter located upstream of the emitters during each irrigation session. During transportation to the laboratories the samples were preserved and stored in an insulated cool box (containing ice). Samples were analyzed according to Standard Methods (1985) for salinity (electrical conductivity, total dissolved solids and principal ions, such as calcium, magnesium, sodium, carbonate, bicarbonate, chloride and sulphate), macronutrients (nitrate, organic nitrogen, potassium and total phosphorus), boron, pH, organic matter (BOD and COD) and heavy metals (Cd, Cr, Co, Fe, Mo, Ni, Pb and Zn).

For microbiological analyses of the effluents grab samples were collected in sterilized flasks once a month during the irrigation period. The microbiological quality of the effluents was determined by enumerating the faecal coliforms and faecal streptococci, and helminth eggs (only for the last out of five tests). Occasionally, *Salmonella* spp. were enumerated.

The potable water was not analyzed microbiologically because it was not expected to contain pathogens.

Physicochemical and Microbiological Analysis of Irrigated Soil

For each subplot, samples from the top 20 cm layer and at 50 cm depth in the soil profile were analyzed for twelve chemical parameters. All the soil samples collected were composite samples, each one comprising a series of subsamples from an individual subplot, as shown in Figure 18.11.

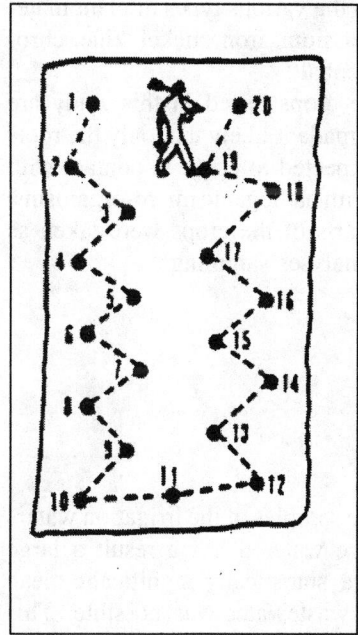
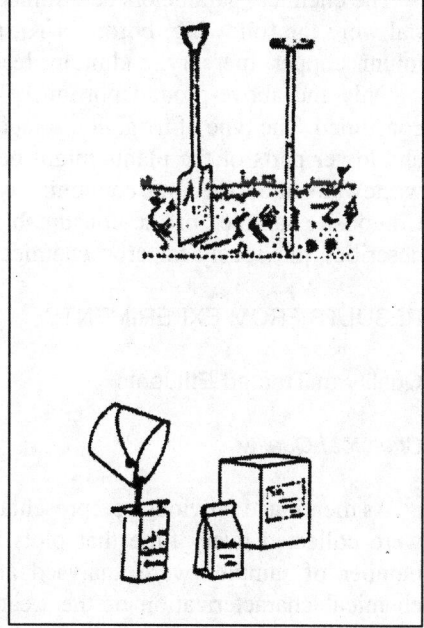

Figure 18.11 Composite soil sampling.

Soil samples were collected before setting up each annual experiment. In practical terms, this was equivalent to sampling before and after each experiment and allowed the assessment of effects from effluent irrigation during the previous season.

Microbiological analysis of soil consisted of faecal coliform enumeration in composite samples of surface soil.

Physicochemical and Microbiological Analysis of Crops

Assuming that the crops irrigated with potable water possess a good chemical quality, their mineral composition was used as the basis of comparison for the same crops irrigated with the different effluents. Samples of the crop material were taken at harvesting for chemical and microbiological analyses. Subsamples made up of two plants were taken at random in each replicate subplot from rows other than two peripheral ones to avoid the so-called "edge effects" and gathered together to make a composite sample. The remaining plants in the central rows were then harvested and weighed for crop yield determination, those in the peripheral rows being discarded for reasons previously explained, to determine crop yield. Sorghum can be harvested up to four times a year. Samples were taken at every harvesting.

The chemical parameters determined in the various types of plant material were the following: boron, N-Kj, potassium, iron, nickel, zinc, chromium, copper, mercury, cadmium, lead, cobalt.

Only the above-ground portion of the crops tested in this study are consumed. The type of irrigation selected made it likely that only the roots and lower parts of the plants might be expected to come in contact with wastewater and so become contaminated with pathogenic micro-organisms. Composite samples of the consumable parts of the crops were taken as described previously for crop chemical analyses sampling.

RESULTS FROM EXPERIMENTS

Quality of Treated Effluents

Chemical Quality

As mentioned previously, representative samples of the irrigation waters were collected every time that plots were watered. As a result a large number of samples were analysed and a statistically significant mean chemical characterization of the treated wastewater was possible. This provided important data in a country where there is not much information regarding water quality. Table 18.7 summarizes the typical concentrations of the three effluents used in the experiments.

TABLE 18.7. Mean Chemical Characterization of Primary Effluent, Secondary Effluent and Facultative Pond Effluent.

	Effluent		
PARAMETER	Primary	Secondary	Facultative Pond
pH	7.4	7.5	8.2
TSS (mg/L)	53.0	29.8	36.2
BOD (mg/L)	178.8	85.8	61.2
COD (mg/L)	358.5	223.5	92.6
Org-N(mg/L N)	12.94	9.48	13.39
NH_4-N(mg/L N)	30.42	20.13	17.92
NO_3-N(mg/L N)	0.97	1.78	1.29
Total-N(mg/L N)	40.77	31.43	30.20
Total-P(mg/L P)	89.64	103.65	14.58
Conduct (μS/cm)	1 236	1 237.5	1 503
Na (mg/L)	118.6	129.7	142.5
K (mg/L)	22.3	24.7	36.8
Ca (mg/L)	39.4	38.9	74
Mg (mg/L)	20.1	19.9	32.2
Cl (mg/L)	144.2	149.6	166.9
SO_4 (mg/L)	50.6	60.2	103.0
B (mg/L)	0.68	0.76	1.53
Hardness (°F)	9.84	9.75	31.9
HCO_3 (mg/L)	400.5		
TDS (mg/L)	662.6	693.95	1 145.3
Aj R_{Na}	7.5	7.91	3.24
As (μg/L)	5.8	5.9	
Cd (μg/L)	2.3	2.1	< 5.0
Co (μg/L)	< 30.0	< 30.0	
Cr (μg/L)	< 20.0	< 20.0	< 5.0
Cu (μg/L)	147.7	125.5	6.4
Fe (μg/L)	400.3	311.5	105.0
Mn (μg/L)	59.9	58.3	13.2
Mo (μg/L)	2.3	2.1	
Ni (μg/L)	68.3	73.1	< 5.0
Pb (μg/L)	22.4	21.5	18.2
Zn (μg/L)	140.3	129.5	570.7

The assessment of the chemical quality of the treated wastewater used for irrigation is important taking into account the impact on the biosystem soil-crop from the agricultural (crop yield) and environmental point of view. The following considerations on the suitability of the effluents referred in Table 18.7 are based on the FAO guidelines for irrigation water (FAO, 1985).

The mean heavy metal contents of the three effluents are clearly less than the maximum levels recommended by FAO; thus heavy metals do not seem to pose a toxicity problem. However, other parameters, such as sodium and chloride ions show mean concentrations which are likely to induce toxicity problems in sensitive plants, mainly through leaf absorption. Therefore, spray irrigation with these effluents must be avoided, especially in the case of the facultative pond effluent. This effluent also is more toxic to plants as it presents higher mean content of boron.

Salinity, assessed by the electrical conductivity, TDS and R_{Na}, showed moderately high values, the facultative pond effluent presenting slightly higher conductivity as proved by its TDS content, which was approximately twice those of the primary and secondary effluents (Figure 18.12).

Apparently the impact of the facultative pond effluent on soil permeability should be expected to be more adverse than the impact of the other effluents. Nevertheless, the ratio R_{Na}/conductivity of the three effluents include them in the same C3-S1 category of the USSL,[5] meaning that they have a high salinity and a low sodium content, so these effluents may be suitable for the irrigation of fairly salt-tolerant crops in soils with good drainage characteristics.

The three effluents present nitrogen content (total-N) slightly greater than 30 mg/L, which is the maximum level recommended by FAO; consequently they are not suitable for crops sensitive to this element, unless the effluents are blended with a water with a lower nitrogen concentration.

The primary effluent shows the higher nitrogen content, but the secondary effluent is the most nitrified, although nitrate concentration is low (Figure 18.13) and similar to the facultative pond effluent.

Concerning other important nutrients, such as phosphorus and potassium, the effluent of the facultative pond contains less phosphorus, but

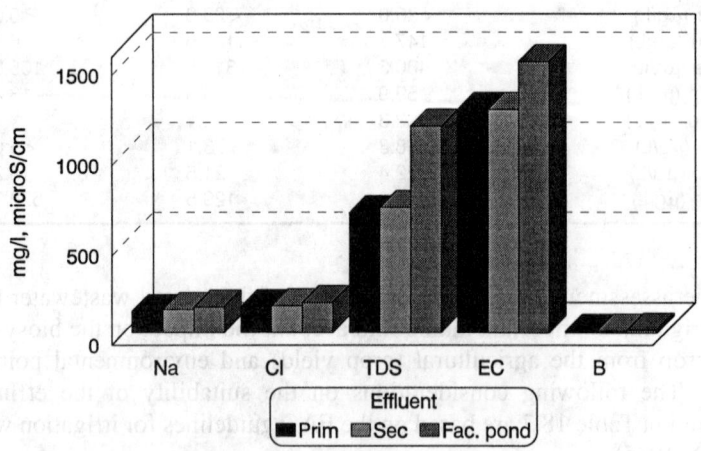

Figure 18.12 Comparison of the salinity in primary, secondary and facultative pond effluents.

[5] USSL—United States Salinity Laboratory.

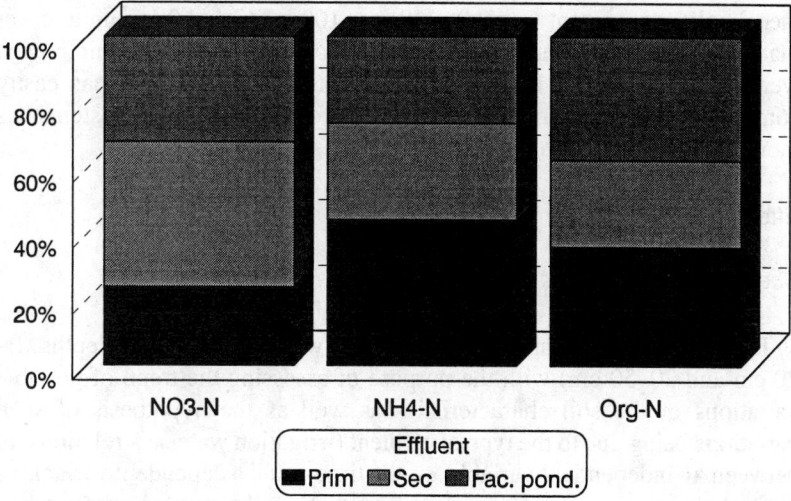

Figure 18.13 Nitrogen compounds in primary, secondary and facultative pond effluents.

more potassium than the other two effluents, which generally can not supply enough potassium to many plants and therefore need complementary chemical fertilisation.

Microbiological Quality of Treated Effluents

The mean contents of pathogenic indicator bacteria in primary and secondary effluents were shown to be very similar, as presented in Table 18.8.

Primary treatment only reduced 1 log unit of faecal coliforms and 62% of helminth eggs. A slightly better efficiency was achieved with secondary treatment, as a 2 log units reduction of faecal coliforms was observed as well as global reduction of 85% of helminth eggs. The results show that restrictions to the use of secondary effluent for irrigation should be the same as for primary treatment.

The microbiologic quality of the effluent of the facultative pond proved to be much better than that of both primary and secondary effluents. Mean

TABLE 18.8. Microbiological Characteristics of Primary and Secondary Effluents.

	FC/100 ml	FS/100 ml	Helminth/L
Raw wastewater	1.3×10^8	4.8×10^7	149
Primary effluent	1.8×10^7	3.1×10^6	56
Secondary effluent	4.7×10^6	4.3×10^5	22

faecal coliform content was 9.9×10^3 per 100 ml (3 and 2 log units lower than primary and secondary effluent, respectively) and no helminth eggs were found. However, it must be stressed that pond effluents can easily contain much lower numbers of faecal coliforms, if the pond systems has more cells and/or if retention times are longer.

Effects on Soil

Chemical Characteristics

In each plot composite samples of soil were taken at two depths (0–20 cm and 20–50 cm) with the purpose of assessing the trend of possible variations of the soil characteristics as well as the hypothesis of such variations being due to the type of effluent (irrigation water). A relationship between an independent variable x—the time—and a dependent variable—the soil chemical parameter y—is assumed. As soil chemical characterisation comprises n such parameters, one will have n dependent variables and consequently n pairs (x_i, y_{ni}), where x_i corresponds to the sampling date and y_{ni} is the mean value of parameter n on date i (mean of the results of the analysis of parameter n in all plots of the same crop watered with the same type of effluent). The means were made up of four values in the case of the experiments with primary and secondary effluents, where the lay-out included four replicates, and of 6 values in the case of the experiments with the facultative pond effluent. The chemical parameters analysed in soil samples were the following: pH, organic matter (%), cationic exchange capacity, saturation degree, assimilable phosphorus, assimilable potassium, and some metals (cadmium, copper, iron, manganese, nickel and zinc). Time was quoted after the first soil sampling (before experiments initiated) and was measured in months, as one study lasted for 5 years and the other for 3 years.

In many cases it is possible to adjust a first order regression equation to sets of data where the dependent variable increases or decreases with time. Frequently this is not possible with biological systems, such as the biosystem soil-plant-effluent. Thus, when the data did not fit a linear equation, progressively more complex statistical models, such as second and third order polynomials, were tested.[6]

Most of the soil chemical characteristics investigated did not show a statistically significant variation along the five years of irrigation with either of the two types of treated wastewater (primary and secondary effluents) or with potable water (control plots). However, in some cases pH, cationic exchange capacity and iron presented some statistically significant varia-

---

[6] The programme Microstat from ECOSOFT was used.

TABLE 18.9. Variation Model of the Cationic Exchange Capacity (meq/100g) in the Upper Soil Layer (0–20 cm) of Sorghum Plots.

Irrigation Water	Regression Equation	R^2 (%)
Potable	$Y = 14.63 - 0.161X$	84.45[a]
Primary effluent	$Y = 13.73 - 0.144X$	78.53[a]
Secondary effluent	$Y = 14.25 - 0.146X$	83.14[a]

[a] $P \leq 0.05$.
Y—Cationic exchange capacity (meq/100g); X—months after the first sampling; R^2—regression coefficient.

tions, and other parameters, such as organic matter (%), saturation degree, assimilable P, assimilable K, copper and zinc showed variations close to a statistically significant model. Generally, the variations observed were less in the lower soil layer (20 to 50 cm depth) than at the surface layer (0–20 cm). As an example, Tables 18.9 and 18.10 present the regression equations of the variation of the cationic exchange capacity in two soil layers in the plots of sorghum irrigated with three types of water (two effluents and control water). The corresponding plots are shown in Figure 18.14.

In those cases where a statistically significant regression model fitted the variation data of soil chemical characteristics, a test of homogeneity of the regression coefficients was used to investigate the hypothesis of the variations being independent of the type of irrigation water versus the hypothesis of these variations being due to the type of water and crop. It was found that the significant variations observed in some soil chemical characteristics were not due to the type of irrigation water. This suggests that the type of crop was the main cause of the variations.

Contamination

Faecal coliforms were enumerated in soil samples to assess contamination effects and consequent health risks for farmers, the group most exposed

TABLE 18.10. Variation Model of the Cationic Exchange Capacity (meq/100g) in the Lower Soil Layer (20–50 cm) of Sorghum Plots.

Irrigation Water	Regression Equation	R^2 (%)
Potable	$Y = 13.11 + 0.362X - 0.0347X^2 + 5.699 \times 10^{-4} x^3$	97.90[a]
Primary effluent	$Y = 11.77 + 0.552X - 0.0418X^2 + 6.405 \times 10^{-4} x^3$	99.99[b]
Secondary effluent	$Y = 12.49 + 0.403X - 0.0342X^2 + 5.426 \times 10^{-4} x^3$	99.89[c]

[a] $P > 0.05$; $P = 0.08$.
[b] $P \leq 0.01$.
[c] $P \leq 0.05$.
Y—Cationic exchange capacity (meq/100g); X—months after the first sampling; R^2—regression coefficient.

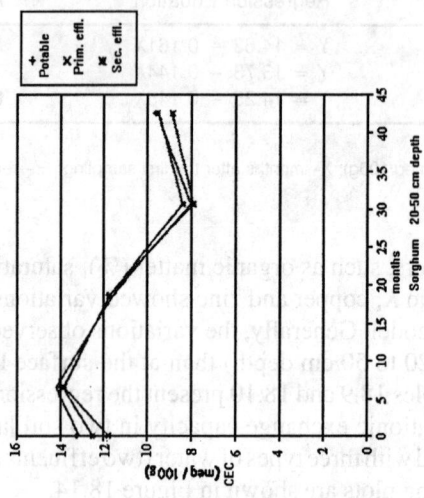

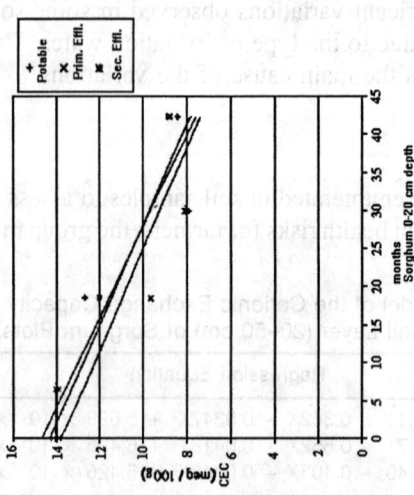

Figure 18.14 Variation pattern of soil cationic exchange capacity in sorghum plots.

to contaminated soils due to irrigation with reclaimed wastewater. Samples were taken before each experiment and after crop harvest. Results from the control plots (irrigated with potable water) were assumed not to have been contaminated.

Between the end of a cultural cycle (harvesting) and the beginning of a new one (preparation of soil for sowing) an almost complete decay of FC was observed (indeed zero contamination was detected before the fifth test), for not only the FC numbers were very low as they were very close to the results of control plots (Figure 18.15).

Primary and secondary effluents presented a similar contamination effect, due to the low pathogen removal efficiency of conventional processes, such as trickling filters.

The time elapsed between the last irrigation session and soil sampling strongly affects the contamination level observed, because most bacteria in soil do not survive longer than 20 days. Due to logistic considerations, it was not possible in this study to keep the time elapsed between last irrigation and sampling constant, but in the last two tests soil sampling was possible immediately after irrigation, when the highest risk to farmers occurs. This explains why the results shown in Figure 18.15 are much higher for plots irrigated with primary and secondary effluents than in the previous three years. These results suggest that the highest soil contamination level is expected to be around 10^5 FC/g of dry soil.

The experiments with facultative pond effluent showed a much lower soil contamination level. This was due to the better microbiological quality

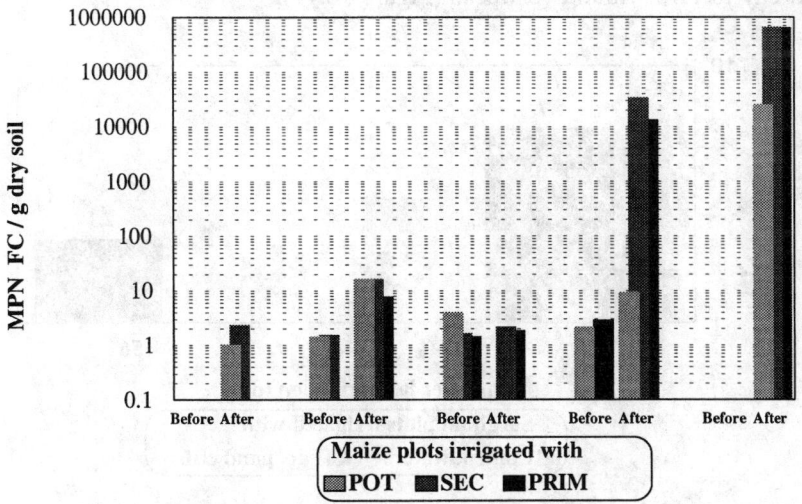

Figure 18.15 Soil contamination in maize plots.

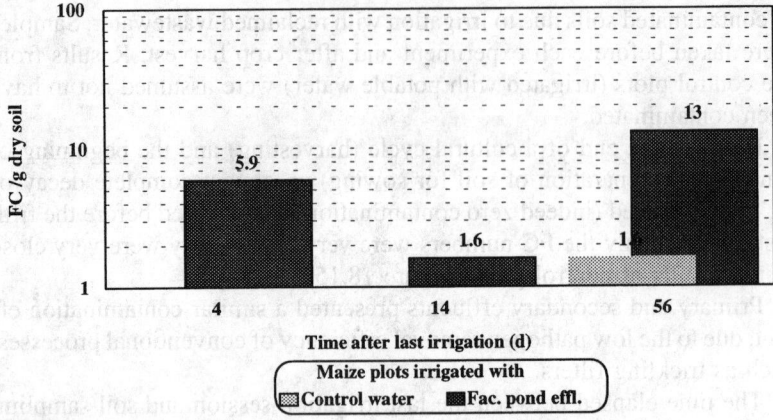

Figure 18.16 Soil contamination decay in maize plots furrow-irrigated with primary and secondary effluents.

of this effluent and to the irrigation method (drip irrigation). An attempt to evaluate faecal contamination decay in soil showed a reduction of 2 log units soil for primary effluent irrigated soil and 3 log units reduction for secondary effluent 4 weeks after irrigation ceased. However, this reduction was insufficient to reach the zero level of contamination of the control plots, as shown in Figure 18.16. On the contrary, soil drip-irrigated with facultative pond effluent does not induce health risk to farmers because contamination levels are very low (<6 FC/g dry soil) and they rapidly decay to zero within two weeks (Figure 18.17).

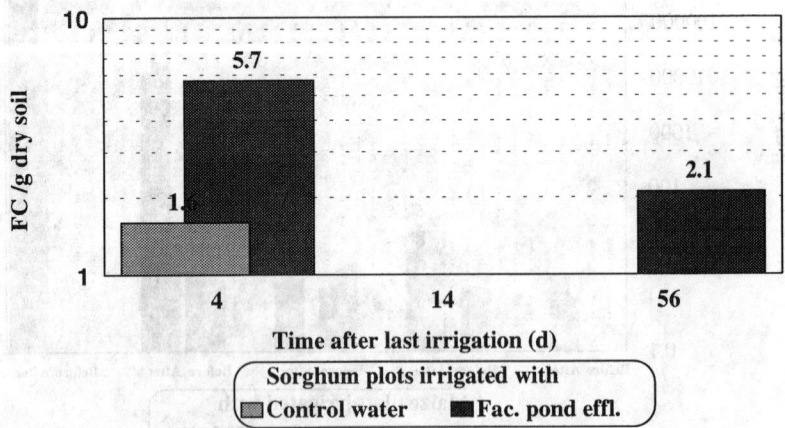

Figure 18.17 Soil contamination decay in sorghum plots drip-irrigated with facultative pond effluent.

Farmers need to be careful so as to avoid direct contact with furrow-irrigated soil, especially when using non-disinfected primary or secondary effluents, because pathogenic bacteria still persist in the soil one month after irrigation has ceased. Furthermore, one must not forget that other pathogens, such as helminths, survive even longer in the soil.

Effects of Irrigation with Treated Wastewater on Crops

The impact of watering the crops with effluents was assessed by the effects on plant quality, evaluated in terms of crop mineral composition, contamination level and yield.

Crop Mineral Composition

The mineral composition of crops irrigated with control water (potable water) was considered good and taken as the basis for comparison. Mean values of five experiments with primary and secondary effluents were compared. Tables 18.11 and 18.12 present the variation (%) of the mineral composition of the three crops studied with reference to the composition of the control crops. Although many parameters show very little or no variation, it is clear that: sorghum composition was more affected than maize and sunflower; sorghum can be harvested three to four times per year and the variation observed in the three harvests did not present a similar trend. For example, the phosphorus content of sorghum irrigated with secondary effluent was identical to the control crops in the first harvest, but increased in the second harvest and decreased in the third harvest. The question is to know whether these variations were due to the type of irrigation water or to experimental error and the answer comes through the analysis of variance (ANOVA) applied to the means of the results.

The values of "F" calculated through ANOVA showed that primary and secondary effluents did not significantly affect the composition of the grains of maize and the seeds of sunflower; only the concentrations of K and B in the sorghum tissues were significantly affected by the type of irrigation water, although B variation was only significant in the second harvest, where increases of 2.5 ppm and 1.9 ppm in sorghum irrigated respectively with primary and secondary effluent were found to be statistically significant. Potassium showed opposite trends in the first and the second harvest of sorghum: while its content decreased in the first harvest (−0.12% and −0.27% for the crop irrigated with primary and secondary effluent, respectively), it increased in the second harvest only for the sorghum irrigated with primary effluent (+0.11%).

The facultative pond effluent appears to induce more variations in crop composition than primary and secondary effluents, because a greater

TABLE 18.11. Variation of Crop Mineral Composition Irrigated with Primary and Secondary Effluents.

| Parameters | Forage Sorghum | | | | | | Maize Grain | | Sunflower | |
| | 1st Harvest | | 2nd harvest | | 3rd harvest | | | | | |
	Prim. Effl.	Sec. Effl.	Prim. Effl.	Sec. Effl.	Prim. Effl.	Sec. Effl.	Prim. Effl.	Sec. Effl.	Prim. Effl.	Sec. Effl.
N (%)	99	101	112	104	99	93	95	101	102	101
P (%)	100	100	112	112	105	95	100	95	100	97
K (%)	98	93	106	99	98	89	103	100	99	100
Ca (%)	—	—	100	100	97	103	100	100	111	105
Mg (%)	—	—	108	97	106	103	100	100	105	103
S (%)	—	—	109	109	87	91	100	100	98	100
Fe (ppm)	100	103	97	90	87	97	113	96	90	98
Mn (ppm)	100	97	84	94	82	87	104	102	124	102
Zn (ppm)	106	104	101	100	76	79	96	97	101	102
Cu (ppm)	—	—	97	99	79	84	109	87	98	101
B (ppm)	129	120	140	117	113	104	123	99	102	95
Ni (ppm)	99	99	106	91	84	116	95	101	93	98
Cd (ppm)	40	100	96	100	100	100	100	100	73	90

TABLE 18.12. Variation of Crop Mineral Composition Irrigated with Facultative Pond Effluent.

Parameter	Maize Grain	Forage Sorghum 1st harvest	Forage Sorghum 2nd harvest	Sunflower
N (%)	114	107	120	115
P (%)	100	133	100	100
K (%)	100	107	121	100
Ca (%)		100	100	50
Mg (%)	100	75	125	125
S (%)	100	100	100	100
Fe (ppm)	102	74	103	103
Mn (ppm)	108	66	85	98
Zn (ppm)	86.7	80	57	80
Cu (ppm)	100	80	77	77
B (ppm)	125	145	153	98
Cd (ppm)	100	100	100	85
Ni (ppm)	93	100	100	126

number of parameters is significantly affected, especially with sorghum, where variations of S, Fe, Mn and Cu were observed. Maize showed significant variations only in two parameters (+0.06% of Ca and +0.58% of N content). The Zn content of sunflower seeds irrigated with facultative pond effluent decreased 19 ppm, but N and K increased 1% and 0.08%, respectively.

In spite of these statistically significant variations observed in some parameters of their mineral composition, the crops did not show any demand or toxicity symptoms.

Crop Contamination

Crop contamination was assessed by the enumeration of faecal coliforms *(E. coli)* on composite samples of the crops' consumable part made up of 2 plants taken at random from the central rows of each of the replicate plots. When possible (only during the first year of experiments), the presence or absence of *Salmonella* spp. and *Campylobacter* spp. was also investigated. Crop contamination was evaluated based on basis of the microbiological limits for market vegetables specified by the International Commission on Microbiological Specifications for Food (1974). Counts of more than 10^3 *E. coli*/100 g would be "undesirable" and more than 10^5/100 g "unacceptable"; salmonellae should not be detected in any instance.

In the first year of experiments with primary and secondary effluents no contamination was found on furrow-irrigated maize grain and sunflower seeds. Sunflower seeds irrigated with the effluents of this study presented

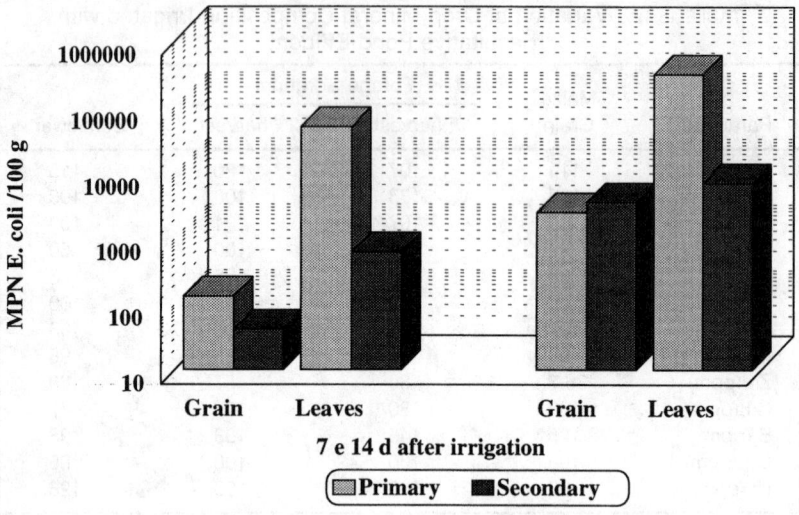

Figure 18.18 Contamination of maize (grain and leaves) furrow-irrigated with primary and secondary effluent.

no health risk, since only very low contamination level were found in all experiments, although higher values were counted in the seeds of the plants irrigated with primary effluent. Indeed, higher levels of contamination were found for all samples of the three crops irrigated with primary effluent.

Similar conclusions can be drawn from the results observed in maize grain, where the highest contamination level was found on the grains of maize furrow-irrigated with primary effluent (1.3×10^2 E. *coli*/100 g), as was to be expected considering the protection that sunflower seeds and maize grain present to direct contact with irrigation water. The contamination tests carried out on maize leaves in the last experimental year indicated that only the leaves of maize plants furrow-irrigated with primary effluent presented undesirable contamination level even one week after irrigation, as illustrated in Figure 18.18. Apparently, the leaves of maize drip-irrigated with the pond effluent reached a suprising unacceptable contamination level 6 days after irrigation (2.1×10^5 E. *coli*/100 g), but the fact is that the leaves of control plants (irrigated with potable water) presented 2.7×10^4 E. *coli*/100 g.

Forage sorghum can be fed to cattle either immediately after harvesting or dried. In the first case,[7] it is important to pay attention to the contamination of the lower parts of the plant, that can be provided especially with

---

[7] When sorghum is dried or ensilaged pathogens are inactivated by solar energy combined with low humidity of the environment; ensilaged sorghum is submitted to a low pH and high temperature during the process.

furrow irrigation, because almost the entire plant is edible. Results of samples taken 7 and 14 days after irrigation showed that sorghum irrigated with primary effluent presented undesirable contamination levels, the same happening with the secondary effluent one week after the irrigation.

The results suggest that two weeks after ceasing irrigation no health risks are likely to cattle fed with sorghum irrigated with primary or secondary effluents.

Crop Yield

The use of reclaimed wastewater for irrigation in agriculture provides many advantages from the environmental and water resources management point of view, but will not become a practice if farmers cannot anticipate good yields. The mean yields (mean of five years) of maize and sorghum irrigated with primary and secondary effluents were statistically significantly greater than the yield of control plots (Figure 18.19). No significant difference was observed in the case of sunflower yield, but it has to be taken in account that the yield measured in the second year was lower than real values, because birds ate a large part of sunflower seeds.

The irrigation with facultative pond effluent provided a clear increase in crop yield as illustrated in Figure 18.20 for sorghum. In spite of the fact that sunflower irrigated with pond effluent presented a greater yield than the control crop, it must be stressed out that the high nitrogen content of the facultative pond effluent (30.2 mg/L N) was the cause of the

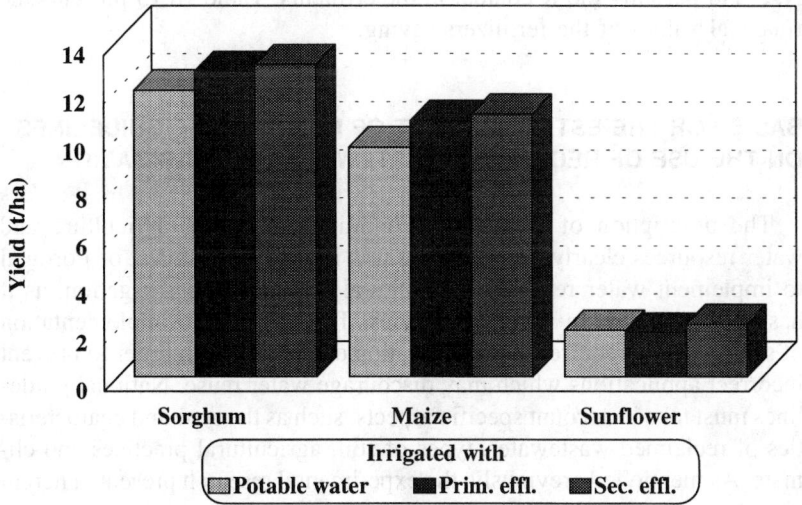

Figure 18.19 Effect of primary and secondary effluent on crop yield.

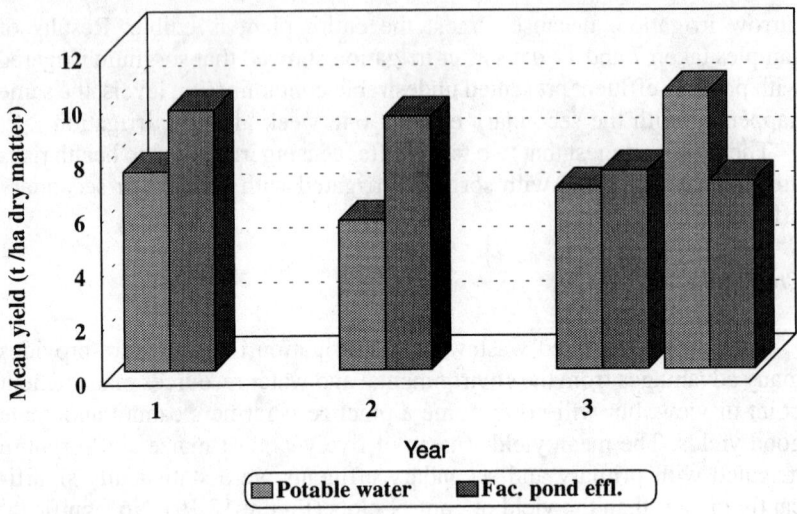

Figure 18.20 Effect of facultative effluent on sorghum yield.

late maturation of sunflower seeds, which can be a serious problem for harvesting if the seeds are only ready in late autumn.

From the agronomic point of view the irrigation of maize, sorghum and sunflower with these effluents is clearly advantageous. Consequently, in addition to the benefit of growing crops during the dry season farmers will also save on commercial fertilizers, as these may be replaced by the N, P and partially the K content in the effluents. Table 18.13 presents the financial values of the fertilizers saving.

BASIS FOR THE ESTABLISHMENT OF PORTUGUESE GUIDELINES ON THE USE OF RECLAIMED WASTEWATER FOR IRRIGATION

The description of Portuguese conditions regarding agriculture and water resources clearly indicates that it is in the clear interest of Portugal to implement water reuse for agricultural and landscape irrigation, as it has been practiced in several countries. The sustainable implementation of such practice requires the specification of guidelines in order to prevent incorrect applications which may discourage water reuse. National guidelines must take in account specific aspects, such as the type and characteristics of reclaimed wastewater, types of soil, agricultural practices and climate. As mentioned previously, the experimental research presented herein aimed at gathering this information.

The guidelines must cover two areas: (a) agronomic aspects, related to the maximization of crop yields and soil and groundwater preservation;

TABLE 18.13. Savings in Commercial Fertilizers
Resulting from Wastewater Irrigation.

Crop	Units	Primary and Secondary Effluents		Facultative Pond Effluent	
		(kg/ha)	(US$./ha)	(kg/ha)	(US$./ha)
Sorghum	N	140	109	140	109
	P_2O_5	50	34	—	—
	K_2O	120	55	60	27
Maize	N	155	121	155	121
	P_2O_5	40	42	—	—
	K_2O	60	27	27	14
Sunflower	N	65	50	65	50
	P_2O_5	30	54	—	—
	K_2O	15	14	15	14

and (b) sanitary aspects, related to public health protection. The document should have the following general characteristics: (a) simplicity, to prevent wastewater reuse being unnecessary descouraging; (b) robustness and reliability, to assure public health protection, good crop production, and to prevent adverse impacts on the environment. Finally, the document must be flexible enough to allow for further improvement in the light of new scientific and technologic developments and acquired experience.

Guidelines can be defined as a simple list of chemical and microbiological parameters whose values should be within a certain range. This simple approach, however, presents a rather serious shortcoming: in most of the Southern European countries, like Portugal, where wastewater reuse for irrigation is more necessary, the monitoring of effluent quality from wastewater treatment plants is not completely assured, due to several factors, including lack of trained staff, available laboratories and also finance. A second simple approach would consist of specifications for minimum treatment levels required for certain irrigation purposes. For instance, not less than primary treatment should be required in all circumstances; and only disinfected effluent would be allowed to irrigate vegetables eaten raw and sport fields. This methodology is not satisfactory, however, because the operation of treatment plants is frequently unreliable in many areas and effluent quality is unlikely to correspond to the presumed level. A combination of both approaches, consisting of a reduction of the number of parameters in the list with some specification concerning the minimum treatment requested for certain irrigation purposes seems to be a better approach.

The selection of parameters and the requests for treatment specifications are now the question. As mentioned previously, the list of parameters for the monitoring of reclaimed wastewater quality for irrigation should include both chemical parameters with major agronomic impact and microbi-

TABLE 18.14. Recommended Parameters for Chemical Quality Monitoring.

Parameter	Monitoring	
	Routine	Periodic
pH	●	◆
TSS		◆
BOD		◆
COD		◆
Org-N		◆
NH$_4$-N		◆
NO$_3$-N		◆
Tot-N	●	◆
Ort-P		◆
Tot-P	●	◆
Conductivity	●	◆
R$_{Na}$	●	◆
Na	●	◆
K	●	◆
Ca	●	◆
Mg	●	◆
TDS		◆
B	●	◆
Cl	●	◆
SO$_4$		◆
As		◆
Heavy metals:		
Cd, Co, Cr,		◆
Cu, Fe, Mn, Ni, Pb, Zn		◆

ological parameters affecting public health. However, these impacts depend on local agricultural practices, such as irrigation methods and important crops in the region, for example. Therefore, the guidelines should bring together the following three factors: specification of limits for some chemical and microbiological quality parameters in combination with treatment requirements and as a function of field practices (crops and irrigation methods).

The first step for the establishment of the guidelines would consist of the selection of the parameters to be included in the list of monitoring requirements. A distinction between routine monitoring and monitoring for licensing and periodic inspection control should be established (Table 18.14): routine monitoring does not need to include a list of parameters as long as the list of parameters for effluent characterization prior to the license being granted for reuse or for periodic inspection.

The second step would be the specification of limiting values for the relevant parameters. The list of chemical parameters should be based on the FAO guidelines (FAO, 1985) adapted to Portuguese conditions regarding climate and agronomic practices including the most interesting crops and usual irrigation methods. Sport fields, such as golf courses, require special attention, as these are large consumers of irrigation water in the tourist resorts of the southern region of the country, and consequently one of the likely potential users of reclaimed wastewater.

The objective of microbiological monitoring must be to assess the level of public health protection. Several factors contribute to a potential health risk in wastewater reusing for irrigation: (a) the type of irrigated crops (for direct or indirect consumption, industrial use or landscape, leisure or sport purposes); (b) the irrigation method; (c) epidemiological characteristics of pathogenic micro-organisms; and (d) wastewater treatment efficiency and effluent storage. Therefore, the selection of the indicator micro-organisms to be used as monitoring parameters must give consideration to the epidemiological characteristics of pathogenic micro-organisms which affect their survival during wastewater treatment and in soils and on crops (persistence and latency), and to their ability to cause illness (infective dose, pathogenicity and virulence), as these characteristics affect their survival after excretion, passing through the wastewater treatment plant, application on crops and infiltration in the soil, and crop marketing and consumption. The highest risk will obviously be caused by the most persistent pathogens, with lower infective dose and greater latency. Helminths have these characteristics, and thus helminths are presently considered to pose the most important sanitary risk in reclaimed wastewater reuse for irrigation (Shuval, 1987; WHO, 1989). Pathogenic bacteria are the second group with a high health risk, especially *Salmonella* spp. In spite of their low infective dose, enteric viruses are only the third important group of risk, because of their low persistence and because their main transmission route is usually direct person to person contact rather than any other path.

A survey of existing microbiological guidelines for reclaimed wastewater use for irrigation shows that total coliforms and faecal coliforms are the most frequently used indicators, although enteric viruses, *Ascaris, Giardia* and *Entamoeba* have been considered in some states, mainly in the USA (WPCF, 1989). Total coliforms have a very limited meaning in the context of effluent reuse for irrigation due to their widespread occurrence in soil and not necessarily from faecal origin. Taking into account the epidemiological characteristics of pathogens and a rational use of laboratory resources, the microbiological parameters to be determined in the quality monitoring of reclaimed wastewater in Portugal should include at least helminths and faecal bacteria (faecal coliforms) and possibly *Salmonella,* especially when vegetables are irrigated.

A third step in the establishment of guidelines should be the requirements for pertinent restrictions to crops and irrigation methods. These should be based on the assessment of the need for irrigating crops in the country and on the agricultural practices, i.e., the most used irrigation methods. The work carried out in Portugal reported herein can provide relevant recommendations which can be incorporated in the establishment of Portuguese guidelines.

Conclusions about the chemical quality of effluents used for irrigation are summarised in Table 18.15. It must be pointed out that, although the pH of the three treated wastewaters, was generally appropriate for irrigation water, sometimes the pH of the facultative pond effluent was higher than the upper limit recommended by FAO. This is due to the diurnal pH variation observed in facultative ponds as a consequence of algal photosynthesis: the pH is often above 8.4 (FAO limit for restricted irrigation), especially in summertime when the need for irrigation is higher. Harmful effects on crop due to high effluent pH can be overcome if irrigation is done outside of the hours of intensive algal CO_2 consumption, i.e., during 4 p.m. to 10 a.m.

Some constituents, like *sodium* and *chloride* ions, show concentrations high enough to warrant some restriction both in crop and irrigation method selection, especially with facultative pond effluents: toxicity can be induced in sensitive plants, the problem being aggravated in the case of spray irrigation. Facultative pond effluents also contain higher concentration of *boron* than primary and secondary effluents, and should only be applied on crops tolerant to this element, such as sunflower, maize and alfalfa and several vegetables, like tomatoes, onions and carrots. However, primary and secondary effluents are appropriate for moderately tolerant crops. The salt content in facultative pond effluents is generally higher than in the primary and secondary effluents, as shown by the mean values of the electrical *conductivity* and TDS (nearly twice as high) as well as of several ions which contribute to salinity (Na, K, Ca, Mg, Cl, SO_4).

No plants sensitive to salinity should be watered with these types of effluents, with especial attention being paid to facultative pond effluents. However, minor impacts on the physical characteristics of soils are to be expected by irrigation with facultative pond effluents. This is due to the relation of conductivity and R_{Na} values which shows a smaller proportion of sodium ions. Nevertheless, these effluents should be applied only in soils with good drainage conditions, and the irrigation method should include techniques for salinity management, such as salts leaching by means of the application of an excessive irrigation water.

The nutrient content is generally considered one of the benefits of using reclaimed wastewater for irrigation. This research has shown that primary effluent contains more total nitrogen, followed by the facultative pond effluent, and that secondary effluent is mostly nitrified, but its nitrate content is only slightly higher than in the facultative pond effluent. More

TABLE 18.15. Comparison of the Agronomic Quality of Primary, Secondary and Facultative Pond Effluent.

Parameter	Effluent		
	Primary	Secondary	Facultative pond
pH	Good	Good	Good, although high occasionally.
N-tot	Effluent with the highest N content, in excess for almost every crop. Dilution may be required.	Lower N content, yet it may be excessive to some crops. It is the most nitrified effluent.	Similar to secondary effluent
P-tot	High	High	Normal
Conductivity and R_{Na}	Class C3-S1	Class C3-S1	Class C3-S1
Sodium	High content for sensitive plants to this element. Spray irrigation must be avoided.	Similar to primary effluent, although slightly higher concentration.	Higher concentration than in primary and secondary effluent. Sodium sensitive crop and spray irrigation must be avoided.
Chloride	High content for sensitive plants to this element. Spray irrigation must be avoided.	High content for sensitive plants to this element. Spray irrigation must be avoided.	Potential toxicity problems for sensitive crops. Spray irrigation must be avoided.
Boron	• Good for sensitive crop • Excellent for semitolerant crop	• Acceptable for sensitive crop • Good for semitolerant crop • Excellent for tolerant crop	• Not appropriate for sensitive crops • Acceptable for semitolerant crops • Good for tolerant crops
Heavy metals	Mean concentration lower than maximum limits recommended for irrigation waters.	Mean concentration lower than maximum limits recommended for irrigation waters.	Mean concentration lower than maximum limits recommended for irrigation waters.

869

than 30 mg/L N is considered excessive even for crops tolerant to an excess of nitrogen (FAO, 1985). The inconvenience for plants may not be visible and they may actually appear to be developing very well. This is due to the fact that nitrogen is an element which can be absorbed by plants beyond their need, provided that it is available. In spite of the usually exuberant plant growth induced by excessive nitrogen, this can have harmful effects for the crop, because it delays fruit maturation sometimes to the point that it may be impossible to harvest in time or be just producing large fruits that are low in taste. In the present study, sunflower seeds tended to mature too late.

The microbiological quality of the effluents presented in Table 18.8 indicates that the removal efficiency of primary treatment was 97.6% for FC and 62% for helminth eggs. However, primary effluent can be considered highly contaminated. The microbiological quality of the secondary effluent is only slightly better, as the mean overall removal of FC and helminth eggs was 99% and 85% respectively. The constraints to the use of secondary effluent for irrigation should be the same as for primary effluent due to their similar microbiological qualities. The effluent of the facultative pond was much better, for it had 3 to 4 log units FC/100 ml less than the secondary effluent and no helminth eggs (counting of helminth eggs was carried out only in the third test of the experiment, when a reliable technique was available (Ayres, 1989)). Normally the effluents of waste stabilization pond systems have an excellent microbiological quality for reuse in irrigation. Effluents with less than 100 FC/100 ml and no helminth eggs per litre are easily achieved by a properly designed pond system, depending principally on the number of ponds in the system, the overall retention time of the wastewater in the pond system and the local mean temperature (Mara and Cairncross, 1989). The effluent of the facultative pond investigated in this study was not as good as possible because this is a system with only two cells, the first being an aerated lagoon.

In spite of the fact that primary and secondary effluents may present a better quality than the ones investigated in this research, these results can be a valid contribution for the production of Portuguese Guidelines, which must be conservative and take into account the fact that many wastewater treatment plants are operating under defective conditions.

In this study the irrigated crops did not show any contamination of their consumable parts even when irrigated with the primary and secondary effluents, except for the lower leaves of forage sorghum. This was due to the precautions taken when selecting the irrigation method and crops. It was found that no health risk occurred when using drip irrigation, because no contamination was found in any crop tested and that sorghum should not be given to cattle before two weeks after furrow irrigation has ceased.

From the agricultural point of view no toxicity or demand problems were detected with the three crops irrigated. However two facts must be pointed out: the effluent of facultative pond caused more deviations (generally not significant) in the crop mineral composition than the other effluents; sorghum proved to be a more sensitive crop than maize and sunflower as it showed a significant variation in a greater number of parameters; although sunflower had a higher yield increase when irrigated with the pond effluent, a delay in maturation was observed, which could be explained by an excess of nitrogen in the irrigation water and its accumulation in the seeds. Further field research is necessary to confirm this hypothesis. The observed significant variations in plant composition must be taken into account when planning the long term effects on crops irrigated with reclaimed wastewater.

Higher yields were even obtained with the crops irrigated with treated wastewater than from the same crops irrigated with potable water supplemented with commercial fertilizers. This indicates that important savings in commercial fertilizers can be made by using treated wastewater for irrigation.

The irrigated soil did not show significant impact on its chemical characteristics during the experimental period, but some contamination was found in the furrow irrigated soil even 4 weeks after irrigation had ceased (only 2 log units reduction of FC per gram of dry soil were observed after this period). This shows that the contamination level of furrow irrigated soil represents a health risk to farmers not only during the irrigation period but also for several weeks afterwards. Therefore, national guidelines should specify that wastewater irrigated soil and subsoil should be monitored at least once per year in order to assess the trend of possible modifications and, if necessary, to modify or cease the irrigation. The surrounding groundwater quality should be also monitored.

The contamination of irrigated crops should be assessed by the competent authorities. The fourth step would, therefore, be the specification of monitoring requirements such as the parameters to be monitored, monitoring frequency and possibly also the analytical methods to be used.

The study concluded that the microbiological characteristics of treated wastewater are the main factor affecting their reuse for irrigation. Severe restrictions concerning both the crops and the irrigation methods must be considered regarding the use of primary and secondary treated wastewater, unless further treatment to improve their microbiological quality can be made. Disinfection is a possibility, but its environmental and agronomic impacts need to be assessed. Maturation ponds and seasonal storage are valid alternatives. The implementation of complementary treatment in a plant in order to improve its effluent quality for irrigation is generally

TABLE 18.16. Recommended Restrictions to the Irrigation with Primary and Secondary Effluents.

	Plants		
Irrigation Method	Edible Crops,[a] Sport Fields, Playgrounds,[b] Parks[b]	Parks, Gardens,[c] Ornamental Trees, Pastures, Forage	Industrial Crops, Nurseries, Forests
Surface	No	No	No
Spray	No	No	No
Localized	No	Yes	Yes

[a]Raw or cooked.
[b]Allowing for users' direct contact.
[c]Without users' direct contact.

expensive, especially considering its chemical characteristics. In many cases blending with natural water will be preferable together with a wise selection of crops taking into account the properties of the effluent used for irrigation. The observed levels of soil contamination suggest that surface (like furrow) and spray irrigation should not be done with primary and secondary effluents without disinfection. In addition, proper salt management is essential for sustainable irrigation with wastewater. Recommended combinations of plant and crop restriction according to the type of effluent are given in Tables 18.16 and 18.17.

The conclusions from this study could be taken into consideration when drafting the technical content of the suggested Portuguese guidelines according to the steps suggested in the methodology. The technical content of any guidelines is clearly important but it must not be forgotten that the institutional framework is very important as well in the application of any legislation. This is an aspect that needs to be observed with great care by the Task Group producing guidelines at a national level.

ACKNOWLEDGEMENTS

The author is grateful for the collaboration of Mr. M. Silva e Sousa, agricultural engineer from Laboratório Químico Agrícola Rebelo da Silva and Ms. Ana Sacramento, environmental engineer from Laboratório Nacional de Engenharia Civil. She is also pleased to acknowledge the financial support from the Junta Nacional de Investigação Científica e Tecnológica (JNICT) and the Direcção Geral da Qualidade do Ambiente (DGQA), as well as the logistic support provided by the Municipality of Évora, the Public Health Laboratory of Évora, the Departamento de Saneamento Básico de Santo André. Special thanks are due to Prof. Takashi Asano, from the University of California, for his guidance, and Prof. D. Mara,

TABLE 18.17. Restrictions to the Irrigation with Facultative Pond Effluent.

		Plants		
Irrigation Method	Edible Sport Playgrounds[b] > 10³ FC/100 mL	Crops[a] and Parks[b] < 10³ FC/100 mL and < 1 egg helm/L	Parks, Gardens,[c] Trees, Pastures, Forage Crops > 10³ FC/100 mL	Industrial Crops, Nurseries, Forest < 10³ FC/100 ml and < 1 egg helm/L
Surface	No	Yes	Yes	Yes
Spray[d]	No	Yes	Yes	Yes
Localized	Yes	Yes	Yes	Yes

[a]Raw or cooked.
[b]Allowing for users' direct contact.
[c]Without users' direct contact.
[d]Minimum distance to houses and roads to be specified.

873

from the University of Leeds, for his support and the cooperation of his Ph.D. students in the laboratory work.

REFERENCES

APHA; AWWA; WPCF (1985). *Standard Methods for the Examination of Water and Wastewater.* 16th Ed., Washington, DC 20005, USA, American Public Health Association.

Asano, T. et al. (1985). Municipal wastewater: treatment and reclaimed water characteristics. In *Irrigation With Reclaimed Municipal Wastewater—A Guidance Manual,* edited by G. S. Pettygrove and T. Asano, Lewis Publishers, Inc., Chelsea, MI, USA.

Asano, T. (1991). Planning and implementation of water reuse projects. In *Wastewater Reclamation and Reuse,* Edited by R. Mujeriego and T. Asano. *Wat. Sci. Tech.* Vol. 24, No. 9, pp. 89–94.

Ayers, R. (1989). Enumeration of parasitic helminths in raw and treated wastewater—A brief practical guide. Leeds, UK, Department of Civil Engineering, University of Leeds.

Centre National du Machinisme Agricole du Génie Rural et des Forêts (1982). *Les Aerosols d'Eaux Residuaires. Stations de Traitement et Dispositifs d'Aspersion.* Bordeaux. Ministère de l'Agriculture, Groupement de Bordeaux, Section Qualité des Eaux.

European Communities Commission (1991). Council Directive regarding the treatment of urban wastewater (91/271/EEC). *Official Journal of the European Communities* No. L 135, of the 91.5.30, pp. 40–50.

Fattal, B. et al. (1984). Community exposure to wastewater and antibody prevalence to several enteroviruses. In *Future of water reuse, Proceedings of the Water Reuse Symposium III,* San Diego, CA, USA, AWWA Research Foundation, pp. 1505–1517.

Feachem, R. G. et al. (1983). *Sanitation and Disease. Health Aspects of Excreta and Wastewater Management.* World Bank Studies in Water Supply and Sanitation 3, World Bank, Bath, Great Britain.

Food and Agriculture Organization of the United Nations (1985). Water Quality for Agriculture. *Irrigation and Drainage,* Paper No. 29 Rev. 1, Rome, FAO.

Henriques, A. G. (1985). *Avaliação dos Recursos Hídricos de Portugal Continental. Contribuição para o Ordenamento do Território.* Caderno 9, Lisboa, Instituto de Estudos para o Desenvolvimento.

Hsiao, J.; Sheikh, B. (1981). Aerosol Generation from sprinkler irrigation with reclaimed wastewater. In *Water Reuse in the Future, Proceedings of the Water Reuse Symposium II,* Washington, D.C., AWWA Research Foundation, pp. 2234–2240.

Ilaco, B. V. (1981). *Agricultural Compendium for Rural Development in the Tropics and Subtropics.* Amsterdam, Elsevier Scientific Publishing Company.

International Reference Centre for Waste Disposal (1985). Health Aspects of Wastewater and Excreta Use in Agriculture and Aquaculture. The Engelberg Report. *IRCWD News* No. 25, Duebendorf, Switzerland.

Mara, D. D.; Cairncross, S. (1989). *Guidelines for the Safe Use of Wastewater and Excreta in Agriculture and Aquaculture.* Geneva, WHO & UNEP.

Marecos do Monte, M. H. (1994). Contributo para a utilização de águas residuais tratadas para irrigação em Portugal. Tese elaborada no LNEC e submetida ao IST da Universidade Técnica de Lisboa para obtenção do grau de Doutor em Engenharia Civil, Lisboa.

Metcalf & Eddy, Inc. (1991). *Wastewater Engineering—Treatment, Disposal, Reuse.* 3rd Ed., Lisbon, McGraw-Hill, Inc.

Ministério da Agricultura, Pescas e Alimentação (1991). *A Agricultura Portuguesa em Números. 1990.* Lisboa, Direcçãp Geral do Planeamento e Agricultura.

Pettygrove, G. S.; Asano, T. (1985). *Irrigation With Reclaimed Municipal Wastewater—A Guidance Manual.* Chelsea, MI, USA, Lewis Publishers, Inc.

Pires, A.; Silva, J. D. (1992). O saneamento básico em Portugal. O Pacto Ambiental. *Informação APESB* No. 45, Lisboa, APESB, pp. 19–27.

Quelhas dos Santos, J.; Cardoso Pinto, F. (1985). O biossistema solovegetação base fundamental da depuração de águas residuais no solo. In *Tratamento e Destino Final de Águas Residuais Municipais e Industriais no Solo,* S 326, Lisboa, Laboratório Nacional de Engenharia Civil.

Sheikh, B. et al. (1990). Monterey Wastewater Reclamation Study for Agriculture. *Research Journal WPCF,* Volume 62, No. 3, Water Pollution Control Federation, Alexandria, VA, USA, pp. 215–226.

Shuval, et al. (1987). Prospective epidemiological study on enteric disease transmission associated with sprinkler irrigation with wastewater: an overview. In *Implementing Water Reuse, Proceedings of the Water Reuse Symposium IV,* Denver, CO, USA, AWWA Research Foundation.

Strauss, M.; Blumenthal, U. (1990). *Human Waste Use in Agriculture and Aquaculture. Utilization Practices and Health Perspectives.* IRCWD Report No. 09/90. Duebendorf, Switzerland, IRCWD.

U.S. Environmental Protection Agency; U.S. Agency for International Development (1992). *Manual—Guidelines for Water Reuse.* EPA/625/R-92/004, Washington, D.C., USA, EPA.

U.S. Environmental Protection Agency et al. (1981). *Process Design Manual for Land Treatment of Municipal Wastewater.* EPA 625/1-81-013, Cincinnati, Ohio, USA, EPA.

Water Pollution Control Federation (1989). *Water Reuse. Manual of Practice SM-3.* WPCF, Alexandria, VA, USA.

World Health Organization (1973). *Reuse of Effluents: methods of wastewater treatment and health safeguard.* Technical Reports Series 517, Geneva, Switzerland, WHO,1973.

World Health Organization (1989). *Health Guidelines for the Use of Wastewater in Agriculture and Aquaculture.* Technical Reports Series 778, Geneva, Switzerland, WHO.

CHAPTER 19

Wastewater Reclamation and Reuse in Tunisia

INTRODUCTION

TUNISIA will likely suffer from water shortages in the next century. Problems of water scarcity may intensify because of population growth, the rise in living standards, and accelerated urbanization which threaten the water supply sector in general and agriculture in particular and lead to both an increase in water consumption and pollution of water resources.

According to forecasts, higher domestic and industrial water consumption by the year 2020 may cause a decrease in the volume of fresh water available for agriculture. It is therefore important to develop additional water resources as well as preserve the existing ones. A water saving policy is being set up and agriculture is already using water of marginal quality for irrigation purposes. Brackish water has been used for several years since most of the water available for agriculture is very often salt-affected and the development of programs of planned reuse of wastewater within Tunisia began in the early 1960s.

Wastewater reuse has become a necessity and even a priority in the Tunisian national water resources strategy. It is an essential component of the policy to integrate all water resources into an effective management plan. Furthermore, Tunisia is among the very few Mediterranean countries which have elaborated and implemented a national wastewater reuse policy for many years [1].

Akissa Bahri, Rural Engineering Research Center, Tunisia, B.P. 10, Ariana 2080, Tunisia.

The objective of this chapter is to present the water resources planning and management, and to evaluate the present status of reclaimed water use in Tunisia. This is done by reviewing the last decades' research results in this field. Institutional, legal, and financial aspects are also screened. We close with giving prospects for the future.

WATER RESOURCES DEVELOPMENT AND MANAGEMENT

In order to overcome the problem of water shortage and marginal quality in Tunisia and to provide most of the country with water for different needs, water planning and management activities have been initiated for conservation of existing water resources, and development of marginal sources. As a first step, considerable efforts have been invested into developing different water resources. The next step is optimal use of the available water resources and increase of the water use efficiency.

Continuing increase in demand by the urban sector has led to higher utilization of fresh water for domestic purposes, on the one hand, and production of greater volumes of wastewater, on the other. Wastewater reuse has been made an integral part of overall environmental pollution control and water management strategy.

An overall water policy has been adopted to strengthen the effort made by the country in the field of water resources development and utilization and to try to ensure a balance between the availability of water resources and needs of water to maintain steady pace in the social and economic development process. A long-term water strategy, in which wastewater reuse is included, is under preparation in Tunisia [2,3] and the outlines will be presented here.

WATER RESOURCES DEVELOPMENT

General

Tunisia extends from the Mediterranean coast in the north to the Sahara desert in the south and its total surface area is 164 150 km^2. Its climate varies from Mediterranean to semiarid and arid; it is characterized by hot and dry summers and mild winters receiving the major part of the annual precipitation. Total rainfall and distribution is highly variable from year to year and from north to south. Average annual evapotranspiration is also high and water deficit is particularly significant from May to October. Annual rainfall is ranging in the northern part from 400 mm to 1000 mm, in the central part from 200 mm to 400 mm, and it is less than 200 mm

TABLE 19.1. Water Supply Sources in Tunisia (Mm^3 yr^{-1}).

	1990	Potential
Large dams	1240	2200
Hillside-dams	18	146
Hillside-lakes	3	50
Floodwater diversion	23	47
Tubewells and springs	858	1146
Open wells	649	669
Reclaimed wastewater	95	275
Total	2886	4533

in the southern part. The agricultural land is only 28% of the total area, i.e., about 4.5 million ha.

The annual precipitation in Tunisia is on average equal to 33 billion m^3 and the water resources of the country are about 4.5 billion m^3 of which 2.7 billion m^3 are from surface water and 1.8 billion m^3 are from groundwater. The volume of water brought on-line during the terms of the Seventh Economic, Social, and Development Plan (1987–1991) [4] was about 2874 million m^3 (Mm^3), i.e., 64% of the potential water resources, from large dams, hillside-dams, open shallow wells, deep tube-wells, and springs.

The water resources management and planning is outlined in the country's five-year development plans. The goals, by the year 2010, are to mobilize most of the surface water through the completion of 42 dams and the construction of 235 hillside-dams, 1000 hillside-lakes, and 2000 floodwater diversion structures (Table 19.1). In addition, the plans emphasize water harvesting and wastewater reuse.

Priority was given the last 25 years to water resources mobilization. Investments in the water sector were about two thirds of the budget of the Ministry of Agriculture during the Seventh Economic, Social, and Development Plan with 59% allocated to large reservoirs or conveying structures and 15% to the water supply sector. As a result, for the next 15 years, water resources seem to be able to satisfy the projected water demand as far as problems relating to location, salinity, rainfall variability, and reservoir management are not taken into account. In the opposite case, water balance may become negative. Water quality and, particularly, salinity are of considerable importance for water resources planning.

The Tunisian population is about 8.8 million of which 50% are concentrated to the coast in urban and industrial areas; 40% of the population live in rural areas (against 60%, 25 years ago) of which 47% live in rural communities and 53% on farms. The Tunisian population is expected to reach 10 million by the year 2000 (the demographic growth rate is 1.9%).

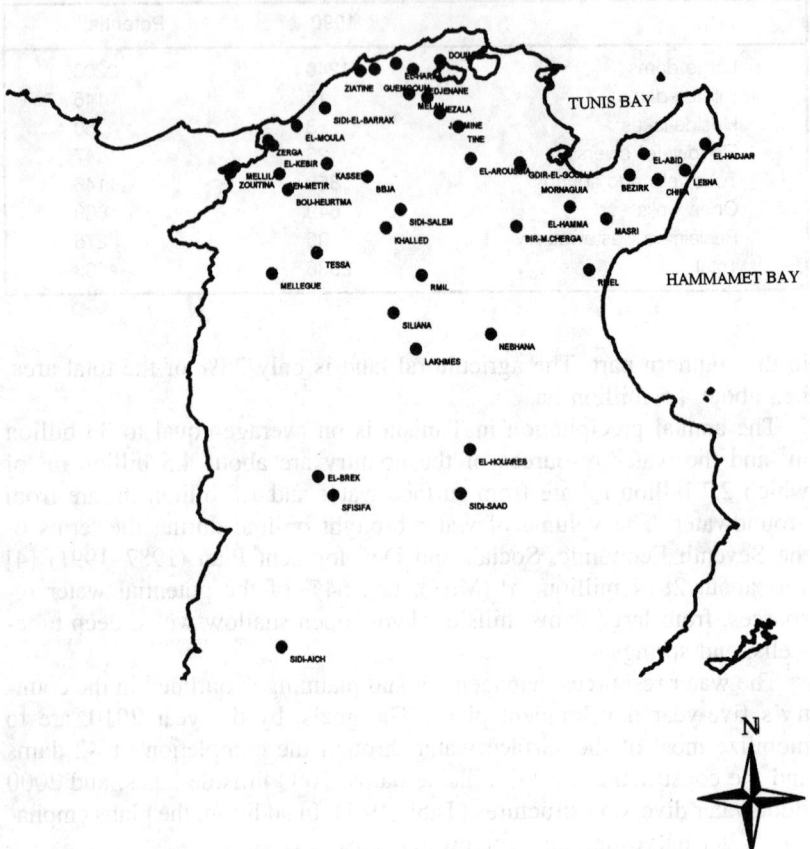

Figure 19.1 Existing and future large reservoirs location.

Surface Water

Large Dams

Total annual rainfall is not sufficient to provide a year-round water source for agricultural crops and to satisfy other requirements. Hence, reservoirs, water transfer systems, and other hydraulic structures such as floodwater diversion are important to store rainwater and fill the gap between rainy and dry years.

Many of these structures have already been completed. Others are under implementation or planned in order to store most of the potential surface water (Figure 19.1). The existing 18 dams currently allow for 1.4 billion m^3 of utilizable surface water per year. The net development rate of surface resources is 53%. With the construction of projected 24 dams, a total useful volume of about 2.2 billion m^3 in year 2010 will allow the storage of 85% of potential surface water resources.

Existing reservoirs are integrated into a complex hydraulic system. Owing to the interconnections (Figures 19.2 and 19.3), interregional transfer and spatial redistribution of water are feasible. Water is piped and conveyed over long distances from inland to the coastal areas (150 km) or from north to south (300 km) through systems of open canals and pipelines, reservoirs, and pumping stations. This is to supply the coastal cities (over 30% of the Tunisian population and a high proportion of touristic, and industrial activities) with drinking water and to preserve some agricultural regions such as the Cap Bon.

Surface water quality is variable; it ranges between 0.5 and 4.5 g L^{-1} with the highest frequency at 2.5 g L^{-1}. The salinity of over a half of surface water resources is less than 1.5 g L^{-1} and may be used without any restriction. Dams' salinity depends upon rainfall and seasons. The salinity of water at some locations may be high (> 2 g L^{-1}); two-thirds of the dams have between 0.5 and 2 g L^{-1}, and ten dams have water whose salinity is less than 0.5 g L^{-1}. Water is stored for various uses, which depend on the water quality. Some of these are multiple purposes dams such as for potable water supply, irrigation, groundwater recharge, and/or flood control.

However, large dams are subject to:

- reduction of their available capacity due to siltation (20–25 Mm^3 are lost each year)
- water losses by evaporation
- eutrophication as in the case of the main surface water reservoir of Tunisia, the Sidi Salem's dam

Different water management measures have been taken and are mainly related to water allocation. Potable water requirements (domestic, tourism, and industry) have to be provided first followed by agricultural requirements, groundwater recharge or production of electrical energy. For environmental protection such as in the case of Lake Ichkeul, specific amounts of fresh water (20 Mm^3 yr^{-1}) are needed to maintain the ecological equilibrium of the lake.

Other Structures

Hillside-dams and hillside-lakes, which are essential to flood control and soil and water conservation, contribute to groundwater recharge. They may also lengthen the expected life of the dams by reducing reservoir siltation. In total, 22 hillside-dams and 50 hillside-lakes (with a capacity

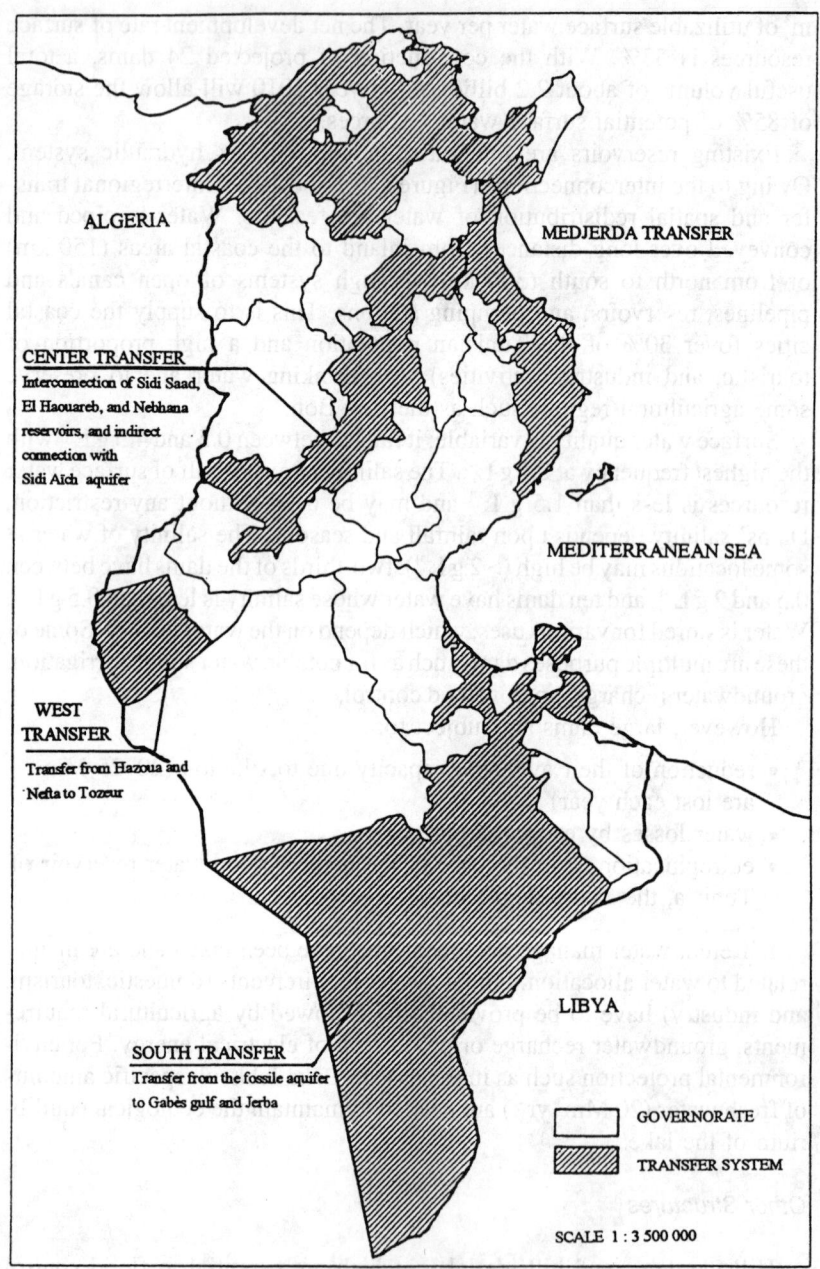

ALGERIA

MEDJERDA TRANSFER

CENTER TRANSFER
Interconnection of Sidi Saad,
El Haouareb, and Nebhana,
reservoirs, and indirect
connection with
Sidi Aïch aquifer

MEDITERRANEAN SEA

WEST
TRANSFER
Transfer from Hazoua and
Nefta to Tozeur

LIBYA

SOUTH TRANSFER
Transfer from the fossile aquifer
to Gabès gulf and Jerba

GOVERNORATE

TRANSFER SYSTEM

SCALE 1 : 3 500 000

Figure 19.2 Existing and planned transfer systems in Tunisia.

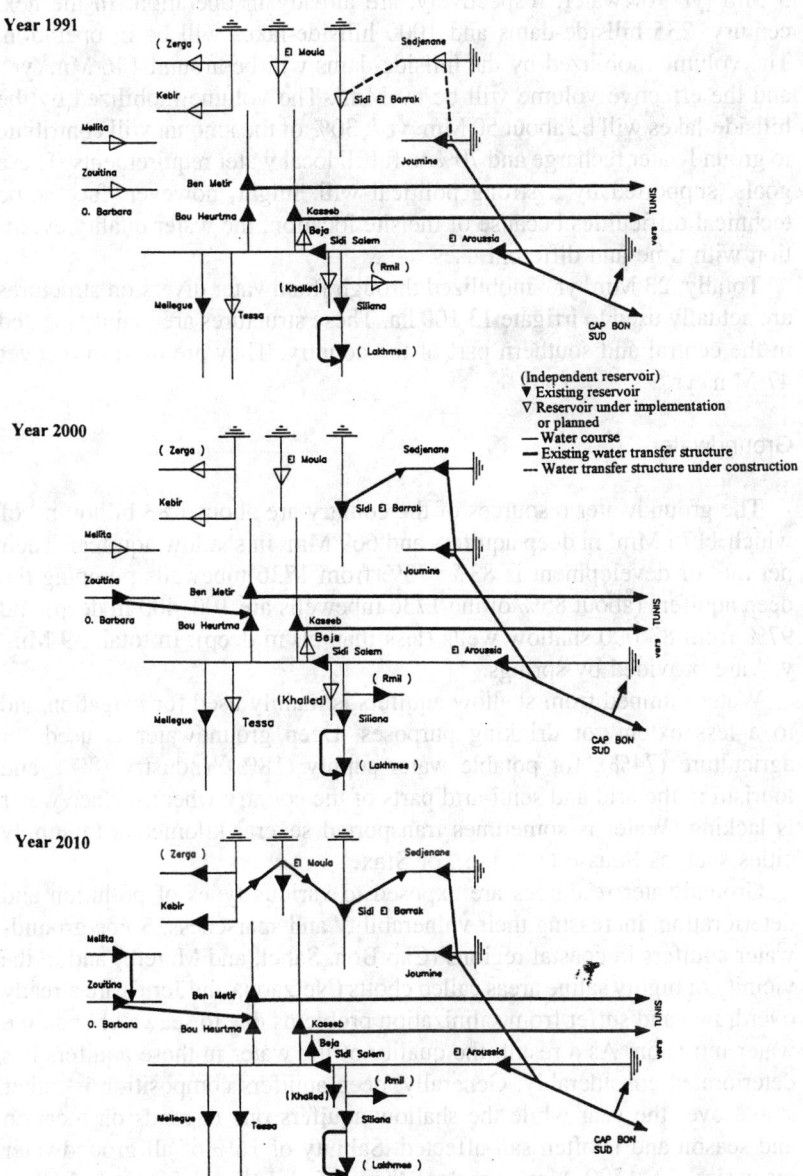

Figure 19.3 Planned interconnections between large reservoirs for the Medjerda transfer system.

The figure contains the following labels:

Year 1991

(Zerga) El Moula Sedjenane
Kebir Sidi El Barrak
Mellita
Zouitina Joumine
O. Barbara Ben Metir
Bou Heurtma Kaseeb
Beja Sidi Salem El Aroussia
(Rmil)
(Khalled) Siliana
Mellegue Tessa
(Lakhmes)
CAP BON SUD
vers TUNIS

Legend:
(Independent reservoir)
▼ Existing reservoir
▽ Reservoir under implementation or planned
— Water course
— Existing water transfer structure
--- Water transfer structure under construction

Year 2000

(Zerga) El Moula Sedjenane
Kebir Sidi El Barrak
Mellita
Zouitina Joumine
O. Barbara Ben Metir
Bou Heurtma Kaseeb
Beja Sidi Salem El Aroussia
(Rmil)
(Khalled) Siliana
Mellegue Tessa
(Lakhmes)
CAP BON SUD
vers TUNIS

Year 2010

(Zerga) El Moula Sedjenane
Kebir Sidi El Barrak
Mellita
Zouitina Joumine
O. Barbara Ben Metir
Bou Heurtma Kaseeb
Beja Sidi Salem El Aroussia
(Rmil)
(Khalled) Siliana
Mellegue Tessa
(Lakhmes)
CAP BON SUD
vers TUNIS

883

varying from 3300 to 500 000 m^3), mobilizing 17.7 Mm^3 yr^{-1} and around 3 Mm^3 yr^{-1} of water, respectively, are already in operation. In the next century, 235 hillside-dams and 1000 hillside-lakes will be in operation. The volume mobilized by the hillside-dams will be around 146 Mm^3 yr^{-1} and the effective volume will be 60 Mm^3. The volume mobilized by the hillside-lakes will be about 50 Mm^3 yr^{-1}: 30% of the amount will contribute to groundwater recharge and 70% to fulfill local water requirements. These goals, supported by a strong political will, might, however, face some technical difficulties because of the site location, the water quality evolution with time and different uses.

Totally, 23 Mm^3 yr^{-1} mobilized through floodwater diversion structures are actually used to irrigate 13 100 ha. These structures are mainly located in the central and southern part of the country. They are used to recover 47 Mm^3 yr^{-1}.

Groundwater

The groundwater resources of the country are about 1.83 billion m^3 of which 1175 Mm^3 in deep aquifers and 669 Mm^3 in shallow aquifers. Their net rate of development is 83%: 73% from 1736 tubewells pumping the deep aquifers (about 85% of the 1736 tubewells are 100–400 m deep) and 97% from 83 000 shallow wells (less than 50 m deep). In total, 59 Mm^3 yr^{-1} are provided by springs.

Water pumped from shallow aquifers is mainly used for irrigation and to a less extent for drinking purposes. Deep groundwater is used for agriculture (74%), for potable water supply (18%), industry (8%), and tourism in the arid and semi-arid parts of the country where surface water is lacking. Water is sometimes transported several kilometers to supply cities such as Sousse (100 km), or Sfax.

Groundwater resources are exposed to various types of pollution and deterioration, increasing their vulnerability and sparseness. Some groundwater aquifers in coastal regions (Cap Bon, Sahel, and Mareth) and in the vicinity of highly saline areas called chotts (Nefzaoua and Jerid) are already overdrawn and suffer from salinization problems due to seawater or saline water intrusion. As a result, the quality of the water in these aquifers has deteriorated considerably. Generally, deep aquifers composition is rather stable over the year while the shallow aquifers one depends on location and season and is often salt-affected. Salinity of 16% of all groundwater resources, i.e., 300 Mm^3, are less than 1.5 g L^{-1} and 30% of shallow groundwater is over 4.0 g L^{-1}. In the alluvial valleys, the water salinity generally varies between 1.5 to more than 5 g L^{-1} and can reach 10 g L^{-1}. The water quality of some deep aquifers (limestone or sandstone) is good (<1–1.5 g L^{-1}) and they are used for potable water supply. In the south, there are three main fossil aquifers with different water qualities (1–7 g L^{-1}) which are already under development.

Some shallow aquifers are already overdrawn and pumping should be reduced. Pollution of some shallow aquifers by nitrates constitutes also a major risk for domestic requirements. Things are different for the deep aquifers and 610 new tubewells have been planned to develop 288 Mm^3 yr^{-1}.

Reclaimed Water

Although wastewater treatment and wastewater reuse are practiced for a long period of time in Tunisia (the first wastewater treatment plant, La Cherguia-Tunis, was built in 1958), only 30% of the urban and rural households are connected to a sewerage network; the remaining amount of wastewater is being discharged in septic tanks (22%), or in the nature (48%). Among the urban population, 2.7 million (51%) are connected to sewers and others to individual septic tanks. The sanitation coverage in the sewered cities is about 75%; this rate, related to the whole urban population, is only 50%. As of 1988, only 17% of the rural population were estimated to have access to on-site or sewered sanitation facilities.

Out of 288.2 Mm^3 of potable water distributed in 1992, 180 Mm^3 were discharged as wastewater. Wastewater is reclaimed in 44 treatment plants with a total design capacity of 130 Mm^3 per year. Several plants are located along the coast to protect the bathing areas and prevent sea pollution (Figure 19.4). The volume of reclaimed water available in 1994 was 106 Mm^3. Sanitation master plans have been designed for several towns. The National Sewerage and Sanitation Office (Office National de l'Assainissement, ONAS) plans to equip other towns, including those in the *Eleven towns* project (some have already been implemented) to protect the Sidi Salem's dam (serving water supply and irrigation of Tunis, Cap Bon, Sousse, and Sfax areas) from eutrophication due to diverse pollution sources. Furthermore, "compact" treatment plants are planned for small towns (<20 000 inhabitants). The annual volume of reclaimed water is expected to reach 175 Mm^3 in year 2000. The expected amount of reclaimed water will then equal approximately 10% of the available groundwater resources. This could be used to replace groundwater currently being used for irrigation in areas where excessive groundwater overdraft is causing salt water intrusion in coastal aquifers.

WATER RESOURCES UTILIZATION—SUPPLY AND DEMAND

The water resources allocation was defined within the framework of three master schemes which cover the entire country and was based on concerns of quantity and quality. High quality water is reserved for drinking water and industry; water allocated to irrigation is generally of a lower quality.

Water Consumption in Urban, Rural, Industrial, and Tourism Sectors

The volume allocated for drinking water supply and industry was, during the Seventh Social, Economic, and Development Plan (1987–1991) [4], 420 Mm^3 yr^{-1} and the overall water consumption for domestic, industry, and tourism was 276 Mm^3 yr^{-1}. This is expected to reach 500 Mm^3 by the year 2010 (73% for domestic purposes, 20% for industry, and 7% for tourism).

In urban areas, water services are almost entirely metered and tariff charges are graduated upward based on consumption. In 1990, the rate varied from 0.11 US\$ (1 DT $\approx$ 1 US\$) per cubic meter for domestic use to 0.53 US\$ per cubic meter for industrial use. The average price was 0.31 US\$ m^{-3}.

In urban areas, 73% of the population have access to piped water supply. In rural areas, 42% of the population are connected to a water supply network through house connections, and 58% rely on public standpipes and wells. Tunisia should achieve full water supply service coverage in urban areas and 78% in rural areas by the end of the century.

The Ministry of Agriculture is responsible for domestic water supply, both piped and unpiped. Piped supplies for urban (3.9 million inhabitants) and rural (1 million inhabitants) communities exceeding 500 inhabitants are constructed and operated by the National Water Supply Company (SONEDE: Société Nationale d'Exploitation et de Distribution des Eaux). Water supply to smaller communities is the responsibility of the Rural Engineering Directorate (1 million inhabitants). The remaining rural population (1.5 million inhabitants) relies on other services such as cisterns, wells, fountains, wadis, or springs.

The volume allocated to domestic (urban and rural) water needs was 240 Mm^3 in 1989. The consumption of the population supplied by SONEDE was 156 Mm^3 in 1989, with 145 Mm^3 for the urban sector and 11 Mm^3 for the rural sector. The average urban specific consumption is 85 L $capita^{-1}$ d^{-1} and varies from 53 L $capita^{-1}$ d^{-1} (Tataouine) to 105 L $capita^{-1}$ d^{-1} (Sfax). In the rural sector, the average specific consumption is 28 L $capita^{-1}$ d^{-1} with a minimum of 5 L $capita^{-1}$ d^{-1} and a maximum of 94 L $capita^{-1}$ d^{-1}. In the remaining part of the rural sector, which is supplied by other offices, this consumption varies between 13 and 18 L $capita^{-1}$ d^{-1}. The whole rural sector consumption is 23 Mm^3.

The salinity of the water available for drinking purposes varies from 0.37 to 2.90 g L^{-1} which is without causing health problems in some cases. Brackish water (3.8 g L^{-1}) desalination using reverse osmosis is in operation

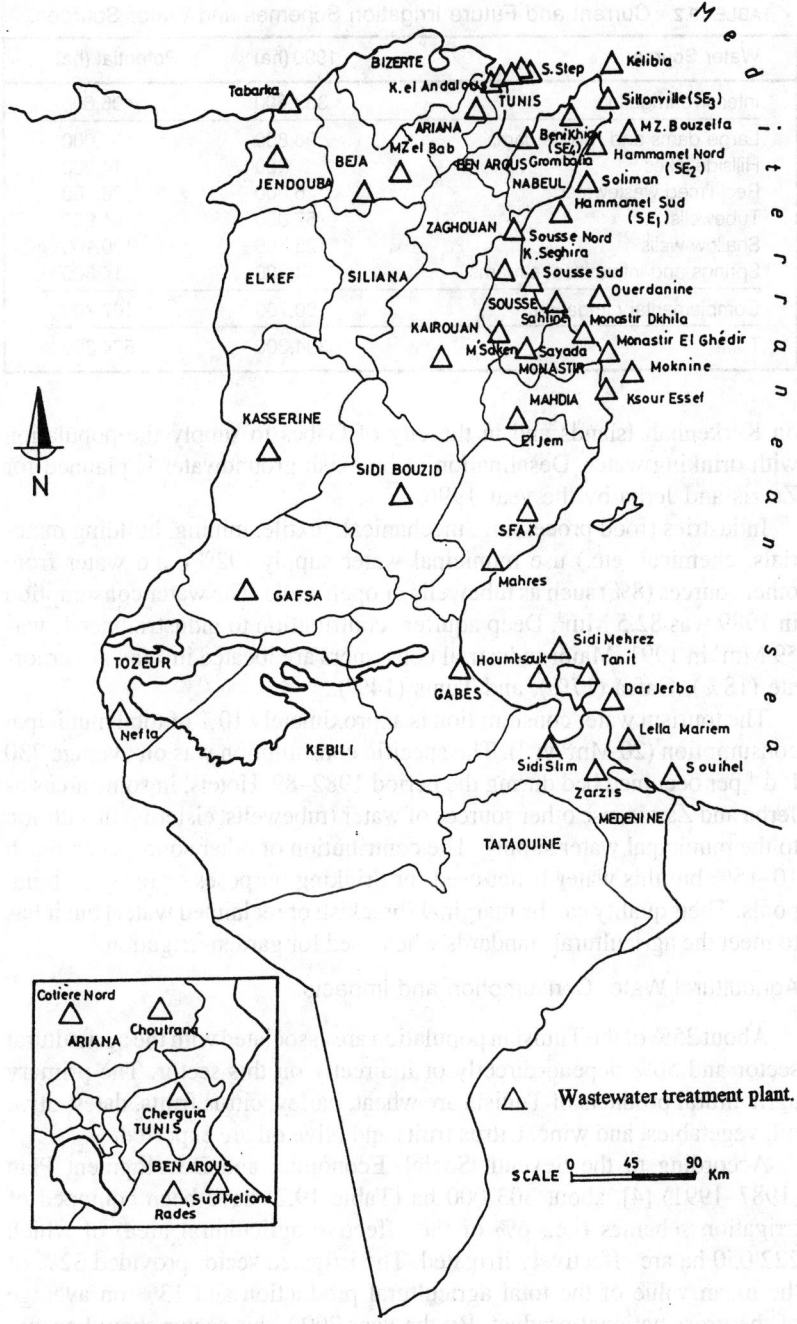

Figure 19.4 Location of the present wastewater treatment plants.

TABLE 19.2. Current and Future Irrigation Schemes and Water Sources.

Water Source	1990 (ha)	Potential (ha)
Intensive irrigation	303,600	406,600
Large dams and hillside-dams	98,800	157,000
Hillside-lakes	900	14,900
Reclaimed wastewater	6,000	28,400
Tubewells	57,900	61,900
Shallow wells	125,700	130,100
Springs and intermittent streams	14,300	14,300
Complementary irrigation	90,700	167,700
Total	394,300	574,300

on Kerkennah Islands and in the city of Gabès to supply the population with drinking water. Desalination of brackish groundwater is planned for Zarzis and Jerba by the year 1996.

Industries (food processing, mechanical, textile, mining, building materials, chemical, etc.) use municipal water supply (92%) and water from other sources (8%) such as tubewells or open wells. The water consumption in 1989 was 82.5 Mm^3. Deep aquifers contribution to industrial needs was 59 Mm^3 in 1991. Major industrial consumers are located in Sfax governorate (18%), Gafsa (17%), and Tunis (14%).

The tourism water consumption is approximately 10% of total municipal consumption (20 Mm^3 yr^{-1}). The specific consumption was on average 730 L d^{-1} per occupied bed during the period 1982–89. Hotels, in some areas as Jerba and Zarzis, use other sources of water (tubewells, cisterns) in addition to the municipal water supply. The contribution of other sources can reach 10–15% but this water is not used for drinking purposes or for swimming pools. Their quality can be marginal (brackish or reclaimed water) but it has to meet the agricultural standards when used for garden irrigation.

Agricultural Water Consumption and Impacts

About 25% of the Tunisian population are associated with the agricultural sector and 50% depend directly or indirectly on this sector. The primary agricultural products of Tunisia are wheat, barley, citrus fruits, dates, olive oil, vegetables, and wine. Citrus fruits and olive oil are exported.

According to the Seventh Social, Economic, and Development Plan (1987–1991) [4], about 303 000 ha (Table 19.2) have been equipped of irrigation schemes (i.e., 6% of the effective agricultural area) of which 222 050 ha are effectively irrigated. The irrigated sector provided 32% of the mean value of the total agricultural production and 13% on average of the gross national product. By the year 2000, this sector should ensure 50% of the agricultural production.

The annual volume allocated, during the Seventh Social, Economic, and Development Plan (1987–1991) [4], for irrigation was 1575 Mm^3. One third of the water used for irrigation comes from surface reservoirs and two-thirds from groundwater. Agricultural water consumption is highly variable from year to year depending on rainfall, complementary irrigation, and difference between the water allocated to irrigation and the water effectively used. In 1993–94, 2400 Mm^3 were allocated to irrigation, but only 1600 Mm^3 were used. About 35 Mm^3 of reclaimed water annually are allocated for irrigation.

Water and soil salinity is a major constraint for Tunisian agricultural development. Looking for a higher agricultural productivity has been compulsory for several years, but it has also led to the development of soil salinization and alcalinization in the irrigated areas. The development of these processes is related to the climatic conditions, the irrigation water quality, the soil characteristics, and the problems of capillary rise from shallow groundwater.

Salt-affected soils in Tunisia cover about 1.5 million hectares that is around 10% of the total area of the country. They are mainly located in the central and southern part of the country. They are most of the time cultivated and almost always irrigated, except in the chott (highly saline) area.

WATER DEMAND FORECAST

Tunisia is rapidly approaching full utilization of its available water resources. A comparison of water resources and requirements shows that, although water requirements will be generally met up to 2015, some areas (coastal and southern) are already suffering from water shortage and groundwater overexploitation. Furthermore, existing water resources often have a certain content of salts and are highly exposed to different pollutants of industrial, domestic, and agricultural origin. The scarcity of water resources is even more acute as the possibilities of developing new sources are limited and their costs may become much higher.

The forecast of water requirements depends on future economic and demographic trends. Therefore, different scenarios are possible. In planning the use and management of water resources in Tunisia [5], the following sub-sectors have been taken into account: domestic, industry, agriculture, and tourism.

The domestic (urban and rural) water requirements will increase with the population growth. The average values may fluctuate from 241 Mm^3 in 1990 to 519 Mm^3 in 2010. According to the development scenarios, they may reach 350 or 574 Mm^3 in 2010 (i.e., 8 to 13% of the conventional resources). According to trends, industrial water requirements may evolve from 82.5 Mm^3 in 1989 to 119 or 123 Mm^3 in 2010 (with Sfax, Gafsa, and the Greater Tunis being the major consumers). If effective water saving measures are

TABLE 19.3. Accessible Water to Cover Water Needs for the Years 1995, 2000, and 2010 (Mm^3 yr^{-1}).

Year	1995	2000	2010
Surface water:			
Large dams[a]	1472	1582	1586
Hillside-dams[b]	22	60	60
Hillside-lakes[c]	1	18	36
Deep aquifers[d]	858	1000	1146
Shallow aquifers[d]	649	660	669
Reclaimed wastewater[e]	35	70	165
Total	3037	3390	3662

[a]Volume technically controlled on an average year assuming 50% useful volume.
[b]Volume technically controllable on an average year.
[c]Excluding groundwater recharge.
[d]Volume technically controllable estimated at 100% for groundwater.
[e]Assuming that 40% of the reclaimed water might be reused in year 2000 and 60% in year 2010.

applied, 10% annually may be saved. The tourism water requirements may, according to the scenarios, be multiplied by a factor equal to 2 or 3. In year 2010, the minimum water requirements could be 18 Mm^3 and the maximum 58 Mm^3. The agricultural sector demand forecast may reach 1900 Mm^3 in 2010. Due to water shortage, agricultural planning is based on the water quantity that may be available to the farmers, after satisfying water demands for domestic and industrial use. In the future, less and less fresh water will be available for agriculture and increasing quantities of marginal (brackish and reclaimed) water will be used.

The accessible water resources and their planned use for years 1995, 2000, and 2010 are shown in Tables 19.3 and 19.4, and Figure 19.5. The planned infrastructure in year 2010, will account for 83% of the potential (4533 Mm^3). Year 2010 forecasts by subsectors as compared with 1995 show that allocation of water to agriculture remains the most important

TABLE 19.4. Projected Water Demand by Category Use for the Years 1995, 2000, and 2010 (Mm^3 yr^{-1}).

	1995		2000		2010	
Year	Min.	Max.	Min.	Max.	Min.	Max.
Domestic (U + R)	257	301	296	373	350	574
Tourism	18	27	18	36	18	58
Industry	92	93	101	103	119	123
Agriculture	1500	1860	1580	1965	1700	2110
Total	1867	2281	1995	2477	2187	2865

Min.: minimum; Max.: maximum.
(U + R): (Urban + Rural).

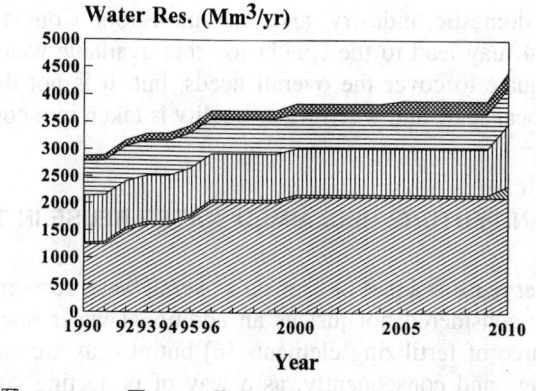

Water Res. (Mm³/yr)

Year

⬚ Large Dams ⬚ Others ▦ Deep GW ⬚ Shallow GW ▦ Recl. WW

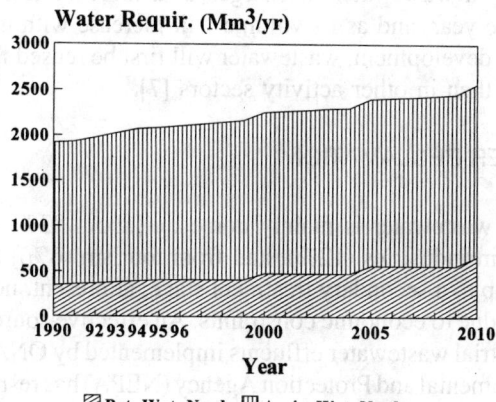

Water Requir. (Mm³/yr)

Year

▨ Pot. Wat. Needs ⬚ Agric. Wat. Needs

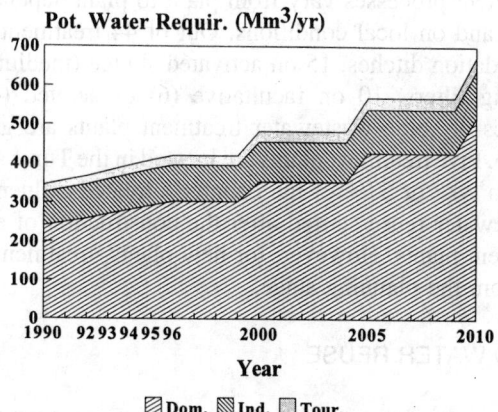

Pot. Water Requir. (Mm³/yr)

Year

▨ Dom. ▦ Ind. ⬚ Tour.

Figure 19.5 Water supply and demand sources.

followed by domestic, industry, and tourism sectors. Comparing Tables 19.3 and 19.4 may lead to the conclusion that available water resources may be adequate to cover the overall needs, but, it is not the case at a regional or local scale and when water quality is taken into consideration.

CURRENT AND FUTURE RECLAIMED WATER REUSE IN TUNISIA

Wastewater reuse is a part of Tunisia's overall water resources balance. It is actually considered not just as an additional water resource and a potential source of fertilizing elements [6] but also as a complementary treatment stage and consequently, as a way of protecting coastal areas, water resources, and sensitive receiving bodies.

As municipal wastewater discharges at a more or less constant rate throughout the year, and as its volume will increase with urban, tourism, and industrial development, wastewater will first be reused for agricultural purposes and then in other activity sectors [7].

WASTEWATER RECLAMATION

Municipal wastewater is mainly domestic [about 82%; the remaining originates from industries (12%), and from tourism (6%)] and processed biologically up to a secondary treatment stage; at present, no further treatment is made due to economic constraints. An effective source control program of industrial wastewater effluents implemented by ONAS and the National Environmental and Protection Agency (NEPA) has resulted in a rather low contribution of industrial flows in municipal wastewater effluents.

The treatment processes vary from plant to plant depending on wastewater origin and on local conditions. Out of 44 treatment plants, 17 are based on oxidation ditches, 15 on activated sludge (medium or low rate), 2 on trickling filters, 10 on facultative (6) or aerated (4) ponds. The characteristics of some wastewater treatment plants are given in Tables 19.5 and 19.6. Five treatment plants are located in the Tunis area, producing about 60 Mm^3 yr^{-1} or 57% of the country's treated effluent.

The wastewater reuse started after the construction of existing wastewater treatment plants. However, for new plants, treatment and reuse are combined from the planning stage.

RECLAIMED WATER REUSE

Irrigation with reclaimed water is a well established practice in Tunisia. Wastewater from Tunis has been used since the early 1960s

TABLE 19.5. Characteristics of Some Wastewater Treatment Plants [8].

WWTP	Capacity (m³ d⁻¹)	Flow (m³ d⁻¹)	Influent Origin (%)	Treatment Process	Reclaimed Wastewater Reuse Rate	Operation Costs (US$ m⁻³)	Non Working Days	Overload (%)
Cherguia	60,000	34,738	D(84) + I(11) + T(5)	AS	28	0.024	0	—
Choutrana	43,000	97,592	D(84) + I(11) + T(5)	OC	34	0.019	7	228
Côtière Nord	15,750	14,566	D(88) + T(12)	SP	—	0.016	0	105
Sud Méliane	37,500	18,310	D(84) + I(16)	OC	20	0.041	6	—
Radès	700	771	D(100)	SP	0	0.113	9	265
Tabarka	5,200	1,762	T(90) + D(9.5) + I(0.5)	AS	0	0.078	5	—
SE_1 Hammamet	4,208	2,828	T(76) + D(15) + I(10)	AS	54	0.073	0	121
SE_2 Hammamet	5,146	3,586	D(67) + T(29) + I(4)	AS	0	0.078	0	108
SE_3 Nabeul	3,500	2,197	T(99.5) + I(0.5)	OC	0	0.095	4	118
SE_4 Nabeul	9,585	7,332	D(93) + T(6) + I(1)	AS	35	0.065	2	115
Kélibia	2,200	2,879	D(99) + I(0.5) + T(0.5)	AS	0	0.071	0	231
Soliman	2,457	1,204	D(89) + I(11)	OC	0	0.092	0	—
Grombalia	3,235	839	D(88) + I(12)	OC	0	0.162	0	—
Menzel Bouzelfa	2,065	1,454	D(95) + I(5)	OC	0	0.067	0	—
Kairouan	12,000	10,412	D(95) + I(4) + I(1)	AS	16	0.039	69	104
Sousse Nord	8,670	12,175	D(76) + T(21) + I(3)	AS	24	0.031	0	190
Sousse Sud	18,700	13,008	D(76) + I(18) + T(6)	AS + TF	24	0.038	24	—
Monastir	2,000	2,774	D(97) + I(3)	TF	77	0.065	1	197
Dkhila	3,100	3,526	T(64) + D(34) + I(2)	AS	17.5	0.060	60	239
Moknine	6,400	3,242	D(74) + I(26)	SP	21	0.030	15	—
Kalaa Seghira	1,450	519	D(100)	OC	10	0.208	7.5	—
Ouerdanine	1,400	764	D(93) + I(3)	OC	0	0.100	5	—

TABLE 19.5. (continued).

WWTP	Capacity (m³ d⁻¹)	Flow (m³ d⁻¹)	Influent Origin (%)	Treatment Process	Reclaimed Wastewater Reuse Rate	Operation Costs (US$ m⁻³)	Non Working Days	Overload (%)
Sahline	2,650	1,490	T(100)	OC	100	0.073	0.5	—
Sayada	1,660	629	D(100)	OC	0	0.091	6	—
Sfax	24,000	21,966	D(85) + I(13) + T(2)	AL	47	0.050	18	236
Gafsa	3,500	3,463	D(94) + I(5) + T(1)	SP	49	0.029	0	146
Nefta	1,335	625	D(95) + T(4) + I(1)	OC	4	0.142	2	—
Tanit	256	141	T(100)	TF	30	0.108	18	—
Houmt Souk	3,500	1,162	D(72) + I(23) + T(5)	AL	6	0.063	0	—
Dar Jerba	1,100	3,512	T(100)	AS	0	0.031	0.7	319
Sidi Slim	1,800	1,478	T(100)	AS	0	0.062	3.5	176
Lalla Meriem	1,724	465	T(100)	AL	33	0.119	0.4	—
Souihel	1,208	366	T(81) + D(19)	AS	10	0.205	0	139
Zarzis	1,355	62	D(100)	OC	0	1.030	0	—
Sidi Mahrez	1,620	2,818	T(97.5) + D(2.5)	AS	40	0.025	43.7	359

D: domestic; I: industrial; T: tourism; AS: activated sludge; SP: stabilization ponds; OC: oxidation ditch; AL: aerated lagoons; TF: trickling filter.

TABLE 19.6. Inflow and Outflow Concentration of BOD_5, COD, and SS (in mg L^{-1}) and Average EC of the Wastewater Discharged from Different Treatment Plants [8,9].

WWTP	Influent			Effluent			Aver. EC
	BOD_5	COD	SS	BOD_5	COD	SS	(μS cm^{-1})
Cherguia	370	812	181	27	66	18	3130
Choutrana	417	987	389	24	124	25	3420
Côtière Nord	327	741	275	47	264	58	8940
Sud Méliane	328	880	351	18	100	13	4490
Radès	245	294	174	34	256	36	—
Tabarka	306	524	315	11	62	28	—
SE$_1$ Hammamet	280	753	207	27	132	27	4100
SE$_2$ Hammamet	433	13	82	13	82	11	2390
SE$_3$ Nabeul	229	634	186	14	85	12	2490
SE$_4$ Nabeul	511	1245	518	12	82	11	3400
Kélibia	388	990	383	80	339	72	—
Soliman	198	731	296	7	79	19	—
Grombalia	276	886	416	12	107	14	—
Menzel Bouzelfa	463	1296	726	9	79	12	—
Kairouan	385	765	413	12	93	13	3780
Sousse Nord	637	1500	750	29	145	32	3060
Sousse Sud	358	830	358	22	108	22	3300
Monastir	375	882	392	24	133	23	4020
Dkhila	237	548	181	34	182	23	—
Moknine	700	2100	1000	90	600	100	5880
Kalaa Seghira	407	1219	562	20	106	16	—
Ouerdanine	250	830	320	20	85	25	—
Sahline	375	900	400	24	90	23	—
Sayada	154	599	308	9	51	12	—
Sfax	668	1019	446	102	301	65	5830
Gafsa	600	1220	460	78	275	17	3280
Nefta	652	800	380	25	110	20	—
Tanit	220	450	170	22	110	15	—
Houmt Souk	463	1542	321	40	210	37	—
Dar Jerba	199	631	278	22	163	22	—
Sidi Slim	209	513	176	15	120	11	—
Lalla Meriem	380	939	347	28	179	50	—
Souihel	163	402	112	16	95	14	—
Zarzis	132	389	250	16	171	43	—
Sidi Mahrez	260	580	220	60	220	40	—
Standards[a]	—	—	—	30	90	30	7000

[a]Standards for reclaimed water reused in agriculture.

895

to irrigate the scheme of La Soukra (600 ha located 8 km North East of Tunis). The reuse has enabled citrus fruit orchards to be saved as it would have been impossible to irrigate by groundwater alone (which had become overdrawn and suffered from intrusion of saline water). Effluents from the treatment plant of La Cherguia (which processes a part of Tunis municipal wastewater) were thus used, mainly during spring and summer, either exclusively or as a complement to ground-water. Wastewater irrigation was restricted to citrus and olive trees since irrigation of vegetables that might be consumed raw was prohibited by the Water Law.

The scheme is still in operation and water from La Cherguia's secondary sewage treatment plant is pumped then discharged into a pond (5800 m³) and finally stored in a reservoir (3800 m³). From there, the water is delivered to farmers' plots, by gravity, through an underground pipe system. Thus, the water is transported over 8 kilometers to the irrigation area. A Regional Department for Agricultural Development (CRDA) supervises the water distribution system (operation and maintenance of all infrastructure) and controls the application of the regulations related to reclaimed water reuse for agricultural purposes.

After the above experience, the wastewater reuse policy was launched at the beginning of the 1980s. The area currently equipped is about 6500 ha (Figure 19.6). Many projects are being implemented or planned that the area irrigated with reclaimed water will be expanded to 20–30000 ha [10]. The main locations are La Soukra, La Cebala, Mornag, Nabeul, Hammamet, Sousse, Monastir, Sfax, and Kairouan. Among these projects, the wastewater irrigated area around Tunis is the most important since it represents 76% of the entire irrigated area (Figure 19.7). The sewage treatment plants supplying irrigation water are La Cherguia, Choutrana, Côtière Nord, and Sud Méliane.

Most of the reclaimed water will be used to irrigate citrus and different kinds of trees (grapes, olives, peaches, pears, apples, grenades, etc.), fodder crops (alfalfa, sorghum, berseem, etc.), industrial crops (cotton, tobacco, sugarbeet), cereals, and golf courses (Tunis, Hammamet, Sousse, Monastir, and Tabarka). Some hotel gardens in Jerba and Zarzis are also irrigated with reclaimed water. Wastewater reuse in agriculture is regulated by the 1975 Water Law and by the 1989 decree (Decree No. 89-1047) [11]. In a separate document, the quality standards as well as the terms and general conditions of reclaimed water reuse have also been set up.

About 35 Mm³ yr⁻¹ of reclaimed water are allocated for irrigation. In some schemes, irrigation with effluents is well-established and most of the volume allocated is used while in new projects, where irrigation has just started, the utilization rate of reclaimed water is slowly increasing. During the irrigation season, 30 to 40% of the overall effluent yield is

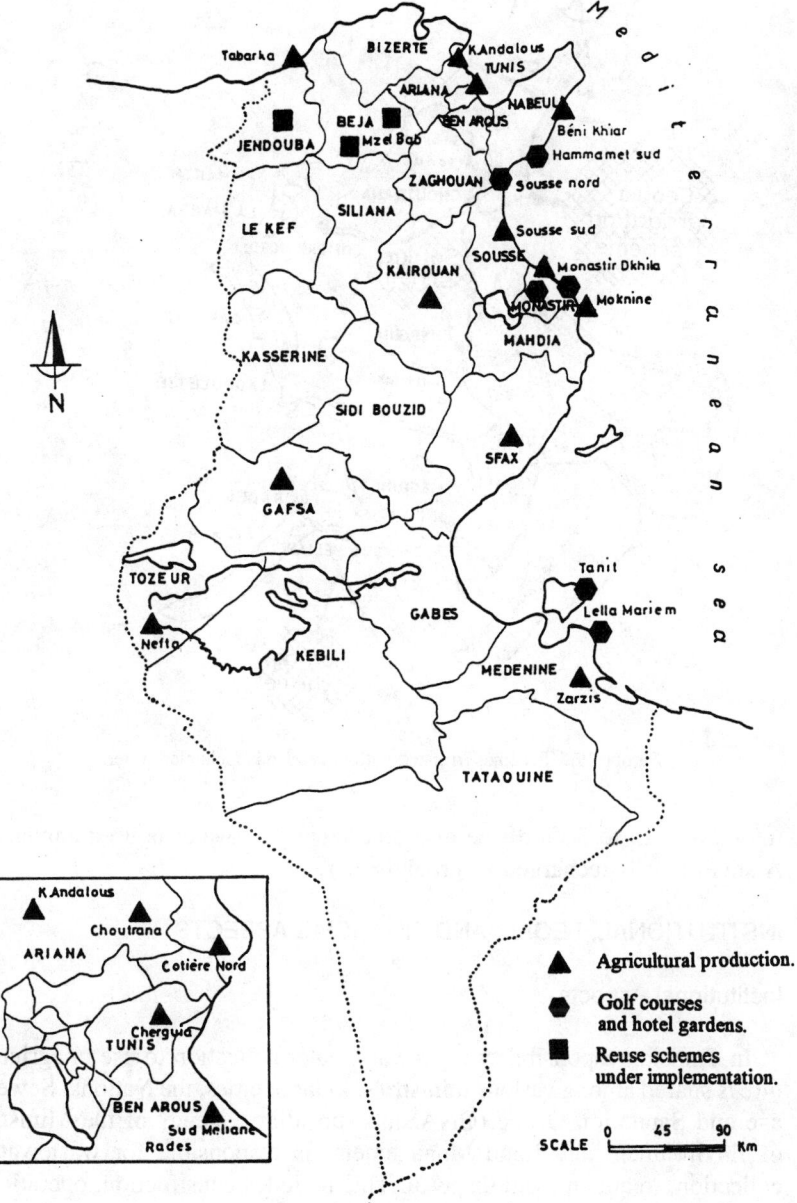

Figure 19.6 Current and future irrigated areas with reclaimed water.

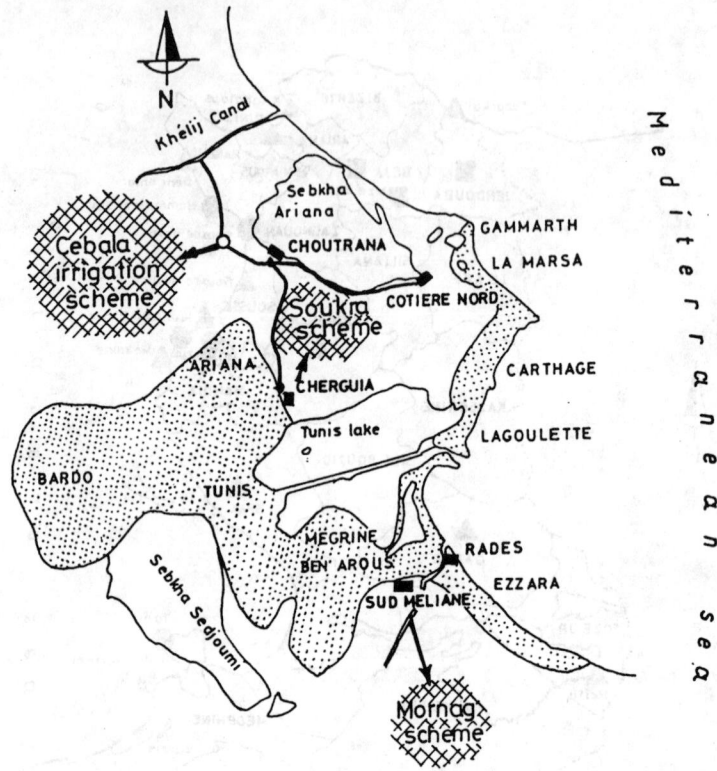

Figure 19.7 Schemes irrigated with Great Tunis reclaimed water.

reused and 25 to 30% of the available reclaimed water is used annually. A small part is recharged to groundwater.

INSTITUTIONAL, LEGAL, AND FINANCIAL ASPECTS

Institutional Aspects

In Tunisia, responsibility from wastewater collection to use in agriculture is shared among various ministries. In large cities, the National Sewerage and Sanitation Office (ONAS), a subsidiary agency of the Ministry of Environment and Land Management, is responsible for wastewater collection, treatment, and disposal. This includes construction, operation, and maintenance of the entire sanitation infrastructure.

The Ministry of Agriculture is responsible for the implementation of wastewater reuse projects, i.e., supply and development of irrigation schemes (pumping stations, reservoirs, pipes, etc.). This work is carried out by different departments: General Directorate of Hydraulic Studies, General Directorate of Rural Engineering, Regional Commissariats for Agricultural Development (CRDA). The latter are in charge of operating the water distribution system, collecting charges and monitoring the application of the Water Law and the decree related to wastewater reuse for agricultural purposes. The Rural Engineering Research Center conducts research on wastewater reuse in agriculture: this means providing data essential in the decision making for the best use of reclaimed water. The Research Center has also to formulate, at agricultural scale and under Tunisian cultural conditions, the standards related to this use, to supervise and follow up the reclaimed water irrigated areas.

The Ministry of Public Health is responsible for the regulation of the hygienic quality of reclaimed water reused for irrigation and of marketed crops, as well as for water pollution monitoring and pollution control enforcement. At a regional level, the different hygiene departments are in charge of periodical monitoring in their laboratories, health education, and prevention campaigns. Non-routine analyses are performed at the Pasteur Institute which is the Ministry's reference and research laboratory.

The Ministry of Tourism and Handicrafts and the Property Agency for Tourism participate in financing wastewater reuse for irrigation of green areas (golf courses, and hotel gardens).

Legal Aspects

Reclaimed wastewater is mainly used for restricted irrigation in Tunisia which means that secondary treated effluent is used for growing all types of crops except vegetables, whether eaten raw or cooked. Wastewater reuse in agriculture is regulated by the 1975 Water Law and by the 1989 decree (Decree No. 89-1047) [11].

The Water Law provides a legal framework for reclaimed water use. This code prohibits use of raw wastewater in agriculture and irrigation with reclaimed water of any vegetable to be eaten raw. This law stipulates that wastewater earmarked for irrigation must be of a quality which does not lead to transmission of diseases.

The 1989 decree specifically regulates reuse of wastewater in agriculture. It specifies that wastewater can be reused only after adequate treatment and the effluent should be used only to irrigate crops which are not directly consumable. It stipulates that the use of reclaimed water must be authorized by the Minister of Agriculture, in agreement with the Minister of Environment and Land Management, and the Minister of Public Health. It sets out the precautions required to protect the health of farmers and consumers, and the environment. Monitoring the physical-chemical and biological quality of reclaimed water (analyses of a set of physical-chemical parameters once a month, of trace elements once every six months, and of helminth eggs every two weeks on 24-hrs composite samples) and of the irrigated crops is planned. In areas where sprinklers are going to be used, buffer areas must be created. Direct grazing is prohibited in the wastewater irrigated areas. In separate documents, reclaimed water quality standards for reuse (Tunisian standard 106.03, 1989, [12]), wastewater disposal standards in receiving waters (Tunisian standard 106.002, 1989, [13]), and a list of crops that can be irrigated have also been set up. Table 19.7 presents common parameters reported on both standards.

Specifications determining the terms and general conditions of reclaimed water reuse [as the precautions that must be taken in order to prevent any contamination (workers, residential areas, consumers, etc.)] have also been published.

The Ministries of Interior, Environment and Land Management, Agriculture, Economy, and Public Health are in charge of the implementation and enforcement of this decree.

Financial Aspects

Wastewater reuse operations are financed entirely or partially from the national budget or co-financed by international organizations or else as part of a bilateral cooperation project.

Sanitation charges are included in the drinking water billing. They are progressive and depend on use (domestic < industry < tourism). They do not cover the treatment costs.

The cost of the reclaimed water sold to farmers varies from one scheme to another. The charges are meant to cover some of the operation costs (operation, maintenance, salaries, and energy). As for schemes irrigated with conventional water, this cost must evolve progressively towards the real cost. However, up to now, the cost of wastewater remained less than that of conventional water. For instance, in the Mornag irrigation scheme, the operation costs amount to 0.065 US$ m^{-3} and the selling price is 0.046 US$ m^{-3}. Water from the Medjerda-Cap Bon canal is sold at 0.058 US$ m^{-3}. This price has already been increased in comparison to previous years. With the creation of new schemes, water is often distributed free of charge at the beginning of the supply to encourage farmers to use it, then at a fixed price per hectare before evolving towards a price per m^3 of water used.

CONSTRAINTS IN RECLAIMED WATER REUSE

Different kinds (technical, sanitary, social, organizational, and institutional) of problems have been encountered, at different levels, in the

TABLE 19.7. Maximum Concentrations for Reclaimed Wastewater Reuse in Agriculture and Discharge in the Receiving Bodies.

Parameter	Agricultural Reuse[a]	Discharge in the environment[b]		Sewers[b]
		Sea	Water bodies	
pH	6.5–8.5	6.5–8.5	6.5–8.5	6.5–9.0
Electrical conductivity (EC) (μS/cm)	7000	—	—	—
Chemical oxygen demand (COD)	90[c,d]	90	90	1000
Biochemical oxygen demand (BOD_5)	30[c,d]	30	30	400
Suspended solids (SS)	30[c,d]	30	30	400
Chloride (Cl)	2000	—	600	700
Fluoride (F)	3	5	3	3
Nitrate (NO_3)	—	50	50	90
Nitrite (NO_2)	—	5	0.5	10
Organic nitrogen + ammonia (N_{org} + NH_4)	—	30	1	100
Total phosphorus or ortho-P (P)	—	0.1	0.05	10
Halogenated hydrocarbons	0.001	0.005	0.001	0.01
Arsenic (As)	0.1	0.1	0.05	0.1
Boron (B)	3	20	2	2
Cadmium (Cd)	0.01	0.005	0.005	0.1
Cobalt (Co)	0.1	0.5	0.1	0.5
Chromium (Cr)	0.1	0.5	0.01	0.5
Copper (Cu)	0.5	1.5	0.5	1
Iron (Fe)	5	1	1	5
Manganese (Mn)	0.5	1	0.5	1
Mercury (Hg)	0.001	0.001	0.001	0.01
Nickel (Ni)	0.02	2	0.2	1
Lead (Pb)	1	0.5	0.1	1

TABLE 19.7. (continued).

| Parameter | Agricultural Reuse[a] | Discharge in the environment[b] | | Sewers[b] |
		Sea	Water bodies	
Selenium (Se)	0.05	0.5	0.05	1
Zinc (Zn)	5	10	5	5
Faecal coliforms (count/100 ml)	—	2000	2000	—
Faecal streptococci (count/100 ml)	—	1000	1000	—
Salmonella and vibrio cholerae (count/5000 ml)	—	None	None	—
Intestinal nematodes (arithmetic mean no. of eggs /l)	≤1	—	—	—

[a]Tunisian standards relative to reclaimed wastewater reuse for agricultural purposes (NT 106.03) [12].
[b]Tunisian standards relative to discharge in the receiving bodies (NT 106.002) [13].
[c]24-hr composite sample.
[d]Except special authorization.
All units in mg/L unless otherwise specified.

TABLE 19.8. Reclaimed Wastewater Utilization Rate in 1994.

Irrigation Scheme	Equipped Area (ha)	Irrigated Area (ha)	Pumped Volume (Mm³)	Pumped Vol./ Reclaimed Vol. (%)
Soukra	615	515	3.6	28
Cebala	3200	1250	12.5	30
Mornag	1047	459	1.3	20
Nabeul	356	266	0.9	35
Hammamet	145	145	0.6	54
Sousse Nord	80	80	1.1	24
Sousse Sud	205	205	1.0	24
Sousse-Monastir	170	170	2.0	62
Moknine	100	60	0.3	21
Kairouan	120	120	0.6	16
Sfax	340	240	3.7	47
Total	6378	3510	27.6	33

schemes irrigated with reclaimed water. Among the most important are:

- the quality of the reclaimed effluents and more particularly their salt content and the variability of their biological composition
- the lack of storage facilities for reclaimed water to meet peak demands, and
- the public acceptance due to the crop restrictions

Some other constraints may be reported such as the inadequate planning of the wastewater reuse projects, the reliability of the distribution system, the insufficient resources to monitor and control reclaimed water effluents and products, the insufficient medical control of the agricultural workers, the land tenure, some equipment and management problems, and the inadequate education and training for farmers and extension services.

As a consequence, the rate of utilization of reclaimed water remain relatively low compared to the potential (Table 19.8). This low percentage can be explained by three main reasons. The first is that wastewater is mainly reused in the agricultural sector which requires water applications only 6 months per year. The second is that, as no reclaimed water can be stored, the irrigation rate is limited to that of the pumps withdrawing treatment plant effluents. Finally, only 55% of the area equipped to use reclaimed water is irrigated. This last point is related to the rate of utilization of irrigation water in general: farmers are taking steps to shift from rainfed to irrigated crops and more particularly to reclaimed water irrigated crops. On the contrary, the utilization rate of reclaimed water is high for golf course irrigation: around 2.7 Mm³ yr⁻¹ are used at La Soukra, Hammamet,

Sousse, and Monastir golf courses. Golf courses are spray irrigated at night, once it is closed to the public.

Assessing the reclaimed water market is a basic step of a project design and may be part of the explanation of the low rate of water reuse. As stressed by Asano and Mills [14], finding potential customers who want and know how to use reclaimed water should be a key task in a planning process. Awareness of this necessity changed the approach to wastewater reuse in Tunisia. In 1992, ONAS launched an analysis of the market of wastewater and residual sludge. It was an attempt at a real commercial prospect to find means, e.g., additional treatments, to satisfy the identified needs under the best possible economic conditions [15].

On the other hand, the remaining amount of effluents not reused is still discharged into the receiving environment (sea, wadis, lakes, and sebkhas). Sea outfalls have already been made for Sousse (1.7 km length, 14–15 m depth), Jerba (1.1 km length, 8–10 m depth), Hammamet (1.6 km length, 23 m depth), and Monastir (1.5 km length, 4.5 m depth). This way of disposal should be considered as a temporary one, since the Mediterranean sea is already under the threat of pollution and because of inland water shortage problems, on the one hand, and of the occurence of coastal marine environmental problems, on the other hand.

So, ways of upgrading wastewater treatment plants should be anticipated in order to improve secondary effluent quality. Different processes such as stabilization ponds, surface or underground storage, wetlands, etc., could be screened depending on the site, the water quality, and other factors, to achieve more economical pollution abatement.

A better coordination among agencies and Ministries may improve the planning, implementation and follow-up of wastewater reuse operations as well as community participation programs at the different steps of project developments.

RESULTS OBTAINED THROUGH APPLIED SCIENTIFIC RESEARCH

A research program co-financed by UNDP (OPE/RAB/80/011) was undertaken from 1981 to 1987 by the Research Center for Rural Engineering (CRGR) within the Ministry of Agriculture. The aim was to determine the conditions to use reclaimed water in agriculture taking into account their composition, soil types, different crops, and sanitary aspects. The results were published in a final report [6]. Research is still continuing. A brief review of the results to date is presented here.

Research is conducted by a multi-disciplinary team (soil physics, water-soil chemistry, plant physiology, water-soil-plant bacteriology, and agronomy). Close contacts and fruitful cooperation have also been set up with offices within various ministries (Agriculture, Public Health, Environment and Land Management).

The methodology which has been adopted is to a major extent studies of real field conditions. Tests carried out on experimental stations have, however, also included systematic laboratory analyses. The experiments have been run on waters, soils, and crops.

A national survey of wastewater composition was conducted for chemical and biological constituents from 15 biological treatment plants located across Tunisia to provide a basis for establishing water quality uses and management for crop production and environment protection.

For experimental purposes, reclaimed water mainly from the activated sludge treatment plants of Tunis (Cherguia) and Nabeul (SE_4) has been used; groundwater has been applied for control experiments; monitoring of the aquifers has been done in order to study the effect of reclaimed water application.

Experimental facilities have been set up at La Soukra and Oued Souhil in order to answer the following questions:

- What are the effects of the use of reclaimed water on soil, crop, and groundwater?
- What are the crops that might be produced under reclaimed water applications and what are the chemical and microbiological levels of contamination?
- What recommendations can be formulated for agricultural use of this resource considering the results acquired?

The achievements of the research program have been twofold: applied research results and training of specialists and technicians. The results concerning the first point are presented here.

The experiments have led to the set-up of specific methodologies, especially for the physical, chemical, and biological analyses and tests carried out in the field. This acquired knowledge, specific to each field, concerns particularly sampling techniques and laboratory analyses. Due to this, many methods used in chemistry, microbiology, and plant physiology have been adopted or partially modified.

PHYSICAL-CHEMICAL AND BIOLOGICAL COMPOSITION OF WASTEWATER

Chemical elements, bacteria, and parasitic content of raw and reclaimed wastewater discharged from different wastewater treatment plants (La Cherguia, Choutrana, Côtière Nord, Sud Méliane, Radès, Nabeul (SE_3 and SE_4), Hammamet (SE_1 and SE_2), Kairouan, Sousse

Nord and Sud, Monastir, Sfax, and Gafsa) and reused in agriculture have been monitored.

Chemical composition of reclaimed effluents varied from one plant to the other depending on the supplied water quality, the seepage of brackish/sea water into the network, the wastewater treatment plant location, the proportion of industrial water in the influent, and finally of the treatment process employed. The composition was characterized by a moderate salinity and sodicity for most of the treatment plants except Côtière Nord, Moknine, and Sfax where they were particularly high. This means, consequently, soil salinization risks. Alkalinization risks may not be important because of the high concentration of Ca and the elevated electrical conductivity of the effluents. Reclaimed wastewater had a high variability of the organic parameters which, for N and P, may be a constraint for fertilization purposes, and a low trace elements content (Table 19.9) [9]. The salt content in wastewater was not affected by the treatment process while total nitrogen removal efficiency was 48% (62% for ammonium) and that of total phosphorus was 63%. For trace elements, the removal averaged: Zn 87%; Cu and Fe 78%; Pb 64%; Cr 50%; Mn 44%; Co and Cd 17%; and Ni 13%. The analyzed reclaimed water samples were consistent with the Tunisian standards regarding water quality required for agricultural reuse and had a high fertilizing content.

The evaluation of the fertilizing value and polluting load of reclaimed water, compared to the crop's uptake, showed that the fertilizing units (N, P_2O_5, and K_2O) brought by 6000 m^3 ha^{-1} (value corresponding to a mean summer irrigation) of reclaimed water may widely exceed the needs of plant growth for nitrogen and potassium. Over application of potassium is not a limiting factor, but, applications of nitrogen exceeding requirements for crop growth may present some risks for crops and/or groundwaters.

The wastewater concentration of faecal microorganisms found in the influents and effluents was of the same order of magnitude independent of treatment process, except for stabilization ponds. Average faecal coliforms and streptococci removal (log units) was about 0–2 in biological wastewater treatment plants such as activated sludge, around 1–2 in aerated lagoons, and from 2 to 5 in stabilization ponds [16]. The different treatment processes did not result in a complete removal of pathogenic bacteria such as salmonella. In La Cherguia (activated sludge), Choutrana (oxidation ditch) and Côtière Nord (stabilization ponds) influents, these pathogens were found in 60% of the samples. In the effluents, they were found in 15% of La Cherguia samples and 19% in the case of Choutrana; Côtière Nord effluents (stabilization ponds) were free of these pathogens. Consequently, stabilization ponds, compared to intensive biological treatment

TABLE 19.9. Descriptive Statistics of Average Element Concentration for Effluents from 15 Wastewater Treatment Plants in Tunisia (in mg L^{-1} unless otherwise indicated).

Parameter	Mean	Median	SD	Min.	Max.	NT
pH	7.6	7.6	0.1	7.5	7.9	6.5–8.5
EC (dS m^{-1})	4.10	3.42	1.68	2.39	8.94	7.0
TDS (g L^{-1})	2.61	2.23	1.08	1.52	5.61	
Ca	168	171	31	121	238	—
Mg	85	72	36	54	188	—
K	52	44	27	18	120	—
Na	537	431	293	293	1438	—
HCO_3	524	473	189	333	1046	—
SO_4	532	484	147	304	922	—
Cl	791	625	526	338	2490	2000
SAR	8.5	7.2	3.8	5.1	17.6	—
SS	42.5	20.0	47.9	14.7	190.9	30[a]
COD	173.6	103.2	152.7	61.4	639.5	90[a]
BOD_5	35.3	26.8	18.4	17.8	69.8	30[a]
Nk	30.0	26.5	11.0	16.9	53.1	—
NH_4-N	26.2	23.7	10.6	14.4	48.3	—
NO_3-N	9.5	7.2	6.0	2.1	23.2	—
NO_2-N	2.48	1.80	2.19	0.52	8.89	—
P	3.60	3.30	1.63	1.61	6.54	—
PO_4-P	2.34	1.90	1.08	1.23	4.34	—
Cd	0.005	0.005	0.001	0.004	0.008	0.01
Co	0.019	0.016	0.006	0.012	0.031	0.1
Cr	0.016	0.016	0.004	0.009	0.023	0.1
Cu	0.017	0.016	0.004	0.011	0.025	0.5
Fe	0.226	0.181	0.115	0.108	0.511	5
Mn	0.054	0.048	0.025	0.022	0.112	0.5
Ni	0.034	0.033	0.009	0.021	0.049	0.2
Pb	0.044	0.042	0.008	0.035	0.066	1
Zn	0.036	0.033	0.011	0.023	0.063	5

[a]Except special authorization.

SD: standard deviation; NT: Tunisian standards; EC: electrical conductivity (25°C); TDS: total dissolved solids; SAR: sodium adsorption ratio; SS: suspended solids; COD: chemical oxygen demand; BOD: biochemical oxygen demand; Nk: Kjeldahl nitrogen; P: total phosphorus.

systems, may produce a higher bacterial effluent quality [17] consistent with WHO guidelines [18].

Eggs and cysts of parasites may be found in raw wastewater. Within the group of intestinal helminths, eggs of *Ascaris lumbricoides, Ancylostoma duodenale, Taenia* (spp.), *Trichuris trichiuria, Enterobius vermicularis,* and *Hymenolepis nana* were present [17]. As to pathogenic protozoa, *Entamoeba histolytica* was often present; *Giardia intestinalis,* on the contrary, has rarely been isolated. Reclaimed wastewater had a certain parasitic load which depended on the treatment process. Stabilization pond effluents were free of parasites.

RECLAIMED WATER IMPACTS

Soil Physical Properties, Chemical Composition, and Bacterial Contamination

Application of reclaimed water on La Soukra and Oued Souhil experimental stations, whose soils were alluvial and sandy-clayey to sandy, has led to little modifications of their physical properties: bulk density of the soil surface layer (0–5 cm) was the same whatever the irrigation water quality (reclaimed water or groundwater) was.

Irrigation with reclaimed water did not affect the soil trace element content but, on the contrary, it affected the chemical composition of the La Soukra clayey-sandy soils [19]. An increase in the electrical conductivity and in the sodium adsorption ratio and a change in the soil solution composition (bicarbonate-calcic → sodium-chloride) were noticed. This increase was, in this case, within the usual range found when using conventional water.

Concerning the bacterial level of contamination, no significant effects of reclaimed water application were recorded on the soil over two years. Microorganisms die-off in the soil was shown to be sensitive to factors such as the climate (temperature, insolation, precipitation, and relative humidity), soil texture, and plant canopy [6].

Salinity Concerns

Because of the extent of the salt problem, salinization and alkalization risks due to the use of brackish water for irrigation purposes have been extensively studied in Tunisia and various but complementary approaches have been adopted to define the range for saline water use and evaluate the consequences for the soil as well as for the plant. Since several years, research has been conducted on how to improve the agricultural productivity in salt-affected environments. The results obtained by the Research Center for Utilization of Saline Water in Irrigation–Rural Engineering Research Center [20] about utilization of saline water in irrigation constitute a major step. This work and its further developments [21,22] especially allowed to formulate, under the cultural conditions of Tunisia and at agricultural scale, standards for the utilization of saline water for irrigation, to study the salinity development and the crop yields on soils irrigated with different water qualities and according to various irrigation regimes taking into account the climate, the soil, the crop type, and the irrigation and drainage management methods. Plant salt tolerance was studied for a certain salinity range (0.2–4.0 g L^{-1}).

When brackish water is used for irrigation, several geochemical reactions occur before chemical composition and accumulation of salt reaching a steady-state. Precipitation and dissolution of mineral salts as well as exchange reactions take place when a new source of irrigation water is introduced. In Tunisia, soil solutions generally follow the neutral saline pathway and reclamation of salt-affected soils may take place through leaching, drainage, and soil tillage operations without gypsum amendment. Transient time to salt steady state may last a certain number of years because of the calcium-saturated conditions of the different solutions. Experiments conducted on silty clay loam soils with a shallow groundwater table, tile drained (1.5 m), and irrigated since 1964 with brackish water (EC = 2.87 dS m^{-1}, SAR = 6.2) from the Medjerda River showed that irrigation water, soil solution, drainage water, and groundwater were in a dynamic equilibrium evolving toward a salt steady state, 26 years after that irrigation started [23].

Plant Growth, Yield, and Chemical Composition, Water Consumption, and Bacterial Contamination

Tests were conducted during four years at La Soukra experimental station on two species, a fodder crop, sorghum (*Sorghum vulgare*) and a vegetable, pepper (*Capsicum annuum*), and were compared to a control irrigated with well water; the first was flood-irrigated, the second was furrow-irrigated. At Oued Souhil (Nabeul), clementine and orange trees were tested. Irrigation with reclaimed water had a favourable effect on the growth of sorghum and led to a significant increase of nitrogen, phosphorus, and potassium contents in plant tissues [24] with, however, a lower efficiency in the use of these elements compared to mineral fertilizing [25]. No effect was noticed for pepper.

Fertilizing element (N, P, and K) uptake differed with the crop species but remained low compared to the soil residual load entailing nitrate groundwater pollution risks. Concerning trace elements, only iron and zinc concentrations in sorghum leaves increased significantly without reaching thresholds values; same was observed for boron [26]. Similar studies were conducted on maize, alfalfa, and barley. The use of reclaimed water compared to irrigation with groundwater resulted in higher annual and perennial crop yields.

In pot experiments, nitrogen, phosphorus, and potassium efficiency of reclaimed water was compared to mineral fertilizing (ammonium nitrate) [25]. Results showed that global effects of effluents were less important than that of mineral fertilizers; however, compared to a non-fertilized control, wastewater application led to a higher dry matter production. Therefore, irrigation with reclaimed water has to be considered as a complementary fertilization that has to be taken into account when the fertilizer amount to apply is to be evaluated. Water consumption of citrus trees did not depend on the water quality [6].

Tests conducted during two years showed that citrus fruits produced on plots irrigated with reclaimed water and not in contact with the soil were free of faecal germs. Contamination level of citrus fruits picked up on the soil of plots irrigated with reclaimed water was significantly higher than that of fruits picked from the trees [6].

As to annual crops, experiments were made first to evaluate the bacterial quality of fodder crops irrigated with effluents and to compare them to non-irrigated fodder crops (winter crops) or to fodder irrigated with groundwater [27]. The second study considered how to lower the contamination level of the fodder irrigated with reclaimed water and improve their microbiological quality. Results showed that faecal contamination of forage crops sampled from control plots was not equal to zero. This contamination, due to natural factors, was higher for summer forage crops. Bacterial quality of forage crops irrigated with reclaimed water depended on the crop specie, the number of days since the last irrigation occured and the climatic conditions. For both sorghum and alfalfa, seven to ten days were required between the last irrigation and the cutting to achieve natural decontamination. The natural die-off of faecal microorganisms on sorghum plants was quicker in summer than in autumn. As to pepper, tests did not show any particular fruit contamination.

Chemical and Bacterial Groundwater Composition

Besides reclaimed water reuse for agricultural purposes, seasonal recharge of the shallow and sandy aquifer of Nabeul has been performed since 1985. Activated sludge effluents that were not used for irrigation during winter season were infiltrated and stored into the aquifer, thus increasing the volume farmers can pump during summer season to irrigate citrus orchards. Artificial groundwater recharge is operated at experimental scale in infiltration-percolation basins and in the river bed in the Cap Bon area. This reuse option allowed an underground storage and an additional treatment stage as wastewater slowly infiltrated through the unsaturated zone [28,29]. However, no clear conclusion could be drawn about the effect of irrigation with reclaimed water on the bacterial and chemical composition of shallow groundwater since the initial contamination level of most of the wells was relatively high and subject to seasonal variations. According to the *state of the art* of soil-aquifer treatment [30,31], improved operation of this facility would lead to a groundwater quality meeting unrestricted irrigation requirement [32]. This subject is still under study and other recharge sites should be included in a survey and monitoring network.

Long-term Effects

Finally, investigations in La Soukra's perimeter which has been irrigated for more than twenty years with reclaimed water have been conducted. The aim of this investigation was long term effects of irrigation with reclaimed water. The results did not show notable effects on soils, crops, or groundwater. A study conducted in this scheme to evaluate health impacts of reclaimed water reuse could not set up a clear cause-effect relationship between the observed diseases and the reuse practice.

PROSPECTS FOR IMPROVING THE WATER REUSE POLICY

Because of the water shortage, enforcement of the wastewater reuse policy will go on. Forthcoming and already implemented projects should allow a higher utilization of reclaimed water. The wastewater market assessment launched by ONAS in 1992 pictured farmers attitudes and wishes with regard to wastewater reuse and helped identifying some measures that would lead to an increase in the rate of wastewater reuse.

The survey pointed out a lack of information amongst farmers about wastewater quality, health risks related to wastewater reuse and impacts on crops and soils. As stated by Wegner-Gwidt [33], winning the support of farmers should be part of the planning and the management of wastewater reuse projects. This means more information and more involvement of the farmers in the decision making process to ensure the success of the projects.

The absence or insufficiency of storage was noticed [34]. Building reservoirs in which a few days effluent production can be stored would allow matching the daily variations of irrigation water demand and increase the reliability of the water supply. As demand for irrigation water is mainly during the dry season, seasonal storage during the non-irrigation period would allow increasing reclaimed water reuse and prevent coastal waters contamination. Reclaimed water can be stored in deep reservoirs or in phreatic aquifers. In both cases, storage can result in a dramatic improvement of reclaimed water quality. Storing secondary effluents in maturation ponds would upgrade the water quality to meet the WHO guidelines for unrestricted irrigation [35]. Effluent storage in deep reservoirs may limit evaporation losses, accumulates water for irrigation and also performs a

TABLE 19.10. **Average Costs of Water Saving Measures.**

Water Saving Measures	(DT m^{-3})
Large dams height raising	0.085–0.660
Interregional transfers (average)	0.149
South	0.234
Center	0.181
West	0.037
Reclaimed wastewater storage	0.167
Brackish water desalination	0.288
Sea water desalination	1.400
Urban water supply network rehabilitation	1.400
Agricultural sector (average)	0.578
distribution network modernization	0.781
equipment rehabilitation	0.155
localized irrigation systems	0.506
other irrigation systems	0.406
Industrial wastewater recycling	0.168

1 DT ≈ 1 US$.

polishing treatment, reducing BOD, COD, and contents in SS, phosphorus, nitrogen, and pathogen indicators. Performing coastal aquifer recharge, where the hydrogeological context is favourable, would make wastewater reuse well accepted by farmers.

Seasonal storage of secondary effluent in deep reservoirs is under consideration as a long term perspective. An economic evaluation of the water saving technical measures which might reduce the water consumption or increase the water production was made (Table 19.10) [3]. The planning period considered for investment and operation costs of the different measures was 20–40 years. As the mobilization rate increases, costs of water saving measures increase. According to Agrar et al. [3], storing reclaimed water is among the cheapest methods for increasing water resources of Tunisia.

Cultivating food crops, particularly market garden crops, in the vicinity of towns is very attractive. Removing the restrictions on irrigated crops is expected to help farmers moving from rainfed to irrigated crops. One major obstacle to the water reuse development would thus be overcome. The removal of restrictions demands that two requirements are fulfilled: new regulations should be layed down and effluent disinfection treatments complying with these regulations have to be set up. A technical-economical analysis concluded that, given the local conditions, maturation ponds should be preferred to ultraviolet light, chlorination and filtration as the treatment to meet the WHO's criteria, as far as land is available.

The wastewater market assessment concluded that farmers would accept prices ranging from 0.014 to 0.040 US$ m^{-3}, according to the type of price rate (per hectare or per cubic meter), the water quality and the water scarcity. These prices are not that different from the current ones, which vary from one scheme to another. Therefore, some of the operation costs are covered.

Golf course irrigation means a high rate of water reuse and a water demand that lasts all the year long through varying climatic conditions. Supplying reclaimed water to golf courses, green belts, and hotel gardens would result in an optimization of both investment and operation costs. These users are never very far from the treatment plants. They are big consumers and customers likely to be able to pay a price that would allow recovering operation and maintenance costs. Polishing secondary effluents through lagooning or seasonal storage would lower health hazards and contribute to increase this demand of reclaimed water.

Research on reclaimed water use has also to develop in order to minimize the risks related to their utilization and to evaluate long-term effects on the soil-plant-groundwater-human system.

CONCLUSION

A lot of achievements have been made in the field of water resources planning and management in Tunisia. Wastewater reuse is considered as a part of Tunisia's overall water resources balance. It has been made an integral part of the environmental pollution control and water management strategy. Most of the reclaimed water produced in Tunisia will be used in agriculture. Many projects expanding the areas irrigated with reclaimed water are under implementation.

The different studies conducted and summarized here have determined the feasibility of the reclaimed water use in irrigation and have confirmed the agronomic advantages of water application. Thus, irrigated soils with reclaimed water may constitute a complementary treatment stage.

Different actions and measures which would contribute to the strengthening of the wastewater treatment and reuse policy have been identified of which: improvement of the water quality (tertiary treatment) and reclaimed water storage; 20 000 ha may be irrigated with 200 Mm^3 of reclaimed water and groundwater recharge operations with reclaimed wastewater may be conducted whenever it is feasible. Sanitary precautions would, however, have to be considered in case of groundwater use for domestic water supply.

The institutional and legal framework of wastewater reuse in agriculture have been set up to organize the treatment and the distribution of reclaimed water for irrigation, to supervise the Water Code and other enactments application and to control sanitary aspects.

During the last decades, Tunisia has gained experience in the field of wastewater reclamation for irrigation of agricultural crops and green areas. The scientific and technical results acquired are of interest to the Maghrebian as well as the Mediterranean climate region. Still, however, research efforts have to be made to improve the quality of the effluents in order to expand the range of crops and irrigation methods. Use of reclaimed water requires follow-up by evaluation of long term effects of the elements brought by the water on soil dynamics, plant and groundwater contamination. Periodical wastewater control must be carried out in all wastewater irrigated areas. Risk assessment studies will also have to be conducted on water-soil-plant-animal-human exposure pathways.

As to the financial and economical feasibility of reclaimed water use, it has still to be evaluated taking into account the cost of the different operations, its fertilizing value, and its effects on crop production and on the environment.

REFERENCES

1 Bahri, A. 1993. ''Wastewater reuse in Tunisia,'' In: *Wastewater Reuse, Synthesis and Mediterranean Experiences, Methodology and Case Studies,* International Office for Water, Sophia Antipolis, pp. 187–204.

2 Direction Générale des Ressources en Eau, 1990. *Stratégie pour le développement des ressources en eau de la Tunisie au cours de la décennie 1991–2000,* Water Resources General Directorate, September 1990, 30 p. + annexes.

3 Agrar- und Hydrotechnik GMBH, GKW, Coyne and Bellier, and Centre National d'Etudes Agricoles. 1993. *Economie d'Eau 2000. Stratégie de Gestion d'Eau,* July 1993, 152 p. + annexes.

4 Seventh Economic, Social, and Development Plan (1987–1991), 1987.

5 Agrar- und Hydrotechnik GMBH, GKW, Coyne et Bellier, and Centre National d'Etudes Agricoles. 1993. *Economie d'Eau 2000.* Rapport de Synthèse, July 1993, 60 p.

6 United Nations Development Programme. 1987. *Reclaimed wastewater reuse in agriculture,* Vol. 1, Technical Report (in French), 3rd Part, 69 p. + annexes, Project RAB/80/011, Water Resources in the North African Countries, Rural Engineering Research Center (Tunisia)—United Nations Development Programme, May 1987.

7 Bahri, A. 1993. *Reclaimed wastewater reuse in Tunisia,* Rural Engineering Research Center Report (in French), May 1993, 33 p. + annexes.

8 National Sanitation Agency. 1993. *Annual report on wastewater treatment plants operation* (in French), Department of Organization and Coordination, Technical Directorate, National Sanitation Agency (ONAS), 55 p.

9 Bahri, A. 1995. ''Fertilizing value and polluting load of reclaimed water in Tunisia,'' (submitted to *Water Research*).

10 Bahri, A. 1988. "Present and future state of treated wastewaters and sewage sludge in Tunisia," *Proceedings of the Regional Seminar on Strengthening the Near East Regional Research and Development Network on Treatment and Reuse of Sewage Effluent for Irrigation,* Food and Agriculture Organization/United Nations Development Programme/The World Bank, December 11–16, 1988, Cairo, Egypt, pp. 53–58.

11 Journal Officiel de la République Tunisienne. 1989. Decree No. 89-1047 regulating reclaimed water reuse for agricultural purposes, July 28, 1989.

12 Institut National de la Normalisation et de la Propriété Industrielle. 1989. Environment Protection—Use of reclaimed water for agricultural purposes—Physical, chemical, and biological specifications (in French), Tunisian standards, INNORPI, NT 106.03.

13 Institut National de la Normalisation et de la Propriété Industrielle. 1989. Environment Protection—Effluent discharge in the water bodies—Specifications relative to discharges in the marine environment, hydraulic environment and in the sewers (in French), Tunisian standards, INNORPI, NT 106.002.

14 Asano, T. and Mills, R. A. 1990. "Planning and analysis for water reuse projects," *Journal, American Water Works Association,* 82 (1), pp. 38–47.

15 National Sewerage and Sanitation Agency. 1992. "Etude de la valorisation et de la commercialisation des eaux usées et des boues résiduaires—Termes de référence." 58 p.

16 Trad-Raïs, M. 1992. "Efficacité bactériologique des principaux procédés de traitement des eaux usées urbaines," *Archs. Inst. Pasteur Tunis,* 69 (3–4), pp. 273–282.

17 Trad-Raïs, M. 1989. "Surveillance bactériologique et parasitologique des eaux usées brutes et traitées de la ville de Tunis," *Archs. Inst. Pasteur Tunis,* 66 (3–4), pp. 275–287.

18 World Health Organization. 1989. *Health guidelines for the use of wastewater in agriculture and aquaculture,* Report of a WHO Scientific Group, Geneva, Switzerland.

19 Bahri, A. 1987. "Utilization of treated wastewaters and sewage sludge in agriculture in Tunisia," *Desalin.* 67: 233–244.

20 Research Center for the Utilization of Saline Water in Irrigation. 1970. *Research and training on irrigation with saline water.* Tech. Rep. 1962–1969. UNDPUNESCO/CRUESI-Tunis 5. Paris. 256 p. + appendixes.

21 Gallali, T. 1980. Transferts sels-matière organique en zones arides méditerranéennes. Contribution à l'étude pédo-biologique des sols salsodiques formés sur matériaux sédimentaires. State Doc. Thesis—Nancy I University. 202 p. + annexes.

22 Bahri, A. 1982. Utilization of saline waters and soils in Kairouan Valley, Tunisia, Doc. Eng. Thesis (in French), INP—Toulouse. 102 p. + annexes.

23 Bahri, A. 1993. Salinity evolution in an irrigated area in the Lower Medjerda Valley in Tunisia (in French), *Science du Sol,* 31 (3):125–140.

24 Rejeb, S. 1992. "Irrigation d'un sudangrass avec des eaux usées traitées. I. Effet fertilisant," *Fourrages,* 130, pp. 171–179.

25 Rejeb, S. 1993. "Comparaison de l'utilisation de l'azote des eaux usées traitées et de l'ammonitrate par le sorgho fourrager (Sorghum sudanense)," *Discovery and Innovation,* 5, 4, pp. 347–353.

26 Rejeb, S. 1992. "Irrigation d'un sudangrass avec des eaux usées traitées. II. Risque de contamination par les micro-éléments," *Fourrages,* 130, pp. 181–190.

27 Trad-Raïs, M. 1991. "Contamination bactérienne des fourrages irrigués avec les eaux usées traitées," *Ann. INRAT,* 64, 12, 16 p.

28 United Nations Development Programme, 1987. *Reclaimed wastewater reuse in agriculture, Artificial groundwater recharge* (in French), Vol. 1, Technical Report, 4th Part, 19 p., Project RAB/80/011, Water Resources in the North African Countries, Water Resources Directorate (Tunisia)—United Nations Development Programme, May 1987.

29 Rekaya, M. 1991. "Role of the General Directorate of Water Resources," Colloquium on "Institutional and organisational aspects of reclaimed water reuse," Ministry of Agriculture— United Nations Development Programme—World Bank, November 1–2, 1991, Tunis, Tunisia, 4 p.

30 Bouwer, H. 1991. "Role of groundwater recharge in treatment and storage of wastewater reuse," *Wat. Sci. Tech.* 24 (9), pp. 95–302.

31 Brissaud, F. and Salgot, M. 1994. "Infiltration percolation as a tertiary treatment," *Hydrotop 94,* April 12–15, 1994, Marseille, France, Vol II. pp. 391–398.

32 Rekaya, M. and Brissaud, F. 1991. "Recharge de la nappe de l'Oued Souhil par des effluents secondaires," *XXI° J. de l'Hydraulique: Les eaux souterraines et la gestion des eaux,* January 29–31, 1991, Sophia Antipolis, France, pp. II.4.1–II.4.8.

33 Wegner-Gwidt, J. 1991. "Winning support for reclamation projects through pro-active communication programs," *Wat. Sci. Tech.* 24 (9), pp. 313–322.

34 Bahri, A. and Brissaud, F. 1995. "Wastewater reuse in Tunisia: Assessing a national policy," *Proceedings of the Second International Symposium on Wastewater Reclamation and Reuse,* October 17–20, 1995, Iraklio, Crete (Greece), pp. 103–110.

35 Pearson, H. W. and Mara, D. D. 1992. "Wastewater treatment and storage for reuse in coastal regions," *Waste Water Management in Coastal Areas,* March 31st–April 2nd, 1992, Montpellier (France), pp. 269–276.

CHAPTER 20

Water Recycling in
Los Angeles County

THE SANITATION DISTRICTS

THE Sanitation Districts of Los Angeles County, formed in 1923 by an act of the California State legislature have a long history as pioneers in the field of water recycling, culminating in one of the most advanced and widespread programs for the treatment, distribution and reuse of reclaimed water. The Districts now provide sewage collection and treatment for over 5 million people and over 5,000 diverse industries in 79 cities and unincorporated areas in Los Angeles County, California (Figure 20.1), essentially all of the populated areas outside of the City of Los Angeles.

In the urbanized areas of the Los Angeles Basin, the Districts have built an integrated network of facilities known as the Joint Outfall System (JOS) which consists of:

- 1,200 miles (2,000 kilometers) of large diameter, trunk sewers.
- the Joint Water Pollution Control Plant (JWPCP), a 385 million gallon per day (16.87 cubic meters per second) wastewater treatment plant. The JWPCP provides advanced primary and partial secondary treatment, with the effluent being disinfected and discharged to the Pacific ocean through a deep ocean outfall. The facility discharged 330.8 MGD (14.49 cms) of wastewater in 1996.
- six water reclamation plants (WRPs) with a total capacity of 190.7 MGD (8.35 cms). Five of these facilities provide filtration and

Earle C. Hartling and Margaret H. Nellor, County Sanitation Districts of Los Angeles County, 1955 Workman Mill Rd., P.O. Box 4998, Whittier, CA 90607.

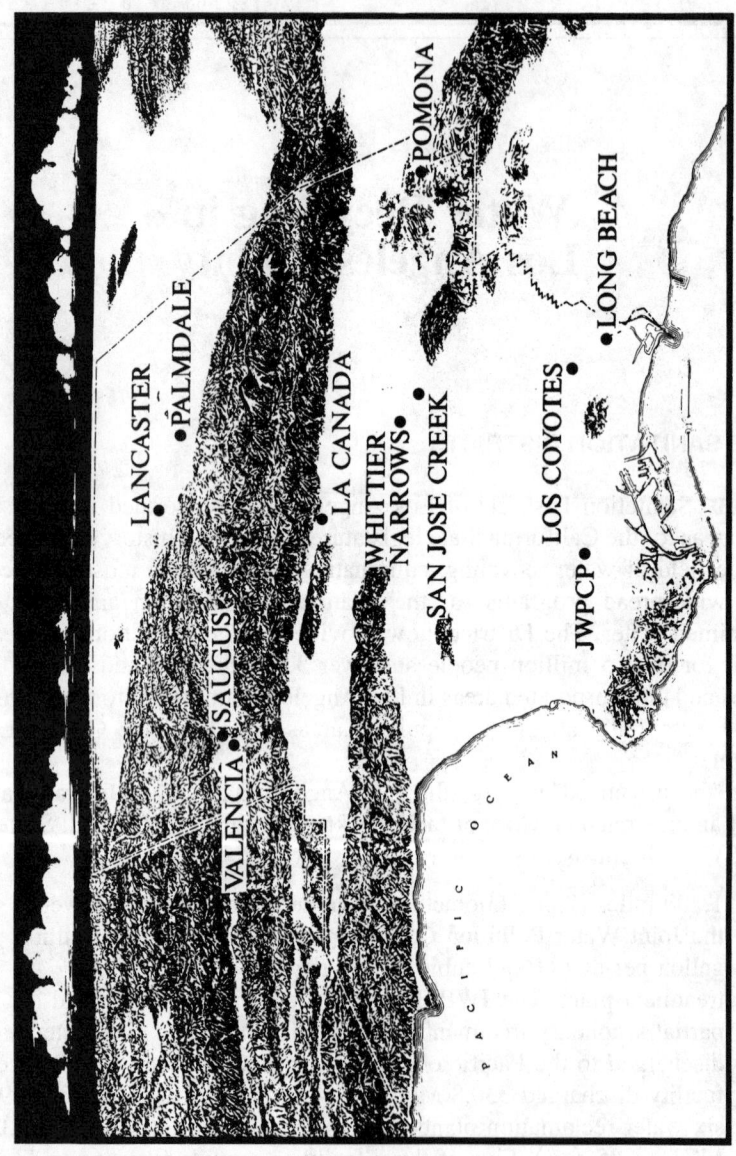

Figure 20.1 Sanitation districts treatment plant locations.

chlorination after secondary treatment. The sixth facility provides secondary treatment and chlorination. During 1996, the plants treated 156.1 MGD (6.84 cms) of wastewater.

Outside of the Los Angeles Basin, sewage treatment is provided for the communities of Santa Clarita, Lancaster and Palmdale. The four WRPs serving these areas produced over 32.2 MGD (1.41 cms) of effluent in 1996.

Within California, the powers and authorities of county sanitation districts are defined under §4700 *et seq.* of the State Health and Safety Code. Accordingly, such districts have the authority to sell reclaimed water, but other statutes have been interpreted to limit their power to provide distribution systems in areas served by other water suppliers. Thus, the Districts have established their role to include producing the reclaimed water, promoting its use, and cooperating with other entities who distribute the water to retail customers. In some cases this role may change slightly. For example, depending on the beneficial use of the receiving water, reclaimed water produced by the Districts is directly provided for ground-water replenishment or maintaining wildlife areas. A few irrigation sites adjacent to Districts' WRPs are directly served. However, the vast majority of irrigation and other reuse sites are served via an intermediary such as a city water department or municipal water district. In practice, the water supply entity must build and operate the transmission and distribution systems, while the Districts contract to produce, sell and, sometimes, pump the reclaimed water to the retailer.

Water reclamation's potential in Los Angeles was recognized as early as 1949 when a Districts' study detailed most of the features incorporated in today's reclamation program:

- the construction of WRPs, incorporating existing, proven treatment technology, along the Districts' sewer system to provide hydraulic relief for the sewerage system and highly treated wastewater suitable for reuse: Economy of scale is achieved by operating these facilities under one agency. For the JOS facilities, economy of scale is also achieved by using the solids handling facilities at the JWPCP as a central processing facility for the WRPs instead of constructing and operating such facilities at each WRP.
- locating the WRPs upstream of the more heavily industrialized areas to treat mostly residential sewage, producing a higher quality effluent: To further improve effluent quality from the reclamation plants, an industrial waste pretreatment program has been implemented to prevent toxic wastes from entering the WRPs. In addition, the Districts' sewer interceptors have been designed to provide sufficient flexibility to bypass industrial waste around the plants as needed.

- the high quality of reclaimed water produced at the WRPs, allowing for broad reuse applications including agricultural and landscape irrigation, manufacturing, construction and industrial cooling, environmental enhancement, recreational activities, and groundwater replenishment.

During 1996, 60.9 MGD (2.67 cms), or 39% of the 156.1 MGD (6.84 cms) of reclaimed water produced in the JOS, was reused. In the outlying areas, 10.1 MGD (0.44 cms), or 31% of the 32.2 MGD (1.41 cms) of reclaimed water produced, was actively reused during the same time period.

WATER SUPPLY CONSIDERATIONS

Perhaps the greatest motivation to use reclaimed water is the fact that Los Angeles is essentially a desert, with a long-term average rainfall of only 15 inches (38 cm) per year and with no major rivers within 100 miles (160 km). Approximately two-thirds of the area's annual water supply is imported through three aqueducts that extend between 200 and 500 miles (320 and 800 km) from the Los Angeles Basin. The delivery capability of each aqueduct is subject to legal, political, operational and climatological factors.

- The Los Angeles City Department of Water and Power's (DWP) groundwater pumping in the Owens Valley has been halted due to the adverse environmental effects resulting from the lowered water table. Diversions from streams feeding Mono Lake have been voluntarily curtailed in order to allow water levels in Mono Lake to rise so that ecosystem can recover. More recently, the air quality authority of Inyo County has approved a plan that will require the DWP to divert some of the Sierra Nevada stream flow feeding its aqueduct to rewet a portion of Owens Lake, to reduce the amount of dust and salt becoming airborne and affecting the respiratory systems of the local inhabitants.
- The Metropolitan Water District of Southern California's (MWD) annual diversion of approximately 1.2 million acre-feet (1.5 billion cubic meters per year) from the Colorado River is expected to be cut by more than half as the Central Arizona Project continues to increase diversions to agricultural and urban areas of Arizona through the 1990's and into the next century. Looming as a future competing interest for this water supply are the increasing water needs of Las Vegas, Nevada, the fastest growing metropolitan area in the country.

- The California State Water Project currently only has facilities sufficient to supply half of its ultimate capacity of water from the state's main watershed, the Sacramento Delta. The defeat of the Peripheral Canal voter initiative in 1982 blocked construction of the remaining necessary facilities that would have brought this system to full capacity. Lack of precipitation and the resulting reduced runoff in 1987–92 prompted reductions in water deliveries to southern California by up to 80 percent. Environmental concerns over effects of water diversions on the delta's wildlife may eventually make such reductions in water diversions permanent. Urban areas, such as southern California, are also in competition with Central Valley agriculture for this water supply.

Existing local groundwater supplies are also limited by the lack of local precipitation, recharge capacities of spreading grounds, basin overdrafting, sea water intrusion in coastal areas and industrial contamination. Compounding these threats to the southern California water supply are anticipated population increases. For example:

- Every year the population in the MWD service area increases by another 400,000 people, equivalent to a city the size of Portland, Oregon.
- The United States Census Bureau estimates that the population of the State of California will increase 52 percent, from 31.4 to 47.9 million, by the year 2020.
- The California State Department of Finance's *Population Projections by Race/Ethnicity for California and Its Counties, 1990–2040,* Report 93 P-1, estimates that the population of the State of California will increase 42 percent, from 30 million in 1990 to 42.5 million, by the year 2010.
- The Southern California Association of Governments' *1994 Regional Comprehensive Plan* estimated population growth in the six county area (Los Angeles, Orange, Riverside, San Bernardino, Ventura and Imperial) to be 40 percent, from 14.6 million in 1990 to 20.5 million in 2010.

This anticipated population growth will prompt increased competition for the State's dwindling water resources.

Within the last 20 years, the State of California has been hit by two serious droughts, in 1976–77 and more recently in 1987–92. Mandatory water rationing of at least 20 percent was instituted by water purveyors throughout the state, and, at one point, the California State Department of Water Resources (DWR) anticipated going to a mandatory 50 percent rationing. Only the extremely wet winters of 1993 and 1995 brought the

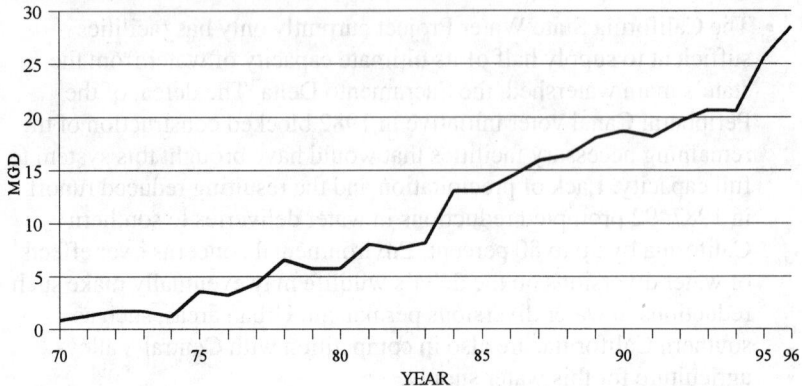

Figure 20.2 Increase in direct nonpotable reuse, 1970–96.

water supply situation in the state out of crisis. However, this relief may be short-lived as tree-ring records indicate that prolonged periods of drought are the historic norm for this region, not the exception.

The State Legislature has long known of the value of water reclamation. Chapter 7 of the California Porter-Cologne Water Quality Act states that, *"the people of the State have a primary interest in the development of facilities to reclaim water containing waste to supplement existing surface and underground water supplies and to assist in meeting the future requirements of the State."* Furthermore, the State Legislature in 1991 officially adopted the goal of reaching one million acre-feet per year of reuse by the year 2010.

Operators of large landscaped areas such as golf courses and parks and water-intensive industries such as those using cooling towers should be well aware of the dire consequences if more draconian conservation measures are ever imposed again. Water shortages have resulted in businesses shutting down and golf courses losing expensive investments in landscaping.

Many heavy water users have since come to the realization that reclaimed water is still a drought-proof supply, despite the drought-induced 10 percent decrease in sewage flow. For example, following the severe 1976–77 drought, several public water purveyors decided to pursue reclaimed water as a supplementary supply to lessen the effects of future water shortages in their service areas. An additional boast for water reclamation came from the most recent drought, which has inspired even more public and private water purveyors to invest in a reclaimed water distribution infrastructure. This is reflected in Figure 20.2, which shows the increase in reclaimed water deliveries through separately constructed distribution systems.

There has been a 40-fold increase in the number of direct non-potable reuse sites, from 9 in 1976, at the beginning of the first drought, to 360 by the end of 1996 (Figure 20.3), with the greatest increases coming just as the second drought began. The total acreage irrigated with reclaimed water has increased nearly nine times, from 940 acres (380 hectares) in 1976 to 8,110 acres (3,282 hectares) by the end of 1996, and usage has increased nearly six times, from 5 MGD (0.22 cms) to 28.49 MGD (1.25 cms).

THE WATER RECLAMATION PROCESS

The treatment process employed by the Districts is essentially the same for seven tertiary treatment plants: five facilities in the Los Angeles Basin JOS and two in the Santa Clarita area (Figure 20.4).

Wastewater entering the plant must first pass through the primary sedimentation tanks, which, over the course of two hours, use gravity and flotation to remove two-thirds of the wastewater settleable and suspended solids. An influent pH meter measures changes in acidity or alkalinity, which could indicate an improperly treated industrial waste discharge, thus allowing the plant operators to take corrective actions before a problem arises in the downstream treatment processes. A chemical polymer is available for dosing the raw wastewater if conditions warrant.

The secondary treatment process is biological in nature, as bacteria aid in the removal of the remaining suspended solids and soluble matter in the primary effluent, converting it to biomass that is subsequently settled out in the final clarifiers. Again, the process is simple in nature. The bacteria are given wastewater as a food source and air is diffused into the aeration tanks to provide oxygen. The plants are equipped with backup air compressors to maintain the flow of air to the tanks and with ceramic bubble diffusers to reduce operating costs and foam production.

A chemical polymer can also be added to the influent end of the final sedimentation tanks to increase solids removal by settling in these tanks. Suspended solids removal by the end of this process is over 95 percent.

This would normally be the end of traditional wastewater treatment, as mandated by the federal Clean Water Act for surface water discharge. However, in order to further protect public health, the equivalent of a raw water treatment plant was constructed at seven of the WRPs. This tertiary treatment process begins when secondary effluent leaving the final clarifiers is dosed with alum (as a coagulant) and chlorine before entering the inert media (either dual-media sand/anthracite coal or mono-media anthracite) gravity filters or pressure filters. These filters are automated

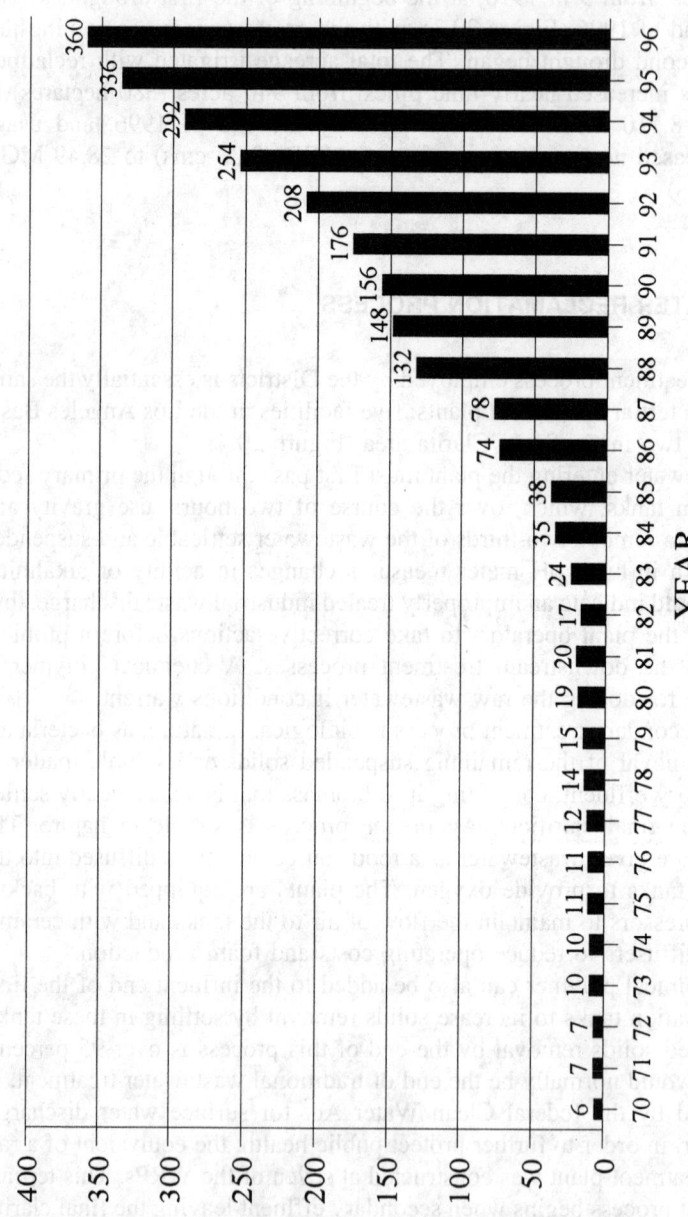

Figure 20.3 Increase in number of reuse sites, 1970–96.

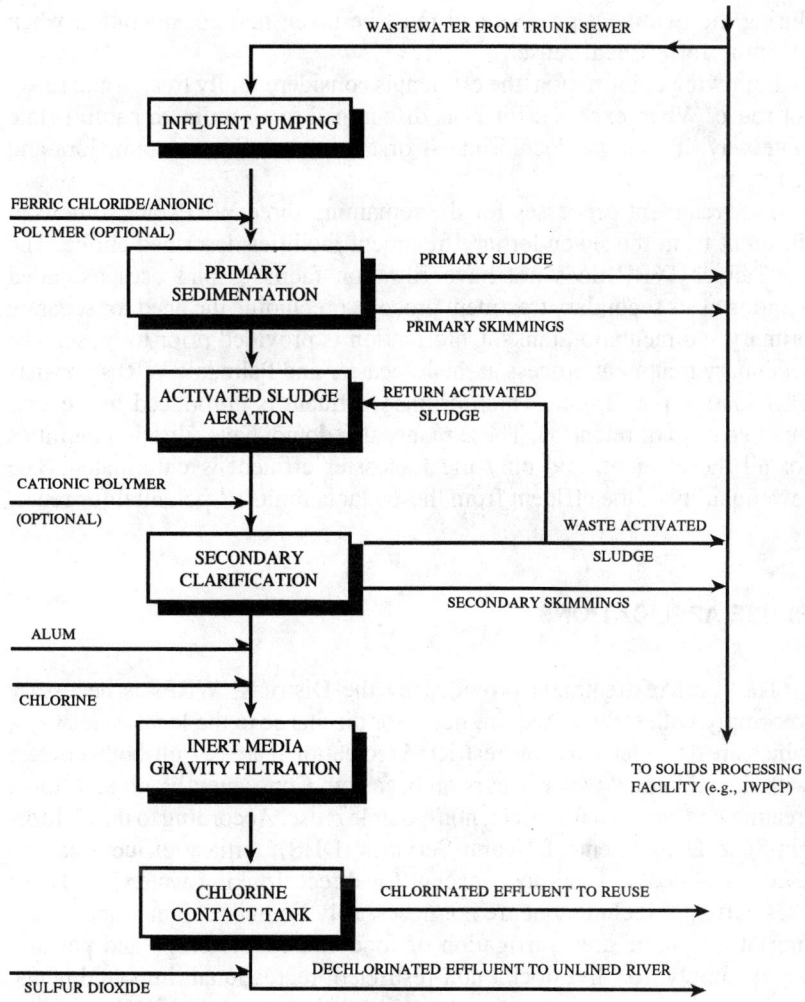

Figure 20.4 Water reclamation process flow schematic.

so that they go into a backwash cycle when they begin to plug with particulate material removed from the wastewater. Filtered effluent is pumped out of the filter underdrain system, then chlorinated a second time before traveling through the chlorine contact tanks for at least 90 minutes. Several residual chlorine analyzers are used throughout the process to ensure that the proper dosage of chlorine is maintained. These three stages of treatment achieve more than 99 percent suspended solids removal. Dissolved solids (i.e., salts), on the other hand, remain relatively unchanged during the treatment process and must be taken into consideration when planning for effluent reuse.

Following chlorination, the effluent is considered fully treated and ready for reuse. When excess effluent is discharged to an unlined, natural-state waterway, it must be dechlorinated first to protect downstream flora and fauna.

The treatment processes for the remaining three WRPs are somewhat different from the seven tertiary treatment facilities described above. The La Cañada WRP does not have filtration facilities and uses extended aeration as its secondary treatment process, precluding the need for separate primary sedimentation tanks. Chlorination is provided prior to reuse. The secondary treatment process at the Lancaster and Palmdale WRPs consists of oxidation ponds, into which primary effluent is introduced for several hundred days of retention. These plants also do not have filtration facilities for all the effluent, and only the Lancaster effluent is chlorinated. The lower quality of the effluent from these plants limits its potential for reuse.

REUSE APPLICATIONS

The level of treatment provided by the Districts' WRPs is necessary to comply with effluent requirements for discharge to the local waterways, which are designated as non-restricted recreational areas. Full-body contact with the reclaimed water occurs on occasion. Consequently, no additional treatment is required for direct, non-potable reuse. According to the California State Department of Health Services (DHS), tertiary effluent can be used for almost any purpose, except for direct drinking water [1]. These uses currently include, but are not necessarily limited to, landscape irrigation of all public areas, irrigation of food and fodder crops and pasture, water supply for livestock, non-restricted recreational impoundments, groundwater recharge, injection into oil-bearing zones and industrial processes.

Almost all of these categories of reuse can be found within the Districts' service areas. By the end of 1996, reclaimed water produced by the Districts was supplied directly to 360 sites for landscape irrigation at 90 parks, 85 schools (from day care to universities), 66 roadway greenbelts, 17 golf courses, 19 nurseries, five cemeteries, and 55 miscellaneous landscaped areas (churches, hospitals, commercial buildings, auto dealerships, landfills, etc.); for agricultural irrigation at 10 sites; and for industrial process water at 12 sites (paper manufacturing, carpet dyeing, concrete mixing, cooling, oil field repressurization and construction applications), as shown in Table 20.1. The Districts also supply water to a 200 acre (81 hectare)

TABLE 20.1. Categories of Reclaimed Water Usage.

Reuse Application	No. of Sites	Area Irrigated		Usage	
		Acres	Hectares	MGD	cms
Parks	90	2,340	947	3.470	0.15
Golf courses	17	2,139	866	3.580	0.16
Schools	85	974	394	1.730	0.08
Roadway greenbelts	66	674	273	1.250	0.05
Nurseries	19	125	50	0.240	0.01
Cemeteries	5	128	52	0.310	0.01
Misc. landscaping	55	168	68	0.650	0.03
Industrial	12	26	10	6.120	0.27
Agriculture[b]	10	1,336	541	4.540	0.20
Wildlife refuge	1	200	81	6.590	0.29
Subtotal	360	8,110	3,282	28.480	1.25
Groundwater recharge	2	689	279	42.443	1.86
Total	362	8,799	3,561	70.930	3.11

[a]1996 data.
[b]The California Polytechnic University, Pomona while technically a school, uses most of its reclaimed water for agricultural purposes and is thus included in this category.

man-made wildlife refuge on Edwards Air Force Base, which is used by migratory birds along the Pacific Flyway.

Approximately two-thirds of the actively reused effluent is used for recharging the Central Groundwater Basin (Figure 20.5). Flood control channels are used to deliver the reclaimed water by gravity flow to existing spreading basins located in the Montebello Forebay. The reclaimed water percolates through the vadose zone of the soil mantle into the groundwater, along with imported and storm water replenishment sources. Since no additional delivery systems need to be constructed and no energy is con-

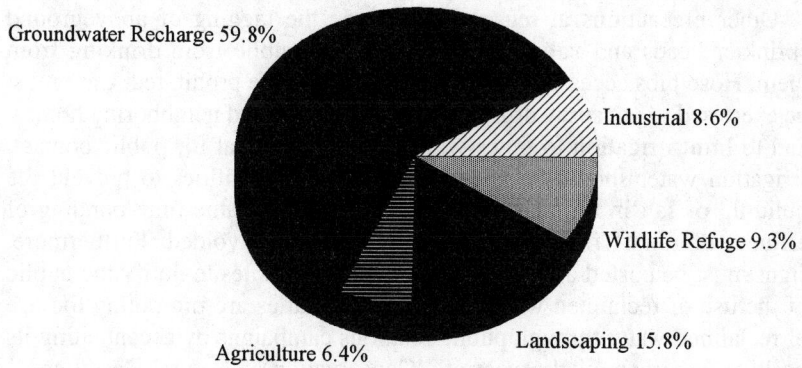

Figure 20.5 Average distribution of reclaimed water usage, 1996.

sumed, groundwater recharge is the simplest and least expensive method of reuse. The remaining one-third of reuse is divided up among the various direct uses listed in Table 20.1.

Revisions to the reuse regulations are expected in the next few years and will formally include additional reuse applications to those already approved: toilet and urinal flushing, cooling towers, fire fighting, commercial laundries, artificial snow making, street cleaning, and various construction uses such as dust control, soil compaction, consolidation of backfill, sewer line flushing and concrete mixing. Many facilities in southern California have already begun to utilize reclaimed water for toilet and urinal flushing, including the Districts' newly constructed administrative office expansion which was dual-plumbed for this use.

WATER REUSE SAFEGUARDS

Even though the quality of reclaimed water produced is safe for full body contact and for the irrigation of food crops, precautions must still be taken to protect the public health, mainly by preventing the accidental ingestion of reclaimed water. For example, the color purple was selected as the identifier of reclaimed water facilities by the California-Nevada Section of the American Water Works Association and adopted by the California DHS and the WateReuse Association of California. This color is also being adopted nationally and internationally (e.g., Australia). Pantone Matching System No. 241 is used for aboveground facilities and No. 512 is used for extruded PVC pipe. New reclaimed water transmission lines must be colored purple and marked ''reclaimed water'' or wrapped with a purple notification tape to prevent an accidental cross-connection to a domestic water supply or outlet. Exposed appurtenances such as air vents and blow-off valves must be labeled and painted purple to differentiate them from potable water facilities.

Other precautions at reuse sites include the tagging of aboveground sprinkler heads and valves to discourage the public from drinking from them. Hose bibs accessible to the general public are prohibited. Care must be exercised to not spray drinking water fountains and neighboring homes, and to limit irrigation to times with the least potential for public contact. Irrigation water should be applied in sufficient quantities to prevent the build-up of salt in the plant root zones, yet at the same time ponding of effluent or runoff from the reuse site should be avoided. Furthermore, signs must be posted at the entrances to the reuse sites to notify the public of the use of reclaimed water. Many communities are promoting the use of reclaimed water through public relations campaigns by accentuating its positive, environmental aspect of ''Conservation of Natural Resources.''

Above all else, the item of most concern is that potable water supplies not be directly connected to the reclaimed water systems, even through a back-flow prevention device. When reclaimed water is provided to a site, such as a park, the existing potable water supply is disconnected from the on-site irrigation system and capped, and a reduced pressure back-flow device is installed downstream of the potable water meter to protect the drinking water supply from inadvertent cross-connections (double check valves are not approved for this use in California). Prior to receiving reclaimed water, each new reuse site must have a cross-connection inspection performed by the local health authorities. This can consist of alternately charging the separate potable and reclaimed water systems while turning off the other, and observing if there is any flow in the inactive system. If a back-up supply to the reclaimed water is desired, the potable water can be air-gapped (no physical connection) into the reclaimed water pump station, storage reservoir or other reclaimed water distribution facility.

EFFLUENT QUALITY CONSIDERATIONS

There are virtually no demonstrable adverse health effects associated with using properly treated reclaimed water for its specified purposes. Reclaimed water consistently meets the EPA's and the State's standards for drinking water in regards to minerals, heavy metals, pesticides, trace organics and radioactivity. Table 20.2 compares current drinking water standards with effluent quality from the San Jose Creek, Whittier Narrows and Pomona WRP's, which together supply reclaimed water for groundwater recharge.

Despite the high quality of water produced by the Districts' WRPs, reclaimed water is not considered to be potable water. Consequently, there are certain water quality aspects that must be taken into consideration for specific reuse applications when dealing with regulatory agencies, water purveyors, end users and the general public.

HUMAN HEALTH EFFECTS STUDIES

In November 1978, the Districts initiated the Health Effects Study to evaluate the health effects of using treated wastewater for groundwater recharge. The focus of the study was the Montebello Forebay Groundwater Replenishment Project located in Los Angeles County, California, where reclaimed water, blended with imported river water (Colorado River and State Project water) and local storm runoff, had been used for replenishment since 1962. At the time the study was conducted, reclaimed water used for recharge averaged 27,000 AFY (1.06 cms), averaging approximately 16% of the inflow to the groundwater basin. The primary goal of the

TABLE 20.2. Health-Related Drinking Water Standards vs. 1996 Tertiary Effluent Quality

Constituent	Units	Limit	San Jose Creek-East	San Jose Creek-West	Pomona	Whittier Narrows
Arsenic	mg/L	0.05	<0.0017	<0.0017	<0.002	<0.0005
Barium	mg/L	1.0	0.05	0.04	0.03	0.04
Cadmium	mg/L	0.010	<0.01	<0.003	<0.003	<0.003
Total Chromium	mg/L	0.05	<0.0003	<0.01	<0.01	<0.01
Fluoride	mg/L	1.6	0.55	0.68	0.39	0.68
Lead	mg/L	0.05	<0.002	<0.002	<0.002	<0.006
Mercury	mg/L	0.002	<0.0001	<0.0001	<0.0001	<0.0001
Nitrate + Nitrite-N	mg/L	10	4.39	4.64	3.47	5.98
Selenium	mg/L	0.01	<0.0011	<0.0022	<0.0010	<0.0011
Silver	mg/L	0.05	<0.01	<0.01	<0.01	<0.01
Lindane	µg/L	4.0	0.05	0.02	0.04	0.04
Endrin	µg/L	0.2	<0.01	<0.01	<0.01	<0.01
Methoxychlor	µg/L	100	<0.01	<0.01	<0.01	<0.01
Toxaphene	µg/L	5	<0.5	<0.5	<0.5	<0.5
2,4-D (acid)	µg/L	100	<0.5	<3.6	<1.7	<9.1
2,4,5-TP (Silvex)	µg/L	10	<0.05	<0.05	<0.05	<0.12
Trihalomethanes	µg/L	100	<6.6	<7.5	<6.3	<12.7
Atrazine	µg/L	3	<3.0	<3.0	<3.0	<3.0
Simazine	µg/L	10	<4.0	<4.0	<4.0	<4.0
Methylene chloride	µg/L	40	<2.4	<1.2	<1.1	<1.6
1,1,1-Trichloroethane	µg/L	200	<0.6	<0.5	<0.5	<0.5
Carbon tetrachloride	µg/L	0.5	<0.3	<0.3	<0.3	<0.3
1,1-Dichloroethene	µg/L	6	<0.3	<0.3	<0.3	<0.3
Trichloroethylene	µg/L	5	<0.3	<0.3	<0.3	<0.3
Tetrachloroethylene	µg/L	5	<0.4	<0.7	<0.3	<0.3

TABLE 20.2. (continued).

Constituent	Units	Limit	San Jose Creek-East	San Jose Creek-West	Pomona	Whittier Narrows
Chlorobenzene	μg/L	30.0	<0.5	<0.5	<0.5	<0.5
Vinyl chloride	μg/L	0.5	<0.5	<0.5	<0.5	<0.5
o-Dichlorobenzene	μg/L	130.0	<0.5	<0.5	<0.5	<0.5
m-Dichlorobenzene	μg/L	130.0	<0.5	<0.5	<0.5	<0.5
p-Dichlorobenzene	μg/L	5	<0.6	<0.6	<0.7	<0.7
1,1-Dichloroethane	μg/L	5	<0.3	<0.3	<0.3	<0.3
1,1,2-Trichloroethane	μg/L	32	<0.3	<0.3	<0.3	<0.3
Benzene	μg/L	1	<0.3	<0.3	<0.3	<0.3
Toluene	μg/L	100	<0.3	<0.3	<0.3	<0.3
Ethyl benzene	μg/L	680	<0.3	<0.3	<0.3	<0.3
o-Xylene	μg/L	1750	<0.5	<0.5	<0.5	<0.5
p-Xylene	μg/L	1750	<0.5	<0.5	<0.5	<0.5
Trans-1,2-dichloroethylene	μg/L	10	<0.3	<0.3	<0.3	<0.3
1,2-Dichloropropane	μg/L	5	<0.5	<0.5	<0.5	<0.5
1,3-Dichloropropene	μg/L	0.5	<0.5	<0.5	<0.5	<0.5
1,1,2,2-Tetrachloroethane	μg/L	1	<0.5	<0.5	<0.5	<0.5
Freon 11	μg/L	150	<1.0	<1.0	<1.0	<1.0
Pentachlorophenol	μg/L	30	<16	<16	<16	<16
Coliform monthly median	#/100 ml	1	<1	<1	<1	<1
Gross alpha radioactivity	pCi/L	15	<3.6	<2.6	<2.6	<2.5
Gross beta radioactivity	pCi/L	50	10.1	7.9	6.7	8.7

931

Health Effects Study was to provide information for use by health and regulatory authorities to determine if the use of reclaimed water for the Montebello Forebay Project should be maintained at the present level, cut back or expanded. Specific objectives were to determine if the historical level of reuse had adversely affected groundwater quality or human health, and to estimate the relative impact of the different replenishment sources on groundwater quality. Research tasks included water quality characterizations of the replenishment sources and groundwater in terms of their microbiological and chemical content; toxicological and chemical studies of the replenishment sources and groundwater to isolate and identify organic constituents of possible health significance; field studies to evaluate the efficacy of soil for attenuating chemicals in reclaimed water; hydrogeologic studies to determine the movement of reclaimed water through groundwater and the relative contribution of reclaimed water to municipal water supplies; and epidemiologic studies of populations ingesting reclaimed water to determine whether their health characteristics differed significantly from a demographically similar control population. The final project report was completed in March 1984 (Nellor et al., 1985) [2]. The results of the study did not demonstrate any measurable adverse effects on either the area's groundwater or the health of the people ingesting the water.

In 1986, the State Water Resources Control Board (SWRCB), the DWR and the DHS established a Scientific Advisory Panel on Groundwater Recharge to review the report and other pertinent information. After its review, the Panel found the research described in the Health Effects Study final report to have been thorough and well-conducted with state-of-the-art methodology. The limitations of the study were also noted, including whether the organic compounds present and of greatest health significance were identified; whether the genotoxicity data were adequate, in the absence of other toxicologic information, to serve as a basis for risk assessment; and whether the ecologic and household survey studies could establish risk since the confounding demographic factors could easily overshadow the results. The Panel concluded that it was comfortable with the continuation of the Montebello Forebay Project and with the safety of the product water. It felt that the risks, if any, were small and probably not dissimilar from those that could be hypothesized for commonly used surface waters.

Based on the results of the Health Effects Study and recommendations of the Scientific Advisory Panel, authorization was given by the Regional Water Quality Control Board in 1987 to increase the annual quantity of reclaimed water used for replenishment from 32,700 AFY (1.28 cms) to 50,000 AFY (1.96 cms). In 1991, the water reclamation requirements for the project were revised to allow for recharge up to 60,000 AFY (2.35

cms) and 50% reclaimed water in any one year as long as the running 3-year total did not exceed 150,000 AF (185 million cubic meters) or 35% reclaimed water. Continued evaluation of the project is being provided by an extensive sampling and monitoring program, and by supplemental research projects pertaining to percolation effects, epidemiology and microbiology.

One of these follow-up studies has been an epidemiologic assessment commissioned by the Water Replenishment District of Southern California. Sloss et al. (1996) [3] studied over 900,000 people in an area receiving reclaimed water and nearly 700,000 people in three demographically similar control areas, resulting in a study of approximately 10 percent of the entire Los Angeles County population. With few exceptions, cancer incidents, cancer deaths and infectious diseases were nearly the same in the reclaimed water and control areas. In the few exceptions, a dose-response relationship was not observed between low reclaimed water exposure areas (one percent reclaimed water in drinking water) and high reclaimed water exposure areas (up to 31 percent reclaimed water). The researchers concluded that the higher observed risk was due *"either to chance or other unmeasured risk factors not necessarily related to reclaimed water exposure."*

MICROORGANISMS

The Districts' tertiary-treated effluent generally contains less than one total coliform bacteria per 100 mL in samples taken daily. Total coliform is used as an indicator organism, because of its extremely high ratio with bacterial pathogens. Infectivity tests for viruses have been performed as often as monthly on the tertiary effluent since 1979 and reported by Yanko (1993) [4]. In the 981 samples, consisting of 268,836 gallons (1,017,544 liters) of tertiary effluent, completed through the end of 1995, only one was found to be positive (*Coxsackie B* at the Valencia WRP in August 1986). These results suggest that the viral risks associated with the use of reclaimed water meeting California water reclamation criteria are within acceptable levels.

NUTRIENTS

Since nutrients such as nitrogen and phosphorous are not removed from the wastewater during treatment, landscape users of reclaimed water can reduce or eliminate fertilizer applications. For example, the operators of the California County Club in Whittier report that they have not fertilized the fairways on that golf course since 1978 when it began receiving effluent from the San Jose Creek WRP, resulting in an estimated annual savings of approximately $10,000. Based on 1995 water quality data, reclaimed water produced at the tertiary treatment plants in the Los Angeles Basin contains approximately 39 pounds per acre-foot (14.5 kilograms per million liters) of nitrogen as N, 18 pounds per acre-foot (6.7 kilograms per million liters) of phosphorous as PO_4, and 45 pounds per acre-foot (16.8 kilograms per million liters) of potassium as K_2O. The constant application of small amounts of fertilizer in the reclaimed water promotes a balanced growth in the vegetation which results in healthier plants, while avoiding the creation of "fertilizer dependence". This also reduces the risk of ground-water contamination from the standard application of large amounts of fertilizer over a short period of time.

PLANT GROWTH

As for the effect of reclaimed water on vegetation, according to a report issued by the SWRCB [5], the chemical constituents (e.g., boron, chloride, total dissolved solids, heavy metals, sodium absorption ratio, etc.) in the effluent produced by the Districts' WRPs should have a slight or no effect on most plants. Experience has borne this out as several nurseries growing very sensitive bedding plants have reported absolutely no problems in using reclaimed water. On the contrary, they have had great success in regards to plant growth and quality.

ALGAE GROWTH

For many end users, perhaps the most inconvenient problem to handle is that of algae growth in lakes holding reclaimed water. Not only do the nutrients in reclaimed water produce lush, green turf, but they can also contribute to heavy algal blooms, especially given the warm, sunny climate in southern California. Various methods to prevent or control algae growth have been experimented with including: introducing algae eating fish, applying algacides such as copper sulfate or chlorine, adding Aqua-Shade to block sunlight and inhibit photosynthesis, aerating and circulating lake water to prevent stagnation, avoiding the introduction of reclaimed water into shallow lakes, reducing the reclaimed water retention time in the lake by using it for irrigation, and covering reservoirs that supply industrial processes. Aside from the last option, not all of these solutions work in all situations all of the time. The Long Beach Water Department has reported excellent results from experiments with a microbial blend called "Bac-Terra" which has significantly reduced algal growth in one of its lakes receiving reclaimed water, at a substantially lower cost than other treatment schemes. However, even this agency is forgoing lake storage

TABLE 20.3. Shallow Well Monitoring Summary.[a]

Well No.	Distance[b] (feet)	No. of Samples	Total N (mg/L)	TDS (mg/L)	TOC (mg/L)	VOCs (μg/L)
2909Y	112	40	<2.08	382	<1.60	<0.19
1590AL	168	41	<2.56	418	<2.21	<0.46
1581P[c]	106	15	<3.43	460	<1.37	<0.13
1582W[c]	61	37	2.17	309	<1.64	<0.40
1620RR	70	51	<3.98	557	<2.27	<0.71
1612T	60	51	<4.81	524	<2.05	<0.26
1613Y	70	51	3.37	520	<2.12	<0.09

[a]Data from August 1988 to December 1996.
[b]Distance is from bottom of spreading basin to top of well perforations.
[c]Well 1581P replaced by Well 1582W in December 1990.

of reclaimed water for covered storage to reduce operating costs and potential water quality problems.

EFFECTS OF NITROGEN ON GROUNDWATER QUALITY

The nitrogen in reclaimed water has also been a concern in groundwater recharge, because of the high levels of ammonia in reclaimed water which potentially could be nitrified by soil bacteria to form nitrates. Nitrates in groundwater have the potential to cause methemoglobinemia ("blue-baby syndrome") and, therefore a Primary Drinking Water Standard of 10 mg/L as nitrogen for nitrate has been established. However, even though reclaimed water has constituted up to 31 percent of the artificial recharge to the Montebello Forebay of the Central Groundwater Basin since 1962, there has been only a slight change in the levels of nitrogen in the groundwater. Bimonthly sampling following construction in 1987 of six shallow wells in and around the spreading grounds (Table 20.3) has also not demonstrated an increase in nitrogen levels attributable to the spreading of reclaimed water.

The nitrification-denitrification process occurring underground was demonstrated quantitatively by Johnson (1993) [6]. For this project, filtered, disinfected effluent from the San Jose Creek WRP was percolated through a 10-foot (3 meter) column of soil collected from the San Gabriel Coastal Spreading Grounds. By mimicking the flooding-drying cycle employed by the Los Angeles County Department of Public Works in the spreading grounds themselves, it was shown that the positively charged ammonium ion was adsorbed by the negatively charged soil particles in the vadose zone during the flooding cycle, where nitrifying bacteria then converted it to negatively charged nitrate ions during the following drying cycle. In micro-environments of anaerobic activity around soil particles in the vadose zone, denitrifying bacteria convert at least one-third of the nitrate to nitrogen gas. The next flooding cycle flushes out the remaining remobilized nitrate ions into the anaerobic zone where denitrifying bacteria could possibly continue the denitrification process. Experimental results also indicate that the rate of denitrification is most likely limited by the availability of organic carbon.

ECONOMIC ADVANTAGES

The Districts' sewerage system is funded primarily through residential and industrial users charges for operation and maintenance costs and capital expenditures for replacement or upgrading of existing facilities. Connection and annexation fees fund capital expenditures for expansions of sewer and treatment plant capacities. The sale of reclaimed water represents a cost recovery for the Districts, partially offsetting the costs of sewage treatment and disposal.

Historically, the price of reclaimed water has been based on approximately one-fifth of the operations and maintenance costs for the WRPs. This resulted in an approximate price of $25 per acre-foot ($20 per million liters). However, since the costs of alternate sources of potable water are increasing rapidly (MWD's rate for treated water is expected to increase to over $800 per acre-foot ($650 per million liters) by 2010), a new pricing policy for reclaimed water sales has been adopted. This new policy is based on "shared savings" whereby the price of the water as a commodity is set as one-half the savings realized by the water purveyor from using reclaimed water to replace a higher priced domestic water supply. To calculate this, the unit cost per acre-foot to build, operate and maintain the reclaimed water system is subtracted from the unit cost of the alternative potable water supply, and half of this amount is set as the price of the reclaimed water.

For example, in Fiscal Year 1995–96, the price for MWD's treated supply was $426 per acre-foot ($345 per million liters). The cost of pumping groundwater (energy plus replenishment fees) is approximately $307 per acre-foot ($249 per million liters) in the Central Basin. (NOTE: These costs are for the water itself and do not take into the consideration the distribution facilities.) The final price of water from local purveyors to the end user can easily exceed $500–$700 per acre-foot ($405 to $568 per million liters). The final cost of reclaimed water to the end user ranges from 85 percent to as low as 22 percent of the domestic water rate (Table 20.4). The low cost of reclaimed water helps offset the high capital costs of constructing a separate reclaimed water distribution system to deliver the water from the WRP to the individual reuse site. Other capital costs

TABLE 20.4. Potable vs. Reclaimed Water Rates[a] ($/AF).

Purveyor	Potable Water	Reclaimed Water	Percent Discount
Long Beach Water Dept.	643.38	319.73	50
City of Cerritos	413.82	217.80	47
City of Lakewood	413.82	370.26	11
Central Basin MWD	429.00	200.00–260.00	39–53
Pomona Water Dept.	276.17–345.43	77.58–119.20	57–78
Walnut Valley Water Dist.	596.77	507.26	15
San Gabriel Valley Water Co.	435.60	236.10–345.87	21–46

[a]Data for Fiscal Year 1995–96. Rate includes distribution system costs.

can include operational storage facilities which might be required to offset diurnal flow variations at the WRP. Since the Districts' WRPs are situated along rivers for effluent discharge, construction costs can be reduced by utilizing the right-of-ways along the banks of these waterways for locating the distribution pipelines.

In 1996, 25.5 billion gallons, or 79,665 AF (98.2 million cubic meters), of reclaimed water were used. Although this represents only about 14 percent of the total annual production of treated effluent at all of the Districts' facilities, it would have been enough water to supply approximately 398,325 people for an entire year, equivalent to a city the size of St. Louis, Missouri, the 34th largest city in the United States. The sale of reclaimed water generates roughly $1 million annually in revenue to the Districts. Also, since the Districts must dechlorinate for river discharge, but can supply chlorinated effluent to its reuse customers (providing a chlorine residual to prevent bacterial regrowth and to help keep the distribution lines clear of slime growths), less sulfur dioxide is required for dechlorination resulting in an annual savings to the Districts of approximately $35,000.

Typically, the Districts do not participate in the construction of the off-site distribution systems for the reclaimed water. This is the responsibility of the municipalities or water purveyors who are sponsoring the projects. However, the Districts generally will use its expertise to design, construct, operate and maintain the pumping plants, which are located on the treatment plant site for convenient access to the reclaimed water, needed to supply the distribution system. The costs of these activities undertaken by the Districts are borne by the water purveyor purchasing the effluent.

In California, funding for reuse projects is available in the form of low-interest loans from the SWRCB. The voters in 1984 and 1988 approved bonds to provide a total of $55 million for such loans, while State Revolving Funds have recently been made available for water reclamation projects. In

1996, the voters approved nearly a billion dollars in bond funding for water projects, of which $60 million will go towards funding water recycling projects. In the southern California area, MWD has established a program to provide funds for local conservation projects, including reclamation. By developing alternative local water supplies, MWD will save not only energy costs by reducing imports of State Project water from the Sacramento Delta area and pumping it over the Tehachapi Mountains, but also the capital costs involved in expanding conveyance, treatment and distribution facilities. The savings are used to provide rebates for reclaimed water projects that would not be economically feasible without this assistance. In 1990, MWD increased its rebate from $84 to $154 per acre-foot ($68 to $125 per million liters), and again in 1995 to $256 per acre-foot ($316 per million liters) to further stimulate water reuse construction activities.

The energy savings realized by using local reclaimed water supplies versus pumping imported water over the mountains provides additional indirect, yet tangible, economic and environmental benefits. It takes approximately 3,000 kilowatt-hours (net) to pump an acre-foot of water over the Tehachapi Mountains. Therefore, in 1996, replacing this imported water with locally produced reclaimed water saved almost 239 million kilowatt-hours of energy and 129,513 barrels (20.6 million liters) of oil. As a result of these energy savings air pollutant emissions were reduced by:

- 1.2 tons (1.1 metric tons) of reactive organic gases,
- 4.8 tons (4.3 metric tons) of particulates,
- 14.3 tons (13.0 metric tons) of sulfur oxides,
- 23.9 tons (21.7 metric tons) of carbon monoxide, and
- 137.4 tons (124.7 metric tons) of nitrogen oxides [7]

Perhaps more importantly, the production of 179,246 tons (162,609 metric tons) of carbon dioxide, a greenhouse gas that is suspected of contributing to global warming, was avoided by the use of locally produced reclaimed water.

RECLAIMED WATER DISTRIBUTION

Since the Districts' WRPs are scattered geographically throughout its service area to better handle locally produced wastewater, they can, therefore, supply reclaimed water to a greater number of communities. It continues to be the Districts' intent to construct additional treatment capacity at the WRPs, instead of for ocean disposal, in order to make additional reclaimed water supplies available for reuse (Figure 20.6).

As the cost of potable water has increased, the cost-effectiveness of reclaimed water has allowed distribution systems to extend further from

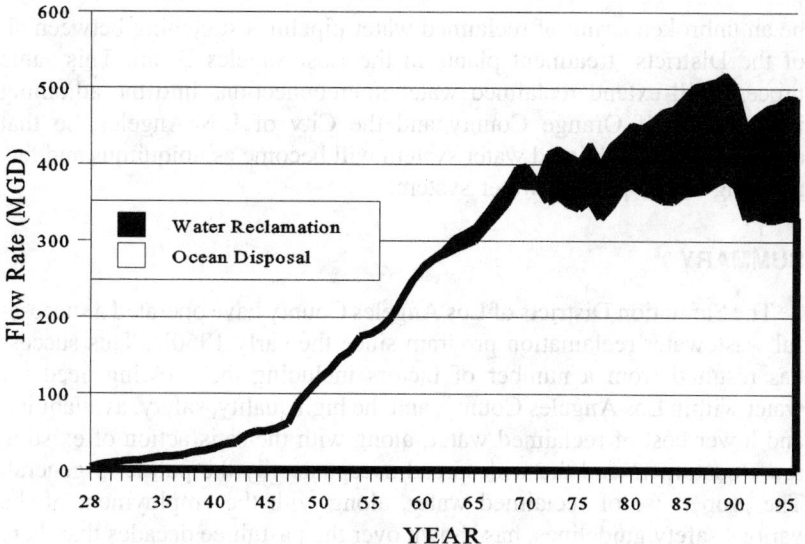

Figure 20.6 Joint outfall system flows, 1928–96.

the WRP source. The first generation of reclaimed water users consisted of individual end users which were located next to a WRP. The second generation was a more extensive municipal distribution system serving many sites within a single city that surrounded a WRP. The third generation is the regional distribution system constructed by a regional water wholesaler which delivers reclaimed water through extensive piping systems to several retail water purveyors which, in turn, sell the reclaimed water to dozens of individual sites. These cooperative, regional systems allow the smaller retail purveyors to participate in water reclamation, despite their lack of resources to accomplish this on their own, allowing for the maximum usage of reclaimed water while keeping the size and cost of the distribution systems to a minimum.

Water recycling is now embarking on the fourth generation of distribution systems, in which the regional systems originating from a single WRP have extended so far that they are merging and interconnecting with regional systems coming from other WRPs. The resulting mega-system provides for a closed loop system, allowing for increased reliability and operational flexibility, and enhanced system pressures and flows, as there is more than one source of water. The Central Basin Municipal Water District's Century and Rio Hondo projects are becoming this kind of mega-system, allowing for reclaimed water from either the Districts' San Jose Creek or Los Coyotes WRPs or both to supply both systems. Over the next few years, several systems will be interconnected so that there will be an unbroken string of reclaimed water pipelines stretching between all of the Districts' treatment plants in the Los Angeles Basin. This same process will extend reclaimed water interconnections into the adjoining jurisdictions of Orange County and the City of Los Angeles, so that eventually, the reclaimed water system will become as ubiquitous and far-reaching as the potable water system.

SUMMARY

The Sanitation Districts of Los Angeles County have operated a successful wastewater reclamation program since the early 1960's. This success has resulted from a number of factors including the growing need for water within Los Angeles County, and the high quality, safety, availability, and lower cost of reclaimed water, along with the satisfaction of existing reuse customers and the widespread acceptance by the public in general. The proper use of reclaimed water, along with the employment of the various safety guidelines, has shown over the past three decades that there have been absolutely no adverse impacts on the health of the general public, on the health and quality of agriculture and urban landscaping, nor on the quality of the groundwater. The use of locally produced reclaimed water is proving to be economically superior to the reliance on costly imported water supplies, with the added economic benefits of reduced energy and fossil-fuel consumption and generation of air pollutants.

REFERENCES

1 California Administrative Code, Title 22, Division 4. 1978. Wastewater Reclamation Criteria. State of California Department of Health Services.

2 Nellor, M. H., R. B. Baird and J. R. Smyth. July 1985. Health effects of indirect potable water reuse. *Journal AWWA.* 88–96.

3 Sloss, E. M., S. A. Geschwind, D. F. McCaffrey and B. R. Ritz. 1996. *Groundwater Recharge With Reclaimed Water: An Epidemiologic Assessment in Los Angeles County 1987–1991.* Rand, Santa Monica, CA.

4 Yanko, W. A. 1993. Analysis of 10 years of virus monitoring data from Los Angeles County treatment plants meeting California wastewater reclamation criteria. *Water Env Fed Research.* 65(3), 221–226.

5 Pettygrove, G. S. and T. Asano. (ed.). 1984. *Irrigation with Reclaimed Municipal Wastewater—A Guidance Manual.* Report No. 84-1 wr. California State Water Resources Control Board, Sacramento, CA.

6 Johnson, P. May 1993. *Experimental Studies of Nitrogen and Organic Carbon Removal from Reclaimed Water by the Soil Vadose Zone.* Sanitation Districts of Los Angeles County, Whittier, CA.

7 *Air Quality Handbook for Preparing Environmental Impact Reports.* Revised, April 1987. South Coast Air Quality Management District, Los Angeles, CA.

CHAPTER 21

Irvine Ranch Water District's Reuse Today Meets Tomorrow's Conservation Needs

HISTORY

INTRODUCTION

ONE of California's youngest and most progressive water agencies, Irvine Ranch Water District (IRWD) was formed in 1961 to provide irrigation and domestic water for a rapidly developing metropolitan community. Over the years, the District has prudently planned programs and designed and operated facilities to ensure a reliable water supply for its customers. To meet the area's unique requirements, IRWD has developed several sources of water. Reclaimed water is just one of those sources, and an important one that supplies approximately 15 percent of the annual water needs of the District.

In 1963, the IRWD Board of Directors made the decision to provide sewage collection and treatment services for the purpose of producing reclaimed water. Within four years of its initial commitment, the District began delivering reclaimed water to its agricultural customers throughout the 6,900-square-mile (1,787,100 hectare) district. Today, IRWD's nationally recognized reclaimed water program encompasses a variety of reclaimed water uses such as landscape irrigation, recreational uses and toilet flushing water for high-rise buildings.

Ronald E. Young, Kenneth A. Thompson, Robert R. McVicker, Richard A. Diamond, Mark B. Gingras, Dave Ferguson, Johnnie Johannessen, Gregory K. Herr, John J. Parsons, Veronica Seyde, Eric Akiyoshi, Jim Hyde, Chris Kinner, and Lars Oldewage, Irvine Ranch Water District, 15600 Sand Canyon Avenue, Irvine, CA 92618.

The cornerstone of IRWD's extensive reclaimed water program is the development of a "dual distribution" system—one set of pipes for potable water and another set for reclaimed water. Thus, the stringent water quality criteria established by the California Department of Health Services have been consistently met and exceeded over a number of years. As a result, the District has been a leader in the design and development of rules and regulations for the use of reclaimed water. The high quality, polished water that leaves the Michelson Water Reclamation Plant (MWRP) in Irvine, California, earned IRWD the first "unrestricted use permit" to be issued in the state. This level of certification allows IRWD's water to be used for almost every purpose except drinking.

RECLAIMED WATER USES

IRWD's reclaimed water system, which contains approximately 200 miles (320 kilometers) of dual distribution pipelines, serves approximately 1,000 acres (400 hectares) of fields and orchards planted with a variety of fruits, vegetables and nursery products. Landscape reclaimed water uses include parks, schools, golf courses, streetscapes, the lake at Mason Park and open space managed by community associations.

In 1991, IRWD became the first water district in the nation to obtain health department permits for the use of reclaimed water from a community system in interior spaces such as office buildings. Reclaimed water is currently used for toilet flushing in IRWD's two facilities as well as in two high-rise office buildings in Irvine, California, constructed with dual piping systems. Domestic water demands in these buildings have dropped by as much as 75 percent due to the reclaimed water use.

Significant landscape irrigation utilizing reclaimed water has been planned for the recently developed Newport Coast area, an exclusive enclave of homes overlooking the Pacific Ocean. In addition to traditional landscape irrigation uses, the District in 1995 began supplying reclaimed water to individual homeowners (rather then homeowners' associations)—another "first" for both the State of California and IRWD.

Working closely with the State of California Department of Health Services, IRWD continues to explore new uses for reclaimed water such as textile manufacturing, which was approved for use in March 1997, and commercial cooling towers.

COMMUNITY RELATIONS PROGRAM

Successful implementation of the "world class" reclamation program developed by IRWD does not occur without a quality community education program. Educating the community about the value of reclaimed water has been a priority of the District for more than three decades. What started out as a simple "show and tell" program has evolved into a very sophisticated, multi-faceted program that reaches virtually every member of the community—from tots to seniors, from students to members of the business community, from government officials to special interest groups. Key components of the program include: the Resident Tour Program, In-School Education Program, Operation Outreach and Community Education Program.

SUMMARY

The remaining sections of this chapter describe the major components of IRWD's water reclamation program: Planning and Implementation, Operations, Water Quality, On-Site Users, Finance, Awards and Recognition and the Appendix. Other sections address the early development of the district, current operational strategies and future directions for the water reclamation program. The authors have presented both the positive and negative aspects of designing and implementing a quality reclaimed water program in the hopes that readers will find useful information to assist them in evaluating and improving their individual reclaimed water program.

PLANNING AND IMPLEMENTATION

PILOT PLANT AND PLANT EXPANSION

In 1967, IRWD constructed a reclamation plant and effluent pipeline connecting to an existing agricultural water distribution system and an existing storage reservoir. Sewage treatment was provided by an activated sludge type treatment plant rated at 2 million gallons per day (mgd) [0.09 cubic meters per second (m³/sec)] capacity. Treated effluent was pumped through a force main to the Sand Canyon Reservoir. The District received a waste discharge permit from the Santa Ana River Basin Regional Water Quality Control Board to operate a 2 mgd (0.09 m³/sec) treatment plant and use the effluent from the treatment plant directly for irrigation of adjacent lands or storage in Sand Canyon Reservoir for eventual irrigation use.

In 1970, construction was completed on tertiary treatment facilities and water reclamation plant expansion, increasing the capacity to 5 mgd (0.22 m³/sec). The Advanced Waste Treatment Plant was designed, constructed and operated to demonstrate nutrient removal employing the Ryckman approach for phosphate removal and biological processes for organic and nitrogen removal. This demonstration project was completed in October 1971 under the partial sponsorship of the Environmental Protection Agency.

From 1970 to 1976, IRWD researched the use of the pressure pipeline treatment (PPT) process and did not expand the plant. (The PPT process is described in the next section.) In 1976, additional clarifiers were added which increased the rated capacity of the treatment plant to 7 mgd (0.31 m^3/sec). In 1979, the plant was expanded to its current rated capacity of 15 mgd (0.66 m^3/sec).

PRESSURE PIPELINE TREATMENT (PPT)

The PPT system utilizes the traditional pipelines in water reclamation systems for the dual purpose of providing both transportation and in-line treatment of the waste water. The in-line treatment in the PPT system is performed by modification of conventional activated sludge treatment. Raw or primary sewage is mixed with activated sludge at the downstream end of the pipeline. Biological treatment then occurs in the pipeline with oxygen being added in the form of air, pure oxygen, or a mixture of air and oxygen. After completion of the biological treatment during conveyance in the pipeline, the activated sludge is then separated at the upstream end of the pipeline and a portion is returned through a sewer to the downstream end of the PPT line.

Theoretical and practical evaluations of the PPT system were conducted at IRWD between 1969 and 1973. The evaluations consisted of three pilot evaluations and computer modeling of the pipeline reactor. The first study was based on the use of an existing 24-inch (0.6-meter) force main. The tests were conducted by mixing primary effluent, return activated sludge and air, and pumping the mixture through the force main. The second study was conducted with a 15,000-foot (4,500-meter-long), 4-inch-diameter (0.1-meter-diameter) steel pipe; a 5-foot-diameter (1.5-meter diameter) flotator; a special raw solids grinder and pump; and a primary screen. This system was designed for a flow rate between 20 and 40 gpm (75 and 150 liters per minute) and was controlled by simple valving. Both of these studies showed that biological oxygen demand (BOD) removals of 85 percent to 90 percent could be achieved at detention times ranging from 10 hours to 6 hours. The computer modeling results showed that the required theoretical treatment time in the PPT system (ideal plug flow) would be only 33 percent of the time required in a conventional (completely mixed) treatment plant. The third study was conducted with a 22,920-foot (6,986-meter-long), 6-inch-diameter (0.15-meter-diameter) polyvinyl chloride (PVC) pipe; a variable speed pump (sewer simulator pump); a Bauer Hydrasieve as the primary treatment facility; a main PPT pump; an 18-foot-diameter

(5.5-meter-diameter) flotator; a 12-foot-diameter (3.7-meter-diameter) sand filter; an 18-foot-square (30 m³/sec) primary solids activation tank; and an aerobic digester [a steel tank 22 feet (6.7 meters) long with a diameter of 8 feet (2.4 meters)]. Operation of this 0.2 mgd (0.009 m³/sec) PPT pilot plant demonstrated 90 percent removal of BOD from raw sewage to secondary effluent. Economic analysis showed an esti- mated capital cost savings of $4,670,000 using the PPT approach for a 12 mgd (0.53 m³/sec) full-scale system.

Although the potential savings of using the PPT system were significant, these potential savings were outweighed by the risk involved in proceeding with an unproven technology. The decision was made to proceed with the conventional treatment plant and water conveyance system.

DUAL DISTRIBUTION SYSTEM

IRWD was a leader in the industry in the design and development of rules and regulations for the use of reclaimed water. As such, District planners had the advantage of building public works infrastructure around the dual distribution reuse system. This concept has been integrated into several recent local developments and promises to become the standard for the future. In the city of Irvine, California, new developments must be built with dual water systems—one set of pipes for drinking water and another for reclaimed wastewater.

The District's dual distribution system began with pipelines to the seasonal storage reservoirs. The first major distribution pipelines were completed between 1975 and 1977. Additional pipelines were constructed between these major distribution pipelines to serve the common area greenbelts, parks and schools in developing areas. As demands increased within the District, additional major distribution pipelines were con- structed.

IRWD's first reclaimed water customers were agricultural users. The Irvine Company, one of the largest landowners and developers in Orange County, California, was the primary customer during the early years of water reclamation and reuse at the District. Other customers used reclaimed water for golf course irrigation (Rancho San Joaquin Golf Course) and park irrigation (William Mason Regional Park). As development progressed throughout the area, the District added the city, the school district, and homeowner associations as reclaimed water customers. Today, reclaimed water is used on crops, golf courses, parks, school grounds, greenbelts, street medians, industrial/commercial properties, churches, construction sites (for dust control), decorative fountains and streams, and freeway landscaping.

In the early 1990s, the District expanded the reclaimed water program to include commercial buildings for uses which do not require potable water quality. Reclaimed water delivered to the buildings via a dual pipe system supplies toilets, urinals, and floor drain trap primers, as well as landscape irrigation requirements. In early 1997, a carpet manufacturing facility joined IRWD's list of reclaimed water customers. The District is currently evaluating the use of reclaimed water in commercial cooling towers, cement manufacturing, carpet manufacturing and commercial car washing.

As reclaimed water demands have increased, the District has developed seasonal and operational storage for reclaimed water. The District has two seasonal storage reservoirs. Rattlesnake Reservoir has a storage capacity of 1,442 acre-feet (1,779,000 m), and Sand Canyon Reservoir has a capacity of 768 acre-feet (947,000 m^3). Reclaimed water is produced at MWRP and stored at these reservoirs during the winter period to meet peak summer demands. Peak demands on the system occur during the hours between 9:00 PM and 6:00 AM. Four operational reservoirs (tanks) are sized up to 5 million gallons (19,000 m^3) to efficiently meet the peak hour demands. Additional reservoirs are planned as development of the District proceeds and reclaimed water demands increase.

All reclaimed water use requires pumping to provide sufficient quantities at adequate pressures to meet customer needs. Initially, the District constructed a pumping station at MWRP to supply water to the Sand Canyon Reservoir and agricultural water distribution system. With the construction of the dual distribution system, additional pumping stations were constructed at MWRP, Sand Canyon Reservoir and Rattlesnake Reservoir. As the use of reclaimed water expanded to the far reaches of the District, additional pumping stations were required and were implemented.

RECLAMATION PLANNING

Reclamation planning for IRWD's largest agricultural customer, the Irvine Ranch, began long before the property was developed and long before IRWD acquired the legal authority to provide sewer and reclamation services. As early as 1961, The Irvine Company (TIC) prepared the "Feasibility Report on Reclamation of Sewage at University City" based on land use planning criteria outlined in the "Second Phase Report—A University Campus and Community Study" prepared by W.L. Periera and Associates. This report outlined the feasibility of constructing a reclamation system which would utilize all anticipated sewage flows from the community for agricultural, landscape irrigation and recreational uses. In 1962, a reclamation master plan was prepared for IRWD's sewer service predecessor, Orange County Sanitation District No. 14. This plan outlined facilities necessary to provide 10 mgd (0.44 m^3/sec) of sewage treatment/reclamation to University City through the year 1990. Both this plan and the 1961 report cited two primary factors benefiting total reclamation: First, the area was undeveloped and under the ownership of a single landowner (TIC), thereby significantly decreasing the jurisdictional hurdles to reclamation; and second, the plan for the University City area included large areas for recreation and covered a relatively small portion of TIC landholdings. Much of the remaining acreage, it was believed, would remain in permanent agriculture and would provide more than ample area for wastewater disposal through reclamation.

In 1968, subsequent to IRWD's acquiring legal authority to provide sewer and reclamation services and the dissolution of Sanitation District No. 14, IRWD completed a Water Reclamation Master Plan. This plan was based on much more extensive development plans proposed by TIC [with ultimate wastewater flows of 70 mgd (3.1 m^3/sec)] and called for equally extensive reclamation through agriculture, landscape irrigation, groundwater recharge and recreational impoundments. However, the elimination of large areas of permanent agriculture, coupled with significant seasonal demand variation, resulted in the need to plan facilities to dispose of excess wastewater flows. This plan recommended a phased construction program for reclamation facilities that primarily consisted of a reclamation plant.

In 1972, the IRWD Water Resources Master Plan introduced the concept of "total water management," whereby complete reclamation and reuse would be pursued through the distribution of reclaimed water to all major agricultural and landscape areas within IRWD. This plan required that excess wastewater, estimated to be 27 mgd (1.2 m^3/sec), be used for groundwater recharge in the vicinity of Santiago Creek/Irvine Lake, or be disposed of through a forestation disposal system.

The two-part Irrigation Water System Plan prepared during 1977–78 focused on total reclamation of all sewage flows and recommended construction of a connection to the Orange County Sanitation District to provide "treatment reliability and winter disposal." These documents also recognized the need for supplemental sources of water to meet irrigation demands within IRWD.

In 1984, IRWD completed the Irrigation/Reclamation System Master Plan. This document used more conservative estimates of future water rates and recommended construction of a more limited reclamation system serving primarily the lower elevation areas of IRWD. The domestic and irrigation documents were again combined in preparation of the 1991 Water Resources Master Plan.

OPERATIONS

COLLECTION SYSTEM

Description

In the early 1960s, the area which was to become the city of Irvine began as an industrial park surrounding the Orange County Airport (now John Wayne Airport), and the newly opened campus of the University of California. Since no treatment plants existed in the area at the time, all sewage was conveyed to the County Sanitation Districts of Orange County, located in Fountain Valley, California, for treatment and disposal. Following the growth of the area's industry, residential development was initiated to satisfy the need for local housing. After the city of Irvine was incorporated in 1971, several early residential developments including the villages of University Park, Turtle Rock, Culverdale, Woodbridge and Deerfield were identified. MWRP was constructed and later expanded to treat and reclaim wastewater from these growing areas and the commercial businesses that supported them. In 1983, the first phases of the Spectrum industrial park were opened, in response to the demand for additional commercial and industrial sites. These areas are tributary to MWRP.

In the early 1980s, IRWD annexed the Foothill Ranch area, a mixed residential, commercial and industrial development similar in character to the city of Irvine. For many years, this area was served by the nearest treatment plant, the Los Alisos Water District, until development in the city of Irvine progressed towards Foothill Ranch. The Foothill Ranch collection system developed from the foothills down to the Irvine plain; the city of Irvine collection system developed up the Irvine plain toward the foothills. Consequently, much of the development in Foothill Ranch is infill development, where small individual development areas tie into a backbone collection system.

In the mid 1980s, IRWD annexed a developed portion of the higher foothills called Portola Hills. This residential area had an established collection system that discharged into the Santa Margarita Water District for treatment at its Chiquita Treatment Plant, which reclaimed water or discharged into the ocean through the Southeast Regional Reclamation Authority outfall in Dana Point. IRWD maintains the collection system and pays treatment and disposal costs by agreement with the Santa Margarita Water District.

SOURCE WATER MANAGEMENT

IRWD has two main sources of domestic water. Domestic water is imported and purchased from the Metropolitan Water District. This water originates from the Sacramento-San Joaquin River Delta and from the Colorado River and is transported by pipeline and open canal to terminal storage reservoirs in Riverside and San Bernardino counties. The total dissolved solids (TDS) content of this water is relatively high, averaging well above 500 mg/L TDS. Even with restrictions on the use of self-regenerating water softeners, the TDS content of the MWRP influent is greater than the basin plan limit of 720 mg/L. Fortunately, IRWD has a well field that produces high quality and low dissolved solids domestic water, which is needed to meet system demands.

Currently, IRWD is limited to a 20,000 acre-foot (25,000,000 m³) annual withdrawal from its well field; however, there are no limitations as to when IRWD can take this water. IRWD modeled and implemented a seasonal water delivery plan to return as much low-dissolved-solids well water for reclamation as possible. The program works as follows:

(1) During the period from late fall through early spring, the demand for reclaimed water is at a minimum due to shorter day length, cooler temperatures, reduced insulation, and adequate rainfall (under normal circumstances). MWRP reduces the quantity of reclaimed water to match the lower system needs. During this period, IRWD maximizes its use of high-dissolved-solids imported water and minimizes the use of its well field. Consequently, only a relatively small quantity of high-dissolved-solids reclaimed water is actually consumed.

(2) During the period from early spring through late fall, the demand for reclaimed water increases, due to longer day length, warmer temperatures, increased insulation, and minimal rainfall (under normal circumstances). MWRP increases the supply of reclaimed water to fill its storage reservoirs and meet the increased demands of urban and agricultural irrigation. IRWD maximizes the use of its well water and directs as much of the well water as is possible into the area tributary to MWRP. Consequently, large volumes of relatively low-dissolved-solids wastewater are reclaimed and reused.

IRWD is able to maximize reclamation of low-dissolved-solids wastewater for a variety of reasons:

(1) During the low reclaimed water demand winter period, wastewater not needed for reclamation is diverted to County Sanitation Districts of Orange County (CSDOC) for treatment and ocean disposal.

(2) The domestic water delivery system has multiple points of connection into the water supply system. Consequently, the flow through these connections is purposefully metered to force water in specific directions.

(3) Forcing water to move in a specific direction may cause water pressure and supply problems in some areas. These problems are minimized because the domestic water distribution system is an extensively looped system, which allows pressure and volume changes to reach a rapid equilibrium. Consequently, the customer does not experience any changes in the delivery of water.

The result of the Michelson Influent Wastewater Quality Improvement Plan has been to improve the quality of reclaimed water without purchasing additional low-dissolved-solids water and without requiring additional treatment to remove salts from reclaimed water. By using this operating scheme, compliance with salt limitations is enhanced, and potential for degradation of the Irvine groundwater basin is lessened.

PRETREATMENT

Industrial Waste Pretreatment Program

Because MWRP has a design capacity greater than 5 mgd (0.22 m^3/sec), it is required to have a pretreatment or industrial waste monitoring and enforcement program as stipulated in Section 40CFR403 of the Code of Federal Regulations. IRWD is also a member of several joint powers authorities, the County Sanitation Districts of Orange County, the Aliso Water Management Authority, and the Southeast Regional Reclamation Authority, all with pretreatment responsibilities in both the Santa Ana and San Diego Regional Boards.

Currently, all of the industrial waste generated within the MWRP service area is also tributary to the CSDOC. To ensure that pretreatment requirements satisfying the needs of both agencies were accomplished, IRWD and CSDOC entered into a Memorandum of Understanding (MOU) to administer industrial discharges. Under the MOU, both agencies retain the ability to enforce, should the other agency decline to do so; however, CSDOC has the initial responsibility to regulate industrial discharges according to the requirements of both agencies. Frequent and direct communication between IRWD and CSDOC has produced a smoothly functioning pretreatment program.

In addition to maintaining direct communication with CSDOC, IRWD also maintains an oversight role over industrial waste activities in its service area. Monthly samples are collected from large users and analyzed for constituents that may affect MWRP operations. Semiannually, the Spectrum, a large industrial park tributary to MWRP, is sampled and analyzed for heavy metals and cyanide. Experience has shown that heavy metals increase as industrial activity increases. To place industrial activity with respect to all discharges to MWRP, five locations representing commercial, industrial, and residential activity are monitored for all the priority pollutants except asbestos. IRWD has developed a 12-year database of monitoring both organic and inorganic toxic and hazardous pollutants.

The city of Irvine is a relatively new city, having been incorporated in 1971. In contrast to many older cities, Irvine is a planned growth development, extending out from the center of the city in discrete planning areas, villages or phases. The Spectrum industrial park, adjacent to the El Toro Marine Corps Air Station, was developed in phases. Each phase has its own collector sewer system which ties into the main interceptor sewers at single points. IRWD established six monitoring locations on the collector sewers when they were constructed. Each location consists of a flow measurement vault, a monitoring vault, and a utility easement at ground level to house monitoring and telemetry equipment. Most of the stations are wired together to allow data to be transferred to a single point for transmission to the central computer. At the present time, only one station is outfitted with monitoring and telemetry equipment. However, as the industrial base develops, new stations will be outfitted with remote sensing and monitoring equipment. To assist in outfitting these stations, students from Harvey Mudd College, a local private engineering college, are determining what equipment best meets the individual needs of each monitoring station. Ultimately, each monitoring station will transmit information specific to each industrial area, unattended and with complete reliability, to the central computer in the Operations Center.

The pretreatment program, managed jointly by IRWD and CSDOC, is effective in maintaining high quality influent into the MWRP. Consequently, the MWRP produces an effluent that meets all Basin Plan objectives. As industrial activity has increased, levels of copper, lead, and zinc have also increased—all planned increases within permit limits. Consequently, there have been no violations of MWRP effluent standards. On the contrary, the levels of organic pollutants has decreased significantly over 12 years from greater than 12 pounds [5.4 kilograms (kg)] per day to 1 pound (0.45 kg) per day. MWRP effluent has decreased to 1 pound (0.45 kg) per day, and the bulk of the priority pollutants detected in the effluent are halomethanes associated with chlorine disinfection. Still, the levels of halomethanes in MWRP effluent are lower than the levels found in domestic water. Presently, thirteen industries are governed by industrial waste permits, and the rate of compliance is high. Overall, the pretreatment program has been effective, because there have been no industry-related plant upset or pass-through events since the inception of the pretreatment program.

Water Softeners

Of particular concern to IRWD are self-regenerating water softeners. These water treatment devices have the potential of contributing large quantities of salt to the sewer, which degrades the quality of reclaimed water. In addition, the Irvine groundwater basin is relatively high in TDS, with no assimilative capacity. To minimize the negative effects of self-regenerating water softeners, IRWD specifically prohibits their installation and discharge into the sewer. IRWD does allow those water softeners that are recharged off site, where brine discharge is accepted. To facilitate enforcement, the city of Irvine has adopted as an ordinance the IRWD prohibition of self-regenerating water softeners, which allows the city of Irvine to exercise its police powers to prohibit the installation of self-regenerating water softeners, and to remove illegally installed units.

PROCESSES

IRWD was formed in January 1961 under provisions of Section 13 of the California Water Code. When IRWD was formed, most of the service area was agricultural with future development in the planning stages. In these early stages, IRWD was able to develop a long-range program of total water management.

In 1965, the District began providing wastewater collection and treatment services and began considering the reuse potential of its treated sewage effluent. By late 1967, the District had completed construction of the first phase of the MWRP, with the plant's effluent being delivered to The Irvine Company for irrigation use. This first phase of the MWRP included: an operations control room; headworks facilities, consisting of one aerated grit chamber and one grit washer; sludge holding tank; sludge pump room; sludge drying beds; three aeration basins, one of which was operated as an aerobic digester; three electric aeration blowers; five secondary clarifiers; and an effluent pumping station, consisting of one gas engine pump and two electric pumps. This original plant was operated as an extended aeration plant with a rated capacity of 2 mgd (0.09 m³/sec) of secondary effluent.

Between 1969 and 1970, the MWRP was expanded to a rated capacity of 5 mgd (0.22 m³/sec). This plant expansion included the addition of: a laboratory; three primary clarifiers, one of which was operated as a holding tank; sludge thickener; two primary sludge pumps; a dissolved air floatation thickener; a 5 mgd (0.22 m³/sec) emergency storage pond; and a chemical feed building. At the time of this expansion, the plant was being operated to remove biological phosphate and organic nitrogen, utilizing the Ryckman process.

Between 1974 and 1975, a new chlorine facility was added. An 8,000 standard cubic feet per minute (SCFM) (4 m³/sec) gas engine air blower and a gas engine effluent pump were also included at this time.

In 1976, continuing IRWD's commitment to total wastewater reclamation, the MWRP was expanded to include tertiary treatment at a rated capacity of 7 mgd (0.31 m³/sec). This expansion included the addition of: two circular tertiary sand filters; four circular secondary clarifiers; filter pump station, consisting of two electric pumps and one gas engine pump; a RAS/WAS pump room; a chlorine contact tank and flash mixer; a storm water pump station; and a 7 mgd (0.31 m³/sec) emergency storage pond.

In 1977, as a continuation of the 7 mgd (0.31 m³/sec) expansion, the following equipment was added: two headwork's bar rakes; two aerated grit chambers; two grit washers; two primary clarifiers; one electric effluent pump; one gas engine effluent pump; and a 4,000 SCFM (2 m³/sec) electric air blower.

Then, in 1979, in order to further meet the growing landscape irrigation demand for reclaimed water and also to comply with Title 22, California Code of Regulations, "Reclamation Criteria," the MWRP was expanded to its present rated capacity of 15 mgd (0.66 m³/sec). This expansion included the addition of the following: a sludge dewatering facility, consisting of three centrifuges; two 2.5 mgd (0.11 m³/sec) flow equalization basins; long-term emergency storage pond; six aeration basins; six aeration blowers; nine secondary clarifiers; a RAS pump station consisting of four RAS pumps; seven dual media tertiary filters; a filter pump station, consisting of three gas engine pumps; a spent backwash tank; a chlorine contact tank; a chlorination facility consisting of three evaporators and two chlorinators; and an effluent pump station consisting of three electric pumps. The plant is operated in a carbonaceous BOD removal mode with continuous pumping of waste activated sludge, which is adjusted daily in order to maintain a 0.4–0.5 F/M ratio.

In 1980, the centrifuges were replaced with belt presses, and in 1982 the headwork's bar rakes were replaced with new mechanical bar rakes.

In 1987, based on the conclusions of the 1984 Irrigation/Reclamation System Master Plan, the IRWD entered into a solids residual disposal agreement with the CSDOC. With the addition of a force main sludge line and pumping station, all solids handling processes were terminated at the MWRP. This included elimination of grit removal, sludge thickening, sludge dewatering and aerobic digestion. Presently, all primary sludge, primary skimmings, waste activated sludge, and secondary skimmings are conveyed, in liquid form, to CSDOC.

In 1993, a new spent backwash tank was constructed, as well as new spent backwash pumps and additional lines for filtration, spent backwash, raw sludge, and waste activated sludge.

From December 1994 until April 1995, the plant operations section experimented by changing from a carbonaceous BOD removal mode to a biological nitrogen removal (BNR) mode of operation. By incorporating major changes in standard operating procedures and utilizing existing plant process equipment, the MWRP successfully produced a nitrified/ denitrified effluent with an average total inorganic nitrogen (TIN) concentration of 4.11 mg/L. This experiment helped to establish what the current MWRP capabilities and limitations for future BNR were and led the way to important design decisions for the upcoming plant upgrade project.

In November 1995, in order to meet the growing demand for reclaimed water and to continue to produce reclaimed water as efficiently as possible, the MWRP began an extensive plant upgrade project. The first phase of the project will replace some badly degraded equipment and include the following: primary clarifier improvements; odor control system; primary clarifiers covers; primary clarifier skimming improvements; flow equalization basin improvements; fine air aeration system; high efficiency aeration blowers; secondary clarifier improvements; secondary clarifier skimming improvements; return activated sludge pump station changes; filter improvements; changes to effluent pumping; and on-site storage of reclaimed water. These upgrades will help to ensure that the MWRP will continue to efficiently produce a high quality reclaimed water product well into the twenty-first century.

SLUDGE DISPOSAL

During the early years of the MWRP, sludge residence times in the aerobic digesters were long enough to allow for on-site storage of sludge in sludge drying beds from 1967 to 1974. Due to the poor design of these early sludge drying beds, odors and flies were a constant problem.

From 1974 to 1979, sludge from the MWRP was hauled by a tanker truck for land disposal. The sludge was sprayed on The Irvine Company land, near a utility easement, and disked into the soil as a soil amendment. No crops were ever grown on this land, which eventually grew over with forage for cattle.

From 1979 to 1982, with the installation of solids handling equipment at the MWRP, the sludge was dewatered using belt presses and trucked to a sanitary landfill.

From 1982 to 1987, the MWRP contracted with a local composting contractor and the dewatered sludge was hauled to a composting site in Ontario, California.

In 1987 the MWRP entered into a solids residual disposal agreement with the CSDOC. Since that time, all sludge processed at the MWRP has been pumped in liquid form to the CSDOC.

COUNTY SANITATION DISTRICT OF ORANGE COUNTY (CSDOC)

Background

IRWD made a commitment to a total wastewater reclamation strategy based on two related factors. First, water quality requirements for ocean disposal were nearly as stringent as the requirements for reuse. Second, total reclamation was less expensive because of reclaimed water sales and reduced imported water costs. However, the cost of reclaimed water as an irrigation source appeared in 1982 to be escalating faster than alternative water sources. Other options for long-term wastewater disposal needed to be evaluated. IRWD and CSDOC approved the concept of investigating treatment and disposal of raw sewage generated in the IRWD service area through facilities of CSDOC.

IRWD retained the engineering firm CH2M HILL to perform a reconnaissance-level Regional Wastewater Management Study (RWMS) to assist IRWD in evaluating various options for wastewater treatment and disposal within the IRWD service area and to recommend the best course of action to implement the selected wastewater management alternative.

Wastewater Management Alternatives

Two basic alternatives to meet IRWD's future wastewater treatment and disposal requirements were identified:

(1) Alternative 1—Wastewater Treatment Expansion (Reclamation/Expand MWRP to 40 mgd (1.75 m³/sec)—Seasonal reclamation of 2.5 to 15 mgd (0.11 to 0.66 m³/sec) average dry weather flows (ADWF) at MWRP in conjunction with direct ocean disposal of MWRP secondary effluent for flows exceeding the reclamation rate [from 25 to 37.5 mgd (1.1 to 1.64 m³/sec) ADWF on a seasonal basis].

(2) Alternative 2—Raw Wastewater Export (Reclamation/Export to CSDOC)—Seasonal reclamation of 2.5 to 15 mgd (0.11 to 0.66 m³/sec) ADWF at MWRP in conjunction with raw wastewater/solids export to CSDOC treatment and disposal facilities for flows exceeding the reclamation rate (from 25 to 37.5 mgd (1.1 to 1.64 m³/sec) ADWF on a seasonal basis).

Alternative 1—Wastewater Treatment Expansion

The existing unit processes and plant layout were evaluated to determine how MWRP could be expanded to accommodate the projected ultimate 40 mgd (1.75 m³/sec) ADWF wastewater flow. It was concluded that a parallel 25 mgd (1.1 m³/sec) ADWF appeared to be the best options for the MWRP capacity. Reconnaissance-level screening of liquid treatment processes was conducted, and the following facilities were selected as the basis of the 25 mgd (1.1 m³/sec) ADWF parallel plant addition at MWRP:

- headworks/influent pumping [80 mgd (3.5 m³/sec) peak wet weather flow (PWWF)]
- circular primary clarifiers
- additional primary effluent flow equalization basins
- plug flow step feed activated sludge basins
- circular secondary clarifiers

Anaerobic digestion was selected as the basis for handling solids in the expansion of the MWRP to 40 mgd (1.75 m³/sec) ADWF capacity based primarily on low present worth costs, energy recovery benefit, and proven technology.

The following facilities are required for transporting up to 40 mgd (1.75 m³/sec) ADWF of MWRP secondary effluent to the ocean:

- secondary effluent pump station
- interceptor pipeline to CSDOC Plant No. 2
- use of IRWD's existing 15 mgd (0.66 m³/sec) PWWF capacity in CSDOC's ocean outfall facilities located in Huntington Beach, California
- new 25 mgd (1.1 m³/sec) PWWF separate IRWD ocean outfall for flows exceeding the existing 15 mgd (0.66 m³/sec) PWWF CSDOC outfall capacity

Alternative 2—Raw Wastewater Export

The following facilities were selected as the basis for transport of up to 80 mgd (3.5 m³/sec) PWWF of raw wastewater from the IRWD service area (MWRP) to CSDOC Plant No. 1:

- raw sewage pump station at MWRP
- dual 42-inch (1.1-meter) force main system
- connection to headworks at CSDOC Plant No. 1
- MWRP operations/maintenance building

Seasonal levels of MWRP solids exported to CSDOC (quantity and quality) over the 50-year planning period were estimated. The following CSDOC improvements were selected as the basis for facilities to handle IRWD exported flows:

- addition of 37.5 mgd (1.64 m³/sec) ADWF headworks, primary, secondary, and solids treatment capacity at CSDOC Plant No. 1

- new 96-inch (2.4-meter) interplant pipeline to handle combined CSDOC and IRWD peak flow increases [IRWD to fund 75 mgd (3.3 m³/sec) PWWF proportionate share]
- new 720 mgd (31.5 m³/sec) PWWF outfall booster pump station to handle combined CSDOC and IRWD flows [IRWD to fund 75 mgd (3.3 m³/sec) PWWF proportionate share]
- extension of existing 78-inch (2.0-meter) CSDOC ocean outfall for combined CSDOC and IRWD peak flow increases [IRWD to fund 60 mgd (2.6 m³/sec) PWWF proportionate share]

Comparative Cost Analysis

Reconnaissance-level capital and operational and maintenance cost estimates were developed for Wastewater Management Alternatives 1 and 2. A "constant dollar" cost comparison of the two alternatives was conducted with inflation equal to the discount rate. The difference in cost of the two alternatives was within the level of uncertainty of the estimates. The total costs of the two alternatives were considered essentially equal.

Optimization of Seasonal Reclamation

A Linear Programming (LP) Model was used to determine the least cost method of meeting landscape/agricultural irrigation demands, treating IRWD's raw sewage, and disposing of effluent from MWRP. In this manner, a broad viewpoint is taken where the wastewater management alternatives are examined in the context of the entire IRWD irrigation/reclamation system.

Three irrigation/reclamation system service area alternatives were run in the LP Model, assuming a dual water system to supply the following demands:

(1) Alternative A Service Area—the existing irrigation/reclamation system landscape service area plus servable agriculture

(2) Alternative B Service Area—a service area composed of existing landscape service, new landscape demands below elevation 300 feet (90 meters) (Zones A and B), potential conversion of areas using domestic water for landscape and servable agriculture

(3) Optimized Service Area—a service area composed of at least the existing landscape service area plus servable agriculture; based on cost-effectiveness, potential of additional service area to include new landscape and conversion areas to the limits of the Alternative B service area

The optimized service area is the most important solution for questions posed by the RWMS. The RWMS focuses on whether to expand treatment capacity at MWRP or to purchase capacity in facilities associated with the CSDOC option. The LP Model was run for four discrete years: 1985, 1995, 2020, and 2032. Each service area is assumed to reach its maximum irrigation demand by year 2011, and ultimate sewage from within the IRWD service area tributary to MWRP is estimated to occur at the year 2032. From the results of the optimizations to determine the most cost-effective dual water system service area, the follow conclusions were derived:

(1) The existing MWRP is always used.

(2) Additional reclamation by increasing MWRP capacity may be feasible.

(3) Any IRWD wastewater not used for reclamation should be exported for treatment and disposal to the ocean via CSDOC facilities.

Based in part on the results of the RWMS, IRWD entered into an agreement with CSDOC, and the necessary facilities were constructed whereby IRWD can divert raw wastewater to CSDOC for treatment and ocean disposal.

TELEMETRY SYSTEM

During the 1970s, the District was dependent upon outside agencies for its water supply, which was comprised of a system consisting of several turnouts. The District maintained a few facilities within the boundaries that supported the water system or boosted to other zones. These sites were operated manually by staff. When an adjustment was necessary, a systems operator would drive to the site, make the appropriate change, and then return to the previous location. The District operated under this procedure until the early 1980s, when telemetry monitoring facilities were installed.

The first facility to be converted to telemetry and control capabilities was the Zone I Reservoir in 1983. Although the District had circular charts indicating the levels of the three other reservoirs and some alarm mechanisms, this facility became the most advanced with upstream and downstream pressures, tank levels, control valves and intrusion alarms all connected to the telemetry system. These features contained manual overriding controls at the facility location, in addition to controls via telemetry, which were located at the Operations Center eight miles (13 kilometers) away. The control panel was made up of large square push buttons on an electrical panel with strip chart recording devices. If a change was required to control the tank level, the systems operator would simply push the increase or decrease button accordingly to achieve the respective change. Within minutes, the valve responded to this telephone-linked command.

In 1980, staff received approval to develop and install a Supervisory Control and Data Acquisition (SCADA) system at the District. This newly developing technology at the time was considered the wave of the future. Although the system could provide a greater level of integrity to the distribution system and still maintain the current labor force, the original projected price was so costly that an independent group was retained to review the design criteria. A study conducted by the Valued Engineering Consulting Group recommended major changes in the district's original contract, including: (1) that leased lines not be used from the telephone company (and that radio signals be pursued as a means of communication); and (2) that a common industry SCADA system be purchased, in addition to a system that meets the needs of other water agencies, not one that is specially designed for the District. In the final analysis, a total of 28 remote terminal units (RTUs) were installed, each independent of the other and designed to communicate by line of sight microwave signals broadcast at 900 MHz. Each RTU was designed to be installed above ground or inside a facility, which eliminated potential damages from vault flooding in the future.

During the year of the RTU installation, IRWD's Information Systems (IS) Director foresaw the need for a dedicated SCADA software staff member, and budgeted additional manpower. This proved to be a significant contribution to orchestrating a smooth startup. After the review of potential software companies, staff determined that the BIF system would meet the District's requirements.

To ensure a smooth start-up, IRWD's IS staff was fully informed and trained in the digital software operating systems prior to the system being delivered to the facility. The system was up and running the same day all of the connections were completed. The initial process of point checking involved placing the contractors in the field to confirm that all wiring and equipment were in place. Signals were transmitted back and forth while the IS system operator and the software contractor monitored the database, or *points*. The total start-up procedure was completed within one month.

The reclaimed system and the collection system was monitored primarily by the SCADA, with no real control capabilities designed for these sites. The original system only had three sites, or RTUs that were dedicated to the reclaimed system. Only one of these sites, Sand Canyon Reservoir pump station, had any type of remote control features via the SCADA system. The control was limited to downloading *start* and *stop* point pressure settings. The remainder of the reclaimed sites were either operating off of hydraulic control valves or local set points not requiring SCADA control. As new facilities were installed, superior features were incorporated into the pump design to allow more control.

The next wave of technology brought in the Micro 84's processors. These microprocessors could accept hard input and soft output, in addition to recognizing program logic. These units increased the reliability of the RTUs and the SCADA system because the solid state components were superior to the original RTUs, which were based on electronic relays. The PLC also enhanced the ease of modifying the current configuration of a system's operating parameters, which was accomplished through the unique programming capabilities.

After years of operating the entire district manually, staff began to feel comfortable with this new computer operating system within a six-month transition period. By the beginning of 1986, staff were operating the distribution system via a portable computer from home after normal business hours. The original portable unit was approximately the same size of a standard CRT in 1995. The unit would connect to *Central* via a modem, giving the operator full access to the CPU and virtual control of the domestic and reclaimed distribution systems.

A new concept in RTUs evolved from the PLC: The main machine interface system, more commonly known as the *Spectrum System,* included a basic key pad access control panel. Attaching directly to the panel face of the RTU's doors, this system allowed any authorized staff to view and modify control variables in real time. Prior to the introduction of this technology, an electrician would have to be scheduled to make minor parameter changes to the PLC prompted by seasonal demands. These changes usually required one to two days before any programming could be performed. With the new key pad access units, the operators were able to eliminate paperwork requirements and the need for additional manpower or equipment such as laptop computers. In 1987, BIF upgraded its software, which made the operation of the SCADA system simpler in some respects and more difficult in others. At the same time, software support became more limited.

In 1990, JPF installed a *dynac* system which emulated the BIF system graphically, but with different key commands. This system was fully operational by January 1991. During the 24-hour period required to implement the switchover, the SCADA system was off line and system operators manually monitored the various facilities and reservoirs. The new DEC VAX 6000 system came with the following components: one tape drive with 290 MB of storage; four hard drives with 280 MB storage each; and one control screen and one printer. The old CRTs were utilized in the new contract package along with the original printers. Although the radio system was changed out to a Micro-wave Data System (MDS 1000), the master unit located at the antenna was also utilized in the package agreement.

A major addition to the new SCADA system was a *Hotback up* system which was located at the headquarters facility. The linkup was accomplished through a T-1 communication link (a high-priority link from the treatment plant to the headquarters facilities via the telephone company's equipment). If switching from the primary system to the *Hotback up* was required, the switch was virtually transparent. The *Hotback up* system consisted of a VAX 6000 with 32 MB of ram and 21.2 MB of hard drives. In addition, the system had a vt320 CRT and a report printer that emulated the primary system reports. This new system included a *true* laptop as part of the hardware package. This unit could be carried in a small briefcase and connected at different locations with little encroachment of room space.

The *Hotback up* and the primary unit were rotated on a monthly basis for a short time to ensure smooth transitioning and reliability of each unit. With this new upgrade, a host of new points were added including several new pressure transducers for the reclaimed system.

In 1994, the SCADA system was moved to a new facility, which prompted the need for an upgrade in the computer's total storage capacity. The new total storage space was 4.5 GB for the primary system, including the *Hotback up* system.

As the District's SCADA system increased, problems arose with communications between the field and the yard. Despite repeated surveys and testing, the District was unsuccessful in polling the 55 RTUs 100 percent of the time. With the collaboration of a consultant, District electricians and a SCADA programmer, polling of the entire system was achieved nearly 100 percent of the time. Additionally, polling time was decreased to 60 seconds. Today, the reclaimed system has 12 dedicated RTU sites and continues to grow with the District.

Future

Currently, IRWD's Operations and IS departments are considering a new PC-based software package, WonderWare, for the treatment plant's SCADA system. The new software is unlimited in the amount of points it can handle as long as the system can handle the storage capacity. Staff will review the MWRP SCADA system and determine if WonderWare is suitable for the distribution system. The alternative is to maintain the existing mainframe system and upgrade this system in cooperation with Trans Dyn Control, Inc. The driving force in IRWD's final decision will be economics and ease of use.

Multiplex

The treatment plant did not receive telemetry until the early 1980s, fifteen years after startup. The original electronic system consisted of local process alarms. The alarms indicated highs, lows, and other variables that required the attention of Plant Operations. These systems consisted of hundreds of relays which were designed based process known as an *If statement* or *conditional parameters*. The relays recognized an input, triggering it to pull in other relays, which combined together in varying configurations to correspond with an alarm condition. This system was later replaced with a microprocessor that performed in a similar fashion, but nonmechanically. The opening and closing of the relays is done through the bits and bytes in the software. The system is programmed in a software system call *ladder logic* which emulates its predecessor, mechanical relays.

Treatment plant operations went to the *Modcon* microprocessor in 1984, which outperformed the relays. Not only could these units alert an operator via audible alarms and visual alarms, but they became local process controllers, turning on and off pumps, aerators, chemical supplies and other crucial treatment process. The microprocessor opened a new door of opportunity— the reduction of the treatment plant staffing, which was accomplished through attrition. As the District developed confidence in the reliability of the equipment, the treatment plant staff went to a schedule of 10-hour days for four consecutive days. The plant was manned seven days a week with two rotating crews, which dwindled down to half the original staff. The plant operators currently work Monday through Friday; during the weekend only one operator surveys the plant and its treatment processes. The microprocessors have been an invaluable addition to the plant, from the viewpoint of economy and ease of operations. This system is tied to a system similar to the distribution system's group computer. The computer allows the staff to monitor the treatment process at the plant including levels, power, chemical levels, dosages, backwashing flows, psi and a host of other significant processes. The next advancement for the plant is to achieve process control via the new system or SCADA system.

In early 1996, the District commissioned a contractor to excavate trenches for a new fiber optic cabling system which would link every process in the plant to the plant Operations Control Center. The existing network of cables connecting all of the processes was not usable, and the contract had to start over. This new cabling system will benefit the integrity of the signal that will be generated and received. The new package will have a laptop similar to the Distribution System's unit. The plant operator on-call will be able to link into central and monitor or adjust the plant accordingly.

RECLAIMED SYSTEM DEMANDS

In the mid-1980s, Southern California experienced a severe drought. This situation caused the District to initiate a program requiring those entities that could use reclaimed water, i.e., agricultural farmers, homeowner associations, etc. to have reclaimed water as the only source available when requesting service.

In the 1990s the District took an even more aggressive approach and began converting to reclaimed water all landscape connections previously connected to domestic water. This conversion was accomplished through low-interest loans issued from the Metropolitan Water District of Southern California. Since the District's reclaimed water is tertiary treated water, customer acceptance was generally not a problem. Tertiary treated water achieves a level of clarity that emulates domestic water, which provides an aesthetic benefit to customers. Also, the price of reclaimed water is a motivator in the conversion from domestic water to reclaimed water, i.e., a service charge of $0.64/ccf ($0.23/m^3) for domestic water and $0.58/ccf ($0.20/m^3) for reclaimed water. All of the above-mentioned factors contributed to the accelerated growth of the reclaimed water system, impacting the demands of the system dramatically.

Seasonal Demands

Seasonal demands are best explained within the context of the annual demands upon the reclaimed water system. During the fall season, the District begins preparing for the next year's high demand season (which occurs during the summer months). The reservoirs are normally at their lowest point at this time of the year. As the reclaimed water demands diminish in direct correlation to the drop in temperature, a gradual filling of the reservoirs occurs. This gradual filling is also directly related to the decreases in temperature and sunlight duration. Typically, fall demands are one-fourth of the summer peaking months and a one-eighth of what the treatment plant is capable of producing.

Occasionally in the fall, a unique local weather phenomena known as the Santa Ana winds occurs. These winds originate in arid desert climates and dry out everything in their path, a condition that increases water demands three-fold. This weather pattern can persist up to several weeks, dramatically changing the requested flow from the treatment plant and the amount of water diverted into storage.

Based on historical data, significant rainfall in the Southern California area [1 inch (2.5 centimeters) or greater] does not usually occur until early December. Once a consistent pattern of rainfall begins, the reclaimed water demand drops off to almost nonexistence. The demands in the system are usually at their lowest between January and March. The treatment plant is then at its lowest operating flow and the reservoirs are close to maximizing their storage capacity.

Beginning in April, the demands on the system are three times greater than the previous month. The demands vary from a daily average of 4.7

cfs to 12.8 cfs (0.13 to 0.36 m³/sec), which indicates the beginning of the peaking season. May through August are considered to be the months of highest demand, with September being the highest demand month. During this time, the weather can range from hot and tropical to dry and windy from the Santa Ana conditions. The average daily demands for this season are 32.3 cfs (0.915 m³/sec). Once summer is past, the seasonal cycle begins again.

In the early 1990s, the District pursued the possibility of providing reclaimed water to urinals and toilets in high-rise buildings to eliminate flushing domestic water into the drains and to find new areas of demand for the reclaimed system's seasonal excess water. These issues are addressed in another section of this chapter.

Historical Demands

The reclaimed water demands for the past 13 years are expected to represent future use unless a significant number of high-rise or industrial users are connected to the system. Historical demands have made an unpredictable change in the total demands (Figure 21.1). The demands are dropping off even though the number of services are increasing annually. This increase is due in part to the conversion of domestic irrigation service to reclaimed water, along with the addition of new transmission lines that reach new areas of potential conversions. The significant decrease in demand seen in Figure 21.1 can be attributed to the implementation of the Operation Outreach, which began in 1992. This program's emphasis is that the correct amount of reclaimed water for irrigation should be based on evapotransporation and acreage, which is enforced by a water rate schedule (tiered rate structure). This schedule provides an incentive rate if the customer manages his system well. If not, the customer is billed according to the excessive usage. The penalties can be as severe as eight times the normal rate. Due to the comprehensiveness of the Operation Outreach's program, it will not be covered in this section. One other factor accounts for the overall reduction in customer reclaimed water usage: IRWD's largest customer, The Irvine Company, has become more self sufficient in supplying its own water and is cutting back on total crops planted and irrigated.

Daily Demands

The monthly demands mentioned earlier represent constant cycles, with the highest demands occurring in September and the lowest in March. Likewise, the daily flows are consistently the same for Sundays, Tuesdays, Wednesdays, and Saturdays (which are typically low demand days). Mon-

Fiscal Year	# of Meter Connection	Total Reclaimed Water Usage (acre-feet)
1977	3	42
1978	80	1120
1979	182	2548
1980	285	3990
1981	328	4592
1982	390	8498.46
1983	411	9375.94
1984	451	9253.09
1985	527	7371.92
1986	651	8519.93
1987	716	9385.53
1988	792	10363.71
1989	863	10929.74
1990	925	10986.03
1991	962	10037.31
1992	1194	12672.28
1993	1291	11017.58
1994	1445	9762.38

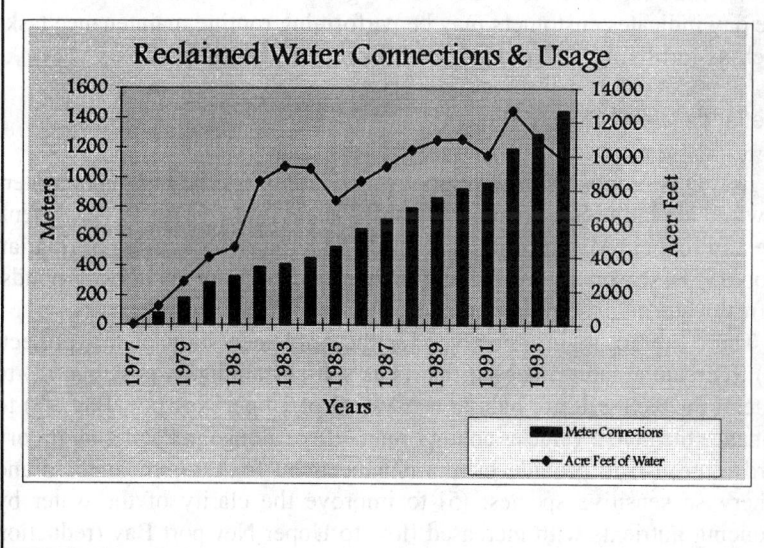

Figure 21.1 Reclaimed water connection and usage.

days, Thursdays, and Fridays, on the other hand, are typically high days of demand. These patterns are influenced by customer acceptance of wet areas the morning after irrigation has occurred (see Figure 21.2). One example is the Irvine Unified School District in Irvine, California, which chooses not to water on Sunday evening so the playing fields will be dry for children's activities during the week.

Hourly Trends

Hourly trends are directly influenced by the District's guidelines, which are based on the County and State Health Departments' standards for reclaimed water usage.

The District enforces strict requirements as to when and how reclaimed water can be used, allowing customers to irrigate landscape between the hours of 9:00 P.M. and 6:00 A.M. without supervision. If the customer wants to survey the system for efficiency, he/she or a representative of the establishment must be present throughout the duration of the irrigation. The intense watering period is between 9:00 P.M. and 6:00 A.M. The maximum peaking occurs between 10:00 P.M. and 2:00 A.M. The actual peak demand drops off at 3:00 A.M. This cut-off is a natural one for the customers as it allows time for water to soak into the ground. A consistent flow during the day indicates customers may be performing routine maintenance tasks such as fertilizing or checking for broken sprinkler heads.

Wetlands Water Supply Project

IRWD is proposing to develop a water supply for the migratory waterfowl ponds in the San Joaquin Marsh adjacent to the MWRP. At present, the San Joaquin Marsh contains a variety of habitats, consisting of riparian woodlands, seasonally wet meadows, managed migratory waterfowl ponds, a freshwater marsh and an upland habitat.

There are six main objectives for the implementation of this project: (1) to create a natural habitat that is as self-sustaining as possible; (2) to extend the rich wildlife habitat ecosystem of Upper Newport Bay; (3) to enhance the marsh value as an important refuge along the Pacific migratory bird route; (4) to provide habitat enhancement for rare, endangered and otherwise sensitive species; (5) to improve the clarity of the water by reducing nutrients with increased flow to Upper Newport Bay (reduction in nutrient concentration in the Bay over time will result in a long-term quality improvement); and (6) to balance water demand and supply, leading to more effective and cost efficient operation.

During the demonstration phase of the project, scheduled to take place during two successive winters from October 1 to March 30 of each year,

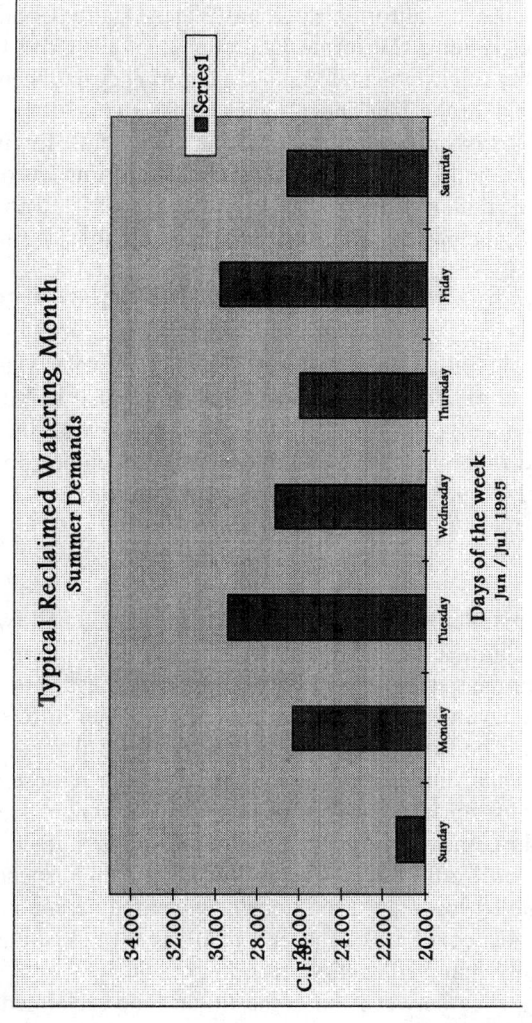

Figure 21.2 Typical reclaimed water month.

the ponds will be filled with reclaimed water from IRWD's distribution system. Approximately three weeks later, the water will be discharged from the ponds into the San Diego Creek. The project will evaluate nitrogen removal from the reclaimed water under various pond operating conditions.

An external monitoring program developed in cooperation with the city of Newport Beach, the California Regional Water Quality Control Board, the UC Reserve System, California Department of Fish and Game and the Orange County Environmental Management Agency has been proposed to measure effects on the San Diego Creek and Newport Bay. Extensive environmental analysis during the preparation of the draft Environmental Impact Report concluded that there will be no adverse impacts on the fish, wildlife or vegetation in and surrounding the Upper Newport Bay. If impacts do occur, the project would be modified quickly to mitigate them.

RECLAIMED WATER DISTRIBUTION SPECIAL REQUIREMENTS

In 1994 and 1995, the District's use of reclaimed water was approximately 23 percent or 13,017 acre-feet (AF) (16,056,000 m³) of the total water used. When reclaimed water usage is combined with the District's use of untreated or raw water, non-potable usage becomes 35.5 percent or 19,545 AF (24,109,000 m³) of the total water used.

Reservoirs

The District maintains four reservoirs (storage tanks) and two earthen dams for the reclaimed water system. In addition, there are two storage tanks in the preliminary planning stages. Design of domestic and reclaimed water reservoirs have very few variations, with the major difference being the floor design. The floors are to be sloped at a 0.25 percent or greater slope than that of a domestic water design. For this reason, reclaimed water is a combination of treatment plant water, raw water from a local lake (Irvine Lake), and strained water from two open surface reservoirs (Sand Canyon and Rattlesnake). These waters contain suspended solids which precipitate by low flow conditions or when they come into contact with varying water conditions. The end result is a thick sludge blanket which is manually removed each year.

Another difference between the domestic and the reclaimed design is the drain line of the reservoir. Since the reclaimed water cannot be poured onto the ground or discharged into a storm drain, special provisions must be applied to accomplish approved disposal of the water and/or sludge. For example, since the water cannot be discharged into the storm drain system, reclaimed water must be pumped from below a buried 30-foot (9.1-meter) tank to a storage or disposal site. This disposal task requires special piping, equipment, electrical sources, and an array of other items just to dewater/desludge a reclaimed water tank. The drain line that attaches to the inside of the reservoir is usually upsized in order to handle the larger volume of sludge.

Reclaimed water tanks are not generally designed using steel, since the higher chlorine levels in reclaimed water make it more aggressive than domestic water on steel coatings. Concrete tanks are typically used for reclaimed water.

Pipelines

Reclaimed water lines are installed in a similar manner to domestic water lines as far as trenching, bedding, and pipe. The same considerations given to the design and installation of domestic water lines are rendered for the reclaimed water. The only exception is when reclaimed water lines come within 10 feet (3.0 meters) of domestic lines or cross a line. In such cases, the domestic line must be encased in concrete for 10 feet (3.0 meters) beyond the interfacing.

All pipeline apparatuses are the same for reclaimed water as domestic water pipeline elements, with three minor exceptions: (1) The pipeline must have a continuous purple banner tape with the words, *Reclaimed water line buried below;* (2) The banner must be laid on top of the pipeline; and (3) The valve can lids are to be of triangular shape, and any item above grade is to be painted either reclaimed purple or beige with a purple sticker with the words, *Reclaimed water.*

Pumps

Pumps and associated parts are virtually the same for reclaimed water as for domestic water systems. All electrical equipment is also identical to domestic pump station equipment and design.

The pumps are subject to similar hydraulic conditions that a domestic system will encounter, with the only major difference being demand. Demands requirements are very low flows during the day and extremely high flows in the evening, so reclaimed pump stations are designed with a jockey pump or a variable speed pump. The pump's hydraulic control features are identical to a domestic design and are maintained the same with the one minor exception, the influence of water quality. Water quality can affect the amount of maintenance and the number of failures that will occur with hydraulic control components. A typical failure is caused by material clogging small sensing tubes or orifices. Another condition is a buildup of scaling material on shafts or stems that cease the moving parts, causing the unit to fail.

Miscellaneous

Any hose bibs located at the reclaimed pump station must be properly tagged with the words, *Domestic Use Only* or *Reclaimed Water Use, Do Not Drink.* This applies to any type of equipment that will transfer or carry these two types of water, i.e., any temporary water tanks, portable pumps, fire hoses, mixing equipment, meters, etc. Repair crews who work on reclaimed water need to exercise good cleaning practices on equipment. The same tool may be used at various sites, but thorough cleaning of the equipment must be performed after each repair job.

The District's pump station facilities have signs posted *(Reclaimed Water Pump Station)* at all points of access and at the general location. When performing weekly bacteria sampling tasks, staff will segregate the days he/she will sample domestic water and reclaimed water as well as segregate the sample carrying container.

WATER QUALITY

Water quality monitoring for IRWD's reclaimed water system is driven by two different motivations: regulatory compliance and internal customer standards. The monitoring program for regulatory compliance is performed at the MWRP and is based on National Pollution Discharge Elimination System and California Administrative Code requirements. Distribution system monitoring has evolved around internal standards for each of the different types of uses. This section will discuss regulatory and internal standards, MWRP and distribution system monitoring programs, reservoir water quality and special projects that are currently underway at the District.

WATER RECLAMATION STANDARDS

During the past thirty years, the State of California Department of Health Services has provided standards in the California Administrative Code for the use of reclaimed water for nonpotable purposes. The evolution of these standards since 1968 will be covered in detail in this section primarily because it closely parallels IRWD's history of operation. As a result, IRWD played a key role in the development of the state regulations that widened the spectrum of acceptable uses of reclaimed water.

Title 17—Public Health (1968)

The May 1968 Water Reclamation Standards were covered in the California Administrative Code, Title 17—Public Health. These basic regulations covered specific definitions and categories of uses. Table 21.1 outlines the approved uses, treatment requirements and water quality standards of the 1968 Title 17 regulations. The regulations included in the 1968 Title 17 are listed below:

- definitions
- irrigation of food crops
- irrigation of fodder, fiber and seed crops
- landscape irrigation
- recreational impoundments

Title 22, Division 4—Environmental Health (1978)

In 1978, the California Administrative Code, Title 22, Division 4—Environmental Health was adopted, which established new standards for water reclamation in the State of California. The 1978 Title 22 regulations included new usage areas, performance and design section, as listed below:

- groundwater recharge

TABLE 21.1. **Title 17 Regulations (1968) Summary.**

Type of Use	Treatment Level	Total Coliform Standard (Median)
Spray irrigation of produce	Disinfected, filtered wastewater	2.2/100 ml
Surface irrigation of produce	Disinfected, oxidized wastewater	2.2/100 ml
Surface irrigation of produce[a] (orchards and vineyards)	Primary effluent	N/A
Fodder, feed, sod	Primary effluent	N/A
Surface irrigation of food crops[b]	Primary effluent	N/A
Spray irrigation of food crops[b]	Disinfected, oxidized wastewater	23/100 ml
Pasture (milking animals)	Disinfected, oxidized wastewater	23/100 ml
Landscape irrigation[c]	Disinfected, oxidized wastewater	23/100 ml
Non-restricted recreational impoundment	Disinfected, filtered wastewater	2.2

[a]Fruit is not harvested that has contacted irrigation water.
[b]Must be processed (chemical or physical) to kill pathogenic bacteria.
[c]Includes golf courses, cemeteries, lawns, parks, playgrounds, freeway landscapes and other landscape areas accessible to the public.

- other methods of treatment
- sampling and analysis
- engineering report and operational requirements
- general requirements of design
- alternative reliability requirements for uses of permitting primary effluent
- alternative reliability requirements for uses requiring oxidized, disinfected wastewater or oxidized, clarified, filtered, disinfected wastewater

Table 21.2 outlines the approved uses, treatment requirements and water quality requirements of the 1978 Title 22 regulations.

R-24-93

R-24-93 is the new draft version of the California Administrative Code,

TABLE 21.2. **Title 22 Regulations (1978) Summary.**

Approved Uses	Treatment Level	Total Coliform Standard (Median)
Spray irrigation of food crops	Disinfected, oxidized, coagulated, clarified and filtered wastewater	2.2/100 ml
Surface irrigation of food crops	Disinfected, oxidized wastewater	2.2/100 ml
Surface irrigation of orchards and vineyards[a]	Primary effluent	N/A
Fodder, fiber and seed crops	Primary effluent	N/A
Pasture for milking animals	Disinfected, oxidized wastewater	23/100 ml
Landscape irrigation[b]	Disinfected, oxidized wastewater	23/100 ml
Landscape irrigation[c]	Disinfected, oxidized, coagulated, clarified, filtered wastewater	2.2/100 ml
Nonrestricted recreational impoundment	Disinfected, oxidized, coagulated, clarified, filtered wastewater	2.2/100ml
Restricted recreational impoundment	Disinfected, oxidized wastewater	2.2/100 ml
Landscape impoundment	Disinfected, oxidized wastewater	23/100 ml
Groundwater recharge	[d]	[d]

[a]No fruit is harvested that has come in contact with irrigating water or the ground.
[b]Includes golf courses, cemeteries, freeway landscapes, and landscapes with similar public access.
[c]Includes parks, playgrounds, school yards, and other landscaped areas with similar access.
[d]Each project is reviewed by the State of California Department of Health Services on an individual basis.

Title 22, Division 4 and Title 17, Division 1—Environmental Health, which is expected to be adopted in 1997. The new regulations have incorporated a number of changes from the 1978 version which are listed below:

- defines three levels of wastewater treatment: Disinfected secondary-2.2 reclaimed water, Disinfected secondary-23 reclaimed water and Disinfected tertiary reclaimed water
- provides an expanded list of acceptable reclaimed water uses for each level of wastewater treatment
- expands the monitoring requirements for using reclaimed water in unrestricted recreational impoundments
- sets standards for the use of reclaimed water in commercial cooling towers
- establishes more explicit definitions for use area requirements
- sets standards for the use of reclaimed water for dual plumbed facilities
- implements language changes for engineering reports and design/ operation and sampling requirements

Table 21.3 outlines the approved uses, treatment level and non-treatment water quality standards of the new Title 22 regulations.

Requirements for treatment have changed significantly during the past twenty-eight years. Table 21.4 lists the treatment process and water quality standards for each treatment level.

INTERNAL CUSTOMER REQUIREMENTS

Distribution system monitoring to meet customer requirements is a major effort at the IRWD. During its thirty years of operation, IRWD has witnessed a steady growth in the diversity of user types, a trend that is expected to continue. IRWD's major categories of customer uses include agricultural, landscape, water features, single family homes, and toilet and urinal flushing. The District is currently working on four new potential uses which are textile manufacturing, cooling towers, cement manufacturing and wetlands enhancement. IRWD has learned from experience that each different type of use has internal requirements required for reclaimed water to be used successfully and/or for the customer to be satisfied. Table 21.5 lists the major areas of concern for the current and future uses of reclaimed water at IRWD.

ENVIRONMENTAL COMPLIANCE

Title 17

California Administrative Code, Title 17, states that water suppliers

TABLE 21.3. **Title 22 Regulations (New) Summary.**

Approved Use	Treatment Level	Non-Treatment Standards
Flushing toilets and urinals, priming drain traps, industrial process water (Worker Contact), structural fire fighting, decorative fountains, commercial laundries, consolidation around potable water pipelines, artificial snow making (commercial outdoor use)	Disinfected tertiary reclaimed water	None
Industrial boiler feed, non-structural fire fighting, consolidation around non-potable piping, soil compaction, mixing concrete, dust control (roads and streets), cleaning roads, sidewalks and outdoor work areas	Disinfected secondary-23 reclaimed water	None
Sanitary sewer flushing	Undisinfected secondary reclaimed water	N/A
Industrial and commercial cooling towers (mist)	Disinfected tertiary reclaimed water	Use of biocide to control growth of legionella and other micro-organisms
Industrial and commercial cooling towers (no mist)	Disinfected secondary-23 reclaimed water	None
Nonrestricted recreational impoundments	Disinfected tertiary reclaimed water (conventional)	None
Nonrestricted recreational impoundments	Disinfected tertiary reclaimed water (non-conventional)	12 consecutive months of monitoring for *Giardia, Cryptosporidium* and enteric viruses; coliform standard met at the entry point into the impoundment
Restricted recreational impoundments	Disinfected secondary-2.2 reclaimed water	None

974

TABLE 21.3. (continued).

Approved Use	Treatment Level	Non-Treatment Standards
Landscape impoundments (no decorative fountains)	Disinfected secondary-23 reclaimed water	None
Irrigation Uses:		
Food crops (water contacts with edible portion of plant), parks and playgrounds, school yards, residential landscaping, unrestricted access golf courses	Disinfected tertiary reclaimed water	None
Irrigation Uses:		
Food crops (water does not contact edible portion of plant)	Disinfected secondary-2.2 reclaimed water	None
Irrigation Uses:		
Cemeteries, freeway landscaping, restricted access golf courses, ornamental nursery stock and sod farms, pasture for milk cows, nonedible vegetation with control access and limited use	Disinfected secondary-23 reclaimed water	None
Irrigation Uses:		
Orchards (water does not contact edible portion), vineyards (water does not contact edible portion), non-food-bearing trees, fodder and fiber crops, food crops (commercial pathogen destruction process), Christmas tree farms (no irrigation within 14 days of allowing public access)	Undisinfected secondary reclaimed water	N/A

must protect the public water supply from contamination occurring through backflow by implementing a cross-connection program. As part of that requirement, the IRWD is required to ensure compliance with the code by periodically reviewing its customers' water systems to determine if compliance is in order.

TABLE 21.4. **Treatment Level Requirements.**

Treatment Level	Treatment Process	Water Quality Requirements
Undisinfected secondary reclaimed water (1996)	Oxidized wastewater	None
Disinfected secondary-23 reclaimed water (1996)	Oxidized and disinfected wastewater	Median total coliform 23/100 ml Maximum of one total coliform Sample in 30 days exceeding 240/100 ml
Disinfected secondary-2.2 reclaimed water (1996)	Oxidized and disinfected wastewater	Median total coliform 2.2/100 ml Maximum of one total coliform Sample in 30 days exceeding 23/100 ml
Disinfected tertiary reclaimed water (1996)	Oxidized, coagulated, filtered and disinfected wastewater	Average turbidity—2 NTU Turbidity does not exceed 5 NTU more than 5% of the time Turbidity never exceeds 10 NTU 90-minute modal contact time at 5 mg/1 chlorine Median total coliform 2.2/100 ml Maximum of one total coliform Sample in 30 days exceeds 240/100 ml 10^5 virus inactivation
Primary effluent (1978)	Solids removal	Maximum of 0.5 milliliter of settleable solids per hour
Disinfected, oxidized wastewater	Oxidized and disinfected	Median total coliform 2.2/100 ml (strictest use) Maximum total coliform does not exceed 240/100 ml in any two consecutive samples (strictest use)
Disinfected filtered wastewater (1978)	Oxidized, coagulated, clarified, filtered and disinfected	Average turbidity of 2 NTU Turbidity does not exceed 5 NTU more than 5% of the time Median total coliform 2.2/100 ml Maximum total coliform does not exceed 23/100 ml (strictest use)
Primary effluent (1968)	Partial solids removal	Maximum 1.0 ml/1 settleable solids
Disinfected oxidized wastewater (1968)	Oxidized and disinfected	Median total coliform 2.2 or 23/100 ml (use specific)
Disinfected filtered wastewater (1968)	Oxidized, coagulated, filtered and disinfected	Turbidity does not exceed 10 NTU Median total coliform 2.2/100 ml

TABLE 21.5. **Water Quality Concerns.**

Type of Use	Concerns
Agricultural	Total dissolved solids, sodium adsorption ratio, boron, chlorides, suspended solids, chlorine
Landscaping	Total dissolved solids, sodium adsorption ratio, boron, chlorides, odors, suspended solids, chlorine
Toilet and urinal flushing	Odors, color, suspended solids
Water features	Nutrients
Single family homes	Total dissolved solids, sodium adsorption ratio, boron, chlorides, odors, suspended solids, chlorine
Cooling towers	Nutrients, TDS, odor, suspended solids, hardness
Textile mill	Color, odor, inorganics
Cement manufacturer	Suspended solids, inorganics
Wetlands enhancement	Nutrients

Prioritizing a list for inspecting customers involves reviewing customers and ranking them according to the probable severity of the health impacts should a backflow incident occur. The likelihood of cross-connections existing on the customer's premises is an important factor for consideration. In general, non-residential customers will account for most of the potential cross-connection problems. Customers that handle either infectious waste or hazardous chemicals should receive top priority.

Title 17 states that at least one person be trained in cross-connection control to carry out the cross-connection program. IRWD has developed an active cross-connection control program with a cross-connection control supervisor to administer the program.

A cross-connection is a potential connection between any part of a potable water system used to supply water for drinking purposes, and a connection to a source not approved as safe for human consumption. Under this condition, contamination of the drinking water system may occur by backsiphonage or backpressure. Backsiphonage is caused by negative or reduced pressure in the drinking water supply system. Backpressure occurs when a pressurized nondrinking water piping system is connected to a drinking water system and exceeds the operating pressure. There are different types of backflow prevention devices required under each of these conditions, depending on the degree of hazard.

Management of dual-plumbed buildings is a key in providing customers with a safe alternative source of water. The California Administrative Code, Title 17, requires that where reclaimed water is used on the same premise and there is no inter-tie between either system, a reduced-pressure-principle backflow assembly is required. Title 17 additionally requires that this backflow device be installed as close as practical to the service connection. This requirement provides protection at the user connection, but does not protect occupants of the building. IRWD inspects both the reclaimed and potable water piping during construction to meet IRWD, state, and county standards specifications. Prior to activating the reclaimed water, and before occupancy, IRWD, state, and county health care agencies work jointly performing a preoccupancy test. Testing procedures requirements must meet the master program report (reclaimed water use in non-residential buildings). Due to the complexity of such a system and the possibility of frequent changes in the plumbing, the health agency and the water supplier may, at their discretion, require that the customer designate a user supervisor. This person will be responsible to ensure that cross-connections are not made during the operation and that the customer's piping and equipment are maintained.

To conserve water, IRWD has expanded the use of reclaimed water to include single-family lots. Each estate lot is provided with reclaimed irrigation water service, and a potable water service for dual purposes. The engineer's report requires a cross-connection survey be performed to approve the dual systems. To identify each system, IRWD inspects the plumbing during construction. The piping system for reclaimed water uses different materials than the domestic piping system. In the engineer's report, all reclaimed piping must be purple PVC. IRWD requires that the customer use copper pipe for all exterior potable water piping. Before activating the reclaimed water service, a thorough cross-connection test and inspection of both the exterior potable and reclaimed water irrigation systems are required. State and county health agencies work together with IRWD to ensure that there are no cross-connections.

National Pollution Discharge Elimination System (NPDES) Permit

IRWD operates the MWRP under an NPDES permit issued by the California Regional Water Quality Control Board, Santa Ana Region. Permit conditions are written to comply with the Water Quality Control Plan, commonly called the Basin Plan, and incorporate requirements of other regulatory entities. This permit regulates MWRP operations, stormwater discharges, site dewatering, reclamation and reuse activities, and reclaimed water distribution system maintenance.

Water reclamation and reuse is governed by Title 22, Section 4, Chapter 3, Water Reclamation Criteria, in the California Administrative Code, which stipulates the quality of reclaimed water required for various uses. MWRP effluent meets the highest standards, permitting reclaimed irrigation of parks, playgrounds, and school yards, and water contact recreation.

To meet this standard, reclaimed water must have completed secondary treatment followed by chemical coagulation and filtration prior to a 90-minute minimum disinfection contact period. Reclaimed water must have an average turbidity less than 2 nephelometric units (NTUs) and a seven day median coliform less than 2.2 organisms/100 ml. In addition to average limitations, reclaimed water must at all times have a turbidity less than 5 NTUs, and a coliform less than 23 organisms/100 ml. MWRP effluent has met these requirements for years without any violations.

AQMD

MWRP is also regulated by the South Coast Air Quality Management District. Some of the critical pumps in the plant are natural gas driven, backed up by propane as an emergency supply. These engines have been fitted with catalytic converters and other equipment to meet current air quality regulations. In addition to critical pumps, IRWD maintains standby engine driven pumps and large standby portable generators in case of emergency. All internal combustion, engine-driven equipment is operated under permit. IRWD does not digest sludge, nor does it recover digested gas. Consequently, it does not require permits for these typical treatment plant activities.

DISTRIBUTION SYSTEM WATER QUALITY

Reclaimed System Sampling and Monitoring Program Design

Monitoring for water quality through sample collection and analysis will yield data that is a precise snapshot of the system at any one given moment in time for that specific location. The accuracy and validity of this data is highly dependent upon the representativeness of the sample collection point.

System Dynamics

When designing a water quality monitoring plan for any system, the dynamics of the system in question is a fundamental component for achieving a beneficial program. The IRWD reclaimed water monitoring program is designed around the general characteristics of the system while allowing for change and ever-increasing utilization. With multiple sources of water, seasonal diversity, operational system changes and constant expansion, the IRWD reclaimed system monitoring plan must be as diversified as the system it is designed to represent.

The IRWD reclaimed water distribution system is highly diversified, with other sources of water besides the MWRP final effluent: the TCE Remediation Well, Irvine Lake, Well 72, and Well 78. The District also has two open reservoirs that are long-term storage and staging sites for the transport of the reclaimed water. Monitoring and understanding the influence of all sources on the system is an important aspect of IRWD's reclaimed water monitoring program. Therefore, each "active" system site is sampled in conjunction with the additional "source" sites during every routine sampling event. If the "source" has any treatment mechanisms prior to the water entering the system, then sampling is conducted both before and after the treatment. This allows the District to evaluate any water quality changes within the storage facility as well as the effectiveness of the treatment process. The TCE Well is the only exception to this rule because the treatment is for TCE only, which does not impact the usability of the water.

Any extensive distribution system such as IRWD's is subject to constant change and variation. Maintenance and repairs can influence the system hydraulics and distribution mechanisms. Operational system manipulations due to unforeseeable circumstances are planned for (within the limits of practicality) by choosing well-distributed locations within the system for sample points on main lines where a need for constant service is assured. As the system expands, additional monitoring sites are added during construction to minimize costs and allow for sampling upon pipeline (or facility) completion.

Sources are in a constant state of flux due to environmental factors such as weather and seasonality. Seasonal demands for reclaimed water limit production in the winter months. When demands are low, some facilities may be out of service for extended periods of time. Thus, IRWD samples throughout the system proper and does not rely on source site monitoring alone. The monitoring sites within the system are chosen to be representational for any one particular zone, independent of the direction of the source feeding it.

Reclaimed Monitoring Program

Why sample, when there are not any regulatory requirements for monitoring within the reclaimed distribution system? IRWD chooses to monitor source and system water to maintain consistency within the system and safeguard the integrity of the water.

IRWD has made a commitment to deliver the finest product possible to its customers. Through continuous weekly sampling, the District can view its reclaimed water product through the users' eyes and thus better maintain the system to meet user needs. The selection of sampling locations, the use of designated sampling devices, the choice of sampling frequency, and a sample collection protocol are all important aspects of the reclaimed monitoring program.

The IRWD monitoring program is progressive yet fundamental in design. In order to facilitate good representative sampling locations, the District has established certain fundamental siting protocols:

- all sources
- on mainlines only (and not laterals)
- all areas subject to blending
- all pressure zones
- away from dead ends

The use of dedicated sample points is an important factor in maintaining sample integrity and thus data validity. The current design incorporates the use of a combination of air vac cans, pressure reducing stations, and hot taps at wellheads. Although these sampling sites have demonstrated themselves to be consistent and a suitable means for sample collection, the District is in the process of retrofitting all locations to a new standardized and dedicated sampling site.

The new standardized sampling devices (or units) bring consistency to the reclaimed water sampling program. These units are the identical counterpart of the domestic water sampling sites except for the purple color with which they are painted. The key features of the ''new'' sampling cans are:

- their ease of sampling (more ergonomic)
- an improved means for isolation and repair
- better sample integrity through dedication
- a continuous standard design throughout both systems

The frequency of the reclaimed monitoring is governed by two factors: analytical turnaround times and the degree of system diversity. The IRWD system is highly diverse; thus, weekly monitoring was chosen. The question of practicality is driven solely by the bacteriological analysis, which can take up to ninety-six hours by the MPN method. Thus, routine sample collection is performed on Tuesdays so as to limit the amount of weekend work necessary.

There are two discrete samples collected upon each sampling event at every location. One container is for the analysis of the physical characteristics and the other is for the bacteriological analysis. The protocols for each of these types are very different and are not interchangeable.

Typically, each site is ''flushed'' for two to three minutes at a high rate of flow. The physical sample container is rinsed once with sample prior to filling. The bacteriological sample is collected according to the very rigid sampling protocol, maintaining the use of aseptic technique so as to forego the possibility of contamination. The sample containers used are governed by the amount of sample required and the specific requirements of the analysis being performed.

Chemical Testing

The monitoring of the chemical quality of reclaimed water begins at the MWRP effluent monitoring station located at the downstream end of the chlorine contact basin. This is the final point of monitoring for the MWRP treatment process and the first location for monitoring reclaimed water delivered to customers. The chemical monitoring performed at this location consists of on-line monitoring, the collection of grab samples and 24-hour composite samples that are flow proportioned and collected by automated samplers. All composite samples are stored in refrigerated sample containers until the end of the sampling period, when they are then brought to the laboratory.

Operational On-line Measurements

On-line monitoring is accomplished by instrumentation designed to measure selected parameters on a continuous basis. One of the most important measurements made in this manner is chlorine residual. The chlorine concentration must be high enough for both effective disinfection and to inhibit re-growth of organisms downstream. The chlorine residual must be low enough to not cause chlorine odors and also not present a potential threat to plants. Turbidity is also a very significant continuous measurement. If turbidity levels are higher than discharge limitations, the water is no longer suitable for all intended uses. Turbidity is also an important factor in chlorination efficiency and treatment process control.

Additional measurements taken on a continuous basis are pH and electrical conductivity. These parameters are of value in determining general characteristics of the water, and the values may fluctuate with the source of water, degree of treatment, or a discharge into the wastewater collection system. Table 21.6 summarizes the methodology used and the results for the 1994 calendar year.

Grab and Composite Samples

Additional chemical analyses performed on grab and composite samples of final effluent include:

(1) Daily monitoring: In addition to the on-line monitoring of the MWRP Final Effluent, the IRWD laboratory also analyzes grab and 24-hour

composite samples on a daily basis. The analyses performed are: nutrients (ammonia-N, nitrite-N, nitrate-N and ortho-phosphate-P), minerals (chloride and sulfate), aggregate organics (BOD5 and COD) and additional chemical and physical constituents (electrical conductivity, pH, total alkalinity, total chlorine residual, suspended solids and volatile suspended solids). These analyses are performed to ensure compliance with water quality standards set by the California Regional Water Quality Control Board and to monitor the potential for degradation within the distribution system. Table 21.6 summarizes the methodology used and the results for the 1994 calendar year.

(2) Monthly monitoring: Monthly analysis of the MWRP Final Effluent is performed by the IRWD laboratory, in addition to the above-mentioned daily monitoring. The analyses tested include: nutrients (total phosphorus—P), minerals (calcium, magnesium, sodium, potassium, fluoride), aggregate organics (TOC), metals (boron) and additional chemical and physical constituents (TDS, MBAS, carbonate, bicarbonate and total hardness). As with the daily monitoring, these analyses are performed to ensure compliance with water quality standards set by the RWQCB and to monitor the potential for degradation within the distribution system. Table 21.6 summarizes the methodology used and the results for the 1994 calendar year.

(3) Quarterly priority pollutants: Some or all of the priority pollutant chemicals are analyzed on either a monthly, quarterly or annual schedule. These analyses include metals, pesticides, base/neutral extractable compounds, acid extractable compounds and volatile organics. As with monthly monitoring, these analyses are performed to ensure compliance with water quality standards set by the Regional Board and to monitor the potential for degradation within the distribution system. Table 21.7 summarizes the methodology used and the results for the 1994 calendar year.

(4) Field sampling: Weekly monitoring of the three reservoirs and twenty sample points within the reclaimed water distribution system is performed by the IRWD laboratory. These analyses include: chlorine residual, color, electrical conductivity and turbidity. Suspended solids and settleable solids are also performed on the reservoir samples only. There is no regulatory requirement for system monitoring beyond the treatment plant. However, since the quality of the water delivered to the customers is of the utmost importance, monitoring is performed beyond the regulatory requirements, allowing the District to track degradation through the system. Table 21.8 summarizes the methodology used and the results for the 1994 calendar year.

TABLE 21.6. MWRP Final Effluent—Methodology and Results.

Parameter	Method	Low Concentration	High Concentration	Average Concentration
Mineral Constituents				
Calcium	200 series	43.8	85.6	68.4
Chloride	300.0	93	238	145
Fluoride	340.2	0.45	0.69	0.56
Magnesium	200 series	14.4	36.2	26.6
Potassium	200 series	10.8	14.9	13.3
Sodium	200 series	127	174	154
Sulfate	300.0	129	338	248
Aggregate Organic Constituents				
Biochemical oxygen demand	405.1	ND	5	ND
Chemical oxygen demand	410.1	16	64	30
Nutrients				
Ammonia (as N)	350.3	1.0	20.3	14.3
Nitrate (as N)	300.0	0.6	16.6	2.8
Phosphate (as P)	365.4	1.2	4.2	2.5
Metals				
Antimony	200 series	ND	ND	ND
Arsenic	200 series	ND	ND	ND
Barium	200 series	0.035	0.091	0.06
Beryllium	200 series	ND	ND	ND
Boron	200 series	0.42	0.53	0.48
Cadmium	200 series	ND	ND	ND
Chromium	200 series	ND	ND	ND
Cobalt	200 series	ND	ND	ND
Copper	200 series	ND	0.015	ND
Iron	200 series	ND	0.13	ND

TABLE 21.6. (continued).

Parameter	Method	Low Concentration	High Concentration	Average Concentration
Lead	200 series	ND	0.035	0.010
Manganese	200 series	0.015	0.041	0.029
Mercury	200 series	ND	ND	ND
Nickel	200 series	ND	ND	ND
Selenium	200 series	ND	0.007	0.002
Silver	200 series	ND	ND	ND
Thallium	200 series	ND	ND	ND
Zinc	200 series	0.024	0.060	0.046
Additional Physical and Chemical Constituents				
Alkalinity as (CaCO$_3$)	310.1	114	288	203
Bicarbonate	310.1	24	288	230
Carbonate	310.1	ND	ND	ND
Cyanide	335.3	ND	0.027	0.011
Chlorine Residual	330.1	2.0	30.8	7.9
Color (CU)	110.2	14	52	28
Electrical Conductivity (umhos/cm)	120.1	977	1610	1300
Foaming Agents (MBAS)	425.1	ND	0.67	0.26
Hardness as CaCO$_3$	130.2	146	340	261
pH (units)	150.1	6.9	7.7	7.5
Suspended Solids	160.2	ND	6.8	3.0
Total Dissolved Solids	160.1	548	941	783
Turbidity (NTU)	180.1	0.7	2.7	1.6

Notes:
Units are milligrams/liter unless noted.
umhos/cm Micromhos per centimeter; NTU Nephelometric turbidity units; ND Not detected; TU Toxicity units; MBAS Methylene blue active substances; CU Color units

TABLE 21.7. Organic Chemicals.

Acenaphthene	2-Chloronaphthalene	4,6-Dinitro-O-Cresol	PCB 1016
Acenaphthylene	2-Chlorophenol	2,4-Dinitrotoluene	PCB 1221
Aldrin	4-Chlorophenyl Phenyl Ether	2,6-Dinitrotoluene	PCB 1232
Alpha BHC	2-Chlorotoluene	Di-N-Octyl Phthalate	PCB 1242
Alpha Endosulfan	4-Chlorotoluene	1,2-Diphenylhydrazine	PCB 1248
Anthracene	Chrysene	Endosulfan Sulfate	PCB 1254
Benzene	cis-1,2-Dichloroethene	Endrin	PCB 1260
Benzidine	cis-1,3-Dichloropropene	Endrin Aldehyde	P-Chloro-M-Cresol
Benzo (A) Anthracene	4,4'-DDD	Ethylbenzene	Phenanthrene
Benzo (A) Pyrene	4,4'-DDE	Fluoranthene	Phenol
Benzo (B) Fluoranthene	4,4'-DDT	Fluorene	Pentachlorophenol
Benzo (K) Fluoranthene	Delta BHC	Gamma BHC	Propylbenzene
1,12-Benzoperylene	1,2,5,6-Dibenzoanthracene	Heptachlor	Pyrene
Beta BHC	1,2-Dibromethane	Heptachlor Epoxide	Styrene
Beta Endosulfan	Dibromochloromethane	Hexachlorobenzene	tert-Butylbenzene
Bis (2-Chloroethoxy) Methane	1,2-Dibromo-3-Chloropropane	Hexachlorobutadiene	1,1,2,2-Tetrachloroethane
Bis (2-Chlorisopropyl) Ether	1,2-Dichlorobenzene	Hexachloroethane	Tetrachloroethylene
Bis (2-Chlorethyl) Ether	1,3-Dichlorobenzene	Hexachlorocyclopentadiene	Toluene
Bis (2-Ethylhexyl) Phthalate	1,4-Dichlorobenzene	Indeno (1,2,3-CD) Pyrene	1,2-trans Dichloroethylene
Bromobenzene	3,3'-Dichlorobenzidine	Isophorone	trans-1,3,-Dichloropropene
Bromochloromethane	1,1-Dichloroethane	Isopropylbenzene	1,2,3-Trichlorobenzene
Bromodichloromethane	1,2-Dichloroethane	4-Isopropyltoluene	1,2,4-Trichlorobenzene
Bromoform	1,1-Dichloroethylene	Methyl Bromide	1,1,1-Trichloroethane
Bromomethane	2,4-Dichlorophenol	Methyl Chloride	1,1,2-Trichloroethane
4-Bromophenyl Phenyl Ether	1,2-Dichloropropane	Methyl Ethyl Keytone	Trichloroethylene
Butyl Benzene Phthalate	1,3-Dichloropropane	Methyl Isobutyl Keytone	Trichlorofluoromethane
N-Butylbenzene	2,2-Dichloropropane	Methylene Chloride	2,4,6-Trichlorophenol

TABLE 21.7. (continued).

Carbon Tetrachloride	1,1-Dichloropropene	Naphthalene	1,2,3-Trichloropropane
Chlordane	Dieldrin	Nitrobenzene	Trichlorotrifluoroethane
Chlorobenzene	Diethyl Phthalate	N-Nitrosodi-N-Propylamine	1,3,5-Trimethylbenzene
Chloroethane	2,4-Dimethylphenol	N-Nitrosodimethylamine	Toxaphene
2-Chloroethyl Vinyl Ether	Dimethyl Phthalate	N-Nitrosodiphenylamine	Vinyl Chloride
Chloroform	Di-N-Butyl Phthalate	2-Nitrophenol	Xylenes, Total
Chloromethane	2,4-Dinitrophenol	4-Nitrophenol	

Detected Organic Chemical	Low	High	Medium
Bromodichloromethane	0.0011	0.0044	0.0023
Carbon tetrachloride	ND	0.0006	ND
Chloroform	0.0061	0.0067	0.0065
Chloroform	0.0061	0.0067	0.0065
Dibromochloromethane	ND	0.0021	ND
Gamma BHC	ND	0.0002	ND
Tetrachloroethylene	ND	0.0066	ND

987

TABLE 21.8. Reclaimed Distribution System.

Constituents	Distribution System				Reservoirs	
	Method	Low	High	Avg.	Low	High
Residual chlorine (mg/L)	330.5	<0.05	34	3	<0.05	32
Color (CU)	Spec 408nm	<5	450	39	<5	320
Electrical Conductivity (umho-cm)	120.1	834	1642	1232	846	1700
Turbidity (NTU)	180.1	<0.1	112	3.7	0.6	76
Suspended solids (mg/L)	160.2	—	—	—	<1	44
Settleable solids (ml/L)	160.1	—	—	—	<0.1	2.5

Notes:
All methods are EPA unless noted.
umhos/cm Micromhos per centimeter.
NTU Nephelometric turbidity units.
CU Color units.

The annual reclaimed water quality report is made available to all IRWD customers and is sent to all reclaimed water users.

Biological Testing

The IRWD maintains a non-potable distribution system for customer uses including crop irrigation, lawn irrigation, flushing toilets in high-rise buildings, decorative fountains, recreational impoundments, and carpet dyeing industries. Potential future uses include cooling tower applications and car washing facilities. The non-potable system includes Title 22 water from a wastewater plant with tertiary treatment, well water, and surface water from Irvine Lake. Two large open reservoirs with carrying capacities of 768 acre-feet (947,000 m^3) and 1,102 acre-feet (1,359,000 m^3/sec) are used to store reclaimed water and local run-off during periods of high demand. The water quality in the non-potable system can vary dramatically depending on the source(s) being used. Inconsistent water quality from the various sources can have a large impact on the users, depending on their specific needs.

In an effort to provide its customers with water quality to meet their individual needs, IRWD has initiated a program to monitor the microbial components of the non-potable distribution system. Unlike domestic water quality, the program is not regulatory driven but designed to meet the needs of customers. The monitoring program consists of total coliform, fecal coliform, heterotrophic bacteria, Legionella organisms, pathogenic protozoa such as *Giardia* and *Cryptosporidium* and enteroviruses. This section will address water quality issues and customer concerns which have been incorporated into the District's non-potable monitoring program.

Total Coliform

An essential part of the process of providing safe, clean water for customer use is the monitoring of water before and after treatment for the presence of potential microbial pathogens. However, the microbial pathogens which originate in the mammalian gut are difficult to isolate from water. This is because they are present in low numbers in feces (and consequently fecally contaminated water) and are susceptible to thermal and chlorine injury. Therefore, microbial pathogens are tested for indirectly by determining the density of total coliform bacteria. Total coliform bacteria include species of several genera within the family Enterobacteriaceae. When monitoring for total coliform, one accepts the basic assumption that the presence of bacterial indicator organisms is associated with the presence of microbial pathogens. The ease of enumeration (Evans, T., et al., 1981), crude relationship to pathogen and disease occurrence (Kabler, P. W., et al., 1960), and lack of better indicators led to incorporation of coliform levels into the state water quality standards. The purpose for total coliform testing is two-fold: to monitor water treatment efficacy and to ensure that the "finished" water is safe and clean for its intended use.

The final effluent at the MWRP is monitored for total coliform on a daily basis as mandated by California Title 22 regulations. Total coliform monitoring of reservoirs, wells and the distribution system is performed on a weekly basis. Table 21.9 summarizes total coliform data for the past two years.

E. Coli

The total coliform group is considered a reliable indicator of treatment efficacy but is, at times, a questionable indication of pathogen occurrence. Part of this shortcoming may be due to poor detection of stressed coliforms and heterotroph interference (Evans, T., et al., 1981). Total coliforms, fecal coliforms and *E. coli* are all used as indicators of fecal pollution. Among these, *E. coli* is often preferred as an indicator because this species is a common inhabitant of the intestines of humans and animals but very seldomly can multiply in the water environment (Linton, A. H., et al., 1988). Also, the presence of *E. coli* indicates recent fecal contamination and the possible presence of enteric pathogens since enteric pathogens often co-exist with fecal coliforms or *E. coli.*

Heterotrophic Bacteria

The IRWD non-potable service area contains a number of high-rise buildings which use water fed by cooling towers to provide comfort cooling for tenants. The use of reclaimed water for cooling tower make-up water would provide a substantial savings of domestic drinking water and increase the baseload demand at the MWRP. The use of reclaimed water for cooling tower make-up presents water quality challenges that will dictate changes from traditional treatment philosophies. One water quality parameter of concern is microbiological activity. Enhanced microbiological activity in a cooling tower can cause excessive slime build-up on internal components of cooling tower systems. It has also been found that with a typical heating, ventilating and air conditioning (HVAC) open loop cooling system, failure is attributable to microbiologically induced problems 50 percent of the time. This is because microbiological upsets can cause increased scale formation, under deposit, corrosion and fouling (Young and Thompson, 1992). These microbiological growth patterns are typically controlled with the use of biocides. Research and field applications have shown biocide dosages approximately double in quantity under recycle conditions. In order to economically evaluate costs associated with the amount of biocide that would be needed, the reclaimed water would first have to be microbiologically characterized. The IRWD water quality department now performs weekly analyses of heterotrophic bacteria using standard methods (APHA, 1992) at sampling points servicing proposed office buildings.

Pathogen Monitoring

The quest for safer, cleaner water has accelerated in late 1995, culminating in IRWD's pathogen monitoring program. Under this program, water quality staff are monitoring various sites to determine levels of pathogenic protozoa, legionella organisms and enteric viruses. With the reuse of highly treated sewage effluent for irrigation, as well as the increase in greenbelt watering and cooling tower applications, it would be beneficial to evaluate the microbial pathogenic composition. Once data and new information are collected, it will be incorporated into a microbial risk assessment model. With this information, staff will be able to assess the population risks of ingestion of reclaimed water and determine rates of daily prevalence for these microorganisms.

GIARDIA AND CRYPTOSPORIDIUM

To determine the distribution of Giardia and Cryptosporidium in the non-potable system, a survey of seven sites was initiated. One sample represented raw sewage entering the MWRP and another sample was collected from the final effluent at the MWRP. This water underwent activated sludge and dual media filtration before receiving chlorine sufficient to obtain a 5–10 mg/L chlorine residual with a minimum two hour contact time. The Coronado site represented tertiary effluent collected

from a sampling valve that supplies water to an industry landscape sprinkling system. The Irvine Center Drive sample represented a sample site
that in the future will distribute water for cooling tower make-up. These
two sample sites represented open reservoirs that receive tertiary effluent.
The final location represented surface water from a lake that receives local
area run off and tertiary effluent. Samples were collected on a monthly
basis, and levels were determined using a combined immunofluorescence
test. Table 21.9 summarizes data from this monitoring program.

LEGIONELLAE

Legionellae are found in human-made environments, e.g., cooling
towers and potable distribution systems. Viable *Legionella pneumophila*
have been found in 63 percent of cooling towers in Japan (Yamamoto,
H., et al., 1991), and the average counts were recorded up to 10^4 CFU/ml.
Although the precise mechanism of legionellosis is not fully understood, it
is generally believed that *L. pneumophila* is inhaled as an aerosol and
causes pneumonia. Cooling towers have often been implicated as a source
for the production of aerosol containing viable legionellae (Dondero, 1980
and Tobin, 1980). Legionella organisms are also highly resistant to chlorine. *Legionella* spp. are ubiquitous in environmental waters and are capable of surviving extreme ranges of environmental conditions (Palmer, C.,
et al., 1993). In the water of cooling towers without biocidal treatments,
heterotrophic bacteria and bacterivorous protozoan first appear, and then
legionellae increase. The efficiency of biocidal treatments by chemicals
has been established in preventing the growth of legionellae. To determine
background levels of legionellae in source waters and sample points servicing proposed office buildings, the water quality staff will initiate a Legionella monitoring program in the near future. *Legionella* will be detected
utilizing plate culture and direct fluorescent antibody techniques.

ENTERIC VIRUSES

During the late 1970s and early 1980s, the District initiated an enteric
virus monitoring project at the MWRP. Samples from the plant influent
and effluent were concentrated and assayed using standard methods
(APHA, 1992). Enteric viruses were detected in 17 of 19 (89.5 percent)
influent samples and 4 of 19 (21 percent) effluent samples. The data
indicated that the enteric virus levels in sewage entering the plant have a
seasonal trend, with higher levels occurring during the summer and fall.
Final recommendations were that the District review pertinent water quality
and plant operation parameters to identify which parameters are associated
with virus breakthrough. It was also suggested that the District initiate an
E. coli phage monitoring program. Reports indicated that *E. coli* phage

TABLE 21.9. Monitoring Program Data.

Location	Total Coliform MPN/100 L (min./max.)	HPC CFU/ml (mean)	E. coli MPN/100 ml (min./max.)	Giardia spp. CYSTS/100 L (min./max.)	Cryptosporidium OOCYSTS/100 L (min./max.)	Somatic Coliphages PFU/100 ml (mean)
1994						
MWRP Filtered Effluent	<2 to 500	17				
Sand Canyon Res.	<2 to 500	1400				
Rattlesnake Reservoir	<2 to 900					
ILP Irvine Park	<2 to 500					
Coronado	<2 to 130					
ICD Spectrum	<2 to 2	69000				
1995						
MWRP Filtered Effluent	<2 to 8	34	<2	<9 to 95	<9 to 17	
Sand Canyon Res.	<2 to 23	760	<2	<4 to 18	<3 to 27	
Rattlesnake Reservoir	<2 to 1600	9300	<2 to 300	4 to 63	4 to 18	
ILP Irvine Park	<2 to 1600	620	<2 to 170	<4 to 7	<4	
Coronado	<2 to 22	20	<2	2 to 29	<2	
ICP Spectrum	<2 to 240	35000	<2	1 to 18	<1	
1996						
MWRP Influent				15000-200000	<5000	170000
MWRP Filtered Effluent	<2 to 50	26	<2 to 50	<3	3	3
Sand Canyon Res.	<2 to 21	80	<2			
Rattlesnake Reservoir	4 to 1600	10000	<2 to 80			
ILP Irvine Park	<2 to 170	130	<2 to 8			
Coronado	<2 to >1600	8900	<2			
ICD Spectrum	<2 to 130	36000	<2 to 4			

levels are higher than enteric virus levels. Because of this, an *E. coli* phage assay would be a better predictor of treatment efficacy. Improved correlation of operating conditions with positive phage results versus enteric virus non-detect levels would result, and improved operations could result. Based on these recommendations and improved methodology, the water quality staff will initiate an enteric virus or *E. coli* phage monitoring program in the 1996.

During the last twenty-eight years, IRWD has demonstrated that water reclamation and reuse is a viable, environmentally safe, and economically feasible form of water conservation. While searching for new areas to expand the use of reclaimed water, IRWD continues to evaluate health risks and water quality issues in order to meet customer needs.

Annual Report

The results of chemical and biological monitoring of the treatment plant effluent are summarized in an annual reclaimed water quality report, available through the IRWD administrative office in Irvine, California. This report lists each constituent that is analyzed along with the minimum, maximum and average values. Where appropriate, detected values are compared to regulatory limitations. The report is also used as a communication tool with the District's systems to provide basic facts about the non-potable system and future projects and activities.

RESERVOIR WATER QUALITY

Watershed Reservoir Management

The District has two major open storage and operational reclaimed water reservoirs, Rattlesnake and Sand Canyon Reservoirs, which receive primarily high-quality treated effluent from the MWRP. Sand Canyon also captures a significant amount of stormwater runoff, which contributes silt to the reservoir. The non-potable system is supplied with MWRP water during low demand periods. During low demand periods, excess treated effluent can be conveyed to Sand Canyon Reservoir and the Zone A/ Rattlesnake Reservoir complex. Water from both of these reservoirs is used to supplement reclaimed water when system demands increase during the summer. Additional sources of supply to the system include two wells and untreated water from Irvine Lake.

The District has experienced several water quality problems characterized as turbidity, odor and algal growth within the two reservoirs. At times, turbidity levels caused by algal growth and natural runoff have exceeded 100 NTUs with apparent color units in excess of 1,000 color units (CUs). Dissolved sulfide levels from the reservoirs have exceeded 20 mg/L,

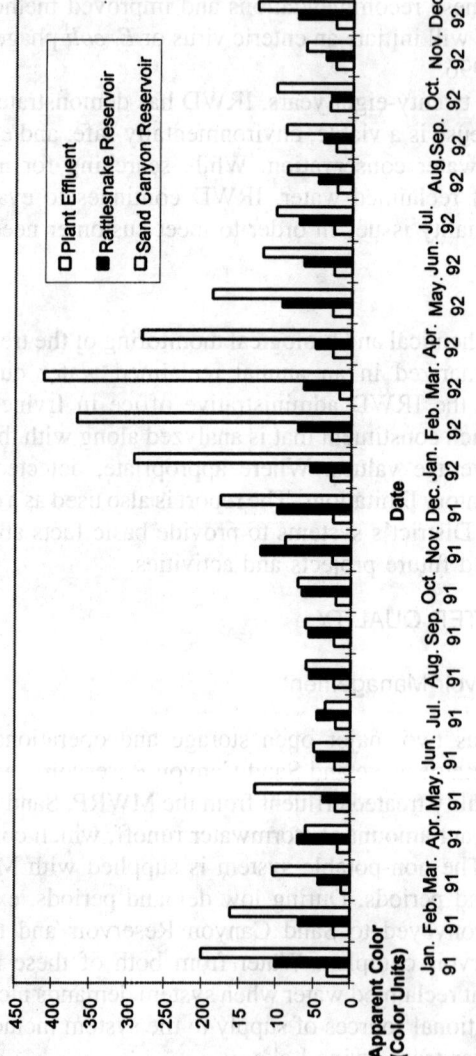

Figure 21.3 Comparison of plant effluent and reservoir color.

resulting in severe odor complaints. Coliform levels have also been found to exceed 23 MPN/100mL in the distribution system. These problems have resulted in: (1) high color levels in the distribution system and, in turn, at high-rise facilities; (2) the continued need for agricultural users to filter water to prevent drip emitter clogging; and (3) odor complaints from homeowner associations (Black & Veatch, May 13, 1993, 3–4, 5 and Black & Veatch, August 19, 1994, 2–3).

Figure 21.3 provides a comparison of average color levels for both reservoirs and the MWRP source water effluent from data collected from 1991 through 1992. Comparing color data of plant effluent and the reservoirs demonstrates that severe increases in color occur within the reservoirs. Color levels in the MWRP effluent are consistently below 35 apparent color units (CUs) while Rattlesnake and Sand Canyon Reservoirs have color levels in excess of 50 CUs and as high as 450 CUs (Harvey Mudd College Engineering Clinic, May 1993, 40–61).

The Harvey Mudd study showed that color and turbidity levels followed a similar pattern. In particular, during winter time storm events and through the following spring, color levels and turbidity levels were highest in Sand Canyon Reservoir. Average turbidity levels for the two year period were 28.4 NTUs and 5.7 NTUs for Sand Canyon and Rattlesnake, respectively. The study also showed a strong correlation between color and turbidity of $r = 0.83$ for the reclaimed water system. Water from Sand Canyon contained high levels of clay and silt, which was verified in a study performed later by District personnel. The source of this problem could be attributed to the runoff and soil characteristics of the reservoir watershed. It was confirmed in the 1994 Black and Veatch study that most of the Sand Canyon watershed is comprised of fine clay sediments which upon entering the reservoir stay in suspension throughout the spring.

The figure above also shows that color levels in both reservoirs still exceed plant effluent levels during the spring, fall and summer months indicating that algal growth was a secondary cause of color and turbidity problems.

In order to take advantage of the operational characteristics of the two reservoirs, the District (Black & Veatch, August 19, 1994) developed a new scheme for resolving water quality problems to satisfy current quality requirements of its customers, as well as requirements for expanded future uses of reclaimed water. Expanded future uses include front yard irrigation, additional high-rise use and industrial applications such as cooling tower water and carpet dye manufacturing process.

The District's water quality objectives or benchmarks for the water discharged from the two reservoirs are based upon the long term performance of the MWRP. These are shown in Table 21.10.

The physical characteristics of a reservoir and its surrounding watershed

TABLE 21.10. Water Quality Objectives for Water Discharged from Rattlesnake and Sand Canyon Reservoirs.

Parameter	Objective
Turbidity	< 2 NTU
Coliform	< 2.2 MPN per 100 mL (7 day average)
Color	< 35 Color units (apparent)
Odor	Not detected

play a major role in determining its water quality. This section summarizes the current and future land uses, rainfall and runoff information, physical characteristics of the two reservoirs and in-reservoir water quality data collected from 1986 through 1991. Table 21.11 summarizes the physical characteristics and water quality data for each reservoir.

Both Sand Canyon and Rattlesnake Reservoirs are very small and would be defined as "shallow lakes" based on classical limnology definitions [mean depth < 66 feet (20 meters)]. Shallow reservoirs tend to eutrophy rapidly when the mean depth is less than 33 feet (10 meters) (Bernhardt and Clasen, 1985).

Temperature and dissolved oxygen profiles collected from 1986–87 for both reservoirs show that direct stratification occurs in early spring with destratification occurring from the fall through the early spring. During the summer, when the reservoirs stratify, anoxic conditions occur which result in hydrogen sulfide formation and subsequent odor production.

Table 21.11 shows that Sand Canyon Reservoir has a much larger

TABLE 21.11. General Physical Characteristics and Water Quality Data.

Parameter	Rattlesnake	Sand Canyon
Physical		
Storage volume (acre-feet)	1,102 (1,359,000 m³)	768 (947,000 m³)
Surface area (acres)	60 (24 hectares)	42 (17 hectares)
Mean depth (feet)	24 (7.3 meters)	18 (5.5 meters)
Watershed area (acres)	1,293 (523 hectares)	4,288 (1,736 hectares)
Soil characteristics	No data	90%+ fine clay sediments
Temperature stratification	Spring-summer	Spring-summer
Dissolved oxygen	< 5 mg/L May-September at 3 meter depth and below	< 5 mg/L May-September at 3 meter depth and below
Nutrients		
Total phosphorous (mg/L)	1.5 to 24.5	5.7 to 15.3
Total nitrogen (mg/L)	2.5 to 6.3	3.1 to 11.4
Chlorophyll a (µg/L)	9.4 to 165	38 to 82
Turbidity (NTU)	1 to 11	1 to 35
Algae (organisms/ml)	3,300 to 77,400	5,500 to 542,000

watershed than Rattlesnake Reservoir. At the time this text was written, the Sand Canyon watershed was used primarily for pasture with a small portion of low to medium density residential uses. Rattlesnake Reservoir watershed area was used for agricultural purposes. The average rainfall is about 12.9 inches (33 centimeters) per year for both reservoirs. However, Sand Canyon watershed is comprised of fine clay sediments that are highly erodible. During storm events, this clay material contributes to turbidity and discoloration of the reservoir water.

The nutrient data in Table 21.11 shows a range of average nutrient concentrations for surface samples taken from both reservoirs. Wide variability in concentration levels occur which is primarily a function of the source of nutrients (reclaimed water and runoff). During 1991 and 1992, reclaimed water *ortho*-phosphate levels averaged 10.8 mg/L and 4.3 mg/L during 1991 and 1992. Total phosphorous levels from runoff to the Sand Canyon Reservoir ranged from 0.3 to 0.9 mg/L. Higher levels in the reservoirs can probably be attributed to the release of nutrients at the water/sediment interface. The average phosphorous levels for both Sand Canyon Reservoir and Rattlesnake Reservoir surface samples ranged from 5.7–15.3 mg/L and 1.5–24.5 mg/L, respectively, which far exceeds the requirements for eutrophic states in natural lakes of 0.1 mg/L. Nitrogen levels follow a similar pattern. Although both of these reservoirs have characteristics of eutrophication, normal theories of nutrient dynamics and morphology relationships cannot be expected to apply directly to these reservoirs because of the significant differences between nutrient concentrations in reclaimed water and natural lakes.

A by-product of excessive nutrients in the water column is high algal or biomass production. Chlorophyll *a* was measured as an indicator of algal production along with algal measurements. In general, chlorophyll levels were higher in the summer for both reservoirs with highest levels in Sand Canyon Reservoir. Higher algal counts were recorded for Sand Canyon Reservoir (5,500–542,000 organisms/mL) than for Rattlesnake Reservoir (3,300–77,400 organisms/mL). The turbidity data also follow the same trend as the algal data, with higher turbidity levels in Sand Canyon Reservoir (1–35 NTUs) than in Rattlesnake Reservoir (1–11 NTUs).

In summary, degradation of reservoir water quality in the two reservoirs was documented. Water quality degradation (i.e., higher turbidity, color and algal levels) was more severe in Sand Canyon Reservoir due to the shallowness of the reservoir, the larger watershed (and the contribution of fine clays during storm events).

A water quality model (BATHTUB) was applied to Rattlesnake and Sand Canyon Reservoirs to aid in the identification and understanding of the water quality problems and to predict impacts on water quality under phosphorous removal alternatives. Data collected over a four-year period beginning in 1988 and ending in 1991 were used to calibrate the model. Major findings of the modeling efforts indicated that phosphorous release rates for Sand Canyon Reservoir are probably higher than Rattlesnake Reservoir. This particular model was not suited for modeling Sand Canyon Reservoir water quality based on calibration of the Sand Canyon Reservoir predicted vs. measured phosphorous data, which varied from 18 to 88 percent. The model provided good results for Rattlesnake Reservoir, with predicted vs. measured phosphorous levels ranging from 1 to 6 percent (Black & Veatch, August 19, 1994, 5-1 to 5-6).

The District took a four-phased approach for determining alternatives to help resolve the water quality problems observed in Rattlesnake and Sand Canyon Reservoirs, which culminated in a final report, "Watershed Management Study for Rattlesnake and Sand Canyon Reservoirs," by Black and Veatch in August 1994.

(1) Phase I: Identification, Screening and Ranking of Twenty Alternatives

(2) Phase II: Technical Evaluation of Ten Feasible Strategies based on Phase I Ranking

(3) Phase III: Feasibility Analysis of Seven Alternatives Based on Phase II Results

(4) Phase IV: Development of the Watershed Reservoir Management Plan

The twenty management strategies that were considered in Phase I, along with their rankings, are shown in Tables 21.12 and 21.13. The alternatives considered were grouped as In-Lake-Management Strategies and Other Strategies.

Rankings were compiled by District personnel and experts from Black & Veatch based on literature surveys, past experience and performance of the various alternatives. A ranking of 1 designated that the strategy would likely be effective in improving water quality. A ranking of 2 indicated uncertainty about the effectiveness of the technique. A ranking of 3 indicated that the technique would not be effective for Rattlesnake and Sand Canyon Reservoirs. Nine strategies received a ranking of 3 and were dropped from additional consideration.

After the preliminary screening phase was completed, the ten alternatives were evaluated based on more specific technical criteria. Phase II presented the problem addressed, case studies, the potential effectiveness of the alternatives and technical feasibility. The results of Phase II determined that six of the ten alternatives and one combined alternative were technically feasible strategies for improving water quality (Table 21.14).

TABLE 21.12. In-Lake Alternative Management Strategies.

Strategy	Problems Addressed	Effectiveness	Negative Impacts	Relative Costs	Ranking
In-line treatment	Turbidity, color	Control external source	Reservoir volume	—	1
Phosphorus removal at MWRP	Reduce TP to 0.1 mg/L	Control external source	—	—	1
MWRP nitrification/denitrification	3 to 5 mg/L total nitrogen	Control external source Moderate effectiveness	—	High	1
Dredging	Phosphorus, Low DO	Control internal sediment sources, Also increased storage	Disposal	—	1
Destratification	PO_4 control, H_2S, possibly algae and turbidity	High for H_2S, questionable for others	Could increase turbidity and algae, Noise, will require sound proofing	Low	1
Copper sulfate	Algae	Moderate, short-term	Toxicity, Cu build-up in sediments	—	2
Biomanipulation	Algae	Variable	—	—	2
Hypolimnetic aeration	H_2S, PO_4	High	Reservoir depth (particularly with drawoff point at bottom)	—	3
Targeted plant growth	Algae	Moderate	Too much plant growth (weed harvesting)	—	3
Alum	Phosphorus	Control internal source (temporary)	Reservoir drawdown operation not compatible	—	3
Harvesting	Algae	Low for nonfilamentous species	—	Moderate	3
Nutri-pod	Algae, nutrients	Unknown	Maintenance	—	3
Aqua Treat	Algae	Unknown	Needs O_2	—	3
Hypolimnetic withdrawal	Nutrients	Moderate to Low	—	—	3

TABLE 21.13. Other Alternative Management Strategies.

Strategy	Problems Addressed	Effectiveness	Negative Impacts	Relative Costs	Ranking
Erosion control	Sediment, color	High in selected (high erosion) areas	Private property	—	1
Wetlands, reclaimed water	Nitrogen	High (90–95%)	Land intensive	High	2
Detention basins	Sediment	Low (for fine grains)	Maintenance	—	2
Detention basins with treatment	Sediment, some nutrient removal	Medium	Maintenance	High cap, low O&M	2
Wetlands, runoff	Runoff nutrients sediments	Low (winter only) Low	—	—	3
Comprehensive watershed management plan	Sediment, toxicants	Moderate	"Harder sell" for reclaimed water reservoir, takes long time to develop	—	3

TABLE 21.14. Summary Comparison of Alternatives.

Management Strategy	Problems Addressed	Success in Previous Projects	Likelihood of Success in Rattlesnake and Sand Canyon Reservoirs	Recommended for Further Evaluation
In-line treatment	Turbidity, color	Has been effective in removing over 90% of turbidity and nutrients in some applications.	Jar test results indicate moderate effectiveness. Chemical requirements are high.	Yes
Phosphorus removal at MWRP	Reduce P concentration	Mixed results. Depends on also controlling internal sources.	Should improve water quality in both reservoirs. Better for Rattlesnake; may require runoff treatment in Sand Canyon	Yes
MWRP nitrification/ denitrification	Reduce N concentration	Very limited information applicable to this project. Requires nitrogen limiting situation.	Success is possible if combined with wetlands treatment.	Yes
Dredging	Phosphorus, Low DO	Has been successful if sufficient sediment is removed and external sources are controlled.	Could be very effective if external sources are controlled.	Yes
Destratification	PO_4 control, H_2S, possibly algae and turbidity	Very effective for increasing DO. Mixed results for algae, turbidity, and phosphorus.	Should improve DO and reduce odors. May also help control algae and turbidity.	Yes
Copper sulfate	Algae	Effective for short-term control. Not as useful in high pH and alkalinity water.	Low to moderate effectiveness for District reservoirs due to high pH and alkalinity. Temporary measure.	Yes

TABLE 21.14. (continued).

Management Strategy	Problems Addressed	Success in Previous Projects	Likelihood of Success in Rattlesnake and Sand Canyon Reservoirs	Recommended for Further Evaluation
Biomanipulation	Algae	Mixed results. Can be effective, but requires on-going maintenance.	District reservoirs do not have some conditions present in other successful projects.	No
Erosion control	Sediment, color	Can be very effective at controlling sediment turbidity and nutrients.	Success depends on locating problem areas and implementing site-specific measures.	Yes
Wetlands, reclaimed water	Nitrogen	Has been very effective in removing nutrients if properly sized and maintained	Land requirements make this impractical without pretreatment.	Yes
Detention basins	Sediment	Has been very effective in other areas.	Presence of colloidal material and size requirements make this impractical.	No

TABLE 21.15. Comparison of Feasible Alternatives.

Management Strategy	Problems Addressed	Effectiveness of Strategy for Controlling				Costs			Other Considerations
		Apparent Color	Odor	Turbidity	Algae	Capital	O&M	Annual	
In-line treatment	Turbidity, color	Moderate	NA	Moderate	Low	$2,777,000	$166,850	$305,700	Permits required; may be incompatible with other planned uses; additional facility site; loss of storage.
Phosphorus removal at MWRP	Nutrients, external P loading	NA	NA	NA	Moderate	$10,934,600	$1,207,000	$1,754,400	Permits required; no additional O&M at reservoirs.
Nitrification/ denitrification[a] at MWRP plus Wetlands Treatment	Nutrients, external N loading	NA	NA	NA	Moderate to high	$265,000	$20,000	$33,300	Permits required; additional facilities at reservoirs; loss of storage.
Dredging	Nutrients, internal P loading, low DO	NA	Low to moderate	NA	Low to high	$2,203,000	—	$110,200	Permits required; disposal options depend on test results; increased storage.
Destratification	Low DO and odor, possibly P, algae and turbidity	NA	High	Unknown	Unknown	$235,000	$34,000	$45,800	Permits required; need to consider noise in the design.
Copper treatment	Algae	Low	NA	Low	Low to moderate	—	$44,000	$44,000	No new permits; concerns about copper build-up in sediments.
Erosion control	Turbidity, color	Moderate	NA	Moderate	NA	Not evaluated			Permits required; other considerations depend on site; coordination with others required.

[a]Costs are for wetlands only.

TABLE 21.16. Comparison of Alternative Management Plans.

Plan	Problems Addressed	Capital Costs	O&M Costs	Annual Costs	Costs per AF[a]
A Destratification, erosion control, and copper treatment	Very effective for low DO and odor; moderately effective for control of turbidity and color; may help alleviate algae problems.	$235,000	$78,000	$89,800	$45
B Destratification, phosphorous removal at MWRP, erosion control, and copper treatment	Very effective for low DO and odor; moderately effective for control of turbidity and color; very effective at controlling external P sources; internal P sources may still cause algae problems.	$11,169,600	$1,285,700	$1,844,200	$922
C Destratification, phosphorus removal at MWRP, dredging, erosion control, and copper treatment	Very effective for low DO and odor; moderately effective for control of turbidity and color; very effective at controlling external and internal P sources.	$13,372,600	$1,285,700	$1,954,400	$977
D Destratification, nitrification/ denitrification at MWRP and wetlands treatment, erosion control, and copper treatment	Very effective for low DO and odor; moderately effective for control of turbidity and color; very effective at controlling external N sources; should control algae problems.	$500,000	$98,000	$123,100	$62
E Destratification, dredging, erosion control, and copper treatment	Very effective for low DO and odor; moderately effective for control of turbidity and color; effective at controlling internal P sources, may help alleviate algae problems; would increase storage.	$2,438,000	$78,000	$200,000	$100

[a]Calculated by dividing the annual yield of the reservoirs [assumed to be 2,000 AF (2,500,000 m³)] into the annual costs.

Each alternative was then evaluated using four criteria: permitting, implementation issues, operation and maintenance (O&M), and costs. A comparison of the individual alternatives can be found in Table 21.15. From this information, five feasible management plans were identified and compared, and an approach was selected. The five alternative management plans are summarized in Table 21.16, which includes a description of the plan, problems addressed and costs for implementation.

Plan D was the management plan adopted by the District to resolve the water quality problems for both reservoirs. This plan addresses all three of the identified water quality problems at a cost that makes the plan feasible. The plan includes installing destratification systems at both reservoirs, implementing nitrification/denitrification at the MWRP, constructing wetlands at the two reservoirs to treat reclaimed water, implementing an erosion control plan in the Sand Canyon watershed, and using copper treatment in each reservoir to temporarily control algal blooms. When fully implemented, this plan should be effective at increasing dissolved oxygen in the reservoirs and controlling odor problems. The plan should also be effective in limiting nitrogen and controlling algal growth. The erosion control plan should be moderately effective at controlling turbidity entering Sand Canyon Reservoir.

The adopted watershed management plan will be implemented over a period of several years. The first stage, scheduled for completion in the spring of 1996, is the installation of the destratification systems, a relatively low-cost measure that will address odor problems. Additional testing is currently being conducted by the District to determine optimum copper testing methodology, i.e., the best form of copper to use and the effectiveness of treatment.

In order to successfully implement the wetlands alternative, the MWRP must be capable of operating in a nitrification/denitrification mode. Upgrades will be made at the MWRP over the next two years to meet this requirement. It is understood, however, that changes in the watershed surrounding Sand Canyon Reservoir could postpone implementation of this alternative and that specific sizes, locations and extensive monitoring are required once this alternative is implemented.

The erosion control plan is being coordinated with The Irvine Company (the major land owner of watershed lands), the city of Irvine, and the county of Orange. With plans in progress to develop a golf course and residential lots near Sand Canyon Reservoir, the adopted watershed management plan is being evaluated on an on-going basis.

To properly measure the performance and effectiveness of the adopted management strategy, a monitoring plan must be put into place. The District adopted a specific monitoring strategy for each element of the

TABLE 21.17. Monitoring Plan for the Recommended Watershed Management Plan.

Plan Element	Sampling Parameter	Station Locations	Sampling Frequency
Destratification	Temperature, DO	Vertical profiles at 3 stations	Weekly for the first month, then 2 times per month
	Chlorophyll a	Sample at 1 meter below the surface within and outside the influence of the air plume	Weekly for the first month, then 2 times per month
	Turbidity	Sample at 1 meter below the surface within and outside the influence of the air plume, also 1 sample near the reservoir outflow	Weekly for the first month, then 2 times per month
Erosion control	Turbidity, color, TSS, estimate of discharge	Major tributaries	During runoff events, 2 or 3 samples each
Copper treatment	pH, alkalinity	Sample at 1 meter below the surface and 1 meter above the bottom within and outside the influence of the air plume	Before treatment, after treatment, 3 days and 6 days after treatment
	Chlorophyll a	Sample at 1 meter below the surface within and outside the influence of the air plume	Before treatment, after treatment, 3 days and 6 days after treatment
Wetlands systems	Nitrogen species, total and orthophosphorus	Sample at inflow and outflow points of the wetlands	Sample at weekly intervals
	Flow	Flow into and out of the wetlands system	Continuous record
	IR photo of the system	Cover the entire wetlands area	Once every two years

watershed management plan. The sampling parameter, location, and sampling frequency are summarized in Table 21.17. As with any plan, it will be modified based on observed results. In most cases, the sampling frequency can be reduced once the system is working properly.

Reservoir Post Treatment

Historical Reservoir Water Quality Data

Regular water quality data monitoring for Sand Canyon Reservoir and Rattlesnake Reservoir was first implemented in 1983 as a means to understand seasonal and watershed related variations. The parameters analyzed included turbidity, apparent color, total coliform and total algae counts (Table 21.18). High algae counts, turbidity and color are of concern to the customers, especially for high-rise building non-potable use, specialty industrial customers, homeowners associations, etc.

Traditionally, algae counts at the Sand Canyon Reservoir are high in spring and early summer, which causes high color in the reservoir outflow and plugging problems for the sand filtration system.

Reclaimed Water Quality Goals

Based on historical water quality data review and customer needs, specific water quality goals were established by IRWD:

- apparent color—less than 30 CU
- turbidity—less than 2
- odor—none

In addition, chlorination is required to remove odors and provide coliform-free water. The odor elimination can be also achieved by other methods described earlier (see previous section on Watershed Reservoir Management).

Reservoir Outflow Treatment Alternatives

Several types of treatment technologies are available for the reservoir post-treatment. Among the most attainable methods are: microstrainers, rapid sand filtration, in-line direct filtration, microfiltration, and dissolved air flotation with filtration.

Evaluation of Treatment Technologies

MICROSTRAINERS

Rattlesnake Reservoir is equipped with 250 μm strainers. The removal

TABLE 21.18. Historical Reservoir Water Quality Data (Summary).

Reservoir	Turbidity (NTU)			Algae Counts (Cells/ml)			Apparent Color (CU)		
	Ave	Min	Max	Ave	Min	Max	Ave	Min	Max
Rattlesnake	5.4	1	25	15,740	400	96,000	600	10	150
Sand Canyon	8.6	1	85	40,200	40	542,000	120	20	500

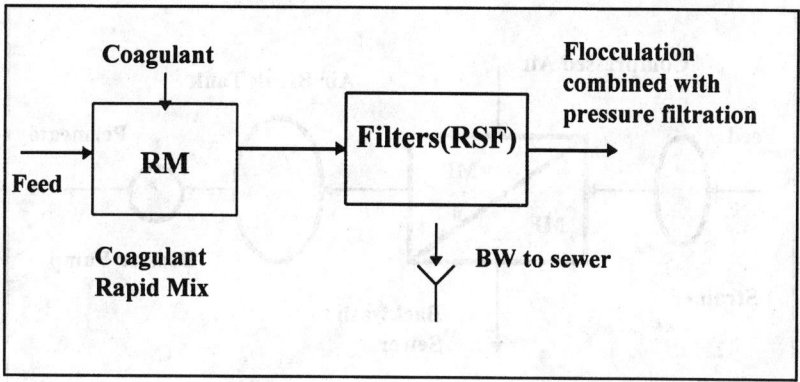

Figure 21.4 IDF block diagram.

principle utilized is based strictly on physical straining. The microstrainer's relatively large mesh size makes it incapable of retaining algae, much less particulates causing turbidity in the outflow. Microstrainers are effective in removing debris only.

RAPID SAND FILTRATION (RSF)

The rapid sand filtration system at the Sand Canyon Reservoir site consists of the two parallel banks of pressure driven sand filters with #16 silica sand [bed depth 16 to 24 inches (0.4 to 0.6 meters)] designed to filter particles larger than 125 μm. The system is designed to treat up to 10 cfs (0.28 m³/sec) of the outflow on a demand basis. Filters are periodically backwashed, based on the headloss increase. Monitoring of the RSF effluent during the Sand Canyon Reservoir Microfiltration Demonstration Study revealed the deficiencies of the existing RSF system: low removal of turbidity, algae and apparent color (only 10 to 20 percent removal). This system also requires a significant amount of free chlorine residual to neutralize pathogenic bacteria (3–7 mg/L).

IN-LINE DIRECT FILTRATION (IDF)

IDF uses the same principle as conventional treatment excluding the sedimentation and flocculation steps (see Figure 21.4). The IDF process involves chemical mixing, coagulation and flocculation in the filter intake header with subsequent rapid sand filtration. This process is known to be effective primarily for the low turbidity waters. Nevertheless, this treatment could be beneficial for seasonal high turbidity and algae laden waters at low dosages of the coagulant.

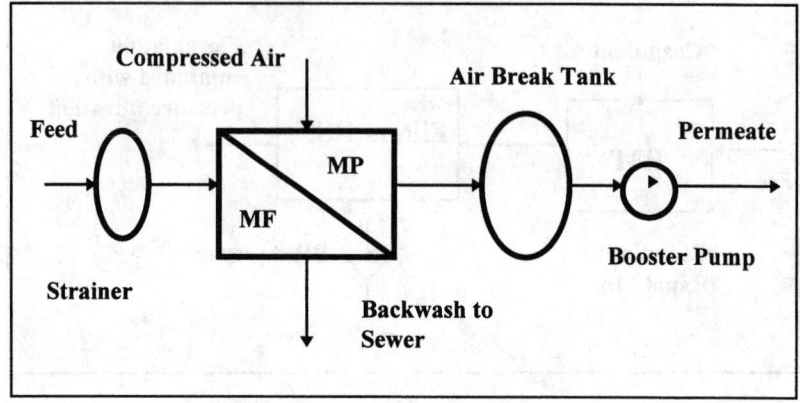

Figure 21.5 Microfiltration block diagram.

MICROFILTRATION (MF)

Microfiltration is a new membrane treatment process based on the continuous filtration of water through hollow fibers with high porosity and small pore size. IRWD conducted several studies at the Sand Canyon Reservoir site to explore the possibility of utilizing this innovative technology. The Memcor (Memtec Corp.) 60M10 unit was used for the Sand Canyon Reservoir effluent treatment during a six-month demonstration scale study. The Memcor process is based on filtering water through polypropylene fibers with the nominal pore size of 0.2 μm. The MF unit has a patented high pressure automated compressed air backwash system and requires only 20 psig (140 kPa) driving pressure for producing up to 30 gpm (110 liters/minute) of high quality effluent. The MF treatment train is shown in Figure 21.5.

The MF unit is not capable of handling high turbidity water with large particulates. Therefore, a 20 mesh strainer was installed on the feed line. The MF unit can be configured in the dead end or crossflow mode. The high pressure backwash system [90 to 110 psig (620 to 760 kPa)] initiates a backwash at a time of interval preset by the timer (usually 15 to 30 minutes). The unit is susceptible to chlorine contained waters; therefore, the feed should be dechlorinated. The results of the IRWD reclaimed water treatment study using the MF Memcor system prove the high reliability of the unit and consistently high quality of the MF permeate.

MF is capable of producing a higher quality effluent than the current existing RSF system (Table 21.19). Other advantages of the MF system include:

TABLE 21.19. Typical WQ Data Comparison Between MF and RSF System.

Range	App. Color (PCU)	Turbidity (NTU)	pH	Algae Count (Org/ml)	T. Coli (MPN/100ml)	TOC, (mg/L)	Susp. Solids (mg/L)
Feed	120–160	9–18	8.3–8.7	$2–3.5 \times 10$	4	12	13–15
MF Filtrate	20–25	0.2	8.5	0	<2	9–10	<0.5
RSF Filtrate	125–225	7–15	7.3	NS	<2	12	8

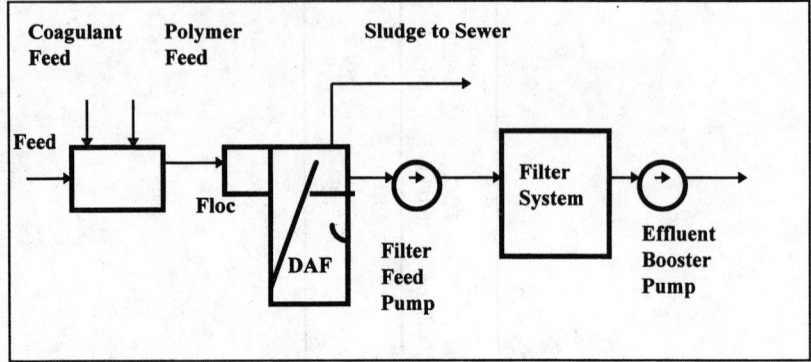

Figure 21.6 DAFF block diagram.

- reduced sensitivity to the changing feed water quality (except to the influxes of the small colloidal particulates in the feed water)
- minimal attention to the treatment process
- removal of total coliform from the feed water (requires much less free chlorine residual in the effluent)

The disadvantages of the MF system for the reclaimed water treatment include:

- relatively high cost of the treated water in comparison to the traditional systems (see below)
- rapid fouling when feed water exhibits high concentration of small (< 0.2 μm) colloids

DISSOLVED AIR FLOTATION WITH FILTRATION (DAFF)

Since high dosages of coagulant make the direct filtration option inappropriate for algae and particulates removal, DAFF with its subsequent filtration may be feasible. DAFF operation principle involves rapid mixing of water with a conventional coagulant (alum or ferric), followed by flocculation and then using air in the flotation tank for binding of the floc particles and creating floating sludge blanket. Air bubbles are released in the flotation tank by dissolving air under pressure (60–70 psig) into the recycle flow (typically 6–10 percent of the raw water) and then lowering the pressure by releasing the compressed air to atmospheric pressure. Either clarified or filtered water is suitable for this purpose. The air bubbles created in the process are of 10 to 80 μm in diameter, and they cannot be obtained by the direct air pumping (see Figure 21.6 below). DAFF tanks are much smaller than conventional sedimentation tanks which significantly reduces coagulant dosages and provides considerable savings in capital costs. Hydraulic loadings in DAFF are about 10 times higher than sedimentation, and detention times are typically 5 to 15 minutes. The DAF effect produces a stable sludge blanket with high dry solids content. The water is then filtered to remove remaining suspended particles and floc. An additional advantage over conventional clarification is that more efficient removal of particles relieves the filter loading, producing longer filter runs. Other potential advantages of DAFF are removal of volatile organics, tastes and odors by air stripping during flotation. On the downside, DAFF requires a lot more attention and constant adjustment in comparison to other treatment methods.

Related Costs and Economic Analysis

KEY FACTORS AFFECTING WATER TREATMENT COST

The key factors affecting water treatment application cost are: capital costs (unit installed, construction, land purchase, contingency, booster pumping); O and M costs (chemicals, power, sludge/concentrate disposal, labor); design configuration (reservoir, regional or point-of-use treatment); seasonal variation of reservoir water quality parameters; and seasonal outflow from the reservoir to the distribution system.

COST ESTIMATES

The cost analysis for the treatment model should be based on the key factors and also on the prospective customer needs. The final cost in dollars per acre-foot of treated water is heavily dependent on the treated effluent outflow and other factors described above. In cases where the data is unavailable or incomplete, assumptions based on industry experience must be made. Most of the cost estimates related to the different water treatment methods are found in detail in the Rattlesnake and Sand Canyon Reservoirs Reclaimed Water Treatment System Study prepared for the IRWD by CH2M HILL. (Rattlesnake and Sand Canyon Reservoirs Reclaimed Water Treatment System Study)

Future IRWD Reclaimed Water Post-Treatment Developments

SCR FILTER MEDIA REPLACEMENT

IRWD carried out a six-month Sand Canyon Reservoir Microfiltration Demonstration Study to explore the possibilities of the regional treatment of the SCR stored reclaimed water. The study showed that MF treatment is capable of meeting and exceeding IRWD water quality goals, but it is costly to implement. A preliminary estimate shows that MF technology would cost \$84 per acre-feet (\$0.068/m³) of treated water compared with \$33 per acre-feet (\$0.027/m³) for the existing Rapid Sand Filtration system. However, MF could be a feasible alternative for the point-of-use application with stringent water quality requirements (cooling towers, textile manufacturing, etc.). IRWD is investigating filter media replacement at Sand Canyon Reservoir as one of the recommendations of the MF demo study. The upcoming project for the filter media replacement will include a two-level test: first, pilot level; and then, a full-scale test.

SPECIAL PROJECTS

Microbial Risk Assessment

IRWD sponsored a project to explore new approaches to microbial risk assessment applicable to assessing the population risks of ingestion of reclaimed water (Eisenberg, 1995). The primary objectives of the project were to: (1) develop an epidemiologically based risk assessment model that allows for the assessment of the public health risk of exposure via ingestion to microbial agents associated with water recycling projects; (2) conduct a formal analysis of uncertainty of key model parameters; and (3) estimate the level of microbial risk associated with several combinations of recycled water treatment methods and uses.

A framework was developed that could integrate available literature data such as immune status, community-specific exposure factors, and dose-response data. A risk assessment model was then developed that was based closely on models used in infectious disease epidemiology. The advantage of this model is that it can be used to: integrate and organize the diverse data bearing on disease risk, account for immunity to disease, model aspects of the transmission dynamics of the agent in the environment, and explicitly acknowledge the uncertainty and variability in the many parameter values characteristic of comprehensive models.

The uncertain and variable nature of environmental systems such as those modeled in this study makes an absolute assessment of risk unreliable. Therefore, a comparative risk approach was developed in which a background simulation scenario was analyzed for a specific pathogen. The risk associated with each of several alternatives for reclaimed water use was then compared with a background prevalence level. All subsequent uncertainty analyses were performed by assessing whether or not a simulation produced an output above or below a mean background level plus one standard deviation.

Introduction

Microbial risk assessment involves evaluating the likelihood that an adverse health effect may occur from human exposure to one or more potential pathogens. Most microbial risk assessments to date have used a framework originally developed for chemicals and defined by four major steps: hazard identification, dose-response identification, exposure assessment, and risk characterization. However, this framework does not explicitly acknowledge the differences between the health effects due to chemical and microbial exposure, such as secondary spread and immunity.

One framework that is consistent with current views on the infectious disease process is based on three components: the microbial agent, the human host, and the environment in which the infection process is mediated. Since it is the interaction of these components that produce human disease, risk is dependent on defining the following:

(1) The exposed population including demographic profile, susceptibility status, and other variables
(2) The microbial characteristics including viability, virulence, etc.
(3) The environmental setting including exposure medium, fate and transport, etc.

Once the population, disease agent, and environment are defined, the corresponding health effects can be described, including such issues as symptoms, severity, and the potential for secondary infection.

This framework, based on host/microbe interaction, makes explicit the mechanistic aspects of the infectious disease process and provides a structure from which data are gathered. Some of the data required to characterize risk may be available in the literature, while some, especially site specific data, may be obtained through surveillance or experimentation. This data collection process is analogous to the analysis phase in the ecological risk assessment framework.

Application of Model

The first application of the model involved the evaluation of the public health risk of exposure through ingestion of microbial agents associated with several combinations of recycled water treatment methods and uses (Eisenberg, 1995). *Giardia lamblia* was selected as the pathogenic agent to consider in detail because the literature was more extensive than for any other agents. In particular, there were reports which provided data on the non-outbreak background incidence of giardiasis (Birkhead, 1989) in a community setting which allowed a "calibration" of some of the biological parameters of the model. RSA was used to establish a set of parameters that reproduced non-outbreak conditions consistent with the data.

This information was then applied to investigate the risk associated with ingestion of reclaimed water containing *Giardia* while swimming in an impoundment. Computer simulations were used to calculate the disease risk for defined ranges of parameter uncertainty and variability. In the *Giardia* case study, the simulations forecast significantly higher prevalence of giardiasis in a community with a public swimming impoundment than in a community without such an impoundment. Filling the impoundment with tertiary treated wastewater did not increase mean prevalence levels above those obtained when the impoundment was filled with water from a pathogen-free source. However, the predicted disease prevalence is highly variable under both scenarios due to the uncertainty and variability of the model's parameter values, and this may have masked differences in predicted public health impact. The results of this simulation study of a hypothetical closed community point towards two finds: (1) An improved understanding of shedding by swimmers is the most critical factor in improving the accuracy of the risk assessment, and (2) depending on the effect of new shedding information on the resultant range of the shedding parameter, then the treatment parameter or the exposure parameter could become the most important factor in predicting the presence of an outbreak.

Conclusions

The original motivation for the study arose from a need to improve methods for carrying out risk assessments for waterborne pathogens by providing a modeling approach that integrates public health information and data in a way that acknowledges the specifics of the situation in particular communities. The analysis requires an explicit statement of what is known about the infectious disease processes involved, the size of the population at risk, and the site-specific details of the exposure scenarios. This conclusion suggests that the differences in particular community and exposure scenarios are likely to be significant and produce widely varying levels of risk. The costs of risk mitigation are generally community or situation-specific which suggests that the approach to risk assessment applied in this study leads naturally to exploring site-specific strategies for risk management.

The results provide a foundation for identifying those parameters that tend to influence the risk assessment results as well as a general sense of the relative relationship and significance, in terms of public health risk, among various alternative uses (i.e., exposure scenarios) and background levels of risk. This information, while not community or situation-specific, can be useful in generally evaluating the relative degree of public health risk posed by different water reclamation alternatives. From the risk management perspective, although risk assessment modeling using generic or site specific input data can be valuable in providing general information pertaining to risk, the results should not be considered definitive. Risk assessment modeling should instead be considered one of many tools providing information to policy makers in the decision making process.

The benefits of this modeling approach are three fold. First, the structure of the model allows for a more realistic analysis of waterborne pathogen dynamics. Second, model simulations can be used to assess the comparative risks of various waterborne pathogens, the difference between exposure scenarios on various demographic populations and the potential effectiveness of different management strategies. Third, the analysis can directly identify dominant sources of uncertainty in the model.

While the study was successful in showing the feasibility and benefits of a new and effective approach to assessing the public health risks of waterborne pathogens, the risk assessment methodology remains in an early stage of development. More work is needed, particularly in studying the dynamic properties of these models in the risk assessment application. It is clear that expanding the scope of the model to incorporate seasonal patterns of exposure and the susceptibility status of different groups within the population will be fruitful directions to pursue. In addition, the uncertainty analysis needs to be refined to include nonlinear multivariate techniques.

System Modeling

Introduction

Since the 1980s, IRWD has used numerous hydraulic and water quality computer models to aid in the development and optimization of the sewer collection system and the reclaimed water distribution system. These models have also helped to develop both the sewer collection master plan and the water resources master plan for the District.

The following are just a few of the models that have produced useful information for IRWD:

(1) The Irvine Lake/Irvine Lake Pipeline (ILP) model provided information for stabilizing the lake elevation.

(2) HYDSIM was used to develop the water resources master plan.

(3) HYDRA hydraulically modeled the sewer collection system while a LOTUS based model analyzed the TDS concentrations in the system.

(4) EPANET is currently being developed to optimize the reclaimed distribution system from both the hydraulic and water quality standpoint.

(5) The ArcInfo GIS package is used to aid both the hydraulic and water quality models that IRWD uses.

In order to further enhance the understanding of the use of modeling for water reuse, the following sections presents a description of the models used by IRWD as well as the benefits each has provided IRWD.

Irvine Lake/ILP Model

The Irvine Lake/ILP Model was developed in the late 1980s. This Fortran-based model helped to determine how much supply to the Irvine Lake was required to keep the elevation stable. Accounting for evaporation, demand and infiltration, the model determined how much untreated surface water purchase was required to supplement the reclaimed water system.

HYDSIM

The HYDSIM model was developed for IRWD by Engineering Sciences (ES) from 1988 to 1989. ES developed and calibrated the model for both the reclaimed and potable water distribution systems.

Running on a VAX computer, the program is capable of modeling flows in the distribution system. Inputs included locations and characteristics of pipes, junctions, reservoirs and pumps. This model enabled IRWD to determine if existing facilities were adequate for ultimate build out. In cases where the system was inadequate to handle the ultimate capacity, HYDSIM helped to evaluate the need for additional facilities.

One drawback IRWD experienced with the HYDSIM was extensive maintenance requirements. A model such as HYDSIM requires periodic calibration checks and updates as the distribution system grows and changes. Also, the accessibility of HYDSIM was limited to a very few people in the District. Because of these drawbacks, and the fact that the hardware required to run the program is no longer in service at IRWD, this model is no longer in use.

HYDRA and the Wastewater Quality Model

Engineering Science (ES) aided IRWD in developing the HYDRA model for the sewer collection system. The firm digitized the IRWD base map and input all the essential land use and sewer system data required for the model.

The sewer system trunks were identified, along with manholes and other pipelines, and the system was calibrated using a wastewater flow monitoring program. The HYDRA model was able to identify capacity levels for the current demand as well as future demand.

The Wastewater Quality Model (WWQM) was developed by ES in 1991. Using data from HYDRA, HYDSIM and a wastewater quality database, the WWQM was able to predict the influent quality for the MWRP. Primarily, the model predicted TDS. At the time, a target of 690 mg/L TDS was necessary to meet the 720 mg/L TDS limit set by the MWRP operating permit.

These models were also used as tools to aid in cost analysis for system upgrades and expansions.

EPANET

Although the EPANET model has not been fully implemented, a portion of the model has been developed for the reclaimed water system. The justifications for developing this model are as follows:

(1) Using a hydraulic/water quality model such as EPANET will enable IRWD to operate the reclaimed distribution system in a way that can ensure a high quality water to sensitive applications (e.g. high-rise users, carpet manufacturing, cooling towers and agricultural users).

(2) EPANET will help to analyze low flow/pressure problem areas in the system.

(3) Water quality problems such as odors, color and other complaints should be reduced due to distribution system optimization.

ArcInfo GIS System

The GIS system can serve as a backbone for both the hydraulic and water quality models. The GIS system contains vital information about the entire distribution system (e.g. pipe locations, sizes, and materials; pump station information; and valve locations).

This information can be used for numerous applications, but specifically provides models with most of the information necessary to accurately emulate the distribution system.

Conclusion

IRWD has utilized computer models to optimize the non-potable system as well as analyze future use requirements. Optimization has included changes in operating conditions as well as suggestions for installing more facilities in the system.

The use of computer models has also enabled IRWD to address problems with both the sewer collection system and the reclaimed distribution system. Although the use of models have been beneficial to IRWD, experience has shown that both the operation and maintenance of the models must be supported at all levels of use. The models have been instruments for meeting regulations as well as providing IRWD customers with high quality reclaimed water sources.

ON-SITE USERS

INTRODUCTION

In the late 1970s, as reclaimed water became more commonly used in agricultural applications, IRWD joined forces with various health agencies to develop requirements and criteria for the on-site use of reclaimed water. In 1976, IRWD chartered its first Irrigation Group to implement these requirements and maintain a degree of control over the on-site use and application of reclaimed water. The group, sectioned in the Engineering and Planning Departments, was commissioned to plan check all reclaimed water systems slated for development within the District and to perform inspections as they were installed. Additionally, specific areas of responsibility were identified to prevent human consumption of reclaimed water and to limit runoff by controlling the installation and operation of irrigation systems. System operation was to be monitored in order to assure complete user compliance.

Over the next decade, the Irrigation Group expanded its scope and purpose to encompass a wider range of duties related to use of reclaimed water. In July 1991, the Irrigation Group was renamed the On-Site Systems Management Group, and the technicians who inspected and plan checked were designated as On-Site Systems Specialists to more closely reflect this wider scope. Due to the nature of the work, and its relationship to potable water on-site, the current group is comprised of American Water Works Association and the University of Southern California Foundation for Cross-Connection Control and Hydraulic Research (AWWA/USC) certified Cross-Connection Program Specialists. This section describes on-site work at IRWD and what lies ahead as innovative uses for reclaimed water continue to be explored.

DESIGN REVIEW

The purpose of design review is to ensure that the intended application of reclaimed water is consistent with the requirements of District procedural guidelines. To achieve this purpose, the intended use must be approved both in concept and design.

The design review process is initiated when plans are received from the irrigation designer. These plans are reviewed and marked with appropriate comments and changes to allow the designer to meet District compliance requirements. At this time, all pertinent information is logged into District files. The designer is then notified to pick up, correct, and re-submit the plan. During all subsequent plan checks, the plans are reviewed to verify that all comments have been addressed, and that required changes or alterations have been integrated into the plan.

The designer is then directed to submit original prints for approval signature. These prints are reviewed to confirm that all final corrections have been completed. The District's official approval signature block is applied to each irrigation print, and the plans are signed. The designer is notified that the plans are approved, and that a final submittal of one sepia mylar copy and a field inspection blueline is required. Once received, mylars and bluelines are filed in the District vault.

FIELD INSPECTION

Inspection of the reclaimed water system is an essential element of the On-Site Systems Specialist's job. From the moment construction begins to the final approval of the site for occupancy, the specialist must be available for consultation and inspection of the water systems. The specialist ensures District specifications are enforced and that prohibitive field conditions are addressed in a manner that fully protects the public health.

While the specifications outlined on the approved plans may be understandable for the contractor or superintendent to follow, the complexity of actual installation requires the visual scrutiny of a trained inspector. Class of pipe and other factors such as color, depth, and identification must be readily noted by the inspector. However, as with the other reclaimed systems and uses discussed in this section, cross-connection emerges as the single most important issue. Often, the proximity of reclaimed to potable piping violates District specifications, so field modifications such as sleeving or, in extreme cases, redesign, may be necessary.

The specialist must facilitate connection to the new reclaimed meter and perform a system evaluation prior to final approval of the project. The evaluation most importantly assures that no connection to other pressure sources has been made. In addition, it confirms the operation of the system as the plan intended. Overspray and runoff are minimized and valves, heads, regulators, and strainers are all in place and operating properly.

Finally, the plans and notes must be logged permanently at the District for reference. Special field conditions which may assist future inspectors are included. This record may be accessed for future operational problems or to add system modifications or additions to the same file.

ON-SITE MONITORING

Monitoring by individual on-site inspectors serves to enforce the District's rules and regulations pertaining to the use of reclaimed water. The focus of the monitoring process involves individual sites that have established plant material and water-use patterns.

The District's rules and regulations permit on-site access to the customer's premises during reasonable hours for the purpose of inspecting on-site reclaimed water facilities and areas of water use. The inspector assures compliance with the District's rules and regulations, including the provision that runoff must be controlled and limited, and the provision that cross-connections between potable and reclaimed water facilities must not exist.

Monitoring is accomplished by focusing on several specific and concurrent functions while routine field work is conducted. And while the emphasis is on established projects and not new construction, District inspectors occasionally patrol their regionally assigned areas investigating systems for compliance as they travel between construction projects.

Each inspector develops an intimate knowledge of the assigned region of the district including where reclaimed water is being used and for what purpose (i.e., irrigation, water features, construction grading, and interior use for toilets and urinals). All of the reclaimed water distribution system is contained in field Atlas books that are updated quarterly. Atlas books held in on-site files are marked to show detailed information about all new systems and meters. All irrigation systems are designed to operate between the hours of 9:00 P.M. to 6:00 A.M., and all control clocks have a label that alerts maintenance personnel in both Spanish and English of the restrictions for reclaimed water. On-site inspectors look for obvious violations and inform the management company or landscaper of specific situations that must be corrected.

On occasion, if complaints regarding overspray or odor are received by customer service, or if a project has demonstrated a record of over-use, an investigation may be performed at night. For systems that are high consumption accounts, monitoring is also achieved during the day by coordinating a thorough site investigation with landscape maintenance personnel. The irrigation system is activated to determine where violations of the District rules and regulations are occurring. Late-night tours provide a visual inspection as systems progress through normal schedules of operation.

When District on-site inspectors are given a specific area of the re-claimed water distribution system to patrol, they are able to maintain this pre-emptive program. The customer and the District are well-served by the program and, most importantly, the health agencies' recommendations are thoroughly enacted in a manner that addresses any concerns of public safety and health.

MAPS AND RECORDS

Record books document relevant information concerning the names and locations of the projects, names and phone numbers of designers or owners, operational criteria and special notes. All correspondence is kept in this file. The permanent mylar plan files contain a complete set of plastic sepia mylars for each reclaimed water irrigation system. Catalogued in the District vault, these plans have been utilized as references for acreage verifications, assessing system operational problems, planning data acqui-sitions, and as a public resource.

SPECIAL USES

In recent years, the District considered several innovative uses for reclaimed water: grading and dust control, dual-plumbed high rises for toilets and urinal waste flushing, water fountains and features, single family full yard irrigation, and industrial uses such as cooling towers and carpet mills for dyeing, and irrigation retrofits. To date, irrigation retrofits, high rises, carpet manufacturing and single family lots have received reclaimed water for approved non-drinking uses. More dual-plumbed high rises have been built and await service, and the commercial/industrial uses are cur-rently being analyzed.

Grading and Dust Control

Grading and dust control provide the opportunity to use reclaimed water in large quantities under controlled conditions, with limited exposure to the public. The District monitors construction sites and enforces the use of reclaimed water in place of potable water, when reclaimed water is available. Control of this program is maintained through site patrols and distribution of District specifications to superintendents and drivers. Trucks and water tanks containing reclaimed water are required to have warning tags and signs installed at every inlet or outlet and are not permitted to be reassigned to new locations using potable water until they have been thoroughly disinfected to the satisfaction of the inspector.

High-Rise Dual-Plumbed Systems

In the late 1980s, a cooperative effort between developers, the city of Irvine, and IRWD resulted in a total of seven high rise buildings being built with dual-plumbed systems. One system serves potable water for drinking purposes; the other is reclaimed for toilet and urinal flushing. Some studies indicate approximately 50 percent of total consumption at commercial office facilities is for toilet and urinal use.

The on-site section must facilitate the project from its conception and plan check phase through to the final connection. Extensive public involvement is required to satisfy the newly created requirements. Together with the On-Site Manager, the Principal Engineer creates an engineer's report detailing the project. Once the plans are approved through the on-site section, inspection of the entire building is conducted during construction. All reclaimed water piping is marked to distinguish it from the potable system. When construction is complete, a cross-connection survey is performed with the regulatory agencies. The burden of care is placed upon the maintenance engineering company servicing the high rises; no reclaimed water system valve may be operated or have work performed upon it without prior notification of the District. This extensive management program assures the District and agencies that cross connections have not occurred between piping systems. To date, two high rises have been connected, while five others await service. IRWD's two newest district buildings, which are low-rises, have been dual-plumbed, and are currently in operation.

Residential Full Yard Irrigation

As development in Southern California moved into areas planned for custom-built homes on large lots, the District identified several tracts that would be suitable candidates for full yard irrigation. As with other new water uses, an engineer's report was necessary to fully describe this use of reclaimed water in all its aspects. Single family custom home lots are unique because each lot is an individual project requiring the attention, inspection, and administration that a much larger project might require. The first development will include two tracts encompassing 150 homes, each having multiple contractors and plans. Entrusting the use and operation of a dual system to a homeowner is unique, and must be administered with caution. However, as this application is accepted over time, other tracts will be considered for full yard irrigation. Likely candidates will be those which are designated as estates or "ranchettes" with larger amounts of irrigable landscape or fodder.

Decorative Uses

Some decorative uses have been designed or retrofitted for reclaimed water. Industrial parks and commercial high rises often employ features such as fountains or lakes. Also, regional parks often install ponds or impoundments as recreational attractions. Open bodies of water differ from other systems in their accessibility to the public and in the likelihood of vaporization due to fountains or wind. Because of this potentially high risk for public contact, signs are posted in these areas. Despite a cautious approach and appropriate warnings, water quality problems have arisen in several cases that required reconnection to a potable source. Each system must be treated as entirely unique, whether it is a pristine pool, whether aquatic life exists, or whether regular public access occurs.

Commercial Applications

Some of the newest and most promising uses for reclaimed water that IRWD is currently exploring occur in industrial and commercial applications. So far, cooling towers and carpet dyeing have been considered for service. Although substantial progress has been made in cooling towers, no retrofits have occurred. The District's first carpet manufacturer was approved in early 1997. Once water quality issues have been addressed and overcome, it is anticipated that significant water savings will be achieved.

After toilets and urinals, it is estimated that cooling towers consume the greatest amount of water—possibly as much as 40 percent in a typical high-rise building. Cooling towers typically do not require significant re-piping to accommodate reclaimed water service. In the carpet mill currently under evaluation by IRWD, it is estimated that 200 to 300 acre-feet (250,000 to 370,000 m^3) per year may be able to serve the dye vats.

IRWD's Retrofit Program

IRWD's most aggressive and successful program by far has been the Irrigation Retrofit Program. Most irrigation accounts are individually metered to the "common area" and are billed based on the ascending block rate structure. Potable water not only costs more than reclaimed water, but the allocations are also more restrictive. Due to the positive economic benefit to the customer, the District has found a favorable response to the program.

The role of the On-Site Specialist in retrofits goes beyond plan check and inspection. The specialist must present a project to the customer that includes funding for the on-site work, specifications and coordination with regulatory agencies. Technically, the specialist must evaluate the premise for existing or potential cross-connections which are not permitted, under any circumstance, between reclaimed and potable systems. As always, it is the issue of cross-connection that remains the highest priority. To date, 550 acres of land (223 hectares) serviced by 180 meter connections have been retrofitted, and many more projects are on the horizon. These projects are estimated to use 2,200 acre-feet (2,700,000 m³) of water per year. Operating concurrently with this program is the agricultural retrofit program. Administered differently, yet still requiring special inspection and caution signage, IRWD's Retrofit Program has successfully delivered reclaimed water service to thirteen users.

FINANCE

At the IRWD, reclaimed water cost allocation decisions are made on a basis consistent with other cost allocation decisions: in short, costs must follow benefits. Although this has been IRWD's underlying principle since its inception, this goal could not be fully realized in the early years because of external price constraints (see section on Rate Setting).

SEPARATION OF CAPITAL AND O&M COSTS

IRWD separates the cost of providing sewage collection, treatment, and disposal into two distinct cost centers: sewage and reclamation. For the purpose of allocating costs between the two, sewer includes collection, treatment costs through secondary levels, and some disposal costs; reclamation includes tertiary treatment and reclaimed water distribution costs. The logic behind this separate approach is based on the theory that sewer customers should assume those costs that would be incurred regardless of whether a reclamation alternative is pursued and reclaimed water customers should assume the incremental cost related to producing and distributing reclaimed water.

Once the sewer/reclamation distinction is made, an additional separation of reclamation costs between capital and operations/maintenance is maintained. Major capital costs, or the cost of building infrastructure, are funded through general obligation bonds, the debt service for which is paid for by property taxes and connection fees. Major capital expenses include such items as tertiary treatment facilities, pump stations, reservoirs, and pipelines four inches (0.1 meters) and larger. In-tract pipelines of less than four inches (0.1 meters) are built and funded by the developer and are donated to the District to become part of its utility plant. The debt service payments attributable to funding major infrastructure costs are paid from property tax revenue and revenues collected through connection fees paid by developers as a condition of plan approval. It is intended that over the long run, property taxes and connection fees will contribute to capital costs in an approximate 50/50 relationship.

Operation and maintenance (O&M) costs, as well as daily costs incurred to operate and maintain the reclaimed water system, are funded through the commodity rate established for reclaimed landscape and agricultural irrigation water. O&M costs include such items as labor, energy, equipment, materials and a proportionate share of overhead costs. The commodity rate set for reclaimed water excludes any capital costs as defined above.

RATE SETTING

Today, virtually all reclaimed water produced by IRWD is sold to either agricultural customers or landscape irrigation customers. As land within the District is converted from agricultural uses to urban development, an increasingly higher percentage of reclaimed water is delivered to landscape irrigation customers so that today approximately 80 percent of all reclaimed water sales is used for landscape irrigation purposes. This relationship is in contrast to the early days of reclamation, where the current apportionment of landscape to agriculture was reversed.

Reclaimed landscape irrigation rates include two components: a commodity charge per hundred cubic feet (2.83 m³) for water used, and a monthly service charge based on meter size. Based on a policy established by the IRWD Board of Directors, the commodity rate for reclaimed landscape irrigation water is set at 90 percent of the District's base commodity rate for potable water. The 10 percent differential is intended to act as an incentive to promote the use of reclaimed water for nonessential uses. The 90 percent indexing is intended to ratchet up the rate for reclaimed water over time until the revenues generated from reclaimed water sales are equal to the cost of producing and distributing reclaimed water. Since the cost of reclaimed water has in the past been greater than the cost of providing potable water, the District was limited in its rate setting ability for reclaimed water by this external constraint (i.e., it was considered impractical to set rates higher for subpotable water than for potable water). In the years when reclaimed water revenues were less than the cost of producing and distributing reclaimed water, the difference was made up from sewer rates based on the sewer system's need for a disposal option.

By Board policy, the District has a uniform agricultural water rate for imported untreated water and for reclaimed water. On an annual basis, the rate set for reclaimed agricultural water is equal to the then-current cost of importing untreated water plus any District O&M costs.

FINANCIAL ASSISTANCE FROM OTHERS

The major source of financial assistance for IRWD's reclaimed water program is from the Metropolitan Water District's (MWD's) Local Projects Program (LPP). The LPP is intended to provide financial assistance for the development of local resources that will reduce demand on MWD's system. In IRWD's case, an LPP agreement was executed and implemented in 1986 for a 25-year term. The agreement was somewhat unique in that, unlike other MWD agreements that provided funds for capital projects, IRWD's agreement provided a per-acre-foot contribution that was used to offset annual O&M costs. The initial LPP rate in 1986 was $75/acre-foot ($0.061/m^3; that rate has been adjusted twice since then and is currently set at $154/acre-foot ($0.125/m^3). MWD is currently exploring an optional LPP that includes a formula which could provide as much as $250/acre-foot ($0.203/m^3), based upon eligible costs. Under the current program, IRWD receives slightly more than $1 million annually through the LPP.

IRWD has also obtained two low interest rate loans from the State of California under a program they have to assist in the development of local resources. Each loan is for $2 million and has been used as partial funding for the construction of two reclaimed water reservoirs.

FINANCIAL ASSISTANCE FROM IRWD TO RECLAIMED WATER USERS

IRWD has implemented several programs aimed at assisting current users of reclaimed water to become more efficient in their water management practices. Among these programs are interest free loans for on-site system upgrades, equipment rebates for hardware such as piping, electrical wiring and spray heads, and rebates to install modern, state-of-the-art irrigation controllers. Additionally, the District conducts seminars periodically throughout the year to educate landscape professionals, homeowners associations, and residential customers in effective irrigation management practices. Through these combined efforts, the District has reduced the average water application rate from approximately 4 acre-feet/acre/year (1.2 m^3/m^2/yr) to 3.5 acre-feet/acre/year (1.1 m^3/m^2/yr). Continued reductions are expected as sound irrigation water management practices are more universally embraced.

IRWD also provides financial assistance to convert irrigation customers currently using potable water to reclaimed water. This assistance usually takes the form of a loan that is paid back over time from the differential between potable and reclaimed water rates as well as from MWD LPP revenue attributable to the yield associated with the converted project.

IRRIGATION RATE STRUCTURE

The base commodity rate established for reclaimed irrigation water is incorporated into IRWD's five-tiered ascending block rate structure. The ascending block rate structure is intended to promote the efficient use of reclaimed water by utilizing the California Irrigation Management Information System (CIMIS) to develop evapotranspiration (Eto) rates that provide adequate water allocations to meet the customer's needs in the first two tiers: the low volume rate and the conservation rate. Water usage beyond the allocated amount is subject to penalty charges, so that as more water is used above the allocated amount, the more costly that water becomes. The three penalty tiers (the penalty rate, the excessive rate, and the abusive rate) double in cost successively from the base conservation rate.

Revenues collected from penalty assessments are used for two primary purposes: first, to fund the low volume rate, which is set at approximately 40 percent of the base conservation rate; and, second, to fund the financial assistance programs offered to IRWD's irrigation customers discussed previously. Funding the financial assistance programs leads to more efficient use of local supplies, reduces the demand for more expensive imported water, and has the net effect of reducing the base conservation charge to all irrigation customers.

APPENDIX: PERTINENT INFORMATION AND SPECIFICATIONS ON IRVINE RANCH WATER DISTRICT OPERATIONS

Demographics:
  Population: 138,000
  Total Area of District: 78,319 acres (31,708 hectares)
  Service Elevations: Sea Level to 1,700 feet (520 meters)
Wastewater Plant Capacity: 15 mgd (0.66 m$_3$)
Number of Open Reclaimed Water Reservoirs: 2 (+1 Untreated Water Reservoir)
  Capacity:
    Sand Canyon Reservoir: 768 AF (947,000 m^3)
    Rattlesnake Reservoir: 1,102 AF (1,359,000 m^3)
    Untreated Water at Irvine Lake: 25,000 AF
    (31,000,000 m^3)/Approx. 19,000 AF (23,000,000 m^3)
    to IRWD
Reclaimed Water Distribution System:
  Amount of Pipe: 189 miles (304 kilometers)
  Booster Pump Stations and Capacity: 10/72,000 GPM (4.5 m^3/sec)

Wells and Capacity: 3/3,400 GPM (0.21 m³)
Number of Pressure Zones: 7 existing/8 near term
Percent of Water Metered: 100%
Number of Services: 1,750
   Irrigation: 1,700
   High Rise/Internal: 4
   Agricultural: 35
   Single Family Lots: 10
Acreage by Type (Reclaimed Accounts Only):
   Irrigation: 5,000 (2,000 hectares)
   Agricultural: 1,000 (400 hectares)
   Single Family Lots: 3 (1.2 hectares)
Current Water Rates:
   Domestic Water: $.64/ccf ($0.23/m₃)
   Reclaimed Water: $.58/ccf ($0.20/m₃)
Staffing Levels:
   Michelson Water Reclamation Plant: 11
   Reclaimed Distribution System Operations: 4
   On-Site Reclaimed Water Management: 4
   Cross Connection Program: 3

IRVINE RANCH WATER DISTRICT TIERED RATES

Tier	Reclaimed Water		
	% Allocation	Increases	Rate
Abusive	121% and up	8 × Base	$4.64
Excessive	111%–120%	4 × Base	$2.32
Penalty	101%–110%	2 × Base	$1.16
Conservation	41%–100%	Base	$0.58
Low Volume	0–40%	75% of Base	$0.44

IRWD Water Use Fiscal Year 94/95

Source	Domestic System	Reclaimed System	Untreated System	Total	Percent %
Imported	20,000			20,000	33
Local Groundwater	20,000	1,000		21,000	33
Local Surface		1,870	8,000	10,500	—
Reclaimed Water		10,000		10,000	16
Total	40,000	12,870	8,000	—	100%

CURRENT PLANT PROCESSES AND EQUIPMENT

1. HEADWORKS

   a. Mechanical Bar Rakes (2)

    b. Grinders (2)

    c. Metering

        1) Raw Influent Flow

        2) PH

        3) Conductivity

2. PRIMARY CLARIFICATION

    a. Primary Clarification Tanks (5)

    b. Flight Drive Systems (5)

        1) Drive Units (4)

    c. Sludge Pumps

    d. Skimming System

        1) Skimming Troughs (5)

        2) Skimming Wetwells (2)

        3) Skimming Pumps (1)

3. FLOW EQUALIZATION

    a. Equalization Basins (2)

    b. Influent Pumps (4)

    c. Return Valves (2)

    d. Recirculation Pumps (2)

    e. Aeration Pumps (3)

    f. Wash Down System

    g. Metering

        1) Secondary Influent Flow

        2) Basin Levels

        3) MPS-#3 Flow

4. ACTIVATED SLUDGE

    a. Aeration Tanks (6)

    b. Aeration Blowers (6)

    c. Jet Clusters (24)

    d. Recirculation Pumps (24)

    e. Dissolved Oxygen Meters (6)

    f. Tank Dewatering Pump (1)

    g. Metering

        1) D.O. Levels

        2) LTES Return Flow

        3) Air Rates

5. SECONDARY CLARIFICATION

    a. Secondary Clarification Tanks (9)

    b. Flight Drive System (9)

        1) Drive Units (9)

    c. RAS Pumps (4)

    d. WAS Pumps (2)

    e. Skimming System

        1) Skimming Troughs (9)

        2) Skimming Wetwell (1)

        3) Skimming Pumps (2)

    f. Metering

        1) RAS Flow

        2) WAS Flow

        3) Wetwell Level

6. TERTIARY TREATMENT

    a. Filters (7)

    b. Filter Pump Station

        1) Filter Influent Pumps (3)

        2) Secondary Turbidity Meter (1)

    c. Valves

        1) ES Valves (4)

        2) SE Valves (2)

        3) Filter Valves (21)

    d. Backwash Supply Pumps (3)

    e. Spent Backwash Tank (1)

f. Spent Backwash Pumps (4)

g. Surface Wash System (7)

h. Metering

1) Filter Influent Flow
2) Backwash Flow
3) Filter Headloss Levels
4) Wetwell Level
5) Secondary Effluent Turbidity

7. DISINFECTION

a. Chlorine Contact Tank (1)

b. 50,000 LB (23,000 kilograms(kg)) Chlorine Storage Tank (1)

c. 4000 LB/day (1,800 kg/day) Chlorinators (2)

d. Chlorine Analyzers (2)

e. Chlorine Leak Detectors (3)

f. Metering

1) Control Chlorine Residual
2) Effluent Chlorine Residual
3) Chlorine Storage Level
4) Chlorine Gas Pressure
5) Contact Tank Level

8. EFFLUENT PUMPING

a. Michelson Pump Station #1

1) Effluent Engine Driven Pumps (3)
2) Effluent Motor Driven Pumps (3)
3) Metering

• Effluent Flow
• Effluent PSI
• Wetwell Level

b. Michelson Pump Station #2

1) Effluent Motor Driven Pumps (3)
2) Surge Tank
3) Metering

• Effluent Flow

- Effluent PSI
- Contact Tank Level
- Secondary Effluent Turbidity
- Final Effluent Turbidity
- Effluent PH
- Effluent Conductivity

D.   SLUDGE DISPOSAL

1.   Michelson Pump Station #3

   a.   Effluent Motor Driven Pumps (2)
   b.   Metering
      1)   Flow to CSDOC
      2)   Total Secondary Sludge Flow

   c.   Force Main
   d.   Pig Launching Station

E.   CSDOC

1.   MPS-3
2.   Force Main

F.   TELEMETRY

1.   SCADA System
2.   Plant Multiplexor System

   a.   Multiplexors (6)
   b.   Auto Dialer (1)

REFERENCES

Allen, M. J. and Geldreich, E. E. 1975. Bacteriological Criteria for Groundwater Quality. *Ground Water* 13:1.

American Public Health Association. 1978. *Standard Methods for the Examination of Water and Wastewater,* 14th edition. American Public Health Association, Inc., New York.

American Public Health Association. 1992. *Standard Methods for the Examination of Water and Wastewater,* 18th edition. American Public Health Association, Inc., New York.

Anderson, R. M., and R. May, *Infectious Diseases of Human Dynamics and Control,* Oxford University Press, New York, 1991.

Auslander, D. M., "Parasol-II: A Laboratory Simulation and Control Tool for Small Computers," *Computers in Mechanical Engineering,* Vol. 1, 1982.

Barbaree, J. M., Gorman, G. W., Martin, W. T., Fields, B. S., and Morrill, W. E. 1987. Protocol for sampling environmental sites for legionellae. *Appl. Environ. Microbiol.* 53: 1454–1458.

Bernhardt, H., and J. Clasen. April 1985. Recent developments and perspectives of restoration for artificial basins used for water supply. In *Conference proceedings from lake pollution and recovery: European Water Pollution Control Association,* Internal Congress, Rome, Italy, 213–27. London: European Water Pollution Control Association.

Birkhead, G., and R. L. Vogt, "Epidemiologic Surveillance for Endemic *Giardia lamblia* Infection in Vermont: The Roles of Waterborne and Person-to-Person Transmission," *American Journal of Epidemiology,* Vol. 129, 1989.

Black & Veatch. August 19, 1994. Final Report to Irvine Ranch Water District, Watershed Management Study for Rattlesnake & Sand Canyon Reservoirs. Irvine, California. Black & Veatch.

Black & Veatch. May 13, 1993. Final Report to Irvine Ranch Water District, Alternative Disinfection Methods Study. Irvine, California: Black & Veatch.

Bois, F. Y., M. T. Smith, and R. C. Spear, "Mechanisms of Benzene Carcinogenesis: Application of a Physiological Model of Benzene Pharmacokinetics and Metabolism," *Toxicology Letters,* Vol. 56, 1991.

Cooper, R. C., A. W. Olivieri, R. E. Danielson, P. G. Badger, R. C. Spear, and S. Selvin, *Evaluation of Military Field-Water Quality, Volume 5: Infectious Organisms of Military Concern Associated With Consumption: Assessment of Health Risks and Recommendations for Establishing Related Standards,* Lawrence Livermore National Laboratory, Livermore, CA 1986.

Dondero, T. J., R. J. Rendtoriff, and G. F. Malluson et al. 1980. An outbreak of Legionnaires' disease associated with contaminated air conditioning cooling tower. *N. Engl. J. Med.* 302:365–370.

Dudely, R. H., K. K. Hekimain, and B. J. Mechalas, "A Scientific Basis for Determining Recreational Water Quality Criteria," *Journal of the Water Pollution Control Federation,* Vol. 48, 2761–2777 (1976).

Eisenberg, J. N., J. Konnan, E. Y. W. Seto, et al., Microbial Risk Assessment for Reclaimed Water, prepared by EOA, Inc. for Irvine Ranch Water District and the National Water Resource Institute, Oakland, CA, 1995.

Evans, T. M., Waarvick, C. E., Seidler, R. J. and LeChevallier, M. W. 1981. Failure of the Most Probable Number Technique to Detect Coliforms in Drinking Water and Raw Water Supplies. *Appl. Environ. Microbiol.* 41:130.

Evans, T. M., LeChevallier M. W., Waarvick, C. E., and Seidler, R. J. 1981. Coliform Species Recovered from Untreated Surface Water and Drinking Waters by Membrane Filter, Standard and Modified Most-Probable Number Techniques. *Appl. Environ. Microbio.* 41:657.

Fuhs, G. W., "A Probabilistic Model of Bathing Beach Safety," *The Science of the Total Environment,* Vol. 4, 1975.

Haas, C. N., "Estimation of Risk Due to Low Doses of Microorganisms: A Comparison of Alternative Methodologies," *American Journal of Epidemiology,* Vol. 55, 1983.

Harvey Mudd College Engineering Clinic. May 1993. Final Report to Irvine Ranch Water District, Reclaimed and Nonpotable Water System Analysis and Recommended Strategies to Water Quality Problems. Claremont, California. Engineering Clinic.

Hornberger, G. M., and R. C. Spear, "Eutrophication in Peel Inlet: 1. The Problem Defining Behavior and a Mathematical Model for the Phosphorous Scenario," *Water Research,* Vol. 14, 1980.

Humphries, R. B., G. M. Hornberger, R. C. Spear, and A. J. McComb, "Eutrophication in Peel Inlet III. A Model for Nitrogen Scenario and a Retrospective Look at the Preliminary Analyses," *Water Research,* Vol. 189, 1984.

Kabler, P. W., and Clark, H. F. 1960. Coliform Group and Fecal Coliform Organisms as Indicators of Pollution in Drinking Waters. *Jour. AWWA* 52:1577.

Kabler, P. W., Clark, H. F., and Geldreich, E. E. 1964. Sanitary Significance of Coliform and Fecal Coliform Organisms in Surface Water. *Publ. Hlth Rpt.* 79:58.

Linton, A. H. and M. H. Hinton. 1988. Enterobacteriaceae associated with animals in health and disease, p. 71–87. *In* B. M. Lund, M. Sussman, D. Jones, and M. F. Stinger (ed.), Society for Applied Bacteriology symposium series no. 17. *Enterobacteriaceae in the environment and as pathogens.* Society for Applied Bacteriology, London.

Palmer, C. J., Tsai, Y. L., Paszko-Kolva, C., Mayer, C. and Sangermano, L. R. 1993. Detection of *Legionella* Species in Sewage and Ocean Water by Polymerase Chain Reaction, Direct Fluorescent-Antibody, and Plate Culture Methods. *Appl. Environ. Microbiol.* 59:3618–3624.

Pipes, W. O. (ed.). 1978. Water Quality and Health Significance of Bacterial Indicators of Pollution, *Workshop Proceedings* Drexel University. National Science Foundation. Washington, D.C. 228 pp.

Regli, S., J. B. Rose, C. N. Haas, and C. P. Gerba, "Modeling the Risk from *Giardia* and Viruses in Drinking Water," *Journal of the American Water Works Association,* Vol. 83, No. 11, 1991.

Spear, R. C., and G. M. Hornberger, "Control of DO level in a River Under Uncertainty," *Water Resources Research* Vol. 19, 1983.

Taubes, G., "Epidemiology Faces Its Limits," *Science* Vol. 269, 1995.

Tobin, J. O., J. Beare, and M. S. Dunnill et al. 1980. Legionnaires' disease in a transplant unit: isolation of the causative agent from shower baths. *Lancet* ii:307–310.

Tsai, K. C., and D. M. Auslander, "A Statistical Methodology for the Design of Robust Process Controllers," *Trans. ASME. Dyn. Sys. Meas. and Control,* Vol. 110, No. 2, 1988.

Yamamoto, H., Ezaki, T., Iced, M. and Yabuuchi, E. 1991. Effects of Biocidal Treatments to Inhibit Growth of Legionellae and Other Microorganisms in Cooling Towers. *Microbiol. Immunol.* 35:795–802.

Young, R. E and K. A. Thompson. 1992. Economic and Water Quality Parameters for Successful Use of Reclaimed Water for Commercial Cooling Towers. Irvine Ranch Water District, Irvine, California.

CHAPTER 22

Wastewater Reclamation and Reuse in the City of St. Petersburg, Florida

HISTORICAL BACKGROUND

THE picturesque City of St. Petersburg is listed among America's most livable cities. Geographically, the city is located at the tip of the Pinellas County peninsula on Florida's west-central coast. The city is somewhat unique in that there is saltwater on three sides of it. It is bordered on its eastern side by Tampa Bay, on its western side by the Gulf of Mexico, on its southern side by the entrance to Tampa Bay from the Gulf and on its northern side by incorporated communities. St. Petersburg is now Florida's fourth largest city with a resident population of over a quarter of a million together with several thousand extra transient winter visitors escaping the harsh cold of the more northerly regions of the continent. The supply of drinking water for the ever increasing population and the treatment of wastewater have thus played dominant roles in the growth and development of this city over the years from its inception in 1880 up to the present day. The state of St. Petersburg's present day drinking water and wastewater disposal systems are depicted in Figure 22.1.

In order to fully understand and appreciate the multiple factors that resulted in St. Petersburg's decision to develop a reclaimed water distribution system, a review of the historical development of the potable water supply and wastewater disposal systems is essential. The following historical accounts of the development of these two separate systems are included

William D. Johnson and John R. Parnell, Department of Public Utilities, City of St. Petersburg, 1635 3rd Ave. North, P.O. Box 2842, St. Petersburg, FL 33731.

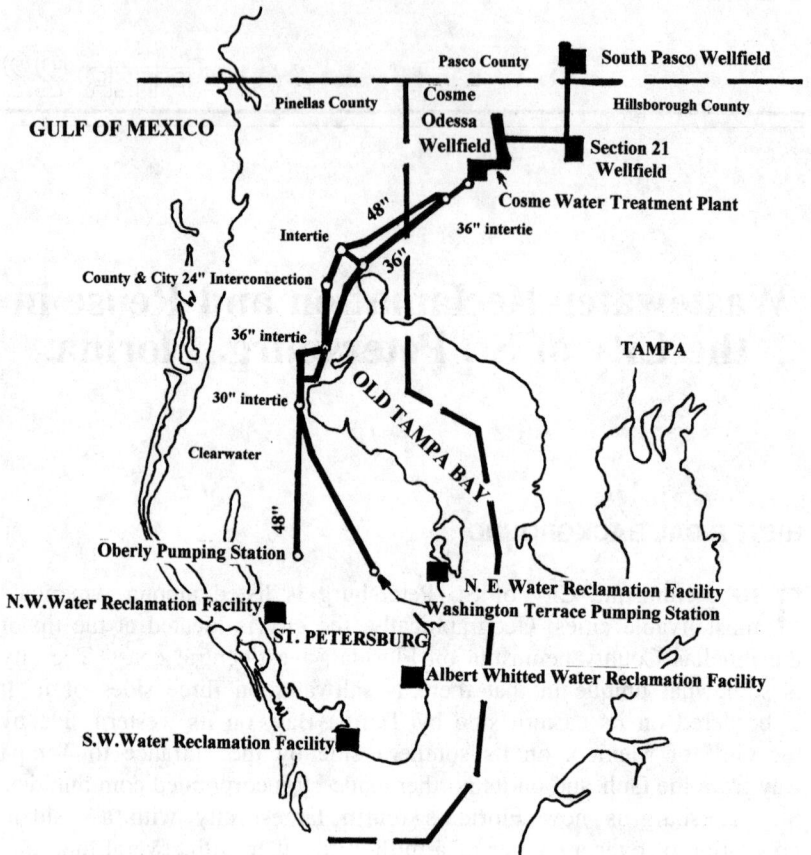

Figure 22.1 St. Petersburg's potable water wellfields and transmission main system (water reclamation facility locations included).

here to illustrate how an ever continuing necessity to conserve rapidly decreasing potable water supplies on one side, coupled with mandated expensive wastewater treatment alternatives on the other, were paramount in persuading the City fathers to pursue the course of action that ultimately led to the development of the nation's largest reclaimed water distribution irrigation system, supplying domestic, commercial and industrial users, that is in place today.

DEVELOPMENT OF ST. PETERSBURG'S POTABLE WATER SUPPLY

In 1880, John Constantine Williams founded the City of St. Petersburg at its present site. The new city's population of 30 was then served by the Orange Belt Railroad which had been extended from Sandford through Tarpon Springs.

Local Water Supplies

With the city in its infancy, most potable water supplies were derived from rain barrels and cisterns, but, as the population increased, motivated entrepreneurs obtained their individual drinking water supplies from small driven wells equipped with pitcher pumps. The first municipal supply was developed when a small pumping station was built on the bank of Reservoir Lake, today known as Mirror Lake, which supplied a tank near the end of the railway. It was not until 1899, after the city had passed a $10,000 bond issue, that municipally owned waterworks machinery was brought in and an elevated tower was constructed. On December 12, 1899, water was pumped into a new network of mains directly supplying downtown city residents.

Increasing demand for fresh water drastically lowered the level of Reservoir Lake and, beginning in 1905, the city drilled several wells to the depth of 450 ft. around both Reservoir and nearby Crescent Lake in a desperate effort to increase the dwindling water supply. Ever increasing withdrawals soon caused the wells to become contaminated by sea water intrusion and the city was forced to begin looking further afield in its quest for water independence. In 1920, the city contracted with the Layne-Bowler Company to study geological formations and suggest a long range cure as well as immediate relief. Their recommendations were to: drill more wells, install gravel filters and a new softening system, build an additional reservoir at Mirror Lake and to look for a new well field and plant site. In the meantime, by 1923, the demand for potable water on the city's wells had increased to nearly 3.5 million gallons per day.

A New Water System

By 1928, the population had grown to 40,000 and a Citizen's Water Committee was set up to report on two sources of water recommended in a report by a New York consultant. Realizing that it was facing a potential water crisis, in 1929, the city entered into its first well field agreement with Layne Southeastern Company (formerly Layne-Bowler), a private developer, to construct a completely new potable water system and deliver wholesale water to the city. The company purchased a section of land in adjacent inland Hillsborough County, developed a well field, constructed a water treatment plant and laid twenty three miles of 36 inch water main from the water plant to a water re-pumping station (Washington Terrace) that the company had constructed north of the city. At this time, this was a massive project and Layne-Southeastern subcontracted with the Lock Joint Pipe Company to locally manufacture and, using local labor, install the 36 inch pipeline through swampy terrain, across bays, in quicksand, muck and water under extremely harsh conditions to bring the water to St. Petersburg. Meanwhile, Layne Southeastern sold the franchise to the Pinellas Water Company, under the same terms. At noon on September 18, 1930, the Pinellas Water Company supplied the first water to St. Petersburg from this pipeline. By 1935, the Pinellas Water Company was supplying potable water to St. Petersburg at approximately 3 million gallons per day (mgd) via a small treatment plant practicing aeration, lime softening and filtration and located near the well field.

In 1940, St. Petersburg was able to exercise its option to purchase the Pinellas Water Company and all of its assets. These included the Cosme Odessa well field, the Cosme Water Treatment Plant, the property where Section 21 well field is located today, the 36 inch water supply pipeline, a water booster station and Weeki Wachee Springs, located in Hernando County. As a point of interest, Weeki Wachee Springs flow averages 103 mgd annually and it was planned as a future water source for St. Petersburg. The purchase of the Pinellas Water Company secured a solution to the city's water needs for many years to come. At the time of the purchase, the delivery rate of fresh water to St. Petersburg was approaching 5 mgd. The biggest drawback to this significant achievement was the fact that the city now became totally dependent on potable water imported from adjacent counties where several of the local municipal governments would eventually also obtain their drinking water.

Expansion of the System

The city continued to improve its water supply system throughout the 1940s and by the end of this decade demand had risen to 15 mgd. Throughout the 1950s a 10 year projection for the system was implemented, several new wells were drilled, new pumps were installed, additional mains were constructed, the treatment plant was upgraded and extra storage capacity was added.

The history and development of the Cosme Water Treatment Plant is further described by Henderson et al. (undated). Two attempts to introduce fluoridation to the system resulted in controversy and the referendums which caused them were eventually abandoned. In the 1960s, six wells were drilled and went on line in the Section 21 well field, a new 48 inch pipeline was constructed from the Cosme Water Treatment Plant to the city to supplement the original 36 inch line and a new pumping station (Oberly Pump Station) went on line. In the late 1960s, the city purchased a third section of land approximately 40 miles from St. Petersburg in

Pasco County. This was developed as the South Pasco well field and a further eight wells were connected to the system and came on line in 1973. These expansions barely kept pace with the rapid population growth experienced by both the city and the surrounding area during the period between 1950 and the mid-1970s. Also, by this time, the city was providing water service to four other growing communities and the water supply capability was once again becoming stressed.

Water Wars

Continued population growth throughout Pinellas, Hillsborough and Pasco Counties in the early 1970s, combined with an overall drop in local rainfall, placed ever increasing demands on all water supply facilities to the extent that inter-governmental cooperation was needed to address critical water issues. The Southwest Florida Water Management District (SWFWMD) had originally been created in 1961 to provide flood control but the Florida Water Resources Act of 1972 brought all waters of the state under regulation and developed five water management districts encompassing the whole state (Baldwin and Carriker 1985).

St. Petersburg now owned substantial blocks of land in Pinellas, Hillsborough and Pasco Counties. These Counties became alarmed when they realized that they might not be able to provide adequate water for their own growing populations because of St. Petersburg's ever increasing water withdrawals. When St. Petersburg joined with Pinellas County in the early 1970s to develop another well field in Pasco County, Pasco, Hillsborough and Hernando Counties joined together to have legislation enacted to block any further water development by municipalities outside of their jurisdiction. As a direct result of this legislation, further development of Weeki Wachee Springs as a water source for St. Petersburg was abandoned. However, the site has been developed into a tourist attraction at the present time.

The period from 1970 onwards subsequently became known as the "water war" years. As a result of these "water wars," in 1974, Hillsborough, Pasco and Pinellas Counties and the cities of St. Petersburg and Tampa agreed to form a separate government entity known as the West Coast Regional Water Supply Authority (WCRWSA) to develop regional water supplies and supply water at wholesale cost to counties and municipalities.

Regional Water Supply

With the formation of the WCRWSA more well fields have been brought on line in recent years and interconnections between well fields have been constructed. At the present time there exists seven interconnected well fields; three owned by the City of St. Petersburg, one owned by Pinellas

County, and three owned by the WCRWSA. All of the well fields are located within a 300 square mile area and have a total annual permitted withdrawal of approximately 145 mgd.

Present Day Problems

In recent times the well fields have become highly stressed and the day was soon to come when the SWFWMD would declare periodic water shortages throughout the Tampa Bay area and limit both lawn irrigation and other non-essential uses to reduce demands on the Floridan aquifer. The SWFWMD also declared St. Petersburg a water short area, and the city reacted to this by prohibiting new customers from using potable water to irrigate parcels of land larger than single residential lots. The potential for further groundwater development in the Tampa Bay area is very limited. Population growth is expected to remain robust so meeting the area's future water needs is of great concern.

New water supplies might come in the form of desalination of brackish and/or sea water and/or the development of groundwater sources in distant locations. There are many environmental and regulatory issues to be resolved before these sources of water can be "harvested" for public use. In any case, the cost of new water will be very expensive and the economic impact to utility customers will be significant.

DEVELOPMENT OF ST. PETERSBURG'S WASTEWATER DISPOSAL SYSTEM

The Initial Years

Records show that as far back as 1894, St. Petersburg had developed a small sewer system which was expanded in the southeast section by 1925 to serve 1,200 acres. A screening system on the site of today's Albert Whitted Water Reclamation Facility was used to process the ever increasing amounts of wastewater that were being produced by the rapidly expanding city. The resulting discharges from this system were discharged directly into Tampa Bay.

New Wastewater Treatment Plants

In the early 1940s when the city was delivering some 5 mgd of potable water to residents from its recently purchased water system, the problems of increasing wastewater disposal began to stress the single wastewater treatment plant. Also, on a national scale, increasing water consumption and wastewater disposal in the 1940s prompted greater recognition of impending water pollution problems by the federal government. The first legislation with broad coverage for the abatement of surface water pollution was the Federal Water Pollution Control Act of 1948.

In St. Petersburg, continued population growth in the 1950s and ever increasing wastewater production soon placed a severe stress on the single treatment plant. The Albert Whitted plant was enlarged and upgraded in 1954 and, in the following two years, construction began on three new plants in the northeast, northwest and southwest sectors of the rapidly expanding city. The Northeast and Southwest plants were constructed as secondary treatment facilities with design capacities of 8 and 9 mgd respectively. The Northwest plant, however, was only a primary facility with a 6 mgd design capacity. Once they became operational, all four plants discharged their partially treated effluent into either Tampa Bay or Boca Ciega Bay.

This method of disposal of wastewater, which depended on dilution and natural purification processes, was quite acceptable at this time. However, increasing quantities of discharge and changing compositions of wastewater due to industrial impacts soon created a threat to the quality of surface waters throughout the United States. The Federal Water Pollution Control Act of 1956 and its amendments in 1966 drew attention to this ever increasing problem, and a major shift in national intent for the management of wastewater was initiated. Federal funds were made available from the 1956 Act for research and development and to assist States and municipalities to construct or upgrade wastewater treatment facilities. Reuse at this time was not encouraged and no funds were made available for any form of wastewater reclamation or recycling.

Tampa Bay Pollution Problems and Plant Upgrades

In the early 1960s, the increasing discharge of partially treated wastewater into adjacent coastal surface waters in the Tampa Bay area by primary treatment plants created a severe water quality problem. To address this problem, St. Petersburg upgraded two of its plants. The Northwest plant became a 6 mgd activated sludge secondary plant in 1966, and the Albert Whitted plant was upgraded to an activated sludge contact stabilization treatment process in 1968. The effluents from both plants were chlorinated prior to disposal into Tampa Bay or Boca Ciega Bay surface waters.

The Wilson-Grizzle Act of 1972

In 1969, the United States Environmental Protection Agency (USEPA) produced a report describing Tampa Bay as one of the most polluted bodies of water in the nation. Following this report, the Florida legislature adopted what was called the "Wilson-Grizzle Act" in 1972. This bill was very short and stated: "No facilities for sanitary sewage disposal constructed after the effective date of this act (March 15, 1972) shall dispose of any wastes into Old Tampa Bay, Tampa Bay, Hillsborough Bay, Sarasota Bay, Boca Ciega Bay, St. Joseph Sound, Clearwater Bay, Lemon Bay and Punta Gorda Bay or any bay, bayou or sound tributary thereto without providing advanced waste treatment approved by the department of pollution control." Some confusion over the meaning of "advanced treatment" led to the State Pollution Control Board setting limits of 5 mg/L of BOD, 5 mg/L of TSS, 1 mg/L of phosphorus and 3 mg/L of nitrogen and requiring a minimum 90% treatment efficiency.

Following the adoption of this bill, in 1972, the City of St. Petersburg evaluated its alternatives and, based on the cost of constructing and operating advanced wastewater treatment (AWT) facilities, and considering the potential potable water supply shortages from the "water wars," the City Council opted to upgrade all plants to advanced secondary treatment and implement a recycling and deep injection well program that would ultimately result in zero-discharge to Tampa and Boca Ciega Bays. This bold decision led to the initiation by the city of what was to become the largest urban reclaimed water irrigation distribution system in the USA.

DEVELOPMENT OF ST. PETERSBURG'S RECLAIMED WATER SYSTEM

Significant Changes in 1972

As well as the Wilson-Grizzle Act, the 1972 decision by City Council to implement a recycling and deep injection well system to achieve zero-discharge to surface waters was also influenced by the results of a pilot study that had been authorized in 1971. This study was designed to determine the feasibility and efficiency of using highly treated wastewater for spray irrigation in an urban environment. The study concluded that spray irrigation using treated wastewater was more feasible and considerably more cost effective than advanced wastewater treatment followed by discharge to Tampa Bay. Also, the construction of a reclaimed water system would benefit the community by reducing the total quantity of water to be imported for potable use.

The approach adopted was to reclaim the wastewater by treating it to a sufficient degree that it would be suitable for the irrigation of parks, schools and golf courses within the city. The City Council also included funds in its 1972 Capital Improvements Program for the upgrading of the regional wastewater treatment facilities to advanced secondary treatment and the development of plans for a reuse distribution system.

The federal government also changed its policies significantly in 1972 by amending the Federal Water Pollution Control Act to develop a sweeping federal/State/local government program to reduce, prevent and eliminate water pollution. Within these amendments, the implementation of innovative reuse technologies was now actively encouraged, unlike the 1956 Act. Federal grants for the development of the "best technology for recycling or elimination of pollutant discharge" were made available. Also, land utilization for the disposal of treated wastewater was made an allowable cost, but land for the construction of treatment plants was disallowed from grant applications. The door was now open for the development of full scale reuse systems with federal approval and funding. The city immediately began to submit applications for these grants to upgrade the four plants and design and install the distribution system and the deep-well disposal system.

Design of the Reclaimed Water Distribution Loop

Plans for a reclaimed water distribution pipeline went forward soon after City Council had approved the scheme in 1972. Also, City Officials spent countless hours between 1973 and 1977 promoting the use of effluent for irrigation to major potential users all over the city. Designs for the system were centered around a centralized loop of large transmission mains that would eventually inter-tie all four treatment plants and would serve golf courses, parks, schools and large commercial areas throughout the city. Because the plants were located at the four corners of the city, the largest pipe in the system only needed to be 42 inches in diameter, and most were less that 36 inches. Thus, in general, large scale excavations were not necessary, a factor that contributed to the cost-effectiveness of the reclaimed water system. The other advantage of having all four treatment plants discharge into one large looped system was to minimize inconveniences to customers if any one or more of the plants had to be taken off-line for some reason.

This is in contrast to the City of Tampa where one very large plant, Hookers Point, serves the entire area. With a central plant, the cost of routing a distribution system through Tampa would have been prohibitive. Thus, Tampa chose to construct a 60 mgd AWT plant, with surface water discharge.

Treatment Plant Upgrading and the Construction of the Distribution System

On July 14, 1974, the USEPA awarded the city a federal grant of $14 million to upgrade its Southwest plant from a 9 mgd secondary facility to a 20 mgd advanced secondary facility, design and construct an effluent spray irrigation system and deep-well injection system. Thus, Phase I of the plan was now under way.

After the expansion to the Southwest plant was completed in 1977, the plant was selected as one of the nation's "Ten Outstanding Engineering Achievements of 1976" by the National Society of Professional Engineers. The first phase of the distribution system was completed to correspond with the Southwest plant expansion, and provided for distribution mains to serve large volume users in an area of the city south of 22nd Avenue. A large transmission main was also constructed along the Pinellas Bayway to a large private condominium and golf course named Isla Del Sol.

The expanded Southwest plant began to discharge some of its effluent into this system in August 1977 at a rate of approximately 7 mgd. Construction of the main loop of the distribution system was not yet completed and excess reclaimed water from the plant was still discharged to surface waters. This plant did not achieve zero-discharge to surface waters for two more years because the three 24-inch injection wells and monitor well system that were completed in 1976 were not permitted for continuous use until September 1979.

In 1975, the USEPA awarded a further grant of $18.8 million to St. Petersburg to upgrade the Northeast plant from 8 mgd secondary treatment to 16 mgd advanced secondary treatment, and construct a distribution system and deep-well injection system similar to the Southwest plant.

In 1977, the USEPA offered further financial incentives for wastewater reuse projects through the reauthorization of the Federal Water Pollution Control Act (now known as the Clean Water Act). This Act further substantiated congressional intent to achieve increased wastewater reuse and recycling of nutrients by giving preference to these programs over conventional wastewater treatment technologies.

The city was successful in obtaining federal funds for the 201-Facilities Plan which was prepared by Black, Crow and Eidsness Inc. (later CH₂M Hill), in 1978. This study concluded that landscape irrigation with treated wastewater was a feasible application and suggested further modifications to the Northwest and Albert Whitted plants and their sewer systems were needed in order to further develop the distribution system.

By early 1980, the expansion of the Northeast plant had been completed together with the distribution system and three deep-wells. The plant immediately began to discharge to the new distribution system and zero-discharge to surface waters was achieved in June 1980 when the three 20-inch injection wells and several monitor wells were permitted for use.

In early 1981, the construction of over 70 miles of the major transmission

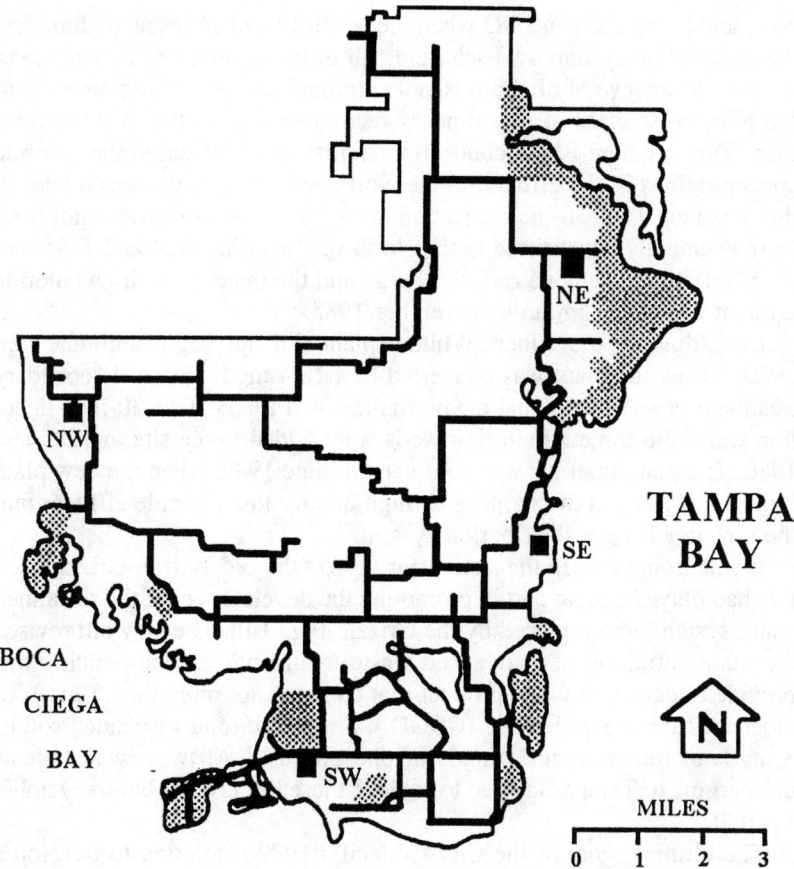

Figure 22.2 St. Petersburg's reclaimed water distribution system and critical water quality areas (shaded).

loops was completed and they were interconnected to form a unified system. This system became the Reclaimed Water Distribution System (Figure 22.2). The completion of this transmission system made St. Petersburg the first community in the nation to develop a dual distribution system to serve urban areas with an alternative water source for irrigation. At this time, however, only the Southwest and Northeast plants were discharging into the system.

A further federal grant of $33.4 million at the end of 1978, enabled St. Petersburg to upgrade the Northwest plant from a 6 mgd primary plant to a 20 mgd facility utilizing advanced secondary processes similar to those at the Southwest and Northeast plants, except that a shallow bed continuous backwash system of filtration was employed. This major upgrade was not completed until early in 1983 when the Northwest plant began to dominate the distribution system by discharging all of its effluent into it. The sewer systems from several offshore island communities were also connected to the Northwest Plant at this time as recommended by the 201-Facilities Plan. This offshore island connection significantly increased the chloride concentrations in the effluent at the Northwest plant as discussed later in this chapter. Two 30-inch injection wells and one 8-inch monitor well were eventually constructed at the Northwest plant in 1984 and 1985, but the plant did not achieve zero-discharge and the injection of high chloride effluent did not begin until November 1985.

Upgrading of the Albert Whitted plant did not begin until the mid 1980s, when the plant was converted to a 12.4 mgd advanced secondary treatment process with dual media shallow bed filters. Two 30-inch injection wells and three monitoring wells were added to the site in 1985 and 1986. The plant upgrade was completed in late 1987, when the new plant immediately began discharging its high quality low chloride effluent into the reclaimed water distribution system.

It was ironic that in the same year (1987) the old Wilson-Grizzle Act, that had played a great part in promoting the development of the reclaimed water system, was replaced by the Grizzle-Figg Bill. The new bill revised the State definition of "advanced waste treatment" as that which "will provide a recovered water product that contains not more than 5 mg/L of biochemical oxygen demand ($CBOD_5$), 5 mg/L of total suspended solids, 3 mg/L of nitrogen and 1 mg/L of phosphorus." All wastewater plants discharging to Tampa Bay had to achieve these limits on or before October 1, 1990.

The ultimate goal of the City Council's 1972 resolution to develop a "zero-discharge system" was finally realized well within the Grizzle-Figg deadline in September 1989, when the Albert Whitted deep-wells were permitted for continuous injection and the plant permanently closed its effluent out-fall to Tampa Bay.

By achieving zero-discharge when it did, the City of St. Petersburg avoided the enormous costs of upgrading its treatment plants to meet the new stringent Advanced Wastewater Treatment (AWT) limits.

At the present time (1998) all four plants are Type I, Category II, advanced secondary, activated sludge treatment facilities. The liquid stream treatment includes: influent pump station, preliminary treatment (screening and grit removal), advanced secondary treatment (activated sludge aeration and clarification), effluent filtration (deep bed filters in the Northeast and Southwest plants, shallow bed filters in the Northwest and Albert Whitted plants), and high level disinfection (chlorination). The solids stream treatment includes: waste activated sludge thickening, anaerobic digestion, and dewatering (sludge belt filter presses) for off-site transport and land disposal (via private contractor).

The Addition of Residential Customers to the Reclaimed System

As soon as the Southwest plant came on line back in 1977, the city realized that it would need a plan to expand the customer base of the distribution system and, in 1981, the city applied for federal funding for this purpose. The study conducted in support of the grant application was called the "Reclaimed Water System Master Plan," and was prepared by Henningson et al. (1981). The study proposed a phased program for expanding the reclaimed water system into selected residential areas. Residential developments designated as "critical water quality areas," were defined as areas where shallow wells produced water that was too saline for irrigation purposes. Residents in these areas were obliged to use potable water for irrigation. A total of four such areas were identified around the coastal regions of the city. In these areas, maximum potable water savings would be realized by the installation of the supplemental non-potable water source for irrigation purposes. This application was successful and various construction grants ranging from 55% to 85% of the costs were received to develop the critical water quality areas (shaded areas on Figure 22.2).

In addition to the critical areas, the 1981 plan recommended that residential customers adjacent to the original reclaimed water distribution mains be permitted to connect. Some residents were allowed to tap into adjacent mains before the city declared a moratorium on further hookups until policies and regulations could be drafted. Policies and regulations for the provision of reclaimed water service were developed and were adopted by the St. Petersburg City Council in November 1981. Once these policies had been adopted, the moratorium on reclaimed water hookups by residential customers was lifted and expansion into residential areas began.

The Present Day Status

As the final plant came on line in 1987, the WCRWSA also prepared a "Needs and Sources" report which projected that by the year 2020, if a 20-year drought occurred, and if reclaimed water was not available for irrigation, the city could anticipate a 23 mgd shortfall on a maximum demand day. However, if reclaimed water were expanded to its full potential by that time, instead of a 23 mgd shortfall, the city would have 1.0 mgd remaining capacity.

At the present time the statistics of St. Petersburg's wastewater treatment plants and the backup disposal deep well system are shown in Table 22.1. Figure 22.3 shows the mean daily potable water consumption of the city since 1977. The mean daily volume of reclaimed water used since startup is superimposed on this figure. It can be seen that the introduction of this

TABLE 22.1.   St. Petersburg's Wastewater Treatment Plant
Capacities to September 1995.

Plant	Design Capacity (mgd)	Avg. Influent Flow (mgd)[a]	Storage (million gals)	Deep Wells (No.)	Total Deep Well Capacity (mgd)
Albert Whitted	12.4	10.0	2	2	47
Northeast	16.0	10.8	8	3	27
Northwest	20.0	13.8	10	2	32
Southwest	20.0	14.6	5	3	27
TOTALS	68.4	49.2	25	10	133

[a]October 1994 to September 1995.

"Third Service" has stabilized the growth of the potable water demand
for the past 19 years. The increase of the customer base and the expansion
of the pipe-line system are also included in Figure 22.3.

The Deep-Well Injection System

In order to maintain a zero-discharge to surface water condition for each
treatment plant, St. Petersburg was permitted to construct a deep-well injection system to act as an alternative effluent disposal procedure. In Florida,
the regulation of these wells, which are classified as Class I, is carried out
by the Florida Department of Environmental Protection (FDEP).

Injection wells and monitoring well networks were constructed at each
of the four plants as described in the previous section so that each plant
has its own independent injection well system. The injection wells are
used during periods when excess reclaimed water is available and storage
tanks are full or when the quality of the product does not meet the high
reuse standards imposed by the city. The injection wells inject excess
reclaimed water through an open borehole into a confined salt water aquifer
between 700 and 1,100 feet below land surface. This salt water aquifer is
located in the Avon Park dolomite and limestone formation beneath St.
Petersburg. The dolomite is highly fractured in the upper layers and allows
wells of large injection capacities to be constructed. The injection zone
out crops in the Gulf of Mexico hundreds of miles west of St. Petersburg.

Originally it was anticipated that the effluent could be stored in freshwater "bubbles" around the injection points. It would thus be available
once again for extraction in times of need. The difference in specific
gravity's between the highly saline water and the effluent caused the latter
to rise and form a lens at the confining layer rather than a bubble at the
injection point so that it proved unfeasible to recover the injected effluent.

Injection well flows and pressures are continuously monitored and the data are periodically examined to determine if a decrease in injection well capacity is occurring due to plugging at the borehole walls. Monthly injectivity tests on each well are required by the FDEP to assess injection well capacity. It has been found that whenever there are increased solids in the effluent, some plugging of the injection well can occur. The city has been able to restore the injection capacity following these events by acidizing the well using concentrated hydrochloric acid.

The monitoring wells vary in depth and are used to monitor for vertical and horizontal movement of the injected effluent. Most of the aquifer

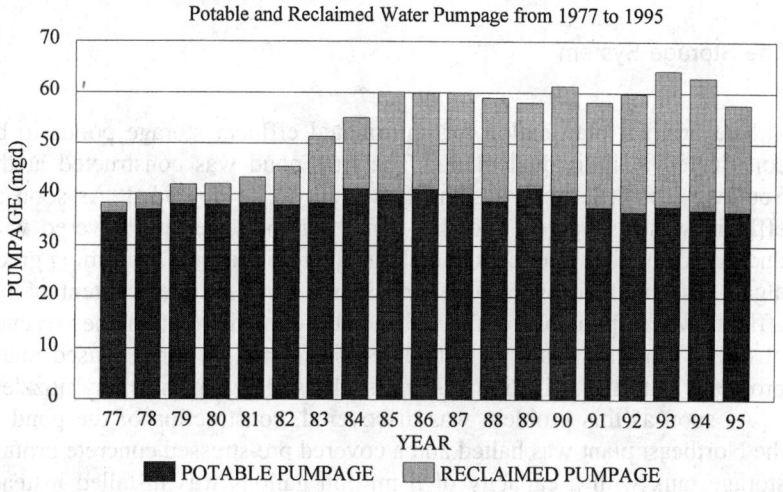

YEAR	'83	'84	'85	'86	'87	'88	'89	'90	'91	'92	'93	'94	'95
RECLAIMED WATER PUMPED (MGD)	13.0	14.0	20.0	18.9	19.8	20.3	16.8	21.4	20.7	23.2	26.7	25.2	21
TOTAL NUMBER OF RECLAIMED WATER CUSTOMERS	610	904	3,100	4,964	5,356	5,871	6,285	6,800	7,175	7,408	7,612	7,924	8,590
RESIDENTIAL	472	761	2,927	4,720	5,107	5,616	6,052	6,547	6,912	7,139	7,307	7,566	8,225
COMMERICAL/ INDUSTRIAL/ OTHERS	138	143	173	244	249	255	259	261	267	269	305	358	365
MILES OF PIPE (ALL SIZES)	94	142	190	228	228	228	229	229	264	264	264	265	275
POTABLE WATER PUMPED(MGD)	38.3	41.1	40.1	41.0	40.2	38.7	41.3	39.9	37.5	36.6	37.7	37	36.7

Figure 22.3 The growth of the reclaimed water system since startup in 1977.

zones monitored at each of the injection sites contain salt water. Each monitor well is regularly sampled to check water quality and water level. Recently, the FDEP has alleged that data from the monitoring wells indicate that there is vertical movement of native groundwaters or reclaimed water into the aquifers overlying the injection zones. Some of these aquifers are classified as underground sources of drinking water (USDWs) as they contain less than 10,000 mg/L of total dissolved solids. State regulations do not allow changes in water quality to occur in USDWs as a result of Class I injection well and the City is at present negotiating with the State to overcome this problem.

The Storage System

The original plan called for chlorinated effluent storage ponds to be constructed at individual plants. The first pond was constructed at the Southwest plant as this was the first plant to be upgraded. As soon as effluent was discharged into this pond, it rapidly became covered with duckweed and the water contained large quantities of free swimming green algae. The high nitrogen, phosphorus and other nutrient content of the effluent was responsible for this enhanced and rapid growth in the presence of direct sunlight. Also, palm tree seeds dropped by birds caused many problems with the irrigation system, such as clogging the spray nozzles.

As soon as this problem was discovered, construction of the pond at the Northeast plant was halted and a covered pre-stressed concrete ground storage tank with a capacity of 8 million gallons was installed instead. The pond at the Southwest plant was also replaced by a similar tank with a 5 million gallon capacity. As the other plants were upgraded, two 5 million gallon tanks were added at the Northwest plant and one 2 million gallon tank was constructed at the Albert Whitted plant.

Reclaimed water is stored in the tanks at each plant when demand for irrigation water is low. When demand exceeds supply, St. Petersburg's inter-connected or "looped" distribution system provides a great deal of flexibility. For example, when storage at one facility is full, effluent from that facility can be discharged into the system to relieve demand in other areas of the city. The total storage capacity of the system (25 million gallons) represents approximately one day's irrigation demand. In extended drought periods, less wastewater is processed at the plants due to reduced infiltration/inflow. This can result in a shortage of reclaimed water for all users if irrigation demands are unusually high. Under these conditions, public education programs and information systems are used to inform the public of the problem, with special emphasis on the responsible use of reasonable quantities of reclaimed water rather than excessive uses.

SPECIAL STUDIES, PUBLIC HEALTH CONSIDERATIONS

St. Petersburg very soon realized that the development of a large re-claimed water distribution system for urban and close public contact residential irrigation increases the risk of transmitting infectious diseases via treated wastewater to the general public. Viral control in the final effluent was a major concern because of the wide variety of viruses (over 100 recorded types) that occur in wastewater.

INITIAL VIRUS STUDIES

The original applications for grant moneys to construct the system and upgrade the treatment plants that were submitted to the USEPA in the early 1970s contained detailed assessments of tertiary treatment methods for viral reduction to a non-detectable level in the effluent. These assessments were based on three years of research studies and testing beginning in 1971 and carried out at the Northwest plant by Dr. Flora Mae Wellings, Administrator of the Epidemiology Research Center, Florida Division of Health and Rehabilitative Services, Tampa, Florida.

In a letter to USEPA, Region IV, in May 1975, with reference to the upgrading of the Southwest and Northeast plants, Dr. Wellings wrote: "There is no longer any doubt that activated sludge treatment alone removes very little, if any, virus. Theoretically, there are four aspects of the proposed treatment which should remove virus to a nondetectable level. The first is precipitation. This technique has been used in the laboratory for years to concentrate virus and is presently used by some investigators to concentrate virus for isolation from surface and/or wastewaters. In practice, this procedure should remove the greatest multiplicity of virus. The second site of removal should occur with the particulate removal during the multimedia filtration if virus present is embedded in these particulates. Rapid mix chlorine contact permitting free chlorine residual would complete the active virus removal treatment. Added to these safeguards is a chlorinated holding pond. Although we do not know the actual survival time of virus under these conditions, we do know that there is a natural die-off over time so if any agents do survive the foregoing, they should be destroyed at this site."

As has already been described, both the Southwest and Northeast plant had alum precipitation prior to filtration, particulate removal by deep bed multimedia filtration and rapid mix chlorine contact permitting a free chlorine residual when they were first connected to the reclaimed water system.

Virus monitoring continued in the influent, clarifiers, post influent, post chlorination, predistribution and distribution areas at the plants after upgrading and prior to and after their connection to the reclaimed water system. This monitoring continued in the early 1980s and virus presence in the chlorinated effluent was rarely detected. When detection did occur, the level was minimal and could usually be linked to a treatment process problem (usually in the filtration system) at the specific plant concerned.

THE PANEL DISCUSSION

The general concern for possible public health problems which might arise from the recycling of treated wastewater in public access areas led St. Petersburg to seek further assistance in obtaining comments from recognized experts on the control of coliform bacteria and viruses in reclaimed water. In December 1986, the consultant firm of Camp Dresser & McKee, Inc., submitted a proposal to assemble a panel of highly experienced and nationally recognized engineers and scientists to review all reports, bacteriological data, design criteria and other associated information dealing with the design, operation and implementation of a virus removal program for the St. Petersburg wastewater reclaimed water system. After this review the experts would make comments and recommendations in a panel discussion with the City of St. Petersburg in early 1987. An expert technical writer would compile the comments and recommendations as a paper directed to assisting in the operations and management of the reclaimed water system. The major issues to be addressed would be:

- What is a reasonable coliform standard for a reclaimed water supply in an urban area?
- What operating conditions would accommodate a "virus free" reclaimed water supply?
- What facility management and data collection procedures are appropriate for a reclaimed water supply in an urban area?

The experts would receive copies of the Colorado Springs Epidemiological Study, the Pomona, CA study of effluent filtration and the St. Petersburg data on coliform and viral counts in the reclaimed water. The final panel of consultants included:

- Dr. Gabriel Bitton, Professor, Department of Environmental Engineering Sciences, University of Florida
- Dr. Kenneth Herrmann, Assistant Director, Division of Viral Diseases, Centers for Disease Control, Atlanta
- Dr. Daniel A. Okun, Kenan Professor Emeritus of Environmental Engineering, University of North Carolina

- Dr. Mark Sobsey, Professor of Environmental Microbiology, University of North Carolina

On February 5, 1987, the panel met St. Petersburg staff for a synopsis of the city's reclaimed water program and visits to the Northwest and Northeast plants were arranged. These plants were selected to provide a view of shallow (Northwest plant) and deep bed (Northeast plant) filtration. The panel discussion took place on February 6, 1987 at the Public Utilities main conference room in St. Petersburg based on questions previously provided to panel members.

THE "WHITE PAPER"

The report (Camp Dresser & McKee, 1987) of the discussion was produced in March 1987. An in depth survey of studies in other parts of the USA and a review of the current standards set for reclaimed water by State regulations led to the following conclusions and recommendations by this report:

(1) There is currently no evidence of increased enteric diseases in urban areas irrigated with treated reclaimed water using coagulation, filtration, and disinfection.

(2) There is no evidence of significant risks of transmission of viral or microbial diseases as a result of exposure to effluent aerosols from spray irrigation with reclaimed water.

(3) Wastewater treatment systems discharging into a reclaimed water system must be designed, constructed, operated, and monitored in the most responsible manner possible to assure public health safety. Items to be considered are: an advanced secondary treatment facility adequate to handle projected hydraulic and organic loadings, and with the capability for chemical addition prior to filtration, Class I reliability, 7 days/week effluent sampling and 24 hours/day, 7 days/week surveillance by qualified operators.

(4) There was overall agreement that the most important operating parameter and measure of performance should be turbidity. Continuous on line turbidity measurements should be made in the effluent from the chlorine contact chamber and it should be feasible to achieve effluent turbidity of 2.0 NTU during any 24 hour period. Additional turbidity readings should be routinely taken of the influent flow to the filters and the effluent from each filter. Turbidity values can and should be correlated with such microbiological data as are collected because turbidity is valuable as a surrogate for more difficult microbiological data.

(5) The recommended standard for fecal coliform is an average of 2.2/ 100 mL with an upper limit of 23/100 mL in not more than 10% of the samples, while maintaining a 4.0 mg/L chlorine residual. [Note: the State requirement at this time for coliform bacteria levels in the effluent was "nondetectable," which was more stringent than the State standard for recreational waters (<1,000/100 mL) including full body contact and drinking water (<1/100 mL)]. The report criticized the "nondetectable" State standard as being unachievable, unrealistic, and contrary to generally accepted concepts for indicator bacteria standards).

(6) Virus standards for reclaimed water are not recommended. First, current virus detection methods are too technically difficult, expensive, time consuming, and unreliable for routine monitoring to determine compliance with standards. Second, current virological methods used to assess reclaimed water will not detect many important virus types (i.e., hepatitis A, rotaviruses) that might be present in treated wastewater. Third, there are not sufficient data to quantitatively relate health risks of viral exposure and resultant infection and illness to the level of viruses in reclaimed water used for specific nonpotable purposes such as landscape irrigation. Until such relationships are established by rigorous epidemiological/virological studies, any virus standards would have to be set by rather arbitrary approaches. Other surrogate parameters such as turbidity provide a good measure of treatment efficiency for virus elimination. If turbidity is being controlled, viral and fecal coliform concentrations are also being controlled. Monitoring for viruses may be continued to provide public assurance of the City of St. Petersburg's concern with virus control and to build public confidence in the reclaimed water product.

(7) In addition to compliance monitoring at the wastewater treatment plant, it is recommended that the distribution system should be randomly monitored (non-compliance) for both fecal coliform and chlorine residual to verify that reclaimed water quality has not become degraded by distribution system colonization, regrowth, stagnation, pipe surface scouring, or loss of integrity of the distribution lines. This procedure would also assure an aesthetic product.

Some extra concern was voiced over the difference in the filtration efficiency between deep bed and shallow bed filters. It was recommended that the installation of deep bed filters was preferable when effluent is to be pumped into a reclaimed water system. However the group agreed that although St. Petersburg has installed shallow bed ABW filters at two of the four plants, the system should not present any public health threat if it is operated in a responsible manner.

A copy of the white paper was sent to the Secretary of the State Department of Environmental Regulation in May 1987. A reply was received in the same month indicating the regulatory agency's would find the report extremely useful to them in evaluating the need for modifying Florida's current standards for reuse of recovered water.

SPECIAL STUDIES, PROJECT GREENLEAF

Wastewater effluent that had previously been discharged to surrounding surface waters with minimal regulation was now being piped directly to domestic homes, parks, schools and golf courses as a new and highly desirable, marketable resource. Once the public health aspects of reclaimed water had been dealt with, St. Petersburg turned its attention to the problems related to continuous supply of high quality reclaimed water which would be beneficial as an irrigant and a dilute fertilizer to the development and maturation of the wide variety of ornamental plant species and turf that abounded in the domestic gardens throughout the city.

ORIGIN OF "PROJECT GREENLEAF"

Between 1983 and 1986, as the reclaimed water system was continuously adding more domestic customers, climatic conditions in the area were highly abnormal. A nineteen month period with extremely low rainfall occurred from October 1983 onwards, severe freezes occurred in December 1983 and February 1985 and a hurricane brushed coastal St. Petersburg in September 1985. These factors resulted in an increase in the demand for reclaimed irrigation water (Figure 22.3), serious damage to many landscape ornamental plants and the death of many delicate species.

Throughout 1985, St. Petersburg experienced a significant increase in the number of complaints received from homeowners regarding damage to ornamental plants and trees. Many of these complaints claimed that leaf damage or plant death was directly caused by reclaimed irrigation water. Klinkenberg (1985) reviewed some of these complaints in a local newspaper article.

To effectively deal with these complaints a Reclaimed Water Committee was formed at the end of 1985. This committee was to identify whether or not a problem truly existed with regard to plant damage caused by reclaimed water and, if so, to develop guidelines for the necessary management required to overcome it. The committee included both staff from the city, the Pinellas County Cooperative Extension Service and members of the public from the Harbor Isle Homeowners Association, Inc. The committee concluded that the unique reclaimed water system developed by the

City of St. Petersburg offered an unprecedented opportunity to conduct original research studies. Approval by city management to fund an investigation of customer complaints and conduct original research on the impact of using reclaimed irrigation water on vegetation was granted early in 1986 and the study which was entitled, "Project Greenleaf" officially came into being. The Pinellas County Cooperative Extension Service participated in the project by offering assistance, producing an educational video and providing free analytical services.

OBJECTIVES OF "PROJECT GREENLEAF"

The overall goals of this original research study were to improve the use of the reclaimed water irrigation system by:

(1) Determining the quality and quantity fluctuations of the influent flow and assessing the resulting reclaimed water quality fluctuations and availability

(2) Assessing the effects of reclaimed water on the growth and development of ornamental plants and trees commonly used for landscaping and decorative purposes within the area

(3) Establishing guidelines for the successful use of reclaimed water on ornamental plants

(4) Listing necessary management practices for more delicate plant species

(5) Producing literature both for public education and scientific publications

(6) Making necessary recommendations to ensure the successful operation of the reclaimed water irrigation system in the future

The study produced a final report (Parnell 1988) in September 1988 and the highlights of this report are included here in the following subsections.

REVIEW OF "PROJECT GREENLEAF"

Influent Wastewater Studies in St. Petersburg

Monthly reports from the Northeast, Northwest and Southwest treatment plants connected to the system showed an 11.5% total increase in influent flow between 1983 and 1987. Census records indicated only a 1% population increase in the city within this period. This discrepancy can be explained by sewer or service lateral deterioration allowing increased infiltration and inflow by groundwater and storm water. The treatment plant reports support this assumption by showing that monthly fluctuations of influent flow to all three plants were positively correlated with monthly rainfall levels, with 22% of variation in influent flow being accounted for by rainfall. Storm water inflow was thus occurring quite extensively. Furthermore, after periods of drought, there was also a positive 29.8% correlation between rainfall in one month and influent flow in the following month. This was due to infiltration from elevated groundwater levels. Previous research by others has shown that these correlations are generally caused by privately owned defective service laterals which are often responsible for over 90% of all infiltration into sewers.

Problems arise at wastewater treatment plants when infiltration/inflow rates climb excessively high or fluctuate violently. On several occasions, influent flows above plant capacities have occurred in St. Petersburg which have resulted in high turbidity levels in the effluent, rendering it unsuitable for use in the irrigation disposal system. For efficient reclaimed water production, an influent wastewater flow not exceeding plant design parameters is highly desirable, if not essential. On-going inspection and maintenance of sewer lines must be carried out to minimize the effects of infiltration/inflow on influent levels. Constant monitoring is also required to determine how much storm water is directly entering the sanitary sewer system from whatever cause. Without these programs, large fluctuations of influent flow will continue to occur and may undermine the concept and integrity of the reclaimed water irrigation distribution system.

St. Petersburg's wastewater treatment plants are primarily designed to treat domestic strength wastewater. The production of industrial wastewater from manufacturing facilities and food processors connected to the system can have a significant effect on influent quality if it is not rigorously regulated. Section 27-206 through 217 of the St. Petersburg code covers the limitations and prohibitions on the quality and quantity of wastewater which may be lawfully discharged into the publicly owned treatment works (POTW) by users. All significant industrial users who produce wastewaters with high pollutant levels are required to pretreat their wastewater to reduce these levels to acceptable concentrations before discharging into the sanitary sewer system. Permits are issued to these users and they are continuously monitored by the city's industrial pretreatment program which enforces the code if violations are detected. Data from the industrial pretreatment program are included here with respect to heavy metals and other pollutants detected in the effluent flows. It is recommended that all wastewater treatment plants that discharge to an effluent irrigation system should have an industrial pretreatment program in place to regulate industrial users of the sewer.

Levels of influent biochemical oxygen demand (BOD), which is an indicator of the amount of organic matter present in the wastewater, and total suspended solids were analyzed for the Northeast and Northwest plants from October 1981 to September 1987. Over this 6 year period, the mean BOD and TSS levels for the Northeast plant were 101 mg/L and 94 mg/L respectively whereas the mean levels for the Northwest plant were 140 mg/L and 147 mg/L. These figures compare well with the national average of (BOD) 181 mg/L and (TSS) 192 mg/L (Culp et al. 1979), and both fall into the weak concentration range (Metcalf & Eddy 1979).

Analysis of the data showed negative correlations between monthly rainfall and levels of both BOD and TSS in the influent. Influent BOD and TSS were also negatively correlated with the previous month's rainfall data so that forecasting was possible. Thus, dilute storm water or ground-water infiltrating/inflowing into the sewers tended to reduce the concentrations of BOD and TSS. Since St. Petersburg's wastewater treatment plants are designed to treat sewage with a mean BOD of 250 mg/L, this dilution by infiltration/inflow is an undesirable factor.

The overall mean total nitrogen at the Northeast Plant was 23.46 mg/L and, at the Northwest Plant, the overall mean was 25.22 mg/L throughout the study period. At both plants, the majority of the total nitrogen was present as ammonium (NH_4^+), with very small amounts of nitrite and nitrate and highly variable amounts of organic nitrogen. This composition is typical of normal wastewaters and the overall mean levels fall well within national ranges of 20 to 85 mg/L (Metcalf & Eddy, 1979).

Phosphorus must be present in minimum quantities for optimum operation of biological treatment systems. It occurs in typical wastewater flows as a mixture of orthophosphate, pyro, poly and meta-phosphate and complex organic phosphates. Orthophosphates are important as they are directly available for biological metabolism. The overall mean total phosphorus level throughout the study at the Northeast plant was 5.45 mg/L and, at the Northwest plant, it was 9.00 mg/L. Orthophosphate levels averaged 65% of the total phosphorus at both plants. National means for total phosphorus levels in typical wastewater vary between 6 and 20 mg/L with inorganic phosphorus making up over 70% of these totals.

Reclaimed Water Quality in St. Petersburg

Effluent quality must depend on influent quality and the degree of treatment that the wastewater receives. The treatment process is designed to remove certain components of the influent such as BOD, TSS and viral and bacterial elements. Other compounds pass through the treatment plant and concentrations in the effluent may be only slightly lowered or un-changed from those in the influent. For instance, total nitrogen and phosphorus levels are reduced somewhat in the effluent, but chlorides are very similar in concentration at both ends of the plant.

The chemical constituents that are of value for the irrigation of landscape vegetation and turf are somewhat different from those that are important to the wastewater treatment process. However, toxic chemicals that affect the biological treatment process such as heavy metals and organic pollutants will also reduce the quality of the irrigation water. The most important compounds in reclaimed wastewater which provide fertilizer benefits to landscape production are nitrogen, phosphorus and potassium. Calcium, magnesium and iron are also essential in smaller quantities and manganese, copper, molybdenum, boron, sulfur and zinc are necessary as trace elements. Table 22.2 shows the mean values for the foregoing and other parameters analyzed in the final effluent from the three plants that were discharging into the irrigation distribution system in 1986 and 1987 when "Project Greenleaf" was in progress.

Nitrogen

Nitrogen is often referred to as the basic building block of all living matter. Plants require large amounts of nitrogen for rapid growth and reproduction. Nitrogen forms part of all protein molecules amino acids and nucleic acids found in living protoplasm and it is essential for many of the physiological processes which occur within green-plants.

Data on total nitrogen concentrations in the effluents produced by the three wastewater treatment plants supplying the distribution system from June 1982 until September 1987 were analyzed. The Northeast plant maintained a fluctuating output with maximum levels not exceeding 14 mg/L and minimum levels as low as 3 mg/L. The Northwest plant showed similar fluctuations to the Northeast plant with slightly higher maximum and minimum values. The Southwest plant showed the greatest fluctuations of all three facilities with maximum levels reaching over 20 mg/L and rarely falling below 6 mg/L. At all plants the effluent total nitrogen levels did not correlate with any external factors, so that the fluctuations were probably a feature of the individual processes at each plant.

Phosphorus

Phosphorus is an essential plant nutrient and its main function is to store and transfer energy within plant cells. It is also associated with the development of the root system in plants. Rhue and Street (1980) describe the fate of phosphorus in Florida soils, and discuss its effect on plants. Nationally, phosphorus occurs in reclaimed water in concentrations ranging from 6 to 15 mg/L, unless some removal occurs at the treatment plant, (Wescot and Ayers 1985).

Effluent phosphorus levels at the three treatment plants from June 1982

TABLE 22.2. Effluent Quality Data from Three Treatment Plants (means of data from January 1986 to September 1987).

Parameter	Units	Northeast Plant	Northwest Plant	Southwest Plant	Westcott & Ayers Recommended Maximum Conc.
B.O.D.	mg/L	2.79	2.94	3.45	—
Suspended solids	mg/L	2.80	2.56	2.77	—
Total dissolved solids	mg/L	1024	1067	580	—
Conductivity	mmhos/cm	1.51	1.61	0.93	—
Turbidity	NTU	0.84	0.90	—	—
Alkalinity	mg/L	154	191	160	—
Kjel-nitrogen	mg/L	3.24	9.77	15.06	—
Organic nitrogen	mg/L	1.03	3.42	3.59	—
Ammonia nitrogen	mg/L	2.21	6.35	11.47	—
Nitrite nitrogen	mg/L	0.18	0.94	1.03	—
Nitrate nitrogen	mg/L	2.33	0.93	0.40	—
Total nitrogen	mg/L	6.22	10.86	15.99	—
Ortho-phosphate	mg/L	3.14	3.12	2.87	—
Total phosphorus	mg/L	3.37	3.45	3.21	—
Total organic carbon	mg/L	9.23	9.14	10.69	—
Foaming agents	mg/L	0.08	0.09	0.11	—
Color	PT/CO	20.00	28.80	25.00	—
Corrosivity	Lang. Index	0.21	0.30	0.09	—
Odor	Threshold	18.33	32.50	28.75	—
Fecal coliform	#/100 mL	<1	<1	<1	—
Aluminium	ug/L	200	200	352	500
Boron	ug/L	410	480	317	—
Calcium	ug/L	80.80	74.40	63.90	—
Chromium	ug/L	<5	20.00	4.30	100
Copper	ug/L	9.20	7.30	8.00	200
Iron	ug/L	127.50	93.10	191.70	5000
Lead	ug/L	<5	<5	7.00	5000
Magnesium	mg/L	22.50	14.80	12.40	—
Manganese	ug/L	19.80	19.30	17.30	200
Molybdenum	ug/L	<100	<100	<50	10
Nickel	ug/L	<5	<5	51.80	200
Potassium	ug/L	14.50	12.00	11.20	—
Silver	ug/L	<10	<1	<10	—
Zinc	ug/L	29.10	19.40	27.80	2000
SAR	units	5.01	4.05	4.04	—

SAR—Sodium Adsorption Ratio.

1062

until September 1987 showed minimal monthly fluctuations and there was a range from 2 to 6 mg/L. Phosphorus is generally present in the effluent as a mixture of orthophosphates, polyphosphates and organic phosphorus. The majority of phosphorus in the reclaimed water from all three plants was present as orthophosphates, with very small proportions of the other forms.

Salinity

Salinity is by far the most important factor as far as wastewater irrigation quality is concerned. The salinity of the effluent is a measure of the total dissolved salts (TDS) including chlorides, sulfates and bicarbonates of calcium, sodium and magnesium. As salinity increases, the availability of water to plant roots decreases and plants will wilt in conditions where adequate irrigant appears to be present, if salinity is exceptionally high.

According to the California agricultural guidelines for the interpretation of water quality for irrigation (Westcot and Ayers 1985), any irrigant with an electrical conductivity between 0.7 and 3.0 mmhos/cm or a TDS between 450 and 2,000 mg/L lies within the slight to moderate "restriction on use" range. Table 22.2 shows that all three effluents lie well within these limits. This indicates that no special procedures or management practices should be necessary in order to use St. Petersburg's reclaimed water for irrigation, especially on landscape ornamental plants where maximum yield is not an essential requirement.

There are many publications on the salinity tolerance levels of a wide variety of green-plant species. Mass (1986) reviews much previous literature and has produced extensive lists of salt tolerance levels for both agricultural and ornamental plants. Barrick (1979) and Johnson and Black (1985) have produced lists of ornamental plants that show varying degrees of salt tolerance in Florida.

Other Beneficial Quality Parameters

Potassium levels in the effluent are rarely, if ever recorded at wastewater treatment plants, as this element has no detrimental effect on treatment processes. It is essential for the development and maturation of all types of landscape vegetation, although its specific role is not clearly understood. According to Rhue and Street (1981), potassium is involved in photosynthesis, plant-water relations, enzyme activation, disease resistance and quality in fruits and vegetables. Potassium concentrations were found to be similar in all three treatment plant effluents and ranged from 11 to 14 mg/L.

Concentrations of magnesium, like potassium, are rarely recorded at wastewater treatment plants. This element is essential for chlorophyll production in green plants and it is generally in short supply in Florida's sandy soils. Usual ranges for this element in wastewaters can fluctuate from zero to 60 mg/L. The "Greenleaf" data in Table 22.2 show that magnesium was present in similar concentrations in the effluents from the Northwest and Southwest plants (12–15 mg/L), but it was somewhat higher at the Northeast plant (22.5 mg/L).

Calcium is almost always present in wastewater in some form and may reach as high as 400 mg/L in some areas. It is essential for vegetative growth and rigidity in landscape plants and often occurs in the soil in high concentrations. Table 22.2 shows calcium levels in the three plant effluents varied between 60 and 80 mg/L which are quite acceptable for irrigation water.

Copper, iron, manganese, molybdenum and zinc are plant micro-nutrients that occur in small quantities in reclaimed water and the levels in μg/L are indicated in Table 22.2. Some of these are discussed further by Street and Rhue (1980a, 1980b).

Boron is an essential trace element for green-plant growth and development but it can accumulate in the tissues and reach levels which become toxic to the plant. This phenomenon has been referred to as specific ion plant toxicity. The sensitivity of plants to boron levels varies greatly. Mass (1986) reviews previous data on this subject and divides several species of ornamental plants into very sensitive, sensitive, moderately sensitive, moderately tolerant and tolerant categories. The threshold level for very sensitive plants is 500 μg/L which was only exceeded once in all of the readings taken at the three treatment plants. It is highly unlikely therefore, that the low concentrations of boron in St. Petersburg's reclaimed water will have toxic effects on any commonly used species of landscape plants.

Westcot and Ayers (1985) have recommended maximum concentrations for other trace elements in irrigation waters for agricultural use in California and these are included in Table 22.2. These authors state that the recommended concentrations are based on water application rates of 4 acre-ft/acre-year. These rates are equivalent to 4 inches per square foot per month (approximately 1 inch per week per square foot of irrigatable green space) for the residential user. It can be seen that none of the elements listed in Table 22.2 occur in concentrations which would have any detrimental effect on plant growth according to Westcot and Ayers's (1985) recommendations.

Sodium

In coastal communities, the greatest source of sodium comes from the infiltration of raw seawater and brackish groundwater into gravity fed sewer lines which are in poor condition. Sodium concentrations are not normally taken in wastewater effluent. A review of the records at the Public Utilities Water Quality Assessment Laboratory showed that, since January 1983, 26 records for the Northeast Plant effluent had a mean value of 197.7 mg/L, 3 records for the Northwest Plant averaged 146.2 mg/L and 10 records for the Southwest Plant averaged 134.6 mg/L. No conclusions can be drawn from these insufficient data.

Sodium ions can affect vegetation in a variety of ways. They often act in conjunction with chloride ions and can be absorbed through the leaves of plants to accumulate in the tissues until they reach toxic concentrations. Also, they can create osmotic problems on wet leaf surfaces when applied by overhead sprinkler irrigation methods. Leaf damage usually begins as a brown patch at the tip of the leaf which spreads upwards until all of the leaf becomes affected. This type of damage is usually referred to as "leaf burn."

Sodium salts in high concentrations can also affect the soil structure which in turn slows down or prevents irrigation water from percolating through the upper layers (Quirk & Schofield 1955, Marshall 1968, Zartman & Gichura 1984). Measurement of the influence of sodium ions on soil properties can be estimated by calculating the Sodium Adsorption Ratio (SAR), which is a measure of the relationships of sodium, calcium and magnesium in the irrigant water. The calculated SARs for the three plants are shown in Table 22.2.

The guidelines for the interpretation of irrigation water quality in California prepared by Westcot and Ayres (1985), suggest that the irrigant SAR should be used in conjunction with the conductivity (EC_w) to evaluate possible soil permeability problems. From the table of values given by these authors, the combination of SAR and conductivity in the effluents from the Northeast and Northwest plants fall into the slight to moderate degree of restriction on use range. The effluent from the Southwest plant, where sodium levels are generally low, falls into the "no restriction on use" range. Thus, the probability of soil damage due to the use of any of the three effluents is minimal. Damage to plants due to contact with sodium ions will only be exhibited by relatively few, highly sensitive species.

Chlorides

Many plants are specifically sensitive to chloride ions (Cl^-), especially species that have woody stems. These ions may be taken up through the plant's root system or directly through the foliage. Toxic accumulations of chloride and/or sodium ions in the leaves rapidly induce leaf burn damage symptoms. Mass (1986) sets tolerance limits for 49 ornamental species of plants. He notes that aesthetic factors such as size and appearance are the most important criteria that must be considered for ornamentals rather than yield factors which are more applicable to agricultural crops.

Walcot and Ayers (1985) set chloride level guidelines for both sprinkler and surface irrigation techniques. For surface or drip irrigation, chloride levels of less than 140 mg/L are not considered to have any damaging effects on plants whatsoever. Levels between 140 and 350 mg/L are considered to be in the slight to moderate degree of restriction on use range and anything over 350 mg/L is considered to fall in the severe restriction on use range. For overhead sprinkler systems, less than 100 mg/L of chloride in the irrigant is considered safe, but restrictions on use begin once chlorides exceed this level. Once again, these levels are suggested for agricultural crops where yield reduction is the most important criterion. Higher chloride levels can be used with ornamentals, since slight growth reduction is not of prime importance provided that no damage symptoms become apparent.

In this study, the chloride levels in the reclaimed water were investigated in depth, as these were the only components of the irrigant that regularly occurred in concentrations high enough to create problems for chloride sensitive vegetation.

Effluent chloride fluctuations from January 1982 to September 1987 for the Northeast, Northwest and Southwest treatment plants are shown in Figure 22.4. The Northeast plant effluent chloride levels fluctuated

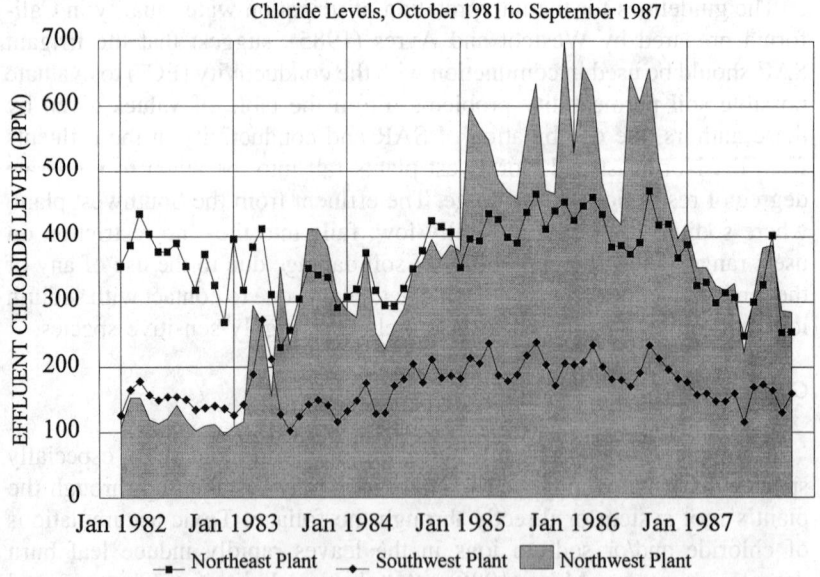

Figure 22.4 Northeast, Northwest and Southwest plant effluent chloride concentration (chloride levels, October 1981 to September 1987).

between 250 and 450 mg/L with a mean of 364 mg/L and a standard deviation of ±58.2, and the Southwest plant chloride levels fluctuated between 100 and 280 mg/L with a mean of 171 mg/L and a standard deviation of ±34.1. The Northwest plant showed large effluent chloride fluctuations in the study period and the computation of means and standard deviations was meaningless for these data.

The effluent chloride concentrations at the Northeast plant were consistently more than double those at the Southwest plant. This was most likely due to saltwater infiltration into the aging and low lying sewer system supplying the Northeast plant.

The fluctuations at the Northwest plant are explainable as follows. In 1981 and in most of 1982, the chloride levels were relatively stable at just over 100 mg/L. In late 1982 the sewer systems from the offshore islands of St. Pete Beach and Treasure Island were connected to the Northwest plant and chloride levels in the effluent immediately jumped to the 300–400 mg/L levels. It was clearly evident that large quantities of saltwater were infiltrating into the old and deteriorating sewer systems on these two islands. A second dramatic rise in chloride levels in the Northwest effluent occurred from November 1984 onwards. Chloride levels rose to the 600 mg/L mark, became extremely erratic, peaked at 700 mg/L in September 1985 and did not begin to significantly decease below the 400 mg/L level again until after November 1986 (Figure 22.4).

The reasons for this second increase were not immediately obvious until the effluent chloride levels were compared for each of the three plants. Figure 22.4 shows that chloride concentrations in the Northeast and Southwest plant effluents fluctuated in the same way as they did at the Northwest plant, but the magnitude of the fluctuations was nowhere near as great at these two plants. There was a positive correlation between all three chloride concentration graphs for the three wastewater treatment plants from September 1984 until April 1987. The statistical coefficient of determination (r^2) showed that the similarity between the chloride fluctuations throughout this time period was 70% between Northeast and Northwest plants, 79% between Northeast and Southwest plants and 66% between Northwest and Southwest plants.

The start of the fluctuations in chloride levels in October 1984 coincided with the beginning of a long drought period which decreased urban runoff inflow rates into the sewer systems and ultimately decreased influent flows to all three wastewater treatment plants. Since there was no significant drop in sea-level, infiltration of sea water into the sewers was not affected by this dry period. Thus, decreased influent flows containing similar amounts of sea water would result in increased concentrations of chlorides at the water treatment plants. This would especially apply to the offshore island's aging sewer systems supplying the Northwest plant, as it has already been established that heavy infiltration occurred in the beach areas.

Environmental Characteristics of the St. Petersburg Area

St. Petersburg's climate is probably one of its greatest natural resources. Summers are warm and relatively humid in this subtropical climate, with daytime high temperatures over 90 degrees F. Winters, although punctuated with invasions of cold air from the north, are mild because of the southern latitude and proximity to the Gulf of Mexico. Approximately 55 inches of rainfall are recorded every year, and 60% of this falls during the summer, usually during afternoon thunderstorms.

It has already been established that rainfall has profound effects on the production and composition of reclaimed water. It was thus essential to evaluate other environmental components to determine which factors directly affected the irrigation of landscape ornamental plants by reclaimed water. The inter relationships of climate with other factors such as evapotranspiration levels and soil composition and chemistry were extensively investigated in the St. Petersburg area and the results that were obtained are summarized below.

Climatic Conditions

Climatic conditions in the St. Petersburg area were extensively studied from 1982 to the end of 1987. Rainfall data showed that six significant drought periods occurred within this time. A drought period was defined as a minimum of a three consecutive month period when less than four inches of rain fell in each month. The important "droughts" were the three that occurred from October 1983 until July 1984, September 1984 until May 1985 and October 1985 until May 1986. The first two of these were almost continuous with each other, so that 19 consecutive months had extremely low rainfall. This unusually long dry period created many problems for the reclaimed water irrigation system. First, it encouraged many more customers to join the system (Figure 22.3), and the overall demand by all users went well over the recommended standard. Second, influent flows to the treatment plants decreased, so less effluent was available to meet the demand. Finally, influent BOD and SS levels rose rapidly, putting extra stress on the treatment processes and effluent chloride concentrations increased to levels that made the irrigation water unsuitable for salt sensitive vegetation.

To further compound these problems, the groundwater levels receded and saltwater intrusion caused many trees to be exposed to highly saline conditions in the deep root zone. Also, percolation rates increased as topsoil dried out, thus reducing the amount of available soil moisture for many shallow rooted ornamental landscape plants and turf.

This extended "drought" was closely followed in 1986 by another long dry period. Many plants had not recovered from the first "drought" before they were exposed to a second eight month period of low rainfall. Effluent chloride levels had not even begun to recover before the 1986 dry period and all previous problems compounded once again.

Temperature fluctuations throughout the period were fairly regular, but severe freezes occurred in late December 1983 and in January 1985. These freezes, coupled with the drought periods, were the main factors responsible for the death of or severe damage to many sensitive ornamental plants and trees recorded throughout 1985. Hurricane force winds in September 1985 and the resulting salt sprays in coastal areas added one more injurious factor to many plants that were already under severe stress conditions.

Evapotranspiration and Irrigation

Evapotranspiration is a measure of the total amount of water evaporated from green plants and from surrounding surfaces. Percolation of water through the soil and runoff are not included in this calculation. If the amount of rainfall were equal to evapotranspiration plus percolation plus runoff, then no supplementary irrigation would be required for adequate vegetative growth to occur.

Estimates of each of these factors for the St. Petersburg area were made and the resulting data showed that supplemental irrigation was required throughout the year. In a "normal" year, January and February require up to 1.5 inches of extra irrigation, March, April and May require increasing amounts up to 6.5 inches per month, June, July and August require 4.5 extra inches per month, September, October and November once again require 6.5 inches per month and December requires 2.0 extra inches.

Soil Characteristics

Extensive soil analyses from 400 sites scattered all over the City showed that soil pH values ranged from highly acidic with pH 3.5 to highly alkaline with pH 9.5. There was a good correlation between the distribution of the pH values and the distribution of the different soil associations as determined by a U.S.D.A study in 1972.

Most areas where reclaimed irrigation water is used have highly alkaline soils which are generally poor for plant growth. Soil amendments such as organic materials, sulfur, and acid forming fertilizers are needed to attain good plant growth in these areas, regardless of the type of irrigation water being used.

The total soluble salts and total chlorides in St. Petersburg's soils were also evaluated and found to be elevated in coastal areas, including those where reclaimed water was used for irrigation.

Effects of Reclaimed Water on Selected Vegetation

Since land treatment has long been used to dispose of municipal waste-water all over the USA, the literature on this subject is also extensive. In Florida, general guidelines on water quality and effluent use in horticulture have been produced by Berry et al. (1980), Hoadley and Ingram (1982), and Fitzpatrick (1983, 1985) and Fitzpatrick et al. (1986).

An investigation of the effects of reclaimed water on the growth and maturation of selected species of commonly used landscape ornamental plants made up the major part of the "Project Greenleaf" study. Eight species of commonly used ornamental plants and two species of sapling trees were used in the experiments. Several complaints about each of these species had previously been received from reclaimed water users concerning damage possibly caused by effluent irrigation.

Ten experimental plots were set up each containing 12 replicates of each of 10 species of plants. These included, Avocado (*Persea americana* Mill.), Boston fern (*Nephrolepis exaltata* "Bostoniensis"), Crape myrtle (*Lagerstroemia indica*), Dwarf azalea (*Rhododendron* sp. cv. Celestine red), Formosa azalea (*Rhododendron* sp. cv. Formosa), Hibiscus (*Hibiscus rosa-sinensis*), Burford holly (*Ilex cornuta* "Burford"), Chinese privet (*Ligustrum sinense* "variegata"), Laurel oak (*Quercus laurifolia*) and Sweet viburnum (*Viburnum odoratissimum*).

All plants were grown in 3 gallon plastic containers. The growing medium consisted of commercial horticultural mix (two thirds peat with not more than 15% nor less than 10% sand, one third cypress chips and dolomite to adjust pH to 6.0). One tablespoon of 14-14-14 Osmocote and one teaspoon of 2% Oxydiazone pre-emergent herbicide were added to each pot at the outset of the experiment.

Three variables were investigated in various combinations as shown in Table 22.3. Irrigant chloride concentration was 15 ppm in potable water, 180 to 200 ppm in reclaimed water supplied to the City Nursery from the Southwest wastewater treatment plant and over 400 ppm in reclaimed water produced at the Northwest wastewater treatment plant. Analyses of reclaimed water from both of these production plants did not vary to any great extent in any other parameters except sodium and chloride levels. Drip or overhead spray irrigation systems were used and the total gallonage applied was metered in each of the plots. Open air (ambient temperature, full sunlight and uncontrollable rainfall) or enclosed greenhouse conditions (12 degrees Fahrenheit mean increase over ambient temperature, 80% sunlight and excluded rainfall) were also included in the experiments to simulate outdoor normal growth or indoor commercial growth conditions.

There was no significant size difference within plant species in all experimental plots at the beginning of the investigation in August 1986.

TABLE 22.3. Treatment Details for the 10 Experimental Plots for "Project Greenleaf" Plant Growth Study (August 1986 to August 1987).

Plot #	Water Type	Chloride Conc. (ppm)	Spray Type	Exposed or Enclosed	Location of Plot
1.	Potable	15	Overhead	Exposed	City nursery
2.	Reclaimed	180+	Overhead	Exposed	City nursery
3.	Reclaimed	400+	Overhead	Exposed	Northwest plant
4.	Potable	15	Overhead	Enclosed	City greenhouse
5.	Reclaimed	180+	Overhead	Enclosed	City greenhouse
6.	Potable	15	Drip	Exposed	City nursery
7.	Reclaimed	180+	Drip	Exposed	City nursery
8.	Reclaimed	400+	Drip	Exposed	Northwest plant
9.	Potable	15	Drip	Enclosed	City greenhouse
10.	Reclaimed	180+	Drip	Enclosed	City greenhouse

Plant growth index determinations were made by summing plant height and average width at monthly intervals from the beginning of September 1986 to August 1987. Observations on the condition of each plant were also recorded monthly. Growth data were entered on a Lotus 123 computer spreadsheet which produced graphical representations and executed analyses of variance and Duncan's multiple range tests (Duncan 1955) at the 5% level.

In August 1987, at the end of the experiment, analyses of variance showed significant differences between the growth indices of all plant species. Significant mean separations of the 10 different species of plants in the 10 different experimental plots were evaluated. Table 22.4 was developed from these data to show trends and significant differences between experimental plots and correlate these with observed differences in the level of vegetative damage sustained by each plant species.

The Boston fern was the only representative of the non flowering plant group referred to as the ferns or Pteridophyta. These plants prefer moist conditions and water is required for their reproductive processes. It is thus not surprising that this plant responded differently from all of the other 9 flowering plant species. Table 22.4 shows that Boston ferns tended to grow better when they were irrigated with overhead sprays, which was the opposite of all other plants. Also, as the chloride content of the irrigation water increased, the growth rate increased, which, again, is the opposite of what was seen in all other species. Inside the greenhouse, in elevated temperatures, Boston ferns grew equally well with drip or overhead irrigation, which, again, differed from all other plant species. Probably the consistently high humidity in the greenhouse created conditions in which the plants could produce good growth, even when they were irrigated by the drip system.

The 9 flowering species of experimental plants in Table 22.4 have been ranked vertically in order of their growth responses to reclaimed irrigation from a good to a poor range. Also, in this table, the columns have been ranked from relatively little significant difference on the left hand side, to mostly significant growth differences on the right hand side.

Burford holly and sweet viburnum both showed no significant growth differences between drip and overhead irrigation systems with either reclaimed or potable water except at the higher temperatures in the greenhouse. Here, they both grew better when irrigated by reclaimed water using a drip application system. Although less growth occurred when reclaimed water was used with the overhead system and when the chloride content of the water increased, both of these plants still increased in size throughout the duration of the experiment. Slight leaf burn was seen on the viburnum plants in the 400+ ppm chloride reclaimed water overhead irrigation plot.

TABLE 22.4. Growth Differences of 10 Ornamental Plant Species under 10 Experimental Irrigation Conditions.

Plot number (z)	1	6	4	9	3	8	5	7	2	10	Damage with 180 ppm Chlorides	Damage with 400 ppm Chlorides
Exposed/Enclosed	Exposed	Exposed	Enclosed	Enclosed	Exposed	Exposed	Exposed	Exposed	Enclosed	Enclosed		
Chloride conc.	15 ppm	15 ppm	15 ppm	15 ppm	400+ ppm	400+ ppm	180+ ppm	180+ ppm	180+ ppm	180+ ppm		
Water type	Potable	Potable	Potable	Potable	Reclaim	Reclaim	Reclaim	Reclaim	Reclaim	Reclaim		
Application type	Overhead	Drip	Overhead	Drip	Overhead	Drip	Overhead	Drip	Overhead	Drip		
Plant Species												
Boston fern	S (w)	inc (v)	<-NS->	<-NS->	<-NS->	<-NS->	S	inc	<-NS->	S	None	None
Burford holly	<-NS->	<-NS-> (y)	<-NS->	<-NS->	<-NS->	<-NS->	<-NS->		inc	S	None	None
Sweet viburnum	<-NS->	<-NS->	<-NS->	<-NS->	<-NS->	<-NS->	<-NS->		inc	S	None	Slight
Hibiscus	<-NS->	<-NS->	inc	S	<-NS->	<-NS->	<-NS->		inc	S	None	Slight
Crape myrtle	<-NS->	<-NS->	inc	T-->	<-NS->	<-NS->	inc	S	inc	S	Slight	Medium
Avocado	inc	T--> (x	inc	T-->	inc	S	inc	S	inc	S	Slight	Medium
Dwarf azalea	inc	T-->	inc	T-->		<-NS->			dec	S	High	V. high
Chinese privet	<-NS->		inc	T-->	dec (u)	S	dec	S	dec	S	High	V. high
Formosa azalea	inc	T-->	inc	T-->	dec	S	dec	S	dec	S	High	V. high
Laurel oak	inc	T-->	inc	T-->	dec	S	dec	S	dec	S	High	V. high

Legend: z Plot number same as shown in Table 22.3.
    y <-NS-> No significant difference in growth between pairs whatsoever.
    x T--> Tendancy to significantly greater growth at arrow point.
    w S Significantly greater growth in this alternative.
    v inc This alternative increasing in size.
    u dec This alternative decreasing in size.

Hibiscus showed similar results to sweet viburnum except that significant growth differences occurred between drip and overhead sprays when potable water was used at high temperatures.

Crape myrtle showed greater leaf burn damage than hibiscus when exposed to overhead sprays of both 180+ ppm and 400+ ppm chloride reclaimed water. This salt sensitivity factor caused significant growth increases to occur in 180+ ppm chloride plots when drip and overhead spray systems were compared. High chloride water reduced the growth rates in both drip and overhead plots to the point where no significant differences were recorded. Although all plants that were sprayed from overhead showed less growth, they did continue to increase in size throughout the duration of the experimental period.

Avocado saplings showed similar results to the crape myrtles except that exposure to 400+ ppm chloride reclaimed water from a drip system produced significantly more growth than exposure to the same water from an overhead system. Thus, avocados can tolerate higher chlorides around the root system better than crape myrtles. Although there was an increase in leaf burn damage as chlorides increased in concentration in the irrigant water, this did not produce a significant reduction in growth when similar application methods in 400+ ppm and 180+ ppm chloride reclaimed water plots at open air temperatures were compared. Throughout the experimental period all plants tended to increase in size regardless of which treatment they were exposed to.

The remaining 4 species of plants all showed a decrease in growth and foliar damage when exposed to reclaimed water by the overhead spray method. Increased concentrations of chlorides also produced increased damage. In both types of azaleas, decreased growth leading to the death of some plants occurred, regardless of the method of application or type of reclaimed water used. This showed a high degree of salt sensitivity in these plants to both foliar and root contact. Chinese privet showed severe defoliation when exposed to 400+ ppm chloride reclaimed water by the overhead method and it showed poor growth, even when the same water was applied by the drip system. This plant can however, tolerate 180+ ppm chloride water around the root system but not on the leaves.

Laurel oak saplings developed leaf burn when exposed to 180+ ppm chloride reclaimed water by the overhead spray method. This especially affected young leaves and the damage increased as chloride concentrations increased. The effects of the leaf burn severely reduced the growth rate of the plants and caused stunting and overall decreased size, which would eventually lead to the death of the plants. Conversely, laurel oaks grown with 400+ ppm or 180+ ppm chloride reclaimed water using the drip system grew well, especially in the elevated temperatures inside the greenhouse where phenomenal growth was recorded. Thus, if foliar contact is avoided in this species, excellent growth can be achieved with the use of reclaimed irrigation water.

The 10 experimental plant species thus showed widely different responses to the experimental treatments. Either 400+ ppm or 180+ ppm chloride irrigation water can be used without problem on the first 4 species shown in Table 22.4 by either irrigation method. High chloride water should be avoided as an irrigant for crape myrtle and avocado and, ideally, 180+ ppm chloride water should be used by a drip irrigation method. Both types of azaleas and chinese privet are highly salt sensitive and the use of these plants in landscapes that have reclaimed water irrigation systems with more than 80 ppm chlorides is not recommended. The leaves of laurel oak trees should not be exposed to contact with any reclaimed water that has a chloride content of more than 80 ppm. Drip irrigation systems are recommended until these plants are large enough to avoid the impact of overhead sprays on a significant percentage of their leaves. Under these conditions, the plants will tolerate high chloride irrigants around their rooting systems.

At the end of the experiment, a sample of the growth medium from each of the plant species in each of the experiments was analyzed by the Greenhouse and Potting Media test. Highest total soluble salts, nitrogen, phosphorus, potassium and magnesium levels were recorded in media irrigated by reclaimed water. Low nutrient levels were recorded in media irrigated by potable water. Reclaimed water was thus determined to contain sufficient nutrients to promote satisfactory growth in the ornamental plants used in the study. Extra applications of micronutrients were still required for optimum growth to occur.

Tissue analyses showed higher nitrogen, phosphorus, potassium and magnesium levels in plants irrigated by reclaimed water, which correlated with their healthier overall appearance when compared with those irrigated by potable water. Levels of the toxic element sodium were also elevated in reclaimed water plants and may have contributed to the damage seen in salt sensitive species such as azaleas and chinese privet.

Studies at 65 Residences

Observations on a total of 203 ornamental plant species were made in 56 private residences randomly selected from all areas served by the reclaimed water system in 1986. Throughout the investigation, these plant species were continuously monitored for damage symptoms which could have been attributed to the use of reclaimed water. Resulting plant lists were divided into 4 major categories with respect to their suitability for use with reclaimed irrigation water. These categories were defined as follows:

(1) *Category A:* Plant species that are completely tolerant to reclaimed irrigation water regardless of its chloride levels.

(2) *Category B:* Plant species that are tolerant to reclaimed water when the chloride concentration is less than 400 ppm.

(3) *Category C:* Plant species that *may* need extra maintenance procedures if reclaimed irrigation water with more than 100 ppm chloride concentration is used.

(4) *Category D:* Plant species that are not recommended for use with reclaimed water.

Plant lists are included in Table 22.5 in alphabetical order of Latin names. These lists are further subdivided into trees, shrubs (including annuals and perennials), palms, ground covers and vines to facilitate easy reference.

Table 22.5 shows that of the 203 plant species found in the 56 residences, 54 species (category A), were highly salt resistant and could tolerate any form of reclaimed irrigation water. Also, 107 plant species (category B), could tolerate reclaimed irrigation water provided that the chloride concentration remained below 400 ppm. A further 39 plant species (category C), would require extra management procedures if reclaimed water with over 100 ppm chloride concentration were used for irrigation. For the 15 tree species in category C, reclaimed water should not be sprayed directly onto the young leaves of saplings. Once the trees grow above spray height, no further management procedures should be necessary. The use of drip systems to prevent reclaimed irrigation water from coming in contact with the foliage of the other plant species in category C is recommended. The 3 species of plants included in category D are not recommended for use with reclaimed water as they have extremely low salt tolerances both to foliar and soil applications of reclaimed irrigation water.

SPECIAL STUDIES, PROJECT RESOURCE MANAGEMENT

ORIGIN AND OBJECTIVES OF PROJECT "RESOURCE MANAGEMENT"

The growth experiments in "Project Greenleaf" were all conducted on plants grown in plastic pots and the amount of irrigation water was not varied throughout the duration of the investigations. It became clear that larger field trials would be required to determine the exact amount of reclaimed irrigation water necessary for optimum plant growth. These

TABLE 22.5. Plant Tolerance to Irrigation by Reclaimed Water, Including Salt and Frost Tolerance.

Latin Name	Common Name	Reclaim Category	Plant Type	Florida Native	Salt Tolerance		Frost Tolerance
					Spray	Soil	
Acalypha wilkesiana	Copper Leaf	B	Shrub	No	Low	Low	Low
Acer rubrum	Red Maple	C	Tree	Yes	Low	Low	Hardy
Acer saccharinum	Silver Maple	C	Tree	No	Low	Low	Hardy
Acoelorrhaphe wrightii	Paurotis Palm	B	Palm	Yes	Medium	Medium	Hardy
Adiantum sp.	Maiden Hair Fern	B	Shrub	No	Medium	Medium	Medium
Agave americana	Century Plant	A	Shrub	Yes	High	High	Hardy
Agave attenuata	Agave	A	Shrub	No	High	High	Medium
Aglaonema sp.	Silver Queen	B	Shrub	No	Medium	Medium	Low
Ajuga repens	Bugle Weed	C	Ground cover	No	Low	Medium	Hardy
Allamanda cathartica	Allamanda	B	Vine	No	Medium	Medium	Medium
Allamanda neriifolia	Allamanda	B	Shrub	No	Medium	Medium	Medium
Allamanda violacea	Purple Allamanda	B	Vine	No	Medium	Medium	Medium
Aloe vera	Aloe	A	Ground cover	No	High	High	Medium
Alternanthera ficoidea	Joseph's Coat	B	Shrub	No	Medium	Medium	Medium
Annona cherimola	Cherimoya	C	Tree	No	Low	Low	Medium
Antigonon leptopus	Coral Vine	B	Vine	No	Low	Medium	Hardy
Aphelandra sp.	Zebra Plant	B	Shrub	No	Medium	Medium	Medium
Araucaria araucana	Monkey Puzzle Tree	A	Tree	No	Medium	Medium	High
Araucaria heterophylla	Norfolk Island Pine	A	Tree	No	High	High	Medium
Arecastrum romanzoffianum	Queen Palm	B	Palm	No	Low	Medium	Medium
Asparagus densiflorus	Asparagus Fern	A	Shrub	No	Moderate	High	Hardy
Averrhoa carambola	Carambola	B	Tree	No	Medium	Medium	Medium
Bambusa sp.	Bamboo	B	Shrub	No	Medium	Medium	Low
Bauhinia purpurea	Orchid Tree	C	Tree	No	Low	Low	Hardy
Beaucarnea recurvata	Ponytail Palm	B	Palm	No	Medium	Medium	Medium

TABLE 22.5. (continued).

Latin Name	Common Name	Reclaim Category	Plant Type	Florida Native	Salt Tolerance Spray	Salt Tolerance Soil	Frost Tolerance
Begonia sp.	Begonia	A	Shrub	No	High	High	Low
Bougainvillea spectabilis	Variegated Bougainvillea	A	Shrub	No	High	High	Medium
Bougainvillea spectabilis	Bougainvillea	A	Shrub	No	High	High	Medium
Brassia actinophylla	Schefflera	B	Tree	No	Low	Medium	Low
Bromeliaceae	Bromeliads	B	Ground cover	No	Medium	—	Hardy
Bucida buceras	Black Olive	A	Tree	No	High	High	Low
Butia capitata	Pindo Palm	A	Palm	No	High	Medium	Hardy
Buxus microphylla	Japanese Boxwood	B	Shrub	No	Medium	Medium	Hardy
Caladium sp.	Caladium	C	Shrub	No	Low	Low	Low
Calliandra haematocephala	Powder Puff Tree	C	Shrub	No	Low	Low	Hardy
Callistemon citrinus	Lemon Bottlebrush	B	Tree	No	Medium	High	Hardy
Callistemon viminalis	Weeping Bottlebrush	B	Tree	No	Medium	Medium	Hardy
Camellia japonica	Camellia	C	Shrub	No	Low	Low	Hardy
Campsis radicans	Trumpet Vine	B	Vine	No	Medium	Medium	Hardy
Canna generalis	Canna Lily	B	Shrub	No	Medium	Medium	Medium
Carica papaya	Papaya	B	Shrub	No	Medium	Medium	Low
Carissa grandiflora	Carissa Boxwood	A	Shrub	No	High	High	Medium
Carissa macrocarpa	Dwarf Carissa Boxwood	A	Ground cover	No	High	High	Medium
Carpobrotus sp.	Hottentot Fig	A	Ground cover	No	High	High	Medium
Carya illinoinensis	Pecan	B	Tree	No	Medium	Medium	Hardy
Caryota mitis	Fishtail Palm	B	Palm	No	Medium	Medium	Medium
Casuarina equisetifolia	Australian Pine	A	Tree	No	High	High	Medium
Catharanthus roseus	Periwinkle	B	Shrub	No	High	Medium	Medium
Cestrum aurantiacum	Orange Cestrum	B	Shrub	No	Medium	Medium	Medium
Chamaerops humilis	European Fan Palm	A	Palm	No	High	High	Hardy

TABLE 22.5. (continued).

Latin Name	Common Name	Reclaim Category	Plant Type	Florida Native	Salt Tolerance Spray	Salt Tolerance Soil	Frost Tolerance
Chlorophytum comosum	Spider Plant	B	Ground cover	No	Medium	Medium	Medium
Chrysalidocarpus lutescens	Areca Palm	B	Palm	No	Medium	Medium	Medium
Chrysanthemum morifolium	Chrysanthemum	B	Shrub	No	Low	Low	Hardy
Cinnamomum camphora	Camphor Tree	C	Tree	No	Low	Low	Hardy
Citrus aurantiifolia	Lime	B	Tree	No	Low	Low	Medium
Citrus lemon	Lemon	B	Tree	No	Low	Low	Medium
Citrus paradisi	Grapefruit	B	Tree	No	Low	Low	Medium
Citrus reticulata	Tangerine	B	Tree	No	Low	Low	Medium
Citrus sinensis	Orange	B	Tree	No	Low	Low	Medium
Clerodendrum thomsoniae	Bleeding Heart	C	Vine	No	Low	Medium	Medium
Coccoloba uvifera	Sea Grape	A	Tree	Yes	High	High	Medium
Codiaeum variegatum	Croton	B	Shrub	No	Low	Low	Low
Cordyline terminalis	Ti Plant	B	Shrub	No	Medium	Medium	Medium
Cortaderia selloana	Pampas Grass	B	Shrub	No	High	High	Hardy
Cryptostegia grandiflora	Palay Rubber Vine	B	Shrub	No	Low	Medium	Medium
Cupaniopsis anacardiodes	Carrotwood	B	Tree	No	High	Medium	Medium
Cuphea hyssopifolia	Heather	B	Shrub	No	Medium	High	Medium
Cupressus sempervirens	Italian Cypress	B	Tree	No	Medium	Medium	Hardy
Cycas circinalis	Queen Sago	B	Shrub	No	Low	Medium	Medium
Cycas revoluta	King Sago	B	Shrub	No	Medium	Medium	Medium
Cyperus alternifolius	Sedge	B	Shrub	No	Medium	Medium	Hardy
Dalbergia sissoo	Indian Rosewood	B	Tree	No	Medium	Medium	Medium
Dietes sp.	African Iris	B	Shrub	No	Low	Medium	Hardy
Diospyros dignya	Black Sapote	B	Tree	No	Medium	Medium	Medium
Diospyros virginiana	Persimmon	C	Tree	No	Low	Low	Medium

TABLE 22.5. (continued).

Latin Name	Common Name	Reclaim Category	Plant Type	Florida Native	Salt Tolerance Spray	Salt Tolerance Soil	Frost Tolerance
Dizygotheca kerchoveana	False Aralia	C	Shrub	No	Low	Low	Low
Dracena deremensis	Dracena	B	Shrub	No	Medium	Medium	Low
Dracena marginata	Dracena	B	Shrub	No	Medium	Medium	Medium
Epipremnum aureum	Potos	B	Ground cover	No	Medium	Medium	Medium
Eriobotrya japonica	Loquat	B	Tree	No	Medium	Medium	Hardy
Ervatamia coronaria	Crape Jasmine	B	Shrub	No	Medium	High	Medium
Eugenia uniflora	Surinam Cherry	C	Shrub	No	Low	Low	Hardy
Euphorbia milli	Crown of Thorns	A	Shrub	No	High	High	Medium
Euphorbia pulcherrima	Poinsettia	C	Shrub	No	Low	Low	Medium
Euphoria longans	Longan	C	Tree	No	Low	Medium	Medium
Ficus benjamina	Weeping Fig	B	Tree	No	Medium	Medium	Medium
Ficus carica	Fig	B	Tree	No	High	High	Hardy
Ficus elastica	Indian Rubber Tree	B	Tree	No	Medium	Medium	Medium
Ficus retusa	Laurel Fig (Cuban Laurel)	B	Tree	No	Medium	Medium	Medium
Fortunella japonica	Kumquat	C	Tree	No	Low	Low	Hardy
Gamolepis chrysanthemoides	African Daisy	A	Shrub	No	High	—	Low
Gardenia jasminoides	Gardenia	C	Shrub	No	Low	Low	Hardy
Gerbera jamesonii	Gerbera Daisy	B	Shrub	No	Medium	Medium	Hardy
Grevilla robusta	Silk Oak	A	Tree	No	High	High	Medium
Hedera helix	English Ivy	B	Ground cover	No	Medium	High	Hardy
Heliconia sp.	Heliconia	B	Shrub	No	Low	High	Medium
Hemerocallis sp.	Day-Lily	B	Ground cover	No	Medium	Medium	Hardy
Hibiscus rosa-sinensis	Hibiscus	A	Shrub	No	Medium	Medium	Hardy
Hibiscus rosa-sinensis	Variegated Hibiscus	A	Shrub	No	High	High	Medium
Hydrangea macrophylla	Hydrangea	C	Shrub	No	Low	Low	Hardy

TABLE 22.5. (continued).

Latin Name	Common Name	Reclaim Category	Plant Type	Florida Native	Salt Tolerance		Frost Tolerance
					Spray	Soil	
Hylocereus undatus	Night Blooming Cereus	B	Vine	No	Medium	High	Low
Ilex cassine	Dahoon Holly	A	Tree	Yes	Medium	Medium	Hardy
Ilex cornuta	Burford Holly	A	Shrub	No	Medium	High	Hardy
Ilex vomitoria	Yaupon Holly	A	Shrub	Yes	High	High	Hardy
Ilex vomitoria	Dwarf Yaupon Holly	A	Shrub	Yes	High	High	Hardy
Ixora sp.	Ixora	C	Shrub	No	Low	Medium	Medium
Jacaranda mimosaefolia	Jacaranda	C	Tree	No	Low	Low	Hardy
Jasminum multiflorum	Downy Jasmine	C	Shrub	No	Low	Low	Medium
Jatropha multifida	Coral Plant	B	Shrub	No	Low	Medium	Medium
Juniperus chinensis	Juniper	B	Ground cover	No	Medium	Medium	Hardy
Juniperus horizontalis	Creeping Juniper	A	Ground cover	No	High	High	High
Juniperus parsoni	Juniper	B	Ground cover	No	Medium	Medium	Hardy
Juniperus procumbens nana	Juniper	B	Ground cover	No	Medium	Medium	Hardy
Juniperus sp.	Torulosa Juniper	B	Tree	No	Medium	Medium	Hardy
Justicia brandegeana	Shrimp Plant	C	Shrub	No	Low	Low	Low
Kalanchoe sp.	Kalanchoe	B	Ground cover	No	Medium	Medium	Hardy
Koelreutaria elegans	Golden Rain Tree	B	Tree	No	Low	Medium	Hardy
Lagerstroemia indica	Crape Myrtle	C	Tree	No	Low	Low	Hardy
Lantana camara	Lantana	A	Shrub	No	High	Medium	Hardy
Ligustrum japonicum	Ligustrum (Privet)	B	Tree	No	Medium	Medium	Hardy
Ligustrum sinense	Chinese Privet	D	Shrub	No	Low	Low	Hardy
Liquidambar styraciflua	Sweetgum	B	Tree	Yes	Medium	Medium	Hardy
Liriope muscari	Liriope	B	Ground cover	No	Medium	Medium	Hardy
Lisianthus nigrescens	Lisianthus	A	Shrub	No	High	High	Medium
Litchi chinensis	Lychee	C	Tree	No	Low	Low	Medium

1081

TABLE 22.5. (continued).

Latin Name	Common Name	Reclaim Category	Plant Type	Florida Native	Salt Tolerance Spray	Soil	Frost Tolerance
Livistonia chinensis	Chinese Fan Palm	A	Palm	No	Medium	Medium	Hardy
Magnolia grandiflora	Southern Magnolia	B	Tree	Yes	Medium	Medium	Hardy
Malus sylvestris	Apple	B	Tree	No	Low	—	Hardy
Mangifera indica	Mango	C	Tree	No	Low	Medium	Medium
Manilkara zapota	Sapodilla	B	Tree	No	High	High	Medium
Melaleuca quinquenervia	Cajeput (Punk) Tree	A	Tree	No	High	High	Hardy
Melia azedarach	Chinaberry	B	Tree	No	Medium	Medium	Hardy
Monstera deliciosa	Monstera	B	Shrub	No	Medium	Medium	Medium
Moraea iridoides	Iris	B	Shrub	No	Medium	Medium	Hardy
Murraya paniculata	Orange Jasmine	C	Shrub	No	Low	Low	Hardy
Musa acuminata	Banana	B	Tree	No	Low	Low	Low
Myrica cerifera	Wax Myrtle	A	Tree	Yes	High	High	Hardy
Nephrolepis exaltata	Boston Fern	A	Shrub	No	High	High	Hardy
Nerium oleander	Oleander	A	Shrub	No	High	High	Hardy
Nerium oleander	Dwarf Oleander	A	Shrub	No	High	High	Hardy
Ophiopogon japonicus	Mondo Grass	A	Ground cover	No	High	High	Hardy
Opuntia sp.	Opuntia Cactus	B	Shrub	Yes	Low	Low	Medium
Pachystachys lutea	Yellow Shrimp Plant	B	Shrub	No	Medium	Medium	Medium
Passiflora sp.	Passion Flower	C	Vine	No	Low	Low	Low
Pelargonium sp.	Geranium	C	Shrub	No	Low	Low	Low
Pentas lanceolata	Pentas	C	Shrub	No	Low	Low	Low
Peperomia obtusifolia	Peperomia	C	Ground cover	Yes	Low	Low	Medium
Persea americana	Avocado	C	Tree	No	Medium	Medium	Low
Petunia sp.	Petunia	B	Shrub	No	Low	Low	Hardy
Philodendron williamsii	Philodendron	B	Shrub	No	Medium	Medium	Medium

TABLE 22.5. (continued).

Latin Name	Common Name	Reclaim Category	Plant Type	Florida Native	Salt Tolerance Spray	Salt Tolerance Soil	Frost Tolerance
Phoenix canariensis	Canary Island Date Palm	A	Palm	No	High	High	Hardy
Phoenix reclinata	Senegal Date Palm	B	Palm	No	Medium	High	Hardy
Phoenix roebelinii	Pygmy Date Palm	B	Palm	No	Medium	Medium	Hardy
Photinia glabra	Red Leaf Photinia	C	Shrub	No	Low	Low	Hardy
Pinus elliottii	Florida Slash Pine	B	Tree	Yes	Medium	Medium	Hardy
Pittosporum tobira	Pittosporum	A	Shrub	No	High	High	Hardy
Pittosporum tobira	Dwarf Pittosporum	B	Ground cover	No	High	High	Hardy
Platycladus orientalis	Oriental Arbor Vitae	A	Tree	No	Medium	Medium	Hardy
Plumbago auriculata	Plumbago	B	Shrub	No	High	High	Medium
Plumeria sp.	Franjipani	B	Tree	No	High	High	Medium
Podocarpus macrophyllus	Yew Podocarpus	B	Tree	No	Medium	Low	Hardy
Polyscias sp.	Aralia	C	Shrub	No	Low	Low	Low
Portulaca grandiflora	Purslane	B	Ground cover	Yes	Medium	High	Low
Prunus caroliniana	Cherry Laurel	B	Tree	No	Medium	Medium	Hardy
Prunus persica	Peach	B	Tree	No	Low	Medium	Hardy
Psidium guajava	Guava	B	Tree	No	Low	Medium	Hardy
Punica granatum	Pomegranate	B	Tree	No	Medium	Medium	Hardy
Pyracantha coccinea	Red Firethorn	B	Shrub	No	Medium	Medium	Hardy
Quercus laurifolia	Laurel Oak	C	Tree	Yes	Low	Medium	Hardy
Quercus nigra	Water Oak	B	Tree	Yes	Low	Medium	Hardy
Quercus virginiana	Live Oak	A	Tree	Yes	High	High	Hardy
Raphiolepis indica	Indian Hawthorn	A	Shrub	No	High	High	Hardy
Ravenala madagascariensis	Travellers Tree	B	Shrub	No	Medium	Medium	Medium
Rhapis excelsa	Lady Palm	B	Palm	No	Medium	Medium	Hardy
Rhododendron sp.	Dwarf Azalea	D	Shrub	No	Low	Low	Hardy

TABLE 22.5. (continued).

Latin Name	Common Name	Reclaim Category	Plant Type	Florida Native	Salt Tolerance		Frost Tolerance
					Spray	Soil	
Rhododendron sp.	Formosa Azalea	D	Shrub	No	Low	Low	Hardy
Rhoeo spathacea	Oyster Plant	B	Ground cover	No	Medium	Medium	Hardy
Rosa sp.	Rose	C	Shrub	No	Low	Medium	Hardy
Rosmarinus officinalis	Rosemary	B	Ground cover	No	Medium	Medium	Hardy
Russelia equisetiformis	Firecracker Plant	B	Shrub	No	Medium	High	Medium
Sabal etonia	Scrub Palmetto	A	Palm	Yes	High	High	Hardy
Sabal palmetto	Cabbage Palm	A	Palm	No	High	High	Hardy
Salix babylonica	Weeping Willow	C	Tree	No	Low	Medium	Hardy
Salvia farinacea	Blue Sage	C	Shrub	No	Low	Low	Hardy
Schefflera arboricola	Dwarf Schefflera	B	Shrub	No	Medium	Medium	Medium
Schinus terebinthifolius	Brazilian Pepper	A	Tree	No	High	High	Hardy
Setcreasea pallida	Purple Queen	A	Ground cover	No	High	High	Medium
Strelitzia reginae	Bird of Paradise	B	Shrub	No	Medium	Medium	Medium
Tagetes sp.	Marigold	C	Shrub	No	Low	Low	Low
Tecomaria capensis	Cape Honeysuckle	A	Shrub	No	High	High	Medium
Trachelospermum jasminoide	Confederate Jasmine	A	Ground cover	No	High	High	Hardy
Tulbaghia violacea	Society Garlic	B	Ground cover	No	Medium	Medium	Hardy
Ulmus parvifolia	Drake Elm	B	Tree	No	Medium	Medium	Hardy
Ulmus parvifolia	Chinese Elm	B	Tree	No	Medium	Medium	Hardy
Verbena sp.	Verbena	C	Shrub	No	Low	Low	Hardy
Vibernum odoratissimum	Sweet Viburnum	A	Shrub	No	Medium	High	Hardy
Viburnum suspensum	Sandankwa Viburnum	B	Shrub	No	Medium	High	Hardy
Washingtonia robusta	Washingtonia Palm	A	Palm	No	High	High	Hardy
Wedelia triloba	Wedelia	A	Ground cover	No	High	High	Hardy
Yucca aloifolia	Spanish Bayonet	A	Shrub	Yes	High	High	Hardy

TABLE 22.5. (continued).

Latin Name	Common Name	Reclaim Category	Plant Type	Florida Native	Salt Tolerance Spray	Salt Tolerance Soil	Frost Tolerance
Yucca elephantipes	Spineless Yucca	A	Shrub	Yes	High	High	Hardy
Zamia floridana	Coontie	A	Ground cover	Yes	High	High	Hardy
Zebrina pendula	Wandering Jew	B	Ground cover	No	Medium	Medium	Medium

trials would use salt tolerant species growing under natural conditions to produce the required supplemental data.

For this type of field investigation, an area known as Colony Point Estates in the southeast corner of St. Petersburg was finally selected as it satisfied the following requirements:

(1) No reclaimed water had been used in the area prior to the beginning of this experiment.

(2) The area was compact, with 46 single family residences, each not more than 28 years old.

(3) Most residences were on waterfront lots with high exposure to salt-water.

(4) The majority of homes had well maintained gardens with a large variety of well established ornamental plants.

(5) Most homes had automatic irrigation systems that were originally professionally installed and were in good condition and well maintained.

(6) Well water in the area was too saline for irrigation so that the majority of irrigation systems were connected to the potable water supply prior to the installation of reclaimed water.

(7) Residents were mostly homeowners who had an interest in maintaining a well kept garden and were thus eager to join the research project.

(8) Records for potable water consumption prior to the beginning of the experiment were easily obtainable from municipal utility accounts.

The name given to the field trial was Project "Resource Management" and it was carried out under the joint financial sponsorship of the Southwest Florida Water Management District (SWFWMD) and the City of St. Petersburg. The grant from the SWFWMD was made possible through a water conservation grant agreement negotiated between the two parties in 1987.

REVIEW OF PROJECT "RESOURCE MANAGEMENT"

The Experimental Procedures

Thirty residential homeowners signed contracts with the City of St. Petersburg allowing access to their landscaped areas to carry out the field trial. Reclaimed water supply was connected to each participant's existing irrigation system and a 1 inch water meter was installed in the delivery line. All irrigation systems were repaired as necessary, defective spray heads were replaced and manual systems were converted to automatic operation.

The number of irrigation zones, and the number, type and position of each sprinkler head within each zone was recorded on a scale drawing of each residence. Areas covered by each zone were subdivided into turf, ornamental, bush, flower bed and tree areas, and included on the scale drawings. Hardcore and any areas that were not irrigated at all were also noted and measured. Where irrigation zones overlapped, total spray areas for each zone were approximated by individual inspection of each occurrence.

Rain gauges from the St. Petersburg "Conservival" promotion were used to determine the uniformity of the spray pattern throughout each irrigation zone. Where variations of more than 50% occurred within one zone, modifications were made to the system until a uniform spray pattern was established. All irrigation systems were calibrated by operating each zone for 15 minutes and noting the total number of gallons that passed through the water meter.

A Lotus 123 spreadsheet was used to compute the number of minutes each irrigation zone should operate when selected numbers of irrigation events per week and required irrigation levels were entered for each residence. The automatic timers in each residence were then set to operate each zone for the calculated amount of time and number of events every week. Meters were read weekly to monitor consumption rates. Irrigation levels of 0.5 in., 0.75 in., 1.0 in., 1.5 in., 2.0 in. and 2.5 in. were used in the plant growth experiment.

Complete residences were used as separate experimental plots and there were five residences in each of the 6 irrigation level groups. Residences were selected at random for each irrigation group.

In order to eliminate the possibility that soil nutrient levels were different in each of the residential areas, known quantities of fertilizer were evenly spread on each garden before the beginning of the experiment in December 1987 and half way through in May/June 1988.

The positions, identifications and quantities of a total of 148 commonly occurring species of ornamental plants in each residential area were mapped. Twenty plant species which occurred commonly in residences in each irrigation group were chosen for the growth experiment as follows:

(1) *Asparagus densiflorus* "sprengeri," Asparagus fern
(2) *Brassaia actinophylla,* Schefflera
(3) *Carissa grandiflora,* Carissa boxwood (Natal plum)
(4) *Chrysanthemum morifolium,* Chrysanthemum
(5) *Citrus sinensis,* Orange
(6) *Codiaeum variegatum,* Croton
(7) *Cupressus sempervirens,* Italian cypress
(8) *Dracaena deremensis,* Dracaena

(9) *Hibiscus rosa-sinensis,* Hibiscus

(10) *Hibiscus rosa-sinensis* (hybrid), Hibiscus variety

(11) *Ilex vomitoria* "schellings," Yaupon holly

(12) *Juniperus procumbens,* Juniper

(13) *Ligustrum japonicum,* Ligustrum (Privet)

(14) *Nephrolepis exaltata* "Bostoniensis," Boston fern

(15) *Philodendron williamsii,* Philodendron

(16) *Pittosporum tobira,* Pittosporum (green)

(17) *Pittosporum tobira* "variegatum," Variegated pittosporum

(18) *Podocarpus macrophyllus,* Yew podocarpus

(19) *Rhododendron* sp., Azalea varieties

(20) *Strelitzia reginae,* Bird of paradise

Plant growth index determinations were made by summing plant height and average width in February, May and August 1988 on twelve replicates of each plant species in each of the 6 experimental irrigation level groups of residences. Plants were selected within a maximum and minimum predetermined size range so that initial measurements would not have an excessively large standard deviation. Total plant measurements were normally made, but where large tree species were used, such as *Citrus, Cupressus* and *Podocarpus,* a branch or particular part of the plant was selected for measurement rather than the complete plant.

Measurements were entered on a Lotus 123 spreadsheet which computed all growth indices, found the mean initial plant size index and calculated the standard deviation around this mean. Initial growth data were corrected to the overall mean value of the 6 individual means and subsequent growth data were also adjusted by the appropriate correction factors. The spreadsheet also produced graphical representations and executed analyses of variance and Duncan's multiple range tests (Duncan 1955) at the 5% confidence level.

Rainfall and maximum and minimum air temperature data were measured daily at the nearby Southwest Wastewater Treatment Plant. Reclaimed water was sampled and analyzed every 14 days from a specially installed faucet located at 200 Colony Point Road and mainline delivery pressure was monitored by a 7 day chart recorder.

Plant Growth at Different Irrigation Levels

During the experiment, the irrigation application rate had to be intentionally increased in the 0.5 in./week group due to severe drying out of the turf areas at these residences. Increases were made in the last three months of the experiment. Also, the 2.5 in./week levels had to be reduced towards the end of the experiment as large areas of the unsightly weed pennywort (*Hydrocotyle umbellata*) were growing in the turf. Thus, the ideal irrigation level fell within the experimental range as the lower level of 0.5 in./week proved to be insufficient and the upper end of 2.5 in./week proved to be excessive.

The 20 experimental plants are listed in Table 22.6 in increasing order of the "F" value obtained from the analysis of variance on the August growth measurements. Values that exceed 2.35 indicate a significant difference between the 6 means. The Duncan's multiple means test (Duncan

TABLE 22.6. **Mean Separation for 20 Plant Species at 6 Irrigation Application Rates by Duncan's Multiple Range Test, 5% Confidence Level. Analysis of August 1988 Growth Indices.**

"F" Plant Species	Value[a]	Irrigation Application Rate (in./week)					
		0.5	0.75	1.0	1.5	2.0	2.5
Grade I: No significant difference							
Cupressus sempervirens[b]	0.5	a[c]	a	a	a	a	a
Asparagus densiflorus	0.6	a	a	a	a	a	a
Podocarpus macrophyllus	0.8	a	a	a	a	a	a
Strelitzia reginae	0.8	a	a	a	a	a	a
Juniperus procumbens	1.0	a	a	a	a	a	a
Philodendron williamsii	1.0	a	a	a	a	a	a
Codiaeum variegatum	1.3	a	a	a	a	a	a
Brassaia actinophylla	1.6	a	a	a	a	a	a
Ilex vomitoria schellings	1.7	a	a	a	a	a	a
Citrus sinensis	1.9	a	a	a	a	a	a
Grade II: Marginally significant difference							
Carissa grandiflora	2.4	a	ab	bc	bc	c	c
Chrysanthemum morifolium	2.5	a	ab	b	a	a	a
Ligustrum japonicum	2.5	a	a	a	b	b	b
Dracaena deremensis	3.8	a	a	a	b	ab	b
Grade III: Distinct significant difference							
Pittosporum tobira	6.4	a	a	ab	bc	c	c
Hibiscus rosa-sinensis	7.7	a	a	a	b	b	b
Hibiscus hybrid	9.7	a	a	a	b	b	b
Pittosporum tobira "var"	10.8	a	a	ab	bc	c	c
Nephrolepis exaltata	18.2	a	a	b	b	b	b
Rhododendron sp.	29.4	a	a	a	b	c	c

[a] "F" value from the analysis of variance of the August 1988 growth indices (a value of 2.35 or greater indicates significant difference between means at $P = <0.05$.).
[b] For each plant species (i.e.) across each row, growth indices at different irrigation application rates are significantly different ($P = <0.05$) if letters are different.
[c] The letter "a" indicates the least growth in each species. Successive letters indicate increased growth indices.

1955) results are included on this table as lower case letters. Where the "F" value is less than 2.35, there is no significant difference between all 6 means and they are represented by the letter "a." Where the "F" value exceeds 2.35, means may be represented by different letters or combinations of letters. Only means which are represented by *different* letters are significantly different from each other. In other words, there is no significant difference between any means that are represented by any similar letter or set of letters.

Table 22.6 shows 10 ornamental plant species which exhibited no significantly different growth responses to different irrigation application levels over the six month period from February to August 1988. Three of these species (*Cupressus, Podocarpus* and *Citrus*) were well established trees with roots that penetrated below the groundwater level, thus making them independent of irrigation application levels except in severe droughts. *Asparagus* has water storage tubers on the root system and can survive long periods without irrigation. *Codiaeum, Ilex* and *Juniperus* are slow growing species that showed no significantly different growth responses in the short experimental period. The other 3 species, *Philodendron, Brassaia and Strelitzia* have uneven and irregular growth habits which resulted in growth indices with high standard deviations, making these plants unsuitable for this type of growth experiment.

Four species showed marginally significant differences in growth responses to different irrigation application rates. *Carissa* responded with significantly increased growth to increasing irrigation levels up to 1.0 in./week. There was no increase in the growth rate once irrigation levels exceeded 1.0 in./week. The soft stemmed *Chrysanthemum* is difficult to measure as it dies back after flowering and grows back from the base of the old plant. Irrigation applications of 1.0 in./week produced maximum growth. Levels above or below 1.0 in./week reduce the growth response. Both *Ligustrum* and *Dracaena* were good experimental species and required irrigation rates of 1.5 in./week to attain optimum growth.

Six ornamental plant species showed highly significant growth responses to different irrigation application levels. Both green and variegated forms of *Pittosporum* showed similar results with irrigation levels between 1.5 and 2.0 in./week producing maximum growth responses. For hedges and well established plants, lower irrigation rates will decrease growth, which is advantageous in many instances, as the need for excessive trimming is avoided. Common and hybrid *Hibiscus* varieties also showed similar results with irrigation levels of 1.5 in./week producing optimum growth responses. Highly significant differences between growth rates were apparent in May 1988, and, in August 1988, this significance increased. An irrigation rate of 1.5 in./week appeared to be ideal for these plants, as no further significant increase in growth rate occurred if watering was increased to 2.5 in./week. *Nephrolepis* showed minimal growth at irrigation applications below 0.75 in./week. This fern also showed no significant increase in growth responses to irrigation rates exceeding 1.0 in./week. *Rhododendron* sp. (azaleas) showed increasing growth responses to irrigation application rates up to a maximum of 2.0 in./week. Above this application level there was no further significant increase in the growth response. "Greenleaf" experiments showed that *Rhododendron* sp. were extremely sensitive to the extra salt levels in reclaimed water and suffered from leaf burn which lead to a slow deterioration and eventual death of the plant. Azaleas are thus not recommended for landscapes that are irrigated with reclaimed water.

Recommendations

This field trial showed that growth responses of half of the selected experimental species appeared to be independent of applied irrigation levels. The other 10 species show a gradation of increased growth responses to different irrigation application levels ranging from very slight to highly significant.

From the results it is recommended that an irrigation rate of 1.5 inches of reclaimed water per week is applied throughout the growing season from March to November to supplement natural rainfall. This rate can be cut back to 1 inch per week from December to February. These application rates will produce adequate growth in most commonly used landscape ornamentals in the Central Florida area.

FLORIDA'S RECLAIMED WATER REGULATIONS

State regulations governing the use of reclaimed water in Florida are to be found in the Florida Administrative Code (F.A.C.) Section 62-610 which is enforced by the FDEP. This code is presently under extensive revision and the date for promulgation of the new law is January 1, 1996. This law applies to all reclaimed water reuse facilities in Florida regardless of the type of reuse involved. All facilities with reuse systems of any type must obtain a permit from the FDEP and Part I of the rule deals with the general regulations related to permit applications. Part III regulates slow rate land application reuse, public access areas, residential irrigation, and edible crops. A brief survey of the MINIMUM State requirements is given in this section. In the following section, St. Petersburg's regulations governing wastewater reuse use are examined in detail to demonstrate the city's intent to continue reclaiming wastewater in a manner that exceeds the requirements of the law and is still environmentally sound and economically feasible.

FLORIDA REGULATIONS FOR PUBLIC ACCESS AREAS AND RESIDENTIAL IRRIGATION USING RECLAIMED WATER

Treatment Plant Capacity

Only treatment plants that have more than 0.5 mgd design average daily flow (to be reduced to 0.1 mgd in 1996) can construct reclaimed water systems to irrigate public access and residential areas. There is no minimum size restriction for toilet flushing and fire protection only.

Reclaimed Water Quality

Reclaimed water must have received at least secondary treatment and high-level disinfection prior to being discharged to the distribution system. The effluent must contain less than 5 mg/L of total suspended solids and continuous on-line monitoring of turbidity is required prior to disinfection. All other limitations and continuous on-line monitoring for chlorine residual shall apply after disinfection. An operating protocol to ensure these conditions are met shall be approved by the FDEP. Reclaimed water that fails to meet the criteria established in the operating protocol (reject water) shall not be discharged to system storage or to the reuse system. Reject water may be discharged to a permitted disposal system.

Treatment Plant Operation

The treatment plant shall have a minimum Class I reliability unless a permitted alternate discharge system exists which has sufficient capacity to handle all reject water flows. The plant shall be staffed by a Class C or higher operator 24 hours per day, 7 days per week. The lead/chief operator shall be a Class A certification for activated sludge process plants exceeding 5 mgd capacity.

Storage System Requirements

System storage shall not be required when an effluent disposal system is incorporated into the system design to ensure continuous facility operation.

Restrictions on Reclaimed Water Use

Where reclaimed water is used in public access areas (e.g. parks), signs, preferably containing the color purple and informing the public of this use shall be posted. Also, in public access areas, application control shall be the responsibility of the management entity or a local ordinance. Only low trajectory spray nozzles may be used within 100 feet of public eating, drinking and bathing facilities to minimize aerosol formation. Above ground hose bibs or faucets are prohibited but special hose connections in locked valve boxes either above or below ground are allowed if they are clearly labeled as non-potable water.

Reclaimed water shall not be used to fill swimming pools, hot tubs or wading pools. Edible crops may only be irrigated with contact reclaimed water if they are peeled, skinned, cooked or thermally processed before consumption is allowed. All others are prohibited unless there is no direct contact between the irrigation water and the edible crop. Reclaimed water can only be used for toilet flushing or fire sprinkler protection in facilities such as hotels, apartment buildings, etc., where guests and residents do not have direct access to the plumbing system for repairs or modifications.

Cross-connection Control, Setback Distances and Color Coding

No cross-connections to potable water systems shall be allowed and permittees are required to submit documentation of inspection procedures for cross-connection control. Separation of potable and reclaimed pipelines shall be a minimum of five feet (center to center) or three feet (outside to outside). All reclaimed water transmission mains shall be at least 75 feet away from potable water wells. All reclaimed water PVC piping, valves and outlets shall be color coded with Pantone Purple 522C using light stable colorants. Concrete and metal pipes may be marked with purple tape.

Inspection Requirements and Procedures

The permittee shall be responsible for conducting inspections within the reclaimed area to ensure proper connections, ensure proper use of reclaimed water, and minimize the potential for cross-connections. Inspections are required when customers first connect to the reclaimed water distribution system. Periodic inspections are required as specified in the permittee's cross-connection control and inspection program.

Groundwater Monitoring

Permittees shall construct groundwater monitoring wells and set up a monitoring program as required by the State. Sampling parameters for different areas will vary depending on the reclaimed water quality, site specific soil and hydrogeologic characteristics, and other considerations.

Permit Requirements

The State will normally issue a reuse permit to the wastewater treatment management facility. This facility is responsible for regulation and management of the individual users of the reclaimed water through binding agreements or by local ordinance.

ST. PETERSBURG'S "RESPONSIBLE RECYCLING" PROGRAM

The key to operating St. Petersburg's highly successful reclaimed water irrigation system for the past twenty four years has evolved around the implementation of, and rigid adherence to, certain basic criteria. These criteria are designed to:

- ensure public health safety and the ability of the wastewater reuse program to deliver a high-quality product water on a continual basis
- to gain increased public acceptance through safe and reliable performance

Collectively, these criteria are referred to by Johnson (1989) under the term, "Responsible Recycling."

For maximum public health safety and continuous effluent quality the items that must be provided for include:

(1) An advanced treatment facility that is adequate to handle all projected hydraulic and organic loadings. It should have the capability for chemical addition at various points in the plant to enhance process control capabilities. The facility should have a high reliability rating to ensure continuous production of an effluent product that consistently meets the stringent quality standards required by the controlling authority.

(2) The plant should meet Class I reliability and have a great deal of operational flexibility to rapidly change operating modes.

(3) It should be staffed at all times with well qualified personnel. Only operators holding advanced certification should be in charge of each shift.

(4) An effective industrial pretreatment program should be established so that harmful chemicals do not enter the wastewater collection system and possibly pass through to the spray sites.

(5) The effluent quality should be monitored twenty-four hours per day, seven days a week. Key parameters must be established that give

instant evaluation of product quality and suitability for use as irrigation water to on-site plant operators.

(6) A storage system must be in place to accumulate excess reclaimed water in low demand periods for use in periods of high demand.

(7) An adequate backup disposal system must be available to discharge effluent that does not meet reclaimed water standards, or excess reclaimed water in times of low demand after storage capacities have been maximized.

To gain increased public acceptance and to ensure that reclaimed water customers connect to and use the system in a responsible manner, St. Petersburg developed a city ordinance and a "Reclaimed Water Policies and Regulations" Manual in 1981. These documents were produced "from the ground up" so to speak, as not even State regulations had been promulgated at this time The manual has recently been updated and contains the current regulations relating to all aspects of the use of the reclaimed water system. Some of the more important sections with respect to domestic hook-ups are listed below:

(1) Rules governing the process of petitioning the city for expansion of the service.
(2) Reclaimed water application procedures
(3) Regulations restricting the use of reclaimed water
(4) The cross-connection and backflow prevention control program
(5) The inspection program
(6) Identification of reclaimed water pipelines
(7) The city maintenance program
(8) Fire protection

Many of these "Responsible Recycling" criteria such as treatment plant design, storage systems and deep-well injection systems have been discussed at length in the foregoing dialogue. Details of those criteria not covered elsewhere are included in the following subsections with specific reference to St. Petersburg's system.

PUBLIC HEALTH AND CONTINUOUS EFFLUENT QUALITY CRITERIA

Treatment Plant Reliability

The basic statistics of St. Petersburg's four plants are shown in Table 22.1 and treatment plant design has been reviewed in a previous section. Reliability studies on these facilities were last prepared in 1993 (Havens & Emerson 1993) and the data presented here is taken from these reports.

This evaluation was performed to establish a reasonable estimate of the ability of the four plants to meet the water quality criteria for the city's reclaimed water system. The criteria for discharging to the reclaimed system in St. Petersburg include: chloride < 600 mg/L, TSS < 5.0 mg/L, turbidity < 2.5 NTU, and fecal coliform < 25 counts/100 ml.

In 1992, some plants exhibited an annual reliability of performance in excess of 99% reliability. Where the few excursions occurred above water quality levels, they were often associated with chloride level problems that cannot be controlled by plant processes and must be regulated by both sewer maintenance and the industrial pretreatment program.

Class I Reliability

Both State and Federal Class I reliability standards are essentially identical, and pertain only to mechanical, electrical and fluid systems and components. These standards are mandatory for facilities which discharge to public access reclaimed water systems such as St. Petersburg and provide the plants with the capacity to rapidly change operating modes. Some of St. Petersburg's plants exceed the Class I criteria which are summarized as follows:

(1) Power source—two separate and independent electric power sources from either two separate utility substations or one substation and one standby generator

(2) Pumps—sufficient capacity with the largest unit out of service to handle peak flow; sludge backup pumps may be installed; electrical redundancy may be restricted to main pumps

(3) Critical lighting and ventilation—all operating stations to have electrical redundancy

(4) Preliminary liquid processing train—minimum facilities: screenings removal and grit removal; mechanically cleaned bar screens must have backup bar screen and power source

(5) Solids processing train—provision for preventing contamination of treated wastewater and alternate method of sludge disposal and/or treatment when unit operations do not contain a backup capability

(6) Liquid processing train redundancy or backup
  - degritting—optional
  - aeration basins—two minimum with equal volume
  - blowers or mechanical aerators—multiple units: remaining capacity with largest unit out of service must be able to achieve design maximum oxygen transfer, back up unit may be uninstalled
  - final sedimentation—multiple units; remaining capacity with

largest unit out of service must be sufficient for at least 75% of design maximum flow
- chemical flash mixer—2 minimum or backup
- disinfection basins—multiple units; remaining capacity with largest unit out of service must be sufficient for at least 75% of design maximum flow

(7) Sludge processing train backup or redundancy
- holding tanks—permissible as alternative to backup capability with adequate capacity for estimated time of repair
- anaerobic digestion—at least two tanks; mixing equipment backup or flexibility should be such that with one piece of equipment out of service, total mixing capability is not lost; backup equipment may be uninstalled
- dewatering—multiple units with capacity to dewater design sludge flow with largest capacity unit out of service

(8) Electrical-power distribution/instrumentation
- no single fault or loss of power source will result in disruption of service to more than 1 MCC.
- Vital components of same type and serving same function shall be divided between at least 2 MCC's.
- Critical electric motors should be isolated against 100 year flood condition.

Operator Certification

In order to ensure that domestic wastewater plants are operated in an effective and safe manner, the FDEP requires all wastewater treatment plant operators to attend training classes and take certification examinations based on this course-work. For domestic wastewater treatment plants, operators first become certified at the "C" level and then, as they increase their experience they can become certified at the "B" and finally the "A" level of competency. Chief operators at St. Petersburg's plant are all required to hold an "A" certification. Shift supervisors have either an "A" or "B" level of certification. The city actively promotes the training of all operators and reimburses tuition fees if examinations are passed.

Industrial Pretreatment Program

St. Petersburg's industrial pretreatment program was approved by the EPA in 1984. The program issues industrial wastewater discharge permits to 45 businesses in St. Petersburg that are identified as significant industrial users. The program also aggressively monitors the wastewater characteristics and regularly inspects these industries to ensure that they remain in compliance with all of the pollutant limits included in USEPA regulations and in the city's ordinance. As a result of this ongoing program, the wastewater influent at each of the four plants contains minimal amounts of pollutants that could either upset the plant processes, be injurious to the health of employees or pass through unchanged into the reclaimed water. The efficacy of this program is further demonstrated and confirmed by the annual influent and effluent priority pollutant scans, submitted to USEPA and FDEP. The program is routinely audited by the USEPA and FDEP legal and technical staff, and has received a National Award as an outstanding program by the USEPA.

Effluent Monitoring

The on-site plant operator must have "now" control information in order to rapidly evaluate product quality and suitability for use as irrigation water. The city uses two key operational parameters for instant evaluation, on line turbidity and chlorine residual monitoring. A turbidity exceeding 2.5 NTU's or a chlorine residual of less than 4.0 mg/L is not discharged into the reclaimed water system. Chloride levels vary more slowly, so monitoring for chlorides is done periodically. A chloride level exceeding 600 mg/L has been determined unsuitable for use as irrigation water. The plant operational permit issued by the State requires that the suspended solids not exceed 5 mg/L. A turbidity level not exceeding 2.5 NTU is typically well below 5.0 mg/L suspended solids, and turbidity provides instantaneous information.

THE "RECLAIMED WATER POLICIES AND REGULATIONS" MANUAL

This comprehensive manual includes regulations for both domestic, commercial and industrial consumers. The regulations included here are mostly applicable to the domestic users which make up the bulk of the customers connected to the system.

Petitioning the City for Expansion of the Service

The city provides reclaimed water service to over 8,000 customers at the present time. Where a reclaimed water transmission main less than 24-inches in diameter is adjacent to a potential customer's residence the customer may apply directly for service. When service is not installed in an area such as a housing development, the property owners in that development may petition the Mayor on an appropriate form demonstrating their interest in obtaining reclaimed water service. If more than 50% of the owners support the petition the city prepares a cost packet consisting of a formal petition cost estimate, a voluntary lien agreement and a hold harmless agreement. If 50% of the property owners approve of the cost estimate and sign and return the legal agreements, the city passes the request to City Council for adoption. The city then designs the reclaimed water system for the approved project area and a "low bid" private contractor installs the distribution mains. Once the main is constructed, the property owners are sent individual bills and various convenient methods of paying these bills are offered to these customers.

Reclaimed Water Application Procedures

Potential customers complete and sign city approved application forms for reclaimed water service, indicating the size of service line required and the location on the form. They are also required to sign a Hold Harmless Agreement as part of the application. Prior to making the application, the potential customer must have an in ground irrigation system with no connections to this system entering the dwelling unit for any purposes. If the customer wishes to have hose connections on the system, only city approved types of connectors are allowed. After inspections by Field Inspectors have been completed, the customer's irrigation system has been approved and the appropriate fees have been paid, the city will connect the customer to the reclaimed water system. Separate permits are required for hose connections.

Regulations Restricting the Use of Reclaimed Water

Reclaimed water is supplied for the irrigation of ornamental flowers and turf only. It shall not be used for: consumption human or animal, interconnecting with another water source, sprinkling of edible crops (gardens), body contact recreation, use through hose bibbs, faucets, quick couplers, etc., filling of swimming pools, sharing a common reclaimed service or connection between properties, augmenting lake or pond levels, filling of decorative pools or fountains, supplying air cooling systems (a/c units), or washing of equipment such as cars, boats, driveways, roofs, structures, etc.

The Cross-Connection and Backflow Prevention Control Program

Cross-connection and backflow go hand in hand as without the one, the other cannot occur. The city cross-connection rule states: "In all premises where there is reclaimed water or other auxiliary water supply, there shall be no physical connection or an air gap of not less than twice the diameter of the potable supply pipe or 1 inch, whichever is greater, between such non-potable supply and the consumer's potable water system." The city backflow regulations state: "In all premises where reclaimed water service is provided, the public potable water supply is fitted with an approved double check valve assembly (1.5 inch and larger services) when applicable, or a dual check device (1 inch and 0.75 inch services)."

Since the reclaim system delivery pressure is approximately 70 psi and the potable pressure is only 50 psi, an illegal cross-connection between the two systems would result in the contamination of the potable water distribution system by reclaimed water. The installation of a backflow prevention device on the customer's side of the potable water meter prevents reclaimed water from entering the potable water distribution system. Illegal cross-connections will still result in the contamination of the potable water supply in the customer's residence however.

The city soon discovered that the installation of backflow devices isolates the potable water system. At some residences, discharges would occur at the water heater relief valve as a result of expansion. To remedy this, the city originally installed a pressure relief device on an outside faucet wherever the problem occurred. Other types of devices such as the thermal expansion tank and the triple purpose toilet tank ball cock valve are now available to overcome this problem.

The Inspection Program

The city has a well established inspection program. All new accounts have to be inspected prior to reclaimed water hook-up. Access to the property to carry out this inspection is given by written consent on the customer's application form. The inspection is comprehensive, and includes cross-connection detection, hose connection inspection, irrigation system inspection and a general survey for any rule infractions. Upon completing this, and all, inspections the inspector prepares an onsite inspection report as to the findings. After hook-up, a re-inspection of the premises is carried out to check for rule infractions. Random inspections after hook-up are carried out at every residence at least once every 18 months. Violations of the rule that are detected lead to various enforcement actions including immediate termination of service until the infractions are corrected.

Identification of Reclaimed Water Pipelines

St. Petersburg color-codes all PVC piping, using blue for potable water, green for sewer force mains and Pantone Purple 522C for reclaimed water. When the reclaimed system was first installed, brown was used for reclaimed water piping. All buried ductile iron piping was affixed with a brown coded tape and fire hydrants located throughout the system had a brown stem and a yellow bonnet. Also, valve box covers were shaped differently than potable water system valve box covers. State regulations require that all new installations since July 1, 1995 be colored purple. PVC pipes are now available that have the Pantone Purple 522C color impregnated into the PVC. Metal and concrete pipes must have purple markings as a predominant color. Retrofitting of brown pipes with purple is not required until these pipes are due to be replaced.

The City Reclaimed Water Maintenance Program

Continuous maintenance is carried out by the city on all sections of the reclaimed water system that belong to the city. This includes leak control, repair of damage to the system by acts of God, vandalism and all other imaginable causes. Subcontractors may also be used by the city for repair work on a contractual basis. The city makes a reasonable effort to keep the facilities in good repair but does not guarantee the uninterrupted supply of reclaimed water to all customers at all times.

Fire Protection

Fire hydrants are installed on reclaimed water mains all over the city at such locations as are deemed appropriate for the purpose of flushing the reclaimed water system to maintain water quality and may be used as an auxiliary source of water for fire protection. Reclaimed water hydrants are distinguished from potable water hydrants as the former are colored purple with a luminescent gray bonnet and caps. The pressure and/or volume of reclaimed water through a reclaimed water hydrant is not guaranteed.

THE IMPACT OF ST. PETERSBURG'S RECLAIMED WATER SYSTEM

The reclaimed water system has continued to expand and change in character since its inception in the late 1970s, from one of an alternative mode of wastewater effluent disposal to one of a fully operational third service encompassing water treatment and distribution, wastewater collection and treatment, and reclaimed water supply. The growth in the reclaimed water system demand since its inception has significantly contributed to the suppression of potable water demands over the past eighteen years. Significant economic and environmental benefits have been derived from the development of this form of water reclamation. Since its inception, annual demand for potable water has been stabilized while the demand for nutrient rich reclaimed water has steadily increased (see Table 22.2).

The reclaimed water system has been an economic benefit to all the City's utility customers in that several potable water system projects such as additional treatment units at the Cosme Water Treatment Plant, a booster station on the 48 inch water transmission main in the Safety Harbor area and the south-side booster station and storage facility have all been delayed indefinitely. The cost avoidance for these projects would have been in the range of $25–$30 million. In addition, economic savings at the treatment plants by avoiding the need to install expensive nutrient removal wastewater processing systems have also been realized.

Furthermore, ongoing expansion of the reclaimed water system will continue to provide an economic benefit to utility customers because the development of new, very expensive, raw water supplies will be delayed.

The expansion of the distribution system into residential areas also required a change in management policy and attitudes. In the early years, the distribution system supplied large land areas such as the parks and golf courses, where landscape management practices were well established. The use of this new resource posed no significant problems to these customers and the city received very few complaints. In 1981, only 18.7% of the total number of customers were residential users, accounting for less than 0.5% of the total acreage under irrigation. In 1987, residential customers represented 96.2% of the total number of users, and the residential acreage under irrigation was 32% of the total area served by the system.

Since 1985, the program has had to change from a non-residential user oriented system to one which must now recognize and respect the needs and deal with the more numerous questions and complaints from the smaller residential homeowner.

Together with the growth of the system, the attitudes of Public Utilities wastewater treatment personnel towards wastewater effluent have had to dramatically change between 1980 and 1987. In St. Petersburg, operators now work in "Water Reclamation Facilities" (WRFs). Signs at plant entrances identify them by this term and "water production," not "sewage treatment," is the theme of a water reclamation program. Plant operations staff members have a "manufacturing" mentality and not a "treat and dispose" attitude. The value of the reclaimed water in helping St. Petersburg meet its total water needs is recognized by all plant employees.

The Reclaimed Water System Master Plan Update produced by Camp Dresser and McKee Inc. (1987), recognized that the system had continued to expand and change in character from an alternate mode of wastewater disposal to a full operation as a third element of the city's Public Utilities

Department. Recommendations on the future development and management of this system were also incorporated in this report.

This final accomplishment has made St. Petersburg the first major municipality in the USA to achieve zero discharge of wastewater effluent into surrounding surface waters. This significant achievement was accomplished over a 15 year period and represents a remarkable example of what can be accomplished by careful planning and notable foresight today, just as the city fathers had foresight in the 1920s concerning potable water issues.

This type of development also provided a workable solution to water supply and water pollution problems in coastal areas. It serves as an ideal example for many other municipalities now striving to attain similar goals.

REFERENCES

Baldwin, L. B. & R. R. Carriker, 1985. Water resource management in Florida. Fla. Coop. Ext. Serv. Bulletin 206, 16 pp.

Barrick, W. E., 1979. Salt tolerant plants for Florida landscapes. State Univ. Sys. Fla., Sea Grant College Program Report 28, 71 pp.

Berry, W. L., A. Wallace and O. R. Lunt, 1980. Utilization of municipal wastewater for the culture of horticultural crops. *Hort. Sci.* 15: 169–71.

Black, Crow & Eidsness Inc./CH2M Hill, 1978. St. Petersburg 201 Facilities Plan, South Pinellas County, Florida. EPA Grant No. C-120631010.

Camp, Dresser & McKee, 1987. A White Paper. Urban Water Reuse in the City of St. Petersburg. Water Quality and Public Health Considerations. City of St. Petersburg Public Utilities Department, 16 pp.

Culp, R. L., Wesner/Culp, 1979. *Water reuse and recycling. Vol 2: Evaluation of treatment technology.* U.S. Dept. of the Interior. Office of Water, Research and Technology. Washington, D.C.

Duncan, D. B., 1955. Multiple range and multiple F tests. *Biometrics* 11: 1–42.

Fitzpatrick, G. E., 1983. Effluent irrigation. (IFAS working for you column). *Fla. Nurseryman* 30(1): 25.

Fitzpatrick, G. E., 1985. Container production of tropical trees using sewage effluent, incinerator ash and sludge compost. *J. Environ. Hort.* 3(3): 123–5.

Fitzpatrick, G. E., H. Donselman and N. S. Carter, 1986. Interactive effects of sewage effluent irrigation and supplemental fertilization on container grown trees. *HortScience* 21(1): 92–3.

Henningson, Durham and Richardson, 1981. Reclaimed Water System Master Plan. City of St. Petersburg, Fl., Pensacola, Florida.

Hoadley, B. & D. Ingram, 1982. Water quality and woody ornamental plant production. Fla. Coop. Ext. Serv. Commercial Fact Sheet OHC-4.

Johnson, W. D., 1989. Responsible Recycling. A practical approach to developing and operating a successful urban irrigation system using recycled wastewater.

Johnson, C. R. & R. J. Black, 1985. Salt tolerant plants for Florida. Fla. Coop. Ext. Serv. Fact Sheet OH-26.

Klinkenberg, J., 1985. Rethinking reclaimed water uses. *St. Petersburg Times* article November 16, 1985.

Marshall, T. J., 1968. Some effects of drag on structure and hydraulic conductivity of soil. *Trans. 9th. Int. Congr. Soil Sci. Adelaide* 1: 213–221.

Mass, E. V., 1986. Salt tolerance of plants. *Appl. Agric. Res.* 1 (1), pp 12–26.

Metcalf & Eddy, 1979. *Wastewater engineering: treatment, disposal, reuse.* McGraw Hill Inc., New York, N.Y.

Parnell, J. R., 1988. Project Greenleaf Final Report. City of St. Petersburg, Public Utilities Department, 500 pp.

Quirk, J. P. & R. K. Schofield, 1955. The effects of electrolyte concentration on soil permeability. *J. Soil Sci.* 6: 163–78.

Rhue, R. D. & J. J. Street, 1980. Phosphorus—A primary plant nutrient. Fla. Coop. Ext. Serv. Fact Sheet SL-35.

Rhue, R. D. & J. J. Street, 1981. Potassium—A primary plant nutrient. Fla. Coop. Ext. Serv. Fact Sheet SL-34.

Street, J. J. & R. D. Rhue, 1980a. Essential micronutrients: Manganese. Fla. Coop. Ext. Serv. Fact Sheet SL-32.

Street, J. J. & R. D. Rhue, 1980b. Essential micronutrients: Zinc. Fla. Coop. Ext. Serv. Fact Sheet SL-32.

U.S. Department of Agriculture, 1972. *Soil survey of Pinellas County, Florida.* U.S. Govt. Printing Press 0-420-436.

Westcot, D. W. & R. S. Ayres, 1985. Irrigation water quality criteria. (in, *Irrigation with reclaimed municipal wastewater—a guidance manual.* G. S. Pettygrove and T. Asano, eds.) Lewis Publishers Inc., Chelsea, Ml.

Zartman, R. E. & M. Gichuru, 1984. Saline irrigation water: Effects on soil chemical and physical properties. *Soil Sci.* 138: 417–22.

Zirschky, J. & A. R. Abernathy, 1985. Land application of wastewater. *J. Water Poll. Contr. Fed.* 57(6): 561–3.

CHAPTER 23

Groundwater Recharge at the Orange County Water District

IN southern California, enhanced groundwater recharge is an increasingly important water resources management tool as previously imported water supplies are reallocated to meet the needs of other southwestern states and endangered species. Orange County Water District (OCWD) has practiced groundwater recharge by both direct injection and surface spreading for over 20 years. These recharge activities maximize the use of local water resources.

Orange County is located in southern California (Figure 23.1) where historic agricultural water use has been replaced with increasing urban water demands. In the early 1950s, groundwater overdraft conditions along the coast caused seawater intrusion into freshwater aquifers. OCWD's implementation of effective groundwater recharge projects, including a seawater intrusion barrier, has controlled the overdraft of the groundwater basin and enhanced water supply reliability.

OCWD was formed in 1933 by a special act of the California Legislature to manage northern Orange County's groundwater supply. Currently over 200 production wells in OCWD's service area supply nearly 75% of the water demand for a population of two million. The remaining demand is met by imported water from the Colorado River and northern California. Most areas in southern Orange County are entirely dependent on imported water supplies, due to the lack of a suitable local groundwater supply.

One of OCWD's goals as the manager of the Orange County groundwater basin is to maximize the availability of high quality groundwater

William R. Mills, Jr., Susan M. Bradford, Martin Rigby, and Michael P. Wehner, Orange County Water District, 10500 Ellis Ave., P.O. Box 8300, Fountain Valley, CA 92728-8300.

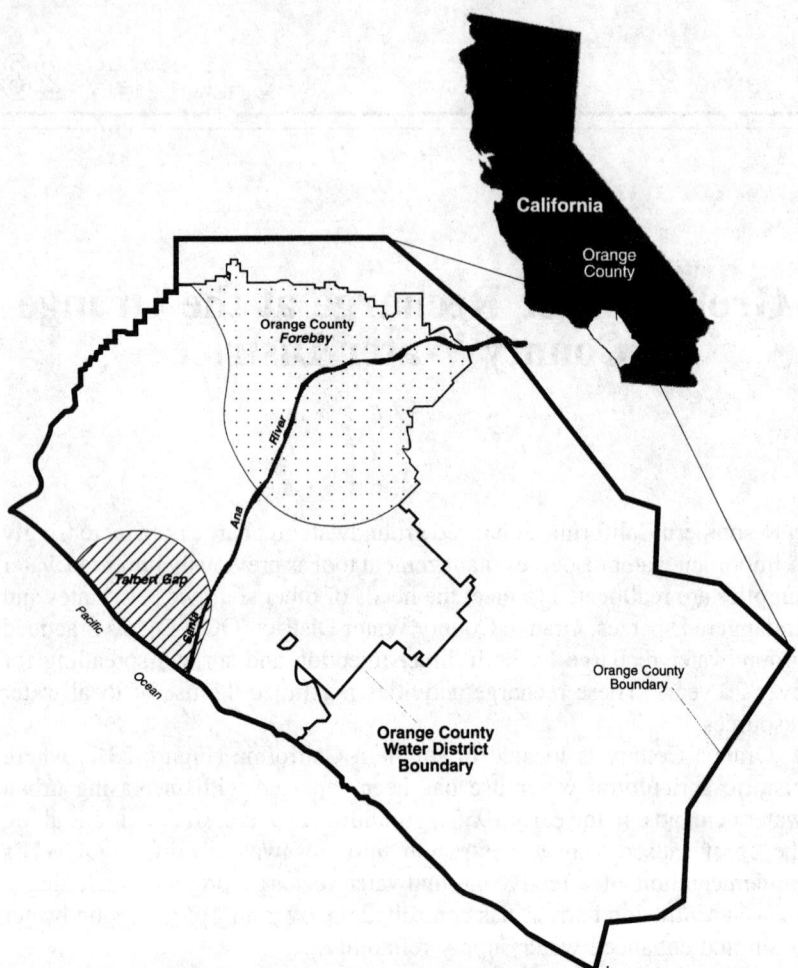

Figure 23.1 Orange County Water District service area.

supplies. Effective groundwater management helps reduce local dependence on less reliable imported water supplies. OCWD has implemented two unique recharge programs as part of its groundwater resources management plan. The first program is Water Factory 21 (WF-21), an advanced wastewater reclamation project that directly injects highly treated reclaimed water into coastal aquifers to prevent seawater intrusion. Mounding of the injected water within the aquifer system provides a hydraulic barrier to seawater intrusion. The second recharge program involves surface spreading of water for groundwater recharge in northeastern Orange County. This program takes advantage of the natural percolation capacity of the Orange County Forebay area where the majority of groundwater recharge in Orange County occurs. Ten spreading basins and in-channel facilities are used to percolate Santa Ana River flows and imported water into the groundwater basin. This chapter discusses the operational parameters, water quality and regulatory considerations for WF-21 and the Orange County Forebay recharge projects.

OCWD is evaluating ways to increase groundwater recharge quantities using existing facilities and spreading basins. It is estimated that an additional 50,000 to 75,000 acre-feet per year (afy) (6.2×10^7 to 9.3×10^7 m³/y) of water could be recharged into the groundwater basin using the existing spreading basins. Development of additional sources of water for recharge is currently being considered. Regulatory considerations for increasing groundwater recharge are discussed herein.

WATER FACTORY 21

WF-21 is an advanced wastewater reclamation facility producing up to 15 million gallons per day (mgd) (57,000 m³/d) of highly treated reclaimed water for injection into coastal aquifers to prevent seawater intrusion into the local groundwater supply. Since 1976, WF-21 has injected over 96,000 acre-feet (af) (1.2×10^8 m³) of highly treated reclaimed water into the aquifer system, creating a freshwater mound or pressure ridge that hydraulically prevents seawater intrusion [1]. Injected water also flows inland and augments potable water supplies.

A brief historical overview of the development of WF-21 is presented in Figure 23.2. OCWD began evaluating the feasibility of injecting reclaimed water to produce a seawater intrusion barrier in 1963, and injected secondary effluent from trickling filters into the Talbert aquifer as part of a pilot study in 1965 [2]. A 1967 report concluded that direct injection of reclaimed water was feasible; however, the injected water had objectionable color, taste, and odor. A final report generated in 1970 recommended additional treatment of the water prior to injection to remove the taste and odor problem. The WF-21 treatment facilities were designed in 1970, and the project was permitted by the Regional Water Quality Control Board (Regional Board) in 1971 following public hearings and recommendations from the California Department of Health Services (DHS) [3]. Construction of WF-21 required three years, and injection of reclaimed water began in 1976. Approximately $10.2 million in state and federal funding was obtained for the $21 million project, including a demonstration seawater desalting facility. A reverse osmosis (RO) component was added to the treatment train in 1977. In 1991, a new permit was issued for WF-21 allowing for injection of up to 100% reclaimed water [4]. Over the

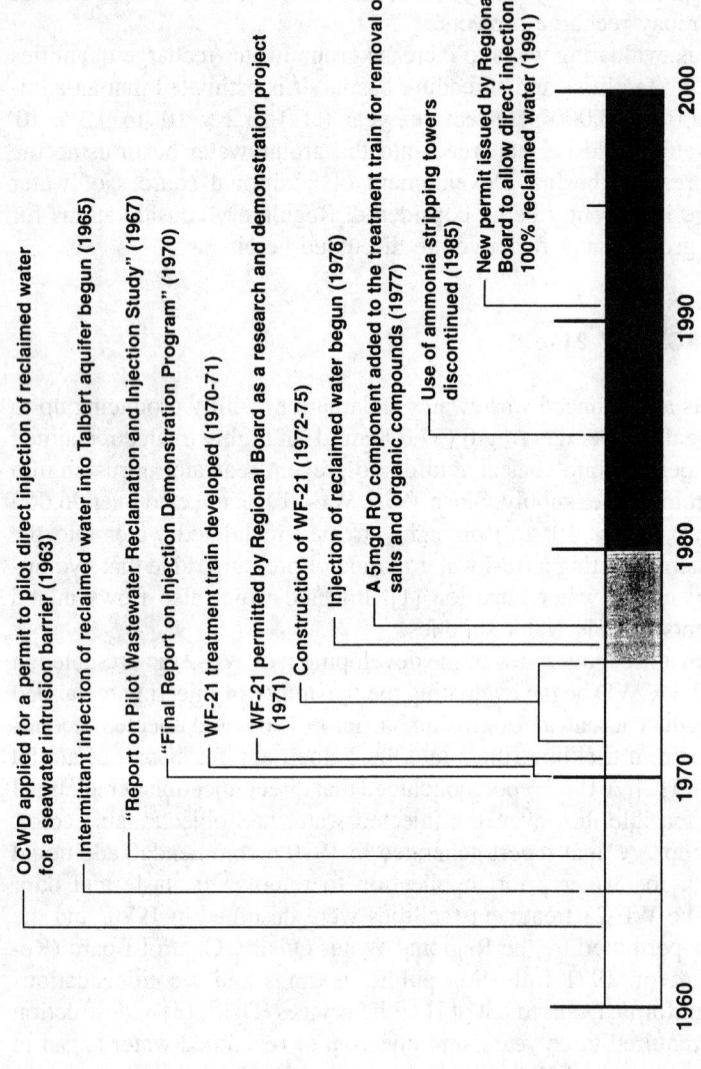

Figure 23.2 A historical overview of the implementation of WF-21.

Source: Wesner, G. M. (December 1987) Historical review of Water Factory 21, Orange County Water District wastewater reclamation and groundwater recharge program

OCWD applied for a permit to pilot direct injection of reclaimed water for a seawater intrusion barrier (1963)

Intermittant injection of reclaimed water into Talbert aquifer begun (1965)

"Report on Pilot Wastewater Reclamation and Injection Study" (1967)

"Final Report on Injection Demonstration Program" (1970)

WF-21 treatment train developed (1970-71)

WF-21 permitted by Regional Board as a research and demonstration project (1971)

Construction of WF-21 (1972-75)

Injection of reclaimed water begun (1976)

A 5mgd RO component added to the treatment train for removal of salts and organic compounds (1977)

Use of ammonia stripping towers discontinued (1985)

New permit issued by Regional Board to allow direct injection of 100% reclaimed water (1991)

1960   1970   1980   1990   2000

past 20 years, OCWD has developed extensive databases on WF-21's advanced wastewater treatment processes and the impact of direct injection of reclaimed water on aquifers used for potable water supplies.

WF-21 PERMIT

The Regional Board, in conjunction with DHS, granted OCWD a discharge permit for WF-21 in 1971. WF-21 was the first groundwater recharge project in California permitted to use reclaimed water to maintain a seawater intrusion barrier. According to the California Water Code, DHS was required to hold a public hearing on WF-21 because the project involved injection of reclaimed water into an aquifer used for domestic water supply [5]. Following the 1971 public hearing, DHS concluded that WF-21 would not impair use of the groundwater basin as a domestic water supply, provided that specific conditions were met. Those conditions became provisions in the permit issued to OCWD by the Regional Board in 1971. A major provision of the 1971 permit was that the injection water be a 50:50 blend of reclaimed and non-reclaimed dilution water (i.e., desalted seawater or deep well water) [3]. After RO treatment was added to the WF-21 treatment train in 1977, the ratio of reclaimed water to blend water was increased to approximately 67% reclaimed and 33% deep well water.

Permit requirements set by the Regional Board and DHS for WF-21 are primarily based on drinking water standards. The total dissolved solids (TDS), chloride, and boron limits are based on water quality objectives set by the Regional Board to protect existing water quality in the groundwater basin [6]. The main concern for DHS with groundwater injection of reclaimed water is the fate and transport of organics of wastewater origin [7]. To minimize concerns with wastewater organics, DHS has proposed to limit the total organic carbon (TOC) of wastewater origin to no more than 1 mg/L at the point of extraction. DHS recognizes TOC is removed in the unsaturated (vadose) zone but not under saturated conditions. Since WF-21 injects directly into the aquifer under saturated conditions the permitted TOC concentration is quite low. In 1994, the TOC in WF-21 injection water ranged from 1.68 to 2.52 mg/L [1]. However extracted groundwater typically had 1 mg/L TOC or less. WF-21 monitoring data has consistently shown that TOC removal occurs under saturated conditions. Hydrogeologic and chemical studies are underway to better characterize TOC removal within the saturated and unsaturated zones.

In 1991, WF-21 became the first project in the State of California permitted to inject 100% reclaimed water, under specific conditions [4]. Much larger RO facilities will be needed to meet requirements for injection of 100% reclaimed water. Currently OCWD injects water containing less

TABLE 23.1. **WF-21 Source Water (secondary effluent) Water Quality.**

Constituent	Minimum (mg/L)	Maximum (mg/L)	Average (mg/L)
Total organic carbon	9.8	12.6	11.4
BOD$_5$	6.2	18.9	11.3
COD	38	45	42
Total dissolved solids	898	1130	1012
Total nitrogen	20.8	34.0	28.0
Ammonia-nitrogen	5.8	30.8	23.8
Nitrate/nitrite nitrogen	0.6	4.4	2.9
Organic-nitrogen	0.8	2.6	1.7
Sodium	207	267	231
Chloride	222	336	265
Sulfate	221	365	276
Phosphate	1.7	3.6	2.9
Boron	0.52	0.68	0.61
Fluoride	0.88	1.39	1.06
Magnesium	21	34	27
Aluminum	16	47	18
Cadmium	< 1.0	2.8	1.4
Lead	< 1.0	6.8	2.0

than 67% reclaimed water while plans are developed to expand the existing RO facilities. To inject 67–100% reclaimed water, WF-21 product water must meet all previous injection requirements plus a 2 mg/L TOC limit. The permit also requires that a research and demonstration Plan of Study and an operating plan be submitted for approval by DHS and the Regional Board. The Plan of Study must include research on the effects, if any, of injecting 100% reclaimed water on the groundwater basin and investigations on the fate of organics of wastewater origin during groundwater recharge [8].

WF-21 TREATMENT FACILITIES

The WF-21 advanced wastewater treatment facility utilizes secondary effluent from County Sanitation Districts of Orange County's (CSDOC) Plant No. 1 as source water. CSDOC's Plant No. 1 is located adjacent to WF-21 in Fountain Valley, California. Plant No. 1 uses both trickling filters and activated sludge to produce secondary effluent. Since 1978 source water to WF-21 has been limited to activated sludge effluent since it is a higher quality effluent containing less suspended solids. An example of WF-21's source water quality is provided in Table 23.1. TOC and turbidity values typically average 11 mg/L and 2.1 nephelometric turbidity units (NTU), respectively [1]. The secondary effluent has very low concentrations of heavy metals due to CSDOC's effective industrial source control program [9].

In 1975, WF-21's advanced wastewater treatment train included chemical (lime) clarification, ammonia stripping, recarbonation, multi-media filtration, granular activated carbon (GAC) filtration, and chlorination. In 1977, a 5 mgd RO facility was added to provide a source of demineralized water for blending. Ammonia stripping was slowly discontinued over the next several years. The current WF-21 treatment train is shown in Figure 23.3. A description of each process is provided in the following paragraphs.

Secondary effluent is first chemically clarified by the addition of lime slurry and polymer to achieve a pH value of approximately 11.2. Clarification involves rapid mixing, flocculation, and sedimentation. WF-21's sys-

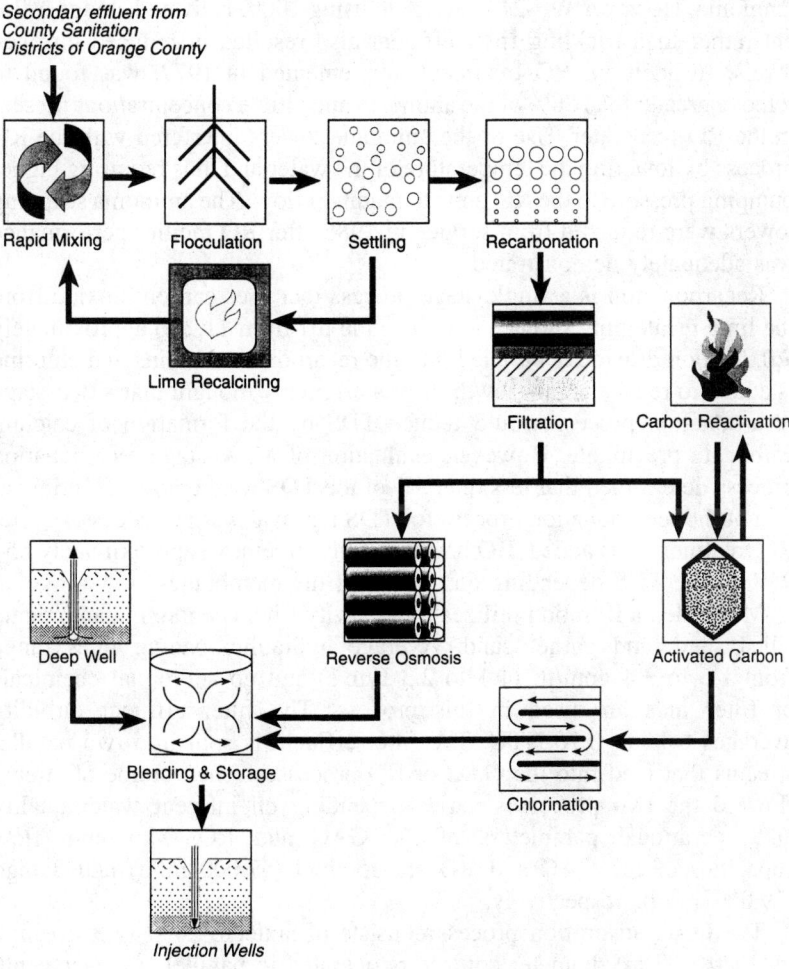

Figure 23.3 WF-21 advanced water reclamation treatment train.

tem uses approximately 1 ton (907 kg) of lime per million gallons of secondary effluent. Over 99% of coliform bacteria are removed, and viruses are effectively inactivated by the high pH. Lime clarification removes approximately 26% of the TOC and significantly reduces the concentration of inorganic constituents such as calcium, magnesium, iron, fluoride, and silica. OCWD operates an on-site recalcining furnace for recycling lime sludge. The recalcination process provides a source of carbon dioxide for recarbonation of the clarified water.

Ammonia stripping was originally used to remove high levels of ammonia from the lime clarified effluent. The ammonia stripping process used packed towers where the high pH values removed nitrogen as gaseous ammonia. However WF-21's switch to using 100% activated sludge effluent rather than trickling filter effluent also resulted in reduced nitrogen levels. In addition, RO treatment, implemented in 1977, was found to remove greater than 80% of the ammonia and nitrate concentrations present in the RO feedwater. Use of the ammonia towers interfered with the RO process by lowering the temperature of the water and thus requiring higher pumping pressure to the RO units to maintain flow. The ammonia stripping towers were removed from service in 1986 after RO facility performance was adequately demonstrated.

Recarbonation is a single-stage process that uses carbon dioxide from the lime recalcining furnace to reduce the pH from 11.2 to approximately 7.0. Carbon dioxide is bubbled into the recarbonation basins and chlorine is added to reduce algal growth. It was originally thought that a two-stage recarbonation process would reduce TDS by the formation of calcium carbonate precipitate. However, evaluation of a two-stage recarbonation process determined that less than 5% of the TDS was removed. Optimization of the recarbonation process for TDS removal was not necessary after RO treatment was added. RO has the ability to remove approximately 85–95% of the TDS depending on the age of the membranes.

Multi-media filtration utilizes four gravity filters containing anthracite, silica sand, and garnet sand. Average hydraulic loading rates range from 1.6 to 3.4 gpm/ft^2 (1.1 to 2.3 L/m^{2}s) and no additional chemicals or filter aids are used in this process. The filter effluent turbidity averages 0.14 to 0.16 NTU. The filter effluent is split into two parallel streams that feed into the GAC or RO processes. The volume of stream flow to the two processes varies depending on influent water quality and operational parameters of the GAC and RO. Maximum flow capacities of the GAC and RO are 15 mgd (57,000 m^3/d) and 5 mgd (19,000 m^3/d), respectively.

The GAC adsorption process consists of sixteen, 24 ft (7.3 m) high by 12 ft (3.7 m) diameter contactors operated in parallel. A seventeenth contactor is used for standby service and carbon storage. Each contactor contains 43 tons (39,010 kg) of activated carbon. The design flow is 0.94 mgd (3,558 m³/d) per contactor with a contact time of 34 minutes. The number of contactors used is dependent on the volume of influent flow. In 1994, no more than three contactors were in service at the same time. Activated carbon removes 30 to 50% of the organics, resulting in an average TOC concentration of approximately 5.5 mg/L. The GAC effluent is pumped into a chlorine contact basin and dosed with approximately 20 mg/L chlorine to achieve a minimum combined chlorine residual of 5 mg/L. Minimum contact time is 30 minutes. Activated carbon has a finite adsorption capacity for organic compounds. OCWD regenerates exhausted carbon at high temperatures using an on-site multiple hearth furnace.

A 5 mgd (19,000 m³/d) RO system was added in 1977 to reduce TDS and TOC concentrations. The RO process consists of six subunits each containing 252 eight-inch (0.2 m) diameter spiral wound, cellulose acetate membranes in a 24-12-6 array. RO removes approximately 85 to 95% of the TDS and 81% of the TOC, resulting in a product water containing 140 mg/L TDS and less than 1.0 mg/L TOC. RO also removes 77 to 97% of copper, chromium, and iron present in the RO feedwater.

RO effluent is pumped into a blending reservoir along with the chlorinated GAC effluent and deep well water. Equivalent volumes of each type of water are blended to achieve a product water containing up to 67% advanced treated wastewater (GAC and RO effluents) and 33% deep well water.

The 1971 WF-21 permit required that the highly treated wastewater be blended with water that is not of wastewater origin, in order to limit the percentage of reclaimed water entering the aquifer. The blend water used by WF-21 is well water pumped from a deep aquifer not impacted by seawater intrusion and not used as a potable supply due to color imparted by subsurface peat deposits.

WF-21 product water is injected into four aquifers prone to seawater intrusion (Figure 23.4), using 23 multi-point injection wells. The injection wells are approximately 500 ft (152 m) apart and span a distance of roughly 2 miles (3.2 km). The location of the injection wells in relation to drinking water wells is shown in Figure 23.5. Although WF-21's permit requires a 2,000 ft (0.6 km) municipal well setback distance from the injection wells, OCWD discourages construction of new wells within 5,000 ft (1.5 km) in order to protect the seawater intrusion barrier.

In addition to WF-21 facilities, OCWD also maintains seven groundwater extraction wells within the Talbert aquifer to remove seawater, if necessary. These extraction wells are located between the injection wells and the coast. Although the extraction wells are not currently needed, they can be used to help prevent seawater intrusion if hydraulic conditions change.

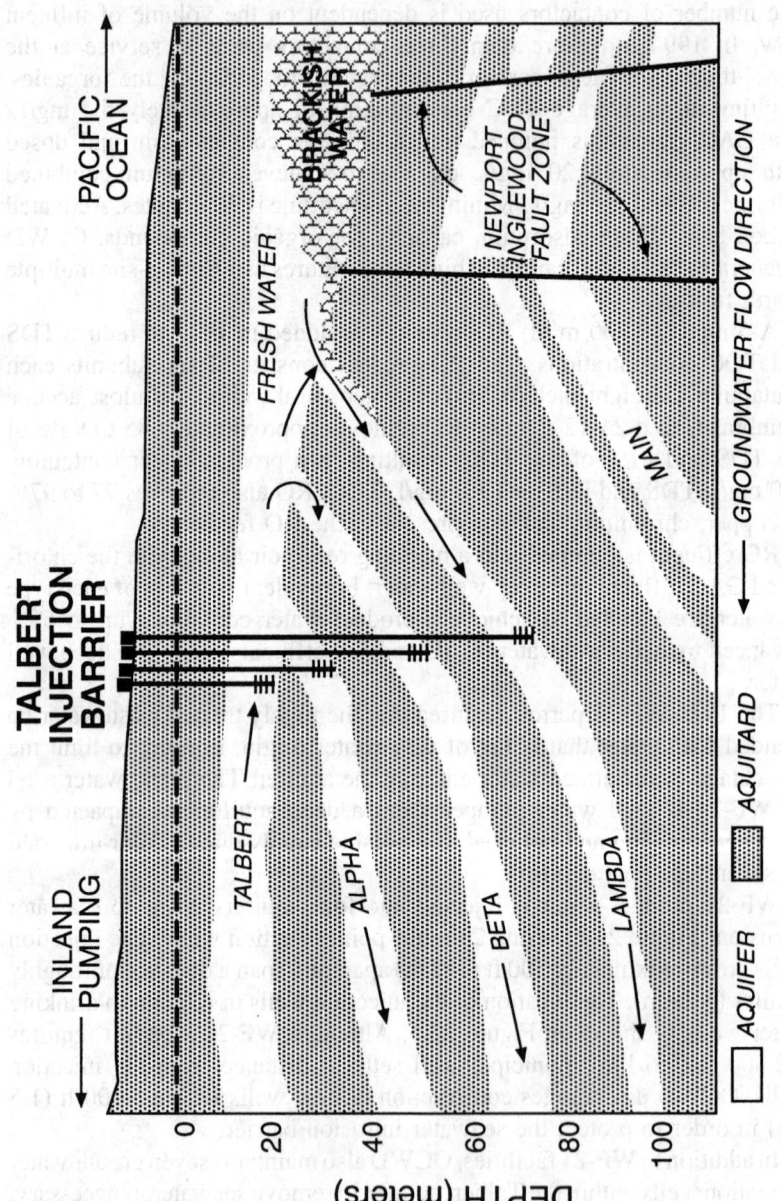

Figure 23.4 Cross-sectional diagram of the Talbert Gap along the southern California coast.

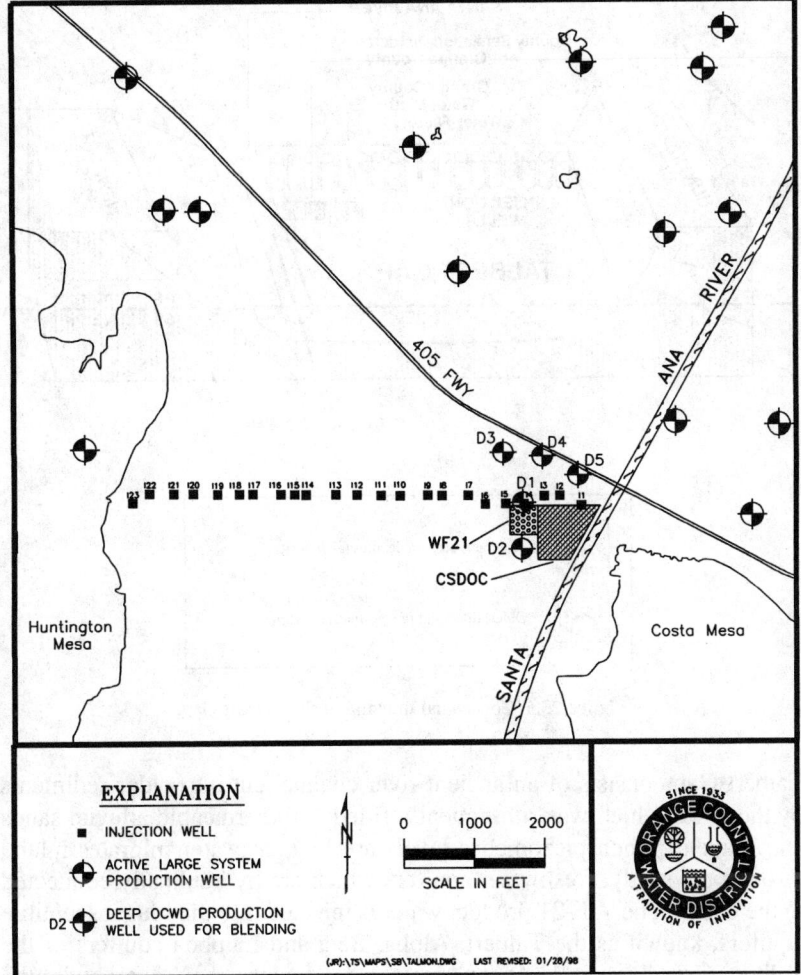

Figure 23.5 Location of WF-21 injection wells and potable (production) wells.

HYDROGEOLOGY OF THE TALBERT GAP

Seawater intrusion into coastal aquifers is not unique to Orange County. Along the California coastline, numerous areas exist where freshwater aquifers are vulnerable to seawater intrusion. Increasing demands on coastal freshwater aquifers for potable water supplies have exacerbated the problem of seawater intrusion. The most significant area of seawater intrusion in Orange County is the 2 mile (3.2 km) wide Talbert (or Santa Ana) Gap at the mouth of the Santa Ana River (SAR, Figure 23.6). The

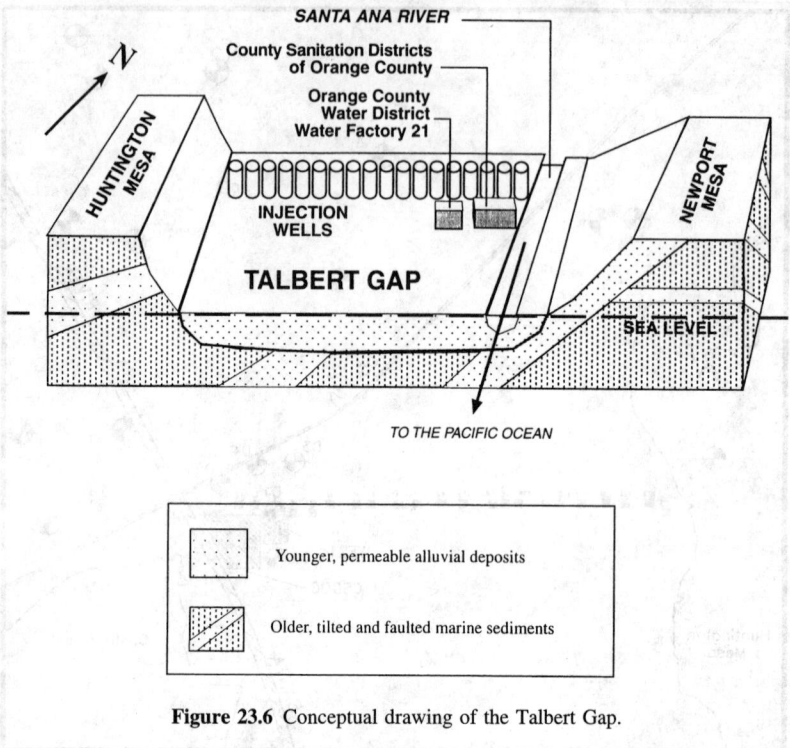

Figure 23.6 Conceptual drawing of the Talbert Gap.

Talbert Gap consists of an ancient river channel cut into older sediments by the SAR, which was subsequently filled with permeable alluvial sands and gravels. When piezometric levels are low, seawater migrates inland through the sandy, freshwater aquifers which are hydraulically connected to the ocean. The WF-21 product water is injected into the four susceptible aquifers, known as the Talbert, Alpha, Beta and Lambda aquifers of the Talbert Gap (Figure 23.4), to create a freshwater mound or hydraulic barrier [10,11]. These aquifers are located under the cities of Huntington Beach and Fountain Valley. Figure 23.5 shows the location of the injection wells, WF-21, and the deep wells used to provide blend water. Since 1976, WF-21 has injected a yearly average of 9,400 af (1.2×10^7 m^3) of water into the aquifers [1].

The total volume of freshwater in the Orange County groundwater basin is not well defined, but is estimated to be several million acre feet (af). There is an estimated 1 million af (1.2×10^9 m^3) of producible water for domestic, industrial, and agricultural uses. However, the operating yield of the basin averages 250,000 afy (3.1×10^8 m^3/y) and varies depending on drought conditions and groundwater recharge supplies. The WF-21 contributes approximately 4% of the total replenishment to the Orange County groundwater basin [11].

OCWD measures water levels monthly at selected monitoring wells to determine groundwater contours and provide information on the effect of direct injection of WF-21 water on the aquifer system. Large fluctuations occur in water levels, especially within the Lambda aquifer, due to seasonal groundwater pumping. The injection barrier increases water levels in the various aquifers by 20 to 40 ft (6–12 m). The effect of the injection barrier on water levels decreases rapidly with increased distance away from the injection wells.

Over 30 monitoring wells around the Talbert Barrier are sampled on a monthly basis to evaluate the effects of WF-21 water on ambient groundwater quality. Monthly monitoring includes measurement of water levels as well as chemical and bacteriological analyses.

Chloride concentrations are used as an indicator of reclaimed water transport or seawater intrusion within the aquifer. Monthly monitoring of chloride concentrations has been used to verify groundwater flow paths that are determined by water level monitoring. Typical injection water has an average chloride concentration of 90 mg/L, in comparison to 20 to 70 mg/L chloride in native groundwater. Chloride concentrations higher than 250 mg/L are indicative of seawater intrusion. In general, chloride concentrations in monitoring wells close to the injection barrier are similar to the injection water quality due to the replacement of pre-existing groundwater with WF-21 derived water.

An isotope tracer study was conducted in 1993 to verify transport time and flow paths of the injected water, as well as determine the age of the groundwater in the various aquifers [12]. WF-21 reclaimed water, deep well water, and ambient groundwater from the different aquifers were tested for isotopes of oxygen, carbon, hydrogen, helium, neon, chlorine, and strontium. Isotopic analyses determined that the travel time for WF-21 injection water to the nearest drinking water well 1,700 ft (518 m) away was approximately 2.3 to 2.7 years [12]. Isotope tracer results were consistent with previous computer modeling studies that estimated injection water travel time to the nearest well at 2 to 3 years.

WF-21 WATER QUALITY MONITORING

Since 1976, WF-21 has produced water that consistently meets or exceeds drinking water standards. The WF-21 multiple barrier treatment process ensures effective removal of various chemicals as well as bacteria, viruses and parasites. Table 23.2 provides selected WF-21 injection requirements and the 1994 average injection water quality values [1,4].

WF-21's monitoring program is summarized in Table 23.3. Extensive testing for inorganic, organic, and bacteriological constituents has been developed not only to meet permit requirements, but also to characterize the performance and reliability of each treatment train process. The WF-

TABLE 23.2.   **WF-21 Permit Requirements and 1994 Average Injection Water Quality.**

Constituent	Selected Permit Requirements[a] (mg/L)	1994 Average Concentrations for Injection Water[b] (mg/L)
Boron	0.7	0.2
Chloride	120	90
Fluoride	1.0	0.5
Total nitrogen	10	3.7
Sodium	115	64
Sulfate	125	40
pH	6.5–8.5	7.4
Total dissolved solids	500	237
Hardness	180	33
Total organic carbon[c]	2.0	1.9
Aluminum	1.00	0.05
Arsenic	0.05	0.002
Barium	1.0	0.006
Cadmium	0.01	0.001
Chromium	0.05	0.001
Cobalt	0.2	0.002
Copper	1.0	0.008
Cyanide	0.2	0.04
Iron	0.3	0.054
Lead	0.05	0.006
Manganese	0.05	0.004
MBAS	0.5	0.06
Mercury	0.002	0.0005
Selenium	0.01	0.005
Silver	0.05	0.001
Zinc	5.0	0.057

[a]Regional Water Quality Control Board Order No. 91–121, adopted November 15, 1991.
[b]Reclaimed water blended with deep well water. Averages based on values above or at the detection limit.
[c]TOC concentration of 2 mg/L becomes effective only when the percentage of reclaimed water injected first exceeds 67 percent and is based on the quarterly average of daily samples.

21 operations have been improved by understanding removal rates for various constituents. For example, the use of double O-rings in the RO process helped reduce leakage through the units.

Studies are underway to increase RO efficiency by using microfiltration (MF) or ultrafiltration (UF) as a pretreatment step. Pilot and demonstration studies indicate that optimizing RO pretreatment will allow higher RO flux rates and improved RO efficiency. OCWD estimates that use of MF as a pretreatment step may increase RO efficiency by as much as 30%.

OCWD conducted a virus monitoring program from 1975 to 1981 as part of the 1971 permit discharge requirements. Virus samples were collected during full-scale operations at WF-21 [13,14]. Activated sludge effluent contained significantly lower levels of enteric viruses than the trickling filter effluent originally used as source water. Therefore, WF-21 receives only activated sludge secondary effluent from CSDOC's Plant No. 1. No viruses have been detected after lime clarification. The high pH achieved during lime clarification (pH 11.2) effectively inactivates viruses. Additional virus removal/inactivation is provided by RO and by chlorine disinfection, establishing a multiple barrier treatment system. No enteric viruses have been detected in the injection water or in any of the monitoring wells. The seven-year monitoring program demonstrated that WF-21 treatment processes effectively remove enteric viruses.

In addition to the virus monitoring program, from 1980 to 1981, the WF-21 treatment processes were also evaluated for the removal of protozoan parasites such as *Giardia*. *Giardia* cysts were detected in 40% of the secondary effluent samples at concentrations ranging from 0.7 to 2 cysts/ 100 gal (380 L) [13]. There were no *Giardia* cysts observed in the RO effluent or the chlorinated GAC effluent. Although parasite cysts tend to be very resistant to disinfection, WF-21's filtration, lime, and RO treatment processes make up an effective multiple barrier system for removal of these organisms.

From 1978 to 1981, an extensive monitoring program was conducted to characterize the organics in WF-21 product water. Effluent from the various treatment processes, as well as groundwater samples from nearby monitoring wells, were evaluated. Analyses included organic priority pollutants, trihalomethane formation potentials, total organic halogens, ultraviolet (UV) absorbance, and specific non-volatile organic compounds. Brominated forms of alkylphenol polyethoxy carboxylates (APECs) were detected at very low concentrations in both chlorinated filter effluent and chlorinated GAC effluent. Additional studies conducted in 1993 determined that APEC compounds are also present at very low concentrations in WF-21 product water and may be a suitable tracer of reclaimed water movement in groundwater. Studies conducted in 1994–1995 have shown that several other groups of anthropogenic compounds, including ethylenediamine tetraacetic acid (EDTA) and nitriloacetic acid (NTA) compounds, may also be useful tracers of organics of wastewater origin [15].

OCWD continues to research removal of organics by the various WF-21 treatment processes. The effect of organics removal by GAC and RO on disinfection byproduct (DBP) formation is being evaluated. Over 18 years of organics research has shown that of the 100 organic priority pollutants analyzed, only 25 are routinely present above detection limits in the secondary effluent feedwater. The organics detected in product water are primarily disinfection byproducts, such as trihalomethanes (THMs).

TABLE 23.3. WF-21 Treatment Train Water Quality Monitoring Schedule.

	Secondary Effluent	Lime Clarified Effluent	Filtered Effluent	GAC Effluent	Chlorinated GAC Effluent	RO Feedwater	RO Effluent	WF-21 Product
Total dissolved solids	WC	WC				WC	WC	WC
pH	MC					MC	MC	7DC
Sodium	MC	MC				MC	MC	WC
Calcium, magnesium, potassium, & trace elements[a]						MC	MC	MC
Ammonia-nitrogen, organic-nitrogen, & total Kjeldahl nitrogen	WG			HWC	HWC	WG	HWC	HWC
Nitrate-nitrogen							MC	MC
Nitrate/nitrite nitrogen	WG			HWC/MC	HWC/MC	MC	HWC/MC	HWC/MC
Carbonate/bicarbonate						WG		WC
Chloride, fluoride, sulfate, hardness, & total alkalinity	MC	MC				MC	MC	MC
Bromide	WC					MC	MC	
Phosphate	MC					MC	MC	WC
Boron	MC						MC	
Silica dioxide								
Total organic carbon	5DC	5DC	5DC	5DC	7DC	MC	7DC	7DC
MBAS	MC						MC	MC
Suspended solids	5DC	5DC						
Color	MC	MC			MC	MC	MC	MC
Cyanide	MG	MG			MG	MG	MG	MC,MG

TABLE 23.3. (continued).

	Secondary Effluent	Lime Clarified Effluent	Filtered Effluent	GAC Effluent	Chlorinated GAC Effluent	RO Feedwater	RO Effluent	WF-21 Product
Residual chlorine					7DG		7DG	7DG
Total coliform	WG	WG	WG		7DG		7DG	7DG
Fecal coliform	WG	WG	WG		4DG		WG	4DG
Turbidity								MC
Radioactivity								QC
Organic chemicals								QG
Inorganic priority pollutants[b]								QC
Organic priority pollutants								QG

[a]Includes: iron, manganese, silver, aluminum, arsenic, barium, cadmium, cobalt, chromium, copper, mercury, nickel, lead, selenium, zinc.

[b]Includes: beryllium, mercury, nickel, antimony, thallium.

C = Composite Sample; 4D = 4-day/week (M-Th); 5D = 5-day/week (Composite: Su-Th, Grab: M-F); 7D = 7-day/week; G = Grab Sample; HMC, HWC = Acid-preserved Composite (collected M or W); Q = Quarterly (January, April, July, and October); M = Monthly; W = Weekly (Wednesday).

In 1994, levels of total THMs in WF-21 injection water were less than 25 μg/L, as compared to the Federal Maximum Contaminant Level (MCL) of 100 μg/L and the proposed MCL of 80 μg/L [1].

FUTURE DIRECTION OF WF-21

As basin groundwater production increases, the injection capacity of the Talbert Barrier, with WF-21 as its supply source, will need to be increased as well. Expansion of WF-21 requires construction of an additional 15 mgd (57,000 m³/d) RO facility to provide for the required organics removal. OCWD continues to evaluate new RO pre-treatment processes and membranes to maximize RO efficiency. These processes are being evaluated as potential alternatives to lime clarification, which has significant space requirements and costs associated with lime sludge disposal and/or recalcining. OCWD is conducting pilot- and demonstration-scale studies on MF and UF as potential RO pre-treatment processes. Use of MF or UF as a pre-treatment step may significantly increase RO flux rates and thereby increase the capacity of RO facilities. Ongoing membrane research by OCWD focuses on methods to verify membrane integrity and the reliability of virus and organics removal. Research is also being conducted on the prevention of biofilm formation to improve membrane performance [16].

OCWD's operation of WF-21 over the past 20 years has demonstrated that highly treated reclaimed water can be successfully used for direct injection into groundwater. Key elements in OCWD's approach to groundwater protection and recharge projects such as WF-21 include monitoring of treatment process performance beyond permit requirements and a commitment to research and development of new treatment technologies. Understanding of treatment process performance and reliability is critical for OCWD's expansion of WF-21 and implementation of new groundwater recharge projects.

GROUNDWATER RECHARGE IN THE ORANGE COUNTY FOREBAY

OCWD operates inland surface spreading basins for groundwater recharge in addition to protection of coastal groundwater by WF-21. Located in the Orange County Forebay, OCWD's surface spreading operations, including the Santiago recharge basins and incidental recharge, are responsible on average for over 95% of the total artificial groundwater replenishment in Orange County [11]. Approximately 250,000 af (3.1×10^8 m³) of water is recharged annually into the groundwater basin by OCWD. During wet years additional storm flows are captured, and the amount of recharge increases to over 300,000 afy (3.7×10^8 m³/y). Management strategies for groundwater recharge include maximizing percolation rates while protecting water quality. Recharge operations allow additional water to be stored within and subsequently pumped from the basin without creating a long-term overdraft condition. OCWD's groundwater recharge operations play a pivotal role in maintaining the groundwater basin as a water resource.

ORANGE COUNTY FOREBAY OVERVIEW

Groundwater recharge has been practiced in Orange County since the early 1900s by maximizing natural infiltration processes along the Santa Ana River (SAR) [17]. Historically, sand dikes were constructed in the SAR to enhance the percolation capacity of the river bottom. Dikes in the river increase the wetted surface area and slowed flow rates to allow for greater infiltration. OCWD instituted a more formalized groundwater recharge program in the 1950s using an off-river channel system and deep spreading basins. Former gravel pits along Santiago Creek were also acquired for groundwater recharge purposes. Pipelines have been constructed to allow for conveyance of water from one basin to another. Two inflatable dams were also constructed in the SAR to allow for additional capture of storm flows. The location of the spreading basins in relation to the SAR is shown in Figure 23.7.

Source water for groundwater recharge in the Orange County Forebay has historically been a blend of water from the Colorado River, northern California, and the SAR. OCWD may purchase imported Colorado River or northern California water, when available, for groundwater recharge. On average 50,000 afy (6.2×10^7 m^3/y) of imported water is purchased by OCWD for groundwater recharge. However, costs for imported water have increased substantially. In addition, over the past 10 years, the availability of imported water has decreased due to reallocation of water for endangered species. In contrast, flow in the SAR has increased. Therefore, the SAR is of critical importance as the primary source of water for replenishment of Orange County's groundwater basin.

The headwaters of the SAR are located in the San Bernardino Mountains, approximately 80 miles (129 km) inland from the coast. Historically, the SAR had perennial flow including snow melt from the San Bernardino and San Gabriel mountains. Diversion of river flows and pumping of upstream groundwaters for agricultural and urban use have significantly reduced natural SAR flows. At present, SAR base flow is primarily derived from tertiary treated wastewater from over eighteen treatment facilities in Riverside and San Bernardino Counties. Without these discharges the SAR would have intermittent flow. The SAR flows are anticipated to continue to increase with further urbanization of Riverside and San Bernardino counties.

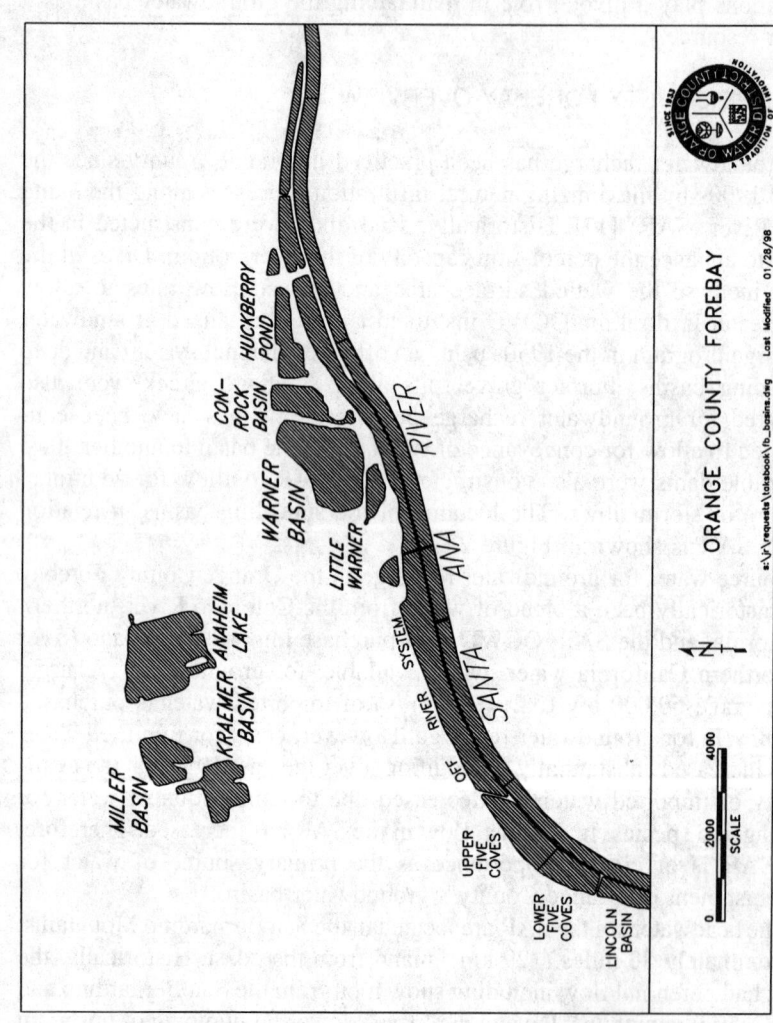

Figure 23.7 Location of Orange County Forebay groundwater recharge basins and the Santa Ana River.

SURFACE WATER SPREADING FACILITIES

OCWD utilizes approximately 1,000 acres (405 ha) of surface water percolation facilities along the SAR in the Orange County Forebay and along Santiago Creek for groundwater recharge. OCWD recharge facilities include sand dikes or levees in the riverbed, an off-river system of shallow basins along the SAR, and deep spreading basins connected by pipelines.

A system of 3 to 5 ft (0.9 to 1.5 m) high sand dikes are constructed by OCWD in the SAR riverbed to enhance percolation. Bulldozers are used to construct the sand dikes in "T" and "L" configurations (Figures 23.8 and 23.9) in approximately six miles (9.7 km) of the SAR riverbed. Sand dikes slow the river flows allowing for greater percolation. However, the "T" and "L" levees wash out during storm events when flows exceed 300 cfs (2,550 m³/s) and must be periodically reconstructed.

The off-river system is a series of shallow ponds and channels used for additional percolation capacity and diversion of river flows to the Burris/Santiago system of spreading basins. The ponds extend parallel to the SAR for approximately four miles (6.4 km). Sand dikes or levees also are maintained in the 300 ft (91 m) wide off-river system. Relatively high percolation rates are maintained in the off-river system as well as in the riverbed, since continuous flows prevent clogging and siltation.

OCWD owns and maintains three deep basin systems for groundwater recharge (Table 23.4). The Warner System includes four basins with storage capacity greater than 4,400 af (5.4 × 10⁶ m³). Selected basins, such as Warner Basin, are stocked seasonally for a fishing concession. The Anaheim/Kraemer System has three of the largest spreading basins, including Miller and Kraemer Basins as well as Anaheim Lake. The Burris/ Santiago System consists of three very deep basins. A series of pipelines allows for flexibility in filling and draining individual basins. For example, SAR water may be pumped to the Santiago basins via a 66 in (1.7 m) diameter pipeline that was constructed in 1990. Deep basins are used to maximize capture of SAR flows for groundwater recharge. The basins can also be filled with imported Colorado River or northern California water via several distribution system turn-outs into the basins or into the SAR.

The spreading basins range in depth from 10 to 120 ft (3 to 37 m) and may hold from 150 to 14,000 af (1.8×10^5 to 1.7×10^7 m³) of water. The total system storage capacity is approximately 25,600 af (3.2×10^7 m³). Estimated percolation rates for individual basins range from 3 to 200 cfs (0.08 to 5.7 m³/s). Percolation rates are dependent on hydrologic and physical conditions such as the depth to groundwater and the accumulation of clogging material on the bottom of the basins. Most basins are cleaned at least once a year to remove the clogging material that eventually reduces

Figure 23.8 "T" and "L" levees in the Santa Ana River.

percolation rates (Figure 23.10). Deep basins must be drained prior to cleaning, due to the use of heavy equipment.

OCWD has conducted research on the nature of the material responsible for clogging the recharge basins [18]. A thin layer of silt, clay and microorganisms forms at the surface/water interface on the bottom of the basins and results in reduced percolation rates. Additional research is underway to evaluate systems for mechanical removal of this low permeability layer without draining the basins.

ORANGE COUNTY FOREBAY HYDROGEOLOGY

The Orange County Forebay is where the majority of natural and artificial groundwater recharge occurs in the Orange County groundwater basin. The Forebay area underlies the cities of Anaheim, Orange, Placentia, and Fullerton. The stratigraphy of the Orange County Forebay

Figure 23.9 Newly constructed "T" and "L" levees in the Santa Ana River.

TABLE 23.4. **Orange County Water District Groundwater Recharge Facilities in the Orange County Forebay.**

Recharge System	Depth (ft)	Area (acres)	Storage (af)	Range of Potential Recharge Rates[a]	
				cfs	afy
Santa Ana River	NA	250	NA	90–200	65,000–145,000
Off-River System	NA	145	NA	15–50	11,000–36,000
Warner System					
Huckleberry Pond	50	21	600		
Con-Rock Basin	50	25	1,000	10–80	7,000–58,000
Warner Basin	50	71	2,600	(system	(system total)
Little Warner Basin	25	11	215	total)	
Anaheim/Kraemer System					
Anaheim Lake	50	72	2,300	40–120	29,000–87,000
Kraemer Basin	50	31	1,030	40–100+	29,000–72,000
Miller Basin	15	21	150	20–60	14,000–43,000
Burris/Santiago System					
Five Coves Basin	10	29	690	3–30	2,000–22,000
Burris Pit	10–80	100	3,000	20–50	14,000–36,000
Santiago Basin	120	200	14,000	100–150	72,000–108,000
Total	NA	976	25,585	340–840	240,000–600,000

[a]Minimum to maximum range of percolation rates under typical conditions.
NA—not applicable.

is characterized by alluvial deposits of unconsolidated sands and gravels with occasional clay and silt lenses. Aquifers in the Forebay area are considered unconfined to semiconfined and have storage coefficients ranging from 10^{-3} to 10^{-1}. Aquifer designations have not been given in the Orange County Forebay since there is a lack of continuous delineating geological features or aquitards.

The aquifer system in the Orange County Forebay deepens and thickens from east to west and ranges from less than 100 ft (30 m) below ground surface (bgs) to over 1,500 ft (457 m) bgs (Figure 23.11). The hydraulic conductivity ranges several orders of magnitude with average values of 200 to 500 ft/d (61 to 152 m/d), as expected for medium to coarse-grained sands. Aquifer transmissivities have been estimated at 10,000 to 100,000 ft²/d (930 to 9,300 m²/d).

Groundwater flow is generally away from the SAR in a westerly direction. Depth to groundwater ranges from 10 to 140 ft (3 to 43 m) bgs and varies seasonally, depending on the volume of water recharged. Recharge operations result in a mounding effect as evidenced by the shallow water levels measured in monitoring wells close to the SAR and spreading basins. Overall, the groundwater elevation and direction of flow are significantly affected by OCWD's recharge operations.

Figure 23.10 Removal of clogging material from groundwater recharge basins.

OCWD maintains and utilizes over 40 monitoring wells in the Orange County Forebay. Screening of wells in specific aquifer zones allows for multi-depth sampling. Many monitoring wells are located adjacent to the spreading basins to evaluate effects of the recharge water on groundwater flow paths and quality.

Numerous production wells in the Orange County Forebay provide potable water for local residents. Production wells are generally screened within the interval from 100 to 1,000 ft (30 to 300 m) bgs, and typical groundwater quality is given in Table 23.5. TDS concentrations exceed 500 mg/L due to groundwater recharge of Colorado River and SAR water, both of which are high in TDS. TOC concentrations in Orange County Forebay groundwater are generally less than 1 mg/L despite recharge of SAR water containing approximately 6–8 mg/L TOC.

An isotopic tracer study was conducted in 1995 by Lawrence Livermore National Laboratory on the Orange County Forebay to determine age of

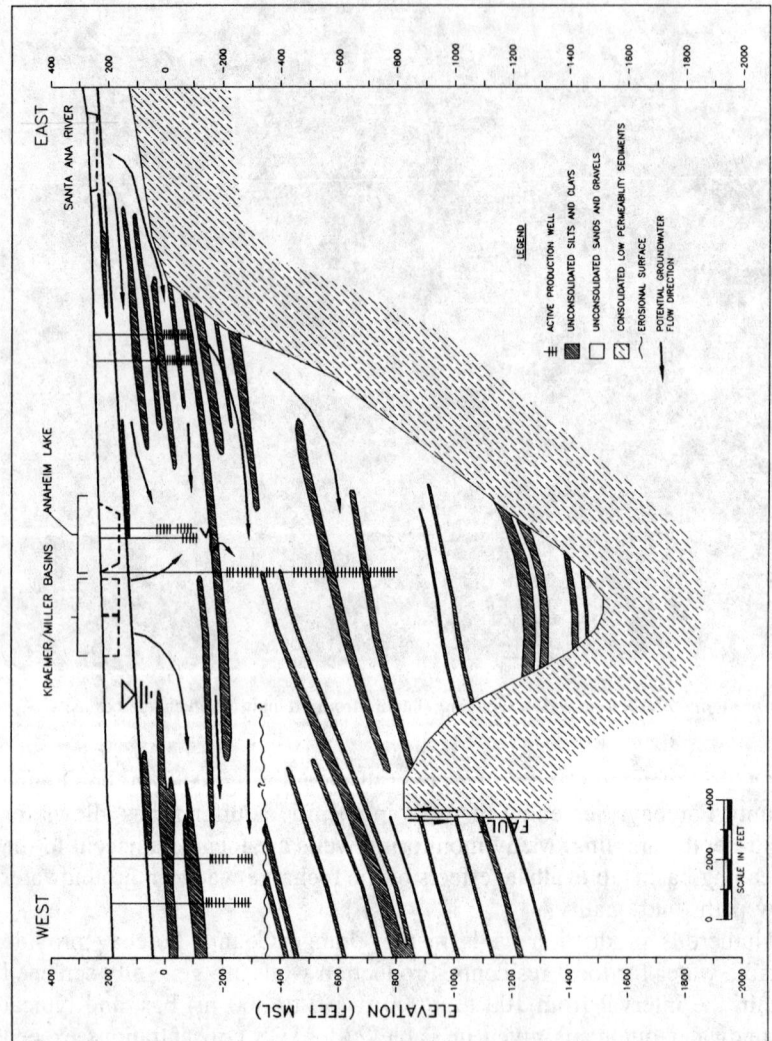

Figure 23.11 Cross-sectional diagram of the Orange County Forebay and groundwater recharge basins.

TABLE 23.5.　Typical Groundwater Quality in the Orange County Forebay.

Parameter	Maximum mg/L	Minimum mg/L	Average mg/L
Turbidity (NTU)	0.7	0.1	0.16
Total organic carbon	1.19	0.70	0.84
Aluminum	0.065	ND	ND
Arsenic	0.008	ND	ND
Barium	0.12	ND	ND
Cadmium	0.002	ND	ND
Cyanide	ND	ND	ND
Mercury	ND	ND	ND
Nickel	0.0065	ND	0.0015
Nitrate-nitrogen	9.27	ND	4.04
Nitrite-nitrogen	ND	ND	ND
Total nitrate/nitrite nitrogen	9.27	ND	4.04
Chloride	117	30	90
Iron	0.13	ND	ND
Manganese	ND	ND	ND
pH	8.5	7.8	8.2
Sulfate	227	77	176
Total dissolved solids	760	302	626
Potassium	9.1	1.8	4.7
Magnesium	32	12	20.5
Sodium	91	34	71
Calcium	148	62	108
Total alkalinity ($CaCO_3$)	221	148	193
Total hardness ($CaCO_3$)	477	204	355

MCL—maximum contaminant level; ND—not detected; NTU—nephelometric turbidity unit.

the groundwater and travel time of recharged water to monitoring and production wells. Monitoring and production wells were tested for isotopes of hydrogen, helium, oxygen, and carbon. Preliminary results showed interference from elevated levels of dissolved gases in the groundwater samples. Given the complexities of the Orange County Forebay aquifer system, a conventional tracer study may be required to define the flow paths and travel times of recharged water.

SANTA ANA RIVER

Water quality of the SAR varies seasonally, depending on the contribution of storm flows and surface runoff. Average water quality for the SAR is given in Table 23.6. SAR water is typically characterized by TDS concentrations above 500 mg/L and nitrogen values ranging from 6 to 8 mg/L. SAR water quality is monitored by the OCWD on a monthly basis for a variety of inorganic and organic compounds.

TABLE 23.6.   Average Santa Ana River Water Quality from 1985–1994.

Constituent	Average Concentration in Santa Ana River Water[a] (mg/L)
Total organic carbon	7.5
Total dissolved solids	629
pH	8.0
Sodium	94
Potassium	10
Magnesium	20
Calcium	81
Ammonia-nitrogen	1.0
Organic-nitrogen	0.8
Total Kjeldahl nitrogen	1.8
Nitrate-nitrogen	6.6
Nitrite-nitrogen	0.1
Total alkalinity	178
Bicarbonate	198
Hardness	286
Fluoride	1.0
Chloride	109
Phosphate	2.5
Sulfate	126
Boron	0.4
Iron	0.046
Manganese	0.111
Aluminum	0.049
Barium	0.063
Silver	0.001
Cadmium	0.0005
Chromium	0.0017
Mercury	ND
Nickel	0.003
Lead	0.0002
Selenium	0.0001
Zinc	0.0014

[a]Orange County Water District, Site No. 5 data, 1985–1994.
ND—not detected.

SAR Base Flows

The majority of the base flows within the SAR originate from discharges in Riverside and San Bernardino Counties. During the summer months over 90% of the flow is derived from tertiary treated wastewater. Over 40 National Pollutant Discharge Elimination System (NPDES) permits are held by wastewater treatment facilities and others that discharge directly or indirectly into the SAR.

In the next 10 to 20 years, SAR flows are expected to increase due to increased wastewater discharges from population growth in Riverside and San Bernardino counties. Wastewater discharged to the SAR is expected to increase from 140,000 afy (1.7×10^8 m³/y), as measured in 1993, to over 150,000 afy (1.9×10^8 m³/y) by the year 2000. Changes in wastewater quality and volume will directly impact OCWD's groundwater recharge project. Approximately 136,000 afy (1.7×10^8 m³/y) of SAR baseflow is recharged into the Orange County groundwater basin.

SAR Storm Flows

Storm flows captured from the SAR by OCWD are estimated to average 34,000 afy (4.2×10^7 m³/y). Depending on the intensity and duration of each storm event, storm flows are captured in the spreading basins. Major storm flows (greater than 500 cfs or 14 m³/s) that cannot be contained in the recharge facilities flow to the Pacific Ocean near Huntington Beach. Unregulated discharges may occur during storm events, including runoff from upstream agricultural areas and high density dairies. These unplanned discharges are of concern to OCWD from a water quality standpoint.

OCWD Wetlands Treatment of SAR Water

Wetlands treatment is becoming increasingly popular for water resources management and improvement of surface water quality. Prado wetlands treatment of SAR water prior to groundwater recharge is advantageous since up to 88% of the nitrate-nitrogen present in the river water is removed (Figures 23.12 and 23.13). In addition, preliminary research has shown that organic carbon compounds may be transformed through biological activities of plants and microorganisms. Currently, OCWD diverts approximately half the non-storm SAR flow into a series of ponds for wetlands treatment [19]. The Prado wetlands consist of a series of 50 shallow ponds ranging in size from several acres to 40 acres (16 ha) and are 1 to 5 ft (0.3 to 1.5 m) in depth. The overall Prado wetlands is about 450 acres and can treat 60 to 85 cfs (1.7 to 2.4 m³/s). Expansion of the wetlands is planned that will increase capacity to 200 cfs (5.7 m³/s).

Removal of nitrate prior to groundwater recharge is important since high nitrate levels in drinking water can cause methemoglobinemia or "blue baby syndrome," which is a serious threat to infants under 6 months of age. The Federal MCL for nitrate-nitrogen in drinking water is 10 mg/L. Proposed California state groundwater recharge regulations also limit total nitrogen to 10 mg/L [7].

Figure 23.12 Prado wetlands.

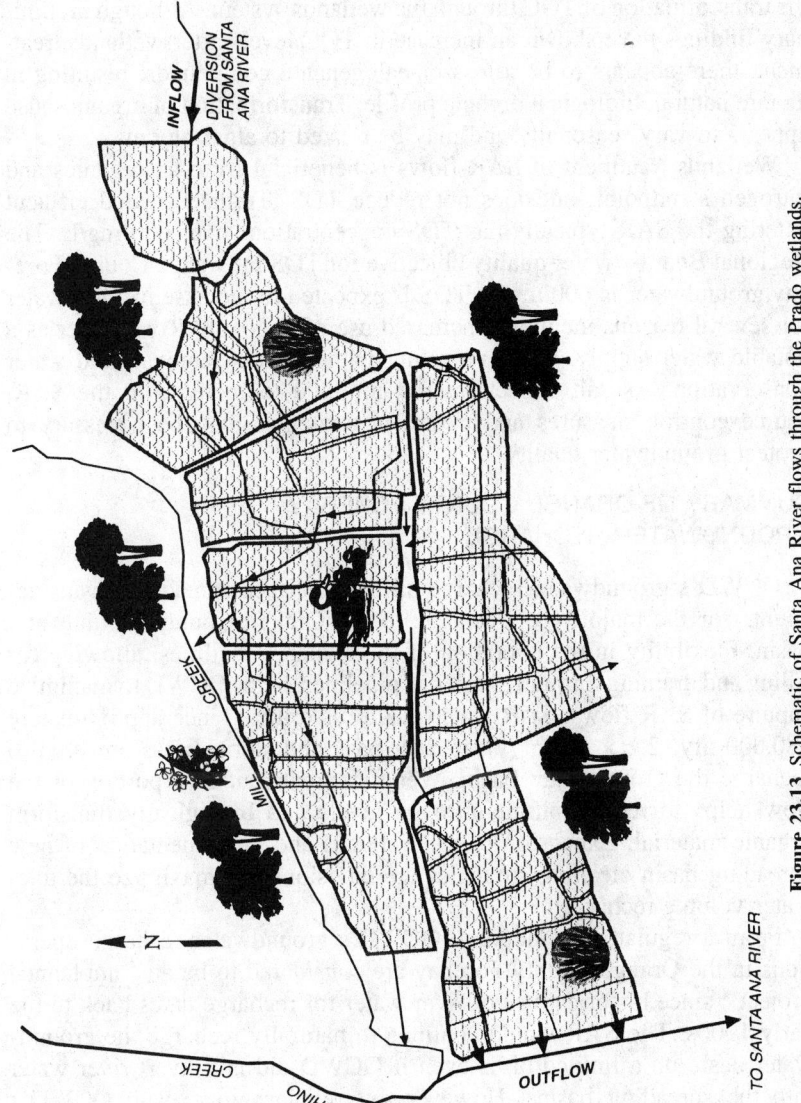

Figure 23.13 Schematic of Santa Ana River flows through the Prado wetlands.

1135

Other constituents such as organic carbon may be biologically transformed within the wetlands system. Research is underway using pyrolysis-gas chromatography/mass spectral analyses (PYGC/MS) to characterize the transformation of TOC through the wetlands system. Although preliminary findings have shown an increase in TOC levels after wetlands treatment, there appears to be a loss of halogenated compounds, resulting in a more natural, biological organic profile. Transformation of organics also appears to vary seasonally and may be related to algal blooms.

Wetlands treatment of SAR flows is beneficial from an organics and nitrogen standpoint, but does not reduce TDS. Tertiary treated effluent entering the SAR typically has TDS concentrations over 600 mg/L. The Regional Board's water quality objective for TDS in Orange County Forebay groundwater is 600 mg/L. TDS is expected to increase in SAR water for several reasons including increased use of Colorado River water as a potable water supply, water recycling without salts removal, and water conservation. As salt concentrations continue to increase in the SAR, source control measures and treatment strategies may be necessary to protect groundwater quality.

SUMMARY OF ORANGE COUNTY FOREBAY GROUNDWATER RECHARGE

OCWD's groundwater recharge utilizing surface spreading basins accounts for the majority of recharge to the Orange County groundwater basin. Flexibility in the operation of the recharge facilities, allowing for filling and draining through a series of pipelines, helps OCWD to maximize capture of SAR flows. SAR water quality is critical since approximately 180,000 afy (2.2×10^8 m^3/y) of river base and storm flows are used to recharge the groundwater basin. Wetlands treatment of a portion of the flow helps to reduce nitrate-nitrogen levels and biologically transform organic material. Expansion of the wetlands and implementation of new spreading basin cleaning strategies are envisioned to maximize the total water volume recharged.

From a regulatory standpoint, OCWD's groundwater recharge operations in the Orange County Forebay are considered to be an ''unplanned project'' since historical use of river water for recharge dates back to the early 1900s. The SAR would continue to naturally recharge the groundwater basin on a limited basis even if OCWD did not divert river water into the spreading basins. However, future plans to expand OCWD's recharge operations may require state and regional permitting.

OPTIMIZATION OF OCWD GROUNDWATER RECHARGE PROJECTS

Groundwater recharge is a critical part of water resources management in southern California and an integral part of OCWD's groundwater management plan. Continued population growth and recurrent drought conditions intensify the need for a reliable, drought-proof water supply. Although OCWD recharges almost 250,000 afy (3.1×10^8 m³/y), additional recharge of local resources needs to be expanded to meet increasing water demands.

The lack of a reliable water supply can have a significant economic impact on Orange County as determined in a 1991 survey of California industries [20]. The survey was conducted to determine the long-term impact of a water shortage on the state's industries by county. It was estimated that prolonged drought conditions in Orange County could result in the loss of 1.5% of manufacturing production, valued at $400 million. If a 30% cut-back in water supplies to Orange County industries occurred, the loss of employment was estimated to be 1.9% or 4,800 jobs. Other losses due to a water shortage include decreased golf course and park revenues, as well as diminished productivity in commercial agriculture. A firm local supply is also critical for disaster preparedness. Imported water supplies rely upon extensive transmission systems that may be damaged during a major earthquake.

OCWD is investigating the feasibility of expanding its groundwater recharge capabilities utilizing existing spreading basins that are underutilized, especially in the summer months. Water resources investigations by OCWD have shown that an average of 182,000 afy (2.2×10^8 m³/y) of unused percolation capacity exists, mostly during Summer and Fall. Use of existing surface spreading basins can be maximized by recharging additional water and improving methods of periodically cleaning the basins.

The benefits of optimizing groundwater recharge include:

(1) Increasing the *reliable* long-term water supply for sustainable economic development

(2) Providing comprehensive water resources management at a local level

(3) Maximizing the use of Orange County's groundwater basin for storage to meet seasonal demands and drought conditions

WATER QUALITY OBJECTIVES FOR FUTURE GROUNDWATER RECHARGE

Future OCWD groundwater recharge projects would require permitting based on a combination of the California Code of Regulations, Title 22 requirements [21], proposed DHS groundwater recharge criteria [7], and the Regional Board's water quality objectives [6]. A comparison of water quality standards set forth by the two regulatory agencies are given in Table 23.7. DHS criteria were developed to protect public health, whereas

TABLE 23.7. **Regulatory Requirements for Groundwater Recharge Projects in California.**

Constituents (mg/L)	DHS Title 22 Reclamation Criteria	Regional Board Groundwater Basin Objectives[a]	Regional Board Groundwater Basin Objectives Plus Mineral Increment[b]
Turbidity	2 NTU	NA	NA
Microbiological	< 2.2 total coliforms per 100 mL	NA	NA
Sodium	NA	60	130
Chloride	250[c]	65	130
Sulfate	250[c]	120	160
Total dissolved solids	500[c]	600	600
Total organic carbon	2–20[d]	NA	NA
Hardness	NA	290	320
Nitrate-nitrogen	10	3	3

[a]Water Quality Control Plan, Santa Ana River Basin, 1994, Regional Water Quality Control Board, Santa Ana Region Number 8, pp. 4–45. Groundwater quality objectives set based on the Porter-Cologne Act (§13050(h)) to "ensure the reasonable protection of beneficial uses and the prevention of nuisance within a specific area."
[b]Water Quality Control Plan, Santa Ana River Basin, 1994, Regional Water Quality Control Board, Santa Ana Region Number 8, pp. 5–14, Section 2b, Mineral Increments.
[c]California Department of Health Services (DHS) drinking water standards, maximum recommended level.
[d]Maximum allowable TOC concentration based on percent contribution of reclaimed water. 1993 Proposed Groundwater Recharge Regulations, California Department of Health Services.
NA—not applicable; NTU—nephelometric turbidity units.

the Regional Board standards are to prevent degradation of the watershed and to protect beneficial uses of surface and groundwaters. Currently, DHS is revising its groundwater recharge requirements.

Current Wastewater Reclamation Criteria were established by the DHS in 1978, and are a part of Title 22, California Code of Regulations [21]. These criteria require case-by-case review and approval of groundwater recharge projects that use reclaimed water. Proposed DHS regulations define the treatment requirements and performance standards for planned groundwater recharge projects (Table 23.8) [7]. The purpose of the proposed regulations is to provide regulatory criteria for all recharge projects, thus reducing the need to evaluate projects on a case-by-case basis. Site requirements in the draft Title 22, Article 5.1 groundwater recharge regulations propose that not more than 50% of the water extracted from a well may be of reclaimed water origin. An underground retention time of at least 6 months is proposed with a minimum 500 ft (152 m) horizontal separation between the spreading area and the nearest well. Minimum depth to groundwater is also specified based on the initial rate of percolation. For

TABLE 23.8. **Proposed DHS Title 22, Article 5.1 Groundwater Recharge Requirements.**[a]

| Project Category | Surface Spreading | | | Direct Injection |
	I	II	III	IV
Maximum percent reclaimed water in potable water well	50%	20%	20%	50%
Minimum depth in feet to groundwater, at initial percolation rate of:				
< 0.2 inches per minute	10	10	20	NA
< 0.3 inches per minute	20	20	50	NA
Retention time underground in months	6	6	12	12
Horizontal separation[2] in feet	500	500	1,000	2,000
Treatment Required:				
Primary	X	X	X	X
Secondary	X	X	X	X
Filtration	X	X		X
Disinfection	X	X	X	X
Organics removal	X			

[a] Draft Proposed Groundwater Recharge Regulations, March 22, 1993, California Department of Health Services, p. 10, California Code of Regulations, Title 22.
[b] Minimum distance between recharge facility and point of extraction.
NA—not applicable.

example, at a percolation rate of greater than 0.2 in/min (5 mm/min) but less than 0.3 in./min (7.6 mm/min), a 20 ft (6 m) depth to groundwater is required for a project recharging 50% reclaimed water.

DHS's concern over the unknown long-term health effects from the consumption of organics of wastewater origin has led to the proposed limit for TOC. Over 85% of the organic constituents in treated wastewater are unidentified and a small percentage of these compounds may be mutagenic and/or carcinogenic. Therefore, organics removal criteria were set so that the concentration of TOC of wastewater origin does not exceed 1 mg/L in groundwater taken from the nearest well. The contribution of organics to groundwater in the proposed regulations is set by limiting the amount of reclaimed water recharged to a groundwater basin and also by limiting the amount of TOC in the reclaimed water. DHS accepts that TOC removal occurs during percolation through unsaturated soil. TOC removal has also been observed under saturated conditions during injection of WF-21 product water. Additional hydrogeological studies and computer modeling of the Forebay are needed to determine the range of TOC removal during groundwater recharge operations.

The Regional Board's water quality requirements for groundwater recharge are based upon the 1995 Santa Ana River Basin Water Quality Control Plan, or Basin Plan [6]. The purpose of the Basin Plan is to protect and enhance quality and beneficial uses of water within the region. The Basin Plan provides physical and chemical limits on discharge water so that projects and activities within the area do not adversely impact groundwater and surface water. At present, constituents of primary concern to the Regional Board are TDS, nitrogen, and metals discharged by wastewater treatment facilities. Factors considered in establishing water quality objectives include the following [6]:

(1) Past, present, and future beneficial uses
(2) Environmental characteristics of the hydrographic unit, including quality of water available
(3) Water quality conditions that could reasonably be achieved through the coordinated control of all factors that affect water quality in the area
(4) Economic considerations
(5) Need for developing housing in the area
(6) Need to develop and use recycled water
(7) Historic and present water quality data
(8) Antidegradation policies

SUMMARY OF OCWD GROUNDWATER RECHARGE PROJECTS AND LONG-TERM PLANNING

OCWD's long-term planning for groundwater recharge includes expansion of WF-21 and maximizing surface spreading capabilities in the Orange County Forebay. As OCWD anticipates implementation of future groundwater recharge projects, source water protection is of primary concern. Increasing TDS concentrations in imported water supplies and wastewater discharges is problematic and may require development of local ordinances to reduce salt contributions from various sources. Ultimately a greater portion of the source water for potable supplies may require demineralization. Enhancing source water quality of the SAR will help provide high quality water for groundwater recharge. Expansion of the Prado wetlands to increase the efficiency of nitrogen removal and transformation of organic compounds is also an integral part of long-term groundwater recharge plans.

OCWD's tradition of innovation has resulted in the successful implementation of WF-21 and groundwater recharge operations in the Orange County Forebay providing nearly 250,000 afy (3.1×10^8 m³/y) of additional recharged groundwater. These projects have provided scientific support for the safe and beneficial use of highly treated reclaimed water for groundwater recharge. OCWD's long-term commitment to research and development has fostered new ideas for improved methods of groundwater recharge and protection of coastal groundwater supplies. Implementation of extensive monitoring programs, beyond regulatory requirements, has proven reliability and performance of advanced wastewater treatment technologies. OCWD's years of experience in groundwater recharge in conjunction with evaluation of new or improved treatment technologies form the basis for development of new projects to maximize groundwater recharge efforts and provide Orange County with a reliable, high quality groundwater supply.

REFERENCES

1 Orange County Water District (1995) Annual Report Orange County Water District Wastewater Reclamation, Talbert Barrier and Recharge Project, Calendar Year 1994.

2 Wesner, G. M. (1987) Historical Review of Water Factory 21 Orange County Water District Wastewater Reclamation and Groundwater Recharge Program. Report to the OCWD.

3 California Regional Water Quality Control Board, Santa Ana Region. Order No. 71-27, Waste Discharge Requirements for Orange County Water District, Water Reclamation and Seawater Intrusion Barrier Project, Orange County. August 26, 1971.

4 California Regional Water Quality Control Board, Santa Ana Region. Order No. 91-121, Waste Discharge Requirements for Orange County Water District, Water Reclamation and Seawater Intrusion Barrier Project, Orange County. November 15, 1991.

5 California Water Code, Division 7, Chapter 7, Article 6, Section 13540.

6 California Regional Water Quality Control Board Region 8 (1994) Water Quality Control Plan, Santa Ana River Basin.

7 State of California (1993) Draft Proposed Groundwater Recharge Regulations. State of California Department of Health Services, Division of Drinking Water and Environmental Management.

8 Crook, J., Herndon, R. L., Wehner, M. P., and M. G. Rigby. (1995) Studies to Determine the Effects of Injecting 100 Percent Reclaimed Water From Water Factory 21. *Conf. Proc. WEFTEC '95,* 68th Annual Conference, Vol. 6. Miami Beach, FL.

9 County Sanitation Districts of Orange County (1993) Annual Report.

10 California Department of Water Resources (1961) Groundwater Basin Protection Projects: Santa Ana Gap Salinity Barrier, Orange County, Bulletin No. 147-1.

11 Wesner, G. M. and R. L. Herndon (1990) Engineering Report. Orange County Water District Reclamation and Seawater Intrusion Barrier Project. Orange County Water District, Fountain Valley, CA.

12 Hudson, G. B., Davisson, M. L., Velsko, C., Niemeyer, S., Esser, B. and J. Beiriger (1994) Preliminary Report on Isotope Abundance Measurements in Groundwater Samples from the Talbert Injection Barrier Area, Orange County Water District. Lawrence Livermore National Laboratory, Nuclear Chemistry Division, Livermore, CA.

13 J. M. Montgomery Engineers (1981) Water Factory 21 Environmental Virus and Parasite Monitoring.

14 J. M. Montgomery Engineers (1979) Water Factory 21 Virus Study.

15 Ding, W. H., Fujita, Y., Wu, J., Semadeni, M. and M. Reinhard. (1996) Annual Report. Behavior and Fate of Organic Contaminants During Groundwater Recharge with Reclaimed Wastewater and Santa Ana River Water—A Field and Laboratory Investigation. Submitted to the OCWD.

16 Ridgway, H. R. and H. C. Flemming. (1996) Chapter 6. "Membrane Biofouling," *Water Treatment Membrane Processes.* McGraw Hill Publishers.

17 Orange County Water District and County Sanitation Districts of Orange County (1995) Orange County Regional Water Reclamation Project, Final Feasibility Study Report.

18 Orange County Water District (1987) Deep Basin Clogging and Infiltration Mechanics Report.

19 Orange County Water District (1995) Engineers Report on the Prado Constructed Wetlands Modification Project.

20 Spectrum Economics Inc. (1991) Cost of Industrial Water Shortages: Preliminary Observations. Report to the California Urban Water Agencies.

21 California Code of Regulations, Title 22, Division 4, Chapter 3, Reclamation Criteria, (1978).

CHAPTER 24

Wastewater Reclamation and Reuse for Cooling Towers at the Palo Verde Nuclear Generating Station

INTRODUCTION

THE Palo Verde Nuclear Generating Station is the energy cornerstone of the Southwest. It is the largest facility for the peaceful use of nuclear power in America. Palo Verde is a standardized triple-unit commercial nuclear power facility. It consists of three identical pressurized water reactors and turbine-generators. Each unit generates 1,270,000 kilowatts of electricity—a total of 3,810,000 kilowatts.

The location selected for the Palo Verde Nuclear Generating Station (PVNGS) is in the arid Sonoran Desert 55 miles west of Phoenix, Arizona. The average rain fall is 7 inches per year. Unlike nearly all power plants, there is not a natural body of water located near PVNGS. There is no ocean, lake, or river in the vicinity.

Every steam-electric plant must have a dependable supply of water to make steam and cool equipment. PVNGS is unique because it's the only nuclear energy facility in the world that uses treated sewage effluent as a source of water for cooling tower operation. PVNGS buys treated wastewater from local cities and then further treats it at the onsite Water Reclamation Facility (WRF). Reuse of wastewater is both environmentally sound and a substantial economical benefit for nearby communities. For reactor coolant and steam, Palo Verde uses demineralized well water.

The PVNGS nuclear units are pressurized water reactors. The process

Daniel E. Blackson and Jerald L. Moreland, Water Reclamation Facility, Arizona Public Service Company, Palo Verde Nuclear Generating Station, P.O. Box 52034, Phoenix, AZ 85072-2034.

depicted in Figure 24.1 is a three loop system: a primary loop, secondary loop, and cooling water loop. The primary loop consists of the reactor, reactor coolant pumps, and the tube side of the steam generators. Major components in the secondary loop are: shell side of the steam generator, turbine, condenser, and pumps. The condenser cooling water loop is composed of the condenser (tube side), cooling towers, and pumps. Make-up water for the cooling water loop is pumped from the reclaimed sewage effluent storage reservoir.

Seven Southwestern utilities own PVNGS and share in the electricity it generates (see Table 24.1). They also share all costs in proportion to their ownership. The largest owner—Arizona Public Service Company—holds the Nuclear Regulatory Commission licenses and operates Palo Verde for itself and the other co-owners.

BACKGROUND

The cooling requirement for the three nuclear generating units is approximately 560,000 gallons per minute per unit. That equates to 1.7 million gallons per minute for all three units operating at 100% power. In all of western Arizona there is not a water source large enough for a once-through cooling system with the possible exception of the Colorado River. The Colorado River water has many different prior appropriations in the western United States. Another option is the Central Arizona Project (CAP). The CAP is a 300 mile canal supplying Colorado River water to the Phoenix and Tucson metropolitan areas. This water is not only totally appropriated, but will not supply enough water to meet the PVNGS operational needs. Groundwater in the arid desert was ruled out because the aquifers in the Phoenix area were already experiencing a decline due to agriculture and municipal usage. Therefore, the designers of the facility had to look elsewhere for a water source.

The City of Phoenix 91st Avenue Wastewater Treatment Plant (WWTP) is located about 38 miles east of the PVNGS project site and has a daily flow rate of approximately 130 MGD. The effluent from this plant was typically discharged into the normally dry Salt River bed and had few prior water rights claims on the flow. As this volume of water is not enough to support a once-through cooling system, it is, therefore, necessary to cycle the water through the system several times. If the water is to be cycled through a cooling tower system, then the make-up water supply must reflect the greatest water demand loss—evaporation. In the summer months as much as 15,000 gallons per minute is lost to evaporation per each nuclear unit in operation at 100% power. The make-up supply of

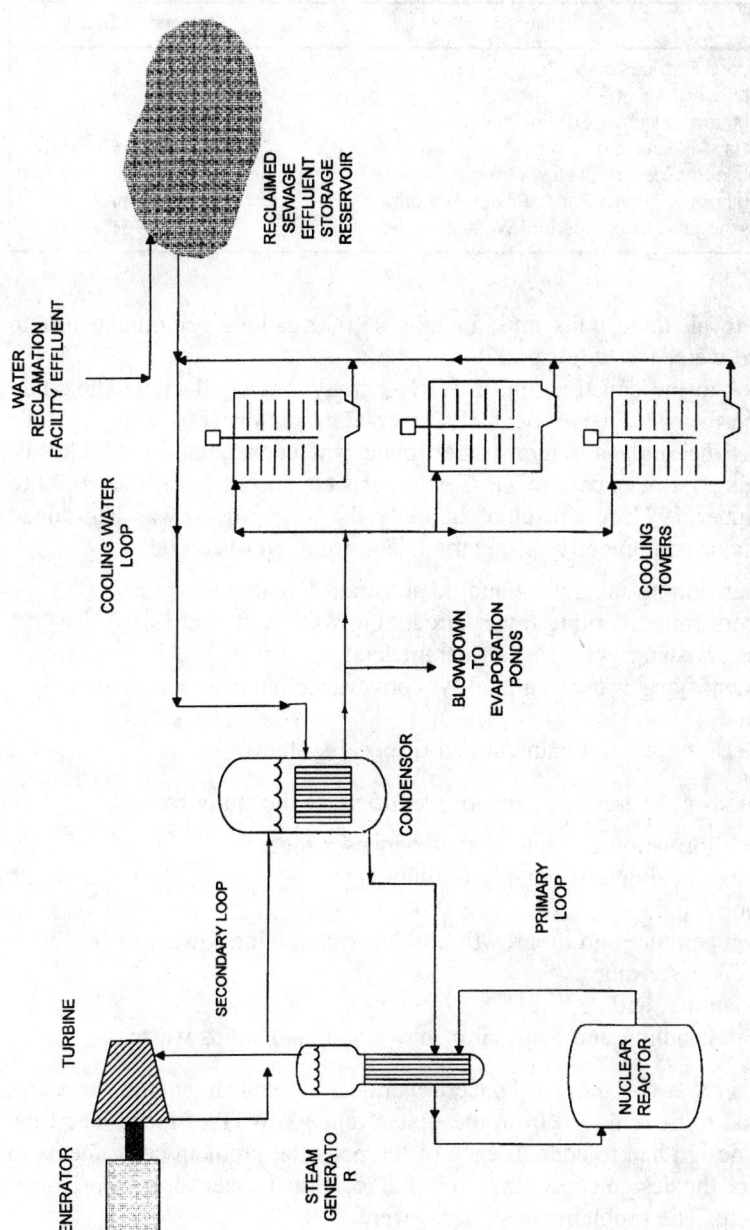

Figure 24.1 PVNGS flow diagram.

1145

TABLE 24.1. **Co-owners of Palo Verde Nuclear Generating Station.**

Utility	Ownership Share
Arizona Public Service Co.	29.1%
Salt River Project	17.5%
Southern California Edison Co.	15.8%
El Paso Electric Co.	15.8%
Public Service Co. of New Mexico	10.2%
Southern California Public Power Authority	5.9%
Los Angeles Department of Water & Power	5.7%

water to all three units must include 45,000 gallons per minute just to replace water lost to evaporation.

Prior to the construction of WRF a study was conducted. The study was performed by the Bechtel Power Corporation (Los Angeles). To support the study a demonstration plant was constructed at the City of Phoenix 91st Avenue WWTP. Testing was conducted from June, 1973 to September, 1974. As a result of the study, the WRF process was determined and major equipment was specified. The study consisted of

- performing laboratory and demonstration plant test studies;
- performing cooling tower circulating water test studies;
- establishing WRF design parameters;
- identifying water availability, conveyance, and storage systems; and
- defining waste treatment and disposal systems.

The specific areas of concern identified by the study were

- the formation of scale in the condenser tubes;
- biological and/or organic fouling;
- corrosion;
- temperature and biogrowth problems in cooling towers or in condenser tubes;
- foaming; and
- the handling and disposal of blowdown and solids waste.

Related to the areas of concern were certain constituents in the water that had to be removed from the 91st Avenue WWTP effluent. The final WRF design had to address each of the potential problem constituents to produce the design concentration of the reclaimed water for cooling tower make-up. The problem constituents were

- ammonia,
- alkalinity,
- biochemical oxygen demand,

- calcium,
- magnesium,
- phosphorus,
- silica,
- sulfate, and
- suspended solids.

The problem constituents were addressed by the study and a process was designed to establish the concentrations for cooling tower operations at 15 cycles. This resulted in defining WRF effluent specifications which became the operating parameters. They are described in Table 24.2.

Typically reuse water treatment includes settleable and suspended solids removal. To meet the cooling demands of the PVNGS it is necessary to cycle the water up to 15 times through the cooling system. This means the reuse treatment system must remove dissolved solids, specifically, calcium, magnesium, silica, and phosphates due to their scaling tendencies.

The most cost effective method of treatment for dissolved solids was a cold lime/soda ash softening system. The pilot plant was equipped with two stage solids contact reactors, gravity filters and a small condenser where the effluent of the treatment system was cycled through a cooling system. This cooling system water was then monitored for its solids concentrations and scaling tendencies to achieve the highest recycle rate without significant damage to the cooling system.

Also included in these studies were treatment schemes for the removal of organic loading and reduction of alkalinity. The treatment systems tested included rotating biological discs, aeration basins and trickling filters. The treatment with the lowest cost turned out to be trickling filters for their ammonia and alkalinity reducing capabilities. Reducing the ammonia content allowed for a lower lime dosage rate and reducing the alkalinity made for easier pH control.

The final design of the Water Reclamation Facility, based on the findings of the pilot studies, was as follows: trickling filters containing plastic

TABLE 24.2. **WRF Effluent Quality Specifications.**

Parameter	Maximum Value	Unit
Suspended solids	10	mg/L
Turbidity	15	T.U.
Orthophosphate	0.5	mg/L
Total calcium (as $CaCO_3$)	70	mg/L
Alkalinity (as $CaCO_3$)	100	mg/L
Silica (as SiO_2)	10	mg/L
Magnesium (as $CaCO_3$)	10	mg/L
Ammonia	5	mg/L

media with 100% recycle capability, first and second stage solids contact clarifiers where lime is injected into the first stage and soda ash and carbon dioxide is used in the second stage, and gravity filters for further suspended solids removal.

The solids generated by this treatment system are gravity thickened and then processed through dewatering centrifuges where the solids are either disposed of in an onsite landfill or recycled through a recalcination furnace and reused as lime.

The cooling tower design includes an induced draft design to take full advantage of heat transfer. Blowdown from the cooling towers is stored in onsite evaporation ponds allowing the entire plant to be zero discharge. Figure 24.2 shows a birds-eye view of the wastewater reclamation plant for the cooling tower operations at the Palo Verde Nuclear Generation Station.

PROCESS DESCRIPTION

The treated municipal wastewater effluent is processed through a multi-stage biochemical treatment operation (see Figure 24.3). First, trickling filters biologically treat the effluent. Second, effluent is chemically treated in the solids contact clarifiers with cold lime, soda ash, and carbon dioxide. Disinfection occurs when chlorine is added prior to the gravity filters where the final step of treatment, filtration occurs. The treated water flows into the 80 acre storage reservoir to be used as cooling tower make-up.

The effluent from the 91st Avenue WWTP is delivered to the Water Reclamation Facility through 36 miles of underground steel reinforced concrete pipe. The first section of the pipe is 114 inches inside diameter and serves as a storage section. This storage section allows for continuous flow to the WRF while the WWTP experiences diurnal flows in a 24 hour period. The majority of the pipe is 96 inches in diameter. The first twenty-eight miles of pipe is gravity flow to the Hassayampa Pump Station where five two speed, multi-stage pumps pump the water the remaining eight miles. The pressure flow section of pipe is 66 inches in diameter.

The first treatment process at the WRF is the trickling filters. The main purpose of the trickling filters is to reduce ammonia concentrations. Water enters the trickling filters through individual rotary distributors located above the top of the trickling filter media. The rotary distributors are driven by water pressure from the Hassayampa Pump Station pumps. The distributors rotate and evenly spray water across the top of the media. As the water trickles down through the plastic media, the natural occurring bacteria will convert the ammonia to nitrate in two steps.

$$NH_4^+ + 3/2\ O_2 \Rightarrow NO_2^- + 2H^+ + H_2O$$

Figure 24.2 On-site wastewater reclamation plant for the cooling tower operations at the Palo Verde Nuclear Generating Station (photo courtesy of Palo Verde Strategic Communications).

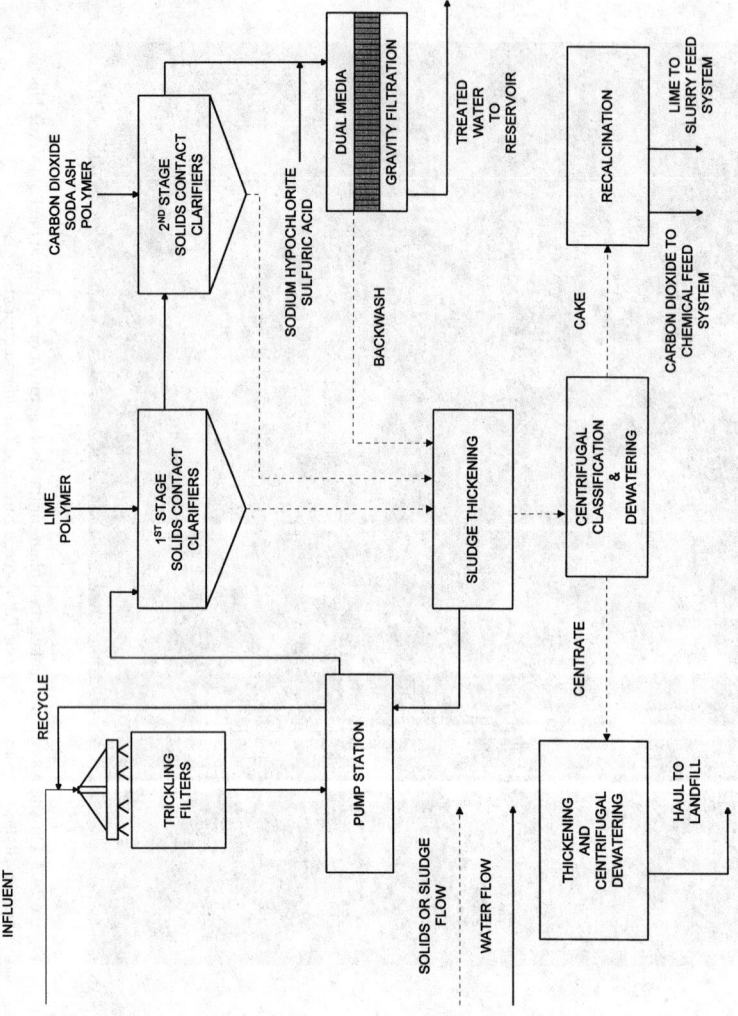

Figure 24.3 WRF flow diagram.

1150

$$NO_2^- + 1/2\ O_2 \Rightarrow NO_3^-$$

The combined reaction is the generation of nitric acid (HNO_3) from the oxidation of ammonia.

$$NH_4^+ + 2\ O_2 \Rightarrow HNO_3 + H_2O + H^+$$

The generation of nitric acid is important for the destruction of alkalinity. Nitric acid will react with bicarbonates in the effluent. The resulting destruction of alkalinity will reduce lime demand in the first stage solids contact clarifiers. If less lime is needed to treat the effluent, less solids will be produced and processed.

$$Ca(HCO_3)_2 + 2(HNO_3) \Rightarrow Ca(NO_3)_2 + 2CO_2 + 2H_2O$$

The trickling filters are 30 feet high and 125 feet across. They are filled with a synthetic media to which the bacteria can attach themselves. Ventilation fans are provided to circulate air down through the media for cooling and to supply oxygen for the bacteria. An additional set of pumps are available to recirculate water back through the trickling filters to achieve and maintain an effluent concentration of 5 mg/L or less of ammonia.

The water is collected at the bottom of the trickling filters and pumped to the first stage solids contact clarifiers. The first stage clarifiers are 140 feet across and about 21 feet deep. An external recirculation pump is provided for a constant hydraulic loading rate during periods of low plant flow. Sludge pumps for solids removal are located on the platform over the clarifier.

Lime slurry is injected into the effluent at the very center of the first stage solids contact clarifier. Provisions for polymer addition are also available at this process point. The mixture then discharges over the top of the mixing zone into the reaction zone. A detention time is designed into the reaction zone to prolong the contact time for chemical reactions. Constant agitation is provided by a gear driven turbine. The water from the reaction zone flows downward. A large volume returns in an upward motion back to the mixing chamber. The remaining water flows outward under the shroud and into the clarification compartment. Here a sludge blanket is formed by solids from the chemical reactions. The water is clarified as it moves up through the sludge blanket. The effluent is then collected in lateral launders and conveyed to the second stage solids contact clarifiers. Precipitated solids (sludge) is raked to the center cone at the bottom of the clarifier and pumped to the solids/liquids system (see Figure 24.4).

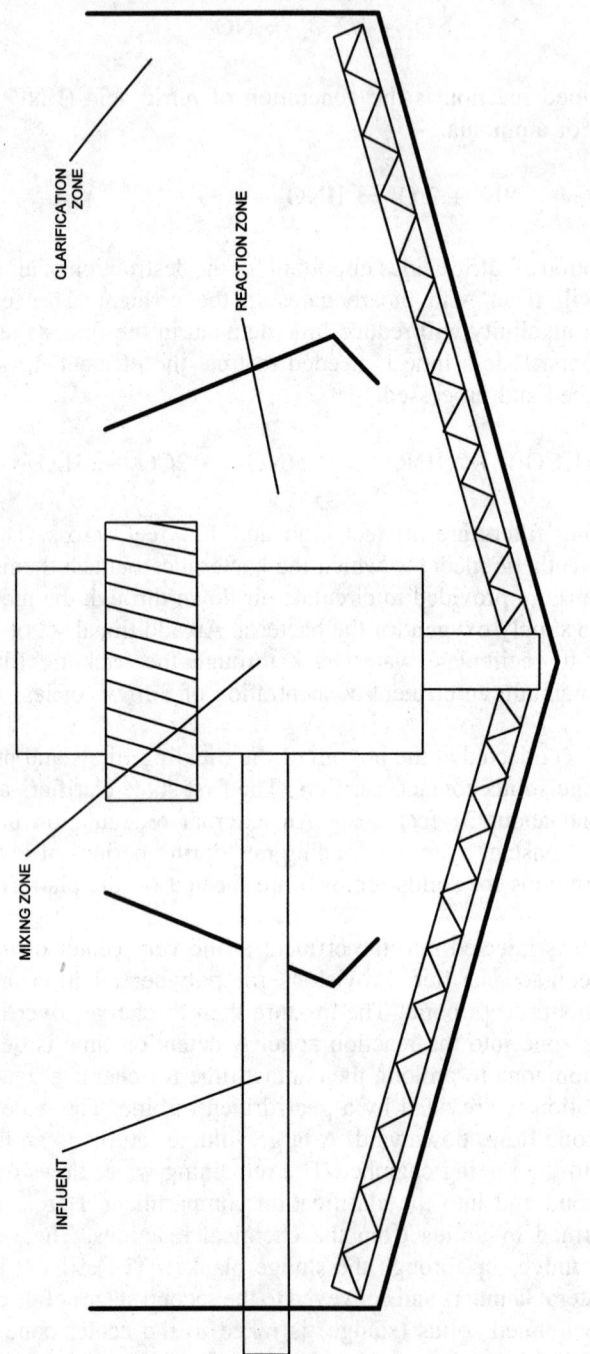

Figure 24.4 Solids contact clarifier.

The desirable chemical reactions in the first stage clarifiers are reactions with lime to remove calcium, phosphates, magnesium, and silica. The first stage clarifiers are maintained at a pH range of 11.0 to 11.3 to encourage the following reactions:

$$Ca(OH)_2 + CO_2 \Rightarrow CaCO_3\downarrow + H_2O$$

$$Ca(OH)_2 + Ca(HCO_3)_2 \Rightarrow 2CaCO_3\downarrow + 2H_2O$$

$$1/2\ Ca(OH)_2 + NH_4^+ \Rightarrow 1/2\ Ca^{+2} + NH_3 + H_2O$$

$$5Ca(OH)_2 + 3HPO_4^{-2} \Rightarrow Ca_5(OH)(PO_4)_3 + 6\ OH^- + 3H_2O$$

$$Ca(OH)_2 + Mg^{+2} \Rightarrow Ca^{+2} + Mg(OH)_2\downarrow$$

$$Mg(OH)_2 + SiO_2 \Rightarrow [Mg(OH)_2]SiO_2\downarrow$$

The next step of treatment occurs in the second stage solids contact clarifiers. The second stage clarifiers are basically identical to the first stage except they are slightly smaller. These clarifiers are 125 feet across and almost 22 feet deep. Soda ash and carbon dioxide are injected at this stage of the process. Provisions for polymer addition are also available. Since a more concentrated, denser sludge is produced, the sludge pumps are located in a tunnel directly under the clarifiers.

Soda ash is added to the center of the clarifier similar to lime slurry in the first stage clarifier. Soda ash supplies excess carbonate to the water to encourage the formation of calcium carbonate. The effluent flows through a mixing zone, reaction zone, and clarification zone within the clarifier.

Carbon dioxide is used to reduce the pH from 11.0 to 10.0. It is injected through diffusers into the effluent in the reaction zone of the clarifiers. At this pH the calcium solubility is greatly reduced, causing the precipitation of calcium carbonate. Available alkalinity is provided by carbon dioxide and soda ash. The reactions are as follows:

$$CO_2 + H_2O \Rightarrow H_2CO_3$$

$$H_2CO_3 + 2OH^- \Rightarrow 2H_2O + CO_3^{-2}$$

$$Ca^{+2} + CO_3^{-2} \overset{pH=10.0}{\Rightarrow} CaCO_3\downarrow$$

The treated water is collected in surface launders and piped to the gravity filters. Sludge is raked to the center cone at the bottom of the clarifier and pumped to the solids/liquids system.

Maintaining the solids inventory in the clarifiers has proven to be an art. The influent from the WWTP tends to change throughout the day or throughout the year. Consequently, the treatment process for the solids contact clarifiers must be adjusted to maintain a quality effluent. In the first stage clarifiers, a sludge blanket level must be maintained to allow a filtering process. Water flows from the reaction zone up through the sludge blanket into the clarification zone. Controlling the sludge blanket level can be difficult. The sludge wasting rate from the clarifier is typically done through batch wasting or by a variable pumping rate. The concentration in the underflow solids can be controlled by varying the rake speed or the wasting rate. The solids concentration is maintained in the reaction zone by increasing or decreasing the turbine speed or by changing the lime slurry feed rate. Constant monitoring of the clarifier influent, pH, and effluent is critical.

During winter months the wastewater effluent from the 91st Avenue WWTP is different than in the summer. The biological process does not work quite as efficiently. Consequently, the influent to the Water Reclamation Plant is more difficult to control. In addition, the cold-lime softening process is also negatively effected by the lower temperatures. For this reason it is necessary to adjust water quality controls for winter and summer.

Just prior to the gravity filters, sulfuric acid is injected to stabilize the water and prevent further precipitation of calcium carbonate. The pH of the effluent is lowered to 9.0. Sodium hypochlorite is also injected to control biological growth. A provision for polymer addition are also available, but has not been necessary.

The final step of treatment is sand and anthracite filtration. Suspended solids content and turbidity are reduced as the water flows through the filter media. Tests have shown that silica and phosphates are measurably reduced by the gravity filters.

There are six gravity filter cell blocks with four filter cells each. Each filter cell has 12 inches of sand and 24 inches of anthracite. The filters are backwashed on time (72 hours) and high water level. The basic backwash operation is to drain the filter, air scour it, and initiate a reverse flow to carry away the suspended solids. The backwashed solids are pumped to thickeners in the solids/liquids system for further processing.

From the gravity filters the treated water flows into a storage reservoir. The reservoir has an 80 acre surface area and is 30 feet deep. It holds 670 million gallons which is at least seven days supply of cooling tower make-up water. The reservoir is lined to prevent percolation into the soil. The bottom is covered with a thick coating of rubberized asphalt and the sides are lined with hypalon sheeting. Biological growth in the reservoir is not controlled with a biocide since most of the nutrients have been removed, the pH is elevated, and sodium hypochlorite was added prior to the gravity filters.

The water is pumped to the cooling towers upon demand. The demand varies according to power levels of the Palo Verde nuclear units and climatic conditions. Peak demands are in July, August, and September with lower demands in December and January.

The solids/liquids system collects the solids (sludge) precipitated by the chemical reactions in the first and second stage clarifiers and solids removed by the gravity filters. The solids/liquids system consists of thickeners, centrifuges, and a recalcination furnace. A thickener concentrates solids mainly by gravity. As the sludge particles become bigger, they drop to the bottom and are pumped to the next thickening process. Any excess water overflows the thickener to the front of the WRF for reprocessing. Polymers may be added to enhance the treatment process.

After the initial thickening of sludge, it is concentrated through one of two processes. The sludge, rich in calcium carbonate, can be diverted to a recalcination furnace to be reclaimed as lime and carbon dioxide. The second option is to process the sludge for onsite landfill disposal.

Sludge destined for the landfill is thickened in a waste thickener and fed to a dewatering centrifuge. The centrifuge spins at a high speed causing the separation of the sludge from the water. These centrifuges are designed for 95% overall solids recovery, a feedrate of 5,499 lb/hr, and at least 49% dry cake solids recovery. The dewatered sludge (cake) collects at one end of the centrifuge and falls down a chute to a truck. It is then hauled to the landfill. The liquid (centrate) collects at the other end of the centrifuge and is returned back to the thickener.

If the decision is made to reclaim the sludge, it is fed from thickeners directly to centrifuges. Centrate is returned to the thickeners and cake is conveyed to the recalcination furnace.

The recalcination furnace is a multiple hearth type. It is rated to handle 6,300 lb/hr of dry solids at 40 to 50% moisture. The cake enters openings at the top of the furnace. The sludge is dried in the upper hearths and drops from level to level as it is raked to openings in the hearth floors. At the higher temperatures (1800°F) in the lower hearths the cake is calcined. Calcium carbonate ($CaCO_3$) breaks down into calcium oxide

(CaO, lime) and carbon dioxide gas (CO_2). Carbon dioxide, along with other products of combustion, is compressed and reused in the second stage solid contact clarifiers. The product lime is pneumatically conveyed and transferred to storage silos for future use in the first stage solid contact clarifiers.

Chemicals to support the treatment process are transported to the facility or manufactured onsite. Lime, soda ash, and sulfuric acid are primarily delivered via rail car. Carbon dioxide and polymer are delivered by tanker trucks. Sodium hypochlorite is manufactured onsite.

Pebble lime is pneumatically unloaded from rail cars or tank trucks into storage silos. Product lime from the recalcination furnace is also pneumatically transferred to storage silos. Upon demand, lime is pneumatically transferred to supply silos located above slakers. The dry, volumetric feeder at the bottom of the supply silo meters lime as it is fed into the slaker and mixed with water. The resulting lime slurry overflows to a storage tank and fed to the first stage clarifiers to maintain a 11.0 to 11.3 pH.

Soda ash is pneumatically unloaded from rail cars or tank trucks into storage silos. Upon demand soda ash is pneumatically transferred to supply silos. A gravimetric feeder and a mixing tank are located under each supply silo. Soda ash is mixed with water and the solution overflows the mixing tanks to holding tanks. The soda ash solution is pumped to the second stage clarifiers.

Carbon dioxide is received in bulk quantities by tank trucks or rail cars. It is pumped into insulated storage tanks. Water bath vaporizers convert the liquid carbon dioxide to a vapor state. Pressure reducing valves regulate the carbon dioxide vapor flow to each second stage clarifier. The carbon dioxide vapor is diffused to the clarifier reaction zone through fine hole diffusers. The pH is controlled between 10.0 and 10.4.

Polymer can be injected into the process by three separate systems. One system services the first and second stage solids contact clarifiers. A second system supplies polymer to the thickeners and centrifuges. The third system provides polymer for injection into the gravity filter influent water.

Each polymer system prepares raw polymer in semi-continuous batch operation before distribution to the application points. Preparation starts with controlled dilution of raw polymer followed by gentle mixing in an aging tank for 30 minutes. Aging allows the polymer to reach full activity. Aged polymer is transferred from the aging tank by a progressing cavity pump to a feed tank. The feed tank provides a reservoir from which the polymer solution is metered by feed pumps. Additional dilution of the aged polymer solution occurs as it is pumped to the process injection points.

Sulfuric acid is delivered to the WRF in rail cars which are unloaded pneumatically into storage tanks. The stored sulfuric acid is metered to the gravity filter influent flow by a positive displacement diaphragm-type

TABLE 24.3. **WRF Effluent Quality Average for 1994.**

Parameter	1994 Annual Average	1995 Annual Average	Unit
Total suspended solids	2.6	2.8	mg/L
Turbidity	1.2[a]	1.1[a]	T.U.
Phosphate (PO^{-3})	0.36	0.53	mg/L
Total calcium (as CaCO$_3$)	67	72	mg/L
Alkalinity (as CaCO$_3$)	73	85	mg/L
Silica (as SiO$_2$)	8.0	8.9	mg/L
Magnesium (as CaCO$_3$)	8.3	6.9	mg/L
Ammonia	0.94*	0.98*	mg/L

[a]Sample location is the reservoir, not the WRF gravity filter effluent

pump. To prevent the concentrated acid from damaging the gravity filter influent pipe and to improve mixing of the acid with plant flow, the sulfuric acid is diluted with process water. The dilute acid is injected into a motionless mixer and dispersed in the pipe by hydraulic turbulence.

Sodium hypochlorite is produced as an 8% solution by two onsite electrolytic cell hypochlorite generators. Sodium hypochlorite is used to control biological growth within the gravity filter media and throughout the cooling tower system. Sodium hypochlorite is transferred from the storage tank at the production facility to a day tank at the gravity filters.

WATER QUALITY

The WRF continues to produce water in the quantity and of the quality necessary to meet the demands of the Nuclear Units. The 1994 and 1995 average water production performance are identified in Table 24.3. Design parameters are listed in Table 24.2.

COOLING WATER SYSTEM

Due to the high concentration of dissolved solids in the cooling water, the equipment used in the cooling water system is designed for sea water conditions. The condenser tubes are made of titanium and the tube sheets are of an aluminum-bronze design. The cooling towers are Type II concrete structures with PVC fill media. The circulating water and the plant cooling water pumps are made of nickel-aluminum bronze. The cooling tower make-up, blowdown and return lines are epoxy lined carbon steel and Type II concrete. Blowdown water is pumped to two onsite evaporation ponds. The ponds cover 485 acres and are lined with a high density PVC lining.

The typical nuclear power plant uses once-through cooling water systems or natural draft cooling towers. A once-through cooling water system typically uses water from a large water supply like an ocean, lake, or river and returns the water back to the source. Natural draft towers work on the principle that there is enough temperature differential between the base surface and the top of the cooling tower to cause air circulation. Natural draft towers will not work in the Arizona desert very efficiently due to the high surface temperatures. The natural draft towers had to be constructed at great heights to achieve the necessary temperature differential. Consequently, they were cost prohibitive.

Palo Verde's cooling tower design is forced draft which means that large fans located on the top of the towers pull air through the sides of the towers and exhaust out of the top. The warm water is pumped to the top of the towers and is distributed around the outside of the tower. This water is then allowed to trickle down through the plastic media where it makes contact with the moving air. In the winter, plumes of water vapor emitted from the towers can be seen across the flat desert for many miles.

Cooling water solids concentration is controlled by a continuous wasting or blowdown rate. Due to the high water loss from evaporation from the towers only a small amount of water needs to be blown down for this control. Feed water or make-up water to the cooling system is further treated with sulfuric acid for pH control and sodium hypochlorite to control biological growth. A dispersant can be added to help inhibit solids precipitation and a defoaming agent is used to control foam.

The cooling water chemistry is sampled daily to monitor the individual scale forming ions. Calcium, magnesium, phosphates and silica are the ions of greatest concern. One way of controlling these is to increase or decrease the blowdown rate as these chemicals reach a particular set point. For example, when the calcium concentration begins to rise near 9,000 ppm the blowdown rate is increased.

A new method to control these concentrations is to closely monitor relationships among these ions. Individual ions are monitored and controlled by comparing their concentrations with a known solubility index. One ion's concentration may be allowed to rise past its set point if the ion with which it would commonly react has a low concentration. For example, if the calcium concentration is high but the sulfate and phosphate concentrations are low, the chances of calcium scaling is low and the blowdown rate may not need to be increased. To initiate this type of control extensive testing and analysis of the cooling water must be done to predict when scaling might occur. This is the future method of control that PVNGS utilizes on cooling tower chemistry.

A test lab was established on site to analyze cooling tower water flows through a model condenser. The primary objective of this test was to limit the amount of water blown down to the evaporation ponds. Cooling water was cycled through the test lab at different concentrations to discover the relationship among the scale forming ions at higher cycles. During the testing periods water was cycled as high as 22 times through the control model before scaling began to form in the test condenser. Continuous monitoring was done on the different ion concentrations. The test lab was operated in both summer (high evaporation rate) and winter (lower evaporation rate) to ensure that the all seasonal conditions were tested.

As a result of the tests, calcium and phosphates are monitored closely during operations due to their affinity to scale at slightly higher pH. Consequently, acid addition to the cooling water is critical and dispersant is added as a backup. Silica is a different player. Because silica does not readily combine with other ions it is still necessary to monitor and control this ion by the traditional set point method.

A potential problem associated with this control method may be at the condenser. This method produces cooling water with high concentrations of ions. The problem occurs when a tube leak in the condenser allows this concentrated water into the secondary loop. The secondary loop contains deionized water and is kept very pure. If even a small leak occurs, ion concentrations in the secondary loop increase dramatically. This normally would not pose much of a problem because the secondary loop is protected by polishing demineralizers which capture these ions. The three ions that cause the most problems in the secondary loop are sodium, chlorides and sulfates. These particular ions are not removed in the Water Reclamation Facility treatment process. In fact, sodium is added in the form of soda ash; chlorides are added in the form of sodium hypochlorites, and sulfates are added with sulfuric acid. If the condensers should leak, high concentrations of these ions may cause problems in the plant steam generators. The current plan is to concentrate monitoring and maintenance efforts on the condensers to ensure leaks are kept to a minimum.

One other method of limiting the amount of blowdown water is to return a portion of the cooling tower blowdown water to the beginning of the Water Reclamation Facility process. This "pump back" system was designed and installed to send as much as 1,000 gpm of blowdown water to the clarifier feed sump at the Water Reclamation Facility. The amount of water returned is determined by the amount of influent received from the 91st Avenue WWTP. Typically, up to 3% of the plant influent is blowdown return water.

To date, the water quality in the cooling tower circulating system has not formed scale on the condenser tubes; created biological and/or organic fouling; generated abnormal corrosion; or caused temperature and bio-growth problems in the condenser tubes. Foaming does occur but is controlled through the use of anti-foam agents and control of the water level in the cooling towers. Biological growth accelerates during the summer months causing high chlorine demands. Although a residual can be maintained a large biomass does grow on the cooling tower deck. This is controlled by adding an additional biocide prior to shutdowns for outages.

A phenomena that was not anticipated is the formation of sludge in the cooling towers. The sludge is mostly comprised of dust with some chemical precipitation. Due to the design of the induced draft cooling towers dust can be pulled into the tower system. During each power plant outage the sludge is removed from the cooling tower basins. It is solar dried in temporary storage locations and sampled for hazardous constituents and radioactivity. Once verified that it is non-hazardous, it is disposed in the onsite WRF sludge landfill.

PVNGS was originally designed and constructed with one evaporation pond. The design also projected a possible need for as much as 1,000 acres of evaporation ponds for the life of the plant. After initial operation, another pond was constructed bringing the total acreage of ponds to 485. As constructing evaporation ponds is very expensive and the methods and equipment needed to treat the blowdown water is still very expensive, the above treatment scheme is working to prevent those expenditures. The current scheme to minimize blowdown water and return a portion of the blowdown water to the WRF influent has been effective. It has prevented the expenses of additional evaporation ponds or alternative treatment. However, cost predictions for blowdown treatment are explored every few years to ensure that current treatment philosophies are the most prudent.

EFFLUENT TREATMENT COSTS

The major components in the operation and maintenance (O&M) cost of treating the 91st Avenue WWTP effluent are for chemicals, manpower, electricity, materials and effluent. A contract for the sale of 91st Avenue WWTP effluent was established in the mid 1970's. A consortium was formed by the City of Phoenix and surrounding cities that share in the use and expenses of the 91st Avenue WWTP. All the cities in the consortium are partners in the contract. The contractual cost of effluent is less than $50.00 per acre-foot.

Chemical costs are the most significant expense for the treatment of effluent water. Chemicals utilized in the treatment process are lime, soda ash, carbon dioxide, sodium hypochlorite, sulfuric acid, polymer, and fuel oil (recalcination furnace operation).

The staff necessary to support the operation and maintenance for effluent water treatment is:

- operations: 35
- maintenance: 26
- laboratory: 5
- engineering: 7
- administration/technical support: 9

The scope of work includes effluent treatment, sludge landfill operations, recalcination furnace operations, chemical handling, and facility maintenance.

The electrical expenses reflect the operation of the Hassayampa Pump Station along the WRF pipeline from the 91st Avenue WWTP. Since the WRF is onsite with the power plant, electricity is directly provided without metering functions.

Material costs include equipment parts, supplies, and consumable materials necessary to maintain equipment and support operations. Examples are replacement of small valves, purchase of personal protective equipment, and computer paper.

The total O&M cost of treating an acre-foot of effluent water for use as cooling tower makeup is approximately $170.00.

CONCLUSION

Treated secondary effluent water has been successfully used as cooling tower make-up water at the Palo Verde Nuclear Generating Station. The effluent from the 91st Avenue Wastewater Treatment Plant has proven to be a reliable source of water that can be successfully treated by the Water Reclamation Facility. The WRF was properly designed to produce economical cooling tower make-up water in sufficient quantities and quality to support the operation of three nuclear power plants. Continuing efforts are underway to optimize the operation and improve water quality production during the winter months, improve controls of the cooling tower water chemistry, and minimize the blowdown water to the evaporation ponds.

REFERENCES

Bechtel Power Corporation, 1978. Palo Verde Nuclear Generating Station System Descriptions Manual. Water Reclamation Facilities, Vol. VI and VII.

Bechtel Power Corporation, 1974. Palo Verde Nuclear Generating Station Water Reclamation Studies, Vol. I.

Arizona Nuclear Power Project, 1985. Water Use Pamphlet.

CHAPTER 25

Wastewater Reuse in South Africa

INTRODUCTION

INDICATIONS are that a lack of sufficient water supplies will become the single most important factor which will limit South Africa's socio-economic development in the twenty-first century. That South Africa is indeed a country with limited water resources, is indicated by the following facts:

- South Africa's mean annual rainfall is 483 mm compared to the world average of 860 mm.
- Rainfall is unevenly distributed, varying from 50 mm per annum on the west coast to 1250 mm per annum in the eastern mountains. Only a narrow region on the eastern and southern coastline is well watered. About 65% of the total land area receives less than 500 mm (considered the minimum for dryland farming) and 21% of the country receives less than 200 mm.
- Mean evaporation varies from 1100 mm per annum in the east to 3000 mm per annum in the west.
- Rainfall is highly variable, with regular occurrence of extended and severe droughts.
- South Africa is poorly endowed with suitable aquifers and only 15% of its water supplies are derived from groundwater. Aquifers

P.E. Odendaal, Water Research Commission, P.O. Box 824, Pretoria 0001, South Africa; J.L.J. van der Westhuizen and G.J. Grobler, Department of Water Affairs and Forestry, Private Bag X313, Pretoria 0001, South Africa.

are generally low-yielding and particularly complex, due to their generally fractured nature.

• Projections indicate that water demand will exceed available supplies soon after the year 2020. Meeting demands beyond that stage will involve considerable incremental cost. Absolute shortages already occur in certain regions, which necessitated the construction of major inter-basin water transfer schemes.

In addition, a high population growth rate and sustained industrial and agricultural development, are posing severe pollution control problems. Very few of the rivers are perennial, and floods occur only intermittently. This means that little dilution capacity is available to ameliorate pollution effects.

Options for supplementing freshwater supplies are limited. The desalination of sea water is, relatively, still very expensive and can, in any case, not supply water to the interior. The towing of icebergs is an unproved option, while the importation of water from neighbouring countries is a politically sensitive issue, and cannot at this stage be stated as a firm option. Research on rainfall stimulation in South Africa has been yielding very promising results, but even if successfully implemented, there will be a limit to what can be achieved.

Against this background, wastewater reclamation is an obvious strategy for extending available water supplies. For this reason, research on water reuse has received considerable attention in South Africa since the late sixties.

Due to geographic considerations, with some of the major cities far from the sea, wastewater reclamation policy in South Africa provides for the planned indirect reuse of treated effluents, which have to be returned to the water course of origin. Therefore, strict control has to be exercised on the treatment of and discharge of effluents. This means that there is a close integration between the country's water reclamation and water quality protection policies.

WATER RECLAMATION AND REUSE POLICY

THE WATER ACT (1956)

Water reuse plays a major role in matching water supply and demand. In fact, the important contribution that water reuse could make to the national water budget was already recognised in the Water Act of 1956 which has become a powerful tool for the implementation of reuse policies. The Act is being administered by the Department of Water Affairs and

Forestry. The provisions of the Act promote water reuse across the total spectrum of wastewater reuse, namely *planned indirect reuse, direct reuse,* and *internal recycling in industry.* The main clauses of the Act that impact on water reuse are discussed below.

PLANNED INDIRECT REUSE

Section 21 of the Act: Water abstracted from a stream and used for industrial or municipal purposes, must be returned to the stream of origin after treatment. Before return, the water must be diminished only by essential consumptive use. For instance, evaporation of tractable waste-water to avoid the expense of treatment, or irrigation of inferior land where there is a possible better form of reuse, would not be acceptable.

DIRECT REUSE AND INTERNAL RECYCLING

Section 12 of the Act: For the use of more than 150 m^3 of water per day, a permit is required from the Department of Water Affairs and Forestry. In granting such permits, the Department has the authority to specify that reuse, recycling and hierarchical use of water should be practised as far as possible.

In certain instances, where water is in short supply, factories that wish to expand production may not be able to obtain permits for additional water supplies and are forced to use reclaimed water or to practice internal recycling and strict water conservation measures.

Section 13 of the Act: A local authority may not construct, enlarge or alter a works in which more than 125,000 m^3 of water may be stored or by which more than 5000 m^3 per day may be diverted, nor enlarge any works to such an extent that more than the above quantities may be abstracted or stored, without a permit from the Department of Water Affairs and Forestry. Through this control of water intake, the Department has the power to exert pressure on local authorities to purify effluent for reuse where water is in short supply.

Section 21 of the Act: Wastewater must be purified to prescribed general standards, while allowance is also made for special standards where required. The Department may grant exemption permits in cases where meeting of the standards would be unduly onerous, in which cases special conditions are laid down.

Where an industry discharges to a sewer owned by a local authority, the responsibility rests with the local authority to meet the standards. The local authority will charge tariffs based on the effluent quality received in its sewers and may impose restrictions on the industrial discharge of certain pollutants.

TABLE 25.1.   Projected Annual Runoff to Hartbeespoort Dam
(million m³ per annum).

Source	1990	2000	2010	2020
Natural runoff	177	177	177	177
Urban storm water runoff	29	46	65	85
Treated wastewaters	100	177	257	335
Upstream abstraction for irrigation	−41	−50	−62	−74
Total	265	350	437	523

While the primary intention of this clause is to control pollution, it does effectively encourage the reuse of wastewater. Where industries and local authorities are forced to spend substantial amounts of money on meeting effluent standards, it becomes likely that they will investigate opportunities for reuse, or may even be prepared to further upgrade effluent for this purpose.

IMPLICATIONS OF PLANNED INDIRECT REUSE

The promotion of planned indirect reuse through the implementation of the Water Act, resulted in a situation where treated effluents constitute a substantial portion of the base flow in many of the rivers and of the content in many dams. A case in point is the Hartbeespoort Dam near Pretoria, which receives considerable quantities of treated wastewater return flows. Projected annual inflow to the dam, up to the year 2020, are presented in Table 25.1. It is clear from the table that effluent return flows will soon contribute as much as the natural runoff, and by the year 2020 will be double the natural runoff.

As a result of the strategies adopted, it has become possible to return approximately 60% of all effluents to the water environment. A further estimated minimum of 5% is *reused directly* by industry and mining.

QUALITY STANDARDS

In view of the rising percentage of effluent return flows in the country's surface waters, quality concerns are of high priority. For this reason, it is vital that the conditions of Section 21 of the Act, controlling the quality of effluent discharges, be strictly applied. In the context of planned indirect reuse, South Africa has adopted a strategy for water quality management which is much more complicated than in countries with abundant water supplies. In such countries, it is possible to adopt water pollution control policies which focus exclusively on the prevention of pollution, without taking quantitative considerations into account (e.g. effluents may not be allowed to be discharged into water courses at all). South Africa has been forced to adopt both a Receiving Water Quality Objectives (RWQO) approach and a Pollution Prevention approach. The latter approach involves the reduction of pollution at source, water recycling, and control over the manner in which wastes are handled and disposed of.

The RWQO approach entails that the water quality requirements of different user categories must be documented and taken into account, when decisions have to be made regarding the quality of the resource as well as the standards that must be imposed upon effluent dischargers. To underpin decision making processes, the Department of Water Affairs and Forestry published a set of water quality guidelines for the recognized water uses: *domestic, industrial, agricultural, recreational* and *aquatic ecosystems.* The target water quality range stated within these guidelines are considered the ideal situation and are accepted as the objective for water quality management purposes.

Central to the RWQO approach is the concept of waste load allocation and assimilative capacity of natural waters. In principle, waste load allocation involves assignments of allowable discharges in such a way that the water quality objectives of the designated water uses can be met. It requires determination of

- water quality objectives for desirable water uses
- understanding of the relationship between pollutant loads and water quality
- economic impacts
- socio-political constraints
- cost/benefit relationships

Assimilative capacity is the concept that water bodies can tolerate the input of some wastes without the deterioration of water quality to the point where water uses are adversely affected. A water body provides many mechanisms to modify, move, or otherwise transform material discharged into it, but this capacity should not be strained.

Assimilative capacity for a constituent differs in fundamental ways, depending on whether the constituent can be considered conservative or non-conservative:

- Conservative constituents are not lost due to chemical reactions or biochemical degradations. Such constituents may include, for example, total dissolved solids and chlorides. Conservative constituents accumulate along the length of a water body in the direction of motion, so that amounts added at the most upstream point are still present at the most downstream point.

Concentrations of conservative constituents can be reduced only by dilution with water with a lower concentration.

- Non-conservative constituents, on the other hand, decay with time due to such mechanisms as chemical reactions, bacterial degradation, radioactive decay, or settling of particulates out of the water column. Many constituents exhibit non-conservative behaviour, including oxidisable organic matter, nutrients, volatile chemicals and bacteria. The amount of a non-conservative constituents decrease with time and/or distance from the point of input.

In line with its precautionary approach, the Department has adopted an hierarchical decision-making procedure in considering applications for effluent discharge to a receiving water body. The steps are the following:

(1) Options for preventing and minimising waste through source reduction, recycling, detoxification and neutralising of wastes must be thoroughly investigated. Caution should be taken that, in this process, one is truly avoiding or minimising waste and not simply shifting it from one environmental medium to another, for example, from water to land, or from water to air.

(2) If, after all the practical options to prevent and/or minimise waste have been exhausted, there is still waste or effluent, such effluent will be required to meet whichever is the strictest: the current general effluent standards (or special effluent standards, where applicable); or standards based on receiving water quality objectives, and taking into account the projected impact of the effluent discharge on the quality of the receiving water.

(3) Exemptions from minimum effluent standards or receiving water quality based standards, will be considered under special circumstances, and as a last resort, but will require sufficient justification on technological, economic and socio-political grounds. Such exemptions may not always be granted, will in most cases be temporary, and will almost certainly be withheld if the effluent discharge investigation shows that the receiving water's fitness for the maintenance of the natural aquatic environment, and for intended water uses, will be significantly reduced.

HEALTH ASPECTS

An expert committee of the World Health Organisation, in 1975, called for the institution of an epidemiological study of any population to be exposed to reclaimed water for potable use, before introduction of the reclaimed water [1]. The South African Department of Health also wrote this requirement into their tentative guidelines for potable reuse [2]. This requirement could not be met in Windhoek, as the urgency to supplement the water supply was too great. However, an epidemiological study was started in 1976 by the South African Institute for Medical Research. This study is reported on in a later section.

There is a possibility that the Cape Town metropolitan area will have to resort to potable reclamation soon after the turn of the century. In anticipation of this eventuality, an epidemiological study of the area was launched in 1982, by the Department of Community Health of the University of Cape Town, on behalf of the Water Research Commission, in order to establish baseline data. This was done by establishing a data bank consisting primarily of information on mortality rates, morbidity as seen by general practitioners, and birth defects [3–5].

Subsequent to establishing the health data base for the Cape Town metropolitan area, the methodologies tested there have been extended to other parts of the country, mainly utilizing routinely collected data. Regional life tables, an atlas of mortality, the national surveillance of birth defects and analysis of morbidity patterns from general practice, have indicated strong geographic differences. Specific analyses correlating mortality from certain cancers with trihalomethane levels in drinking water, and another comparing the incidence of cardiovascular mortality and water hardness have not, however, shown a significant relationship.

TECHNOLOGY DEVELOPMENT

In this section a review is given of technologies which have been developed, evaluated and implemented for reuse applications in South Africa.

It should be stated, at the outset, that in many (if not most) instances of direct reuse, a biologically well treated effluent, sometimes followed by disinfection and/or sand filtration, is more than adequate to meet the quality requirements for a wide range of reuse applications. This is clearly valid for the irrigation of treated municipal wastewaters, but applies equally well for many cases of industrial reuse, as is demonstrated in Table 25.2. Technology development per se had, obviously, not been required for these applications. On the other hand, certain technologies have been developed or evaluated specifically for direct reuse applications. Other technologies were evaluated to cope with problems derived from the broad-based policy of planned indirect reuse. Membrane technology, in particular, has been the subject of wide-ranging research and evaluation in South Africa, and is beginning to have a significant impact on industrial recycling.

CATCHMENT QUALITY CONTROL

Although, strictly speaking, catchment quality control (quality of wastewaters entering the sewerage network) cannot be considered as technology

TABLE 25.2. Examples of Industrial Use of Secondary Effluent with No or Little Tertiary Treatment.

Industry	Purpose	Treatment	Quantity
AECI (Chemicals manufacturer), Modderfontein [15]	Gas washing in ammonia from coal process. Thereafter used on a cascade basis and finally for ash slurrying	• Orbal activated sludge plant • Chlorination • Sand filtration	1200 m³/d
SASOL (oil from gas), Sasolburg [16]	Ash transport	• Biofiltration	12000 m³/d
General Hide (tanning), Harrismith [17]	Hide processing and dilution of final, high TDS effluent for irrigation	• Activated sludge nutrient removal	200 m³/d
Galvanizing Techniques, Port Elizabeth [18]	Cooling and rinsing of steel fabrications	• Activated sludge • Sand filtration • Chlorination	15 m³/d
Beatrix Goldmine, Welkom [19]	Gold reduction works and slurrying to slimes dams	• Activated sludge • Chlorination	3000 m³/d
Kelvin power station, Johannesburg [20]	Cooling	• Biofiltration and nutrient removal activated sludge*	35000 m³/d
Cape Fellmongering Port Elizabeth [18]	Process water (completely replaced potable water for this purpose)	• Activated sludge • Sand filtration • Chlorination	360 m³/d
Fishwater Flats wastewater treatment plant Port Elizabeth [18]	Compressor cooling, bearing cooling, bearing lubrication, cleaning and irrigation	• Activated sludge • Sand filtration • Chlorination	4000 m³/d
Nampak (Tissue paper manufacturer), Bellville [21]	Process (replaced 85% of potable water)	• Orbal activated sludge • Sand filtration	800 m³/d

1170

development, it should conceptually be regarded as a first step in, and an integral part of, water reclamation technology. The incoming wastewater may contain intractable constituents of industrial origin which could complicate or adversely affect technology for achieving defined quality criteria. In such cases, an analysis of the sewer network should be conducted in order to identify polluters, and steps be taken to remove or divert the discharge of the offending pollutants. Four examples of catchment quality control, to meet reuse objectives, are cited:

(1) The well-known instance of direct reclamation for potable reuse at Windhoek, the capital of Namibia, will be dealt with in a later section. The application of catchment quality control was crucial to the implementation of this, the world's first potable reclamation system. For example, an analysis of wastewater composition during the period 1961–65, indicated that only 10% of the wastewater flow, contributed 50% of the nitrogenous load in the total wastewater stream. The wastewater from the industrial area was then permanently diverted to a separate, small treatment plant. All new industries are now located on the basis of expected wastewater composition [6].

(2) There is a possibility that the Cape Town metropolitan area will have to resort to potable reclamation soon after the turn of the century. In planning for this scenario, the city already took steps in the early seventies to segregate industrial effluents as far as possible from the wastewater flow to the Cape Flats treatment plant which is earmarked as the site for a possible reclamation system [7].

(3) In the City of Port Elizabeth, the use of reclaimed water is recognised as a possible future strategy for the augmentation of potable water supplies. The sewerage network collects separately from the predominantly domestic and industrial areas of the city, and the separation is maintained throughout at the Fishwater Flats activated sludge treatment plant [8].

(4) The Mondi Paper Company near Durban has been upgrading secondary effluent as process water since 1972. Intermittently, problems have been experienced due to the presence of colour in the secondary effluent, deriving from textile dye-house effluents in the catchment of the municipal wastewater treatment plant. This prompted the municipality to take the matter up with the textile manufacturers to deal with the problem at source.

UPGRADING OF BIOLOGICAL EFFLUENT TREATMENT

In a situation of extensive indirect reuse through the discharge of treated effluents to surface waters, it is essential to maintain a high standard of effluent treatment. The biological removal of nitrogen and phosphorus from sewage effluents through the modified activated sludge process, had been pioneered in South Africa, and has been the subject of intensive research. Research is, in fact, still continuing to refine the process.

Nitrification has long been a requirement for sewage treatment in South Africa, while an orthophosphate standard of 1 mg/L, as P, is applicable in sensitive catchments since 1985. The new generation of modified activated sludge plants not only reduces phosphates to meet the phosphate standard, but also reduces COD, nitrates which can have adverse health effects, and ammonia which increases the cost of disinfection. Some 57 of the new generation activated sludge plants are now operating in South Africa.

TREATMENT OF ALGAL WATERS

Eutrophication inevitably followed the increasing discharge of treated effluents to the country's surface waters. Since the middle of the seventies, many water treatment plants experienced problems such as floating floc, high chemical dosages, short filter runs and off-tastes and -odours in chlorinated water. The ortho-phosphate limit of 1 mg/L, as P, referred to in the previous paragraph, is unlikely to achieve a permanent reversal of eutrophication. This situation prompted wide-ranging research into the treatment of eutrophied waters in various parts of the country, and various technologies are now in full-scale application.

Dissolved Air Flotation (DAF)

DAF proved an effective technology in the treatment of eutrophied waters, and soon resulted in the modification of six existing treatment plants to incorporate DAF technology. Since then DAF features in most new plants treating eutrophied waters. Figure 25.1 schematically depicts a typical DAF plant. Normal ranges of operating and performance data are shown in the figure.

Combined DAF/Filtration (DAFF)

A novel development in DAF is DAFF technology which integrates DAF with gravity filtration. A typical DAFF system is depicted in Figure 25.2. Scrapers to remove the float layer are normally not used on DAFF systems in South Africa. The layer of sludge is removed during backwash and thickened separately.

During the development phase, it was shown that the conversion of one full-scale sand filter at Richards Bay water works to implement DAFF, doubled its capacity. The first application was in Pretoria, where existing

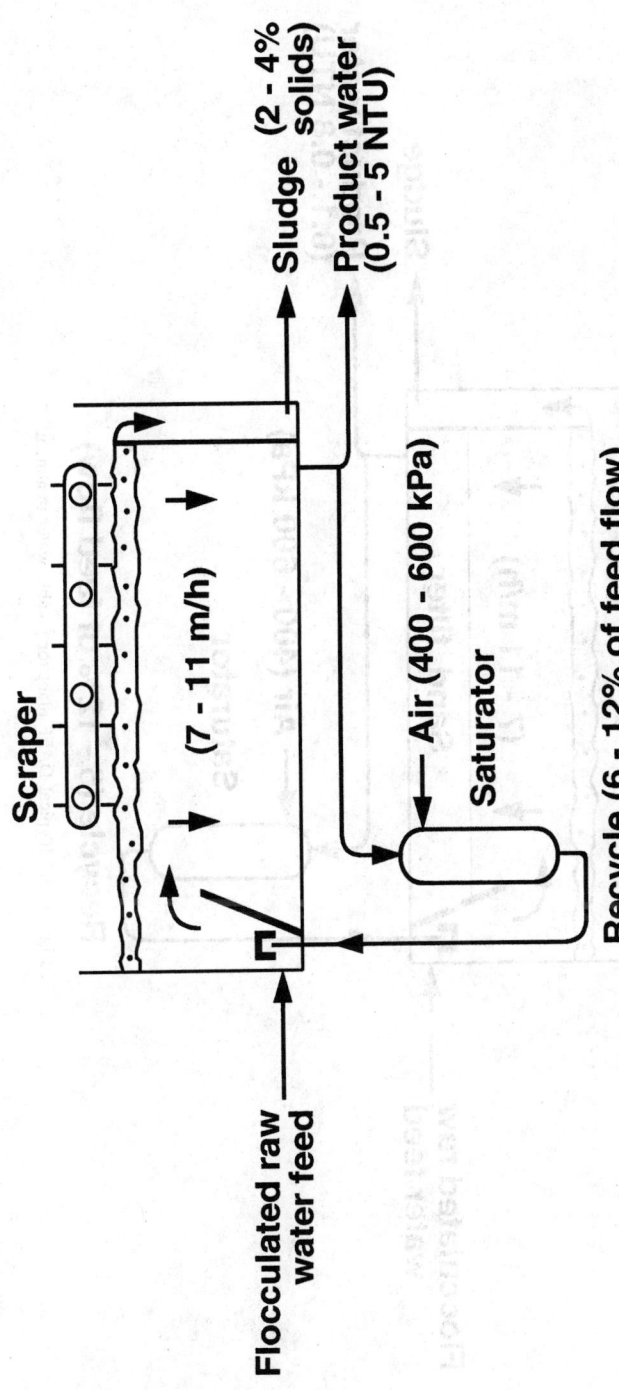

Figure 25.1 Typical DAF plant for potable water treatment.

Scraper

(7 - 11 m/h)

Sludge (2 - 4% solids)

Product water (0.5 - 5 NTU)

Air (400 - 600 kPa)

Saturator

Recycle (6 - 12% of feed flow)

Flocculated raw water feed

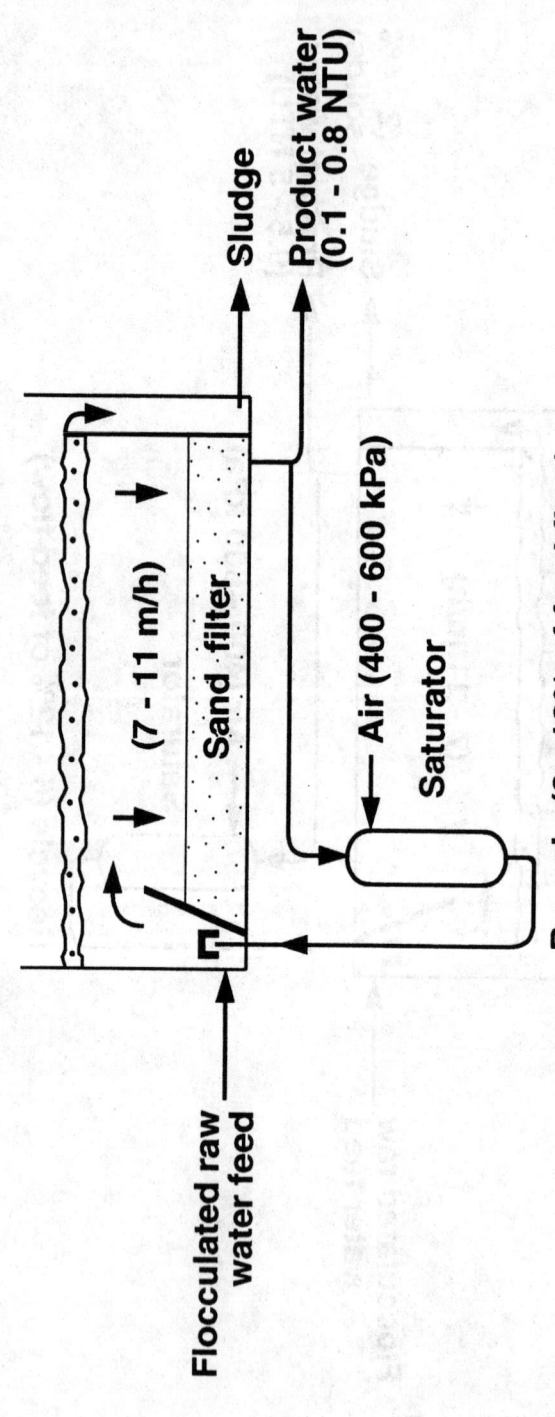

Figure 25.2 Typical DAFF plant for potable water treatment.

Flocculated raw water feed

Sludge

Product water (0.1 - 0.8 NTU)

(7 - 11 m/h)

Sand filter

Air (400 - 600 kPa)

Saturator

Recycle (6 - 12% of feed flow)

facilities were upgraded and a new plant, incorporating DAFF (36,000 m³/d), commissioned at Rietvlei in 1988 [9].

A 120,000 m³/d DAFF plant, treating mixed eutrophic and high turbidity water is under construction near East London. The plant incorporates pre-settlement in clarifloculators, PAC addition with 4 minute contact time before flocculation, and DAF on top of the sand filters. Raw water feed may contain algae of up to 400 μg/L chlorophyll *a,* and turbidities of up to 500 NTU [10].

Powdered Activated Carbon (PAC) with DAF (PACDAF)

The use of PAC was researched to deal with taste and odour problems that may occur during certain periods of the year. Because of the advantages of PAC over granular activated carbon (GAC) and because dosing was only required during certain periods of the year, the dosage of PAC before DAF was investigated at the Schoemansville plant [11,12].

Following successful application there, PAC was used on two other DAF plants using eutrophied waters. An interesting laboratory finding was that the adsorption capacity of the PAC is not impaired by the presence of ferric chloride flocs [13].

In the PACDAF system, PAC is used with DAF on a counter-current basis in two contact stages. Fresh PAC is added to the second reactor or contact tank to polish the partially treated effluent from the first reactor's DAF separator. PAC with a relatively low adsorbate loading is then removed from the process stream by coagulation, flocculation and DAF, and the product water leaves the system to be treated by final filtration. The flocculated PAC is recirculated to the first reactor and brought into contact with algae-laden raw water. The exhausted PAC and flocculated algal biomass is then separated by DAF and discharged from the system, whilst the clear product water passes to the second reactor. In this way, product water and PAC pass through the system in a counter-current mode.

The first full-scale PACDAF plant was commissioned in 1992 at Brits [10]. In this instance, it was more cost-effective to employ a lamella settler in place of the second DAF. A schematic diagramme of the process is shown on Figure 25.3 and some operating data are the following:

Flow	60,000 m³/d
Hydraulic loading with 10% recycle	7 m/h
Recycle, adjustable	5–15%
Saturator pressure (adjustable)	300–500 kPa
Standard cupped South African nozzles are used	
Primary PAC contact time alone	3.5 min
Primary PAC contact time with coagulant (FeCl₃)	25 min
Primary PAC separation time in lamella settler	40 min

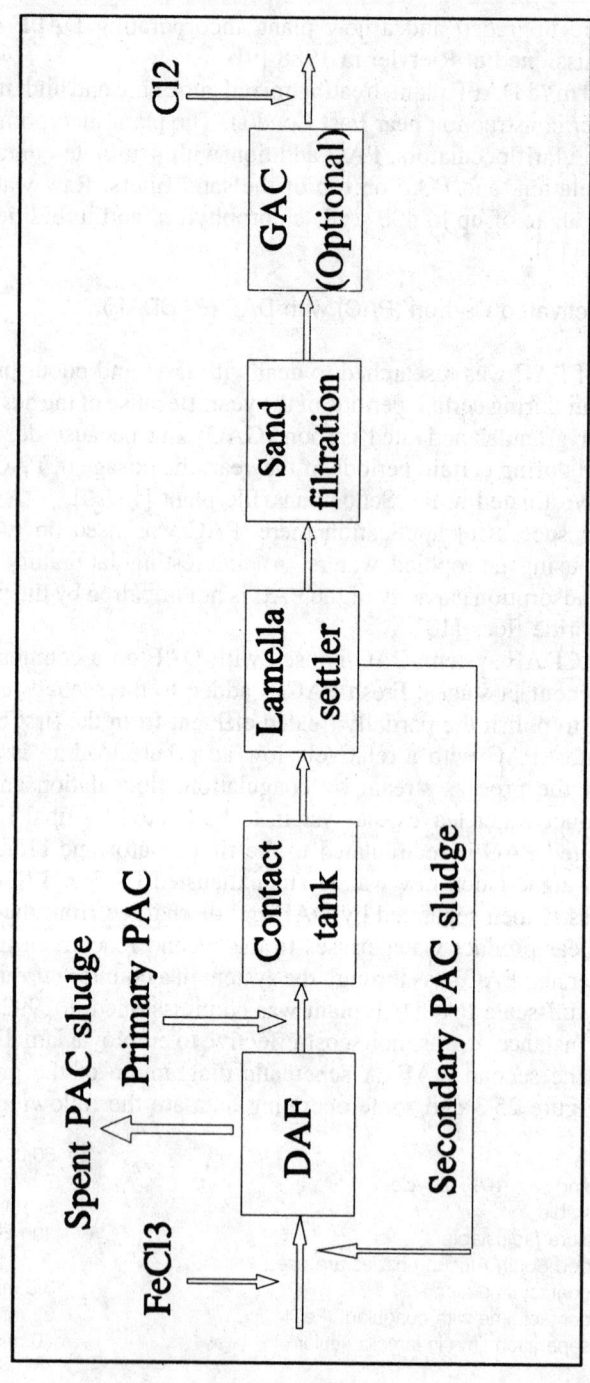

Figure 25.3 Process diagram for Brits PACDAF plant.

Secondary PAC contact time with coagulant	20 min
Secondary PAC separation time in DAF	15 min
PAC (primary) dosage required	5–6 mg/L

Counter-current DAF and DAFF

In this system, which is also finding full-scale application, flocculated raw water feed enters near the top of the tank, below the float layer, and saturated recycled water enters near the bottom of the tank through the aeration nozzles. Flow is thus counter-current, and clarified water is withdrawn below the recycle inlets. These plants are being used for both drinking water and industrial effluent treatment. They are operated with or without high rate presettlement and a later innovation includes a DAFF version, with dual media filtration within the counter-current DAF unit [10].

MEMBRANE TECHNOLOGY

Membrane technology is certainly not novel, but its application in the treatment and recycling of industrial wastewaters does not have a long history. In South Africa membrane research has been proceeding since the middle seventies, aiming primarily at the development of technology for industrial water management. This research resulted in the development of a South African membrane manufacturing industry, and in novel applications which clearly demonstrated the potential of membrane technology for the treatment and recycling of effluents in a wide range of industries. Case studies are reported in a later section.

Pilot plant studies to investigate the viability of reclaiming a potable quality water from activated sludge effluent, using a tubular reverse osmosis system, were successfully undertaken at the Fishwater Flats plant, Port Elizabeth. The investigation was done in two phases, first with a 30 m^3/d pilot plant, followed by a demonstration plant of 400 m^3/d capacity.

Three considerations, in particular, favour the application of membrane processes in water reclamation:

(1) Over the last two decades or so, there has been a steady improvement in the robustness, versatility, support systems and flux properties of these systems, coupled to a significant reduction in real cost.

(2) The TDS levels in water increases by 300–400 mg/L for each cycle of domestic use, and in many cases by higher values for each cycle of industrial use, thereby rendering the wastewater unsuitable for certain reuse applications without some degree of desalination.

(3) Membranes offer a positive barrier to microorganisms and a variety of pollutants.

WATER REUSE APPLICATIONS IN SOUTH AFRICA

POTABLE REUSE OF MUNICIPAL WASTEWATER

Potable reuse research in South Africa commenced during the late sixties. At the time, severe water shortages faced Windhoek, the capital of Namibia (then a mandate territory administered by South Africa), and reclamation of municipal wastewater, for potable reuse, presented itself as the only short-term solution to the problem. Following pilot-scale research by the South African Council for Scientific and Industrial Research, a potable reclamation plant (4800 m³/d capacity) was inaugurated in 1968. The plant has now consistently produced water of acceptable quality over a period of more than 25 years. It was upgraded on several occasions, incorporating new research findings, and its capacity is now being expanded to 21,000 m³/d.

The public had been kept fully informed at all stages—from conceptualization, through the pilot testing and construction phases, until final implementation. Although there had been initial opposition, this gradually faded, and the fact that the reclamation plant had been in use for more than 25 years, demonstrates that the public had fully accepted this option for water supply augmentation.

Although Windhoek is not in South Africa, the case is presented here fairly extensively, in view of the uniqueness of this plant (the world's first, and still the only, full-time direct potable reuse facility) and the historic involvement of South African scientists and engineers in the research and in the design and construction of the reclamation plant.

An intensive research programme on potable research continued after the inauguration of the Windhoek plant, involving *inter alia* three experimental plants: The Stander reclamation plant in Pretoria, and the Athlone and Cape Flats demonstration plants in Cape Town. The Stander plant, in particular, served as testing ground for many of the process modifications which were made to the Windhoek system over the years.

In a paper entitled *Twenty-five years of wastewater reclamation in Windhoek, Namibia,* Haarhoff and Van der Merwe gave a comprehensive review of the Windhoek reclamation system. The information presented in this section is mainly drawn from that source [6].

The Windhoek project, from the start, was based on the premise that successful reclamation for potable reuse could only succeed if three equally important elements can be controlled:

- diversion of industrial and potentially toxic wastewater from the main wastewater stream
- secondary wastewater treatment to produce an effluent of adequate and consistent quality

• advanced treatment to produce acceptable potable water

Diversion of Industrial Wastewaters

This was a key issue in the planning of the reclamation system, and has been dealt with previously.

Secondary Wastewater Treatment

The secondary treatment of wastewater at the time of commissioning the reclamation plant, consisted of a biofiltration system followed by a series of maturation ponds (retention time 6 days, later extended to 14 days). An activated sludge module of 9000 m³/d capacity was commissioned in 1979, which effectively reduced the ammonia content of the feed water to the reclamation plant. The effluent from the activated sludge plant discharged into a separated set of maturation ponds, and became the sole source of supply to the reclamation plant. The wastewater treatment plant was again enlarged in 1992/93 when the activated sludge system was extended to 17,000 m³/d. The process configuration was also modified to incorporate biological phosphate removal. The effluent is excellent with COD concentrations of 33–37 mg/L, ammonia-N concentrations of 0.3 to 0.5 mg/L and nitrate-N concentrations of 3 mg/L. Biological phosphate removal has not been optimized and phosphate levels are between 6 and 8 mg/L.

Advanced Treatment

The reclamation plant underwent a steady process of evolution and improvement since its commissioning, and four distinct process configurations have been used since 1968. These are indicated in Table 25.3.

The most serious problem experienced during the initial phase (Configuration 1) was the ammonia carried over from the maturation ponds, which steadily became worse due to overloading of the wastewater treatment plant. This problem, and the fact that foam fractionation was no longer required (due to the phasing out of hard detergents), prompted a major modification of the reclamation plant in 1977. Most important was the introduction of excess lime dosing and ammonia stripping (Configuration 2). Modifications were based on extensive testing at the Stander demonstration reclamation plant in Pretoria.

Technical problems during the second phase centred around the high pH lime system: the lime dosing equipment failed regularly, and severe scaling occurred in the ammonia stripping towers. Treatment costs also escalated due to the high use of chemicals. When the activated sludge effluent became available in 1979, with substantially reduced ammonia levels, the reclamation system was again adapted in 1980. Ammonia stripping was eliminated and alum dosage introduced (Configuration 3). Again, the changes were only introduced after extensive testing at the Stander demonstration reclamation plant.

Problems were encountered with poor settling in the primary settling tank due to the presence of algae in the maturation pond effluent. This led, in 1986, to the commissioning of dissolved air flotation which replaced the primary settling step (Configuration 4).

Monitoring

From the start of the reclamation project, a comprehensive battery of tests were conducted on samples taken at different points in the treatment sequence, in the storage reservoirs, at key points in the distribution system and at consumer off-takes. This was supplemented by monitoring every patient who visited a hospital, clinic or general practitioner's consulting room since 1968. Originally, analyses were carried out by the City of Windhoek, the Department of Water Affairs, the South African Institute of Medical Research and the National Institute for Water Research in South Africa. Testing and sampling were initially duplicated, but due to rigorous quality control measures the duplication proved to be superfluous and abandoned after 1982.

To ensure independent monitoring of system performance, a technical committee, representing experts from five independent professional bodies, convened three times a year for a detailed review of water quality, and conidering technological and operational improvements at the plant. In addition, a larger group of 28 experts met once a year for an annual analysis of system performance. This formal procedure was discontinued and since 1988 expert reviews are invited on an ad hoc basis.

At present the water produced by the reclamation plant is monitored by three independent laboratories. The treated wastewater, before reclamation, is also continuously monitored to ensure a consistent, high quality maturation pond effluent.

The monitoring programme involves testing for chemical quality, toxicity (water flea lethality, urease enzyme activity, and bacterial growth inhibition), virological monitoring, somatic coliphages, bacteriological monitoring, algae, and mutagenicity (Ames test).

An epidemiological study was conducted over the period 1976 to 1983. The total population varied between 75,000 and 100,000 during the study period, making the numbers too small for statistical significance. However, when classified by cause and race, the profile corresponded to earlier global analyses by the WHO. Similarly, when only cancer deaths were classified by site and race, they were consistent with other studies on ethnic differences in cancer prevalence. Within the limitations of the low numbers, this analysis seems to confirm that the patterns of mortality and cancer were not affected by reclamation. It would have been preferable to compare the Windhoek statistics to overall Namibian statistics rather than to global statistics, to eliminate local factors other than the water reclamation scheme, but such data were not available [14].

Contribution to Windhoek's Water Supply

The production cost of the reclaimed water was consistently higher than for water derived from other sources, mainly because of the high cost of activated carbon treatment. In addition, the water supplied from state-run conventional sources is subsidised in respect of capital cost. As a result, the reclamation plant was operated on an intermittent basis to supplement supplies during times of peak summer demand or during emergencies. For this reason, the production of the plant was, on average, not as high as it could have been. For the period 1968 to 1991, the average production was 27.3% of the plant's capacity, while the contribution of reclaimed water to the total Windhoek supply was only 4.0%. In 1982, due to a preceding series of exceptionally dry years, the plant produced 80.1% of its capacity—its maximum output since commissioning.

Up to the present, the reclaimed water was blended in two steps. The first blending took place at the Goreangab treatment plant, where the reclaimed water was blended with conventionally treated surface water from the Goreangab Dam, which was treated at the same site. The treatment capacities of the reclamation and conventional treatment plants were about the same, which ensured a minimum 1:1 dilution of reclaimed water. At most times, the reclamation section was operated below full capacity or not at all, which caused a much higher dilution factor. For the period 1968 to 1991, the average blend of reclaimed water to conventional water was 1:3.5.

The second blending step took place in the bulk water system of Windhoek, where the blend from Goreangab mixed with treated water from other sources. The water from Goreangab was only dispersed over a limited number of water supply zones, which meant that the reclaimed water was not fully blended with all the water from conventional sources. Reclaimed water was only supplied to 5 of the 12 supply zones, with a maximum of 25% in any zone at any period.

At present the monitoring and testing of the system represent about 20% of the total production cost. In future, with an anticipated increase in plant capacity, these costs should drop to about 8% of the total costs.

The Future

Despite the commissioning of a long-distance water carrier from the north in 1971, which coped with increased water demand since that time, demand was again outstripping supply by 1995. The only affordable option was to enlarge the reclamation system capacity to the maximum—from the existing 4800 to 21,000 m³/d.

When the reclamation plant was commissioned in 1968, the Goreangab Dam provided a good quality surface water which was conventionally treated in a parallel treatment plant alongside the reclamation plant. In recent years, the dam water deteriorated sharply due to urbanization of the area immediately surrounding the impoundment, to the point where the conventional treatment plant became inadequate. In future, the parallel treatment of two separate streams will be abandoned and the new, upgraded treatment plant will be designed for the integrated treatment of dam water and maturation pond effluent.

The proposed process train was derived at after careful analysis of the historical process performance, site constraints and maximum integration of existing structural elements into the new upgraded and extended treatment plant. The initial phase separation steps will continue to be dissolved air flotation, sedimentation and rapid sand filtration. After filtration, ozone will be added to the water and allowed contact for twenty minutes before granular activated carbon adsorption. Chlorination at the plant will follow as well as at intermediate pumping stations to maintain a residual of 1.0 mg/L at the inlet to the service reservoirs.

REUSE OF WASTEWATER BY IRRIGATION

In South Africa, policy relating to the reuse of wastewater for irrigation differs from that of many other countries. In general the disposal of purified effluent by means of irrigation is discouraged. The view is that in most instances, more beneficial applications of treated wastewater are possible. Exceptions are: the irrigation of public parks and sport fields which would otherwise require abstraction of water from the available freshwater supplies; at the coast, where wastewater would otherwise be lost to sea; instances where the user is willing to trade existing water rights for treated effluent; and temporarily, where the producers of wastewater find it economically and technically impossible to meet the effluent standards for discharging to water courses.

The South African Department of Health formulated guidelines with respect to the public health effects of irrigation with treated municipal wastewater. In general, no irrigation with sewage effluent is allowed on crops that may be eaten raw. Surface irrigation of fruit trees, trellised vines and crops not eaten raw, is allowed with effluents that have undergone at least tertiary treatment and contain less than 1000 *E. coli* type 1 per 100 cm³. The same quality requirements apply to the irrigation of pastures for beef cattle, provided no grazing is permitted during irrigation. Dairy cattle may not graze on these pastures. Similarly, recreation facilities such as golf courses, parks and sports fields may be irrigated when not in use. There are limited restrictions on the quality of effluent for irrigation of crops from which dry fodder is to be produced, provided contact by humans and animals is prevented.

Applications for the irrigation of treated effluents are considered on merit, taking into account the specific circumstances pertaining. Some general policies bear on decisions concerning individual applications. For instance, in the Vaal River system no extensions of present irrigation with wastewater or any other water is allowed. Irrigation with very saline effluents are not acceptable, due to the impact on receiving soils and the risk of leachate polluting surface or groundwater sources.

An instance of promoting the indirect reuse of treated wastewater for irrigation, concerns the Zeekoegat sewage treatment works. Although not the cheapest option, the works was sited where its effluent could flow into the Roodeplaat Dam, which is primarily used for irrigation by downstream farmers.

An interesting example where beneficial use is made of effluent for crop production, is at Messina where dates are grown on the banks of an oxidation pond system. The quality of the water is poor, but this has no negative impact on the quality of the dates produced.

REUSE FOR THE ENVIRONMENT

In terms of its policy, the Department of Water Affairs and Forestry considers aquatic ecosystems to be ''the base from which the water resource is derived.'' The natural water environment is not considered to be a ''user'' of water, in competition with other users, but as the resource base, without which no development or water use can be sustained. Its protection is, therefore, seen as vital and essential.

In addition to the protection of aquatic ecosystems generally, wetlands are regarded as particularly important. There is also international pressure for their conservation, in view of the fact that South Africa in 1975 became the fifth contracting party to the Convention on Wetlands of International Importance, with special reference to Waterfowl Habitat (''Ramsar Convention''). The convention requires that wetlands be selected for the ''List of Wetlands of International Importance'' on account of their international significance in terms of ecology, botany, zoology, limnology or hydrology.

Because South Africa is a semi-arid country, with seasonal variability causing some rivers to completely dry up for parts of the year, aquatic ecosystems are already stressed under natural conditions. The situation is severely compounded by structures such as dams and management interventions by man. In this context, the controlled release of treated effluents to surface waters is making a significant contribution towards maintaining and protecting the natural environment, although quality considerations have to be carefully monitored.

The Blesbokspruit was listed as a Ramsar site in 1986. The Blesbokspruit is a typical highveld wetland, but the water levels are artificially maintained by embankments and by mining, industrial and municipal effluents.

Specific forms of environmental enhancement by reuse of wastewater is not widespread, but some interesting cases do exist. Several cases occur where treated effluent is discharged into a system of ponds, or a wetland, before flowing into a river or dam, creating environmentally attractive areas which attract numerous birds and sustain viable, albeit artificial, ecosystems.

INDUSTRIAL REUSE OF MUNICIPAL WASTEWATER

As stated previously, in many (if not most) instances of industrial reuse of municipal wastewaters, a biologically well treated effluent, sometimes followed by disinfection and/or sand filtration, is more than adequate to meet the quality requirements for a wide range of industrial reuse applications. This is illustrated in Table 25.3, which presents a selection of instances where municipal wastewaters are reused in industry. Two case studies, from the pulp and paper industry, where more advanced treatment is required, are dealt with below.

Sappi Pulp and Paper Group, Enstra [22]

The Enstra Mill, an integrated pulp and fine paper operation, and part of the Sappi pulp and paper group, situated near the city of Springs, uses secondary sewage effluent as the major part of its water supply. The present rate of water usage is made up of 15,000 m³/d fresh water (Rand Water Board) and 15,000 m³/d treated wastewater from the McComb and Ancor works of the Springs Municipality. These volumes have reduced substantially over the years as water saving measures have been implemented.

As the residual organic matter, plus colour bodies and heavy metal content of the secondary wastewater, are not compatible with the production of high quality fine paper, the effluent is further upgraded by an advanced treatment plant that was commissioned in July 1970.

The design of the reclamation plant was based on investigations carried out by Sappi and the National Institute for Water Research and makes use of alum coagulation, dissolved air flotation and a combination of chlorine

TABLE 25.3. Process Configurations and Modifications Since Commissioning [6].

Configuration 1 (1969)	Configuration 2 (1977)	Configuration 3 (1980)	Configuration 4 (1986)
Gammams Wastewater Treatment Plant			
Primary settling	Primary settling	Primary settling	Primary settling
Biological filters	Biological filters	Activated sludge†	Activated sludge
Secondary settling	Secondary settling	Secondary settling	Secondary settling
Maturation ponds	Maturation ponds	Maturation ponds	Maturation ponds
Goreangab Water Reclamation Plant			
Carbon dioxide‡	Lime†‡	Chlorine†‡	
Alum‡	Settling†	Alum‡	Alum†
	Ammonia stripping†	Lime‡	
Autoflotation	Primary carbon dioxide‡		Dissolved air flotation†
Foam fractionation	Chlorine‡	Settling	Chlorine*†‡
Breakpoint chlorination‡	Settling	Breakpoint chlorination‡	Lime†‡
Settling	Secondary carbon dioxide‡	Settling	Settling
Sand filtration	Sand filtration	Sand filtration	Sand filtration
	Breakpoint chlorination†‡	Chlorine‡	Breakpoint chlorination‡
Carbon filtration	Chlorine contact†	Chlorine contact	Chlorine contact
	Carbon filtration	Carbon filtration	Carbon filtration
Chlorine‡	Chlorine‡	Chlorine‡	Chlorine‡
Blending	Blending	Blending	Blending

Notes: † = Process modifications. ‡ = Chemical addition. Chlorine marked * only for intermittent shock dosing. Modifications to Configuration 4 (adapted in 1995) entail the use of ferric chloride in the place of alum, carbon filtration moved to immediately follow sand filtration and with breakpoint chlorination after carbon filtration.

plus activated bromine as used for bio-control. Alum is controlled via an on-line charge analyzer and bromine by an on-line redox potential unit, automating the plant and allowing it to adjust rapidly to incoming waste-water quality and demand variations.

Mondi Paper Mill, Durban [23]

The Mondi paper mill, consisting of a groundwood pump mill, a thermo-mechanical pulp mill, a recycled fibre plant and five paper machines for the production of approximately 1300 t/d of newsprint, supercalendered magazine paper and fine paper grades, was designed to use up to 8000 m^3/d of treated municipal wastewater from the nearby Durban southern sewage works as process water.

The advanced treatment plant required for upgrading the quality of the secondary effluent to that compatible with the production of the above grades of paper, was designed on the basis of result obtained from a pilot plant study and was commissioned in October 1972.

The plant consists of chemical coagulation with aluminium sulphate, activated silica and polyelectrolyte, followed by clarification, foam frac-tionation, sand filtration and activated carbon adsorption.

The mill has also instituted a mill effluent recovery scheme.

INTERNAL RECYCLING BY INDUSTRY

Under the pressures of limited water supplies, several South African industries have adopted intensive water conservation and recycling prac-tices. This approach is further promoted by the fact that when effluent has to be upgraded to meet discharge requirements, opportunities for reuse often arise. Some case studies are presented here.

Sasol Chemical Industries and Sasol Synthetic Fuel [16]

Sasol operates the world's largest fuel and chemicals from coal plants at its facilities in Sasolburg and Secunda. The Sasol Chemical Industries plant at Sasolburg has been in production since 1955. This factory has a daily water intake of ca 75,000 m^3/d and returns, on average, 50,000 m^3/d to the Vaal River. Through intensive recycling of treated raw water and effluents, large volumes of water are saved which otherwise would have to be abstracted from the river.

At Sasol Synthetic Fuel in Secunda, ca 200,000 m^3/d of raw water is taken in, used as utility cooling water and recovered for boiler feed water purposes. In addition, 120,000 m^3/d of industrial effluent is recovered through secondary treatment and used as make-up water in the process cooling water systems. No industrial effluent from this plant is discharged to surface water systems, making it a zero effluent discharge plant.

A tubular reverse osmosis plant (cellulose acetate membranes) is also being installed at Sasol Synthetic Fuel, in Secunda. The objective is to recover and recycle water from saline effluent dams, thereby maintaining the zero effluent discharge status without having to provide costly new storage capacity in the future. The full-scale computer controlled plant will consist of 8800 modules, and will process up to 5500 m^3 of effluent per day. The process was first tested in a 10-module plant and subsequently in a 400-module demonstration plant.

Water Conservation and Recycling in the Chemical Industry—AECI [24]

In the early eighties, severe drought conditions occasioned close scrutiny of water consumption and use in industry. This was true too for the chemical industry and indeed for AECI. Total water consumption at the major manufacturing facility at the time was 44,000 m^3/d, comprising 37,000 m^3/d from Rand Water, and 7000 m^3/d purified sewage effluent.

An extensive water conservation programme was introduced in 1982, which embraced reviews of technology capabilities in respect of water use, employee awareness and incentives, technical improvements and the hierarchical reuse of water. By 1984 total water consumption at the site had reduced to 40,000 m^3/d, comprising 30,700 m^3/d of water from Rand Water and 9300 m^3/d low grade water.

Water conservation continued to be a substantial issue for the AECI Group since that time, and between 1984 and 1995 total water consumption was reduced by a further 25%. Water consumption at the site in 1995 was 30,000 m^3/d: 27,400 m^3/d from Rand Water, and 2600 m^3/d low grade water. The further reductions in water consumption was mainly effected through in-plant recycling and the hierarchical reuse of water. These additional water savings since 1984 have been achieved in spite of the fact that the output of the factory has increased by about 15%. The savings in water cost is about $470,000 per year.

Desalination at Eskom Power Generation Plants [25]

Eskom generates the bulk of South Africa's electric power, through a network of coal-fired power generation plants and one atomic energy plant. In addition to the use of dry cooling systems at some of its plants, Eskom has also introduced desalination systems at two of its plants in order to increase the degree of recycling in the cooling systems, and to achieve zero effluent discharge.

At the Tutuka Power Station, cooling tower blowdown is recovered, using an electrodialysis reversal (EDR) system, with a capacity of 13,000 m³/d. This enables the implementation of a zero effluent discharge philosophy. It also allows the recovery of some 2500 m³/d of underground mine water with a high content of sodium chloride, which is introduced into the cooling system. Some 12 to 14 tons of salts are removed from the cooling water system in the brine. The volume of the brine is reduced through evaporation. The final brine is used in the ash conditioning process.

Zero effluent discharge is also pursued at the Lethabo Power Station. A tubular reverse osmosis system (cellulose acetate membranes) with a capacity of 9000 m³/d, is used for the recovery of cooling water blowdown. The system is operated in three streams of 3000 m³/d each. Normally two of the streams are in operation, with the third on maintenance/standby.

There is no mine water recovery in this plant, so that only 4 to 7 tons of salt are removed from the system per day. The final brine is used for ash conditioning and due to the higher ash content of the coal—as compared to Tutuka—evaporation is not required, as more brine, by volume, is needed in the ash plant.

MONDI PAPER MILL, PIET RETIEF [26]

In 1994 Mondi Kraft, a leading international supplier of pulp, paper and board products, installed a world-first system for the treatment of ''black liquor'' effluent streams generated in the production of pulp. The system was designed for the integrated pulp and liner-board facility at Piet Retief. The decision was motivated by a need to conserve water, to reduce operating costs, and to minimise the impact of the mill's effluent stream on the environment. The effluent treatment process was the result of more than three years of on-site pilot plant and laboratory studies.

The treatment plant currently treats 1700 m³/d, with facilities to upgrade to 2400 m³/d. The treatment process consists of four stages:

(1) Tubular ultrafiltration for the removal of suspended solids and organic compounds with high molecular weight
(2) Ion exchange for the removal of low molecular weight compounds
(3) Reverse osmosis, whereby the majority of remaining organic materials and dissolved salts are removed

The effluent stream is treated until the water is sufficiently clean for reuse in the factory. The final concentrated waste, in the form of salt cake, is produced in saleable form.

Columbus Stainless Steel Plant [26]

The new Columbus Stainless Steel factory at Middelburg produces various effluent streams during the manufacturing and beneficiation process, and has taken a proactive stance in order to comply with the discharge requirements laid down by the Department of Water Affairs and Forestry.

The effluent streams have been categorised into two broad streams, namely strong and weak effluents. After pilot trials, it was decided to treat the weak effluents stream with the aid of reverse osmosis. This effluent stream derives mainly from contaminated site run-off water and ion exchange regeneration effluents, and has a flow rate of ca 3300 m^3/d. The first reverse osmosis plant with a capacity of 1500 m^3/d was commissioned during July 1994, and a second with a capacity of 1800 m^3/d in May 1995.

Pretreatment consists of screening, coagulation, sand filtration, pH adjustment with sulphuric acid, and cartridge filtration. The permeate is reused in the factory as process water and the brine is further concentrated in an evaporator-crystallizer.

Abakor Abattoir, Johannesburg [27]

Abakor owns and manages 11 abattoirs throughout South Africa. Annually it processes some 1 million cattle, 4 million sheep and 0.75 million pigs. Its largest abattoir is situated in Johannesburg and can process up to 2000 head of cattle (or equivalent) daily. The fresh water consumption at the plant is about 2500 m^3/d and the effluent generated has a COD concentration of about 5000 mg/L. The cost of fresh water is $0.83/m^3$ and the cost of discharging untreated effluent into the municipal sewer is $2.74/m^3$.

A process was developed in which effluent is treated by using single cell bacteria. This reduces the effluent COD by 90–96% and the associated cost of discharging effluent is reduced to $0.21/m^3$. The single cell bacteria is harvested and used as high value animal feed with 43% protein.

The pretreated effluent is treated further by a process of sedimentation, filtration and ozonation to achieve potable water standards for general reuse in processing areas. The plant, commissioned in February 1996, has a capacity of 1000 m^3/d.

Iscor Steel Mills [28]

The Iscor Mills at Pretoria and at Vanderbijlpark both operate on a water quality cascading system, where the best quality water is used where there is a definite need for it, and the bleed-off or blow-down from one system feeds the next system and so on.

Steel production at the Pretoria works started in 1934 and since water is relatively scarce, the reuse of water was already planned at the design stage, but further improvements to the reuse systems have been made over the years. The following figures give an outline of water recirculating rates and consumption, for a daily production of 42,000 t of steel (in m^3/d):

Daily water intake	21,300
Daily circulating rate	730,000
Water circulated per ton of steel	175
Water consumed per ton of steel (evaporation plus effluent)	5.1
Water discharged as effluent	4100

Application of Counter-Current Dissolved Air Flotation [29]

Two cases are mentioned which illustrates the applicability of counter-current dissolved air flotation for industrial water recovery and recirculation.

South African Breweries at Alrode, uses two DAF units of 36 m^3/d and sand filtration to recover and reuse bottle wash water. In the case of Spoornet, Durban, train and wash water containing oil and suspended solids is recovered, employing sedimentation, foam fractionation, DAF and sand filtration.

CONCLUSION

Approximately 25% of wastewaters in South Africa is still being directly discharged to sea. This remains one of the areas where reuse strategies must be further investigated and developed.

Whereas planned indirect reuse is extensively applied in the interior of the country, this is generally not possible in coastal areas. Although the infiltration of treated wastewater into aquifers and subsequent withdrawal is an obvious route for planned indirect reuse in coastal areas, South Africa has very few aquifers that are suitable for this purpose. An exception is the large sandy aquifers that occur near Cape Town, and work has already been done towards exploiting their infiltration potential. In the south western Cape, surface runoff is being infiltrated together with treated wastewater of mainly domestic origin, while hydrogeological and other background studies, over a number of years, were conducted on a large sandy aquifer in the Cape peninsula to establish its potential for infiltration.

During 1995, the Minister of Water Affairs and Forestry launched a national water conservation campaign which will, during the course of 1997 culminate in formal policies by central government to promote water conservation in the various sectors of water use. These policies will, undoubtedly, encourage further application of water reuse practices, particularly in the industrial and municipal environment.

Pollution control policies too are being tightened, and as the "polluter pays" principle becomes more strictly applied, this will also become a driving force towards the reuse of wastewater which has to be upgraded, in any case, at high cost.

REFERENCES

1   World Health Organization. (1975). Health effects relating to direct and indirect reuse of waste water for human consumption. Technical paper series No. 7. World Health Organisation, International Reference Centre for Community Water Supply. The Hague, 1975.

2   Viviers, F. S. (1982). Proposed directives for the reclamation of treated effluents for direct recycling and reuse as domestic water. *Wat. Sci. Tech.* 14:1458–1463, 1982.

3   Bourne, D. E., Sayed, A. R. and Klopper, J. M. L. (1987). The Cape Town epidemiological base line and its sensitivity to detect changes in health patterns following the implementation of potable reuse. *4th Reuse Symp.,* Denver, 1987.

4   Bloom, B., Bourne, D. E., Sayed, A. R. and Klopper, J. M. L. (1988). Morbidity patterns from general practice in Cape Town. *S. Afr Med J,* 73:224–226.

5   Sayed, A. R., Bourne, D. E., Nixom, J. H. M., Klopper, J. M. L. and Opt Hof, J. (1989). Birth defects surveillance: a pilot system in the Cape Peninsula. *S. Afr Med J,* 76:5–7.

6   Haarhoff, J. and Van der Merwe, B. (1995). Twenty-five years of wastewater reclamation in Windhoek, Namibia. Paper presented at the *2nd International Symposium on Water Reclamation and Reuse,* Iraklio, Crete, Greece. 17–20 October 1995.

7   Beekman, H. G., Klopper, D. N., Fawcett, K. S. and Novella, P. H. (1991). Construction and operation of the Cape Flats water reclamation plant and the surveillance of the reclaimed water quality. Final Report to the Water Research Commission by the City Engineer's Department, Cape Town City Council. Pretoria 1991.

8   Vail, J. W. and Barnard, J. P. (1986). Reclamation of secondary sewage effluent by reverse osmosis: a pilot plant study. *Water SA,* 12(1):37–42, 1986.

9   Haarhof, J. and Fouche, L. (1989). Evaluation of combined flotation/filtration at Rietvlei water treatment plant. *First Bienn. Conf. Wat. Inst. Southern Africa.* Cape Town, 28–30 March 1989.

10   Offringa, G. (1995). Dissolved air flotation in South Africa. *Wat. Sci. Tech.* Vol. 31, No. 3–4, pp. 159–172, 1995.

11   Le Roux, J. D. (1989). The treatment of odorous algae-laden water by dissolved air flotation and power activated carbon. *Water Quality Bull.* 13(203), 72–77, 100.

12   Sontheimer, H., Crittenden, J. C. and Summers, S. (1988). *Activated Carbon for Water Treatment* (2nd ed. in English) DVGW Forschungstelle, Engler-Bunte-Institut, Universität Karlsruhe.

13   Le Roux, J. D. and Van der Walt, C. J. (1991). Development of a two-stage counter-current powdered activated carbon-dissolved air flotation system for the removal of organic compounds from water. Water Research Commission, 244/1/91. PO Box 824, Pretoria, 0001.

14 Isaäcson, M., Sayed, A. R. and Hattingh, W. (1987). Studies on health aspects of water reclamation during 1974 to 1983 in Windhoek, South West Africa/Namibia. Report WRC 38/1/87 to the Water Research Commission, Pretoria.

15 Breyer-Mencke, C. J. (1991). Private communication. AECI, Modderfontein. 1991.

16 Stegmann, P. (1996). Private communication. Sasol, Sasolburg. 1991.

17 Bester, I. (1991). Private communication. General Hide, Harrismith. 1991.

18 Slim, J. A. (1991). Private communication. Port Elizabeth Municipality, Port Elizabeth. 1991.

19 Du Preez, J. (1991). Private communication. Beatrix Gold Mine, Welkom. 1991.

20 Pitman, A. R. (1991). Private communication. Johannesburg City Council. Johannesburg. 1991

21 McAteer, D. (1996). Private communication. Nampak, Bellville. 1996.

22 Davies, C. (1996). Private communication. Sappi, Enstra, Springs. 1996.

23 Reid, J. C. (1996). Private communication. Mondi Paper, Merebank. 1996.

24 Breyer-Menke, C. J. (1996). Private communication. AECI Limited, Johannesburg 1996.

25 Aspden, D. (1996). Private communication. Eskom, Megawatt Park. 1996.

26 Rencken, G. E. (1995). Private communication. Debex Desalination, Booysens. 1995.

27 Muller, J. (1996). Private communication. Abakor, Pretoria. 1996.

28 Van Tonder, G. (1996). Private communication. ISCOR, Pretoria 1996.

29 Marais, S. W. (1996). Private communication. Aquatek (Pty) Ltd, Pretoria. 1996.

CHAPTER 26

The French Wastewater Reuse Experience

F RANCE has a temperate climate and is privileged for its water resources: it has no arid lands. Public authorities, of course, think and act on water resources management; however, these thoughts and actions are essentially geared towards a better protection of aquifers and surface waters.

The direct reuse of wastewater, in particular for irrigation, remains very marginal. Over the last 20 to 30 years, industry has made many efforts to save water and, therefore, within the development of "clean technologies," to recycle it. The driving forces behind this trend were the increase in water prices and the beginning of pollution taxes collected by the Water Agencies (Agences de l'Eau).

In contrast, there is no incentive for the agricultural sector to consider the potential new water resource constituted by treated wastewaters. Historically, in France, the spreading of wastewaters has only been used as a wastewater treatment technique. Today, the idea of preserving the receiving sensitive surface environments (bathing waters, fish farming sites, sites protected for ecological reasons) from pollution is very prevalent in the development of projects for wastewater reuse through irrigation.

The domestic reuse of wastewater, as practiced in Japan, has not been considered to date in France in spite of a few exploratory experiments, both because of its cost and because of the fear the sanitary authorities have about the development of double networks and the risks of cross-connections. In addition, the inclusion of wastewater treatment facilities in the utility rooms of housing units is considered undesirable.

J. Bontoux, Département Sciences de l'Environnement et Santé Publique, Université Montpellier I, Faculté de Pharmacie, av. Ch. Flahault—34060 Montpellier Cedex, France; G. Courtois, Direction Départementale des Affaires Sanitaires et Sociales, Montpellier, 615, boulevard Antigone—34000 Montpellier Cedex, France.

As a consequence, the interest in France for wastewater reuse, in particular for irrigation, often arose because French water experts were involved in development programs for southern Mediterranean countries. This interest also increased in the south of France in the 1990s, which was linked to the creation of golf courses, water guzzlers during the summer in coastal areas where water demand is also high. Finally, wastewater reuse also turned out to be interesting for a few particular sites, such as small islands, where water resources are a problem.

LOOKING BACK ON WASTEWATER REUSE IN FRANCE: THE SPREADING FIELDS NEAR PARIS

The first sewers of the Paris urban area started to be built in 1857 and were discharging into the river Seine. Soon, people realized the negative impact on river water quality and studied the possibility of spreading wastewater on sandy agricultural land.

In 1889, a law declared the creation of the spreading fields, a public utility project; several regulatory texts in 1899, 1905, and 1910 defined the protection perimeters of urban areas, within which spreading was forbidden, and they codified usage conditions. In 1910, the city of Paris had close to 5000 ha of land and could spread-treat almost all of its sewage.

Of course, wastewater treatment in Paris has changed since the beginning of the century. Today, the Acheres Biological Treatment Plant alone treats more than 2 million m^3 wastewater per day. However, the spreading fields still exist, but they have shrunk over the last few decades under the pressure of urbanization. In 1995, they formed a 2000-ha unit, receiving approximately 40 million m^3 screened raw wastewater per year, with flow peaks in the summer of several hundred thousand m^3 per day. More than half of this land belongs to the city of Paris, the rest belonging to private owners linked by contract to the city of Paris.

The crops grown there are maize, fruit, and vegetables. There are also poplar plantations, nurseries, etc. It is forbidden to grow vegetables susceptible to being eaten raw. The technical services of the city of Paris and the sanitary authorities are in charge of ensuring that the farmers respect their obligations concerning water usage.

The wastewaters are distributed on the fields by a network of reinforced concrete conduits totalling almost 200 km. In principle, the irrigated lands are prepared in billows so that the water can penetrate the soil directly at root level. The percolated waters are collected by open or underground channels carrying the (very satisfactorily) treated water to the Rivers Seine and Oise.

It is worth mentioning that, since 1947, the private landowners benefiting from the irrigation must make a yearly payment of 150 kg wheat per ha. Additionally, no serious sanitary problem was ever detected.

These spreading fields survive from a historical approach to the sanitation of a large urban area. They are not the result of a programmed wastewater reuse policy. In fact, they will eventually disappear; however, none are playing an appreciated role by avoiding an excessive charge of the River Seine during the low water periods. They also allowed the development of a local form of agriculture, still active, which will have to be reckoned with if the city of Paris decides to stop the distribution of wastewater.

THE 1991 RECOMMENDATIONS OF THE "CONSEIL SUPERIEUR D'HYGIENE PUBLIQUE DE FRANCE" (CSHPF; SUPERIOR PUBLIC HYGIENE COUNCIL OF FRANCE)

In 1990 in France, attention was drawn to wastewater reuse because the WHO had just published its recommendations (1989) and the country was experiencing a drought. Therefore, the Ministry of Health gave the CSHPF the responsibility of drawing up "Sanitary Recommendations for the Use, after Treatment, of Municipal Wastewaters for the Irrigation of Crops and Landscaped Areas," which were published in July 1991.

These recommendations are strictly adapted to the French national territory and rely in particular upon sanitary structures that are specific to the French administration. They are not yet of compulsory application. Their objective is to prepare forthcoming legislation to be enacted after an evaluation, within 5 years of their publication, of the wastewater reuse projects put in practice.

They link the approval of each project to an authorization, taking into account

- the protection of ground and surface waters
- the restrictions of use according to the quality of treated effluents
- the building of specific distribution networks for treated wastewaters
- the chemical quality of treated effluents
- the control of the hygiene regulations applicable to treatment and irrigation facilities
- the training of operators and supervisors

As for biological contaminants, the French recommendations are similar to those of the WHO, defining analogous water categories, A, B and C, with their associated usage.

The opinion of the CSHPF is given in Appendix 1. It would be wrong to sum it up by its similarities to the WHO recommendations because it goes much further, by maintaining an administrative authorization on a case by case basis, enabling the authorities to impose any restriction estimated necesssary for each project. The CSHPF did not want to fix drastic water quality limits once and for all. It wished to preserve an appreciation margin for the competent authorities within the frame of the WHO recommendations.

Appendix 2 provides, in substance, the composition of a dossier for the petition of an authorization for wastewater reuse by irrigation. Figure 26.1 summarizes the procedure followed by such a dossier. It highlights the preliminary consultation step between the administration and the project leader. It also shows the need for the parallel involvement of the administrative authorities from the Public Health and Environment administrations.

AUDIT OF THE CURRENT OPERATIONS

The Ministry of Health is currently carrying out a survey of the sites in operation and of the projects under consideration. The results of this survey are not yet final, but the following has been determined:

- There seems to be approximately 20 sites in France where wastewater reuse is being practiced. They appear to concern less than 500 ha if the historical and very particular cases of Paris and Reims are omitted.
- Only four or five of these projects appear to use category A water quality, which is required for the irrigation of golf courses or sports fields.
- About 10 projects appear to be in preparation for a total surface area of around 1500 ha. The most important is the Clermont-Ferrand project where 50 ha of crops could be extended to 500 or 600 ha in a second phase. In Clermont-Ferrand, the sanitary authorities have imposed the use of category A water in case maize is cultivated. This particular type of maize requires a manual detasseling. The frame of an epidemiological study has just been put in place to monitor the development of the project.

Once the national survey is completed, more precise data will be available; however, the order of magnitude of the surface areas irrigated by direct reuse of wastewaters is unlikely to change. These surface areas are extremely small when compared to the total French agricultural surface area.

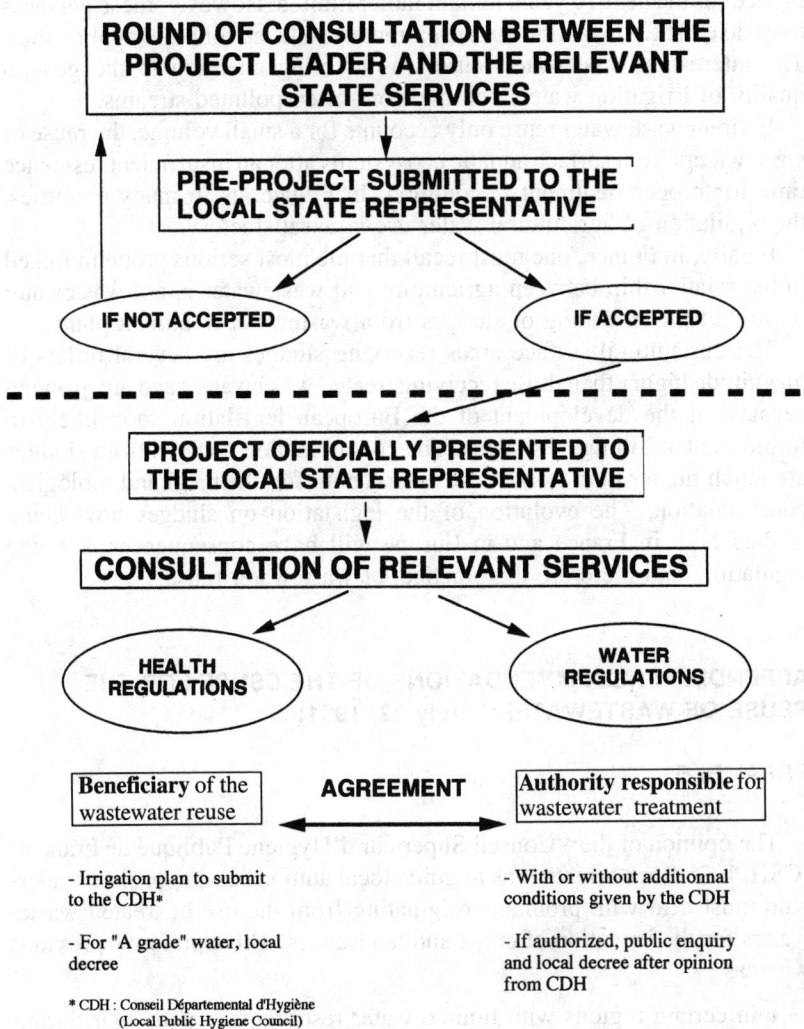

ROUND OF CONSULTATION BETWEEN THE PROJECT LEADER AND THE RELEVANT STATE SERVICES

PRE-PROJECT SUBMITTED TO THE LOCAL STATE REPRESENTATIVE

IF NOT ACCEPTED

IF ACCEPTED

PROJECT OFFICIALLY PRESENTED TO THE LOCAL STATE REPRESENTATIVE

CONSULTATION OF RELEVANT SERVICES

HEALTH REGULATIONS

WATER REGULATIONS

Beneficiary of the wastewater reuse	**AGREEMENT**	**Authority responsible** for wastewater treatment
- Irrigation plan to submit to the CDH*		- With or without additionnal conditions given by the CDH
- For "A grade" water, local decree		- If authorized, public enquiry and local decree after opinion from CDH

* CDH : Conseil Départemental d'Hygiène
(Local Public Hygiene Council)

Figure 26.1 Procedures followed by a dossier for the petition of an authorization for wastewater reuse by irrigation.

CONCLUSIONS

The perspectives for the development of direct wastewater reuse in France for the next 10 years remain rather limited. However, these perspectives do exist for irrigation but are extremely reduced for other applications. The interest in wastewater reuse was to bring attention to the general quality of irrigation waters coming from often-polluted streams.

If direct wastewater reuse only accounts for a small volume, the reuse of wastewaters from surface aquatic ecosystems after an insufficient residence time for proper treatment is common. In France, as in many countries, the regulation of agricultural water use is unsatisfactory.

Finally, in France, one must recall that the most serious problem linked to the relationship between agriculture and wastewater is not wastewater reuse, but the spreading of sludges from wastewater treatment plants.

The agricultural surface areas receiving sludges are several orders of magnitude higher than those receiving treated wastewaters and are growing because of the development of the European legislation soon likely to forbid the landfilling of sludges. The contamination charges from sludges are much higher than from wastewaters, both for chemical and biological contamination. The evolution of the legislation on sludges now being studied both in France and in Europe will have consequences over the regulation, and therefore the practice, of wastewater reuse.

APPENDIX 1: RECOMMENDATIONS OF THE CSHPF FOR THE REUSE OF WASTEWATERS (July 22, 1991)

PREAMBLE

The opinion of the ''Conseil Superieur d'Hygiene Publique de France'' (CSHPF) defines orientations to guide local authorities and project leaders who must deal with problems originating from the use of treated wastewaters for the irrigation of crops and landscapes. This opinion is provided whereas

- In certain regions with limited water resources, in particular during the summer, the use of properly treated wastewaters can be of large interest.
- The improvement in certain cases, of ground and surface water quality (land and marine), and consequently the safe uses of these waters (production of water for human consumption, bathing waters, aquaculture) can be expected; the irrigation with properly treated wastewaters allows, during certain periods, the avoidance

of discharges likely to contribute to the deterioration of surface waters.

This opinion only deals with the use of properly treated municipal wastewaters for the sole irrigation of crops and landscapes, i.e., to rationally satisfy the water needs of plants; the use of the soil as a way to dispose of and treat municipal wastewaters is not taken into consideration by these recommendations. This opinion has been drawn from international recommendations and from the national situation according to the available knowledge.

The first applications will have to provide a support to the development of additional research, to be promoted by competent authorities and organizations, aiming at improving the estimation of sanitary risks and at developing new techniques increasing the safety of the facilities. The following points will be worth a particular effort:

- the monitoring of the chemical and microbiological (especially intestinal helminth eggs, *Cryptosporidium* and *Giardia*) parameters in water (wastewater, irrigation water and aerosols), soil, and food products
- an epidemiological approach, in particular of the exposed personnel as well as the populations visiting the spray irrigated areas (golf courses, public gardens, etc.)
- the research and development of new treatment, storage, and irrigation processes aiming at improving the decontamination of effluents and decreasing the transmission risks of pathogenic agents

Additionally, the CSHPF requests the examination, at both national and international levels (and in particular at the European Union level) and comparatively to the requirements for wastewater reuse, of the risks linked to

- the irrigation of crops with surface waters that would not meet the quality requirements of treated wastewaters
- the consumption of imported vegetables irrigated with wastewater, knowing that, in some countries, the sanitary constraints on wastewater treatment, type of irrigation, and choice of crops are far from being respected

Finally, the CSHPF suggests

- that a guide be prepared at the national level for the technicians of both the local authorities and the project leaders, drawing expertise from the projects already implemented both in France and abroad
- that the elected representatives be at some point informed in order

to raise their awareness about the different aspects (sanitary, social, etc.) of wastewater reuse
- that an evaluation of the recommendations presented in this opinion be carried out within five years

OPINION

In the current state of knowledge, as presented to the World Health Organization (brochure nr. B0 778-1989) and after the examination of the risks to which operation personnel in daily contact with wastewaters, populations that could consume produce cultivated with treated effluents and people living near zones so irrigated are potentially exposed, the Superior Public Hygiene Council of France (Conseil Superieur d'Hygiene Publique de France, CSHPF) is favorable to the principle of using treated municipal wastewaters, under the conditions thereafter defined.

The conditions apply essentially to

- the protection of ground and surface water resources: the restrictions of use according to the quality of treated effluents, the distribution networks of treated wastewaters, the chemical quality of treated effluents, the control of the hygiene regulations applicable to treatment and irrigation facilities
- the training of operators and supervisors

However, the CSHPF draws attention to the (uncontrolled) risk to which the public could be exposed if all the restrictions defined in this opinion were not strictly respected. This concerns in particular the choice of crops to irrigate, the quality levels expected from wastewater treatment, and the potential impact on water resources in general.

PROTECTION OF GROUND AND SURFACE WATER RESOURCES

Preliminary Study

A prefectoral authorization, delivered in application of the decree nr. B0 73-218, dated February 23, 1973, taken in application of the law nr. B0 64-1245, dated December 16, 1964, relative to the flow and distribution of waters, is required for any municipal wastewater reuse project. This procedure allows for checking the compatibility of the project with the protection and vocation of the receiving environment, superficial or underground.

However, one should tailor the preliminary study required by this regulation by introducing the following information:

- the general characteristics of the site (topography, geology, surface or underground hydrology, soil condition)
- the local climatic conditions, the characteristics of the irrigation project: characteristics of the wastewaters (origin, quantity, main physico-chemical characteristics), water needs of the crops, frequency and conditions of water deliveries according to the absorption and exchange capacities of the soils, fate of the wastewaters when they are not used for irrigation
- other possible disposal options

In particular, this study must allow the evaluation of the risk of contamination of water resources used for human consumption as well as the risk of alteration of water resources that are especially vulnerable (karstic zones, land with insufficient natural protection, etc.).

Beyond the need for irrigation, the objective of a wastewater use project may be the improvement of the protection of surface water resources, e.g., by temporarily diverting an existing discharge. The impact study must in that case contain all the information needed to evaluate the desired improvements: receiving stream flow; quality of the environment, particularly in view of its uses; quality objectives, and so on.

The authorization document will set, on top of the daily flows permitted, the periods of irrigation, the quality levels of the discharges, the monitoring frequency, and the distances to be kept from the banks of streams, lakes, and ponds.

Adaptation of the Overall Regulation

In some particularly sensitive zones (notably karstic areas), the so-called "negligible adverse effect" threshold [set at 500 eq. inhabitants by the decree ("arrete") of May 15, 1975], below which the above-mentioned authorization is not required, will have to be lowered by decision of the prefect.

One is at least requested to forbid any water reuse project within the "immediate" (immediat) and "near"(rapproche) protection perimeters of points of water uptake for human consumption. This is due to the possible risks of contamination of groundwater that may be used to produce water for human consumption.

Irrigation Plan

Below the "negligible adverse effect" threshold, i.e., when no authorization is required, the irrigation facilities must still respect the requirements for the protection of water resources, as defined in the Departmental

Sanitary Regulation (Reglement Sanitaire Departemental, article 159) for wastewater spreading.

Thus, in conformity with this regulation, when an irrigation plan will be established and approved by the departmental sanitary authority, the requirements of this plan (quantity and quality of the effluents used, mode and periods of irrigation, lots irrigated) will be the only ones applicable. In the absence of an officially declared irrigation plan, spray irrigation will be forbidden and agricultural use will only be allowed on ploughable lands if:

- it is not practiced on lands that are or will be dedicated to fruit and vegetable crops within one year
- it is performed at more than 200 m of streams in case the slope is steeper than 7%

USAGE RESTRICTIONS ACCORDING TO THE QUALITY OF TREATED EFFLUENTS

In order to ensure the protection of public health and, in particular, that of the personnel in contact with wastewaters because of their occupation, that of the final consumer, and that of the populations living near irrigated areas, sanitary constraints on crop types and treated wastewater quality must be strictly followed. The type of irrigation also plays a very significant role on the possible long distance transport of pathogenic agents.

In general, the favored projects for the reuse of treated wastewaters will be based on a rigorous management plan:

- eliminating or severely restricting the possible contacts between people and water, as well as the risks of food chain contaminations
- limiting the dispersion of effluents, spray irrigation being merely tolerated in the absence of alternatives for reasons of hydrology

In reference to the work of the WHO (1989), three sanitary constraints categories are proposed, C, B and A, expressing increasing risks from the planned types of use and modes of irrigation.

Type C Constraints

No limit is set on the microbiological quality of wastewaters because the techniques used and the types of crops irrigated provide a break in the trasmission of risks from water. This is mainly dealing with underground or local (micro-irrigation) irrigation techniques, for cereal, industrial, and fodder crops, orchards, and forested zones, but also for green and landscaped areas not open to the public.

Previous treatment of the effluents is required for technical reasons (hydraulics, clogging, etc.).

Type B Constraints

Constraints level: Intestinal helminth eggs (taenia, ascaris) ≤1/liter

The respect of the type B constraint level aims at protecting the populations against parasites. This is especially the case for personnel of irrigated farms. This level is required for gravity or channel irrigation of orchards, cereal and fodder crops, nurseries, and vegetables eaten cooked (potatoes, beets, cabbages, carrots, etc.).

Spray irrigation of these crops, pastures, or fodder grass fields, as well as landscaped areas out of the reach of the public, are tolerated with this quality level under the following conditions:

- Spraying must be carried out far enough from houses and sports and leisure areas, taking local climatic conditions into account (this distance cannot be less than 100 m).
- Obstacles or screens (trees) must be set up to limit the propagation of aerosols, and the direct irrigation of the public paths and other ways of communication must be avoided.
- Operation and maintenance personnel must be sufficiently protected against the risks of inhalation.

Sports fields, used several weeks after irrigation, can be irrigated with wastewaters meeting the type B constraint level. The type B constraint level can be met, for example, by a series of settling tanks with a residence time of around ten days or any other process of comparable efficiency.

Type A Constraints

Constraint level: Intestinal helminth eggs (taenia, ascaris) ≤1/liter
Thermotolerant coliforms <10,000/liter

By introducing an additional bacteriological constraint, the type A constraint level aims at ensuring, beyond the protection of farm personnel and cattle, the safety of products that can be eaten raw. This quality requirement must be complemented by the use of irrigation techniques, avoiding wetting the fruit and vegetables: gravity irrigation, watering under canopy, etc.

This level will also be tolerated for the irrigation of sports fields (golf courses) and landscaped areas open to the public if the following constraints are simultaneously respected:

- Spray irrigation must be performed outside of the opening hours for the public.

- The sprays must have a short reach.
- The same distance requirements from houses as those for type B constraint level must also be respected.

The type A constraint level can be attained, for example, by treatment in a facultative pond system or any other system providing an equivalent level of treatment. A 30-day residence time in well-designed, properly operated, and sufficiently illuminated basins can provide the required bacteriological quality level.

DISTRIBUTION NETWORKS OF TREATED WASTEWATER

The dispositions of the "Departmental Sanitary Regulations" about nonpotable water distribution networks must be applied to the special case of the distribution networks of wastewater under pressure. Thus, any interconnection between the drinking water network and the treated wastewater network is forbidden. The treated wastewater network must remain out of the reach of the public or any person other than the operation personnel.

CHEMICAL QUALITY OF THE TREATED WASTEWATERS FOR THE IRRIGATION OF CROPS

Effluents that are mainly municipal, as defined by the NFU 44041 standard, can be used, after treatment, for the irrigation of crops and landscapes. The use of nonmunicipal effluents remains dependent on a specific examination of their chemical quality because of the possible presence (in excessive quantities) of mineral or organic chemical micropollutants. In some cases, this use may be forbidden.

Whatever the case, the application for discharge authorization requested by the water police must contain

- precise information about the nature and quantities of the products released upon discharge of industrial effluents to the sewers
- at least one analysis of the treated effluent covering the general pollution parameters (SS, BOD5, COD, TKN), the heavy metals mentioned in the NFU 44041 standard and the organic substances likely to be present in large quantities
- one analysis of the sludges generated by the treatment plant (NFU 44041 standard)

When the concentrations measured in the sludge exceed the levels set by the standard for at least one trace element parameter (Cd, Cr, Cu, Hg, Ni, Pb, Zn), a more in-depth examination of the treated wastewater will be required, particularly if this water is destined for the irrigation of vegetable, cereal, industrial, and fodder crops, as well as for grazing.

The discharge authorization will have to be reexamined if

- the reused wastewaters have been significantly enriched in toxic substances
- the limit values of the yearly amounts of heavy metals that can be added to cultivated soils, introduced by the NFU 44041 standard, are not respected

It is also important to know and monitor the concentrations of nutrients (N, K, P) in the treated wastewaters. These data will allow the tailoring of the possible agronomical use of fertilizers and avoiding an excessive use of nitrogen.

CONTROL OF THE HYGIENE REGULATIONS APPLICABLE TO TREATMENT AND IRRIGATION FACILITIES

Administrative Procedures

Beyond the procedures considered previously for the protection of water resources, the CSHPF proposes that any project for the use of wastewaters in which the water must reach a type A constraint level should be submitted to the approval of the prefect after advice from the Departmental Hygiene Council.

This procedure will allow

- checking that the above-mentioned crop restriction conditions are effectively respected in the projects
- properly performing the control of the hygiene rules in operating facilities
- informing and advising the operators

This procedure could be introduced by a decree taken in application of article L1 of the Code of Public Health.

For discharges submitted to authorization in the frame of application of the law of December 16, 1964, the instruction of both procedures could be performed simultaneously in order to avoid the repetition of administrative constraints.

Monitoring and Control of the Facilities

One must distinguish the monitoring of the operations of the facilities providing the required quality level; it will be performed by the Departmental Directorate of Sanitary and Social Affairs (DDASS), either directly or with the help of the Service for Technical Assistance to the Treatment Plant Operators (SATESE). The monitoring frequency of these services should be increased, in particular when a type A constraint level is required.

Microbiological analyses and chemical analyses of the treated effluent covering the nutrients must be performed on a regular basis (at least once every three months). When the treated wastewaters are used for the irrigation of vegetables that can be consumed by humans or cattle, these determinations will be complemented by the detection of specific micropollutants, in particular nickel and cadmium; the detection of other heavy metals and organic substances will be made if justified by the nature and quantity of discharges upstream from the facility.

The control of hygiene rules applicable to the irrigation of crops must also be enforced. This control must be performed by the Departmental Directorate of Sanitary and Social Affairs, particularly in the case of crops to which the type A constraint level is applicable. The control of plants (cadmium measurement) to be made by the competent services.

Periodical Audit

For five years, a periodical audit will be performed by the Departmental Directorate of Sanitary and Social Affairs and presented to the Departmental Hygiene Council and, if necessary, to the CSHPF. After this audit, and if the recommendations of this opinion are not respected, the withdrawal of the authorizations must be considered.

TRAINING OF OPERATORS AND SUPERVISORS

The operating personnel, the agents in charge of the control or of the technical assistance to the treatment plant operators, and, if applicable, the agents from official laboratories, must be properly trained (hygiene, sanitary risk, water quality, self-control, analysis, etc.).

ADDITIONAL RECOMMENDATIONS FOR THE USE, AFTER TREATMENT, OF MUNICIPAL WASTEWATER FOR THE IRRIGATION OF CROPS AND LANDSCAPES (August 3, 1992)

Whereas a number of elements have already been taken into account by the CSHPF for the estimation of the sanitary risks linked to the use of properly treated municipal wastewater (Report from July 1991) for the irrigation of crops and landscaped areas;

Whereas, in particular, an efficient protection of, on the one hand, consumers of produce susceptible to being eaten raw and, on the other hand, of the personnel of wastewater reuse facilities, as well as the users of landscaped areas, needs to be ensured;

Whereas certain limit values must not be exceeded by treated wastewaters in order to comply with the type A constraint level (intestinal helminth eggs ≤1/liter, thermotolerant coliforms ≤10,000/liter);

Whereas a sufficient number of analytical results are needed in order to assess the efficiency of treatment processes;

The CSHPF, according to the proposal from the task force on the "Use of Wastewater," complements its previous opinion by introducing the following precisions:

(1) The limit values accepted for the intestinal helminth eggs and for the thermotolerant coliforms must be considered as imperative values, to be respected under all circumstances by treated wastewater used for the irrigation of crops and landscaped areas.

(2) The sampling frequency must have a rhythm of at least one sample every two weeks, for a period of at least one year before the beginning of the effective use of the treated wastewater for irrigation and during the first period of use.

(3) In the case when the limit value is exceeded, a new analysis will immediately be performed in order to confirm the previous result. If the breach persists and after enquiry from the sanitary authorities, the use of wastewater must cease, either temporarily or permanently.

(4) The sampling frequency can be reduced by half when all the analysis results from the previous period of operation have been judged to comply and no new event could affect the performance of the treatment facility.

APPENDIX 2: TYPICAL DOSSIER COMPOSITION OF A WASTEWATER REUSE PROJECT

(1) Name and address of the applicant

(2) General objectives
  - irrigation, protection of the receiving environment
  - describing the plants to be irrigated, the structure and characteristics of the farm or urban facilities (size, operators, likely evolution, etc.)
  - describing the surrounding environment and its uses

(3) Characteristics of treatment facilities and of treated effluents (existing facilities and plans for future development)
  - available effluent volume, current outfall, outfall in case of partial reuse of the water or in case of interruption of irrigation, preserved stream flow

- fate of the out-of-specification effluents
- characteristics of the sludge produced (representative analysis of the sludge: AFNOR U 44-041 standard)
- characteristics of the effluents before treatment (nature and type of effluent, connected industries)
- characteristics of the effluents after treatment (representative analyses of the usual parameters but also of any other parameter that may be involved in the various types of protection to be considered for the surface environment, the soil, the crop, the would-be human or animal consumer, the operating personnel and the neighboring communities). The content frequency and periodicity of the sampling for these analyses will be determined during the first project meeting. They can be complemented by other analyses of interest for the operator (fertilizing capacity of the effluent, phytotoxic compounds, etc.).

(4) Characteristics of the storage facilities
- justification and conditions of the storage (according to the characteristics of the treated wastewater)
- storage management (according to the crop calendar: immediate and delayed needs)

(5) Environmental constraints
- 1/25000 map showing the location and layout of the plots involved. The data and constraints that can be schematized will be indicated on this map.
- general geological and hydrogeological data from the study of the available data (vulnerability of the aquifers)
- existing urban zoning documents (taking into account zoning plans and rules when they exist, as well as the protection perimeters for public water collection sites)
- map of the plots at an adequate scale (e.g. 1/2500) identifying:
  —the local topology (slope, relief, etc.)
  —the hydrographic network
  —the inhabited buildings, the paths and roads, the water uptake points (wells, boreholes, springs), the ditches or small streams located less than 200 m from the irrigated zones or that could be influenced by the irrigation
  —the pedological characteristics of the plots based on the available knowledge from soil analysis drill sampling. Any hydro-morphic soils detected must be noted. Soil pH, soil metal content, slope, useful soil thickness, stone load and available water holding capacity, cation exchange capacity, and texture are also determined.

—the geological and hydrogeological data, in particular the depth of the aquifer under the plot and the special features (faults, etc.)

—the irrigated surface areas and the corresponding mode of irrigation (in the case of spray irrigation, the reach of the sprays must be drawn as well as the protection zones for third parties)

(6) Climatic and crop characteristics
- the climatic factors (temperature, frost days, precipitation, winds); the crop factors (evapotranspiration, water needs, water holding capacity, capacity available for the plant, etc.) will be taken into account to define the irrigation water needs, the volume of water used and discharged to the environment according to the periods and the various types of crops (case of crop rotation)
- An impact document will provide, according to the climatic and seasonal variations, the impacts of the irrigation, storage and transport included, on water resources, on the aquatic environment, on water quality, including run off, so as to satisfy the public health and well-being requirements.

(7) Management
- The irrigation calendar per type of crop will be completed by diagrams of distribution of effluent volumes:
  —used by the farmer
  —not used (direct discharge to the environment or long-term storage) (volume, kg BOD_5/y, kg nitrogen/y, etc.)
- An agreement between the treated effluent producer and the user will be established, which will provide the recommendations for the use of the treated effluents and the mutual agreements on a minimum duration (5 years would be desirable).
- person in charge of the treatment of the effluents
- person in charge of the management of the wastewater reuse plan

(8) Design of the monitoring
- As a complement to the monitoring as required by the "arrete" of December 22, 1994, concerning the monitoring of the wastewater collection and treatment facilities, a log book will be kept available for the competent supervisory services and containing at least
  —the analyses results (according to the predefined content and frequency)
  —the quantities of effluent used
  —the dates of irrigation
  —the inventory of the irrigated plots

—the crops on each plot
—the incidents occurred
- Soil monitoring specifications: on average, one complete physical, chemical and trace element analysis every three years

ACKNOWLEDGEMENTS

We thank Laurent Bontoux of the Institute for Prospective Technological Studies (European Commission, Spain) for the advice he gave during the writing of this text and his translation into English.

CHAPTER 27

Indirect Potable Reuse of Reclaimed Water

INTRODUCTION

THE purpose of this chapter is to present the state of the art of indirect potable reuse. Indirect potable reuse is the recovery of water from wastewater for the purposeful reintroduction into either a surface water or groundwater body that ultimately serves as a drinking water supply. This chapter does not focus on direct potable reuse which is the purposeful reintroduction of reclaimed water directly into either a water treatment plant or potable water distribution system. The following topics will be covered in this chapter:

- What is indirect potable reuse?
- Why are water utilities increasingly considering indirect potable reuse for augmenting raw water supplies?
- History of indirect potable reuse.
- Indirect potable reuse water quality and public health safeguards.
- Indirect potable reuse case studies.

WHAT IS INDIRECT POTABLE REUSE?

The concept of indirect potable reuse can be described by what it is, as well as, by what it is not. Figures 27.1 through 27.3 help to distinguish

Brock McEwen, CH2M Hill, P.O. Box 241325, Denver, CO 80224-9325.

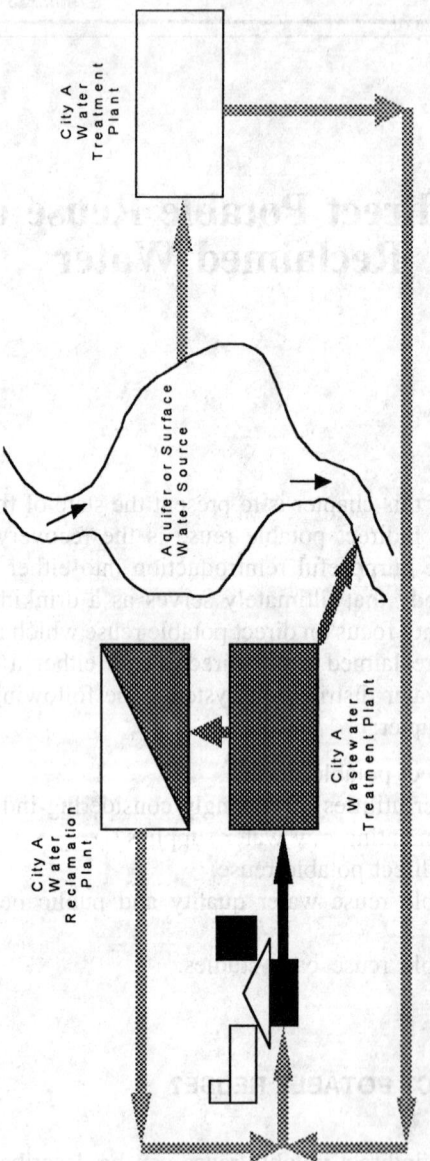

Figure 27.1 Direct potable reuse.

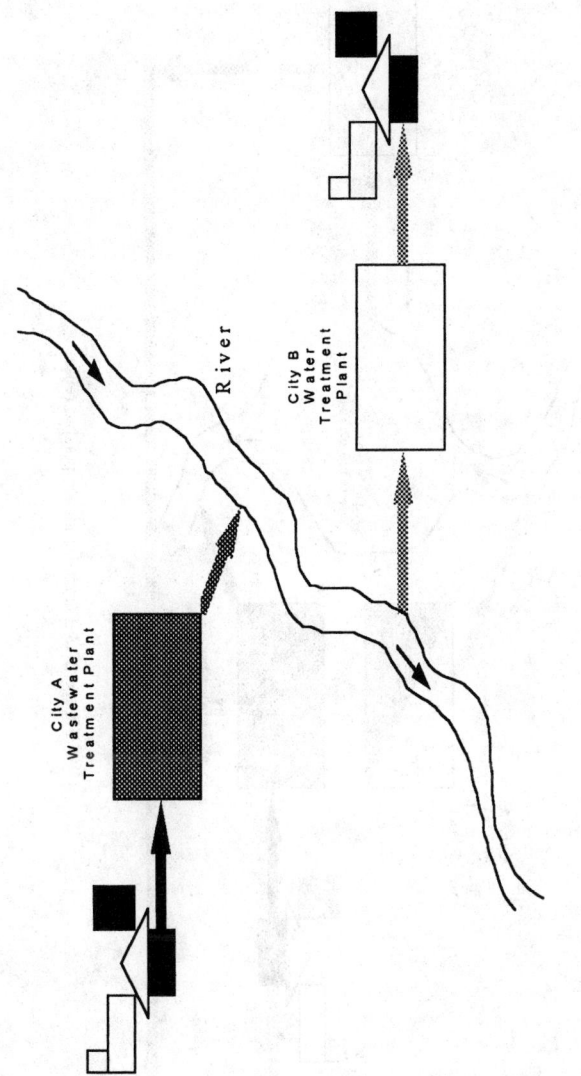

Figure 27.2 Unplanned, indirect potable reuse.

1213

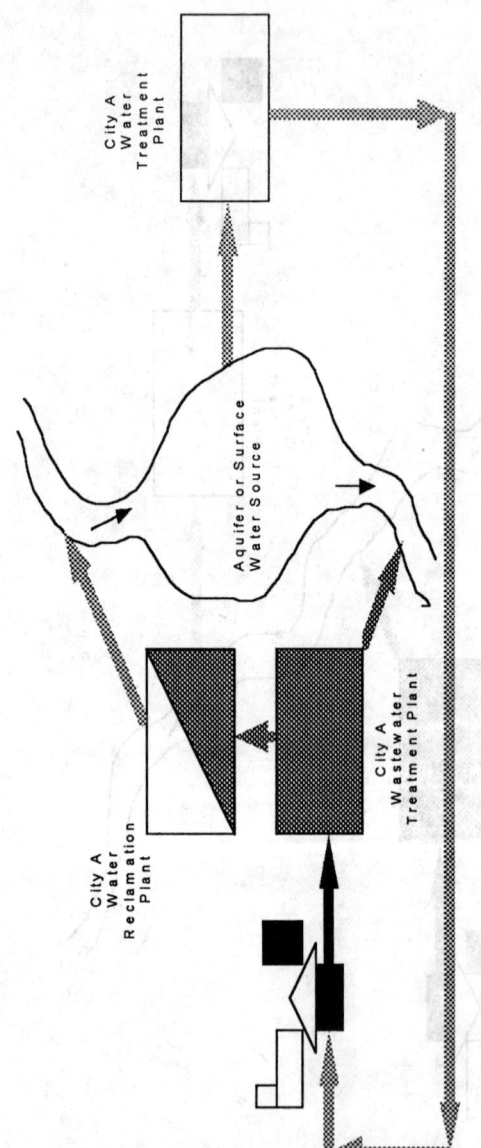

Figure 27.3 Planned, indirect potable reuse.

the concept of indirect potable reuse as a unique water resource management tool.

Figure 27.1 presents the concept of *direct potable reuse*. Direct potable reuse involves the direct conveyance of water recovered from wastewater to either a potable water distribution system or a water treatment plant. Currently, there are no full-scale applications of direct potable reuse in the United States.

Figure 27.2 presents the concept of *unplanned or incidental, indirect potable reuse*. Unplanned, indirect potable reuse occurs whenever a water supply is withdrawn for potable purposes from a natural surface or underground water source that is fed in part by the discharge of a wastewater effluent. The wastewater is discharged to the water source as a means of disposal and therefore subsequent reuse is an unplanned or incidental byproduct of the wastewater disposal practice.

Figure 27.3 presents the concept of *planned, indirect potable reuse,* which is the subject of this chapter. Planned, indirect potable reuse is the purposeful augmentation of a surface water source or recharge of an underground water source with a water recovered from wastewater with the intent of reusing the water resource.

WHY ARE UTILITIES CONSIDERING INDIRECT POTABLE REUSE?

Water reuse is becoming an increasingly common component of water resource planning as opportunities for conventional water supply development dwindle and costs for wastewater disposal climb. Indirect potable reuse of reclaimed municipal wastewater offers means to extend and maximize the utility of limited water resources.

The following are common factors that often contribute to the serious consideration of indirect potable reuse as an alternative water supply.

HIGH COST AND ENVIRONMENTAL IMPACTS OF CONVENTIONAL APPROACHES

Locally available conventional water supplies are limited or exhausted in many large southwestern cities in the United States leading them to rely on, or plan to use, indirect potable reuse as part of their overall water resource management program to satisfy growing water demands. For instance, both the cities of Los Angeles and San Diego, California are planning for indirect potable reuse to supply 10 percent of their water demands by the year 2010.

Clearly, the more remote a source of supply from the water demand, the higher the costs involved in construction of raw water transmission and storage facilities. But, beyond the increased infrastructure cost, the cost to permit such remote supply development projects is becoming prohibitive. Major water resources projects involving trans-basin water conveyance, such as California's State Water Project, the Central Arizona Project, the Central Utah Project, and much of the water supply along the front range of the Rocky Mountains must satisfy a large number of local, state, and federal permits. These permits cover the legislative authority that oversees water rights, environmental, public health, and public interest issues that are either directly or indirectly affected by the various components of a water supply development project. Most major water supply development projects involving regional reallocation of water resources must meet the requirements of the National Environmental Policy Act and adequately address the needs of a number of other federal laws including, but not limited to, the Endangered Species Act, the Migratory Bird Act, the Golden Eagle Protection Act, the Fish and Wildlife Coordination Act, the Historical Preservation Act, and the Clean Air Act. In short, the entire programmatic policy of federal agencies traditionally involved with water supply development has shifted from dams, reservoirs, and aqueducts to water conservation and reuse.

The EPA disapproval of the Denver Water Department's proposed Two Forks Dam and Reservoir Project is an excellent example of growing public dissatisfaction with traditional water supply development practices in the United States. The Denver Water Department spent tens of millions of dollars preparing environmental impact statements for this project which would have served the Denver metropolitan area's water needs well into the twenty-first century. Ultimately, however, pressure applied by environmentalists to preserve the South Platte River trout fishery and allegations of poor water conservation practices in Denver led to the project's failure.

Similar environmental concerns fortified by the Endangered Species Act, have thwarted further expansion of California's State Water Project based on preservation of the environmental integrity of the San Joaquin/Sacramento River Delta in Northern California.

Beyond environmental concerns, remote water development to support urban demands is in increasing competition with agriculture. In some instances agricultural water rights are sold outright or exchanged with urban water users, but in other instances intense court battles ensue that further complicate and increase the cost of remote water supply development. In Florida, agriculture is the second largest industry in the state behind tourism. Agriculture is geographically located inland, and tourism is geographically located along the coastlines. There is pressure on the urban coastal cities to improve water management to reduce pressure on inland water supplies that support agriculture.

For the previously stated reasons, indirect potable reuse, in many cases, offers a water supply development alternative more tractable than remote, conventional water supply development.

WASTEWATER DISPOSAL STANDARDS ARE BECOMING MORE STRINGENT

Throughout the U.S., National Pollutant Discharge Elimination System (NPDES) permits, issued to wastewater treatment plants under the Clean Water Act, are becoming more restrictive in the area of both nutrient control and aquatic life metals criteria.

In Florida, for instance, protection of coastal waters and the inland Everglades ecosystem is resulting in extremely low nutrient limits with total nitrogen less than 2 mg/L and total phosphorus less than 0.1 mg/L. The need to provide nutrient removal, in some cases, begs the question as to whether it might be more advantageous to recover water for indirect potable reuse rather than discharge the highly treated wastewater. For this reason, the City of Tampa has been pursuing an indirect potable reuse project to augment surface water supplies during their annual summer dry period.

Protection and restoration of both freshwater and coastal fisheries are, in some instances, resulting in aquatic life metals criteria that are an order-of-magnitude lower than drinking water metals standards. These criteria often require intense chemical treatment and/or membrane treatment for discharge compliance. For this reason, it may be more advantageous to reclaim water from wastewater for indirect potable reuse than discharge to the aquatic system. For example, San Jose, California; Boise, Idaho; Phoenix, Arizona; and Gwinnett County, Georgia have all investigated the feasibility of indirect potable reuse as an alternative to complying with strict NPDES proposals.

WATER RIGHTS FAVOR REUSE RATHER THAN DISPOSAL

In some instances, water rights encourage reuse rather than disposal. For example, water rights that the Denver Water Department acquired and import to the front range from west of the continental divide are contingent on frugal usage including successive reuse. In many cases where water is imported from one basin to another, the importer has no obligation to discharge their wastewater for downstream users. If, however, the water importer does not diligently practice and communicate their imported water reuse plans, then downstream users will likely grow accustomed to the wastewater discharge augmentation of their raw water supply, and possibly contest withdrawal of the importer's discharge. This was the case in Phoenix, Arizona, where a downstream user on the Salt River litigated the right of the city to sell wastewater effluent to a nuclear power plant for cooling water reuse rather than discharge to the river. The case was ultimately decided in favor of the city by the state supreme court. The court declared that once the water passed through the city's water management infrastructure, ownership of the water right became the city's. Water rights and case law vary regionally throughout the country. In addition, case law is in a continuing state of flux due to changing priorities, issues, and concerns. Therefore, the impact of indirect potable reuse practices on a regional basis must be considered for the protection of impacted parties.

ECONOMICALLY VIABLE, NONPOTABLE REUSE OPPORTUNITIES ARE EXHAUSTED

The highest quality water supplies should be prioritized for drinking purposes. Consequently, it is unlikely that indirect potable reuse will gain wide support unless supply of nonpotable demands with water of less pristine origin have been fully investigated and exhausted. Urban nonpotable demands, however, are typically as widely dispersed geographically as are potable demands. Consequently, a nonpotable dual distribution system must often nearly duplicate the breadth and scope of current potable distribution systems to take full advantage of the available nonpotable supply. Furthermore, in many instances wastewater treatment facilities are strategically located for effluent discharge as opposed to redistribution to users. As a result, in a developed urban area, it is often more cost effective to implement more treatment intensive indirect potable reuse and use the existing potable water supply, treatment, storage, and distribution network than to implement a dual nonpotable system. Generally, nonpotable reuse is most economical for serving large users, such as parks, schools, golf courses, and industries that are geographically located in reasonable proximity of the nonpotable, treated wastewater effluent source of supply. Dual distribution systems are also more cost effective when masterplanned in advance of development, thus avoiding the costly installation associated with distribution system construction in existing and often congested roadways. Lastly, urban demand for nonpotable water fluctuates both diurnally and seasonally which often precludes full usage of the reclaimed water supply, or requires costly storage systems that often require near drinking water quality to permit.

POLITICALLY VIABLE, DEMAND REDUCTION MEASURES ARE EXHAUSTED

Similar to nonpotable reuse, most water professionals and the public alike philosophically agree that conservation of potable water should be a priority, particularly when the next increment of raw water supply may be of wastewater origin. Certainly, an aggressive leak detection and repair program should be in place to minimize losses. In addition, water use restrictions and water rate schedules can be modified to encourage wise water use. Although it is difficult to argue against water conservation, we live in a competitive society where water and its availability impact economic development. Consequently, city policy makers must balance water resource management practices with socioeconomic costs and benefits.

U.S. HISTORY OF INDIRECT POTABLE REUSE

Unplanned, indirect, potable reuse has been in practice since man first began disposing of wastewaters into watersheds that are hydrologically connected to raw water supplies. As populations have increased, so too has the quantity of wastewater and the technology to manage these increased volumes of wastewater. Indirect potable reuse is one of the developing strategies to both manage wastewater and recover and reuse water resources.

The following is a summary of some of the historical milestones marking the development of planned potable reuse as a viable component of a water resource management plan.

1956–1957: CHANUTE, KANSAS [1]

In 1956 and 1957, severe drought conditions forced the City of Chanute, Kansas to practice direct potable reuse. Secondary treated water was impounded and recycled back through the city WTP where it received super chlorination to inactivate the greater pathogen concentrations associated with the recycled effluent. No adverse health effects were noted, however, after several cycles, the water became pale yellow in color and was aesthetically unappealing.

1960: USPHS

In 1960, the Advanced Waste Treatment Research Program of the United States Public Health Service (USPHS) was directed to develop new treatment technology for renovating wastewater to allow more direct and deliberate water reuse.

1962: WHITTIER NARROWS GROUNDWATER REPLENISHMENT PROJECT, CALIFORNIA [2]

Since 1962, the County Sanitation Districts of Los Angeles County (Figure 27.4) have been surface spreading disinfected secondary effluent

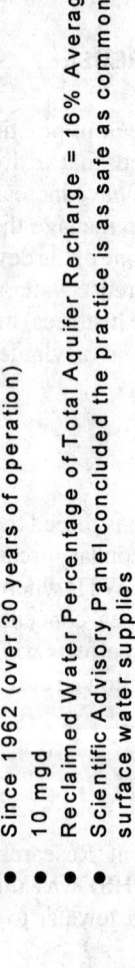

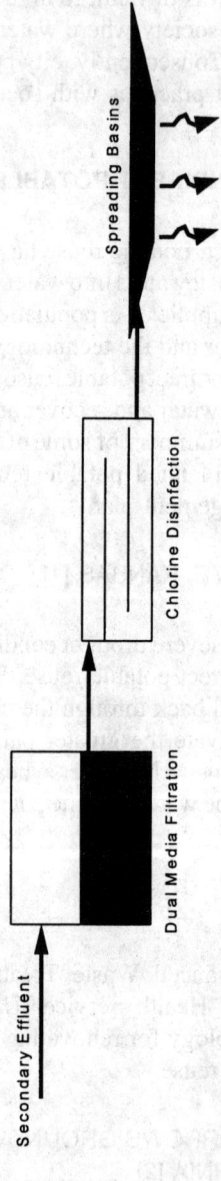

- Since 1962 (over 30 years of operation)
- 10 mgd
- Reclaimed Water Percentage of Total Aquifer Recharge = 16% Average
- Scientific Advisory Panel concluded the practice is as safe as commonly used surface water supplies

Secondary Effluent

Dual Media Filtration

Chlorine Disinfection

Spreading Basins

Aquifer Recharge

Figure 27.4 Whittier Narrows Groundwater Replenishment Project.

(dual media filtration was added later in 1978) from a 10-million-gallon-per-day (mgd) water reclamation plant for infiltration to an underground potable water supply. This operation continues and the amount of reclaimed water recharged annually averages 16 percent of the total inflow to the groundwater basin.

Depending on the physical characteristics, location, and pumping history of a given well, the population drawing potable water from the groundwater basin is estimated to be exposed to a reclaimed water percentage ranging from 0 to 23 percent. After extensive data acquisition, evaluation, and statistical analysis by an independent scientific advisory panel to the state of California, the panel concluded that the Whittier Narrows groundwater replenishment project was as safe as commonly used surface water supplies.

1965: SOUTH LAKE TAHOE WATER RECLAMATION PLANT [3]

In 1965, the South Lake Tahoe Water Reclamation Plant was the first large-scale facility to use advanced wastewater treatment processes to remove nutrients and organics from wastewater (Figure 27.5).

1969: WINDHOEK, NAMIBIA [4]

In 1969, the Windhoek (Figure 27.6) experimental, 1 mgd direct potable reuse plant was commissioned. Since 1971, finished water from the Windhoek plant has occasionally been used to directly augment potable water supplies. Based on epidemiological studies, no increased incidence of illness or disease has been associated with this water recovery and reuse practice. Founded on these positive results, the plant is slated to be expanded to 4 mgd in the near future.

1972: FEDERAL WATER POLLUTION CONTROL ACT [5]

In 1972, the Federal Water Pollution Control Act stated that discharge of pollutants into all navigable waters will be eliminated thereby encouraging water recovery and reuse.

1976: ORANGE COUNTY, CALIFORNIA WATER DISTRICT [6,8]

In 1976, the Orange County California Water District's Water Factory 21 began operation (Figure 27.7). The 15-mgd facility reclaims unchlorinated secondary effluent to drinking water quality and recharges it into a heavily used groundwater to prevent salt water intrusion. The water recovery treatment includes lime clarification, air stripping, recarbonation, filtra-

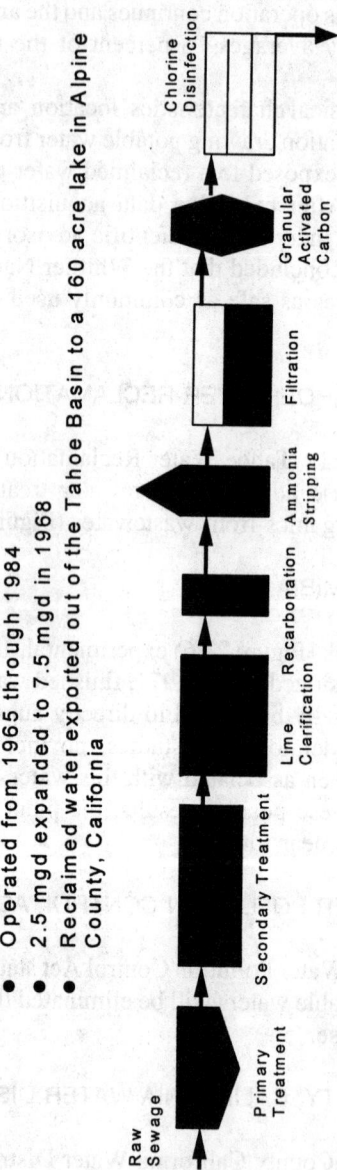

Figure 27.5 South Lake Tahoe Water Reclamation Plant.

- Operated from 1965 through 1984
- 2.5 mgd expanded to 7.5 mgd in 1968
- Reclaimed water exported out of the Tahoe Basin to a 160 acre lake in Alpine County, California

Raw Sewage

Primary Treatment

Secondary Treatment

Lime Clarification

Recarbonation

Ammonia Stripping

Filtration

Granular Activated Carbon

Chlorine Disinfection

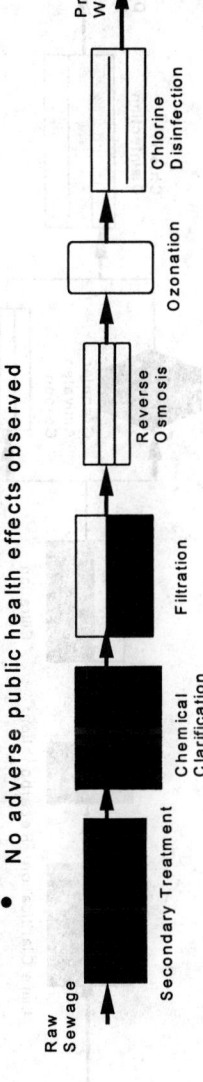

- Since 1971 (over 20 years of operation)
- No adverse public health effects observed

Figure 27.6 Windhoek, Namibia Direct Potable Reuse Plant.

Raw Sewage → Secondary Treatment → Chemical Clarification → Filtration → Reverse Osmosis → Ozonation → Chlorine Disinfection → Product Water

- Since 1976 (20 years of operation)
- 15 mgd
- No more than 5% of the reclaimed water actually comprises the domestic supply
- No observed water quality degradation that constitutes a public health concern

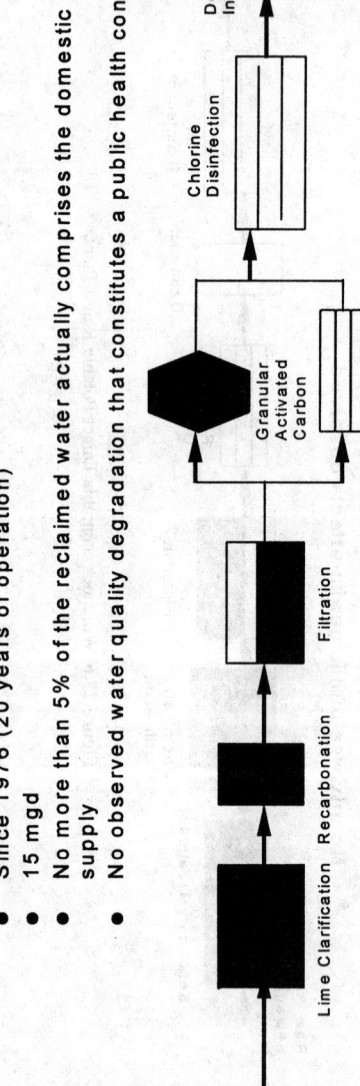

Figure 27.7 Water Factory 21.

1224

tion, carbon adsorption, slip-stream reverse osmosis, and disinfection. Estimates project that no more than 5 percent of the recovered water actually comprises the domestic supply. The Orange County Water District has found no evidence that indicates that this indirect potable reuse practice poses a significant risk to users of the groundwater.

1978: UOSA WATER RECLAMATION PLANT [7]

In 1978, the 15-mgd Upper Occoquan Sewage Authority (UOSA) Water Reclamation Plant (Figure 27.8) began reclaiming wastewater for subsequent discharge to the 11 billion gallon Occoquan Reservoir. The Occoquan Reservoir is a critical source of drinking water for about 1 million people in Northern Virginia. During extended droughts, the plant discharge has accounted for as much as 90 percent of the flow into the reservoir. The reclamation treatment includes primary treatment, secondary treatment, biological nitrification, lime clarification and recarbonation, filtration, activated carbon adsorption, and disinfection. Due to the positive reservoir response to the reclaimed water inflow, the plant was expanded to 27 mgd and will be further expanded to 54 mgd by the year 2000. No negative health effects attributable to the plant or effluent discharges have been reported since the plant has been in operation.

1978: TAHOE-TRUCKEE SANITATION AGENCY WATER RECLAMATION PLANT [8]

Also in 1978, the 5-mgd Tahoe-Truckee Sanitation Agency Water Reclamation Plant (Figure 27.9) began operation. This plant also uses advanced wastewater reclamation processes to recover water suitable for release to the Truckee River that is used as a water supply by the City of Reno, Nevada.

1981–1983: POTOMAC ESTUARY EXPERIMENTAL WATER TREATMENT PLANT [9]

From 1981 to 1983, the 1-mgd Potomac Estuary Experimental Water Treatment Plant (Figure 27.10) was operated with a plant influent blend of Potomac Estuary water and nitrified secondary effluent to simulate the influent water quality expected during drought conditions when as much as 50 percent of the estuary flow would comprise treated wastewater. Treatment included aeration, coagulation, clarification, predisinfection, filtration, carbon adsorption, and post disinfection. An independent National Academy of Science/National Academy of Engineering panel reviewed the extensive testing performed by the Army Corps of Engineers.

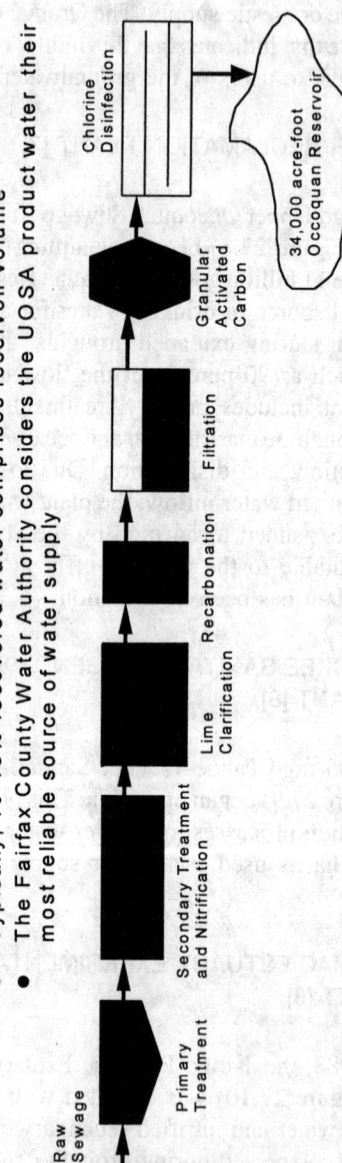

- Since 1978 (18 years of operation)
- 15 mgd expanded to 27 mgd in 1987, & expanded to 54 mgd by 2000
- Typically, 10-15% recovered water comprises reservoir volume
- The Fairfax County Water Authority considers the UOSA product water their most reliable source of water supply

Figure 27.8 UOSA Water Reclamation Plant.

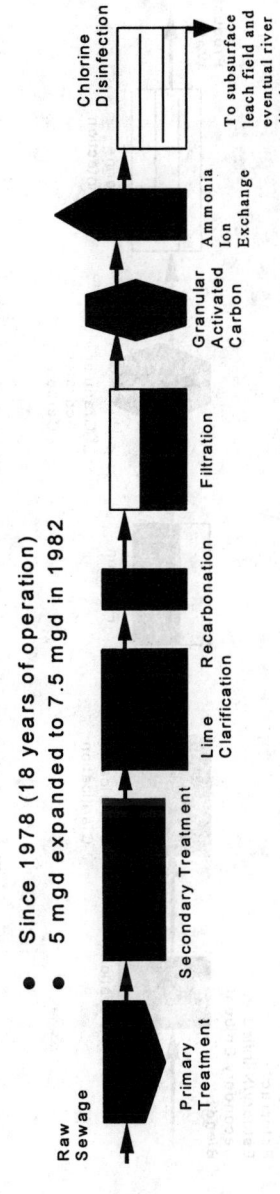

Figure 27.9 TTSA Water Reclamation Plant.

- Since 1978 (18 years of operation)
- 5 mgd expanded to 7.5 mgd in 1982

Raw Sewage

Primary Treatment

Secondary Treatment

Lime Clarification

Recarbonation

Filtration

Granular Activated Carbon

Ammonia Ion Exchange

Chlorine Disinfection

To subsurface leach field and eventual river discharge

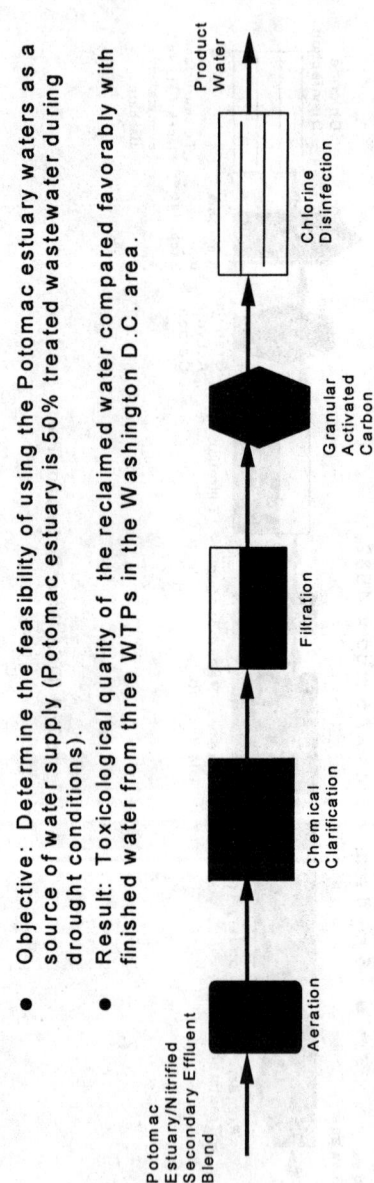

- Objective: Determine the feasibility of using the Potomac estuary waters as a source of water supply (Potomac estuary is 50% treated wastewater during drought conditions).

- Result: Toxicological quality of the reclaimed water compared favorably with finished water from three WTPs in the Washington D.C. area.

Potomac Estuary/Nitrified Secondary Effluent Blend

Aeration

Chemical Clarification

Filtration

Granular Activated Carbon

Chlorine Disinfection

Product Water

Figure 27.10 Potomac Estuary Experimental Water Treatment Plant.

The panel concluded that the advanced treatment could recover water from a highly contaminated source that is similar in quality to three major water supplies for the Washington, D.C. metropolitan area.

1980: USEPA REVIEW OF SURFACE WATER COMPOSITION

To gain a better understanding of the prevalence of unplanned, indirect potable reuse in our nation's surface water supplies, the U.S. EPA conducted a study and published a report in 1980 called Wastewater in Receiving Waters at Water Supply Abstraction Points. The purpose of the project was to determine how much wastewater and wastewater-derived material from discharges are present in the surface water supplies of U.S. cities with populations greater than 25,000. The study identified 1,246 municipal water supply utilities using surface water from 194 basins serving 525 cities with populations greater than 25,000. From the nearly 80 million users of surface water included in this study, about 33 percent of this population withdrew their water supplies from sources that contain from 5 to 100 percent wastewater during low flow periods.

1983: DENVER POTABLE REUSE DEMONSTRATION PLANT PROJECT, COLORADO [10]

In 1983, the 1-mgd Potable Reuse Demonstration Plant in Denver, Colorado began operation (Figure 27.11). This plant was designed to evaluate the feasibility of direct potable reuse of secondary treated municipal wastewater. After seven years of testing and evaluating alternative treatments, a potable water recovery system comprised of lime clarification, recarbonation, filtration, ultraviolet light intermediate disinfection, carbon adsorption, reverse osmosis, air stripping, ozone primary disinfection, and chloramine secondary disinfection was selected for comprehensive health effects testing. The results of this $4 million whole animal health effects testing program were integrated with chemical, physical, and microbiological examinations to provide a basis for determining the suitability of recovered water as a drinking water supply. The final report for this $30 million project was issued in 1993. The findings of this extensive research effort unequivocally verified the ability of advanced water treatment processes to reliably remove a broad spectrum of pollutants and render a product which satisfies every currently known measure of drinking water safety. Denver does not currently need to implement potable reuse, but will consider it among other water supply alternatives for meeting future needs.

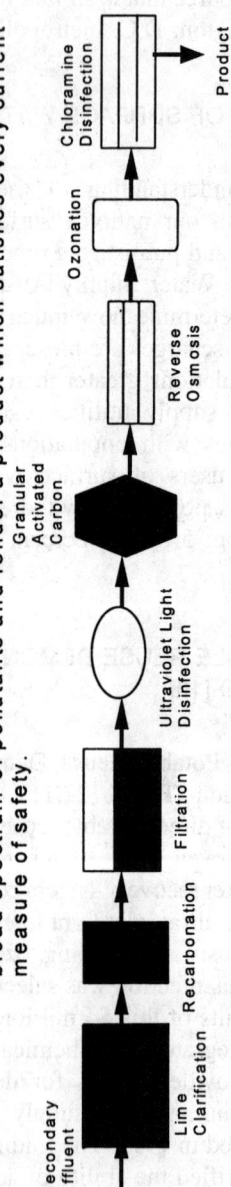

- Objective: Determine the feasibility of reliably converting secondary effluent to potable water quality ($30 million program).
- Result: The most extensive water treatment technology research project to date verified the ability of advanced water treatment processes to reliably remove a broad spectrum of pollutants and render a product which satisfies every current measure of safety.

Figure 27.11 Denver Potable Reuse Demonstration Plant.

1983: SAN DIEGO TOTAL RESOURCE RECOVERY PROJECT, CALIFORNIA [11]

Also in 1983, the San Diego, California 1-mgd potable water recovery demonstration facility was commissioned as part of a total resource recovery program established in San Diego (Figure 27.12). The treatment system included: primary treatment, a water hyacinth aquaculture system, coagulation, clarification, filtration, ultraviolet disinfection, reverse osmosis, aeration, carbon adsorption, and disinfection to reclaim raw water from raw sewage. The program included an extensive chronic toxicity risk analysis to determine the potential health effects resulting from reuse of the recovered water and to compare the recovered water to current raw water supplies used by the City of San Diego. Results of the health effects showed that the risk associated with use of the recovered water as a raw water supply is less than or equal to that of the existing raw water entering the City's Miramar Water Treatment Plant. Based in large part on these positive results, the City is planning on reclaiming up to 20 mgd of secondary effluent for augmentation of their 90,000 acre foot San Vicente Reservoir where it will blend with imported water prior to passage through the City's Alvarado Water Treatment Plant and on to customers.

1985: EL PASO, TEXAS FRED HERVEY WATER RECLAMATION PLANT [12]

In 1985, the 10-mgd Fred Hervey Water Reclamation Plant began operation in El Paso, Texas (Figure 27.13). Recovered water is recharged to the Hueco Bolson drinking water aquifer where over a 2-year period, the water travels to one of El Paso's potable water well fields to become part of the potable water supply. The treatment of raw wastewater to recharge quality water includes: primary treatment, activated sludge/powdered activated carbon treatment, lime treatment, recarbonation, filtration, ozonation, and granular activated carbon adsorption. No negative health effects have been correlated with this practice, however, some increase in the total dissolved solids content of the aquifer has occurred. Future plant expansions will include slip-stream demineralization to address this concern.

1986: TAMPA WATER RESOURCE RECOVERY PROJECT, FLORIDA [13]

In 1986, the City of Tampa, Florida's Water Resource Recovery Pilot Plant began operation (Figure 27.14). The pilot project was designed to evaluate the feasibility of reclaiming denitrified secondary effluent to a

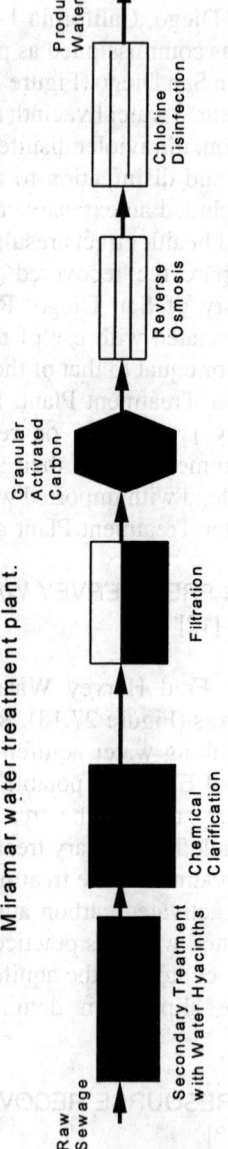

- Objective: Determine the feasibility of reliably converting raw sewage to a quality commensurate with existing raw drinking water supplies.
- Result: The health risk associated with using reclaimed water as a raw water supply is less than or equal to that of the existing City raw water entering the Miramar water treatment plant.

Figure 27.12 San Diego Total Resource Recovery Project.

Raw Sewage → Secondary Treatment with Water Hyacinths → Chemical Clarification → Filtration → Granular Activated Carbon → Reverse Osmosis → Chlorine Disinfection → Product Water

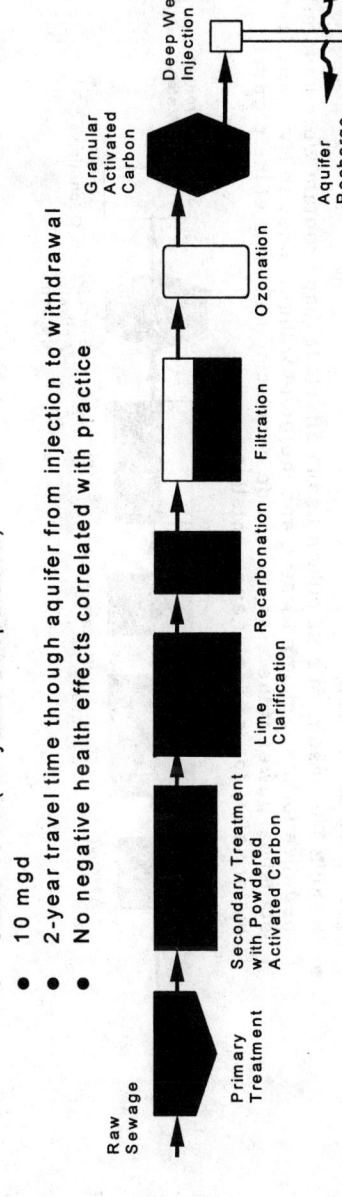

- Since 1985 (11 years of operation)
- 10 mgd
- 2-year travel time through aquifer from injection to withdrawal
- No negative health effects correlated with practice

Raw Sewage

Primary Treatment

Secondary Treatment with Powdered Activated Carbon

Lime Clarification

Recarbonation

Filtration

Ozonation

Granular Activated Carbon

Deep Well Injection

Aquifer Recharge

Figure 27.13 El Paso, Fred Hervey Water Reclamation Plant.

- Objective: Determine the feasibility of reliably converting denitrified, secondary effluent to a quality suitable for blending with existing surface water and groundwater sources.

- Result: The results of a $2 million health effects testing program were uniformly negative for the reclaimed water, and therefore within the capability, limits, and statistical power of the assays, provide convincing evidence of the product water safety for raw water augmentation.

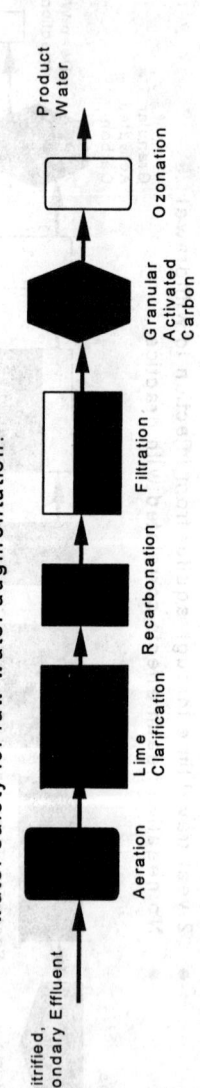

Denitrified, Secondary Effluent

Aeration

Lime Clarification

Recarbonation

Filtration

Granular Activated Carbon

Ozonation

Product Water

Figure 27.14 Tampa Water Resource Recovery Project.

quality suitable for blending with existing surface water and groundwater sources for indirect potable reuse. Several alternative treatments were evaluated and one was selected for health effects testing after 2 years of evaluation. The treatment selected included: aeration, high pH lime clarification, recarbonation, filtration, granular activated carbon adsorption, and ozonation. Final results of the study were documented in 1993. The results of the $1.5 million whole animal health effects testing coupled with the microbiological and chemical analyses performed revealed that the quality of the reuse product water is equivalent to or exceeds the quality of the Hillsborough River raw water supply. The City of Tampa is planning on implementation of a 20 to 50 mgd Water Resource Recovery Plant in the near future.

1988: PHOENIX, ARIZONA POTABLE REUSE FEASIBILITY STUDY [14]

In 1988, the City of Phoenix, Arizona conducted a potable reuse feasibility study to evaluate the cost-effectiveness, institutional constraints, and social constraints associated with direct potable reuse. The feasibility study results suggested that potable reuse is cost competitive with other alternative water supply development projects for this desert-based city. The state, county, and local health departments participated in project workshops to develop the most desirable potable reclamation approach. Consequently, the concept of potable reuse gained regulatory favor. Phoenix now recognizes potable reuse among other alternatives as a viable future water source. Neighboring Scotsdale, Arizona is planning on implementing indirect potable reuse via groundwater recharge in the near future.

1990: CALIFORNIA DEPARTMENT OF HEALTH SERVICES PROPOSED GROUNDWATER RECHARGE REGULATIONS [15]

Based primarily on the success of Orange County Water District's Water Factory 21, the California Department of Health Services (CDHS) proposed regulations to govern future indirect potable reuse projects via groundwater recharge. These regulations are the first of their kind in the United States. The regulations apply to both surface spreading and direct injection of reclaimed water into potable aquifers. The regulations include the following primary stipulations: (1) water of reclaimed origin shall not comprise more than 20 percent (direct injection) and 50 percent (surface spreading) of the total flow at any domestic well based on the average of the past 5 years of calculations; (2) reclaimed water shall be retained underground for a minimum of 12 months (direct injection) and 6 months (surface spreading) prior to being withdrawn at a domestic water supply well; and (3) the maximum total organic carbon (TOC) concentration in the reclaimed water shall be 2 mg/L (direct injection) and 3 to 5 mg/L (surface spreading). This last stipulation regarding TOC content essentially requires carbon adsorption and/or reverse osmosis treatment.

1990–1995: WEST BASIN WATER RECYCLING PROGRAM [16]

From 1990 through 1995, the West Basin Municipal Water District conceived, designed, constructed, and began operation of their West Basin Water Recycling Program, which includes reclaiming 5 mgd expandable to 20 mgd of secondary effluent from the City of Los Angeles's Hyperion Treatment Plant for injection into the West Coast Basin Barrier Project (Figure 27.15). The West Coast Basin Barrier Project was constructed in the 1950's and 1960's to inject imported water into the coastal reaches of local South Bay aquifers for mitigation of saltwater intrusion. The Barrier has historically received an average of about 20 mgd of potable water. Substitution of reclaimed water for potable water provides substantially greater water use efficiency in southeast Los Angeles county. Reclamation treatment includes pre-decarbonation, lime clarification, recarbonation, filtration, reverse osmosis, post-decarbonation, and final disinfection. A baseline groundwater monitoring program was conducted in advance of the recycling project to allow assessment of reclaimed water impacts on the aquifer water quality. Based on hydrogeologic investigation and modeling of the West Coast Basin, it is anticipated that the reclaimed water will improve groundwater quality along the Barrier due to the high quality of the reclaimed water relative to the imported water and native groundwater. The system has been designed to meet California Department of Health Services proposed regulations for groundwater recharge presented previously.

1993: CITY OF COLORADO SPRINGS [17]

Current projections of population growth and future water usage estimates indicate that by the year 2012, the current local water sources and the transmountain water delivery systems will not be adequate to meet all of the City's projected water needs. New water supply options are under consideration including several concepts to facilitate exchange of wastewater return flows for additional water supplies in the Arkansas River Basin and potable reclamation of water in Fountain Creek that is largely comprised of the City's treated wastewater (Figure 27.16). In 1993, the City of Colorado Springs conducted a water reclamation assessment that included development, screening, and recommendations for potable water recovery from the City's treated wastewater effluent. The concept ulti-

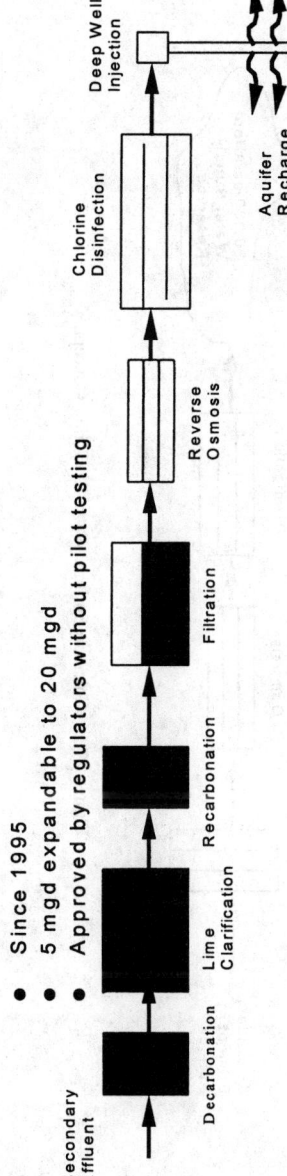

- Since 1995
- 5 mgd expandable to 20 mgd
- Approved by regulators without pilot testing

Secondary Effluent

Decarbonation

Lime Clarification

Recarbonation

Filtration

Reverse Osmosis

Chlorine Disinfection

Deep Well Injection

Aquifer Recharge

Figure 27.15 West Basin Water Reclamation Plant.

- Planning concept for future water supply
- 28 to 58 mgd capacity

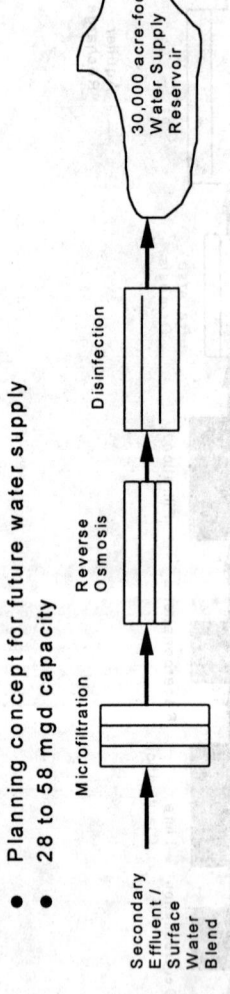

Figure 27.16 Colorado Springs Water Recycling Concept.

mately selected includes diversion of Fountain Creek flows 16 miles downstream of the City's wastewater effluent discharge point; reclamation in a new plant including microfiltration and reverse osmosis; discharge of recovered water to a new 30,000 acre-foot recreational/storage reservoir; followed by a new drinking water plant; and ultimate water redistribution to the City's potable system. The present value unit cost of this indirect potable reuse concept was found to be competitive with the other water supply concepts involving further development of the Arkansas River Basin. The City is currently evaluating the comparative socioeconomic impacts of the alternative water supply concepts. Results to date indicate greater public acceptance for water reuse than for concepts involving additional development of upland Arkansas River Basin water. Exchange of wastewater return flows for additional Arkansas River water via storage in the Pueblo Reservoir located about 60 miles south of Colorado Springs and indirect potable reuse are respectively the leading candidates for the City's next increment of raw water supply.

1994: CALIFORNIA DEPARTMENT OF HEALTH SERVICES CONCEPTUAL APPROVAL OF THE CITY OF SAN DIEGO WATER REPURIFICATION PROJECT [18]

The CDHS conceptual approval of the City of San Diego's indirect potable reuse project (described previously) expands existing CDHS policy to allow repurified water to supplement drinking water supplies stored in local reservoirs (Figure 27.17). Currently, repurified water for drinking and other domestic purposes may be stored only in groundwater basins. Based on this action, it is anticipated that the CDHS will ultimately have regulations in place to govern indirect potable reuse via both groundwater recharge and surface water augmentation. In the past, indirect potable reuse projects have been evaluated on a case-by-case basis. Regulations will foster more expeditious evaluation and approval of indirect potable reuse in California which will set a standard for other states to consider.

1996: THE CALIFORNIA POTABLE REUSE COMMITTEE PROPOSES A FRAMEWORK FOR REGULATING THE INDIRECT POTABLE REUSE OF ADVANCED TREATED RECLAIMED WATER BY SURFACE WATER AUGMENTATION [19]

The California Potable Reuse Committee was formed in May 1993 to look into the feasibility and safety of potable reuse of advanced treated reclaimed water. The Committee, commissioned by the Departments of Health Services and Water Resources, concluded after two years of meetings and examination of relevant data that planned, indirect potable reuse

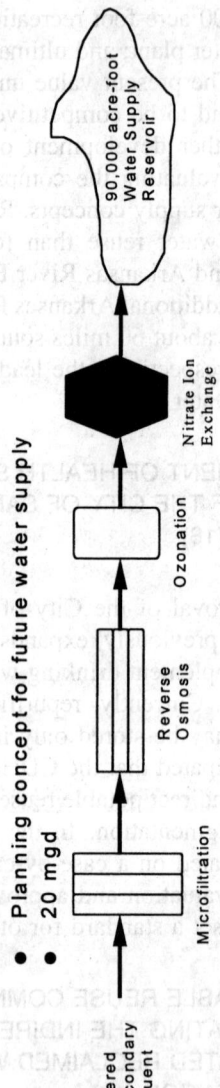

Filtered Secondary Effluent

Microfiltration

Reverse Osmosis

Ozonation

Nitrate Ion Exchange

90,000 acre-foot Water Supply Reservoir

● Planning concept for future water supply
● 20 mgd

Figure 27.17 San Diego Water Repurification Project.

of advanced treated reclaimed water using surface water reservoirs is feasible under specific criteria. The Committee recommended that the following six criteria be met before indirect reuse is allowed:

(1) Application of best available technology in advanced wastewater treatment with the treatment plant meeting operating criteria
(2) Maintenance of appropriate retention times based on reservoir dynamics
(3) Maintenance of advanced wastewater treatment plant operational reliability to consistently meet primary microbiological, chemical, and physical drinking water standards
(4) Surface water augmentation projects using advanced treated reclaimed water must comply with applicable State of California criteria for groundwater recharge for direct injection with reclaimed water
(5) Maintenance of reservoir water quality
(6) Provision for an effective source control program

INDIRECT POTABLE REUSE WATER QUALITY AND PUBLIC HEALTH SAFEGUARDS [20,21]

MULTIPLE BARRIER APPROACH

In typical drinking water supply systems, state and federal drinking water standards are used as a measure to determine if a given supply has been adequately treated prior to distribution to the community. However, drinking water standards were established under the assumption that water supplies would be derived from only the highest quality raw water sources. Since the existing standards were not originally developed with the goal of regulating drinking water derived from a wastewater origin, extra care must be incorporated into a water supply system that serves a water of wastewater origin.

Due to this lack of specific standards covering indirect potable reuse applications, the concept of "multiple barriers" has been adopted by the water supply industry to achieve the appropriate level of safety and reliability. In this concept, multiple unit processes and other mechanisms are relied upon to remove or inactivate various water quality parameters that are of concern, primarily pathogenic organisms. For example, an indirect potable reuse application may include two or three unit treatment processes designed to remove or inactivate viruses and parasites that may be present in the supply. Should one process fail at the task, backup mechanisms are available to do the job. Although this multiple barrier approach is particularly necessary for pathogenic organisms, the approach also can be used to provide protection against trace organics or metals.

As depicted in the previous section on history, there are several advanced water treatment unit processes typically incorporated into an indirect potable reuse system. These unit processes and their relative capabilities to remove constituents within the gross contaminant categories are shown in Table 27.1. The removal of a contaminant is considered to be a relative measure of the ability of a unit process to act as a barrier to that contaminant. Unit processes identified in Table 27.1 are expected to effect at least 50 percent removal within the contaminant categories specified.

Indirect potable reuse, by definition, includes blending, dilution, and retention of reclaimed water with conventional raw water supplies prior to passage through a water treatment plant and distribution to the public. In this fashion, indirect potable reuse offers several additional barriers to contaminant passage into the potable distribution system.

Diligent and responsive monitoring is also paramount to assuring proper operation and performance of the reclaimed water treatment system. Therefore, well designed and operated reclamation systems are required to provide reliable and safe reclaimed water production via the multiple barrier approach.

Planned, indirect potable water recovery systems include considerably more contaminant barriers than an incidental indirect potable reuse system. These additional barriers primarily reside in the advanced treatment provided downstream of secondary biological treatment and upstream of raw water supply blending and dilution.

COMPARATIVE SAFETY STANDARD

Drinking water supply sources are subjected to rigorous monitoring to insure that the sources represent a safe water supply. As an overall goal, indirect potable reuse projects should provide a degree of safety at least equal to that of a community's current water supply.

Until specific standards are developed that can be directly used to measure the safety and reliability of an indirect potable reuse project, indirect potable reuse projects are judged in comparison to existing local water supplies. For instance, in San Diego, California, Denver, Colorado, and Tampa, Florida, three communities that have conducted health effects testing to measure the safety of indirect potable reuse systems, water supplied by indirect potable reuse was compared with the existing local water supply. Water managers and the regulatory community have insisted that these health effects studies must demonstrate that water developed by an indirect potable reuse project must be at least as safe as the water that is currently available and in use by that community. In each of these

TABLE 27.1. Advanced Water Treatment Multiple Unit Process Barriers.

Gross Contaminant Category	Biological Treatment	Biological Treatment w/Nutrient Removal	Biological Nitrogen Removal	High Lime w/ Recarbonation	Chemical Oxidation/ Disinfection
Suspended solids	X	X		X	
Dissolved solids					
Biological oxygen demand	X	X	X	X	
Total organic carbon	X	X			
Volatile organic chemicals	X	X	X		X
Heavy metals	X	X		X	
Nutrients		X	X	X	
Radionuclides	X	X			
Microbial Factors	X	X		X	X

Gross Contaminant Category	Filtration	Granular Activated Carbon	Membrane Demineralization	Air Stripping	Membrane Particle Removal
Suspended solids	X	X	X		X
Dissolved solids			X		
Biological oxygen demand		X	X		X
Total organic carbon		X	X		
Volatile organic chemicals		X	X	X	
Heavy metals			X		X
Nutrients			X		X
Radionuclides			X	X	
Microbial factors					X

1243

three instances, comprehensive health effects studies revealed that water could be reliably and safely recovered from wastewater to a quality commensurate with the safety of existing comparative local water supplies.

ADDITIONAL SAFEGUARDS PROVIDED BY RECEIVING WATER

Using a raw water supply as a recycle conduit and buffer between the reclaimed water and drinking water offers system reliability, redundancy, and psychological satisfaction.

Blending and Dilution

One of the primary safeguards provided by both incidental and planned indirect potable reuse is the blending and dilution provided to the reclaimed water from the conventional raw water supply. The appropriate level of blending and dilution to provide for remains largely a site specific exercise. In California, for instance, proposed requirements limit the blending ratio of both reclaimed groundwater recharge and surface water augmentation to no more than 50 percent. In other states, the capacity of the water reclamation plant is permitted, and cannot be increased without raw water supply monitoring data that supports the safety of a capacity increase. With respect to incidental, indirect potable reuse, the NPDES permitting process essentially hinges on the assimilative capacity of the receiving stream to handle the discharge without experiencing deleterious effects. Blending and dilution stream assimilation issues are key to the NPDES permitting process.

Retention Time

Retention time is the time that elapses from the time reclaimed water enters the raw water supply to the time it is withdrawn for potable water treatment and redistribution. Retention time provides time for blending and dilution, time for natural treatment processes to occur, and time for water quality monitoring and potential corrective actions. Provision of retention time generally requires geographic separation between the point of reclaimed water augmentation and withdrawal. This geographic separation in turn provides a psychological comfort that a "natural" barrier is present. Like blending and dilution, retention time provisions are site specific. California's proposed groundwater recharge regulations require 6 months retention when reclaimed water is introduced to the aquifer via spreading basins, and 12 months retention when reclaimed water is directly injected into the aquifer. At the Fred Hervey Plant in El Paso, Texas, the retention time in the Hueco Bolson aquifer is approximately 2 years.

Water movement underground is slow and dependent on the transmissivity of the aquifer materials and hydraulic gradient. Aquifer hydraulics are relatively stable in comparison to reservoir hydrodynamics. Retention time in a surface water reservoir is dependent on the reservoir geometry, inflow location, outflow location, water temperature, density currents, and wind currents. Preliminary guidelines in California require an average hydraulic retention time of 12 months. In addition, short circuiting must be avoided. The UOSA project includes about 1 month of average retention time in the Occoquan Reservoir.

Natural Treatment

Beyond dilution and retention, the receiving raw water supply also can provide natural treatment. Typically, in planned indirect potable reuse, this natural treatment is viewed as a redundant system, above and beyond the engineered treatment systems. Spreading basins for aquifer recharge provide vadose zone percolation treatment that can offer suspended solids removal, nutrient removal, metals removal, organics removal, and pathogen removal via soil and aquifer material physical filtration, biological degradation, ion exchange, and pathogen mortality. These same factors, but to a lesser degree, are available to direct injection aquifer recharge applications. Surface waters offer nutrient removal, metals removal, organics removal, and pathogen removal via aeration, biological degradation, photodecomposition, adsorption, and sedimentation.

CONVENTIONAL WATER TREATMENT

Indirect potable reuse, by definition, includes a drinking water plant as the final contaminant barrier in the reuse system. To date, the approach to planned, indirect potable reuse has been based on assuring no net treatment burden increase on the water treatment plant. Therefore, the contaminant burden has been purposefully shifted to the water reclamation plant in advance of raw water blending and dilution. Although this approach may be overly conservative, it has been viewed as necessary to gain both regulatory and public acceptance.

INDIRECT POTABLE REUSE CASE STUDIES

In this section, the following planned, indirect potable reuse projects will be presented in more detail relative to the impetus for the project and performance of the project:

- Upper Occoquan Sewage Authority Water Reclamation Plant

- Tampa Water Resource Recovery Project
- West Basin Municipal Water District Groundwater Augmentation Project
- Fred Hervey Water Reclamation and Groundwater Recharge Project

UPPER OCCOQUAN SEWAGE AUTHORITY WATER RECLAMATION PLANT [7]

Project Impetus

The 11 billion gallon Occoquan Reservoir is a critical source of water for about 1,000,000 people in Northern Virginia. In 1960, most of the Occoquan Watershed, including historic Manassas and Centreville, had changed little since the Civil War. However, the decade of the 1960s marked the beginning of its rapid transformation from a largely rural to a predominantly urban/suburban region. Stimulated by the opening of a new interstate highway and Washington, DC's suburban expansion, Northern Virginia grew by some 500 people a week during the 1960s, and the residential development radiating into the Occoquan Watershed began to adversely affect the water quality in the Reservoir. In response to this deterioration, the Virginia Water Control Board, after extensive studies and public hearings (including consideration of exporting all wastewater out of the Occoquan Watershed), adopted a comprehensive policy for water quality management in the Watershed. The Occoquan Policy mandated the construction of a highly sophisticated regional wastewater reclamation plant to replace the Watershed's 11 small secondary treatment plants. In 1971, the Upper Occoquan Sewage Authority (UOSA), a regional agency serving four jurisdictions, was created to comply with the state's mandate. The project was initiated because unplanned, indirect potable reuse was insufficient to protect the Occoquan Reservoir water supply, so proper planning and institution of more stringent treatment standards were required (Figure 27.18).

Project Description and Performance

Table 27.2 presents the plant effluent limits. To meet these strict limitations, the UOSA treatment system consists of primary and secondary treatment processes, followed by five advanced treatment processes, as shown below.

Primary treatment includes screening, comminution, grit removal and primary clarification. Screenings and grit are dewatered and conveyed to a sanitary landfill. Primary sludges are conveyed to anaerobic digestion.

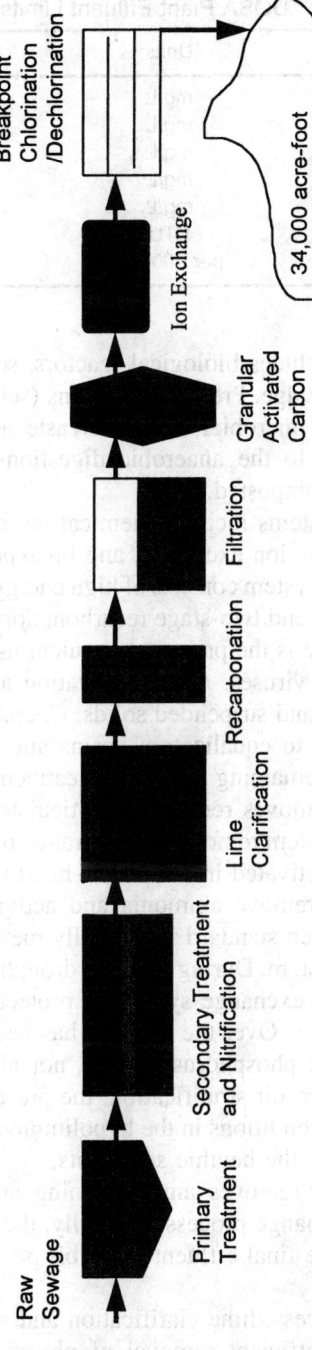

Figure 27.18 UOSA Water Reclamation Plant.

Raw Sewage

Primary Treatment

Secondary Treatment and Nitrification

Lime Clarification

Recarbonation

Filtration

Granular Activated Carbon

Ion Exchange

Breakpoint Chlorination /Dechlorination

34,000 acre-foot Occoquan Reservoir

1247

TABLE 27.2. **UOSA Plant Effluent Limits.**

Parameter	Units	Concentration
COD	mg/L	10.0
Suspended solids	mg/L	1.0
Unoxidized nitrogen	mg/L	1.0
Total phosphorus	mg/L	0.1
MBAS	mg/L	0.1
Turbidity	NTU	0.5
Coliform Bacteria	per 100 mL	< 2

Secondary treatment includes biological reactors, secondary clarification and return activated sludge. Preaeration basins (selectors) are used to promote a more efficient microbial culture. Waste activated sludge is thickened and transferred to the anaerobic digestion system. Digested sludge is dewatered and composted.

Advanced treatment systems include chemical clarification, filtration, activated carbon adsorption, ion exchange, and breakpoint chlorination.

The chemical treatment system consists of high energy mixing, flocculation, chemical clarification and two-stage recarbonation with intermediate settling. Calcium hydroxide is the primary coagulant used to raise the pH to 11.3 for destruction of viruses, and precipitation and coagulation of phosphorus, heavy metals and suspended solids. Chemical treatment system effluent is discharged to equalization basins and then pumped at a uniform rate through the remaining advanced treatment processes.

Multimedia filtration removes remaining particulate matter. The activated carbon adsorption system removes a wide range of synthetic organic compounds. Carbon is reactivated in a multiple-hearth furnace.

Ion exchange reactors remove ammonia and activated carbon fines. UOSA's unoxidized nitrogen standard is normally met by nitrification in the secondary treatment system. During extreme droughts, however, nitrogen is removed in the ion exchange system to protect the water supply from excessive nitrate levels. Over the years, it has been determined that the Occoquan Reservoir is phosphorus limited, not nitrogen limited. In fact, during periods of reservoir stratification, the presence of nitrate has curbed onset of anaerobic conditions in the hypolimnion which are known to release phosphorus from the benthic sediments.

Breakpoint chlorination removes any remaining ammonia following nitrification or the ion exchange processes. Finally, the water is dechlorinated and discharged to the final effluent reservoir prior to entry into the Occoquan Reservoir.

With respect to the process, lime clarification and recarbonation continue to be used due to efficient removal of phosphorus, metals, and viruses. Due to relatively low total dissolved solids in the wastewater, the use of reverse osmosis is not required to match the ambient reservoir solids content. Provisions to add final ozone disinfection are being included in the latest plant expansion should concern over *Cryptosporidium* necessitate future ozonation.

Table 27.3 shows typical UOSA treatment performance over its history of operation. The plant consistently meets the effluent requirements.

A testimonial to the success of the UOSA project is the change in opinion regarding the value of the reclaimed water. At first, the Fairfax County Water Authority and the Virginia Department of Health had reservations about the proposed plan, but establishment of strict recovered water standards and plant process reliability features persuaded both agencies to allow construction of an initial 15 mgd facility that started up in 1978.

Since 1978, commercial and residential development in the watershed has continued and necessitated expansion of the UOSA plant to 27 mgd. Originally, the plant was not to exceed 15 mgd in capacity, but both the Water Authority and the Health Department agreed to the expansion based on the exceptional performance of the plant and the vast improvement to the reservoir water quality.

To further demonstrate the success of the UOSA program, the plant is now currently being expanded from 27 mgd to 54 mgd. Construction is scheduled for completion by the turn of the century.

Project Costs

The original 15 mgd plant that began operation in 1978 was constructed for $60 million (1975 dollars). The expansion from 15 to 27 mgd completed in 1987 was constructed for $18 million (1986 dollars). The expansion from 27 to 54 mgd slated for completion by the turn of the century

TABLE 27.3. Typical UOSA WRP Performance.

Flow Stream	COD mg/L	TSS mg/L	TKN mg/L	TP mg/L	MBAS mg/L	Turb. NTU
Plant influent	400	170	34	9.0	5.3	—
Primary effluent	240	65	29	7.2	—	—
Secondary effluent	40	16	2.5	5.9	—	4.5
Chem. clarifier	18	11	1.2	0.15	—	1.5
2nd stage recarb.	17	10	1.2	0.09	—	0.85
Filter effluent	13	0.2	0.75	0.03	0.08	0.3
Carbon effluent	11	1.0	0.6	0.03	0.05	0.6
Ion exchange effluent	—	0.3	0.55	0.03	—	0.25
Final effluent	8	0.1	0.46	0.03	0.04	0.25
Effluent limits	10	1	1	0.1	0.1	0.5

has a projected construction cost of about $200 million (1994 dollars). Operational costs are approximately $0.85 per one thousand gallons ($280 per acre-foot) of reclaimed water produced.

UOSA's FUTURE

Expansion of the plant from 27 mgd to 54 mgd began in 1991 and will be finished by the year 2000. The new facilities will essentially replicate the existing processes, but will also incorporate the latest equipment available and include some process modifications as follows:

- modification of the biological reactors to allow full-time nitrification with denitrification abilities during drought conditions
- expansion with gravity filters as opposed to the existing pressure filters
- two-stage, upflow/downflow carbon contacting versus single stage downflow contacting

The UOSA plant is the only full-scale facility that has practiced indirect potable reuse via surface water augmentation successfully for nearly 20 years. Recovered water from the plant has historically accounted for 10 to 15 percent of the reservoir volume, but during extended droughts has accounted for as much as 90 percent of the inflow to the reservoir. The Occoquan Reservoir is the principal water supply for nearly 1 million people in Northern Virginia, projected to reach 1.5 million people by the end of the century.

TAMPA WATER RESOURCE RECOVERY PROJECT [13]

Project Impetus

The City of Tampa, Florida has long recognized that the high quality effluent from their Hookers Point Advanced Wastewater Treatment Facility represents a valuable water resource to the metropolitan area. The Hookers Point Facility discharges to the Hillsborough Bay and has a rather strict NPDES permit to protect the quality of the bay as follows:

- BOD_5 = 13 mg/L
- total suspended solids = 15 mg/L
- total nitrogen = 5 mg/L
- total phosphorus = 7.5 mg/L
- fecal coliform < 200 per 100 mL
- total residual chlorine = 0.5 mg/L

The Hookers Point Facility is designed to process an average flow of 60 mgd of wastewater while meeting the previous standards. The treatment system used at the facility includes:

- high-purity oxygen activated sludge
- two-stage nitrification
- high-rate, deep-bed denitrification filters

The effluent from the plant meets or exceeds the treatment requirements with near-perfect consistency. Table 27.4 presents typical annual average performance. Currently, only a fraction of this high quality effluent is reused.

Population projections for the Tampa area, project that during the annual summer dry season, the year 2020 water demand will exceed currently available sources by anywhere from 15 to 30 mgd, depending on the level of conservation measures instituted in the interim.

Based on these water supply projections and the availability of the high quality Hookers Point AWT Facility effluent, the City of Tampa has been pursuing the investigation and ultimate development of an indirect potable reuse project to augment their Hillsborough River raw water supply during annual dry periods.

Project Description and Performance

Figure 27.19 is a schematic of the proposed water supply plan to augment the City of Tampa's raw water supplies.

Essentially the proposed plan includes recovery of water suitable for augmentation of the Hillsborough River during low flow periods. The existing Tampa Bypass Canal would serve as a mechanism to convey reclaimed water to the river. The Bypass Canal was constructed to divert Hillsborough River flood waters around the City of Tampa, and

TABLE 27.4. Hookers Point AWT Facility Performance.

Parameter	Influent	Effluent	NPDES Limit
BOD$_5$	255 mg/L	5 mg/L	13 mg/L
Suspended solids	180 mg/L	3 mg/L	15 mg/L
Total nitrogen	27.5 mg/L	2.8 mg/L	5 mg/L
Total phosphorus	8.0 mg/L	5.0 mg/L	7.5 mg/L
Fecal coliform		1.6 per 100 mL	200 per 100 mL
Chlorine residual		0.9 mg/L	0.5 mg/L
pH		6.9	
Alkalinity		224 mg/L	
Total organic carbon		15.3 mg/L	

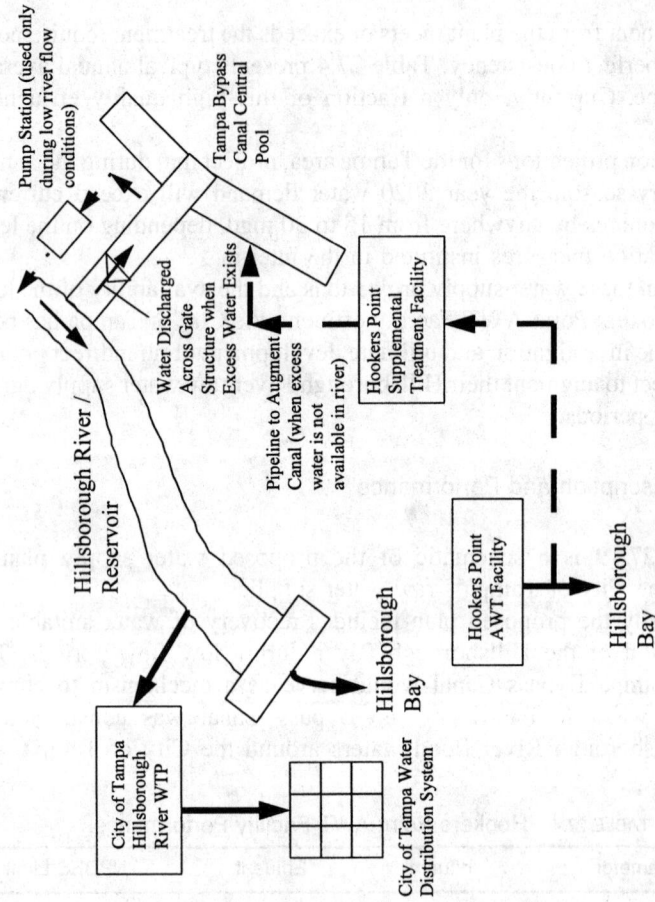

Figure 27.19 Tampa Water Resource Recovery Project.

is otherwise available to serve in the capacity proposed for the water resource recovery project.

The supplemental treatment element of the water resource recovery project was determined to be most critical to the project's feasibility. Therefore, it was decided that a pilot plant should be constructed and operated to test the viability of the project, as well as gain regulatory and public support for the concept.

In late 1985, the design of a 50 gpm supplemental treatment pilot plant was completed. Construction and start-up of the pilot facility was completed in 1986. The pilot plant was operated for 2-1/2 years. Toxicological testing continued beyond the pilot plant operational period, and the final results were documented in 1992. The primary goal of the pilot testing was to select the best supplemental treatment process alternative considering the ability to meet or exceed the quality of the existing Hillsborough River source water, the total treatment cost, the operational simplicity, and the overall system reliability.

Various combinations of the following unit operations were evaluated: preaeration, lime clarification/recarbonation, filtration, activated carbon adsorption, reverse osmosis, ultrafiltration, ozone disinfection, and chlorine disinfection. Ultimately, a unit process train consisting of preaeration, lime clarification/recarbonation, filtration, activated carbon adsorption, and ozone disinfection was selected for in-depth toxicological evaluation in comparison to Hillsborough River source water (Figure 27.20).

The following briefly summarizes the average performance of each of the unit processes in the selected treatment train over the course of the pilot testing with respect to the primary process control parameters:

- **preaeration:** approximately 75 percent carbon dioxide removal
- **lime treatment/recarbonation:** 96 percent total phosphorus removal; 40 percent total hardness removal; 36 percent total organic carbon removal; 16 percent total dissolved solids removal; trace metals reduced to levels below federal and state drinking water standards
- **filtration:** 80 percent turbidity removal
- **lime treatment/recarbonation and filtration:** approximately 3-log reduction of fecal and total coliform bacteria; 1-log reduction of heterotrophic plate count bacteria; 1.8-log reduction of viruses; protozoans (*Giardia* and *Cryptosporidium*) not detected in any samples after lime treatment and filtration
- **GAC adsorption:** 71 percent total organic carbon removal; 0.5-log reduction of fecal and total coliform bacteria; 0.3-log reduction of heterotrophic plate count bacteria
- **Ozone disinfection:** approximately 1-log reduction of both fecal

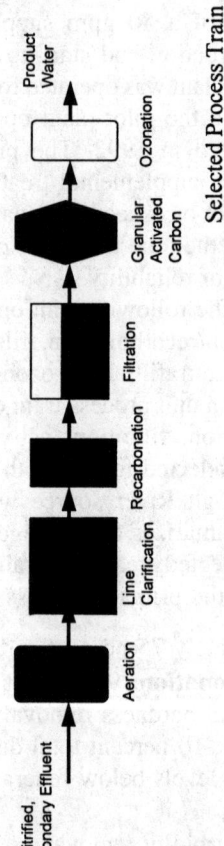

Figure 27.20 Tampa Advanced Water Reclamation Pilot Plant.

and total coliform bacteria; approximately 2-log reduction of heterotrophic plate count bacteria

Concentration of both the recovered water and Hillsborough River organic compounds was required to support conduct of a $1.5 million toxicological testing program to evaluate the health effects potentially associated with complex mixtures of unidentified organic contaminants in both waters. The non-volatile organic compounds were collected, concentrated, and tested at up to 1,000 times the potential human exposure of a 70 kilogram person consuming two liters of water per day for the following:

- **mutagenicity:** Ames *Salmonella* reverse mutation assay
- **genotoxicity:** sister chromatid exchange and micronuclei assay in mouse splenocytes
- **subchronic toxicity:** 90-day subchronic toxicity assay in mice and rats
- **carcinogenicity:** SENCAR mouse skin initiation-promotion assay and strain A mouse lung adenoma assay
- **reproductive effects:** two-generation reproductive toxicity assay in mice
- **teratogenicity:** developmental toxicity in rats

The results of these toxicological studies were uniformly negative for the reclaimed water and provided convincing evidence of the acceptability of the water produced by supplemental treatment of the Hookers Point AWT Facility effluent.

Project Costs

Additional facilities required to augment the City of Tampa's raw water supply include the proposed Supplemental Treatment Facility, as well as a pump station and a pipeline to the Tampa Bypass Canal. Additional operational costs for the pump station at the gate structure connecting the Canal to the River would also be required. Three-phase construction of an ultimate 50 mgd water resource recovery system is proposed as follows:

(1) The first phase would include construction of a new 12.5 mgd Supplemental Treatment Facility at Hookers Point, and a transmission line and pump station to transport recovered water from Hookers Point to the south end of the central pool of the Tampa Bypass Canal.

(2) The second phase would include expansion of the Supplemental Treatment Facility from 12.5 mgd to 25 mgd, and an additional 12.5 mgd of pumping capacity from Hookers Point to the Tampa Bypass Canal.

(3) The third phase would include expansion of the overall system from 25 mgd to 50 mgd.

The estimated capital cost for the entire 50 mgd system is $81 million. The annual operation and maintenance cost is estimated to be about $4 million for summer only operation, and about $12 million for year-round operation.

Project Future

In the 1995–1996 time frame, the City of Tampa is working through the project permitting issues and educating the public about this new source of drinking water supply. It is anticipated that the project will go into design shortly thereafter, and construction will begin around the year 2000.

WEST BASIN MUNICIPAL WATER DISTRICT GROUNDWATER AUGMENTATION PROJECT [16]

Project Impetus

The West Basin Municipal Water District (WBMWD) was formed in 1948 under the California Municipal Water District Act to bring supplemental imported water supplies from the Metropolitan Water District of Southern California (MWD) to the South Bay area because the West Coast groundwater basin was in serious overdraft condition and incapable of meeting the area's needs. WBMWD is the regional wholesaler of imported water within the South Bay area and its primary mission is to ensure a reliable water supply is maintained in sufficient quantity to meet their customers' needs. WBMWD provides supplemental water for 18 cities and unincorporated areas of southwest Los Angeles County. As one of the 27 member agencies of the MWD, WBMWD purchases State Water Project and Colorado River water from MWD for resale to its clients.

Imported water used in the West Coast Basin to supplement the local water supplies represents about 90 percent of the basin's water supply. The decreasing availability and increasing cost of imported water have heightened the need to develop new sources of water for the basin, including the use of reclaimed water. WBMWD has implemented the West Basin Water Recycling Program, which includes reclaiming secondary effluent from the City of Los Angeles's Hyperion Treatment Plant and distributing the reclaimed water for beneficial uses, including injection into the West Coast Basin Barrier Project (WCBBP). Substitution of recycled water for imported water in the WCBBP helps to maximize the regional benefits offered by the water resource imported into Southern California.

The WCBBP was constructed in the 1950s and 1960s to inject imported MWD water into the coastal reaches of local aquifers for mitigation of saltwater intrusion. These aquifers serve as a major source of water for many coastal communities. For example, in the early 1990s groundwater accounted for 40,000 ac-ft of the 190,000 ac-ft consumed. The WCBBP consists of two sections of pressurized pipeline connecting 144 injection wells. The injection wells have casing perforations at selected depths ranging from 280 to 700 feet to allow water injection into three different water bearing strata. An average of approximately 20,000 to 25,000 ac-ft/yr is injected into the aquifers through the WCBBP, however, historically up to 40,000 ac-ft/yr of potable water has been injected into the barrier system.

WBMWD initiated a baseline groundwater monitoring program along the WCBBP in December 1991. Baseline groundwater quality monitoring in advance of recycled water injection will enable clear evaluation of the impacts (positive and negative) of substituting recycled water for potable water. Unfortunately, most other reclaimed water injection projects in Southern California have not had the benefit of a baseline water quality data base for comparison purposes.

Several sampling events have been completed. The first sampling event included sampling of 56 wells for 258 chemical, physical, and biological constituents. The baseline sampling program indicates a wide variation in water quality in the three principal aquifers along the barriers as shown in Table 27.5. In addition, state and federal primary drinking water standards have been exceeded for several volatile organic compounds and metals.

Project Description and Performance

The WBMWD began supplying 5 mgd of highly treated recycled water to blend with Metropolitan's imported surface water for injection in June 1995. The source of wastewater is secondary effluent from the Hyperion Treatment Plant. Secondary effluent is conveyed to the WBWRP in El Segundo, treated, conveyed from the WBWRP to a blend facility, blended

TABLE 27.5. **West Coast Basin Barrier Project Aquifer Baseline Water Quality.**

Constituent	Units	200-Foot Sand	Silverado	Lower San Pedro
Total dissolved solids	mg/L	260–5,640	330–1,570	280–13,400
Chlorida	mg/L	68–2,250	35–670	22–7,700
Sulfate	mg/L	51–1,500	83–260	ND–1,050
Boron	mg/L	0.095–1	0.13–0.24	0.11–1
Nitrate	mg/L	ND–18	ND–0.3	ND–8
Total organic carbon	mg/L	ND–31	1.2–6.9	0.9–7.1

with Metropolitan water, then distributed to injection wells through the existing WCBBP distribution pipeline. The WCBBP consists of north and south reaches. The blend ratio at the blend station can be varied from 0 to 100 percent of either recycled water or imported surface water.

Treatment of the secondary effluent includes decarbonation, lime/soda ash softening, recarbonation, filtration, and reverse osmosis followed by lime stabilization and chlorine disinfection as shown schematically in Figure 27.21.

At the time of this writing, the plant has been in operation for only a short period of time; therefore, projections of the quality of water produced by the WBWRP for selected constituents are presented in Table 27.6.

Water produced from the WBWRP will be blended with imported water and injected. The average blend ratio during Phase 1 will be approximately 25 percent based on the average quantity of water injected over the last decade. It is possible to inject up to 100 percent of recycled water into the north reach of the WCBBP by adjusting the blend ratio. The WBMWD proposes to inject up to 50 percent recycled water based on a 5-year average, which could include injection of up to 100 percent recycled water

TABLE 27.6. **Average Reclaimed Water Quality Projections.**

Parameter/Units	Barrier Injection Water
pH	8.3
TSS, mg/L	0
TDS, mg/L	168
COD, mg/L	3
TOC, mg/L	1.5
Ammonia-N, mg/L	4
Nitrate-N, mg/L	1
Alkalinity, mg/L	92
Hardness, mg/L	92
Silica, mg/L	3
Calcium, mg/L	37
Phosphorus, mg/L	0
Chloride, mg/L	35
Sulfate, mg/L	2
Sodium, mg/L	18
Oil and Grease, mg/L	0
Aluminum, μg/L	< 100
Iron, μg/L	6
Manganese, μg/L	< 1
Copper, μg/L	< 5
Zinc, μg/L	1
Fluoride, μg/L	80
Barium, μg/L	< 10

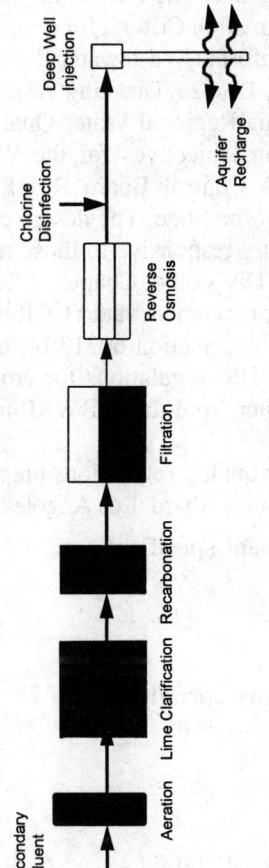

Secondary Effluent · Aeration · Lime Clarification · Recarbonation · Filtration · Reverse Osmosis · Chlorine Disinfection · Deep Well Injection · Aquifer Recharge

Figure 27.21 West Basin Water Reclamation Plant.

in some years in the north reach of the WCBBP. This will provide flexibility in operation, allow better coordination with the Los Angeles County Department of Public Works, who operates the barrier, and allow observation of performance of the injection system using higher percentages of reclaimed water, which is a goal for future phases.

The regulatory requirements governing development of the West Basin Water Recycling Project included the following: proposed CCR, Title 22, Division 4, Chapter 3 Reclamation Criteria for groundwater recharge using reclaimed water; CCR, California Wastewater Reclamation Criteria, Title 22, Section 60313(b); CCR, Title 22, Drinking Water Quality and Monitoring Requirements; California Regional Water Quality Control Board-Los Angeles Region Basin Plan Objectives for the West Coast Basin; and the State Water Resources Control Board Resolution 68-16 regarding nondegradation of waters of the State. The design criteria for the conveyance and treatment is directly responsive to those requirements contained in proposed CCR, Title 22, Division 4, Chapter 3 Reclamation Criteria for groundwater recharge using reclaimed water; CCR, California Wastewater Reclamation Criteria, Title 22, Section 60313(b); and California Department of Health Services (DHS) regulations for cross connection control. In addition, the product water from the WBWRP meets federal and state drinking water standards.

Specific reclaimed water quality regulations prepared by the California Regional Water Quality Control Board, Los Angeles Region are as follows:

(1) **Reclaimed water influent specifications**

- BOD_5 <30 mg/L
- SS <30 mg/L
- TOC <20 mg/L

(2) **Reclaimed water quality specifications**

- BOD_5 <1 mg/L
- SS <1 mg/L
- turbidity <2 NTU
- TOC <2 mg/L
- total coliform < or = 2.2/100 mL
- pH = 6.5–8.5
- oil and grease < or = 1 mg/L
- meet drinking water standards

(3) **Blended reclaimed water quality specifications**

- TDS < or = 800 mg/L
- sulfate < or = 250 mg/L
- chloride < or = 250 mg/L
- boron < or = 1.5 mg/L

- total nitrogen < or = 5 mg/L
- chlorine residual > or = 0.2 mg/L

The proposed DHS regulations for groundwater recharge using reclaimed water also require the following: (1) the injected reclaimed water be retained in the aquifer for 12 months prior to extraction, (2) the maximum percentage of reclaimed water within 2,000 feet of an extraction well shall not exceed 50 percent, and (3) the injected reclaimed water travel at least 2,000 feet prior to extraction. The first two requirements are based on a 5-year average of yearly determined values of retention time, and percentage of reclaimed water in production wells, respectively.

Hydrogeologic investigations were conducted to address the regulatory requirements listed above. These investigations included (1) evaluation of retention time of reclaimed water in the basin prior to being extracted in production wells, (2) evaluation of dilution of reclaimed water in the basin including chemical constituents such as TOC and nitrate introduced to the basin by injection of reclaimed water, and (3) evaluation of the potential impact of reclaimed water on production wells. A three-dimensional groundwater flow and solute transport model of the West Coast Basin was used during these hydrogeologic investigations to further assess the retention time of reclaimed water in the basin, and dilution of reclaimed water.

Based on the model projections, injection of a 50 percent blend of reclaimed water and imported water will not adversely affect the quality of groundwater in the West Coast Basin; instead, use of reclaimed water for injection is anticipated to improve groundwater quality along the WCBBP because of the projected high quality of the reclaimed water relative to imported water and groundwater. Conclusions from the three-dimensional groundwater flow and solute transport modeling included the following:

- Reclaimed water is estimated to reside in the aquifer for at least 15 years and travel more than 5,000 feet prior to being extracted.
- The dilution of reclaimed water prior to injection and in the aquifer prior to extraction is estimated to limit the reclaimed water contribution to less than 15 percent at the nearest extraction well after 20 years of injection.

Project Costs

The construction cost for the first 5 mgd phase of the West Basin Barrier Plant was about $22 million. This cost included infrastructure to readily accommodate expansion to an ultimate capacity of 20 mgd.

Expansion to the ultimate 20 mgd capacity is estimated to cost about an additional $40 million in construction.

Annual operation and maintenance costs are projected to be about $4.5 million for the first 5 mgd phase.

Project Future

The treatment plant will likely be built out in 5 mgd increments to the ultimate 20 mgd capacity. The next 5 mgd increment will likely be constructed by the year 2000. Based on continued monitoring of the aquifer water quality, the WBMWD's goal is to substitute reclaimed water for all the treated, imported surface water currently used to operate the West Coast Basin Barrier Project.

FRED HERVEY WATER RECLAMATION AND GROUNDWATER RECHARGE PROJECT [12]

Project Impetus

The Fred Hervey Water Reclamation and Groundwater Recharge Project is a unique recharge system, being a full scale operating program in which sewage is treated to potable water quality and injected directly into the major drinking water source for the City of El Paso, Texas, Hueco Bolson aquifer. The project concept was developed in response to two factors. First, studies performed by the United States Geological Service (USGS) showed that the Hueco Bolson aquifer was being depleted at rates which could exhaust the 10 million acre-feet of fresh water in the Hueco Bolson aquifer by early in the 21st century. Second, there was a need to upgrade and enlarge the sewage treatment facility that serviced the Northeast area of El Paso from 5 to 10 mgd. With these two driving factors at play, the concept of recharging the Hueco Bolson aquifer with reclaimed water was developed, along with other water resource management strategies. In the final analysis, there were three major considerations that led to a consensus for implementing the recharge project.

(1) Conventional wastewater treatment and disposal alternatives were limited and expensive because the project area is distant from a surface water discharge point.

(2) Alternative, conventional raw water supply sources were expensive. The lowest cost source was groundwater in New Mexico, but New Mexico state law prohibits export of groundwater. Of supplies in Texas, the lowest cost alternative involved developing water over one hundred miles away.

(3) The USEPA Environmental Impact Statement analysis, in concert with an extensive public participation program, determined that the public health risks of potable water recycling were acceptable to the public.

Project Description and Performance

Table 27.7 presents the influent water quality and effluent water quality criteria used to guide development of the Fred Hervey Water Reclamation Plant. The table also presents the quality of the Hueco Bolson aquifer for comparison, as well as the Texas Water Commission (TWC) Wastewater Discharge Permit Requirements. The discharge permit from the TWC requires monitoring of the 23 parameters, with reporting of 30-day average values. The permit also requires that each 8-hour batch of reclaimed water have less than 10 mg/L nitrate and turbidity less than 5 NTU.

Based on the water quality criteria presented in Table 27.7 below, a treatment train including the following unit processes was selected: primary treatment; two-stage bio-physical PACT process; lime treatment; recarbonation; sand filtration; ozonation; granular activated carbon adsorption; and clearwell storage prior to aquifer injection.

A schematic of the 10 mgd Fred Hervey Water Reclamation Plant is depicted in Figure 27.22.

The two-stage bio-physical PACT process combines a conventional aerated biological treatment system with the use of powdered activated carbon (PAC). This process provides the majority of the organics removal and all of the nitrogen removal. The first stage provides carbonaceous BOD removal and nitrification. The second stage uses methanol and denitrifying bacteria to remove nitrates, and continue the PAC organics adsorption.

Lime clarification, recarbonation, and sand filtration provide heavy metals removal, virus inactivation and removal, phosphorus removal, and suspended solids removal.

Ozone disinfection provides pathogen kill and conditions remaining organics for biological removal on downstream activated carbon.

Granular activated carbon filters remove remaining synthetic organic chemicals, assimilable organic carbon, taste and odor, and suspended solids.

Three 3.3 million gallon clearwells are operated sequentially. While one clearwell is filling, another is undergoing quality control monitoring, while the third is emptying. Composite samples are taken for each 8-hour batch of water from the clearwell. This water is analyzed for turbidity, nitrates, total organic carbon, pH, alkalinity, ozone residual, and chlorine residual prior to releasing the water for injection.

The reclaimed water is conveyed from the Fred Hervey Water Reclamation Plant to eleven injection wells. The injection wells are located within an area of the Hueco Bolson aquifer expected to provide much of the future water supply for the El Paso area. Specific locations were chosen by locating the injection wells approximately three-fourths of a mile up gradient and one-fourth of a mile down gradient from the exisitng El Paso

TABLE 27.7. Fred Hervey Water Reclamation Plant Water Quality Criteria.

Parameter	Units	Influent Quality Design Criteria	Effluent Quality Design Criteria	Hueco Bolson	TWC Permit
Alkalinity	mg/L	250	150	129	
Ammonia-N	mg/L	25	<0.1		
Boron	mg/L	<1	<0.8		
Calcium	mg/L	35	50	45	
Chloride	mg/L	122	140	64.7	300
COD	mg/L	400	<10		
E. Coli	#/100 mL		0		
EC	umho/cm	1,200	1,000	510	
Fluoride	mg/L	0.9	0.9	1	
Magnesium	mg/L	7.9	3	10.6	
pH	units	7.1	7.5–8	8.1	
Phosphate-P	mg/L	20	0.1	<0.5	
Sodium	mg/L	192	145	96	
Sulfate	mg/L	125	125	53.7	300
Silica	mg/L	38	38	38	
Total hardness	mg/L	120	140	102	
Total nitrogen	mg/L	36	2		
Nitrate-N	mg/L	0	1.5	<2.1	10
TOC	mg/L		<2		
Turbidity	NTU	97	<0.5	0.14	1
TSS	mg/L	150	<1		
Cyanide	mg/L		0.02		
Arsenic	mg/L	0.008	0.05		0.05
Barium	mg/L	0.46	0.1		1
Cadmium	mg/L	<0.01	<0.01	<0.01	0.01
Chromium	mg/L	0.01	0.01	<0.05	0.05
Copper	mg/L	<0.05	<0.05	<0.1	1
Iron	mg/L	<0.31	0.05	<0.1	0.3
Lead	mg/L	<0.05	<0.05	0.05	0.05
Manganese	mg/L	<0.05	<0.05	0.01	0.05
Mercury	mg/L	0.002	0.0014		0.002
Selenium	mg/L	<0.0035	0.01		0.01
Silver	mg/L	<0.01	<0.01		0.05
Zinc	mg/L	<0.1	<1	0.1	5
Aluminum	mg/L	0.28	<0.15		
Color	units	52	<10		
MBAS	mg/L	3.5	0.2		
Corrosivity			noncorrosive		
Hydrogen sulfide	mg/L	1	0.01		
Odor	TON		1		
TDS	mg/L	770	1	391	1,000
Dissolved oxygen	mg/L		1		
Phenol	mg/L	0.004	0.001	0.001	
Chlorine residual	mg/L		0.25 free		
Virus			0		
Endrin	mg/L		<0.0002		0.0002

TABLE 27.7. (continued).

Parameter	Units	Influent Quality Design Criteria	Effluent Quality Design Criteria	Hueco Bolson	TWC Permit
Lindane	mg/L		< 0.004		0.004
Methoxychlor	mg/L		< 0.1		0.1
Toxaphene	mg/L		< 0.005		0.005
2,4-D	mg/L		< 0.1		0.1
2,4,5-TP	mg/L		< 0.01		0.005
TTHMs	mg/L		< 0.1		
Radium 226 & 228	pCi/L	32	< 5		
Gross alpha	pCi/L		< 15		
Beta particle	mrem/yr		< 4		

Water Utility production wells in the area. This was done to provide a minimum 2-year residence time for injected water prior to withdrawal for potable reuse.

Based on monitoring to date, the Fred Hervey Water Reclamation Plant is very effective in removing the priority pollutants that enter the plant.

Results of aquifer modeling project that by the year 2005, the artificial recharge project will have provided two or more feet of water level benefit over an area of more than 175 square miles, effectively reversing the aquifer water table decline.

Project Costs

Construction of the 10 mgd Fred Hervey Water Reclamation Plant was completed in 1984 at a cost of $26.7 million. The direct recharge facilities— a holding tank, ten injection wells, and seven observation wells each completed to depth of approximately 750 feet—cost $1.79 million. Total capital costs for the project were approximately $33 million. While operating at 50 percent capacity, the project operational cost for treatment and injection is $1.35 per 1,000 gallons. While operating at 75 percent capacity, the operational cost declines to $1.14 per 1,000 gallons.

Project Future

Although there are no firm plans for project expansion at this time, one water supply manangement scenario has the project expanding to 50 mgd with two 20 mgd increments by the year 2015. This increase in water recycling would be accompanied by increased use of surface water in the region. Under this alternative, the Hueco Bolson aquifer would reach a

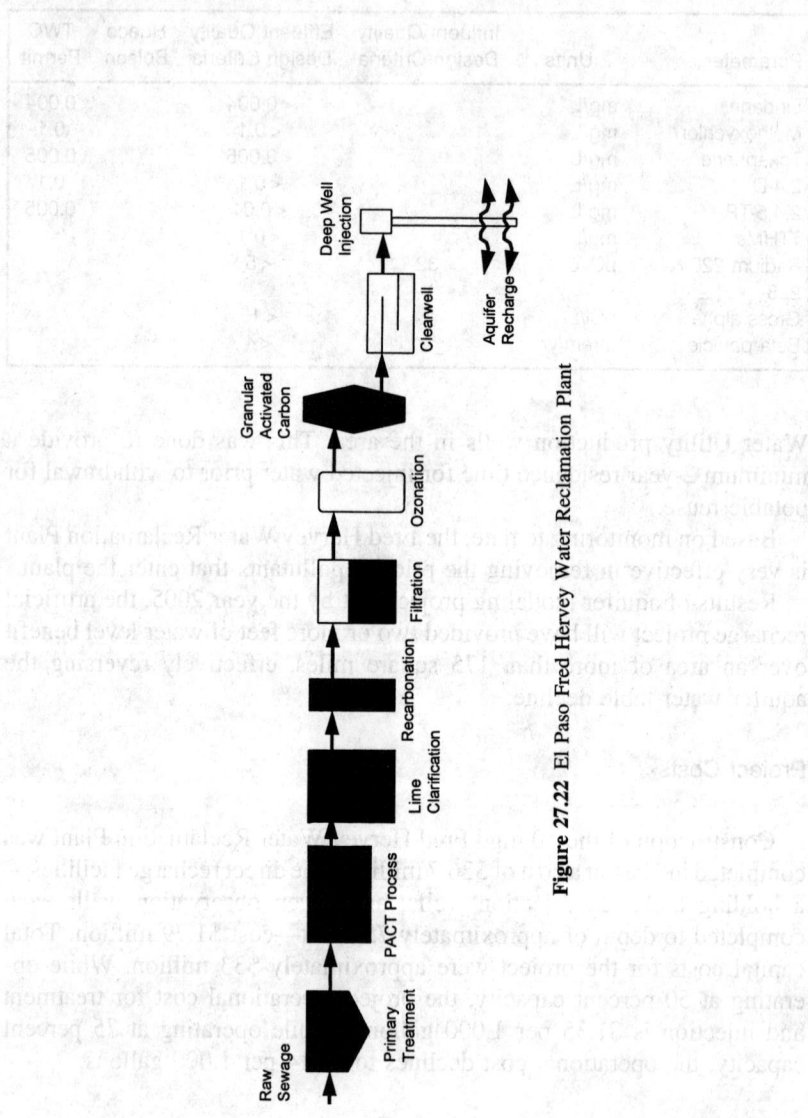

Figure 27.22 El Paso Fred Hervey Water Reclamation Plant

steady-state condition (withdrawal equals recharge), so long as neighboring Mexico does not over pump the aquifer.

REFERENCES

1 Metzler, D. F., et. al., Emergency Use of Reclaimed Water for Potable Supply at Chanute, Kansas, *JAWWA*, 50:1021, 1958.
2 Report of the Scientific Advisory Panel on Groundwater Recharge with Reclaimed Wastewater, prepared for State of California Water Resources Control Board, Department of Water Resources, and Department of Health Services, November 1987.
3 Culp, R., Water Reclamation at South Tahoe, *Water and Wastes Engineering*, Vol. 6, No. 4, April 1969.
4 Offringa, G., 20 Years of Direct Potable Reuse, *OWR News*, Volume 5, Issue 2, June 1994.
5 Federal Water Pollution Control Act (Clean Water Act) as amended by the Water Quality Act of 1987, P.L. 100-4, approved February 4, 1987.
6 Argo, D. G., Water Reuse: Where are We Headed?, *Environ. Sci. Technol.*, Vol. 19, No. 3, 1985.
7 CH2M HILL, UOSA, Upper Occoquan Sewage Authority Overview and Progress Report. December 1989.
8 The National Research Council, *Quality Criteria for Water Reuse*, National Academy Press, Washington, D.C., 1982.
9 The National Research Council. *The Potomac Estuary Experimental Water Treatment Plant*. National Academy Press, Washington, D.C., 1984.
10 Lauer, W. C., Denver's Direct Potable Water Reuse Demonstration Project, Final Report, Executive Summary, April 1993.
11 Western Consortium for Public Health, The City of San Diego Total Resource Recovery Project, Health Effects Study, Final Summary Report, September 1992.
12 Bureau of Reclamation, PSC, EPWU, Hueco Bolson Recharge Demonstration Project, Project Overview. February 1993.
13 CH2M HILL, Tampa Water Resource Recovery Project. Summary Report. January 1993.
14 CH2M HILL and Black & Veatch, Potable Reuse Feasibility Study, Final Report, Prepared for the City of Phoenix, November 1988.
15 Proposed Draft Regulation, Title 22, California Code of Regulations, Division 4. Environmental Health, Chapter 3. Reclamation Criteria, Article 5.1 Groundwater Recharge, May 8, 1992.
16 CH2M HILL, WBMWD, West Basin Municipal Water District Groundwater Augmentation Project. AWWA Conference Presentation, June 1995.
17 Bostrom, G., McEwen, B., Potable Reuse—Becoming a Necessary Component of Water Resources Management Plans, presented at the *1994 AWWA/WEF Water Reuse Symposium*, Dallas, Texas, March 1994.
18 Montgomery Watson, et al., San Diego County Water Authority, Final Water Repurification Feasibility Study, June 1994.
19 The California Potable Reuse Committee, A Proposed Framework for Regulating the Indirect Potable Reuse of Advanced Treated Reclaimed Water by Surface Water

Augmentation in California, for the California Department of Health Services and Department of Water Resources, January 1996.

20 McEwen, B., Richardson, T., Indirect Potable Reuse: AWWA Committee Report, 1995.

21 Hamann, C. L. & McEwen, B., Potable Water Reuse, *Water Environment & Technology,* pp. 75–80, January 1991.

CHAPTER 28

The Demonstration of Direct Potable Water Reuse: Denver's Landmark Project

INTRODUCTION

D ENVER'S Direct Potable Water Reuse Demonstration Project was developed to determine if potable water of comparable quality to Denver's existing supply could be continuously and reliably produced from secondary treated wastewater. The project investigated the interrelated issues of product safety, technical and economic feasibility, and public acceptance. A 1.0 mgd (44L/s) demonstration treatment plant, which served as the main testing facility for the project, was constructed at a cost in excess of $18 million. The $30 million project, of which $7 million was contributed by the U. S. Environmental Protection Agency (EPA) (Table 28.1) included integral health effects testing as well as comprehensive analytical studies, and was completed in 1992. Public information programs were carried out concurrently with the scientific studies. The project provided information necessary to consider direct potable reuse as one possible alternative which may be used to satisfy Denver's future water supply needs.

Although the research described in this article has been funded wholly or in part by the United States Environmental Protection Agency contract CS-806821 to the Denver Board of Water Commissioners, it has not been subject to the Agency's review and therefore does not necessarily reflect the views of the Agency, and no official endorsement should be inferred.

William C. Lauer, Montgomery Watson, 10065 East Harvard Avenue, Suite 201, Denver, CO 80231; Stephen E. Rogers, Camp Dresser & McKee Inc., 1331 17th Street, Suite 1200, Denver, CO 80202. William C. Lauer was Project Manager and Stephen E. Rogers was Plant Supervisor for the Denver Water Department during the Reuse Project.

TABLE 28.1.   Reuse Demonstration Project Costs (1979–1992).

Activity	Total Cost	EPA Contribution
Treatment plant design	$ 1,319,059	$ 346,958
Treatment plant construction	$18,501,107	$4,150,000
Treatment plant operation	$ 8,419,591	$ 600,000
Scientific studies		
Analytical	$ 3,085,363	$ 750,000
Health effects	$ 3,022,053	$1,200,000
Project cost	$34,347,173	$7,046,958

HISTORICAL BACKGROUND

The study of direct potable water reuse in Denver has been investigated since 1968. It was started in part to satisfy the requirements of the court decree which allowed Denver to divert water from the Blue River on the west slope of the Continental Divide, but also required Denver to investigate ways to maximize the use of this trans-basin water [1]. It was recognized that in order to satisfy the demands of the growing Denver metropolitan area, all water resource possibilities including additional trans-basin diversions and various forms of successive water use would have to be employed.

The Successive Use Project, as it was known, began in 1968 by acquiring a grant from the Federal Water Quality Administration to build a 5 gpm (19 L/min) pilot plant on the site of the Metropolitan Denver Sewage Disposal District Plant No. 1 (renamed Metropolitan Denver Reclamation District). In 1970 the construction was completed and 10 years of research conducted by the Denver Water Department and the University of Colorado's Environmental Engineering Department followed. Over thirty (30) advanced degrees were supported by this effort funded by the Denver Water Department. Many processes were investigated to convert secondary treated wastewater to higher quality effluent. All of the processes eventually included in the demonstration plant design were investigated at this facility [2].

The Successive Use Project has investigated many possibilities other than potable reuse. Alternatives such as river exchange, industrial reuse, agricultural reuse, and dual distribution systems were studied in addition to potable reuse. Due to various legal, economic, and technical considerations these options were discarded in favor of potable reuse [3]. Two options which still hold promise for at least part of the water which is available for reuse are exchange and dual distribution systems. Exchange is presently utilized to the maximum extent possible. Dual distribution systems require a new area of development of at least 10,000 population to be economically feasible. Therefore, the potable water reuse option surfaced as the only viable reuse alternative capable of utilizing the volume of water which would be available from the trans-mountain diversion.

Plans were initiated to construct a demonstration facility to study the reliability and cost of potable reuse [4]. This intermediate step between pilot and full scale treatment was necessary due to the pioneer nature of this conversion and to answer the many technical and nontechnical issues which would need to be addressed prior to a full scale implementation.

A project advisory committee of national experts was formed to review the progress of the project and to act as consultants to the Department. The membership evolved since its formation in 1979 and in its final form was composed of the following:

Project Advisory Committee	Affiliation
Dr. Elmer W. Akin	U.S. EPA—National Regional Office, Atlanta, Georgia
Dr. Fred Kopfler	U.S. EPA—Gulf of Mexico Program Office, Stennis Space Center, Mississippi
Dr. Franklyn N. Judson	Director, Denver Public Health, Denver, Colorado
Dr. Robert Neal	Center in Molecular Toxicology, Vanderbilt University School of Medicine, Nashville, Tennessee
Dr. Harold Walton	Chemistry Department, University of Colorado, Boulder, Colorado
Dr. Mark D. Sobsey	Professor of Environmental Microbiology, University of North Carolina, Chapel Hill, North Carolina
Dr. I. H. Suffet	Professor of Environmental Engineering and Science, University of California, Los Angeles, California
Dr. Joel S. Cohen	Professor of Mathematics and Computer Science, University of Denver, Denver, Colorado
Dr. Richard J. Bull	College of Pharmacy, Washington State University, Pullman, Washington
Dr. Alfred Dufour	U.S. EPA—Microbiology Section, Cincinnati, Ohio
Dr. Joseph F. Borzelleca	Department of Toxicology and Pharmacology, Medical College of Virginia, Richmond, Virginia
Dr. John Doull, M.D.	Department of Pharmacology and Toxicology, University of Kansas Medical Center, Kansas City, Kansas
Mr. David Argo	Black & Veatch Consulting Engineers, Santa Ana, California
Mr. Franklin D. Dryden	Environmental Engineer, Pasadena, California
Dr. K. Daniel Linstedt	Black & Veatch Consulting Engineers, Aurora, Colorado
Mr. Kip Cherotes	Legislative Assistant to the Honorable Patricia Schroeder—House of Representatives, Denver, Colorado

Mr. Carl Hamaan	CH2M Hill, Inc., Denver, Colorado
Mr. Jack Hoffbuhr	American Water Works Association, Denver, Colorado
Mr. Jerry Biberstine	Colorado State Health Department, Denver, Colorado
Dr. Lyman Condie	U. S. Army Proving Ground, Dugway, Utah
Mr. Carl Brunner	Project Officer, Water Engineering, U.S. EPA—Research Laboratory, Cincinnati, Ohio
Dr. Joe Cotruvo	U.S. EPA—Office of Toxic Substances, Health & Environmental Review Division, Washington, D.C.
Dr. Raymond S. H. Yang	Professor of Toxicology, Department of Environmental Medicine and Biomedical Sciences, Colorado State University, Fort Collins, Colorado

PROJECT OBJECTIVES

A proposal for direct potable reuse raises many questions regarding safety, reliability, public acceptance, and technical and economic feasibility. The objectives for the demonstration project involved addressing each of these questions. The major objectives [5] of the project were:

- establish the product water safety
- demonstrate the dependability of the process
- generate public awareness
- establish a foundation for future regulatory acceptance
- provide data for a large scale implementation

The key objective upon which all the others depended was the establishment of the safety of the product. Since the health standards thus far established for drinking water were not intended to apply to treated waters which were derived from a polluted source, additional criteria had to be used to establish that the product water is suitable for human consumption [6]. The criteria used in this project were as follows:

(1) The product was compared with parameters included in the National Primary and Secondary Drinking Water Regulations [7] and the Colorado State Drinking Water Regulations [8].

(2) The product was compared with potential or proposed Federal or State regulated parameters [7,8] and World Health Organization (WHO) standards [9], and other international standards [7].

(3) Concentrations of parameters in the product water were compared with Denver's current drinking water in those areas where there are no existing or proposed standards.

(4) Whole animal health effects testing including chronic toxicity and carcinogenicity and reproductive toxicity was conducted on the product water using Denver's current drinking water as a comparison standard.

Denver's current drinking water was selected for use as a comparison since it is derived from a relatively protected source and there is no reason to believe that it will fail to satisfy any future health standards that are developed. This insured a margin of safety necessary to apply this technology many years hence. Also, surveys have shown that if water were produced which would be comparable to Denver's existing supply, the customers would be more inclined to accept it as an alternate source [10,11].

PROCESS SELECTION FOR HEALTH EFFECTS STUDIES

In a 1974 conceptual design report commissioned by the Denver Water Department, CH2M-Hill, Inc. [4] described a sequence of processes which could be utilized to convert secondary treated wastewater directly to potable water quality. After five more years of pilot study with the University of Colorado Environmental Engineering Department, the Denver Water Department proposed a treatment sequence for a Potable Water Reuse Demonstration Plant in its proposal to the U.S. EPA for a cooperative funding agreement [3]. A $7 million funding ceiling was approved and the final treatment train was agreed upon after meetings with the EPA in 1979. The Reuse Demonstration Plant treatment train is shown in Figure 28.1.

The plant was designed to allow experimentation by changing inter-process modes of operation and also intra-process impacts [12]. For example, in the carbon adsorption system, both first- and second-stage could be operated either up-flow or down-flow. Changes could be made in one process which affect the operation of the following processes, perhaps even making some processes unnecessary so that they could be bypassed.

These alternatives provided an almost endless variety of operational modes [13]. During plant start-up, many of these options and combinations were evaluated in order to establish the best operating conditions for the first phase of the project [14]. Subsequently, laboratory tests and controlled plant scale experiments were conducted to achieve lower cost and higher quality treatment. Promising treatments were incorporated into the plant operational criteria for a full scale trial.

During the first two years of operation, the processes underwent numerous changes to provide the least-cost, best-quality treatment [15]. The routine sampling and analysis program documented the quality of the effluent and the plant operational records revealed the cost.

A schedule was established in the project experimental design for the optimization period (Figure 28.2); a description of each alternative process configuration which was examined during this period is listed below:

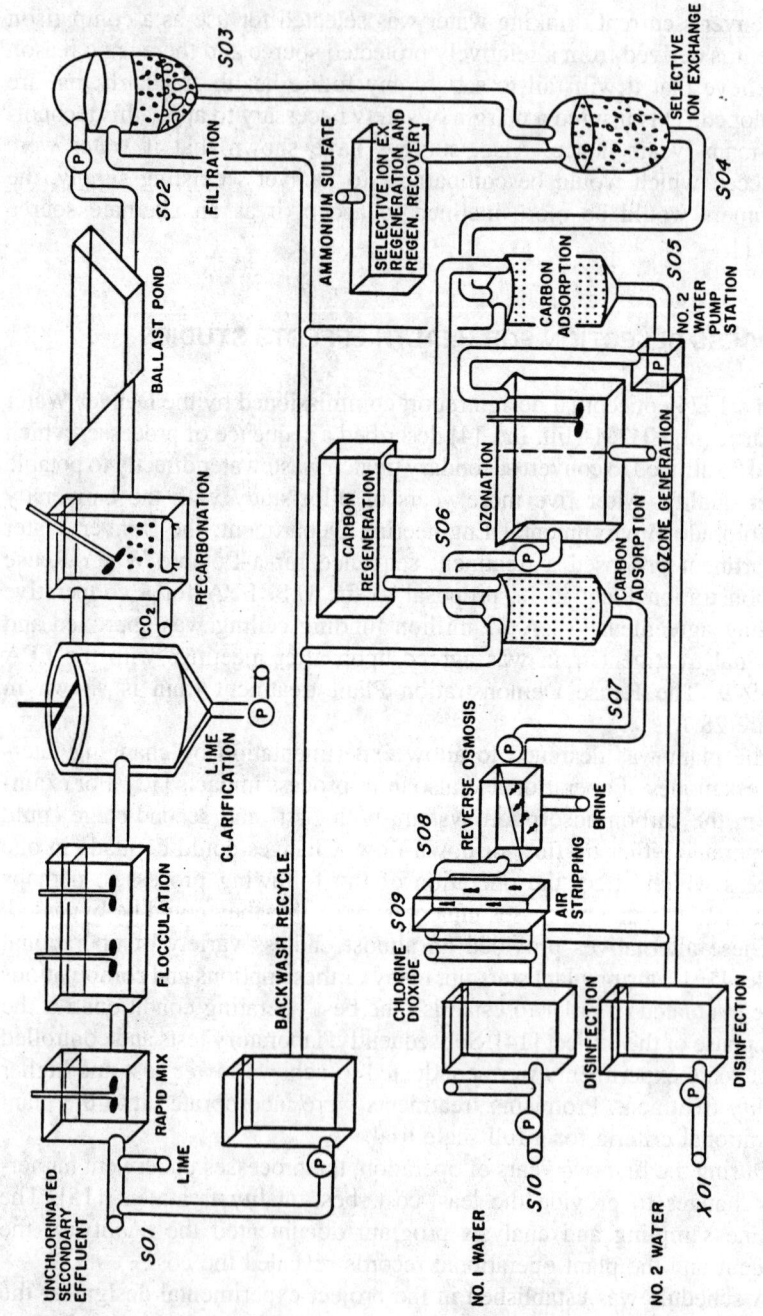

Figure 28.1 Water reuse treatment process (showing sample locations).

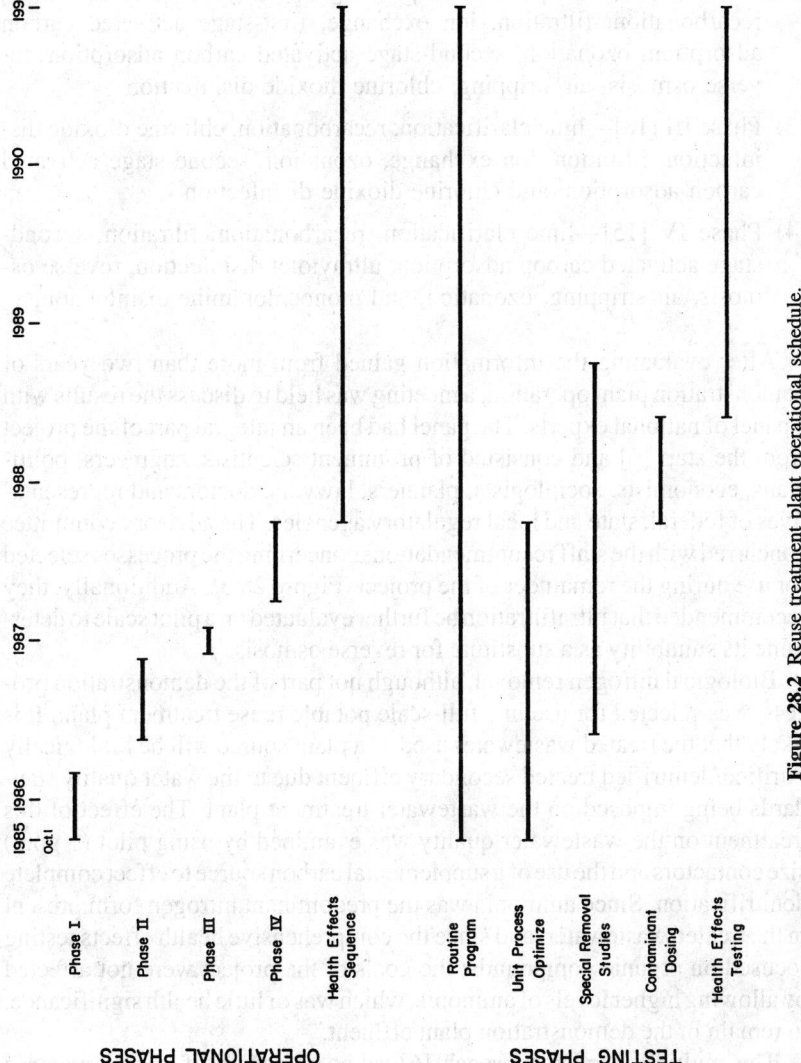

Figure 28.2 Reuse treatment plant operational schedule.

1275

(1) Phase I [16]—lime clarification, recarbonation, filtration, first-stage activated carbon adsorption, ozonation, second-stage activated carbon adsorption, reverse osmosis, air stripping, chlorine dioxide disinfection

(2) Phase II [17]—all processes operating at design: lime clarification, recarbonation, filtration, ion exchange, first-stage activated carbon adsorption, ozonation, second-stage activated carbon adsorption, reverse osmosis, air stripping, chlorine dioxide disinfection

(3) Phase III [18]—lime clarification, recarbonation, chlorine dioxide disinfection, filtration, ion exchange, ozonation, second-stage activated carbon adsorption, and chlorine dioxide disinfection

(4) Phase IV [15]—lime clarification, recarbonation, filtration, second-stage activated carbon adsorption, ultraviolet disinfection, reverse osmosis, air stripping, ozonation, and monochloramine disinfection.

After evaluating the information gained from more than two years of demonstration plant operation, a meeting was held to discuss the results with a panel of national experts. The panel had been an integral part of the project from the start [6] and consisted of prominent scientists, engineers, politicians, economists, sociologists, planners, lawyers, doctors and representatives of federal, state and local regulatory agencies. The advisory committee concurred with the staff recommendations concerning the processes selected for use during the remainder of the project (Figure 28.3). Additionally, they recommended that ultrafiltration be further evaluated on a pilot scale to determine its suitability as a substitute for reverse osmosis.

Biological nitrogen removal, although not part of the demonstration process, was selected for use in a full-scale potable reuse treatment plant. It is likely that the treated wastewater used as a plant source will be biologically nitrified/denitrified treated secondary effluent due to the water quality standards being imposed on the wastewater treatment plant. The effect of this treatment on the wastewater quality was examined by using pilot (1 gpm) size contactors and the use of a supplemental carbon source to effect complete denitrification. Since ammonia was the predominant nitrogen form present in the treated wastewater, and since the comprehensive health effects testing focused on organic compounds; the goals of the project were not affected by allowing higher levels of ammonia, which was of little health significance, to remain in the demonstration plant effluent.

The multiple barrier approach [6] relies on layers of complementary processes, several of which may be effective for the removal of any single contaminant. This design concept was adopted early in the planning stages of the project and was retained in the final process selection procedure. The demonstration plant health effects treatment process (Figure 28.3) provided multiple barriers of protection against all primary contaminants.

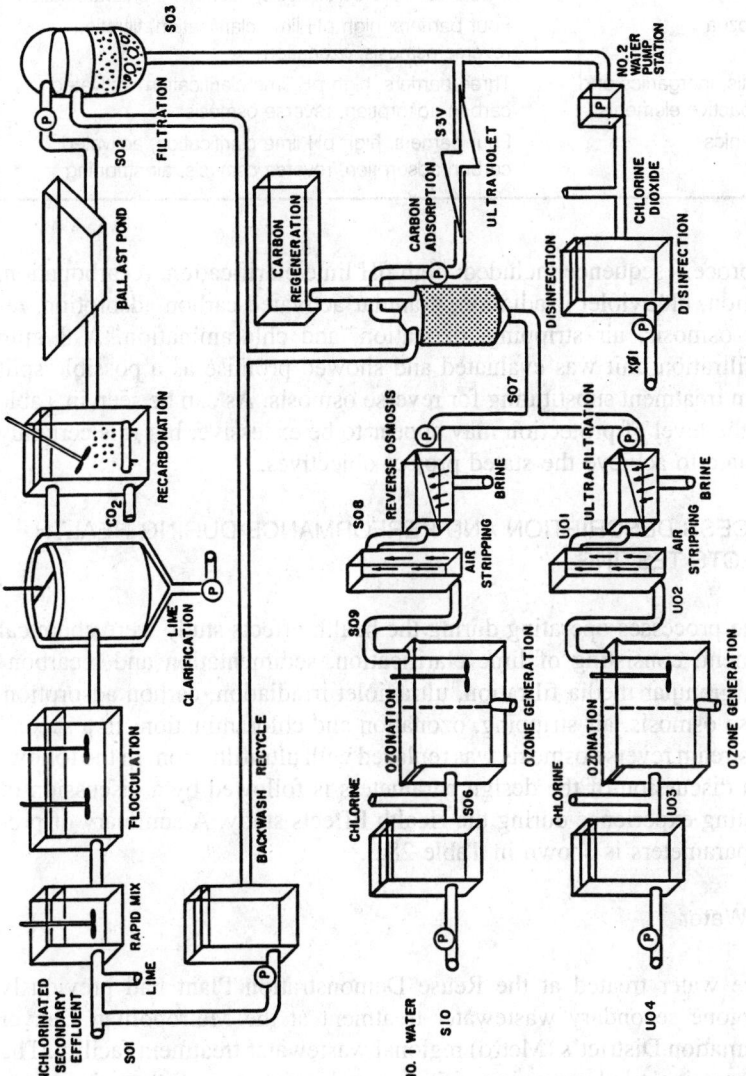

Figure 28.3 Water treatment process (health effects with ultrafiltration sidestream).

TABLE 28.2. **Primary Contaminant Grouping and Reuse Treatment Process Barriers.**

Viruses and bacteria	Five barriers: high pH lime clarification, ultraviolet irradiation, reverse osmosis, ozonation, chloramination
Protozoa	Four barriers: high pH lime clarification, filtration, reverse osmosis, ozonation
Metals, inorganics and radioactive elements	Three barriers: high pH lime clarification, activated carbon adsorption, reverse osmosis
Organics	Four barriers: high pH lime clarification, activated carbon adsorption, reverse osmosis, air stripping

The process sequence included: high pH lime clarification, recarbonation, filtration, ultraviolet irradiation, granular activated carbon adsorption, reverse osmosis, air stripping, ozonation, and chloramination. A 3 gpm ultrafiltration unit was evaluated and showed promise as a possible split stream treatment substituting for reverse osmosis. As can be seen in Table 28.2 the level of protection may appear to be excessive, but it is certainly adequate to achieve the stated project objectives.

PROCESS DESCRIPTION AND PERFORMANCE DURING HEALTH EFFECTS TESTING

The processes operating during the health effects study were chemical treatment, consisting of lime clarification, sedimentation and recarbonation, granular media filtration, ultraviolet irradiation, carbon adsorption, reverse osmosis, air stripping, ozonation and chloramination. In a second flow stream reverse osmosis was replaced with ultrafiltration. In the following, a discussion of the design parameters is followed by a discussion of operating experience during the Health Effects study. A summary of process parameters is shown in Table 28.3.

Raw Water

The water treated at the Reuse Demonstration Plant had previously undergone secondary wastewater treatment at the Metropolitan Denver Reclamation District's (Metro) regional wastewater treatment facility. The treatment included screening, grit removal, primary sedimentation, pure oxygen activated sludge and secondary sedimentation. The treatment at Metro was designed to meet a 30 mg/L total suspended solids and a 17 mg/L BOD_5 discharge permit [19]. Although a portion of Metro's flow had been nitrified, none of the flow entering the Reuse Plant had undergone nitrogen removal. Additionally, none of this flow had been disinfected.

TABLE 28.3. **Selected Operational Parameters (geometric means).**

Location	Parameter	Value	
Plant influent	Flow rate	0.90	mgd
	Turbidity	9.2	NTU
	pH	7.0	pH
	Backwash equalization flow rate	0.067	mgd
Rapid mix basin	Flow rate	0.967	mgd
	Detention time	2.6	min.
	Velocity gradient	326	sec^{-1}
	Lime dose	471	mg/L
	Ferric chloride dose	20.9	mg/L
	pH value	11.2	Units
Flocculation basin	Detention time	25	min.
	Velocity gradient 1	125	sec^{-1}
	Velocity gradient 2	100	sec^{-1}
	Velocity gradient 3	100	sec^{-1}
Chemical clarifier	Flow rate	0.967	mgd
	Detention time	82	min.
	Overflow rate	763	gpd/ft^2
	Flow rate of waste sludge	0.022	mgd
	Avg. concentration sludge	5.2	wt.%
	Lime sludge wasted	9,561	lb/day
	Turbidity of overflow	1.3	NTU
Recarbonation basin	Flow rate	0.945	mgd
	Detention time	13	min
	Velocity gradient 1	533	sec^{-1}
	Velocity gradient 2	188	sec^{-1}
	Carbon dioxide dose	172	mg/L
	pH Value	7.7	pH
Filters	Hydraulic loading rate	4.4	gpm/ft^2
	Average filter run length	22.9	hr
	Length of filter backwash	25	min.
	Backwash flow rate	1500	gpm
	Length of surface wash	20	min.
	Surface wash flow rate	56	gpm
	Backwash loading rate	19	gpm/ft^2
	Terminal pressure drop	11	ft
	Influent turbidity	1.4	NTU
	Effluent turbidity	0.33	NTU
Ultraviolet irradiation	Flow rate	0.082	mgd
Second stage carbon	Flow rate	0.082	mgd
	Throughput rate	1.37	BV/hr
	Hydraulic loading rate	4.5	gpm/ft^2
	Empty bed contact time	42	min.
	Length of backwash	20	min
	Backwash flow rate	200	gpm
	Backwash loading rate	15.9	gpm/ft^2
	Terminal pressure drop	15–35	ft
	Effluent turbidity	0.25	NTU

TABLE 28.3. (continued).

Location	Parameter	Value	
Reverse osmosis	Feed flow rate	0.082	mgd
	Feed pressure	359	psig
	Feed conductivity	1,072	mmhos/cm
	Feed turbidity	0.22	NTU
	Product conductivity	69.4	mmhos/cm
	Product water recovery	86.3	%
	Rejection based on conductivity	93.3	%
	Hydrochloric acid dose	121	mg/L
Air stripping	Flow rate	0.079	mgd
	Gas/liquid ratio	100:1	unitless
Ozone basin	Flow rate	0.079	mgd
	Detention time	86.5	min.
	Ozone residual	0.14	mg/L
	Concentration ozone off-gas	0.028	wt.%
	Applied ozone dose	0.67	mg/L
	Absorbed dose	0.62	mg/L
	Generator power consumption	4.06	kwh/day
Plant Effluent	Chlorine dose	0.97	mg/L
	Chlorine residual	0.56	mg/L
	Chlorine contact time	14.6	min.
	Turbidity	0.06	NTU

The pumping facilities were designed to deliver a nominal 1.0 mgd to the Reuse Plant.

Chemical Treatment

The chemical treatment system consisted of three individual processes: high pH lime treatment, sedimentation, and recarbonation. Upon reaching the Reuse Plant, the secondary effluent was aerated to remove dissolved carbon dioxide. Approximately one volume of air per volume of water was used for stripping [16]. Removing the dissolved carbon dioxide was shown to reduce the lime dosage required to reach the desired pH in the rapid mix basin by approximately 10 percent [16]. The air was sparged through a fine bubble diffuser in the influent channel.

After aeration the raw water was supplemented with backwash waste flow from the equalization basin and then hydrated lime was added in the rapid mix basin. The purpose of the lime addition was to raise the pH of the solution to 11 and exceed the saturation concentration for calcium carbonate. This treatment was expected to remove suspended solids, precipitate heavy metals and phosphates, reduce the total organic carbon, and reduce bacterial populations. Virus inactivation was also achieved under the high pH conditions. Ferric chloride was added to the flow stream as it exited the rapid mix basin to aid in flocculation and sedimentation.

Although the facilities for adding polymer, alum and seed sludge were available, none was necessary and these systems were not used. The design detention time in the rapid mix basin was approximately 2.2 minutes [20].

The mixer installed in the rapid mix basin provided a mixing intensity (G value) of 326 sec^{-1} at a circulation rate of approximately 4.4 times the throughput rate [16], and baffling was used to reduce vortex formation in the basin. The pH of the basin was measured continuously and control provided through a closed loop, pH paced system by varying the amount of lime added. Redundant systems allowed for cleaning and maintenance without downtime.

Flocculation was carried out in three serially connected basins. The basins were separated with baffling that created a serpentine flow and mitigated short circuiting. In the first basin the installed mixer provided a G value of 125 sec^{-1} at a circulation rate of 15.6 times the throughput rate. The mixers installed in the second and third basins provided a G value of 100 sec^{-1} at a circulation rate 12.4 times the throughput rate [16]. Although the facilities for adding polymer to the flocculation basin were available, none was deemed necessary. The total detention time in the flocculation basins was approximately 27 minutes.

Sedimentation occurred in a circular, center-feed clarifier with a nominal surface loading rate of 796 gpd/ft^2. Settled solids were removed in the underflow as a thickened sludge with a solids concentration of 3 to 5 wt%. The detention time in the clarifier was approximately 110 minutes.

Following clarification, the pH of the process flow was lowered to approximately 7.8 by the addition of gaseous carbon dioxide. The dispersion and dissolution of the carbon dioxide was accomplished by sparging the gas beneath a radial flow turbine mixer that provided a G value of 533 sec^{-1} at a circulation rate of approximately 11.7 times the throughput rate [16]. Additional mixing was carried out in a secondary mixing chamber with a mixer that provided a G value of 188 sec^{-1}. The theoretical detention time in the basin was approximately 9 minutes [20]. The pH of the basin was measured continuously and control provided through closed loop instrumentation that regulated carbon dioxide flow.

During the phases prior to the Health Effects testing phase, the chemical treatment system was extensively monitored to determine contaminant removal efficiency [21]. During the Health Effects testing, however, only parameters that would give an indication that the chemical treatment system was performing as expected were measured. The results of these analyses are shown in Table 28.4.

With the exception of the membrane heterotrophic plate count (mHPC), the contaminant removals shown in Table 28.4 are consistent with the removals seen during previous phases [16]. The increased mHPC values through chemical treatment may be a result of increased growth within the ballast ponds located upstream of the sample point. The results in

TABLE 28.4. **Contaminant Removal in Chemical Treatment.**

Parameter	Effluent Concentration	Percent Removal
Total coliform, count/100 mL	1,084	99.8
m-HPC, count/mL	1.2×10^6	-14.2
Coliphage E. coli 137, PFU/100 mL	485	99.1
Coliphage E. coli B, PFU/100 mL	84	99.6
Total organic carbon, mg/L	9.4	42.0
Turbidity, NTU	1.4	84.5

Table 28.4 combined with the fact that the high pH environment was consistently maintained in the rapid mix basin suggest that the system performed as expected throughout the Health Effects phase.

Sludge Handling

The sludge produced in the chemical treatment system was disposed by two methods. The first method involved thickening the sludge in an off line clarifier and then dewatering it with a vacuum filter. The dewatered sludge cake was then hauled for final disposal at a lined landfill. The second method was thickening within the clarifier and then direct disposal to the headworks at Metro.

Filtration

Following chemical treatment the process water from the recarbonation basin flowed into a ballast pond which served as a hydraulic buffer for upstream or downstream flow interruptions. Connected to the ballast pond was the filter pump wetwell.

The granular media pressure filters were supplied by 3-stage vertical turbine pumps and operated at a pressure of approximately 70 psig. The nominal loading rate to the filters was 4.4 gpm/ft². The primary function of the filters was to remove turbidity. The goal in operating this process is to achieve a product water turbidity below 0.5 NTU at all times.

Backwashing of the filters was accomplished using a programmable logic controller at a loading rate of 20 gpm/ft². It was initiated automatically at a terminal headloss of 10–12 ft and lasted for a minimum of 10 minutes. Hydraulic surface washing was used to improve backwashing and was done at flow rate of 56 gpm. At the start of each service cycle, the filter product water was wasted for 10 minutes. This allowed time for the product water quality to improve to quality goals.

The average length of the filter runs during the Health Effects study was 22.9 hours. The mean effluent turbidity was 0.33 NTU. The removal of other contaminants are shown in Table 28.5. As can be seen in Table 5, the removal of turbidity across the filters was significant. The average effluent turbidity of 0.33 NTU was well below the goal of 0.5 NTU in the effluent. Although some excursions beyond the goal were experienced, they were short duration and did not measurably impact plant effluent water quality.

The removal of microbiological contaminants and total organic carbon across the filters was modest. However, as stated above, the removal of these contaminants was not the primary function of the filters.

A loss of dissolved oxygen (D.O.) seen across the filters, approximately 42 percent, was indicative of the growth of nitrifying bacteria within the filters. It was determined that the D.O. loss could be controlled by disinfecting the filters with chlorine dioxide. As a result, process control testing included measuring the nitrite concentrations across the filters. A standard operating procedure established the criteria and procedures for disinfecting the filters. The filters required disinfection approximately twice per month.

Ultraviolet Irradiation

After filtration, the flow stream was split and only 0.1 mgd was given further treatment. This began with an ultraviolet irradiation (UV) process that was used to reduce the bacterial loading on the downstream carbon adsorption process. The UV contact tank consisted of a stainless steel cylindrical shell fitted with ten 60 watt mercury vapor lamps that provided a majority of the spectral energy at a wavelength of 254 nm. The contact time within the contactor was approximately 13 seconds. In the contactor the process water flowed around the quartz jacket surrounding each lamp. The system was cleaned when the intensity of the UV light, measured with a photocell, dropped to 75 percent of that of a clean unit. Cleaning was done with a 2 wt% citric acid solution.

The UV system proved effective in reducing the coliform bacteria

TABLE 28.5. Contaminant Removal in Filters.

Parameter	Effluent Concentration	Percent Removal
Total coliform, count/100 mL	785	27.7
m-HPC, count/mL	9.5×10^5	23.6
Coliphage *E. coli* 137, PFU/100 mL	443	8.6
Coliphage *E. coli* B, PFU/100 mL	65	22.7
Total organic carbon, mg/L	8.9	5.6
Turbidity, NTU	0.33	76.7

populations in the flow stream. Overall bacteria populations were less affected by the treatment but reductions were still significant. Coliphage virus inactivation was greater than 99 percent. Table 28.6 shows the removals across this process.

The UV system was inserted into the treatment configuration in attempt to reduce bacterial loadings on the carbon column. Time constraints left a very limited testing period prior to the start of the Health Effects study. Subsequent evaluations, however, suggest that the UV system is of limited value as an intermediate process and probably would not be included in the design of a full scale facility.

Carbon Adsorption

Granular activated carbon adsorption was used to remove organics from the flow stream. The column was operated in a downflow mode and was filled with regenerated carbon prior to the start of the Health Effects testing. The columns were backwashed with plant product water. After backwashing, the product flow stream was wasted for a period of 20 minutes to flush the low ionic strength water from the column and mitigate transient effects on the reverse osmosis system.

The carbon in the contactors was 8 × 30 mesh Filtrasorb 300 (Calgon Corporation, Pittsburgh, Pennsylvania) at an effective depth of 26-feet 8-inches. The nominal bed volume was 340 cubic feet (11,300 pounds). The design hydraulic loading rate to the columns was 5.5 gpm/ft^2, or 1.64 bed volumes per hour, and the empty bed contact time was 36 minutes [20]. The carbon column was backwashed at a 20 gpm/ft^2 loading rate for 15 minutes at a terminal headloss of 15 ft.

The carbon column effluent was sampled for both organic and inorganic contaminants and the results of these analyses are discussed below. A parameter used to monitor process performance during the study was total organic carbon (TOC). Figure 28.4 shows the TOC concentrations in the carbon effluent. As can be seen, the concentrations are low at the beginning of the study and increase steadily until September 1989. Regenerated carbon was then added to the bed to replace the carbon that had been

TABLE 28.6. Microbial Removals by Ultraviolet Irradiation.

Parameter	Effluent Concentration	Percent Removal
Total coliform, count/100 mL	N.D.	>99.9
mHPC, count/mL	1.2 × 10^4	98.7
Coliphage E. coli 137, PFU/100 mL	N.D.	>99.98
Coliphage E. coli B, PFU/100 mL	0.35	99.5

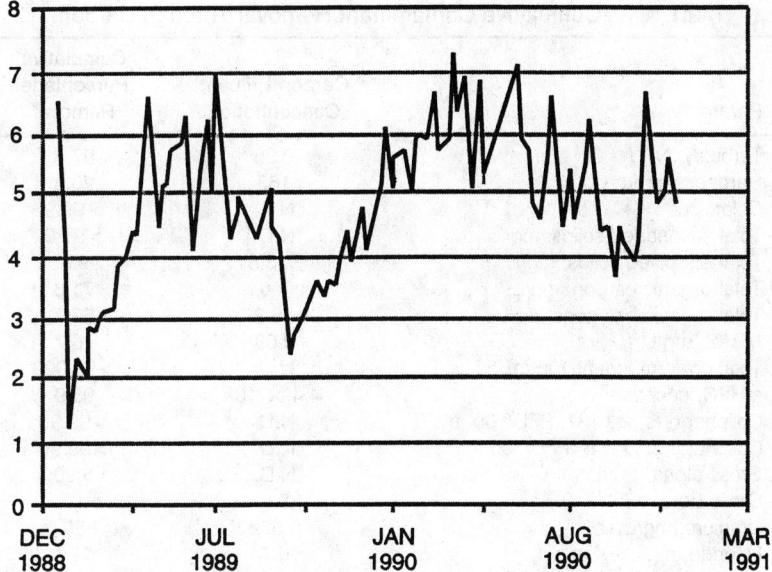

Figure 28.4 Reuse plant carbon effluent, total organic carbon (mg/L).

removed during skimming operations. This resulted in lower TOC concentrations for a short period and increasing concentrations with time. Over the course of the study, the mean TOC removal across the carbon column was 48 percent.

The TOC removals across the carbon column remained relatively constant after the initial increase in removal seen with virgin or regenerated carbon. The consequence of this is that the extended service between regeneration greatly reduces the cost of operating a carbon system.

The carbon column effluent was sampled and tested for a wide variety of contaminants. The influent flow to the carbon columns was not tested to the same degree or frequency. Therefore, the test results from the carbon effluent are best discussed in context of cumulative removal through the first portion of the plant. This is important in that it shows the degree to which the water has been treated prior to the use of the membrane processes. Table 28.7 shows selected results from the analyses of the carbon effluent.

The parameters listed in Table 28.7 were selected because of their aesthetic value or because they are a regulated compound found in significant concentrations in the influent. The cumulative removals point out the effectiveness of the multiple barrier approach used in the plant design. The compounds shown in the table are further reduced by the remaining treatment.

Not shown in Table 28.7 is that previous testing [18] indicated that

TABLE 28.7. Cumulative Contaminant Removal Through Carbon.

Parameter	Carbon Effluent Concentration	Cumulative Percentage Removal
Turbidity, NTU	0.25	97.3
Hardness, mg/L CaCO₃	163	20.8
Color, pcu	N.D.	>99.9
Total suspended solids, mg/L	N.D.	>99.9
Total dissolved solids, mg/L	523.8	9.9
Total organic carbon, mg/L	4.6	71.8
Total organic halogens, μg/L	35.2	67.7
MBAS, mg/L	0.09	76.7
Total coliform, count/100 mL	N.D.	>99.999
mHPC, count/mL	4.4×10^4	95.9
Coliphage E. coli 137, PFU/100 mL	N.D.	>99.99
Coliphage E. Coli B, PFU/100 mL	N.D.	>99.99
Gross alpha, pCi/L	N.D.	>99
Gross beta, pCi/L	8.1	19
Uranium, mg/L	0.002	50
Fluoride, mg/L	0.7	41.7
Nitrate, mg/L as N	0.03	78.6
1,1,1 Trichloroethane, mg/L	N.D.	>99
Chloroform, mg/L	N.D.	>99

N.D. = not detected or 50 percent of samples below detection limit.

free living nematodes were found in the effluent of the carbon columns. The presence of these organisms was instrumental in the decision to use a membrane barrier downstream of the carbon columns.

Reverse Osmosis

The reverse osmosis (RO) system consisted of three units each containing seven vessels. Each vessel housed six elements. The three pass system was configured in a 4-2-1 array. At approximately 50 percent recovery per pass, the system was designed to operate at a total recovery of 90 percent and an average salt rejection of 95 percent, based on conductivity. The membranes were Fluid Systems® Inc. model 4600, thin film composite, spiral wound polyamide membranes.

Pretreatment for the reverse osmosis system included pH adjustment to 5.8 with hydrochloric acid and the addition of a scale inhibitor, sodium hexametaphosphate. The acid was used to control carbonate scaling within the system while the hexametaphosphate was used to mitigate noncarbonate scaling. Cartridge filters with a nominal 5 micron rating were used to remove particulate matter prior to the flow entering the pumps and membranes.

During conceptual design, reverse osmosis was considered in part for its ability to remove ammonia [22]. Later, it was determined necessary because of the project goal to meet or exceed Denver drinking water quality for all parameters. The concentration of Total Dissolved Solids in Denver drinking water could only be met in the reuse plant product water by removing dissolved solids from the wastewater. In addition, the reverse osmosis system provided a final barrier for removing any particulate matter, including living organisms and viruses, from the flow stream.

The feed flow rate through the reverse osmosis system averaged 0.082 mgd during the Health Effects study. The mean specific conductance of the product water was 69 mmhos/cm. The average rejection based on specific conductance was 93.3 percent and the mean recovery was 86.6 percent. Hydrochloric acid was added to the feed flow at a dosage of 121 mg/L and sodium hexametaphosphate was added at a dosage of approximately 5 mg/L.

The product water leaving the RO system was consistently of high quality as measured by specific conductance. The removal of specific contaminants by RO was not studied during the Health Effects phase but typical removals seen earlier in the project are documented elsewhere [16].

In a separate study, the effect of operating the reverse osmosis system at 95 percent recovery was examined. The study was performed because of the difficulty of disposing of brine in the Denver area. By increasing the recovery, the volume of brine would be reduced. The results of this evaluation indicated that a 95 percent recovery RO system could provide significant benefits.

Air Stripping

The air stripping tower was located immediately downstream of the reverse osmosis system and served to remove dissolved carbon dioxide and volatile organic compounds from the water. The design gas to liquid ratio was approximately 100:1 by volume. The air stream was filtered with a standard industrial air filter to remove airborne particulates.

High concentrations of dissolved carbon dioxide in the reverse osmosis product water depressed the pH below 5. The removal of the dissolved carbon dioxide with the air stripping process raised the pH to near neutral.

The influent and effluent pH values for the air stripper flow averaged 4.8 and 6.4, respectively. On average there was a net increase in both the coliform bacteria concentrations and mHPC concentrations across the air stripper. The air stripper was routinely disinfected with chlorine dioxide to mitigate bacterial growths.

Ozonation

Ozone was used as the primary disinfectant for the reuse plant product water. It was generated on site using generators sized to produce 14 lbs/day at 1.3 wt%. Subordinate systems included air compressors and dryers, concentration monitors and a thermal destruction unit for the ozone in the off-gas.

The ozone was fed to the flow stream in a plug flow reactor that was baffled to provide serpentine flow. The air stream carrying the ozone gas was added at a single location in the contactor countercurrent to the liquid flow. The theoretical detention time at design flow was 59 minutes.

The applied ozone dosage was 0.67 mg/L and the absorbed dosage was 0.62 mg/L. This relates to a mean transfer efficiency of over 92 percent. The detention time in the ozone contactor was 86.5 minutes and the average residual at the end of the contactor was 0.14 mg/L. The power consumption for the generation of ozone averaged 4.06 kwh/day. The mean mHPC concentration in the effluent was approximately <0.01/mL. Coliform bacteria and coliphage viruses were not detected in any samples.

Chloramination

Chlorine was used to provide a residual disinfectant in the process flow following ozonation. Chlorine gas was mixed with water and then added as a solution to the flow stream. Residual ammonia in the process water was used in the formation of chloramines. A residual set point of 0.5 mg/L total chlorine was established to match the residual disinfectant concentrations of Denver drinking water.

The average chlorine dosage during the Health Effects phase was 0.97 mg/L. The average residual was 0.56 mg/L total chlorine and the contact time was 14.6 minutes.

Ultrafiltration Sidestream

The ultrafiltration sidestream was designed to provide a sample for Health Effects evaluation of a product water that had been filtered with an ultrafilter (UF) in the place of a reverse osmosis unit. The pilot scale plant was manually operated. It contained the ultrafiltration units, an ozone contactor and a chlorine contact tank. An air stripper originally installed following the ultrafilter was removed when hydraulic problems could not be resolved. The system was not engineered to provide optimum treatment but instead to provide information on possible byproducts that might be formed in a system that contained residual organics, and to provide a sample for the Health Effects Studies.

The ultrafiltration system consisted of three parallel units each containing three spiral wound elements. The elements were Desalination Systems® Inc. model G-10, thin film membranes on a polysulfone base and were rated at a molecular weight cutoff of 2500. The system was manually controlled and monitored. The flow entering the UF system had undergone the treatment of the entire demonstration plant through carbon. No acid addition or cartridge filtration was required for the ultrafilters.

The ozone contactor was constructed from 12 inch diameter stainless steel pipe. Ozone was diffused into the bottom of the contactor through a fine bubble diffuser. The dosage was controlled by adjusting the rate at which the air flow containing the ozone entered the tank. The control was rudimentary and dosages difficult to ascertain.

The chlorine contactor consisted of an unbaffled, stainless steel tank. A chlorine solution made with product water from the reverse osmosis plant was added to the pilot plant flow stream prior to it entering the contact tank. Again, control and measurement of the added chlorine were imprecise.

At the beginning of the Health Effects study, the ultrafiltration system consisted of two parallel units. A third unit was added during the study to increase capacity. The average recovery for unit 1 during this time was 78.8 percent. The average recoveries for units 2 and 3 were 78.5 percent and 81.1 percent, respectively. The feed pressure varied with degree of fouling but was as low as 60 psig for new elements and as high as 175 psig for older, more fouled elements.

The ultrafiltration system was effective in removing a variety of contaminants. Table 28.8 shows the removals for selected contaminants.

The use of the pilot sidestream provided a means of comparing the treatment provided an ultrafilter process with that provided by a reverse osmosis process. In Table 28.9 the removal efficiencies for the UF stream are compared with the RO stream. The efficiencies are calculated based on loadings taken from the carbon effluent flow stream common to both.

The data in Table 28.9 corroborate the expectation that the reverse osmosis system had superior removal capabilities to the ultrafiltration

TABLE 28.8. Contaminant Removal In Ultrafiltration.

Parameter	Effluent Concentration	Percent Removal
Turbidity, NTU	0.05	80.0
Total organic carbon, mg/L	1.0	78.2
Total coliform, count/100 mL	N.D.	—[a]
mHPC, count/mL	924	97.9
Coliphage *E. coli* 137, pfu/100 mL	N.D.	—
Coliphage *E. Coli* B, pfu/100 mL	N.D.	—

[a]—Not calculated because none present in process influent.
N.D. = not detected or 50 percent of samples below detection level

TABLE 28.9. Removal Efficiency, RO vs. UF.

Parameter	Percent Removal	
	Reverse Osmosis	Ultrafilter
Turbidity, NTU[a]	74.8	80.9
Hardness, mg/L as CaCO₃	97.5	28.6
Total dissolved solids, mg/L	96.8	34.7
Total organic carbon, mg/L	97.0	80.5
Total organic halogens, mg/L	83.3	29.6
Gross beta, Pci/L	>99.0	31.2
Uranium, mg/L	37.8	15.0
Fluoride, mg/L	>99.0	12.4
Chloride, mg/L	84.8	9.4

[a]UF turbidity measured prior to ozone contactor.
N.D. = not detected or 50 percent of samples below detection limit.

system. Of particular interest, however, is the organics removal in the ultrafiltration sidestream. The 80.5 percent removal of Total Organic Carbon (TOC) and 29.6 percent removal of Total Organic Halogens (TOX) show the ultrafilter's effectiveness in organics removal and, by inference, the precursors for disinfection byproduct formation.

OPERATIONS AND MAINTENANCE

One of the principal concerns in the early stages of the reuse demonstration project was providing process reliability. The multiple barrier approach used in the design of the plant has been well documented [6] and was the first step in achieving the reliability goals set for the plant. During the operations phase many other factors contributed to process reliability.

The design of the plant incorporated redundant equipment for all processes with the exception of air stripping. This allowed backup equipment to be put into service in the event of equipment failure or maintenance.

The design of the plant included a limited backup power supply. The on site generator was not intended to meet the electrical requirements of the entire plant but only critical equipment. During power outages, the plant shut down. During the Health Effects study, electrical outages occurred on 19 different days in 1989 and on 18 different days in 1990. These shut downs were generally of short duration and always unexpected.

Primary to a successful project was the collection and storage of process data. The computer system at the reuse plant was used to monitor each of the plant processes and store the data. In total, 250 analog and 550 digital points were scanned by the computer system every 4 seconds and the data stored. The integrity of the data base was assured by ensuring that the information entering the computer from the on-line instrumentation was correct. This was done by calibrating the on-line instrumentation to manufacturers' specifications with calibration equipment that had been certified as NBS traceable. Additionally, flow rates to the analyzers were checked and adjusted at least once per day. To ensure that the on-line instrumentation was operating properly between calibrations, the plant operations staff would routinely sample process effluent flow and perform analyses in the plant laboratory. Comparisons would then be made with values from the on-line instrumentation. If the comparison was not favorable, corrective action would be taken. This could include cleaning, calibration or removing the equipment from service until repairs could be made.

The tests performed by the operations technicians would also indicate whether or not the system was operating within prescribed ranges. In the event that a test result indicated that the system was operating beyond the prescribed range, the operations staff would take corrective action. This could include adjusting flows or dosages until the system was brought into compliance with the prescribed parameters. The tests were run every four hours.

Maintenance records for all of the plant instrumentation and mechanical equipment were kept in a PC based, commercial maintenance program. The program stored nameplate data, maintenance procedures and a parts inventory for each piece of plant equipment. Entered into the data base was information for over 80 pumps, 100 motors and 500 instrumentation items. The program was used for the scheduling of preventive and predictive maintenance and storing the maintenance history for the equipment.

The daily operation of the reuse plant was conducted by two operators assigned to a 12-hour shift. The lead technician was required to be certified by the State of Colorado at the "A" level in either the water or wastewater classification and at the "D" level in the other discipline. The assistant technician was required to be certified at the "C" level in either discipline and at the "D" level in the second. Many of the technicians achieved certification levels above those required. The State of Colorado also offers certification in an industrial classification for operations personnel required to operate the types of processes in use at the reuse plant. At one point during the project, seven members of the operating staff were certified at the "A" level in all three of the classifications.

The operators were responsible for process monitoring and adjustments. Technicians normally assigned to maintenance duties replaced the operators in their absence. These substitute personnel met the same certification requirements as the shift technicians.

The training of the operations and maintenance staff was done in house. Each technician was trained in plant operations through the use of a series of classroom sessions which included the use of video tapes and training manuals. The manuals and video tapes were produced by the operations and maintenance staff, at the beginning of the project. Each technician was assigned a unit process and was responsible for preparing an operations manual and video tape for that process. The information covered in the manuals and tapes ranged from text book information on the treatment to information specific to the facility. In preparing the training material, the technician became an in-house expert on the process. Video tapes were also made when manufacturers representatives were on site to start up the equipment they supplied. The combination of training materials and on-site experts proved to be valuable resources during the operations phases of the project.

SUMMARY

The treatment provided during the Health Effects study met expectations. The work in previous phases provided a basis for comparison of system performance and the analyses performed during the Health Effects phase verified that each system was operating as expected.

The chemical treatment system provided removals of suspended solids, coliform bacteria, coliphage viruses, and organic carbon. The filtration system removed turbidity and suspended material. The ultraviolet irradiation process reduced the coliform bacteria counts to near the detection limit and carbon provided additional removal of organics. The reverse osmosis system removed dissolved solids while providing a final physical barrier for particulate and microbial contaminants. The air stripper removed hydrogen sulfide and carbon dioxide and increased the pH of the process flow. Ozone provided primary disinfection for the water and the chlorine ensured the presence of a residual disinfectant. In the pilot stream, the ultrafiltration system provided the same physical barrier as the reverse osmosis system did in the demonstration plant.

The design and subsequent operations of the treatment plant ensured that a high quality product could be produced reliably. The quality assurance program for on-line instrumentation and the process testing program ensured that reliable data were collected. Staff training and certification provided a means of utilizing the available equipment to its capability.

The treatment plant and its' personnel were challenged many times during the Health Effects testing phase of the project. However, it was demonstrated that this highly complex treatment facility could be managed by water treatment operators with normal qualifications supplemented by extensive on-the-job training to continuously produce potable drinking water quality from secondary treated wastewater.

WATER QUALITY RESULTS

The Health Effects sampling protocol required that, for the duration (2 years) of the animal feeding portion of the study, a continuous supply of product be available. The treatment plant was therefore, operated under steady-state conditions. The health effects treatment sequence was unaltered for the entire period. Comprehensive sampling and analysis of the plant influent and various effluents were conducted [5] to document the ability of the treatment plant to continuously process secondary wastewater effluent.

Virtually every known water contaminant was examined in this test program. The data are summarized in Table 28.10. The results are so copious that a discussion of the data has been divided into groupings. Most results were obtained from 24 hour composite samples taken with specialized automatic sampling devices from the treatment plant effluent locations. The primary objective was to determine the product water safety, therefore these sample locations received the most scrutiny. The plant influent was used to establish the variability and extent that contaminant removal was required. The water following activated carbon adsorption was of interest itself but also identified contaminants entering the split stream treatments: reverse osmosis and ultrafiltration. Sample locations are shown on Figure 28.3. Reuse RO is indicated by the location of S10 and Reuse UF is indicated by the location of U04.

GENERAL PARAMETERS

Table 28.10 lists the results for general parameters. The mineral content of the plant influent is somewhat high (hardness—206 mg/L as $CaCO_3$, TDS—581 mg/L) for drinking water but many communities now consume water with values in this range. Ultrafiltration unexpectedly reduced hardness by 28.6 percent and total dissolved solids by 35 percent. Since the intention was to blend the reverse osmosis and ultrafiltration effluents before distribution, this removal may not be necessary. Alternate ultrafiltration membranes may reduce TOC without removing minerals.

The process flow from the reverse osmosis process downstream received only air stripping to adjust the pH. A value of 6.4 was obtained by this method. A small amount of alkaline reagent could be added if further adjustment is required. One parameter which might require some adjustment is temperature. The mean values expressed in Table 28.10 are quite different from Denver's current drinking water supply (Reuse = 22°C, Denver Drinking Water = 6°C). However, when seasonal variations are considered, differences (greater than 5°C) occur only during relatively short periods and much of this is during the winter when Denver Drinking Water is too cold to use without hot water augmentation. In any case,

TABLE 28.10. Health Effects Reuse Plant Performance.

Parameter	Reuse Plant Influent	Reuse Plant Product		Denver Drinking Water
		RO	UF	
GENERAL				
Total alkalinity—CaCO₃	247	3	166	60
Total hardness—CaCO₃	203	6	108	107
TSS	14.2	a	a	a
TDS	583	18	352	174
Specific conductance—μmho/cm	907	67	648	263
pH—units	6.9	6.6	7.8	7.8
Turbidity—NTU	9.2	0.06	0.2	0.3
Particle size (count/50 mL)				
>128μ	—b	a	a	a
64–128μ	—	a	a	1
32–64μ	—	1.2	18	18
16–32μ	—	58	100	168
8–16μ	—	147	448	930
4–8μ	—	219	1290	3460
Radiological (pCi/L)				
Gross Alpha	2.9	<0.1	<0.1	1.3
Gross Beta	10.0	<0.4	5.6	2.3
Radium 228	<1	<1	<1	<1
Radium 226	<0.3	<0.3	<0.3	<0.3
Tritium	<100	<100	<100	<100
Radon 222	<20	<20	<20	<20
Plutonium—total	<0.02	<0.02	<0.02	<0.02
Uranium—total	0.004	<0.0006	<0.0006	0.002
Microbiological				
m-HPC (count/mL)	1.3 × 10⁶	a	350c	3.3
Total coliform (count/100 mL)	7.7 × 10⁵	a	a	a
Fecal coliform (count/100 mL)	6.3 × 10⁴	a	a	a
Fecal strep (count/100 mL)	9.3 × 10³	a	a	a
Coliphage B (count/100 mL)	1.7 × 10⁴	a	a	a
Coliphage C (count/100 mL)	4.8 × 10⁴	a	a	a
Giardia (cysts/L)	0.8	a	a	a
Endamoeba coli (cysts/L)	0.5	a	a	a
Nematodes (count/L)	3.8	a	a	a
Enteric virus (count/L)	—	a	a	a
Entamoeba histolytica (cysts/L)	a	a	a	a
Cryptosporidium (oocysts/L)	0.1	a	a	a
Algae (count/mL)	1.1	a	a	1.9
Clostridium perfringens (count/ 100 mL)	8.5 × 10³	<0.2	<0.2	<0.2
Shigella	Present	Absent	Absent	Absent
Salmonella	Present	Absent	Absent	Absent
Campylobacter	Present	Absent	Absent	Absent
Legionella	Present	Absent	Absent	Absent

TABLE 28.10. (continued).

Parameter	Reuse Plant Influent	Reuse Plant Product		Denver Drinking Water
		RO	UF	
Inorganic[d]				
Aluminum	0.051	a	a	0.2
Arsenic	0.001	a	a	a
Boron	0.4	0.2	0.3	0.1
Bromide	a	a	a	a
Cadmium	a	a	a	a
Calcium	77	1	38	27
Chloride	98	19	96	22
Chromium	0.003	a	a	a
Copper	0.024	0.009	0.01	0.006
Cyanide	a	a	a	a
Fluoride	1.4	a	0.8	0.8
Iron	0.025	0.02	0.07	0.03
Potassium	13.5	0.7	9.1	2.0
Magnesium	12.6	0.1	1.8	7.2
Manganese	0.103	a	a	0.008
Mercury	0.0001	a	a	a
Molybdenum	0.019	a	0.004	0.02
TKN	29.5	5	19	0.9
Ammonia-N	26.0	5	19	0.6
Nitrate-N	0.2	0.1	0.3	0.1
Nitrite-N	a	a	a	a
Nickel	0.007	a	a	a
Total Phosphorus	5.6	0.02	0.05	0.01
Selenium	a	a	a	a
Silica	15.0	2.0	8.8	6.4
Strontium	0.44	a	0.13	0.2
Sulfate	158	1	58	47
Lead	0.002	a	a	a
Uranium	0.003	a	a	0.001
Zinc	0.036	0.006	0.016	0.006
Sodium	119	4.8	78	18
Lithium	0.018	a	0.014	0.007
Titanium	0.107	a	0.035	0.005
Barium	0.034	a	a	0.04
Silver	0.001	a	a	a
Rubidium	0.004	a	0.003	0.001
Vanadium	0.002	a	a	a
Iodide	a	a	a	a
Antimony	a	a	a	a
Organic (μg/L)				
Total organic carbon (mg/L)	16.5	a	0.7	2.0
Total organic halogens	109	8	23	45
Methylene blue active substances	400	a	a	a

TABLE 28.10. **(continued).**

Parameter	Reuse Plant Influent	Reuse Plant Product RO	Reuse Plant Product UF	Denver Drinking Water
Total trihalomethanes	2.9	a	a	3.9
Methylene chloride	17.4	a	a	a
Tetrachloroethene	9.6	a	a	a
1,1,1-Trichloroethane	2.7	a	a	a
Trichloroethene	0.7	a	a	a
1,4-Dichlorobenzene	2.1	a	a	a
Formaldehyde	a	a	12.4	a
Acetaldehyde	9.5	a	7.2	a
Dichloroacetic Acid	1.0	a	a	3.9
Trichloroacetic Acid	5.6	a	a	a

[a] Below detection limit.
[b] (−) indicates not analyzed.
[c] Disinfection considered to be non-optimal at pilot scale.
[d] Additional parameters which were tested but concentrations were below the detection limit or only very limited data are available.

Hafnium	Holmium	Cesium	Zirconium
Holmium	Neodymium	Palladium	Gadolinium
Tellurium	Thulium	Dysprosium	Rhodium
Beryllium	Bismuth	Platinum	Galium
Iridium	Niobium	Ytterbium	Germanium
Terbium	Tin	Erbium	Ruthenium
Cobalt	Cerium	Praseodymium	Gold
Lanthanum	Osmium	Yttrium	Samarium
Lutetium	Tungsten	Europium	Scandium

Geometric Mean Values 9 January 1989–20 December 1990.
All concentrations in mg/L unless otherwise indicated.

blending in the distribution system would result in virtual elimination of this aesthetic concern.

RADIOLOGICAL PARAMETERS

All radiological parameters were below regulatory standards in the plant influent. Therefore, removal was not required. Several unusual isotopes were analyzed (radon-222, total plutonium, and uranium) due to the presence of sources in the area. Low levels of uranium were detected in the influent, but none was detected following lime treatment. As part of the plant contaminant spike studies [15], uranium was introduced at a dosage more than twice the highest background amount and none reached the plant effluent. More than 80 percent was removed by lime treatment alone.

Gross beta activity was the only radiological parameter which showed persistence through any part of the treatment process. Low values were present in the plant influent (less than 20 percent of the Colorado guidance standard of 50 pCi/L) and a 44 percent removal was realized through the ultrafiltration process sequence. The reverse osmosis sequence removed this parameter to below detection limit values.

MICROBIOLOGICAL PARAMETERS

Microbiological contamination in the unchlorinated secondary wastewater effluent used as the demonstration plant supply was quite high. Virtually all of the microbes that were tested were found to be present. However, nearly complete elimination of these contaminants was achieved prior to membrane treatment and final disinfection. Due to limited resources, testing of carbon effluent for many of these parameters was omitted. Instead, analysis focused on the microbiological integrity of the plant effluents (ultrafiltration and reverse osmosis). Previous work [16–18] and plant contaminant spiking studies [15] had established the removal capabilities of lime treatment for microbiological parameters [e.g., 100 percent removal of resistant coliphage [JJ and MS-2], attenuated polio virus, and 3μ latex spheres (to simulate parasite cysts) introduced into the treatment plant at levels ten to twenty times the amount normally encountered].

Viruses were examined by enteric virus analysis and coliphage testing utilizing two bacterial hosts. Enteric virus testing focused on the plant effluents due to the extremely labor intensive nature of this analysis method. Previous work [2] had established the presence of enteric viruses in the plant influent ranging in concentration from 20 to 500 MPNIU/ 10L. No enteric virus was ever detected in any effluent sample in more the five years of testing. Additionally, attenuated polio virus was introduced as a plant challenge [15] at a level at least 1,000 times higher than ever encountered and none was detected in the plant effluent or even the carbon effluent. Coliphage was routinely determined to establish the ability of the treatment plant to meet the challenge of variable virus concentrations under changing conditions. While high levels were routinely encountered in the influent, coliphage was not detected in the plant effluents.

The membrane heterotrophic plate count provided the only positive results after carbon treatment. At this point in the treatment process, disinfection had not been introduced with the exception of UV irradiation which was provided to possibly extend the run time of the carbon column operation by reducing bacterial colonization of the columns. While there was a two log removal of m-HPC bacteria and total coliform were eliminated by UV treatment, a substantial concentration of m-HPC remained. Some m-HPC bacteria were present in the UF effluent. However, this pilot system was inefficient and a full-scale properly mixed disinfection contactor should not have this problem. Chlorine was added to the UF effluent primarily to produce chlorinated organic by-products for inclusion in the animal health effects sample, if they were formed. Even with these limitations the concentration of m-HPC bacterial was lower than many drinking water supplies.

Generally the microbiological parameters were removed and reduced by many logs to undetectable levels. Parasites and protozoan organisms were also eliminated. Reuse treatment provided unprecedented protection from microbiological contamination. This seemingly excessive treatment is certainly sufficient to assure safety.

INORGANIC PARAMETERS

Inorganic parameters include the metals, cations, and anions. None of these parameters were detected in concentrations in excess of existing or proposed regulatory standards at any sample location, including the plant influent. Metals present in the influent at measurable concentrations were: aluminum, barium, boron, calcium, chromium, copper, iron, lead, lithium, magnesium, manganese, mercury, molybdenum, nickel, potassium, rubidium, silver, sodium, strontium, titanium, vanadium, and zinc.

Fluoride is naturally occurring at about 1 mg/L in Denver South Platte Supply System. Diversions from other sources at times lower this optimum amount and the concentration must be augmented as directed by the Colorado Department of Public Health and Environment. Reverse Osmosis (RO) treatment removes fluoride to less than detectable values. If RO were employed to provide reclaimed water, fluoride may be added to produce water conforming with health department guidelines.

Virtually all of the reuse plant influent nitrogen was in the form of ammonia (a t-test used to compare the significance of two means resulted in no difference when looking at the Total Kjeldahl Nitrogen and ammonia nitrogen results). Ammonia was reduced by about 30 percent by ultrafiltration and 84 percent by reverse osmosis. The demonstration project health effects studies were designed to evaluate the organic materials which might remain. Therefore, a small ammonia residual would not affect this analysis. Also, new stream standards being developed which will require the wastewater utility to nitrify and at least partially denitrify its effluent led to the expectation that a full-scale reuse treatment plant built several years in the future would receive water where nitrogen would be significantly reduced [15]. Therefore, the presence of ammonia in demonstration plant effluent is of no consequence and removal by biological processes is assumed for large-scale implementation [15].

Sodium is one of the most difficult inorganic elements to remove. Reuse pretreatment processes (high pH lime clarification, recarbonation, filtration and activated carbon adsorption) do not remove sodium. Ultrafiltration is responsible for a 28 percent (similar to potassium) reduction in sodium

TABLE 28.11. Trace Organic Test Summary (January 9, 1989–December 20, 1990).

	Number of Tests Performed				
Test Method	Influent	Reuse Carbon	Reuse UF	Reuse RO	Denver Drinking Water
Volatile organics (EPA 502.2)	36	35	97	103	27
Grob closed loop stripping— GC/MS (EPA 8270)	38	32	63	67	29
Pesticides (EPA 508)	6	7	7	7	7
Herbicides (EPA 515.1)	11	11	12	12	13
Carbamate pesticides (EPA 531.1)	6	6	6	6	6
Polychlorinated biphenyls (EPA 505)	7	6	7	6	6
Polynuclear aromatic hydrocarbons (EPA 610)	7	6	7	6	6
Base, neutral and acid extractables (EPA 625)	9	7	9	9	8
Disinfection byproducts[a]	8	8	7	8	8
Haloacetic acids[a]	8	8	7	8	8
Aldehydes[a]	5	5	5	5	5

[a]Montgomery Laboratory Methods, Pasadena, California.

concentration. A 96 percent rejection of sodium is achieved by reverse osmosis. A 50/50 blend of reverse osmosis and ultrafiltration would result in a geometric mean of 18 mg/L which is comparable to Denver drinking water. No EPA limit has been established for sodium. However, the WHO set a guideline value of 200 mg/L for aesthetic quality and EEC set a maximum advisable concentration of 150 to 175 mg/L. The sodium levels found in the reuse treatment plant carbon effluent satisfy these limits.

ORGANIC PARAMETERS

A great deal of the analysis effort was directed toward the detection and enumeration of organic compounds. Table 28.11 lists the eleven test methods routinely used to examine reuse project samples for the presence of trace organic contaminants. Two methods (Volatile Organics—EPA 502.2, and Grob Closed Loop Stripping GC/MS—EPA 8270) were used as primary screening procedures and were more frequently performed than some of the other methods and targeted a broad spectrum of trace organic compounds. The remaining methods were used to determine the purity of the samples with regard to a complete list of toxic substances. Several methods were added during the final two years (disinfection by-products,

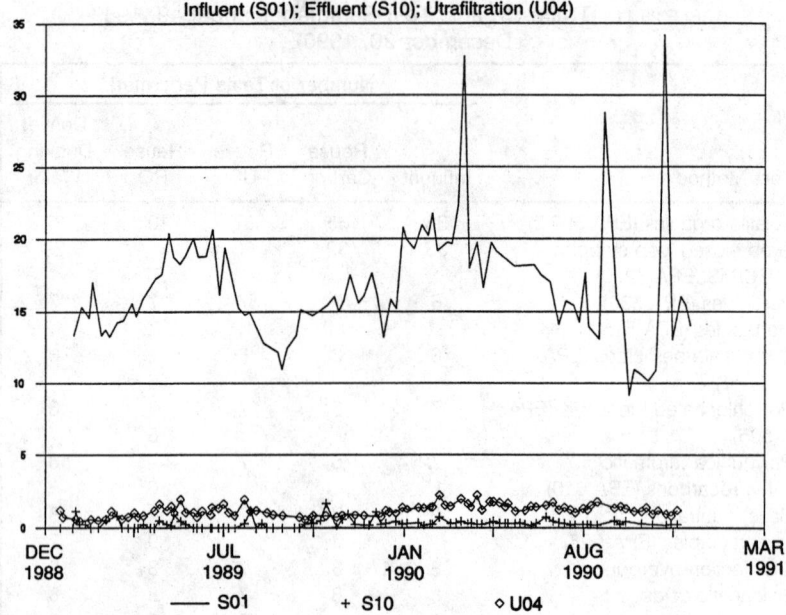

Figure 28.5 Reuse plant comparisons for total organic carbon (mg/L): influent (S01); effluent (S10); ultrafiltration (U04).

haloacetic acids, and aldehydes) when preliminary findings from EPA research [23] indicated that these compounds may be important and further work might lead to regulatory limits being established. A total of 1487 organic test procedures were performed over a two year period characterizing the organic content of the samples examined to an unprecedented degree.

Total organic carbon (TOC) was used to monitor activated carbon performance and served as measure of general organic removal by reuse treatment processes. The geometric mean TOC concentration following carbon treatment was 4.6 mg/L. Previous work [15—18] had established TOC removal of high pH lime clarification at about 40–50 percent of the influent value. Activated carbon adsorption can remove very high amounts of TOC for a relatively short time (1–2 months) [23]. After this initial period, a so called "steady state" removal of about 50 percent is reached [17]. Membrane and air stripping processes were utilized for reuse treatment to reduce the remaining organic concentration to "acceptable" levels. In this case, Denver drinking water was used as a comparison and the acceptable level established as 2 mg/L TOC.

The variability of TOC in the plant influent is shown in Figure 28.5. Generally the concentration was stable around 16 mg/L. However, several large pulses (35 mg/L) of short duration occurred and considerable data below 15 mg/L should be noted. Some of the general trends in the influent appear to be reflected in the activated carbon adsorption product and to a lesser degree, in the ultrafiltration effluent (Figures 28.4 and 28.5). The ultrafiltration effluent TOC never exceeded 2.2 mg/L in spite of these influences. Reverse osmosis results are at the detection limit of the instrumentation.

Another screening parameter which found use in determining the relative process performance for remaining organics from water and for comparisons with drinking water samples was total organic halide (TOX). Large reductions (68 percent) by the pretreatment processes through activated carbon adsorption. Further reductions by the membrane processes (ultrafiltration—31 percent and reverse osmosis—83 percent) lower the TOX concentration to a value well below that found in Denver's current drinking water supply. Figure 28.6 illustrates the TOX variability over time. Considerable fluctuation in influent values does not appear to be reflected in the effluent. Denver drinking water values were related to seasonal chlorination practices with spring elevation and winter decreases.

Trihalomethanes (THM) include chloroform, bromodichloromethane,

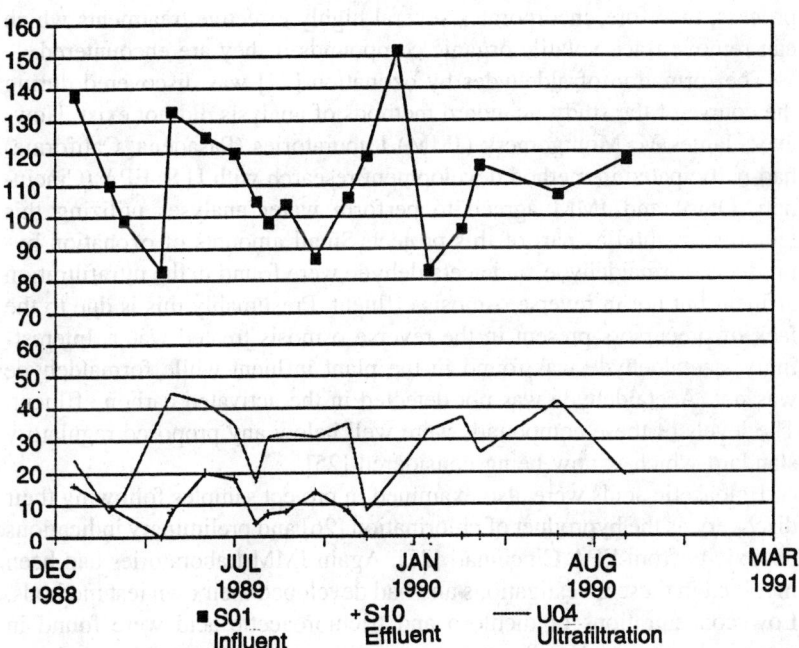

Figure 28.6 Reuse plant comparisons for total organic halogen (mg/L): influent (S01); effluent (S10); ultrafiltration (U04).

chlorodibromethane, and bromoform. The U.S. EPA has established an MCL for total trihalomethanes of 100 µg/L. Chloroform is the predominant THM in most waters resulting from chlorination for disinfection practices. Low levels of chloroform are present in the reuse plant influent and in Denver drinking water. Small concentrations of bromodichloromethane are also present in drinking water samples. None of the concentrations encountered in reuse process samples approach even one tenth of the regulatory standard. Denver drinking water at times reaches levels about one third of the standard. THMs were not detected in either reuse plant effluent sample.

The predominant trace volatile organic compounds found in reuse plant were: methylene chloride, tetrachloroethene, 1,1,1-trichloroethane, trichloroethene, and 1,4-dichlorobenzene. Erratic results were experienced for all of these compounds. However, concentrations were reduced below detectable levels following activated carbon adsorption. None of these compounds were detected in plant effluent samples. Additionally, methylene chloride was injected into the plant at more than 15 times the amount normally found and none was detected following carbon treatment [15]. Certainly the membrane processes and air stripping would provide additional removals if residual methylene chloride were to remain. The reuse process, therefore, incorporates several highly effective treatments which can remove trace volatile organic compounds if they are encountered.

The formation of aldehydes by ozonation [24] was discovered during the course of the study. Standard methods of analysis did not exist. However, James M. Montgomery (JMM) Laboratories (Pasadena, California) had participated in method development research with U.S. EPA (Cincinnati, Ohio) and JMM agreed to perform water analyses utilizing this tentative method as part of this project. Small amounts of ozonation byproducts, formaldehyde, and acetaldehyde were found in the ultrafiltration effluent but not in reverse osmosis effluent. Presumably this is due to the lack of precursors present in the reverse osmosis treated water. Interestingly, acetaldehyde was found in the plant influent while formaldehyde was not. Acetaldehyde was not detected in the activated carbon effluent. The levels of these compounds were well below any proposed regulatory standard which is now being considered [25].

Haloacetic acids were also examined in project samples following their discovery as the byproduct of chlorination [26] and preliminary indications of toxicity from EPA Cincinnati [27]. Again JMM Laboratories had been involved in these investigations and had developed their own test methods. Low concentrations of dichloro and trichloroacetic acid were found in plant influent samples. These compounds were not detected in any reuse plant effluents. Dichloroacetic acid was routinely found at low concentrations in Denver drinking water samples.

Many compounds were detected only occasionally by the various methods employed and at some of the sample locations. The infrequent nature of these compounds detection prohibited the use of statistics (means, etc.). Most of these compounds were detected from Grob Closed Loop Stripping samples utilizing broad spectrum GC/MS techniques [30].

A total of 54 infrequently detected compounds were identified in the plant influent by all the organic test methods over the two year health effects study period. Most of these compounds (45) were determined by Grob Closed Loop Stripping and broad spectrum GC/MS analysis. Twenty-two compounds were detected in the activated carbon effluent and only seven compounds were found in the ultrafiltration and reverse osmosis effluents. Eight compounds of this type were found in the Denver drinking water samples. None of these compounds at any of the sample locations were present in concentrations approaching any existing or proposed regulatory standard. A comparison of these compounds was complicated by their diversity and inconsistency of detection. Three compounds were found at times in all reuse sample locations (2-methyl 1-1-1,[1,1-dimethylethyl] propanoic acid, 2,2-dimethyl decane, and dimethyl undecane) generally decreasing in maximum concentration from influent to reverse osmosis effluent. However, two of these same compounds were also found in Denver drinking water (2-methyl-1-1,[1,1-dimethlethyl]) propanoic acid and dimethyl undecane) in the same amount. This result raises the specter of a systematic contaminant. Table 28.12 lists two possible comparison methods for infrequently detected compounds. It is possible to reject compounds which are detected in less than five percent of the samples tested. When this is done, testing anomalies may be reduced. In this case, the ultrafiltration, reverse osmosis, and Denver drinking water samples would compare favorably with few or no compounds being detected. The carbon effluent contained 17 compounds which were detected in more than five percent of the samples tested. Another possible way of eliminating spurious results is to list those compounds which were detected at sufficiently high concentration (>1 µg/L) so as to make their

TABLE 28.12.   Infrequently Detected Organic Compounds
Sample Location Comparison.

Sample Location	# Compounds Detected in More Than 5% of Samples	# Compounds Detected at Max. Concentration >1 µg/L
Reuse influent	30	19
Reuse carbon effluent	17	10
Reuse UF effluent	1	2
Reuse RO effluent	0	0
Denver drinking water	2	0

TABLE 28.13. Reuse Plant Contaminant Challenge Study Organic Compounds Cumulative Percent Removal.

Compound	Initial Dosage (mg/L)	Lime	Carbon	Reverse Osmosis	Plant Effluent
Acetic acid	5054	100	—	—	—
Anisole	23	100	—	—	—
Benzothiazole	86.2	63	100	—	—
Chloroform	229.6	26	99.7	99.9	100
Clofibric acid	17.1	0	100	—	—
Ethyl benzene	25.1	100	—	—	—
Ethyl cinnamate	67.8	100	—	—	—
Methoxychlor	44.6	84	100	—	—
Methylene chloride	230	8	100	—	—
Tributyl phosphate	69.4	51	100	—	—
Gasoline (1st trial)	97.8	100	—	—	—
Gasoline (2nd trial)	2115	—	—	—	—
Toluene	—	25	97	100	—
Benzene	—	40	100	—	—
Ethylbenzene	—	36	100	—	—
Xylene	—	32	100	—	—

detection and identification more certain. When this is done, the UF, RO, and Denver drinking water samples again compare favorably. The carbon effluent contained only ten compounds with maximum concentrations in excess of 1 µg/L. The plant influent had 19 compounds which exceeded this criterion.

A selection of organic compounds was used in a special study to challenge the reuse treatment plant with high concentrations in order to define the extent of protection provided by these advanced processes [15]. A list of the organic compounds used in this evaluation is given in Table 28.13. Most of these compounds were removed completely or substantially by lime treatment. The remainder of the compounds were removed by activated carbon. The only exception was chloroform. A very small residual concentration (<1 µg/L) survived reverse osmosis treatment, but was eliminated by air stripping prior to final product water sampling. As with most of these compounds chloroform was added to the treatment plant in an amount 100 times the concentration normally found in the plant influent.

This stress testing augmented the ambient organic analysis further illustrating the capability of the treatment processes to remove organic contaminants to a degree never achieved by conventional water treatment plants.

CORROSION

It is difficult to predict the degree of corrosion which may occur when water comes in contact with metals. There are many factors which affect the rate and capacity of this reaction. One of the most common indices used to indicate the corrosion of water is the Langelier Index (also known as the Langelier Saturation Index). This index reflects the equilibrium pH of a water with respect to calcium and alkalinity. If the pH is raised from the equilibrium point, the water becomes scale forming depositing calcium carbonate. Lower pH values result in corrosive water. There are several problems with this approach [28,29] which may lead to incorrect evaluation of the corrosion potential of the water tested. In fact, Denver drinking water should be moderately aggressive according to this measure when, in fact, corrosion is not observed in the Denver distribution system.

Even though the Langelier Index (LI) has proved unreliable in accurately predicting corrosion tendencies in several situations, it is an accepted measure and may be useful in determining the relative aggressiveness of water samples.

The Langelier Index was calculated [30,31] for several reuse project waters and the results are listed in Table 28.14. The column headed "Reuse UF and RO" lists the values for a theoretical blended water that would result from a 50 percent ultrafiltration sequence mixed with 50 percent reverse osmosis treatment sequence. The values used in the table are arithmetic mean values of actual samples taken over the two year reuse Health Effects study. Negative values of LI would indicate that the water would be unstable with respect to calcium carbonate saturation and therefore corrosive. Using this measure, reuse plant water treated through activated carbon and the ultrafiltration split stream would be relatively non-corrosive (LI = 0.0 and −0.1 respectively). Reuse plant influent, Denver drinking water and theoretical reuse UF and RO blended water would be moderately aggressive. Reuse plant reverse osmosis treated water would be highly aggressive. As mentioned previously, even though Denver drinking water would be predicted to be moderately aggressive, due to a negative

TABLE 28.14. Langelier Index Parameters and Calculated Values.

Parameter	Influent	Reuse Carbon	Reuse UF	Reuse RO	Reuse UF and RO	Denver Water
Calcium (mg/L CaCO$_3$)	133	134	84	3	44	65
Alkalinity (mg/L CaCO$_3$)	251	204	157	3	80	65
TDS (mg/L)	583	525	344	19	182	186
Temperature (°C)	20	21	22	22	22	6
pH$_a$ (units)	6.8	7.5	7.7	6.7	7.0	7.8
pH$_s$ (calculated)	7.5	7.5	7.8	11.1	8.3	8.4
Li (calculated)	−0.7	0.0	−0.1	−4.4	−1.3	−0.6

LI, corrosion is not found in practice. Similarly, the reuse plant influent should be aggressive, but again corrosion was not experienced. Reverse osmosis treated effluent was highly aggressive as predicted and required special materials (stainless steel or fiberglass) to ensure long life. The theoretical blended UF and RO water may be somewhat aggressive. However, the addition of only 5–10 mg/L of lime would be all that is necessary to adjust the pH and result in an LI comparable to Denver drinking water. Since lime treatment equipment would be included in a large scale reuse treatment plant as part of the lime clarification process, the additional cost for pH adjustment would be negligible in comparison (probably increasing the cost by less than $0.01/1000 gal).

ULTRAFILTRATION/REVERSE OSMOSIS BLENDED WATER QUALITY

One possible configuration for a full-scale direct potable water reuse treatment plant could utilize a split process sequence. Perhaps 50 percent of the water could be treated through ultrafiltration while the remaining water would receive reverse osmosis processing. Figure 28.7 illustrates this treatment train. There would be several advantages in employing this type of treatment. One important outcome would be the mitigation of highly corrosive water that would result from reverse osmosis treatment if used alone. The production of relatively stable water would provide flexibility when delivering the water to customers and eliminate the need to transport large quantities of hard water from other sources to the treatment plant location for blending prior to distribution. A second reason for split treatment would be lower utility costs due to the lower pressures required by ultrafiltration. Also, water of various qualities could be produced from the same plant. For example, filter effluent could be disinfected and supplied as non-potable water for irrigation. Carbon effluent would be available for industrial non-potable use and ultrafiltration or reverse osmosis products could be used in the plant or provided to special customers who require higher quality water.

The quality of a blended supply can be calculated from the results of the two separate water streams. Means of selected parameter test results are included in Table 28.15. The calculated values for the blended water, Denver drinking water, and regulatory standards are listed in this table.

The water produced by a theoretical 50/50 blending of the ultrafiltration and reverse osmosis effluent would be a very high quality water. General water parameters would indicate that the blended water would be roughly equivalent to Denver drinking water. The hardness would be about half that of current drinking water while the alkalinity, TDS, and specific conductance would be in the same range. Radiological parameters are

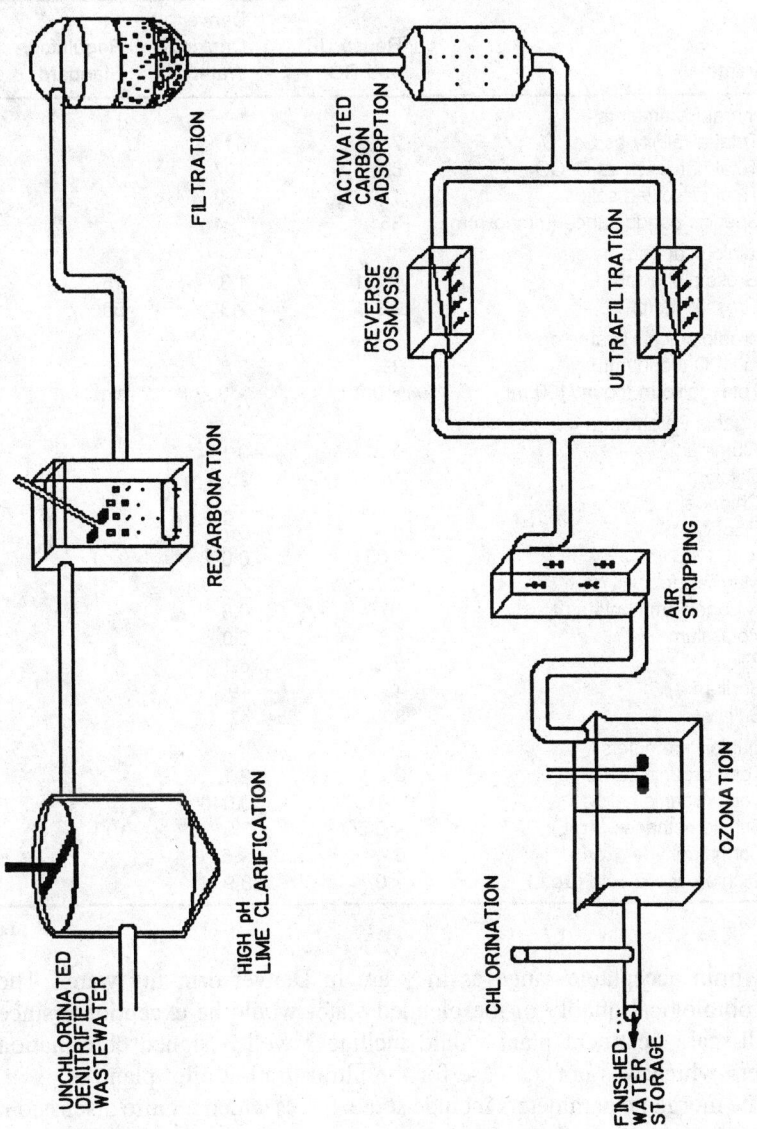

UNCHLORINATED
DENITRIFIED
WASTEWATER

HIGH pH
LIME CLARIFICATION

RECARBONATION

FILTRATION

ACTIVATED
CARBON
ADSORPTION

REVERSE
OSMOSIS

ULTRAFILTRATION

AIR
STRIPPING

OZONATION

CHLORINATION

FINISHED
WATER
STORAGE

Figure 28.7 Full-scale split treatment sequence.

TABLE 28.15. **UF and RO Blended Water Quality Comparison—Selected Parameters Means Values January 9, 1989–December 20, 1990 (mg/L unless noted).**

Parameter	Reuse UF and RO	Denver Drinking Water	Regulatory Standard
General Parameters			
Total alkalinity as CaCO$_3$	78	64	
Total hardness as CaCO$_3$	53	107	
Total dissolved solids	180	183	
Specific conductance (mmhos/cm)	361	294	
Radiological Parameters			
Gross alpha (pCi/L)	<0.1	1.3	15
Gross beta (pCi/L)	2.8	2.3	50
Microbiological Parameters			
m-HPC (count/mL)	91	2.8	
Total coliform (count/100 mL)	<0.2	<0.2	1
Inorganic Parameters			
Aluminum	0.011	0.144	
Calcium	16.7	25.9	
Chloride	55	25	
Fluoride	0.4	0.7	4
Iron	0.034	0.028	
Magnesium	0.9	7.9	
Nitrogen-ammonia as N	10.6	0.6	
Potassium	4.5	2.0	
Silica	5.2	6.1	
Sodium	42	19	
Sulfate	30	47	
Organic Parameters			
Total organic carbon	0.7	2.1	
Total organic halide	0.015	0.046	
Trihalomethanes (μg/L)	<0.5	3.9	100
Formaldehyde (μg/L)	6	<5	
Dichloroacetic acid (μg/L)	<0.1	3.9	

all within acceptable range as they are in Denver drinking water. The microbiological quality of the blended water would be exceptional since a full-scale treatment plant would include a well designed chlorination system which was not the case for the ultrafiltration pilot plant.

The inorganic parameters include some values which require discussion. None of the regulated inorganic substances would be present in the blended water in an amount approaching a regulatory standard. Calcium and magnesium would be lower while sodium and potassium would be higher than Denver drinking water values. However, these higher values are for non-toxic elements and are well within guidelines set by the WHO (sodium =

200 mg/L) or EEC (sodium = 150–175 mg/L, potassium = 12 mg/L) for aesthetic quality. Ammonia nitrogen for the blended water is listed as 10.6 mg/L as N. However, as discussed earlier, for large scale implementation a denitrified plant influent is now assumed and, therefore, this value would be reduced below the detection level [32].

The low level of organic material in the blended water would be superior to Denver drinking water. One possible exception would be formaldehyde which apparently formed in ozonation. The concentration remaining in the blended water would be near the detection limit. Although this amount is well below any health concern [25], if it were desirable to lower it further this may be accomplished by treatment with air stripping after ozonation instead of before.

Although a blended effluent would contain higher levels of a few compounds it would be lower in most and far lower in organic content than Denver drinking water. A blended reuse effluent would provide a suitable supply lacking in corrosiveness and at a lower cost than 100 percent reverse osmosis treated water. The blended reuse effluent would also compare favorably with Denver drinking water.

SUMMARY

Two years of extensive comprehensive testing have documented the quality of waters produced from the treatment of unchlorinated secondary wastewater at Denver's Direct Potable Water Reuse Demonstration Plant. More than 10,000 samples were collected to verify plant operation and examine thousands of possible contaminants which may be found in water. The variability of the wastewater influent to the reuse plant and the treatment process response was established by routine sampling and analysis. The ambient water quality was augmented by plant-scale contaminant dosing studies of a wide variety of potential contaminants which further established the incredible degree of protection provided by the reuse treatment system.

The reuse plant effluents from ultrafiltration and reverse osmosis membrane systems satisfied all existing and proposed U.S. EPA, WHO, and EEC standards. Additionally, these product waters compared favorably with Denver's existing drinking water. A 50/50 blend of the two reuse plant effluents would also satisfy all standards and provide a less corrosive water at a lower cost than reverse osmosis alone. The blended supply would also compare favorably with Denver drinking water. Lower concentrations of calcium, magnesium, and sulfate would be offset by higher (but not in excess of any standard) concentrations of sodium, potassium and chloride making this possible supply relatively soft and quite desirable.

The water quality evaluations which have paralleled the whole animal health effects studies have established the chemical, physical, and microbiological characteristics of reuse project samples at a level of completeness and duration never before established for any drinking water supply. Although the source water was highly contaminated, product waters never exceeded any water quality standard for any substance. The treatment processes provided a level of protection not experienced in conventional water purification facilities. All of the water quality results establish reuse plant effluent as a high quality water supply satisfying all health standards and equal to or better than Denver drinking water in every important way.

ANIMAL HEALTH EFFECTS TESTING

The NAS Panel on Quality Criteria for Water Reuse [23] stated that the ultimate evaluation of the potential adverse health effects from reused water must come from chronic toxicity studies in whole animals. Lifetime feeding studies to detect carcinogenicity using a maximum tolerated dose with even a single chemical are often difficult to interpret with regard to anticipated risks in humans. The problems of interpretation of single chemical testing approaches can be greatly magnified when the test material consists of a complex, undefined mixture of compounds that would each require testing and assessment of total impact. Therefore, they concluded that the problem can be best approached by exposing the subject animals to concentrates of the chemical constituents found in the water under evaluation.

Several studies of the possible harmful effects of water constituents on mammalian reproduction and embryonic growth development have been reported. However, the NAS Panel on Quality Criteria for Water Reuse felt that the studies were few and inadequate [33]. Borzelleca and Carchman investigated the effects of selected organic drinking water contaminants on male mouse and rat reproduction [34]. The organic compounds tested (multichloro compounds; 2,4-dinitrotoluene, 1,2,3,4-tetrabromobutane, hexachlorobenzene, dibromochloromethane, chloroform) all had very severe effects at very high levels on adult male mouse and rat reproductivity. Five other studies were reviewed by the NAS Panel on Quality Criteria for Water Reuse. This panel concluded that there were short-comings in these studies, and they did not yield definitive information on the potential effects of substances in water on prenatal development [33].

An important conclusion of the NAS Panel on Quality Criteria for Water Reuse was that results from animal bioassays of single chemicals do not provide an adequate basis for evaluation of impacts in drinking water. As such, simple approaches whereby individual constituents are identified may not provide a realistic basis for determining risks. The limited studies that have been performed suggest a relationship to water quality that needs to be explored in greater depth for complex mixtures. Similar conclusions regarding animal bioassays on reclaimed wastewater have been reached by others [35,36].

The National Toxicology Program (NTP) conducted a subchronic toxicity study on a chemical mixture simulating contaminated groundwater [36]. The study investigated the health effects of repeated intake of a mixture of 25 chemicals representing frequently occurring groundwater contaminants. These 25 chemicals include 19 organic and 6 inorganic compounds. The NTP protocol outline noted that most previous studies involved only two chemicals. The outline did cite several drinking water-related studies using chemically defined mixtures [37–39] and specifically dealing with toxicity of complex mixtures in contaminated water [40,41]. Although these studies provide guidance on how toxicity studies evaluated complex mixture should be conducted, they do not attempt to examine the possible toxic effects of the organic compounds present in a reclaimed water.

SAMPLE CONCENTRATION PROCEDURE RATIONALE

The acquisition of a suitable sample is complicated by several factors: the spectrum of potential compounds that could be present, the volume of water which must be concentrated to provide the necessary quantity of test article, the lack of sufficient previous work on the collection of environmental samples for this purpose to allow the adoption of a "standard method," and the probable change in the type and concentration of organic compounds over time. These and other concerns have led to a series of trade-offs when evaluating and selecting a method for the isolation and concentration of organic compounds from water samples to be used in toxicity studies.

Lauer et al. [42] described the tasks performed to review, analyze, and select a procedure for concentrating the water samples. The primary target for organic sample concentration is the compounds that cannot be analyzed by existing methods and which may separately or in combination cause long or short term health effects. The methods of concentrating organic chemicals eliminate certain classes of compounds, such as volatile organic compounds. These compounds are susceptible to analysis, however, and those compounds which are reliably identified can be reconstituted to the test article.

The volume of water that was processed in order to produce the required quantity of 500× concentrate for the toxicology studies necessitated the construction of apparatus on a scale never before attempted. Considerations limiting the volume of solvents and simplifying the procedures were required to ensure success by minimizing areas that might introduce contaminants or interrupt the sample flow. Many methods that could possibly be considered for analytical sample preparation and for the limited sample requirements for Ames mutagenicity testing are unsuitable for a two year whole animal chronic study. On the other hand, since the objective was to produce samples for health effects testing rather than analytical evaluation, the importance of artifacts which may interfere with chemical analysis may be introduced as long as they do not produce toxic effects.

An alternative approach to concentrating water samples would be to prepare a defined chemical mixture simulating the organics in the test water. This approach was taken by the NTP for a toxicology study on contaminated groundwater [36]. A concern with adopting this approach for the Denver reuse project is that only 10 percent of the organic compounds in the test waters are known and the types of compounds present and their concentrations change over time. Thus, it would be impossible to make up a typical mixture of organic compounds for a biological testing program since the target water changes over time. Therefore, a concentrate of a composite sample representing the water over a period of time seemed best for this biological toxicity study.

The organic concentrate should be representative of chemical species and concentration ratios of all the organic substances present in the water (i.e., broad spectrum analysis). Currently, no single concentration procedure exists that is capable of concentrating all the organic materials present in the water samples in this manner. Such a procedure may never be possible and the fallback position that is usually taken is that a representative sample of organic matter be collected which includes the efficient extraction of potential toxicants. The concentration method must be chosen on the basis of the chemical and physical properties of the organic constituents being tested. Other criteria for the development of a meaningful method include that the method must be quantitative, reproducible, and should not alter the chemicals.

The alternative methods suggested for concentrating samples larger than 100 L are:

- adsorption on XAD resins
- liquid-liquid extraction by continuous processing
- membrane processes (ultrafiltration or reverse osmosis) followed by resin adsorption or solvent extraction

Based on a review of the literature of isolation and concentration methods used to collect organic chemicals for toxicity testing, the primary methods were XAD resins and liquid-liquid extraction. Bench scale studies were conducted at the reuse plant on both a continuous liquid-liquid extractor and an XAD resin concentrator.

The continuous liquid-liquid (helical tube coil contactor) extractor appeared to be suited for the isolation and concentration of gas chromatographable compounds. However, naturally occurring compounds found in Denver's drinking water were not isolated by this method. This is evident because only approximately 1 percent of the dissolved organic carbon (DOC) content was extracted during the testing sequence.

After it became apparent that the continuous liquid-liquid extractor would not isolate a significant fraction of the DOC content of actual water samples, the only remaining demonstrated method for preparing samples for health effects testing was XAD resin concentration. Bench scale tests were begun to confirm the ability of XAD resins to isolate organic compounds from water.

Initial testing examined the effect of pH on XAD-2 resin capacity. Low pH increased the percent removal and apparent resin capacity for DOC. All further trials were run at low pH. The importance of pH on the recovery of mutagens has also been demonstrated by Ringhand et al. [43]. The pH adjustment was accomplished with the use of hydrochloric acid.

Further testing examined the specific XAD resins and solvents for elution. Several resin and solvent elution systems were evaluated; a dual column system of a XAD-8 column followed by a mixed XAD-2 and -4 column, utilizing acetone as an eluant, was the best of the systems tested. The resins have been used by other researchers for health effects testing and acetone is a low toxicity solvent with good elution characteristics. The choice of a mixed XAD-2 and -4 secondary column was due to the similarity of the two resins and the increased pore volume of XAD-4 which may be an advantage when attempting to isolate organics from reuse plant samples which have undergone reverse osmosis treatment.

Results from the bench scale evaluation of the XAD resin concentrator indicated that it was capable of isolating and concentrating approximately 40 percent of the DOC content. Because of the relatively high fraction of DOC concentrated and the fact that these DOC compounds are not those detectable by standard analytical procedures, the XAD resin concentrator was used to produce samples for the toxicity study.

HEALTH EFFECTS TESTING PROTOCOL

The testing protocols were developed over a period of several years with the advice of numerous experts in toxicology, chemistry, microbiology, and engineering. The final protocols were prepared by Joseph F. Borzelleca (Medical College of Virginia) and Lyman W. Condie (U.S. EPA) and were reviewed by an advisory panel and final modifications were incorporated following their input [44].

The following members comprised the panel:

- Richard Bull, Washington State University
- Joseph Cotruvo, U.S. EPA, Washington, D.C.
- Fred Kopfler, U.S. EPA, Stennis Space Center, Mississippi
- I. H. Suffet, Drexel University
- Lyman W. Condie, U.S. Army Dugway Proving Ground, Dugway, Utah (Formerly, U.S. EPA Cincinnati)
- Joseph F. Borzelleca, Medical College of Virginia
- Robert Neal, Vanderbilt University (Formerly, C.I.I.T)
- Paul Ringhand, U.S. EPA, Cincinnati, Ohio
- Raymond Yang, National Toxicology Program at Colorado State University (new affiliation)
- Carl Brunner, Project Officer, U.S. EPA, Cincinnati, Ohio
- John Doull, University of Kansas Medical Center

The protocol provided for the comparative testing of concentrates obtained from Denver drinking water, reuse demonstration plant product treated with reverse osmosis, and reuse demonstration plant product treated with ultrafiltration. The program included 104-week chronic toxicity and carcinogenicity studies in Fischer 344 rats and $B_6C_3F_1$ mice. Reproductive toxicity studies were performed with Sprague-Dawley rats.

The National Toxicology Program (NTP) general statement of work for the Conduct of Toxicity and Carcinogenicity Studies in Laboratory Animals [45] was used as a basis for the work requirements. The study was also conducted in accordance with the United States Environmental Protection Agency (U.S. EPA—TSCA) Good Laboratory Practice (U.S. Environmental Protection Agency 40 CFR Part 792) regulations [46]. Some of the more unique modifications of these experimental guidelines which were required for this complex mixture study are described below.

Dosage Setting Rationale

The chronic toxicity and carcinogenicity study used two dosage groups per water sample (reclaimed water from the reuse demonstration plant with reverse osmosis treatment and Denver drinking water from the Foothills Water Treatment Plant) administered to rats and mice (as described below). The dosages were set at 150 and 500 times the concentration in the original water samples. Due to sample availability and funding constraints the ultrafilter sample (reclaimed water from the reuse demonstration plant, but with ultrafiltration in place of reverse osmosis treatment) received testing at the high dose (500×) in the rat only. Distilled water controls were used for both the rat and mouse studies. The high dose (500×) was set based on 10× for interspecies factors, 10× for individual factors, and 5× for inherent variabilities of the water sample and incomplete recovery of total organic carbon (TOC). A lower dose of approximately one-third the upper value was determined to be adequate to provide a dose response relationship should the high dose prove to be toxic.

The reproductive toxicity study used one dosage concentration per water sample (reclaimed water using reverse osmosis, Denver drinking water, and reclaimed water substituting ultrafiltration for reverse osmosis) plus a control (distilled water). If overt reproductive toxicity was observed at any dose, one would be skeptical of using the water for drinking purposes. A no-effect result at a lower dose would do little to ameliorate the conclusion. Therefore, only the highest practical attainable dose (500×) was used for the reproductive toxicity study.

Animals

Chronic Toxicity and Carcinogenicity

Fischer 344 rats and $B_6C_3F_1$ mice were selected as the test animals because of the large data base available for these animals from the NTP. All animals were housed in an environmentally controlled room (20–25°C, 40–70 percent relative humidity, and lighting 12 hours per day). All animals were individually caged.

Reproductive Toxicity

Sprague-Dawley rats were selected as the test animals because this strain is commonly used in reproductive toxicity studies. Animals were individually caged until mating, and females were individually caged after mating. Pups were caged with their mothers for 21 days. Housing was in an environmentally controlled room (20–25°C, 40–70 percent relative humidity, and lighting for 12 hours per day).

Administration and Handling of Test Article

Routes of Administration

The test articles were concentrated drinking water samples. Dosed drinking water on an *ad libitum* basis was the route of administration. The concentrated samples were solubilized in distilled water (described below). The solubility and palatability of the water types was investigated by the U.S. EPA prior to the start of the health effects testing. No taste aversion problems were noted. Control animals were administered distilled water containing 0.025 percent/w Emulphor (EL-620, GAF Corp., Wayne, New

Jersey). All drinking water solutions were supplied in amber glass bottles fitted with Teflon stoppers and stainless steel sipper tubes.

Sample Handling

The samples used for the chronic and reproductive toxicity studies were obtained by isolating organic compounds from Denver drinking water and reclaimed water on XAD resins. The organic substances were eluted from the XAD resin columns with acetone. The acetone was removed from the water residue by rotary vacuum distillation. A concentrated water sample was shipped to the testing laboratory packaged under nitrogen and stored at 4°C. The testing laboratory diluted the concentrate with distilled water to the dosage level, adding in the volatile organic compounds (described below) as required, and re-sealing the concentrate under nitrogen.

The procedures for obtaining the water concentrates required large quantities that could not be prepared in advance of the study. Therefore, samples were prepared continually during the chronic and reproductive toxicity studies.

Vehicle (Distilled Water)

The sample and controls were administered via drinking water. The only substances which contacted the water were stainless steel, amber glass, and Teflon. Due to the minute quantities of organic compounds which may be present in tap water, only distilled water was used to dilute samples. The distilled water was obtained from Polar Water Company (Savage, Maryland).

The control water was distilled water as described above with an appropriate amount of Emulphor added to reach a concentration of 0.025 percent and acetone added (100 mg/L $\pm$ 50 mg/L) to equal the residuals remaining in the organic water concentrates obtained from the sample preparation method.

Test Article Preparation

The water concentrate (5000x) obtained from the elution of XAD resins was received by the health effects laboratory as an emulsion (0.25 percent Emulphor) shipped in amber glass bottles at 4°C and stored under nitrogen in the dark. The water concentrate was diluted, with distilled water, to the proper concentration factor (150x or 500x) following agitation to ensure sample homogeneity. Only the amount necessary for immediate use was diluted and the remaining water concentrate was resealed in the original container after filling the air space with nitrogen and storing at 4°C in the

TABLE 28.16. Volatile Organic Chemicals Added to Denver Drinking Water Test Article (mg/L).

Compound	Concentration in 500 × Dose	Concentration in 150 × Dose
Chloroform	1.5	0.45
Bromodichloromethane	0.5	0.15
1,1-Dichloropropanone	0.2	0.06

dark. The volatile organic compounds (VOCs) listed (Table 28.16) for Denver Drinking water was then added to the diluted sample prior to administering the test article. These VOCs were lost during the concentration procedure and represent mean concentrations determined from chemical tests of the original water. No VOCs were consistently identified in either the reclaimed water or the ultrafiltration treated reclaimed water samples, therefore, none were added to these test article concentrates. The VOCs were obtained from Aldrich Chemical Company (Milwaukee, Wisconsin). All test articles and controls contained equal amounts of acetone (100 mg/L ± 50 mg/L) and Emulphor (0.025 percent).

Samples collected for rodent health effects testing were prepared by Denver's reuse project staff and shipped to Hazleton Laboratories America (Vienna, Virginia) for use. The samples consisted of organic concentrates of the three water samples (Denver drinking water, reclaimed water treated by ultrafiltration, and reclaimed water treated by reverse osmosis) continuously prepared over a two year period. Nearly 5 million gallons of water (Table 28.17) was processed to produce the needed test materials.

104-Week Chronic Toxicity and Carcinogenicity Study Protocol

Purpose

The objective of the chronic toxicity and carcinogenicity study was to evaluate potential adverse effects on growth and development and potential carcinogenic effects during a 104 week (2 year) study. All observations and examinations were made according to the NTP Protocol (April 1987).

Test Animals

Seventy males and 70 females per species per dose level per article were used (50 males and 50 females for terminal sacrifice and 10 males and 10 females for each interim sacrifice at 6 and 15 months). Fischer 344 rats and $B_6C_3F_1$ mice were the test animal species. The animals were six to seven weeks of age at start of study. The animals

TABLE 28.17. Health Effects Studies Sample Preparation Summary.

	Number of Isolation Runs	Gallons Sampled	Gallons[a] Sample Produced	Average Kilograms TOC Removed	Average Solids[b] Recovered (mg/L)
Denver drinking water	52[c]	1,846,000	337	10.9	4500
Ultrafiltration sample	65	1,249,000	249	6.4	1100
Reverse osmosis sample	38[c]	1,864,000	354	< DL[d]	75

[a] 5000:1 concentrate.
[b] 5000:1 concentrate Blank Solids = 22 mg/L.
[c] Most runs were 2 column sets.
[d] < DL: less than instrument detection limit.

were randomly selected for each test group by formal randomization procedures. The total number of animals at the start of the study is shown in Table 28.18.

Animals were observed twice daily for mortality and moribundity. Physical examinations for clinical signs of toxicity were recorded weekly. Animal body weights and food consumption were recorded weekly through Week 13, then every other week thereafter. Drinking water was measured twice weekly for the duration of the study when the water bottles were changed. Evaluation of hematology, clinical chemistry and urinalysis was performed prior to the two interim sacrifices and prior to the terminal sacrifice of surviving animals. Analysis of serum for diseases was performed on the sentinel animals after 27, 52 or 78 weeks of study.

Complete gross necropsy, including organ weights and preservation of protocol-specified tissues and any observed gross lesion, for all animals in the treatment groups was performed at scheduled sacrifice times or if the animals were found moribund or dead. Histopathological evaluations were made with the preserved tissues for all animals in the 500× treatment

TABLE 28.18. Test Animals In Study.

	Animals	Sex	Species	Levels	Total
Reclaimed water	70 ×	2 ×	2 ×	2 =	560
Drinking water	70 ×	2 ×	2 ×	2 =	560
Ultrafilter water	70 ×	2 ×	1 ×	1 =	140
Control	70 ×	2 ×	2 ×	1 =	280
					1,540

and control groups. If a treatment-related effect was seen in any tissue of the high dose group of animals, the corresponding tissue would be evaluated in the low dose group.

Reproductive Toxicity Study Protocol

Purpose

The objective of the reproductive toxicity study was to identify potential adverse effects on reproductive performance, intra-uterine development, and growth and development of the offspring during a two generation study. A teratology phase was included to identify potential embryo toxicity and teratogenicity of the test article.

Test Animals

Fifty male and 50 female Sprague-Dawley rats per 500x water sample were used for the F_0 generation. The animals were 12 to 15 weeks of age at start of study. The animals were randomly selected to each test group by computer-generated randomization procedures.

Breeding Procedures

For the F_0 parental generation, 50 male and 50 female Sprague-Dawley rats were supplied with one of three kinds of water (DW, RO and UF) plus distilled water for controls. After ten weeks of exposure, each male was cohabitated with one female of the same dose group for 21 days or until evidence of mating was observed. Prior to mating, physical examinations were performed weekly, as were measurements of body weight and food consumption. Water consumption was measured twice weekly. Animals were observed at least twice daily for mortality and moribundity. During the breeding period, only mortality and moribundity observations were conducted. After confirmation of mating, the males were sacrificed via CO_2 asphyxiation and necropsied with preservation of selected reproductive tissues for histopathological examination.

Observation and Examination

During gestation females were weighed and physical examinations were performed on Days 0, 7, 14 and 20. Food consumption was measured at each body weight interval, and water consumption was measured on Days 3, 7, 10, 17, and 20 of gestation. Females were observed four times daily at the first possible Day 21 of gestation for any female. During lactation females were weighed and physical examinations performed on Days 0, 4, 7, 14, and 21. Food consumption was measured on Days 4, 7, 10, 14, 16, 18, 20, and 21, and water consumption was measured on Days 2, 4, 7, 10, 12, 14, 16, 18, 20, and 21 of lactation.

As soon as possible after birth of the F_1 generation pups, litter size (number born live and dead) was recorded; pups were sexed, weighed and examined for external abnormalities. Dead pups were examined grossly and preserved in 95 percent ethanol. On Day 4 pups were weighed and examined for external abnormalities. Litters of more than eight pups were culled to that number to produce litters of four males and four females if possible. Culled pups were sacrificed, examined grossly, and preserved in 95 percent ethanol. On Days 7, 14, and 21 each remaining pup was weighed and examined for external abnormalities. On Day 21 of lactation the pups were weaned and selected for F_1 parental generation. Thirty-five pups per sex per group were randomly selected to comprise the F_1 parental generation and continued to receive the same type of drinking solution. Unselected pups were discarded without necropsy. After weaning of the F_1 pups, the F_0 females were sacrificed and necropsied with selected reproductive tissues preserved for histopathology.

Reproductive indices and gross necropsy and histopathology of reproductive organs of F_0 and F_1 parental animals were performed. The following tissues and organs were examined microscopically: clitoral glands; gross lesions; mammary gland with adjacent skin; ovaries; reputial gland; prostate; seminal vesicles; testes epididymus and vaginal tunics of testes; uterus; and vagina.

Gross necropsies were performed on all parental animals dying during the course of treatment, moribund animals sacrificed during the course of the study and all F_{2a} and F_{2b} offspring. Microscopic examination were made of all tissues showing gross pathological changes.

Pups selected for the F_1 parental generation were housed individually after weaning; they were mated at 12 to 17 weeks of age to produce the F_{2a} litters. Following weaning of the F_{2a} pups, the F_1 females were rested for at least 10 days before mating to produce the F_{2b} litters. Following of the F_{2b} litters, the females were rested again for a minimum of 10 days, after which they were mated to produce the F_{2c} fetuses for teratologic evaluation. Mortality and morbidity checks, clinical observations, and measurements of body weights and food and water consumption for the F_1 parental animals were the same as that of the F_0 parental animals. Observations and handling of the F_{2a} and F_{2b} pups were the same as that of the F_1 pups through weaning. After weaning, the F_{2a} and F_{2b} pups were sacrificed via CO_2 asphyxiation and subjected to gross examination of cervical, thoracic and abdominal viscera with preservation of gross lesions.

On gestational Day 20 of the F_{2c} breeding, each F_1 female was sacrificed for cesarean section and gross necropsy. Abnormal viscera and protocol-designated tissues and organs were preserved for histopathology. The uterus from each gravid female was excised, weighed and examined for the number and placement of implantation sites, number of live and dead fetuses and the number of early and late resorptions. The ovaries were examined for the number of corpora lutea. Each fetus was sexed, weighed, examined externally, tagged with the maternal identification number and a number indicating its position in the uterus, and sacrificed via intraperitoneal injection of sodium pentobarbital. Approximately one third of the fetuses were processed for visceral examination by the Wilson technique for assessing soft tissue development. The remaining fetuses were processed for skeletal examination using the Alizarin Red S staining method.

HEALTH EFFECTS STUDY RESULTS

Two Year Chronic Toxicity/Carcinogenicity Rat Study

Administration of RO water, UF water, or Denver's present drinking water at up to 500 times the concentration in the original water samples and supplied to F344 rats for at least 104 weeks did not result in any toxicologic or carcinogenic effects. Survival ranged from 52 percent to 70 percent in the males and from 64 to 84 percent in the females. These ranges of survival in each group and sex were within the ranges normally observed. All animals received a complete gross necropsy examination. No treatment-related gross lesions were observed at necropsy. Many gross lesions were noted in unscheduled death animals; however, these lesions were noted sporadically across the groups.

The incidence and type of clinical signs were comparable in all groups of the same sex. There were statistically significant differences between control and treated groups in the body weight, food consumption, and water consumption data, but all differences were small in magnitude and not consistent throughout the study. Therefore, these differences were not considered treatment related. The slight taste aversion noted in the groups receiving the drinking water groups was most likely the result of the addition of the VOCs. These were added to the drinking water dosing solutions since they are naturally present in the Denver drinking water but were removed as a result of the concentration process. No VOCs are present in the RO water or the UF water prior to the concentration process, therefore, it was not necessary to add VOCs to these water types.

Clinical pathology (hematology, clinical chemistry and urinalysis) and gross pathology conducted at Weeks 26, 52, and termination did not reveal any findings that could be considered to be treatment related. No treatment-related histomorphologic alterations were observed in unscheduled deaths or animals sacrificed by design. A wide variety of spontaneously occurring incidental lesions was observed in all groups and this appeared to be without relationship to treatment. A wide variety of spontaneously occurring neoplasms was observed in all groups, and these, generally, were of the type and frequency anticipated in this age and strain of rat. Table 28.19 summarizes the incidence of representative neoplasms, benign and malignant, which occurred at a rate greater than five percent in a group. A somewhat higher incidence of "C" cell adenoma of the thyroid gland was observed in drinking water males and RO females than in controls. This higher incidence, however, is well within the range anticipated for this type of neoplasm. Furthermore, these are small neoplasms that are often undetected at necropsy, and their observation may be due, in part, to fortuitous sectioning. Additional sections of the thyroid gland may increase the number of these types of neoplasms observed in all groups. The "C" cell tumors in the thyroid are not considered treatment-related. It can be concluded that the variety, frequency and severity of spontaneously occurring incidental lesions and neoplasms were within the anticipated range for this age and strain of rat.

Two-Year Chronic Toxicity/Carcinogenicity Mouse Study

Similar negative findings of the rat chronic study were also observed in the mouse chronic study. Administration of RO water or Denver's present drinking water at up to 500 times the concentration in the original water samples and supplied to $B_6C_3F_1$ mice for at least 104 weeks did not result in any toxicologic or carcinogenic effects. Survival ranged from 52 percent to 70 percent in the males and from 64 to 84 percent in the females. These ranges of survival in each group and sex were within the ranges normally observed. The incidence and type of clinical signs were comparable in all groups of the same sex. There were numerous statistically significant differences between control and treated groups in the body weight, food consumption, and water consumption data, but all differences were small in magnitude and not consistent throughout the study. The slight taste aversion (also seen in the rat study) noted in the groups receiving the Denver drinking water was most likely the result of the addition of the VOCs.

Clinical pathology, gross pathology, and microscopic pathology conducted at Weeks 26, 65, and at termination did not reveal any findings that could be considered to be treatment related. An increased incidence of minimal to slight renal tubular regeneration was observed in male $B_6C_3F_1$ mice which received either DW water or RO water for at least 26 weeks. This increased incidence of renal tubular regeneration was not confirmed at Week 65 or at terminal sacrifice.

TABLE 28.19. Number of Rats with Representative Neoplasms at Terminal Sacrifice (week 104) and Unscheduled Deaths.

	Sex							
	Male				Female			
Group	Con	DW	RO	UF	Con	DW	RO	UF
Pituitary								
Number examined	50	49	49	50	50	50	50	50
Adenoma	19	11	22	19	19	22	19	20
Carcinoma	0	1	0	0	0	0	0	0
Adrenal medulla								
Number examined	50	49	50	50	50	50	48	50
Benign pheochromocytoma	8	4	8	9	3	3	2	0
Malignant pheochromocytoma	0	0	0	0	0	1	1	0
Thyroid								
Number examined	50	49	50	50	50	50	48	50
Follicular cell adenoma	5	1	3	3	2	0	2	2
Follicular Cell Carcinoma	0	1	2	3	1	0	0	0
"C" cell adenoma	2	9	3	5	4	4	8	2
"C" cell carcinoma	3	4	1	5	2	3	1	2
Hematopoietic neoplasia								
Number examined	50	49	49	49	50	49	50	50
Leukemia, mononuclear	27	21	22	21	16	11	18	11
Pancreas								
Number examined	50	49	49	50	50	49	48	50
Islet cell adenoma	3	1	4	3	0	2	0	1
Islet cell carcinoma	1	1	0	2	0	1	0	0
Testis								
Number examined	49	50	50	49	—	—	—	—
Benign interstitial cell tumor	46	47	38	44	—	—	—	—
Malignant mesothelioma	1	1	2	4	—	—	—	—
Uterus								
Number examined	—	—	—	—	50	49	49	50
Endometrial stromal polyp	—	—	—	—	7	4	7	6
Endometrial stromal sarcoma	—	—	—	—	1	0	1	0
Mammary gland								
Number examined	47	44	49	49	50	50	50	50
Fibroadenoma	1	0	1	0	7	10	5	1

Con = Distilled water control.
DW = Denver drinking water.
RO = Reverse osmosis treatment.
UF = Ultrafilter treatment.

TABLE 28.20. **Distribution of Common Neoplasms by Organ Observed During Mouse Chronic Study.**

	Sex					
	Male			Female		
Group	Con	DW	RO	CON	DW	RO
Number in group[a]	70	70	70	70	70	70
Hematopoietic system	9	7	10	20	21	25
Liver	28	24	21	2	7	5
Lung	10	11	12	4	2	2
Pituitary	0	0	0	10	9	12

[a]Number in group at beginning of study; does not represent number of tissues examined for each organ in any group.
Con = Distilled water control.
DW = Denver drinking water.
RO = Reverse osmosis treatment.

A total of 357 neoplasms, of which 95 percent occurred after the Week 66 interim sacrifice, were observed. The most commonly affected organs were the hematopoietic system, liver, lung, and pituitary (see Table 28.20). The neoplasms observed are common in aging mice, and no relationship to either Denver drinking water or RO water was evident. The remaining microscopic changes observed in mice which received either Denver drinking or reclaimed water were considered consistent in type and severity with common spontaneous processes in mice and were similar to those observed in the concurrent distilled water control group.

Reproductive Study

The most notable result of the multigeneration reproductive study was the absence of any demonstrable treatment-related effects on reproductive performance, growth, mating capacity, survival of the offspring or fetal development in any of the treatment groups. In fact, the F_1 generation was exposed to the treatment regimen for a substantial portion of its lifespan (minimum of 48 weeks from weaning to terminal sacrifice).

The survival in the F_0 generation was 100 percent except for the death of one female rat in the DW dose group due to a difficult delivery. There were no differences of body weight gain during the growth period of the F_0 generation. Mean daily water intake was consistently lower in the DW dose group when compared to the other dose groups. This is attributable to a possible taste aversion resulting from the volatile organic compounds added to the DW dosing solutions. The reduced water consumption was not sufficient to reduce the body weights or food intake. F_1 pup survival was very good throughout the lactation period; pup body weights were similar in all groups throughout lactation.

There were three unscheduled deaths among the male groups and six unscheduled deaths among the female groups during the treatment period of the F_1 parental generation. As expected, pregnancy rates were reduced with each successive breeding of the F_1 parental animals. There were no adverse effects on pup survival or growth in either the F_{2a} or F_{2b} pups. A summary of the teratology evaluations is described in Table 28.21. There was only one malformed, tailless fetus in the study that also had skeletal malformations of the vertebrae and ribs. The observation of a single malformed fetus does not indicate a teratogenic effect but rather is considered a spontaneous event. Skeletal and visceral variations in development did not occur in a pattern that would be indicative of an experimental effect on fetal development.

There were no clinical signs or gross tissue alterations noted at necropsy in either parental generation which were attributed to any of the dose water exposures. There were no treatment-related histopathologic findings in parental animals of either generation.

TABLE 28.21. Summary of Results from F_{2c} Generation Teratogenicity Study.

	Control Group	RO Group	UF Group	DW Group
Fetuses evaluated	185	187	121	166
Fetal incidence	0	0	0	0
Fetal External Malformations				
Fetuses evaluated	185	187	121	166
Fetal incidence	0	1	0	0
Fetal Soft Tissue Variations				
Fetuses evaluated	7 (12.1%)	63	4 (9.5%)	58
Fetal incidence		12 (19.0%)		20 (34.5%)
Soft Tissue Malformations				
Fetuses evaluated	58	63	42	58
Fetal incidence	0	0	0	0
Fetal Skeletal Variations				
Fetuses evaluated	127	124	79	108
Fetal incidence	92 (72.4%)	79 (63.7%)	50 (63.3%)	82 (75.9%)
Fetal Skeletal Malformations				
Fetuses evaluated	127	124	79	108
Fetal incidence	0	1	0	0

RO = Reverse osmosis treatment.
UF = Ultrafilter treatment.
DW = Denver drinking water.

COST ESTIMATES

The feasibility of direct potable water reuse depends on several factors: product water safety, regulatory agency approval, public acceptance, and cost. The financial viability of this proposal depends on the cost to convert secondary wastewater effluent to potable water quality and the availability and relative cost of alternate traditional water supply development projects. Before proceeding with the multi-million dollar health effects testing program designed to evaluate the long-term effects of drinking reclaimed water, cost estimates for this complex treatment system were developed and compared to other future water supply projects [15]. Although accurate estimates for distant projects suffer from numerous deficiencies, they can be used to determine relative merit and to support decisions regarding the advisability of proceeding with further study. The previous evaluation [15] established that the cost of wastewater conversion was within acceptable limits and the health effects study proceeded. The cost estimates listed below utilize the knowledge gained from more than two years of continuous operation of the reuse demonstration plant in its final configuration. Also these cost figures have been revised to reflect January 1994 conditions.

The Army Corps of Engineers completed (May 1988) a massive (13-volume) environmental impact study [49] on Metropolitan Denver's water supply alternatives. This document examined the many issues surrounding the augmentation of metropolitan Denver's water supply. Among the items evaluated were the cost of various alternative traditional water supply projects which could be used to satisfy Denver's projected water demand. Some of the projects (e.g., Two Forks Dam) could be needed in less than twenty years while others were identified as future projects which may satisfy demands 20 to 50 years in the future. Direct potable water reuse costs are compared to these projects, since implementation of this new technology will most likely not be required until after the year 2010. The final environmental impact study [49] identified these longer-term project costs ranging in value from $250/ac-ft to $960/ac-ft. In order to be competitive, reuse costs should fall within this range.

Table 28.22 contains the cost estimates for the RO plus UF treatment sequence. The individual processes, as well as the major treatment groupings, are represented, along with the data for capital cost, amortized capital costs, O&M costs, and total costs. The capital costs are presented in a total dollar format. The amortized capital costs are presented in cents per 1000 gal of finished product water. These estimates were determined by amortizing the capital cost over a twenty year period at a 7.5 percent interest rate. The O&M costs were presented in two formats which included a total dollar amount, as well as the costs in cents per 1,000 gal of finished product water. The total costs are presented in cents per 1,000 gal of

TABLE 28.22. Engineering Table Cost Estimate for a 100 mgd 50% RO, 50% UF Reuse Plant (January 1994).

Process	Total Cost
Biological	0.16
High pH lime classification	0.47
Filtration	0.05
Activated carbon	0.31
Reverse osmosis	0.55
Ultrafiltration	0.33
Ozonation	0.09
Chloranation	0.005
Miscellaneous	0.03
Total	$2.00/1,000 gal

treated water, and include the amortized capital cost, O&M costs, and the project overhead costs.

DISCUSSION

In previous reports the operation and maintenance (O&M) costs for the 1.0 mgd demonstration facility have been computed [16–18]. These costs figures have been used to compare different treatment sequences within the plant, aiding in process evaluations leading to the selection of the health effects treatment system.

The O&M costs for the reuse demonstration plant are higher than the estimates obtained from the engineering tables. In the case of sludge removal, the reuse plant was dependent on other entities for disposal which impacted the cost. For a full scale reuse facility, sludge handling from the plant would employ a recalcination furnace that would reduce this cost significantly. Carbon regeneration costs for the reuse plant were derived from operating the furnace during phases that only covered six month periods [16–18]. The regeneration costs obtained from the engineering tables were calculated on the basis of operating the furnace twice a year. Experience gained during the health effects phase, revealed that the operation of the furnace was not required. Therefore, regeneration may not be required for two years or longer, reducing the cost of regeneration. The difference in cost for reverse osmosis and air stripping is due primarily to the assumption of obtaining membranes that operate at lower osmotic pressures. The cost derived for the ultrafiltration system at the reuse demonstration plant was based on a pilot system operating at 2.3 gpm. This cost should decrease at a full scale facility. Other factors which may reduce the O&M costs include Metro Wastewater Reclamation District's capability to denitrify, which will eliminate

TABLE 28.23. Cost Estimate for a 100-mgd Reverse Osmosis Plant (using demonstration plant operational values) $/1,000 Gal (January 1991).

Process	Amortized Capital	O&M Costs	Total Costs
Biological nitrogen removal	0.09	0.05	0.14
High pH lime clarification (including sludge disposal)	0.10	0.32[a]	0.42
Sludge disposal	0.05	0.10	0.15
Filtration	0.03	0.02[a]	0.05
Activated carbon contact and regeneration (including regeneration and replacement)	0.09	0.19[a]	0.28
Reverse osmosis (including brine disposal)	0.46	0.72[a]	1.18
Ozonation	0.008	0.07[a]	0.08
Chloramination	0.002	0.003[a]	0.005
Miscellaneous plant	0.01	0.017	0.027

[a]O&M Cost derived from the Reuse Demonstration Plant [16–18].
Total treatment cost estimate = $2.34/1000 gal = $762/ac-ft.

the need for a denitrification process in a full-scale reuse facility, and a lower cost method for brine disposal. The ultraviolet radiation system was not included in the 100 mgd reuse facility. From experience gained through the health effects period, it was determined that its use would not significantly impact water quality.

Table 28.23 incorporates amortized capital costs including construction related expenses determined for the 100 mgd plant, along with the pertinent operating experience from the reuse demonstration plant. By substituting the higher O&M cost estimates from demonstration plant experience with available capital construction costs estimates an upper cost limit may be calculated ($2.34/1,000 gal or $762/ac-ft).

The engineering table cost estimates contained in Table 28.22 exhibit lower values both for the reverse osmosis plant option as well as the split stream process incorporating reverse osmosis and ultrafiltration ($1.73/1,000 gal or $563/ac-ft and $1.64/1,000 gal or $534/ac-ft respectively). These values may be used to establish the minimum value for the cost estimate range.

The several cost estimates for a full scale reuse facility utilizing the engineering tables, and actual reuse demonstration plant operations values range from $534/ac-ft to $762/ac-ft. These estimates compare favorably with those of equal uncertainty for future water supply augmentation projects described in the Final Environmental Impact Study [49] which evaluates various Denver metropolitan water supply alternatives (estimates range from $250/ac-ft–$960/ac-ft). Therefore, cost does not appear to be a barrier to the future implementation of direct potable reuse in Denver.

PUBLIC INFORMATION PROGRAM

One of the objectives of the project was to "generate public awareness" of direct potable water reuse and its role as one possible component to satisfy metropolitan Denver's future water supply needs. A stated measure of success was a five percent increase in the number of Denver residents who had knowledge about potable reuse [16]. This seemingly modest goal was established since the implementation of direct potable reuse would only be considered following the completion of the demonstration project assuming that absolute reliability and product safety could be assured. These issues are critical to gaining public acceptance.

A public information program which was designed to educate as many Denver area residents as possible about potable reuse was implemented as part of the demonstration project, since a more knowledgeable public would be more receptive when and if this technology became a reality impacting their daily lives [10,50,51]. The program was comprised of a multi-media approach centered around escorted tours of the demonstration plant. The plant tour which included a professional audio-visual orientation, and escorted explanation of the treatment system, and a full- color brochure, was found to have the greatest impact on public attitudes [51]. More than 7,000 visitors were exposed to this educational program (Table 28.24). Included in this number were foreign visitors from more than 40 countries representing six continents. The majority, however, were Denver area residents which were the primary focal point.

In addition to plant tours, several other forms of information transfer were utilized. A newsletter was published periodically to inform interested parties of various project milestones. Newsletter circulation increased from 400 to more than 2,000 by the time the final issue (#21) was distributed. Informational bill stuffers were used to reach more than 200,000 house-

TABLE 28.24. **Reuse Plant Visitor Summary (1980–1991).**

Visitor Category	Number
Metropolitan Denver Residents	
Students	3,732
Public	1,052
Foreign	486
Technical	2,184
Total	7,454

holds on several occasions. Numerous newspaper and television reports documented project progress. A full-length (26-minute) video documentary was produced (''Pure Water . . . Again'') and distributed to community access channels and local public television for their use. Based upon audience estimates, more than 50,000 area residents were exposed to this educational program.

CONCLUSIONS

The demonstration of direct potable water reuse was conducted over a period of 13 years to illustrate the capability of reliably producing potable water quality from unchlorinated secondary treated wastewater. The 1-mgd (44 L/s) research treatment facility was unique in the world and provided valuable information regarding the effectiveness of numerous treatment processes for the removal of natural and man-made contaminants from water. Major project findings are as follows:

- The treatment processes (high pH lime clarification, recarbonation, filtration, activated carbon adsorption, reverse osmosis, or ultrafiltration, air stripping, ozonation, and chloramination) reliably produce a product from secondary treated wastewater which easily satisfies *all* current and proposed U.S. EPA drinking water standards.
- A complete two-year chronic toxicity and carcinogenicity study was conducted for the first time on the reclaimed water and compared to currently used drinking water. No adverse health effects were detected from lifetime exposure to any of the samples.
- Reproductive studies were conducted on the reclaimed water and compared to Denver's current drinking water. No adverse health effects were detected during a two generation reproduction study.
- Unprecedented physical, chemical and microbiological testing of the reclaimed water revealed a purity not normally found in domestic water supplies. *No* compound (organic or inorganic) or organism (bacteria or virus) was found in any of the samples which even approached regulatory limits.
- Public attitudes were found to be cautiously optimistic with the majority expressing a willingness to accept potable water reuse if the need were demonstrated and the safety assured.
- Although regulatory agency approval was not sought as part of this project; the analyses and health effects test results provide a strong foundation for acceptance.
- The estimated cost of a full-scale reuse treatment plant, based upon the demonstration plant experience, would compare favorably ($534–$762/ac-ft vs $250–$960/ac-ft) with that of Denver's projected future conventional water supply augmentation projects.

COMMENTARY BY THE AUTHORS

The most extensive water treatment technology research project ever undertaken has been successfully completed. The findings verify the ability of advanced water treatment processes to reliably remove a broad spectrum of pollutants and render a product which satisfies every current measure of safety. Direct potable water reuse has now been successfully demonstrated and is available as one possible alternative to be considered along with more conventional water supply development projects for satisfying future water needs.

REFERENCES

1 Work, S. W. and Hobbs, N., "Management Goals and Successive Water Use." *Journal AWWA,* 68:2:86, (1976).

2 Linstedt, K. D., and Bennett, E. R., *Evaluation of Treatment for Urban Wastewater Reuse.* U.S. EPA, EPA-R2-73-122, (1973).

3 Rothberg, M. R., Work, S. W., Linstedt, K. D., and Bennett, E. R., "Demonstration of Potable Water Reuse Technology, The Denver Project." *Proceedings, Water Reuse Symposium I,* AWWA Research Foundation, Denver, Colorado (1979).

4 CH$_2$M Hill, Inc., "Conceptual Design Report—Potable Water Reuse Plant—Successive Use Program." Report to the Denver Board of Water Commissioners, Denver, Colorado (1975).

5 Lauer, W. C., and Work, S. W., "Denver's Analytical Studies Program." *Proceedings Water Reuse Symposium II,* AWWA Research Foundation, Denver, Colorado (1982).

6 Work, S. W., Rothberg, M. R., and Miller, K. J., "Denver's Potable Reuse Project: Pathway to Public Acceptance." *Journal AWWA,* 72:8:435, (1980).

7 Sayre, I. M. International Standards for Drinking Water. *Journal AWWA,* 80:1:53 (January 1988).

8 Colorado Primary Drinking Water Regulations, Colorado Department of Health (1991).

9 World Health Organization. *Guidelines for Drinking Water Quality, Vols. I–III* (1984).

10 Carley, Robert L., "Attitudes and Perceptions of Denver Residents Concerning Reuse of Wastewater." *Proceedings, 92nd AWWA Annual Conference,* Denver, Colorado (1972).

11 Milliken, J. G., and Lohman, L., "Public Attitudes About Denver Water and Wastewater Reuse." Denver Research Institute, University of Denver, DRI Project No. 5-31597, Denver, Colorado (1985).

12 Lauer, W. C., Work, S. W., "The Denver Potable Water Reuse Demonstration Project: Initial Plant Start-up." *Proceedings, 104th AWWA Annual Conference,* Denver, Colorado (1984).

13 Rogers, S. E., and Lauer, W. C., "Start-up and Operation of the Denver Potable Water Reuse Demonstration Plant." presented at *WPCF Annual Conference,* New Orleans, Louisiana (1984).

14 Lauer, W. C., Ray, J. M., and Rogers, S. E., "The Denver Potable Water Reuse Project—Project Update." presented at *AWWA Water Quality Technology Conference,* Denver, Colorado (1984).

15 Lauer, W. C., et al., "Process Selection for Potable Reuse Health Effects Studies." *Journal AWWA,* 83:11:52 (1991).

16 Lauer, W. C., et al., "Preliminary Process Evaluations Phase I Report," Denver Water Department, Denver, Colorado (1987).

17 Lauer, W. C., et al., "Preliminary Process Evaluations Phase II Report," Denver Water Department, Denver, Colorado (1987).

18 Lauer, W. C., et al., "Preliminary Process Evaluations Phase III Report," Denver Water Department, Denver, Colorado (1988).

19 E.P.A. N.P.D.E.S. Permit No. CO-0026638, 1991.

20 "Operations Manual—Potable Water Reuse Demonstration Plant," CH_2M Hill, Denver, Colorado, 1983.

21 Nealey, M. K., et al., "Multicomponent Water Treatment Plant Hydraulic Model and Applications," Denver Water Department, unpublished.

22 Arber, R. P., et al., "Process Evaluation and Selection in the Design of a Potable Water Demonstration Plant," *Proceedings, AWWA Reuse Symposium II.*

23 McGuire, M J., and Suffet, I. H., "Treatment of Water by Granular Activated Carbon," *Advances in Chemistry 202,* American Chemical Society, Washington, D.C., (1983).

24 Jacangelo, J. G., et al., "Ozonation: Assessing Its Role in the Formation and Control of Disinfection By-Products," *Journal AWWA,* 81:8:74 (August 1989).

25 Glaze, W. H., Koga, M., and Cancilla, D., "Ozonation By-Products. 2. Improvement of an Aqueous-Phase Derivation Method for the Detection of Formaldehyde and Other Carbonyl Compounds Formed by the Ozonation of Drinking Water." *Environmental Science and Technology,* 23:7:838 (July 1989).

26 Krasner, S. W., et al., "The Occurrence of Disinfection By-Products in U.S. Drinking Water," *Journal AWWA,* 81:8:41 (August 1989).

27 Stevens, A. A., Moore, L. A., and Miltner, R. J., "Formation and Control of Non-Trihalomethane Disinfection By-Products." *Journal AWWA,* 81:8:54 (August 1989).

28 "Internal Corrosion of Water Distribution Systems," American Water Works Association Research Foundation, ISBN 0-915295-07-5 (1985).

29 Kerri, Kenneth D., Project Director, "Water Treatment Plant Operation," Volume 1, California State University, Sacramento, School of Engineering (1989).

30 *Standard Methods for the Examination of Water and Wastewater,* 17th Edition, APHA, AWWA, WPCF, American Public Health Association, 1015 15th Street, NW, Washington, D.C. (1989).

31 Larson, T. E., and Buswell, A. M., "Calcium Carbonate Saturation Index and Alkalinity Interpretations," *Journal AWWA,* Volume 34, pp.1667 (1942).

32 Rogers, S. E., "Biofilm Nitrogen Removal for Potable Water Reuse." Masters Thesis, Department of Civil, Environmental, and Architectural Engineering, University of Colorado, Boulder, Colorado (1989).

33 National Academy of Sciences, *Quality Criteria for Water Reuse.* Washington, D.C., National Academy Press (1982).

34 Borzelleca, J. F., and Carchman, R. A., "Effects of Selected Organic Drinking Water and Contaminants on Male Reproduction." National Technical Information Service Report No. PB 82-259:847 (December 1982).

35 National Academy of Sciences, *The Potomac Estuary Experimental Water Treatment Plant.* Washington, D.C., National Academy Press (1984).

36 Yang, R. S. H., and Rauckman, E. J., "Toxicological studies of chemical mixtures of environmental concern at the National Toxicology Program: Health effects of groundwater contaminants." *Toxicology* 47:15–34 (1987).

37 Chu, I., Villeneuve, D. C., Becking, G. C., and Lough, R., "Subchronic study of a mixture of inorganic substances present in the Great Lakes eco-system in male and female rats. *Bulletin Environmental Contamination Toxicology,* 26:42–45 (1981).

38 Cote, M. G., Plaa, G. L., Valli, V. E., and Villeneuve, D. C., "Subchronic effects of a mixture of persistent chemicals found in the Great Lakes. *Bulletin Environmental Contamination Toxicology,* 34:285–290 (1985).

39 Wester, P. W., Van Der Heijden, C. A., Bisshop, A., Van Esch, G. J., Wegman, R. C. C., and DeVries, T., "Carcinogenicity study in rats and mixture of eleven volatile halogenated hydrocarbon drinking water contaminants. *Science Total Environment,* 47:427–432 (1985).

40 Bull, R. J., "Toxicological of natural and man-made toxicants in drinking water, *Proceedings, 14th Annual Conference Environmental Toxicology,* AFAMRL Technical Report 83-099:259–266.

41 Kool, H. J., Kuper, F., Fan Haeringen, H., and Koeman, J. H., "A carcinogenicity study with mutagenic organic concentrates of drinking water in the Netherlands." *Food Chem. Toxicol.,* 23:79–85 (1985).

42 Lauer, W. C., LaChance, A., Shuck, D., Suffet, I. H., and Croy, R., "Water Sample Preparation for Comprehensive Health Effects Testing," Published report for the Denver Water Department Potable Water Reuse Demonstration Plant (1988).

43 Ringhand, H. P., and Neier, J. R., Kopfler, F. C., Schentz, K. M., Kaylor, W. H., and Mitchell, D. R., "Importance of sample pH on the recovery of mutagenicity from drinking water by XAD resins." *Environ. Science Technology* 21:382 (1987).

44 Lauer, W. C., Johns, F. J., Wolfe, G. W., Myers, B. A., Condie, L. W., Borzelleca, J. F., "Comprehensive Health Effects Testing Program for Denver's Potable Water Reuse Demonstration Project." *Journal of Toxicology and Environmental Health,* 30:305–321 (1990).

45 National Toxicology Program. "General Statement of Work for the Conduct of Toxicity and Carcinogenicity Studies on Laboratory Animals." April 1987 Revision, NIEHS, NTP, Research Triangle Park, N.C. (1987).

46 U. S. Environmental Protection Agency, "Good Laboratory Practices." Toxic Substances Control Act (TSCA), 40 CFR Part 792.

47 Aikin, G. R., Thurman, E. M., Malcolm, R. L., and Walton, H. F., "Comparison of XAD Macroporous Resins for the Concentration of Fulvic Acid from Aqueous Solution," *Anal. Chem.* 51, 1799, (1979).

48 Greff, R. A., Setzkorn, E. A., and Leslie, W. D., "A Colorimetric Method for the Determination of Parts/Million of Nonionic Surfactants," *J. Amer. Oil Chem. Soc.,* 42, 180, (1965).

49 "Metropolitan Denver Water Supply Environmental Impact Statement," U.S. Army Corps of Engineers, Omaha District (1988).

50 Lohman, L. C., and Milliken, J. G., "Informational/Educational Approaches to Public Attitudes on Potable Reuse of Wastewater." DRI (1985).

51 Lohman, L. C., "Potable Water Reuse can win Public Support." *Proceedings, Water Reuse Symposium IV,* AWWARF, Denver, Colorado (1988).

CHAPTER 29

The Cost of Wastewater Reclamation and Reuse

W ASTEWATER reclamation can be defined as the treatment or processing of wastewater to make it reusable. In many areas of the United States wastewater reclamation is viewed increasingly as a means to augment existing water resources against the specter of continued droughts and water supply shortages as well as to provide improved reliability to operating systems. Unfortunately, wastewater reclamation costs are not well-documented. This chapter is presented because of the increasing interest in wastewater reclamation and the need for reliable cost data for meaningful economic comparisons. To facilitate subsequent data use and manipulation, the chapter is divided into six distinct sections: (1) Background, (2) Selection of Wastewater Reclamation Facilities, (3) Design Criteria for Wastewater Reclamation Facilities, (4) Cost Methodology for Wastewater Reclamation Facilities, (5) Wastewater Reclamation Facility Costs, and (6) Conclusions.

BACKGROUND

Reclaimed wastewater can be used for a number of options including agricultural irrigation, landscape irrigation, groundwater recharge, and industrial processes. Water quality requirements for reuse alternatives vary depending on the extent of potential public exposure. Background information is presented below including a discussion of multiple reuse options, water quality considerations, and reclamation facility components.

David Richard, Nolte and Associates, Inc., 1750 Creekside Oaks Drive, Suite 200, Sacramento, CA 95833.

DEFINITION OF RECLAMATION

As defined in the Porter-Cologne Water Quality Control Act (Section 13050, Chapter 2, Division 7 of the California Water Code), reclaimed water "means water which, as a result of treatment of waste, is suitable for a direct beneficial use or a controlled use that would not otherwise occur" [1]. In California, reclamation refers commonly to a wastewater treatment system that produces an effluent of such quality to allow its beneficial "reuse" by a water consumer. The planning and design of a reclamation program, therefore, is intended to maximize the benefits to end users within a water supply network. Although reclamation may represent an attractive alternative to costly wastewater disposal options, the principal objective for a reuse project remains to augment or supplement the water resources inventory of the area.

DESCRIPTION OF REUSE OPTIONS

In the past, the use of reclaimed water has been evaluated in terms of the following reuse categories:

(1) Agricultural
(2) Landscape irrigation
(3) Impoundments
(4) Groundwater recharge
(5) Industrial
(6) Livestock, wildlife, and fisheries

Other potential reuse options within each of the above categories are also considered in the following discussion.

Reuse Alternative 1— Agricultural

Historically, agricultural irrigation has constituted more than 50 percent of all reuse activities [2,3]. Within the agricultural reuse alternative, irrigation with reclaimed water may be utilized for the following:

(1) Food crops (spray or surface irrigation)
(2) Fodder, fiber, and seed crops
(3) Pasture for milking animals

Reuse Alternative 2—Landscape Irrigation

Urban irrigation of landscaped areas using reclaimed water represents the fastest growing reuse option in the United States. Because residential and commercial landscape watering comprise more than 40 percent of the total water consumption in arid and semi-arid regions [3], substitution of reclaimed water for potable water in a dual distribution system can generate significant long-term benefits to a community's water supply sources. Based on the potential for public exposure to reuse activities, reclaimed water irrigation of landscaped areas can be divided into the following sub-categories:

(1) Golf course, cemetery, freeway median, and greenbelt irrigation

(2) Parks, playgrounds, and schoolyard irrigation

Reuse Alternative 3— Impoundments

Man-made ponds, lakes, or reservoirs constructed to store or hold reclaimed water are referred to as impoundments. Depending upon public access limitations or use restrictions, impoundments may be grouped under the following sub-headings [4]:

(1) Restricted recreational impoundment (recreation limited to fishing, boating, and other non-body contact water recreation activities)

(2) Nonrestricted recreational impoundment (no limitation imposed on body contact water sport activities)

(3) Landscape impoundment (no public contact allowed)

Reuse Alternative 4—Groundwater Recharge

The use of reclaimed water for groundwater recharge and the control of saltwater intrusion may be accomplished through either of the following:

(1) Injection wells

(2) Surface spreading basins

Recharge which represents an indirect potable water reuse option is employed to: reduce, stop, or reverse declines in groundwater levels due to aquifer overdrafting; provide a means to store treated effluent for future beneficial purposes and protect underground freshwater in coastal aquifers against salt water intrusion [5,6]. Examples of large groundwater recharge projects include Water Factory 21 operated by the Orange County Water District, San Jose Creek-Whittier Narrows Reclamation Plants operated by the Los Angeles County Sanitation District, and the Fred Hervey Water Reclamation Project in El Paso, Texas [6].

Reuse Alternative 5— Industrial

Reclaimed municipal wastewater has been utilized by industry for the following:

(1) Cooling water
(2) Process water
(3) Boiler feed water

To date, reclaimed municipal wastewater used as cooling water constitutes 99 percent of the total volume of industrial reuse water [7]. Industries
with potential process water reuse requirements include primary metal
production, petroleum and coal products, tanning, lumber, textiles, chemicals, pulp and paper, food canning, and soft drinks. Although the use of
reclaimed water for boiler feed water is technically feasible, it has proven
to be operationally difficult to achieve because of severe problems with
scaling.

Reuse Alternative 6—Livestock, Wildlife, and Fisheries

Within this category, reclaimed water may be used for:

(1) Livestock and wildlife watering
(2) Fisheries habitat (warm water and cold water fisheries)

Regulatory standards and water quality requirements for the identified
reuse alternatives are described in the following subsection. End use alternatives are referred to subsequently by numerical designations (e.g. 1a,
1b, etc.)

DISCUSSION OF WATER QUALITY REQUIREMENTS

In view of the lack of federal guidelines or regulations concerning
wastewater reclamation, detailed criteria governing water reuse have been
developed by state and local agencies. As an example, the use of reclaimed
water in California is regulated by the California Regional Water Quality
Control Board predicated on guidelines established by the California Department of Health Services (DHS) under Title 22 [4]. The basic objective
of the Wastewater Reclamation Criteria promulgated as Title 22 is to
assure public health protection without discouraging unnecessarily water
reuse [8]. The potential for disease transmission through pathogenic organisms (bacteria, protozoa, helminths, and viruses) is the critical concern
addressed in the regulations and thus, those end uses that create the greatest
opportunities for public exposure are subject to the strictest requirements.

Provisions enumerated under Title 22 and accompanying guidelines for
reclaimed water use [9] include requirements for specific treatment process
configurations, effluent quality, process redundancy, and facility reliability.
A summary of treatment and reclaimed water quality requirements for
specific end use categories described earlier is presented in Table 29.1.

TABLE 29.1. General Treatment and Effluent Quality Requirements for Various Water Reuse Alternatives as Established under Title 22 [7,10].

Reuse Alternative Designation	Title	Treatment and Effluent Quality Requirement[a]
1a	Agricultural food crop	Secondary treatment, filtration and disinfection, total effluent coliform ≤2.2/100 milliliter (mL)
1b	Fodder, fiber, and seed crops, orchards and vineyards	Primary treatment
1c	Pasture for milking animals	Secondary treatment and disinfection, total effluent coliform < 23/100 mL
2a	Golf course, cemetery, freeway median, and greenbelt irrigation	Secondary treatment and disinfection, total effluent coliform < 23/100 mL
2b	Parks, playgrounds and schoolyard irrigation	Secondary treatment, filtration, and disinfection, total effluent coliform ≤2.2/100 mL
3c	Restricted recreational impoundment	Secondary treatment and disinfection, total effluent coliform ≤2.2/100 mL
3b	Nonrestricted recreational impoundment	Secondary treatment, filtration, and disinfection, total effluent coliform ≤2.2/100 mL
3c	Landscape impoundment	Secondary treatment and disinfection, total effluent coliform < 23/100 mL

[a]Effluent coliform requirements refer to a 7 day median value.

From a review of Table 29.1 it can be concluded that treatment and quality requirements are specifically included to insure the removal or inactivation of pathogens, including viruses. The highest standard or greatest level of treatment is intended to produce essentially a pathogen-free effluent [10].

Following the Title 22 guidelines, a reclamation facility must also incorporate the following features:

(1) Each individual treatment process shall consist of multiple units capable of producing the required effluent quality with one unit out of service.
(2) The facility shall include sufficient alarms to indicate failure of individual unit processes and loss of plant power supply.
(3) Reliability provisions at the facility shall include either a standby power source or automatically actuated emergency storage or disposal alternative.

Guidelines for Groundwater Recharge

Although not addressed specifically under Title 22, the California DHS has proposed general guidelines for groundwater recharge with reclaimed municipal wastewater using surface spreading basins [11]. Suggested requirements have included design guidelines for the following project characteristics:

(1) Depth to groundwater
(2) Retention time underground for reclaimed water
(3) Horizontal distance from the recharge basin to the extraction well
(4) Percent reclaimed water extracted

Although the recommended California DHS water quality requirements vary depending upon specific applications for evaluation purposes the following treatment process configuration will be assumed for the groundwater recharge alternatives (injection wells and surface spreading basins): secondary treatment, filtration, disinfection (total coliforms $\leq 2.2/100$ mL), and organics removal.

Standards for Industrial Reuse

Although no specific standards exist for the use of reclaimed water for industrial purposes, the California DHS has indicated the need to produce an effluent quality equal to that required for agricultural food crops (i.e., total coliforms $\leq 2.2/100$ mL) in historical correspondence [12]. Interestingly, water quality requirements for industrial use (cooling water, process water, and boiler feed) are typically more stringent than DHS restrictions. Potable water quality or greater may be necessary for various reuse applications in the food processing industry while dissolved metals and organic impurities in reclaimed water are critical concerns in the textile, paper, and photographic industries because these components may create staining problems [7]. Hardness and alkalinity removal also must be considered in cooling water or boiler feed water reuse alternatives. Scale formation (from insoluble calcium and magnesium salts) and excessive localized corrosion due to the release of carbon dioxide gas are the principal problems when treated effluent is used without additional processing. Because of these potential negative impacts associated with water reuse for industrial purposes, a high degree of advanced wastewater treatment (organics removal, water softening, and some demineralization) will be included in the cost analyses for industrial reuse systems.

Water Quality Considerations for Animal Watering and Fisheries Habitat

Water quality considerations for the livestock/wildlife watering and fisheries habitat alternatives are primarily a function of potential reclaimed water toxicological effects on animal or marine organisms. The sensitivity of livestock, various species of wildlife, and fish to specific impurities present in the reclaimed water will determine the required level of wastewater treatment. Because of the need for high water quality in fisheries, more advanced stages of reclaimed water "purification" will be assumed, including extensive nutrient removal. In contrast, secondary wastewater treatment only will be required for the livestock/wildlife watering reuse option due to the greater tolerance of domesticated animals or wildlife to fluctuations in water quality [13].

Summary of Water Quality Requirements

Predicated upon the recommended levels of treatment for the respective reuse alternatives, the corresponding effluent quality criteria for each end-use option are summarized in Table 29.2. Data presented in the referenced table reflects the following assumptions:

(1) For effective disinfection (total coliforms $\leq 2.2/100$ mL), the reclaimed water total suspended solids (TSS) and biochemical oxygen demand, 5-day (BOD_5) concentrations must be equal to or less than 10 mg/L.
(2) Nutrient removal will be required for the groundwater recharge, industrial, and fisheries habitat reuse alternatives.
(3) Reclaimed water used for cooling water, industrial process water, and boiler feed must undergo processes to reduce concentrations of total dissolved solids (TDS), total organic carbon (TOC), and hardness [14].

Specific treatment trains designed to achieve the desired water quality are discussed in the following section of this chapter.

DESCRIPTION OF WASTEWATER RECLAMATION FACILITIES

In evaluating the feasibility of reclamation versus other water supply alternatives, it is important to recognize the various elements that comprise a reclamation system. Wastewater reclamation facilities include treatment, storage, and delivery system components. Treatment facilities consist of unit processes designed to produce a specific water quality required for various reuse options. Short-term storage of reclaimed water is provided typically to balance user requirements versus production capabilities. Delivery system elements include pipelines and pump stations. Design criteria

TABLE 29.2. **Summary of Effluent Quality Requirements for Various Reuse Alternatives.**

Reuse Designation	Maximum Constituent Concentration[a]								
	TSS	BOD$_5$	NH$_3$-N[b]	NO$_3$-N[c]	PO$_4$[d]	TOC	TDS	Hardness[e]	Coliform[f]
1a	15	15	NR[g]	NR	NR	NR	NR	NR	2.2
1b	100	120	NR	NR	NR	NR	NR	NR	NR
1c	30	30	NR	NR	NR	NR	NR	NR	23
2a	30	30	NR	NR	NR	NR	NR	NR	23
2b	15	15	NR	NR	NR	NR	NR	NR	2.2
3a	15	15	NR	NR	NR	NR	NR	NR	2.2
3b	15	15	NR	NR	NR	NR	NR	NR	2.2
3c	30	30	NR	NR	NR	NR	NR	NR	23
4a	10	10	10	10	2	5	NR	NR	2.2
4b	10	10	10	10	2	5	NR	NR	2.2
5a	10	10	10	10	2	5	250	100	2.2
5b	10	10	10	10	2	5	250	100	2.2
5c	10	10	10	10	2	5	100	5	2.2
6a	30	30	NR	NR	NR	NR	NR	NR	NR
6b	10	10	5	5	2	NR	NR	NR	NR

[a] mg/L unless indicated otherwise.
[b] Ammonia nitrogen.
[c] Nitrate nitrogen.
[d] Phosphate.
[e] mg/L as calcium carbonate (CaCO$_3$).
[f] Total coliforms/100 mL.
[g] No specific requirement.

and cost information for these reclamation elements are provided later in this chapter.

SELECTION OF WASTEWATER RECLAMATION FACILITIES

The selection of wastewater reclamation facilities is dependent upon specific treatment objectives and desired effluent quality. Water quality objectives have been established previously for various reuse options (see Table 29.2). The basis for treatment process selection is described below as a function of final effluent quality characteristics.

SELECTION OF TREATMENT PROCESSES

Based on a review of the literature [14–17], treatment flow diagrams have been developed for various process configurations that will produce a specific effluent quality. The treatment processes selected for analysis

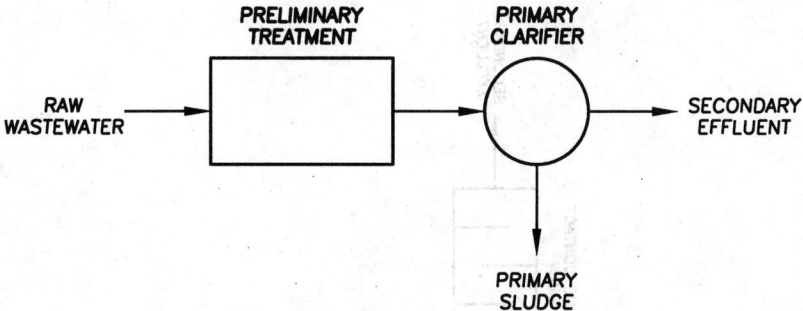

Figure 29.1 Treatment flow diagram 1: primary treatment.

are summarized in Table 29.3. In general, wastewater reclamation treatment flow diagrams were selected based on their ability to produce reclaimed water of the quality required to meet the end-use criteria enumerated in Table 29.2. The flow diagrams were also developed based on an increasing level of treatment, with the lowest quality effluent produced from Treatment Flow Diagram 1, Primary Treatment and the highest quality effluent available from Treatment Flow Diagram 12, Secondary Treatment–Lime Treatment–Reverse Osmosis.

Treatment Process Flow Diagrams

Detailed flow diagrams for the treatment processes summarized in Table 29.3 are presented in Figures 29.1 through 29.12. Within each flow diagram, unit processes were configured based on their inter-

TABLE 29.3. Summary of Reclamation Treatment Processes.

Flow Diagram Number	Title
1	Primary treatment
2	Conventional activated sludge
3	Combined trickling filter and activated sludge
4	Extended aeration
5	Secondary treatment plus full title 22 facility
6	Secondary treatment plus direct filtration
7	Secondary treatment plus contact filtration
8	Secondary treatment—contact filtration–phosphorus removal
9	EIMCO bardenpho process
10	Secondary treatment—contact filtration–carbon adsorption
11	Secondary treatment—contact filtration–carbon adsorption–reverse osmosis
12	Secondary treatment—lime treatment–reverse osmosis

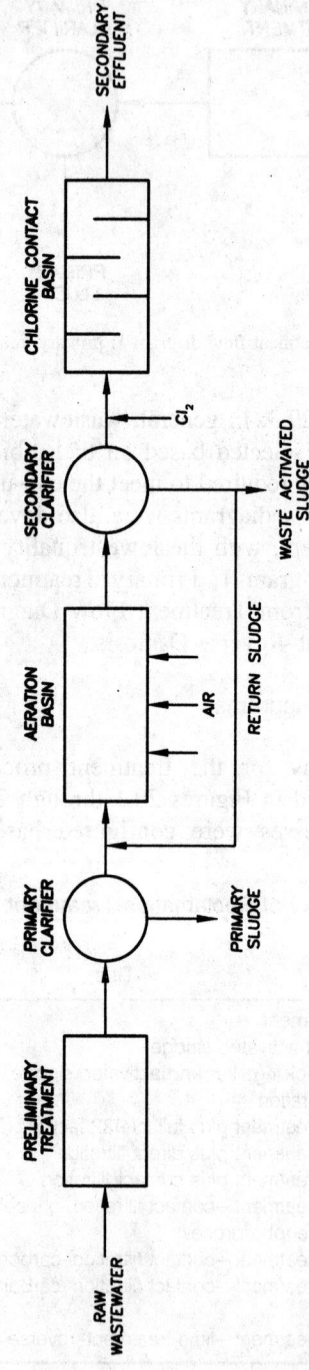

Figure 29.2 Treatment flow diagram 2: conventional activated sludge.

1344

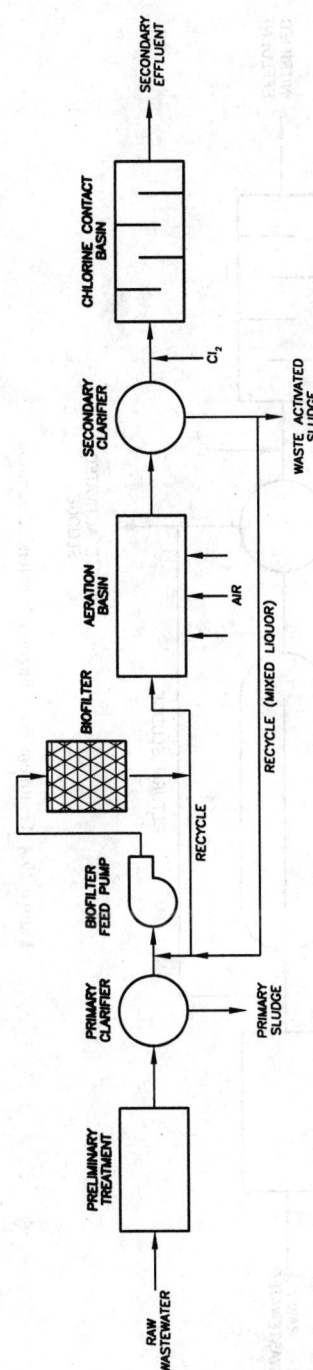

Figure 29.3 Treatment flow diagram 3: combined biofilter—activated sludge process.

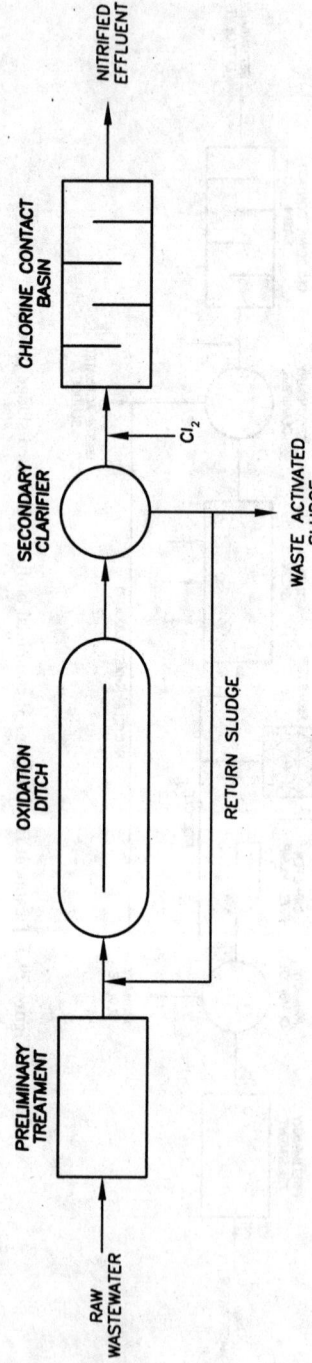

Figure 29.4 Treatment flow diagram 4: extended aeration.

1346

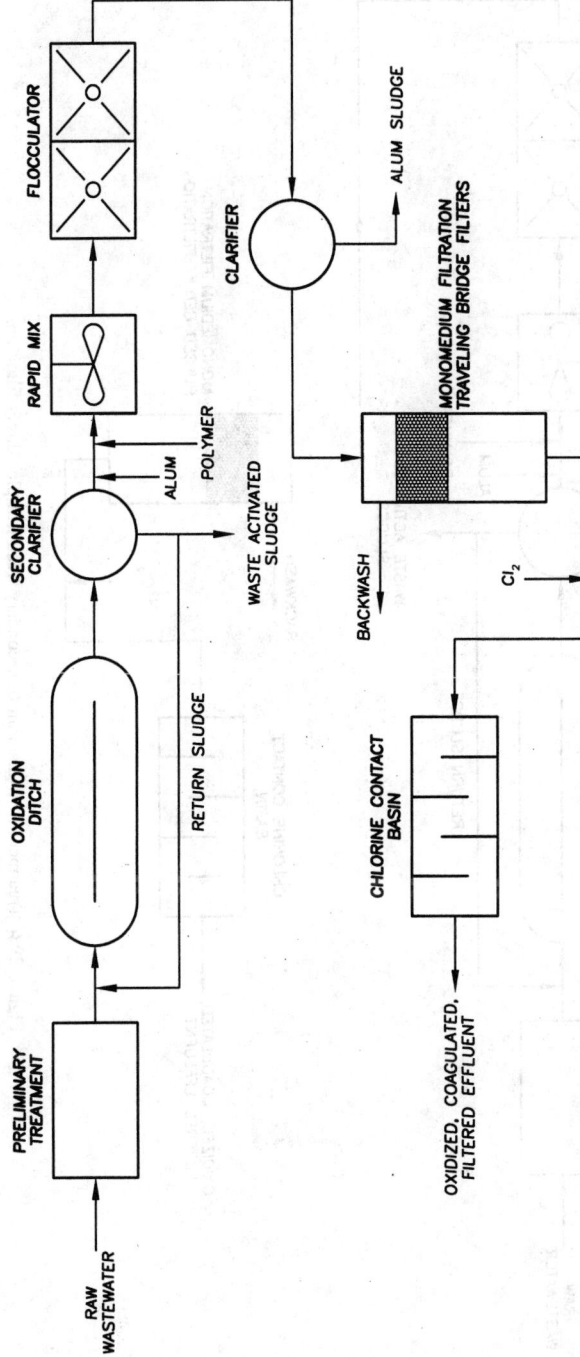

Figure 29.5 Treatment flow diagram 5: secondary treatment plus full Title 22 facility.

1347

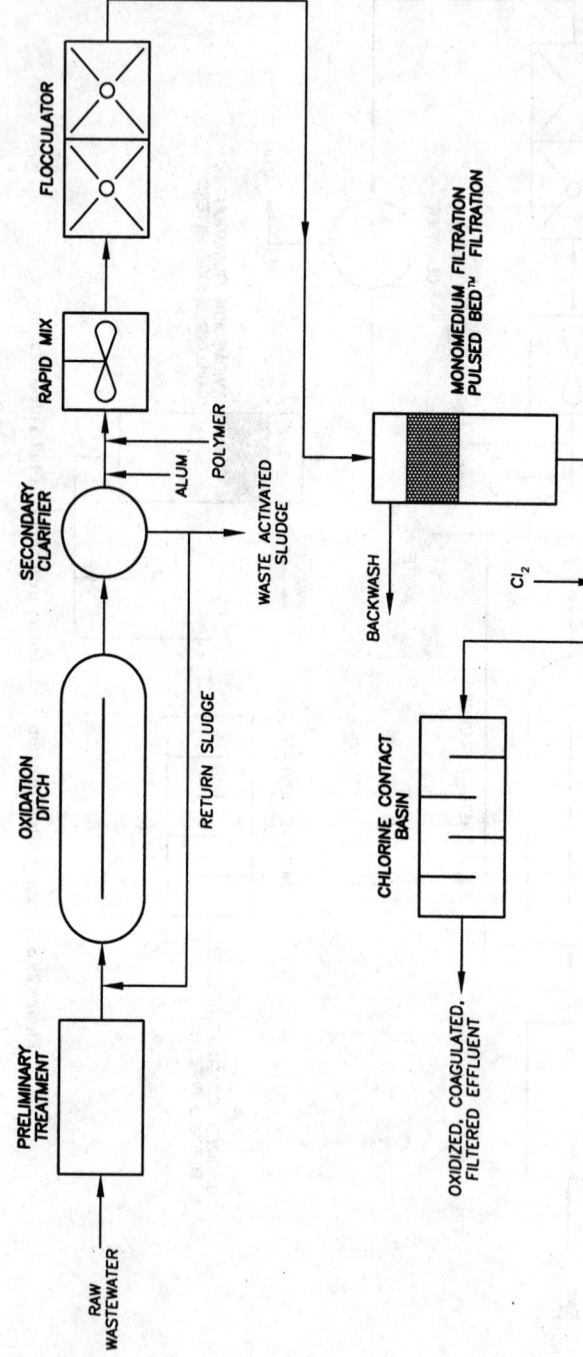

Figure 29.6 Treatment flow diagram 6: secondary treatment plus direct filtration.

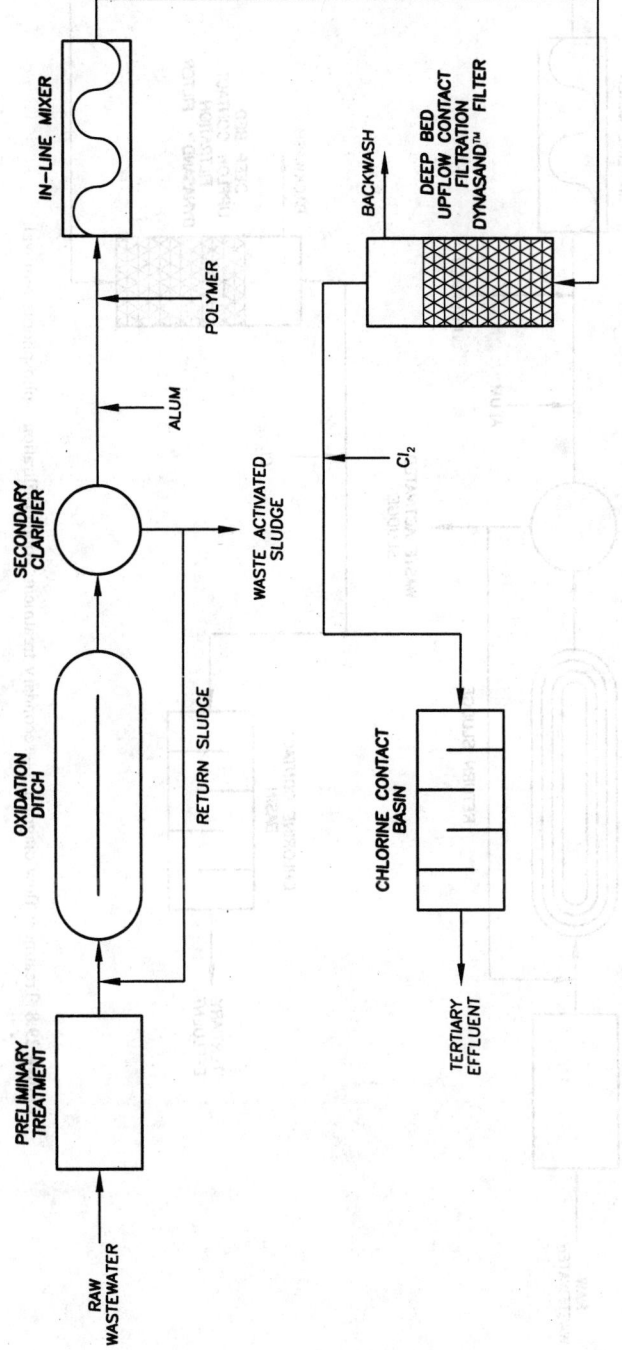

Figure 29.7 Treatment flow diagram 7: secondary treatment plus contact filtration.

1349

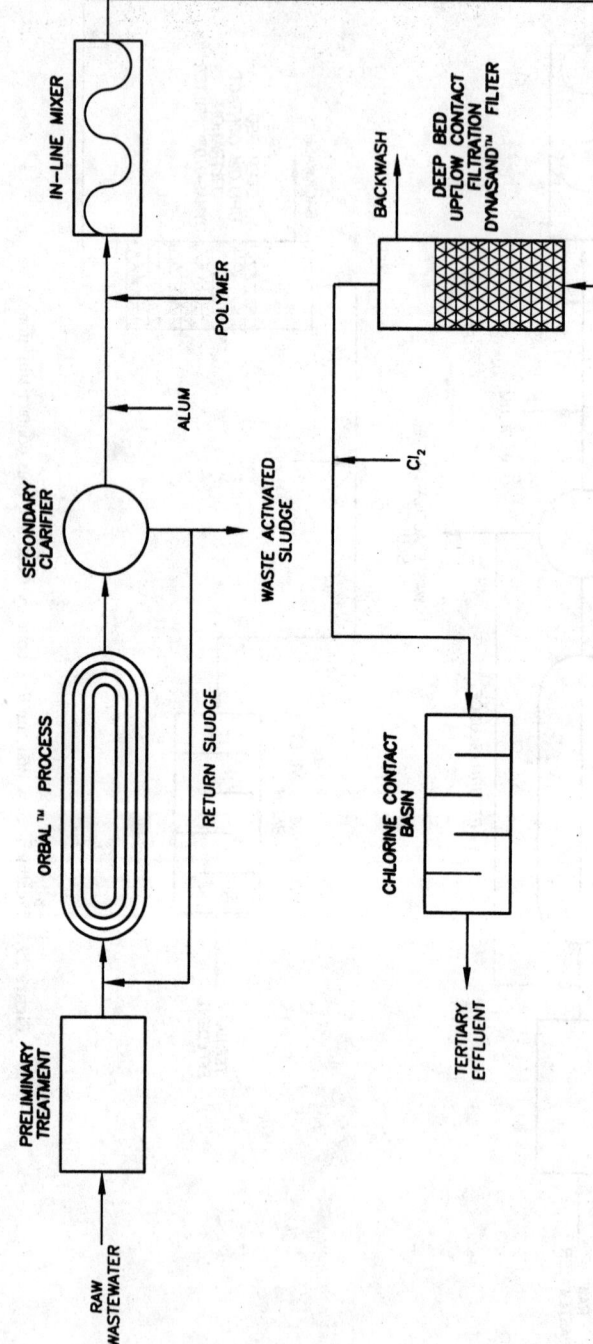

Figure 29.8 Treatment flow diagram 8—secondary treatment; contact filtration—phosphorus removal.

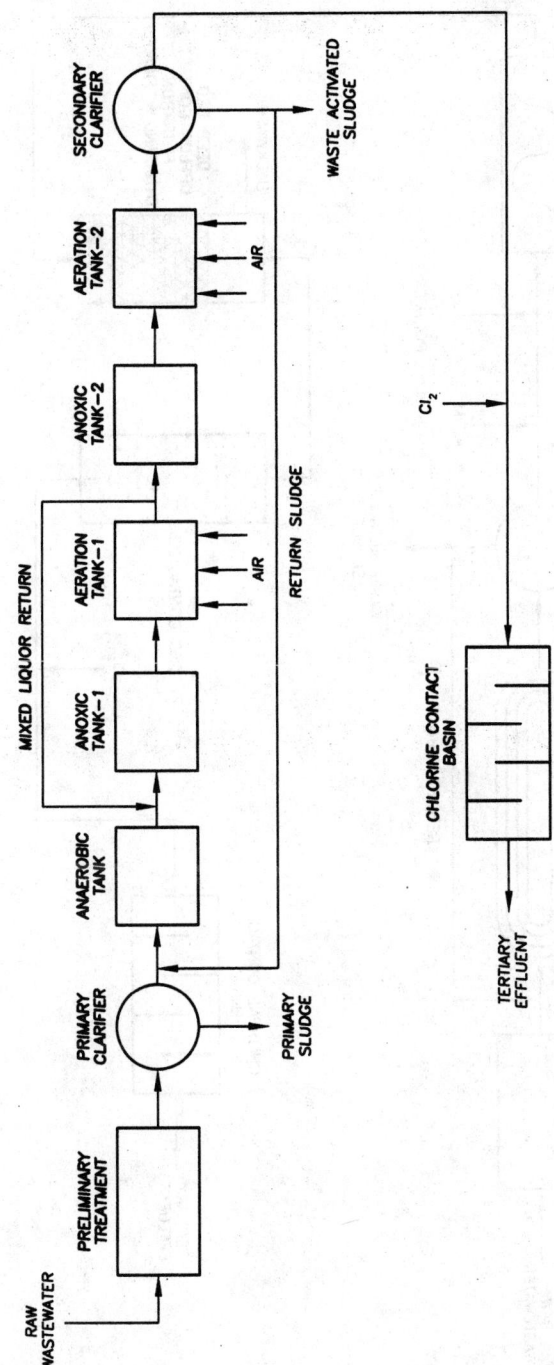

Figure 29.9. Treatment flow diagram 9: EIMCO five-stage Bardenpho™ process.

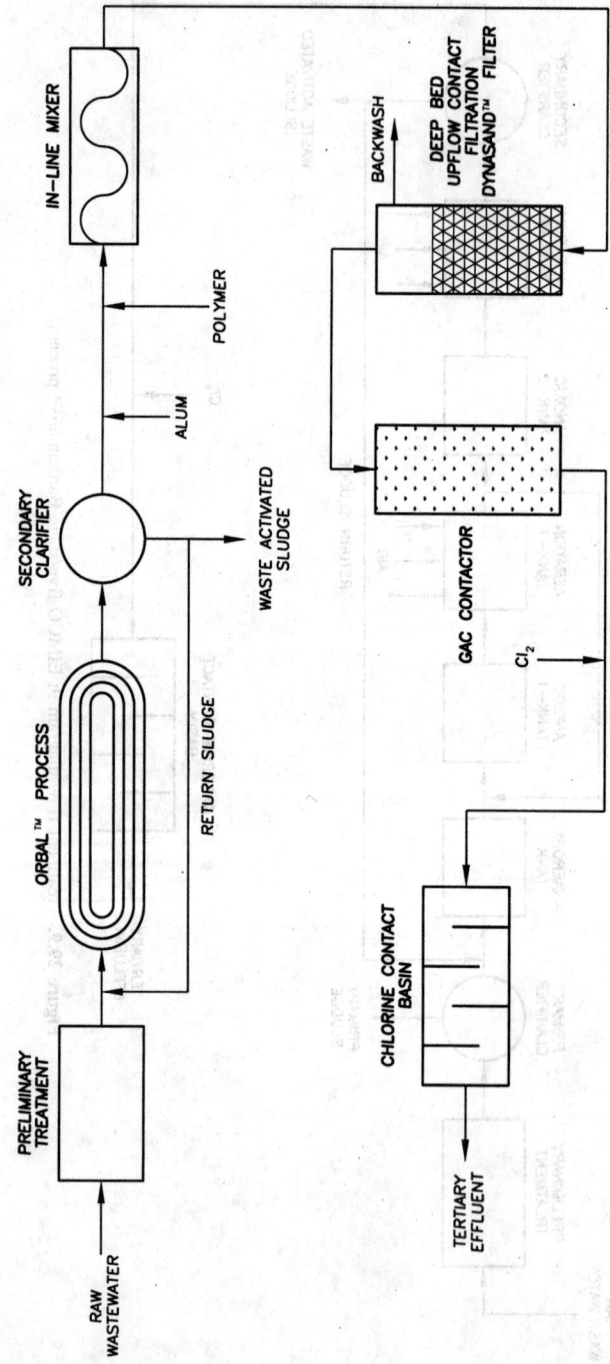

Figure 29.10. Treatment flow diagram 10—secondary treatment; contact filtration—carbon adsorption.

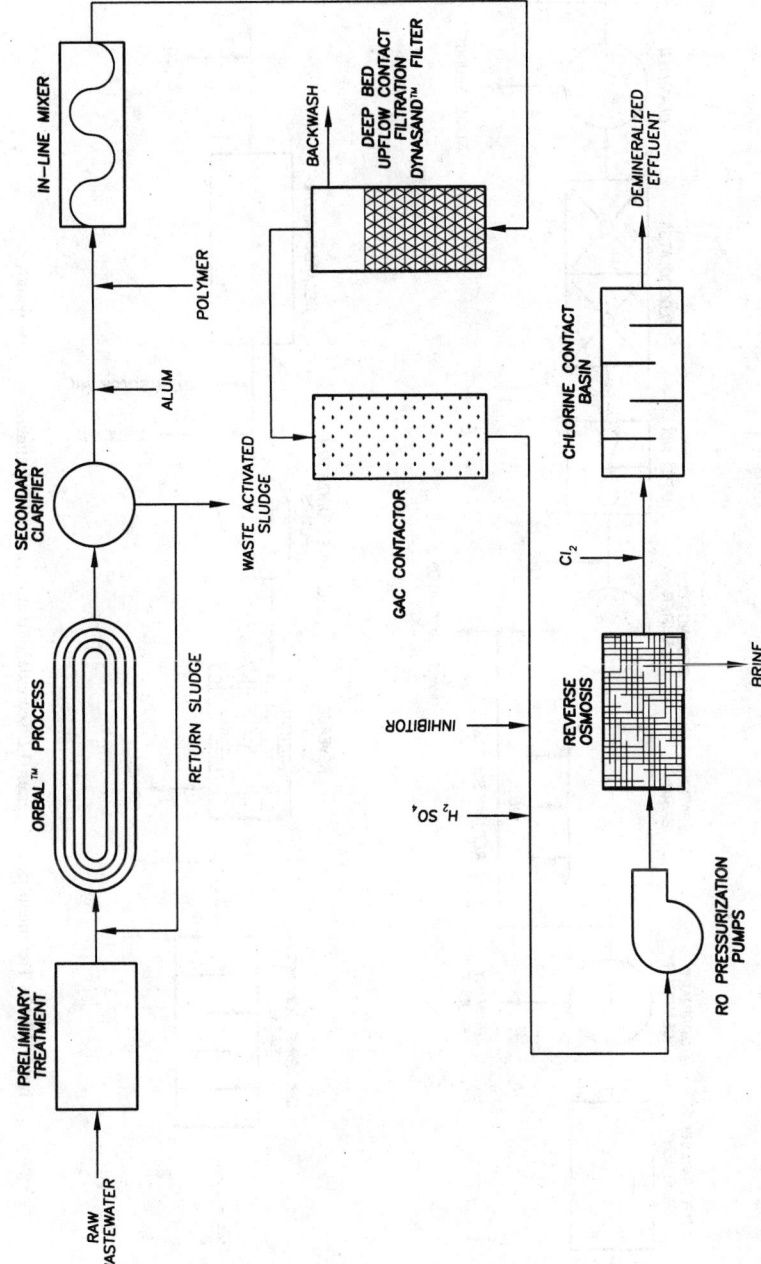

Figure 29.11. Treatment flow diagram 11: secondary treatment—contact filtration—carbon adsorption—reverse osmosis.

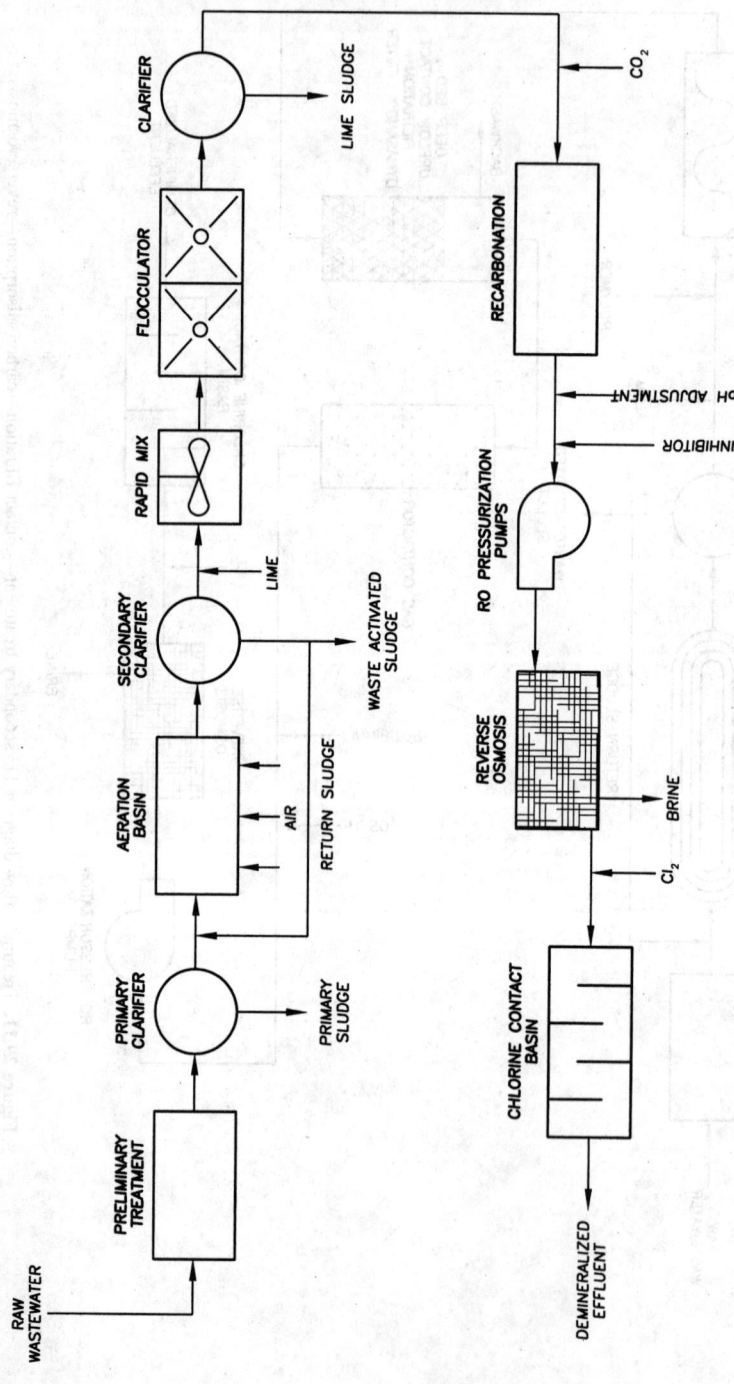

Figure 29.12. Treatment flow diagram 12: secondary treatment—lime treatment—reverse osmosis.

TABLE 29.4. Projected Effluent Quality from Alternative
Treatment Process Flow Diagrams.

Flow Diagram Number	Constituent Concentration[a]								
	TSS	BOD$_5$	NH$_3$-N	NO$_3$-N	PO$_4$	TOC	TDS	Hardness[b]	Coliform[c]
1	80	120	NA[d]	NA	NA	NA	NA	NA	NA
2	20	20	NA	NA	NA	NA	NA	NA	< 23
3	25	25	NA	NA	NA	NA	NA	NA	< 23
4	10	10	5	NA	NA	NA	NA	NA	< 23
5	10	10	5	NA	NA	NA	NA	NA	< 2.2
6	10	10	5	NA	NA	NA	NA	NA	< 2.2
7	10	10	5	NA	NA	NA	NA	NA	< 2.2
8	10	10	1	2	2	NA	NA	NA	< 2.2
9	10	10	1	2	2	NA	NA	NA	< 2.2
10	< 2	< 2	1	2	2	< 5	NA	NA	< 2.2
11	< 1	< 1	< 1	< 1	2	< 2	< 50	< 10	< 2.2
12	< 1	< 1	< 1	< 1	2	< 2	< 50	< 10	< 2.2

[a]mg/L unless indicated otherwise.
[b]mg/L as CaCO$_3$.
[c]Total coliform/100 mL.
[d]Not applicable, treatment process not designed typically for specific constituent removal.

changeability in other flow diagrams. As an example, conventional activated sludge (Treatment Flow Diagram 2) is compared to extended aeration (Treatment Flow Diagram 4) and therefore, can be substituted if desired, for extended aeration in the various filtration alternatives. In this manner, by creating a matrix of individual comparable components and developing a basic cost comparison of unit processes, both the cost of overall treatment flow diagrams and incremental increases in treatment can be evaluated.

Treatment Process Performance

The effluent quality that can be expected from the flow diagrams presented in Figures 29.1 through 29.12 is repeated in Table 29.4. Projected constituent concentrations are based upon full scale operating experience at wastewater treatment facilities with similar process configurations.

TREATMENT PROCESS DESCRIPTION AND APPLICATIONS

A brief discussion of the treatment flow diagrams and the appropriate reuse options in which the individual process configurations could be utilized is presented below. A more specific description of the individual unit processes and design criteria is included in Table 29.5.

TABLE 29.5. Typical Raw Wastewater Characteristics Used in the Sizing of Reclamation Unit Processes.

		Concentration		
Contaminants	Unit	Weak	Medium	Strong
TSS	mg/L	100	220	350
BOD$_5$	mg/L	110	220	400
Nitrogen (total as N)	mg/L	20	40	85
Phosphorus (total as P)	mg/L	4	8	15

Source: Adapted from Reference [18].

Treatment Flow Diagram 1, Primary Treatment

For Reuse Alternative 1b (fodder, fiber, and seed crops; orchards and vineyards), primary treatment will consist of screening, grit removal (preliminary treatment), and sedimentation.

Treatment Flow Diagram 2, Conventional Activated Sludge

To satisfy fundamental secondary treatment requirements, a conventional activated sludge system coupled with chlorine disinfection will be employed. Reuse Alternatives 1c (pasture for milking animals), 2a (golf course, cemetery, freeway median, and greenbelt irrigation), 3a (restricted recreational impoundment), 3c (landscape impoundment), and 6a (livestock and wildlife watering) will require as a minimum secondary wastewater effluent.

Treatment Flow Diagram 3, Combined Trickling Filter and Activated Sludge

As a secondary treatment alternative, a combined biological process that links a high-rate biofilter and a suspended-growth aeration basin will be considered in a similar application to Treatment Flow Diagram 2.

Treatment Flow Diagram 4, Extended Aeration

Nitrification is typically the initial unit process in any nutrient management scheme. Although ammonia conversion alone is generally not a requirement for a specific reuse option, an extended aeration system will be analyzed nonetheless, because of its importance in a total nitrogen removal system. Also, extended aeration represents a popular derivative of the high-rate activated sludge process, particularly in the oxidation ditch mode. Because of the number of installations nationwide, extended aeration systems to be evaluated will include the ENVIREX Orbal™ process and EIMCO Carrousel™ process.

Treatment Flow Diagram 5, Secondary Treatment Plus Full Title 22 Facility

In accordance with the California DHS Title 22 criteria, wastewater oxidation (extended aeration), chemical addition, coagulation, sedimentation, filtration, and disinfection will be provided to produce essentially a pathogen-free effluent. Reuse Alternatives 1a (agricultural food crops), 2b (parks, playgrounds, and schoolyards irrigation), and 3b (nonrestricted recreational impoundments) will require the described level of treatment.

Treatment Flow Diagram 6, Secondary Treatment Plus Direct Filtration

As an alternative to a Full Title 22 Facility (Treatment Flow Diagram 6), direct filtration following secondary treatment (extended aeration) will be considered for Reuse Alternatives 1a, 2b, 3a, and 3b. In direct filtration, the tertiary sedimentation process is deleted with coagulation- flocculation occurring in a separate reactor immediately upstream of the filters.

Treatment Flow Diagram 7, Secondary Treatment Plus Contact Filtration

As a third option for satisfying the Title 22 criteria for the production of a pathogen-free effluent, upflow contact filtration (Parkson Dynasand™ deep bed sand filter) following secondary treatment (extended aeration) and chemical addition/coagulation will be analyzed also for Reuse Alternatives 1a, 2b, 3a, and 3b. Similar to the direct filtration alternative, the contact filtration process configuration does not include an intermediate clarification step. Unlike Treatment Flow Diagram 6, chemical addition/coagulation is accomplished through an in-line mixer mechanism with flocculation and aggregation occurring subsequently in the lower layers of the deep bed Dynasand™ filter.

Treatment Flow Diagram 8, Secondary Treatment—Contact Filtration–Phosphorus Removal

Combining a single tank nitrification-denitrification Orbal™ process with chemical addition and upflow contact filtration, complete nutrient management (nitrogen and phosphorus) will be assessed for Reuse Alternative 6b (fisheries habitat).

Treatment Flow Diagram 9, EIMCO Bardenpho™ Process

Biological nutrient removal in a compartmentalized treatment process will be evaluated as an alternative to contact filtration (Treatment Flow Diagram 8) for nitrogen and phosphorus reductions.

Treatment Flow Diagram 10, Secondary Treatment—Contact Filtration–Carbon Adsorption

For the groundwater recharge alternatives 4a and 4b (injection wells and surface spreading basins), nutrient and organics removal will be required. Addition of carbon adsorption to the contact filtration flow scheme will provide the desired effluent polishing.

Treatment Flow Diagram 11, Secondary Treatment—Contact Filtration–Carbon Adsorption-Reverse Osmosis

To produce the high quality demineralized water required for industrial reuse (Alternatives 5a, 5b, and 5c), reverse osmosis will be added to the contact filtration-carbon adsorption flow diagram (Treatment Flow Diagram 10).

Treatment Flow Diagram 12, Secondary Treatment—Lime Treatment–Reverse Osmosis

Patterned after the unit process configuration at Water Factory 21, lime treatment-reverse osmosis will be presented as an alternative to Treatment Flow Diagram 11 for the industrial reuse options.

DESIGN CRITERIA FOR WASTEWATER RECLAMATION FACILITIES

Following the selection of a treatment process flow diagram, the sizing of individual unit process components may be accomplished. Depending upon the specific unit process, criteria governing design for reclamation unit processes will vary. Recommended design parameters and numerical values for reclamation unit processes based upon typical raw wastewater characteristics are presented below. In addition, requirements for overall facility reliability, redundancy, and standby equipment are discussed in view of California DHS guidelines and desired operational control. An example is provided to illustrate the suggested design considerations involved in the preliminary sizing of reclamation treatment facilities. A detailed summary of design criteria is also provided to the reader to allow for "customization" in the development of facility plans and costs. Finally, guidelines are presented regarding the design of reclaimed water delivery systems.

DESIGN PARAMETERS AND NUMERICAL CRITERIA

In the design of a wastewater treatment and reclamation plant, raw wastewater characteristics and hydraulic-organic loading parameters must be considered. The potential deterioration in process performance under minimum and maximum conditions will determine the likely governing criteria to be utilized in the sizing of individual treatment components. A summary of typical wastewater composition and suggested criteria—numerical values for reclamation unit processes is presented in Tables 29.5, 29.6, and 29.7. Peaking factors referenced in Tables 29.6 and 29.7 are plotted as a function of average daily wastewater flow (ADWF) in Figure 29.13. Use of the data listed in Tables 29.5 and 29.6 is discussed subsequently in Example 1.

RECLAMATION FACILITY RELIABILITY PROVISIONS

For a reclamation facility, reliability is defined as the probability of successful performance of the facility in satisfying specific operating conditions and producing a minimum effluent quality. Reliability can be increased typically by improving reclamation plant influent quality, increasing the contaminant removal efficiencies of individual unit operations and processes, or adding supplemental treatment mechanisms in series [29]. Reclamation unit processes designed to produce a desired effluent quality consistently were described earlier in the previous section of this chapter. Once a treatment plant configuration has been selected, the reliability of the facility will then be a function of the number of treatment units, emergency power provisions, and the ability of an operator to monitor/manage the functions of the plant through instrumentation or alarms. A discussion of these reliability provisions follows below.

Standby Units

Because mechanical equipment should be serviced and overhauled routinely while concrete or steel structures must be drained and cleaned annually, individual treatment processes should include a sufficient number of units or standby equipment to compensate for system down-time. Facilities must be sized to accommodate periodic maintenance while maintaining average design loading conditions with one unit out of service. For reference, standby replacement equipment is defined as equipment that can be

TABLE 29.6. Summary of Governing Criteria Used in the
Sizing of Reclamation Unit Processes.

Unit Process or Operation	Governing Criteria in Process Sizing
Preliminary treatment	Performance during peak hour wastewater flow
Primary treatment (sedimentation)	Hydraulic loading rate during peak hour wastewater flow
Secondary treatment	
High-rate activated sludge	
Aeration basin volume	Detention time at ADWF
	Food to microorganisms (F/M) ratio at average daily organic loading
Aeration equipment	Maximum daily organic loading
Secondary clarification	Hydraulic loading rate during peak hour wastewater flow
	Solids loading rate at ADWF with 100% recycle
Oxidation ditch	
Ditch volume	Detention time at ADWF
	F/M ratio art average daily organic loading
Aeration equipment	Maximum daily organic loading
Secondary clarification	Hydraulic loading rate during peak hour wastewater flow
	Solids loading rate at ADWF with 100% recycle
Trickling filter	
Filter medium volume	Maximum daily organic loading
Coagulation, flocculation, sedimentation	
Chemical addition-flash mix	Detention time at peak hour wastewater flow
Flocculation	Detention time at peak hour wastewater flow
Sedimentation	Hydraulic loading rate at peak hour wastewater flow
Filtration	Filtration rate at peak hour wastewater flow with one unit out of service
Carbon adsorption	Hydraulic loading rate at peak hour wastewater flow
Lime treatment	Hydraulic loading rate at peak hour wastewater flow
Reverse osmosis	Hydraulic loading rate at peak hour wastewater flow
Chlorination	
Disinfection	Detention time at peak hour wastewater flow
Ultraviolet light	UV dose at maximum day wastewater flow

Source: Adapted from Reference [18].

1360

TABLE 29.7. Design Criteria for Reclamation Unit Operations and Processes.

Design Parameter	Unit	Value Range	Typical
Preliminary treatment:			
Aerated grit tank			
Detention time during peak hour wastewater flow	minute	3–5	3
Air delivered	ft^3/ft-minute	3–8	5
Primary treatment (sedimentation):			
Hydraulic loading rate during peak hour wastewater flow	gal/ft^2-d	1,500–2,500	2,000
Clarifier sidewater depth	ft	8–12	12
Secondary treatment:			
High-rate activated sludge			
F/M ratio	lb BOD$_5$/lb MLVSS-d	0.15–0.40	0.25
Aeration tank detention time	hr	4–8	6
Sludge age	day	2–8	6
Mixed liquor volatile suspended solids (MLVSS, aeration tank)	mg/L	1,000–3,500	2,500
Oxygen required	lb O$_2$/lb BOD$_5$ removed	0.8–1.2	1.0
Sludge production	lb solids/lb BOD$_5$ removed	0.4–0.8	0.6
Recirculation rate	%	30–100	50
Clarifier hydraulic loading rate during peak hour wastewater flow	gal/ft^2-d	800–1,000	1,000
Clarifier solids loading rate—ADWF with 100% recycle	lb solids/ft^2-d	10–30	20
Clarifier sidewater depth	ft	12–16	14
Oxidation ditch			
F/M ratio	lb BOD$_5$/lb MLVSS-d	0.05–0.15	0.10
Oxidation ditch detention time	hr	16–30	24
Sludge age	day	20–30	25
Organic loading rate	lb BOD$_5$/10^3 ft^3-d	10–15	15
MLVSS (oxidation ditch)	mg/L	1,500–5,000	4,000

TABLE 29.7. (continued).

Design Parameter	Unit	Value	
		Range	Typical
Oxygen required			
BOD₅ removal	lb O_2/lb BOD_5 removed	1.5–2.2	1.7
Nitrification	lb O_2/lb ammonia oxidized	4.6–6.0	4.7
Sludge production	lb solids/lb BOD_5 removed	0.65–1.20	0.80
Recirculation rate	%	100–300	100
Clarifier hydraulic loading rate during peak hour wastewater flow	gal/ft²-d	800–1,000	800
Clarifier solids loading rate—ADWF with 100% recycle	lb solids/ft²-d	10–30	20
Clarifier sidewater depth	ft	16–20	18
Trickling filter			
Organic loading rate			
Low rate	lb BOD_5/10³ ft³-d	25–40	30
Roughing filter	lb BOD_5/10³ ft³-d	100–200	150
Coagulation, flocculation, sedimentation:			
Chemical addition-flash mix			
Hydraulic detention time	s	0.5–5	1
Flocculation			
Hydraulic detention time	min	10–30	20
Velocity gradient (G)	s⁻¹	20–100	40
Mixing energy-detention time (Gt)		20,000–150,000	50,000
Sedimentation			
Clarifier hydraulic loading rate during peak hour wastewater flow	gal/ft²-d	800–1,000	800
Filtration:			
Filtration rate with one filter out of service or in the backwash mode	gal/ft²-min	4–6	5

TABLE 29.7. (continued).

Design Parameter	Unit	Value Range	Value Typical
Carbon adsorption:			
Hydraulic loading rate	gal/ft²-min	4–6	5
Empty bed contact time	min	20–30	30
Lime Treatment:			
Clarifier hydraulic loading rate during peak hour wastewater flow	gal/ft²-d	1,500–2,000	1,800
Reverse Osmosis:			
Hydraulic loading rate	gal/ft²-d	5–10	7.5
Disinfection:			
Chlorination			
Rapid mix detention time	s	1–2	1
Chlorine contact basin—contact time during peak wastewater flow[a]	min	120	120
UV disinfection			
UV dose during maximum day wastewater flow[b]	mW-s/cm²	100–120	100

[a]Contact time is based upon satisfying a discharge requirement of 2.2 total coliforms/100 mL. The contact time may be reduced to 30 min during peak hour wastewater flows if the discharge requirement is 23 total coliforms/100 mL.
[b]UV dose is based upon satisfying a discharge requirement of 2.2 total coliforms/100 mL. A minimum UV dose of 140 mW-s/cm² during maximum week flow is recommended to satisfy the coliform requirement for reclaimed water intended for unrestricted use.
Source: Adapted from References [7,10,14–15,18–28].

1363

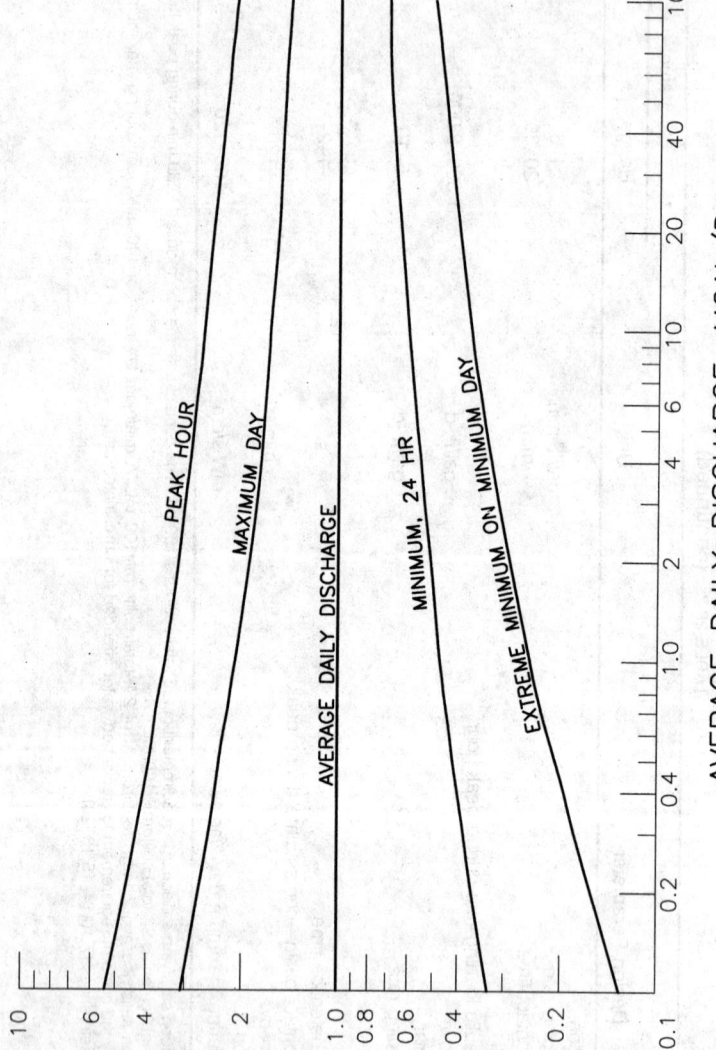

Figure 29.13. Wastewater peaking factors.

AVERAGE DAILY DISCHARGE, MGAL/D

RATIO OF INDICATED PARAMETER TO AVERAGE DAILY DISCHARGE

PEAK HOUR

MAXIMUM DAY

AVERAGE DAILY DISCHARGE

MINIMUM, 24 HR

EXTREME MINIMUM ON MINIMUM DAY

placed in operation within a 24 hr period [4]. Reclamation treatment systems for various reuse alternatives, described later in this chapter, include provisions for redundant or standby facilities that will allow for preventative maintenance without comprising effluent quality.

Emergency Power

Unless an influent wastewater stream can be diverted to an alternate treatment facility and reclaimed water deliveries may be curtailed temporarily, provisions for emergency electrical power must be included in a water reclamation facility [10]. A standby diesel generator and automatic transfer switch should be incorporated into the electrical power and control systems of the reclamation facility. Depending upon the required effluent quality and level of reliability desired, standby power systems should be sized only to serve essential electrical loads. Load shedding should be considered for non-critical areas with the ultimate objective to only maintain reclamation plant performance and reclaimed water quality at some minimum standard during a power outage. Economy of scale, however, should be considered in diesel generator selection and operation. Cost data for emergency power systems are included in a later section of this chapter.

Instrumentation and Alarms

To operate a reclamation facility reliably, plant personnel must be able to monitor process flows and characteristics conveniently. The design engineer in developing instrumentation and control strategies for a reclamation plant must, therefore, identify the key process parameters and optimum methods for monitoring. Alarms must be included to alert an operator to emergency conditions or poor effluent quality. Depending on operator preference, alarms and parameter measurement can be installed locally (at the respective equipment or unit process) or linked to a central monitoring station. Because chemical addition is an integral component of typical reclamation plants, sufficient instrumentation must be provided to allow for automatic chemical dosing either through flow pacing or in response to other system variables (e.g., chlorine residual, wastewater turbidity). Also in accordance with California DHS guidelines, automatic activation of supplemental treatment processes (chemical pretreatment) and flow by-passing or re-routing for further treatment should be possible in the event of unsatisfactory reclaimed water quality [10]. A typical list of water quality parameters, controls, and alarms for a reclamation plant designed to satisfy an effluent requirement of ≤ 2.2 total coliforms/100 mL is presented in Table 29.8 [30].

TABLE 29.8. **Recommended Process Parameters and Alarm Conditions to Be Monitored in a Water Reclamation Plant [28,30].**[a]

Category	Characteristic Monitored
Water quality parameters	Filter influent turbidity
	Filter effluent turbidity (composite and individual filters)
	Effluent chlorine residual
	Wastewater transmittance (UV systems)
Treatment process characteristics	Influent flow
	Effluent flow
	Filter backwash rate
	Filter surface wash rate
	UV intensity
	UV dose
Alarm conditions	High filter influent turbidity
	High filter effluent turbidity
	Chlorine contact tank bypass (high filter effluent turbidity)
	Low chlorine residual
	Low chemical supply
	High chlorine supply
	Multiple UV lamp failure
	Low UV intensity
	Low UV transmittance
	Pump/equipment failure
	Power failure
	Chemical feeder malfunction

[a]Limited to chemical addition-flocculation, sedimentation, filtration, and disinfection unit processes in reclamation facility.

From a review of Table 29.8, the number and potential complexity of alarms and controls required in a reclamation plant is apparent. Because of the emphasis placed on reliability and product quality, costs for instrumentation and alarms in a reclamation facility are generally twenty percent higher than in a conventional wastewater treatment plant.

EXAMPLE OF RECLAMATION PLANT SIZING (EXAMPLE 1)

A water reclamation plant is proposed to serve the master planned community of Lakeborough in western Stanislaus County in Northern California. Reclaimed water will be utilized throughout the development for landscape irrigation, including golf courses, street medians, parks, and playgrounds. The estimated wastewater flow to the reclamation plant is 2.0 Mgal/d (ADWF). The peak hour wastewater flow is projected at 7.0 Mgal/d using a peaking factor of 3.5 from Figure 29.13. Influent wastewater characteristics are assumed as 220 mg/L TSS, 220 mg/L BOD_5, and 40 mg/L total Kjeldahl nitrogen (TKN) (medium strength wastewater, Table 29.5). The maximum daily to average daily organic and nitrogen loading factors are assumed as 2.5 and 2.25 respectively, based on a historical analysis of similarly-sized treatment plants. To satisfy an effluent discharge requirement of 15 mg/L TSS, 15 mg/L BOD_5, and ≤ 2.2 total coliforms/ 100 mL, Treatment Flow Diagram 6 (secondary treatment plus direct filtration) has been selected for the reclamation facility configuration. Sizing of individual unit processes at the Lakeborough facility using criteria listed in Tables 29.6 and 29.7 is as follows:

(1) Preliminary treatment elements:
   (a) Influent pump station, wastewater grinders, mechanically-cleaned bar screen, vortex type grit removal tank, parshall flume
   (b) All units sized based on a peak hour wastewater flow = 7.0 Mgal/d

(2) Secondary treatment:
   (a) Extended aeration oxidation ditch design criteria:
   - hydraulic detention time at ADWF = 28 hr
   - organic loading rate = 15 lb $BOD_5/10^3$ ft^3-d
   - sludge production = 0.8 lb solids/lb BOD_5 removed
   - oxygen requirement = 1.7 lb O_2/lb BOD_5 removed and 4.7 lb O_2/lb ammonia oxidized
   (b) Extended aeration oxidation ditch sizing:
   - Provide two ditches, each capable of treating 1.0 Mgal/d. The hydraulic detention time in each ditch will be 28 hr.
   - Each ditch will have a volume of 1.2 Mgal and a mixed liquor concentration of 3,500 mg/L. Based on an average organic loading of 1,835 lb BOD_5/d per ditch, the F/M ratio is 0.05 lb BOD_5/lb MLVSS-d. Sludge production in each ditch will be 1,470 solids/d with a sludge age in the biological system of 24 days.
   - The oxidation ditch aeration system must be capable of responding to maximum daily organic loading. For BOD_5 removal and nitrification, the maximum daily loading to the system is 4,590 lb BOD_5/d and 1,500 lb TKN/d. The oxidation ditch aeration system should be sized to deliver 14,860 lb O_2/d (620 lb O_2/hr).
   (c) Secondary clarification design criteria:
   - hydraulic loading rate during peak hour wastewater flow = 900 gal/ft^2-d
   - solids loading rate during ADWF with 100% recycle = 15 lb solids/ft^2-d
   (d) Secondary clarifier sizing
   - Provide two 70 ft diameter clarifiers.

- The hydraulic loading rate at the clarifiers during the peak hour wastewater flow (assuming no recycle) will be 910 gal/ft²-d. The clarifier's maximum solid loading rate during peak hour wastewater flows will be 15.2 lb solids/ft²-d.

(3) Coagulation-flocculation:
   (a) Maximum flows through the tertiary treatment components will be limited to 2.0 Mgal/d. Excess secondary effluent flows will be routed to storage ponds or a flow equalization facility prior to additional treatment.
   (b) Rapid mix chamber hydraulic detention time at maximum flow = 1 s, volume = 25 gal.
   (c) Flocculator hydraulic detention time at maximum flow = 20 min. Provide two units each with a volume of 14,000 gal.

(4) Filtration:
   (a) Filtration rate with one filter out of service = 5.0 gal/ft²-min.
   (b) Provide four filter cells each with a surface area of 100 ft².

(5) Chlorination:
   (a) Rapid mix chamber hydraulic detention time flow = 1 s, volume = 25 gal.
   (b) Chlorine contact basin hydraulic detention time at maximum flow = 120 min. Provide two basins each with a volume of 85,000 gal.

DELIVERY SYSTEM DESIGN CONSIDERATIONS

Following treatment, reclaimed water is conveyed to potential reuse sites or customers through a distribution network. The delivery system consists of pumping stations, pipelines, and operational storage tanks, typically. Depending upon the application, reclamation demands may vary daily as well as seasonally. As illustrated in Table 29.9, irrigation requirements for landscape are greatest during the summer months when plant evapotranspiration rates are high and precipitation amounts are negligible. As an example, during the months of July and August, maximum day irrigation demands can be 1.5–2.0 times average day demands. If the delivery of reclaimed wastewater is also restricted to certain hours each day (10 P.M.–6 A.M.) because of the desire to limit public exposure, more water must be conveyed over a shorter period of time to satisfy the daily demand. Combining maximum day demands with delivery restrictions can generate peaking factors in the range of 4–6 times the average daily demand. A delivery system must be designed to accommodate this range of demands as summarized in Table 29.10.

TABLE 29.9. Seasonal Variations in Landscape Irrigation Requirements in Northern California [31].

Month	Percentage of Total Annual Requirement
January	0
February	0
March	0
April	5.2
May	14.5
June	20.4
July	22.4
August	20.4
September	13.6
October	3.5
November	0
December	0
Total	100.0

COST METHODOLOGY FOR WASTEWATER RECLAMATION FACILITIES

To maximize the usefulness of the data presented in this chapter and to allow for future use of this information by engineers and planners in evaluating the economics of water reclamation, a clear understanding of the methodology employed in the development of reuse costs is required. As a comparison to current predicted costs, previous cost studies and technical reports describing unit process parameters and examples of actual reclamation system costs are also highlighted in this section. Using the

TABLE 29.10. Design Criteria for Reclaimed Water Delivery Systems [32].

Delivery System Component	Design Criteria
Pipelines	
Governing criteria	Peak delivery conditions
Maximum velocity	10 ft/sec
Desirable velocity	3–5 ft/sec
Minimum velocity	1 ft/sec
Design head loss	10 ft/1,000 ft of pipe
Pump stations	
Governing criteria	Peak delivery conditions[a]
Minimum number of pumps	3
Minimum number of standby pumps	1
Operational storage	
Governing criteria	Peak delivery conditions
Minimum capacity	2/3 of peak day delivery demand

[a]Criteria may be revised to maximum day demand depending on the availability of operational storage.

historical information and an expanded data base, the rationale for the development of various facility component capital and operation and maintenance costs is presented. Finally, the method of analyzing both total reclamation systems linking a series of treatment processes as well as incremental reuse costs are described below.

BASIS FOR COST ESTIMATES

The development of a cost estimate includes projections of capital costs, annual operation and maintenance costs, and life cycle costs. Capital or construction costs are presented typically based on a standard cost index. In this chapter, life cycle costs are computed by combining amortized capital costs with annual operation and maintenance costs and converting the figure to $/ac-ft. Life cycle costs are useful in comparing the economic feasibility of alternative water resources projects over a specific time period (20 yrs as an example). In conjunction with other economic evaluations and revenue projections, an analysis of life cycle costs may be helpful in identifying the break-even point for a project or the requirement for various subsidies.

Description of Data Base

Wastewater reclamation system costs are developed in this chapter as a function of facility capacity, end-use option, and treatment process configuration. Costs are identified by estimating facility construction costs, equipment purchases, and operation and maintenance fees. Initially, reclamation systems are analyzed in terms of individual components based on design criteria presented previously. Cost data are derived for each element of a reclamation system at various capacity levels and unit sizes. The information is summarized in a series of cost tables found in the next section. Data are developed for reclamation systems ranging in production capacity from 1.0 Mgal/d to 10.0 Mgal/d. Additional descriptions of estimated capital and operating cost data are presented below.

Description of Reclamation System Capital Costs

Capital costs for reclamation system components are derived based on recent treatment plant and pipeline construction contract bid results, contractor quotations, vendor information, and published material prices. Major equipment costs reflect data supplied by several diversified manufacturers including ENVIREX, Walker-Process, Westech, EIMCO, Infilco-Degremont, DAVCO, Passavant-Zimpro, Trojan Technologies, Parkson, Lightnin, Wallace & Tiernan, Milton-Roy, Fairbanks-Morse, and

TABLE 29.11. Summary of Unit Costs Utilized in the Development of Reclamation System Capital Costs.

Item	Basis	Unit Cost
Grading	$/yd^3	5.00
Structural excavation/backfill	$/yd^3	10.00
Structural concrete		
Foundations	$/yd^3	400.00
Walls	$/yd^3	450.00
Slabs on grade	$/yd^3	350.00
Elevated slabs	$/yd^3	500.00
Grating	$/ft^2	40.00
Handrail	$/ft	50.00
Building cost	$/ft^2	150.00
Asphalt concrete paving	$/ft^2	2.00
Piping		
8 in. diameter	$/ft	35.00
10 in. diameter	$/ft	40.00
12 in. diameter	$/ft	45.00
18 in. diameter	$/ft	60.00
24 in. diameter	$/ft	80.00
30 in. diameter	$/ft	100.00
36 in. diameter	$/ft	120.00
42 in. diameter	$/ft	140.00
48 in. diameter	$/ft	160.00
Fencing	$/ft	10.00

WEMCO. Where appropriate, specific references to particular equipment are included in the cost tables referenced previously. Common unit costs for construction materials and activities employed in the development of system capital costs are also summarized in Tables 29.11 and 29.12. Site development and electrical system costs are assumed as 10 and 15 percent of the total facility costs, respectively.

Development of Reclamation System Operation and Maintenance Costs

Reclamation system annual costs are comprised of treatment and distribution facility personnel salaries, operating fees (recurring power and chemical costs), and maintenance costs (equipment repairs and replacements). Personnel requirements are a function of facility size and complexity. Operating costs depend upon energy usage and chemical consumption. Maintenance costs (spare parts, replacement) are estimated generally as a percentage of equipment first costs (i.e., 5 percent). For pipelines and storage tanks, maintenance costs are projected as two percent of capital costs. Each of these annual cost components are analyzed for various reclamation options based on the guidelines presented in the technical literature and actual operating experi-

TABLE 29.12. **Summary of Pump Station and Storage Tank Costs Utilized in the Development of Reclamation System Capital Costs [31].**

Delivery System Component		Capital Cost, $
Pump stations		
1.0 Mgal/d	TDH[a] = 20–50 ft	100,000
	TDH = 50–300 ft	200,000
	TDH > 300 ft	400,000
5.0 Mgal/d	TDH = 20–50 ft	250,000
	TDH = 50–300 ft	525,000
	TDH > 300 ft	1,100,000
10.0 Mgal/d	TDH = 20–50 ft	375,000
	TDH = 50–300 ft	850,000
	TDH > 300 ft	1,750,000
Reclaimed water storage tanks		
500,000 gal capacity		750,000
1,000,000 gal capacity		950,000
2,000,000 gal capacity		1,250,000
4,000,000 gal capacity		2,000,000

[a]TDH = total dynamic head.

ence [14,32–34]. Typical unit costs employed in the evaluations of these annual costs are enumerated in Table 29.13.

Use of Cost Indexes

Because construction costs vary over time, it is important to reference cost data to a specific cost index. In this chapter all capital costs are presented in June 1996 dollars, using an Engineering News Record Construction Cost Index (ENRCCI) of 5,600 (20 cities average). The ENRCCI is employed because of its general availability and frequent use by the United States Environmental Protection Agency (USEPA) as a cost estimating reference for wastewater treatment facilities. The use of the 20 cities average is recommended unless a specific city ENRCCI can be obtained. Costs summarized in the next section may then be revised as follows:

$$\text{Revised cost} = \frac{\text{Revised ENRCCI}}{5,600} \times \text{Cost cited in chapter}$$

Analysis of Total Reclamation System Life Cycle Costs

As noted previously, total reclamation system life cycle costs are estimated by combining amortized capital costs with annual operation and maintenance costs and converting the figure to $/ac-ft. The $/ac-ft figure is computed by dividing the estimated life cycle costs ($/yr) by the reclamation facility capacity (ac-ft/yr). The life cycle analysis is based on a 20-year facility life and a return rate of ten percent.

Capital costs include costs incurred for wastewater treatment, storage and distribution. Capital costs for wastewater treatment and water reclamation are estimated by summing individual unit process costs with required supporting facilities fees. Standby units are assumed for each unit process to ensure redundancy and reliability. Ancillary facility costs will reflect the following:

(1) Site development costs including grading, paving, fencing, and landscaping
(2) Process piping and liquid stream conduit costs linking reclamation facility unit processes
(3) Administration—operations building costs
(4) Standby generator costs

Operation and maintenance costs for a total reclamation system are analyzed in a similar manner to capital costs. Incremental costs are accrued and adjusted to reflect additional administrative and monitoring costs. Solids management costs are also added to the total reclamation system annual costs. Solids refer to primary sludges, waste secondary sludges, and tertiary chemical sludges. Because an analysis of sludge handling and disposal alternatives is beyond the scope of this chapter, a solids management cost of $150/dry ton of solids produced is assumed based on the results of a recent sludge management master plan study in Southern California [35]. The stated sludge "handling" cost reflects some type of

TABLE 29.13. Summary of Unit Costs Utilized in the Development of Reclamation System Operation and Maintenance Costs.

Item	Basis	Unit Cost
Plant operator	$/hr	18.00[a]
Utility maintenance worker	$/hr	15.00[a]
Electrical power	$/kWh	0.10
Chemicals		
Alum	$/lb	0.45
Polymer	$/lb	1.50
Chlorine	$/lb	0.20
Sulfuric acid	$/lb	0.50
Lime	$/lb	0.20
Sodium hexametaphosphate	$/lb	1.00
Sodium hypochloride	$/gal	0.80
Granular activated carbon replacement	$/lb	1.10

[a]Labor costs do not include fringe benefits, insurance, and administrative overhead.

solids digestion, sludge pumping, dewatering, and beneficial agricultural reuse application. The $150/dry ton figure includes amortized capital costs and operation and maintenance fees.

Analysis of Incremental Treatment Costs

In many instances, planning for major reclamation systems including complete advanced wastewater treatment—reclamation facilities is limited to new, master planned communities. More frequently, the upgrading of existing wastewater treatment plants with the addition of AWT components to produce reclaimed water for specific end-uses must be analyzed. These "incremental" treatment costs are evaluated by isolating the specific additional components required to ensure a certain water quality and analyzing their likely impacts on facility life cycle costs. Impacts to be considered include the need for additional chemical handling equipment, increased process monitoring requirements, incremental increases in energy and chemical consumption, and greater sludge handling costs. A typical cost analysis of a treatment plant upgrade and the associated increase in incremental facility costs is illustrated in Example 2.

Example of Incremental Cost Analysis (Example 2)

A 1.5 Mgal/d extended aeration, oxidation ditch treatment plant was constructed in 1985 to serve the residential community of Dos Ranchos in the San Joaquin Valley. Chlorinated secondary effluent from the treatment plant was conveyed to a series of holding ponds prior to pumping to a local farmer's spray irrigation system. The agricultural site was then used as pastureland for milking animals. The farmer recently sold the agricultural land to a developer who has planned a regional golf course and subdivision estates to be served entirely by reclaimed water produced by the Dos Ranchos wastewater treatment plant. The local Regional Water Quality Control Board (RWQCB) in conjunction with the California DHS has mandated a treatment plant upgrade however, prior to implementation of the proposed reuse program. In accordance with Title 22 criteria, a required plant effluent quality of ≤2.2 total coliforms/100 mL has been established by the RWQCB.

Following completion of a detailed pre-design study and pilot plant program, the feasibility of satisfying the plant discharge requirements with contact filtration using a Dynasand™ upflow sand filter has been demonstrated. The incremental life cycle treatment costs at the Dos Ranchos plant are estimated as follows:

(1) Advanced wastewater treatment (AWT) components:
    (a) Chemical addition, in-line mixing, coagulation

    (b) Three Dynasand™ filter cells, 100 ft^2 each, hydraulic loading rate = 5 gal/ft^2-min with one cell out of service

    (c) Expand chlorine contact tank, total detention time = 120 min

(2) Chemical usage:
    (a) Alum = 30 mg/L
    (b) Polymer = 1 mg/L
    (c) Chlorine = 10 mg/L

(3) Additional electrical loads:
    (a) Filter compressors, filter reject pumps = 50 horsepower (hp)
    (b) Chlorine mixer, chemical metering pumps = 5 hp

(4) Additional personnel-power requirements = 1/2 - time operator

(5) The treatment plant site was originally planned for the addition of tertiary treatment components. There is sufficient hydraulic fall through the plant to allow for placement of the in-line static mixer-filter-chlorine contact tank without additional wastewater pumping. Secondary wastewater effluent stored in the holding ponds can flow by gravity through the AWT plant elements to the existing plant effluent pumping station. Filtered, reclaimed water will be pumped to the proposed golf course.

(6) Incremental capital and operation and maintenance costs are summarized in Table 29.14.

(7) By combining the amortized capital costs and the annual operation and maintenance fees presented in Table 29.14, the annual incremental life cycle cost of the Dos Ranchos reclamation plant upgrade is computed. Using a capital recovery factor of 0.11745 (design life of 20 years, return rate of 10%), the amortized capital costs are $221,650/y ($1,887,000*0.11745) [36]. The total annual life cycle cost is then $333,000/yr ($221,650/yr + $111,350/yr). Based on an annual reclaimed water production rate of 1,680 ac-ft/yr (1.5 Mgal/d*365 d/yr), the incremental unit life cycle cost is $198/ac-ft ($333,000/yr/1,680 ac-ft/yr) for the proposed project.

(8) The $198/ac-ft cost can then be compared to costs associated with the development of a potable water source for golf course irrigation to determine, in general, the economic feasibility of the reclamation program.

SUMMARY OF PREVIOUS WATER RECLAMATION COST DATA

In the past, comprehensive literature has been presented documenting historical wastewater treatment and disposal costs. During the early years of the Clean Water Grant Program, the USEPA published capital

TABLE 29.14. Estimated Incremental Capital and Operation
and Maintenance Costs Associated with
Dos Ranchos Reclamation Plant Upgrade, Example 2.

Item	Cost, $
Capital costs:	
1. Chemical addition, flash mix, coagulation	125,000
2. Dynasand™ filter, air compressors, reject pumps	550,000
3. Chlorine contact tank, chlorine mixer, injector	350,000
4. Chemical handling improvements[a]	265,000
5. Process piping and site work	175,000
6. Electrical improvements	250,000
Subtotal	1,715,000
Contingency, 10%	172,000
Total estimated capital cost	1,887,000
Operation and maintenance costs:	
1. Energy	
a. Filter compressors, reject pumps[b]	13,060/yr
b. Chlorine mixer, chemical pumps[c]	3,270/yr
2. Chemicals	
a. Alum[d]	61,640/yr
b. Polymer[e]	6,850/yr
c. Chlorine[f]	9,130/yr
3. Personnel[g]	15,000/yr
4. Chemical sludge[h]	2,400/yr
Total annual operation and maintenance costs	111,350/yr

[a]Chemical handling improvements include enclosed alum storage tanks, alum metering pumps, polymer solution batching equipment, polymer metering pumps, chlorine ton cylinders, chlorinators.
[b]Filter: 50 hp $\times$ 0.746 kW/hp $\times$ 3,500 hr/yr $\times$ $0.10/kWh = $13,060/yr.
[c]Mixer: 5 hp $\times$ 0.746 kW/hp $\times$ 8,760 hr/yr $\times$ $0.10/kWh = $3,270/yr.
[d]Alum: 1.5 Mgal/d $\times$ 8.34 $\times$ 30 mg/L $\times$ $0.45/lb $\times$ 365 day/yr = $61,640/yr.
[e]Polymer: 1.5 Mgal/d $\times$ 8.34 $\times$ 1 mg/L $\times$ $1.50/lb $\times$ 365 day/yr = $6,850/yr.
[f]Chlorine: 1.5 Mgal/d $\times$ 8.34 $\times$ 10 mg/L $\times$ $0.20/lb $\times$ 365 day/yr = $9,130/yr.
[g]Personnel: 1,000 hr/yr $\times$ $15/hr = $15,000/yr.
[h]Chemical sludge: 16 ton/yr $\times$ $150/ton = $2,400/yr.

and operation and maintenance cost estimating curves for various unit processes and treatment facilities [37,38]. The wastewater treatment cost data base has since been embellished by several technical studies including those undertaken by Tchobanoglous [39–41], Benjes [42], and Scroggs [33] that have explored economies of scale and treatment costs as a function of facility capacity. In contrast, historically, the costs of water reclamation systems have not been the subject of extensive research. A brief description of earlier engineering investigations in this subject area follows.

Although water reuse has been practiced extensively in California only during the last two decades, reclamation costs were estimated in the late 1950s as part of planning studies for the California State Water Project.

Orlob and Lindorf projected reclamation costs at $10–$20/ac-ft for large plants and $25–$90/ac-ft for smaller facilities in 1958 [43]. Twenty years later one of the first efforts to develop a rationale basis for projecting reuse costs was technical information published in the report, "Wastewater Reuse and Recycling Technology" [14]. Eighteen beneficial reuse options were identified ranging from agricultural irrigation to groundwater recharge. Thirteen levels of wastewater treatment comprising twenty-four different unit process configurations were developed and analyzed subsequently in terms of life cycle costs. Assuming certain water quality requirements for each reuse alternative, specific treatment schemes were then linked with the respective beneficial use. A summary of the estimated reclamation costs from "Wastewater Reuse and Recycling Technology" [14] is presented in Table 29.15.

The wide variation in projected reuse costs both as a function of treatment level and facility capacity is shown clearly in Table 29.15. Life cycle costs (including sludge disposal but excluding storage and distribution costs) ranged from $110–$3,300/ac-ft depending upon specific reclamation requirements.

During the last ten years site specific costs have also been presented in the technical literature for several major reuse systems in California. Argo noted that treatment costs for groundwater recharge at Water Factory 21 (WF 21) in Orange County averaged $428/ac-ft for a 5.0 Mgal/d lime clarification reverse osmosis process and $480/ac-ft for a comparable filtration-carbon adsorption-reverse osmosis treatment configuration in 1983 [27]. Similarly, Young et al. analyzed reclamation costs for the Irvine Ranch Water District (IRWD) in 1987 [44]. Incremental tertiary treatment operational costs (chemical addition, filtration, solids treatment) were estimated to be only $74/ac-ft while distribution costs, administrative charges (accounting, monitoring, overhead), and replacement reserve fees were projected at $132, $50, and $47/ac-ft respectively for a total water reuse cost of $303/ac-ft at IRWD. A comparison of these site specific costs versus estimated treatment costs is presented in Table 29.16.

As indicated in Table 29.16, actual treatment costs have been somewhat lower than those predicted earlier, reflecting certain process optimizations and greater treatment efficiencies created by improved equipment. Process optimization in water reclamation plants is discussed in a later section of this chapter.

In evaluating reclamation system costs in Southern California, Hoover indicated that operation and maintenance (O&M) fees represented typically between 25–40 percent of overall reuse life cycle costs [45]. For reference, a break-down of operation and maintenance costs for the IRWD is included as Table 29.17. Based on a review of the data, the critical importance of

TABLE 29.15. Summary of Estimated Water Reclamation Treatment Process Life Cycle Costs, 1996 Dollars [14].[a,b]

	Reuse Alternative	Recommended Treatment Process	Annual Cost ($/ac-ft)[c,d]
(1)	Agricultural irrigation	Activated sludge	245–682
(2)	Livestock and wildlife watering	Trickling filter	268–711
(3)	Power plant and industrial cooling, once-through	Rotating biological contactors	379–728
(4)	Industrial water supply, primary metals		
(5)	Urban irrigation—landscape	Activated sludge, filtration of secondary effluent	291–903
(6)	Industrial water supply—petroleum and coal products		
(7)	Recreation—secondary contact		
(8)	Power plant and industrial cooling-recirculation	Tertiary lime treatment	404–1334
(9)	Industrial water supply—paper and allied products		
(10)	Recreation—primary contact		
(11)	Industrial boiler make-up—low pressure fisheries	Tertiary lime, nitrified effluent	412–1474
		Tertiary lime plus ion exchange	524–1525
(12)	Industrial water supply—chemicals and allied products	Activated sludge, filtered secondary effluent, carbon adsorption	388–1200
(13)	Industrial water supply—food and kindred products	Tertiary lime, carbon adsorption, ion exchange	623–1957
(14)	Industrial boiler make-up—intermediate pressure	Infiltration—percolation	108–260
(15)	Groundwater recharge—spreading basins	Activated sludge, filtration of secondary effluent, carbon adsorption, reverse osmosis of advanced wastewater treatment (AWT) effluent	1166–3271
(16)	Groundwater recharge—injection wells		

[a]Selected values from Summary Tables 1 and 2 in "Water Reuse and Recycling Technology" [14].
[b]Summarized costs reflect September 1977 dollars (ENRCCI = 2700) escalated to June 1996 dollars.
[c]Costs are estimated for facility capacities ranging from 1 Mgal/d to 50 Mgal/d. Lower cost figure within each treatment process category represents cost for a 50 Mgal/d reclamation plant while the upper cost limit is presented for a 1 Mgal/d facility.
[d]Annual costs include amortized capital costs based on a facility life of 20 years and a return rate of 7 percent.

TABLE 29.16. Comparison of Historical Estimated Life Cycle Reclamation Costs Versus Site Specific Treatment Costs.

Site	Treatment System	Life Cycle Water Reclamation Costs, $/ac-ft	
		Estimated[a]	Actual[b]
WF21	Filtration—carbon adsorption—reverse osmosis	921–2,589[c]	674
IRWD	Direct filtration	46–221[d]	157[e]

[a] Historical estimated water reclamation costs taken from Table 29.15.
[b] Actual costs presented in June 1996 dollars.
[c] Treatment cost derived by noting difference in costs between "Reverse Osmosis of AWT Effluent" and "Activated Sludge" in Table 29.13.
[d] Treatment cost derived by noting difference in costs between "Filtration of Secondary Effluent" and "Activated Sludge" in Table 29.15.
[e] Life cycle cost derived by combining operational cost and replacement reserve allocation.

labor and energy costs in a reuse treatment system is clear. Reclamation system operation and maintenance costs are explored further in the next two sections of this chapter.

WASTEWATER RECLAMATION FACILITY COSTS

Following the methodology presented previously, reclamation facility costs are developed for the twelve process flow diagrams described earlier. Capital, operation and maintenance, and life cycle costs are summarized in tabular and graphical form for reclamation plants with capacities of 1.0, 5.0, and 10.0 Mgal/d. Incremental treatment costs associated with the upgrade of existing wastewater treatment plants to satisfy the Title 22 criteria are highlighted in this section. Distribution system costs are also described along with a comparison of alternative disinfection systems.

TABLE 29.17. Breakdown of Reclamation System Operation and Maintenance Fees for the Irvine Ranch Water District [44].

Cost Component	Percent of Total O&M Cost	
	Treatment	Distribution
Wages and benefits	45	22
Energy	17	44
Chemicals	15	—
Maintenance	—	14
Monitoring	—	9
Miscellaneous	5	11
Total	100	100

RECLAMATION FACILITY TREATMENT COSTS

Capital and operation and maintenance costs for the following treatment process flow diagrams, described previously, are presented in Tables 29.18 and 29.19.

(1) Primary treatment
(2) Conventional activated sludge
(3) Combined biofilter—activated sludge
(4) Extended aeration
(5) Secondary treatment plus full Title 22 facility
(6) Secondary treatment plus direct filtration
(7) Secondary treatment plus contact filtration
(8) Secondary treatment—contact filtration—phosphorus removal
(9) Bardenpho™ process
(10) Secondary treatment—contact filtration—carbon adsorption
(11) Secondary treatment—contact filtration—carbon adsorption—reverse osmosis
(12) Secondary treatment—lime treatment—reverse osmosis

Facility capital costs include construction costs for individual unit processes, Operations-Laboratory Building, Maintenance Building, site development, process-yard piping, instrumentation, electrical distribution-controls (Motor control centers (MCC), power and control wiring), and electrical service (switchgear and standby power systems). Operation and maintenance costs include personnel charges, power costs, spare parts, chemicals, and sludge handling fees.

LIFE CYCLE COSTS FOR RECLAMATION TREATMENT FACILITIES

Life cycle costs are computed by combining amortized capital costs with operation and maintenance costs. Amortization of capital costs is based upon a design life of 20 yr and a return rate of 10 percent. The combined life cycle annual costs ($/yr) are converted to a $/ac-ft figure through normalization (i.e., dividing by the reclamation facility output in ac-ft/yr). Life cycle costs for wastewater reclamation facilities are summarized in Table 29.20.

INCREMENTAL TREATMENT COSTS

Reuse Alternatives 1a (agricultural food crops), 2b (parks, playgrounds, and schoolyard irrigation), and 3b (nonrestricted recreational im-

TABLE 29.18. Estimated Capital Costs for Reclamation Treatment Facilities [46].

	Treatment Flow Diagram	Capital Cost, $[a]		
		1.0 Mgal/d	5.0 Mgal/d	10.0 Mgal/d
1.	Primary treatment	2,950,000	5,300,000	7,550,000
2.	Conventional activated sludge	6,100,000	14,400,000	24,900,000
3.	Combined biofilter-activated sludge	6,500,000	15,200,000	26,100,000
4.	Extended aeration	5,700,000	13,200,000	24,950,000
5.	Secondary treatment plus full Title 22 facility	8,400,000	18,400,000	35,300,000
6.	Secondary treatment plus direct filtration	6,900,000	15,700,000	30,000,000
7.	Secondary treatment plus contact filtration	7,050,000	16,650,000	30,900,000
8.	Secondary treatment—contact filtration—phosphorus removal	7,100,000	18,100,000	34,500,000
9.	Bardenpho™ process	7,600,000	20,800,000	38,200,000
10.	Secondary treatment—contact filtration—carbon adsorption	9,050,000	25,550,000	49,350,000
11.	Secondary treatment—contact filtration—carbon adsorption—reverse osmosis	13,450,000	43,800,000	84,150,000
12.	Secondary treatment—lime treatment—reverse osmosis	12,100,000	35,450,000	65,450,000

[a]Costs presented in June 1996 dollars (ENRCCI = 5,600).

TABLE 29.19. Estimated Operation and Maintenance Costs for Reclamation Treatment Facilities [46].

	Treatment Flow Diagram	Operation and Maintenance Cost, $/yr[a]		
		1.0 Mgal/d	5.0 Mgal/d	10.0 Mgal/d
1.	Primary treatment	150,000	530,000	960,000
2.	Conventional activated sludge	270,000	930,000	1,730,000
3.	Combined biofilter-activated sludge	300,000	1,060,000	1,990,000
4.	Extended aeration	300,000	1,030,000	1,950,000
5.	Secondary treatment plus full Title 22 facility	520,000	1,960,000	3,810,000
6.	Secondary treatment plus direct filtration	350,000	1,200,000	2,290,000
7.	Secondary treatment plus contact filtration	340,000	1,200,000	2,280,000
8.	Secondary treatment—contact filtration—phosphorus removal	660,000	2,730,000	5,340,000
9.	Bardenpho™ process	280,000	1,040,000	2,120,000
10.	Secondary treatment—contact filtration—carbon adsorption	820,000	3,400,000	6,680,000
11.	Secondary treatment—contact filtration—carbon adsorption—reverse osmosis	1,240,000	4,960,000	9,600,000
12.	Secondary treatment—lime treatment—reverse osmosis	1,060,000	4,230,000	8,100,000

[a]Costs presented in June 1996 dollars (ENRCCI = 5,600).

TABLE 29.20. Summary of Reclamation Facility Life Cycle Costs.

		Life Cycle Costs, $/ac-ft[a]		
	Flow Diagram Number—Title	1.0 Mgal/d	5.0 Mgal/d	10.0 Mgal/d
1.	Primary treatment			
	Capital[a]	309	112	80
	Operation and maintenance	134	96	86
	Total	443	208	166
2.	Conventional activated sludge			
	Capital[a]	644	303	262
	Operation and maintenance	242	168	154
	Total	886	471	416
3.	Combined biofilter-activated sludge			
	Capital[a]	687	319	275
	Operation and maintenance	273	191	179
	Total	960	510	454
4.	Extended aeration			
	Capital[a]	600	279	262
	Operation and maintenance	269	184	175
	Total	869	463	437
5.	Secondary treatment plus full Title 22 facility			
	Capital[a]	886	388	371
	Operation and maintenance	465	351	342
	Total	1,351	739	713
6.	Secondary treatment plus direct filtration			
	Capital[a]	726	331	316
	Operation and maintenance	314	215	206
	Total	1,040	546	522

1383

TABLE 29.20. (continued).

Flow Diagram Number—Title	Life Cycle Costs, $/ac-ft[a]		
	1.0 Mgal/d	5.0 Mgal/d	10.0 Mgal/d
7. Secondary treatment plus contact filtration			
Capital[a]	742	350	326
Operation and maintenance	310	215	205
Total	1,052	565	531
8. Secondary treatment—contact filtration—phosphorus removal			
Capital[a]	748	382	363
Operation and maintenance	594	489	479
Total	1,342	871	842
9. Bardenpho™ process			
Capital[a]	802	439	403
Operation and maintenance	244	179	181
Total	1,046	618	584
10. Secondary treatment—contact filtration—carbon adsorption			
Capital[a]	953	539	529
Operation and maintenance	731	610	600
Total	1,684	1,149	1,129
11. Secondary treatment—contact filtration—carbon adsorption—reverse osmosis			
Capital[a]	1,415	922	886
Operation and maintenance	1,109	889	859
Total	2,524	1,811	1,745
12. Secondary treatment—lime treatment—reverse osmosis			
Capital[a]	1,273	745	690
Operation and maintenance	945	757	726
Total	2,218	1,502	1,416

[a]Capital costs are amortized based on a facility life of 20 years and a return rate of 10 percent.

poundment), described in the first section, require essentially a pathogen-free effluent as defined under Title 22 (i.e., reclaimed wastewater suitable for unrestricted use). To upgrade secondary wastewater treatment plants to produce the desired product quality, chemical addition, coagulation, filtration, and chlorination facilities must be provided. The advanced wastewater treatment (AWT) elements included in Treatment Flow Diagrams 5 (Full Title 22 Facility), 6 (Direct Filtration), and 7 (Contact Filtration) will satisfy the Title 22 criteria when constructed downstream of typical secondary process components. Upgrading costs for each filter alternative are summarized in Table 29.21.

Full Title 22 Facility

Following secondary treatment, chemical addition, coagulation, sedimentation, filtration, and disinfection would be provided under the Full Title 22 AWT alternative. Filtration would be accomplished utilizing traveling bridge filters. Capital costs and operation and maintenance costs for three facility capacities are summarized in Tables 29.21 and 29.22. Combining the initial capital investment with annual expenses, incremental life cycle costs are computed and presented in Figure 29.14. As depicted in Figure 29.14, upgrading costs for the Full Title 22 alternative range from $274–$490/ac-ft.

Direct Filtration

As an equivalent AWT process, assuming good secondary effluent quality (i.e., turbidity less than 10 NTU), the direct filtration option would include chemical addition, coagulation, filtration with Hydro Clear Pulsed Bed™ filters, and chlorine disinfection. Capital costs and operation and maintenance costs for 1.0, 5.0, and 10.0 Mgal/d plants are enumerated in Tables 29.21 and 29.22. Life cycle costs for the direct filtration option as summarized in Table 29.23 vary from $96–$217/ac-ft.

Contact Filtration

Chemical addition, in-line mixing/coagulation, filtration with Parkson Dynasand™ upflow sand filters, and chlorine disinfection is presented as an equivalent second alternative to the Full Title 22 facility. A summary of the estimated capital investment and annual operating and maintenance costs for the contact filtration option is presented in Tables 29.21 and 29.22. Life cycle costs for the alternative range from $101–$221/ac-ft as illustrated in Figure 29.14.

TABLE 29.21. Estimated Incremental Capital Costs Associated with a Wastewater Treatment Plant Upgrade Designed to Produce Reclaimed Water Suitable for Unrestricted Use per Title 22 [46].

AWT Process Beyond Secondary Treatment	Capital Cost, $[a]		
	1.0 Mgal/d	5.0 Mgal/d	10.0 Mgal/d
Full Title 22	2,650,000	5,050,000	9,550,000
Direct filtration	1,450,000	2,950,000	5,450,000
Contact filtration	1,550,000	3,650,000	6,100,000

[a]Costs presented in June 1996 dollars (ENRCCI = 5,600).

Distribution System Costs

Dual distribution system costs will vary considerably depending upon site specific conditions such as proposed or existing land use, reuse area density, (i.e., are the reuse sites together), proximity of the reclamation plant to the application areas, and topography. To illustrate, amortized capital costs for non-potable water distribution networks for proposed master planned communities in the Central Vally range from $150–$175/ac-ft [47]. However, capital costs for dual distribution system retrofits in urbanized areas in Sacramento County are estimated at $400–$600/ac-ft [32]. Because of these potential variations as illustrated in Table 29.24, the development of standardized cost data for dual distribution systems is difficult. For order of magnitude estimating purposes, therefore, the following life cycle costs are offered for reclaimed water distribution systems: (1) new systems—$175/ac-ft and (2) retrofits—$350/ac-ft.

COMPARISON OF ALTERNATIVE DISINFECTION COSTS

Historically, gaseous chlorine has served as the principal disinfectant in wastewater reclamation plants. The low cost of gaseous chlorine coupled

TABLE 29.22. Estimated Operation and Maintenance Costs Associated with a Wastewater Treatment Plant Upgrade Designed to Produce Reclaimed Water Suitable for Unrestricted Use per Title 22 [46].

AWT Process Beyond Secondary Treatment	Operation and Maintenance Cost, $/yr[a]		
	1.0 Mgal/d	5.0 Mgal/d	10.0 Mgal/d
Full Title 22	240,000	970,000	1,930,000
Direct filtration	70,000	210,000	420,000
Contact filtration	65,000	215,000	400,000

[a]Costs presented in June 1996 dollars (ENRCCI = 5,600).

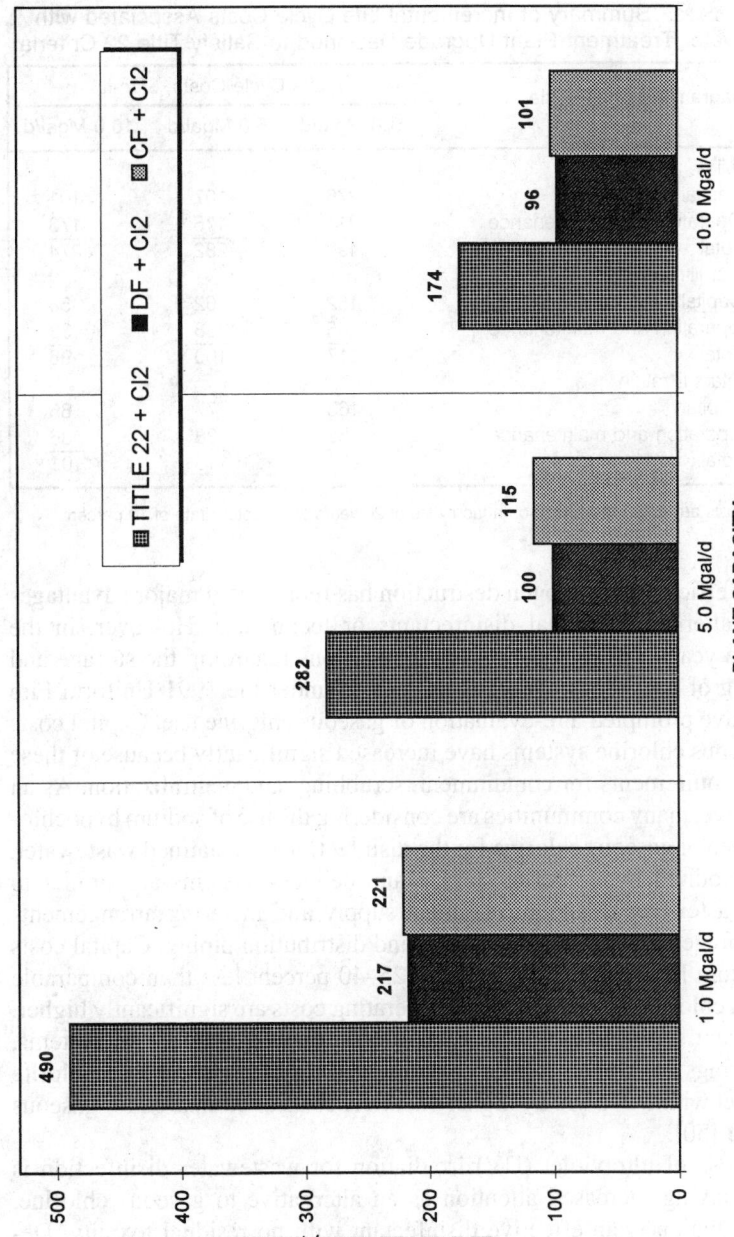

Figure 29.14. Incremental life cycle costs for tertiary treatment systems designed to satisfy Title 22.

TABLE 29.23. Summary of Incremental Life Cycle Costs Associated with Wastewater Treatment Plant Upgrade Designed to Satisfy Title 22 Criteria.

Flow Diagram Number—Title	Life Cycle Costs, $/ac-ft[a]		
	1.0 Mgal/d	5.0 Mgal/d	10.0 Mgal/d
5. Full Title 22 facility			
Capital[a]	276	107	101
Operation and maintenance	214	175	173
Total	490	282	274
6. Direct filtration			
Capital[a]	152	62	58
Operation and maintenance	65	38	38
Total	217	100	96
7. Contact filtration			
Capital[a]	163	77	65
Operation and maintenance	58	38	36
Total	221	115	101

[a]Capital costs are amortized based on a facility life of 20 years and a return rate of 10 percent.

with its efficacy in pathogen destruction has represented major advantages versus alternate chemical disinfectants or techniques. However, in the last five years, safety concerns and restrictions regarding the storage and handling of hazardous materials as defined under the 1991 Uniform Fire Code have prompted a re-evaluation of gaseous chlorine use. Capital costs for gaseous chlorine systems have increased significantly because of these recent requirements for containment, scrubbing, and neutralization. As an alternative, many communities are considering the use of sodium hypochlorite in lieu of gaseous chlorine for the disinfection of reclaimed wastewater. Liquid sodium hypochlorite storage and delivery systems are similar to other wastewater treatment chemical supply and metering arrangements with storage tanks, metering pumps, and distribution piping. Capital costs for sodium hypochlorite systems are 25–40 percent less than comparable gaseous chlorine systems, however operating costs are significantly higher. In a recent study of life cycle costs for alternative disinfection systems, any savings in capital costs associated with liquid sodium hypochlorite are offset within 3–5 yrs by higher chemical costs as compared for gaseous chlorine [50].

The use of ultraviolet (UV) irradiation for wastewater disinfection is also receiving increased attention as an alternative to gaseous chlorine. UV irradiation is an effective disinfectant with no residual toxicity. Depending on wastewater characteristics and effluent quality requirements, significant capital and operating cost savings could be realized with a UV system compared to a conventional chlorination-dechlorination process [51]. Additional full-scale operating experience with UV irradiation pro-

TABLE 29.24. Examples of Projected Reclaimed Water Distribution System Costs for Urbanized Areas in California.

Location	Projected Reclaimed Water Deliveries (ac-ft/yr)	Projected Distribution System Costs, $/ac-ft			Comment
		Capital[a]	Operation and Maintenance	Total	
Manteca, California[b]	2,700	305	40	345	Flat terrain
Sacramento, California[c]	5,700	325	120	445	Scattered users
Moulton Niguel, California[d]	6,800	85	125	210	System expansion

[a] Capital costs are amortized based on a facility life of 20 years and a return rate of 10 percent.
[b] Adapted from Reference [48].
[c] Adapted from Reference [32].
[d] Adapted from Reference [49].

ducing various qualities of reclaimed water is required, however, to confirm trends in long-term cost comparisons with gaseous chlorine.

CONCLUSIONS

The primary objective of this chapter has been to present a detailed methodology for estimating reclamation costs. Twelve process flow diagrams have been identified and evaluated as a function of potential reuse alternatives. Design criteria have been summarized for specific unit processes used in reclamation plants. Reclamation costs have been developed for various production capacities and end-use options. A summary of these reclamation costs is presented as Table 29.25. Specific conclusions are highlighted below.

(1) The production of reclaimed water for Reuse Alternative 1b (fodder, fiber, and seed crops; orchards and vineyards) requires the lowest level of treatment and generates the least reclamation costs. Life cycle costs for primary treatment range from $443 to $166/ac-ft for facilities with a capacity of 1 and 10 Mgal/d respectively.

(2) Secondary effluent suitable for Reuse Alternative 1c (pasture for milking animals), 2a (golf course, cemetery, freeway median, and greenbelt irrigation), and 3c (landscape impoundment) is produced from either the conventional activated sludge, activated biofilter, or extended aeration process configurations. Capital costs for the extended aeration alternative are 10–15 percent lower than similar costs for the conventional activated sludge and activated biofilter options. Life cycle costs are comparable for extended aeration and conventional activated sludge systems.

(3) Effluent from three advanced wastewater treatment configurations can satisfy the Title 22 criteria for reclaimed water suitable for unrestricted use. Secondary effluent that is treated through either a Full Title 22 facility, direct filtration plant, or contact filtration process can be utilized for agricultural food crops; parks, playgrounds, and schoolyard irrigation; and nonrestricted recreational impoundments.

(4) Incremental capital costs for a Full Title 22 facility are 1.5–2.0 times as great as those associated with either a direct filtration or contact filtration alternative. The higher capital costs are the result of the requirement for tertiary sedimentation equipment and additional chemical handling facilities inherent with the Full Title 22 option.

(5) Incremental operation and maintenance costs for a direct filtration

TABLE 29.25. Summary of Reclamation Treatment Costs
for Various Reuse Alternatives.

Reuse Designation	Flow Diagram Number	Reclamation Treatment Cost, $/ac-ft[a,b]		
		1.0 Mgal/d	5.0 Mgal/d	10.0 Mgal/d
1a	5	1,351	739	713
	6	1,040	546	522
	7	1,052	565	531
1b	1	443	208	166
1c	2	886	471	416
	3	960	510	454
2a	2	886	471	416
	3	960	510	454
2b	5	1,351	739	713
	6	1,040	546	522
	7	1,052	565	531
3a	5	1,351	739	713
	6	1,040	546	522
	7	1,052	565	531
3b	5	1,351	739	713
	6	1,040	546	522
	7	1,052	565	531
3c	2	886	471	416
	3	960	510	454
4a	10	1,684	1,149	1,129
4b	10	1,684	1,149	1,129
5a	11	2,524	1,811	1,745
	12	2,218	1,502	1,416
5b	11	2,524	1,811	1,745
	12	2,218	1,502	1,416
5c	11	2,524	1,811	1,745
	12	2,218	1,502	1,416
6a	2	886	471	416
	3	960	510	454
6b	8	1,342	871	842
	9	1,046	618	584

[a]Life cycle costs include amortized capital costs based on a facility life of 20 years and a return rate
of 10 percent.
[b]Reclamation treatment costs for Flow Diagram 4 are:
1. 1.0 Mgal/d: $869/ac-ft
2. 5.0 Mgal/d: $463/ac-ft
3. 10.0 Mgal/d: $437/ac-ft

or contact filtration process configuration are 20–30 percent of
the annual costs incurred with a Full Title 22 facility. The large
difference in operation and maintenance costs is due primarily to
higher chemical usages associated with a Full Title 22 plant.

(6) Certain economics of scale may be realized for reclamation plants
in the 5 to 10 Mgal/d capacity range. Incremental unit capital

costs for small reclamation plants (capacity less than 1.0 Mgal/d) are 2–2.5 times as great as larger facilities. Because ancillary equipment costs do not increase linearly with plant size, the cost of supporting facilities is a major component for small reclamation plants, thereby contributing to the disproportionately high unit costs.

(7) Life cycle costs for the direct filtration and contact filtration alternatives vary less than 10 percent. Specific equipment characteristics and other factors (filter head loss, reliability under varying influent loads, operator preference, compatibility with existing equipment, flexibility, and aesthetics) will govern ultimate system selection.

(8) Incremental treatment costs for the additional treatment processes beyond secondary treatment required to produce reclaimed water suitable for unrestricted use will range from $96–$221/ac-ft depending upon facility capacity for either a contact filtration or direct filtration process configuration.

(9) Nutrient management (nitrogen and phosphorus removal) may be accomplished through utilization of either Treatment Flow Diagram 8 (Contact Filtration—Phosphorus Removal) or 9 (EIMCO Bardenpho™ Process). Capital costs for the Contact Filtration—Phosphorus Removal alternative are 10–20 percent less than comparable costs for the Bardenpho™ process. Life cycle costs for the Bardenpho™ alternative, however, are 30 percent lower than the Contact Filtration option because of high chemical costs associated with Treatment Flow Diagram 8.

(10) The production of reclaimed water for groundwater recharge will require a high degree of treatment including nutrient removal, filtration, and carbon adsorption. Reclamation costs for the contact filtration–carbon adsorption process configuration vary from $1,129–$1,149/ac-ft.

(11) Reclamation costs for demineralized wastewater vary from $1,416–$2,524 ac-ft using either the Contact Filtration—Carbon Adsorption—Reverse Osmosis or the Lime Treatment—Reverse Osmosis alternatives.

REFERENCES

1 California Water Code, Division 7, Chapter 2, Section 13050, Definitions.
2 Crook, J. and D. A. Okun, 1988. "The Place of Nonpotable Reuse in Water Management," *Journal Water Pollution Control Federation,* 59:5, 237–241.

3 Miller, K. J., June 1987. "Water Reuse Practices in the United States," Current Status and Future Trends presented at the *Annual Conference of the American Water Works Association,* Kansas City, Missouri.

4 State of California Department of Health Services, Sanitary Engineering Section, California Administrative Code, Title 22, Division 4, Environmental Health, 1978. "Wastewater Reclamation Criteria."

5 Nichols, A. B., 1988. "Water Reuse Closes Water-Wastewater Loop," *Journal Water Pollution Control Federation,* 60:11, 1931–1937.

6 Asano, T. and K. L. Wassermann, 1980. "Groundwater Recharge Operations in California," *Journal American Water Works Association,* 72:7, 380–385.

7 Water Pollution Control Federation, 1983. "Water Reuse, Manual of Practice, SM-3, Systems Management."

8 Crook, J., June 1987. "Wastewater Reuse for Potable and Nonpotable Purposes: A Regulatory Viewpoint," presented at the *Annual Conference of the American Water Works Association,* Kansas City, Missouri.

9 State of California Department of Health Services, "Guidelines for Use of Reclaimed Water."

10 State of California Department of Health Services, Environmental Management Branch, June 1988. "Policy Statement for Wastewater Reclamation Plants with Direct Filtration."

11 State of California Department of Health Services, May 1988. "Proposed Policy and Guidelines for Groundwater Recharge with Reclaimed Municipal Wastewater."

12 Correspondence from Harvey F. Collins, Chief, Environmental Health Division, Department of Health Services to Roger J. Dolan, General Manager-Chief Engineer, Central Contra Costa Sanitary District, May 1988. "Use of Filtered Reclaimed Water for Industrial Cooling."

13 Williams, R. B., 1982. "Wastewater Reuse—An Assessment of the Potential and Technology," *Water Reuse,* Middlebrooks, E. J., editor, Ann Arbor Science Publishers, Inc., Ann Arbor, Michigan.

14 Culp, G., Wesner, G. M., Williams, R., and M. Hughes, 1980. "Wastewater Reuse and Recycling Technology," Noyes Data Corporation, Park Ridge, New Jersey, 1980. (Office Water Research & Technology, OWRT/RU-79-1,2).

15 Metcalf and Eddy, Inc., 1979. *Wastewater Engineering, Treatment, Disposal, Reuse,* Second Edition, revised by G. Tchobanoglous and F. L. Burton, McGraw-Hill Book Company, New York City, New York.

16 Culp, R. L., Wesner, G. M., and G. L. Culp, 1978. *Handbook of Advanced Wastewater Treatment,* Second Edition, Von Nostrand Reinhold Company, New York City, New York.

17 Sanitation Districts of Los Angeles County, June 1977. "Pomona Virus Study," prepared for the California State Water Resources Control Board.

18 Metcalf and Eddy, Inc., 1990. *Wastewater Engineering, Treatment, Disposal, Reuse,* Third Edition, revised by G. Tchobanoglous and F. L. Burton, McGraw-Hill Book Company, New York City, New York.

19 Tchobanoglous, G., 1974. "Wastewater Treatment for Small Communities," *Public Works,* 105:7, 61–68.

20 Lakeside Equipment Corporation, "Suggested Design Guidelines for the Oxidation Ditch Modification of the Extended Aeration Process."

21 Harrison, J. R., Daigger, G. T., and G. T. Filbert, October 1983. "A Survey of Combined Trickling Filter and Activated Sludge Processes" presented at the *Annual Conference of the Water Pollution Control Federation,* Atlanta, Georgia.

22 Asano, T., Tchobanoglous, G. T., and R. C. Cooper, 1984. "Significance of Coagulation-Flocculation and Filtration Operations in Wastewater Reclamation and Reuse," *Proceedings of Water Reuse Symposium III, Future of Water Reuse,* American Water Works Association Research Foundation, San Diego, California, pp. 1,185–1,204.

23 Kirkpatrick, W. R. and T. Asano, 1986. "Evaluation of Tertiary Treatment Systems for Wastewater Reclamation and Reuse," *Water Science Technology,* 18:10, 82–95.

24 Treweek, G. P., 1979. "Optimization of Flocculation Time Prior to Direct Filtration," *Journal Water Pollution Control Federation,* 71:2, 96–101.

25 Horne, F. W., Anderton, R. L., and F. A. Grant, 1980, "Water Reuse: Markets and Project Costs in Southern California" presented at the *Annual Conference of the American Water Works Association,* Atlanta, Georgia.

26 Report of the Scientific Advisory Panel on Groundwater Recharge with Reclaimed Wastewater, prepared for the State of California State Water Resources Control Board, Department of Water Resources, and Department of Health Services, 1987.

27 Argo, D. G., 1984, "Use of Lime Clarification and Reverse Osmosis in Water Reclamation," *Journal Water Pollution Control Federation,* 56:12, 1,236–1,246.

28 National Water Research Institute, September 1993. "UV Disinfection Guidelines for Wastewater Reclamation in California and UV Disinfection Research Needs Identification," prepared for the State of California Department of Health Services.

29 McCarty, P. L., Argo, D. G., and M. Reinhard, 1979. "Reliability of Advanced Wastewater Treatment," *Proceedings of Water Reuse Symposium I, Water Reuse— From Research to Application,* American Water Works Association Research Foundation, Washington, D.C., pp. 1,249–1,268.

30 Siemak, R. C., 1984, "Tertiary Filtration: Practical Design Considerations," *Journal Water Pollution Control Federation,* 56:8, 944–949.

31 Brown and Caldwell in association with Nolte and Associates, July 1994. "Water Reclamation Master Plan," prepared for the El Dorado Irrigation District, Placerville, California.

32 Nolte and Associates in association with HYA Consulting Engineers, Bookman Edmonston Engineering, Inc., August 1994. "Sacramento County Water Reclamation Study," prepared for the Sacramento Regional County Sanitation District.

33 Scroggs, J., 1979. "Capital and Operation and Maintenance Costs for Small Municipal Wastewater Collection, Treatment, and Disposal Systems," Unpublished M.S. Thesis, University of California, Davis, California.

34 Department of Industrial Engineering and Engineering Research Institute, Iowa State University, June 1973. "Estimating Staffing and Cost Factors for Small Wastewater Treatment Plants Less Than 1 MGD," EPA Grant No. 5P2-WP1950452.

35 Metcalf & Eddy in association with Nolte and Associates, August 1986. "San Diego Sludge Management Program Master Plan Report," prepared for the City of San Diego Water Utilities Department.

36 Grant, E. L., Ireson, W. G., and R. S. Leavenworth, 1976. *Principles of Engineering Economy,* Sixth Edition, The Ronald Press Company, New York City, New York.

37 Dames & Moore, February 1978. *Analysis of Operations and Maintenance Costs for Municipal Wastewater Treatment Systems,* EPA 430/9-77-015.

38 Sage Murphy & Associates, June 1983. *Construction Costs for Wastewater Treatment Plants: 1973–1982,* EPA 430/9-83-004.

39 Mills, R. A. and G. Tchobanoglous, 1976. "Energy Consumption in Wastewater Treatment," *Proceedings for the Seventh National Agricultural Waste Management Conference,* Ann Arbor Science Publishers, Ann Arbor, Michigan.

40 Tchobanoglous, G., 1974. "Wastewater Treatment for Small Communities," *Public Works,* 105:7, 61–68.

41 Tchobanoglous, G., 1974: "Wastewater Treatment for Small Communities, Part II— Special Design Considerations," *Public Works,* 105:8, 58–62.

42 Benjes, H. H., 1980. *Handbook of Biological Wastewater Treatment, Evaluation, Performance, and Cost,* Garland Publishing, Inc., New York City, New York.

43 Orlob, G. T. and M. R. Lindorf, 1958. "Cost of Water Treatment in California," *Journal American Water Works Association,* 50:1, 45–55.

44 Young, R. E., Lewinger, K., and R. Zenk, 1988. "Wastewater Reclamation—Is It Cost Effective? Irvine Ranch Water District—A Case Study," *Proceedings of Water Reuse Symposium IV, Implementing Water Reuse,* American Water Works Association Research Foundation, Denver, Colorado, pp. 55–64.

45 Hoover, M. G., October 13, 1988. "The Cost of Water Reclamation" presented at *Symposium III—Water Reclamation,* San Diego, California.

46 Richard, D., Tchobanoglous, G., and T. Asano, November 1992. "The Cost of Wastewater Reclamation in California," Department of Civil and Environmental Engineering, University of California—Davis, Davis, California.

47 Mingee, T. J. and Richard, D., April 1991. "Planning for Reclamation in the Central Valley" presented at the *Annual Conference of the California Water Pollution Control Association,* Pasadena, California.

48 Nolte and Associates, January 1996. "Water Reuse Investigation," prepared for the City of Manteca Department of Public Works.

49 Nolte and Associates, 1991. "Reclaimed Water Market Assessment and Facility Analysis" prepared for the Southeast Regional Reclamation Authority, Laguna Hills, California.

50 Richard, D., Scroggs, L., Anderson, M., and T. J. Mingee, March 1996. "Using Sodium Hypochlorite for Disinfection of Reclaimed Wastewater: Design and Operational Issues" presented at the *Water Environment Specialty Conference,* Disinfecting Wastewater for Discharge and Reuse, Portland, Oregon.

51 Water Environment Federation, 1995. "Comparison of UV Irradiation to Chlorination: Guidance for Achieving Optimal UV Performance."

CHAPTER 30

Legal Aspects of Water Reclamation

INTRODUCTION

RESOURCE managers across the nation are turning to water reclamation as a means of either stretching limited supplies or lessening
the cost of compliance with waste discharge requirements. Water
reclamation and reuse, as we know it today, is a highly regulated
activity. The California Legislature set the stage for regulation of
water reuse in 1969 and has alone enacted over 100 statutes pertaining
to reclaimed water [1]. State administrative agencies have complemented this body of law with supporting regulations, plans, policies,
and guidelines directed toward the administration and management
of this resource.

The basic structure for federal regulation of reclaimed water has
been in place since the adoption of the Federal Water Pollution
Control Act of 1972. Since that time, however, the water industry
has gained a better appreciation of the finite nature of the nation's
water resources. Periodic drought conditions coupled with an enlight-
ened understanding of the impact water management decisions have
on the aquatic environment have helped shape the state and federal
water reclamation policies. A growing trend today among the states
is to include within the state's plan for the management of its water
resources legislative policies favoring the beneficial use of reclaimed
water.

Gordon Cologne, Justice (ret.), California Court of Appeal, 48511 Via Amistad, La
Quinta, CA 92253; Peter M. MacLaggan Esq., WateReuse Association of California,
4021 Liggett Drive, San Diego, CA 92106.

Legislation and regulations have a major impact on many aspects of a water reclamation program, particularly important in the areas of discharges into water ways, public health, environmental protection, and permitting of various aspects of these activities. Additionally, local ordinances play an important role in managing the use of reclaimed water. The purpose of this chapter is to provide the reader with a general overview of the legal considerations associated with water reclamation. While specific aspects of the materials that follow may not be applicable in all jurisdictions, the discussion will provide the reader with a basic understanding of the primary legal and regulatory authority, with particular emphasis on California law.

HISTORICAL PERSPECTIVE

Concern for the quality of water dates back to the Roman Empire when aqueducts were constructed to bring water into areas devoid of water acceptable for public use. The notion of protecting water quality was not developed until the 1800's when disease was found to be associated with bacteria most easily detected in water. Massachusetts established a Board of Health in 1869 to respond to a finding that water-borne bacteria were contributing to epidemic diseases. Shortly thereafter, other diseases were traced to water pollution, including cholera, typhoid, and dysentery.

The United States (U.S.) government's role in exerting some control over water pollution began in the early 20th century. Initial efforts were primarily to control water-borne disease related to disposal of human wastes. In 1901, the U.S. Public Health Service (PHS) was created. In 1912, the PHS's authority was broadened to include control over pollution of surface water. However, the states were given primary enforcement over water quality standards.

In 1915, California created the Bureau of Sanitary Engineers and required all suppliers of drinking water to obtain permits for their operation. The Bureau had no enforcement powers and no meaningful standards were put in place. The only protection the public had at that time was through the Department of Fish and Game, which had the authority to prohibit discharges that adversely affected fish. If there was any real control, it was with the local agencies.

As late as 1940 many cities were discharging untreated or minimally treated wastewater into the streams, bays, and ocean. The Navy was discharging wastewater from ships directly into the bays. Even after World War II the Navy refused to stop the practice pointing to the expense of taking the vessels out to sea for the disposal.

After the War, the public became more concerned about pollution. There was a public outcry on finding that the shellfish in San Francisco

Bay were contaminated and quarantined. Miles of coastline near El Segundo were shut down because of the grease on the beaches.

While wastewater treatment had been used for years and was practiced in the United States as early as 1880, it was primarily a pretreatment process incidental to discharge. But the concept of treatment and reuse began to gain support not only as a new water source but also for its nutrient value to agriculture. By 1926, the federal government had constructed a wastewater reclamation plant at Grand Canyon National Park in Arizona and used the reclaimed water to irrigate the lawns, cooling water and boiler-feed water at a power plant. In 1929, the City of Pomona, California, began providing treated wastewater to irrigate lawns, gardens, and agriculture. Orange County, California, was using primary effluent for irrigation of agricultural crops in the 1930s.

In 1949, the California Legislature passed the Dickey Act, which created nine regional boards, to establish specific water quality standards under the direction of a central board. In 1967, the Department of Public Health was directed to "establish statewide standards for each direct use of reclaimed water where such use involves the protection of public health." While the use of reclaimed water was growing and reclamation was a common practice, real enforcement of the standards was wanting.

After a two-year study by the California Legislature and various state and local agencies, a need was established for a more comprehensive program for setting standards and enforcement of the standards for the public protection. Taking the lead in legislation was Carley Porter, chairman of the Assembly Water Committee, and Gordon Cologne, chairman of the Senate Water Resources Committee. In 1969, the Legislature enacted the statute which bears their names, the Porter Cologne Water Quality Act [1]. The Act expanded the supervisory and appellate powers of the water quality control boards and required the establishment of specific water quality objectives and the plans for achieving those goals. Most importantly, it strengthened the roles of the state and regional boards by requiring any agency reclaiming, or proposing to reclaim, water to file a report with the appropriate regional board. The regional boards are responsible for enforcement of the criteria established by the Department of Public Health now contained in the Title 22 regulations.

Congress followed California's lead by enacting the 1972 amendments to the Federal Water Pollution Control Act, which recognized water reclamation as a valuable adjunct to wastewater treatment. Today, water reclamation is now an important component of the water management plans and policies of many states. Federal, state, and local governments are very much aware of the importance of pollution control and the role each has in its development and regulation.

FEDERAL AUTHORITY GOVERNING RECLAIMED WATER

A fundamental principle of water policy in the U.S. is that the federal government must respect the primary role that individual states have in shaping and controlling their own policies regarding water use and allocation. Federal statutes, as interpreted by the U.S. Supreme Court, recognize the plenary authority of the states to enact laws governing the use of waters within the state in accordance with the needs of its citizens [2]. The rationale for a passive federal policy with respect to water is that the individual states are in a better position to assess the needs of their citizens and balance competing interests. However, the federal government's deference to state law on water policy matters is not without limitations. State policies on water use and allocation are subject to prohibitions against interference with powers reserved to the federal government under the U.S. Constitution.

Congress, on the other hand, may enact laws in furtherance of federal powers which substantially interfere with state water policy. For example, Congressional acts directed at protecting the environment frequently interfere with state primacy over water. Congress derives its power to enact far-reaching environmental legislation from the Constitutional provision granting the federal government the power to regulate navigable waters and interstate commerce: "Congress shall have the power to . . . regulate Commerce . . . among the several States . . . and to make all laws which shall be necessary and proper for carrying into execution the foregoing powers" [3].

Federal laws governing water reclamation are rather broad and general in nature. Significant congressional acts directly concerned with water reuse are currently limited to the Federal Water Pollution Control Act, the Reclamation Wastewater and Groundwater Studies and Facilities Act of 1992, and the Reclamation Recycling and Water Conservation Act of 1996.

FEDERAL CLEAN WATER ACT

The principal federal law pertaining to water reclamation is the Federal Water Pollution Control Act, commonly referred to as the Clean Water Act (CWA) [4]. The legislative origin of the CWA stretches back to the Rivers and Harbors Act of 1899. The 1972 CWA dramatically changed the balance of federal and state responsibilities for water quality management. Ten major bills subsequently revised the 1972 statute. Today, the CWA is one of the United States most sweeping environmental laws. It represents a major federal policy and financial commitment to make all surface waters "fishable and swimmable."

The CWA regulates discharge of pollutants into navigable waters through permits issued pursuant to the National Pollution Discharge Elimination System (NPDES). Primary jurisdiction under the CWA is with the United States Environmental Protection Agency (EPA), but in all but a few states, the CWA is administered and enforced by the state water pollution control agencies. Under the CWA, the term ''navigable waters'' means waters of the United States [5]. The federal courts follow the Tenth Circuit's conclusion that this definition is an expression of congressional intent ''to regulate discharges made into every creek, stream, river or body of water that in any way may affect interstate commerce'' [6].

The goal of the CWA is to ''restore and maintain the chemical, physical and biologic integrity of the nation's waters.'' The CWA sets forth specific goals to conserve water and reduce pollutant discharges and directs the U.S. EPA Administrator to assist with the development and implementation of water reclamation plans which will achieve those goals. Major objectives of the CWA are to eliminate all discharges of pollutants into navigable waters, stop discharges of toxic pollutants in toxic amounts, develop waste treatment management plans to control sources of pollutants, and to encourage water reclamation and reuse.

Several bills to amend the CWA were introduced in Congress between 1993 and 1995. Most of these bills included provisions to strengthen national policy with respect to water reclamation. An important provision for the western states was the proposed amendment directing the Administrator to establish water quality standards for ephemeral streams supplied primarily by reclaimed water. For reasons unrelated to the water reclamation provisions, none of the proposed amendments have been passed by Congress.

FEDERAL FUNDING PROGRAMS

The federal interest in the funding of water reclamation programs stems from a long history of federal participation in flood control and supply development projects. More recently, federal interest stems from a recognition of the importance of water reclamation as part of a long-term, comprehensive approach to management of the nation's water resources and environmental protection. The growing federal interest in water reclamation prompted Congress to authorize federal participation in feasibility studies, research, demonstration projects, and construction programs directed at furthering water reclamation.

State Revolving Fund

The 1987 amendments to the CWA eliminated the federal construction grants program, replacing it with a $20 billion revolving fund loan program.

The State Revolving Fund (SRF) loan program is administered by the states under general U.S. EPA guidance and regulations. The SRF is jointly funded by the federal government (83%) and state matching money (17%). The SRF is intended to provide funding in perpetuity for the design and construction of eligible facilities. States may establish eligibility criteria within the limits of the CWA. Water reclamation and reuse facilities which further the goals of the CWA are eligible for SRF funding under the SRF [7].

Under the SRF, states may enter into low-interest loans to local agencies. Interest rates are set by the states at or below market rates and may be as low as 0%. The amount of such loans may be up to 100% of the cost of eligible facilities. Loan repayment begins one year after completion of construction and the debt must be retired within twenty years. Repayment of principal and interest is returned to the state SRF for reissuance to eligible projects. California participates in this program [8].

Water Reclamation Grant Program

In response to the growing national interest in water reclamation, Congress directed the U.S. Bureau of Reclamation (BuRec) to promote water reclamation as a viable option to expand western water resources [9]. Under Title XVI of Public Law 102-575 (Reclamation Wastewater and Groundwater Studies and Facilities Act of 1992), the BuRec, through the Secretary of Interior, is authorized to conduct appraisal and feasibility studies on water reclamation and reuse projects and has general authority to conduct research and demonstration to further water reclamation technologies. Title XVI also gives the BuRec authority to participate in the design and construction of five specific water reclamation projects in California and Arizona, with federal contributions of up to 25% of total eligible cost. Federal appropriations to carry out the purposes of Title XVI have been on the order of $35 million per year.

In 1996, Congress amended Title XVI authorizing BuRec participation in the design and construction of an additional 15 water reclamation projects in California, Nevada, Utah, New Mexico and Texas [10]. Congress maintained the 25% maximum federal contribution and established a $20 million cap for federal contributions to each of the newly authorized projects. This legislation requires completion of an appraisal investigation and a feasibility study, completed by the Secretary of the Interior or the non-federal project sponsor, prior to appropriation of funds for construction.

Passage of the additional authorizations has increased the pressure on BuRec to fulfill funding expectations of the local project participants. BuRec is developing a methodology for establishing funding priorities for the limited federal resources available to advance water reclamation [11].

STATE LAW GOVERNING RECLAIMED WATER USE

The production and use of reclaimed water is carefully controlled through state laws and administrative regulations. In most jurisdictions a state agency, such as the State Engineer, Water Resources Control Board or Department of Health Services, is responsible for the adoption of regulations for the use of reclaimed water. The reuse of municipal wastewater is typically limited by state regulation to areas that can safely accommodate reclaimed water without adversely affecting public health or environmental quality. Chapter 16 provides an in-depth review of state regulation of water reclamation.

STATE POLICIES ENCOURAGING WATER RECLAMATION AND REUSE

Many state legislatures have enacted laws designed to encourage water reclamation and reuse. Collectively, these laws create a favorable climate for undertaking the water reclamation projects necessary to carry the state into the 21st century. In setting forth their water management strategy, the states contemplated that local public agencies would be responsible, in part, for implementing the water reclamation programs. Accordingly, numerous provisions were added to the state codes to encourage local entities to consider water reclamation as an alternative means of offsetting the state's growing demand for water.

Consider for example the following statute enacted by the State of Florida [12]:

> The Legislature finds that, due to a combination of factors, vastly increased demands have been placed on natural supplies of fresh water, and that, absent increased development of alternative water supplies, such demands may increase in the future. The Legislature also finds that potential exists in the state for the production of significant quantities of alternative water supplies, including reclaimed water, and that water production includes the development of alternative water supplies, including reclaimed water, for appropriate uses. It is the intent of the Legislature that utilities develop reclaimed water systems, where reclaimed water is the most appropriate alternative water supply option, to deliver reclaimed water to as many users as possible through the most cost-effective means, and to construct reclaimed water system infrastructure to their owned or operated properties and facilities where they have reclamation capability. It is also the intent of the Legislature that the water management districts which levy ad valorem taxes for water management purposes should share a percentage of those tax revenues with

water providers and users, including local governments, water, wastewater, and reuse utilities, municipal, industrial, and agricultural water users, and other public and private water users, to be used to supplement other funding sources in the development of alternative water supplies.

The California Legislature also has made a practice of enacting laws designed to encourage water reclamation and reuse [13]. The Legislature has declared that the people of the state have a primary interest in developing water reclamation facilities to supplement existing surface and groundwater supplies [14]. A substantial portion of the future water requirements of the state may be economically met by beneficial use of reclaimed water [15]. It is the intent of the Legislature and the state to undertake all possible steps to encourage development of water reclamation facilities [16]. One such step taken was the adoption of goals for the beneficial reuse of water. The aim of the Legislature is to beneficially reuse 700,000 afy (2,400,000 m³/d) by the year 2000 and 1 million afy (3,400,000 m³/d) by the year 2010 [17]. About 500,000 afy (1,750,000 m³/d) of reclaimed water is now used every year in California.

PROHIBITION AGAINST WASTE AND UNREASONABLE USE OF WATER

Seven western states (Arizona, California, Colorado, Idaho, Montana, New Mexico, and Wyoming) have codified provisions within their state constitutions which impose a strict requirement of beneficial use as a condition on the exercise of water rights [18]. Other states have enacted similar provisions within their statutory scheme for managing water [19]. California's prohibition of waste or unreasonable use of water expressly precludes the use of potable water where reclaimed water is determined to be available for that use [20]. The California law requires water users, water suppliers, and waste dischargers to assume responsibility for the development of reclaimed water. Through a combination of statutes and regulations the state requires consideration be given to water reclamation opportunities anywhere there is a significant demand for non-potable water or there is a substantial discharge of once-used water to the ocean.

Accommodating Changing Needs for Water

The California Constitution imposes dual requirements of reasonable use and beneficial use on the exercise of all water rights in California [21]:

> It is hereby declared that because of the conditions prevailing in the State the general welfare requires that the water resources of the State be put to beneficial use to the fullest extent of

which they are capable, and that the waste or unreasonable use or unreasonable method of use of water be prevented, and that the conservation of such waters is to be exercised with a view to the reasonable and beneficial use thereof in the interest of the people and for the public welfare.

Under this rule of "reasonable use," it is not enough that a use of water serves a beneficial purpose. The use must also be reasonable. Further, what constitutes a reasonable use of water is always subject to reconsideration. Since the 1930's and continuing through the present, the California courts have allowed public entities and public interest groups to use the reasonable use mandate to compel water rights holders to re-examine and modify their uses in order to accommodate competing interests for water.

The California courts have construed Article X, Section 2, to require that all water rights be exercised in accordance with contemporary economic conditions and social values [22]. Reasonableness is not a static concept. As changes in economic conditions and societal needs give rise to new demands for water, the California Water Resources Control Board (State Board) is under an obligation to balance competing uses [23].

Water Reuse Mandate

Responding to drought, increasing population, and environmental constraints on development of traditional water supplies, the California Legislature enacted Water Code Sections 13550–13555.3 to clarify that water suppliers and individual water users are not guaranteed the right to use potable water for non-potable applications. Such use is not to be tolerated if reclaimed water meeting certain conditions is available. The list of prohibited uses of potable water includes all forms of irrigation, non-residential toilet flushing, industrial supply, and make-up water for cooling towers and air conditioning units. Private property owners may be required to plumb new construction to accommodate future availability of reclaimed water.

Administrative Enforcement

The State Board is vested with jurisdiction to conduct proceedings to adjudicate claims of waste and unreasonable use of potable water. To establish a waste and unreasonable use of potable water the State Board must find: (1) reclaimed water of adequate quality is available; (2) at a reasonable cost to the user; (3) the use will not be detrimental to public health; (4) downstream water rights will not be adversely affected; (5) water quality will not be degraded; and (6) plants, fish and wildlife will not be injured [24]. Where reclaimed water is available meeting these criteria, suppliers and individual users of water are constitutionally and statutorily mandated to use reclaimed water for their nonpotable applications in lieu of potable water.

The rule of reasonable use has fostered the promulgation of regulations by the State Board which require a periodic review of permits and licenses. The purpose of the review is to make a determination whether changing conditions may have converted an otherwise beneficial use of water into a waste of water. The periodic review applies water rights permits and wastewater discharge permits.

Water right permittees and licensees are required to report periodically on the potential to use reclaimed water for all or part of their needs [25]. The State Board, when acting on a water right permit application, may reduce the amount of water requested and require the applicant to adopt a water reclamation program [26]. Waste discharge permits are subject to renewal every five years. Where an applicant in a water-short area proposes a discharge of once-used wastewater to the ocean, it is the policy of the State Board to require the applicant to submit to the Regional Water Quality Control Board a complete report on the potential for water reuse with an explanation as to why the water is not being recycled for further beneficial use [27].

Case Studies

The State Board has reviewed claims of waste and unreasonable use involving reclaimed water. While these decisions have not been reviewed by the courts, they are instructive as to the obligations of water customers, suppliers, and waste dischargers to participate in the development and use of reclaimed water. As a general rule, inherent in the concept of obtaining maximum beneficial use of the state's waters is the idea that a user may be required to incur some reasonable costs or incur some inconvenience to prevent waste of water. The rulings of the State Board which follow illustrate what reasonable costs or inconveniences must be incurred to prevent waste of water. In reviewing these decisions, however, it is important to recognize that what constitutes a reasonable use of water will change as the water supply situation may change.

Consumer's Obligation to Use Reclaimed Water

The petition of the City of Santa Barbara brought against the Montecito country club clarified the consumer's obligation to use reclaimed water under Section 13550 [28]. This case involved a situation where a municipality was attempting to compel one of its larger consumers, a country club, to use reclaimed water for irrigation of a private golf course. Principal issues in dispute were: (1) the suitability of the water for the intended use; and (2) the City's refusal to indemnify the country club from liability arising from the use of reclaimed water. The State Board relied on expert testimony in making a determination that the use of reclaimed water would not adversely affect the fairways on the golf course. Less certain about the sensitivity of the greens, the State Board instructed the parties to conduct a demonstration test to determine whether the greens could be irrigated with reclaimed water. As for the country club's demand for indemnification, the State Board found that the statute does not expressly require a supplier of reclaimed water to give indemnification assurances to users. Since no evidence was offered demonstrating that the cost of defending, settling or paying adjudicated claims arising out of the use of reclaimed water would make the cost of using reclaimed water greater than the cost of using potable water, the State Board was unable to find any legal or factual basis for allowing the country club to continue to receive potable water for landscape irrigation.

The Water Supplier's Obligation to Make Reclaimed Water Available

In 1987, the San Gabriel Water Company initiated an action against the Sanitation District No. 2 of Los Angeles County in Los Angeles Superior Court pursuant to the Service Duplication Law [29]. The Company sought to recover alleged service duplication losses claimed to result from the District's provision of reclaimed water within the Company's potable water service area. Subsequently, the District filed the complaint against the Company with the State Board alleging that, due to the availability of reclaimed water, the Company's use of potable water for greenbelt irritation is a waste and unreasonable use of water.

The State Board concluded in 1990 that reclaimed water, satisfying the conditions of Section 13550, was available for greenbelt irrigation. In reaching this decision, the State Board found that the District was supplying, and could supply additional reclaimed water, at a cost that is less than, or comparable to, the cost of potable water from the Company. The Company's potable water rate at the time ranged from $144 to $280 per acre foot. The District was charging between $9 and $17 per acre foot for reclaimed water and users bear the cost of transporting the water to their places of use. The State Board concluded that the cost of reclaimed water is likely to be comparable to or less than the cost of potable water even if the Company were to assume responsibility for reclaimed water distribution within its service area.

In 1993, the Superior Court held that the service of reclaimed water by the District to customers within the Company's service area constitutes a taking of property of the Company for a public purpose which is compensable service duplication under the Service Duplication Law. The court went on to find that the Water Reclamation Law does not provide the District with any defense to the claim of the Company for compensable service duplication under the Service Duplication Law.

This decision has not been reviewed by an appellate court. Thus, the questions as to whether the Service Duplication Law applies to reclaimed water remains open. However, in 1995, the California Legislature enacted a statute providing that payment of just compensation will not apply where the duplication of service involves an entity's self use of reclaimed water [31]. Self use is limited to the premises of a water reclamation plant or a landfill owned by the entity operating the water reclamation plant. The scope of this change in the law is limited to Los Angeles County.

Wastewater Discharger's Obligation to Make Reclaimed Water Available

In 1984, the State Board considered a complaint filed by the Sierra Club to enjoin an unreasonable use of water by a wastewater discharger [32]. At issue was a permit issued by the Regional Board authorizing the Fallbrook Sanitary District to discharge up to 1.6 mgd (6000 m³/d) of treated wastewater to the ocean. The Sierra Club alleged that under the circumstances, the discharge of the District's wastewater to the ocean where it cannot be recovered for beneficial use was a waste of water.

Before a wastewater discharger can be required to reclaim water, a determination must be made whether the particular discharge constitutes a waste or unreasonable use of water. Water Code Section 13550, with its focus on prohibiting the use of potable water for nonpotable applications, provided no guidance to the State Board in this instance. Thus, in making its determination, the State Board sought guidance from the state's constitutional prohibitions on waste and related case law.

In keeping with the case law which indicates that a reasonable use of water today may be a waste of water at some time in the future, the State Board ordered the District and all future applicants proposing a discharge of once-used water into the ocean to evaluate the feasibility of reclaiming its wastewater. The State Board insisted that water reclamation be carefully analyzed as an alternative, or partial alternative, to the discharge of once-used wastewater to the ocean in all water-short areas of the state. In adopting its order, the State Board recognized the requirements were consistent with the Board's authority to conduct investigations and prevent waste of water [33].

WATER QUALITY CONTROL

The CWA gives states the responsibility for developing water quality standards. However, the U.S. EPA reserves the right to approve or disapprove state standards. Water quality standards adopted by the states and approved by the U.S. EPA become the basis for setting permit limits for discharges of reclaimed water. Generally, state and federal laws are similar in that the fundamental purpose of both is to protect the beneficial uses of water. An important distinction between the two is that the states typically address both ground and surface waters while the Clean Water Act addresses surface water only. The state laws direct various administrative agencies to adopt regulations, plans, policies, and procedures to carry out the legislative intent of the state and federal laws pertaining to water reclamation [34].

LOCAL LAWS AND ORDINANCES

Typically, the state legislature, through enabling statutes, will authorize its political subdivisions responsible for water supply to acquire, store, deliver and sell reclaimed water for any authorized beneficial use [35]. Additionally, it may be necessary, or desirable, for the local water district to adopt ordinances to carry out its responsibilities under state law and to facilitate the implementation of its water reclamation program. Under its basic police powers, the local government has broad authority to regulate water reclamation activity in order to protect the public health, safety, and general welfare.

RIGHT TO CONTROL RECLAIMED WATER

An important consideration arising early on in the development of any water reclamation plan is the establishment of the right to use the reclaimed water. A water right is a legal entitlement authorizing water to be diverted from a specified source and put to beneficial use. Water rights are property rights but their holders do not own the water itself, they possess the right to use it. The exercise of some water rights requires a permit or license from the state. The objective of the state permitting system is to ensure that water resources are put to the best possible use and that the public interest is served. Water right law in the western states is markedly different from the laws governing water use in the eastern states.

APPROPRIATIVE WATER RIGHTS

The appropriative right system evolved as the preferred method of allocating water rights among the western states because of constraints on the availability of water. The "first in time, first in right" principle is an important feature of the modern appropriative water right. Under this system, water is allocated by the state on a first come, first serve basis. The first party to use the water has the most senior claim to that water. The senior user has a continued right to non-wasteful beneficial use of the water that is not to be diminished in quantity or quality by a junior user. This assures the senior right holder an adequate supply of water under almost any climatic condition. The junior right holder may be subject to a reduction in supply during times of shortage.

RIPARIAN WATER RIGHTS

The riparian water rights system is found primarily in the eastern states. Riparian rights usually attach to a parcel of land that is adjacent to a source of water. A riparian right entitles the landowner to use a correlative share of the water flowing past his or her property. In times of shortage, landowners along the stream receive a proportional share of the available supply. Generally, riparian rights do not require permits, licenses, or government approval and apply only to water which would naturally flow in the stream. A riparian right does not entitle a water user to divert water to storage in a reservoir for use in the dry season or to use water on land outside of the watershed. Riparian rights remain with the property when it changes hands, however, parcels severed from the adjacent water source generally lose their right to the water.

Seasonal, geographic, and quantitative differences in precipitation across California was the motivating factor causing that state to adopt a system of water rights that includes a unique blend of both appropriative and riparian water rights [36].

OWNERSHIP AND CONTROL OF RECLAIMED WATER

As the competition for limited water resources increases, urban, agricultural, and environmental interests are turning to water reclamation as a means of shoring up reliability. In many areas of the western U.S., treated wastewater represents by far the largest block of unappropriated water. Consequently, it is important to understand whom among the discharger, water supplier, other appropriators, and environmental interests retains the right to ownership and control of the reclaimed water.

The actual right of the downstream water user of reclaimed water will depend on the nature of the state's water allocation system. Some states issue permits to the owners of reclaimed water or to appropriators of it when discharged into a natural water course. These states granting permits to the appropriators of reclaimed water do so treating such discharges into a reclaimed water course as if it has been abandoned and thus available for appropriation. Other states issue appropriation permits containing a provision that clarifies that the permit does not, in itself, give the permittee a right against a party discharging water upstream who may cease to discharge the water to the water course in the future [37].

Discharger's Right to Reclaimed Water

Some states provide that the owner of the wastewater treatment plant has the right to the wastewater that is superior to anyone who has supplied the water discharged into the wastewater collection system unless there is an agreement to the contrary [38]. The owner of a water reclamation plant must obtain approval from the state prior to initiating a change in the point of discharge, place of use, or purpose of use of the reclaimed water [39]. The state may not approve such a change in use if it will operate to the injury of any legal user of the reclaimed water [40]. In reviewing the merits of a petition for change in use of reclaimed water, the state would consider whether anyone has established a right to the water adverse to that of the treatment plant owner. Other states simply provide that the appropriator does not have the right to rely on the same point of discharge or return of reclaimed water [41].

CASE STUDIES

The review of California, Arizona, New Mexico and Wyoming water rights decisions that follows illustrates the practical application of many of these legal theories in the resolution of actual disputes over ownership of reclaimed water.

Deer Creek Decision

The El Dorado Irrigation District (EID), operator of the Deer Creek Wastewater Treatment Plant in northern California, petitioned the State Board to change the point of discharge, place of use, and purpose of use of the reclaimed water.

EID had contracted to sell up to 2.5 million gallons per day (mgd) (9,500 m³/d) of reclaimed water to El Dorado Hills Development Company. The Development Company constructed a pipeline from the treatment plant to the place of use in the El Dorado Hills. The reclaimed water was to be used to irrigate two golf courses, school grounds, and a park. The Development Company began taking 1.2 mgd (4,500 m³/d) reclaimed water in 1994 when the first of the two golf courses was planted. The second golf course was to be built when more water was available from the treatment plant.

In 1994, the average dry weather discharge from the treatment plant to Deer Creek was 1.9 mgd (7,200 m³/d). Absent the discharge, the creek at some distance below the treatment plant would become dry during the summer months. The discharge has created a lengthy reach in which the creek flows year around, supporting downstream diversions for irrigation and domestic use. The creek provides some recharge to the local groundwater basin. The discharge also supported well-developed riparian vegetation and populations of fish, birds, mammals, reptiles, and amphibians.

While the State Board's discussion in the Deer Creek decision is the subject of considerable debate among legal scholars, the language of the decision bears noting [42]. The State Board held downstream users rights to the continuous flow in the water course are secondary to those of the dischargers of reclaimed water. It also held, however, that fish and wildlife had gained status as a "legal user" and would be entitled to a continued release by the generators of this reclaimed water. The decision was not reviewed by the courts.

Arizona Public Service v. Long

In *Arizona Public Service v. Long,* several municipalities contracted to sell reclaimed water to a group of electric utilities for use as cooling water for the Palo Verde nuclear power plant [43]. The agreement was structured as four supply options, one for each of the planned generators, totaling 120 mgd (450,000 m³/d). The utilities exercised the first two options for approximately 60 mgd (225,000 m³/d). They spent some $290 million to construct a pipeline to transport the reclaimed water 50 miles to the west of Phoenix and a water treatment plant to further treat the reclaimed water so that it could be used as a coolant.

Downstream appropriators and property development interests within the contracting cities brought suit to invalidate the contract. Two questions were posed by the plaintiffs: (1) can the cities contract to sell reclaimed water for use on lands outside the area subject to the original appropriation, and (2) once the cities abandon the reclaimed water in the stream and it is appropriated by downstream users, are they obligated to continue the discharge? The parties agreed that the reclaimed water the cities contracted to sell to the utilities originates from a combination of both groundwater and surface water.

The development interests argued that the contract was in conflict with restrictions on the transport of groundwater contained in the Arizona Groundwater Code. The downstream appropriators asserted the right to continue to divert the reclaimed water from the stream to satisfy their appropriative rights. The removal of the discharge of reclaimed water from the stream by the cities in order to execute the contract with the utilities left the downstream appropriators with an inadequate water supply. In previous years, most of the water diverted by the appropriators was treated effluent. The cities and utilities argued that the reclaimed water was not subject to regulation under the Surface Water or Groundwater Code because it had lost its original character and that the reclaimed water is the property of the entity providing for the collection and treatment of the wastewater.

In its 1989 decision, the Supreme Court of Arizona concurred that the reclaimed water was neither surface water nor groundwater. The Court validated the contract, holding that the reclaimed water could be put to any reasonable use the cities saw fit, including selling it for use on lands other than those involved in the original appropriation. The Court found that the reclaimed water is subject to appropriation by downstream appropriators, but the cities did not permanently abandon the reclaimed water through previous discharges to the water course. Further, the Court determined that the cities were not obligated to continue to discharge reclaimed water to satisfy the needs of the downstream appropriators.

In 1991, the Arizona Legislature amended its Surface Water Code and Groundwater Code to conform with the *Long* Court's holding that those statutes are not intended to regulate reclaimed water [44].

Roswell-Artesian Waters and Users' Association v. City of Roswell

In *Roswell-Artesian Waters and Users' Association v. City of Roswell*, the City of Roswell sought to reserve its right to change the place of use of its reclaimed water [45]. An application by the city to change the place of use of an appropriative water right was protested by Roswell-Artesian Waters and Users' Association on behalf of its members. The State Engineer approved the city's application subject to a condition that it not change the place of use of its reclaimed water in the future. At issue was whether the city could reclaim and reuse its water for municipal purposes free of the restriction.

In 1968, the City of Roswell acquired the Walker Air Force Base from the United States government. Included in the transfer was the right to 2,500 acre-feet per year (afy) (8,500 m^3/d) of groundwater underlying the Air Base. The city also acquired the Air Base wastewater treatment plant. Reclaimed water generated from this facility was used to irrigate near-by agricultural land and a golf course. In 1974, the city abandoned the Air Base wastewater treatment plant, preferring to pipe the wastewater to its municipal wastewater treatment plant. This action resulted in a change in the place of use of the reclaimed water. The new use involved the sale of the reclaimed water for irrigation of farm land and a golf course east of the city. Excess flows were discharged to the Hondo River which runs through the city.

Having lost the benefit of the return flows from the reclaimed water usage, the production from the city's wells declined. The city applied to the State Engineer for a permit to drill supplemental wells so that it could continue to pump its appropriative right of 2,500 afy (8,500 m³/d). Additionally, the city sought to change the place of use of the groundwater from the Air Base to the entire city. The State Engineer granted the city's applications upon finding that the proposed actions would not impair or detrimentally affect any existing water rights. However, the State Engineer attached conditions to the permit which required that the city continue the present uses of the reclaimed water.

At issue on appeal was whether the State Engineer, in granting a permit for change of place of use, may limit the city's discretion regarding the place of use of its reclaimed water. The New Mexico statutes provide that no point of diversion can be changed if it would impair existing rights [46]. In deciding the issue of impairment, the State Engineer has the authority to approve an application subject to conditions if the change in use will impair the rights of others.

In deciding the case, the New Mexico Supreme Court cited the independent findings of the State Engineer that the change in the place of use of the Air Base water will not impair the existing rights of others. The Court held that absent an impairment of existing rights, the State Engineer may not impose return flow requirements as a condition of a permit authorizing a change in the place of use of water rights.

Thayer v. City of Rawlins

In *Thayer v. City of Rawlins,* the Supreme Court of the State of Wyoming addressed the downstream appropriators' right to continued diversion of reclaimed water derived from imported water sources [47]. The City of Rawlins proposed to change the location of its reclaimed water discharge point in Sugar Creek. Several parties had been diverting the reclaimed water from Sugar Creek for production of agriculture and livestock. Such diversions, commencing in 1914, were made pursuant to certificates of appropriation issued by the state. The appropriators sought compensation from the city when it proposed to change the location of its discharge point downstream below the point of diversion authorized by the state. The Wyoming Supreme Court held that since the water in question was imported into the watershed of Sugar Creek and since a priority relates only to the natural supply of the stream at the time of appropriation, the downstream users had no priority of use and no right to compensation for the loss of such waters.

REFERENCES

1 Porter-Cologne Water Quality Control Act, California Code § 13000 et seq.

2 *California Oregon Power Co. v. Beaver Portland Cement Co.*, 295 U.S. 142 (1935).

3 U.S. Constitution, Article I, § 8.

4 Federal Water Pollution Control Act, Pubic Law 92-500, 33 U.S.C. 1251-1387.

5 33 U.S.C. § 1362(7).

6 *United States v. Earth Sciences, Inc.*, 599 F.2d 368, 375 (1979).

7 33 U.S.C. § 1383(c).

8 California Water Code § 13,475 et seq.

9 Reclamation Wastewater and Groundwater Studies and Facilities Act of 1992, Pubic Law 102-575, Title XVI.

10 Reclamation Recycling and Water Conservation Act of 1996, Public Law 104-266 (amended Title XVI of P.L 102-575).

11 United States Bureau of Reclamation. December 1994. "Draft Report of the Bureau of Reclamation's Water Recycling Team to the Commissioner."

12 Florida Statute § 373.1961(2). See also Washington Revised Code § 90.46.005.

13 California's Water Reclamation Law, Water Code §§ 13500–13556.

14 California Water Code § 13510.

15 California Water Code § 13511.

16 California Water Code § 13512.

17 Water Recycling Act of 1991, California Water Code § 13577.

18 Arizona Constitution article XVII, § 2; California Constitution article X, § 2; Colorado Constitution article XVI, § 6; Idaho Constitution article 15, § 3; Montana Constitution article IX, § 3; New Mexico Constitution article XVI, § 2; Wyoming Constitution article 8, § 3.

19 Florida Statutes § 373.016; Hawaii Revised Statutes § 174C-2; Texas Water Code § 11.024.

20 California Water Code § 13550–13555.3.

21 California Constitution, article X, § 2.

22 *Joslin v. Marin Municipal Water District*, 67 Cal. 2d 132, 60 Cal. Rptr. 277 (1967).

23 *United States v. State Water Resources Control Board*, 182 Cal. App. 3d 82 (1986).

24 California Water Code § 13550.

25 California Code of Regulations. Title 23, § 848.

26 California Code of Regulations, Title 23, § 780.

27 "In the Matter of the Sierra Club, San Diego Chapter" State Water Resources Control Board Order 84-7, 1984.

28 In the Matter of the Complaint by the City of Santa Barbara Against the Use of Potable Water by the Tsukamoto Sogyo Company, LTD. for the Irrigation of the Montecito Country Club When Reclaimed Water Meeting the Requirements of Water

Code Section 13550 Is Available, California State Water Resources Control Board Decision 1625, 1990.

29   *San Gabriel Valley Water Company v. County Sanitation Districts of Los Angeles County,* Case No. C661402 Superior Court of California, Los Angeles County (1993).

30   In the Matter of Availability of Reclaimed Water for Greenbelt Irrigation in the San Gabriel Valley Water Company Service Area in the Vicinity of the San Jose Creek Reclamation Plant of the County Sanitation Districts of Los Angeles County, California State Water Resources Control Board Decision 1623, February 16, 1989 (Amended By Order 90-1, January 18, 1990).

31   California Public Utilities Code § 1507.

32   "In The Matter Of The Sierra Club, San Diego Chapter" State Water Resources Control Board Order 84-7, 1984.

33   California Water Code §§ 174, 275, 13142.5(e) and California Code of Regulations Title 23, §§ 856–859.

34   California Water Code § 13521; Oregon Revised Statutes § 537.132; Washington Revised Code §§ 90.40.030–90.40.44.

35   California Water Code § 13556; Arizona Revised Statute § 45-494.

36   State Water Resources Control Board. 1994. "The Water Rights Process."

37   California State Water Resources Control Board water right permit.

38   California Water Code § 1210.

39   California Water Code § 1211.

40   California Water Code § 1211; New Mexico Statutes § 75-5-24.

41   *Metropolitan Denver Sewage v. Farmers Res. & I. Co.,* 179 Colo. 36, 499 P.2d 1190 (1972).

42   In The Matter Of Treated Wastewater Change Petition WW-20 Of El Dorado Irrigation District—Order Reconsidering Approval of Changes In Point Of Discharge, Purpose Of Use, And Place Of Use Of Treated Wastewater, California State Water Resources Control Board (1995).

43   *Arizona Public Service v. Long,* 160 Ariz. 429, 773 P.2d 988 (1989).

44   A.R.S. § 45-467(D), 45-576(L) (Current version at A.R.S. § 45-576(M) (1994)].

45   *Roswell-Artesian Waters and Users' Association v. City of Roswell,* 99 N.M. 84, 654 P.2d 537 (1982).

46   New Mexico Statutes § 75-5-24.

47   *Thayer v. City of Rawlins,* 594 P.2d 951 (1979).

CHAPTER 31

Public Support and Education for Water Reuse

INTRODUCTION

T HE importance of public attitudes in the adoption of reclamation projects has been recognized for some time. We know that people's perceptions and opinions are forces that can mean the difference between success or failure—survival or extinction. This is a reality that must be recognized in the planning and implementation of every water reuse program.

While there are no fool-proof guarantees of success, a sound and proactive communication and education program is essential. The first part of this chapter will tell you how to plan a communication program that will help you reach technical objectives, obtain public support, and win acceptance for your program.

The second part of this chapter presents the expertise and experience of public relations experts in the western United States, professionals who almost daily must meet the press and the public to answer questions about their operations. The sheer size and complexity of their agencies demand that they plan and execute effective community relations programs to educate the public; network and champion to gain advocates for their operations; and effectively communicate with and respond to the press— in good times as well as bad times. By meeting these challenges in a proactive manner, they are investing in programs that will reap many benefits for years to come.

Joyce Wegner-Gwidt, Irvine Ranch Water District, 15600 Sand Canyon Ave., Irvine, CA 92618.

From battling big business for the public's favor to establishing and rolling out media and community relations programs in highly charged political environments, these case studies illustrate the trials, tribulations and triumphs of developing and promoting water programs in a diverse, complex and constantly changing society.

WINNING SUPPORT FOR RECLAMATION PROJECTS

As the 1990s draw to a close, water resource professionals are facing a major transition from the technical side of operations to the human side. The consumer today is more responsive to public issues than in the past, and all too often this responsiveness is the result of unfounded emotion and prejudice. When planning a reclamation project, it is important to recognize that consumerism is "in"—citizens are more interested and involved in the decision-making process. This transition in the way we do business necessitates a change in the way we communicate with our publics.

Traditionally, the technical community has done a good job communicating among ourselves. We conduct research and produce studies and reports about how water and wastewater systems function, why they are needed, how they are designed and maintained, and how much it costs to install and repair them. But unless we start talking to the ultimate decision makers—those who vote on and fund these projects—as effectively and as often as we talk to one another, we could find ourselves and our projects without support.

The challenge is to share our understanding with the public and to communicate with them as well as we communicate within our own industry. We have to recognize that a successful project requires more than a balance sheet and cost benefit analysis. We need to deal with the perceptions and expectations of people, community and special interest groups, government and the media.

To be responsive to all these groups, both public and private, it is necessary to develop a planned communication program. Information and education are the solutions. If people understand what you understand, they will probably agree with what you are doing.

Today, communication networks are dependent not so much on communication products, but on process—on each individual's being a part of the communication system. This takes negotiation, facilitation and consensus. The challenge is to become expert at adapting effective communication processes to a more complex environment.

Reaching the public requires creative, proactive outreach programs. The community relations program is the cornerstone upon which all other elements can be built. A sound community relations program is essential for the success of any communication effort.

When a project is successfully accepted by the community, it is invariably because a commitment has been made to include citizens in the decision-making process. It's important to recognize that people's attitudes are genuinely important and to believe that a fundamental respect for each individual's opinion is essential. The public you serve needs and deserves to understand what you are doing. Informing the public is as much a part of your normal work activity as providing the designated service.

An important first step is to find out what people think about your organization. There are many ways to gather this information. Which ones you choose to implement will depend on your budget and resources. Public opinion surveys are always an excellent first step. While opinion surveys can provide extremely useful information, they are best left in the hands of professionals who specialize in designing and conducting such polls.

Once you have identified your audience, you need to gain the support of opinion leaders, media contacts and third-party experts. It is important to start the communication process well in advance; in fact, in the earliest planning stages. Make people part of the process—meet with stakeholders, form an advisory committee and utilize your employees. Employees can be the best resource you have.

The Citizens' Advisory Committee is an excellent way to make that vital connection between government and citizens. The committee serves not only to provide in-depth input on the project but also to reinforce to the public that their input is important and utilized. The public has more faith in a program that utilizes a Citizens' Advisory Committee of community leaders than it would in a program that was strictly formulated by the agency.

A Citizens' Advisory Committee should be made up of people from the following groups:

(1) Private citizens who are not likely to incur financial gain or loss greater than the average homeowner
(2) Representatives of public interest groups
(3) Public officials
(4) Citizens or representatives of groups with substantial economic interest in the project or plan

A dynamic, well-organized Citizens' Advisory Committee will:

(1) Provide information from a group that is politically active in the community
(2) Raise questions that serve as valuable insight into questions that will be asked at other meetings

(3) Provide a sounding board for ideas

(4) Test market the public participation process to determine its success

(5) Lend credibility to the process in the eyes of elected officials

Dissemination of information can produce a "ripple effect." The first rule is to always fill the vacuum of information; because if you don't, someone else will. Accept the established fact that the public responds to the information it has. If it has no information, it reacts without it. High-level briefings move outward from key opinion leaders to civic groups, neighborhood groups and service clubs. If you have a well-informed community when the media seeks reaction interviews, responses are likely to be reasonable and favorable. Today, people want to be involved in decisions that affect them; consequently, effective community relations is responding to your critics, not stonewalling them.

When working with the community today, the public relations practitioner serves as an orchestrator rather than a presenter. Effectively responding to the community goes beyond renting out a local auditorium, placing an ad in the newspaper and letting the audience barrage the experts with questions—a process that generally deteriorates into mass hysteria, rather than mass communication. A far better technique is moving from one-on-one meetings to workshops or open houses and tours. Good displays, maps and exhibits visually demonstrate the salient points of the project.

One of the very best ways to convince people of the merits of your project is to take them to visit a similar successful project. IRWD has invited many groups from around the world to visit our facilities and share our experiences with water reclamation. If a site visit is not economically feasible, the next best option would be a presentation of the various projects with testimony about how successful they've been. Remember, one picture (or slide or video) is worth a thousand words.

Community relations is a successful approach to building consensus. A successful program combines positive internal management actions with energetic and well-focused external communication efforts. It can be used to build a reputation, evidence a commitment, and favorably influence targeted public attitudes.

One caveat: Effective community relations cannot be built overnight. The most successful community relations programs are ones that establish long-term support and goodwill within the community, so that when you need allies, they are there. They don't have to be sold on you—they already are. They know you, believe you, and trust you. It's like a "trust bank" in which you make deposits over time to draw upon if a crisis does occur. If people realize you're in it for the long haul, they're inclined to be forgiving and give you the benefit of the doubt if you make a mistake (and that can happen to anyone).

Imagine you're a juror. The accused is charged with disrupting traffic and creating a nuisance. If the prosecutor's evidence is persuasive and you have only the facts to go on, you'll surely convict. But if you know the accused, if he's helped you personally and you know him to be a valuable member of the community who has done many good things, you surely will give him the benefit of the doubt. That's the goal of preventive public relations: the benefit of the doubt, the advantage that accrues from being known favorably for your many good works instead of being substantially unknown except for the current accusations.

Public relations is a continuing requirement through the ups and downs of the business cycle. It is not a luxury and should never be abandoned.

Communication is the heart of this response—the thread that links growing accountability with increasing demands for information. Proactive communication programs will help educate citizens, establish priorities and build consensus for projects.

Communication, like the rest of your operations, should be a planned, programmed, continuous professional function of business. If it isn't, you run the same risk of breakdown, failure and public criticism that any other unplanned, unsupervised office or plant activity would deserve. Unlike the relative permanence of pipelines or a fine new facility, a communication program is a pursuit that changes constantly. An agency's relation with the public can never stand still.

The need is so urgent and the number of things which we do to deserve public understanding is so great—and time is so short—that we can no longer be satisfied with damning the press, the environmental extremists, an apathetic public or the politicians. Our job now is to convert any unbelievers in the media, the environmental movement, the politicians and the general public. We can't withdraw from these people—we've got to become related to them. And the quality of that relationship is largely up to us.

If we take this approach and insist that we be seen for what we are—responsive to the public's needs and providing a needed service—then we have told the public who we are, what we do and how people benefit from it.

The new imperatives are cooperation, consent, consensus and communication—but without communication, none of the other three is possible!

FOUR REASONS TO INITIATE THE COMMUNICATION PROCESS

(1) To inform and educate the public: This process is not as easy as it may appear. Initially, all information must be provided to a "base case" audience. Although some may feel patronized while others feel the material is above their comprehension, it is necessary to ensure everyone is speaking the same language and understands the concepts and terminology.

(2) To add public input to the development of the final approach: All too frequently, the planners of a project discover that members of the public are aware of hidden pitfalls. They may say, "That was tried already and . . ." or "They had a problem with that back when. . . ." To avoid such problems, it's wise to involve the public early in the process.

(3) To raise issues early and avoid surprises: The public may find some technologically sound solutions to be unpalatable. Knowing this early in the process avoids locking into a solution that ultimately will fail the test of public acceptability.

(4) To identify the naysayers and their pet issues: There are always community opinion leaders—not necessarily the elected officials—who will become the driving forces in stopping a project. Knowing who these people are will enable the project coordinators to speak directly with them and seek a common ground for discussions.

HOW TO IMPLEMENT THE COMMUNICATION PROCESS

(1) Actively solicit public input. Prior to an official public comment period, awareness of sensitivities helps in planning for a successful project. Form advisory committees for gauging reactions to the project. Form work groups for educating the committee. Keep the members aware of changes in strategy or in a willingness to negotiate certain aspects of the design or content of a project.

(2) Develop and maintain a series of small group educational/informational activities. This approach avoids the large arena that is the most likely place for disruption while allowing concerned citizens to seek information in a setting that encourages the open exchange of concerns.

(3) Be prepared to share both the decision-making power and the problem-solving responsibility. This is a two-way street. If the community expects the right to raise issues, it must be prepared to offer solutions.

(4) Focus energies on winning the support of the community and then keeping it by providing responses to new questions or concerns. Once you have developed supporters, keep their confidence by keeping them in the information loop. Don't let them read about the project in the papers. Encourage project supporters to attend meetings, and persuade them to speak up. Encourage supporters to convince their network of contacts—or anyone who will listen!

GUIDELINES FOR DEVELOPING EDUCATION PROGRAMS

Every public agency has a responsibility in the formation and maintenance of a desirable living environment for the population within its service area. An increased consciousness on the part of the general public has created a demand to know more about our natural resources and how they are being used. Additional concerns have been generated by the widespread publicity about environmental concerns.

The interest of an agency and its customers can best be served by developing and maintaining a comprehensive program of education in the local schools. Such a program will provide an increased level of awareness of water and wastewater issues affecting operations. A higher awareness level will also be developed in the students' families which will assist in maintaining a high level of communication with the public.

Water education in the schools is the key to grooming future voters and citizens who will be knowledgeable about water, will use water wisely, and will make rational voting decisions related to water.

Why is water education so important? Today, water resources planning and management is a more complex task than ever before. The gamut of water issues facing local water agencies can range from water quality and supply/demand to groundwater overdraft and building new facilities. The solutions to these problems are not simple, and often require public approval. Yet many times, the solutions to these problems are seen as simplistic when viewed by an uninformed public. The complex interrelationships that exist in resources planning and management need to be brought into public awareness. Educating the community on the wise use of resources is not something that can be done overnight; but these concepts can and should be introduced to children throughout their years as students, while habits and attitudes are being formed. Only through education about the basic principles of water, its occurrence, development, use and administration will children grow into citizens whose response to water issues will be reasonable, justifiable and socially beneficial, rather than a reaction to misunderstanding.

The important components of a successful education program include:

(1) Know the education system. In order to have a successful water education program, particularly at the elementary and high school level, you must know what the educational system is and how it operates in your local area.

(2) Provide a quality curriculum. We work closely with the schools to ensure that materials complement both grade level and course requirements. Every school district has guidelines or frameworks that tell educators what they must teach for each grade level. To be effective, an education program must offer value to the recipient in terms of the world they live in. The questions you must ask yourself are: What expertise can I offer that will enable people to enjoy the best quality of life available? How can I best communicate so that people will be able to use this expertise?

(3) Plan what to say. Young students are not too interested in acre-feet, hydraulics, the miles of pipe your agency has, or how your engineer solved a very difficult pressure differentiation problem. What the student is interested in, if you make it interesting, is where water comes from. How is it brought to his house? What happens to the water that goes down the drain?

Generalize. Make your information basic. You don't have to give a course in aqueducts, hydraulic pumping or laminar flow to describe how water is transported from where Mother Nature left it to where it can be put to use. Keep it simple. Keep it interesting. That's the name of the game; don't try to make it something else.

You can experience the joy of watching faces light up when, for the first time, a student really understands the hydrologic cycle. You can witness this "lighting up" in the engineering classes every bit as much as in the fourth grade. You will find many who have no idea how water is distributed in the world or how much is available for human use.

You want to talk about how to treat wastewater? The biggest problem you are going to encounter is the fact that people do not want to talk about sewage. It's a cultural problem. But don't let that defeat you, or mislead you into talking about everything except sewage, as usually happens. Don't get trapped into telling students about cement tanks, the size of pipes, the cost of projects and next year's budget, thinking you are getting through.

Wastewater treatment is fascinating! All those little bugs eating all those organics, that's what sewage treatment is about. Sewage is real, it's something everyone contributes to. Sewering agencies don't make sewage, people do. Sewage is a waste-bearing water system, no more. It's a tool to carry away what would otherwise be bothersome, and potentially dangerous, materials in a society living as close together as ours does.

Removing the wastes from sewage and making the water fit to return to nature—that's the job of wastewater treatment plants. It's a necessary job, an important one, and an exciting one. Tell them about it—about the bugs eating organics, about the testing, about the reuse. It's exciting. Make it so!

(4) Decide how to say it. Once you decide *what* you're going to say, decide *how* you're going to say it. You have three choices: generic presentations designed for your industry, individually designed audio-visual presentations, or live presentations. Depending on your resources and preferences, you can use all or any of the three to develop an effective educational program.

Using videos produced for your general field is the least desirable of your options. There are many very fine videos available on many subjects, but too many of these contain too much information. Any journalist will tell you that one of the first rules of good reporting is to tell one story at a time. Teachers tell me that students are distracted from the main theme frequently, because the videos tell a multitude of stories in the same presentation. These videos are useful if you have the luxury of more than one presentation, or if you can leave them with the teacher for follow-up learning activities.

Next, you have the opportunity to create special audio-visual presentations for your individual agency. It is important to find out what equipment is available for using the programs you develop.

Personal presentations are by far the most effective means of communication. When a personal presentation is designed properly, and augmented with audio-visuals, you really have a winner. The best way to communicate is to entertain—make it a performance. Use experiments, maps and charts. Draw a lot of pictures. Tell it from your experience. There is nothing so dull as a speaker who can't take his nose out of his notes.

(5) Train your teachers. Teachers and administrators get lots of mail, and much of it lands up in the wrong file or on a shelf somewhere to be read when time permits. As we all know, time never permits. Teachers are the consumers and we are there in a sense to sell them a product, our educational materials. Teacher training or in-servicing is one way to get their individual attention and actively involve them in the subject of water. A teacher training session can be anything from a 45-minute presentation to a two-day course given for college credit, depending on the level of commitment or involvement you want to make.

Use a variety of presenters with various backgrounds. Show a video or slides. Take a field trip to your facilities so teachers can see first-hand what goes on. Have your key person or a staff member give demonstrations and involve the teacher in activities from the curriculum.

(6) Participate at educational events. Participating at educational events is another key element to a successful education program. It's a great way to promote your program not only to the educational community, but to the general public as well. Most areas have environmental or science fairs and workshops that will allow you to display your materials, and provide guest speakers. IRWD has been very successful in contributing to our local science fair and selecting water-related projects for awards.

(7) Develop an ongoing public relations program with the schools. Before you begin any education program, get the approval and support of the highest ranking school official. The most important

element of developing good public relations with schools is to find a key person in the school system who can help inaugurate and carry out the program. This key individual might be a science coordinator, curriculum specialist, principal or teacher with a special interest in water or environmental education.

As we all know, follow-up is important for almost any task to ensure that the job gets done, and it is particularly important when dealing with another system that has its own unique methods of operation. It is very easy for your contact person, however interested they may be, to get sidetracked on another equally important project. Your job is to keep that enthusiasm ever present and work closely with the person to make the program a joint effort.

It has been our experience that the members of a community who are knowledgeable of the many resources that determine the ecology of their living environment are better able to make decisions regarding that environment. We believe that water is one of the most important of these resources.

HOW TO DEAL WITH THE MEDIA

One of the first steps in any public involvement/education process is the formulation of a good media relations program. The purpose of this program is to educate the public and increase their awareness of your agency and your project. To be effective, media relations must begin well in advance of the project.

For years, public agencies avoided potential confrontation with the media. They felt safer keeping a low profile. The price for that hands-off attitude continues to be paid today in terms of a general skepticism and atmosphere of distrust, even confrontation, between the media and the agencies.

Many people fear reporters and feel that they are out to get them. A friendly reporter, however, can be a great asset. It takes understanding and cooperation on both sides to achieve this give-and-take relationship.

Members of the media are professionals, just like bankers, lawyers, engineers and other specialists. They are skilled in the business of gathering pertinent information and presenting it in an appealing and understandable fashion to the public.

As in any profession, you will find those who are cantankerous, belligerent and take a slip-shod approach to their jobs. These, happily, represent a very small minority. Most journalists are objective, conscientious and responsible. It is, after all, in their best interest to present the facts accurately. Their success depends on maintaining a good reputation, just as ours does.

Another point to realize is that a reporter regularly faces two demands that most us don't. First, each product of his efforts is examined by great numbers of people. Accuracy becomes essential, and the good reporter is always pursuing and presenting precise information. Second, a reporter is usually working against a deadline, sometimes only minutes away. Therefore, urgency often characterizes the reporter's efforts to fulfill an assignment. It is important to recognize this and return calls promptly.

Recognize that these demands are usually present and that they color the manner in which the reporter does his job, sometimes to the point of the reporter's pressing hard for additional details and elaboration and insisting upon an immediate response.

When a prominent national journalist was asked to define the role of the press, he replied, ''Our job is to report the uncommon, unique and unusual—it's not news if people are doing what they're supposed to be doing.''

Consequently, you won't hear or read a lot in the news about ordinary, routine, positive things. You won't read the headline, ''Sewage Treatment Plant Ran Well Today.'' However, if the plant fails, you can expect to see two inches on the front page—or maybe six inches of sewage in your living room.

The media serves as our professional watchdog. It keeps us honest, keeps us on our toes and reminds us that we are here to serve the public. Here are some guidelines to help you deal with the press and get better coverage of your positive stories:

(1) Read the publication in which you wish to place the material. Determine the kinds of stories they run. Find out who writes what before trying to get particular ''beat'' reporters to cover your stories. Find out if there is a special reporter who covers environmental issues.

(2) Look for unusual angles in the things your organization does. Reporters often seek atypical stories that competing publications might overlook. Provide a human interest aspect to your story.

(3) Use courtesy in your approach to reporters. Many media people are overworked—don't add to their burden. Drop them a short note on the kind of story you're plugging. Follow it up with a phone call. If there is no interest, drop the subject. *Don't push too hard.*

(4) React quickly if a call comes expressing interest. Supply all the requested information; set up any interviews with dispatch. Make absolutely sure that you interview your client or subject before the reporter does. This helps get the person accustomed to the interviewing process and prepared to anticipate any questions that might catch him or her off guard.

(5) Don't try to sell non-stories. Reporters keep a mental list of people who suggest worthwhile subjects to cover. They also retain another list of the people who waste their time with irrelevant and useless information. "Dog Bites Man" is *not* news. "Man Bites Dog" *is* news!

(6) Educate the reporter. Be prepared and have understandable, concise, and interesting explanations. Put technical jargon into layman's terms. Prepare fact sheets for statistics.

(7) Remember to follow up. If the reporter bought your idea and wrote a story that pleases everyone, drop the journalist a short note commending him or her on the handling of the story—especially on its accuracy. Everyone likes to be complemented, even a journalist.

(8) Take the initiative to tell your story. Invite reporters to tour your facilities. Accompany them as your guest to professional meetings to hear an important speaker. Be a source of technical information to them with related issues.

(9) Be honest. Honesty is not just the best policy; it's the only policy. If you have deficiencies, or make mistakes—admit them. Journalists will respect you for it. If you lie and they print it, you'll both be caught in a difficult situation.

(10) Assist with deadlines. Do what you can to help reporters get their story. Be accessible. Respond to questions in a timely manner.

(11) Don't exert pressure. News is competitive. If submitted copy is cut or revised, or if the end product contains *minor* errors, don't complain. If you don't like a story as printed, grin and bear it. Don't argue and threaten. Headlines are typically written by a copy editor, not the reporter.

(12) Don't cover up. It isn't ethical, and the story you want to kill will probably get out anyway. If they think you are covering up, the media's natural reaction will be to play the story up bigger than ever. Never answer a reporter with a curt, "No comment." If you can't divulge information, tell them why (i.e., litigation). Most reporters will understand your position.

(13) Choose a qualified spokesperson. Have this person trained and ready. In a crisis, reporters need answers fast. That is especially true of radio and television newscasters, who must meet frequent deadlines. As much as possible, it is advisable to be cooperative—but do so only on the condition that you are confident your response is accurate, consistent, and the best possible representation of your agency's position or viewpoint. The best way to ensure quick, consistent and factual responses to media inquiries is to select an individual to speak for your company in a period of crisis. The individual you choose

should be articulate, cool under pressure, and have the respect of the reporters who are assigned to cover the story.

(14) Improve yourself. You can't promote what doesn't exist. You can't expect favorable press coverage if you don't deserve it. Take action now to correct deficiencies and improve operations, and then let the press know what you are doing. They'll be more tolerant of your shortcomings if they know you are doing everything you can to overcome them.

(15) Don't overreact. The relationship between agencies and reporters can be an adversarial one, but adversarial need not mean hostile. Do not aggravate the situation by overreacting if a reporter does something that appears unfair.

(16) Be candid and open. Exhibit a spirit of cooperation. Be willing to volunteer information that will help them understand the situation. If a mistake is made, or if your agency is in the wrong, the best response you can make is an open admission of error, an apology and expression of compassion and caring for those hurt by your mistake. The public will forgive a mistake if it is admitted openly and followed by a sincere apology and accompanied by obvious sympathy for those adversely affected. In this kind of situation, of course, it is wise to discuss the response in advance with your legal counsel.

These guidelines can be summarized by the Three C's of Good Media Relations: Courtesy, Cooperation and Candor. I'd like to add two more: Class and Courage (see Figure 31.1).

Even under the best circumstances, the media should be only one of the tools used in an effective communication program. Keep all your communication lines open. You'll be on the way to turning confrontation into communication—or at least cooperative confrontation.

CASE STUDIES

BUILDING SUPPORT FOR URBAN WATER RECYCLING—SAN FRANCISCO'S MASTER PLAN[1]

Efficiently using local water resources and providing maximum flexibility regarding water quality and quantity has been an aggressive goal of the City of San Francisco, California, for over five years. The San Francisco Department of Public Works has been working to comply with the City's

---

[1] Reported by Karen Kubick, Project Manager, San Francisco Department of Public Works and Bonnie Nixon, Public Outreach Consultant, Public Affairs Management, San Francisco, California.

THE EIGHT MAXIMS OF SUCCESSFUL MEDIA RELATIONS

- LACKING INFORMATION, PEOPLE ASSUME THE WORST.
- NEWS THRIVES ON CONFLICT.
- BACKGROUND MATERIALS PREPARED IN ADVANCE ARE FAR SUPERIOR TO MATERIALS PREPARED AFTER THE FACT.
- RELEASE GOOD NEWS QUICKLY AND DISPEL RUMORS PROMPTLY.
- SIMPLICITY REIGNS OVER COMPLEXITY.
- REFUSAL TO COOPERATE *WILL* BACKFIRE.
- KEEP A RECORD OF ALL CONTACTS.
- GOOD MEDIA RELATIONS LEADS TO GOOD RELATIONS WITH EMPLOYEES AND THE COMPANY.

Figure 31.1 The eight maxims of successful media relations.

1991 Reclaimed Water Use Ordinance, which was adopted by the Board of Supervisors during the most recent drought. The initial Phase I, $140 million program to be voted on in November of 1996 will assist in drought measures for the City and meeting future water needs by developing non-potable water for fire fighting, landscape irrigation, industrial uses and office cooling systems. To implement this program, the City needs to construct recycled water treatment plants, storage reservoirs and a 15-mile distribution pipeline system.

A Draft Recycled Water Master Plan was issued in September of 1995, the result of the efforts of several City departments: Public Works, Water, Recreation and Parks, Fire and the School District. The recommended project described in the Master Plan is an integrated treatment storage and distribution system that will reliably supply recycled water to City irrigation, residential and industrial users, as well as to the San Francisco Fire Department (SFFD) to augment its emergency firefighting capabilities.

During recent droughts, water policies have resulted in mandatory water use restrictions. Since 1992 residential and business water conservation legislation has been passed. However, the City has no alternative water supplies to the Hetch Hetchy system. Through the Bay Delta hearings, the City's pre-1914 water rights to the Hetch Hetchy system are being evaluated. Environmental, urban and agricultural water needs are being thoroughly examined in those hearings, and the need for alternative water supplies is becoming clearly evident. Recycled wastewater can provide a reliable alternative for non-potable uses. The City currently treats 87 million gallons per day (mgd) of wastewater that is discharged into either the San Francisco Bay or the Pacific Ocean. A strong potential exists for using this effluent to supply non-potable water demands. The City's Department of Public Works (DPW) and the San Francisco Water Department (SFWD) teamed up to begin studying reclamation in 1989. This union was necessary and functional because DPW operates the wastewater treatment plants and SFWD distributes the City's water.

The challenge for the Department of Public Works, as the lead project agency in developing the project plan, was determining where needed facilities and infrastructure for treatment, storage and distribution could be built. Open space in San Francisco is primarily limited to schools and parks, and it is desired by the residents of San Francisco that open space remain undeveloped. As a result, facilities are sited based on a "joint use" basis where open space can be maintained. All of the three reservoirs proposed for the system are below ground and the areas would be restored and improved so they remain a neighborhood park, a playing field, and a parking lot. The treatment plant will be constructed with a facade that matches an existing building along the ocean. Distribution lines will utilize and extend the Fire Department's non-potable fire-fighting system.

Neighborhoods where the facilities will be located have been a major focus of the public outreach and education program. Efforts were made to get the addresses of all adjacent properties early on. The Water Recycling Master Plan is currently in its environmental review phase, which will produce a comprehensive Draft Environmental Impact Report, written in accordance with the California Environmental Quality Act, and due for distribution in early 1996.

In the initial stages of the program, it became apparent that to get the support of the public, the public needed to see a benefit. Fire protection and the use of integrated water supplies provided the public with tangible goals for the recycled water program. The project obtained the approval of the Public Utilities Commission, the Chief Administrative Officer and the Board of Supervisors. Now the challenge was how to reach and get voter support from San Francisco's diverse community.

Developing and implementing a program that will engender public support for water recycling has been a key goal from the start. The public includes potentially affected residential and business interests, and local and regional governmental entities, as well as the general voting public. To increase public awareness of the proposed water recycling program and build support and momentum for the November, 1996, ballot measure, the program focuses on: (1) informing and educating the public by disseminating information, (2) identifying and addressing issues of concern to the public by soliciting their participation and comments at public meetings, events and presentations, and (3) encouraging their support by involving them in the planning and development process. The major program elements include communication materials, general and targeted outreach efforts, legislative relations, media relations and agency partnerships. These elements are discussed below.

(1) Communication materials: An extensive database of potentially affected, influential and interested individuals and organizations was compiled. A general brochure discussing the need, safety, costs and features of the overall program was produced and distributed to this mailing list. Quarterly newsletters, fact sheets, briefing packets, utility bill inserts, common questions and answer sheets, and specialized brochures are also being prepared on a regular basis. All of the communication materials are available at community events and presentations.

(2) General outreach: Helping San Francisco citizens understand the need to pay for a water recycling program is the foundation of the general outreach program. The Institute for Participatory Planning once stated that "the biggest single obstacle to broad citizen participation and support is citizen apathy." To offset this type of apathy an aggressive outreach program with a sophisticated slide presentation and handout materials are used weekly to reach civic, trade, environmental, business, educational and multi-cultural organizations. San Francisco's diverse neighborhoods and ethnic organizations demand that the program's messengers be knowledgeable and sensitive to cultural differences, and possess bilingual capabilities. The recently completed public awareness survey was translated and administered in Spanish and Cantonese as well as English. The City has a regular tradition of street fairs and outdoor special events that the outreach staff attend. In addition, staff are available by telephone during business hours.

(3) Targeted community outreach: If directly impacted residents and businesses understand the goals of a project in their neighborhood and help shape the program's implementation and mitigation, they are likely to be allies and support the recommended changes. Tailored, organized and well-run neighborhood meetings have helped to explain complex technical issues and solicit opinions and concerns. A Joint Use Task Force has been established to directly involve residents in the project planning phase. A Neighborhood Monitoring Committee will be established during the construction phase. The staff is trained in facilitation and public presentation skills and prepares for anticipated questions well in advance of the meetings. Community liaisons and

spokespersons are identified and kept informed and involved on an ongoing basis. Community interests are asked to help determine the most effective times and vehicles for disseminating and receiving information (i.e., existing newsletters, community newspapers, phone trees, flyers and door hangers). Monitoring and responding to feedback in a timely manner helps to establish ongoing trust with the community. Creating a climate where community members feel consulted and involved has helped to foster the sense of ownership that is critical to successful program implementation.

(4) Legislative relations: The objective of the legislative relations program is to keep the ballot measure sponsors and other elected officials apprised of the outreach effort and the public's response. The 600-resident public awareness survey was an invaluable tool for this effort. Regular briefings with the Board of Supervisors, special commissioners and other elected officials helped to maintain the momentum of the program. Political winds have been known to change the course of many a program. Keeping the decision makers aware and informed of voter concerns has been necessary for the program team to stay the course.

(5) Media relations: While community relations professionals search for agreement in the midst of disagreement, media professionals often look for disagreement in the midst of agreement so they can get a good story. In order to gain voter approval, it is absolutely necessary to raise general awareness of the overall program and its benefits. The media undoubtedly reaches the largest city-wide audience. Understanding how to work with media representatives has been central to the program's success to date. Delivering tailored pitches to neighborhood and city-wide media and highlighting significant issues has helped to generate many articles. One timely article appeared in the *San Francisco Chronicle* showing the fire dangers associated with low water quantity and pressure.

　　Early on, the team identified all the target media including business and association newsletters, neighborhood publications, city-wide publications, trade journals, and local radio and television outlets. Establishing relationships with key editors and media contacts has been essential to receiving balanced coverage. Meeting their deadlines and ascertaining the timing, format, circulation and audience considerations are all important factors. A press packet, press releases, articles, public service announcements and media advisories are prepared in a timely fashion and updated regularly.

(6) Interagency coordination: One of the more unique aspects of the program is the solid partnership that has been created between the Department of Public Works, the Water Department, the Fire Department, the Recreation and Parks Department, the Zoo, the School District and several

other local government entities. Regular consultation, meetings, presentations and communications help to keep this network cohesive. The program's project manager has been responsive, persistent, sensitive and, most importantly, positive in all her communications with agency counterparts. Along with staff and consultant support, the agency partnerships have definitely helped to define the overall success of San Francisco's Water Recycling Program.

CLOWNING AROUND WITH BEER IN A RECLAIMED WATER DEBATE[2]

It all began with a full-page newspaper ad that would be any public works manager's nightmare. The bold headlines blared, "Do You Want Reclaimed Sewer Water in Your Underground Drinking Water?" and then went on to contend "Your Health Risk . . . Cancer, Dementia, Birth Defects, Hormone Deficiency, Cardiac Disease—Chemicals." The ad was paid for by a self-proclaimed grassroots organization called "Citizens for Clean Water." Who were these guys, and why were they attacking a water reclamation project?

Full-page ads seemed to pop up every couple of weeks in the local media, castigating the proposed project for percolating 15 million gallons per day of reclaimed water into the groundwater table in the San Gabriel Valley. Similar programs had put an average of 30 million gallons per day into the groundwater in a basin south of the San Gabriel basin since the early 1960s. In addition, 80 million gallons were already being reclaimed and reused in the same vicinity as the proposed project. No one had protested the use of reclaimed water before. What was this all about?

Citizens for Clean Water was the creation of Forrest Tennant, M.D., who had extensive political connections and aspirations in the San Gabriel Valley. It was difficult to determine Dr. Tennant's motives for organizing the alleged grassroots movement—or whether he truly believed his far-fetched allegations. But his group kept banging away, speaking out at Chamber of Commerce meetings and other public forums. Citizens for Clean Water also distributed a newsletter on a regular basis that was purported to reach 8,000 households.

The attacks were not only unexpected and unwarranted, they were full of falsifications and half-truths meant to scare and inflame the public. In trying to establish a community relations program to address the concerns of the group (and thus diminish the impact on the general public), it became clear that scientific data would not be well received. The opinions

[2] Reported by Joe Haworth and Earle Hartling, Sanitation Districts of Los Angeles County, Los Angeles, California.

of the opposition were set in stone, despite an abundance of evidence to the contrary.

The Upper San Gabriel Valley Municipal Water District was the original proposer for the groundwater recharge project. The source of the water was going to be the San Jose Creek Water Reclamation Plant, operated by the Sanitation Districts of Los Angeles County. Districts' Information and Operations personnel became a part of the information program that was pulled together in response to the negative ad campaign. Because the main San Gabriel Groundwater Basin was already designated a Superfund clean-up site (the result of pollution from aerospace and dry cleaner-type solvents), the Superfund Working Information Group (SWIG) of the U.S. Environmental Protection Agency was part of the information resources of the community. Member organizations to the SWIG were also involved, such as the League of Women Voters and the Sierra Club.

Earle Hartling was the Sanitation Districts' main representative to the community on the reclamation issue. He would often be asked to speak at meetings to discuss the project, which had become high profile in the local press. He spoke before organizations such as the Kiwanis, Rotary, city commissions and even the Irwindale Chamber of Commerce (home of the Miller Brewery Co.).

It was early in the sequence of these presentations that Earle first encountered E.T. the Clown (yes, he was dressed like a circus clown). E.T. was what can best be described as a local gadfly to many city councils and government organizations. He was cynical about any kind of government program or proposal. Determined to bring this project down by force of personality, he constantly harassed Earle during presentations and generally created a ruckus. He also hosted a public access television show in the Valley. Though no one is sure of his viewership, he was certainly able to amplify his clownish presence on the subject of water reuse. He stayed on the topic for well over a year until he was finally diverted by other interests just as his audience was fading away.

Fairly early in the response program, it became clear that Miller Brewing Co. was deeply involved in the opposition to the project. Claiming interest for the public health, Miller alleged that serious environmental problems would result from the reclaimed water project. To the casual observer, it appeared that Miller was less fearful of the public interest and more fearful of the potentially negative impact percolated reclaimed water might have on consumers if a competing beer company used it against them. In an industry driven by images of sparkling clear mountain water, Miller seemed to equate reclaimed water with the kiss of death (this fear was not alleviated by journalists, who repeatedly made reference to Miller having to "put up with the sewage" if the project progressed).

The irony of the situation was that Miller drew 860 million gallons of water each year from the local groundwater to make more than 5 million barrels of beer at its Irwindale plant. Yet to date, none of the competition had hit the industry giant with the Superfund issue; i.e., showing images of EPA lab technicians with test tubes.

Nevertheless, Miller hired a group of local attorneys to do everything they could to undercut or overturn the project proposal. Of course, one of the major opportunities would be in the Environmental Impact Report process. Their efforts also included trying to replace directors on the Upper Basin's Water Board through elections. The *Los Angeles Times* reported that Miller Brewing Co. had contributed $11,000 to a political action committee that spent that same sum the very next day to help two candidates who opposed the project. In addition, Miller filed a lawsuit in an attempt to stop the project dead in its tracks—or at least bog down its progress in a tangle of accusations and rebuttals (which they did).

Before all was said and done, two years worth of legal work was rumored to have cost the opponents several million dollars. It's hard to estimate what the proponents had to spend to "defend" the project, because it was such a diverse working group, including volunteers as well as paid staff from the various organizations involved.

Because most of the battle was taking place in a local newspaper, the *San Gabriel Valley Daily Tribune,* the project proponents and other supporters of water reclamation met with the editorial board of the newspaper. This was a newspaper that had a clearly declared opposition to the Sanitation Districts' Puente Hills Landfill, a large regional landfill within the newspaper's distribution area. However, the editorial board did not drag old baggage into the water reuse proposal; instead, they found the merits of the proposal far outweighed any alleged environmental problems and would be helpful to both the economy and the public health and welfare in the communities.

Organizations like the California WateReuse Association stepped in to help. WateReuse took out a full-page ad in the Tribune in support of the concept of water reclamation which defined what water reuse was in lay terms and why it was so important in southern California. Additional media coverage in support of the project was gained from journalists reporting on the WateReuse Association's condemnation of the Miller Brewing Co.'s fight against the plan to replenish the valley's underground water reservoir.

One of the strongest allies of water cleanup and reclamation in southern California, U.S. House of Representatives member Esteban E. Torres, came to the rescue of the project on the political level. He even made an impassioned plea before the Irwindale Chamber of Commerce, an organization that had already taken the side of the opposition at the request of Miller Brewing Co. Representative Torres clarified the reclaimed water issue in southern California even though he had friends and associates who worked with Miller Brewing Co. He pulled no punches in a fiery speech on the value of water and water reuse. His aggressive support was essential, as the attorneys for Miller had already started actions against other groundwater recharge projects and even Title 22, the State's Administrative Code on reclaimed water programs. They were ready to bring down the house for the sake of destroying this one project.

Even the Los Angeles County Medical Association had to leap into the fray to defend reclamation and reuse. They crafted a resolution that clearly stated their level of support and served as an excellent public relations tool from an organization that carries the value of excellence in health care on its masthead.

What was most amazing was the resiliency of the public's acceptance of water reclamation. From the press and other media to public institutions, from community groups to environmental organizations, it became clear that enough grassroots communication had taken place over the years that the public was strongly on the side of reclaimed water utilization.

The support of third-party, non-profit organizations like Heal the Bay and the Sierra Club was essential to gain credibility with the public. The Board members of the Upper San Gabriel Valley Municipal Water Districts stayed solidly behind the water reclamation project, although they had to ultimately compromise as a result of the Miller Brewing Co. litigation. They moved the project farther away from the Miller Brewing Co. plant and reduced the project size by about 40 percent. A research program was also established to determine any potential health effects from both the reduced project and a possible full-size project in the future.

All in all, the reclaimed water debate garnered substantial media coverage—mostly in support of the reclaimed water issue or at least demonstrative of fair and accurate reporting. The emotional (verging on hysterical) approach used by Citizens for Clean Water eventually worked against the group, which received negative coverage in the press from both journalists and members of the water resources community. In a debate fraught with exaggerations and misrepresentation, opinion-editorial pieces provided an excellent opportunity for setting the record straight.

What resulted from almost three years of legal wrangling—played out before the public in the media and through an aggressive community participation program—was compromise. Miller won space and a diminished threat to its corporate image. The proponents won a "demonstration project," using a smaller pipe that will take in enough water annually for 25,000 households, instead of the 60,000 households originally planned. Once in place, the project as it stands holds the promise of expansion to create an ample and reliable future water supply for the San Gabriel Valley.

In the time between now and then, proponents will continue to work to educate the public about wastewater recycling—perhaps with the assistance of Miller Brewing Co. and other companies whose livelihoods depend on water—in an attempt to assuage fears and lead the state to self-sufficiency and a plentiful water supply.

WASTEWATER AS A SOURCE OF DRINKING WATER . . . IS THE PUBLIC READY?[3]

Imagine that you live in a desert, where it rarely rains and water is scarce—so scarce that 90 percent of your water comes many miles from the north and the east. Recognizing that you and your fellow desert residents could be in grave danger if that imported supply becomes unavailable, the city planners come up with a new idea. They propose treating reclaimed water—which isn't scarce—to standards so high that it can be used as a potable supply.

It doesn't take a rocket scientist to recognize that this new source of water is really highly treated wastewater. Knowing where it comes from, will *you* drink it?

This is the question that many residents in San Diego, California, are asking themselves. The scenario of a new water source for the desert isn't just a far-fetched story. In fact, the San Diego County Water Authority and the City of San Diego are pursuing such a project right now. The question these water managers are asking themselves is not whether water repurification is possible, but rather, "Is the public ready?"

Is the public *really* ready to drink treated wastewater?

Based on extensive research, the Authority and the City of San Diego think the concept of repurified water is something county water users are ready for and a project they will support.

Coming to this conclusion has taken a lot of thought and research, both from a technological and public affairs standpoint. Once the technological aspects of the project were ironed out, the Authority took the lead on gauging the public's acceptance of the concept of repurified water.

Determining how to go about such research was a challenge for the Authority. Although repurified water projects exist in several locations throughout the world, they began for other reasons such as groundwater protection. San Diego's project is the first to propose the use of repurified water as a supplement to the drinking water supply. Therefore, no public outreach model for such a project exists.

The Authority recognized that without public acceptance, a project of this magnitude cannot go far. "What is repurified water?" "Is it safe?" "Why do we need to use this source of water?" are important questions

[3] Reported by Patricia A. Tennyson and Sara M. Katz, Katz & Associates, San Diego, California.

that must be answered in order to gain the public's understanding and acceptance.

To address these questions, and many more, in the fall of 1993 the Authority and the City embarked on a comprehensive research project to better gauge the public's willingness to accept repurification and to identify specific issues that would need to be addressed. Information was gathered through public opinion research, focus groups and individual interviews with community leaders and policy makers.

With the help of San Diego State University, more than 300 City of San Diego residents were questioned by telephone. The surveys were designed to determine the respondents' level of knowledge and personal opinions about general water issues, how they felt about various water recycling options, their response to terminology such as "would you be more likely to drink reclaimed, recycled or repurified water?" and the trustworthiness of potential message carriers with regard to water quality and safety.

Survey findings indicated there was a significant level of interest and concern about water supply, quality and treatment options. A surprisingly high number of respondents indicated support for the concept of repurifying water and many responded favorably when asked if they would use "repurified water." Research also indicated that the University of California, San Diego Medical School, the Environmental Protection Agency and the State Department of Health were all viewed favorably as independent sources of information about this subject.

Overall, those surveyed were supportive of repurified water once it was explained, and indicated they would be willing to drink, wash with and cook using repurified water.

Four focus groups also were conducted with participants of all ages and ethnic backgrounds to assess community perceptions of repurified water. The first two groups were presented the concept of repurified water from a water supply perspective. The second two groups were presented several "toilet-to-tap" visuals designed to underscore the wastewater origin of the water. The goal of the focus groups was to identify how citizens felt about various water recycling options and what their level of knowledge was regarding water supply in general. The sessions also sought to identify participant thresholds at which individuals would accept repurified water for potable use.

Focus group findings confirmed the perception that San Diego is a region of water scarcity. Although toilet-to-tap visuals made participants feel uncomfortable, the majority still responded positively to the concept of water recycling for potable reuse. All respondents thought it was possible to technologically recreate nature's water cycle.

Project team leaders from the Authority visited more than 70 community leaders individually to find out how they felt about repurified water. These one-on-one interviews indicated that there is a high comfort level with repurified water when the process is explained. The main concerns from the interviews were affordability of repurified water as a supply portion and public perception.

Although the results of the Authority's research were favorable, telephone surveys, focus group input and individual stakeholder's interviews were not enough. What else could the Authority do to ensure that San Diegans would be comfortable with repurified water?

The Authority and the City submitted their water repurification project proposal to the scrutiny of an Independent Advisory Panel (IAP). Made up of nationally recognized experts in the fields of water treatment and public health, the IAP reviewed the proposal with a fine-toothed comb and provided insight into technical aspects of the project that would eventually be reviewed by the California Department of Health Services (DHS).

Input from the water and health experts increased the Authority's comfort level, but there were other stakeholders to consult. Who better to assess public acceptance of the project than members of the San Diego community?

A citizens' review committee composed of representatives from a broad range of community, environmental, medical, business, biotech and recreational groups was convened to review the project in detail. The Repurified Water Review Committee (RWRC) met over a period of five months in mid- to late-1994 to examine San Diego's water supply issues, the costs and economics of various supply alternatives, available technology, and public health and safety, as well as public perception issues associated with repurified water.

Committee members spent hours gaining expertise on water quality and supply issues, as well as contributing their individual experience to assist in studying the proposed project. Representatives of the business community were concerned about the cost of the project. "How will it affect the price of doing business in San Diego?" asked one member. "Can we afford it?" and "Will it bring more business to the region?" asked another.

A representative of the regional Parent-Teachers Association wanted to know what the schools could do to educate the public about the project, while another member wanted to know how the Authority could garner the medical community's support.

In the RWRC's final report, committee members concluded that "there is sufficient information available to determine the suitability of water repurification as a supplement to the San Diego region's water supply. Additional planning, economic analysis and environmental studies should proceed."

The RWRC's findings, as well as the recommendations of the Independent Advisory Panel and the DHS, have helped the Authority and the
City of San Diego decide to proceed with detailed planning, design and
feasibility studies.

The timing of the committee's conclusion could not have been better.

The Authority and City's research was put to the test in January 1995
when a consumer reporter at San Diego's ABC affiliate decided the proposed repurified water project "smelled rotten."

Trying to uncover a scandal for "sweeps week" in late February, the
reporter traveled across the country grilling water quality and health experts
on their assessment of the project. After speaking with a representative
of the Environmental Protection Agency in Washington, D.C, officials at
the American Water Works Association in Denver, and officials at the
Upper Occoquan Sewerage Authority in Virginia, the reporter found only
one water quality expert who spoke "on the record" in opposition to the
project.

The Authority worked aggressively with the reporter, providing facts
and figures, offering briefings and arranging interviews with additional
water quality experts. As a result, what could have been a project-stopping
news report became an excellent opportunity for public education.

In advance of the news report, the Authority held a media briefing,
providing local and regional reporters with a one-hour technical briefing
on the project. Although the briefing was intended only as a source of
information, the project received several news articles and extensive coverage on the local CBS affiliate. Now reporters are better informed about
who the project players are and who they can turn to for answers.

Although the Authority was forced to speed up its outreach program,
the results of the February media blitz were positive. The Authority outlined
an aggressive plan to educate the public and is now coordinating these
efforts in tandem with preliminary environmental studies.

Traditional methods of milestone news releases, media briefings and
the development of an information kit have been helpful in initiating the
public outreach efforts. A project information line is available 24 hours
a day for callers to record their comments and ask questions about the
project. A project video and slide show are in final production to use in
presentations to community organizations, businesses and other interested
parties. As the project progresses, an aggressive speakers' bureau program
will provide many opportunities to educate the public on the feasibility
of repurified water.

The Authority and the City continue to interview community leaders
to discuss project issues with the members of the RWRC to ensure that
San Diegans feel comfortable with the concept. Because of the steps that
have been taken to improve public understanding of repurified water issues,

the agencies have confidence the public is ready to accept repurified water as a supplementary source of drinking water, and to support efforts by the Authority and City to ensure that the county has a highly reliable water supply now and in the future.

WHEN THOSE WHO CONFLICT HELP SOLVE THE PROBLEM[4]

Developing a solution to San Diego County's emergency water supply problems has been fraught with conflict in the past. Recently, however, those who were on opposite ends of the spectrum on the issue have become part of the solution. The conflict, which began in the 1980s, has slowly begun to dissipate as community members have become more involved in hands-on project activities.

There is a significant need for emergency storage in San Diego County, California. Since it lacks an adequate water supply of its own, the county relies on water imported from the north and east. The aqueducts supplying water to the county cross the San Andreas, San Jacinto and Elsinore earthquake fault systems. Should an earthquake sever the imported water pipelines, the region lacks sufficient water storage to meet the demand. As a result, several locations throughout the county would be without water within 24 hours. The region's water supply would be severely curtailed for up to six months, and parts of the county would be completely without water for up to two months.

In addition to lessening the effects of a disaster, an emergency storage system is necessary to meet the demands of the existing population. Much of the region's water storage system was established before the Authority was organized to import water to the county. As a result, most of the reservoirs are not connected to the Authority's pipelines and can neither receive imported water nor provide water to other parts of the county. Lake Sutherland, the last major reservoir to be built in the county, was completed in 1953. Since then, the county's population has grown by nearly 500 percent. The county has a substantial shortage of available water storage capacity, especially in the area of emergency storage.

In 1980, the San Diego County Water Authority and the City of San Diego agreed to pursue construction of a dam and reservoir in the Pamo Valley. When completed, this reservoir would provide emergency water storage for much of the county. For many reasons, including the way in which the project was conducted, it became the center of a war between the agency and special interest groups and was never completed.

By deciding on the project and then informing the community about

---

[4] Reported by Patricia A. Tennyson and Sara M. Katz, Katz & Associates, San Diego, California.

their plans, the Authority and the City lost support for the project. Had they established two-way communication between the project team and the public, the project may have been successful.

The Corps of Engineers certified environmental documents related to the project and announced its intention to approve the 404 permit application. However, the Environmental Protection Agency (EPA), which has veto power over 404 permits, objected and called for further study. The intensity of the project increased and it became the center of a large public debate. Despite a county-wide vote to fund the project, the EPA's indication that it planned to veto the 404 permit application brought the project to a halt in 1988. The EPA did not believe the Authority had demonstrated that the Pamo Valley alternative was the least environmentally damaging and practical solution to the emergency storage problem because they did not study several viable options in detail.

So in 1989, the Authority went back to the drawing board and began an extensive alternatives study to select the most appropriate site for an emergency storage system. In beginning the second round of planning, the project team had several strikes against it: the knowledge of an unsuccessful first attempt, growing opposition from opponents of the defunct Pamo Valley project and the usual conflicts associated with accusations of a "growth-inducing," "expensive," large-scale project.

In planning a large-scale water resources project, one of the toughest responsibilities is answering to a variety of diverse interest groups. What the business community may think is a great project, the environmental community might fight tooth and nail. The project team also needs to factor in the opinions of homeowners, politicians and other public agencies. All of these groups—and others with interests just as vital—can place major roadblocks in the path of completing a project. But this is not inevitable.

Lessons learned from the Pamo Valley project have enabled the Authority to jump the initial hurdles in the second round of planning, and it is currently in the environmental evaluation phase of siting and developing the Emergency Storage Project (ESP).

To combat potential conflict in planning the project, the Authority discarded the traditional decide, announce, defend policy of decision-making. The Authority has implemented a thorough alternatives analysis that will yield an environmentally sound, cost-effective project that meets all legal requirements. The analysis involves a full range of possible solutions, including new pipelines to improve the use of existing reservoirs, use of groundwater basins, new or expanded reservoirs, and the use of alternative water supplies such as desalination or reclamation.

In addition to the alternatives analysis, the Authority launched an aggressive public information program. Each alternative raises a unique set of issues and is subject to the scrutiny of community interest groups that have differing views on the importance and acceptability of the systems being considered. Questions that the public information program addresses include the amount of storage required, the community's ability to fund the project, environmental and social impacts, and the engineering reliability of the project.

To address these issues, the Authority created the following programs with the help of a community and media relations firm:

(1) A Speakers' Bureau, available to make presentations to any community group: It comprises ESP team members and the Authority's directors. The speakers are trained regularly to ensure the accuracy of the information being disseminated

(2) An ESP Hotline, which allows citizens to ask questions and voice concerns to project experts

(3) Newsletters, media releases and a video, which are produced, updated and distributed to the media, elected officials, environmental groups, private citizens and other interested parties

(4) Tours of alternative sites, which are conducted to educate Authority Board members, project staff, member water agencies, the public and the media on issues associated with each site

All of these measures have succeeded in keeping county residents informed about the project, but conflict is inevitable. Generating the usual one-way communications tools is just not enough.

The Authority raised the amount of public input to a new level by forming the Emergency Storage Working Committee (ESWC), which has brought stakeholders into the planning process and has incorporated community values and interests into the project. What makes the ESWC different from many community advisory boards is that the committee was given a specific job to complete. While the group provided the usual community input on a project, it actually became part of the decision-making process.

The SWC is made up of 27 representatives of diverse interests in San Diego County: those who live and work in the site-affected areas, business and environmental leaders, educators, disaster preparedness experts, agricultural and biotech industry leaders, and political representatives. Working side by side with Authority staff and consultants, the group reached a consensus agreement on the priority of various environmental, social, technical and economic issues that must be addressed before an emergency storage system can be developed.

At first, many of the groups asked to participate in the ESWC were suspicious of the Authority's motivation for bringing their representatives into the process. However, after the Authority demonstrated it wanted honest opinions on the project, trust developed between the committee members and the Authority. The potential for a shared-decision outcome increased, and resolutions evolved among those who once stood at opposing ends of the spectrum. Through the ESWC, committee members were able to put their concerns into a bigger picture perspective.

Education was an important tool in forming a larger project perspective for ESWC members. In initial committee meetings, the Authority staff and technical consultants presented project information to the committee in a thorough and easy-to-understand manner. This preliminary work was important in establishing a more uniform understanding of the technical issues associated with the project. A day-long tour of surface sites also allowed ESWC members to learn more about the alternatives and question technical consultants on the project. Discussion at each committee meeting helped to increase the committee members' level of understanding about the project and broadened the staff's and consultants' perspectives on the community's interests. What was created throughout this information-sharing process was a more even playing field in determining weights for the system alternatives.

The group was given a time frame of six monthly meetings to determine what weights should be given to various factors used to screen potential emergency storage systems. They were encouraged to develop additional commentary on other aspects of the project. Their final product was a report, endorsed by all members of the committee, which was presented to the Authority's Board of Directors and explained the results of their meetings.

To ensure that the Authority did not have a hidden agenda in forming the ESWC or put pressure on members to agree on a certain outcome, an outside facilitator joined the process to lead the group. The facilitator's being from outside the county reinforced the Authority's contention that there was no bias towards any particular alternative system. The use of a facilitator proved to be effective in garnering participation from all ESWC members.

"The difference between the first working committee meeting and the last was amazing," said Sara Katz, a public information consultant to the Authority. "Most of the committee members who had opposed each other on other projects sat with arms crossed, glaring at their one-time foes during the first meeting. When the committee's work was done, these people who had come from different directions were friends."

To tackle the pre-existing attitudes, the committee's first task was to adopt ground rules for its meetings. Approved rules included respecting the personal integrity and values of each member (including the avoidance of personal attacks and stereotyping). Disagreements were to be regarded as problems to be solved rather than battles to be won. Each member was responsible for keeping his or her respective group aware of committee decision-making processes and time lines. Commitments by members were to be taken seriously and to be kept.

Consistent participation by all members was expected. Each member was responsible for stating his or her views; this was an essential part of the process. If a committee member had a special interest in a considered site or system configuration, he or she was to continue to participate in all meetings for the greater good of the committee. Members had an obligation to provide pertinent, public information for items under discussion. Also, the committee agreed that contact with the media regarding committee decisions was not allowed during the decision-making process, in order to preserve the group's ability to modify draft proposals.

The ESWC's task was not easy. Initially, the committee considered how much weight (out of 100 percent) to attach to each of five broad goals: environmental impacts, project costs, operational effectiveness, social impacts and system implementability. The second step was to recommend weights for subgoals. For instance, under environmental impacts, the ESWC evaluated biological and cultural impacts, and under system implementability, technical concerns and jurisdictional constraints were weighted.

Using the ESWC's criteria, the Authority's staff pared down a list of alternative systems from thirteen to four. The four systems will be included in a draft Environmental Impact Report/Environmental Impact Statement.

Because of their specific knowledge and experiences, ESWC members shed new light on project issues. But more importantly, they helped to solve a problem that public agencies fear most when developing a large-scale project: communication with the public.

"Most of us, though we entered the process somewhat skeptically, went away feeling it was a fair process," said Linda Michael, the Sierra Club's representative on the ESWC. "The Authority truly has reached out to try to educate the community on emergency storage."

In fact, the committee members were so pleased with the ESWC process that they asked to continue working on the ESP. Having worked together successfully for more than six months, the members are eager to take part in the recreation and mitigation phases of the project.

The committee members' satisfaction with the ESWC process is one reason they wanted to continue to participate in the project. Their satisfaction was due in part to the ESWC's democratic voting procedure. Each committee member's vote carried equal weight, and the confidentiality of the procedure allowed committee members to express their true thoughts on the project.

The most evident sign of the committee's effectiveness lies in a final consensus report. Signed by all ESWC members, the report outlined the committee's recommendations and conclusions. The final report stands as a testimony that in an arena where polarization on views often seems inevitable, a broadly based community group can discuss issues openly and find areas of real consensus.

Because of the ESWC's success, the Authority hasn't had to take on the role of conflict resolution manager with the ESP as it has with other projects in the past. Although the project is not conflict-free, all of the usual voices of dissension are now on the same team—the ESP team.

"I would never have believed such a diverse committee could ever have come together with a unified consensus, but we have achieved a strengthened unanimity because of and through our diversity," said Jere Lien, an ESWC member.

The Authority's use of the ESWC represents a valuable model for future environmental decision-making in San Diego County and elsewhere in the country. With most projects of this magnitude, those who create conflict do not feel involved in the resolution of the problem, making further conflict inevitable. The committee allowed an unprecedented amount of interaction to take place between the project team and the community. What the Authority has discovered is that to truly resolve conflict, it is essential to have those who clash become part of the solution.

REUSE: EIGHT CRITICAL STEPS TO BUILDING COMMUNITY SUPPORT[5]

According to some fairly credible folks, the reuse of municipal wastewater is one of the greatest challenges of the next century. But the idea *is* catching on. Surveys indicate the use of reclaimed water has increased fairly rapidly during the past fifteen years.

One thing we know for certain: the more up-front public awareness and community involvement, the better the chances of public acceptance. For a reuse program to be successful, it must first be based on a clear understanding of public values and concerns, and it must be based on a partnership with the stakeholders most affected by the plan. The perceived or real risk factors must be dealt with early on and head-on—and that's the difficult part.

We have a lot to learn about dealing with perceived and/or real risk or any other issues of such concern that citizens refuse to allow a project to occur. Savvy managers have long recognized the value of public participation in advance or wherever and whenever it is necessary to accomplish a project and they are directing their organizations toward public process. The really good ones are doing so with a belief that the public's input

[5] Reported by Linda Kelly, Unified Sewerage Agency, Washington County, Oregon.

broadens the range of possibilities for projects and deepens society's understanding of our difficulties in serving the public.

Eight critical steps are required to take a proactive approach to building community support:

(1) Be up-front and proactive.
(2) Develop a basic information campaign.
(3) Work with your local media.
(4) Use credible third-party testimony.
(5) Show successful projects elsewhere.
(6) Be visible and creative.
(7) Go for more public awareness rather than less.
(8) Use demonstration projects.

It wasn't so long ago that citizens looked to elected officials and business leaders to make the decisions that affected their health and the environment. In general, citizens presumed that the public and private sector leadership had strong public protection and environmental ethics. As the perceived or real risks get more attention, the average citizen is less likely to trust that the government knows how or will do its best to protect public health and the environment. They are beginning to feel not just global risks, but personal risks. Now they believe they have no choice but to intervene when their health or quality of life is likely to be impacted by agency/government decisions.

Public opinion matters, and it's no wonder the public has doubts. Technical limitations and disagreements among experts inevitably affect communication in the adversarial climate that surrounds many risk issues. If the experts can't agree, how can citizens trust that government organizations are providing "risk-free" or at least acceptable solutions?

What usually happens is that agencies feel secure in the realm of technology and science; they believe they have an acceptable, low-risk project. The citizens learn of the so-called "low-risk" project, then become concerned that *any* risk exists. The battle begins when agencies ignore risks that are considered important by the public.

Most groups have legitimate concerns, and they have every right to information and to be involved in the decision-making process. Remember, the public includes doctors, chemists and teachers, as well as persons with less scientific background who understand more than we might expect. Ignoring the issues that influence public perception—or worse yet, labeling the citizens as irrational and then discounting their opinions—is guaranteed to raise the level of hostility between your agency and the community. What many agencies don't realize is that these hostilities will ultimately stand in the way of a project. The following case study of a community uprising resulting from a planned recycled wastewater project illustrates these principles in action—as well as the results and the lessons learned.

The Unified Sewerage Agency (USA) of Washington County, Oregon, is a 26-year-old service district which collects urban wastewater from 370,000 residents and businesses from twelve cities and unincorporated areas. USA operates four treatment plants collecting an average of 50 million gallons of wastewater per day. Once that water is cleaned to advanced tertiary levels, it is recycled onto land or discharged to the Tualatin River.

Almost eight years ago, a lawsuit regarding wastewater permit violations resulted in USA creating technology for reducing the phosphorous levels to .07. The technology is in place and the water is being cleaned to near drinking water standards. In 1990, USA developed a Wastewater Facilities Plan to meet regulatory requirements for water quality in the Tualatin River Basin. A related plan was developed for wastewater recycling. Several intergovernmental and public committees were involved in a most extensive public involvement process for the development of both plans.

At one of the wastewater facilities, application of the effluent on USA-owned land exceeded regulatory guidelines. In an effort to reduce the application rate, USA purchased 362 acres of farmland adjacent to the treatment facility. The land-use application for construction of the irrigation system was requested and notices of the intended use were posted/mailed to notify adjacent property owners.

When the notification process began, a concern was raised by the staff at the Regional Drinking Water Treatment Facility because the intake for this facility is located on the opposite side of the river just upstream from the potential reuse site. The concern was whether or not applying levels II and IV effluent on this property would have any adverse effects on the drinking water for three of the large cities the District serves. Neighbors and representatives of the three cities receiving the water chimed in. Even though USA's staff and technical specialists considered the risk to be practically nonexistent, the Agency agreed to implement significant safeguards and adopted management practices for additional protection of the quality of the surface water that could migrate to the river upstream of the intake. The safeguards included additional land buffers along riparian areas, construction of cut-off drainage trenches and normal farming operation which includes irrigation of non-food crops at the regulatory rates.

USA discontinued the permit process and formed a task force composed of regulatory and municipal representatives and concerned citizens to address the issues. Over six months, USA held three public meetings and addressed more than 100 questions in writing in addition to providing vast amounts of technical data. The concerns covered a wide range of topics including cryptosporidium, giardia and bacterial contamination; aerosols; and groundwater contamination.

The safeguards were considered secure and the site was approved by all appropriate regulatory agencies, which included the State Department of Environmental Quality, and the State Health Division. However, the citizen opposition based on perceived risk was overwhelming. The citizens demanded zero risk. Because the agency could not guarantee zero risk, the project was halted.

Some level of public involvement is necessary from the beginning when seeking the public's acceptance of controversial projects, especially if they pose real or perceived risks. It is always better to take a proactive approach and become part of the solution to potential problems by working with the public to create a project that meets community needs, solves community problems and provides community services. If the public feels their participation hasn't been adequate, putting a project on hold until the public has had genuine participation will engender trust and very likely prevent your project from being stopped.

The case referred to was costly in a number of respects. The property purchased for the recycled wastewater project is being partially leased and partially used for mitigation of wetlands related to the construction of a water reservoir nearby. The reactive process took a tremendous amount of time and energy for staff.

Philosophical discussions ensued about "giving in" to public pressure, especially if the public was only a few vocal participants and the risk was too small to even consider from a technical standpoint. We also discussed if the final decision to drop the reuse project was really the best decision for the good of the community. Staff had concerns about spending more public funds on an alternative project when they were sure the first one was a safe and smart long-term investment. They also had concerns about future projects. If a project can never be declared "risk-free," would future projects pass the public's scrutiny?

Overall, however, much was gained. The public was not as suspicious of follow-on projects, because the Agency had demonstrated a commitment to engage in a genuine public participation process—despite the investment of time and money. And yes, even if the chosen alternative cost more than the least-costly alternative. Projects are being built, and the Agency has recouped much of its credibility. Perhaps the greatest lesson of all was that the Agency *needed* to know the answers to the questions generated by the public.

Recycled water programs, sometimes referred to as "reuse," are imminent in the future of water management, but they have the potential for conflict by the very nature of the product. Public acceptance may be the major obstacle to these programs' becoming a regular part of life. Yet the successful programs are proof that when public awareness and involvement are enhanced, public acceptance may be on the horizon.

The incentives for a reuse program make perfect sense to the technical and financial experts—water conservation, economic advantages, environmental benefits, government support, a new water source, and the fact that the cost of wastewater treatment makes the product too valuable to "throw away." So why hasn't the concept been embraced and supported by the community?

The main obstacle to reuse programs is public acceptance. Without public acceptance, your project won't leave the drawing board—even if you're armed with the most technologically advanced safety/environmental data.

To build a successful recycled wastewater program, start with a clear understanding of public values and concerns. Form a partnership with the stakeholders most affected by the plan, including:

(1) Citizens in general: These groups may have several different levels of interest in your organization. There is generally a group of opinion leaders in your community who participate in and contribute to your decisions; there are those who read the headlines and keep track of you, but rely on others to watch out for how you might impact their safety or quality of life; and there are those who barely know of you or care about your business until you impact their lives in some way.

(2) Potential users: These individuals may be farmers, industries, parks, schools, and cemeteries.

(3) Potential users' neighbors: These individuals live or work near the users.

(4) Agricultural agencies/extension agencies: These individuals may carry forth information on your behalf.

(5) Water distributors

(6) Food processing industry

(7) Regulatory agencies

(8) Resource agencies

(9) Environmental groups

(10) Community and civic organizations

(11) Ratepayers

Once you have identified your potential stakeholders, surveys are a useful tool for identifying the various perceptions about your project. Most of those surveyed for USA's water reuse project showed interest in the project and expressed concerns about safety, costs and public acceptance. Those who indicated strongest opposition were concerned about exposure of recycled wastewater to the public (i.e., schools and parks). Representatives from the cemeteries were concerned about negative public perceptions associated with using "sewer water on graves."

While most of these fears are unfounded, they show the need for public education as an integral part of the public participation process. Think about it: the average citizen has never toured a treatment facility and has little knowledge of wastewater processes or operations. When they think of "sewage," what images come to mind? Unclean, smelly, disease-laden, AIDS carrying, filthy . . . industrial waste, metals, toxins, unknowns. It's no wonder apprehension and oftentimes panic prevail.

The average citizen has little or no knowledge of how a wastewater treatment plant "cleans" wastewater. The do not understand the biological processes, the chemical analysis, pretreatment, TMDLs, Class IV water, etc. Perhaps they have heard a few horror stories about E. coli or cryptosporidium killing innocent people. They may find it hard to trust scientists and government officials when these are the same officials that tested and approved the use of hazardous and sometimes deadly materials such as DDT, asbestos, nuclear waste, alar, and red food coloring. Even the once-touted oatmeal has turned suspect. It's no wonder the public is skeptical.

Oftentimes, the public exerts pressure on the business and governmental community. A food processing company owner, a farmer, a school superintendent—all share the same concerns. What if I use recycled wastewater and the public finds out? Will media attention and misinformation drive me out of business? Prevent people from buying my crops? Outrage parents and cause a public outcry?

That's where a comprehensive public information program comes into play. Eight strategies your organization should consider to be critical follow:

(1) Be up-front and proactive. As an organization, make a deliberate decision to explore the *possibility* of implementing rather than the *certainty* of implementing a recycled wastewater program. Go into the community and disseminate enough information so that the public can understand your project thoroughly and communicate their concerns. And remember—credibility is your ally.

(2) Develop a basic information campaign. Line up all your information and get your facts straight *before* you present them to the public. Prepare an internal question-and-answer (Q&A) document. Brainstorm with staff about every possible question that might be asked about your program. Develop responses that offer options. Avoid policy decisions at this point because you want to be sure your policies encompass the community. Update your Q&A as new questions arise and policies develop. In addition, develop (or borrow) informative, exciting materials such as pamphlets and videos to illustrate your point. Use questionnaires and surveys to measure the community's willingness to consider a recycled wastewater program. Develop an

aggressive speakers' bureau and attend every civic event. Make it your mission to explain your cause and explore the public's thinking about whether a recycled wastewater program makes sense for the community. At the end of every presentation, ask for input. Provide additional forums for people to air their thoughts and feelings, such as hot-lines, small meetings and one-on-one communication. Watch for trends in their response, and note whether they are related to values, interests, relationships, perceived or real risks—or a combination.

(3) Work with your local media. This is a good time to educate reporters/editorial staff members and to seek opportunities for sharing information. Headlines at this stage should read something like "County Government Asks Citizens What They Think About Using Recycled Wastewater for Irrigating Parks." Can you see how this differs from "County Government Imposes Recycled Wastewater on Parks"? Guest columns will help the community talk about the pros/cons and get the discussion going. Your goal here is not to guarantee that the program is 100 percent the answer to everything, but that it is safe and makes sense for the community. Now is the time to prepare a summary of all education activities including every related press clipping. Keep these records current from the formative stages to implementation. They will become valuable tools when the few who oppose the program claim that you never asked for input and they were never told about such a program.

(4) Use credible third party testimonials. Few citizens will believe a government agency like they will believe an objective bystander. Approach credible people in the environmental and farming communities, perhaps the extension agent, to see if they are willing to share their experiences, understanding and knowledge of such programs. They may be willing to discuss benefits such as lower cost alternatives, the bonus of phosphorus and nutrients for crops which will reduce the need for fertilizers. They have a valuable story to tell that will be much more interesting and real to the community than the one you're telling. They may be willing to talk on a local cable forum, to participate in a video, or simply to be quoted.

(5) Show successful programs elsewhere. No one has to be a guinea pig nowadays. There are plenty of successful recycled wastewater programs to highlight. Sponsor a seminar on the topic and bring in experts to show and tell what they know and what they've learned. Charter a bus and take folks on tours of areas where the product is being used. Look for long-term, proven programs where excellent management has fostered community support.

(6) Be visible and creative. Name the program and incorporate the name

in a logo. Use this logo on every education piece in order to foster recognition. Take some creative risk to make the program visible. We printed the slogan, "It's safe, it's clean, it makes sense" on ping-pong balls and our general manager and Board chair used golf clubs to "chip" them off the back of a parade float decorated like a golf course. The children caught the souvenirs in the air, and it was a big hit. That gimmick was so popular, we did it two years in a row. (I wouldn't do it again, however, because balls bounced back near the wheels of the float and we had to have walkers alongside the float to make sure children were safe. But, you get the idea.) Consider billing inserts, bumper stickers, cable forums and any other visible way you can acquaint your community with this concept.

(7) Go for more public awareness and involvement than less. Don't under-estimate the extent to which the public education and involvement effort will be needed. Spend time on this element from concept to implementation. "Awareness" will help people get acquainted with the concept; "involvement" is critical in shaping the program. Identify stakeholders, develop citizen involvement committees (perhaps a tech-nical committee and a policy committee or some combination), identify potential users, establish goals, define the public education needs, and develop and implement policies and strategies. Even with the best public education and involvement presentation, people will come in late and their concerns will need to be addressed just as if you were at the beginning. Every time a policy issue is presented, a new round of questions must be asked to see if its options are acceptable to your community.

(8) Use demonstration projects. Demonstration projects should be a major component of your discussions because they increase public awareness of the recycled wastewater benefits, techniques and uses. The more often people see signs of a project, the more likely they will become used to the concept and more trusting of it. We used signs to increase public understanding that water conservation is a big part of recycling wastewater. Demonstrate every facet of the program and the range of benefits: farm fields, golf courses, athletic fields, etc.

Most wastewater agencies do not expect the community to accept a recycled wastewater program wholeheartedly. It will take years of excellent resource management before the water (and sometimes the organization) can be trusted. Don't expect to resolve problems and answer all questions quickly or easily. Neither should you expect speedy resolution of difficult communication problems with citizens. A spirit of respect and concern should be maintained throughout the process by all the organization's staff. This means timely response to questions, sufficient lead time for meetings and public review of policies. It means listening to citizens' ideas and concerns and providing the information they request. Even if it takes a long time and consumes valuable resources, it will be well worth the effort if your community leaders and citizens believe in the value of your program. Then, you must maintain excellent management of the program because your credibility is at stake—and credibility is everything.

HOW COMMUNITY RELATIONS HELPED WIN APPROVAL FOR BOLSA CHICA[6]

For more than twenty-five years, one of the most controversial land use issues in Orange County, California, involved Bolsa Chica—a 1,700-acre beach-front oil field in Huntington Beach. The area included severely degraded (dry) lowland wetlands bordered by two upland mesas. In 1989, the landowner, state and local government, and the area's largest environmental group adopted a historical "coalition agreement" for a land use plan including 1,100 acres of restored wetlands funded by development on 300 acres of the most severely degraded lowlands and one of the mesas. The plan actually included a net gain of wetlands with 75 percent of the property restored and development on only 25 percent. Although the coalition agreement was widely praised, delays in the approval process gave those who wanted no development the time to establish a vocal opposition group.

The challenge was to build support for the most highly charged, controversial land entitlement project in Orange County, in spite of efforts to block the plan by powerful local activist groups, an adversarial city council and unsympathetic local media.

Koll Real Estate Group, a publicly held company with stockholders nationwide, is the owner of Bolsa Chica and is seeking approval for a wetlands restoration project and a 3,300-home residential development. Gladstone International provided community, media and government relations services to support the Bolsa Chica Plan.

To understand the present-day dynamics of the Bolsa Chica area, one has to look to the area's place in history. The relevant history of the Bolsa Chica Wetlands begins at the turn of the century, when the area was in full bloom. According to a local historian, the sun was eclipsed by birds flying overhead on their journey along the Pacific flyway. Salt water flushed over the land, cleansing it and bringing it to life. Food was abundant and animals and birds flocked to the land for the feast.

Such a grand supply of wildlife attracted the Bolsa Chica Gun Club, who took ownership of the property. Duck blinds were "enhanced" by

[6] Reported by Joan Gladstone and Jill Kanzler, Gladstone International, Irvine, California.

damming the natural tidal inlets that fed the wetlands and gun shots rang out in sport. Bolsa Chica was a hunter's paradise, the supply seemed never-ending. At the time, the hunters were unaware that Bolsa Chica carried with it another plentiful resource—oil.

In the 1920s, oil was discovered in Huntington Beach and the boom began. Bolsa Chica offered vast quantities of oil pumping from beneath its skin. Hundreds of oil derricks dotted the Huntington Beach landscape while oil roads, pipelines, sumps and tanks crisscrossed back and forth, cutting off Bolsa Chica's wetlands from the sea and creating isolated "cells" of habitat including semi-stagnant ponds and dusty weed patches.

In 1970, the property passed from the Bolsa Chica Gun Club's descendants to Signal Landmark Corporation, now a subsidiary of Koll Real Estate Group. Soon thereafter, the State of California made its claim on Bolsa Chica. Citing a California law that places tidal waters in the public domain, the state declared that a large portion of the Bolsa Chica lowland was subject to the ebb and flow of the tide and, therefore, was public property. After much negotiation, then-Governor Ronald Reagan signed a land exchange agreement in 1973 which settled the dispute between Signal and the State of California.

The agreement gave approximately 300 acres, now known as the Bolsa Chica State Ecological Reserve, to the state. Signal received the title to the balance of the property, which included more than 1,100 acres of lowland and 600 acres of upland area. In exchange, the state agreed to help Signal implement its plan to construct an ocean entrance for a marina development at Bolsa Chica. In 1978, the state began to restore some of those degraded lands acquired from Signal Landmark. This is what the public sees today—a strip of 163 acres of restored wetlands between Pacific Coast Highway and a field of degraded wetlands that continues to this day in oil production.

Reviewing these actions closely was the League of Woman Voters, who later formed the citizens' group, the Amigos de Bolsa Chica, which would become one of the Bolsa Chica wetlands' most avid protectors. The Amigos took issue with the proposed marina development of the remaining lowland area. Their goal was to preserve as much open space as possible. To that end, the Amigos disputed the terms of the state's agreement with Signal Landmark and filed a lawsuit in 1979 to set aside the agreement, as well as foreclose the possibility of development at Bolsa Chica.

Meanwhile, Signal Landmark proceeded with its vision for the property—to build a 1,300-slip marina with two 2,000-foot jetties, space for ocean-front hotels, seaside shops and restaurants, and the development of 5,700 houses. By 1986, the land-use plan was approved by the County of Orange and received certification, subject to conditions, from the California

Coastal Commission. Although the marina land-use plan included the expansion and permanent preservation of 915 acres of wetlands, local citizens were outraged. What began as a rumble from the Amigos de Bolsa Chica, soon became a roar. These activists would not stand for the building of a marina and vowed to fight the landowner until the end.

Fight they did. Bolsa Chica became a major campaign platform for local city and county candidates, and the Amigos de Bolsa Chica emerged as a powerful citizens' group with considerable political clout and membership swelling to 1,000.

In response to the community's outcry, Orange County Second District Supervisor Harriet Wieder and former Huntington Beach Mayor John Erskine made an unprecedented move—they brought the interested parties together and retained a mediator to resolve the issues regarding Bolsa Chica.

The landowner, Amigos de Bolsa Chica, City of Huntington Beach, State of California and the County of Orange embarked on a long negotiation process with the formation of an innovative public/private partnership called the Bolsa Chica Planning Coalition.

For six intense months, the five main parties and sixteen agencies and organizations that provided guidance and technical support, worked to map out a land-use plan that would satisfy all concerned. The top two concerns were how to guarantee the restoration of the Bolsa Chica wetlands and how to protect the landowner's property rights.

On May 22, 1989, the Bolsa Chica Planning Coalition Concept Plan, a guideline for developing the Bolsa Chica property, was announced to the public. For the first time, the community was supportive. Details of the plan included more than 75 percent of the property to remain as open space with a 1,100-acre contiguous wetlands ecosystem. In fact, that public ownership and expansion of restored wetlands would increase from 163 acres to over 1,100 acres—a 600 percent increase!

Development of 3,300 homes would be limited to only 400 acres, or one-fourth of the site, and would guarantee the funding of the restoration and other improvements at no cost to taxpayers. The plan called for 2,400 homes on the upland mesa and 900 homes on lowland, degraded wetlands.

The Coalition succeeded because the agreement met the needs and concerns of all participants. They had charted a plan to create a minimum of 1,000 acres of wetlands with quality feeding, nesting and breeding grounds for six endangered species and other animals that use the site throughout the year. They respected the landowner's rights to profitable use of its property, allowing for residential development along the periphery of the site in exchange for the restoration and preservation of the wetlands.

This exchange would create the largest real gain of wetlands in Southern California and increase the quality of wetlands at Bolsa Chica, a restoration project of particular ecological significance to California, which has lost

90 percent of its coastal wetlands over the years. In addition, the plan provided for a 106-acre linear park overlooking the 1,100 acres of the restored wetlands ecosystem.

Unfortunately, even with the cooperation of the many groups involved, the public process was not easy and many lessons were learned along the way. During the planning stages of the land-use plan, changing regulations and regulatory agencies' inconsistencies often caused unforeseen delays and a lack of certainty about the process. In addition, because the regulatory process took so long, it encouraged opposition and splinter groups to form. The purpose behind creating the Coalition was to stop the fighting among opposing groups which was impeding the process. However, despite the Amigos' best efforts to be fair and build consensus, a vocal minority was adamant that absolutely no development whatsoever take place.

By the spring of 1993, four years after the Coalition Agreement, Bolsa Chica had still not undergone planning commission hearings, and the once-heralded plan was being distorted by the opposition. The Bolsa Chica Land Trust, a splinter group founded in May 1992 by former Amigos de Bolsa Chica members and others, had the stated goal of purchasing Bolsa Chica from the landowner to halt the entitlement and implementation of the land-use plan. The Land Trust maintains its goal to raise public funds to purchase the land to create a bio-diversity park.

Fundraising efforts over two years have included selling honorary deeds and book sales, but to date they have not exceeded $50,000. The property itself, at 1,700 acres, has been valued with similar coastal property in the $1 million per acre range.

Members of the Bolsa Chica Land Trust were able to position their ''no development'' agenda by claiming that Koll Real Estate Group (Signal Landmark was acquired by Koll Real Estate Group) was ''building on the wetlands.'' This was complicated by the fact that the State Ecological Reserve, located along a stretch of Pacific Coast Highway, was most visible to the public. The public never saw the dry and dusty interior with its oil wells, dry brush and sand, and assumed that all of Bolsa Chica was a beautiful, thriving wetlands.

With local hearings slated for the fall of 1994, it was critical to change the focus of the debate back to the issue of wetlands restoration with a guaranteed funding source. A series of meetings with local community leaders including members of the Chamber of Commerce and City Council, City staff, realtors and seniors determined that videos, brochures and slide shows could not convey the extent of degradation at Bolsa Chica. The first challenge was to get people out to the actual site to see the real Bolsa Chica first hand.

The second challenge was to educate the community through media relations. Gladstone International began media relations efforts to bring the developer's story to the public in 1992. The following year was marked by several major milestones in the land-use entitlement process that dramatically increased the hostile and active opposition to the project by environmentalist groups. The main local opposition group established alliances with the Sierra Club, National Audubon Society and Surfrider Foundation. The local media was focused on the escalated debate, resulting in crises scenarios and negative coverage that could ultimately affect the attitudes of political decision makers, public officials, opinion leaders, stockholders and local residents unless responded to and challenged.

First, Gladstone International analyzed the results of focus group sessions and a telephone survey. It was determined that print media was by far the most effective way to reach the target publics: Huntington Beach residents, county business leaders and county public officials. The research was also used to develop the basic message in the program. Throughout the years, news articles and letters to the editor about the project were analyzed for message content and strategies were developed to respond to the press. Editors, opinion-editorial writers and reporters at all the local papers were investigated to determine the best people to work with in a given situation. Newsletters by opposition groups were analyzed to anticipate organized protests and demonstrations against the project and environmental group meetings were attended to learn more about their position. Research efforts also included contact with twelve expert consultants to gather information on particular issues.

The goal was to take local and regional opinion leaders, plus anyone with an interest in the project, on a one and one-half hour tour of Bolsa Chica. A tour script was written to keep messages consistent. Active community groups were targeted for tours, so their word-of-mouth support of the plan would reach a larger audience. The community relations plan included production of mementos and literature which underscored the project's messages (i.e., cups with graphics that turned from an oil field to a wetlands, binoculars and expandable travel bags). Those who became supporters would receive recognition pins and T-shirts and be included on a mailing list to receive bulletins, information and follow-up materials after the tours.

One major milestone in 1993 for Koll Real Estate Group was their decision to switch the entitlement process from the City of Huntington Beach to the County of Orange in mid-March after a local election changed the sentiment of the City Council. The development proposal had been under City jurisdiction for the previous three years and Koll expected that this change would create a huge controversy with City officials, residents and the media, requiring careful justification.

In December, the County of Orange released the Environmental Impact Report that was heavily scrutinized by opposition groups and the City of

Huntington Beach. At the same time, Koll Real Estate Group announced the results of the first report on the projected economic benefits of Bolsa Chica. The study predicted that 17,000 jobs would be created and $1.2 billion of economic activity would be generated when the project was approved. The positive information helped to offset some of the opposition's claims about the EIR. This strategy was a result of formal strategy meetings that were held once a week to review public relations issues and discuss the status of the entitlement process. In addition, there was daily contact with Koll and members of its team of specialists to discuss new coverage and response strategies.

The goal throughout the campaign was to increase community support for the project. To support that goal, the media relations objective was to attain balanced news coverage (an equal number of pro and con paragraphs, measured by content analysis) for the project throughout the year. The strategy was to educate reporters covering the project in advance of major milestones, and provide feedback of the coverage to reporters and editors either directly or with letters to the editor. Strategy included having third-party opinion leaders deliver messages where possible.

The most likely supporters were contacted, such as the Chamber of Commerce, every real estate office in the city, the Senior Center, elected officials and their staff, media and others to offer site tours. The tours were publicized by press releases, notices in the Chamber newsletter, Senior Center flyers and Association of Realtor newsletters, and through a billboard with an 800-line. Later, more than 125 local business owners put information/tour pamphlets in their stores. At the conclusion of each tour, guests were given support cards and encouraged to sign up on a support list. The names collected were entered into a database to track their affiliation and level of support. People who signed on as supporters received a thank-you letter from Koll Real Estate Group, plus "Bolsa Chica Countdown," a newsletter designed to be quickly produced by computer, and distributed within a day to keep supporters informed as quickly as events occurred. Regarding education through media relations, program elements included the following: crises response; 35 meetings or tours of the project site for reporters, opposition-editorial writers, editors and TV crews for local television stations to film B-roll footage; re-education of reporters when new issues arose; education of new reporters; maintaining contact list; preparing and distributing news releases to announce key milestones; ghost-writing of opinion articles for Koll senior management and third-party opinion leaders; booking Koll senior management on local cable TV and radio talk shows; and preparing fact sheets for background use. By reacting quickly in crisis situations, we were able to prevent coverage of publicity stunts by opposition groups. For instance, one group was able to attract a TV crew to the site by claiming that mice on the property carried a deadly virus. Gladstone International staff convinced the producer that this was a hoax, and no story appeared on the network newscast as a result.

In one year, more than 150 tours were organized, bringing 2,000 people to the site. Over 1,200 signed on as active supporters. In addition, a Chamber member and a real estate professional came forward to form their own grass-roots organization—Business for Bolsa Chica, with 125 members and a realtor support group of 170 members. Regional leaders were impressed by the environmental and economic benefits. They formed another organization, Californians for Bolsa Chica, which included a blue-chip roster of leaders from business, community, education and public utilities.

By late August 1994, supporters became just as vocal as the opposition. At the first planning commission hearings, more than 600 people attended, with half or more of the audience in support of the project. Through all five commission hearings, supporters outnumbered opponents. Their speeches at the hearings referred to the site tour, and the fact that Bolsa Chica was not a thriving wetlands. They urged approval of the project so that funds from development could be used to restore the oil field into a functional wetlands. On December 14, 1994, at the Orange County Board of Supervisors hearings, the project received unanimous approval.

The objective of attaining balanced media coverage was met in spite of the constant pressure and publicity stunts the opposition groups held at the project site (including a bluegrass concert, a Native American religious ceremony, a candlelight vigil, an event to sell fake grant-deeds to the property, and a winter solstice celebration). The most dramatic evidence of success was in the tremendous growth in local support. In January 1993, there was a formal rally of approximately 100 supporters. By December 1993, this number had grown to more than 5,000.

Having positive messages in the print media about economic and environmental benefits aided the efforts to mobilize citizens to combat negative articles and letters by writing opinion articles and letters supporting the project. Having a level playing field in the media also allowed the team to gain the support of the Orange County business community, resulting in written endorsements by trade and commerce associations.

Efforts in community and media relations continued after the Orange County Board of Supervisors' approval because the plan still needed additional approvals. Continuing the program throughout 1995 required a very proactive campaign because many people believed the Board of Supervisors' hearing was the only approval necessary. In fact, it was only one of many.

The California Coastal Commission was the next hurdle and was expected to be the most complex. Since the California Coastal Commission serves the entire coast of California, the hearing could take place anywhere along the coast, making it very difficult for supporters to attend a hearing. Therefore, two regional strategies were devised to influence the Coastal Commission. The first included a major letter writing campaign by the supportive constituency to the twelve Coastal Commissioners urging their support of the Bolsa Chica Plan. The other plan included seeking endorsements from statewide organizations to demonstrate widespread support.

The letter writing strategy included educating supporters on issues important to the commissioners and encouraging them to send letters as often as possible. Letters were also sent by the leadership of the Bolsa Chica Alliance, Business for Bolsa Chica and Californians for Bolsa Chica.

Another regional strategy included seeking endorsements for the plan from statewide organizations. This was helpful to influence commissioners representing areas of the state who were not as familiar with the project. We targeted business-oriented organizations as well as environmental organizations that supported a balance of nature and the economy. Several of these organizations endorsed the Bolsa Chica Plan and sent letters to the Coastal Commissioners encouraging approval of the plan.

Finally, more than a year after the Orange County Board of Supervisors' approval, in January of 1996, the California Coastal Commission held a hearing on Bolsa Chica in Los Angeles.

Although the hearing was held an hour's drive away from the project, more than 600 supporters and opponents attended the hearing to testify on behalf of Bolsa Chica. After hundreds of people spoke to the Coastal Commissioners, it was apparent that there were more supporters than opponents.

After an exhausting fourteen-hour public hearing, the California Coastal Commission approved the proposed Orange County Bolsa Chica Plan.

As of December 1996, Koll Real Estate Group is in negotiations with State and Federal wildlife officials to sell the lowland or wetlands portion of the property, representing approximately 75 percent of Bolsa Chica. The agreement includes using $67 million in funds provided by the Ports of Long Beach and Los Angeles to pay for restoration of the wetlands. The ports, in return, will receive mitigation credits that will allow them to expand their ports. Koll Real Estate Group will still own and build 2,400 homes on the mesa portion of their property, but will not build the 900 homes in the degraded wetlands section.

The successful community and media relations program helped Koll Real Estate Group to win necessary government approvals that put the company and Bolsa Chica in a favorable position with either outcome.

The Regulatory Status of Wastewater Reuse in the European Union

THIS book presents many examples of the potential benefits of wastewater reuse in the world. Naturally, they are most obvious for the arid areas but the increasing pressures on water resources in industrialized countries in Europe should also make wastewater reuse attractive there. The increasing concern for sustainable development, as increasingly recognized in the policies of the European Union, is a strong promoter of sound water management.

Wastewater reuse can be a matter of choice in general water management strategy. Worldwide, wastewater reclamation and reuse is estimated to represent a potential extra water resource amounting to approximately 15% of existing water consumption. On a local basis, this proportion can be significantly higher (e.g., 30% of agriculture irrigation water and 19% of total water supply in Israel in the future). In view of the increasing pressure on all water resources, both in industrialized and in developing countries, supplementing water resources with reclaimed wastewater can no longer be neglected.

CALIFORNIA: A PIONEER WITH RESPECT TO WASTEWATER REUSE REGULATION

Within the industrialized countries, California has been a pioneer in the systematic approach to wastewater reuse. The first Californian regulations on wastewater reuse for irrigation date back to 1918 (Asano and Levine, 1996; see also Chapter 14 by J. Crook). Progress in wastewater

Laurent Bontoux, European Commission, Joint Research Center, Institute for Prospective Technological Studies, W.T.C., Isla de la Cartuja s/n, E-41092 Seville, Spain.

reuse has been strongly linked to the development of wastewater treatment technologies: primary settling, activated sludge, ponds, chemical treatment, filtration, disinfection, nutrient removal, etc. Since the 1960s, California has been actively promoting wastewater reuse by drafting regulations and promoting research for all types of applications (irrigation, industrial and municipal reuse, groundwater recharge, potable reuse, etc.). Generally, the requirements of the application define the technological means of treatment. Today, California has a number of large-scale applications and has generated many invaluable experimental results, making it a reference the world around. The Californian approach is a high tech, "better safe than sorry" approach designed for the needs of a rich society.

Today, the available experience has spread beyond California, and wastewater reuse schemes are being developed in a variety of contexts worldwide: wastewater reuse technology and practice have come of age, even if some public health concerns persist.

There are many examples of the potential benefits of wastewater reuse in the world, but irrigation is by far the largest application. While these benefits are most obvious for the arid areas, the increasing pressures on water resources in "wet" industrialized countries, both in a quantitative and a qualitative perspective, are also making wastewater reuse attractive in such places. The rising general concern for sustainable development, as increasingly recognized in the policies of the European Union, is a strong promoter of sound water management.

WASTEWATER REUSE IS ON THE RISE IN SOUTHERN EUROPE

Since the beginning of the 1990s, several droughts have hit Europe. The most severe was probably the recent one (1991–1995) experienced in Spain, where its impact on agriculture and on the population was considerable, but the 1995 drought in England, while much more limited in time, also left crops thirsty and seriously disrupted municipal water supply.

Additionally, various long-term climate change models predict an increase in rainfall in the wetter areas of the world and a decrease in the drier areas (Anonymous, 1992; Von Storch et al., 1993), together with changes in weather patterns. This highlights the increased risks of water shortages in regions such as the Mediterranean (Shelef and Azov, 1996; Anonymous, 1992) and gives an early warning for immediately improving the management of water resources in the threatened areas.

In this context, wastewater reuse is increasingly attractive in southern Europe, but also in France and even in the UK (Stuart and Chilton, 1995). Wastewater reuse for crop irrigation is already being practiced at large scale in Spain and Italy and is being seriously considered in Greece and Portugal where wastewater reclamation and reuse experiments are being carried out. Most of these countries are large exporters of agricultural products. Reclaimed wastewater is also increasingly used to irrigate golf courses and resorts, a significant international tourist attraction and money earner. This is the case in southern Spain. Southern Portugal is also implementing wastewater reuse. Cyprus and Tunisia also practice this type of irrigation.

Finally, an overall improvement of the quality of treated wastewater in line with the implementation of the European Wastewater Directive makes the development of wastewater reuse a certainty (Marecos do Monte et al., 1996). However, this trend is taking place in a very uneven regulatory context.

THE WHO GUIDELINES

At the international level, the 1989 World Health Organization (WHO) "Health guidelines for the use of wastewater in agriculture and aquaculture" are the only existing guidelines for wastewater reuse. While reviewing the health risks and the limited epidemiological evidence available at the time, the only specific criteria the WHO proposes are microbiological (see Table A.1). Work has started on chemical guidelines since then (Chang et al., 1995).

The main justification for a standard of less than 1000 fecal coliforms per 100 ml is in contrast to the less than 2000 fecal coliforms per 100 ml used as the European standard for bathing waters. Protozoa are not included in the WHO guidelines because the technologies effective in achieving the nematode standard arguably also provide a certain removal of protozoa. Enteric viruses are not considered. These guidelines were intended to guide wastewater treatment design engineers in the choice of treatment and management technologies that will reliably achieve these standards. Since these guidelines have a worldwide scope, they were also designed to stand realistic chances of being applied in developing countries, where an unnecessarily stringent stance would most probably result in them being ignored (Mara and Cairncross, 1989).

These guidelines essentially cover the microbiological quality of treated wastewater and are accompanied by recommendations on the wastewater treatment technology to be applied. They have become a point of reference for many existing legislations in Europe but remain widely criticized in some industrialized countries, particularly in California, where the "Title 22" regulation (see Appendix II of this book) requires less than 2.2 fecal

TABLE A-1. Recommended Microbiological Quality Guidelines for Wastewater Use in Agriculture[a] (WHO, 1989).

Category	Reuse Conditions	Exposed Group	Intestinal Nematodes[b] (arithmetic mean number of viable eggs per liter)[c]	Faecal Coliforms (geometric mean, number per 100 ml[3])	Wastewater Treatment Expected to Achieve the Required Microbiological Quality
A	Irrigation of crops likely to be eaten uncooked, sports fields, public parks[d]	Workers, consumers, public	≤1	≤1000[d]	A series of stabilization ponds designed to achieve the microbiological quality indicated, or equivalent treatment
B	Irrigation of cereal crops, industrial crops, fodder crops, pasture and trees[e]	Workers	≤1	No standard recommended	Retention in stabilization ponds for 8–10 days or equivalent helminth and faecal coliform removal
C	Localized irrigation of crops in category B if exposure of workers and the public does not occur.	None	Not applicable	Not applicable	Pretreatment as required by the irrigation technology, but not less than primary sedimentation

[a] In specific cases, local epidemiological, socio-cultural and environmental factors should be taken into account and the guidelines modified accordingly.
[b] Ascaris, Trichuris, and hookworms.
[c] During the irrigation period.
[d] When edible crops are always consumed well-cooked, this recommendation may be less stringent.
[e] In the case of fruit trees, irrigation should cease two weeks before the fruit is picked, and no fruit should be picked off the ground. Sprinkler irrigation should not be used.

1466

coliforms per 100 ml for food crops and high exposure landscape and less than 23 per 100 ml for pasture and landscape impoundment (State of California, 1978). The California standard does not have a value for nematodes because of the low public health significance of the issue in the USA.

EXISTING LEGISLATION AND GUIDELINES ON WASTEWATER REUSE IN MEMBER STATES OF THE EUROPEAN UNION

In general, the adoption of standards for wastewater reclamation and reuse follows the problems encountered in each country. As a result, across Europe, the legal status of wastewater reuse is not uniform. Many European countries and most northern European countries (e.g., the United Kingdom, Belgium, the Netherlands) do not have any specific legislation on the matter. France has national recommendations, Italy a national law and Spain various regional regulations. Portugal and Greece are considering developing national guidelines.

FRANCE

In line with its administrative tradition, France has enacted a comprehensive national code of practice under the form of recommendations from the Conseil Supérieur d'Hygiène Publique de France (CSHPF, 1991). The 1991 "Sanitary Recommendations on the Use, after Treatment, of Municipal Wastewaters for the Irrigation of Crops and Green Spaces" use the WHO guidelines as a basis, but complement them with strict rules of application. In general, the approach is very cautious and the main restrictions given by the CSHPF are

- the protection of the ground and surface water resources
- the restriction of uses according to the quality of the treated effluents
- the piping networks for the treated wastewaters
- the chemical quality of the treated effluents
- the control of the sanitary rules applicable to wastewater treatment and irrigation facilities
- the training of operators and supervisors

The CSHPF calls for a strict observation of these restrictions to ensure the best possible protection of the public health of the populations concerned. The WHO guidelines are introduced in the second point, but all the points covered are accompanied by very precise lists of requirements, such as the performance of hydrogeological studies, the characterization of the waters to be reused, the respect of distances from inhabited areas, the delivery of administrative authorizations, strict monitoring, and the like. In fact, the authorizations for wastewater reuse are attributed on a case by case basis after review of a very detailed dossier. There are no explicit strict standards for minerals or trace organics, but the experts providing their advice before delivery of the permits follow recommendations and usual practices and can in every case refuse the authorization.

ITALY

In Italy, a national water legislation exists (law 319 of May 10, 1976), complemented for wastewater reuse in agriculture by the "Criteria, Methodologies and General Technical Standards" of February 4, 1977 (Ministero dei Lavori Pubblici, 1977). These standards aim at protecting the soil used for agriculture and the crops. It gives limits on certain minerals, such as Na, Mg and Ca, by ways of ratios and tables of values. For the irrigation of crops that can be eaten raw (unrestricted irrigation), municipal wastewater effluents must go through secondary treatment and disinfection, in order for the level of total coliforms not to exceed 2 per 100 ml. For pastures (restricted irrigation), the level of total coliforms must be lower than 20 per 100 ml. In the case of crops that do not come in contact with the water (restricted irrigation) and in all the other cases, only primary treatment is required. However, "chemicals that may leave undesirable residues" in the crops must be absent.

For the soils not used for agriculture, the only interest of the law is to preserve the landscape, or natural value, with possible limitations due to the local hydrogeology. Precautions against the contamination of water resources and monitoring are also foreseen.

In Italy, the regions benefit from a certain autonomy in the regulatory area, and the three regions where wastewater reuse is most practiced (Puglie, Emilia Romagna, and Sicilia) have enacted comprehensive standards, without necessarily following the line set by the national legislation. Puglie takes a single value of 10 total coliforms per 100 ml; Emilia Romagna takes a value of 12 total coliforms per 100 ml for unrestricted irrrigation and of 250 total coliforms per 100 ml for restricted irrigation. Sicilia takes a radically, and probably more realistic, stance. It forbids the irrigation of fodder crops and of food crops that come in direct contact with treated wastewater. For the other cases (restricted irrigation) the applicable standard is 3000 total coliforms per 100 ml and 1000 fecal coliforms per 100 ml, simultaneously. It also requires the absence of salmonella and less than 1 helminth egg per liter.

SPAIN

Spain, a country composed of autonomous regions, also has a national legislation and a number of regional regulations in the Autonomous Provinces. The national water law (Ley de Aguas, 29/1985) merely foresees that the government will "establish the basic conditions for the direct reuse of wastewaters," according to the treatment processes, water quality, and foreseen uses. There are so far no standards.

A draft Royal Decree to extend this existing law was prepared (Ministerio de Obras Públicas, 1996) but was put on hold by the change of government in Spain in 1996. The proposed decree foresees a standard of 1 nematode egg per liter for all types of irrigation and 10 fecal coliforms per 100 ml for unrestricted irrigation. For restricted irrigation, the fecal coliform standard becomes 200 per 100 ml and in the case of the irrigation of cereals, industrial crops, fodder crops, and pastures, it becomes 500 fecal coliforms per 100 ml. Limits on Cl_2 are also foreseen. Specific standards for heavy metals must be respected for the reuse of industrial wastewaters. The draft law would also allow the water agencies to set stricter standards for sanitary reasons. As in the French case, the draft law requires the persons who ask for a permit for the reuse of treated wastewater to constitute a complete dossier for the water agency.

There is an institutional debate in Spain as to whether the central government can legislate in this area in view of the existence of competent autonomous regions. This places doubt over the future of the draft decree but shows the interest raised by the issue of wastewater reuse standards.

A few regional legislations and standards do exist (in Andalusia, Baleares, and Catalonia, and planned in Canarias). In the Balearic Islands, wastewater reuse is regulated by a 1992 decree (B. OC. A. I. B.) with legal value. The approach taken is strictly that of the WHO. Two other pieces of Balearic legislation favor the reuse of wastewater. One prescribes the irrigation of golf courses with water other than for domestic consumption or agricultural irrigation, and the other recognizes agricultural irrigation with reused wastewater as being of public utility (Salgot and Pascual, 1996).

Catalonia has guidelines with a de facto legal value containing limit values for boron, cadmium, molybdenum, and selenium, all relevant for the health of irrigated crops (Salgot et al., 1994). The microbiological standards are those of the WHO.

Andalusia also has recommendations dating from 1994, largely following the French approach with a case by case authorization. However, these guidelines specifically exclude the reuse of wastewater for potable water, street cleaning, municipal heating and cooling, and the cleaning of urban premises, as well as for the washing and transport of materials. Ground-

TABLE A.2.   Quality Guidelines for the Various Applications
of Wastewater Reuse in Andalusia.

Type of Standard	Application	Fecal Coliforms per 100 ml	Nematode Eggs/L
1	Irrigation of sports fields and parks with public access	< 200	< 1
2	Vegetables to be consumed raw	< 1,000	< 1
3	Production of biomass intended for human consumption and refrigeration in open circuits	< 1,000	None
4	Recreational lakes	< 2,000	< 1
5	Refrigeration in semi-closed circuits	< 10,000	None
6	Industrial crops, cereals, dry fodder seeds, forests and conserved or cooked vegetables	None	< 1
7	Irrigation of green areas with no public access, production of biomass not intended for human consumption and recreational lakes with access prohibited	None	None

Source: Adapted from Castillo et al., 1994.

water recharge is also restricted. Overall, the permitted types of reuse fall into seven categories. Table A.2 summarizes the guidelines (Castillo et al., 1994).

WASTEWATER REUSE, TRADE AND CONSUMER PROTECTION

While this review shows that the WHO guidelines have provided the basis for many national or regional standards, this has not always been the case, such as in Italy. In any case, this degree of homogeneity disappears completely for the parameters (mostly chemical) not covered by the WHO guidelines, and many member states of the European Union do not even have *any* wastewater reuse standards. What are the reasons for these differences? Local conditions vary strongly between countries and even between regions. Who can say any value is more valid than another one? Besides, the public health relevance of the indicator microorganisms such as the fecal coliforms is being increasingly challenged. Therefore, some sort of internationally validated risk assessment approach is needed. This is certainly an issue in the case of the export of crops irrigated with reused wastewater, mainly for produce that can be consumed raw.

Unequal enforcement of the standards, training of the operators, or monitoring of the installations are also issues. Any news of health problems rightly or wrongly attributed to "dirty" fruit or vegetables, because of standards or practices without pan-European recognition, or their improper application, could have devastating consequences on the European market. The recent crisis of consumer confidence and subsequent market crashes due to the presence of the bovine spongiform encephalopathy (BSE) prion in British beef is a good example of what can happen. As a consequence, there is a clear need for a better epidemiological justification and a wider international acceptability of standards, even beyond the borders of the European Union, in the area of wastewater reuse.

There have already been calls for the establishment of European guidelines for the Mediterranean countries of the European Union (Marecos do Monte et al., 1996), accompanied by detailed proposals. Guidelines for wastewater reuse for irrigation must cover the chemical quality, the physico-chemical quality, and the microbiological quality of the treated wastewater in order to protect, on the one hand, the soils and the crops, and on the other hand, the agricultural workers and the consumers. Broadly, the chemical quality guidelines must be suitable for the local soil and climatic conditions. The microbiological quality guidelines, by contrast, must be universally suitable and guarantee the safety of the local field workers, as well as that of the potential consumers wherever they are, since crops may be exported far from their place of production. In the case of landscape or golf course irrigation, the microbiological standards must also cover the health of international tourists. These last two points do require a wide acceptability of the standards, well beyond the Mediterranean regions.

The increasing concern about undesirable chemical residues (e.g., pesticide residues, surfactant residues, etc.) may even require widely acceptable chemical standards. This has not yet been stressed very much but is likely to become an increasingly relevant issue as the debate about the use of wastewater sludges in agriculture expands.

The existence of European guidelines for wastewater reuse, beyond ensuring an adequate level of consumer protection in Europe, would also offer an advantage for non-European Mediterranean countries (a sensitive point in the context of the future Euro-Mediterranean Free Trade Area to be set up within 15 years). On the one hand, the existence of clear guidelines allows them to know the standards that must be met. This offers them a clear opportunity for benchmarking the quality of their production and assures them they will encounter no bad surprise on the European market. On the other hand, wastewater reuse is a valid water resource for these countries, and their position as exporters of fruit and vegetables to the European Union is a clear incentive to develop this resource locally and adopt the same guidelines.

There is therefore clearly a need for harmonized European regulation on wastewater reuse, or at least a common regulatory approach. In the event that such work would start, the economic aspects of implementing more or less stringent standards should be considered from the start. Answers to questions such as "is it economically justifiable to implement standards as stringent as in California?" should be provided.

WASTEWATER REUSE: AN OPPORTUNITY FOR BETTER PROTECTING THE EUROPEAN ENVIRONMENT

An extra benefit of the development of European guidelines for wastewater reuse would be to promote the practice in Europe with the objective to protect the environment. This can make wastewater reuse attractive, even in areas where water resources are still sufficient. Pilot schemes have already been set up in Belgium (Guillaume and Xanthoulis, 1996), and this application is already spreading in France. It was actually the main perspective present in the mind of the experts when the French recommendations were drawn.

This aspect could be tackled by the development of "good wastewater reclamation and reuse practices." This could be an alternative approach to the painstaking elaboration of precise standards where many debates come to light. In Israel, for example, besides the existence of standards, the concept of "required treatment" has been introduced. This can be related to the introduction of the concept of "Best Available Technique" in the recent European directive for the Integrated Pollution Prevention and Control (96/61/EC) [European Commission (2), 1996]. This approach can be viewed as agreeing on the use of the technology or technique able to deliver a satisfactory water quality for the foreseen type of reuse.

LEGISLATION AND GUIDELINES ON WASTEWATER REUSE AT EUROPEAN LEVEL

So far, no regulation of wastewater reuse exists at the European level. The only reference to it is article 12 of the European Wastewater Directive (91/271/EEC) [European Commission (1), 1991] stating: "Treated wastewater shall be reused whenever appropriate." In order to make this statement reality, common definitions of what is "appropriate" are needed.

The new Communication of the European Commission on the future European Community Water Policy (European Commission, 1996) does not specifically mention the desirability of wastewater reuse, but it introduces a quantitative dimension to water management, on top of the usual qualitative dimension, which may stimulate the consideration of wastewater reuse. It also states that "water resources should be of sufficient quality and quantity to meet other economic requirements." With wastewater reuse being a water resource often mobilized for economic reasons, such a statement does have economic implications.

In parallel, a "Task Force Environment-Water" has been set up by the European Commission (Commission Européenne, 1996), largely with an advisory role, in order to set R&D priorities and improve the coordination of the various actions of the European institutions in the domain of water research. One of its declared areas of concern is the promotion of the reuse and recycling of water in the various branches of agriculture and industry (irrigation and cooling in particular), through the development of standards for reuse, the development of techniques for on-site treatment and storage of wastewater, and awareness campaigns.

CONCLUSION

The European Commission is currently drafting a new framework directive encompassing all existing European regulations dealing with water. This work aims at providing a coherent regulatory approach compatible with the concept of "Integrated Pollution Prevention and Control" and promotes the application of "Best Available Techniques."

This is clearly an opportunity to start a reflection at European level on suitable criteria (chemical, microbiological, etc.) and practices for wastewater reuse. In order to enhance the acceptability and applicability of these criteria, such a reflection should be coupled to R&D and regulatory work on the development of a set of Best Available Techniques, suitable for a variety of cases. At the same time, epidemiological research on this topic in Europe must be stimulated, potentially contributing important results for the debate around the WHO guidelines. The end result could be a revision of the WHO guidelines with an improved worldwide acceptability.

AKNOWLEDGEMENTS

I would like to express my thanks to the following experts for their useful contributions: H. Blöch, EC DG XI, Brussels (B); J. Bontoux, Univ. Montpellier (F); A. El Bahri, CRGR Tunis (TN); G. Lawrence, EC DG XI, Brussels (B); H. Marecos do Monte, Ministerio da Obras Públicas, Lisbon (P); and M. Salgot, Univ. Barcelona (E).

REFERENCES

Anonymous, The Hadley Centre Transient Climate Change Experiment, The Hadley Centre for Climate Prediction and Research, UK, August, 1992.

Asano, T. and Levine, A. D., Wastewater reclamation, recycling and reuse: past, present and future, *Wat. Sci. Tech.*, Vol. 33 N°10–11, pp. 1–14, 1996.

Castillo, A., Cabrera, J. J., Fernández Artigas. M. P., García-Villanova, B., Hernández-Ruíz, J. A., Laguna, J., Nogales, R. and Picazo, J., *Criterios para la evaluación sanitaria de proyectos de reutilización directa de aguas residuales urbanas depuradas,* Consejería de Salud, Junta de Andalucía, Ed. A. Castillo, Granada, España, (255 p.), 1994.

Chang, A. C., Page, A. L. and Asano, T. *Developing human health-related chemical guidelines for reclaimed wastewater and sewage sludge applications in agriculture,* WHO, Geneva, Switzerland, 114 pp., 1995.

Commission Européenne, Brochure "La Task Force Environnement-Eau," Centre Commun de Recherche, Ispra, 1996.

Conseil Supérieur d'Hygiène Publique de France, Recommandations sanitaires concernant l'utilisation, après épuration des eaux résiduaires urbaines pour l'irrigation des cultures et des espaces verts, Ministère chargé de la Santé, Paris, July 1991.

European Commission (1), Communication de la Commission au Conseil et au Parlement Européen, La Politique Communautaire de l'Eau, COM(96) 59 final, Feb, 21, 1996.

European Commission (2), Council Directive concerning integrated pollution prevention and control, 96/61/EC of September 24, 1996, OJ N°L 257/26 of October 10, 1996.

European Commission, Council Directive concerning urban wastewater treatment, 91/271/EEC of May 21, 1991, OJ N°L 135/40 of May 30, 1991.

Guillaume, P. and Xanthoulis, D., Irrigation of vegetable crops as a means of recycling wastewater: applied to Hesbaye Frost, *Wat. Sci. Tech.*, Vol. 33, N°10–11, pp. 317–326, 1996.

Mara, D. and Cairncross, S., *Guidelines for the unsafe use of wastewater and excreta in agriculture and aquaculture, Measures for public health protection,* WHO, Geneva, Switzerland, 1989.

Marecos do Monte, H., Angelakis, A. and Asano, T., Necessity and basis for establishment of European guidelines for reclaimed wastewater in the Mediterranean region, *Wat. Sci. Tech.*, Vol. 33, N°10–11, pp. 303–316, 1996.

Ministerio de Obras Públicas, Transportes y Medio Ambiente, Secretaría General Técnica, Proyecto de Real Decreto por el que se establecen las condiciones básicas para la reutilización directa de las aguas residuales depuradas, Madrid, Spain, 1996.

Ministerio dei Lavori Pubblici, Criteri, metodologie e norme tecniche generali di cui all'art.2, lettere b), d) ed e), della legge 10 maggio 1976, n. 319, recante norme per la tutel delle acque dall'inquinamento, Allegato 5, Supplemento ordinario alla "Gazzetta Ufficiale" n. 48 del 21 febbraio 1977.

Salgot, M., Cortés, A., Gomá, P. and Pascual, A., Prevenció del risc sanitari derivat de la reutilizació d'aigües residuals depurades com a aigües de reg, Generalitat de Catalunya, Direcció General de Salut Publica, Barcelona, 1994.

Salgot, M. and Pascual, A., Existing guidelines and regulations in Spain on wastewater reclamation and reuse, *Wat. Sci. Tech.*, Vol. 34, No. 11, pp. 261–267, 1996.

Shelef, G., and Azov, Y., The coming era of intensive wastewater reuse in the Mediterranean region, *Wat. Sci. Tech.*, Vol. 33, N°10–11, pp. 115–126, 1996.

State of California, Wastewater reclamation criteria, an excerpt from the California Code of Regulations, Title 22, Division 4, Environmental Health, Department of Health Services, Sacramento, California, 1978.

Stuart, M. E. and Chilton, P. J., Wastewater reuse, in *Groundwater in the UK: a strategic study,* Groundwater Issues, FR/GF 2, Foundation for Water Research, UK, September 1995.

Von Storch, H., Zorita, E. and Cubasch, U., Downscaling of climate change estimates to regional scales: an application to Iberian rainfall in wintertime, *J. of Climate,* Vol. 6, N°6, pp. 1160–1171, June 1993.

WHO, *Health guidelines for the use of wastewater in agriculture and aquaculture,* World Health Organization Tech. Rep. series 778, WHO, Geneva, Switzerland, 1989.

WASTEWATER
RECLAMATION CRITERIA

An Excerpt from the

CALIFORNIA ADMINISTRATIVE CODE
TITLE 22, DIVISION 4

ENVIRONMENTAL HEALTH

1978

STATE OF CALIFORNIA
DEPARTMENT OF HEALTH SERVICES
SANITARY ENGINEERING SECTION
2151 Berkeley Way, Berkeley 94704

INTENT OF REGULATIONS

The intent of these regulations is to establish acceptable levels of constituents of reclaimed water and to prescribe means for assurance of reliability in the production of reclaimed water in order to ensure that the use of reclaimed water for the specified purposes does not impose undue risks to health. The levels of constituents in combination with the means for assurance of reliability constitute reclamation criteria as defined in Section 13520 of the California Water Code.

As affirmed in Sections 13510 to 13512 of the California Water Code, water reclamation is in the best public interest and the policy of the State is to encourage reclamation. The reclamation criteria are intended to promote development of facilities which will assist in meeting water requirements of the State while assuring positive health protection. Appropriate surveillance and control of treatment facilities, distribution systems, and use areas must be provided in order to avoid health hazards. Precautions must be taken to avoid direct public contact with reclaimed waters which do not meet the standards specified in Article 5 for nonrestricted recreational impoundments.

CHAPTER 3.   RECLAMATION CRITERIA

Article 10.   Alternative Reliability Requirements for Uses Requiring Oxidized,
Disinfected Wastewater or Oxidized, Coagulated, Clarified, Filtered,
Disinfected Wastewater

CHAPTER 3. RECLAMATION CRITERIA

Article 1. Definitions

60301. Definitions. (a) **Reclaimed Water.** Reclaimed water means water which, as a result of treatment of domestic wastewater, is suitable for a direct beneficial use or a controlled use that would not otherwise occur.

(b) **Reclamation Plant.** Reclamation plant means an arrangement of devices, structures, equipment, processes and controls which produce a reclaimed water suitable for the intended reuse.

(c) **Regulatory Agency.** Regulatory agency means the California Regional Water Quality Control Board in whose jurisdiction the reclamation plant is located.

(d) **Direct Beneficial Use.** Direct beneficial use means the use of reclaimed water which has been transported from the point of production to the point of use without an intervening discharge to waters of the State.

(e) **Food Crops.** Food crops mean any crops intended for human consumption.

(f) **Spray Irrigation.** Spray irrigation means application of reclaimed water to crops by spraying it from orifices in piping.

(g) **Surface Irrigation.** Surface irrigation means application of reclaimed water by means other than spraying such that contact between the edible portion of any food crop and reclaimed water is prevented.

(h) **Restricted Recreational Impoundment.** A restricted recreational impoundment is a body of reclaimed water in which recreation is limited to fishing, boating, and other non-body-contact water recreation activities.

(i) **Nonrestricted Recreational Impoundment.** A nonrestricted recreational impoundment is an impoundment of reclaimed water in which no limitations are imposed on body-contact water sport activities.

(j) **Landscape Impoundment.** A landscape impoundment is a body of reclaimed water which is used for aesthetic enjoyment or which otherwise serves a function not intended to include public contact.

(k) **Approved Laboratory Methods.** Approved laboratory methods are those specified in the latest edition of "Standard Methods for the Examination of Water and Wastewater", prepared and published jointly by the American Public Health Association, the American Water Works Association, and the Water Pollution Control Federation and which are conducted in laboratories approved by the State Department of Health.

(l) **Unit Process.** Unit process means an individual stage in the wastewater treatment sequence which performs a major single treatment operation.

(m) **Primary Effluent.** Primary effluent is the effluent from a wastewater treatment process which provides removal of sewage solids so that it contains not more than 0.5 milliliter per liter per hour of settleable solids as determined by an approved laboratory method.

(n) **Oxidized Wastewater.** Oxidized wastewater means wastewater in which the organic matter has been stabilized, is nonputrescible, and contains dissolved oxygen.

(o) **Biological Treatment.** Biological treatment means methods of wastewater treatment in which bacterial or biochemical action is intensified as a means of producing an oxidized wastewater.

(p) **Secondary Sedimentation.** Secondary sedimentation means the removal by gravity of settleable solids remaining in the effluent after the biological treatment process.

(q) **Coagulated Wastewater.** Coagulated wastewater means oxidized wastewater in which colloidal and finely divided suspended matter have been destabilized and agglomerated by the addition of suitable floc-forming chemicals or by an equally effective method.

(r) **Filtered Wastewater.** Filtered wastewater means an oxidized, coagulated, clarified wastewater which has been passed through natural undisturbed soils or filter media, such as sand or diatomaceous earth, so that the turbidity as determined by an approved laboratory method does not exceed an average operating turbidity of 2 turbidity units and does not exceed 5 turbidity units more than 5 percent of the time during any 24-hour period.

(s) **Disinfected Wastewater.** Disinfected wastewater means wastewater in which the pathogenic organisms have been destroyed by chemical, physical or biological means.

(t) **Multiple Units.** Multiple units means two or more units of a treatment process which operate in parallel and serve the same function.

(u) **Standby Unit Process.** A standby unit process is an alternate unit process or an equivalent alternative process which is maintained in operable condition and which is capable of providing comparable treatment for the entire design flow of the unit for which it is a substitute.

(v) **Power Source.** Power source means a source of supplying energy to operate unit processes.

(w) **Standby Power Source.** Standby power source means an automatically actuated self-starting alternate energy source maintained in immediately operable condition and of sufficient capacity to provide necessary service during failure of the normal power supply.

(x) **Standby Replacement Equipment.** Standby replacement equipment means reserve parts and equipment to replace broken-down or worn-out units which can be placed in operation within a 24-hour period.

(y) **Standby Chlorinator.** A standby chlorinator means a duplicate chlorinator for reclamation plants having one chlorinator and a duplicate of the largest unit for plants having multiple chlorinator units.

(z) **Multiple Point Chlorination.** Multiple point chlorination means that chlorine will be applied simultaneously at the reclamation plant and at subsequent chlorination stations located at the use area and/or some intermediate point. It does not include chlorine application for odor control purposes.

(aa) **Alarm.** Alarm means an instrument or device which continuously monitors a specific function of a treatment process and automatically gives warning of an unsafe or undesirable condition by means of visual and audible signals.

(bb) **Person.** Person also includes any private entity, city, county, district, the State or any department or agency thereof.

NOTE: Authority cited: Section 208, Health and Safety Code and Section 13521, Water Code. Reference: Section 13521, Water Code.

History: 1. New Chapter 4 (§§ 60301–60357, not consecutive) filed 4-2-75; effective thirtieth day thereafter (Register 75, No. 14).
2. Renumbering of Chapter 4 (Sections 60301–60357, not consecutive) to Chapter 3 (Sections 60301–60357, not consecutive), filed 10-14-77; effective thirtieth day thereafter (Register 77, No. 42).

Article 2. Irrigation of Food Crops

60303. Spray Irrigation. Reclaimed water used for the spray irrigation of food crops shall be at all times an adequately disinfected, oxidized, coagulated, clarified, filtered wastewater. The wastewater shall be considered adequately disinfected if at some location in the treatment process the median number of coliform organisms does not exceed 2.2 per 100 milliliters and the number of coliform organisms does not exceed 23 per 100 milliliters in more than one sample within any 30-day period. The median value shall be determined from the bacteriological results of the last 7 days for which analyses have been completed.

60305. Surface Irrigation. (a) Reclaimed water used for surface irrigation of food crops shall be at all times an adequately disinfected, oxidized wastewater. The wastewater shall be considered adequately disinfected if at some location in the treatment process the median number of coliform organisms does not exceed 2.2 per 100 milliliters, as determined from the bacteriological results of the last 7 days for which analyses have been completed.

(b) Orchards and vineyards may be surface irrigated with reclaimed water that has the quality at least equivalent to that of primary effluent provided that no fruit is harvested that has come in contact with the irrigating water or the ground.

60307. Exceptions. Exceptions to the quality requirements for reclaimed water used for irrigation of food crops may be considered by the State Department of Health on an individual case basis where the reclaimed water is to be used to irrigate a food crop which must undergo extensive commercial, physical or chemical processing sufficient to destroy pathogenic agents before it is suitable for human consumption.

Article 3.   Irrigation of Fodder, Fiber, and Seed Crops

60309.   Fodder, Fiber, and Seed Crops.   Reclaimed water used for the surface or spray irrigation of fodder, fiber, and seed crops shall have a level of quality no less than that of primary effluent.

60311.   Pasture for Milking Animals.   Reclaimed water used for the irrigation of pasture to which milking cows or goats have access shall be at all times an adequately disinfected, oxidized wastewater. The wastewater shall be considered adequately disinfected if at some location in the treatment process the median number of coliform organisms does not exceed 23 per 100 milliliters, as determined from the bacteriological results of the last 7 days for which analyses have been completed.

Article 4.   Landscape Irrigation

60313.   Landscape Irrigation.   (a) Reclaimed water used for the irrigation of golf courses, cemeteries, freeway landscapes, and landscapes in other areas where the public has similar access or exposure shall be at all times an adequately disinfected, oxidized wastewater. The wastewater shall be considered adequately disinfected if the median number of coliform organisms in the effluent does not exceed 23 per 100 milliliters, as determined from the bacteriological results of the last 7 days for which analyses have been completed, and the number of coliform organisms does not exceed 240 per 100 milliliters in any two consecutive samples.

(b) Reclaimed water used for the irrigation of parks, playgrounds, schoolyards, and other areas where the public has similar access or exposure shall be at all times an adequately disinfected, oxidized, coagulated, clarified, filtered wastewater or a wastewater treated by a sequence of unit processes that will assure an equivalent degree of treatment and reliability. The wastewater shall be considered adequately disinfected if the median number of coliform organisms in the effluent does not exceed 2.2 per 100 milliliters, as determined from the bacteriological results of the last 7 days for which analyses have been completed, and the number of coliform organisms does not exceed 23 per 100 milliliters in any sample.

NOTE: Authority cited: Section 208, Health and Safety Code and Section 13521, Water Code. Reference: Section 13520, Water Code.

History: 1. Amendment filed 9-22-78; effective thirtieth day thereafter (Register 78, No. 38).

Article 5.   Recreational Impoundments

60315.   Nonrestricted Recreational Impoundment.   Reclaimed water used as a source of supply in a nonrestricted recreational impoundment shall be at all times an adequately disinfected, oxidized, coagulated, clarified, filtered wastewater. The wastewater shall be considered adequately disinfected if at some location in the treatment process the median number of coliform organisms does not exceed 2.2 per 100 milliliters and the number of coliform organisms does not exceed 23 per 100 milliliters in more than one sample within any 30-day period. The median value shall be determined from the bacteriological results of the last 7 days for which analyses have been completed.

60317. Restricted Recreational Impoundment. Reclaimed water used as a source of supply in a restricted recreational impoundment shall be at all times an adequately disinfected, oxidized wastewater. The wastewater shall be considered adequately disinfected if at some location in the treatment process the median number of coliform organisms does not exceed 2.2 per 100 milliliters, as determined from the bacteriological results of the last 7 days for which analyses have been completed.

60319. Landscape Impoundment. Reclaimed water used as a source of supply in a landscape impoundment shall be at all times an adequately disinfected, oxidized wastewater. The wastewater shall be considered adequately disinfected if at some location in the treatment process the median number of coliform organisms does not exceed 23 per 100 milliliters, as determined from the bacteriological results of the last 7 days for which analyses have been completed.

Article 5.1. Groundwater Recharge

60320. Groundwater Recharge. (a) Reclaimed water used for groundwater recharge of domestic water supply aquifers by surface spreading shall be at all times of a quality that fully protects public health. The State Department of Health Services' recommendations to the Regional Water Quality Control Boards for proposed groundwater recharge projects and for expansion of existing projects will be made on an individual case basis where the use of reclaimed water involves a potential risk to public health.

(b) The State Department of Health Services' recommendations will be based on all relevant aspects of each project, including the following factors: treatment provided; effluent quality and quantity; spreading area operations; soil characteristics; hydrogeology; residence time; and distance to withdrawal.

(c) The State Department of Health Services will hold a public hearing prior to making the final determination regarding the public health aspects of each groundwater recharge project. Final recommendations will be submitted to the Regional Water Quality Control Board in an expeditious manner.

NOTE: Authority cited: Section 208, Health and Safety Code and Section 13521, Water Code. Reference: Section 13520, Water Code.

History: 1. New Article 5.1 (Section 60320) filed 9-22-78; effective thirtieth day thereafter (Register 78, No. 38).

Article 5.5. Other Methods of Treatment

60320.5. Other Methods of Treatment. Methods of treatment other than those included in this chapter and their reliability features may be accepted if the applicant demonstrates to the satisfaction of the State Department of Health that the methods of treatment and reliability features will assure an equal degree of treatment and reliability.

NOTE: Authority cited: Section 208, Health and Safety Code and Section 13521, Water Code. Reference: Section 13520, Water Code.

History: 1. Renumbering of Article 11 (Section 60357) to Article 5.5 (Section 60320.5) filed 9-22-78; effective thirtieth day thereafter (Register 78, No. 38).

Article 6. Sampling and Analysis

60321. Sampling and Analysis. (a) Samples for settleable solids and coliform bacteria, where required, shall be collected at least daily and at a time when wastewater characteristics are most demanding on the treatment facilities and disinfection procedures. Turbidity analysis, where required, shall be performed by a continuous recording turbidimeter.

(b) For uses requiring a level of quality no greater than that of primary effluent, samples shall be analyzed by an approved laboratory method of settleable solids.

(c) For uses requiring an adequately disinfected, oxidized wastewater, samples shall be analyzed by an approved laboratory method for coliform bacteria content.

(d) For uses requiring an adequately disinfected, oxidized, coagulated, clarified, filtered wastewater, samples shall be analyzed by approved laboratory methods for turbidity and coliform bacteria content.

Article 7. Engineering Report and Operational Requirements

60323. Engineering Report. (a) No person shall produce or supply reclaimed water for direct reuse from a proposed water reclamation plant unless he files an engineering report.

(b) The report shall be prepared by a properly qualified engineer registered in California and experienced in the field of wastewater treatment, and shall contain a description of the design of the proposed reclamation system. The report shall clearly indicate the means for compliance with these regulations and any other features specified by the regulatory agency.

(c) The report shall contain a contingency plan which will assure that no untreated or inadequately-treated wastewater will be delivered to the use area.

60325. Personnel. (a) Each reclamation plant shall be provided with a sufficient number of qualified personnel to operate the facility effectively so as to achieve the required level of treatment at all times.

(b) Qualified personnel shall be those meeting requirements established pursuant to Chapter 9 (commencing with Section 13625) of the Water Code.

60327. Maintenance. A preventive maintenance program shall be provided at each reclamation plant to ensure that all equipment is kept in a reliable operating condition.

60329. Operating Records and Reports. (a) Operating records shall be maintained at the reclamation plant or a central depository within the operating agency. These shall include: all analyses specified in the reclamation criteria; records of operational problems, plant and equipment breakdowns, and diversions to emergency storage or disposal; all corrective or preventive action taken.

(b) Process or equipment failures triggering an alarm shall be recorded and maintained as a separate record file. The recorded information shall include the time and cause of failure and corrective action taken.

(c) A monthly summary of operating records as specified under (a) of this section shall be filed monthly with the regulatory agency.

(d) Any discharge of untreated or partially treated wastewater to the use area, and the cessation of same, shall be reported immediately by telephone to the regulatory agency, the State Department of Health, and the local health officer.

60331. Bypass. There shall be no bypassing of untreated or partially treated wastewater from the reclamation plant or any intermediate unit processes to the point of use.

Article 8.   General Requirements of Design

60333. Flexibility of Design. The design of process piping, equipment arrangement, and unit structures in the reclamation plant must allow for efficiency and convenience in operation and maintenance and provide flexibility of operation to permit the highest possible degree of treatment to be obtained under varying circumstances.

60335. Alarms. (a) Alarm devices required for various unit processes as specified in other sections of these regulations shall be installed to provide warning of:

(1) Loss of power from the normal power supply.
(2) Failure of a biological treatment process.
(3) Failure of a disinfection process.
(4) Failure of a coagulation process.
(5) Failure of a filtration process.
(6) Any other specific process failure for which warning is required by the regulatory agency.

(b) All required alarm devices shall be independent of the normal power supply of the reclamation plant.

(c) The person to be warned shall be the plant operator, superintendent, or any other responsible person designated by the management of the reclamation plant and capable of taking prompt corrective action.

(d) Individual alarm devices may be connected to a master alarm to sound at a location where it can be conveniently observed by the attendant. In case the reclamation plant is not attended full time, the alarm(s) shall be connected to sound at a police station, fire station or other full-time service unit with which arrangements have been made to alert the person in charge at times that the reclamation plant is unattended.

60337. Power Supply. The power supply shall be provided with one of the following reliability features:

(a) Alarm and standby power source.
(b) Alarm and automatically actuated short-term retention or disposal provisions as specified in Section 60341.
(c) Automatically actuated long-term storage or disposal provisions as specified in Section 60341.

Article 9.   Alternative Reliability Requirements for
Uses Permitting Primary Effluent

60339.   Primary Treatment. Reclamation plants producing reclaimed water exclusively for uses for which primary effluent is permitted shall be provided with one of the following reliability features:

(a) Multiple primary treatment units capable of producing primary
effluent with one unit not in operation.

(b) Long-term storage or disposal provisions as specified in Section
60341.

Article 10.   Alternative Reliability Requirements for Uses Requiring
Oxidized, Disinfected Wastewater or Oxidized, Coagulated,
Clarified, Filtered, Disinfected Wastewater

60341.   Emergency Storage or Disposal. (a) Where short-term retention or disposal provisions are used as a reliability feature, these shall
consist of facilities reserved for the purpose of storing or disposing of
untreated or partially treated wastewater for at least a 24-hour period.
The facilities shall include all the necessary diversion devices, provisions for odor control, conduits, and pumping and pump back equipment. All of the equipment other than the pump back equipment shall
be either independent of the normal power supply or provided with a
standby power source.

(b) Where long-term storage or disposal provisions are used as a
reliability feature, these shall consist of ponds, reservoirs, percolation
areas, downstream sewers leading to other treatment or disposal facilities or any other facilities reserved for the purpose of emergency storage or disposal of untreated or partially treated wastewater. These
facilities shall be of sufficient capacity to provide disposal or storage of
wastewater for at least 20 days, and shall include all the necessary
diversion works, provisions for odor and nuisance control, conduits, and
pumping and pump back equipment. All of the equipment other than
the pump back equipment shall be either independent of the normal
power supply or provided with a standby power source.

(c) Diversion to a less demanding reuse is an acceptable alternative
to emergency disposal of partially treated wastewater provided that the
quality of the partially treated wastewater is suitable for the less demanding reuse.

(d) Subject to prior approval by the regulatory agency, diversion to
a discharge point which requires lesser quality of wastewater is an
acceptable alternative to emergency disposal of partially treated wastewater.

(e) Automatically actuated short-term retention or disposal provisions and automatically actuated long-term storage or disposal provisions shall include, in addition to provisions of (a), (b), (c), or (d) of
this section, all the necessary sensors, instruments, valves and other
devices to enable fully automatic diversion of untreated or partially
treated wastewater to approved emergency storage or disposal in the
event of failure of a treatment process, and a manual reset to prevent
automatic restart until the failure is corrected.

60343. Primary Treatment. All primary treatment unit processes shall be provided with one of the following reliability features:

(a) Multiple primary treatment units capable of producing primary effluent with one unit not in operation.

(b) Standby primary treatment unit process.

(c) Long-term storage or disposal provisions.

60345. Biological Treatment. All biological treatment unit processes shall be provided with one of the following reliability features:

(a) Alarm and multiple biological treatment units capable of producing oxidized wastewater with one unit not in operation.

(b) Alarm, short-term retention or disposal provisions, and standby replacement equipment.

(c) Alarm and long-term storage or disposal provisions.

(d) Automatically actuated long-term storage or disposal provisions.

60347. Secondary Sedimentation. All secondary sedimentation unit processes shall be provided with one of the following reliability features:

(a) Multiple sedimentation units capable of treating the entire flow with one unit not in operation.

(b) Standby sedimentation unit process.

(c) Long-term storage or disposal provisions.

60349. Coagulation.

(a) All coagulation unit processes shall be provided with the following mandatory features for uninterrupted coagulant feed:

(1) Standby feeders,

(2) Adequate chemical stowage and conveyance facilities,

(3) Adequate reserve chemical supply, and

(4) Automatic dosage control.

(b) All coagulation unit processes shall be provided with one of the following reliability features:

(1) Alarm and multiple coagulation units capable of treating the entire flow with one unit not in operation;

(2) Alarm, short-term retention or disposal provisions, and standby replacement equipment;

(3) Alarm and long-term storage or disposal provisions;

(4) Automatically actuated long-term storage or disposal provisions, or

(5) Alarm and standby coagulation process.

60351. Filtration. All filtration unit processes shall be provided with one of the following reliability features:

(a) Alarm and multiple filter units capable of treating the entire flow with one unit not in operation.

(b) Alarm, short-term retention or disposal provisions and standby replacement equipment.

(c) Alarm and long-term storage or disposal provisions.
(d) Automatically actuated long-term storage or disposal provisions.
(e) Alarm and standby filtration unit process.

60353. Disinfection.

(a) All disinfection unit processes where chlorine is used as the disinfectant shall be provided with the following features for uninterrupted chlorine feed:

(1) Standby chlorine supply,
(2) Manifold systems to connect chlorine cylinders,
(3) Chlorine scales, and
(4) Automatic devices for switching to full chlorine cylinders.

Automatic residual control of chlorine dosage, automatic measuring and recording of chlorine residual, and hydraulic performance studies may also be required.

(b) All disinfection unit processes where chlorine is used as the disinfectant shall be provided with one of the following reliability features:

(1) Alarm and standby chlorinator;
(2) Alarm, short-term retention or disposal provisions, and standby replacement equipment;
(3) Alarm and long-term storage or disposal provisions;
(4) Automatically actuated long-term storage or disposal provisions; or
(5) Alarm and multiple point chlorination, each with independent power source, separate chlorinator, and separate chlorine supply.

60355. Other Alternatives to Reliability Requirements.

Other alternatives to reliability requirements set forth in Articles 8 to 10 may be accepted if the applicant demonstrates to the satisfaction of the State Department of Health that the proposed alternative will assure an equal degree of reliability.

Health Guidelines Reported by a WHO Scientific Group

A World Health Organization Report, entitled "Health guidelines for the use of wastewater in agriculture and aquaculture," *Technical Report Series 778,* considers "the health implications of the reuse of treated wastewater and . . . review[s] and evaluate[s] the health safeguards necessary in reusing effluents for irrigation and aquaculture" (p. 7).

The report discusses the background of wastewater reuse, reviews its use concerning health effects, recommends guidelines for controlling the spread of infectious disease, and identifies needs for additional research and development.

The remainder of the appendix is reprinted with permission from "Health guidelines for the use of wastewater in agriculture and aquaculture," Report of a WHO Scientific Group, *World Health Organization Technical Report Series 778,* World Health Organization, Geneva,© 1989, pp. 36–44.

7.2 EFFLUENT QUALITY GUIDELINES FOR AGRICULTURE

Removal of pathogens is the prime objective in treating wastewater for reuse. However, as previously pointed out, wastewater quality guidelines and standards for reuse are often expressed in terms of the maximum permissible number of faecal coliform bacteria. Since the faecal origin of wastewater is not in question, the implication is that these faecal indicator organisms can be used as pathogen indicators, and that there is at least a semi-quantitative relationship between pathogen and indicator concentrations. In practice, faecal coliforms can be used as reasonably reliable indicators of bacterial pathogens, as their environmental survival characteristics and rates of removal or die-off in treatment processes are broadly

1491

similar. The "total coliform" group is less reliable as an indicator since not all coliforms are exclusively faecal in origin and, especially in warm climates, the proportion of non-faecal coliforms is often very high. Faecal coliforms are less satisfactory as indicators of excreted viruses and are of very limited use in relation to protozoa and helminths, for which no reliable indicators exist.

Standards or guidelines for the quality of wastewater to be used for unrestricted crop irrigation, including that of salad and vegetable crops eaten raw, have generally specified both explicit standards (e.g., maximum numbers of coliforms) and minimum treatment requirements (primary, secondary or tertiary) according to the class of crop to be irrigated (consumable or non-consumable). The standards developed over the past 50 years have tended to be very strict, as they were based on a theoretical evaluation of the potential health risks associated with pathogen survival in wastewater and soil and on crops, rather than on firm epidemiological evidence of actual risk. To some extent, those early standards were based on a "zero risk" concept, with the aim of achieving an "antiseptic" or pathogen-free environment. At that time, the method of choice for pathogen removal, as judged by coliform removal, was secondary biological treatment followed by carefully controlled effluent chlorination. Since this could, at least theoretically, achieve very low residual coliform concentrations, the maximum permissible number of coliforms was set correspondingly low. For example, the standards of the California State Department of Public Health (4) permit a total of only 23 or 2.2 coliforms per 100 ml, depending on the crop being irrigated and on the irrigation method.

In 1971, the WHO Meeting of Experts on the Reuse of Effluents (1) recognized that the extremely strict California standards for effluent reuse were not justified by the available epidemiological evidence and recommended a microbial guideline for the unrestricted irrigation of vegetables eaten cooked of not more than 100 total coliforms per 100 ml, which was in effect a significant liberalization. The Meeting felt that there was a need for wastewater irrigation guidelines to be given a sounder epidemiological basis, and recommended that this matter be fully investigated.

Since that time, major efforts have been made by WHO, the World Bank, the United Nations Development Programme, the United Nations Environment Programme, the International Development Research Centre, Canada, the International Reference Centre for Waste Disposal, Switzerland, the Food and Agriculture Organization of the United Nations, the US Environmental Protection Agency and many academic institutions throughout the world to provide a more rational epidemiological basis for wastewater irrigation guidelines.

Extensive new epidemiological evidence has been accumulated and earlier studies and reports have been evaluated. The findings of these studies have been carefully reviewed by leading public health experts, environmental scientists and epidemiologists at meetings in Engelberg (2) and Adelboden (3) in 1985 and 1987, respectively, and at numerous national and international meetings and consultations. The consensus view of the epidemiologists and public health experts who have reviewed these data is that the actual risk associated with irrigation with treated wastewater is much lower than previously estimated and that the early microbial standards and guidelines for effluent to be used for unrestricted irrigation of vegetables and salad crops normally consumed uncooked were unjustifiably restrictive, particularly in respect of bacterial pathogens. The epidemiological evidence is summarized in this report; for more detailed information, readers are referred to the original reports by Shuval et al. (6) and Blum & Feachem (22) and to the Engelberg report (2).

On the basis of this new evidence, the Engelberg report recommended new guidelines containing less stringent standards for faecal coliforms than those previously suggested. However, they were stricter than previous standards in respect of numbers of helminth eggs, which were recognized to be the main actual public health risk associated with wastewater irrigation in those areas where helminthic diseases are endemic. The Engelberg recommendations were subsequently reviewed and confirmed at the Adelboden meeting. After consideration of this preparatory work and the epidemiological evidence currently available, the Scientific Group now recommends the guidelines shown in Table A.3. These are based on the fact that in many developing countries the main actual health risks, as pointed out above, are associated with helminthic diseases and that the safe use of wastewater in agriculture or aquaculture will therefore require a high degree of helminth removal. Thus, these guidelines introduce a new, stricter approach concerning the need to reduce numbers of helminth eggs (*Ascaris* and *Trichuris* species and hookworms) in effluents to a level of one or less per litre. This means that some 99.9% of helminth eggs must be removed by appropriate treatment processes in areas where helminthic diseases are endemic and present actual health risks (field studies indicate that helminth concentrations are rarely greater than 1000 per litre, even in endemic areas). Stabilization ponds with a retention time of 8–10 days are particularly effective in achieving this but other technologies are also available. While not all helminths and protozoa of public health importance are referred to specifically in the guidelines (for example, *Amoeba* and *Giardia* species are not mentioned), the intestinal nematodes covered should serve as indicator organisms for all of the large settleable pathogens (including amoebic cysts); other pathogens of interest apparently become non-viable in long-retention pond systems. It is thus implied by the guide-

TABLE A.3. Recommended microbiological quality guidelines for wastewater use in agriculture[a]

Category	Reuse conditions	Exposed group	Intestinal nematodes[b] (arithmetic mean no. of eggs per litre[c])	Faecal coliforms (geometric mean no. per 100 ml[c])	Wastewater treatment expected to achieve the required microbiological quality
A	Irrigation of crops likely to be eaten uncooked, sports fields, public parks[d]	Workers, consumers, public	≤1	≤1000[d]	A series of stabilization ponds designed to achieve the microbiological quality indicated, or equivalent treatment
B	Irrigation of cereal crops, industrial crops, fodder crops, pasture and trees[e]	Workers	≤1	No standard recommended	Retention in stabilization ponds for 8–10 days or equivalent helminth and faecal coliform removal
C	Localized irrigation of crops in category B if exposure of workers and the public does not occur	None	Not applicable	Not applicable	Pretreatment as required by the irrigation technology, but not less than primary sedimentation

[a] In specific cases, local epidemiological, sociocultural and environmental factors should be taken into account, and the guidelines modified accordingly.
[b] *Ascaris* and *Trichuris* species and hookworms.
[c] During the irrigation period.
[d] A more stringent guideline (≤200 faecal coliforms per 100 ml) is appropriate for public lawns, such as hotel lawns, with which the public may come into direct contact.
[e] In the case of fruit trees, irrigation should cease two weeks before fruit is picked, and no fruit should be picked off the ground. Sprinkler irrigation should not be used.

TABLE A.4. Faecal coliforms in rivers[1]

Number of faecal coliforms per 100 ml	No. of rivers in each region			
	North America	Central and South America	Europe	Asia and the Pacific
< 10	8	0	1	1
10–100	4	1	3	2
100–1,000	8	10	9	14
1,000–10,000	3	9	11	10
10,000–100,000	0	2	7	2
>100,000	0	2	0	3
Total number of rivers	23	24	31	32

Source: reference 23. No data for African rivers are available.

lines that all helminth eggs and protozoan cysts will be removed to the same extent.

Based on current epidemiological evidence, a *bacterial guideline* of a geometric mean of 1000 faecal coliforms per 100 ml for unrestricted irrigation of all crops is recommended. This is considered to be technologically feasible. The Group concluded that no bacterial guideline need be recommended in cases where farm workers are the only exposed population, since there is little or no evidence indicating a risk to such workers from bacteria; nevertheless, some degree of reduction in bacterial concentration is desirable in wastewater used for any purpose.

The natural die-off of pathogens in the field constitutes a valuable additional safety factor in reducing potential health risks. Pathogen inactivation by ultraviolet irradiation, by desiccation and by natural biological predators when effluent is applied to crops and soil can often provide an additional 90–99% reduction of pathogens within a few days after application. In addition to this important factor, field and laboratory studies which indicated that wastewater effluent with 1000 faecal coliforms per 100 ml contained few, if any, detectable pathogens were taken into account by the Scientific Group in formulating the guidelines.

The new bacterial guidelines are in line with the actual quality of river water used for the unrestricted irrigation of all crops in many countries without known ill effects. Table A.4 shows typical faecal coliform concentrations in rivers throughout the world, based on data gathered from 1979 to 1984; in about 45% of the rivers, such concentrations were 1000 per 100 ml or greater, while nearly 15% had faecal coliform levels of 10 000

[1] Table A.4 is reprinted from GEMS: Global Environmental Monitoring System. *Global Pollution and Health. Results of Health-Related Environmental Monitoring.*© 1987, with permission from World Health Organization/United Nations Environment Programme.

per 100 ml or more. Waters from such rivers are widely used outside the USA for irrigation, without any legislative restrictions on their use. In the USA, the US Environmental Protection Agency, together with the US Academy of Sciences, recommended in 1973 that the acceptable guideline for irrigation with natural surface water, including river water, be set at 1000 total coliforms per 100 ml (24).

The Scientific Group also compared earlier wastewater irrigation standards and guidelines for the irrigation of vegetables eaten raw (2.2–100 total coliforms per 100 ml) with existing microbial guidelines and standards for bathing-water quality developed by the Mediterranean Pollution Monitoring and Research Programme of the United Nations Environment Programme and WHO (1000 faecal coliforms per 100 ml, 25) and by the European Economic Community (less than 10 000 total coliforms per 100 ml and less than 2000 faecal coliforms per 100 ml, 26). Finally, the Group concluded that it was not reasonable or rational to retain the earlier wastewater irrigation guidelines, which were close to those for drinking-water, when health authorities consider natural river waters used for irrigation and waters used for bathing to be acceptable with faecal coliform concentrations of 1000 per 100 ml and more.

The irrational application of unjustifiably strict microbial standards for wastewater used for irrigation has undoubtedly led to some anomalous situations. Standards are often not enforced and serious public health problems have resulted from totally unregulated, and often illegal, irrigation of salad crops with raw wastewater, as widely practised in many developing countries. The recommended approach now calls for the introduction of realistic revised national standards which are strict for helminth egg removal but less so with regard to allowable bacterial levels. The Group considered that this new approach would increase public health protection for a greater number of people while at the same time setting targets which were both technologically and economically feasible.

The guideline values given in Table A.3, however, must be carefully interpreted and, if necessary, modified in the light of local epidemiological, sociocultural and environmental factors. Greater caution may be justified where there are significant exposed groups that are more susceptible to infection than the population at large, such as people lacking immunity to the local endemic infections. On the other hand, some degree of flexibility may sometimes be justified. For example, where intestinal helminths are not endemic, an egg removal efficiency of 99.9% is not necessary. Edible crops such as tomatoes for canning and peanuts for roasting might also be properly considered as industrial crops, and sports fields which will not be used for many weeks after irrigation might be regarded as belonging to Category B.

Where members of the public have direct access to lawns and parks which are irrigated with treated wastewater, the potential public health risk may then be greater than that associated with the irrigation of vegetables eaten raw. The Scientific Group took note of the epidemiological investigation of the health effects of landscape irrigation with reclaimed wastewater at Colorado Springs (27), which indicated that people who visited parks irrigated with non-potable water derived from wastewater did not report gastrointestinal symptoms with greater relative frequency than those who visited parks irrigated with either potable or non-potable water of runoff origin. Nevertheless, an effluent standard for park lawn irrigation of 200 faecal coliforms per 100 ml was recommended in the report of the study (27), and the Scientific Group felt that it would be prudent to accept this more stringent guideline for public lawns. This bacteriological effluent guideline can normally be achieved only by means of secondary biological treatment (ponds or conventional treatment) followed by effective disinfection. Additional treatment would be required for helminth egg removal, if relevant.

The *helminth egg guideline* value in Table A.3 is intended as a design goal for wastewater treatment systems, and not as a standard requiring routine testing of effluent quality. The most sensitive techniques currently available for the detection of helminth eggs in wastewater are able to detect a minimum of the order of one egg per litre. However, these are not practicable for field monitoring purposes, for which the procedures described in Annex 2 (capable of detecting of the order of 10 eggs per litre) are more suitable. These procedures are designed to detect eggs of *Ascaris* and *Trichuris* species, the absence of which can be used in most circumstances as an indication of effective helminth removal. However, in regions where the prevalence of these parasites is so low that their eggs are outnumbered by hookworm eggs in raw sewage, procedures for the detection of hookworm eggs should be used instead.

7.3 EFFLUENT QUALITY GUIDELINES FOR AQUACULTURE

A number of infections caused by excreted pathogens are of concern in connection with waste-fed aquaculture. Aquatic snails are intermediate hosts of several helminth parasites, including *Schistosoma* species. Transmission can occur when people wade in fish-ponds in which infected snails are present, and the larval schistosome penetrates the human skin. Certain species of fish are the secondary intermediate hosts of several helminth parasites, for example *Clonorchis* species (liver fluke). Transmission can occur when fish are eaten raw or undercooked, and the cysts in the fish flesh hatch out in the human gut. With some helminth infections, cysts are formed on edible aquatic plants (for example, *Fasciolopsis* species encyst on water caltrop), and transmission can occur when the fruit of the plant is eaten. Fish grown in excreta-fertilized or wastewater ponds may also become contaminated with bacteria and viruses. These are passively carried on the scales, or in the gills, intraperitoneal fluid, digestive tract or muscle of the fish. If fish are eaten raw or undercooked, transmission of bacterial or viral infections may then occur.

Strauss (*18*) reviewed the literature on the survival of pathogens in and on fish and concluded that:

(1) Invasion of fish muscle by bacteria is very likely to occur when the fish are grown in ponds containing concentrations of faecal coliforms and salmonellae greater than 10^4 and 10^5 per 100 ml, respectively, the potential for muscle invasion increasing with the duration of exposure of the fish to the contaminated water.

(2) Some evidence suggests that there is little accumulation of enteric organisms and pathogens on, or penetration into, edible fish tissue when the faecal coliform concentration in the fishpond water is below 10^3 per 100 ml (*28*).

(3) Even at lower contamination levels, high pathogen concentrations may be present in the digestive tract and the intraperitoneal fluid of the fish.

There are, in general, only limited experimental and field data on the health effects of sewage-fertilized aquaculture. Further work is needed, therefore, before a definitive bacteriological quality standard can be established for pisciculture. A tentative *bacterial guideline* of a geometric mean number of faecal coliforms of $\leq 10^3$ per 100 ml is recommended for fishpond water. In view of the dilution of wastewater which occurs in most fish-ponds, this ambient bacterial indicator concentration can normally be achieved by treating the wastewater feed water so as to give a level of 10^3–10^4 faecal coliforms per 100 ml. The same faecal coliform standard should be applied to pond water in which aquatic vegetables (macrophytes) are grown, because they are eaten raw in some areas.

This bacterial guideline, which is based on the present state of knowledge regarding wastewater use in aquaculture, should ensure that invasion of fish muscle is prevented. However, research to date shows that pathogens may accumulate in the digestive tract and intraperitoneal fluid of fish. These pathogens may then pose a risk through cross-contamination of the fish flesh or other edible parts and transmission to consumers if standards of hygiene in fish preparation are inadequate. A further necessary public health measure, therefore, is to ensure that high standards of hygiene are maintained during fish handling and especially gutting. This is easier to achieve in commercial operations than in subsistence aquaculture, for which sustained health education programmes will often be required. Cooking of fish, which is a common practice in many areas where waste-fed aquaculture exists, is an important health safeguard.

Transmission of the helminth infections clonorchiasis and fasciolopsiasis is known to occur only in restricted geographical areas in eastern Asia. Given the cultural preference in some of these areas for eating fish and aquatic vegetables uncooked, transmission can be prevented only by ensuring that no eggs enter the pond or by snail control. The latter is unlikely to be achieved at all times in practice, especially in the small subsistence ponds common in Asia, so that the only feasible means of control is to remove all viable trematode eggs from the wastewater before it enters ponds. *All* eggs must be rendered non-viable because the parasites multiply asexually on an enormous scale within their first intermediate host. Similar considerations apply to the control of schistosomiasis, a disease that is endemic over a much wider geographical area. The appropriate *helminth quality guideline* for all aquacultural use of wastewater is thus the absence of viable trematode eggs. This is readily achieved by stabilization pond treatment.

Index

1501